Calcul

An Integrated Approach to Functions and Their

P R E L I M I N A R Y   E D I T I O N

Calculus

An Integrated Approach to Functions and Their Rates of Change

PRELIMINARY EDITION

ROBIN J. GOTTLIEB

HARVARD UNIVERSITY

Boston San Francisco New York
London Toronto Sydney Tokyo Singapore Madrid
Mexico City Munich Paris Cape Town Hong Kong Montreal

Sponsoring Editor: Laurie Rosatone
Managing Editor: Karen Guardino
Project Editor: Ellen Keohane
Marketing Manager: Michael Boezi
Manufacturing Buyer: Evelyn Beaton
Associate Production Supervisor: Julie LaChance
Cover Design: Night and Day Design
Cover Art: The Japanese Bridge by Claude Monet; Suzuki Collection, Tokyo/Superstock
Interior Design: Sandra Rigney
Senior Designer: Barbara Atkinson
Composition: Windfall Software

Library of Congress Cataloging-in-Publication Data

Gottlieb, Robin (Robin Joan)
    Calculus: an integrated approach to functions and their rates of change / by Robin
    Gottlieb.—Preliminary ed.
      p. cm.
    ISBN 0-201-70929-5 (alk. paper)
    1. Calculus. I. Title.
QA303 .G685 2001   00-061855
515—dc21

1 2 3 4 5 6 7 8 9 10—CRS—04 03 02 01

To my family,
especially my grandmother,
Sonia Gottlieb.

PREFACE

The concepts of calculus are intriguing and powerful. Yet for a learner not fluent in the language of functions and their graphs, the learner arriving at the study of calculus poorly equipped, calculus may become a daunting hurdle rather than a fascinating exploration. The impetus to develop a course integrating calculus with material traditionally labeled "precalculus" emerged from years of working with a sequential college system.

Few students regard the prospect of taking a precalculus course as inspiring. In the eyes of many students it lacks the glamour and prestige of calculus. For some students, taking precalculus means retaking material "forgotten" from high school, and, bringing the same learning skills to the subject matter, such a student may easily "forget" again. At many colleges, students who successfully complete a precalculus course subsequently enroll in a calculus course that compresses into one semester what their better prepared fellow students have studied back in high school over the course of a full year. Yet any lack of success in such a course is bemoaned by teachers and students alike.

The idea behind an integrated course is to give ample time to the concepts of calculus, while also developing the students' notion of a function, increasing the students' facility in working with different types of functions, facilitating the accumulation of a robust set of problem-solving skills, and strengthening the students as learners of mathematics and science. An integrated course offers freedom, new possibilities, and an invigorating freshness of outlook. Freshness in particular is valuable for the student who has taken some precalculus (or even calculus) but come away without an understanding of its conceptual underpinnings.

This text grew out of an integrated calculus and precalculus course. Three general principles informed the creation of both the course and the text.

- Developing mathematical reasoning and problem-solving skills must not be made subservient to developing the subject matter.

- Making connections between mathematical ideas and representations and making connections between functions and the world around us are important to fostering a conceptual framework that will be both sturdy and portable.

- Generating intellectual excitement and a sense of the usefulness of the subject matter is important for both the students' short-term investment in learning and their long-term benefits.

Developing Mathematical Reasoning

Mathematical reasoning skills are developed by learning to make conjectures and convincing mathematical arguments. Mathematics, like any other language, must be spoken before being spoken well. Students initially need to learn to make mathematical arguments at their own level, whatever that level may be. The instructor can model logically convincing arguments, but if students are not given the opportunity to engage in discussion themselves, they are not likely to acquire discussion skills.

The structure of this text is intended to facilitate learning deductive and inductive reasoning, learning to use examples and counterexamples, and learning to understand the usefulness of a variety of perspectives in devising an argument. Students are encouraged to seek patterns and connections, to make conjectures and construct hypotheses. The reflective thinking this fosters helps students develop judgment and confidence.

Making Connections

Ideas are presented and discussed graphically, analytically, and numerically, as well as in words, with an emphasis throughout on the connections between different representations. There is an emphasis on visual representations. Topics are introduced through examples and, often, via applications and modeling, in order to build connections between mathematics, the students' experience outside mathematics, and problems in other disciplines, such as economics, biology, and physics.

Generating Enthusiasm

When the storylines of mathematics get buried under technicalities and carefully polished definitions, both those storylines and the enthusiasm new learners often bring to their studies may well be lost. When all the technical details and theory are laid out in full at the start, students may become lost and, not understanding the subtleties involved, simply suspend judgment and substitute rote memorization. The intrepid learner has more potential than the timid, self-doubting learner. For these reasons, answers to questions students are unready to ask are often omitted. Definitions may be given informally before they are provided formally; likewise, proofs may be given informally or not given in the body of the text but placed in an appendix. In this text, the presentation is not always linear, not all knots are tied immediately, and some loose ends are picked up later. The goal is to have students learn material and to have them keep concepts solidly in their minds, as opposed to setting the material out on paper in a neat and exhaustive form.

About the Problems

Problems are the heart of any mathematics text. They are the vehicles through which the learner engages with the material. Certainly, they consume the bulk of students' time and energy. A lot of class time can be constructively spent discussing problems as well. To do mathematics requires reflection, and discussion both encourages and enriches reflection.

The first 16 chapters offer "Exploratory Problems." These are integral to the text, and some are referred to in later sections. Exploratory problems can be incorporated into the course in many ways, but the bottom line remains that they need to be worked and discussed by students. Exploratory problems can be done as in-class group exercises, given as group homework problems, or given as homework to be discussed by the class during the following class meeting. Many of these problems combine or encourage different viewpoints and require the student to move between representations. Some exploratory problems call for conjectures that will subsequently be proven; some call for experimentation. The problems attempt to exercise and stretch mathematical reasoning and the process of discussing them is meant to constitute a common core experience of the class.

This preliminary edition includes many problems that require basic analytic manipulation. In the sense that mathematics is a language, these problems are analogous to vocabulary drills. They are exercises designed to support the less routine problems. But doing only these warm-up exercises would mean missing the spirit of the text and circumventing the goals laid out in this preface. Because problem solving involves determining which tools to use in a given situation, sometimes a few problems at the end of a section may best be solved by using tools from a previous section.

The text assumes that students have access to either a graphing calculator or a computer. Technology may be incorporated to a greater or lesser extent depending upon the philosophy and goals of the instructor.

Structuring the Content

The text covers the equivalent of a precalculus course plus one year of one-variable calculus. Parts I through VII meld precalculus and first-semester calculus. A yearlong course might cover Parts I through VIII and sections of Part IX, although the composition of the syllabus is, of course, at the instructor's discretion.

Part I provides an introduction to functions and their representations with an emphasis on the relationship between meaning and symbolic and graphic representations. From the outset the study of functions and the study of calculus are intertwined. For example, although the first set of exploratory problems requires no particular mathematical knowledge, the ensuing discussion inevitably involves the notion of relative rates of change. Similarly, in extracting information about velocity from a graph of position versus time, or extracting information about relative position from a graph of velocity versus time, students explore the relationship between a function and its derivative without being formally introduced to the derivative.

Part II focuses on rates of change and modeling using linear and quadratic functions. Linearity and interpretation of slope precede the derivative and its interpretation. Knowing about lines and the relationship between a function and its derivative provides a new window into quadratics. A chapter devoted to quadratics allows students to work through issues of sign and the relationship between a function and its graph as well as tackle optimization problems both with and without using calculus.

Traditionally applied optimization problems appear in a course after all of the formal symbolic derivative manipulations have been mastered. Taking on these problems incrementally permits the topic to be revisited multiple times. The most difficult aspect of optimization involves translating the problem into mathematics and expressing the quantity to be optimized as a function of a single variable. Part I, Chapter 1 and Part II, Chapters 4 and 6 address these skills.

Once students are able to appreciate the usefulness of computing derivatives, the notions of limits and continuity can be addressed more thoroughly in Chapter 7. Chapter 8 builds on that basis, revisiting the idea of local linearity and introducing the Product and Quotient Rules.

Part III introduces exponential functions through modeling. These functions are treated early on, because students of biology, chemistry, and economics need facility in dealing with them right away. The derivatives of exponentials are therefore discussed twice, first before the discussion of logarithms and then, more completely, after it. This order leaves some loose ends, but the subsequent resolution after a few weeks is quite satisfying and makes the natural logarithm seem natural. The number e is introduced as the base for which the derivative of b^x is b^x. Part III also takes up polynomials and optimization.

Part IV deals more fully with logarithmic and exponential functions and their derivatives. The number e is revisited here. By design, the Chain Rule is delayed until after the differentiation of the exponential and logarithmic functions. Differentiating these functions without the Chain Rule gives students a lot of practice with logarithmic and exponential manipulations. For instance, to differentiate $\ln(3x^7)$ the student must rewrite the expression as $\ln 3 + 7 \ln x$. Chapter 15, which introduces differential equations via the exponential function, can be postponed if the instructor prefers.

Part V revisits differentiation by addressing the Chain Rule and implicit differentiation.

Part VI provides an excursion into geometric sums and geometric series. A mobile chapter, it can easily be postponed to immediately precede Part X, on series. If, however, the class includes students of economics and biology who will not necessarily study Taylor series, then the students will be well served by studying geometric series. Part VI emphasizes modeling, using examples predominantly drawn from pharmacology and finance.

Part VII presents the trigonometric functions, inverse trigonometric functions, and their derivatives. From a practical point of view, this order means that trigonometry is pushed to the second semester. The rationale is twofold. First, some traditional precalculus material must be delayed to make room for the bulk of differential calculus in the first semester. Second, delaying trigonometry has the benefit of returning students to the basics of differential calculus. Too often in a standard calculus course students think about what a derivative is only at the beginning of the course, but by mid-term they are thinking of a derivative as a formula. This text looks at the derivative of $\sin t$ from graphic, numeric, analytic, and modeling viewpoints. By the time students have reached Part VII, they are more sophisticated and can follow the more complicated analytic derivations, if the instructor chooses to emphasize them. Delaying trigonometry presents the opportunity to revisit applications previously studied. Students should now have enough confidence to understand that the basic properties of trigonometric functions can be easily derived from the unit circle definitions of sine and cosine; they will be capable of retrieving information forgotten or learning it for the first time without being overwhelmed by detail.

Part VIII introduces integration and the Fundamental Theorem of Calculus. There is a geometric flavor to this set of chapters, as well as an emphasis on interpreting the definite integral. Part IX discusses applications and computation of the definite integral, with an emphasis on the notion of slicing, approximating, and summing.

Part X focuses on polynomial approximations of functions and Taylor series. (Convergence issues are first brought up in Part VI, in the context of a discussion of geometric series.) In Part X the discussion of polynomial approximations motivates the subsequent series discussion.

Differential equations are the topic of Part XI. Although the emphasis is on modeling and qualitative behavior, students working through the chapter will come out able to solve separable first order differential equations and second order differential equations with constant coefficients, and they will understand the idea behind Euler's method. Some discussion of systems of differential equations is also included.

The first few sections of Chapter 31 may easily be moved up to follow an introduction to integration and thereby be included at the end of a one-year course. Ending the year with these sections reinforces the basic ideas of differential calculus while simultaneously introducing an important new topic.

Certain sections have been made into appendices in order to give the instructor freedom to insert them (or omit them) where they see fit, as determined by the particular goals of the course. Sections on algebra, the theoretical basis of calculus, including Rolle's Theorem and the Mean Value Theorem, induction, conics, l'Hôpital's Rule for using derivatives to evaluate limits of an indeterminate form, and Newton's method of using derivatives to approximate roots constitute Appendices A, C, D, E, F, and G, respectively. Certain appendices can be transported directly into the course. Others can be used as the basis of independent student projects.

This book is a preliminary edition and should be viewed as a work in progress. The exposition and choice and sequencing of topics have evolved over the years and will, I expect, continue to evolve. I welcome instructors' and students' comments and suggestions on this edition. I can be contacted at the addresses given below.

Robin Gottlieb
Department of Mathematics
1 Oxford Street
Cambridge, MA 02138
gottlieb@math.harvard.edu

Acknowledgments

A work in progress incurs many debts. I truly appreciate the good humor that participants have shown while working with an evolving course and text. For its progress to this point I'd like to thank all my students and all my fellow instructors and course assistants for their feedback, cooperation, help, and enthusiasm. They include Kevin Oden, Eric Brussel, Eric Towne, Joseph Harris, Andrew Engelward, Esther Silberstein, Ann Ryu, Peter Gilchrist, Tamara Lefcourt, Luke Hunsberger, Otto Bretscher, Matthew Leerberg, Jason Sunderson, Jeanie Yoon, Dakota Pippins, Ambrose Huang, and Barbara Damianic. Special thanks to Eric Towne, without whose help writing course notes in the academic year 1996-1997 this text would not exist. Special thanks also to Eric Brussel whose support for the project has been invaluable, and Peter Gilchrist whose help this past summer was instrumental in getting this preliminary edition ready. Thanks to Matt Leingang and Oliver Knill for technical assistance, to Janine Clookey and Esther Silberstein for start-up assistance, and to everyone in the Harvard Mathematics department for enabling me to work on this book over these past years.

I also want to acknowledge the type-setting assistance of Paul Anagnostopoulos, Renata D'Arcangelo, Daniel Larson, Eleanor Williams, and numerous others. For the art, I'd like to acknowledge the work of George Nichols, and also of Ben Stephens and Huan Yang. For their work on solutions, thanks go to Peter Gilchrist, Boris Khentov, Dave Marlow, and Sean Owen and coworkers.

My thanks to the team at Addison-Wesley for accepting the assortment of materials they were given and carrying out the Herculean task of turning it into a book, especially to Laurie Rosatone for her encouragement and confidence in the project and Ellen Keohane for her assistance and coordination efforts. It has been a special pleasure to work with Julie LaChance in production; I appreciate her effort and support. Thanks also to Joe Vetere, Caroline Fell, Karen Guardino, Sara Anderson, Michael Boezi, Susan Laferriere, and Barbara Atkinson. And thanks to Elka Block and Frank Purcell, for their comments and suggestions.

Finally, I want to thank the following people who reviewed this preliminary edition:

Dashan Fan, University of Wisconsin, Milwaukee
Baxter Johns, Baylor University
Michael Moses, George Washington University

Peter Philliou, Northeastern University

Carol S. Schumacher, Kenyon College

Eugene Spiegel, The University of Connecticut

Robert Stein, California State University, San Bernardino

James A. Walsh, Oberlin College

To the Student

This text has multiple goals. To begin with, you should learn calculus. Your understanding should be deep; you ought to feel it in your bones. Your understanding should be portable; you ought to be able to take it with you and apply it in a variety of contexts. Mathematicians find mathematics exciting and beautiful, and this book may, I hope, provide you with a window through which to see, appreciate, and even come to share this excitement.

In some sense mathematics is a language—a way to communicate. You can think of some of your mathematics work as a language lab. Learning any language requires active practice; it requires drill; it requires expressing your own thoughts in that language. But mathematics is more than simply a language. Mathematics is born from inquiry. New mathematics arises from problem solving and from pushing out the boundaries of what is known. Questioning leads to the expansion of knowledge; it is the heart of academic pursuit. From one question springs a host of other questions. Like a branching road, a single inquiry can lead down multiple paths. A path may meander, may lead to a dead-end, detour into fascinating terrain, or steer a straight course toward your destination. The art of questioning, coupled with some good, all-purpose problem-solving skills, may be more important than any neatly packaged set of facts you have tucked under your arm as you stroll away from your studies at the end of the year. For this reason, the text is not a crisp, neatly packed and ironed set of facts. But because you will want to carry away something you can use for reference in the future, this book will supply some concise summaries of the conclusions reached as a result of the investigations in it.

We, the author and your instructors, would like you to leave the course equipped with a toolbox of problem-solving skills and strategies—skills and strategies that you have tried and tested throughout the year. We encourage you to break down the complex problems you tackle into a sequence of simpler pieces that can be put together to construct a solution. We urge you to try out your solutions in simple concrete cases and to use numerical, graphical, and analytic methods to investigate problems. We ask that you think about the answers you get, compare them with what you expect, and decide whether your answers are reasonable.

Many students will use the mathematics learned in this text in the context of another discipline: biology, medicine, environmental science, physics, chemistry, economics, or one of the social sciences. Therefore, the text offers quite a bit of mathematical modeling—working in the interface between mathematics and other disciplines. Sometimes modeling is treated as an application of mathematics developed, but frequently practical problems from other disciplines provide the questions that lead to the development of mathematical ideas and tools.

To learn mathematics successfully you need to actively involve yourself in your studies and work thoughtfully on problems. To do otherwise would be like trying to learn to be a good swimmer without getting in the water. Of course, you'll need problems to work on. But you're in luck; you have a slew of them in front of you. Enjoy!

CONTENTS

Preface   vii

PART V | Adding Sophistication to Your Differentiation   513

PART VI | An Excursion into Geometric Series   559

Appendices   1051

C H A P T E R

1

Functions Are Lurking Everywhere

■ 1.1 FUNCTIONS ARE EVERYWHERE

Each of us attempts to make sense out of his or her environment; this is a fundamental human endeavor. We think about the variables characterizing our world; we measure these variables and observe how one variable affects another. For instance, a child, in his first years of life, names and categorizes objects, people, and sensations and looks for predictable relationships. As a child discovers that a certain phenomenon precipitates a predictable outcome, the child learns. The child learns that the position of a switch determines whether a lamp is on or off, and that the position of a faucet determines the flow of water into a sink. The novice musician learns that hitting a piano key produces a note, and that *which* key is hit determines *which* note is heard. The deterministic relationship between the piano key hit and the resulting note is characteristic of the input-output relationship that is the object of our study in this first chapter.

Mathematical modeling involves constructing mathematical machines that mimic important characteristics of commonly occurring phenomena. Chemists, biologists, environmental scientists, economists, physicists, engineers, computer scientists, students, and parents all search for relationships between measurable variables.[1] A chemist might be interested in the relationship between the temperature and the pressure of a gas, an environmental scientist in the relationship between use of pesticides and mortality rate of songbirds, a physician in the relationship between the radius of a blood vessel and blood pressure, an economist in the relationship between the quantity of an item purchased and its price, a grant manager in the relationship between funds allocated to a program and results achieved. A thermometer manufacturer must know the relationship between the temperature and the volume of a gram of mercury in order to calibrate a thermometer. The list is endless. As human beings trying to make sense of a complex world, we instinctively try to identify relationships between variables.

We will concern ourselves here with relationships that can be structured as input-output relationships with the special characteristic that the input completely determines the output. For example, consider the relationship between the temperature and the volume of a gram of mercury. We can structure this relationship by considering the input variable to be temperature and the output variable to be volume. A specific temperature is the input; the output is the volume of one gram of mercury at that specific temperature. The temperature determines the volume. As another example, consider a hot-drink machine. If your inputs are inserting a dollar bill and pressing the button labeled "hot chocolate," the output will be a cup of hot cocoa and 55¢ in change. In such a machine the input completely determines the output. The mathematical machine used to model such relations is called a function.

Mathematicians define a function as a relationship of inputs and outputs in which each input is associated with exactly one output. Notice that the mathematical use of the word "function" and its use in colloquial English are not identical. In colloquial English we might say, "The number of hours it takes to drive from Boston to New York City is a function of the time one departs Boston." By this we mean that the trip length *depends on* the time of departure. But the trip length is not *uniquely determined by* the departure time; holiday traffic, accidents, and road construction play roles. Therefore, in a mathematical sense the length of the trip is *not* a function of the departure time.

Think about the task of calibrating a bottle, marking it so that it can subsequently be used for measuring. The calibration function takes a volume as input and gives a height as output. For any particular bottle we can say that the height of the liquid in the bottle is a function of the volume of the liquid; that is, height (output) is completely determined by volume (input). We use pictures to illustrate the relationship. Tracking the input variable along the horizontal axis and the output variable along the vertical axis is a mathematical convention for displaying graphs of functions.

---

[1] While physicists hope to uncover physical laws, economists and other social scientists often aim for some working understanding that can be applied appropriately.

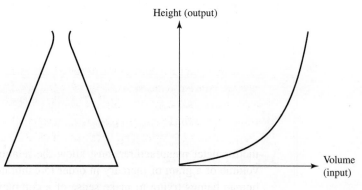

Figure 1.1   Calibration of conical flask

The concept of a function is important and versatile. We can attach many different mental pictures and representations to it; our choice of representation will often depend upon context. Upon first exposure, the notation and representations may be confusing, but as you use the language of functions the nuances will become as natural as nuances in the English language.

Aside: In any discipline accurate communication is critical; in mathematics a great deal of care goes into definitions because the field is structured so that the validity of arguments rests on logic, definitions, and a few postulates. Definitions arise because they are needed in order to make precise, unambiguous statements. Sometimes a formal definition can initially seem unnatural to you, usually because the person who made the definition wanted to be sure to include (or exclude) a certain situation that hasn't yet crossed your mind. To circumvent this problem, in this text we will sometimes begin with an informal definition and formalize it later.

Just in case you are not kindly disposed to learning the language of functions, we'll take a brief foray into English word usage to put mathematical language in perspective. Consider the word "subway," meaning an underground railway. The word conveys a meaning; we associate it with the physical object. A Bostonian, a New Yorker, and a Tokyo commuter might each have a slightly different mental image—but the essence is similar. We have convenient shorthand notations for subway. In Boston, people refer to the subway as the "T." If there is a "T" symbol with stairs going downward, that indicates a subway stop; if you see a "T" symbol on a street sign, with no stairs in sight, chances are that you're at a bus stop. The symbol takes on a life of its own when a Bostonian says, "I'll take the 'T' downtown." On the other hand, New Yorkers look for an "M" (for Metropolitan Transit Authority) when they want to find a subway. But a New Yorker never says, "I'll take the 'M' downtown." To make matters muddier, in London a subway refers to an underground walkway, while the underground rail is referred to as "the tube." Adding to the general zaniness of usage is a chain of "subway shops" selling submarine (hero, or grinder) sandwiches. And if you think you can clarify everything by switching to the term "underground railroad," think again: Harriet Tubman's underground railroad was something altogether different. Compared with this murky tangle, mathematical notation and usage may provide lucid relief.

Exploratory Problems for Chapter 1

Calibrating Bottles

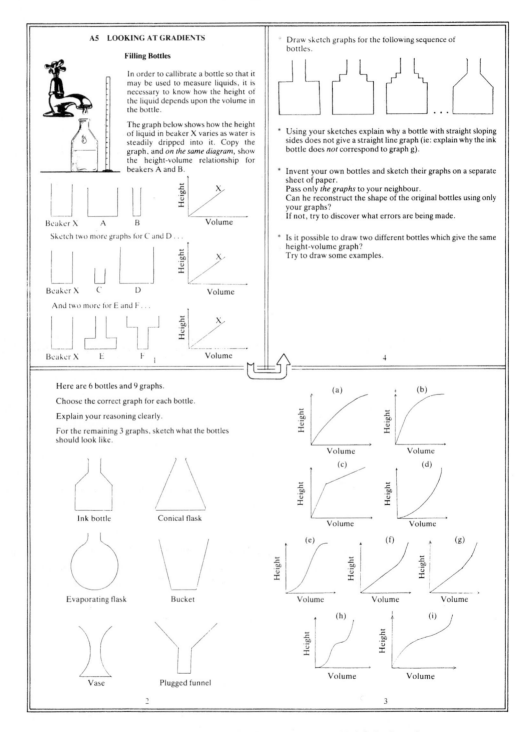

A5   LOOKING AT GRADIENTS

Filling Bottles

In order to callibrate a bottle so that it may be used to measure liquids, it is necessary to know how the height of the liquid depends upon the volume in the bottle.

The graph below shows how the height of liquid in beaker X varies as water is steadily dripped into it. Copy the graph, and *on the same diagram,* show the height-volume relationship for beakers A and B.

Beaker X   A   B

Sketch two more graphs for C and D . . .

Beaker X   C   D

And two more for E and F . . .

Beaker X   E   F

Draw sketch graphs for the following sequence of bottles.

* Using your sketches explain why a bottle with straight sloping sides does not give a straight line graph (ie: explain why the ink bottle does *not* correspond to graph g).

* Invent your own bottles and sketch their graphs on a separate sheet of paper.
Pass only *the graphs* to your neighbour.
Can he reconstruct the shape of the original bottles using only your graphs?
If not, try to discover what errors are being made.

* Is it possible to draw two different bottles which give the same height-volume graph?
Try to draw some examples.

Here are 6 bottles and 9 graphs.

Choose the correct graph for each bottle.

Explain your reasoning clearly.

For the remaining 3 graphs, sketch what the bottles should look like.

Ink bottle   Conical flask

Evaporating flask   Bucket

Vase   Plugged funnel

From *The Language of Functions and Graphs: An Examination Module for Secondary Schools,* 1985, Shell Centre for Mathematical Education.

■ 1.2   WHAT ARE FUNCTIONS? BASIC VOCABULARY AND NOTATION

◆ **EXAMPLE 1.1**   The following table describes three possible designs for soda machines.

Machine A			*Machine B*			*Machine C*	
Button #	**Output**		**Button #**	**Output**		**Button #**	**Output**
1	Coke		1	Coke		1	Coke/Sprite
2	Diet Coke		2	Coke		2	Orange Crush/Diet Coke
3	Sprite		3	Diet Coke		3	Ginger Ale
4	Orange Crush		4	Diet Coke		4	Fresca
5	Ginger Ale		5	Coke			
			6	Coke			

You've seen machines similar to machines A and B, but you've probably never seen one similar to machine C. If you press button #1 of machine C you don't know whether you'll be getting a Coke or a Sprite; your input does not uniquely determine the output of the machine.   ◆

D e f i n i t i o n s

■   A **function** is an input-output relationship with the characteristic that for each acceptable input there is exactly one corresponding output. In other words, the input completely determines the output. The input variable is referred to as the **independent variable**. The output generally depends on the input, and therefore the output variable is called the **dependent variable**.[2]

■   The **domain** of a function is the set of all acceptable inputs.

■   The **range** of a function is the set of all possible outputs.

Functions are typically given short names, like f or g.[3] We typically call our generic function f (for function). For each input, x, in the domain of the function there is a corresponding output. This output is called **the value of the function at x** and is denoted $f(x)$. Using set language, we say that a function f assigns to each element in its domain a specific element in its range.

---

[2] We use the word "generally" because it is possible that all inputs result in the same output, i.e., that the function is a constant function. (A soda machine may be stocked only with Ginger Ale so that no matter what button you press—input— you'll get the same output).

[3] Some functions have longer names, like the cosine function, with the tag cos, or, to be more esoteric, the hyperbolic cosine function, with the tag cosh.

Mental Picture:
Function as a Machine

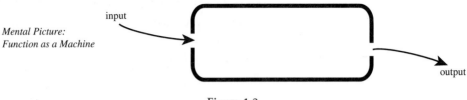

input

output

Figure 1.2

You can think of a function as a machine (or box) that accepts an input and produces a predictable outcome. By "predictable outcome" we mean that if we input "button #2" into the machine and it produces a Diet Coke, we can predict, with certainty, that whenever we press button #2 the output will be a Diet Coke.

- Machine A can be modeled by a function because the button pressed (the input) completely determines the soda received (the output). This machine has an additional feature—each button corresponds to a different soda. There is a one-to-one correspondence between inputs and outputs. A function having this characteristic is called **1-to-1**. Notice that for machine A if you know the type of soda (the output), you know precisely which button was pressed (the corresponding input).

- Machine B can be modeled by a function because the button pressed (the input) completely determines the soda received (the output). The fact that more than one button will give the same type of soda does *not* prevent this relationship from being a function; it simply means that the function is not 1-to-1. When using machine B, if you know the type of soda (the output), you cannot determine which button was pressed.

- Machine C cannot be represented by a function because some buttons do not give a unique output. Selecting button #1 sometimes results in a can of Coke and other times in a can of Sprite. The input does not uniquely determine the output.

Let's make diagrammatic models of the three machines. We'll call the mappings of inputs to outputs A, B, and C, corresponding to machines A, B, and C, respectively. Assign button #1 the number 1, button #2 the number 2, etc. We'll use the first letter (small case) of each type of soda to represent the soda type, a convention that works because no two sodas listed begin with the same letter. Then we can represent the input-output relationships (mappings) as follows:

A	B	C
$1 \longrightarrow c$	$1 \longrightarrow c$	$1 \longrightarrow c$ or s
$2 \longrightarrow d$	$2 \longrightarrow c$	$2 \longrightarrow o$ or d
$3 \longrightarrow s$	$3 \longrightarrow d$	$3 \longrightarrow g$
$4 \longrightarrow o$	$4 \longrightarrow d$	$4 \longrightarrow f$
$5 \longrightarrow g$	$5 \longrightarrow c$	
	$6 \longrightarrow c$	

A is a function with domain $\{1, 2, 3, 4, 5\}$ and range $\{c, d, g, o, s\}$.[4]

B is a function with domain $\{1, 2, 3, 4, 5, 6\}$ and range $\{c, d\}$.

C is not a function.

To emphasize that mapping A is 1-to-1, mapping B is a function but not 1-to-1, and mapping C is not a function, we can represent the maps as follows:

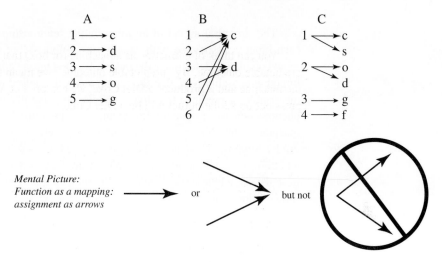

Figure 1.3

You can think of a function as a mapping or association rule with no "split arrows" allowed. Multiple arrows can hit the same target; this simply indicates that the function is not 1-to-1.

EXERCISE 1.1   Which of the following rules is a function? Are any of the functions 1-to-1?

Q	R	S
$0 \longrightarrow 13$	$0 \longrightarrow \pi$	$0 \longrightarrow 0$
$1 \longrightarrow 14$	$1 \longrightarrow \pi$	$1 \longrightarrow \pm 1$
$3 \longrightarrow 16$	$3 \longrightarrow \pi$	$3 \longrightarrow \pm 3$
$6 \longrightarrow 11$	$6 \longrightarrow \pi$	$6 \longrightarrow \pm 8$

Answer

Q and R are both functions, since every number in the domain has a unique output assigned to it; for each input there is one and only one output. S is not a function, because the output is not uniquely determined by the input. Q is 1-to-1; R is not 1-to-1.

CAVEAT It is perfectly fine for many different inputs of a function to be assigned the same output. In fact, *all* inputs may be assigned to the same output, as illustrated by mapping R in the previous exercise. The "every-student-is-an-honors-student" rule, H, that assigns

---

[4] Curly brackets, { }, are used to enclose items in a set.

to each student in our class the grade of "*A*" is a function. *H* is a function because *H* associates with each input (student) a single output (grade). Like the mapping *R* in the previous exercise, it is a constant function. Likewise, if the Red Queen's procedure after her tea party is to assign to every guest the "off-with-his-head" command, her procedure can be modeled by a constant function, albeit a grisly one. A **constant function** is a function whose range consists of only one element. The output is fixed; it does not vary with the input.

The structural aspect of an input-output relationship that makes it a function is that each input is associated with exactly one output. More casually, we could say that the input completely determines the output. By the phrase "completely determines" we are *not* implying causality (i.e., that the input *causes* the output to occur), nor are we implying a discernible or predictable pattern. (While we generally *look* for patterns, sometimes the phenomenon we are examining exhibits no pattern.) For example, let's return to drink machine A, but this time cover up the key that tells us which buttons correspond to which drinks. Suppose we've never been to this machine before. We put in our money and press a button; we obtain a drink as output. Were we to put in the same amount of money and push the same button, we would get the same output. Before seeing the output for the first time we don't know what it will be—but we do know that it is completely determined by the input. Although we cannot predict the output before using the machine, the machine operates like a function. Here the key is hidden.

Now consider the input-output relationship that takes as input a date (day, month, year) and gives as output the highest temperature on the top of the Prudential Center, a skyscraper in downtown Boston. Is it a function? There is no causality between the date and the temperature, nor is there a formula to determine the temperature from the date, and yet, because each date is associated with exactly one high temperature, this *is* a function. We say that the temperature is a function of the date. Conversely, the date is *not* a function of the temperature. Why not?

Functional Notation

We use a very efficient shorthand notation to give information about the input-output relationship of a function. Let's work with the function *Q* from Exercise 1.1. Suppose we want to indicate that the function *Q* assigns to the input of 3 the output of 16. We write

$$Q(3) = 16$$

name of
function   input   output

We read this aloud as "*Q* of 3 is 16."[5]

---

[5] You can think of the notation $Q(3) = 16$ using the mental model of a function machine. Q is the function machine that has received the input 3 and spits out the output 16. To be anthropomorphic about it, think of the set of parentheses ( ) as a pair of hands cupped to receive the input. The output spurts out of the equal sign.

Definition

The output of a function f for a particular input is called the **value** of f at that input.

The following statements are equivalent.

- The function f associates with the input 2 the output 7.

- The *value* of f at 2 is 7.

- $2 \overset{f}{\mapsto} 7$

- $f(2) = 7$

The expression $y = f(x)$ is shorthand for "y is the output of f corresponding to an input of x" and it is read "y equals f of x." When we say $y = f(x)$, we mean that y depends on x and is uniquely determined by x.

◆ **EXAMPLE 1.2**   *Functional notation and the associated meaning.* The Bee Line Trail in the White Mountains of New Hampshire is a hiking trail that goes directly up to the summit of Mount Chocorua. Let $T(a)$ give the temperature (in degrees Fahrenheit) as a function of the altitude a (in feet above sea level) at a particular time on a particular day along the Bee Line Trail.

$$\text{altitude} \xrightarrow{T} \text{temperature}$$

Suppose that right now you are at an altitude of H feet above sea level and that temperature is a function of altitude.

Answer the questions that follow.

Q: What is the temperature at your altitude?

A: Use your altitude, H, as the input of the function T; $T(H)$ is the temperature.

Q: What is the temperature if you double your current altitude?

A: Use the input $2H$, twice your current altitude; the temperature will be $T(2H)$.

Q: What is the temperature if you double your original altitude and then ascend 1000 more feet up the mountain?

A: Use the input $2H + 1000$, which is 1000 feet above an altitude of $2H$; the temperature will be $T(2H + 1000)$.

Q: Interpret the meanings of $T\left(\frac{H}{2}\right)$ and $\frac{T(H)}{2}$.

A: $T\left(\frac{H}{2}\right)$ is the temperature at one-half of your current altitude; the *input* (the altitude) has been halved. $\frac{T(H)}{2}$ is one-half of the temperature at your current altitude; the *output* (temperature) has been halved. Notice that in the first expression the new altitude is specified, and in the second expression the new temperature is specified. Also note that the two expressions will generally not be equal. If it is 60 degrees at 2000 feet, then it probably won't be 30 degrees at 1000 feet.

Q: Interpret $T(H + 10)$ and $T(H) + 10$.

A: $T(H + 10)$ is the temperature at an altitude 10 feet higher than your current altitude. $T(H) + 10$ is a temperature 10 degrees higher than your current temperature. Notice that the 10 inside the parentheses represents 10 more feet while the 10 *outside* the parentheses represents 10 more degrees.

Q: At what altitude will the temperature be 40 degrees?

A: Solve the equation $T(a) = 40$ for a. (There may not be a value for a that satisfies this equation. At the moment in question it is possible that at no height on the mountain is the temperature 40 degrees.)

Q: How high must you go for the temperature to drop 10 degrees?

A: The temperature at your current height is $T(H)$ degrees, thus we need to find the altitude that gives a temperature of $T(H) - 10$ degrees. This means solving for a in the equation $T(a) = T(H) - 10$. (*Note:* H is a constant and a is the variable.)   ◆

PROBLEMS FOR SECTION 1.2

1. Below are pictures of three different train trips. In each case, the train is far enough from its station of origin to have reached a steady cruising speed. One picture corresponds to a trip on Amtrak in America; the second corresponds to a train trip in the People's Republic of China, where trains have a tendency to be slow and train travel is a very leisurely affair; and the third corresponds to a trip on the Japanese bullet train (*shinkansen*), a very fast train. Which picture corresponds to which trip?

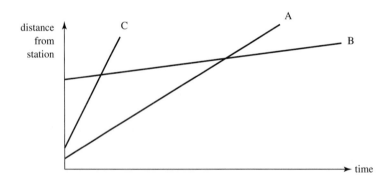

2. Which of the following rules can be modeled as a function? If a rule is not a function, explain why not.

   (a) For a particular flask, the rule assigns to every volume (input) of liquid in the flask the corresponding height (output).

   (b) For a particular flask, the rule assigns to every height (input) the corresponding volume (output).

   (c) The rule assigns to every person his or her birthday.

   (d) The rule assigns to every recorded singer the title of his or her first recorded song.

   (e) The rule assigns to every state the current representative in the House of Representatives.

   (f) The rule assigns to every current member of the House of Representatives the state he or she represents.

(g) The rule assigns to every number the square of that number.

(h) The rule assigns to every nonzero number the reciprocal of that number.

3. (a) Which of the following maps are maps of functions?

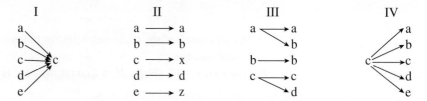

(b) For each function, determine the range and the domain.

4. Which of the functions in Problem 3 are 1-to-1?

5. There are infinitely many prime numbers. This has been known for a long time; Euclid proved it sometime between 300 B.C. and 200 B.C.[6] Number theorists (mathematicians who study the theory and properties of numbers) are interested in the distribution of prime numbers. Let $P(n) =$ number of primes less than or equal to n, where n is a positive number. Is $P(n)$ a function? Explain.

6. *Writing:* We would like to tailor this course to your needs and interests; therefore we'd like to find out more about what these needs and interests are. On a sheet of paper *separate* from the rest of your homework, please write a paragraph or two telling us a bit about yourself by addressing the following questions.

   (a) What are you interested in studying in the future, both in terms of math and otherwise?

   (b) Are there things you have found difficult or confusing in mathematics in the past? If so, what?

   (c) What was your approach to studying mathematics in the past? Did it work well for you?

   (d) What are your major extracurricular activities or interests?

   (e) What do you hope to get out of this course?

For Problems 7 through 9 determine whether the relationship described is a function. If the relationship is a function,

(a) what is the domain? the range?

(b) is the function 1-to-1?

---

[6] The proof went like this: Assume there are finitely many primes: $2, 3, 5, 7 \ldots, N$, where N is the largest prime. Let $M = 2 \cdot 3 \cdot 5 \cdot 7 \cdot \ldots N + 1$; i.e., let M be 1 bigger than the product of *all* the primes. Then M is not divisible by any of the primes on the list. But that means the list doesn't include all of the prime numbers. Whoops! We have a contradiction. Therefore, our assumption that there are finitely many primes is false. So there must be infinitely many primes.

7. Input	Output
0	2
1	3
2	2
3	3
4	2

8. Input	Output
$\sqrt{2}$	2
$\sqrt{3}$	3
$\sqrt{5}$	5
$\sqrt{6}$	6

9. Input	Output
$\sqrt{2}$	0
$2\sqrt{2}$	0
$3\sqrt{2}$	0
$4\sqrt{2}$	0

10. Express each of the following rules for obtaining the output of a function using functional notation.

   (a) Square the input, add 3, and take the square root of the result.

   (b) Double the input, then add 7.

   (c) Take half of 3 less than the input.

   (d) Increase the input by 10, then cube the result.

11. Let C be a circle of radius 1 and let $A(n)$ be the area of a regular n-gon inscribed in the circle. For instance, $A(3)$ is the area of an equilateral triangle inscribed in circle C, $A(4)$ is the area of a square inscribed in circle C, and $A(5)$ is the area of a regular pentagon inscribed in circle C. (A polygon inscribed in a circle has all its vertices lying on the circle. A regular polygon is a polygon whose sides are all of equal length and whose angles are all of equal measure.)

   (a) Find $A(4)$.

   (b) Is $A(n)$ a function? If it is, answer the questions that follow.

   (c) What is the natural domain of $A(n)$?

   (d) As n increases, do you think that $A(n)$ increases, or decreases? This is hard to justify rigorously, but what does your intuition tell you?

   (e) Will $A(n)$ increase without bound as n increases, or is there a lid above which the values of $A(n)$ will never go? If there is such a lid (called an upper bound) give one. What is the smallest lid possible? Rigorous justification is not requested.

12. Some friends are taking a long car trip. They are traveling east on Route 66 from Flagstaff, Arizona, through New Mexico and Texas and into Oklahoma.

   Let f be the function that gives the number of miles traveled t hours into the trip, where $t = 0$ denotes the beginning of the trip. For instance, $f(7)$ is the mileage 7 hours into the trip. If the travelers set an odometer to zero at the start of the trip, the output of f would be the reading on the odometer.

   Let g be the function that gives the car's speed t hours into the trip, where $t = 0$ denotes the beginning of the trip. For instance, $g(7)$ is the car's speed 7 hours into the trip. The output of g corresponds to the speedometer reading.

   Suppose they pass a sign that reads "entering Gallup, New Mexico," h hours into the trip.

   (a) Write the following expressions using functional notation wherever appropriate.
      i. The car's speed 1 hour before reaching Gallup
      ii. 10 miles per hour slower than the speed of the car entering Gallup
      iii. Half the time it took to reach Gallup

iv. Their speed 6 hours after reaching Gallup

v. The distance traveled in the first 2 hours of the trip

vi. The distance traveled in the second 2 hours of the trip

vii. Half the distance covered in the second 3 hours of travel

viii. The average speed in the first 5 hours of travel (Average speed is computed by dividing the distance traveled by the time elapsed.)

ix. The average speed between hour 6 of the trip and hour 12 of the trip

(b) Interpret the following in words.

i. $f(h+2)$

ii. $\frac{1}{2}f(h)$

iii. $f(\frac{h}{2})$

iv. $f(h-2)$

v. $f(h)-2$

vi. $f(h)+2$

vii. $g(h+2)$

viii. $g(h)+2$

ix. $g(h)-2$

x. $\frac{1}{2}g(h)$

xi. $\frac{1}{2}g(h-1)$

13. Let $C(w)$ be the amount (in dollars) it costs you to mail your grandmother a first-class package weighing w ounces. Suppose you just mailed her a birthday present that weighed A ounces. Describe in words the practical meaning of each of the following expressions.

(a) $C(A)$ (b) $C(2A)$ (c) $2C(A)$ (d) $C(A+1)$ (e) $C(A)+1$

14. If $f(x) = \sqrt{\frac{1}{x+1}}$, find the following. Simplify your answer where possible.

(a) $f(0)$ (b) $f(3)$ (c) $f(-\frac{1}{4})$ (d) $f(b)$ (e) $f(b-1)$

(f) $f(b+3)$ (g) $[f(7)]^2$ (h) $f(b^2)$ (i) $[f(b)]^2$

15. If $g(x) = \frac{\sqrt{x^2+4}}{2}$, find the following. Simplify your answer where possible.

(a) $g(0)$ (b) $g(2)$ (c) $g(\sqrt{5})$ (d) $g\left(\frac{1}{\sqrt{2}}\right)$ (e) $-g(3t)$ (f) $g(\sqrt{t-4})$

16. If $h(x) = \frac{x^2}{1-2x}$, find

(a) $h(0)$ (b) $h(3)$ (c) $h(p+1)$ (d) $h(3p)$ (e) $2h(3p)$ (f) $\frac{1}{h(2p)}$

17. If $j(x) = 3x^2 - 2x + 1$, find the following. Simplify your answer where possible.

(a) $j(0)$ (b) $j(1)$ (c) $j(-1)$ (d) $j(-x)$ (e) $j(x+2)$ (f) $3j(x)$

(g) $j(3x)$

18. If $P(x) = 5 - 2x$, find the following.

(a) $P(1)$ (b) $P(-1)$ (c) $P(2W)$ (d) $P(2W+1)$ (e) $P(W^2)$

(f) $[P(W)]^2$ (g) $P(W^2+1)$

In Problems 19 through 21

(a) find the value of the function at $x = 0$, $x = 1$, and $x = -1$.

(b) find all x such that the value of the function is (i) 0, (ii) 1, and (iii) -1.

19. $f(x) = \frac{3x+5}{2}$

20. $g(x) = x^2 - 1$

21. $h(x) = \frac{x^2+2x}{3x+2}$
    (For a review of quadratic equations, refer to the Algebra Appendix.)

For each function in Problems 22 through 27, determine the largest possible domain.

22. (a) $f(x) = \frac{1}{5x+10}$ (b) $g(x) = \sqrt{5x+10}$

23. (a) $f(x) = \frac{1}{x^2-1}$ (b) $g(x) = \sqrt{x^2-1}$
    For part (b), factor the quadratic. The product must be positive. For more assistance, refer to the Algebra Appendix.

24. (a) $f(x) = \frac{3}{x^2+3x-4}$ (b) $g(x) = \sqrt{x^2+3x-4}$
    Factoring will help clarify the solution.

25. (a) $f(x) = \frac{1}{x^2+2x+1}$ (b) $g(x) = \sqrt{x^2+2x+1}$
    Factoring will help clarify the solution.

26. (a) $f(x) = \frac{1}{x+2} - \frac{1}{x-1}$ (b) $g(x) = \sqrt{x+2} - \sqrt{x-1}$

27. (a) $f(x) = \frac{3}{x} + \frac{2}{3-x} - \frac{x}{2x+2}$ (b) $g(x) = \sqrt{x} + 2\sqrt{3-x}$

28. The volume of a sphere and the surface area of a sphere are both functions of the sphere's radius. The volume function is given by $V(r) = \frac{4}{3}\pi r^3$ and the surface area function is given by $S(r) = 4\pi r^2$.

    (a) If the radius of a sphere is doubled, by what factor is the volume multiplied? The surface area?

    (b) Which results in a larger increase in surface area: increasing the radius of a sphere by 1 unit or increasing the surface area by 12 units? Does the answer depend upon the original radius of the sphere? Explain your reasoning completely. (It may be useful to check your answer in a specific case as a spot check for errors.)

    (c) In order to double the surface area of the sphere, by what factor must the radius be multiplied?

    (d) In order to double the volume of the sphere, by what factor must the radius be multiplied?

29. Let $A = A(S)$ be the area function for a square of side S. A takes as input the length S of the side of a square and gives as output the area of the square.

    (a) Find the following.
        i. $A(4)$ ii. $A(W)$ iii. $A(\sqrt{2}+3)$ iv. $A(4+h)$ v. $A(x-1)$

(b) Suppose that S is bigger than 1. Which is larger, $A(S - 1)$ or $A(S) - 1$? (Does the answer depend on the size of S? If so, how?)

(c) Explain in words the difference between $A(S - 1)$ and $A(S) - 1$. Which one of the expressions corresponds to the shaded area in the accompanying figure? What area corresponds to the other expression?

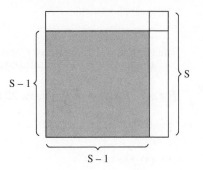

1.3 REPRESENTATIONS OF FUNCTIONS

Presenting a Function Using a Table

◆ EXAMPLE 1.3 Let N be the function that assigns to each year the number of applicants to Harvard College for the fall of that year. N is a function, because each input completely determines the output. For a restricted domain, this function can be expressed in the form of a table.

Year	Number of Applicants
1999	18,164
1998	16,818
1997	16,597
1996	18,183
1995	17,852
1994	15,261
1993	13,865
1992	13,029
1991	12,589
1990	12,190
1989	12,843

We may not be able to express this function in the form of an equation or formula that will correctly predict future numbers. In fact, we cannot determine the output associated with an input of 2020 or 2021 at the present time. Nevertheless, N is a function because each year a certain number of students apply. ◆

Presenting a Function in Words; Presenting a Function Using a Formula

◆ EXAMPLE 1.4   *Tariffs.* Consider the following two schemes of tariffs on the sale of automobiles. State 1 collects a fixed tax of $500 on every car sold; state 2 levies a 4% tax on the list price of every car sold. Let x be the list price of a car and let f and g be the tax functions in states 1 and 2, respectively. We can describe the input-output relationship of the functions f and g using formulas.

$$f(x) = 500 \qquad g(x) = 0.04x$$

The graphs of the functions f and g are given below.
What is the significance of the x-coordinate of the point of intersection of the lines?

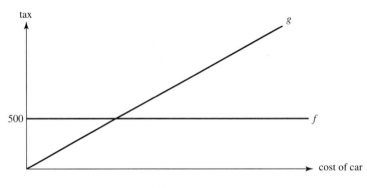

Figure 1.4

REMARKS

i. The function f is a constant function, assigning the output 500 to every acceptable input. When thinking about functions it is handy to have a little stash of specific functions at your fingertips to contemplate as case studies. Stash the set of constant functions in your mind for future reference.

ii. $g(x) = 0.04x$ tells us that the tax in state 2 is 4% of the list price of the car.

We translate "$A\%$ of B" as $\frac{A}{100} \cdot B$.

If the tax is kx, where x is the price and k is a constant, we say that the tax is **directly proportional** to the price, with **proportionality constant** k.

We translate "C is proportional to D" as $C = kD$ for some constant k.

The idea of proportionality is important; you will see it come up repeatedly in the physical and biological sciences and in economics.

iii. The function g is 1-to-1 because every output is used only once. The function f is not 1-to-1 because one output corresponds to more than one input (infinitely many in this case).   ◆

Chances are that many of the functions you have seen so far in your mathematics courses have been given by formulas. This is the type of function with which you are probably most familiar.[7] For example, because a circle's area is uniquely determined by its radius,[8] we can say that the area A of a circle is a function of the radius r: $A = \pi r^2$. In functional notation we can write $A = f(r)$, where $f(r) = \pi r^2$; we read this as "A equals f of r." On the other hand, we might decide to name our area function "A" instead of "f," in which case we could write $A = A(r) = \pi r^2$. To indicate that a circle of radius 3 has an area of 9π we write $A(3) = 9\pi$; 9π is the value of the function at 3. When we read $A = A(r)$ we think "the area is uniquely determined by the radius."

NOTE In the example above we've let "A" denote both the area of the circle *and* the function whose input is the radius of a circle and whose output is the area. Recall that a "T" sign may sometimes be a landmark indicating a subway station and other times may indicate a bus stop; we can discern the meaning by the context. Similarly, $A(r)$ may sometimes be used to indicate the area *function* and other times be used to indicate the *output* associated by the area function with the input r; we figure out which is implied by the context.

Domain and Range

When a function is modeling a real-life situation, the conditions of the situation determine the appropriate domain. Such is the case for the area function where the domain (the set of all possible radii) is the set of positive numbers. When a function is simply given by a formula without context then, unless otherwise specified, the domain is the largest set of inputs for which the function is defined. We'll refer to this as the **natural domain** of the function.

When looking for the natural domain of a function, keep in mind the following.

■   Division by zero is undefined; $\frac{h(x)}{g(x)}$ is undefined wherever $g(x) = 0$.[9]

■   The square root is undefined for negative numbers; $\sqrt{g(x)}$ is undefined for $g(x) < 0$.

Equality of Functions.   The functions f and g are **equal** if:

■ f and g have the same domain, and

■ $f(x) = g(x)$ for every x in the domain.

For example, the functions $f(x) = \frac{x^2 - x}{x}$ and $g(x) = x - 1$ are *not* equal; the domain of g is all real numbers, while the domain of f does not include $x = 0$.

◆ **EXAMPLE 1.5**   Suppose $f(x) = \frac{\sqrt{x+1}}{x-2}$. Find the natural domain of f.

SOLUTION   In order for $\sqrt{x+1}$ to be defined, we need $x + 1 \geq 0$. Therefore, $x \geq -1$.

---

[7] In fact, it was not until the nineteenth century that the mathematician Lejeune Dirichlet formulated the modern definition of a function as an input-output mapping such as we gave in the previous section. (Eli Maor, *e: The Story of a Number*, Princeton University Press, 1994.) Earlier mathematicians, including such giants as Euler and Johann Bernoulli, thought of functions only as formulas, just as you perhaps did before starting this chapter. (Howard Eves, *An Introduction to the History of Mathematics*, 6th Edition, Saunders College Publishers, 1990.)

[8] For a review of geometric formulas, see Appendix B: Geometric Formulas. Do you confuse the formulas for circumference and area? Circumference is one-dimensional; the circumference formula involves an r to the first power: $C(r) = 2\pi r$. Area is two-dimensional; the area formula involves an r^2: $A(r) = \pi r^2$.

[9] What does it mean when we write $\frac{A}{B} = C$? It means $A = C \cdot B$. Can you see why $\frac{A}{0}$ is undefined? Suppose $A \neq 0$ and $\frac{A}{0} = C$. But $C \cdot 0 = 0$ and $A \neq 0$, giving a contradiction. Suppose $A = 0$. Then $0 = C \cdot 0$ regardless of the value of C, so we still can't define division by zero.

The denominator of the fraction cannot be zero, so $x \neq 2$.

The domain of f is $\{x: -1 \leq x < 2 \text{ or } x > 2\}$.[10]

We can express this using interval notation by writing

$$x \text{ is in } [-1, 2) \text{ or } (2, \infty). \quad \blacklozenge$$

Interval Notation.   Interval notation is useful for describing an interval on the number line.

■ [2, 5] is the set of all numbers between 2 and 5 *including* both 2 and 5. We can write $x \in [2, 5]$, where the symbol "$\in$" is read "in"; this expression is equivalent to $2 \leq x \leq 5$. [2, 5] is called a **closed interval** because the numbers 2 and 5 are included in the interval. We indicate this on a number line by using filled circles at 2 and 5.

■ (2, 5) denotes the set of all numbers greater than 2 and less than 5 but *excluding* 2 and 5 themselves. $x \in (2, 5)$ is equivalent to $2 < x < 5$. (2, 5) is called an **open interval** because neither of its endpoints are included in the interval. We indicate this on a number line by using open circles at 2 and 5.

Notice that there is neither a smallest number nor a largest number in the interval (2, 5). The number 5 is an upper bound (ceiling) for the interval; in fact, it is the smallest upper bound. The number 2 is a lower bound (floor) for the interval; it is the greatest lower bound.

■ $(0, \infty)$ is the set of all positive real numbers.[11] Writing $x \in (0, \infty)$ is equivalent to writing $x > 0$. The symbol "∞" is read "infinity" and denotes the absence of an upper bound for this interval. "∞" is a symbol, *not* a number!

■ $[0, \infty)$ denotes the set of all nonnegative numbers.

■ $(-\infty, \infty)$ denotes the set of all real numbers.

We can combine "[" and "(" to include or exclude the endpoints of an interval as appropriate. However, we **never** use a square bracket "]" with the symbol for infinity, ∞. It is *very* important to keep in mind that ∞ is *not* a number.[12]

Suppose that y is a function of x and that we would like to present this function using a formula. If we have an equation relating x and y, then solving this equation unambiguously for y in terms of x allows us to express y as a function of x.

For example, suppose that a person has $2000 to invest. He decides to spend a certain amount on slow-growth stocks, twice that amount on riskier but potentially high-return

---

[10] Braces, {}, are used because this is a set. The colon, :, is read "such that." The expression $x:$ is read aloud as "the set of x such that . . . ."

[11] Real numbers will be discussed in Chapter 2.

[12] Because ∞ is not a number it cannot be treated like a number: You can't put it on a number line, and you can't add, subtract, multiply, or divide with ∞ as if it were a number.

stocks, and the rest on bonds. Let s be the amount he puts into slow-growth stocks, $2s$ be the amount he puts into riskier stocks, and b be the amount he puts into bonds. Then

$$s + 2s + b = 2000.$$

We can solve this equation for b unambiguously in terms of s; therefore b is a function of s:

$$b = f(s) = 2000 - 3s.$$

We can solve this equation for s unambiguously in terms of b; therefore s is a function of b:

$$s = g(b) = \frac{2000 - b}{3}.$$

On the other hand, suppose that we are interested in a point on the rim of a 13-inch bicycle tire. We choose the hub of the wheel to be the origin of our coordinate system, letting v denote the vertical height of the point (v is positive when the point is above the hub and negative when it is below) and h denote the horizontal coordinate of the point. The distance between the hub, located at $(0, 0)$, and the point on the rim, located at (h, v), is 13; therefore,

$$v^2 + h^2 = (13)^2.$$

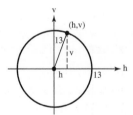

Figure 1.5

We *cannot* solve this equation unambiguously for either variable. For instance, $v = \pm\sqrt{169 - h^2}$, so $v = \sqrt{169 - h^2}$ or $v = -\sqrt{169 - h^2}$. There is ambiguity here. Knowing h does not uniquely determine v. In this example v is not a function of h. Similarly, h is not a function of v.

In the preceding example we used the distance formula. This formula will be useful to us, so we review it here. The distance between the points (x_1, y_1) and (x_2, y_2) in the plane is given by

$$d = \sqrt{(\Delta x)^2 + (\Delta y)^2} = \sqrt{(x_2 - x_1)^2 + (y_2 - y_1)^2},$$

where the capital Greek letter Δ (read: delta) is used as a shorthand for the words "change in." This is a direct consequence of the Pythagorean Theorem.

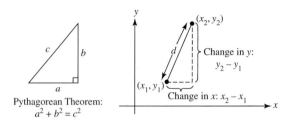

Figure 1.6

EXAMPLE 1.6   *Bottle calibration.* When calibrating a bottle, the input is the volume of liquid poured into that bottle. For a particular bottle the calibration function assigns to each volume a unique height; this is the function's output. We can give the calibration function a name, like C. If we want to specify that we are referring to the calibration function for a *beaker* we might even name the function C_b, which can be read as "C sub b." The calibration function for a given bottle assigns to each volume a unique height; therefore, we say that "height is a function of volume."

Consider the calibrating function C_b for a cylindrical beaker with radius 3 inches and height 7 inches.

    i. Find the domain and range of the function.

   ii. Find the coordinates of the point marked "P."

  iii. Find a formula for the function. Is height directly proportional to volume?

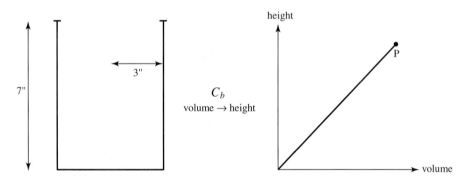

Figure 1.7

SOLUTION   i. The domain is the set of all acceptable volumes: $[0, B]$, where B is the volume of the full beaker. The volume of a cylinder of height h and radius r is the product of the height and the area of the disk that is its base, i.e. $\pi r^2 h$. Since the radius is 3 inches, the volume is $9\pi h$ cubic inches. (Notice that the height varies with the amount of liquid but the radius does not.) The maximum height of liquid that the beaker can hold is 7 inches, so the maximum volume is $\pi(3^2)(7) = 63\pi$ cubic inches. Therefore, the domain is $[0, 63\pi]$. The range, the set of all possible heights, is $[0, 7]$.

ii. Using the information from part (i) we label the point P on the graph $(63\pi, 7)$.

iii. The volume of liquid in this beaker, V, is $9\pi h$. We want to calculate height given volume, so we solve for h in terms of V:

$$h = \frac{V}{9\pi} \quad \text{so} \quad C_b(V) = \frac{V}{9\pi}.$$

The height *is* proportional to the volume. We can see this from the function formula, $h = C_b(V) = \frac{1}{9\pi} \cdot V$, or from the physical situation itself. The walls of the beaker are perpendicular to the base; therefore all cross-sections are equal in area and the height is proportional to the volume.   ◆

In the examples that follow we will work on expressing one variable as a function of another. These examples are chosen to highlight relationships that will arise repeatedly throughout our studies. They will also serve as a review of some geometry including similar triangles and the Pythagorean Theorem. (For a summary of some useful geometric formulas, refer to Appendix B.)

REMARK Examples in a mathematics text are meant to be read actively, with a pencil and paper. A solution will have more impact, and stay with you longer, if you have spent a bit of time tackling the problem yourself. Read the problems that follow and try each one on your own before reading the solutions. The problem-solving strategies highlighted below should help you out. Think of them as a way of coaching yourself through a problem.

Portable Strategies for Problem Solving

- Draw a picture whenever possible.

- Label known quantities and unknown quantities so you can refer to them. Make your labeling clear and explicit.

- Take the problem apart into a series of simpler questions. Plan a strategic approach to the problem.

- Make the problem more concrete (or simpler) in order to get started. Either solve the problem with the concrete numbers and try to generalize, or spot check your answer by making sure it works in a concrete example or two.

◆ **EXAMPLE 1.7**   *Functioning around the house.* An 8-foot ladder is leaning against the wall of a house. If the foot of the ladder is x feet from the wall, express the height of the top of the ladder as a function of x.

SOLUTION   We begin with a picture, labeling all known and unknown quantities. Let y be the height of the top of the ladder.

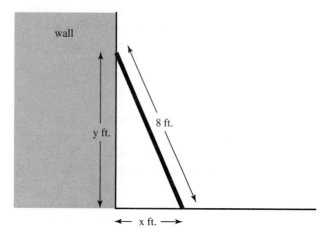

ladder against wall

Figure 1.8

The Pythagorean Theorem tells us how x and y are related.

$$x^2 + y^2 = 8^2$$
$$y^2 = 64 - x^2$$
$$y = \pm\sqrt{64 - x^2},$$

but height is never negative, so

$$y = \sqrt{64 - x^2}.$$

Therefore, $y = f(x) = \sqrt{64 - x^2}$, where $x \in [0, 8]$.   ◆

◆ EXAMPLE 1.8   *Functioning around the house.* A 13-foot ladder is leaning against the wall of a house. The foot of the ladder is 5 feet from the house and the top has a height of 12 feet. The distance between the wall of the house and a point on the ladder is a function of the height, h, of the point on the ladder. Write a formula for this function.

SOLUTION   Let d be the distance between the point on the ladder and the wall.

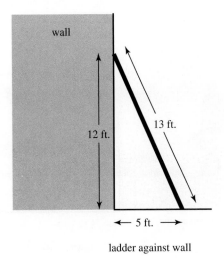

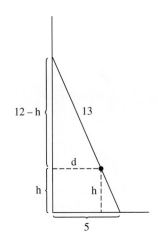

ladder against wall

Figure 1.9

We must relate d and h. We can do this by using similar triangles. We'll then solve for d in terms of h.

$$\frac{d}{12 - h} = \frac{5}{12}$$

$$d = \frac{5}{12} \cdot (12 - h)$$

$$= 5 - \frac{5h}{12}$$

Therefore, $d = f(h) = 5 - \frac{5h}{12}$.   ◆

Thinking ahead can save energy. There were several different algebraic options when using similar triangles to relate d and h. Since our goal is to solve for d, our work is simplified by using a relationship that is written so that d is in the numerator of the ratio.

EXERCISE 1.2   *Functioning in the evening*   Late in the evening an elegant $5\frac{1}{2}$-foot-tall woman is standing by a 14-foot-high street lamp on a cobbled road. The length of the shadow she casts is a function of her distance from the lamppost. Write a formula for this function, where x is the distance between the woman and the lamppost.

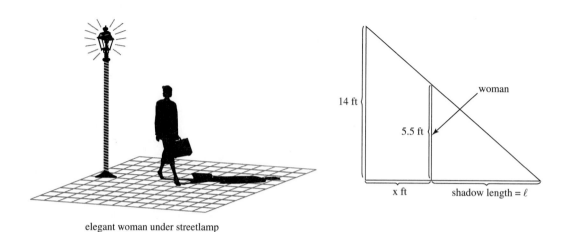

elegant woman under streetlamp

Figure 1.10

Answer

The length of her shadow is $f(x) = \frac{11}{17}x$.

◆ **EXAMPLE 1.9**   *Functioning in the morning.* We're making a pot of coffee using a conical coffee filter. The coffee filter holder is a right circular cone with a radius of 6 centimeters at the top and a height of 12 centimeters. Express the amount of liquid in the filter as a function of the height of the liquid.

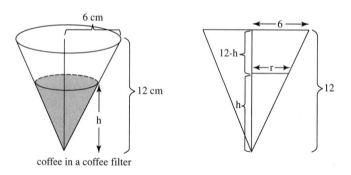

coffee in a coffee filter

Figure 1.11

SOLUTION   We can use the volume of a cone to relate the amount of liquid, V, and the height, h, of the liquid.

$$V = \frac{1}{3}\pi r^2 h$$

Here r is the radius of the top of the cone of liquid, *not* of the conical filter itself, so r varies with the amount of liquid. But now V is expressed as a function of *two* variables, r and h. We would like the volume expressed as a function of h only. Since r depends on h, we must express r in terms of h. We can do this using similar triangles.

$$\frac{r}{h} = \frac{6}{12}$$

$$r = \frac{1}{2} \cdot h$$

Therefore,

$$V = \frac{1}{3}\pi \left(\frac{h}{2}\right)^2 h$$

$$= \frac{1}{3}\pi \frac{h^2}{4} h$$

$$= \frac{1}{12}\pi h^3. \quad \blacklozenge$$

◆ **EXAMPLE 1.10**   *Functioning in the kitchen.* Chocolate pudding is being served in hemispherical bowls with a radius of 2 inches. The top skin of the pudding forms a disk whose radius, r, depends upon the height, h, of the pudding. Express r as a function of h.

SOLUTION   First, be sure you understand the question. The radius of the pudding skin and the radius of the bowl are generally *different*. They are the same only if the bowl is filled all the way to the top.

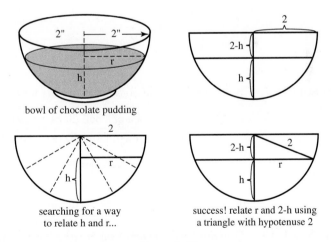

bowl of chocolate pudding

searching for a way
to relate h and r...

success! relate r and 2-h using
a triangle with hypotenuse 2

Figure 1.12

We can relate r and h by looking at a cross-sectional slice and using a right triangle. While it may be tempting to draw a triangle whose legs are h and r, this is *not* useful, since the hypotenuse of the triangle is unknown. Instead, we must draw a triangle that involves the radius of the bowl itself. This radius *must emanate from the center of the sphere*.

$$(2-h)^2 + r^2 = 2^2$$
$$r^2 = 4 - (2-h)^2$$
$$r^2 = 4 - (4 - 4h + h^2)$$
$$r^2 = 4h - h^2$$
$$r = \pm\sqrt{4h - h^2}$$

but $r \geq 0$, so

$$r = \sqrt{4h - h^2}$$
$$r = f(h) = \sqrt{4h - h^2}. \quad \blacklozenge$$

◆ **EXAMPLE 1.11**   *Functioning with friends.* Javier goes to a pizza shop intending to order a small pizza and eat it. When he enters the shop he sees some of his friends and they decide to split a large pizza. If the radius of a large pizza is twice the radius of a small pizza, what fraction of the large pizza should be allocated to Javier to give him the amount of food he originally intended to eat?

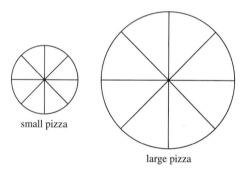

small pizza

large pizza

Figure 1.13

SOLUTION   We'll use functional notation in solving the problem.

$$\frac{\text{Area of small pizza}}{\text{Area of large pizza}} = \frac{A(r)}{A(2r)} = \frac{\pi r^2}{\pi(2r)^2} = \frac{\pi r^2}{\pi 4r^2} = \frac{1}{4}$$

Allocate a quarter of the large pizza to Javier.   ◆

Notice that we did not need to know the actual radius of either the small or the large pizza. We were able to determine the answer simply by calling the radius of the small pizza r.

◆ **EXAMPLE 1.12**   *Functioning on a diet of sugar cane.* Giant pandas eat sugar cane. S pounds of sugar cane provide N calories. On average, each day a panda needs to take in C calories for every pound of the panda's weight.

    i. The number of pounds of sugar cane it takes to support a panda for a day is a function of the weight, x, of the panda. Express this function with a formula. The formula will involve the constants S, N, and C.

    ii. The number of pounds of sugar cane it takes to support a pair of pandas, one weighing P pounds and the other weighing Q pounds, for a period of W weeks is a function of

W, the number of weeks they are to be supported. Express this function as a formula. The formula will involve the constants S, N, C, P, and Q.

SOLUTION   i. Let's use the strategy of taking the problem apart into a series of simpler questions. We'll use unit analysis to help us.

We're looking for the number of pounds of sugar cane it takes to support a panda weighing x pounds for one day. In order to use unit analysis constructively we'll need to be very precise; we don't want to confuse pounds of sugar cane with pounds of panda. *Question:* How many calories does the panda need for one day?

$$C \frac{\text{calories}}{\text{lbs of panda}} \cdot x \text{ lbs of panda} = Cx \text{ calories}$$

Question: How many pounds of sugar cane will provide Cx calories?

If we know the number of pounds of sugar cane per calorie, then multiplying by Cx, the number of calories needed, will give the answer. S pounds of sugar cane provide N calories. We can express this as

$$\frac{S \text{ lbs of sugar cane}}{N \text{ calories}} \quad \text{or} \quad \frac{S}{N} \frac{\text{lbs of sugar cane}}{\text{calorie}}.$$

Therefore, to support the panda for one day we need

$$\frac{S}{N} \frac{\text{lbs of sugar cane}}{\text{calorie}} \cdot Cx \text{ calories} = \frac{SC}{N} x \text{ lbs of sugar cane}.$$

Number of pounds of sugar cane $= f(x) = \frac{SC}{N} x$.

ii. Again, let's break this down into a series of simpler questions.

How many pounds of sugar cane are needed to support a pair of pandas, one weighing P pounds and the other weighing Q pounds for one day?

$$\text{Amount to support the pair for a day} = \text{amount to support one panda}$$
$$+ \text{ amount to support the other}$$

$$\text{Amount to support the panda weighing } P \text{ lb} = \frac{SCP}{N}$$

$$\text{Amount to support the panda weighing } Q \text{ lb} = \frac{SCQ}{N}$$

$$\text{Amount to support the pair for a day} = \frac{SCP}{N} + \frac{SCQ}{N}$$

$$= \frac{SC}{N} \cdot (P + Q)$$

Whatever it takes to feed the pandas for a day, it takes 7 times that amount to feed them for a week and $7W$ times that amount to feed them for W weeks. To feed the pandas for W weeks it takes

$$f(W) = 7W \cdot \frac{SC}{N} \cdot (P + Q)$$

pounds of sugar cane.   ◆

COMMENT *Working with Rates*   Consider the statement "S pounds of sugar cane provide N calories." From this we can determine the amount of sugar cane equivalent to one calorie (pounds of sugar cane per calorie):

$$\frac{S \text{ lbs of sugar cane}}{N \text{ calories}} \quad \text{or} \quad \frac{S}{N} \frac{\text{lbs of sugar cane}}{\text{calorie}}.$$

Similarly, we can determine the number of calories per pound of sugar cane:

$$\frac{N \text{ calories}}{S \text{ lbs of sugar cane}} \quad \text{or} \quad \frac{N}{S} \frac{\text{calories}}{\text{lbs of sugar cane}}.$$

In other situations we may be given analogous sorts of *rate* information. For instance, if a typist types W words in H hours then he types at a rate of

$$\frac{W \text{ words}}{H \text{ hours}} \quad \text{or} \quad \frac{W}{H} \frac{\text{words}}{\text{hour}}.$$

If we know the number of hours the typist worked, we can estimate the number of words he typed. Unit analysis gives

$$\frac{\text{words}}{\text{hour}} \cdot \text{hours} = \text{words}.$$

Multiplying rate by time gives the amount done.

Alternatively, we can calculate the time per word.

$$\frac{H \text{ hours}}{W \text{ words}} \quad \text{or} \quad \frac{H}{W} \frac{\text{hours}}{\text{word}}.$$

If we know the number of words the typist must type, we can estimate the time it will take him.

$$\frac{\text{hours}}{\text{words}} \cdot \text{words} = \text{hours}.$$

Presenting a Function Graphically

While you may be accustomed to seeing functions described by formulas, frequently information about a function is obtained directly from its graph. The function itself may be presented graphically, without a formula; think about a seismograph recording the size of tremors in the earth, an electrocardiogram machine measuring electric activity in the heart, or the output of many standard measuring instruments of meteorologists. These machines give us pictures, not formulas. Alternatively, consider the function that takes time as input and gives as output the temperature at the top of the Prudential Center in downtown Boston. The function can be quite easily presented in the form of a graph, while arriving at a formula to fit past data is a much more complicated task. (Even if such a formula is found, it will require frequent alteration as data comes in that does not fit the existing formula.) The function that gives the number of applicants to Harvard College as a function the calendar year can also be presented much more simply by a graph than by trying to fit a formula to past data.

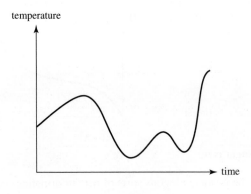
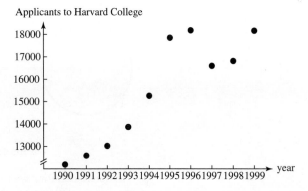

Figure 1.14

When graphing functions we use the convention that the independent variable (input) is plotted along the horizontal axis and the dependent variable (output) is plotted along the vertical axis.[13] The coordinates of points on the graph are of the form (input, corresponding output). Given the function f, for every x in the domain of f the point with coordinates $(x,$ output of f corresponding to $x)$ is a point on the graph of f. Recall that "the output of f corresponding to x" can be written using the shorthand $f(x)$. Thus the points on the graph are of the form

(input, corresponding output)

$(x,$ output of f corresponding to $x)$

$(x,$ value of f at $x)$

$(x, f(x))$.

All four expressions carry the same meaning; we usually use the last one.

> The height of a graph at a point is the *value* of the function at that point.

◆ **EXAMPLE 1.13** C_E is the calibrating function for a particular evaporating flask. The graph of C_E is drawn below. Use the graph to identify the domain and range of C_E.

---

[13] When constructing a function to model a situation, you must decide which variable you consider input and which you consider output. For instance, when calibrating bottles, we consider the input (or *independent*) variable to be the volume and the output (*dependent*) variable to be the height. When using a calibrated bottle for measuring, we consider the input variable to be the height; the output is the volume.

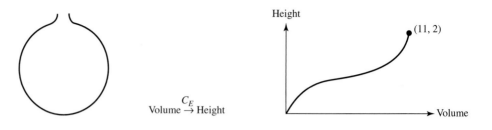

Figure 1.15

SOLUTION   By looking at the graph of C_E we can tell that the domain of the function, the set of all possible volumes, is $[0, 11]$ and the range, the set of all possible heights for these volumes, is $[0, 2]$. ◆

It is useful to be able to obtain information about a function from its graph, as we will see below.

If x is in the domain of f then the point $(x, f(x))$ is on the graph of f.

If $(3, 16)$ lies on the graph of some function named g, then 3 is an acceptable input of the function, and 16 is the associated output, so $g(3) = 16$. If an input of 8 produces an output of 3, then the point $(8, 3)$ will be on the graph. More generally, if (x_1, y_1) is on the graph of f, then $y_1 = f(x_1)$. The converse[14] is also true. If $y_1 = f(x_1)$, then the point (x_1, y_1) is on the graph of f.

If a function is given by an equation $y = f(x)$, then a point (x, y) lies on the graph of the function if and only if it satisfies the equation $y = f(x)$.

The "if and only if" construction comes up frequently in mathematical discussions, so we'll pause for a moment to clarify the meaning of an "if and only if" statement.

Language and Logic: An Interlude.

"A if and only if B" means "A and B are equivalent statements." Using symbols we write $A \Leftrightarrow B$. Specifically, "P is a square if and only if P is a rectangle with sides of equal length" says that the two statements "P is a square" and "P is a rectangle with sides of equal length" are equivalent. They carry the same information.

$A \Leftrightarrow B$ means $A \Rightarrow B$ and $B \Rightarrow A$.

A if and only if B means A implies B and B implies A.

---

[14] The word "converse" has a precise meaning in math. Suppose that if A is true, then B is also true. The converse reverses the two pieces, saying if B is true, A is also true. The converse of a true statement is often not true. For example, the statement "if a shape is a square, then it is also a rectangle" is true, but the converse ("if a shape is a rectangle, then it is also a square") is false.

	Specific Example		General Statement	Symbolic Representation
P is a square	if and only if	*P* is a rectangle with sides of equal length	*A* if and only if *B*	*A* ⇔ *B*
P is a square	if	*P* is a rectangle with sides of equal length	*A* if *B*	
	equivalently		*equivalently*	
If *P* is a rectangle with sides of equal length	then	*P* is a square	if *B* then *A*	*B* ⇒ *A*
P is a square	only if	*P* is a rectangle with sides of equal length	*A* only if *B*	
	equivalently		*equivalently*	
If *P* is a square	then	*P* is a rectangle with sides of equal length	if *A* then *B*	*A* ⇒ *B*

The graphs of some functions are given in the figure below.

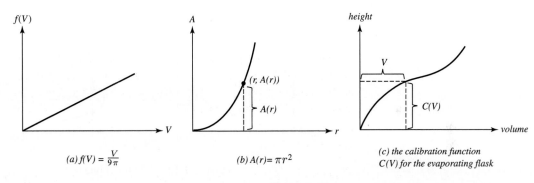

(a) $f(V) = \frac{V}{9\pi}$ (b) $A(r) = \pi r^2$ (c) the calibration function $C(V)$ for the evaporating flask

Figure 1.16

Graphs of mappings Q, R, and S from Exercise 1.1 are given in Figure 1.17. You should now be able to check visually that Q and R are functions while S is not a function.

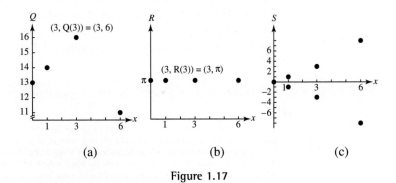

(a)   (b)   (c)

Figure 1.17

EXERCISE 1.3   Two of the four graphs given in Figure 18 are the graphs of functions.[15] Which two are they? Can you come up with a rule for determining whether or not a given graph is the graph of a function?

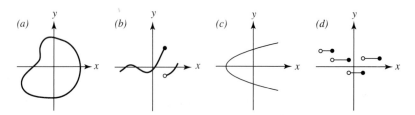

Figure 1.18

Answer

We can tell that the relationships represented in Figures 1.16(a)–(c), 1.17(a) and (b), and 1.18(b) and (d) are, in fact, functions. The test for a function is that every input must have only one output assigned to it; graphically, this means if we draw a vertical line through any point on the horizontal axis, this line cannot cross the graph in more than one place. (The vertical line will not cross the graph at all if it is drawn through a point on the horizontal axis that is not a valid input, i.e., not in the domain of the function.) This test for deciding if a graph represents a function is called the **vertical line test**.

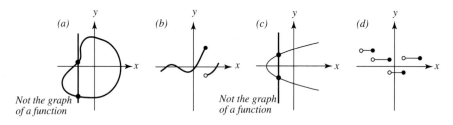

Figure 1.19   Vertical line test

Let's return to the problem of calibrating bottles. When we calibrate a bottle we put in a known volume of liquid and the calibration function $C(v)$ produces a height. Once we have calibrated a bottle we can then use it as a measuring device. We turn the procedure around so that the height becomes the input while the volume is the output. The key ingredient that makes the original volume-height assignment useful is that not only does each volume correspond to one height but each height corresponds to only one volume. The calibration function is 1-to-1. A function f that is 1-to-1 has an **inverse function** that "undoes" f. The domain of the inverse function is the range of f; if f maps a to b then its inverse function maps b to a.[16]

EXERCISE 1.4   Given the graph of a function, how can we determine whether the function is 1-to-1?

---

[15] The conventions about hollow and filled circles are the same as they are for interval notation. A hollow circle denotes a point that is not on the graph, while a filled circle indicates that the point is on the graph.

[16] Sometimes people *mistakenly* think of the words "inverse" and "reciprocal" as being the same. The inverse of the function f undoes f. If f adds 3 to its input then its inverse function subtracts 3; if f multiplies its input by 3 then its inverse function divides by 3. If f maps 7 to 11 then its inverse function maps 11 to 7. On the other hand, the reciprocal of N is simply $1/N$.

Answer

By the horizontal line test: If any horizontal line intersects the graph in more than one place, then the function is not 1-to-1; if no horizontal line intersects the graph in more than one place, then the function is 1-to-1.

Functions: The Grand Scheme

In this text we will be looking at functions of one variable—but not all functions *are* functions of one variable. For instance, suppose you were to calculate the value, V, of the change in your pocket. V is a function of q, d, n, and p, where q, d, n, and p are the number of quarters, dimes, nickels, and pennies, respectively, in your pocket. We can write this function as

$$V = f(q, d, n, p) = 25q + 10d + 5n + p.$$

As another example, suppose we denote by V the volume of a fixed mass of gas, by T its temperature on the Kelvin scale, and by P its pressure. Then the combined gas law tells us that

$$\frac{PV}{T} = k, \text{ where } k \text{ is a constant,}$$

or, equivalently,

$$V = \frac{k}{P} \cdot T.$$

We see that the volume of the gas will depend upon both the pressure *and* the temperature. V is a function of two variables, $V = f(P, T)$. If pressure is held constant, then volume is directly proportional to temperature; as the temperature of a gas goes up, the volume goes up as well.

$$V = \text{ (a constant) } \cdot T \qquad \text{is the gas law of Charles and Gay-Lussac.}$$

In other words, if pressure is held constant, then V can be expressed as a function of one variable.

Think back to Example 1.9 where we were expressing the volume of water in a conical coffee filter as a function of the height of the water. We began by writing $V = \frac{1}{3}\pi r^2 h$, where both r, the radius of the liquid, and h, the height of the liquid are variables. We wanted to express V as a function of only h. In this example we cannot simply say "let r be a constant," since h cannot vary without r varying. Instead, we found a relationship between h and r that allowed us to express the former in terms of the latter, resulting in a function of only one variable.

Because our focus will be on functions of one variable, when we do modeling we need to determine one independent variable and one dependent variable, for a total of two variables. If we appear to have more variables, then it is necessary to do one of two things:

- Figure out how we can express one of the variables in terms of another, or

- be less ambitious with our model, treating some quantities as constants when it is reasonable to do so.

PROBLEMS FOR SECTION 1.3

1. (a) Consider the following graphs. For each graph decide whether or not y is a function of x. If it is a function, determine the range and domain.

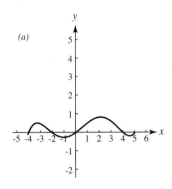

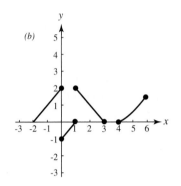

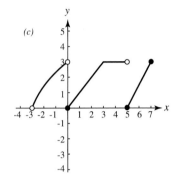

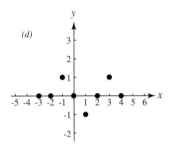

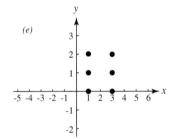

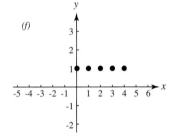

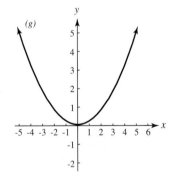

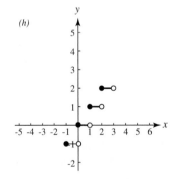

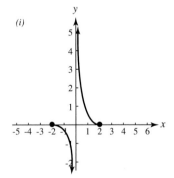

(b) Are any of the functions from part (a) 1-to-1? If so, which ones?

In Problems 2 through 4 find the domain of each of the functions.

2. (a) $f(x) = \frac{1}{x+2}$ (b) $g(x) = \frac{5}{x^2+4}$

3. (a) $f(x) = \sqrt{x}$ (b) $g(x) = \sqrt{x-3}$ (c) $h(x) = \sqrt{x^2-4}$

4. (a) $f(x) = \sqrt{\frac{1}{x+1}}$ (b) $g(x) = \sqrt{\frac{x}{x+1}}$

   (*Hint:* Keep in mind that if the numerator and denominator of a fraction are both negative, then the fraction is positive.)

5. Find the range of the functions in Problems 2 through 4.

6. Let $g(x)$ be the function graphed below with domain $[-5, 4]$. Use the graph to answer the following questions.

   (a) What is the range of g?
   (b) Is $g(x)$ 1-to-1?
   (c) Where does g take on its highest value? Its lowest value?
   (d) What is the highest value of g?
   (e) What is the lowest value of g?
   (f) For what x is $g(x) = 0$?
   (g) Approximate $g(0)$.
   (h) Approximate $g(-4)$.
   (i) Approximate $g(\pi)$ and $g(-\pi)$.
   (j) For approximately what values of x is $g(x) = 1$?
   (k) For approximately what values of x is $g(x) < -1$?

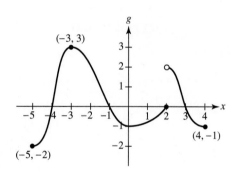

7. Two functions are equivalent if they have the same domain and the same input/output relationship. The first function listed on each line below is called f. Which of the functions listed on each line are equivalent to f? The domain of each function is the set of all real numbers. (Be careful to think about the sign of each function.)

   (a) $f(x) = -2x^2$ $g(w) = (-2w)^2$ $i(t) = 2(-t^2)$ $j(x) = -\sqrt{2x^2}\sqrt{x^2}\sqrt{2}$

   (b) $f(x) = (2x-1)^2$ $g(c) = (1-2c)^2$ $h(t) = 1-2t^2$ $j(x) = 1-(2x)^2$

   (c) $f(x) = \sqrt{x^2}$ $\lambda(m) = m$ $\mathcal{T}(x) = \sqrt{(-x)^2}$ $\varphi(s) = |s|$

8. Below is a graph of the function f. Use it to approximate the following.
   (a) The value of f at $x = -1$, $x = 0$, and $x = 1$
   (b) All x such that $f(x) = 0$
   (c) All x such that $f(x) = 2$
   (d) $-f(0) + 2f(3)$

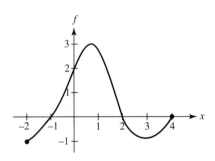

9. Below is a graph of the function g. Use it to approximate the following.
   (a) $g(-2)$, $g(0)$, $g(1)$
   (b) All t such that $g(t) = 0$
   (c) All t such that $g(t) = 1$
   (d) $\frac{g(-1) + g(1)}{2}$
   (e) $g(3) + 3g(1)$

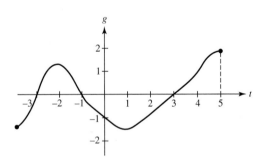

10. The graph of h is given at the top of page 37.
    (a) Give the domain and range of h.
    (b) Evaluate $h(0)$, $h(1)$, $h(2)$.
    (c) Fill in the output values on the table.

x	$h(x)$
-0.5	
-0.01	
0	
0.1	
0.5	
1.5	
1.95	
2	
2.01	

(d) Find all x such that $h(x) = 3$. Use inequalities in your answer.

(e) Find all x such that $h(x) = 1$. Use interval notation, paying attention to [versus (.

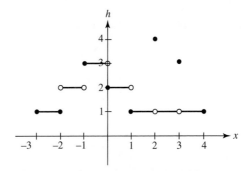

11. The graph of k is given below.
    (a) Give the domain and range of k.
    (b) Find $2k(-1) + [k(-1)]^2 + k((-1)^2)$.
    (c) Find all t such that $k(t) = 2$.

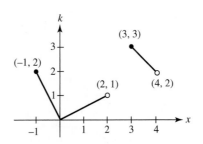

12. Consider the functions f, g, h, and k presented as graphs in Problems 8 through 11. Which, if any, of these functions are 1-to-1?

13. The graphs of f and g are given below. Approximate the following.
   (a) All x such that $f(x) = g(x)$
   (b) All x such that $g(x) > f(x)$

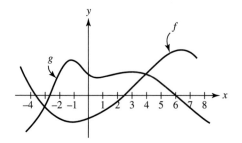

14. Determine whether or not the given graph is the graph of a function.

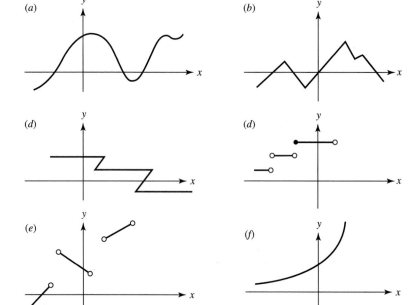

15. Draw the graph of a function f that is 1-to-1 and a function g that is not 1-to-1.

The graph of a function is given in Problems 16 through 22. Determine the range and domain. Is the function 1-to-1?

16.

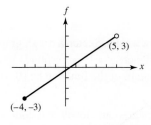

17.

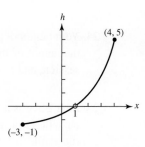

18.

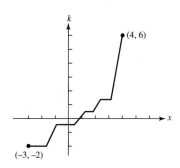

19.

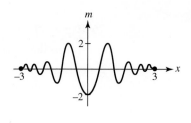

20.

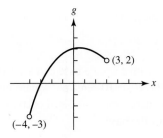

21.

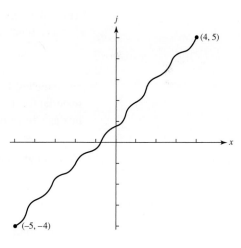

22.

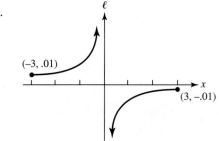

23. You make yourself a hot cup of tea. The telephone rings, distracts you, and when you return the tea is at room temperature. Sketch a rough graph of the function $T(t)$, where t is the time, in minutes, from when the phone rings, and $T(t)$ is the temperature of the tea at time t.

24. Your neighbor prunes his hedge once every three weeks. Sketch a rough graph of $h(t)$, the height of the hedge as a function of time, where $t = 0$ corresponds to a pruning session, and t is measured in weeks. Let the domain of the function be [0, 10].

25. First Slice is a "come from behind" horse. He starts out slowly and comes on at the end. Tass is a sprinter, starting out fresh and tiring at the end. First Slice and Tass both run a mile in 1:12. On the same set of axes, graph $F(t)$ and $T(t)$, the distances traveled by First Slice and Tass, respectively, as a function of time t, t in seconds.

26. Express the area of a circle as a function of:
    (a) its diameter, d.   (b) its circumference, c.

27. Express the surface area of a cube as a function of the length s of one side.

28. A closed rectangular box has a square base. Let s denote the length of the sides of the base and let h denote the height of the box, s and h in inches.
    (a) Express the volume of the box in terms of s and h.
    (b) Express the surface area of the box in terms of s and h.
    (c) If the volume of the box is 120 cubic inches, express the surface area of the box as a function of s.

29. You are constructing a closed rectangular box with a square base and a volume of 200 cubic inches. If the material for the base and lid costs 10 cents per square inch and the material for the sides costs 7 cents per square inch, express the cost of material for the box as a function of s, the length of the side of the base.

30. A rectangle is inscribed in a semicircle of radius R, where R is constant.
    (a) Express the area of the rectangle as a function of the height, h, of the rectangle, $A = f(h)$.
    (b) Express the perimeter of the rectangle as a function of the height, h, of the rectangle, $P = g(h)$.

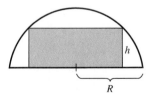

31. Two sisters, Nina and Lori, part on a street corner. Lori saunters due north at a rate of 150 feet per minute and Nina jogs off due east at a rate of 320 feet per minute. Assuming they maintain their speeds and directions, express the distance between the sisters as a function of the number of minutes since they parted.

32. Below are graphs of $f, g, h,$ and j. Use the following clues to match the function with its graph.

   - $h(0) = g(0)$
   - $h(x) > g(x)$ for $x < 0$
   - $h(2) = f(2) = j(2)$
   - $j(x) = 2f(x)$ for $x > 0$

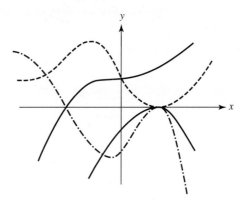

For Problems 33 through 35, if the interval is written using inequalities, write it using interval notation; if it is expressed in interval notation, rewrite it using inequalities. In all cases, indicate the interval on the number line.

33. (a) $-1 \le x \le 3$ (b) $(-2, -1]$

34. (a) $-7 \le x < -5$ (b) $(-\pi, \pi)$

35. (a) $-1 < x$ (b) $(-\infty, 3)$

36. Let $f(x) = 2x^2 + x$. Find the following.
   (a) $f(3)$ (b) $f(2x)$ (c) $f(1 + x)$
   (d) $f(\frac{1}{x})$ (e) $\frac{1}{f(x)}$

37. A right circular cylinder is inscribed in a sphere of radius 5.
   (a) Express the volume of the cylinder as a function of its radius, r.
   (b) Express the surface area of the cylinder as a function of its radius, r.

38. The height of a right circular cone is one third of the diameter of the base.
   (a) Express its volume as a function of its height, h.
   (b) Express its volume as a function of r, the radius of its base.

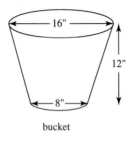

39. A vitamin capsule is constructed from a cylinder of length x centimeters and radius r centimeters, capped on either end by a hemisphere, as shown at left.

    Suppose that the length of the cylinder is equal to three times the diameter of the hemispherical caps.

    (a) Express the volume of the vitamin capsule as a function of x. *Your strategy should be to begin by expressing the volume as a function of both x and r.*

    (b) Express the surface area of the vitamin capsule as a function of x.

40. *Calibration function for a bucket:* Consider the bucket drawn below. The bucket walls can be thought of as a slice of a right circular cone, as shown on the right.

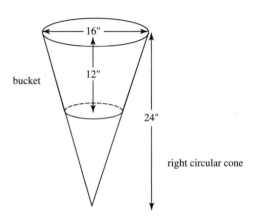

    Let C be the calibration function. C takes as input a volume of liquid and produces as output the height of the liquid in the bucket. You will need to know how to find the volume of a cone in order to do this problem. Refer to the geometry summary in Appendix B: Geometric Formulas if you do not know the formula.

    (a) What is the domain of C?

    (b) What is the range of C?

41. Late at night a caped man is standing by a 14-foot-high street lamp on a highway. Denote his height by H. The length of the shadow he casts is a function of his distance from the lamppost. Write a formula for this function, where s is the distance between the man and the lamppost.

42. Assume that f is a function with domain $(-\infty, \infty)$. Which of the following statements is true for every such function f and all p, w, and z in the domain of f? If a statement is not true for every function, find a function for which it is false. (*Hint:* The constant functions are good functions to use as a first check.)

    (a) $f(2)f(3) = f(6)$

    (b) $f(p) + f(p) = 2f(p)$

    (c) $f(4+5) = f(4) + f(5)$

    (d) $f(w)f(w) = \left[f(w)\right]^2$

    (e) $f(z)f(z) = f(z^2)$

    (f) $\dfrac{f(x)}{5[f(x)]^2} = \dfrac{1}{5f(x)}$, where $f(x) \neq 0$

    (g) $f(1)f(7) = f(7)$

43. Translate each of these English sentences into a mathematical sentence, i.e., an equation. Then comment on the validity of the statement, making qualifications if necessary. You will need to define your variables. The first example has been worked for you.

Example: The cost of broccoli is proportional to its weight.

Answer: Let C = the cost of broccoli and B = the weight of broccoli.

$$C = kB \qquad \text{for some constant } k.$$

We can emphasize that cost is a function of weight by writing

$$C(B) = kB.$$

Validity: In general this is true, because broccoli is usually sold by the pound. The proportionality constant k is the price per pound; at any given time and store it is fixed. The proportionality constant may vary with the season and the specific store.

   (a) The cost of a piece of sculpture is proportional to its weight.
   (b) The rate at which money grows in a savings account is proportional to the amount of money in the account.
   (c) The rate at which a population grows is proportional to the size of the population.
   (d) The total distance you travel is proportional to the time you spend traveling.

44. A gardener has a fixed length of fence that she will use to fence off a rectangular chili pepper garden. Express the area of the garden as a function of the length of one side of the garden.

   If you have trouble, reread the "Portable Strategies for Problem Solving" listed in this chapter. We've also included the following advice geared specifically toward this particular problem.

   • Give the length of fencing a name, such as L. (We don't know what L is, but we know that it is fixed, so L is a constant, not a variable.)
   • Draw a picture of the garden. Call the length of one side of the fence s. How can you express the length of the adjacent side in terms of L and s?
   • What expression gives the area enclosed by the fence?

45. Cathy will fence off a circular pen for her rabbits. Express the area of the rabbit pen as a function of the length of fencing she uses.

46. A commuter rides his bicycle to the train station, takes the subway downtown and then walks from the subway station to his office. He bikes at an average speed of B miles per hour and can walk M miles in H hours. The subway ride takes R minutes. The commuter bikes X miles and walks Y miles to get to work. Assume that X, Y, B, H, M, and R are all constants.

   The amount of time it takes the commuter to get to work varies with how long he has to spend at the subway station locking his bike and waiting for the next train. Denote this time by w, where w is in hours. Express the time it takes for his commute as a function of w. Specify whether your answer is in minutes or in hours.

47. A rectangular piece of thick cardboard measuring 10 inches by 6 inches is being used to make an open-top box for raspberries. The box is constructed by cutting out squares

x inches by *x* inches from each of the corners of the cardboard sheet and then folding up the sides, as shown in the acccompanying figure. Express the volume of the box as a function of *x*.

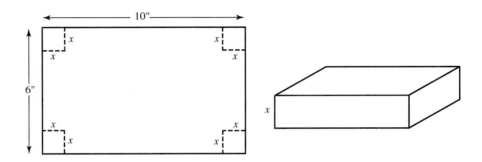

48. A new mosque is being built in the Turkish town of Iznik. Under construction is an archway whose structure will be a rectangle capped by a semicircle, as shown in the figure below. The distance from the highest point of the arch to the floor is 7 meters. Denote the width of the archway by w and the height of the vertical wall of the rectangle y. The width and height are given in meters.

  (a) Express y, the height of the side wall, as a function of w, the width of the archway.

  (b) Express the area enclosed by the archway as a function of w.

  (c) Express the perimeter of the archway (excluding the floor) as a function of y.

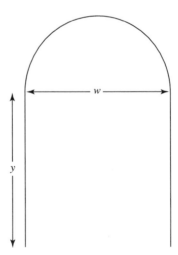

49. During road construction gravel is being poured onto the ground from the top of a tall truck. The gravel falls into a conical pile whose height is always equal to half of its radius. Express the amount of gravel in the pile as a function of its

  (a) height.

  (b) radius.

50. In Durham, England, the circular plots of land at the center of the roundabouts[17] are often meticulously planted. Along the 8-meter diameter of one such circular plot is a line of yellow tulips. Rows of purple tulips, pansies, and marigolds are neatly planted parallel to the center row of yellow tulips. Each row of flowers extends from one side of the plot to the other. Express the length of a row of flowers as a function of its distance from the center line of yellow tulips.

51. (a) A bead maker has a collection of wooden spheres 2 centimeters in diameter and is making beads by drilling holes through the center of each sphere. The length of the bead is a function of the diameter of the hole he drills. Find a formula for this function.

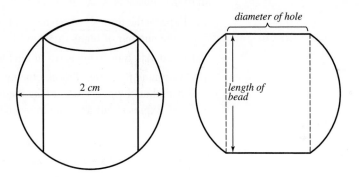

   If you are stuck, begin by trying to express half the length of the bead as a function of the radius of the hole drilled.

   (b) More generally, suppose he works with spherical beads of radius R. Again, the length of the bead is a function of the diameter of the hole he drills. Find a formula for this function.

In the following problems, demonstrate your use of the portable strategies for problem solving described in this chapter. What simpler questions are you asking yourself? What concrete example can you give to convince your friends and relatives that you are right? Write this up clearly, so a reader can follow your train of thought easily.

52. At the Central Perk coffeehouse in Manhattan, Rachel serves c cups of coffee and d desserts per hour. The coffee costs a dollars per cup, and the desserts cost b dollars each. She averages a tip of k cents per dollar of the customers' bills (excluding taxes). In addition, she makes a fixed wage of F dollars per hour. Consider c, d, a, b, k, and F as constants. Express Rachel's earnings as a function of h, the number of hours she works. (In actuality, Rachel's earnings are *not* a function of the hours she puts in. Other considerations complicate the situation. For instance, business is slow at certain times of the day, and some customers tip more generously than others. Nevertheless, by using the information provided, we can make a mathematical model of the situation that gives us a reasonably accurate picture.)

---

[17] "Roundabout" is a British term for traffic circle.

53. I work an h-hour day. I spend $1/w$ of these h hours on the road and the remainder in consultation. I receive A dollars per hour as a consultant. I receive no money when I'm on the road. In fact, each day I pay G dollars in gas and tolls and I estimate that each day costs C cents in wear and tear on the car. I have no other expenses. Express my daily profit as a function of h, the number of hours I work. (A, w, G, and C are all constants.) (*Note:* If $w = 5$, I spend 1/5 of my workday on the road and the rest of my workday in consultation.)

54. Two bears, Bruno and Lollipop, discover a patch of huckleberries one morning. The patch covers an area of A acres and there are X bushels of huckleberries per acre. Bruno eats B bushels of huckleberries per hour; Lollipop can devour L bushels of huckleberries in C hours. Express your answers to parts (a) and (b) in terms of any or all of the constants A, X, B, L, and C.

   (a) Express the number of bushels of huckleberries the two bears eat as a function of t, the number of hours they have been eating.

   (b) In t hours, how many acres of huckleberries can the two bears together finish off?

   (c) Assuming that after T hours the bears have not yet finished the berry patch, how many hours longer does it take them to finish all the huckleberries in the patch?
      Express your answers in terms of any or all of the constants A, X, B, L, C, and T.
      If you are having difficulty, use this time-tested technique: Give the quantity you are looking for a name. (Avoid the letters already standing for something else.)

55. A manufacturer is packaging oatmeal in cylindrical containers. She needs the volume of the container to be 88 cubic inches in order to hold 18 ounces of rolled oats. Given this requirement, the height of the cylinder will depend upon the radius, r, selected.

   (a) Express the height, h, of the cylindrical oatmeal container with volume 88 cubic inches as a function of r.

   (b) Suppose that the lid, bottom, and sides of the container are all made of cardboard. (The lid will be attached to the container with tape.) Express the number of square inches of cardboard used in the container as a function of r, the radius of the container. (To figure out how much cardboard is used in the sides of the container, imagine cutting along the height and rolling the sheet out into a rectangle. The height of the rectangle will be h. What will its length be?)

   (c) When making the containers, the lid and the base are cut from squares of cardboard, $2r$ by $2r$, and the excess cardboard is tossed into a recycling bin. Assume that the company must pay full price for the excess cardboard it uses.
      If cardboard costs k cents per square inch, express the cost of the cardboard for the container as a function of r. (k is a constant.)

   (d) Suppose the manufacturer decides to switch to plastic lids and bottoms to eliminate the taping problems. Assume that custom-made plastic lids and bottoms cost $7k$ cents per square inch. Express the cost of the container as a function of r.

56. A typist can type W words per minute. On average, each computer illustration takes C minutes to create and I minutes to insert.

   (a) What is the estimated amount of time it will take for this typist to create a document N words long and containing Z illustrations?

(b) The typist is paid $13 per hour for typing and a flat rate of $10 per picture. The cost of getting a document typed is a function of its length and the number of pictures. Write a function that gives a good estimate of the cost of getting a document of x words typed, assuming that the ratio of illustrations to words is 1:1000.

(c) Given the document described in part (b), express the typist's wages per hour as a function of x.

57. Filene's Basement regularly marks down its merchandise. A discounted item now costs D dollars. This is after a p-percent markdown. Express the initial price of the item in terms of p and D. Try out your answer in the concrete case of an item that now costs $100 after a 20-percent markdown. Why should your answer not be $120?

58. Amir and Omar are tiling an area measuring A square meters. They lay down N tiles per square meter. Omar can put down q tiles in r hours while it takes Amir m minutes to lay one tile. A, N, q, r, and m are constants.

(a) Give the number of tiles Amir and Omar can put down as a function of t, the number of hours they work together.

(b) How many square meters can they tile in t hours?

(c) After H hours of working with Omar, Amir leaves. The job is not yet done. How many hours will it take Omar to finish the job alone? Express the answer in terms of any or all of the constants A, N, q, r, m, and H.

59. Sam Wright plays the role of Sebastian the crab in the Disney film "The Little Mermaid." He spent H hours working on the production. P percent of this time was spent on the taping; of the remaining time, $1/n$ was spent on rehearsal and the rest on dubbing and looping. Sam was paid D dollars per hour for each of the H hours spent on the production.

(a) If Disney had changed the contract so that they paid for taping, dubbing, and looping but not rehearsal, how much would Sam's pay have been?

(b) If Disney were paying for taping only, and Sam wanted to earn the same thing he would have under the original contract, how much would he need to charge per hour?

2

Characterizing Functions and Introducing Rates of Change

■ 2.1 FEATURES OF A FUNCTION: POSITIVE/NEGATIVE, INCREASING/DECREASING, CONTINUOUS/DISCONTINUOUS

Just as you might characterize a fish by type, behavior, and habitat, so too can you characterize a function by type and behavior. Throughout this book we will build up a "type" categorization of groups of functions (polynomial, rational, exponential, logarithmic, trigonometric); for now we will focus on some basic behavioral descriptions that can be used to characterize functions.

In characterizing a function we direct our attention to the *values* of the function—the outputs.

Positive/Negative

We say

■ f is **positive** at a if the output value $f(a)$ is positive;

■ f is **negative** at a if the output value $f(a)$ is negative.

Graphically, this means that

■ f is positive wherever its graph lies above the horizontal axis;

■ f is negative wherever its graph lies below the horizontal axis.

The graph of f can change sign from positive to negative (or vice versa) by either cutting through the horizontal axis or leaping over the horizontal axis. Frequently in our study of calculus we will uncover a wealth of information behind simple sign information.

Suppose y is a function of x. We label the horizontal (input) axis the x-axis and the vertical (output) axis the y-axis. Wherever the graph of f meets with the x-axis the value of the function is zero. The x-values at which the graph of f meets the x-axis are the zeros of f.

If $f(a) = 0$, then a is called a **zero of f** or an **x-intercept** of f. We say that a is a **root** of the equation $f(x) = 0$. For example, if $f(x) = (x - 1)^2$ as in Figure 2.1(a), $x = 1$ is a root of the equation $f(x) = 0$. Each of the functions in Figure 2.1 has a zero at $x = 1$. As illustrated in Figure 2.1, the sign of f may change on either side of a zero or it may remain the same on either side of a zero. (There is a lack of consensus on terminology; sometimes one speaks of roots of functions and zeros of equations.)

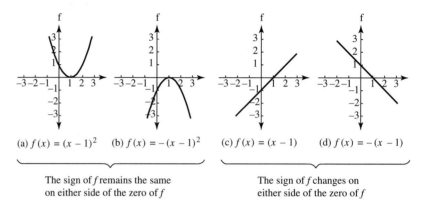

(a) $f(x) = (x - 1)^2$ (b) $f(x) = -(x - 1)^2$ (c) $f(x) = (x - 1)$ (d) $f(x) = -(x - 1)$

The sign of f remains the same on either side of the zero of f The sign of f changes on either side of the zero of f

Figure 2.1

If $x = 0$ is in the domain of the function f, then the graph of f will intersect the y-axis. The y-value (output) corresponding to $x = 0$, denoted $f(0)$, is called the **y-intercept** of f.[1] If f is a function, then it can have at most one y-intercept. (Why?) There can be any number of x-intercepts because several different inputs can be assigned an output of 0.[2]

The question of whether $f(x)$ (the output) is positive or negative is completely different from the question of whether the x (the input) is positive or negative.

EXERCISE 2.1   The following two graphs give a "happiness index" as a function of outdoor temperature (in degrees Celsius). When the index is positive the person is happy, and when the index is negative the person is unhappy. Can you guess which person is a skier and which is a bicyclist?

---

[1] Keep x-intercepts and y-intercepts straight in your mind:

   i. x-intercepts are x-values (inputs) where the output y is zero. A point where the graph crosses the x-axis is of the form $(a, 0)$ where the number a is the x-intercept.

   ii. the y-intercept is the y-value (output) where x is zero. A point where the graph crosses the y-axis is of the form $(0, b)$ where the number b is the y-intercept.

[2] A function that is 1-to-1 can cross the x-axis at most once.

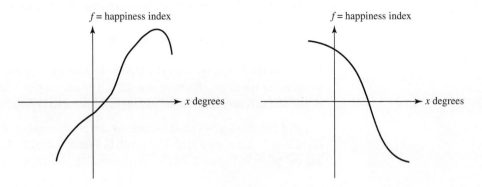

Figure 2.2

Increasing/Decreasing

Let I be an interval in the domain of f. We say that

- f is **increasing** on the interval I if the output of f increases as the input increases, i.e., $x_2 > x_1$ implies $f(x_2) > f(x_1)$ for all x_1, x_2 in I;

- f is **decreasing** on the interval I if the output of f decreases as the input increases, i.e., $x_2 > x_1$ implies $f(x_2) < f(x_1)$ for all x_1, x_2 in I.

Graphically, this means that f is increasing wherever the graph of f is going uphill from left to right, and decreasing wherever the graph of f is going downhill from left to right.

The functions graphed in Figure 2.3 are all increasing.

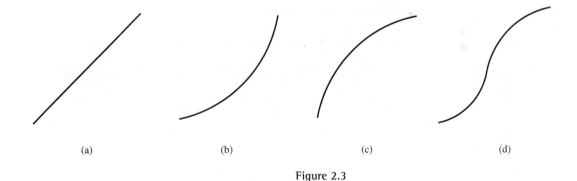

(a)   (b)   (c)   (d)

Figure 2.3

Observations

- The graph in Figure 2.3(a) is increasing at a constant rate over the interval shown; i.e., the steepness is constant.

- The graph in Figure 2.3(b) is increasing at an increasing rate; i.e., the steepness is increasing as x increases.

  We say f is increasing and **concave up**.

■   The graph in Figure 2.3(c) is increasing at a decreasing rate; i.e., the steepness is decreasing as x increases.

We say f is increasing and **concave down**.

In fact, *all* the functions in Figure 2.3 are increasing. The rate of increase is related to the steepness of the slope. (Rates of change will be taken up in Section 2.3; the main point here is that all the functions graphed in Figure 2.3 are increasing.)

The functions graphed in Figure 2.4 are all decreasing. The graph in Figure 2.4(a) is decreasing at a constant rate. The graph in Figure 2.4(b) is decreasing and concave down, and the graph in Figure 2.4(c) is decreasing and concave up.[3]

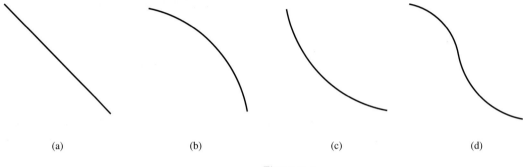

(a)   (b)   (c)   (d)

Figure 2.4

As an informal working definition, we say that a curve is **concave up** if it could be a part of a bowl holding water and **concave down** if it could be part of a bowl held upside-down, spilling its contents. See Figure 2.5.

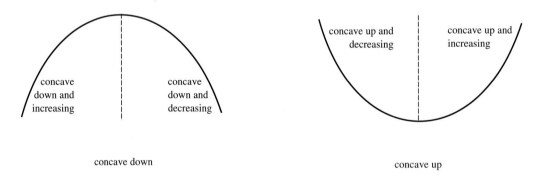

concave down   concave up

Figure 2.5

---

[3] We will *not* use the phrases "decreasing at an increasing rate" and "decreasing at a decreasing rate" because the meaning is ambiguous. Consider Figure 2.1(b), for instance. The function is definitely decreasing. Although the steepness of the *decline* is decreasing, the number we use to describe the steepness is becoming less negative; as a value becomes less negative, it is increasing.

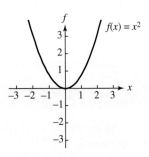

Figure 2.6

As a concrete example, consider the function $f(x) = x^2$, which is graphed in Figure 2.6.

f is positive on $(-\infty, 0)$ and $(0, \infty)$.

f is increasing on $[0, \infty)$. f is decreasing on $(-\infty, 0]$.[4]

f is concave up on $(-\infty, \infty)$.

◆ EXAMPLE 2.1   A city depends on a nearby reservoir for its drinking water. Let $f(t)$ be the function that assigns to each time t the amount of water in the reservoir in comparison to a benchmark amount.[5]

Suppose the benchmark chosen is the average amount of water in the reservoir over the 10-year period 1987–1996. Figure 2.7 is a graph of f on the domain $[-3, 6]$, where t is measured in months and $t = 0$ corresponds to January 1, 1997.

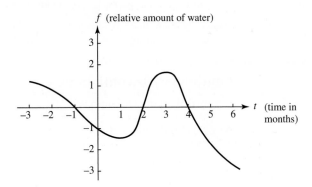

Figure 2.7

---

[4] Note that 0 is included in both the interval on which f is increasing *and* the interval on which f is decreasing. This works according to our definitions of increasing and decreasing.

[5] The idea of a benchmark is critical to making a mathematical model of real-world phenomena. As an example, consider the civil calendar predominately in use today, the Gregorian calendar introduced by Pope Gregory XIII in 1582 to replace the Julian calendar. The year 2000 has meaning only in reference to a benchmark year, the year in which at one point it was thought that Christ was born. (In fact, according to the Encyclopedia Britannica, historical evidence has shown that the benchmark was off its mark by about six years.) Other calendar systems have other benchmarks. For example, in China between 1911 and 1949 the Chinese government adopted a calendar taking the year of that government's founding (1911) as year zero. The Gregorian calendar sets January 1 as the first day of the year, replacing a previous benchmark that made March 25 the first day of the year.

■ f is positive on the intervals $[-3, -1)$ and $(2, 4)$. On these time intervals the reservoir had more water than the benchmark amount.

■ f is negative on the intervals $(-1, 2)$ and $(4, 6]$. On these time intervals the reservoir had less water than the benchmark amount.

■ The zeros of f are at $t = -1$, $t = 2$, and $t = 4$. (These are the t-intercepts.) At these times the water level is equal to the benchmark level.

■ f is increasing on $[1, 3]$. On this time interval the amount of water in the reservoir is increasing.

■ f is decreasing on $[-3, 1]$ and $[3, 6]$. On these time intervals the amount of water in the reservoir is decreasing.

EXERCISE 2.2   For each of the four parts of this problem, sketch the graph of a function f with domain $[-3, 1]$ that has the characteristics indicated.

(a) f is positive and increasing   (b) f is negative and increasing

(c) f is positive and decreasing   (d) f is negative and decreasing

EXERCISE 2.3   Repeat Exercise 2.2, expanding the domain to all real numbers. Make sure the characteristics hold throughout the interval $(-\infty, \infty)$.

EXERCISE 2.4   Refer to the function in Example 2.1.

(a) On what time interval(s) is the amount of water in the reservoir below the benchmark and decreasing?

(b) On what time interval(s) is the amount of water above the benchmark and decreasing?

(c) If the benchmark for the water level was taken to be the level on October 1, 1996, what would the graph of the function look like?

Continuous/Discontinuous

Informal Working Definitions

■ f is **continuous** at $x = a$ if a is in the domain of f, $f(a)$ is defined, and $f(x)$ is arbitrarily close to $f(a)$ whenever x is sufficiently close to a.[6] In other words, the output values for f near $x = a$ are close to $f(a)$; the graph has no gap or hole at a. Very informally, we can say that its graph can be drawn without lifting pencil from paper.

■ f is **discontinuous** about $x = a$ if its graph has a hole, a gap, or a vertical asymptote there.

If f is continuous on $(-\infty, \infty)$ then its graph will not have holes, gaps, or vertical asymptotes. In this text, if we say "f is continuous" with no qualifiers we mean that f is continuous on $(-\infty, \infty)$. In order to determine whether or not f is continuous at $x = a$ it

---

[6] Continuity at a point will be formally defined in Chapter 7.

is necessary to investigate the values of f around a. We will arrive at a formal definition of the term continuous after discussing limits; for the time being, we want you to have an intuitive notion of continuity. Study Figure 2.8.

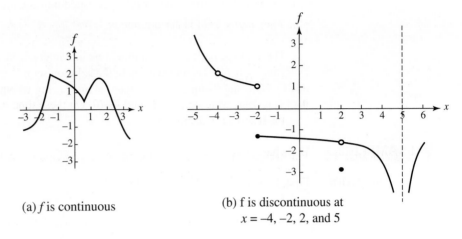

(a) f is continuous

(b) f is discontinuous at
$x = -4, -2, 2,$ and 5

Figure 2.8

Observations

■   If f is a continuous function, then f can change sign (from positive to negative or vice versa) only on either side of a zero of f. This simply means that the graph of a continuous function must cross the x-axis in order to make the transition from lying above the axis to lying below it (or vice versa). If f is continuous on $[a, b]$ and has no zeros in $[a, b]$, then its sign cannot change on $[a, b]$.

This "observation," sometimes called the **Zero Value Theorem**, is equivalent to the **Intermediate Value Theorem**, which tells us that if a continuous function takes on the values m and M it also takes on all values between m and M.[7] (These theorems may seem quite reasonable to you, but in fact quite a bit of work is required to prove them. For starters, a proof requires a precise definition of continuity, which will be presented in Chapter 7.)

■   If f is a continuous function, then wherever f changes from increasing to decreasing or vice versa it must have a "turning point" (a maximum or minimum value).

If you are dubious about either of the points above, try to draw counterexamples. For instance, try to draw a function that changes sign but doesn't cross the horizontal axis—and remember that you can't lift your pencil off the paper.

Inequalities

The Intermediate Value Theorem can be applied not only to determine the sign of a function but also to solve inequalities. Suppose we want to find all x such that $f(x) > g(x)$ (or,

---

[7] To see the equivalence, notice that we can consider any value between m and M to be the benchmark, or zero value.

equivalently, $f(x) - g(x) > 0$). The inequality can tip either by passing through a balance where $f(x) = g(x)$ or by "hopping over" the balance. Therefore, plot on a number line the values of x for which

 i. $f(x) = g(x)$ (the roots of the "corresponding equation"), or
 ii. $f(x)$ and/or $g(x)$ is discontinuous.

These values partition the number line into intervals in which the balance does *not* tip. Determine whether each interval is in the solution set by testing a point in the interval. Do this by choosing a number in the interval and seeing whether or not that number satisfies the inequality.

◆ **EXAMPLE 2.2**   For what x is $x^2 > x$?

SOLUTION   First, look at $x^2 = x$. Solving yields:

$$x^2 - x = 0$$
$$x(x - 1) = 0$$
$$x = 0 \quad \text{or} \quad x = 1.$$

Second, note that both x^2 and x are continuous functions.
Plotting the roots of the corresponding equation on a number line and testing points in each interval yields:

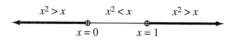

Figure 2.9

So $x \in (-\infty, 0)$ or $(1, \infty)$.   ◆

◆ **EXAMPLE 2.3**   *Postage costs: First-class mail.* The cost of sending a first-class letter inside the United States through the U.S. Postal Service depends on the weight of the letter. According to the system in use in 1997, the cost was $0.32 for letters weighing less than or equal to 1 ounce, and $0.23 for each additional ounce, up to a maximum of 11 ounces. Thus, a letter weighing more than 10 and less than or equal to 11 ounces would cost $2.62 ($= 0.32 + 10 \cdot 0.23$). For mail weighing more than 11 ounces and less than 2 pounds, the cost is $3.00. The cost then increases by $1.00 for each additional pound, up to a maximum of 5 pounds, which would cost $6.00. The cost for packages with weights greater than 5 pounds depends on the place of origin and destination. Therefore, the cost of mailing a parcel is a function of weight for parcels weighing up to 5 pounds.
    Below is a graph that represents this situation.

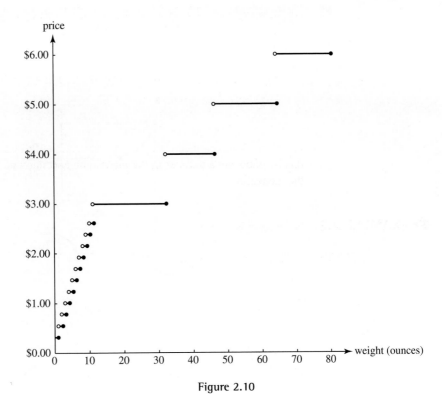

Figure 2.10

i. Explain how the description of the pricing system determines where the graph should have hollow circles and filled circles.

ii. What is the cost function's domain and range?

iii. Is this function continuous?

SOLUTION

i. A hollow circle denotes a point that is not part of the graph. For example, the point $(11, 3)$ should not be part of the graph because an 11-ounce letter costs $2.62. However, if a letter weighs just a tad more than 11 ounces the price jumps to $3.00. To represent this situation graphically, we draw a hollow circle at the point $(11, 3)$ and a filled circle at $(11, 2.62)$.

ii. The domain is any weight between 0 and 80 ounces, not including 0, i.e., $(0, 80]$. The range is a set of discrete points: $\{0.32, 0.55, 0.78, 1.01, 1.24, 1.47, 1.70, 1.93, 2.16, 2.39, 2.62, 3.00, 4.00, 5.00, 6.00\}$. These are all the possible outputs (prices) of the function; in 1997 it was not possible, for example, to have a first-class domestic letter costing $0.42 to mail.

iii. The function is discontinuous at every point where the price "jumps" to a higher level.

◆

PROBLEMS FOR SECTION 2.1

1. Below are the graphs of functions.

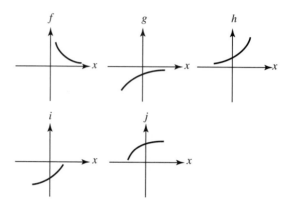

(a) Which functions are positive and increasing?

(b) Which functions are negative and increasing?

(c) Which functions are increasing at an increasing rate?

(d) Which functions are increasing at a decreasing rate?

(e) Which functions have graphs that are concave up?

2. A function f is decreasing throughout its domain $[-8, -1]$. Can we determine where f takes on its largest value? Does your answer depend upon whether or not f is continuous?

3. The graph of $f(t)$ is the line drawn below. Let $A(x)$ be the area between the line and the t-axis from $t = -2$ to $t = x$, where $x \in [-2, 8]$.

(a) Is $A(x)$ a function?

(b) If $A(x)$ is a function, is it increasing, or decreasing?

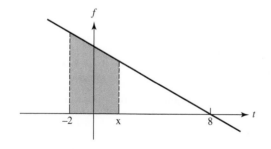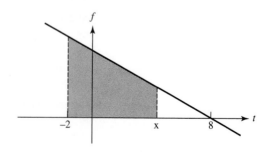

4. Consider the graph of f drawn below. The domain of f is $[-6, 10]$.

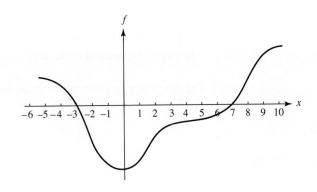

(a) On what intervals is f negative?

(b) On what intervals is f decreasing?

(c) On what intervals is f concave down?

(d) On what intervals is f concave up and decreasing?

(e) On what intervals is f concave down and increasing?

5. In the peak of apple season in rural Vermont, migratory workers are often hired to pick apples. In this problem, model the situation in one orchard. *Depending upon your interpretation of the situation, your model may differ from those of your classmates. Discuss your answers with others.*

(a) Sketch a graph of the time it takes to harvest the crop of apples as a function of the number of people picking apples. Label your axes.

(b) Is the graph you drew a straight line? Why or why not?

(c) Is the graph you drew continuous or discontinuous? Explain.

(d) Does the graph you drew intersect the vertical axis? The horizontal axis? If so, where? If not, why not?

6. Sketch a graph of a continuous function defined for all real numbers with all three characteristics listed. If it is impossible to do this, say so, and draw the graph of a function with the three characteristics on the domain $[-1, 1]$.

(a) f is positive, increasing, and concave up

(b) f is positive, increasing, and concave down

(c) f is negative, increasing, and concave up

(d) f is negative, increasing, and concave down

(e) f is positive, decreasing, and concave up

(f) f is positive, decreasing, and concave down

(g) f is negative, decreasing, and concave up

(h) f is negative, decreasing, and concave down

For each of the functions in Problems 7 through 9 estimate:

(a) *the zeros of the function,*

(b) *where the function is positive,*

(c) *where the function is increasing,*

(d) *where the function is negative and increasing.*

7. The function f is given by the graph below.

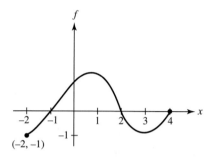

8. The function g is given by the graph below.

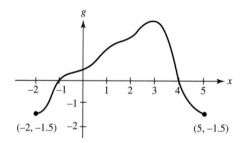

9. The function h is given by the graph below.

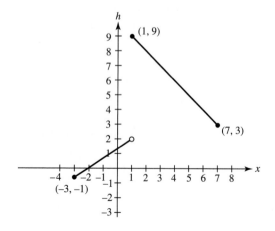

10. Which of the following functions is continuous at $x = 2$?

  (a) $f(x) = x + 3$

  (b) $f(x) = \frac{x+3}{x-2}$

  (c) $f(x) = \frac{x^2+x-6}{x-2}$

11. Which of the following functions is continuous at $x = 1$?

  (a) $f(x) = 5$

  (b) $f(x) = \frac{5x}{x}$

  (c) $f(x) = \frac{5x(x-1)}{x-1}$

  (d) $f(x) = \frac{x^2(x-1)}{x-1}$

2.2   A POCKETFUL OF FUNCTIONS: SOME BASIC EXAMPLES

Throughout this text we will be looking at functions analytically and graphically. Consider any equation relating two variables. Every point that lies on the graph of the equation satisfies the equation. Conversely, every point whose coordinates satisfy the equation lies on the graph of the equation. The correspondence between pairs of x and y that constitute the coordinates of a point on a graph and that satisfy an equation was a key insight of Rene Descartes (1596–1650) that unified geometry and algebra.[8] As a consequence of this insight, the fields of geometry and algebra became irrevocably intertwined, opening the door to much of modern mathematics, including calculus.

  We have been discussing functions and their graphs in a very general way, but it is always strategically wise to include a few simple, concrete examples for reference. In this section you will become familiar with a small sampling of functions, functions that you can carry about and pull out of your back pocket at any moment. As we go along in the text this collection will grow; you will learn that these functions belong to larger families of functions sharing some common characteristics, and you'll also be introduced to a greater variety of families of functions. But for the time being, we will become familiar with a few individual functions.

A Few Basic Examples

Consider the following function:

$$f(x) = x \quad g(x) = x^2 \quad h(x) = |x| \quad j(x) = \tfrac{1}{x}$$

■ The function f is the **identity function**; its output is identical to its input.

■ The function g is the **squaring function**; its output is obtained by squaring its input.

■ The function h is the **absolute value function**; its output is the magnitude (size) of its input. In other words, if the input is positive (or nonnegative), then the output is identical to the input; if the input is not positive, then change its sign to obtain the

---

[8] We have already discussed this correspondence when looking at the graph of a function. In fact, it holds for the graph of any equation, whether defining a function or not.

output. We can define the absolute value function $|x|$ analytically as follows.

$$|x| = \begin{cases} x & \text{if } x \geq 0; \\ -x & \text{if } x < 0. \end{cases}$$

■   The function j is the **reciprocal function**; the output is the reciprocal of the input.

$$\text{output} = \frac{1}{\text{input}}.$$

The graphs of these functions are given below. Familiarize yourself with these functions by doing the exercises that follow.

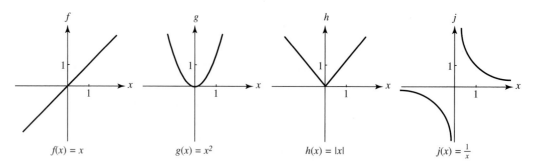

Figure 2.11

EXERCISE 2.5   Look at each of the functions $f(x) = x$, $g(x) = x^2$, $h(x) = |x|$, and $j(x) = \frac{1}{x}$ one by one and answer the following questions.

(a) What are the domain and range of the function?

(b) Where is the function positive? Negative?

(c) Where is the function increasing? Decreasing?

(d) Is the function continuous? If it is not continuous everywhere, where is it discontinuous?

(e) Is the function 1-to-1?

Answers are provided at the end of the section.

EXERCISE 2.6   Consider the function $j(x) = \frac{1}{x}$.

(a) Why is $j(x)$ never equal to zero?

(b) Why is zero not in the domain of j?

(c) As x increases without bound what happens to the values of j? As x decreases (becomes more and more negative) without bound what happens to the values of j? How is this information displayed on the graph of j? How is it displayed on your graphing calculator's rendition of the graph?

(d) Find $j(x)$ for $x = 0.01$, 0.001, and 0.0001. If x is a positive number, as x gets increasingly close to zero, what happens to the values of j? How is this information displayed on the graph?

(e) Find $j(x)$ for $x = -0.01$, -0.001, and -0.0001. If x is a negative number, as x gets increasingly close to zero, what happens to the values of j? How is this information displayed on the graph? How is this displayed on your calculator's rendition of the graph?

(f) Fill in the blanks with $>$, $<$, or $=$ as appropriate.

    i. If $x > 0$ then $j(x)$ ____ 0.

    ii. If $x < 0$ then $j(x)$ ____ 0.

    iii. If $|x| > 1$ then $|j(x)|$ ____ 1.

    iv. If $0 < |x| < 1$ then $|j(x)|$ ____ 1.

(g) Translate the statements from part (f) into plain prose. Your translation should be accessible to someone who knows nothing about functions and absolute values.

EXERCISE 2.7   Consider the function $g(x) = x^2$.

(a) Fill in the blanks with $>$, $<$, or $=$ as appropriate.

    i. If $|x|$ ____ 1, then $g(x) > |x|$.

    ii. If $|x|$ ____ 1, then $g(x) < |x|$.

(b) Translate the statements from part (a) into plain prose. Your translation should be accessible to someone who knows nothing about functions and absolute values.

(c) Consider $h(x) = |x|$.

    i. For what values of x is $h(x) > g(x)$?

    ii. For what values of x is $h(x) < g(x)$?

REMARKS

■   The graph of $f(x) = x$ given in Figure 2.11 has the same scale on the x- and y-axes and therefore the line cuts the angle between the axes evenly in two. If different scales were used on the axes this would not be true.[9]

■   For $x \geq 0$ the functions $f(x) = x$ and $h(x) = |x|$ are identical. The graph of $h(x)$ has a sharp corner at $x = 0$. No matter how much you magnify the portion of the graph around that point, the corner remains sharp.

■   The graph of the squaring function $g(x) = x^2$ is called a parabola. Notice that $g(-3) = (-3)^2$, not -3^2. Therefore $g(3) = g(-3)$ and, more generally, $g(x) = g(-x)$. The graph of g is **symmetric** about the y-axis; the graph is a mirror image on either side of the y-axis. You could fold the graph along the y-axis and the function would exactly match itself on either side of the y-axis. The characteristic $g(x) = g(-x)$ guarantees this symmetry.

■   The graph of $j(x) = \frac{1}{x}$ has a vertical asymptote at $x = 0$ and a horizontal asymptote at $y = 0$.

■   If $y = \frac{1}{x}$, then y is inversely proportional to x.

---

[9] If you use a graphing calculator to look at the graph of a function you can set the range and domain so that the scales are very different. You should be aware that your choice of domain and range can make the line $f(x) = x$ look almost horizontal or almost vertical.

A is inversely proportional to *B* if

$$A = \frac{k}{B}$$

for some constant k. As *B* increases *A* decreases, and vice versa.

This type of relationship arises frequently in the world around us. For example, the amount of time required to complete a trip is inversely proportional to the rate at which one travels.

Informal Introduction to Asymptotes

The function $Q(x)$ has a **vertical asymptote** at $x = a$ if Q is undefined at $x = a$ and as x gets closer and closer to a the magnitude of $Q(x)$, written $|Q(x)|$, increases without bound. Because Q is undefined at its vertical asymptotes, the graph of a function will never cross its vertical asymptote. Around a vertical asymptote, the graph will resemble one of the four options shown below. If Q has a one-sided vertical asymptote at $x = a$ then as x approaches a from one side $|Q(x)|$ increases without bound.

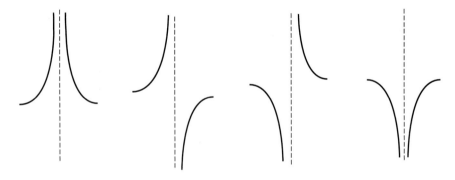

Figure 2.12

For the function $j(x) = \frac{1}{x}$, as x approaches 0 from the right, the value $j(x)$ grows without bound. We can write this symbolically:

$$\text{as } x \to 0^+ \quad j(x) \to \infty.$$

"As $x \to 0^+$" means "as x approaches zero from the right"; "$j(x) \to \infty$" means "$j(x)$ grows without bound." On the other hand, as x approaches 0 from the left, the value of $j(x)$ gets increasingly negative without bound. We can write this symbolically:

$$\text{as } x \to 0^- \quad j(x) \to -\infty.$$

The function Q has a **horizontal asymptote** at $y = c$ if, as the magnitude of x increases without bound, the value of $Q(x)$ gets closer and closer to c. A horizontal asymptote tells us only about the behavior of the function for $|x|$ very large; therefore, it can be crossed. A one-sided horizontal asymptote tells us what happens to Q as x either increases or decreases without bound. It can be crossed. As $|x|$ increases without bound, $j(x) = \frac{1}{x}$ approaches zero. We can express this symbolically by writing

$$\text{as } x \to \infty \quad j(x) \to 0 \quad \text{and} \quad \text{as } x \to -\infty \quad j(x) \to 0.$$

Even and Odd Symmetry

A function may be even, odd, or neither.

Definition

A function f is **even** if $f(-x) = f(x)$ for all x in the domain of f. This means that the height of the graph of f is the same at x and at $-x$. The graph of an even function is a mirror image about the y-axis.

A function is **odd** if $f(-x) = -f(x)$ for all x in the domain of f. An odd function is said to be symmetric about the origin. For further description see the answers to Exercise 2.8.

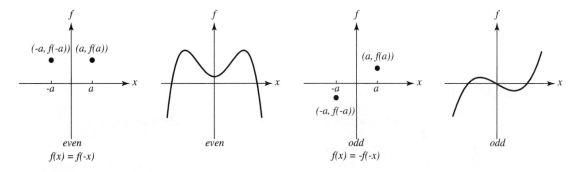

Figure 2.13

EXERCISE 2.8   In this exercise we return to the four functions we introduced at the beginning of this section:

$$f(x) = x, \quad g(x) = x^2, \quad h(x) = |x|, \quad and \quad j(x) = \frac{1}{x}$$

(a) Which of functions f, g, h, and j are even?

(b) Which of functions f, g, h, and j are odd?

(c) Describe in words the characteristic of the graph of an odd function. (The graph of an odd function is *not* a mirror about the x-axis.)

Answers to Exercise 2.8 are provided at the end of the section.

Working with Absolute Values

To look at the absolute value of a quantity is to look only at its magnitude, to make it nonnegative.

Analytic Principle for Working with Absolute Values

To deal with an absolute value, *remove* it by breaking the problem up into two cases.

Case (1): If the expression inside the absolute value is nonnegative, then simply remove the absolute value signs.

Case (2): If the expression inside the absolute value is negative, then change its sign and remove the absolute values. It follows that:

$$|x - 3| = \begin{cases} x - 3 & \text{for } x \geq 3, \\ -(x - 3) & \text{for } x \leq 3. \end{cases}$$

Geometric Principle for Working with Absolute Values

Equivalent to the preceding analytic definition is the "geometric" definition: The absolute value of x is the distance between x and 0 on the number line. More generally, this geometric definition tells us that $|x - a|$ is the distance between x and a on the number line. It follows that $|x - 3|$ is the distance between x and 3, while $|x + 2|$ (which equals $|x - (-2)|$) is the distance between x and -2.

We can use this geometric interpretation to solve equations and inequalities involving absolute values. For example, we can interpret the inequality

$$|x - 100| \leq 0.1$$

as "x differs from 100 by no more than 0.1," or $x \in [99.9, 100.1]$.

◆ EXAMPLE 2.4   Solve:

(a) $|x - 1| = 2$ (b) $|x + 1| \leq 2$ (c) $|2x + 3| > 1$

SOLUTIONS

(a) Geometric Approach:

$$|x - 1| = 2$$

(the distance between x and 1) $= 2$

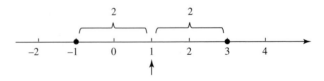

Figure 2.14

$$x = -1 \quad \text{or} \quad x = 3$$

Analytic Approach:

Case (1): $x - 1 \geq 0$	Case (2): $x - 1 < 0$
$x - 1 = 2$	$-(x - 1) = 2$
$x = 3$	$-x + 1 = 2$
	$x = -1$

Check that when $x = 3$, $x - 1 \geq 0$ and when $x = -1$, $x - 1 < 0$. Alternatively, check the answers in the original problem.

(b) Geometric Approach:

$$|x + 1| \leq 2$$

$$|x - (-1)| \leq 2$$

(the distance between x and -1) ≤ 2

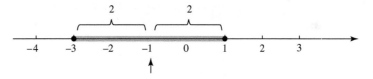

Figure 2.15

x is between -3 and 1, including both endpoints

$$x \in [-3, 1]$$

Analytic Approach: Solve the corresponding equation: $|x + 1| = 2$.

Case (1): $x + 1 \geq 0$	Case (2): $x + 1 < 0$
$x + 1 = 2$	$-(x + 1) = 2$
$x = 1$	$-x - 1 = 2$
	$x = -3$

These two solutions partition the number line into three intervals. To determine whether each of these intervals contains acceptable values of x, substitute one number from each of these intervals into the original equation. This technique gives us $-3 \leq x \leq 1$.

(c) Geometric Approach:

$$|2x + 3| > 1$$

$$|2x - (-3)| > 1$$

(the distance between $2x$ and -3) > 1

We'll focus on $2x$ and solve for x later.

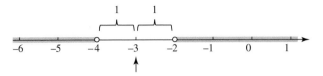

Figure 2.16

$$2x < -4 \quad \text{or} \quad 2x > -2$$
$$x < -2 \quad \text{or} \quad x > -1$$

Analytic Approach: Solve the corresponding equation $|2x + 3| = 1$.

Case (1): $2x + 3 \geq 0$	**Case (2): $2x + 3 < 0$**
$2x + 3 \;=\; 1$	$-(2x + 3) \;=\; 1$
$2x \;=\; -2$	$2x + 3 \;=\; -1$
$x \;=\; -1$	$2x \;=\; -4$
	$x \;=\; -2$

These two solutions once again divide the number line into three intervals. Check to see if a number inside each of these intervals is a solution to determine whether all numbers within the interval are solutions. Here we discover $x < -2$ or $x > -1$.   ◆

EXERCISE 2.9   Solve for x. Do these problems in two ways, first using the analytic definition of the absolute value and then using the geometric definition of the absolute value.

(a) $|3 - x| = 2$
(b) $|2x + 1| = 4$ ◆

Answers to Exercise 2.9 are provided at the end of the section.

Answers to Selected Exercises

Answers to Exercise 2.5

■   The linear function $f(x) = x$
(a) The domain of f is all real numbers. Likewise, the range of f is all real numbers.
(b) f is positive for x positive, negative for x negative.
(c) The graph of f is a straight line. The graph is always increasing.
(d) f is a continuous function.
(e) f is 1-to-1.

■ $g(x) = x^2$
(a) The domain of g is all real numbers. The range of g is all nonnegative real numbers.
(b) g is positive for all x except $x = 0$. $g(0) = 0$.
(c) g is decreasing on $(-\infty, 0]$ and increasing on $[0, \infty)$.
(d) g is a continuous function.

(e) g is *not* 1-to-1. By restricting the domain to nonnegative real numbers the function can be made 1-to-1.

■   The absolute value function $h(x) = |x|$

(a) The domain of h is all real numbers. The range of h is all nonnegative real numbers.

(b) h is positive for all x except $x = 0$. $h(0) = 0$.

(c) h is decreasing on $(-\infty, 0]$ and increasing on $[0, \infty)$.

(d) h is a continuous function.

(e) h is *not* 1-to-1. By restricting the domain to nonnegative real numbers the function can be made 1-to-1.

■ $j(x) = \frac{1}{x}$

(a) The domain of j is all nonzero real numbers. The range of j is all nonzero real numbers.

(b) j is positive for x positive, negative for x negative.

(c) j is decreasing on $(-\infty, 0)$ and decreasing on $(0, \infty)$.

(d) j is undefined and discontinuous at $x = 0$.

(e) j is 1-to-1.

Answers to Exercise 2.8

(a) g and h are even functions.

(b) f and j are odd functions.

(c) The graph of an odd function is said to have symmetry about the origin. What does this mean? Graph the function for all nonnegative x in the domain. The other half of the graph can be obtained by reflecting this portion of the graph across the y-axis and then across the x-axis. (This is equivalent to rotating the selected portion of the graph $180°$ around the origin.)

Answers to Exercise 2.9

(a)   i. Analytic Approach: Remove the absolute value signs by breaking the problem into two cases—one in which the expression inside the absolute value is positive and the other in which the expression inside the absolute value sign is not positive.

Case (1): The expression inside the absolute value is positive. Because $3 - x > 0$, we replace $|3 - x|$ by $3 - x$.

$$3 - x = 2$$
$$-x = -1$$
$$x = 1$$

Case (2): The expression inside the absolute value is not positive. Because $3 - x \leq 0$, we replace $|3 - x|$ by $-(3 - x) = -3 + x$.

$$-3 + x = 2$$
$$x = 5$$

Check that these answers satisfy the original equation.

ii. Geometric Approach:

$$|3 - x| = 2$$

(the distance between 3 and x) $= 2$

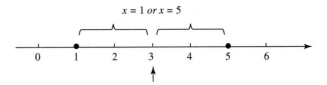

Figure 2.17

(b)   i. Analytic Approach: Remove the absolute value signs by breaking the problem down into two cases.

Case (1): The expression inside the absolute value is positive. Because $2x + 1 > 0$, we replace $|2x + 1|$ by $2x + 1$.

$$2x + 1 = 4$$
$$2x = 3$$
$$x = \frac{3}{2}$$

Case (2): The expression inside the absolute value is not positive.
Because $2x + 1 \leq 0$, we replace $|2x + 1|$ by $-(2x + 1) = -2x - 1$.

$$-2x - 1 = 4$$
$$-2x = 5$$
$$x = -\frac{5}{2}$$

Check that both of these answers satisfy the original equation.

ii. Geometric Approach:

$$|2x + 1| = 4$$
$$|2x - (-1)| = 4$$
(the distance between $2x$ and -1) $= 4$

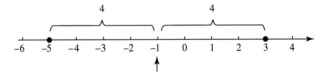

Figure 2.18

$$2x = -5 \quad \text{or} \quad 2x = 3$$
$$x = -\frac{5}{2} \quad \text{or} \quad x = \frac{3}{2}$$

■ PROBLEMS FOR SECTION 2.2

1. This problem applies the definitions of proportionality (also referred to as direct proportionality), and inverse proportionality.

If P is the pressure of a gas, T is its temperature, and V is its volume, then the combined gas law tells us that for a fixed mass of gas

$$\frac{PV}{T} = \text{a constant.}$$

From this statement, demonstrate the following.

(a) If temperature is held constant, then pressure is inversely proportional to volume (Boyle's Law). (This should make sense intuitively; as pressure increases, volume decreases.)

(b) If pressure is held constant, then volume is directly proportional to temperature (Law of Charles and Gay-Lussac). (This too should make sense intuitively; as the temperature of a gas goes up, volume goes up as well.)

2. Physicists define the work done by a force on an object to be the magnitude of the force, F, times the distance, d, that the object is moved. Notice that this definition is different from our conversational use of the word "work." For instance, if you stand holding a 50-pound object stationary for an hour, then according to the physicists' definition you have done *no* work because the object has not moved. According to the physicists' definition, if force is kept constant is the work done proportional to the distance the object moves. Explain.

3. The following problems are warm-up exercises for absolute values.

(a) Express the following without using absolute value signs.
   i. $\big|\, |-5| - |-3|\, \big|$
   ii. $|x - 3|$ (You need two cases.)

(b) Express the following using absolute values.
   i. The distance between -5 and π
   ii. The distance between $\sqrt{3}$ and π

(c) Fill in the blanks in the following table.

	Algebraic Statement	Geometric Statement
i.	$|x - c| > 6$	x is more than 6 units from c
ii.	$|x - 3| > 2$	_____
iii.	_____	c is closer to 0 than b is
iv.	$|x - 3| \leq 4$	_____
v.	_____	w is 6 units from d
vi.	_____	q is at most 18 units from 5

4. Use the geometric interpretation of absolute value to solve the following equations and inequalities. Display the solutions on a number line.

(a) $|x + 3| < 2$

(b) $|x - 5| \leq 3$

(c) $|x - a| = b$,   where a and b are positive

(d) $|x + a| \leq a$,   where a is positive

5. Use the algebraic interpretation of absolute value to solve each of the following. Please display your answers on a number line.

(a) $|2x + 1/2| \leq 6$ (b) $|3x - 4| > 8$

6. Solve the following inequalities. Display your answers on a number line using interval notation.

(a) $-2x - 7 < -8$ (b) $|-2x - 8| \geq 2$ (c) $|-2x - 8| < 2$

7. Solve these inequalities and explain your answers: *Think carefully.*

(a) $|3x - 4| > -4$ (b) $|3x - 4| > 0$ (c) $|3x - 4| < -4$

8. Which of the following statements are true and which are not always true? For a statement to be true, it must *always* be true. If a statement is not always true, give a counterexample. To give a counterexample is to give an example of values for x and y for which the statement is false.

(a) $|x^2| = |x|^2$ (b) $|x| = |-x|$

(c) $|x - y| = |x| - |y|$ (d) $|x + y| = |x| + |y|$

(e) $\frac{|x|}{|y|} = \left|\frac{x}{y}\right|$ for $y \neq 0$ (f) $|x||y| = |xy|$

(Four of the statements are true.)

9. Use absolute values to write the following statement more compactly: Whenever x is within 0.02 of 7, $f(x)$ differs from 19 by no more than 0.3.

10. Refer to the accompanying figure. Given the graph of $f(x)$ below, for which values of x is:

(a) $f(x) = 0$?   (b) $f(x)$ discontinuous?   (c) $f(x) < 0$?

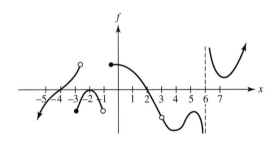

11. Let $h(x) = |x|$. Solve the following. Do parts (a) and (b) twice—once using an analytic approach and once using a geometric approach.

(a) $h(x + 2) \leq 3$ (b) $h(x - 1) = 5$ (c) $h(x + 3) \geq 0.1$ (d) $h(3x + 1) > 4$

12. Let $h(x) = |x|$. Solve the following.

(a) $2h(x) > 4$ (b) $h(2x - 1) \leq 3$ (c) $h(x^2 - 1) \geq 0$

In Problems 13 through 18, determine whether the function is even, odd, or neither.

13. (a) $f(x) = x^2 + 3x^4$ (b) $g(x) = \frac{1}{x^2 + 3x^4}$

14. (a) $f(x) = 2x^3 + 3x$ (b) $g(x) = 2x^3 + 3x + 1$

15. (a) $f(x) = \frac{x^2-1}{x^3}$ (b) $g(x) = \frac{x^2-1}{x^4+1}$

16. (a) $f(x) = |x| + 3$ (b) $g(x) = -2|x|$

17. (a) $f(x) = \frac{1}{x^2}$ (b) $g(x) = \frac{2}{x^3}$

18. (a) $f(x) = x + \frac{1}{x}$ (b) $g(x) = 1 + \frac{1}{x}$

19. Let $f(x) = \frac{1}{x}$. Solve the following.

   (a) $f(x^2) = 1$ (b) $-f(x) = f(x-1)$ (c) $2f(x-2) = f(x+3)$

20. A function can be neither even nor odd. For example, consider $f(x) = x^3 + x^2$. Can a function be both even and odd? If your answer is yes, give an example. Can you give two examples?

2.3   AVERAGE RATES OF CHANGE

The problem of how to determine the rate at which the output of a function is changing is the fundamental question that gives rise to differential calculus. In Section 2.1 we discussed characterizing a function by describing where it is increasing and where it is decreasing. Suppose we want to be more specific and determine the function's *rate* of increase or *rate* of decrease. We first explored this question when looking at the calibration of a bottle and discussing how a change in the volume of water changes the height. In the case of a cylindrical beaker we saw that the ratio

$$\frac{\text{change in height}}{\text{change in volume}}$$

is constant. We refer to this ratio as the rate of change of water level "with respect to volume." In the case of the conical flask, we saw that this ratio is not constant. We continue our exploration of rates of change in this section.

◆ EXAMPLE 2.5   Below is a graph of $y = T(t)$, which gives the temperature in Green Bay, Wisconsin, as a function of time between 6 A.M. and 11 A.M. on a cold day. We can see from the graph that the temperature is increasing throughout this time interval (sometimes negative and sometimes positive, but always increasing).

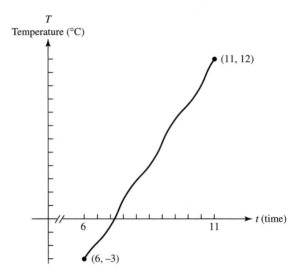

Suppose we want to determine how fast the temperature is increasing between 6 A.M. and 11 A.M. ("Why?" you may ask. Perhaps a football game is starting at noon and you would like to guess the temperature at the start of the game.) Although the temperature is not increasing at a constant rate, we know that it has increased 15 degrees (from -3 to $+12$) in a 5-hour span, so it is increasing at an average rate of 3 degrees per hour. In mathematical notation, we calculate the average rate of change of temperature with respect to time as follows:

$$\text{Average rate of change of temperature} = \frac{\text{change in temperature}}{\text{change in time}}$$
$$= \frac{\text{final temperature} - \text{initial temperature}}{\text{final time} - \text{initial time}}$$
$$= \frac{15 \text{ degrees}}{5 \text{ hours}}$$
$$= 3\frac{\text{degrees}}{\text{hour}}$$

(Therefore, to answer the question posed above, you might guess that the temperature will increase about three more degrees in the next hour before the game starts, making the kickoff temperature a balmy 15 degrees—perfect weather for going to a Packers game.)   ◆

We'll use the next example to distinguish between three different notions:

   i. Change in price is (final price − initial price).

   ii. Percent change in price is $\frac{\text{change in price}}{\text{initial price}}$.
   (This gives percent as a decimal: $0.05 = 5\%$.)

   iii. Average rate of change of price with respect to time is $\frac{\text{change in price}}{\text{change in time}}$.

◆ **EXAMPLE 2.6**   The *Boston Globe* (summer, 1999) reports that average apartment-rental prices in the Allston-Brighton neighborhood of Boston have gone up 53 percent in the past six years. For instance, a two-bedroom apartment that was being rented for $800 in 1993 could be renting for $1224 in 1999. Find:

    i. the change in price,

    ii. the percent change in price,

    iii. the average rate of change in price over the six-year period specified.

SOLUTION   i. The change in price is $424. It is measured in dollars.

    ii. The percent change in price is $\frac{\$424}{\$800} = 0.53 = 53\%$. It is unitless.

    iii. The average rate of change of the price with respect to time is $\frac{\$424}{6 \text{ years}} \approx 70.67$ $/year. It is measured in dollars per year.   ◆

In this section we will be focusing on average rate of change.

Definition

The **average rate of change** of a function $y = f(x)$ over the interval from $x = a$ to $x = b$ is given by

$$\text{average rate of change of } f \text{ on } [a, b] = \frac{f(b) - f(a)}{b - a}.$$

In other words,

$$\text{average rate of change of } f = \frac{\Delta f}{\Delta x} = \frac{\Delta y}{\Delta x}.$$

Geometrically, the average rate of change represents the slope of the line between the two points used. This is because the average rate of change of f on $[a, b]$ is given by

$$\frac{\text{change in output}}{\text{change in input}} = \frac{\text{change in } y}{\text{change in } x} = \frac{\text{rise}}{\text{run}},$$

which gives the slope of a straight line through the points $(a, f(a))$ and $(b, f(b))$.

A line through two points on a curve is called a **secant line**. Therefore, the average rate of change of f on $[a, b]$ is the slope of the secant line through the points $(a, f(a))$ and $(b, f(b))$.

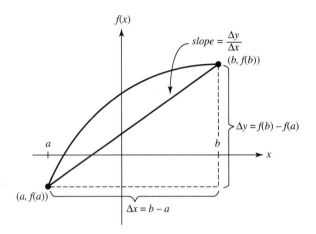

Figure 2.19

Average velocity over a time interval is the average rate of change of position with respect to time:

$$\text{average velocity} = \frac{\Delta \text{position}}{\Delta \text{time}}.$$

Suppose, for example, that we're analyzing the motion of a cheetah. If $s(t)$ gives the cheetah's position at time t, then the cheetah's average velocity from time $t = 1$ to time $t = 5$ is given by:

$$\frac{\Delta \text{position}}{\Delta \text{time}} = \frac{\Delta s}{\Delta t} = \frac{s(5) - s(1)}{5 - 1}.$$

In colloquial English, to accelerate is to pick up speed. Mathematically, acceleration refers to the rate of change of velocity with respect to time:

$$\text{average acceleration} = \frac{\Delta \text{velocity}}{\Delta \text{time}}.$$

◆ **EXAMPLE 2.7**   Suppose we are filling a bucket with water. We have the following information, where time is measured in seconds, amount of water in liters, and height of the water in centimeters.

Time	Amount of Water	Height of Water
0	0	0
1	1.5	4
2	4	9

The average rate of change of volume with respect to time over the time interval [0, 1] is given by

$$\frac{\Delta \text{volume}}{\Delta \text{time}} = \frac{1.5 - 0}{1 - 0} \frac{\text{liters}}{\text{sec}} = 1.5 \frac{\text{liters}}{\text{sec}}.$$

The average rate of change of volume with respect to time over the time interval [1, 2] is given by

$$\frac{\Delta \text{volume}}{\Delta \text{time}} = \frac{4 - 1.5}{2 - 1} \frac{\text{liters}}{\text{sec}} = 2.5 \frac{\text{liters}}{\text{sec}}.$$

The average rate of change of height with respect to time over the time interval [2, 0] is given by

$$\frac{\Delta \text{height}}{\Delta \text{time}} = \frac{9 - 0}{2 - 0} \frac{\text{cm}}{\text{sec}} = 4.5 \frac{\text{cm}}{\text{sec}}.$$

The average rate of change of height with respect to volume as the volume increases from 1.5 to 4 liters is given by

$$\frac{\Delta \text{height}}{\Delta \text{volume}} = \frac{9 - 4}{4 - 1.5} \frac{\text{cm}}{\text{liter}} = \frac{5}{2.5} \frac{\text{cm}}{\text{liter}} = 2 \frac{\text{cm}}{\text{liter}}. \quad \blacklozenge$$

◆ **EXAMPLE 2.8**   A trucker drives west a distance of 240 miles stopping only once to get gas. He begins the trip parked at a truck stop and ends the trip parked at another truck stop. The trip takes him 4 hours. What is his average velocity? Did he ever exceed the 60-mile-per-hour speed limit?

SOLUTION   His average velocity for the trip is given by

$$\frac{\text{change in position}}{\text{change in time}} = \frac{240 \text{ miles}}{4 \text{ hours}} = 60 \text{ mph}.$$

He stopped once for gas and began and ended the trip with zero velocity; therefore he wasn't traveling at 60 mph all the time. There must have been some times when his speed exceeded the 60 mph speed limit. While verifying this mathematically takes some work, this conclusion should make logical sense to you.[10] His velocity is varying; he is not always traveling at 60 mph.

Notice that his average velocity is *not* the average of his final and initial velocities. His final and initial velocities are both zero, but his average velocity is *certainly* positive. We will return to rates of change in Chapter 5.   ◆

PROBLEMS FOR SECTION 2.3

1. The average price of an 8-ounce container of yogurt in upstate New York was 35 cents in 1970. In 2000 the average price had risen to 89 cents.
   (a) What is the price increase?
   (b) What is the percent increase in price?
   (c) What is the average rate of change in price from 1970 to 2000?

2. The average price of a 12-ounce cup of coffee in Seattle is modeled by the function $p(t)$, where t is the number of years since 1950 and $p(t)$ is price in dollars. Express the following using functional notation.

---

[10] The theorem assuring us that this is true is called the Mean Value Theorem. It is proven in Appendix C.

(a) The increase in the price of a cup of coffee in Seattle from 1970 to 2000

(b) The percent increase in the price of a cup of coffee in Seattle from 1970 to 2000

(c) The average rate of change of the price of a cup of coffee from 1970 to 2000

3. The graph of a function f is given below.

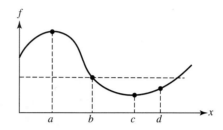

(a) Put in ascending order (smallest to largest) $f(a)$, $f(b)$, $f(c)$, $f(d)$.

(b) Determine which of the expressions listed below are positive. Then put the expressions in ascending order, with < or = signs between them.

$$f(b) - f(a) \qquad f(b) - f(c) \qquad f(d) - f(c)$$

(c) Determine which of the expressions listed below are positive. Then put the expressions in ascending order, with < or = signs between them.

$$\frac{f(b) - f(a)}{b - a} \quad \frac{f(a) - f(b)}{a - b} \quad \frac{f(c) - f(b)}{c - b} \quad \frac{f(d) - f(c)}{d - c} \quad \frac{f(d) - f(b)}{d - b}$$

4. Consider the statement "Colleges report that their tuition increases are slowing down." Suppose we set $t = 0$ to be three years before this statement was made and measure time in years. If we let $T(t)$ be the average college tuition in year t, put the following expressions in ascending order (the smallest first), assuming the statement is true.

$$T(3) - T(2) \quad 0 \quad T(2) - T(1) \quad \frac{T(3) - T(1)}{3 - 1}$$

5. The *Boston Globe* (August 31, 1999, p.1) reports: "While the number of AIDS deaths continues to drop nationally, the rapid rate of decline that had been attributed to new drugs is starting to slow dramatically." The newspaper supplies the following data.

Year	Number of Deaths Attributed to AIDS
1995	149,351
1996	36,792
1997	21,222
1998	17,047

Let $D(t)$ be the number of deaths from AIDS in year t, where t is measured in years and $t = 0$ corresponds to 1995.

(a) Is $D(t)$ positive or negative? Increasing or decreasing?

(b) What is the average rate of change of $D(t)$ from $t = 0$ to $t = 1$? What is the percent change in $D(t)$ over that year?

(c) What is the average rate of change of $D(t)$ from $t = 1$ to $t = 2$? What is the percent change in $D(t)$ over that year?

(d) What is the average rate of change of $D(t)$ from $t = 2$ to $t = 3$? What is the percent change in $D(t)$ over that year?

6. The average rate of change of a function f over the interval $a \le x \le b$ is defined to be

$$\frac{\Delta f}{\Delta x} = \frac{f(b) - f(a)}{b - a}.$$

Let $P(t)$ be the size of a population at time t. The graph of $P(t)$ is given below.

(a) Using functional notation, write an expression for the average rate of change of the population over the interval $t_1 \le t \le t_2$.

(b) Using functional notation, write an expression for the average rate of change of the population over the interval $t_2 \le t \le t_3$.

(c) Which expression is larger, your answer to part (a) or your answer to part (b)?

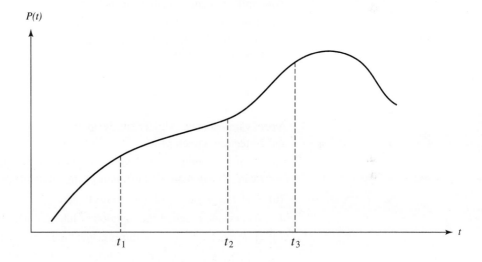

7. Let $h(t)$ denote the height of a rocketship t seconds after takeoff.

(a) Express the average rate of change of height of the rocket betweeen 2 and 2.01 seconds after takeoff in terms of the function h.

(b) Express the average rate of change of height (average vertical velocity) of the rocket on the time interval $[a, a + 0.001]$ in terms of h.

(c) Express the average vertical velocity of the rocket on the time interval $[a, a + k]$.

8. Find the average rate of change of $f(x) = \frac{1}{x^2 + 1}$ on each interval. Simplify your answer.

(a) $[1, 3]$

(b) $[1, 1.5]$

(c) $[1, 1.01]$

(d) $[1, 1 + h]$ (Show that your answer agrees with the answers you obtained in parts (a), (b), and (c).)

(e) Illustrate your answers to parts (a), (b), and (c) by sketching $f(x)$ and drawing secant lines whose slopes correspond to these average rates of change.

9. Answer the previous question using $g(x) = 2x^2 + 3x$ instead.

10. Find the average rate of change of $f(x) = x^2$ over each of the following intervals.

(a) $[0, 3]$

(b) $[1, 4]$

(c) $[2, 5]$

(d) $[a, a + 3]$

(e) $[a, a + h]$

11. Find the average rate of change of $g(t) = \frac{t}{t^2+2} + 3t$ over the intervals $[-1, 1]$, $[0, 2]$, $[1, 1 + p]$.

12. A bicyclist does a one-mile climb at a constant speed of 12 miles per hour followed by a one-mile descent at a constant speed of 30 miles per hour.

(a) Sketch a graph of distance traveled as a function of time. Assume the cyclist starts at $t = 0$ minutes, and be sure to label the times at which he reaches the top and bottom of the hill.

(b) What is his average speed for the two miles? Is this the same as the average of 12 mph and 30 mph? Explain why or why not.

13. A backyard pool is a cylinder sitting above the ground and measuring 3.5 feet in height and 20 feet in diameter.

(a) Express the volume of water in the pool as a function of the height h of the water. (*Note:* The domain of this function, the set of all acceptable inputs, is $0 \leq h \leq 3.5$.)

(b) Sketch a graph of volume versus height.

(c) What is the range of the function? Make sure this is indicated on your graph.

(d) How much additional water is needed to increase the depth of water in the pool by 1 foot? By 1/2 foot? Is $\frac{\Delta V}{\Delta h}$ constant? If so, what is it?

(e) You've expressed the rate of change of volume with respect to height in terms of ft³/ft, but the volume of water is more likely to be measured in gallons or liters. Knowing that 1 gallon ≈ 0.16054 ft³, convert your answer to gallons/ft.

(f) Is the volume of water in the pool directly proportional to the height of water?

14. During the 1996 Summer Olympics in Atlanta, track and field world records were set in both the men's 100 meters dash and the men's 200 meters dash. Donovan Bailey won the 100 in 9.86 seconds, and Michael Johnson won the 200 in 19.32 seconds.

(a) What was the average speed of each runner? Which race had the higher average speed? Explain why you think this might be so. Please use graphs to illustrate your answer.

(b) In 1996 the record for 400 meters was 43.29, set by Butch Reynolds. How does this average speed compare to the two given above? Does the "pattern" of the longer race having a higher average speed continue if we include the 400 meter? Give a possible reason for this.

15. A and B are points on the graph of $f(x) = \frac{1}{x^2}$. The x-coordinate of point A is 3 and the x-coordinate of point B is $(3 + h)$. Which of the expressions below correspond to the average rate of change of f on the interval $[3, 3 + h]$?

(a) $\dfrac{\frac{1}{9} + h - \frac{1}{9}}{h}$

(b) $\dfrac{\frac{1}{(3+h)^2} - \frac{1}{9}}{h}$

(c) $\dfrac{\frac{1+h}{9} - \frac{1}{9}}{3}$

(d) $\frac{1}{9} + \frac{1}{h} - \frac{1}{9}$

(e) $\dfrac{\frac{1}{(3+h)^2} - \frac{1}{9}}{3}$

Exploratory Problems for Chapter 2

Runners

1. Early one morning three runners leave Washington State University in Pullman and run the Palouse Path[11] joining their town to the University of Idaho. Alicia, Bertha, and Catrina run for 1 hour. The graphs below give position, $s(t)$, as a function of time, for each of the runners. The position function gives position measured from the start of the path from time $t = 0$ to t. The runners' position graphs are labeled A, B, and C, respectively.

   (a) Narrate the run, comparing and contrasting the running styles of the three women.

   (b) Compare the average velocities of the three runners.

   (c) Who is ahead after 1 hour of running?

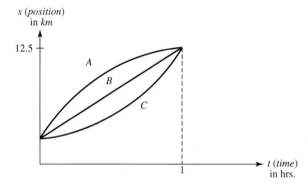

2. At lunchtime Amir, Baboucar, and Carlos leave Washington State University and head for the University of Idaho along the same trail. The graphs on page 83 give velocity, $v(t)$, as a function of time for each of the runners. Each man runs for an hour. Their position graphs are labeled A, B, and C, respectively.

---

[11] This path, officially called the Bill Chipman Palouse Path, is a rails-to-trails conversion completed in the late 1990s. It joins the small college towns of Pullman, Washington, and Moscow, Idaho, along a defunct railway track.

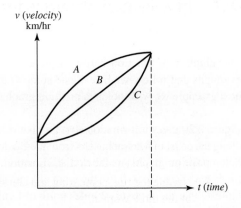

(a) Narrate the run. Be sure to mention who is the frontrunner and who is behind after half an hour and then after 1 hour. Compare and contrast the runners' running styles.

(b) Compare the average velocities of the runners for the 1-hour time block.

(c) Compare the average accelerations of the runners for the 1-hour time block.

2.4   READING A GRAPH TO GET INFORMATION ABOUT A FUNCTION

The graph of a function carries a great deal of information about the function. In the examples that follow we investigate some graphs and the information they contain. In the next example we will look at a position graph; we follow this by looking at a rate graph.

◆ EXAMPLE 2.9   Figure 2.20 gives information about a bike ride along Route 1A, a road going north-south along the coast of Massachusetts from the Boston area, through a town called Lynn, and past Salem. The graph tells us the cyclist's position, p, along the road (in miles) as a function of time: $p = f(t)$. We will use the town of Lynn as our benchmark for measuring position; we indicate that the cyclist is 2 miles north of Lynn by setting position equal to 2 and indicate that the cyclist is 2 miles south of Lynn by writing position equals -2. We let the benchmark time of 12 noon correspond to $t = 0$.

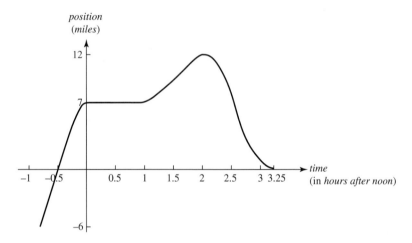

Figure 2.20

Use the graph in Figure 2.20 to answer the following questions.

i. Where (in relation to Lynn) and at what time does the bike trip start?

ii. Interpret the intercepts on the horizontal axis and the intercept on the vertical axis in terms of the trip.

iii. When is the cyclist traveling north? South?

iv. Salem is about 7 miles north of Lynn. Does the cyclist ever reach Salem? If so, approximate the time she gets there. If she comes back through Salem, indicate that time.

v. What is the cyclist doing between noon and 1:00 P.M.?

vi. We define **speed** to be the absolute value of velocity. Speed is always nonnegative; it gives no information about direction.

What is the cyclist's average speed between 2:00 P.M. and 3:15 P.M.?
What is the cyclist's average velocity between 2:00 P.M. and 3:15 P.M.?

vii. Does the cyclist ever turn around and change direction? When?

viii. What is the farthest north of Lynn the cyclist ever gets? What is the farthest south?

ix. Does she begin and end the trip in the same place?

SOLUTION   First we will answer these question colloquially, and then we will use the notation of functions introduced in Chapter 1.

i. The trip starts 6 miles south of Lynn at 11:00 A.M. $f(-1) = -6$, where $t = -1$ corresponds to 11:00 A.M. and $p = -6$ corresponds to a place 6 miles south of Lynn.

ii. The horizontal intercepts, solutions to the equation $f(t) = 0$, indicate when she passes through Lynn. The vertical intercept, $f(0)$, indicates her position relative to Lynn at time $t = 0$. $f(0) = 7$, so she is 7 miles north of Lynn at noon.

iii. She is going north when the graph is increasing; this is from 11:00 A.M. to noon and again from 1:00 P.M. to 2:00 P.M. She is going south when the graph is decreasing; this is from 2:00 P.M. to 3:15 P.M. (Note that $t = 3.25$ corresponds to 3:15 P.M.)

iv. Yes. She reaches Salem for the first time at noon, when her position is $+7$. She returns through Salem the next time her position is $+7$, or at about 2:30 P.M. Answering this question is equivalent to finding t such that $f(t) = 7$. Notice that the position function is not 1-to-1.

v. She is not moving. Perhaps she is taking a break in Salem or changing a flat tire.

vi. During this time period she travels 12 miles. The time period is 1.25 hours long, so her average speed is (12 miles)/(1.25 hours) = 9.6 miles per hour.

At 2:00 P.M. $t = 2$ and at 3:15 P.M. $t = 3.25$. Therefore, her average velocity can be expressed as

$$\frac{f(3.25) - f(2)}{3.25 - 2} = \frac{0 - 12}{1.25} = -9.6$$

The sign of velocity indicates direction. The negative sign indicates that the cyclist is going south. A positive velocity would indicate that she is biking north (because position, p, increases as she goes north).

When her velocity is -9.6 mph her speed is 9.6 mph.

vii. Yes, she turns around when the graph changes from increasing (going north) to decreasing (going south). This happens at 2:00 P.M.

viii. The farthest north corresponds to the maximum of the graph, which is 12 miles. The farthest south is represented by the minimum, which is -6, or 6 miles south of Lynn.

ix. No. She starts 6 miles south of Lynn, but she finishes in Lynn.   ◆

◆ EXAMPLE 2.10   Figure 2.21 is the graph of a different bike ride; now the graph represents velocity, v, as a function of time. $v = f(t)$. When velocity is negative the rider is moving south, and when velocity is positive she is moving north. Here again, time $t = 0$ represents noon.

CAVEAT While this graph is similar in *shape* to Figure 2.20, the two graphs represent completely different bike rides. A common error is to be drawn in by the shape of a graph and leap headlong into a tangled web of incorrect conclusions without stopping to look at the labeling of the axes of the graphs. You can be misled by not reading graphs carefully! The story told by a graph hinges very much on the labeling of the axes.

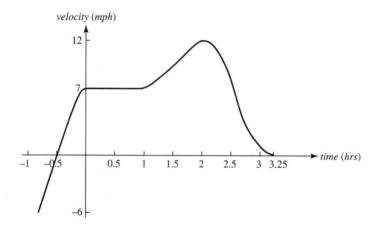

Figure 2.21

Use the graph to answer the following questions.

i. During what time period(s) is the cyclist riding south? North?

ii. Interpret the intercepts on the horizontal axis and the intercept on the vertical axis in terms of the trip.

iii. When is the cyclist traveling most rapidly south? Most rapidly north?

iv. Does the cyclist ever change direction? If so, when?

v. Does the cyclist ever stop? If so, when?

vi. Can you determine where the cyclist started her trip?

vii. What is the cyclist's maximum speed? Minimum speed?

SOLUTION   i. She is moving south when velocity is negative (below the horizontal axis)—from 11:00 A.M. until 11:30 A.M.. She is moving north when velocity is positive (above the horizontal axis)—from 11:30 A.M. until 3:15 P.M.

ii. The horizontal intercepts (at -0.5 and 3.25) indicate times when her velocity is zero. Her velocity is zero at 11:30 A.M. and 3:15 P.M. The vertical intercept gives her velocity at noon. She is traveling 7 mph north at noon.

iii. She is traveling most rapidly south when the graph lies furthest below the t-axis—at 11:00 A.M. She is moving most rapidly north when the graph lies furthest above the t-axis—at 2:00 P.M.

iv. She turns around when her velocity changes sign (either positive to negative or vice versa). This occurs at 11:30 A.M.

v. She stops for an instant to turn around at 11:30 A.M. and stops again at 3:15 P.M., when her velocity hits zero.

vi. We cannot determine where she started because this is only a graph of velocity. We have absolutely no idea where any of this takes place. The biker could have begun her trip in Lynn, in Seattle, in Bangkok—anywhere there is a north-south route.

vii. Her maximum speed is 12 mph, and her minimum speed is 0 mph. (*Note:* When velocity is -6 mph, speed is $+6$ mph.)   ◆

These last three examples are designed to help you read information from graphs and to keep you aware that the story told by a graph cannot be determined without carefully reading the labeling of the axes. Notice that we were able to get some information about velocity from the position graph and some information about relative position from the velocity graph. As we proceed, we will develop a deeper understanding of the connections between position and velocity, and, more generally, between amount functions and their corresponding rate functions.

EXERCISE 2.10   Oil is leaking from a point and spreading evenly in a thin, expanding disk. We can measure the radius of the disk and want to know the rate of change of the area of the disk with respect to the radius. The area is a function of the radius: $A(r) = \pi r^2$ square feet.

    i. When the radius increases from 1 foot to 3 feet, by how much has the area changed? What is the average rate of change of area with respect to radius?

    ii. When the radius increases from 3 feet to 5 feet, by how much has the area changed? What is the average rate of change of area with respect to radius?

Answers to Exercise 2.10 are provided at the end of Chapter 2.

Information About a Function Not Readily Available From Its Graph

You can obtain a tremendous amount of information about a function from its graph. The increasingly sophisticated technology of graphing calculators puts a mountain of information in the palm of your hand and access to it at your fingertips. However, there are instances in which it can be difficult to get useful information about a function from its graph. In this subsection we will explore facets of functions not readily available from their graphs and not readily accessible via a calculator.

◆ EXAMPLE 2.11   We define the function f as follows:

$$f(x) = \begin{cases} -1 & \text{for } x \text{ a rational number,} \\ 1 & \text{for } x \text{ an irrational number.} \end{cases}$$

To make sense out of this function let's clarify the terms "rational number" and "irrational number."

Definitions

A **rational** number is a number that can be expressed in the form p/q, where p and q are integers.[12] This is equivalent to saying that a rational number can be expressed as either a finite decimal expansion (e.g., $1/4 = 0.25$) or an infinitely repeating decimal (e.g., $1/6 = 0.1\bar{6}$).[13]

An **irrational** number is a number that cannot be expressed as a ratio of integers. The decimal expansion of an irrational number is infinite and nonrepeating. Some examples of irrational numbers are π, $\sqrt{2}$, and $\sqrt{5}$.[14]

The set of **real** numbers is the set of all rational and all irrational numbers. The real numbers correspond to all the points on the number line.

Because we cannot write a complete decimal expansion of any irrational number, sometimes people are left with the incorrect idea that there are only "a few" of them. The truth is that irrational numbers are more slippery than fish, and much more numerous. There are infinitely many of them.

Let's return to our function f. If we try to graph f, the best we can do is along the lines of the figure below. The domain of this function is the set of all real numbers, and the range is $\{-1, 1\}$. The horizontal lines at $y = 1$ and $y = -1$ have infinitely many holes, but we have no way of indicating this. In fact, since we cannot indicate the holes in the lines, this graph doesn't even look like the graph of a function; it doesn't appear to pass the vertical line test, although in actuality it does. Every real number is either rational, in which case the function maps it to -1, or irrational, in which case the function maps it to $+1$.   ◆

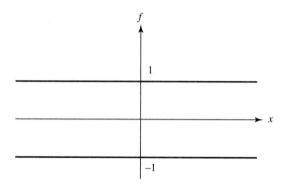

Figure 2.22

A computer or graphing calculator can be an extremely useful tool, but, like any tool, it has limitations. Try using your machine to do the following exercise.

EXERCISE 2.11   Let $f(x) = x^2 - 2\sqrt{2}x + 1$. See if you can use your calculator to find the exact values of the zeros and turning points of this function. (Some calculators can do this.)

The answers are:

zeros: $x = \sqrt{2} + 1$ and $x = \sqrt{2} - 1$

turning point: $x = \sqrt{2}$

The exact answers *cannot* be expressed as decimals.

---

[12] The set of integers is $\{\ldots -3, -2, -1, 0, 1, 2, 3, \ldots\}$.

[13] See Reference Section A, Algebra, for details.

[14] It is not obvious that these numbers are irrational, but it can be proven. The standard proof that $\sqrt{2}$ is irrational is a proof by contradiction. We begin by assuming $\sqrt{2}$ can be written in the form p/q and show that this assumption leads to a contradiction. While $\pi \approx 3.14159\ldots$, its decimal representation cannot be presented in its entirety; it is a sequence that never repeats yet goes on forever. Refer to Section 2.5 where reals and rationals are discussed.

Historically, irrational numbers were perplexing and disturbing, particularly to the Greek Pythagoreans around 500 B.C. They have also had a history of being perplexing to students in this course. Section 2.5 includes a discussion of irrational numbers and exact answers as they pertain to our work.

EXERCISE 2.12   Let $g(x) = \frac{1}{x^{10}}$. Look at the graph of g to see what type of information is accessible from its graph.

Graph this function on the standard viewing window of your calculator (with a domain and range of $[-10, 10]$). The graph almost looks like two vertical lines. Why? See if you can adjust your viewing window to get a "good" picture of g. Investigate this function numerically in order to get a feel for why, regardless of the viewing window used, it is difficult to get a handle on g from its graph.

Here is another function that is difficult to get a handle on by looking at a single graph.

EXERCISE 2.13   Let $h(x) = x^3 + 999x^2 - 1000x$. What are the zeros of this function? How many turning points does it have?

Answers to Exercise 2.13 are provided at the end of Chapter 2.

Exercises 2.12 and 2.13 are designed to make you think about the window through which you view a function on your calculator or computer. A bug and a bird have very different views of the world. Flexibility and the ability to view the world through different lenses and from different perspectives build problem-solving power.

PROBLEMS FOR SECTION 2.4

1. Use the graph of f on the following page to give approximate answers to the following questions. Assume that the domain of f is $-5 \leq t \leq 5$.

   (a) For what value of t between 0 and 5 is $f(t)$ largest?

   (b) How many solutions are there to the equation $f(t) = 0$? Approximate these solutions.

   (c) How many solutions are there to the equation $f(t) = 3$? Approximate these solutions.

   (d) Approximate $f(0)$.

   (e) For what values of t is $|f(t)| < 1$?

   (f) What are the values of k between -3 and 5 for which $f(t) = k$ has exactly one solution?

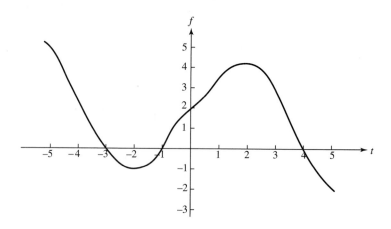

2. Refer to the graph for Problem 1. Suppose that this graph gives information about a mule's travels along the Nile River in Egypt between the hours of 7:00 A.M. and 5:00 P.M. The mule travels along the Corniche, the road following the Nile. $t = 0$ corresponds to noon; $t = -5$ indicates five hours before noon, i.e., 7 A.M.

   (a) Suppose $f(t)$ gives the mule's position (in miles) at time t, where t measures time in hours. Karnak Temple is used as a benchmark for position; $f = 0$ when the mule is at the Karnak Temple. Positions north of the Karnak are assigned positive values and positions south of the Karnak are assigned negative values. Your job is one of translation. Translate each of parts (a)–(e) in the previous problem to the corresponding questions about the mule's trip. Your answers should contain neither "t"s nor "$f(t)$"s.

   (b) Now suppose $f(t)$ gives the mule's *velocity* at time t (measured in miles per hour). When the velocity is positive the mule is traveling north along the Nile, and when the velocity is negative the mule is traveling south along the Nile. Translate each of parts (a)–(e) in the previous problem to the corresponding questions about the mule's trip, again avoiding "t"s and "$f(t)$"s.

3. Suppose f is a continuous function with domain $[-4, 7]$; f is decreasing on $[-4, 4]$ and increasing on $[4, 7]$.

   (a) Can you determine where f takes on its lowest value? If this can be done, do it. If not, can you narrow down the selection to a few possibilities ?
   Can you determine the lowest value of f?

   (b) Can you determine where f takes on its highest value? If this can be done, do it. If not, can you narrow down the selection to a few possibilities ?
   Can you determine the highest value of f?

4. Below are graphs giving information about trips. The trips are all taken along Interstate Route 90, a road that runs east-west through Sturbridge, Massachusetts. The top four graphs give position versus time while the bottom four give velocity versus time.

   We use as our benchmark location Sturbridge itself. We will indicate that we are 10 miles east of Sturbridge by writing position $= 10$; we will indicate that we are 10 miles west of Sturbridge by writing position $= -10$. Positive velocity will indicate that we are traveling from west to east; negative velocity will indicate that we are traveling from east to west. We use noon as our benchmark time; noon corresponds to time $t = 0$. Therefore time $t = -2$ is 10:00 A.M.

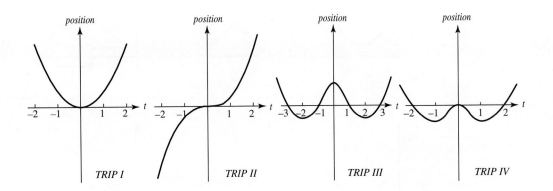

TRIP I   TRIP II   TRIP III   TRIP IV

Answer parts (a), (b), and (c) for each of the trips corresponding to graphs I, II, III, and IV.

(a) For what values of t is velocity positive? When is travel from west to east?

(b) For what values of t is velocity negative? When is travel from east to west?

(c) To which trip do each of the following velocity graphs correspond? (Be sure your answer to part (c) agrees with your answers to parts (a) and (b).)

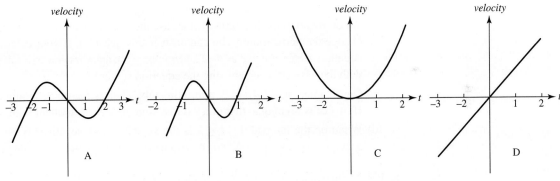

A   B   C   D

When you have completed your work on this problem, compare your answers with those of one of your classmates. If you disagree about an answer, each of you should discuss your reasoning and see if you can come to a consensus on the answers.

5. Look back at the figures for Problem 4. What characteristic of the graph of position versus time determines the sign of the velocity?

6. An ape with budding consciousness throws a bone straight up into the air from a height of 2 feet.[15] From the seven graphs that follow pick out the one that could be the bone's

(a) height versus time,   (b) velocity versus time,   (c) speed versus time.

---

[15] Problem by Eric Brussel.

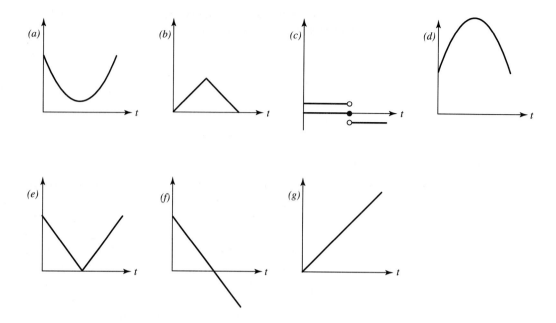

7. The velocity of an object is given in miles per hour by $v(t) = 2t^5 - 6t^3 + 2t^2 + 1$ over the time interval $-2 \le t \le 2$, where t is measured in hours. Use your graphing calculator to answer the following questions.

   (a) Sketch a graph of the velocity function over the time interval $-2 \le t \le 2$.

   (b) Approximately when does the object change direction? Please give answers that are off by no more than 0.05. (Either use the "zoom" feature of your calculator or change the domain until you can answer this question. If your calculator has an equation solver, use that as well and compare the answers you arrive at graphically with the answers you get using the equation solver.)

   (c) On the interval $-2 \le t \le 2$, approximately when is the object going the fastest? How fast is it going at that time? (Give your answer accurate to within 0.1.)

   (d) When on the interval $0 \le t \le 2$ is the velocity most negative? (Give an answer accurate to within 0.1.) When you zoom in on the graph here, what do you observe?

8. The displacement of an object is given by $d(t) = 2t^5 - 6t^3 + 2t^2 + 1$ miles over the time interval $-2 \le t \le 2$ where t is measured in hours.

   (a) Approximately when does the object change direction? Please give answers accurate to within 0.1. When you zoom in on the graph here, what do you observe?

   (b) Approximately when is the object's velocity positive? Negative?

   (c) Approximate the object's velocity at time $t = 0$.

9. At time $t = 0$ three joggers start at the same place and jog on a straight road for 6 miles. They all take 1 hour to complete the run. Jogger A starts out quickly and slows down throughout the hour. Jogger B starts out slowly and picks up speed throughout the hour. Jogger C runs at a constant rate throughout. On the same set of axes, graph distance traveled versus time for each jogger. Clearly label which curve corresponds to which jogger. Be sure your picture reflects all the information given in this problem.

10. A baseball "diamond" is actually a square with sides 90 feet long. Several of the fastest players in history have been said to circle the diamond in approximately 13 seconds.

   (a) Sketch a plausible graph of speed as a function of time for such a dash around the bases. Label the point at which the player touches each of the bases on your graph. (Keep in mind that your player will probably need to slow down as he approaches each base in order to make the necessary 90-degree turn.)

   (b) Sketch a graph of his acceleration as a function of time. Again, label the point at which he touches each base.

11. Before restrictions were placed on the distance that a backstroker could travel underwater in a race, Harvard swimmer David Berkoff set an American record for the event by employing the following strategy. In a 100-meter race in a 50-meter pool, Berkoff would swim most of the first 50 meters underwater (where the drag effect of turbulence was lower) then come up for air and swim on the water's surface (at a slightly lower speed) until the turn. He would then use a similar approach to the second 50 meters, but could not stay underwater as long due to the cumulative oxygen deprivation caused by the time underwater.

   Assume that Berkoff is swimming a 100-meter race in Harvard's Blodgett pool (which runs 50 meters east to west). He starts on the east end, makes the 50-meter turn at the west end, and finishes the race at the east end. Sketch a graph of his velocity, taking east-to-west travel to have a positive velocity and west-to-east a negative velocity.

12. Below are graphs of position versus time corresponding to three trips. To be realistic the graphs ought to be drawn with smooth curves; to make things simpler the situation is approximated by a model using straight lines. The trips are all taken along the Massachusetts Turnpike (Route 90) a road running east-west. Positions are given relative to the town of Sturbridge. Positive values of position indicate that we are east of Sturbridge, and negative values indicate positions west of Sturbridge. Thus, positive velocity indicates that we are traveling from west to east; negative velocity indicates that we are traveling from east to west. For each trip do the following:

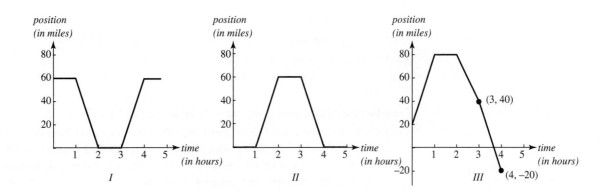

(a) Describe the trip in words. Include where the trip started and ended and how fast (and in what direction) we traveled.

(b) Graph velocity as a function of time.

(c) Graph speed as a function of time. (*Note:* Speed is always nonnegative (zero or positive); velocity may be zero, positive, or negative depending on direction.)

13. The annual Ironman triathlon held in Hawaii consists of a 2.4-mile swim followed by a 112-mile bicycle ride, and finally a 26.2-mile run. One entrant can swim approximately 3 miles per hour, bike approximately 18 miles per hour, and run about 9 miles per hour. In addition, during each portion of the event, she slows down toward the end as she gets tired. Sketch a possible graph of her distance as a function of time.

14. Below is a graph that gives information about a boat trip. The boat is traveling on a narrow river. The trip begins at 7:00 A.M. at a boathouse. To be realistic the graph ought to be drawn with smooth curves; to make things simpler the situation is approximated by a model using straight lines.

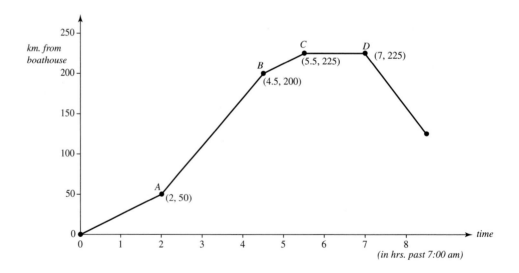

(a) How fast is the boat going between 7:00 A.M. and 8:00 A.M.?

(b) At what time do you think that the boaters stopped to go fishing?

(c) What happens after they fish?

(d) How can you tell that the boat is going at a steady speed between 9:00 A.M. and 11:30 A.M.?

(e) How fast is the boat going between 10 A.M. and 11 A.M.?

(f) Using the information given above, sketch a graph to show how the speed of the boat varies with time. Label your vertical axis speed (in kilometers per hour) and your horizontal axis time (in hours past 7:00 A.M.). Consider the portion of the trip beginning at 7:00 A.M. and ending at 2:00 P.M.

2.5   THE REAL NUMBER SYSTEM: AN EXCURSION

The Development of the Real Number System

The concept of the real number system did not emerge fully formed from the ancient world; it had a very long and tumultuous gestation period over tens of thousands of years, beginning with early counting systems.[16] Tallying systems using notches in bones and sticks (or knots in ropes) gave way to symbolic number systems, beginning with counting numbers: 1, 2, 3, ... We have ten fingers to count on; the ancient Egyptians,[17] along with people of many other ancient civilizations, counted by tens—as we do today. The Babylonians, on the other hand, had a well-developed place-value number system based on 60.[18] Well over a thousand years elapsed between the use of the Babylonians' symbolic number system and the first evidence of the use of zero. Zero was first introduced *not* as a number in its own right but simply as a positional place-holder.[19] It took about another thousand years for zero to gain acceptance as a number.

The Chinese mathematician Liu Hui used negative numbers around 260 A.D.[20], but the famous Arab algebraists of the 800s such as al-Khowarizmi, whose work laid the foundations of Western Europe's understanding of algebra, avoided negative numbers. While the Hindu mathematician Bhaskara put negative and positive roots of equations on equal footing in the early 1100s, Europeans were skeptical.[21] As late as the mid-1500s, when the Italian mathematicians Cardan and Tartaglia were battling over the solution of cubic equations, Cardan, while farsighted in terms of recognizing negative roots, referred to them as fictitious.[22]

The use of fractions dates back to the ancient Egyptians, although their fractions were always reciprocals of counting numbers (1/4, but not 3/4). Positive rational numbers gained acceptance early, but irrational numbers were a cause of great consternation for well over a millennium. The Greek Pythagorean school (around 530 B.C.) held the mystical belief that the universe is governed by ratios of positive integers. Yet it was Pythagoras (or one of his followers) who proved the Pythagorean Theorem, which tells us that the diagonal of a square with sides 1 has length $\sqrt{2}$. Not only that, but the Pythagoreans proved that $\sqrt{2}$ is irrational: that is, they proved that $\sqrt{2}$ cannot be expressed in the form p/q where p and q are integers. One possibly apocryphal story says that the Pythagoreans punished with

---

[16] For some fascinating details and a very readable account, consult *The History of Mathematics: An Introduction*, by David Burton, McGraw-Hill Companies, Inc. 1997.

[17] Due in part to their methods of record keeping and their hot dry climate, the ancient Egyptians and Babylonians have left modern historians more evidence of their mathematical development than have the people of other ancient civilizations. The writings of the Greek historian Herodotus (around 450 B.C.) have helped establish records of life in this part of the world. The climate of regions such as China and India has contributed to the disintegration of evidence. In addition, much destruction of ancient work throughout the world has been deliberately carried out in the course of political and religious crusades.

[18] Perhaps the base of 60 was selected because it has so many proper divisors (a position advocated by Theon of Alexandria, father of Hypatia, the first famous woman mathematician), or because a year was thought to be 360 days. For more information about the Babylonian system, see Burton's book, Section 1.3, or read Howard Eves' *An Introduction to the History of Mathematics*, Saunders College Publishing, 1990.

[19] The Mayans used zero in this manner in the first century A.D. (*The Story of Mathematics,* by Lloyd Motz and Jefferson Weaver, Plenum Press, 1993, p. 33). Circa 150 A.D. the astronomer Ptolemy used the symbol "o" as a place-marker in his work. The symbol came from the first letter of the Greek word for "nothing" (Burton, p. 23). The Hindus used a dot as a zero place-holder in the fifth century A.D.; this dot later metamorphosized into a small circle. Arabic uses a dot for zero.

[20] Burton, pp. 157-58.

[21] Burton, p. 173.

[22] Burton, p. 294.

death one of their members who revealed to the outside world this dreadful contradiction in beliefs.[23]

While rational numbers (positive, zero, and negative) were fully accepted by the late 1500s, irrational numbers were still viewed with some confusion. For example, in 1544 the algebraist Michael Stifel wrote " . . . just as an infinite number is not a number, so an irrational number is not a true number, but lies hidden in some sort of cloud of infinity."[24]

It wasn't until the 1870s that Dedekind, Cantor, and Weierstrass put irrational numbers on solid ground. As previously mentioned, the set of real numbers is the set of all rational and all irrational numbers; the real numbers correspond to all the points on the number line.

What Does it Mean to Give An Exact Answer When Your Answer Is Irrational?

Question: What exactly is the square root of 2? Is it 1.414213562?

In our math class this issue has been hotly debated. Tempers have flared; voices have been raised; we barely escaped pistols at dawn. Let's take a moment to think about this calmly and rationally. In fact, *irrationality* is the root of the problem. The number $\sqrt{2}$, the positive square root of 2, is the positive number such that, when you square it, you get 2. Try computing $\sqrt{2}$ on your calculator. Your calculator spits out 1.414213562, or something very similar. Is this really the square root of two? Our class annals show that Ted claims yes, exactly, while Kevin claims no. Both are adamant. We'll listen in.

Ted: "Yes, I now press $\boxed{x^2}\,\boxed{=}$ and my calculator gives me 2. This shows that 1.414213562 is $\sqrt{2}$."

Kevin: "That's your calculator covering up for itself. x^2 and $\sqrt{x}$ share the same key on many calculators so the calculator covers up for its inaccuracies."

Ted: "Will you listen to this? Give me a break. A calculator is a machine; it won't cover up its mistakes."

Kevin: "Try this: Press in 1.414213562 × 1.414213562. What do you get?"

Ted: "1.999999999. But it's 9's forever, so it's essentially two."

Kevin: "No way. You see nine 9's, not 9's forever. Think about multiplying out by hand; for sure you'll get a 4 as your last decimal place on the right."

Ted: "All right, all right—so how do I get $\sqrt{2}$ exactly?"

Kevin: "You already **have** it exactly."

Ted glares. The tension builds. What is going on here? Ted is desperately seeking a finite decimal expansion that is exactly $\sqrt{2}$ and Kevin is trying, without success, to convince him that there simply isn't one. The problem is that Ted is locked into thinking that any number can be written as a finite decimal or an infinitely repeating decimal. In fact, any *rational* number can be expressed this way. The irrational numbers are numbers that cannot

---

[23] E. Maor, *To Infinity and Beyond: A Cultural History of the Infinite*, Boston, Birkhauser, 1987. p. 46.
[24] Burton, p. 170.

be expressed in this way. Their decimal expansions are infinite and nonrepeating. So we can never write one down exactly as a decimal, and our calculators can never tell us the decimal expansion exactly. Ted believes, like the ancient Pythagoreans, that the world ought to be ruled completely by rational numbers. Historically, this dilemma caused even greater consternation for the Pythagoreans than it is causing for Ted.

Question: Between any two rational numbers, how many rational numbers are there?
Answer: Infinitely many.

Question: Between any two rational numbers, how many irrational numbers are there?
Answer: Infinitely many.

Question: So between 1.414213562 and 1.414213563 how many irrational numbers are there?
Answer: Infinitely many. You can string infinitely many different infinite sequences of numbers behind that final 2 of the former without reaching beyond the final 3 of the latter.

Question: How many of these does your calculator indicate to you?
Answer: None. However, some calculators do store away more digits than they show you. For example, the *T I*-85 stores three digits more of the decimal representation of $\sqrt{2}$ than it displays. Look in your calculator's instruction book to find out what your calculator really "knows."

Question: Can your calculator ever "give" you an irrational number?
Answer: Not as a decimal. It can approximate an irrational number by a rational one.

Question: Can straightforward problems lead to irrational answers?
Answer: Yes. Think about the diagonal of a square with sides of length 1 (or with sides of any rational length, for that matter). Or consider the ratio of the circumference of a circle to the diameter:

$$\frac{\text{circumference}}{\text{diameter}} = \frac{2\pi r}{2r} = \pi.$$

This is in fact taken to be the definition of π.[25] π is irrational, although it wasn't until the 1700s that it was proven to be so, about 2000 years after Pythagoras had proven the irrationality of $\sqrt{2}$.[26]

Are you still feeling a little queasy about irrational numbers, still feeling in the pit of your stomach that you could really sweep them under the table and no one would notice? While it is true that in any practical application a "good enough" approximation is all we

---

[25] Cuneiform tablets from the Babylonians indicate that they took π to be 3. The ancient Egyptians essentially took it to be 3.16. Archimedes showed that $3.14103 < \pi < 3.14271$. (Eli Maor, *e: The Story of a Number*, p. 43.) In the United States a state legislature actually tried to "legislate" the value of π, forcing it to be rational. They felt they were fighting for the underdog, the poor engineers, rocket scientists, doctors, and students whose lives were being unnecessarily complicated by an irrational π.

[26] Johann Lambert proved π is irrational in 1768. (Eli Maor, *e: The Story of a Number*, p. 188.)

need, let's think for a moment about the problems that would arise if we were to toss out irrational numbers. Consider this: Suppose we were to graph only rational points—meaning points with both the x- and y-coordinates rational numbers. While we could still make out the shape of the unit circle, it would have infinitely many holes in it. For example, the line $y = x$ would pass right through without intersecting it because the points where this line intersects the circle are $\left(\frac{1}{\sqrt{2}}, \frac{1}{\sqrt{2}}\right)$ and $\left(-\frac{1}{\sqrt{2}}, -\frac{1}{\sqrt{2}}\right)$, both of which have irrational coordinates.

The graph of $x^4 + y^4 = 1$ looks like a deformed circle,[27] but graphing only rational points gives us just four points, as shown in Figure 2.23.

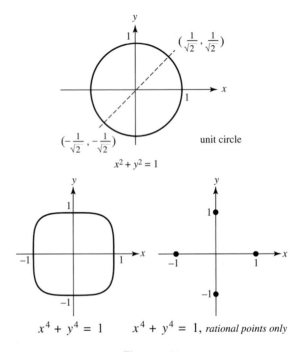

Figure 2.23

Similarly, the graph of $x^6 + y^6 = 1$ looks like a radically deflated circle, but graphing rational points only gives us four single points.

The graph of $x^n + y^n = 1$, for any $n \geq 3$ would look like the figure on the left in Figure 2.24 for n even and the figure on the right for n odd.[28]

---

[27] To plot these using your calculator you'll probably need to construct pairs of functions: $y_1 = \sqrt{1 - x^2}$ and $y_2 = -\sqrt{1 - x^2}$, i.e., $y_1 = \left(1 - x^2\right)^{1/2}$ and $y_2 = -\left(1 - x^2\right)^{1/2}$ for the unit circle, $y_1 = \left(1 - x^4\right)^{1/4}$ and $y_2 = -\left(1 - x^4\right)^{1/4}$ for $x^4 + y^4 = 1$.

[28] The fact that $x^n + y^n = 1$ has no rational solutions other than 0's and 1's is a consequence of Fermat's Last Theorem, which states that $x^n + y^n = z^n$ has no nontrivial integer solutions for $n \geq 3$. This longstanding conjecture had challenged mathematicians for hundreds of years. It was proven in 1994 by Andrew Wiles with some last-minute assistance from Richard Taylor. Those of you who frequent the theater might be interested to know that Fermat's Last Theorem is discussed in Tom Stoppard's play *Arcadia*, and those of you who watch *Star Trek: The Next Generation* know that Captain Picard is working on this still "unsolved" problem many centuries from now.

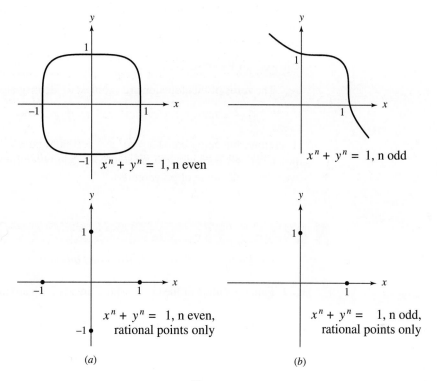

Figure 2.24

An even more basic though less dramatic example is obtained by simply thinking of the number line. Between any two rationals there are infinitely many irrationals; therefore, if we were to ignore irrationals the line would be full of holes—to put it mildly!

Why does the idea of having infinitely many holes in the graph of a function cause us such great consternation? Soon we will move from looking at an average rate of change of a quantity to inquiring about an instantaneous rate of change; for example, instead of asking about the average velocity of a biker over a five-minute period we will be interested in her velocity at a certain instant. In order to find the instantaneous rate of change of f it is necessary that f be a continuous function, that its graph have no holes or jumps.

Answers to Selected Exercises

Answers to Exercise 2.10

i. Change in area: $A(3) - A(1) = 9\pi - \pi = 8\pi$ (This is a little more than 25 square feet.)

$$\text{Rate of change}: \frac{\text{change in area}}{\text{change in radius}} = \frac{\Delta A}{\Delta r} = \frac{A(3) - A(1)}{3 - 1} = \frac{8\pi}{2} = 4\pi$$

ii. Change in area: $A(5) - A(3) = 25\pi - 9\pi = 16\pi$

$$\text{Rate of change}: \frac{\text{change in area}}{\text{change in radius}} = \frac{\Delta A}{\Delta r} = \frac{A(5) - A(3)}{5 - 3} = \frac{16\pi}{2} = 8\pi$$

Notice that as r increases the rate of change of area with respect to the radius also increases; increasing the radius from 3 to 5 adds more area than does increasing the radius from 1 to 3.

Answers to Exercise 2.13

The zeros of this function are simplest to identify algebraically.

$$x^3 + 999x^2 - 1000x = x(x^2 + 999x - 1000) = x(x + 1000)(x - 1)$$

Therefore, the zeros are at $x = 0$, $x = -1000$, and $x = 1$. The function has two turning points. How we know there are not more than two turning points will be discussed when we take up the topic of polynomials.

PROBLEMS FOR SECTION 2.5

For Problems 1 through 7, give exact answers, not numerical approximations.

1. Find the radius of the circle whose area is 2 square inches.

2. Find the diameter of the circle whose circumference is 7 inches.

3. How long is the diagonal of a square whose sides are 5 inches?

4. A rectangle is 3 meters long and 2 meters high. How long is the diagonal?

5. Solve: $x^2 + 1 = 6$.

6. Solve: $(\pi x)^2 = \pi x$. (There are two answers.)

7. Solve: $\pi^2 x^3 = \pi x^2$.

8. (a) How many rational numbers are in the interval [2, 2.001]?
   (b) How many irrational numbers are in the interval [2, 2.001]?

9. The number π lies between 3.141592653489 and 3.141592653490. How many other irrational numbers lie between these two?

10. How many points on the graph of $f(x) = x^2$ have at least one irrational coordinate?

11. (a) Is it possible for the graph of a function f with domain [0, 2] to have at most finitely many points with an irrational coordinate? If so, give such a function.
    (b) Is it possible for the graph of a function g with domain $\{0, 1, 2, \ldots\}$ to have no points with an irrational coordinate? If so, give an example of such a function.

C H A P T E R

3

Functions Working Together

In this chapter we'll look at ways of combining functions in order to construct new functions.

3.1 COMBINING OUTPUTS: ADDITION, SUBTRACTION, MULTIPLICATION, AND DIVISION OF FUNCTIONS

The sum, difference, product, and quotient of functions are the new functions defined respectively by the addition, subtraction, multiplication, and division of the *outputs* or values of the original functions.

Addition and Subtraction of Functions

◆ EXAMPLE 3.1   Suppose a company produces widgets.[1] The revenue (money) the company takes in by selling widgets is a function of x, where x is the number of widgets produced. We call this function $R(x)$. We call $C(x)$ the cost of producing x widgets. Producing and selling x widgets results in a profit, $P(x)$, where profit is revenue minus cost; $P(x) = R(x) - C(x)$.

The height of the graph of the profit function is obtained by subtracting the height of the cost function from the height of the revenue function. Where $P(x)$ is negative the company loses money. The x-intercept of the $P(x)$ graph corresponds to the break-even point where revenue exactly equals costs, so there is zero profit. (See Figure 3.1 on the following page.)

---

[1] A widget is an imaginary generic product frequently used by economists when discussing hypothetical companies.

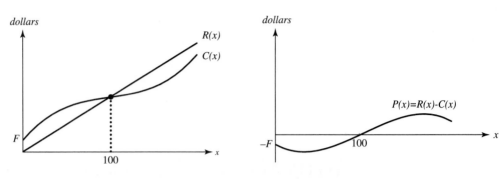

Figure 3.1   ◆

More generally, if h is the sum of functions f and g, $h(x) = f(x) + g(x)$, then the output of h corresponding to an input of x_1 is the sum of $f(x_1)$ and $g(x_1)$. In terms of the graphs, the height of h at x_1 is the sum of the heights of the graphs of f and g at x_1. An analogous statement can be made for subtraction. The domain of h is the set of all x common to the domains of both f and g.

◆ **EXAMPLE 3.2**   Let $f(x) = x$ and $g(x) = \frac{1}{x}$. We are familiar with the graphs of f and g. We can obtain a rough sketch of $f(x) + g(x) = x + \frac{1}{x}$ from the graphs of f and g by adding together the values of the functions as shown in the figure below.

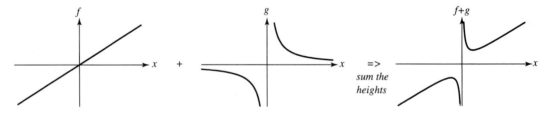

Figure 3.2

◆

EXERCISE 3.1   The following questions refer to Example 3.2, where $f(x) + g(x) = x + \frac{1}{x}$.

(a) What is the domain of $f + g$?

(b) For $|x|$ close to zero, which term of the sum dominates (controls the behavior of) the sum?

(c) For $|x|$ large, which term of the sum is dominant?

EXERCISE 3.2   When a company produces widgets they have fixed costs (costs they incur regardless of whether or not they produce a single widget, such as renting some space for widget production) and they have variable costs (costs that vary with the number of widgets they produce, such as the cost of materials and labor). Figure 3.3 shows graphs of the fixed cost function, FC, and the variable cost function, VC, for widgets.

Total cost = fixed costs + variable costs

Graph the total cost function, TC, where $TC = FC + VC$.

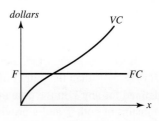

Figure 3.3

EXERCISE 3.3   A patient is receiving medicine intravenously. Below are two graphs. One is the graph of $R_I(t)$, the rate at which the medication enters the bloodstream. The other is a graph of $R_O(t)$, the rate at which the medication is metabolized and leaves the bloodstream. [2]

(a) Let $R(t)$ be the rate of change of medication in the bloodstream. Express $R(t)$ in terms of $R_I(t)$ and $R_O(t)$.

(b) Graph $R(t)$.

(c) At approximately what value of t is $R(t)$ minimum?

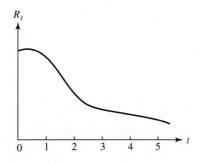
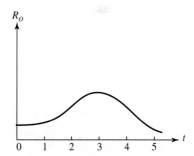

Figure 3.4

Multiplication and Division of Functions

Let's suppose a consultant wants to construct a function to model the amount of money he spends on gasoline for his automobile. The price of gasoline varies with time. Let's denote it by $p(t)$, where $p(t)$ is measured in dollars per gallon. The amount of gasoline he uses for his commute also varies with time; we'll denote it by $g(t)$. Then the amount of money he spends is given by

$$p(t)\frac{\$}{\text{gal}} \cdot g(t) \text{ gal.}$$

This is the product of the functions p and g.

A demographer is interested in the changing economic profile of a certain town. If the population of the town at time t is given by $P(t)$ and the total aggregate income of the town at time t is given by $I(t)$, then the per capita income is given by

---

[2] R_I for "rate in"; R_O for "rate out."

$$\frac{I(t)}{P(t)}.$$

This is the quotient of the functions I and P.

If $h(x) = f(x) \cdot g(x)$, then the output of h is the product of the outputs of f and g. If $j(x) = \frac{f(x)}{g(x)}$, then the output of j is the quotient of the outputs of f and g. The product is defined for any x in the domains of both f and g; the quotient is defined for any x in the domains of both f and g provided $g(x)$ is not equal to 0.

EXERCISE 3.4 $h(x) = f(x) \cdot g(x)$

(a) If $h(a) > 0$, what can be said about the signs of f and g at $x = a$?

(b) If $h(a) < 0$, what can be said about the signs of f and g at $x = a$?

(c) How are the zeros of h related to the zeros of f and g?

EXERCISE 3.5 $j(x) = \frac{f(x)}{g(x)}$

(a) If $j(a) > 0$, what can be said about the signs of f and g at $x = a$?

(b) How are the zeros of j related to the zeros of f and g?

◆ **EXAMPLE 3.3**   The number of widgets people will buy depends on the price of a widget. Economists call the number of widgets people will buy the demand for widgets. Thus, demand for a widget is a function of price. Let's suppose that the number of widgets demanded is given by $D(p)$, where p is the price of a widget. If a company has a monopoly on widgets, then it can fix the price of a widget to be whatever it likes. The revenue, R, that this company takes in is given by (price of a widget) · (number of widgets sold), so

$$R(p) = p \cdot D(p).$$

Below is the graph of the demand function, where quantity demanded is a function of price.[3]

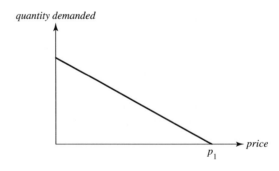

Figure 3.5

(a) What prices will yield no revenue? Why?

(b) Sketch a rough graph of $R(p)$.

---

[3] Economists would reverse the axes.

SOLUTION   (a) Prices of $0 and p_1 will yield no revenue. (In fact, any price above p_1 will yield no revenue.) If widgets are free there is no revenue, and likewise if the price of a widget is p_1 or more then nobody will buy widgets, so the revenue is also zero.

(b) The graph of $R(p)$ is given below.

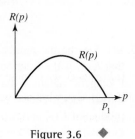

Figure 3.6   ◆

EXERCISE 3.6   Let $C(q)$ be the total cost of making q widgets. Assume that $C(0)$ is positive. How would you compute the average total cost of making q widgets? When graphing the average total cost function, what units might be on the coordinate axes? Why is it that average total cost curves never intersect either of the two coordinate axes?

EXERCISE 3.7   Let $f(x) = x$ and $g(x) = 1/x$. Sketch the graph of $h(x) = f(x) \cdot g(x)$. Where is $h(x)$ undefined? How can you indicate this on your graph? How does your graphing calculator deal with the point at which h is undefined?

The answer to Exercise 3.7 is supplied at the end of the chapter.

PROBLEMS FOR SECTION 3.1

1. Let $f(x) = x^2$ and $g(x) = 1/x$. Use your knowledge of the graphs of f and g to sketch the graph of $h(x) = f(x) \cdot g(x)$. Where is $h(x)$ undefined? How can you indicate this on your graph? How does your graphing calculator deal with the point at which h is undefined?

2. Let $f(x) = |x|$ and $g(x) = 1/x$. Use your knowledge of the graphs of f and g to sketch the graph of $h(x) = f(x) \cdot g(x)$. Where is $h(x)$ undefined? *Note:* You must deal with the cases $x > 0$ and $x < 0$ separately. This is standard protocol for handling absolute values.

3. Let $f(x) = |x|$ and $g(x) = x$. Use your knowledge of the graphs of f and g to sketch the graph of $h(x) = f(x) + g(x)$.

4. Let $B(t)$ denote the birth rate of Siamese fighting fish as a function of time and $D(t)$ denote the death rate. Then the total rate of change of the population of Siamese fighting fish, $R(t)$, is given by subtracting the death rate from the birth rate; thus,

$R(t) = B(t) - D(t)$. Graphs of $B(t)$ and $D(t)$ are shown below. Sketch a graph of $R(t)$.

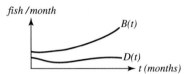

5. Let $F(t)$ be the number of trout in a given lake as a function of time and suppose that $K(t)$ is the fraction of these fish in the lake at time t that are "keepers" if caught ("keepers" meaning that they are above a certain minimum length—smaller ones are thrown back). Then the total number of keepers in the lake at any time is given by the product of $F(t)$ and $K(t)$. Below are graphs of $F(t)$ and $K(t)$. Sketch a graph of $N(t)$, the total number of keepers as a function of time.

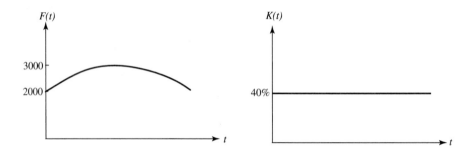

6. A town draws its water from the town reservoir. The town's water needs vary throughout the day; the rate of water leaving the reservoir (in gallons per hour) is shown on the graph below. Also recorded is the rate at which water is flowing into the reservoir from a nearby stream.

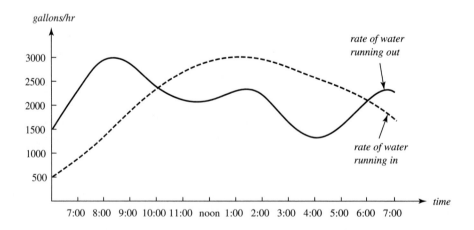

(a) At 6:00 A.M., at what rate is the water being used by the town? At what rate is water flowing in from the stream? Is the water level in the reservoir increasing or decreasing at 6:00 A.M.? At what rate?

(b) At approximately what time(s) is the rate of flow of water into the reservoir equal to the rate of flow out of the reservoir?

(c) During what hours (between 6:00 A.M. and 7:00 P.M.) is the water level in the reservoir increasing?

(d) At approximately what time is the water level in the reservoir increasing most rapidly? How can you get this information from the graph?

7. Let's return to the city reservoir in Problem 6. We'll denote the rate at which water is flowing into the reservoir by $R_I(t)$ and the rate that it is flowing out by $R_O(t)$. Then the total rate of change of water in the reservoir is given by

$$\text{Total rate of change} = \text{Rate in} - \text{Rate out}$$

or

$$R(t) = R_I(t) - R_O(t).$$

Graph $R(t)$.

8. Let $f(x) = x(x + 1)$, $g(x) = x^3 + 2x^2 + x$.

(a) Simplify the following.

   i. $f(x) + g(x)$ ii. $\frac{f(x)}{g(x)}$ iii. $\frac{g(x)}{f(x)}$ iv. $\frac{[f(x)]^2}{g(x)}$

(b) Solve $xf(x) = g(x)$.

9. The Cambridge Widget Company is producing widgets. The fixed costs for the company (costs for rent, equipment, etc.) are $20,000. This means that before any widgets are produced, the company must spend $20,000. Suppose that each widget produced costs the company an additional $10. Let x equal the number of widgets the company produced.

(a) Write a total cost function, $C(x)$, that gives the cost of producing x widgets. (Check that your function works, e.g., check that $C(1) = 20,010$ and $C(2) = 20,020$.) Graph $C(x)$.

(b) At what rate is the total cost increasing with the production of each widget? In other words, find $\Delta C / \Delta x$.

(c) Suppose the company sells widgets for $50 each. Write a revenue function, $R(x)$, that tells us the revenue received from selling x widgets. Graph $R(x)$.

(d) Profit = total revenue − total cost, so the profit function, $P(x)$, which tells us the profit the company gets by producing and selling x widgets, can be found by computing $R(x) - C(x)$. Write the profit function and graph it.

(e) Find $P(400)$ and $P(700)$; interpret your answers. Find $P(401)$ and $P(402)$. By how much does the profit increase for each additional widget sold? Is $\Delta P / \Delta x$ constant for all values of x?

(f) How many widgets must the company sell in order to break even? (Breaking even means that the profit is 0; the total cost is equal to the total revenue.)

(g) Suppose the Cambridge Widget Company has the equipment to produce at maximum 1200 widgets. Then the domain of the profit function is all integers x where

$0 \le x \le 1200$. What is the range? How many widgets should be produced and sold in order to maximize the company's profits?

10. A photocopying shop has a fixed cost of operation of $4000 per month. In addition, it costs them $0.01 per page they copy. They charge customers $0.07 per page.

   (a) Write a formula for $R(x)$, the shop's monthly revenue from making x copies.
   (b) Write a formula for $C(x)$, the shop's monthly costs from making x copies.
   (c) Write a formula for $P(x)$, the shop's monthly profit (or loss if negative) from making x copies. Profit is computed by subtracting total costs from the total revenue.
   (d) How many copies must they make per month in order to break even? Breaking even means that the profit is zero; the total costs and total revenue are equal.
   (e) Sketch $C(x)$, $R(x)$, and $P(x)$ on the same set of axes and label the break-even point.
   (f) Find a formula for $A(x)$, the shop's average cost per copy.
   (g) Make a table of $A(x)$ for $x = 0, 1, 10, 100, 1000, 10000$.
   (h) Sketch a graph of $A(x)$.

3.2   COMPOSITION OF FUNCTIONS

Whereas the addition, subtraction, multiplication, and division of functions is simply the addition, subtraction, multiplication, and division of the *outputs* of these functions, another way of having functions work together is to have the output of one function used as the input of the next function.

Suppose you are blowing up balloons for a celebration. The surface area S of the balloon is a function of a, the amount of air inside of the balloon. Let's say $S = f(a)$. The amount of air inside the balloon is a function of time. Let's say $a = g(t)$. Then

$$S = f(a) = f(g(t)).$$

We can say that S is equal to the composition of f and g.

Composition of functions is analogous to setting up functions as workers (or machines) on an assembly line. If the output of g is handed over to f as input, we write $f(g(t))$, where the notation indicates that f acts on the output of g. This can be represented diagrammatically by

$$t \xrightarrow{g} g(t) \xrightarrow{f} f(g(t)).$$

The expression $f(g(t))$ says "apply f to $g(t)$"; that is, use $g(t)$ as the input of f. This is called the **composition of f and g}** and is also sometimes written as $f \circ g$. The expression $(f \circ g)(t)$ means $f(g(t))$: Start with t; apply machine g and then apply machine f to the result.

$$t \longrightarrow \boxed{g} \longrightarrow g(t) \longrightarrow \boxed{f} \longrightarrow f(g(t)) = (f \circ g)(t)$$

Caution: Do not let the notation mislead you into thinking that f should be applied before g; $(f \circ g)(t)$ *means* apply g, then apply f to the result. The domain of $f \circ g$ is the set of all t in the domain of g such that $g(t)$ is in the domain of f.

Notice that the order in which machines are put on an assembly line is generally critical to the outcome of the process. Suppose machine W pours a liter of water in a specified place and machine L places a lid on a bottle. We send an open empty bottle down the assembly line. Putting machine W first on the assembly line, followed by machine L, results in the production of bottled water, while reversing the order results in the production of sealed, washed, empty bottles.

Open empty bottle $\xrightarrow{W}\xrightarrow{L}$ bottled water   corresponding to $L(W(\text{bottle}))$

Open empty bottle $\xrightarrow{L}\xrightarrow{W}$ washed bottle   corresponding to $W(L(\text{bottle}))$

From this example we see that generally $f(g(t)) \neq g(f(t))$.

When unraveling the composition of functions, *always start from the innermost parentheses and work your way outward*. This will assure the correct order on the assembly line of functions.

◆ **EXAMPLE 3.4**   Let $f(x) = x^2$, $g(x) = 2x + 3$. Find the following.

   i. $(f \circ g)(x)$, i.e., $f(g(x))$ ii. $(g \circ f)(x)$, i.e., $g(f(x))$

SOLUTION   i. $(f \circ g)(x) = f(g(x))$. To evaluate, replace $g(x)$ by its output value, which is $2x + 3$. Next, treat $(2x + 3)$ as the input of the function f; f squares the input.

$$f(g(x)) = f(2x + 3) = (2x + 3)^2 = 4x^2 + 12x + 9$$

   ii. $(g \circ f)(x) = g(f(x)) = g(x^2) = 2x^2 + 3$ ◆

Notice that in Example 3.4 for almost all values of x, $f(g(x)) \neq g(f(x))$.[4] The order in which the functions are composed determines the result. In this case, doubling the input, adding 3 to it, and then squaring the sum is different from squaring the input, doubling the result, and then adding 3. When we write mathematics, we indicate the order of operations in an expression through a combination of parentheses and conventions for orders of operations.[5]

◆ **EXAMPLE 3.5**   Let $f(x) = x^2$, $g(x) = 2x + 3$, as in Example 3.4. Find

   i. $f(g(g(x)))$ ii. $g\left(\frac{1}{f(x)}\right)$

---

[4] There are only two values of x for which $2x^2 + 3$ is the same as $4x^2 + 12x + 9$. See if you can find them. If you need a refresher, refer to Appendix A: Algebra, under Solving Quadratic Equations.

[5] For a review of conventions for order of operations, please refer to Appendix A: Algebra.

SOLUTION   i.

$$f(g(g(x))) = f(g(2x + 3))$$
$$= f(2(2x + 3) + 3)$$
$$= f(4x + 6 + 3)$$
$$= f(4x + 9)$$
$$= (4x + 9)^2$$
$$= 16x^2 + 72x + 81$$

ii. $g\left(\frac{1}{f(x)}\right) = g\left(\frac{1}{x^2}\right) = \frac{2}{x^2} + 3$ ◆

◆ **EXAMPLE 3.6**   Let $g(x) = 2x + 3$, $h(x) = \frac{x-3}{2}$.   Find

i. $h(g(x))$ ii. $g(h(x))$

SOLUTION   i. $h(g(x)) = h(2x + 3) = \frac{(2x+3)-3}{2} = \frac{2x}{2} = x$

ii. $g(h(x)) = g\left(\frac{x-3}{2}\right) = 2\left(\frac{x-3}{2}\right) + 3 = x - 3 + 3 = x$ ◆

Observation

In this example diagrammatically we have $x \xrightarrow{g} g(x) \xrightarrow{h} x$ and $x \xrightarrow{h} h(x) \xrightarrow{g} x$. Thus, whether g is followed by h or h is followed by g, the result is not only the same, but it is the original input. The functions h and g undo one another. If $h(g(x)) = x$ and $g(h(x)) = x$, then h and g are called **inverse functions**. (We first introduced the topic of inverse functions in Section 1.3 and will discuss it in detail in Chapter 12.)

Notice that in order to perform the composition of functions you need to be comfortable evaluating a function even when the input is rather *messy*. You must distinguish in your mind the difference between the functional rule itself and the input of the function. The following exercise may be helpful.

EXERCISE 3.8   Let f be the function given by $f(x) = \frac{x}{x-1} + 2x$. Find the following.

i. $f(3)$ ii. $f(y + 1)$ iii. $f(1/x)$ iv. $f(x + h)$ v. $\frac{f(2h)}{h}$

To do this exercise, it is important to keep in mind that whatever is enclosed in the parentheses of f is the input of f. What does the function do with its input? f divides the input by a number that is one less than the input and then adds twice the input to that to the quotient.

For a silly but fail-proof way to find $f(mess)$, run through the following questions:

What is $f(2)? f(3)? f(\pi)? f(mess)$?

This serves to get the functional rule firmly established in your mind.

Solutions to Exercise 3.8 are given at the end of the chapter.

EXERCISE 3.9   Let $f(x) = x^2$ and $g(x) = x + 1$.

(a) Find $h(x) = f(g(x))$ and sketch the graph of h. (Use any means at your disposal.)

(b) Find $j(x) = g(f(x))$ and sketch the graph of j.

(c) Now look at parts (a) and (b) as variations on the theme $f(x) = x^2$.

i. In part (a) the input of f is increased by 1. What is the effect on the graph?

ii. In part (b) the output of f is increased by 1. What is the effect on the graph?

EXERCISE 3.10   Let $f(x) = 1/x$ and $g(x) = x + 2$.

(a) Find $h(x) = f(g(x))$ and sketch the graph of h.

(b) Find $j(x) = g(f(x))$ and sketch the graph of j.

(c) Now look at parts (a) and (b) as variations on the theme $f(x) = 1/x$.

i. In part (a) the input of f is increased by 2. What is the effect on the graph?

ii. In part (b) the output of f is increased by 2. What is the effect on the graph?

◆ **EXAMPLE 3.7**   *An oil spill on a large lake.* Let's return to our model of an oil spill where oil spreads out into a thin expanding disk. In Chapter 2 we looked at how the area of the disk of oil varies with the radius. Of more practical importance is how the area of the spill varies with time. Is the leak steady, worsening, or slowing down? Suppose the cannister from which the oil is leaking is opaque so we can't take our measurement directly from there. Instead, we measure the radius of the spill (in yards) at various times. The idea is to express the radius r as a function of time, $r = g(t)$. Then, since we can express the area of the spill as a function of the radius, $A = f(r) = \pi r^2$, we will be able to replace r by $g(t)$, obtaining $A = f(g(t))$. The composition $f(g(t))$ allows us to express the area as a function of time.

Suppose we measure the radius of the spill at one-hour intervals. The information is recorded below.[6]

Time (hr)	0	1	2	3	4	5	6	7	8	9	10
Radius	0.00	1.00	1.41	1.73	2.00	2.23	2.44	2.65	2.83	3.00	3.16

A plot of the results of our measurements is given in Figure 3.7. The graph of r versus t should be modeled by a continuous function because r is continuously increasing.

---

[6] Note that if measurements had been taken just for $t = 4, 5, 6, 7, 8$, and 9 and if the radius of the spill was recorded to only one decimal place, it would appear that each hour the radius increased by two tenths of a yard, and that the radius was a linear function of time. You can see this by looking at the graph in Figure 3.7 on the following page and seeing that the points above $t = 4, 5, 6, 7, 8$, and 9 *almost* appear to lie on a single line. In fact, it is not linear, but locally (on small enough intervals) it is approximately linear. We say that it is locally linear. The notion of local linearity is central to differential calculus and will be discussed in more detail in Chapter 4.

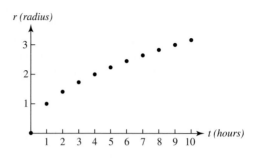

Figure 3.7

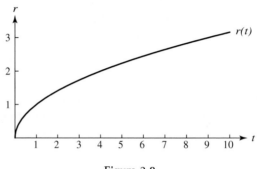

Figure 3.8

Answer the following questions.

   i. How does the radius vary with time? Give a possible model for r as a function of t. In other words, find a function $g(t)$ that would have a graph similar to that in Figure 3.8.

  ii. Using your answer to part (i), express area as a function of time.

 iii. How is the area changing with respect to time? What does this tell us about the oil leak?

SOLUTION   i. $r = g(t) = \sqrt{t}$ looks like an excellent model for t for $0 \le t \le 10$.

  ii. Area of a circle $= f(r) = \pi r^2$. Using part (i), we have $r = g(t) = \sqrt{t}$, so

$$A = f(r) = f(g(t)) = f(\sqrt{t}) = \pi(\sqrt{t})^2 = \pi t.$$

t	0	1	2	3	4	5	6	7	8	9	10
r	0.00	1.00	1.41	1.73	2.00	2.23	2.44	2.65	2.83	3.00	3.16
A	0.00	3.14	6.28	9.42	12.57	15.71	18.85	21.99	25.13	28.27	31.42

$$t \xrightarrow{g} r \xrightarrow{f} A$$

(The bottom row in this table is πt rounded to two decimal places.)

iii. Once we know A as a function of r and know r as a function of t we can find A as a function of t using composition of functions. $A(t) = \pi t$, so each hour the area of the spill increases by π square yards. This indicates that the leak is steady. It shows no signs of diminishing or speeding up; oil is leaking out at a steady rate. The rate at which the area is increasing is π square yards/hour; this is approximately 3.14 sq yd/hr.

Mental Picture: Assembly Line Model

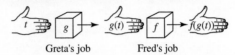

Greta's job   Fred's job

In this assembly line, g stands for Greta's job. At each time t, Greta determines the radius of the spill, $g(t)$. Greta hands her output, the radius, over to Fred, who uses it to compute the area of the spill. Fred's job, f, is to take what he is given, square it, and multiply by π.

$$t \to g(t) \to f(g(t)) = f(\sqrt{t}) = \pi(\sqrt{t})^2 = \pi t \quad \blacklozenge$$

PROBLEMS FOR SECTION 3.2

1. To find $f(g(x))$, apply g to x and then use the output of g as the input of f. Work from the inside out. Let $f(x) = x^2$, $g(x) = 1/x$, and $h(x) = 3x + 1$.
   Worked example:

$$f(g(h(x))) = f(g(3x + 1)) = f\left(\frac{1}{3x + 1}\right) = \left(\frac{1}{3x + 1}\right)^2$$

   Find the following.
   (a) $f(g(x))$ (b) $g(f(x))$
   (c) $xh(f(x))$ (d) $f(h(g(x)))$
   (e) $g(g(w))$ (f) $h(h(t))$
   (g) $g(f(1/x))$ (h) $g(2h(x - 1))$
   (i) Show that $g(g(x)) \neq [g(x)]^2$ (j) Show that $[h(x)]^2 \neq h(x^2)$.

For Problems 2 through 10, f and g are functions with domain $[-3, 4]$. Their graphs are provided below. Use the graphs to approximate the following.

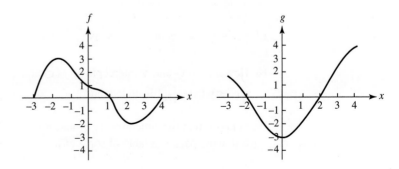

2. (a) $f(g(1))$ (b) $f(g(0))$

3. (a) $g(f(1))$ (b) $f(g(0))$

4. (a) $f(f(2))$ (b) $f(f(1))$

5. (a) $g(g(0))$ (b) $f(f(0))$

6. Find x such that $2f(x) = 0$.

7. Find all x such that $f(g(x)) = 0$.

8. Find all x such that $g(f(x)) = 0$.

9. (a) Find all x such that $f(x) = 3$.
   (b) Find all x such that $f(g(x)) = 3$.

10. Solve for x: $f(0)x + f(1) = g(0)x + g(1)$.

For Problems 11 through 13, find $f(g(h(x)))$ and $g(h(f(x)))$.

11. $f(x) = \frac{1}{x}$, $g(x) = \sqrt{x}$, $h(x) = x - 3$

12. $f(x) = 3x^2 + x$, $g(x) = x + 1$, $h(x) = \frac{2}{3x}$

13. $f(x) = x + 2$, $g(x) = x^2$, $h(x) = \frac{x}{2-x}$

For Problems 14 and 15 let $f(x) = x - 3$ and $g(x) = x^2 - 6x$.

14. Evaluate and simplify each of the following expressions.
   (a) $f(x) + g(x)$ (b) $f(x) - g(x)$
   (c) $f(x)g(x)$ (d) $f(g(x))$
   (e) $g(f(x))$ (f) $\frac{f(x)}{g(x)}$

15. Find the x- and y-intercepts of the following.
   (a) $f(x)$ (b) $g(x)$ (c) $f(x)g(x)$ (d) $\frac{f(x)}{g(x)}$

16. How do the x- and y-intercepts of $f(x)$ and $g(x)$ affect the intercepts of $f(x)g(x)$? State this as a general rule.

17. For what values of x (if any) is $\frac{f(x)}{g(x)}$ undefined?

18. How do the x- and y-intercepts of $f(x)$ and $g(x)$ affect the intercepts of $\frac{f(x)}{g(x)}$ and the places where $\frac{f(x)}{g(x)}$ is undefined?

19. Suppose that the functions f, g, and h are defined for all integers. At the top of the following page is a table of some of the values of these functions.

x	-2	-1	0	1	2	3	4
$f(x)$	0	2	1	3	4	-2	5
$g(x)$	2	3	4	1	3	-1	0
$h(x)$	3	4	-3	2	8	1	2

Evaluate the following expressions. If not enough information is available for you to do so, indicate that.

(a) $f(-1) \cdot g(-1)$

(b) $f(g(-1))$

(c) $g(f(-1))$

(d) $h(g(f(2)))$

(e) $\frac{f(0)+2}{g(0)}$

(f) $5h(3) + f(f(1))$

(g) $f(f(f(0)))$

20. The graphs of $f(x)$ and $g(x)$ are given below.

   (a) Approximate all the zeros of the function $h(x) = f(g(x))$.

   (b) Approximate all the zeros of the function $j(x) = g(f(x))$.

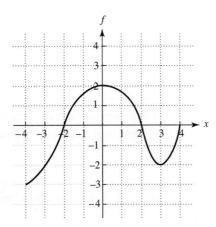 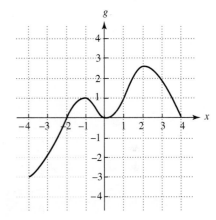

21. You put $300 in a bank account at 4% annual interest compounded annually and you plan to leave it there without making any additional deposits or withdrawals. With each passing year, the amount of money in the account is 104% of what it was the previous year.

   (a) Write a formula for the function f that takes as input the balance in the account at some particular time and gives as output the balance one year later. Write this formula as one term, not the sum of two terms.

   (b) Two years after the initial deposit is made, the balance in the account is $f(f(300))$ and three years after, it is $f(f(f(300)))$. Explain.

   (c) What quantity is given by $f(f(f(f(300))))$?

   (d) *Challenge:* Write a formula for the function g that takes as input n, the number of years the deposit of M dollars has been in the bank, and gives as output the balance in the account.

22. $f(x) = \frac{1}{2-x}$ and $g(x) = x^2 + 1$. Find the following. Simplify your answers. If simplifying is difficult, consult Appendix A: Algebra.

(a) $2f(x+1)$ (b) $f(2x-2)$ (c) $g(\sqrt{x}+1)$ (d) $f(g(x))$

(e) $g(f(x))$ (f) $f(f(x))$ (g) $g\left(\frac{1}{f(x)}\right)$ (h) $\frac{g(x)}{f(x)}$

23. Most of the time, when a store provides coupons offering $5 off any item in the store they include the clause "except for sale items." Suppose that clause were omitted and you found an item you wanted on a "30% off" rack. There would be some ambiguity; should the $5 be taken off the reduced price, or off the price before the 30% discount?

Let C be the function that models the effect of the coupon, S be the function that models the effect of the sale, and x be the original price of the item.

(a) Which situation corresponds to $C(S(x))$?

(b) Which situation corresponds to $S(C(x))$?

(c) Which order of composition of the functions is in the buyer's favor?

24. Two brothers, Max and Eli, are experimenting with their walkie-talkies. (A walkie-talkie is a combined radio transmitter and receiver light enough to allow the user to walk and talk at the same time.) The quality of the transmission, Q, is a function of the distance between the two walkie-talkies. We will model it as being inversely proportional to this distance.

At time $t = 0$ Max is 100 feet north of Eli. Max walks north at a speed of 300 feet per minute while Eli walks east at a speed of 250 feet per minute. All the time they are talking on their walkie-talkies.

(a) Write a function f such that $Q = f(d)$, where d is the distance between the brothers. Your function will involve an unknown constant.

(b) Write a function g that gives the distance between the brothers at time t.

(c) Find $f(g(t))$. What does this composite function take as input and what does it give as output?

25. If $h(x) = f(g(x))$, then x is in the domain of h if and only if x is in the domain of g and $g(x)$ is in the domain of f. In other words, x must be a valid input for g and $g(x)$ must be a valid input for f.

(a) If $h(x) = f(g(x))$, where $g(x) = \sqrt{x}$ and $f(x) = x^2$, what is the largest possible domain of h? For all x in its domain, $h(x) = x$. Why is the domain *not* $(-\infty, \infty)$?

(b) If $h(x) = f(g(x))$, where $g(x) = \frac{1}{x-1}$ and $f(x) = \frac{1}{x+3}$, what is the largest possible domain of h? (There are *two* numbers that must be excluded from the domain.)

26. Let $f(x) = \frac{2x}{x+3}$ and $g(x) = \frac{1}{x+1}$.

(a) Find $f(g(2))$.

(b) Find $f(g(x))$ and simplify your answer. Be sure that your answer is in agreement with the concrete case from part (a).

27. Let $f(x) = \frac{x}{x+3}$ and $g(x) = \frac{3x}{1-x}$.

(a) Find $f(g(2))$ and $g(f(2))$.

(b) Find $f(g(x))$ and $g(f(x))$.

(c) What does part (b) suggest about the relationship between f and g?

28. Below are graphs of f and g.

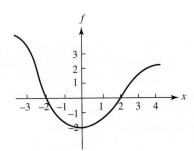 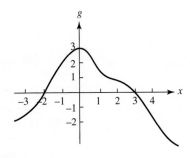

(a) Approximate $f(g(2))$.

(b) Approximate $g(f(2))$.

(c) For what values of x is $f(g(x)) = 0$?

29. If the function $m(t) = \frac{1}{t+2}$ and $h(t) = t - 2$, then is it ever true that $m(h(t)) = h(m(t))$?

30. The functions $R(x)$, $K(x)$, $D(x)$, and $L(x)$ are defined as follows:

$$R(x) = \frac{1}{x^2}, \qquad K(x) = |x|, \qquad D(x) = x + 3, \qquad L(x) = -5x.$$

Evaluate the following expressions. (Be sure to give simplified expressions whenever possible.)

(a) $R(K(L(x)))$ (b) $R(L(R(x)))$

(c) $R(K(x))$ (d) $R(D(R(x)))$

In Problems 31 through 33, let $f(x) = |x|$, $g(x) = \sqrt{x}$, and $h(x) = x - 2$. Find the domain for each of the following.

31. (a) $j(x) = g(h(x))$ (b) $k(x) = h(g(x))$

32. (a) $l(x) = g(f(x))$ (b) $m(x) = g(h(f(x)))$

33. (a) $p(x) = h(g(h(x)))$ (b) $q(x) = f(h(g(x)))$

34. Let $f(x) = x^2 + 9$, $g(x) = \sqrt{x}$, and $h(x) = g(f(x))$. Find the average rate of change of h over the following intervals.

   (a) $[-4, 4]$ (b) $[0, 4]$ (c) $[4, 4 + k]$

In Problems 35 through 38, find $h(x) = f(g(x))$ and $j(x) = g(f(x))$. What are the domains of h and j?

35. $f(x) = x^2 + 9$ and $g(x) = \frac{1}{\sqrt{x}}$

36. $f(x) = \frac{2}{x+2}$ and $g(x) = x - 2$

37. $f(x) = x^2$ and $g(x) = -2x + 3$

38. $f(x) = \frac{x}{x-3}$ and $g(x) = \frac{2}{x}$

In Problems 39 through 43, find $(f + g)(x)$, $(fg)(x)$, and $\left(\frac{f}{g}\right)(x)$, and find their domains.

39. $f(x) = ax + b$ and $g(x) = cx + d$

40. $f(x) = 3x + 2$ and $g(x) = 5x - 1$

41. $f(x) = 2x + 3$ and $g(x) = x^2 - 1$

42. $f(x) = \frac{3}{x+1}$ and $g(x) = \frac{2x}{x-5}$

43. $f(x) = \sqrt{x}$ and $g(x) = \sqrt{x - 3}$

In Problems 44 through 49, let $f(x) = \frac{1}{x} + x$ and $g(x) = \frac{2x}{x^2+1}$. Evaluate the following expressions.

44. (a) $g(f(2))$ (b) $f(g(2))$

45. (a) $g(f(\frac{1}{3}))$ (b) $f(g(\frac{1}{3}))$

46. (a) $g(f(1))$ (b) $f(g(1))$

47. (a) $f(f(2))$ (b) $g(g(-1))$

48. (a) $f(g(x))$ (b) $g(f(x))$

49. (a) $(f \circ f)(x)$ (b) $(f \circ f \circ f)(x)$

50. (a) Suppose f and g are both even functions. What can be said about $(f + g)(x)$? $(fg)(x)$?

   (b) Suppose f and g are both odd functions. What can be said about $(f + g)(x)$? $(fg)(x)$?

   (c) Suppose f is an even function and g is an odd function. What can be said about $(f + g)(x)$? $(fg)(x)$?

Algebraic calisthenics: Let $f(x) = 2x^2$, $g(x) = x + 1$, and $h(x) = \frac{1}{x}$. In Problems 51 through 58, if what is written is an expression, simplify it. If it is an equation, solve it.

51. $h(f(x)) + h(g(x))$

52. $g(x)h(f(x)) + f(x)h(g(x))$

53. $g(x)h(f(x)) = 1$

54. $2h(f(x)g(x)) + h(3g(x))$

55. $h(f(x) + 3g(x)) = h(2)$

56. $f(g(x)) = 10$

57. $f(g(f(x))) = 8$

58. $f(g(-5 + f(x))) = 8$ (There are four solutions.)

■ 3.3   DECOMPOSITION OF FUNCTIONS

To decompose a function means to express it as the composition of two or more functions. Using our assembly line analogy, the process of decomposition corresponds to breaking up a task into a sequence of simpler jobs to be done in succession on an assembly line, the output of one operation constituting the input for the next. Just as there may be different ways of setting up an assembly line to accomplish a given task, there are often different ways to decompose a single function. We give some examples below.

◆ **EXAMPLE 3.8**   If $h(x) = \sqrt{x^2 + 7x}$, find f and g such that $f(g(x)) = h(x)$.
Diagrammatically we have the following situation:

$$x \xrightarrow{g} g(x) \xrightarrow{f} f(g(x))$$
$$\underbrace{\qquad\qquad}_{h}$$

Think of g as the first worker on the assembly line; the output of g is passed on for f to act upon. There are different ways we can set up the assembly line. One possibility is that the first worker, g, can produce $x^2 + 7x$ and, to finish off the job, f can take a square root. Then $g(x) = x^2 + 7x$ and $f(x) = \sqrt{x}$. Sometimes this is written $g(x) = x^2 + 7x$ and $f(u) = \sqrt{u}$, this notation (i.e., the different variable) reminding us that f works on the output of another function.   ◆

EXERCISE 3.11   There are many ways of decomposing functions. For instance, suppose

$$h(x) = \frac{x^2 + 1}{x^2 + 2}$$

and we want to find f and g such that $f(g(x)) = h(x)$. Determine which of the following pairs of functions produce the appropriate result.

$$x \xrightarrow{g} g(x) \xrightarrow{f} f(g(x))$$
$$\underbrace{\qquad\qquad}_{h}$$

i. $g(x) = x^2$, $\qquad f(x) = \frac{x+1}{x+2}$

ii. $g(x) = x^2 + 1$, $\qquad f(x) = \frac{x}{x+1}$

iii. $g(x) = x^2 + 2$, $\qquad f(x) = \frac{x-1}{x}$

iv. $g(x) = \frac{x^2}{x^2+2}$, $\qquad f(x) = \frac{x+1}{x}$

v. $g(x) = x^2 + 1$, $\qquad f(x) = \frac{x}{x^2+2}$

Answers are provided at the end of this chapter.

◆ **EXAMPLE 3.9** $f(x) = \sqrt{(x^2+2)^2 + 3}$. Find g, h, and j such that $g(h(j(x))) = f(x)$.

SOLUTION

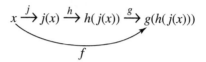

There are many different sets of functions g, h, and j that satisfy $g(h(j(x))) = f(x)$. For example,

$$j(x) = x^2$$
$$h(x) = (x+2)^2 + 3$$
$$g(x) = \sqrt{x}$$

or

$$j(x) = (x^2+2)^2$$
$$h(x) = x + 3$$
$$g(x) = \sqrt{x}$$

or

$$j(x) = x^2$$
$$h(x) = (x+2)^2$$
$$g(x) = \sqrt{x+3} \quad ◆$$

Why decompose? Decomposing a function can give us insight into its underlying structure and aid us in dealing with more complex functions. A basic strategy when solving any complex problem is to break the problem down into simpler subproblems and to construct a solution by appropriately building it up from solutions to the subproblems. Determining the skeletal structure of the problem is critical when searching for a strategy for solving a complex problem.

◆ **EXAMPLE 3.10** Suppose we want to solve the equation

$$x^4 - 5x^2 + 4 = 0.$$

In general, fourth degree equations are difficult to tackle, but this particular equation can be viewed as a "quadratic in disguise." By this we mean that the underlying structure is that of a quadratic; that is, if $h(x) = x^4 - 5x^2 + 4$ then $h(x) = f(g(x))$, where $g(x) = x^2$ and $f(u) = u^2 - 5u + 4$. To solve the original equation we can let $u = x^2$ and solve the equation $u^2 - 5u + 4 = 0$ for u. Once we have found u, we can solve[7] for x easily because we know $u = x^2$. Breaking the original (hard) problem into two (easier) subproblems makes solving the original problem possible.

$$u^2 - 5u + 4 = 0$$
$$(u - 4)(u - 1) = 0$$
$$u = 4 \quad \text{or} \quad u = 1$$

But $u = x^2$.

$$x^2 = 4 \quad \text{or} \quad x^2 = 1$$
$$x = \pm 2 \quad \text{or} \quad x = \pm 1.$$

The solutions to $x^4 - 5x^2 + 4 = 0$ are $x = -2, -1, 1, 2$.   ◆

Calculus involves the study of rates of change of functions. Later in this text, when we are looking for the rate of change of a composite function, we will use the same approach of decomposition to decompose the function and find its rate of change from our knowledge of the rates of change of these simpler functions.[8]

PROBLEMS FOR SECTION 3.3

1. Let $h(x) = f(g(x))$ and suppose that $h(x) = \dfrac{1}{\sqrt{x^2+6}}$. Write possible formulas for $f(x)$ and $g(x)$.

2. For each of the functions given below, give possible formulas for $f(x)$ and $g(x)$ such that $h(x) = f(g(x))$. Do not let $g(x) = x$; do not let $f(x) = x$.
   (a) $h(x) = \sqrt{x^2 + 3}$
   (b) $h(x) = \sqrt{x} + \dfrac{5}{\sqrt{x}}$
   (c) $h(x) = \dfrac{3}{3x^2+2x}$
   (d) $h(x) = 5(x^2 + 3x^3)^3$

3. Let $j(x) = \dfrac{2}{3\sqrt{4x^2+3x}}$. Suppose that $j(x) = h(g(f(x)))$. Write possible formulas for $f(x)$, $g(x)$, and $h(x)$. None of f, g, and h should be the identity function.

4. Let $j(x) = 10(x^{-2} + 2x^2)^3$. Give two possible decompositions of $j(x)$ such that $j(x) = f(g(h(x)))$. None of the functions f, g, and h should be the identity function.

5. Let $h(x) = f(g(x))$, where $f(x) = x^2$. If

$$h(t - 2) = \left(\frac{1}{t - 2} + 1 \right)^2,$$

   what is $g(x)$?

---

[7] For more on solving quadratics, see Appendix A: Algebra.
[8] The method of constructing the rate of change function of a composite function is called the Chain Rule.

Decompose the functions in Problems 6 through 9 by finding functions $f(x)$ and $g(x)$,
$f(x) \neq x$ and $g(x) \neq x$, such that $h(x) = f(g(x))$.

6. $h(x) = (x^2 + 7x + 1)^3$

7. $h(x) = \frac{1}{x^2 + 4}$

8. $h(x) = \sqrt{x^2 + 1}$

9. $h(x) = (\sqrt{x})^3 - 2\sqrt{x} + 3$

In Problems 10 through 17, find functions f and g such that $h(x) = f(g(x))$ and neither
f nor g is the identity function, i.e., $f(x) \neq x$ and $g(x) \neq x$. Answers to these problems
are not unique.

10. $h(x) = \frac{3}{\sqrt{x+2}}$

11. $h(x) = 3x^4 + 2x^2 + 3$

12. $h(x) = 5(x - \pi)^2 + 4(x - \pi) + 7$

13. $h(x) = 2|3x - 4|$

14. $h(x) = \frac{x+1}{x+2}$

15. $h(x) = \sqrt{5x^2 + 3}$

16. $h(x) = 3^{2x} + 3^x + 1$

17. $h(x) = 4\pi^{2x} + 3\pi^x + 2$

In Problems 18 through 20, find functions f, g, and h such that $k(x) = f(g(h(x)))$
and $f(x) \neq x$, $g(x) \neq x$, and $h(x) \neq x$.

18. $k(x) = \frac{3}{\sqrt{x^2 + 4}}$

19. $k(x) = \frac{1}{(\sqrt{x} + 1)^9}$

20. $k(x) = \sqrt{(x^2 + 1)^3 + 5}$

Exploratory Problems for Chapter 3

Flipping, Shifting, Shrinking, and Stretching: Exercising Functions

In these exercises you will experiment with altering the input and output of functions and draw conclusions about the effects of these alterations on the graph of the function. Part of being a scientist is learning how to design an experiment. A graphing calculator or a computer will be an invaluable tool. We'll suggest some experiments, but you are expected to come up with most of them on your own. Experiment until you have confidence in your conclusions. Look for patterns, test your discoveries, and then try to generalize. When you generalize, think about *why* these generalizations make sense. Compare your conclusions with those of your classmates and resolve any discrepancies.

1. Modifying the output of f

   (a) How does the graph of $y = kf(x)$ (where k is a constant) relate to that of $y = f(x)$?

   You might want to break your answer into cases depending upon the sign of k and whether $|k|$ is greater than, less than, or equal to 1.

       i. Does multiplying $f(x)$ by a constant change the location of its zeros? If so, how?

       ii. Does multiplying $f(x)$ by a constant affect where the graph has turning points (peaks and valleys)? Does it affect the value of the function at these peaks and valleys?

   (b) How does the graph of $y = f(x) + k$ (where k is a constant) relate to that of $y = f(x)$?

       i. Does adding a constant to $f(x)$ change the location of its zeros?

       ii. Does adding a constant to $f(x)$ affect where the graph has turning points (peaks and valleys)? Does it affect the value of the function at these peaks and valleys?

   (c) How does the graph of $y = cf(x) + k$ (where c and k are constants) relate to that of $y = f(x)$?

   In particular, let $f(x) = x^2$. On the same set of axes sketch the graphs of $f(x)$ and $h(x)$, where $h(x) = -f(x) + 2$. Basic order of operation rules tell us to multiply by -1 first and then add 2. Do we follow the same procedure when analyzing the function graphically? In other words, to obtain the graph of h do we first flip the graph of f across the x-axis and then shift it up 2 units? Demonstrate that if we first shift the graph of f up 2 units and then flip the result across the x-axis that we will obtain the graph of $y = -[f(x) + 2]$, *not* the graph of h.

2. Altering the function's input

(a) How is the graph of $y = f(x - k)$ (where k is a constant) related to that of $y = f(x)$? Distinguish between the cases where k is positive and where k is negative.

   i. How are the zeros of $f(x - k)$ related to the zeros of $f(x)$? In particular, if $x = r$ is a zero of $f(x)$, what can we say about a zero of $f(x - k)$?

   ii. Does replacing x by $(x - k)$ affect where the graph has turning points (peaks and valleys)? Does it affect the value of the function at these peaks and valleys?

   iii. Given the generalizations you made, how would you expect the graph of $y = \frac{1}{x-k}$ to look? Will the position of the vertical asymptote be affected by k? Will the position of the horizontal asymptote be affected by k? Explain your reasoning. Check your expectations by looking at some concrete cases.

(b) How is the graph of $y = f(kx)$ (where k is a constant) related to that of $y = f(x)$? Try looking at a function such as $f(x) = (x - 1)(x - 2)(x + 4)$ and experimenting with various values of k.

   Break your answer into cases: k > 1, k < 1, k = −1, k < −1, *and* −1 < k < 0.

   i. What is the effect on the x-intercepts?

   ii. Do the locations of the peaks and valleys change?

   iii. Do the heights of the peaks and valleys change?

3. Alterations to both input and output

(a) Let $p(x) = -\frac{1}{2}(x + 3)^2 - 1$. Interpret the graph of p in terms of transformations of a familiar function. *Note: $p(x)$ can be viewed as a variation of the familiar function $f(x) = x^2$.* Unraveling a function is just like unraveling a very complicated algebraic expression; you need to start from the innermost part and remember the order of operations rules.

(b) Let $q(x) = -\frac{2}{x-4} + 3$. Interpret the graph of q in terms of transformations of a familiar function. Where are the vertical and horizontal asymptotes?

(c) Below is the graph of $f(x)$. Graph the following.

    i. $y = f(2x)$

    ii. $y = 3f(\frac{x}{2})$

    iii. $y = -f(x-1)$

    iv. $y = 2f(x-1) + 3$

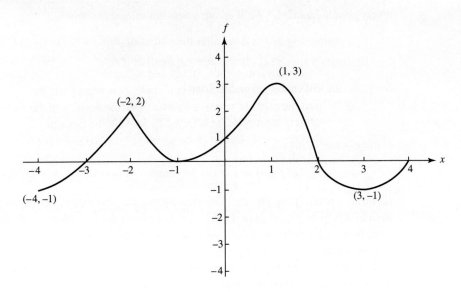

3.4   ALTERED FUNCTIONS, ALTERED GRAPHS: STRETCHING, SHRINKING, SHIFTING, AND FLIPPING

Suppose we are quite familiar with a certain function f and know the shape of its graph. Now suppose we alter this function f by composing it with another function, g, where g is a very simple function.[9] We'll create a new but related function.

By constructing $g(f(x))$ we alter the output of f.

By constructing $f(g(x))$ we alter the input of f.

Suppose, for instance, $g(x) = 3x$. Then

$$g(f(x)) = 3f(x) \quad \text{and} \quad f(g(x)) = f(3x).$$

If $g(x) = x - 2$, then

$$g(f(x)) = f(x) - 2 \quad \text{and} \quad f(g(x)) = f(x - 2).$$

If you are a good sleuth, you can determine a great deal about these new but related functions, such as $3f(x)$ or $f(x - 2)$ or $f(3x) - 2$. The exploratory problems preceding this section were designed to help you figure out how your knowledge of f can lead you to understand its close relatives. Being able to construct functions that stretch, shrink, shift, and flip graphs with which you are familiar is an extremely powerful tool. Not only can it brighten a rainy day, it can also help you model real-life phenomena by enabling you to alter functions you know. Conversely, being able to take what you know about f and apply it to other related functions can save you a great deal of time because you will not have to start from scratch with every new function you encounter. For example, by the time you are done with this section, you should be able to look at a function like $r(x) = \frac{-1}{x-1} + 2$ and break it down into a sequence of operations performed on a function that you know well. This will help you determine what the graph of $y = r(x)$ will look like.

Variations on the Theme of $y = f(x)$

Let k be a constant. [10]

- $y = kf(x)$: vertical stretching and shrinking plus flipping across the x-axis

   If $k > 1$, the graph of $kf(x)$ is the graph of $f(x)$ stretched vertically away from the x-axis, stretched by a factor of k.

   If $0 < k < 1$, the graph of $kf(x)$ is the graph of $f(x)$ shrunk vertically toward the x-axis, rescaled by a factor of k.

   If $-1 < k < 0$, the graph of $kf(x)$ is the graph of $f(x)$ flipped across the x-axis and rescaled vertically by a factor of $|k|$.

   If $k = -1$, the graph of $kf(x)$ is the graph of $f(x)$ flipped across the x-axis.

   If $-1 < k$, the graph of $kf(x)$ is the graph of $f(x)$ flipped across the x-axis and stretched by a factor of $|k|$.

---

[9] We will begin by looking at g where $g(x)$ is of the form $ax + b$, a linear function.

[10] A constant, such as k, that can vary from equation to equation but can be used to represent a family of related equations is often referred to as a **parameter**.

Suppose that a is a zero of $f(x)$. This means that $f(a) = 0$. Therefore $kf(a) = k \cdot 0 = 0$ as well. $kf(x)$ and $f(x)$ have the same roots.

Multiplying $f(x)$ by a constant does *not* affect the location of its zeros.

- $y = f(x) + k$: vertical shifts

  If $k > 0$, the graph of $f(x) + k$ is the graph of $f(x)$ shifted up k units.

  If $k < 0$, the graph of $f(x) + k$ is the graph of $f(x)$ shifted down |k| units.

- $y = cf(x) + k$, where c and k are constants

  Just as $2a - 3$ is not the same as $2(a - 3) = 2a - 6$, the function $2f(x) - 3$ is not the same as $2(f(x) - 3)$. Following order of operations rules, we do the multiplication (i.e., the stretching or shrinking and possible flipping around the x-axis) before we do the addition or subtraction (i.e., the shifting up or down).

  Aside: Notice that because $2(f(x) - 3) = 2f(x) - 6$, we can say that shifting down 3 units then stretching vertically by a factor of 2 is the same as stretching vertically by a factor of 2 then shifting down 6 units.

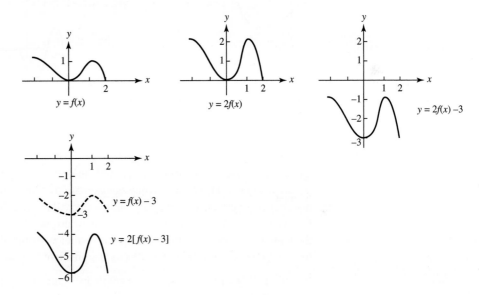

Figure 3.9

- $y = f(x - k)$: horizontal shifts

  If $k > 0$, the graph of $f(x - k)$ is the graph of $f(x)$ shifted k units to the *right*.

  For example, replacing x by $x - 2$ shifts the graph of f to the right 2 units.

  If $k > 0$, the graph of $f(x + k)$ is the graph of $f(x)$ shifted k units to the *left*.

  For example, replacing x by $x + 3$ shifts the graph of f to the left 3 units.

  Be careful here. This is easy to get mixed up; although $f(x + 3)$ has a plus sign, the graph is shifted to the *left*.

- $y = f(kx)$: horizontal shrinking and stretching plus flipping across the y-axis

  If $k > 1$, then the graph of $f(kx)$ is the graph of $f(x)$ *compressed* horizontally.

  If $0 < k < 1$, then the graph of $f(kx)$ is the graph of $f(x)$ *stretched* horizontally.

If $-1 < k < 0$, then the graph of $f(kx)$ is the graph of $f(x)$ flipped across the y-axis and *stretched* horizontally.

If $k = -1$, then the graph of $f(kx)$ is the graph of $f(x)$ flipped across the y-axis.

If $k < -1$, then the graph of $f(kx)$ is the graph of $f(x)$ flipped across the y-axis and *compressed* horizontally.

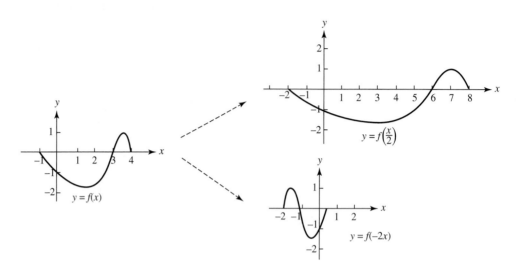

Figure 3.10

While the effects on the graph of $f(x)$ of altering the output, for instance, $y = 2f(x) + 3$, strike most people as natural, the effects of altering the input of f, such as $f(2x)$ or $f(x + 1)$, often provoke consternation. This tends to be the case even if you discover these effects on your own. Below we will rewrite the altering of the output so that the symmetry between the two cases is more transparent. We begin with $y = f(x)$. To alter the input is to alter x; to alter the output is to alter y. Notice that $y = f(x) + 2$ is equivalent to $y - 2 = f(x)$. Similarly, $y = 2f(x)$ is equivalent to $\frac{y}{2} = f(x)$.

■ Replacing y by $y - 2$ shifts the graph up 2 units; replacing y by $y + 3$ shifts the graph down 3 units.

■ Replacing x by $x - 2$ shifts the graph right 2 units; replacing x by $x + 3$ shifts the graph left 3 units.

■ Replacing y by $\frac{y}{2}$ stretches the graph vertically by a factor of 2; replacing y by $2y$ compresses the graph vertically. The new height is half the old one.

■ Replacing x by $\frac{x}{2}$ stretches the graph horizontally by a factor of 2; replacing x by $2x$ compresses the graph horizontally. The x-coordinates are half the old ones.

Taking Control: A Portable Problem-Solving Strategy

Suppose you forget what happens when x is replaced by $x + 2$. Do you just say, "I used to know this but I've forgotten. If I study it I'll remember"? No! If you want your knowledge to be portable you've got to find ways of reconstructing it so the knowledge that you've worked hard to gain doesn't fly off in the wind or wear thin with the passing of time.

Even though the details may fade, chances are you'll remember some broad outlines. For instance, you might recall that alterations of the input variable x will be manifested horizontally. Or maybe you'll remember that this is a shift, not a stretch. Probably you'll have some question like "Is this a shift right or a shift left?" Make a hypothesis (for instance, that it is a shift right) and test the hypothesis out on a simple function like $y = x^2$.

Does the point $(2, 0)$ satisfy the equation $y = (x + 2)^2$?   No.

Does the point $(-2, 0)$ satisfy the equation $y = (x + 2)^2$?   Yes.

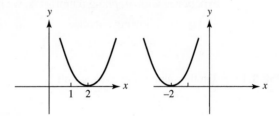

Figure 3.11

Learning to ask and answer questions that help you retain information is a large part of being a mathematician or scientist.

More Transformations of Output

In addition to shifting, stretching, and flipping, we can do a few other interesting things to functions to obtain new functions.

Absolute Values: $|f(x)|$

Suppose that $v(t)$ is a function that gives velocity as a function of time. Then the speed function is given by $|v(t)|$. Similarly, if $s(t)$ is a position function, then $|s(t)|$ gives the distance from a benchmark position.

How is the graph of $y = |f(x)|$ related to the graph of $y = f(x)$? Where the graph of $f(x)$ lies above or on the x-axis the graph of $y = |f(x)|$ is identical; where the graph of f lies below the x-axis the graph of $y = |f(x)|$ is obtained by flipping the graph of f over the x-axis, corresponding to multiplying the y values by -1. Notice that sharp corners are created.

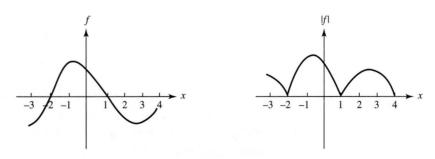

Figure 3.12

Reciprocals: $\frac{1}{f(x)}$

Suppose a function $f(t)$ gives the price of gasoline in $\frac{\text{dollars}}{\text{gallon}}$ at time t. We may be interested in the function $\frac{1}{f(t)}$ that gives the number of gallons of gas that can be purchased for a dollar. Or suppose a function tracking an employee's training gives the amount of time it takes her to produce an item, with units of $\frac{\text{minutes}}{\text{item}}$. We may be interested in the reciprocal function that gives us the number of items she can produce per minute.

How is the graph of $y = \frac{1}{f(x)}$ related to the graph of $y = f(x)$? We already have some experience with graphing reciprocals; we know how to graph $f(x) = x$ and $g(x) = \frac{1}{f(x)} = \frac{1}{x}$. Try the following exercise.

EXERCISE 3.12   On the same set of axes, graph

$$y = m(x) = x - 3$$

and its reciprocal,

$$y = \frac{1}{m(x)} = \frac{1}{x-3}.$$

Label all intercepts and asymptotes. Notice that you can graph these two functions by simply applying shifting principles to the functions $f(x) = x$ and $g(x) = \frac{1}{x}$. Instead of starting out that way, try to think in terms of reciprocals and use the shifts to check your work.

General Principles: Sign, Magnitude, and Asymptotes (See Figure 3.13 for examples.)

■   Where $f(x)$ is positive, $\frac{1}{f(x)}$ is positive; where $f(x)$ is negative, $\frac{1}{f(x)}$ is negative.

Where $f(x)$ is zero, $\frac{1}{f(x)}$ is undefined.

■   Where $|f(x)| = 1$, $|\frac{1}{f(x)}| = 1$.

Where $|f(x)| > 1$, $|\frac{1}{f(x)}| < 1$.

$$\frac{1}{\text{LARGE positive number}} = \text{small positive number.}$$

Where $|f(x)| < 1$, $|\frac{1}{f(x)}| > 1$.

$$\frac{1}{\text{small positive number}} = \text{LARGE positive number.}$$

■   Suppose that $f(x) \to \infty$ as $x \to \infty$. Then as $x \to \infty$, $\frac{1}{f(x)} \to 0$.

Suppose that $f(x) \to L$ as $x \to \infty$, where L is a nonzero constant. Then as $x \to \infty$, $\frac{1}{f(x)} \to \frac{1}{L}$.

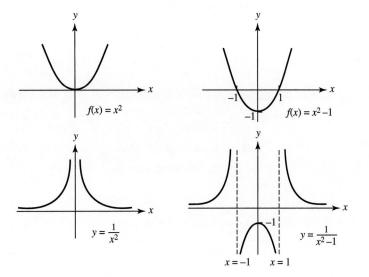

Figure 3.13

Answers to Selected Exercises

Answer to Exercise 3.7

$$h(x) = x \cdot \frac{1}{x} = \frac{x}{x} = \begin{cases} 1 & \text{for } x \neq 0 \\ \text{undefined} & \text{for } x = 0 \end{cases}$$

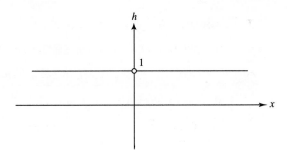

Figure 3.14

Your calculator probably won't indicate the pinhole in the graph.

Answer to Exercise 3.8

    i.

$$f(3) = \frac{3}{3-1} + 2(3)$$

$$= \frac{3}{2} + 6 = 7.5$$

ii.

$$f(y+1) = \frac{y+1}{(y+1)-1} + 2(y+1)$$

$$= \frac{y+1}{y} + 2y + 2$$

$$= 1 + 1/y + 2y + 2$$

$$= 2y + 3 + 1/y$$

iii.

$$f(1/x) = \frac{1/x}{(1/x)-1} + 2(1/x)$$

$$= \frac{\frac{1}{x}}{\frac{1}{x}-\frac{x}{x}} + \frac{2}{x}$$

$$= \frac{\frac{1}{x}}{\frac{1-x}{x}} + \frac{2}{x}$$

$$= \left(\frac{1}{x}\right)\left(\frac{x}{1-x}\right) + \frac{2}{x}$$

$$= \frac{1}{1-x} + \frac{2}{x}$$

If you want to add the fractions, get a common denominator.

$$\frac{x}{(1-x)x} + \frac{2(1-x)}{x(1-x)} = \frac{x+2-2x}{(1-x)x} = \frac{2-x}{(1-x)x}$$

iv.

$$f(x+h) = \frac{x+h}{x+h-1} + 2(x+h) = \frac{x+h}{x+h-1} + \frac{(2x+2h)(x+h-1)}{x+h-1}$$

$$= \frac{x+h+2x^2+2xh-2x+2hx+2h^2-2h}{x+h-1}$$

$$= \frac{2x^2+4xh-x+2h^2-h}{x+h-1}$$

v.

$$\frac{f(2h)}{h} = \frac{\frac{2h}{2h-1}+2(2h)}{h}$$

$$= \frac{\frac{2h}{2h-1}+4h}{h}$$

$$= \frac{\frac{2h}{2h-1}+\frac{4h(2h-1)}{2h-1}}{h}$$

$$= \frac{\frac{2h+8h^2-4h}{2h-1}}{h}$$

$$= \left(\frac{8h^2 - 2h}{2h - 1} \right) \left(\frac{1}{h} \right)$$

$$= \left(\frac{(8h - 2)(h)}{2h - 1} \right) \left(\frac{1}{h} \right)$$

$$= \frac{8h - 2}{2h - 1} \quad \text{or} \quad \frac{2(4h - 1)}{2h - 1}$$

Solution to Exercise 3.11

All but the last two pairs of functions give a decomposition.

PROBLEMS FOR SECTION 3.4

1. The zeros of the function $f(x)$ are at $x = -4, -1, 2,$ and 8. What are the zeros of
   (a) $m(x) = 5f(x)$?
   (b) $g(x) = f(x + 2)$?
   (c) $h(x) = f(2x)$?
   (d) $j(x) = f(x - 1)$?
   Verify your answers analytically.

2. The zeros of the function $f(x)$ are at $x = -5, -2, 0,$ and 5. Find the zeros of the following functions. If there is not enough information to determine this, say so.
   (a) $g(x) = 3|f(x)|$
   (b) $h(x) = w(f(x))$, where $w(x) = -2x^2$
   (c) $p(x) = 3f(x) + 1$
   (d) $q(x) = 4f(x + 1)$
   (e) $m(x) = 4f(-x)$
   (f) $n(x) = -f(x)$

3. The graph of $y = f(x)$ is symmetric about the y-axis. Which of the following functions is equal to $f(x)$?
   (a) $g(x) = -f(x)$
   (b) $h(x) = f(-x)$
   (c) $j(x) = -f(-x)$

4. Using what you know about shifting, flipping, and stretching, match the graphs on page 135 with the equations.
   (a) $y = \frac{-3}{x}$
   (b) $y = \frac{1}{x - 3}$
   (c) $y = \frac{1}{x + 1} - 1$
   (d) $y = 2 \left(\frac{1}{x + 1} + 1 \right)$
   (e) $y = \frac{-2}{x + 1} - 1$

(i)

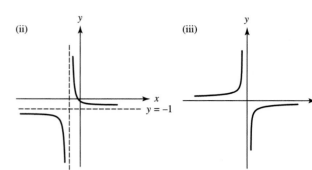

(ii)

(iii)

(iv)

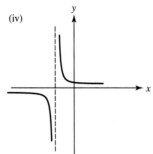

(v)

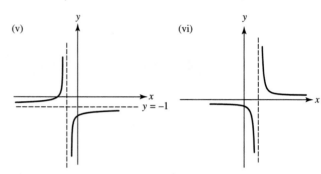

(vi)

5. Below is the graph of $y = 2^x$.

As $x \to \infty$, $y \to \infty$.

As $x \to -\infty$, $y \to 0$.

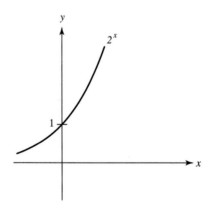

Using what you know about shifting, flipping, and sliding, match the graphs on the following page with the equations.

(a) $y = -2^x$

(b) $y = 2^{-x}$

(c) $y = 2^x + 1$

(d) $y = 2^{-x} - 1$

(e) $y = -2^{-x} + 1$

(f) $y = -2^x + 1$

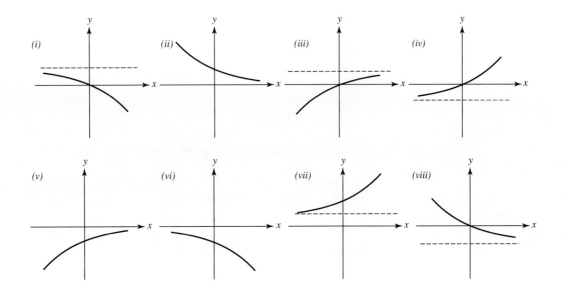

6. Using what you know about shifting, flipping, and stretching, match the graphs below with the equations.

   (a) $y = |x - 2|$

   (b) $y = -3|x + 1|$

   (c) $y = |x + 1| + 2$

   (d) $y = -|x + 1| - 1$

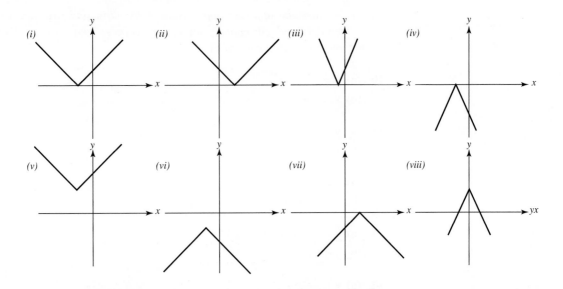

7. Applying what you learned in the last section of this chapter to the pocketful of functions you've been introduced to (the identity, squaring, reciprocal, and absolute value functions), graph the following functions. Label any asymptotes and x- and y-intercepts.

(a) $f(x) = \frac{1}{x^2}$

(b) $g(x) = |(x-1)^2|$

(c) $h(x) = |x^2 - 1|$

(d) $j(x) = \frac{1}{x+1} + 2$

(e) $m(x) = \frac{-1}{x-2} + 1$

(f) $p(x) = \left|\frac{1}{x}\right|$

8. Which of the functions in the previous problem are even?

9. f is a function with domain $[-3, 4]$. The graph of f is given below. Sketch $g(x) = f(x+2)$. What is the domain of $g(x)$?

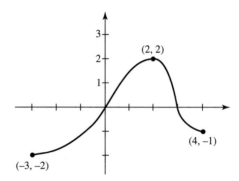

Graph the functions in Problems 10 through 18 by starting with the graph of a familiar function and applying appropriate shifts, flips, and stretches. Label all x- and y-intercepts and the coordinates of any vertices and corners. Use exact values, not numerical approximations.

10. (a) $y = (x-1)^2$

    (b) $y = -x^2 - 1$

11. (a) $y = |x+2|$

    (b) $y = -|x| + 2$

12. (a) $y = -(x+3)^2 - 1$

    (b) $y = (x-3)^2 + 1$

13. (a) $y = -2(x+1)^2 + 3$

    (b) $y + 3 = 7(x+1)^2$

14. (a) $y = \frac{2}{x+4} + 1$

    (b) $y = \frac{-1}{x-\pi}$

15. (a) $y = -x + \pi$

    (b) $y = -(x+\pi)$

16. (a) $y - \pi = (x - 2\pi)^2$

(b) $y - \pi = -(x - 2\pi)^2$

17. (a) $y = \frac{x+3}{x+2}$ (rewrite $x + 3$ as $x + 2 + 1$)

(b) $y = \frac{x+1}{x-1}$

18. (a) $y = -1 - 2|x + 1|$

(b) $y = -1 - 2(x + 1)^2$

For Problems 19 through 21, let $f(x) = |x|$. Graph the functions on the same set of axes.

19. $g(x) = (x - 3)^2 - 4$ and $f(g(x))$

20. $g(x) = -(x + 2)^2 + 1$ and $f(g(x))$

21. $g(x) = |x - 2| - 3$ and $f(g(x))$

22. Let $f(x) = \frac{1}{x}$ and $g(x) = x^2$. Using what you've learned in Section 3.4, graph the following equations.

(a) $y = f(g(x))$

(b) $y = |g(x - 1) - 4|$

(c) $y = |f(x)| - 1$

23. The graph of $y = f(x)$ is given below.

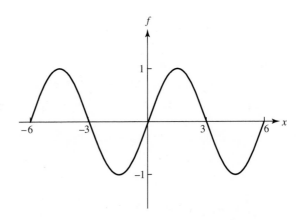

Sketch:

(a) $y = f(x - 3)$

(b) $y = 2f(x)$

(c) $y = -.5f(x)$

(d) $y = f(x/2)$

24. A and B are points on the graph of $k(x)$. The x-coordinate of point A is 6 and the x-coordinate of point B is $(6 + h)$. Write mathematical expressions, using functional notation, for each of the following.

(a) The change in value of the function from point A to point B

(b) The average rate of change of the function k over the interval $[6, 6 + h]$

(c) Suppose that the average rate of change of the function k over the interval $[6, 6 + h]$ is -5. The functions f, g, and h are defined as follows:

$$f(x) = k(x) + 2, \qquad g(x) = k(x + 2), \qquad h(x) = 2k(x).$$

i. Which of the following must also be equal to -5?
   A. The average rate of change of the function f over the interval $[6, 6 + h]$
   B. The average rate of change of the function g over the interval $[6, 6 + h]$
   C. The average rate of change of the function h over the interval $[6, 6 + h]$

ii. One of the functions f, g, and h has an average value of -10 on the interval $[6, 6 + h]$. Which is it? Explain briefly.

C H A P T E R

4

Linearity and Local Linearity

■ 4.1 MAKING PREDICTIONS: AN INTUITIVE APPROACH TO LOCAL LINEARITY

Every day we are bombarded with predictions; weather forecasters, sports announcers, financial consultants, demographers, experts, and self-proclaimed experts alike are constantly letting us know what they think will happen in the future. And, based on certain assumptions, each of us makes his or her own projections and acts accordingly. Let's look at a couple of examples.

◆ **EXAMPLE 4.1** It is Thursday, November 6, 1997, and a student is sitting at her desk making a schedule for herself. She likes to have completed her daily afternoon run by sunset, so she has been checking the local newspaper for sunset times. She has recorded the following information.[1]

---

[1] These times (and all that follow in this problem) are given in Greenwich mean time for a north latitude of 30°. Source, *1998 World Almanac.*

Date	Sunset Time
Tuesday, November 4	5:11 P.M.
Wednesday, November 5	5:10 P.M.
Thursday, November 6	5:09 P.M.

She estimates that on Sunday, November 9, the sun will set at 5:06 P.M. Upon what assumptions is her projection based? Is it reasonable to use the same assumptions to predict the time of sunset on November 20, 1997? On December 25, 1997? One year later, on November 6, 1998?

SOLUTION   Over the past few days the hour of sunset had been getting earlier at a rate of 1 minute/day.[2] In other words, the sunset time has been changing at a rate of -1 minute/day. Assuming that this rate of change remains constant for the next few days, then three days later the sunset will be three minutes earlier than it was on November 6.

$$\text{Change in sunset time} = (\text{rate of change per day}) \cdot (\text{days})$$

$$= -1\frac{\text{minute}}{\text{day}} \cdot (3 \text{ days}) = -3 \text{ minutes}$$

The predicted sunset time is

$$5:09 \text{ P.M.} + \left(-1\frac{\text{minute}}{\text{day}}\right) \cdot (3 \text{ days}) = 5:09 \text{ P.M.} - 3 \text{ minutes} = 5:06 \text{ P.M.}$$

Keep in mind that this prediction is based upon the assumption that the time of sunset is changing at a constant rate of -1 minute/day throughout the six-day period. *If* this were correct, then the graph of sunset time plotted versus the date would be a straight line.

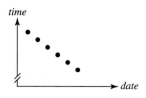

Figure 4.1

If we used the same assumptions to predict the hour of sunset on November 20, 1997, we'd get

$$5:09 \text{ P.M.} + \left(-1\frac{\text{minute}}{\text{day}}\right) (14 \text{ days}) = 5:09 \text{ P.M.} - 14 \text{ minutes} = 4:55 \text{ P.M.}$$

In fact, on Sunday, November 9, sunset was at 5:07 P.M., while on November 20 it was at 5:02 P.M.

Common sense tells us that it is unreasonable to assume that the sun will continue to set a minute earlier every day over a long period of time. By December 25 the days are getting longer; sunset is at 5:07 P.M., not at 4:20 P.M., the prediction based on a constant rate of

---

[2] This is the best we can say given the degree of accuracy in the newspaper accounts. The newspaper has rounded off to the nearest minute.

change of −1 minute/day. Certainly it is ridiculous to estimate that on November 6, 1998, the sun will set at 11:04 A.M. (where 5:09 P.M. −(1 minute/day)(365 days) = 5:09 P.M. − 365 minutes = 5:09 P.M. − (6 hours 5 minutes) = 11:04 A.M.)!

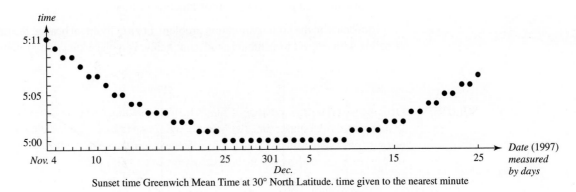

Sunset time Greenwich Mean Time at 30° North Latitude. time given to the nearest minute

Figure 4.2   Data from the *1998 World Almanac*, pp. 463–474.[3]

As we can see by looking at the graph in Figure 4.2, over small enough intervals, the data points either lie on a line or lie close to some line that can be fitted to the data. However, the line that fits the data best varies with the interval chosen. When looked at over the entire interval from November 4 to December 25, the graph does *not* look linear. ◆

In the last example we looked at a discrete phenomenon and made predictions based on the assumption of a constant rate of change over a small interval. In the next example we'll look at a continuous model.

◆ **EXAMPLE 4.2**   Brian Younger is a high-caliber distance swimmer; in competition he swam approximately 1 mile, 36 laps of a 25-yard pool.[4] If he completes the first 24 laps in 12 minutes, what might you expect as his time for 36 laps?

SOLUTION   Knowing that Brian is a distance swimmer, it is reasonable to assume that he does not tire much in the last third. Assuming a constant speed of 12 laps every 6 minutes (or 120 yards/minute) we might expect him to finish 36 laps in about 18 minutes. We would feel less confident saying that he could swim 4 miles if given an hour and 12 minutes, or 8 miles if given 2 hours and 24 minutes. ◆

A quantity that changes at a constant rate increases or decreases **linearly**.

If the rate of change of height with respect to time,

$$\left(\frac{\Delta \text{ height}}{\Delta \text{ time}} \right),$$

is constant over a certain time interval, then height is a linear function of time on that interval.

---

[3] These times might look suspect to you; the sun begins to set later by December 10, well before the winter solstice. Do not be alarmed: Sunrise gets later and later throughout December and continues this trend through the beginning of January.

[4] One lap is 50 yards.

If the rate of change of position with respect to time,

$$\left(\frac{\Delta \text{ position}}{\Delta \text{ time}} \right),$$

is constant, then position is a linear function of time.

Think back to the bottle calibration problem. For a cylindrical beaker, the rate of change of height with respect to volume is constant; height is a linear function of volume.

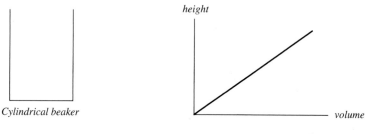

Cylindrical beaker

Figure 4.3

Many of the functions that arise in everyday life (in fields like biology, environmental science, physics, and economics) have the property of being **locally linear**. What does this mean?

"Local" means "nearby" and "linear" means "like a line."[5] So a function is locally linear if, in the immediate neighborhood of any particular point on the graph, the graph "looks like a line." This is not to say that the function *is* linear; we mean that near a particular point the function can be well approximated by a line. In other words, over a small enough interval, the rate of change of the function is approximately constant.

Graphically this means that f is locally linear at a point A if, when the graph is sufficiently magnified around point A, the graph looks like a straight line.[6] The questions of *which* line best approximates the function at a particular point and just what we mean by "nearby" are very important ones, and we will examine them more closely in chapters to come. First, let's look back at Examples 4.1 and 4.2.

In Example 4.1 the runner estimating the time of sunset is assuming local linearity; her assumption leads to a prediction that is only 1 minute off when she predicts just a few days ahead. The function is only *locally* linear; the idea of "locality" does not extend from the first three readings in early November all the way to Christmas day. In Example 4.2 the assumption that the swimmer's pace is maintained for another 5 minutes is an assumption of local linearity.

You probably have made many predictions of your own based on the assumption of local linearity without explicitly thinking about it. For instance, if you buy a gallon of milk and you have only a quarter of a gallon left after three days, you might figure that you'll be out of milk in another day. Here you're assuming that you will consume milk at a constant rate of $\frac{1}{4}$ gallon/day. Or, suppose you come down with a sore throat one evening and take three throat lozenges in four hours. You might take six lozenges to work with you the next day, assuming that you'll continue to use them at a rate of $\frac{3}{4}$ lozenge/hour for eight hours.

---

[5] By "line" we mean *straight* line.

[6] A computer or graphing calculator can help you get a feel for this if you "zoom in" on point A.

Clearly you wouldn't expect to pack six lozenges with you every day; you're assuming only *local* linearity.

Many examples of the use of local linearity arise in the fields of finance and economics. Investors lay billions of dollars on the line when they use economic data to project into the future. The question of exactly how far into the future one can, within reason, linearly project any economic function based on its current rate of change, and by how much this projection may be inaccurate, is a matter of intense discussion.

Local linearity plays a key role in calculus. The problem of finding the best linear approximation to a function at a given point is a problem at the heart of calculus. In order to work on this keystone problem, one must be very comfortable with linear functions. So, before going on, let's discuss them.

PROBLEMS FOR SECTION 4.1

1. Lucia has decided to take up swimming. She begins her self-designed swimming program by swimming 20 lengths of a 25-yard pool. Every 4 days she adds 2 lengths to her workout. Model this situation using a continuous function. In what way is this model not a completely accurate reflection of reality?

2. Cindy quit her job as a manager in Chicago's corporate world, put on a backpack, and is now traveling around the globe. Upon arrival in Cairo, she spent $34 the first day, including the cost of an Egyptian visa. Over the course of the next four days, she spent a total of $72 on food, lodging, transportation, museum entry fees, and baksheesh (tips). She is going to the bank to change enough money to last for three more days in Cairo. How much money might she estimate she'll need? Upon what assumptions is this estimate based?

3. It is 10:30 A.M. Over the past half hour six customers have walked into the corner delicatessen. How many people might the owner expect to miss if he were to close the deli to run an errand for the next 15 minutes? Upon what assumption is this based?

   Suppose that between 9:30 A.M. and 11:30 A.M. he had 24 customers. Is it reasonable to assume that between 11:30 A.M. and 1:30 P.M. he will have 24 more customers? Why or why not?

4.2   LINEAR FUNCTIONS

The defining characteristic of a linear function is its constant rate of change.

◆ EXAMPLE 4.3   For each situation described below, write a function modeling the situation. What is the rate of change of the function?

(a) A salesman gets a base salary of $250 per week plus an additional $10 commission for every item he sells. Let $S(x)$ be his weekly salary in dollars, where x is the number of items he sells during the week.

(b) A woman is traveling west on the Massachusetts Turnpike, maintaining a speed of 60 miles per hour for several hours. Her odometer reads 4280 miles when she passes the Allston/Brighton exit. Let $D(t)$ be her odometer reading t hours later.

SOLUTION   (a)

$$\text{salary} = \text{base salary} + \text{commission}$$

$$\text{salary} = \text{base salary} + \left(\frac{\text{dollars}}{\text{item}}\right)(\text{items})$$

$$S(x) = 250 + 10x$$

The rate of change of $S = \frac{\Delta S}{\Delta x} = \$10$ per item.

(b)

$$\text{odometer reading} = (\text{initial odometer reading}) + (\text{additional distance traveled})$$

$$\text{odometer reading} = (\text{initial reading}) + \left(\frac{\text{miles}}{\text{hour}}\right)(\text{hours})$$

$$D(t) = 4280 + 60t$$

The rate of change of $D = \frac{\Delta D}{\Delta t} = 60$ miles per hour.   ◆

Definition

f is a **linear function of** x if f can be written in the form $f(x) = mx + b$, where m and b are constants.

The graph of a linear function of one variable is a straight nonvertical line; conversely, any straight nonvertical line is the graph of a linear function. As we will show, the line $y = mx + b$ has slope m and y-intercept b. The slope corresponds to the rate of change of y with respect to x. Every point (x_0, y_0) that lies on the graph of the line satisfies the equation. In other words, if (x_0, y_0) lies on the graph of $y = mx + b$, then $y_0 = mx_0 + b$. Conversely, every point whose coordinates satisfy the equation of the line lies on the graph of the equation. As discussed in Chapter 1, this is what it means to be the graph of a function.

The following are examples of equations of lines and their graphs.

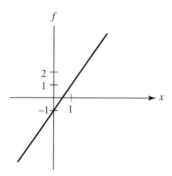

Figure 4.4 $f(x) = 2x - 1$
$m = 2, \ b = -1$

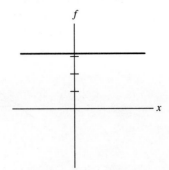

Figure 4.5 $f(x) = \sqrt{10}$
$m = 0, b = \sqrt{10}$

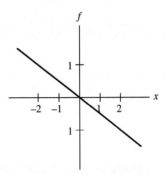

Figure 4.6 $f(x) = -0.5x$
$m = -0.5, b = 0$

The slope of a line is the ratio

$$\frac{\text{rise}}{\text{run}}, \quad \text{or} \quad \frac{\text{change in dependent variable}}{\text{change in independent variable}}.$$

If y is a linear function of x, then the slope is

$$\frac{\text{change in } y}{\text{change in } x}, \quad \text{or} \quad \frac{\Delta y}{\Delta x}.$$

We will now verify that if $y = mx + b$, then the constant m is the slope of the line.

Verification: Suppose (x_1, y_1) and (x_2, y_2) are two distinct points on the graph of $y = mx + b$. We want to show that the rate of change of y with respect to x is m, regardless of our choice of points.

Since $y = mx + b$, the points (x_1, y_1) and (x_2, y_2) can be written as $(x_1, mx_1 + b)$ and $(x_2, mx_2 + b)$.

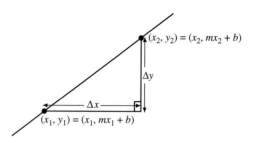

$(x_2, y_2) = (x_2, mx_2 + b)$

Δy

Δx

$(x_1, y_1) = (x_1, mx_1 + b)$

Figure 4.7

Therefore,

$$\frac{\text{change in } y}{\text{change in } x} = \frac{\Delta y}{\Delta x}$$

$$= \frac{y_2 - y_1}{x_2 - x_1}$$

$$= \frac{(mx_2 + b) - (mx_1 + b)}{x_2 - x_1}$$

$$= \frac{mx_2 - mx_1}{x_2 - x_1}$$

$$= \frac{m(x_2 - x_1)}{x_2 - x_1}$$

$$= m$$

We have shown that the slope of the line $y = mx + b$ is indeed m. The slope of a line is a fixed constant, regardless of how it is computed.

To verify that b is the y-intercept of the line, we set $x = 0$. Then $y = m \cdot 0 + b = b$. (Remember that the y-*intercept* is the value of the function on the y-axis, i.e., when $x = 0$.)

◆ **EXAMPLE 4.4** $3x + 2y = 7$ is the equation of a line. Find the slope and the x- and y-intercepts.

SOLUTION   Put the equation into the form $y = -\frac{3}{2}x + \frac{7}{2}$. We can then read off the y-intercept as $\frac{7}{2}$ and the slope as $-\frac{3}{2}$.

Find the x-intercept by setting $y = 0$:

$$3x + 2(0) = 7$$

$$3x = 7$$

$$x = \frac{7}{3}.$$

Alternatively, begin with $3x + 2y = 7$ and find the x-intercept by setting $y = 0$ and solving for x: $(\frac{7}{3}, 0)$. Find the y-intercept by setting $x = 0$ and solving for y: $(0, \frac{7}{2})$. (The x- and y-intercepts can be useful in graphing the line.) Given any two points, you can find the slope by computing $\frac{\Delta y}{\Delta x}$. For instance, given the points $(\frac{7}{3}, 0)$ and $(0, \frac{7}{2})$, the slope is computed as follows.

$$\frac{\Delta y}{\Delta x} = \frac{\frac{7}{2} - 0}{0 - \frac{7}{3}} = \frac{\frac{7}{2}}{-\frac{7}{3}} = \left(\frac{7}{2}\right)\left(-\frac{3}{7}\right) = -\frac{3}{2}$$

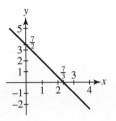

Figure 4.8

It is not necessary to use the x- and y-intercepts in order to calculate the slope; any two points will do. ◆

EXERCISE 4.1   Verify that any equation of the form $ax + cy = d$, where a, c, and d are constants and $c \neq 0$ is a linear equation. In other words, verify that it can be written in the form $y = mx + b$. If $ax + cy = d$, what is $\frac{\Delta y}{\Delta x}$?

Lines and Linear Equations

EXERCISE 4.2   Graph the following lines. Use three sets of axes, one for each of the Parts I, II, and III.

I.   a) $y = x$ b) $y = 2x$ c) $y = 3x$ d) $y = 0.5x$ e) $y = -2x$ f) $y = -0.5x$

II. a) $y = 2x$ b) $y = 2x + 1$ c) $y = 2x - 2$

III. a) $y = 3$ b) $x = 4$ c) $x = 0$ d) $y = 0$

Notice that the results of this exercise are consistent with the principles of stretching, shrinking, flipping, and shifting discussed in Section 3.4.

From this exercise you can observe how the constant m corresponds to the steepness of the line.

$m = 0$ ⇒   The line is horizontal.

$m > 0$ ⇒   The line rises from left to right, so the function is increasing.
The more positive m is, the steeper the rise of the line.

$m < 0$ ⇒   The line falls from left to right, so the function is decreasing.
The more negative m is, the steeper the fall of the line.

The closer m is to zero, the less steep the line.

Vertical Lines

A vertical line is the graph of an equation of the form $x = $ constant. Notice that a vertical line is not the graph of a function of x. One x-value is mapped to infinitely many y-values; a vertical line certainly fails the vertical line test!

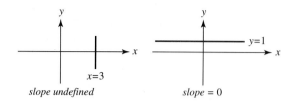

Figure 4.9

Slope $= \frac{\Delta y}{\Delta x}$, so the slope of a vertical line is undefined, while the slope of a horizontal line is zero.

Parallel and Perpendicular Lines

If $\mathcal{L}_1$ and $\mathcal{L}_2$ are nonvertical lines with slopes m_1 and m_2, respectively, then

$\mathcal{L}_1$ is parallel to $\mathcal{L}_2$ if and only if $m_1 = m_2$, i.e., their slopes are equal.

$\mathcal{L}_1$ is perpendicular to $\mathcal{L}_2$ if and only if $m_1 = \frac{-1}{m_2}$, i.e., their slopes are negative reciprocals.

Fundamental Observation

A line is completely determined by either

i. the slope of the line and any one point on the line, or

ii. any two points on the line.

This geometric observation can be translated as follows. If there is a linear relationship between two variables, then that relationship can be completely determined provided

i. one data point is known and the rate of change of one variable with respect to the other is known, or

ii. two data points are known.

Finding the Equation of a Line

i. Suppose we know the slope of the line, m, and a point on the line, (x_1, y_1). We can find the equation of the line by using either of the following two methods.

- *Method 1:* The equation of any nonvertical line can be put into the form $y = mx + b$. If we know the slope, we know m. Our job is to find b. Because (x_1, y_1) is a point on the line, (x_1, y_1) satisfies the equation $y = mx + b$. Therefore, $y_1 = mx_1 + b$, where x_1, y_1, and m are constants. We can solve for b, the only unknown, and then write $y = mx + b$.

- *Method 2:* The slope $= \frac{\Delta y}{\Delta x}$. Point (x_1, y_1) is a fixed point on the line. Therefore, if (x, y) lies on the line,

$$m = \frac{y - y_1}{x - x_1}.$$

This is written

$$y - y_1 = m(x - x_1)$$

to make it clear that the domain is all real numbers. Solving for y will put this into the form $y = mx + b$.

The equation $y - y_1 = m(x - x_1)$ is called the **point-slope form** of the equation of a line.

ii. Suppose we know two points on a line: (x_1, y_1) and (x_2, y_2). We can find the slope of the line:

$$\frac{\Delta y}{\Delta x} = \frac{y_2 - y_1}{x_2 - x_1} = m.$$

Knowing the slope and a point (two points, actually), we can continue as described in part i.

◆ **EXAMPLE 4.5**   Find the equation of the line passing through points $(2, -3)$ and $(-4, 5)$.

SOLUTION   Calculate the slope.

$$m = \frac{-3 - 5}{2 - (-4)} = \frac{-8}{6} = -\frac{4}{3},$$

so

$$y = -\frac{4}{3}x + b.$$

Use the fact that point $(2, -3)$ lies on the line to find b.

$$-3 = -\frac{4}{3}(2) + b \Rightarrow b = -3 + \frac{8}{3} = -\frac{9}{3} + \frac{8}{3} = -\frac{1}{3}$$

$$y = -\frac{4}{3}x - \frac{1}{3}.$$

Alternatively, after finding $m = -\frac{4}{3}$ we can use Method 2 above.

$$-\frac{4}{3} = \frac{y + 3}{x - 2}$$

$$y + 3 = -\frac{4}{3}(x - 2)$$

We leave it up to you to verify that the equations are equivalent.   ◆

The Equation of a Line

Slope-intercept form of a line	Point-slope form of a line
$y = mx + b$	$y - y_1 = m(x - x_1)$
where m is the slope and b is the y-intercept	where m is the slope and (x_1, y_1) is a point on the line

Scientists and social scientists often make general models in which constants or parameters are represented by letters.[7] Cases in which the slope and/or points of a line are not given as numbers but as letters representing constants typically are stumbling blocks for students trying to determine the equation of the line. Unless you think carefully and clearly, it is easy to lose track of the main features of the equation:

■ the two variables,

■ the constants that are given, and

■ the unknown constants that you are trying to find.

If you plan to consider slope, you must decide which variable will play the role of the independent variable and which will be treated as the dependent variable. (If you're accustomed to x's and y's, then determine which variable will play the role of x and which will play the role of y.) By taking inventory, keeping your goal firmly in mind, and determining the steps you'll take to reach that goal, you ought to be able to proceed fearlessly. Try the following exercises.

EXERCISE 4.3

   i. Find the equation of the line through the points $(2\pi, \pi^2)$ and $\left(\pi, \frac{1}{\pi}\right)$.

   ii. Find the equation of the line through the point (a, c^2) having slope p.

   iii. Find the equation of the line through (m, n) with slope b.

CAUTION Here m and b are playing nonstandard roles. To avoid confusion, try a strategic maneuver; use something such as $y = Ax + C$ as your "reference" equation to avoid getting the m's and b's in the problem confused with the slope and the y-intercept.)

EXERCISE 4.4   Over the interval $[0, T]$ a horse's velocity, v, is a linear function of time, t.

(a) At time $t = 0$, the horse's velocity is v_0. At time $t = T$, his velocity is v_T. Find $v(t)$ on the interval $[0, T]$. Note: v_0, T, and v_T are all constants.

(b) What can you say about the horse's acceleration?

Answers to Selected Exercises

Answers to Exercise 4.3

   i. $y = \frac{\pi^3 - 1}{\pi^2}x - \pi^2 + \frac{2}{\pi}$

   ii. $y = px + c^2 - pa$

   iii. $y = bx + n - mb$

Answers to Exercise 4.4:

(a) $v(t) = mt + b$ because v is a linear function of t. Our job is to find m and b. We are given two points, $(0, v_0)$ and (T, v_T). (Points are written in the form (time, velocity) because time is the independent variable.) In other words, t plays the role of x and v

---

[7] Physicists use g to denote acceleration due to the force of gravity. Biologists might denote the carrying capacity of an ecosytem for a certain species by a letter, such as C, without specifying the value of C. Chemists use r for Avogadro's number, the number of molecules in a mole.

plays the role of y.

$$m = \frac{\Delta \text{velocity}}{\Delta \text{time}} = \frac{v_T - v_0}{T - 0} = \frac{1}{T}(v_T - v_0)$$

The v-intercept is v_0, so $b = v_0$.

$$v(t) = \frac{v_T - v_0}{T}t + v_0.$$

(b) Acceleration $= \frac{\Delta v}{\Delta t} = \frac{v_T - v_0}{T}$.

The horse's acceleration is constant on $[0, T]$.

PROBLEMS FOR SECTION 4.2

1. The graph of f is given in the figure below.

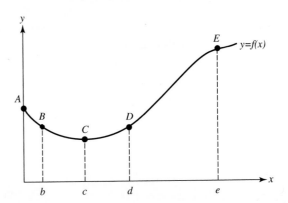

Lines $\mathcal{L}_1, \mathcal{L}_2$, and $\mathcal{L}_3$ are as shown. $\mathcal{L}_1$ is horizontal and $\mathcal{L}_3$ is vertical. A, B, C, D, and E are points on the graph of f. Express your answers to the following questions in terms of b, c, d, e, and the function f.

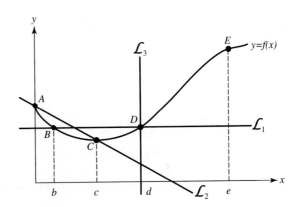

(a) Write the coordinates of points A, B, C, D, and E using functional notation. For instance, point A has coordinates $(0, f(0))$.

(b) What is the slope of $\mathcal{L}_1$?

(c) What is the equation of $\mathcal{L}_1$?

(d) What is the length of the line segment joining point B and $(b, 0)$?

(e) What is the length of the line segment joining points B and D?

(f) What is the slope of $\mathcal{L}_2$?

(g) What is the equation of $\mathcal{L}_2$?

(h) What is the equation of $\mathcal{L}_3$?

(i) What is the slope of the line $\mathcal{L}_3$?

2. Find the slope of the line through the two points given.

(a) $(3, -1), (-2, -3)$

(b) $(\pi, 2\pi), (0, -\pi)$

(c) $(\sqrt{2}, 3), (\sqrt{2}, 5)$

(d) $(\sqrt{2}, \sqrt{3}), (1, \sqrt{3})$

3. On the same set of axes, sketch lines through point $(0, 1)$ with the slopes indicated. Label the lines.

(a) slope $= 0$

(b) slope $= \frac{1}{2}$

(c) slope $= 1$

(d) slope $= 2$

(e) slope $= -\frac{1}{2}$

(f) slope $= -1$

(g) slope $= -2$

For Problems 4 through 13, find the equation of the line with the given characteristics.

4. Slope $-\frac{1}{2}$, passing through $(-2, -3)$

5. Slope π, passing through $(3, 5)$

6. Passing through points $(0, a)$ and $(b, 0)$

7. Passing through points $(\pi, 3)$ and $(-\pi, 5)$

8. Passing through point $(\sqrt{3}, \sqrt{2})$ and parallel to $3x - 4y = 7$

9. Passing through the origin and perpendicular to $\pi x - \sqrt{3}y = 12$

10. x-intercept of $\sqrt{\pi}$ and parallel to the y-axis

11. Perpendicular to $y - \pi = \pi(x - 1)$ with a y-intercept of 3

12. Horizontal and passing through $(-\sqrt{\pi}, \pi^2)$

13. Vertical and passing through $(-\sqrt{\pi}, \pi^2)$

For Problems 14 through 16, find the slope of the secant line passing through points P and Q, where P and Q are points of the graph of $f(x)$ with the indicated x-coordinates.

14. $f(x) = \frac{3}{x} + 2x$; the x-coordinates of P and Q are $x = 3$ and $x = 3 + k$ $(k \neq 0)$, respectively.

15. $f(x) = x^2 + 3x + 1$; the x-coordinates of P and Q are $x = b$ and $x = b + h$ $(h \neq 0)$, respectively.

16. $f(x) = ax^2 + bx + c$; the x-coordinates of P and Q are $x = k$ and $x = k + h$ $(h \neq 0)$, respectively.

17. There is a proliferation of telephone-call billing schemes. According to one scheme, a call to anywhere in the United States is billed at 50 cents for the first three minutes and 9.8 cents per minute after that. Express the cost of a call as a function of its duration.

18. From the early 1500s to nearly 1700, the Turkish town of Iznik was famous for its beautiful colored tiles. In the 1990s, tile-making was pursued with renewed vigor in the town. In the late 1990s, a new mosque was built, and the walls, both inside and outside, are currently being covered with the blue and red tiles for which the town is known. If the mosque cost C dollars to construct with an additional T dollars for each tile used, find the total cost as a function of x, where x is the number of tiles used.

4.3 MODELING AND INTERPRETING THE SLOPE

Interpreting slope as a rate of change is the key to many applications of calculus to other disciplines. Given the line $y = 5x + 5$ we can interpret the slope 5 as $\frac{5}{1} = \frac{\Delta y}{\Delta x}$; for each increase of 1 unit in x, y increases by 5 units. As a second example, consider the line

$$y = -\frac{3}{2}x + 7.$$

The slope is $-\frac{3}{2}$. We can interpret this as follows.

$$-\frac{3}{2} = \frac{-3}{2} = \frac{\Delta y}{\Delta x}. \qquad \text{If } x \text{ increases by 2, } y \text{ will decrease by 3.}$$

Alternatively,

$$-\frac{3}{2} = \frac{3}{-2} = \frac{\Delta y}{\Delta x}. \qquad \text{If } x \text{ decreases by 2, } y \text{ will increase by 3.}$$

Or,

$$-\frac{3}{2} = \frac{-\frac{3}{2}}{1} = \frac{\Delta y}{\Delta x}. \qquad \text{If } x \text{ increases by 1, } y \text{ will decrease by } \frac{3}{2}.$$

To interpret slope in context, we must first clarify which variable is independent (the input) and which variable is dependent (the output).

The slope is

$$\frac{\Delta \text{dependent variable}}{\Delta \text{independent variable}} = \frac{\text{change in variable plotted on vertical axis}}{\text{change in variable plotted on horizontal axis}}.$$

◆ **EXAMPLE 4.6**   In this example we consider position to be a function of time, quantity demanded to be a function of price, and velocity to be a function of time.

i. Suppose s is position, in miles, and t is time, in hours. If

$$s = -\frac{3}{2}t + 7,$$

then the slope is

$$\frac{\Delta s}{\Delta t} = \frac{\text{change in position (in miles)}}{\text{change in time (in hours)}}$$

$$= \text{velocity (measured in miles per hour).}$$

Here the velocity is $-\frac{3}{2}\frac{\text{miles}}{\text{hours}}$, the negative sign indicating direction.

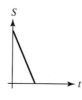

Figure 4.10

ii. Suppose D is quantity demanded, in bushels, and p is price, in dollars. If

$$D = -\frac{3}{2}p + 7,$$

then the slope is

$$\frac{\Delta D}{\Delta p} = \frac{\text{change in bushels demanded}}{\text{change in price (in dollars)}}.$$

For every increase of \$2 in price, 3 fewer bushels will be demanded.

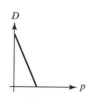

Figure 4.11

Equivalently, for every $2 decrease in price, 3 more bushels will be demanded. Notice that a negative slope is exactly what we would expect in this kind of situation. Students of economics should check the footnote below. [8]

iii. Suppose v is an object's velocity, in meters/second, and t is time measured in seconds. If

$$v = -\frac{3}{2}t + 7,$$

then the slope is

$$\frac{\Delta v}{\Delta t} = \frac{\text{change in velocity}}{\text{change in time (in seconds)}}.$$

Every two seconds, the object's velocity decreases by 3 meters/second. The object is decelerating.

Figure 4.12

PROBLEMS FOR SECTION 4.3

1. A photocopying shop has a fixed cost of operation of $6000 per month. In addition, it costs them $0.01 per page they copy. They charge customers $0.07 per page.

   (a) Write a formula for $R(x)$, the shop's monthly revenue from making x copies.

   (b) Write a formula for $C(x)$, the shop's monthly costs from making x copies.

   (c) Write a formula for $P(x)$, the shop's monthly profit (or loss if negative) from making x copies. (Profit is computed by subtracting total costs from the total revenue.)

   (d) How many copies must they make per month in order to break even? (Breaking even means that the profit is zero; the total costs and total revenue are equal.)

   (e) Sketch $C(x)$, $R(x)$, and $P(x)$ on the same set of axes and label the break-even point.

   (f) Find a formula for $A(x)$, the shop's average cost per copy.

   (g) Make a table of $A(x)$ for $x = 0, 1, 10, 100, 1000, 10000$.

   (h) Sketch a graph of $A(x)$.

---

[8] Economists tend to put price on the vertical axis and quantity on the horizontal axis. We can solve the equation above for p.

$$p = -\frac{2}{3}D + \frac{14}{3}$$

The slope is $-\frac{2}{3}$, but the interpretation in words is identical to that in part (ii).

2. An item costs $1000 this year. This is a 10% increase over the price last year. What was the price last year?

(Caution: it was not $900. It would be wise to give last year's price a name—like "x," or "P," or some other labeling of your choice.)

3. According to a study done by Chester Kyle, Ph.D. (*Long Distance Cycling*, Rodale Press, Emmaus, PA, 1993), adding 6 pounds to a bicycle slowed the rider down by 22 seconds on a certain 2-mile course. Assume that riding the course without the extra weight took K seconds (actual time not specified).

(a) Assuming that this relationship is linear, find an equation for $T(w)$, the time needed to complete the course as a function of the amount of extra weight added.

(b) What is the rate of change of $T(w)$? Interpret this rate of change in practical terms.

4. Economists use demand curves to express the relationship between the price of an item and the number of items demanded by consumers. Below is a demand curve for a certain good. q is the quantity of the good demanded (i.e., the number of items demanded). p is the price per item.

(a) Write an equation for the line in terms of p and q. Express p as a function of q.

(b) In words, interpret the meaning of the slope of the line.

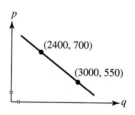

5. The three most commonly used temperature scales are the Fahrenheit (°F), the Celsius (°C), and the Kelvin (K, an absolute temperature) scales. One interval on the Kelvin scale is equal to one degree Celsius. The freezing point of water is 0°C, which is 32°F and 273.15 on the Kelvin scale. The boiling point of water is 100°C and 212°F. On the Celsius scale, the interval between the freezing and boiling points of water is divided into 100 degrees while on the Fahrenheit scale it is divided into 180 degrees.

You have been given *more* than enough information to answer the following questions! You'll have to select the information you will use.

(a) Write a formula for a function that takes as input degrees Celsius and gives as output degrees Fahrenheit.

(b) Write a formula for a function that takes as input degrees Fahrenheit and gives as output degrees Celsius. Do this as efficiently as possible! The function you've arrived at is the inverse of the function from part (a).

(c) Write a formula for a function that takes as input degrees Celsius and gives the temperature on the Kelvin scale as output.

(d) Write a formula for a function that takes as input degrees Fahrenheit and gives the temperature on the Kelvin scale as output. Express this function as the composition of two functions from previous parts of this problem.

6. Economists use indifference curves to show all combinations of two goods that give the same (fixed) level of satisfaction to a household. Generally an indifference curve is nonlinear, but for certain combinations of goods it is possible to have a straight-line indifference curve. The following is a linear indifference curve. Let $R =$ the number of units of item 1 and $S =$ the number of units of item 2.

(a) Write an equation for the line in terms of S, R, a, and c.
(b) Interpret the meanings of the intercepts.
(c) Optional (but suggested for those studying economics): Give an example of two items for which the indifference curve could reasonably be linear.

7. Stories are told about some of the less fair-minded teams of early baseball (e.g., the Baltimore Orioles of the 1890s) freezing baseballs until shortly before game time so that although the cover would feel normal, the core of the ball would be much colder. Then, they would attempt to introduce these balls into play when the opposing team was at bat, working on the assumption that the frozen balls would not travel as far when hit. Experiments have shown that a ball whose temperature is $-10°$F would travel 350 feet after a given swing of the bat, while a ball whose temperature is $150°$F would be hit 400 feet by the same swing. Assume this relationship is linear.

Let $B(T)$ be the distance this swing would produce, where T is the temperature in degrees Fahrenheit.

(a) Find an equation for $B(T)$.
(b) What is the B-intercept? What is its practical meaning?
(c) What is the slope of $B(T)$? What is its practical meaning?

8. A horseman has some ponies of his own and boards horses for other people. For his own ponies, he orders 9 bales of hay from the supplier. The total number of bales he orders increases linearly with the number of horses he boards. When he boards 6 horses, he orders a total of 36 bales of hay (for these horses and his ponies). Express the number of bales of hay he orders as a function of the number of horses he boards.

Exploratory Problem for Chapter 4

Thomas Wolfe's Royalties for The Story of a Novel

Thomas Wolfe was an author born in Asheville, North Carolina, in 1900. His first novel, *Look Homeward Angel*, was a thinly veiled fiction that gave such scathingly accurate portrayals of well-known townspeople that despite the novel's enthusiastic reception nationwide, Wolfe felt he received a cool welcome in his hometown. He proceeded to write *You Can't Go Home Again*. The problem you are asked to consider involves one of Wolfe's lesser-known novels.

Exploratory Problem:

In lieu of a straight royalty percentage on sales of his book, *The Story of a Novel*, Thomas Wolfe agreed to accept a sliding scale of 10% on the first 3000 copies sold, 12.5% on the next 4500 copies, and 15% on all copies after that. The book was priced at $1.50. (Elizabeth Nowell, *Thomas Wolfe: A Biography*, Doubleday, Garden City, NY, 1960.)

(a) Write a function $R(x)$ that gives Wolfe's royalties as a function of x, the number of books sold. You will have to write the formula in pieces, because the formula varies depending upon the number sold. Make sure your formula works as you want it to by checking it out on some concrete cases. Does your formula work if Wolfe sells 3001 copies? 7501 copies?

(b) Graph $R(x)$. Compare your answers to parts (a) and (b) with those of some of your classmates.

(c) Solve $R(x) = 1000$. What does this equation mean?

(d) Graph $R'(x)$, where $R'(x)$ is the slope function—giving the slope at every point on the graph of R.

(e) What is the practical meaning of $R'(x)$? Include units in your answer.

4.4   APPLICATIONS OF LINEAR MODELS: VARIATIONS ON A THEME

Linear models are used in many contexts. Sometimes they are used to express linear or piecewise linear relationships and other times to deduce information about relationships that, while not actually linear, are locally linear. The examples in this section direct our attention to each of these topics:

- simultaneous linear equations,

- piecewise linear functions, and

- approximating nonlinear behavior locally using linear approximations.

◆ EXAMPLE 4.7

    i. Salesman A gets a base salary of $250 per week plus an additional $10 for every item he sells. Salesman B gets a base salary of $200 per week plus a commission of $20 for each item sold. At what number of sales does salesman B's salary scheme yield a higher weekly income than salesman A's salary arrangement?[9]

SOLUTION   Let $A(x)$ and $B(x)$ be the weekly wages (in dollars) of salesmen A and B, respectively, where x denotes the number of items sold in the week.

$$A(x) = 250 + 10x \quad \text{and} \quad B(x) = 200 + 20x$$

Graphs of $A(x)$ and $B(x)$ are drawn below.

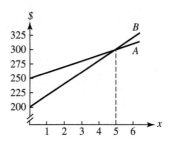

Figure 4.13

    We are most likely interested in an integer answer; if we draw very accurate graphs using appropriate scales this information is readily available. If one assumes that x can take on only positive integer values, then it may be reasonable to check salaries for various x-values numerically. However, the quickest way to answer the question is to find out at what value of x the salaries are equal.

---

[9] Because these items are discrete (let's say they are selling radios, for instance), we could specify that x must be a nonnegative integer and draw our graph as a discrete set of points. In practice it is often simpler to use a continuous model to study phenomena that are actually discrete. (Economists, biologists, and others do it all the time.) Calculus deals with continuous functions, so in this course most of our models will be continuous; however, when solving a problem you will want to interpret your continuous model in a way that makes sense in context.

$$250 + 10x = 200 + 20x$$
$$50 = 10x$$
$$5 = x$$

So, if a salesperson can sell more than five items per week, then salesman B's salary scheme pays more.

ii. Saleswoman C has been offered the following arrangement. Her base salary is $220 per week. For each of the first six items she sells during the week she will get $5 commission; for each additional item after that she will earn $25 commission. Let x = the number of items she sells each week. Write a function $C(x)$ giving her weekly salary. Make a continuous model with domain $[0, \infty)$. Before looking at the solution, spend some time trying this problem on your own. As a spot-check for errors, check your answer for $x = 0$, $x = 1$, $x = 6$, and $x = 7$ to make sure that your function "works." If it does not work, see where it goes astray.

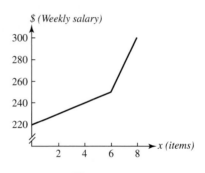

Figure 4.14

SOLUTION

$$C(x) = \begin{cases} 220 + 5x & \text{for } 0 \le x \le 6, \\ 250 + 25(x - 6) & \text{for } x > 6, \end{cases}$$

i.e.,

$$C(x) = \begin{cases} 220 + 5x & \text{for } 0 \le x \le 6, \\ 100 + 25x & \text{for } x > 6. \end{cases}$$

For $0 \le x \le 6$ we have the standard scheme. At $x = 6$, the saleswoman is making $250. In a sense, that is the woman's new base rate if she sells more than six items. She gets $25 commission on the number of items over six. The number of items over six is $(x - 6)$, not x. Therefore, for $x > 6$ we have $250 + 25(x - 6)$.

Common Errors

(a) Suppose you tried

$$C(x) = \begin{cases} 220 + 5x & \text{for } 0 \le x \le 6, \\ 220 + 25x & \text{for } x > 6. \end{cases}$$

This model erroneously indicates that once the seventh item is sold, each item, including the first six, earns a $25 commission.

(b) Suppose you tried

$$C(x) = \begin{cases} 220 + 5x & \text{for } 0 \leq x \leq 6 \\ 250 + 25x & \text{for } x > 6. \end{cases}$$

For $x > 6$ this model correctly counts the $5 commission for the first six items in with the base rate, but then it gives an *additional* $25 for each of those first six items. This amounts to giving a $30 commission for each of the first six items sold. For instance, when $x = 7$ this model gives

($220) + ($5/item)(6 items) + ($25/item)(7 items)

instead of the correct

($220) + ($5/item)(6 items) + ($25/item)(1 item).

REMARK The function $C(x)$ is defined piecewise: It is defined as one function on one interval and as another function on a second interval. Because each piece is linear, the function is called **piecewise linear**.

The slope of $C(x)$ corresponds to the commission rate, the rate of change of salary with respect to the number of items sold. The commission rate changes at $x = 6$. Below is a sketch of the **slope function**, typically denoted by C'.

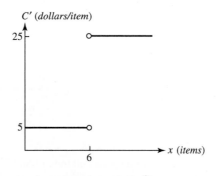

Figure 4.15 ◆

◆ EXAMPLE 4.8 The resale value of a used TI-81 calculator is a function that decreases with time; as newer and more advanced models come out there is less demand for the older TI-81. Let's call $P(t)$ the price, in dollars, of a used TI-81 at time t, where t is measured in years, with $t = 0$ corresponding to January 1, 1992. Suppose that on January 1, 1992, the resale value was $75 and was decreasing at a rate of $10 per year and that on January 1, 1995, the resale value was $51. Furthermore, let's assume that although the value is always going down, it is going down less and less steeply as time passes. (The rationale behind this assumption might be that inflation tends to drive prices up over time, and that the calculator will always have some positive value.)

    i. Sketch a possible graph of $P(t)$ incorporating all the information given.

    ii. What is the average rate of change of the calculator's value between $t = 0$ and $t = 3$?

    iii. What can we say about the value of the calculator on January 1, 1994, at $t = 2$? Give good upper and lower bounds for the price of the calculator on that date.

SOLUTION

i. $P(t)$ is decreasing and concave up.

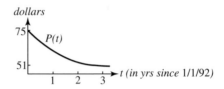

Figure 4.16

ii. The average rate of change is the change in the price divided by the change in time. This is

$$\frac{P(3) - P(0)}{3 - 0} = \frac{51 - 75}{3} = -8\frac{\text{dollars}}{\text{year}},$$

or a decrease of $8 per year.

Graphically, this is the slope of the line connecting the points $(0, 75)$ and $(3, 51)$.

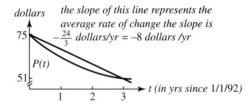

Figure 4.17

iii. To find upper and lower bounds for $P(2)$ means that we must find a price that is *greater* than $P(2)$ and a price that is *less* than $P(2)$. For instance

$$51 < P(2) < 75,$$

so 51 is a lower bound and 75 is an upper bound. We could do worse (for instance, using 0 as a lower bound and 75 as an upper bound), but we could do better! To "do better" means to find a larger lower bound and a smaller upper bound so we can pin down the price at $t = 2$ as much as possible.

First, let's find a good lower bound. We know that the value was dropping at a rate of $10 per year at $t = 0$ and that after this time the value dropped at a slower and slower rate. So, we know that in the two years between $t = 0$ and $t = 2$, the value dropped by some amount *less* than (2 years) · ($10/year) = $20. If the value dropped by less than $20, then at $t = 2$ it must have been *more* than $P(0) - \$20 = \$75 - \$20 = \$55$.

How can we visualize this graphically? We know that at $t = 0$ the price is $75. If we assume that the price drops by $10 per year, this corresponds graphically to the line through $(0, 75)$ with slope -10. The value of the calculator is going down less and less steeply as time passes, therefore this line lies *below* the graph of *P*. Therefore, the point on this line with the *t*-coordinate of 2 lies below the point (2, *P*(2)).

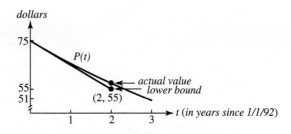

Figure 4.18

Now, let's find an upper bound for the price at $t = 2$. We know that the average rate of change of price between $t = 0$ and $t = 3$ is represented by the line we drew in Figure 4.17. We can see that the point on this line at $t = 2$ is an overestimate for the actual value of the calculator at that time since this secant line lies above the curve. The slope of the secant line is -8, so the point on the secant line with a t-coordinate of 2 corresponds to a price of $P(0) - 2 \cdot \$8 = \$75 - \$16 = \$59$. Thus, the value of a calculator at $t = 2$ must be *under* \$59.

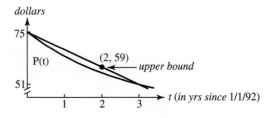

Figure 4.19

In summary, our lower bound is \$55 and our upper bound is \$59. Notice that the difference between these two estimates for $P(2)$ is \$4. To obtain our two estimates, we used two different linear approximations, both using the point (0, 75) corresponding to a price of \$75 in 1992. The difference in the slopes of the lines we used was \$2 per year. We were looking at a period of 2 years; therefore it makes sense that the difference between the estimates is (2 years) · (\$2/year) = \$4. If we were to make estimates closer to $t = 0$, this difference would be smaller; this reflects the fact that estimates based on the idea of local linearity are more accurate the nearer we are to the point at which we have definite information.

Figure 4.20 shows both linear approximations. Notice that the line through (0, 75) with slope -10 is a much better linear approximation of the curve near (0, 75) than is the line through (0, 75) and (3, 51). In fact, we will soon find that the former line is the best linear approximation of the curve at (0, 75).

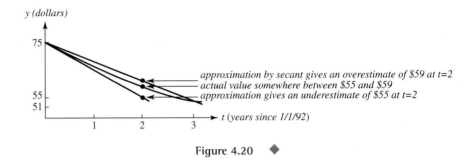

Figure 4.20   ◆

REMARK The key ideas in Example 4.8 are the geometric ones.

■   The average rate of change of a function f on the interval $[a, b]$ can be represented by the slope of the secant line through $(a, f(a))$ and $(b, f(b))$.

■   Where a curve is concave up, its secant lines lie above the curve.
Where a curve is concave down, its secant lines lie below the curve.

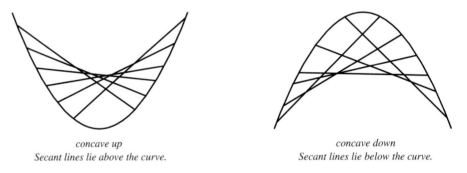

concave up
Secant lines lie above the curve.

concave down
Secant lines lie below the curve.

Figure 4.21

The particular information we were given in this problem determined the approach we took to solving it. Our problem-solving strategy involved drawing pictures to represent the information available in a visual way.

PROBLEMS FOR SECTION 4.4

1. A social worker gets paid $\$D$ per hour up to 40 hours per week. If he puts in more than 40 hours, the hours over 40 count as overtime, which pays an additional 50% per hour. Express his weekly wages as a function of x, where x is the number of hours he has worked that week. (You'll have to write a function in two pieces since the pay equation is described in two different ways depending upon the value of x.)

2. Inflation in Turkey has caused prices of small everyday items to be measured in tens of thousands of lira. One day I went to a market and purchased one container of yogurt and two packets of honey for 180,000 Turkish lira. Two days later I returned to the

same market and purchased two containers of yogurt and three packets of honey for 310,000 Turkish lira.

(a) Assuming that the price remained constant over this two-day period, what is the price of a yogurt? What is the price of a packet of honey?

(b) The figures given in this problem are accurate for the summer of 1998 in the town of Iznik, a beautiful, tiny lakeside town founded nearly 3000 years ago. The exchange rate at the time was 258,000 Turkish lira per dollar. Convert the prices of yogurt and honey into dollars.

3. (a) Determine the equation of the supply and demand curves shown in the figure below.

   (b) What are the equilibrium price and quantity? Assume price is measured in dollars and quantity in thousands of units. (The equilibrium occurs when supply and demand are equal.)

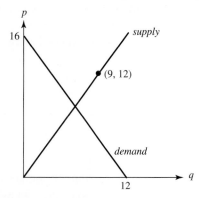

4. A moving company charges a minimum of $250 for a move. An additional $100 per hour is charged for time in excess of two hours. Write a function $C(t)$ that gives the cost of a move that takes t hours to complete.

5. The graph that follows indicates the salary scheme at Company A for a certain job. The pay scheme at Company B for the analogous job is as follows: Workers get paid $80 per week plus an additional $10 for each item sold. How many items must a worker sell over the course of a week in order to have the job at Company B to be to her advantage? Please give all possible answers. We would like you to answer this question in three different ways.

   (a) First approach it numerically. Make a salary table for each job.

   (b) Now approach the problem graphically. Use numerical methods only for fine-tuning.

   (c) Finally, approach the problem algebraically: Let S_A be the salary scheme at Company A and S_B be the salary scheme at Company B. Write S_A and S_B each as functions of x, the number of items sold per week. Solve the required inequality by solving the corresponding system of equations.

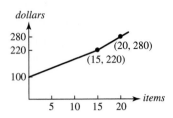

6. You've been presented with two different pay plans for the same job. Plan A offers $12 per hour with overtime (hours above 40 per week) paying time and a half. Plan B offers $14 per hour with no overtime. Let x denote the number of hours you work each week. Let $P_A(x)$ give the weekly pay under plan A and $P_B(x)$ give the weekly pay under plan B.

   (a) What is the algebraic formula for $P_B(x)$?

   (b) What is the algebraic formula for $P_A(x)$?
       Note that you must define this function differently for $x \leq 40$ and for $x > 40$. Check your answer and make sure that the pay for a 50-hour work week is $660.

   (c) i. For what value(s) of x are the two plans equivalent?
       ii. For what values of x is plan B better?
           (*Hint:* A good problem-solving strategy is to draw a graph so you can really see what is going on.)

   (d) True or False:
        i. $P_B(x + y) = P_B(x) + P_B(y)$
       ii. $P_A(x + y) = P_A(x) + P_A(y)$
           (*Hint:* If you are not sure how to approach a problem, a good strategy—frequently used by mathematicians everywhere!—is to try a concrete case. If the statement is false for this special case, then you know the statement is definitely false. If the statement holds for this special case, then the process of working through the special case may help you determine whether the statement holds in general.) *Caution:* Since the rule for $P_A(x)$ changes for $x > 40$, you need to check several cases. If *any* case doesn't hold, then the statement is false.

7. You've written a book and have two publishers interested in putting it out. Both publishers anticipate selling the book for $20. The first publisher guarantees you a flat sum of $8000 for up to the first 10,000 copies sold and will pay 12% royalty for any copies sold in excess of 10,000. For instance, if 10,001 copies were sold, you would receive $8002.40. The second publisher offers a royalty of 10%.

   Let x be the number of books sold.

   Let $A(x)$ give the income under plan A and $B(x)$ give the income under plan B.

   (a) What is the algebraic formula for $A(x)$?

   (b) What is the algebraic formula for $B(x)$?

   (c)   i. For what value(s) of x are the two plans equivalent?
        ii. For what values of x is plan B better?
       iii. For what values of x is plan A better?

8. An investment fund has two different investment options. Option C, the more conservative option, puts 70% of the investor's money into slow-growing reliable stocks and 30% of the money into high-risk stocks with high growth potential. Option R, the riskier option, puts 60% of the money into high-risk stocks and 40% into low-growth stocks. If a client has $2 million invested in high-risk stocks and $3 million in low-risk stocks, how much of the client's money is in option C and how much in option R?

9. Below is a graph of temperature, T, plotted as a function of time, t. The temperature function is increasing on $[0, 21]$. It is concave down on $[0, 14]$ and concave up on $[14, 21]$.

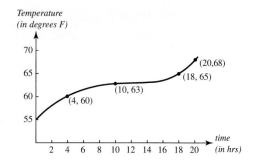

(a) On average, between hours 4 and 10, what is the rate of increase of temperature with respect to time? In other words, what is the average rate of change of temperature between hour 4 and hour 10?

(b) What is the average rate at which temperature is increasing between hour 18 and hour 20?

(c) Draw the secant line through the points on the graph where $t = 4$ and $t = 10$. Find the equation of this line.

(d) Draw the secant line through the points on the graph where $t = 18$ and $t = 20$. Find the equation of this line.

(e) Using your answer to part (d) and approximating the graph by the secant line, estimate the temperature at hour 21. Would you guess that the T-coordinate of the point on the secant line is slightly higher than the temperature at hour 21 or slightly lower?

10. After 3 miles of difficult climbing in the morning, a group of hikers has reached a plateau and they are confident they can maintain a steady pace for the next 10 miles. After covering a total of 13 miles, they'll set camp. Twenty minutes after reaching the plateau, they've covered $1\frac{1}{3}$ miles. Express the total daily mileage as a function of t, where t is the number of hours spent hiking since they reached the plateau. What is the domain of the function?

11. At 8:00 A.M., a long-distance runner has run 10 miles and is tiring. She runs until 9:00 A.M. but runs more and more slowly throughout the hour. By 9:00 she has run 16 miles.

(a) Sketch a possible graph of distance traveled versus time on the interval from 8:00 to 9:00. What are the key characteristics of this graph?

(b) Suppose that at 8:00 A.M. she is running at a speed of 9 miles per hour. Find good upper and lower bounds for the total distance she has run by 8:30 A.M. Explain your reasoning with both words and a graph.

12. This problem focuses on the difference between being piecewise linear (made up of straight lines) and being locally linear (being approximately linear when magnified enough). Consider the functions f, g, and h below.

$$f(x) = |x + 2| - 3$$

$$g(x) = \begin{cases} x & \text{for } x \leq 0 \\ x^2 & \text{for } x > 0 \end{cases}$$

$$h(x) = (x - 1)^{10} + 1$$

(a) Graph f, g, and h.

(b) Specify all intervals for which the given function is linear (a straight line.)
   i. f
   ii. g
   iii. h

(c) Specify the point(s) at which the given function is *not* locally linear (that is, where it does not look like a straight line, no matter how much you zoom in).
   i. f
   ii. g
   iii. h

13. As part of a conservation effort we want to buy a monogrammed mug for every student, staff, and faculty member in the mathematics department. We check with several companies and get the following price quotes.

Great Mugs will charge $20 just to place the order and then they charge an additional $6 for each mug that we order.

Name It will only charge $10 to process the order and has a varying scale depending upon the number of mugs ordered. For the first 20 mugs we order, the cost is $7 per mug; for the next 50 mugs, the cost is $6 per mug; and for all mugs after that, the cost is $5 per mug.

Let $G(x)$ be the cost of ordering x mugs from *Great Mugs*.

Let $N(x)$ be the cost of ordering x mugs from *Name It*.

(a) Graph $G(x)$ and $N(x)$.

(b) Write functions for $G(x)$ and $N(x)$.

(c) For which values of x is it cheaper to order from *Great Mugs* as compared to ordering from *Name It*?

(d) How much can the difference in prices between the two companies ever be if we place an order for the same number of mugs from each company?

CHAPTER 5

The Derivative Function

■ 5.1 CALCULATING THE SLOPE OF A CURVE
AND INSTANTANEOUS RATE OF CHANGE

In Chapter 4 we looked at linear functions, functions characterized by a constant rate of change. This characteristic is unique to linear functions; the rate of change of the output of any nonlinear function varies with the value of the independent variable.

Consider, for example, the bucket calibration problem for a bucket as drawn in Figure 5.1. The change in height produced by adding one gallon of water to an empty bucket is greater than the change in height produced by adding the same amount of water to a partially filled bucket. The more water in the bucket the less impact an additional gallon of water will have on the height. For the bucket, the change in height produced by the addition of water is a function of the volume of water already in the bucket. Similarly, for the conical flask the $\frac{\text{change in height}}{\text{change in volume}}$ ratio depends upon volume.

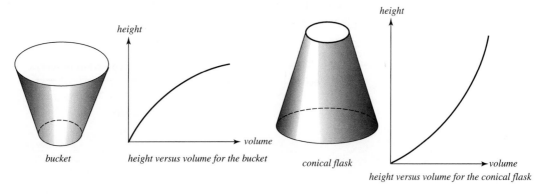

bucket   height versus volume for the bucket   conical flask   height versus volume for the conical flask

Figure 5.1

We have looked at several examples in which a rate of change is a function of the independent variable. While we have not yet defined the *instantaneous* rate of change of a function, we already have some notion of it. For instance:

■   Velocity is the instantaneous rate of change of position over change in time. When riding in a car, we look at the speedometer to determine our velocity at an instant.

■   In Chapter 2 we analyzed a graph of a cyclist's velocity plotted as a function of time. Implicit in this graph is the idea that at each instant the cyclist's velocity can be determined.

■   In Chapter 3 we looked at graphs of the rates of flow of water in and out of a reservoir plotted as a function of time. Again, implicit in these graphs is the idea that at each instant such a rate of change can be determined.

For a given function we know how to calculate the average rate of change over an interval. We'll now tackle the problem of calculating an instantaneous rate of change. To find an average rate of change we need two data points; an instant provides us with only one data point. This is the fundamental challenge of differential calculus. We need a strategy for approaching this problem.

Problem-Solving Strategies

■   Look at a concrete problem and determine what methods of attack can be applied to the more abstract problem.

■   Use the method of successive approximations. Approximate what you're looking for, determining upper and lower bounds if possible. Then improve on the approximation. Repeat this process until the approximation is good enough for your purposes or until you arrive at an exact answer.

Let's consider a rock dropped from a height of 256 feet. From the moment the rock is dropped the fundamental forces acting on it are the force of gravity and the opposing force of air resistance. (We will consider only the force of gravity because air resistance in this situation is negligible.) The force of gravity results in a downward acceleration of the rock. The rock's speed will increase as it falls. Its position, s, is a function of time. Let's set $t = 0$ to be the moment the rock is dropped and measure time in seconds. Suppose we are given the following data:

t (time in seconds)	s (position in feet)
0	256
1	240
2	192
3	112
4	0

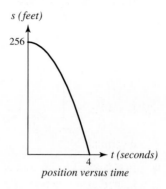

Figure 5.2   Position vs. time

The rock hits the ground in 4 seconds. What is its average velocity in the first second? In each consecutive second? Recall that the average velocity is given by

$$\frac{\Delta \text{ position}}{\Delta \text{ time}} = \frac{\Delta s}{\Delta t}.$$

average velocity in the 1st second $= \frac{\Delta s}{\Delta t} = \frac{s(1)-s(0)}{1-0} = \frac{240-256}{1} = -16$ ft/sec

average velocity in the 2nd second $= \frac{\Delta s}{\Delta t} = \frac{s(2)-s(1)}{2-1} = \frac{192-240}{1} = -48$ ft/sec

average velocity in the 3rd second $= \frac{\Delta s}{\Delta t} = \frac{s(3)-s(2)}{3-2} = \frac{112-192}{1} = -80$ ft/sec

average velocity in the 4th second $= \frac{\Delta s}{\Delta t} = \frac{s(4)-s(3)}{4-3} = \frac{0-112}{1} = -112$ ft/sec

Why are these velocities negative? Velocity carries information about both speed and direction; the sign indicates direction, while the magnitude (size, or absolute value) gives the speed. A positive velocity indicates that the height of the rock was increasing and a negative velocity indicates that the rock is falling. We see that the speed itself is increasing as the rock is falling; on the other hand, the velocity is decreasing because it is becoming more and more negative.

Suppose we are interested in the rock's velocity at $t = 2$. Simply knowing that at $t = 2$ the rock's height is 192 feet doesn't help us determine the instantaneous velocity; a snapshot of a rock in midair gives us virtually no clue as to its speed and direction. However, from the computations above we see that the velocity should lie between the average velocity on the interval [1, 2] and that on the interval [2, 3]. The velocity at $t = 2$ is between -48 ft/sec and -80 ft/sec.

Graphically, the average velocity on the interval [1, 2] can be represented by the slope of the secant line through the points $(1, s(1))$ and $(2, s(2))$. Similarly, the average velocity on the interval [2, 3] can be represented by the slope of the secant line through the points $(2, s(2))$ and $(3, s(3))$.

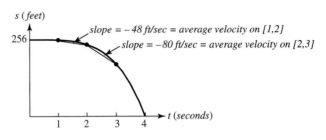

Figure 5.3

We'd like to find better approximations to the instantaneous velocity at $t = 2$ by finding better upper and lower bounds, but without more information we can make little progress. It would be helpful to have more information about the position of the rock in the neighborhood of time $t = 2$. The average velocity of the rock over the interval $[2, 2.1]$, for instance, will give us a better estimate of the rock's velocity at $t = 2$ than did the average velocity over the interval $[2, 3]$. The average speed of the rock will still be a little greater than its speed at $t = 2$ because speed increases with time (its average velocity will be more negative than the instantaneous velocity), but we will have a better approximation. Looking at the interval $[2, 2.01]$ will give us an even better estimate, although again the average speed will be greater on the interval than at $t = 2$ because the rock is traveling faster as time goes on. From a graphical perspective we are asserting that we can get better and better approximations to the instantaneous velocity at $t = 2$ by looking at the slopes of secant lines passing through $(2, 192)$ and a second point on the graph of s that gets closer and closer to the point $(2, 192)$, as shown in Figure 5.4 below.

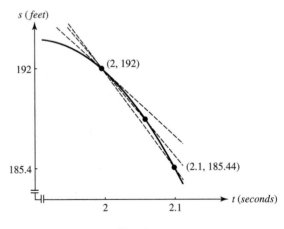

Figure 5.4

Suppose we ask for more data about the position of the rock around $t = 2$ and we are supplied with the following:

t (in seconds)	s (position in feet)
1.9	198.24
1.99	192.6384
2	192
2.01	191.3584
2.1	185.44

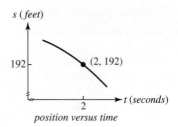

position versus time

Figure 5.5

The average velocity on [2, 2.1] is

$$\frac{\Delta s}{\Delta t} = \frac{s(2.1) - s(2)}{2.1 - 2} = \frac{185.44 - 192}{2.1 - 2} = -65.6 \text{ ft/sec,}$$

so the instantaneous velocity at $t = 2$ is less negative than -65.6. We can improve on this. The average velocity on [2, 2.01] is

$$\frac{\Delta s}{\Delta t} = \frac{s(2.01) - s(2)}{2.01 - 2} = \frac{191.3584 - 192}{2.01 - 2} = -64.16 \text{ ft/sec,}$$

so the instantaneous velocity at $t = 2$ is less negative than -64.16.

We've put a higher floor on the instantaneous velocity at $t = 2$. First we said that the instantaneous velocity was greater than -80 ft/sec, then we said that it was greater than -65.6 ft/sec, and now we have determined that it is greater than -64.16 ft/sec.

As a lid on the value of the instantaneous velocity we can use -48 ft/sec, the average velocity on [1, 2]. We can lower the lid and get a better upper bound by computing the average velocity on [1.9, 2] or get an even better answer using [1.99, 2]. The average velocity on [1.99, 2] is

$$\frac{\Delta s}{\Delta t} = \frac{s(2) - s(1.99)}{2 - 1.99} = \frac{192 - 192.6384}{2 - 1.99} = -63.84 \text{ ft/sec.}$$

We now know that the instantaneous velocity at $t = 2$ is between -63.84 ft/sec and -64.16 ft/sec. That's a much higher degree of accuracy than we had without the additional data. To close in further on the instantaneous velocity of the rock at $t = 2$ we need more data. If only there were an algebraic formula to give us position as a function of time . . . Indeed, there is such a formula:

$$s(t) = -16t^2 + 256.$$

This formula can be derived using a combination of physics and mathematics; you will be able to derive it on your own by the end of the course.[1]

Armed with a formula for position as a function of time, we can find the average velocity over the intervals, say, [2, 2.0001] and [1.9999, 2] to get lower and upper bounds. We could get better and better approximations by taking smaller and smaller time intervals.

Let's save ourselves a bit of work. We'll find the average velocity on the interval between 2 and $2 + h$, where $h \neq 0$. For the interval [2, 2.0001] we have $h = 0.0001$; for [1.9999, 2] we have $h = -0.0001$. The smaller the magnitude (absolute value)[2] of h, the better the average velocity will approximate the instantaneous velocity at $t = 2$.

$$\text{Average velocity on the interval between 2 and } 2 + h = \frac{\Delta s}{\Delta t}$$

$$= \frac{s(2 + h) - s(2)}{2 + h - 2} \qquad \text{Since } s(t) = -16t^2 + 256,$$
$$\qquad\qquad \text{we can find } s(2 + h).$$

$$= \frac{-16(2 + h)^2 + 256 - [-16(2)^2 + 256]}{h} \qquad \text{Simplify.}$$

$$= \frac{-16(4 + 4h + h^2) + 256 + 16 \cdot 4 - 256}{h}$$

$$= \frac{-16 \cdot 4 - 64h - 16h^2 + 16 \cdot 4}{h}$$

$$= \frac{-64h - 16h^2}{h}$$

$$= \frac{h(-64 - 16h)}{h} \qquad \text{As long as } h \neq 0, \text{ we know } h/h = 1.$$

$$= -64 - 16h$$

In particular, the average velocity on [2, 2.0001] is -64.0016 ft/sec and the average velocity on [1.9999, 2] is -63.9984 ft/sec.

Now we can see what will happen as h approaches 0.

■   If h is positive we are looking at the average velocity on $[2, 2 + h]$. This number, $-64 - 16h$, will be more negative than -64 and will provide a lower bound for the velocity at $t = 2$. The smaller the magnitude of h, the closer $-64 - 16h$ gets to -64.

■   If h is negative we are looking at the average velocity on $[2 + h, 2]$. This number, $-64 - 16h$, will be less negative than -64 and will provide an upper bound for the velocity at $t = 2$. Again, the smaller the magnitude of h, the closer $-64 - 16h$ gets to -64.

We have constructed a vise, an upper bound (ceiling) and a lower bound (floor) between which the velocity at $t = 2$ is trapped. As h gets increasingly close to zero both the upper bound and the lower bound approach -64. The instantaneous velocity of the rock is -64 ft/sec.

---

[1] This equation can be derived from the facts that:
  i. The downward acceleration due to gravity is -32 ft/sec^2.
  ii. The rock was dropped from a height of 256 ft and given no initial velocity.
[2] $|h|$ is a handy notation for "the size of h" or "the magnitude of h."

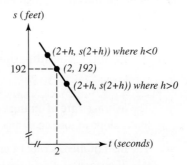

Figure 5.6

Graphically, we have a set of secant lines all going through (2, 192); we choose a second point on the graph of s to be $(2 + h, s(2 + h))$. This point lies to the right of (2, 192) if h is positive and to the left if h is negative. As we take values of h closer and closer to zero, this second point slides along the graph of s closer and closer to (2, 192). The slope of the secant line corresponds to the average velocity of the rock on the interval between 2 and $2 + h$.

We make the following observations about h.

- h must be nonzero. If we try to calculate the average velocity on [2, 2] we will get $\frac{\Delta s}{\Delta t} = \frac{0}{0}$; we can't make sense of this. Furthermore, in our calculations above we said $h/h = 1$; this is true only if $h \neq 0$.

- There is no *smallest* possible magnitude for h. If you think you might have the smallest h, just divide it by two and you'll get something half the size.

We'll introduce some notation that allows us to summarize what we've done less verbosely.[3] The average rate of change of the position of the rock on the interval between 2 and $2 + h$, where $h \neq 0$, is given by $\frac{s(2+h)-s(2)}{(2+h)-2}$ or $\frac{s(2+h)-s(2)}{h}$. As h approaches zero, $\frac{s(2+h)-s(2)}{h}$ approaches the instantaneous rate of change of position with respect to time at $t = 2$. We can write this in shorthand as follows:

$$\text{As } h \to 0, \frac{s(2+h)-s(2)}{h} \to \text{(the instantaneous velocity at } t = 2).$$

Or, equivalently,

$$\lim_{h \to 0} \frac{s(2 + h) - s(2)}{h} = \text{(the instantaneous velocity at } t = 2).$$

Note: Built into this $\lim_{h \to 0}$ notation is the notion that $h \neq 0$.

Unraveling the notation (an intuitive approach):

$\lim_{x \to 5} f(x) = 7$ is read "the limit as x approaches 5 of $f(x)$ is 7."

As x approaches 5, $f(x)$ approaches 7, is equivalent to saying or writing

$$\text{as } x \to 5, f(x) \to 7.$$

---

[3] Modern mathematicians relish being concise and precise. We'll start working on the former now and postpone the latter for Chapter 7.

In words this means that $f(x)$ can be made *arbitrarily* close to 7 provided x is *close enough* to 5 *but not equal* to 5.[4]

Summary and Generalization

If $y = f(x)$, then

■   The **average rate of change of** f **on** $[a, b] = \frac{f(b)-f(a)}{b-a}$. Graphically, this corresponds to the slope of the secant line through $(a, f(a))$ and $(b, f(b))$.

■   The **instantaneous rate of change of** f **at** $x = a$ can be approximated by the average rate of change of f on $[a, a + h]$ for h very close to zero. As h gets closer and closer to zero, the average rate of change over the tiny interval approaches the instantaneous rate of change at $x = a$.

$$\text{The instantaneous rate of change of } f \text{ at } x = a \approx \frac{f(a+h)-f(a)}{(a+h)-a} \text{ for } h \text{ small.}$$

$$\text{The instantaneous rate of change of } f \text{ at } x = a = \lim_{h\to 0} \frac{f(a+h)-f(a)}{h},$$

if this limit exists.

Graphically the instantaneous rate of change of f at $x = a$ corresponds to the slope of the line tangent to the graph of f at $x = a$. The **tangent line to** f **at** $x = a$ is *defined* to be the limit of the secant lines through $(a, f(a))$ and $(a + h, f(a + h))$ as h approaches zero, provided this limit exists.

A typical misconception is that a tangent line can touch the graph only at one point. The picture below should dissuade you of this notion.

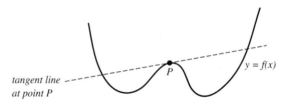

Figure 5.7

Taking Inventory

The strategy of successive approximation combined with a limiting process has given us a method of tackling (and defining) two important and closely linked problems:[5]

1. How can we find the instantaneous rate of change of a function?

2. How can we define the slope of a curve at a point? Or, equivalently, how can we find the slope of the line tangent to a curve at a point?

---

[4] We will take up limits later and make this more precise. Loosely speaking, "arbitrarily close" is "as close as anyone could conceivably insist upon." And how close is "close enough"? Close enough to satisfy whatever conditions have been insisted upon.

[5] The strategy of successive approximation in combination with a limiting process will continue to be a keystone in calculus. This is only the beginning.

Both problems encompass the same computational challenge. We need two data points in order to compute a rate of change, and we need two points in order to compute the slope of a line. A snapshot tells us nothing about an average rate of change, and a single point gives no information about slope. Our strategy allowed us to see our way through this dilemma.

We found that

$$\lim_{h \to 0} \frac{f(a + h) - f(a)}{h} = \text{slope of the tangent line to } f \text{ at } x = a$$

$$= \text{the instantaneous rate of change of } f \text{ at } x = a.$$

We introduce some shorthand notation for this expression.

Definition

Let c be in the domain of the function f. We define $f'(c)$ to be

$$f'(c) = \lim_{h \to 0} \frac{f(c + h) - f(c)}{h}$$

if the limit exists. We refer to this as **the limit definition of the derivative of** $f'(c)$. $f'(c)$ is called the **derivative of** f **at** $x = c$. We read $f'(c)$ aloud as "f prime of c." $f'(c)$ can be interpreted as the rate of change of f at $x = c$ or as the slope of the tangent line to the graph of f at $x = c$.

Recall that informally speaking, "locally linear" means that if the graph were magnified enormously and you were a miniscule bug positioned on the curve, the curve would appear, from your bug's-eye perspective, to be a straight line. If f is locally linear and nonvertical at $x = c$, then $f'(c)$ exists and gives the slope of the tangent line, the best linear approximation to the curve, at $x = c$.

A function f is **differentiable at** c if $f'(c)$ is defined. We say f is **differentiable on the open interval** (a, b) if f' is defined for all x in the interval.

Since the derivative is defined as the limit of a quotient and the quotient is the quotient of differences, sometimes the derivative is referred to as the limit of a difference quotient.

Cases in Which $f'(c)$ Is Undefined

If $f'(c)$ is undefined, then the limit of the difference quotient does not exist. $f'(c)$ is undefined if the graph of f has a vertical tangent at the point $(c, f(c))$ or if the graph of f is not locally linear at $x = c$. f is not locally linear at $x = c$ if f is not continuous at $x = c$ or if there is a sharp corner at $x = c$. If the graph of f has a sharp corner (like the "V" of the absolute value function as opposed to the "U" of the squaring function), then the corner will look sharp regardless of how greatly it is magnified.

Notice that $f'(c)$ is undefined if c is not in the domain of f; we cannot write the difference quotient, $\frac{f(c+h)-f(c)}{h}$.

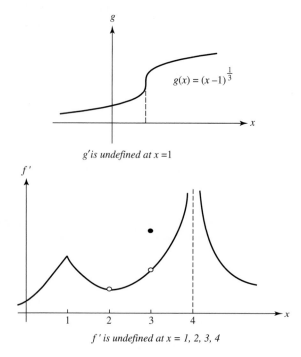

g' is undefined at $x = 1$

f' is undefined at $x = 1, 2, 3, 4$

Figure 5.8

At present we do not have this on very solid ground. We'll need to discuss the notion of limits to get a more sturdy foundation. But historically, neither Isaac Newton nor Gottfried Leibniz, both of whom developed the fundamentals of calculus in the decade from 1665 to 1675, had a rigorous notion of limit and, nevertheless, they made enormous mathematical strides. Following the spirit of the historical development, we will proceed with our informal notion for the time being. In Chapter 7 we will make our notion of limit more precise.

Below we work a few examples. We'll reiterate the entire line of reasoning in the first example and concentrate on the mechanics of the computation in the second one.

◆ EXAMPLE 5.1   Find the slope of the tangent line to the graph of $f(x) = x^2 + 3x$ at $x = 2$.

SOLUTION   Let's run through the line of reasoning again. Choose a point on the graph of f near $(2, 10)$. We choose $(2 + h, f(2 + h))$, i.e., $(2 + h, (2 + h)^2 + 3(2 + h))$, where h is very small. The slope of the secant line through $(2, 10)$ and $(2 + h, (2 + h)^2 + 3(2 + h))$ approximates the slope of the tangent line to f at $x = 2$. We will denote by m_{sec} the slope of the secant line and by m_{tan} the slope of the tangent line.

$$m_{\text{sec}} = \frac{f(2+h) - f(2)}{(2+h) - 2}$$

$$= \frac{(2+h)^2 + 3(2+h) - (2^2 + 2 \cdot 3)}{h}$$

$$= \frac{4 + 4h + h^2 + 6 + 3h - 10}{h}$$

$$= \frac{7h + h^2}{h}$$

$$= \frac{h(7+h)}{h}$$

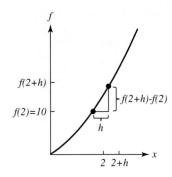

Figure 5.9

Since $h \neq 0$, we can cancel to get $m_{\text{sec}} = 7 + h$.

As h tends toward zero, the point $(2 + h, (2 + h)^2 + 3(2 + h))$ slides along the graph of f toward $(2, 10)$.

As h tends toward zero, the slope of the secant line tends toward the slope of the tangent line.

As h tends toward zero, the slope of the secant line tends toward 7: $\lim_{h \to 0} 7 + h = 7$.

Notice that we've computed $\lim_{h \to 0} \frac{f(2+h) - f(2)}{h}$ by first simplifying the difference quotient and then computing the limit. ◆

The next example is a wonderful way to practice your algebra skills. Work it out on your own and then compare your work with the solution given.

◆ **EXAMPLE 5.2** Find the equation of the tangent line to the graph of $y = \frac{5}{x^2}$ at $x = -3$.

SOLUTION First let's find the *slope* of the tangent line at $x = -3$. Let $y = f(x)$. We want $f'(-3)$.

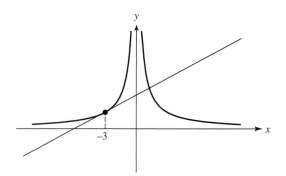

Figure 5.10

$$f'(-3) = \lim_{h \to 0} \frac{f(-3+h) - f(-3)}{h}$$
We know $f(x) = \frac{5}{x^2}$; use this.

$$= \lim_{h \to 0} \frac{\dfrac{5}{(-3+h)^2} - \dfrac{5}{(-3)^2}}{h}$$

$$= \lim_{h \to 0} \frac{\dfrac{5}{(-3+h)^2} - \dfrac{5}{9}}{h}$$
Get a common denominator.

$$= \lim_{h \to 0} \frac{\dfrac{45}{9(-3+h)^2} - \dfrac{5(-3+h)^2}{9(-3+h)^2}}{h}$$
Simplify.

$$= \lim_{h \to 0} \frac{\dfrac{45}{9(-3+h)^2} - \dfrac{5(9 - 6h + h^2)}{9(-3+h)^2}}{h}$$
We must multiply out the numerator in order to combine like terms.

$$= \lim_{h \to 0} \frac{45 - 45 + 30h - 5h^2}{9(-3+h)^2} \cdot \frac{1}{h}$$
We pull out the $1/h$ to make this easier to read.

$$= \lim_{h \to 0} \frac{30h - 5h^2}{9(-3+h)^2} \cdot \frac{1}{h}$$
Notice that we have not multiplied out the denominator. It is rarely in our interest to do so.[6]

$$= \lim_{h \to 0} \frac{h(30 - 5h)}{9(-3+h)^2} \cdot \frac{1}{h}$$
Factor out an h.

$$= \lim_{h \to 0} \frac{(30 - 5h)}{9(-3+h)^2}$$
Since $h \neq 0$, $\frac{h}{h} = 1$.

$$= \frac{30}{9(-3)^2} = \frac{10}{3(-3)^2} = \frac{10}{27}$$

---

[6] Many beginners waste a lot of time and energy multiplying out denominators that are factored. Yet, when getting a common denominator or simplifying an expression, one factors. Therefore the effort expended is not a badge of virtue; rather, it is counterproductive.

Therefore the slope of the tangent line is 10/27; the equation of the tangent line is of the form $y - y_1 = \frac{10}{27}(x - x_1)$. The tangent line goes through the point $(-3, f(-3)) = (-3, 5/9)$, so when $x = -3$, y must be $5/9$.

$$y - \frac{5}{9} = \frac{10}{27}(x + 3) \quad \text{or} \quad y = \frac{10}{27}(x + 3) + \frac{5}{9}.$$

Alternatively, the equation of the tangent line can be written

$$y = \frac{10}{27}x + \frac{5}{3}. \quad \blacklozenge$$

In the next section we look at f', the derivative *function*.

PROBLEMS FOR SECTION 5.1

The limiting process enables us to get a handle on the slope of a curve and the instantaneous rate of change. Problems 1 through 4 ask for approximations before arriving at an exact answer; this is to remind you of the process.

1. Suppose a ball is thrown straight up from a height of 48 feet and given an initial upward velocity of 8 ft/sec. Then its height at time t, t in seconds, is given by $h(t) = -16t^2 + 8t + 48$, for $t \in [0, 2]$.

    In this problem we will look at the ball's velocity at $t = 1$.

    (a) At $t = 1$, is the ball heading up, or down? Explain your reasoning.

    (b) By calculating the average rate of change of height with respect to time, $\frac{\Delta h}{\Delta t}$, on the intervals $[0.9, 1]$ and $[1, 1.1]$, give bounds for the ball's velocity at $t = 1$.

    (c) Improve your bounds by using the intervals $[0.99, 1]$ and $[1, 1.01]$.

    (d) Use the limit definition of $f'(1)$ to find the ball's instantaneous velocity at $t = 1$.

2. Let $f(x) = \frac{1}{x}$. In this problem we will look at the slope of the tangent line to $f(x)$ at point $P = (\frac{1}{2}, 2)$.

    (a) Is the slope of the tangent line to f at P positive, or negative?

    (b) By calculating the slope of the secant line through P and a nearby point on the graph of f, approximate $f'(\frac{1}{2})$. First choose the point with an x-coordinate of 0.49. Next choose the point with an x-coordinate of 0.501. Now produce an approximation that is better than either of the previous two.

    (c) By calculating the limit of the difference quotient, find $f'(\frac{1}{2})$.

    (d) Find the equation of the tangent line to f at P.

3. Let $f(x) = x^3$ and P be the point $(1, 1)$ on the graph of f.

    (a) Approximate the slope of the line tangent to f at P by looking at the slope of the secant line through P and Q, where $Q = (1 + h, f(1 + h))$. Calculate the difference quotient for various values of h, both positive and negative. See if your calculator or computer will produce a table of values.

    (b) Calculate $f'(1)$ by computing the limit of the difference quotient.

(c) For what values of h is the difference quotient greater than $f'(1)$? For what values of h is the difference quotient less than $f'(1)$? Make sense out of this by looking at the graph of x^3.

4. Let $f(x) = x^3$ and P be the point $(0, 0)$ on the graph of f.

   (a) Approximate $f'(0)$ by looking at the slope of the secant line through P and a nearby point Q on the graph of f. Use a calculator or computer to get a sequence of successive approximations corresponding to allowing the point $Q = (h, f(h))$ to slide along the graph of f toward P. Choose both positive and negative values of h.

   (b) Use the results of part (a) to guess the slope of the line tangent to x^3 at $x = 0$.

   (c) Calculate $f'(0)$ by computing the limit of the difference quotient $\frac{f(0+h)-f(0)}{h}$.

   (d) Challenge question: In the previous three problems, by varying the sign of h in the difference quotient $\frac{f(c+h)-f(c)}{h}$ we obtain upper and lower bounds for $f'(c)$. In this problem, the difference quotient computed in part (a) is always larger than $f'(0)$. Explain what is going on by looking at the graphs of the various functions.

In Problems 5 through 8, estimate $f'(c)$ by calculating the difference quotient $\frac{f(c+h)-f(c)}{h}$ for successively smaller values of $|h|$. Use both positive and negative values of h.

5. $f(x) = \sqrt{x}$. Approximate $f'(9)$.

6. $f(x) = \frac{1}{\sqrt{x}}$. Approximate $f'(4)$.

7. $f(x) = x^4$. Approximate $f'(1)$.

8. $f(x) = \sqrt[3]{x}$. Approximate $f'(8)$.

9. Let $f(x) = \frac{6}{x+1}$. Let P and Q be points on the graph of f with x-coordinates 3 and $3 + h$, respectively.

   (a) Sketch the graph of f and the secant lines through P and Q for $h = 1$ and $h = 0.1$.

   (b) Find the slope of the secant line through P and Q for $h = 1$, $h = 0.1$, and $h = 0.01$.

   (c) Find the slope of the tangent line to f at point P by calculating the appropriate limit.

   (d) Find the equation of the line tangent to f at point P.

10. A pool is being drained. The amount of water in the pool at time t, t in hours, is given by $w(t) = -16t^2 + 256$ gallons.

   (a) Find the average rate of change of water in the pool over the time interval $[2, 2.5]$. Include units in your answer. Why is the value $\frac{\Delta w}{\Delta t}$ negative?

   (b) Use the dropped-rock problem results from this section to determine the rate of change of water in the pool at time $t = 2$. Include units in your answer.

In Problems 11 and 12, estimate the slope of the tangent line at point P on the graph of $f(x)$ by choosing a nearby point Q on the graph of f, $Q = (a, f(a))$ and finding

*the slope of the secant line through P and Q. By choosing Q successively closer to P,
guess the slope of the tangent line at P.*

11. $f(x) = \sqrt{x}$. $P = (4, 2)$. Fill in the table.

x-coordinate of Q	Slope of PQ
3	
3.8	
3.95	
3.998	
4.001	
4.02	
4.3	
5	

12. $f(x) = x^5$. $P = (2, 32)$. Use your calculator to construct a table.

x-coordinate of Q	Slope of PQ

13. The point $P = (3, f(3))$ lies on the graph of $f(x)$. Suppose $Q = (w, f(w))$ is another point on the graph of $f(x)$.

   (a) Write an expression for the slope of the secant line through P and Q.

   (b) Which one of the following corresponds to the slope of the tangent line to f at point P?

   i. $\frac{f(3) - f(w)}{3 - w}$

   ii. $\lim_{w \to 0} \frac{f(3) - f(w)}{3 - w}$

   iii. $\lim_{w \to 3} \frac{f(3) - f(w)}{3 - w}$

   (c) Explain, using pictures, why the answer you chose in part (b) corresponds to the process of taking point Q closer and closer to P and computing the slope of PQ.

14. If we have a formula for f we can get quite good numerical estimates of the slope of the tangent line to f at a particular point. In this exercise we will do that.

Below is the graph of $y = f(x) = x^2 - 4$.

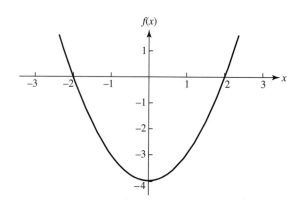

(a) *Goal:* We want to estimate the value of $f'(2)$, i.e., to approximate the slope of the tangent line to the graph at the point $(2, 0)$. To this end, do the following:

   i. Sketch the tangent line to the graph at $(2, 0)$.

      Finding the slope of this line is problematic, since we need two points on a line to find its slope. The only point we know for sure is on the tangent line is $(2, 0)$.

   ii. Draw a line through the points $(2, 0)$ and $(2.1, f(2.1))$.

      The slope of this line is close to the slope of the tangent line. Is it bigger than the slope of the tangent line, or less than the slope of the tangent line? Compute the slope of this line through points $(2, 0)$ and $(2.1, f(2.1))$.

   iii. Draw a line through the points $(2, 0)$ and $(1.9, f(1.9))$.

      The slope of this line is close to the slope of the tangent line. Is it bigger than the slope of the tangent line, or less than the slope of the tangent line? Compute the slope of this line through points $(2, 0)$ and $(1.9, f(1.9))$.

      We are now in pretty good shape: We have an upper and lower bound for the slope of the tangent line. If you have done the problem correctly, you should now have the slope of the tangent line to within 0.2. Suppose we want to get an even better estimate. Do the following:

   iv. Find the slope of the line through $(2, 0)$ and $(2.01, f(2.01))$. Is this slope bigger than the slope of the tangent line at $(2, 0)$, or smaller?

   v. Find the slope of the line through $(2, 0)$ and $(1.99, f(1.99))$. Is this slope bigger than the slope of the tangent line at $(2, 0)$, or smaller?

   vi. We will ask you a few more questions of this sort. You can save yourself time and energy by computing the slope of the line through $(2, 0)$ and $(2 + h, f(2 + h))$. Then you will be able to simply evaluate your answer for different values of h. Make your life easier by computing this slope.

   vii. Find the slope of the line through $(2, 0)$ and $(2.002, f(2.002))$.

   viii. Find the slope of the line through $(2, 0)$ and $(2.0001, f(2.0001))$.

ix. Find the slope of the line through $(2, 0)$ and $(1.998, f(1.998))$.

x. Find the slope of the line through $(2, 0)$ and $(1.9999, f(1.9999))$.

xi. At this point, you should have a fairly accurate idea of the slope of the graph of f at $(2, 0)$. Give the best bounds you can for $f'(2)$ given the work you've done in parts vii through x. If you simplified your answer to part vi, then you can see what happens as h gets arbitrarily small. What do you think is the value of $f'(2)$?

We now turn our attention to another point.

(b) *Goal:* We want to estimate the value of $f'(1)$; that is, we want to approximate the slope of the tangent line to the graph at the point $(1, -3)$.

i. Sketch the tangent line to the graph at $(1, -3)$.

ii. Find the slope of the line through $(1, -3)$ and $(1 + h, f(1 + h))$. Use this to answer parts iii through vi. (The more you simplify your answer the less time this problem will take.)

iii. Find the slope of the line through $(1, -3)$ and $(1.1, f(1.1))$.

iv. Find the slope of the line through $(1, -3)$ and $(1.001, f(1.001))$.

v. Find the slope of the line through $(1, -3)$ and $(0.9, f(0.9))$.

vi. Find the slope of the line through $(1, -3)$ and $(0.99, f(0.99))$.

vii. What happens to the slope of the secant line through $(1, -3)$ and $(1 + h, f(1 + h))$ as h gets arbitrarily small? What do you think is the value of $f'(1)$?

(c) *Goal:* We want to estimate the value of $f'(0)$; i.e., we want to approximate the slope of the tangent line to the graph at the point $(0, -4)$.

i. Sketch the tangent line to the graph at $(0, -4)$.

ii. Find the slope of the line through $(0, -4)$ and $(0 + h, f(0 + h))$.

iii. Use your answer to part ii to estimate the slope of the tangent line.

(d) *Goal:* To estimate the value of $f'(c)$ for c a constant.

i. Find the slope of the line through $(c, f(c))$ and $(c + h, f(c + h))$.

ii. Use your answer to the previous question to find the slope of the tangent line to f at $x = c$.

iii. Sketch the graph of $f'(x)$.

15. Let $f(x) = 2x^2 - 5x - 1$. Let P and Q be points on the graph of f with x-coordinates 3 and $3 + h$, respectively.

(a) Sketch the graph of f and the secant lines through P and Q for $h = 1$ and $h = 0.1$.

(b) Find the slope of the secant line through P and Q for $h = 1$, $h = 0.1$, and $h = 0.01$.

(c) Find the slope of the tangent line to f at point P by calculating the appropriate limit.

(d) Find the equation of the line tangent to f at point P.

16. Find the equation of the line tangent to $y = \frac{2}{x+1}$ at the point $(1, 1)$.

17. Let g be a function of t. Assume g is locally linear.

(a) Label the coordinates of each of the six points shown on the following page.

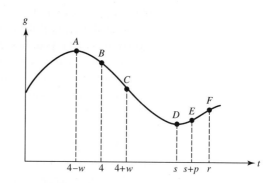

(b) Match the following expressions on the left with one of those on the right. The expressions on the right can be used more than once. (We use m_{sec} as shorthand for "the slope of the secant line" and m_{tan} as shorthand for "the slope of the tangent line.")

i) $\dfrac{g(4+w)-g(4)}{w}$

ii) $\dfrac{g(4)-g(4-w)}{w}$

iii) $\lim_{\Delta t \to 0} \dfrac{g(4+\Delta t)-g(4)}{\Delta t}$

iv) $\dfrac{g(r)-g(s)}{r-s}$

v) $\lim_{t \to s} \dfrac{g(t)-g(s)}{t-s}$

vi) $\lim_{t \to r} \dfrac{g(r)-g(t)}{r-t}$

vii) $\dfrac{g(s+p)-g(s)}{p}$

viii) $\lim_{h \to 0} \dfrac{g(s+h)-g(s)}{h}$

A. m_{sec} through points D and F

B. m_{sec} through points B and C

C. m_{sec} through points A and B

D. m_{sec} through points D and E

E. $g'(4) = m_{\text{tan}}$ at point B

F. $g'(r) = m_{\text{tan}}$ at point F

G. $g'(s) = m_{\text{tan}}$ at point D

18. Let g be a function that is locally linear. We know that $g'(a)$ is the slope of the tangent line to the graph of g at point $A = (a, g(a))$.

Let $Q = (t, g(t))$ be an arbitrary point on the graph of g, Q distinct from A.

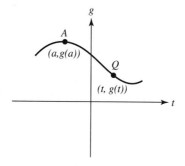

(a) Write a difference quotient (i.e., an expression of the form $\frac{?-?}{?-?}$, the quotient of two differences) that gives the slope of the secant line through points A and Q.

(b) Take the appropriate limit of the difference quotient in part (a) to arrive at an expression for $g'(a)$.

19. Let $f(x) = \frac{3}{x-5}$.

(a) Using the limit definition of derivative, find $f'(2)$.

(b) Find two ways of checking whether or not your answer is reasonable. These methods should *not* involve simply checking your algebra. They can be numerical or graphical—use your ingenuity.

20. Let f be a function that is locally linear. Let $A = (5, f(5))$ and $B = (5 + h, f(5 + h))$. A is a fixed point on the graph of f while B is mobile; as h varies, B moves along the graph of f.

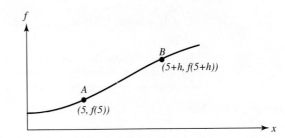

The following statements are equivalent. Replace each question mark with the appropriate expression. (We use m_{sec} as shorthand for "the slope of the secant line" and m_{tan} as shorthand for "the slope of the tangent line.")

As $B \to A$, (m_{sec} through points A and B) $\to$ (m_{tan} at point A).

(a) As $h \to \boxed{?}$, $\dfrac{\boxed{?}}{\boxed{?}} \to f'\left(\boxed{?}\right)$

(b) $\lim\limits_{h \to \boxed{?}} \dfrac{\boxed{?}}{\boxed{?}} = f'\left(\boxed{?}\right)$

For Problems 21 through 24 use the limit definition of $f'(a)$ to find the derivative of f at the point indicated.

21. $f(x) = \dfrac{3}{2-x}$ at $x = 1$

22. $f(x) = x(x + 3)$ at $x = 2$

23. $f(x) = \dfrac{x}{2} + \dfrac{2}{x}$ at $x = 1$

24. $f(x) = (x - 3)^2$ at $x = 3$

5.2   THE DERIVATIVE FUNCTION

Definition

The **derivative function** $f'(x)$ is given by

$$f'(x) = \lim_{h \to 0} \frac{f(x + h) - f(x)}{h}.$$

The function $f'(x)$ can be thought of as the **slope function of** $f(x)$. An Alice-in-Wonderland exercise is useful. Imagine the smooth curve given by $y = f(x)$ enlarged enormously. Shrink yourself to the size of a tiny bug and crawl along the curve from left to right. At each point you should feel as if you are crawling along a line.[7] As you crawl along the curve the slope changes, but from your bug's perspective, it looks like a line with a given slope at each point. The slope is a function of x. The derivative gives a bug's-eye view of the world.

Let's start by considering two simple cases.

If f is a linear function, $f(x) = mx + b$, then $f'(x) = m$; the slope function is the slope.

If f is a constant function, $f(x) = k$, then $f'(x) = 0$; the slope of a horizontal line is zero.

EXERCISE 5.1

(a) Use the limit definition of $f'(x)$ to show that if $f(x) = mx + b$, then $f'(x) = m$. In other words, verify that the derivative of a linear function gives the slope of the line.

(b) Conclude that if f is the constant function, $f(x) = k$, then $f'(x) = 0$.

Answers are provided at the end of the section.

In the example of the falling rock in Section 5.1, the rock's velocity is a function of time. This is equivalent to saying that the slope of the position versus time graph is a function of t. From this graphical perspective you might observe that the slope is always negative; the graph is decreasing. As t increases, the slope (corresponding to the instantaneous velocity) gets more and more negative. The derivative function, $f'(t)$, which is both the velocity function and the slope function, is negative and decreasing.

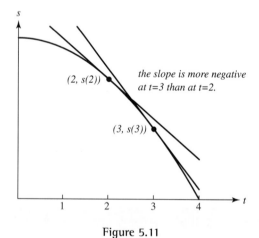

Figure 5.11

---

[7] As an analogy for those uncomfortable with the metamorphosis into a bug, if you were standing in the middle of a huge cornfield in a Midwestern plain, you would feel as if you were standing on a plane—completely oblivious to the roundness of the earth.

Given a function $f(x)$, we can interpret f' in two ways: as the slope function that associates to each value of x the slope of the graph of f at $(x, f(x))$, or as the function whose output is the instantaneous rate of change of f with respect to x.

If $y = f(x)$ we can think of $f'(x)$ both as $\lim_{h \to 0} \frac{f(x+h)-f(x)}{h}$ and as $\lim_{\Delta x \to 0} \frac{\Delta y}{\Delta x}$ where $\Delta x = (x + h) - x = h$ and Δy is defined to be $f(x + h) - f(x)$.

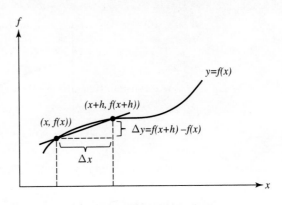

Figure 5.12

◆ **EXAMPLE 5.3**   Consider the graph of the function $h = f(V)$ drawn below. This is a height versus volume graph for the bucket in the first set of exploratory problems.

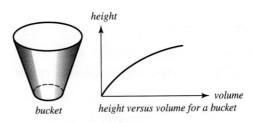

bucket   height versus volume for a bucket

Figure 5.13

Characterize the function $f'(V)$ and explain the implications of the answers in plain English.

   i. Is f' positive, zero, or negative?

   ii. Is f' increasing, or decreasing?

   iii. Graph f'.

SOLUTION   i. $f'(V)$ is positive.

Graphical Reasoning: $f'(V)$ is the slope of the graph and the graph always has positive slope.

Implication: $f'(V) = \lim_{\Delta V \to 0} \frac{\Delta h}{\Delta V} = \lim_{\Delta V \to 0} \frac{\Delta \text{height}}{\Delta \text{volume}}$.

It is the instantaneous rate of change of height with respect to volume.

As the volume increases, the height increases.

ii. $f'(V)$ is decreasing.

Graphical Reasoning: The slope of the graph of f decreases as V increases.

Implication: Because the bucket is widest at the top, as the volume increases the rate of change of the height with respect to volume decreases.

iii.

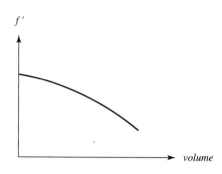

Figure 5.14   ◆

Calculating the Derivative Function

The limit definition of the derivative function is

$$f'(x) = \lim_{h \to 0} \frac{f(x+h) - f(x)}{h}.$$

In the examples below we use this definition to calculate derivative functions.

◆ **EXAMPLE 5.4**   Suppose $f(x) = x^2 + 3x.$

(a) Find $f'(x)$.

(b) Where does the graph of f have a horizontal tangent line?

SOLUTION

$$f'(x) = \lim_{h \to 0} \frac{f(x+h) - f(x)}{h}$$

$$= \lim_{h \to 0} \frac{(x+h)^2 + 3(x+h) - (x^2 + 3x)}{h} \qquad \text{We know } f(x) = x^2 + 3x.$$

$$= \lim_{h \to 0} \frac{x^2 + 2hx + h^2 + 3x + 3h - x^2 - 3x}{h} \qquad \text{Multiply out in order to simplify.}$$

$$= \lim_{h \to 0} \frac{2hx + 3h + h^2}{h} \qquad \text{Simplify.}$$

$$= \lim_{h \to 0} \frac{h(2x + h + 3)}{h} \qquad \text{Again, since } h \neq 0, \tfrac{h}{h} = 1.$$

$$= \lim_{h \to 0} 2x + 3 + h$$

$$= 2x + 3$$

Notice that knowing $f'(x) = 2x + 3$ we now know that $f'(1) = 5$, $f'(2) = 7$ (as shown in Example 5.1), $f'(7) = 17$, etc.; by finding $f'(x)$ we have a formula for determining the slope of the graph of f at *any* x-value.

Where $f'(x) = 0$ the graph of f has a horizontal tangent line.

$$2x + 3 = 0$$
$$2x = -3$$
$$x = -1.5$$

Below we graph f and f'. Notice that the answers make sense in terms of our intuitive ideas about slope.

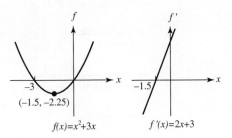

$f(x)=x^2+3x$ $f'(x)=2x+3$

Figure 5.15   ◆

EXERCISE 5.2   Use the limit definition of derivative to find the derivatives of x^2 and $3x$. Conclude that the derivative of $x^2 + 3x$ is the sum of the derivatives of x^2 and $3x$.

Notice that so far the strategy that has been working for computing derivatives is to simplify the difference quotient to the point that we can factor an h from the numerator to cancel with the h in the denominator (assuming that $h \neq 0$) and, once done, the limit becomes apparent. Sometimes the simplification allowing cancellation is a bit more complicated, as illustrated in the next problem. You should compute some simpler derivatives on your own before reading the next example. It is designed for your *second* reading of the chapter.

◆ **EXAMPLE 5.5**   Let $f(x) = \sqrt{x}$. Find $f'(x)$.

SOLUTION

$$f'(x) = \lim_{h \to 0} \frac{f(x+h) - f(x)}{h}$$

$$= \lim_{h \to 0} \frac{\sqrt{x+h} - \sqrt{x}}{h}$$

We have a dilemma here; to make progress we must cancel the h in the denominator. This means getting rid of the square roots. We're working with an expression, so our options are to multiply by 1 (multiply numerator and denominator by the same nonzero quantity) or to add zero. The former will be most productive. Multiplying the expression $\sqrt{A} - \sqrt{B}$ by $(\sqrt{A} + \sqrt{B})$ will eliminate the square roots.[8] $(\sqrt{A} - \sqrt{B})$ is called the *conjugate* of $(\sqrt{A} + \sqrt{B})$.

$$(\sqrt{A} - \sqrt{B})(\sqrt{A} + \sqrt{B}) = A - \sqrt{AB} + \sqrt{AB} - B = A - B$$

---

[8] Multiplying $(\sqrt{A} + \sqrt{B})$ by itself does not eliminate the square roots. $(\sqrt{A} + \sqrt{B})(\sqrt{A} + \sqrt{B}) = A + 2\sqrt{AB} + B$, which leaves the square root in the mixed term.

This algebraic maneuver is worth stashing away in your mind for future reference.

$$f'(x) = \lim_{h \to 0} \frac{\sqrt{x+h} - \sqrt{x}}{h} \quad \text{Multiply numerator and denominator by} \sqrt{x+h} + \sqrt{x}.$$

$$= \lim_{h \to 0} \frac{(\sqrt{x+h} - \sqrt{x})}{h} \cdot \frac{(\sqrt{x+h} + \sqrt{x})}{(\sqrt{x+h} + \sqrt{x})}$$

$$= \lim_{h \to 0} \frac{x+h-x}{h(\sqrt{x+h} + \sqrt{x})}$$

$$= \lim_{h \to 0} \frac{h}{h(\sqrt{x+h} + \sqrt{x})} \quad \text{Since} \quad h \neq 0, \quad \frac{h}{h} = 1.$$

$$= \lim_{h \to 0} \frac{1}{\sqrt{x+h} + \sqrt{x}}$$

$$= \frac{1}{\sqrt{x} + \sqrt{x}}$$

$$= \frac{1}{2\sqrt{x}}$$

So the derivative of $x^{1/2}$ is $\frac{1}{2}x^{-1/2}$.   ◆

Answers to Selected Exercises

Answers to Exercise 5.1

(a) $f'(x) = \lim_{h \to 0} \dfrac{f(x+h) - f(x)}{h}$

$$= \lim_{h \to 0} \frac{m(x+h) + b - (mx + b)}{h}$$

$$= \lim_{h \to 0} \frac{mx + mh + b - mx - b}{h}$$

$$= \lim_{h \to 0} \frac{mh}{h}$$

$$= m$$

(b) $k = mx + k$ where $m = 0$, so the derivative of k is zero.

PROBLEMS FOR SECTION 5.2

1. Use the limit definition of derivative to show that the derivative of the linear function $f(x) = ax + b$ is a. Why is this exactly what you would expect? You have shown that the derivative of a constant is zero. Explain, and explain why this is exactly what you would expect.

2. Use the limit definition of derivative to find the derivative of $f(x) = kx^2$.

3. Let $f(x) = x^2$. Find the point at which the line tangent to $f(x)$ at $x = 2$ intersects the line tangent to $f(x)$ at $x = -1$.

4. Use the limit definition of derivative to find the derivative of $f(x) = x^3$.

5. Using the limit definition of the derivative, find $f'(x)$ if $f(x) = (x - 1)^2$.

6. Let $g(x) = \frac{x}{2x+5}$. Using the limit definition of derivative, find $g'(x)$.

For Problems 7 through 13, find $f'(x)$, $f'(0)$, $f'(2)$, and $f'(-1)$.

7. $f(x) = 3x + 5$

8. $f(x) = \pi x - \sqrt{3}$

9. $f(x) = \frac{2x-5}{3}$

10. $f(x) = \pi(x + 7) - 2$

11. $f(x) = x^2$

12. $f(x) = \frac{1}{x}$

13. $f(x) = \frac{x+\pi}{2}$

14. Suppose $f(x) = x^2$. For what value(s) of x is the instantaneous rate of change of f at x equal to the average rate of change of f on the specified interval? Illustrate your answers with graphs.
    (a) the interval $[0, 3]$ (b) the interval $[1, 4]$

15. Let $g(x) = \frac{1}{x}$. For what value(s) of x is the slope of the tangent line to g equal to the average rate of change of g on the interval indicated? Illustrate your answer to parts (a) and (b) with pictures.
    (a) $[\frac{1}{2}, 2]$ (b) $[1, 4]$ (c) $[c, d]$ $d > c > 0$

16. Let $f(x) = \frac{1}{x^2}$. Let P and Q be points on the graph of f with coordinates $(x, f(x))$ and $(x + \Delta x, f(x + \Delta x))$, respectively.
    (a) Find the slope of the secant line through P and Q. Simplify your answer as much as possible.
    (b) By calculating the appropriate limit, find the slope of the tangent line to $f(x)$ at point P.

17. We showed that the derivative of $\sqrt{x}$ (or $x^{\frac{1}{2}}$) is $\frac{1}{2}\frac{1}{\sqrt{x}}$ (or $\frac{1}{2}x^{-\frac{1}{2}}$). Here we focus on
    $f(x) = \sqrt{x - 1}$.

(a) How is the graph of $\sqrt{x-1}$ related to that of $\sqrt{x}$?

(b) How is the graph of the derivative of $\sqrt{x-1}$ related to that of the derivative of $\sqrt{x}$? Illustrate with a rough sketch.

(c) Given your answer to part (b) explain why $\frac{d}{dx}\sqrt{x}\Big|_{x=4} = \frac{d}{dx}\sqrt{x-1}\Big|_{x=5}$. In other words, explain why the derivative of $\sqrt{x}$ at $x=4$ is equal to the derivative of $\sqrt{x-1}$ evaluated at $x=5$.

(d) Show that $f'(5) = \frac{1}{4}$ using the limit definition of derivative:

$$f'(5) = \lim_{x\to 5}\frac{f(x)-f(5)}{x-5}.$$

(You'll need to rationalize the numerator.)

18. Show that $\frac{d}{dx}\sqrt{x+8} = \frac{1}{2\sqrt{x+8}}$ using the limit definition of derivative. You'll use different versions of the definition in parts (a) and (b). In both cases it will be necessary to rationalize the numerator in order to evaluate the limit.

(a) $f'(x) = \lim_{h\to 0}\frac{f(x+h)-f(x)}{h}$

(b) $f'(x) = \lim_{b\to x}\frac{f(b)-f(x)}{b-x}$

19. Let $f(x) = x^{-\frac{1}{2}}$. Use the limit definition of derivative to show that $f'(x) = -\frac{1}{2}x^{-\frac{3}{2}}$.

5.3   QUALITATIVE INTERPRETATION OF THE DERIVATIVE

Key Notions

In this section and the following, we'll interpret the derivative function f' as the slope function. Assertions will be made that should make sense intuitively. However, we will not prove these assertions and "facts" here; proofs will be delayed until after Chapter 7.

Fact:

Where f is increasing, the slope of its graph is nonnegative; where f is decreasing, its slope is nonpositive.[9]

---

[9] You may wonder why we say that f is increasing implies that the slope is nonnegative rather than just saying that the slope is positive. Consider a function like $f(x) = x^3$. It is increasing everywhere, yet its slope at $x=0$ is zero; locally it looks like a horizontal line around $(0, 0)$.

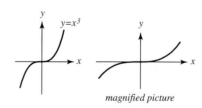

magnified picture

Where f is increasing, $f' \geq 0$; where f is decreasing, $f' \leq 0$.

It follows that

where f' is positive, the graph of f is increasing (from left to right);

where f' is negative, the graph of f is decreasing (from left to right);

where f' is zero, the graph of f locally looks like a horizontal line.

Concavity is determined by whether f' is increasing or decreasing.

Where $f' > 0$ and f' is increasing, the graph of f looks like Figure 5.16(a).

Where $f' > 0$ and f' is decreasing, the graph of f looks like Figure 5.16(b):

Where $f' < 0$ and f' is increasing, the graph of f looks like Figure 5.16(c):

Where $f' < 0$ and f' is decreasing, the graph of f looks like Figure 5.16(d):[10]

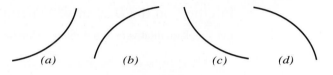

Figure 5.16

Where the slope of f is increasing, we say the graph of f is **concave up**. In other words, where f' is increasing the graph of f is concave up. Figures 5.16(a) and (c) are examples of concave-up graphs.

Figure 5.17

Where the slope of f is decreasing, we say the graph of f is **concave down**. In other words, where f' is decreasing, the graph of f is concave down. Figures 5.16(b) and (d) are examples of concave-down graphs.[11]

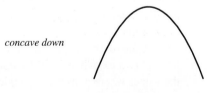

Figure 5.18

---

[10] If $f' < 0$ and f' is decreasing, then f' is becoming increasingly negative as the independent variable increases.

[11] Concave up "holds water"; concave down "spills water."

COMMON ERROR Consider the graph in Figure 5.16(c). A common error is to think that the slope is decreasing, because as x increases the drop is becoming more gentle. This is incorrect.

As the slope changes from, say, -2 to -1, the slope is *increasing* (becoming less negative); similarly, if the temperature goes from -10 degrees to -5 degrees, we say that the temperature is increasing. As the slope changes from -2 to -1, the *steepness* of the line decreases, but the *slope* of the line increases. Should you ever become confused, label the slopes at a few points and put these slopes in order on a number line.

Interpreting the Derivative Function Graphically

Let's look qualitatively at the relationship between a function and its derivative function.

Sketching the Derivative Function Given the Graph of f

The problem of sketching f' when we are given the graph of f is equivalent to the problem of sketching the velocity graph for a trip when we know the graph of position versus time. When sketching the derivative function, begin with the most fundamental questions:

Where is f' zero? undefined? positive? negative?

Note that, like any other function, f' cannot change sign without passing through a point at which it is either zero or undefined.

COMMON ERROR A standard mistake of the novice derivative sketcher is to get mesmerized by an internal conversation about where the slope is increasing and decreasing without first considering the question of whether the slope is positive or negative. Keep your priorities in order! Observations about the concavity of f are fine tuning and are *not* the first order of business.

On the following page are several worked examples. Where f has a horizontal tangent line, f' is zero. Where f is discontinuous or has a sharp corner, f' is undefined. We mark the values where f' is zero or undefined on a number line. At the bottom we have tracked the sign of f'; above the number line we have used arrows to indicate where the graph of f is increasing and decreasing. Where f is increasing, $f' \geq 0$; the region above the x-axis on the graph of f' is shaded to indicate this. Where f is decreasing, $f' \leq 0$; the region below the x-axis on the graph of f' is shaded. The graph of f' must lie in the shaded regions.

◆ **EXAMPLE 5.6**   Given the graph of f, produce a rough sketch of f'.

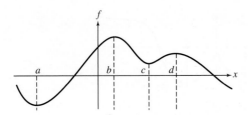

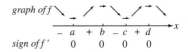

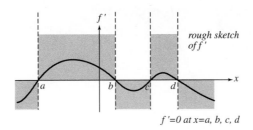

$f'=0$ at $x=a, b, c, d$

◆ **EXAMPLE 5.7**   Given the graph of f, produce a rough sketch of f'.

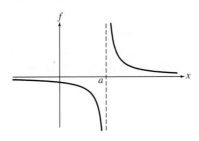

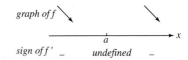

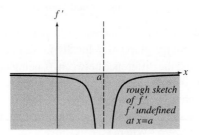

rough sketch
of f'
f' undefined
at $x=a$

◆ **EXAMPLE 5.8**   Given the graph of f, produce a rough graph of f'.

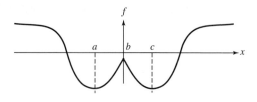

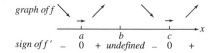

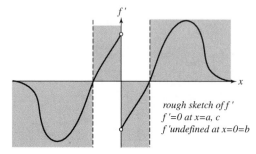

rough sketch of f'
$f'=0$ at $x=a, c$
f' undefined at $x=0=b$

◆

Getting Information About f Given the Graph of f'

The problem of sketching f given the graph of f' is equivalent to the problem of sketching the position versus time graph for a trip given the graph of velocity as a function of time. Unless we know either where the trip began or the location at some particular time, we cannot produce the actual graph of the trip; we can only get the general shape of the trip graph without information as to position at any time.

For instance, if we are told that a cyclist travels 12 miles per hour for 2 hours, we know that he has traveled 24 miles, but we have no idea at all as to where he started or finished his trip. If we are given his position at some particular time, then we can piece together a picture of his position throughout the trip.

Analogously, because f' gives information only about the slope of f, we can simply obtain information about the shape of f, not its vertical position.

To make a rough sketch of the shape of f given f', begin again with the most important features of f':

Where is f' zero? undefined? positive? negative?

Where f' is positive, the graph of f is increasing; where f' is negative, the graph of f is decreasing; where f' is zero, the graph of f locally looks like a horizontal line.

Keep clear in your mind the difference between f' being positive and being increasing. When you look at a graph of f' it is easy to get swept away by the shape of the graph. Hold back! Sometimes the graph of f' gives you more information than you can easily process. We suggest organizing the sign information on a number line. Begin by asking yourself (in a calm voice) "Where is f' zero? Where is f' positive?"

As pointed out in Chapter 2, sign information is a fundamental characteristic of a function; we see now that the sign of f' gives vital information. Begin by constructing a number line highlighting this information. This method is illustrated in the following example.

◆ **EXAMPLE 5.9**   Use the graph of f' to answer the following questions.

    i. Where on the interval $[-3, 10]$ is the value of f the largest?

    ii. Which is larger, $f(2)$ or $f(5)$?

    iii. Where on the interval $[-3, 10]$ is f increasing most rapidly?

    iv. Where on the interval $[-3, 10]$ is the graph of f concave up?

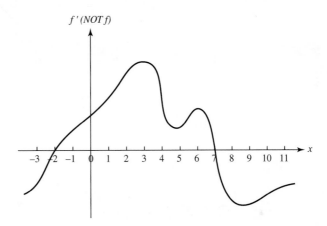

Figure 5.19

SOLUTION   i. Begin by constructing a number line with information about the sign of f' and its implications for the graph of f.

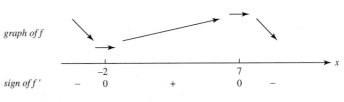

Figure 5.20

From this analysis we see that f is largest at either $x = -3$ or $x = 7$.

Since f' is positive on $[-2, 7]$ the graph of f is increasing on the entire interval. The increase on the interval $(-2, 7)$ is substantially greater than the decrease on $(-3, -2)$ so $f(7) > f(-3)$.

f is largest at $x = 7$.

    ii. $f(5) > f(2)$ since f is increasing on $[-2, 7]$.

    iii. f is increasing most rapidly where f' is greatest. This is at $x = 3$.

iv. f is concave up where f' is increasing. This is on the intervals $[-3, 3]$, $[5, 6]$, and $[9, 10]$. ◆

◆ **EXAMPLE 5.10**   For each of the following, given the graph of f', sketch possible graphs of f. Because f' gives us only information about the slope of f, there are infinitely many choices for f; each is a vertical translate of graphs given.

   The most constructive way to "read" this example is to do so actively. Do the problems on your own and then look at the solutions.

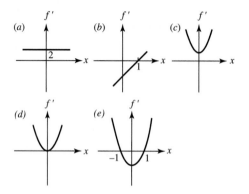

SOLUTION

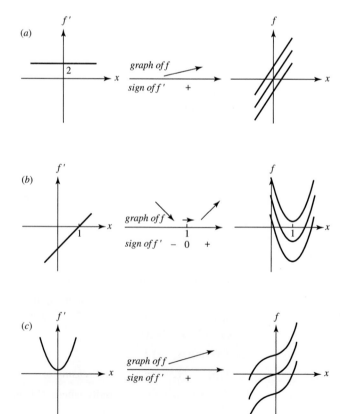

(d)

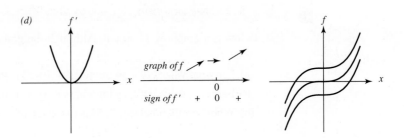

(e)

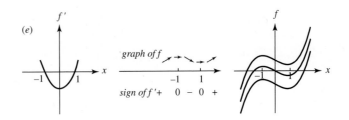

As you can see, vertically shifting the graph of f doesn't change its slope.

EXERCISE 5.3   Given the graph of f, sketch the graph of f'. Include a number line indicating both the sign of f' and the direction of the graph of f.

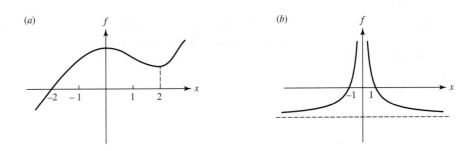

EXERCISE 5.4   Given the graph of f', sketch two possible graphs of f. Make one of these graphs pass through the origin.

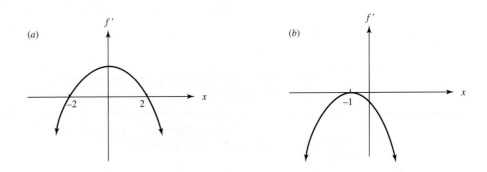

PROBLEMS FOR SECTION 5.3

1. Below is a graph of f', not f. Although the graph is of f', the questions we ask are about f.

    (a) For what values of x in the interval $[0, 5]$ is the graph of f increasing?

    (b) For what values of x in the interval $[0, 5]$ is the graph of f decreasing?

    (c) Where on the interval $[0, 5]$ is the value of $f(x)$ the smallest?

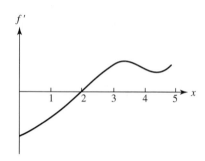

2. Below is a graph of g', not g. Although the graph is of g', the questions asked are about g.

    (a) For what values of x in the interval $[-2, 6]$ is the graph of g increasing?

    (b) For what values of x in the interval $[-2, 6]$ is the graph of g decreasing?

    (c) Where on the interval $[-2, 6]$ is the value of $g(x)$ the smallest?

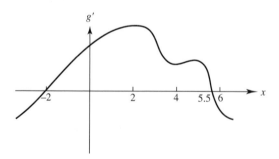

3. On the following page, match each of the eight graphs $(a–h)$ with the correct graph $(A–H)$ of its slope function, i.e., its derivative. When you have completed your work on this problem, compare your answers with those of one of your classmates. If you disagree about an answer, each of you should discuss your reasoning to see if you can come to a consensus on the answers.

(a)	(b)	(c)	A	B	C
(d)	(e)	(f)	D	E	F
(g)	(h)		G	H	

4. The graph of $g(x) = x^2 - 4$ looks just like the graph of $f(x) = x^2$ shifted vertically downward 4 units. Will f' and g' be the same or different? Explain your reasoning.

For Problems 5 through 8, sketch the graph of f'.

5.

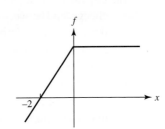

6.

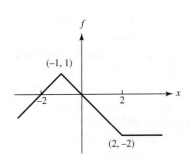

7.

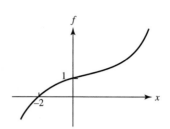

8.

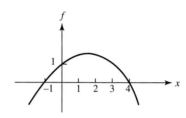

9. How are the graphs of f' and g' related in each of the following situations? Explain your reasoning.

   (a) $g(x) = f(x) + 3$

   (b) $g(x) = f(x + 3)$

   (c) $g(x) = 3f(x)$

10. *Writing Assignment:* You have formed a study group with a few of your friends. One of the people in your study group has been ill for the past $2\frac{1}{2}$ weeks and is concerned about the upcoming examination. She needs to understand the main ideas of the past few weeks. Your essay should be designed to help her.

   (a) Explain the relationship between average rate of change and instantaneous rate of change, and between secant lines and tangent lines. Your classmate is unclear why the definition of derivative involves some limit with h going to zero. She wants to know why you can't just set $h = 0$ to begin with and be done with it. Why does calculus involve this limit business?

   (b) She also has one specific question. She is not clear about how you get the graph of f' from the graph of f. Just before she got sick we were doing that. She believes that if the graph of f' is increasing, then the graph of f is also increasing. Her Course Assistant says this is wrong, but she claims that sometimes she gets the right answer using this reasoning. What's the story?

   How to write this essay: First, think about your friend's position. Particularly in part (a) see if you can understand what is confusing her and how to clarify it for her. Outline your answer. Then write your essay. Use words precisely. Try to avoid pronouns. For example, do not say "it" is increasing; be specific—what is increasing? Use words to say precisely what you mean.

   The real purpose of this essay: These are issues that are important for you to understand. We want you to put together what you have learned in your own words. We also want you to learn to write about mathematics by using words precisely to say exactly what you mean.

11. In many problems you have been either looking at graphs of trips (position versus time) and producing velocity graphs or, equivalently, looking at graphs of $y = f(x)$ and producing graphs of $f'(x)$, the slope function. In this problem, we give you a velocity graph and ask you to produce the position function given the starting point of the trip. We'll denote the starting point of the trip by $s(0)$. For each velocity graph, produce three position graphs, one so that $s(0) = 0$, one with $s(0) = 1$, and a third with $s(0) = -1$. (Disclaimer: Note that these graphs of trips are models; that is, it is physically impossible to change from a velocity of 10 units to a velocity of 0 units without traveling at every speed between 10 and 0. In the real world, these graphs would be continuous. But for our purposes, we will work with simplified models.)

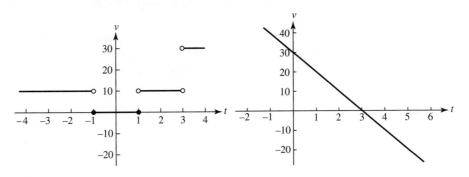

12. Below is the graph of a velocity function. Our benchmark time $(t = 0)$ is, as usual, noon. Sketch a graph of position versus time (use our usual conventions: positive position corresponding to east of our benchmark location, negative position corresponding to west) if

(a) at 8:00 A.M. $s = 0$ (b) at 8:00 A.M. $s = -20$

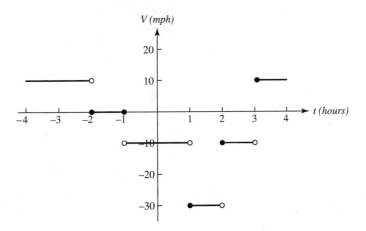

Exploratory Problems for Chapter 5

Running Again

1. Three runners run along the Palouse Path from joining towns in Washington and Idaho. (We discussed these runners in the Exploratory Problems for Chapter 2.)

   Alicia, Bertha, and Catrina run for one hour. The graphs below give position, $s(t)$, as a function of time for each of the runners. The position function gives the distance run from time $t = 0$ to t. Their position graphs are labeled A, B, and C, respectively.

   • Sketch velocity graphs for the three women. Label your graphs clearly.

   • Relate what you've done to the graphs of functions and their derivatives. Can you give labels to the graphs making the connections between function and derivative clear?

   • Identify relationships between the graph of s and the corresponding graph of v. What characteristic of s assures you that v is positive or zero? What characteristic of s assures you that v is increasing?

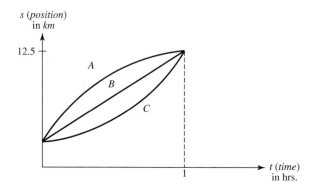

2. In this problem we ask you to conjecture about how alterations to f affect the slope function. Here we suggest you arrive at your conjecture by thinking about the relationship between the new and old graphs. At the end of Chapter 7 we will ask you to check your conjectures using the limit definition of derivative.

   Let $f(x)$ be differentiable on $(-\infty, \infty)$ and let k be a constant.

   (a) How is the slope of $f(x) + k$ at $x = a$ related to the slope of $f(x)$ at $x = a$?

   If $g(x) = f(x) + k$, express $g'(x)$ in terms of f' and k.

   (b) How is the slope of $kf(x)$ at $x = a$ related to the slope of $f(x)$ at $x = a$?

   If $g(x) = kf(x)$, express $g'(x)$ in terms of f' and k.

   (c) If $g(x) = f(x + k)$, express $g'(x)$ in terms of f' and k.

(d) *Challenge:* If $g(x) = f(kx)$, express $g'(x)$ in terms of f' and k.

3. Let $f(x) = 1/x$.

   (a) Sketch $f(x)$. For what values of x is $f(x)$ increasing, and for what values of x is $f(x)$ decreasing?

   (b) Sketch the graph of $f'(x)$.

   (c) Use the limit definition of derivative to find the derivative of $f(x) = 1/x$.

   (d) Are your answers to parts (b) and (c) consistent?

   (e) At which point(s) on the graph of f is the slope of the tangent line equal to $\frac{1}{4}$?

■ 5.4   INTERPRETING THE DERIVATIVE: MEANING AND NOTATION

Getting Perspective

We began this chapter with the problem of calculating the velocity of a falling rock at a single instant. Using average velocity to approximate the instantaneous velocity, we applied the strategy of successive approximations and thus narrowed in on the desired quantity. Using analytic methods and a limiting process, we allowed the second data point to get arbitrarily close to the instant in question and found the instantaneous velocity.

Adopting a graphical perspective provided us with a new insight; the problem of finding the instantaneous rate of change of f at $x = 2$ corresponds to finding the slope of the tangent line to f at $x = 2$. The fundamental challenge is the same, we have one data point but need two. The function itself gives us information about a second data point, Q, which we can slide along the curve to get arbitrarily close to point $P = (a, f(a))$. By taking the limit of the slopes of the secant lines through P and Q as Q approaches P along the curve, we arrive at the slope of the tangent line.

We started with the problem of calculating the instantaneous rate of change at a specific instant ($t = 2$) and found that the same methods could be applied to a generic time, t, to produce a rate function. From a graphical perspective we produced a slope function that can be evaluated at a specific point in the domain of f to get the slope at that point. In Section 5.2 we distinguished between the derivative evaluated at a point and the derivative function.

Meaning and Notation

Because the limit definitions of $f'(c)$ and $f'(x)$ have been given the dignity of boxes, it may be tempting to memorize them. But there are *many* different versions of these definitions; if you try to memorize each new version you may find yourself with an annoying abundance of definitions floating around anchorless in your head. It is simpler to focus on the *meaning* and help yourself along with a picture. Then your knowledge will be portable and robust.

Regardless of the form in which it is written, the derivative is simply the limit of the slope of a secant line, $\frac{\Delta y}{\Delta x}$, as Δx tends toward zero. In other words, the slope of the tangent line is $\lim_{\Delta x \to 0} \frac{\Delta y}{\Delta x}$.

EXERCISE 5.5   Which of the following are equal to $f'(1)$?

You may find it helpful to interpret these quotients as slopes of secant lines, labeling the relevant points on the figure on the next page.

a) $\displaystyle\lim_{\Delta x \to 0} \frac{f(1 + \Delta x) - f(1)}{\Delta x}$

b) $\displaystyle\lim_{x \to 1} \frac{f(x) - f(1)}{x - 1}$

c) $\displaystyle\lim_{h \to 0} \frac{f(1 + h) - f(1)}{h}$

d) $\displaystyle\lim_{q \to 1} \frac{f(1) - f(q)}{1 - q}$

e) $\displaystyle\lim_{\Delta x \to 0} \frac{f(\Delta x) - f(1)}{\Delta x}$

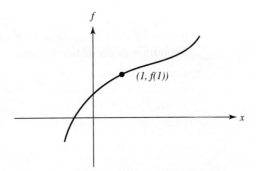

Answers are given at the end of this section.

Notation

Thinking of f' as $\lim_{\Delta x \to 0} \frac{\Delta y}{\Delta x}$ led Leibniz to introduce the notation $\frac{dy}{dx}$ or, equivalently, $\frac{df}{dx}$, for the derivative. These latter two notations are helpful in part because they are suggestive of a rate of change. The notation reminds us of the limit we're computing. Keep in mind that $\frac{dy}{dx}$ is not a fraction; it is simply another way of denoting f'. We read $\frac{dy}{dx}$ aloud as "*dy dx*." If we want to write $f'(2)$ in Leibniz notation, we write $\left.\frac{df}{dx}\right|_{x=2}$ and read this as "*df dx* evaluated at $x = 2$."

If $y = f(x)$, then the derivative function can be denoted in any of the following ways:

$$f', \qquad f'(x), \qquad y', \qquad \frac{dy}{dx}, \qquad \frac{df}{dx}.$$

The derivative evaluated at $x = 3$ can be denoted by:

$$f'(3), \qquad y'(3), \qquad \left.\frac{dy}{dx}\right|_{x=3}, \qquad \left.\frac{df}{dx}\right|_{x=3}.$$

We will sometimes find it useful to use the notation $\frac{d}{dx}$ to indicate that a derivative is to be taken. $\frac{d}{dx}$ is called "operator notation" for the derivative, since it operates on some function and has no independent meaning. For instance

$$\frac{d}{dx}(x^2 + 3x)$$

means "the derivative of $x^2 + 3x$."

Interpreting the Derivative as an Instantaneous Rate of Change

Leibniz's notation, $\frac{df}{dx}$, is very useful when interpreting the derivative as an instantaneous rate of change. In Example 5.11 we interpret the statement $f'(2) = 3$ in different contexts.

◆ **EXAMPLE 5.11**

(a) *Bottle Calibration:* Suppose $H(v)$ gives height (measured in centimeters) as a function of volume v (measured in liters). Then $H'(v)$, also written $\frac{dH}{dv}$, is the function that gives the instantaneous rate of change of height with respect to volume for any volume in its domain.

$$\frac{dH}{dv} = \lim_{\Delta v \to 0} \frac{\Delta H}{\Delta v}; \text{ the units for } \frac{dH}{dv} \text{ are cm/L.}$$

$H'(2) = 3$ tells us that when the volume in the bottle is 2 L, the instantaneous rate of change of height with respect to volume is 3 cm/L. In plain English, at the instant when there are 2 liters of liquid in the bottle, the height of the liquid is increasing at a rate of 3 centimeters per additional liter.

(b) *Population Growth:* Suppose $P(t)$ gives population (measured in thousands of people) as a function of time t (time measured in years, where $t = 0$ corresponds to January 1, 1950). Then $P'(t)$, also written $\frac{dP}{dt}$, is the function that gives the instantaneous rate of change of population with respect to time.

$$\frac{dP}{dt} = \lim_{\Delta t \to 0} \frac{\Delta P}{\Delta t}; \text{ the units for } \frac{dP}{dt} \text{ are thousands of people/year.}$$

$P'(2) = 3$ tells us that when $t = 2$, the instantaneous rate of change of population with respect to time is 3000 people/year. In plain English, on January 1, 1952, the population is increasing at a rate of 3000 people per year.

(c) *Position versus Time:* Suppose $s(t)$ gives position (measured in miles) as a function of time t (measured in hours). Then $s'(t)$, also written $\frac{ds}{dt}$, is the function that gives the instantaneous rate of change of position with respect to time.

$$\frac{ds}{dt} = \lim_{\Delta t \to 0} \frac{\Delta s}{\Delta t}; \text{ the units for } \frac{ds}{dt} \text{ are miles/hour.}$$

$s'(2) = 3$ tells us that at time $t = 2$, the instantaneous rate of change of position with respect to time is 3 miles/hour. In plain English, the velocity at time $t = 2$ is 3 mph.

(d) *Velocity versus Time:* Suppose $v(t)$ gives velocity (measured in miles per hour) as a function of time t (measured in hours). Then $v'(t)$, also written $\frac{dv}{dt}$, is the function that gives the instantaneous rate of change of velocity with respect to time.

$$\frac{dv}{dt} = \lim_{\Delta t \to 0} \frac{\Delta v}{\Delta t}; \text{ the units for } \frac{dv}{dt} \text{ are mph/hour.}$$

$v'(2) = 3$ tells us that at time $t = 2$, the instantaneous rate of change of velocity with respect to time is 3 mph/hour. In plain English, the acceleration at time $t = 2$ is 3 mph/hr.

(e) *Production Cost versus Amount Produced:* Suppose $C(x)$ gives cost (measured in dollars) as a function of x, the number of pounds of material produced (measured in pounds). Then $C'(x)$, also written $\frac{dC}{dx}$, is the function that gives the instantaneous rate of change of cost with respect to number of pounds produced.

$$\frac{dC}{dx} = \lim_{\Delta x \to 0} \frac{\Delta C}{\Delta x}; \text{ the units for } \frac{dC}{dx} \text{ are \$/lb.}$$

$C'(2) = 3$ tells us that when 2 pounds are being produced, the cost is increasing at a rate of 3 \$/lb. When 2 pounds are being produced, the *additional* cost for producing an *additional* pound is *approximately* \$3.

Economists refer to $C'(x)$ as the marginal cost. The term "marginal cost" is used by economists to mean the additional cost for producing an additional unit *or* to be $C'(x)$.[12] ◆

---

[12] For Δx small, $\frac{\Delta C}{\Delta x}$ and $\frac{dC}{dx}$ are approximately equal, but they are not exactly equal. Economists often tend not to bother with the distinction; on the other hand, mathematicians do distinguish between the two.

◆ **EXAMPLE 5.12**   *Burning Calories While Bicycling:* Suppose $C(s)$ gives the number of calories used per mile of bicycle riding (measured in calories/mile) as a function of speed s (measured in miles/hour).[13] Then $C'(s)$, also written $\frac{dC}{ds}$, is the function that gives the instantaneous rate of change of calories/mile with respect to speed.

$$\frac{dC}{ds} = \lim_{\Delta s \to 0} \frac{\Delta C}{\Delta s} \text{ so the units for } \frac{dC}{ds} \text{ are } \frac{\text{calories/mi}}{\text{mi/hr}}.$$

$C'(12) = 7$ tells us that when the speed of the cyclist is 12 mph, the number of calories used per mile is increasing at a rate of 7 calories/mile per mph. Practically speaking, this means that if you currently ride at a pace of 12 mph, increasing your speed by 1 mph will result in your burning approximately 7 *more* calories for every mile you ride.[14] Notice that $C'(12) = 7$ does not give us any information about the calories being burned by riding 1 mile at 12 mph. $C(12)$ would tell us that.   ◆

EXERCISE 5.6   Suppose we know not only that $C'(12) = 7$ but also that $C(12) = 22$. Then we can estimate that $C(13) \approx 29$. Explain the reasoning behind this.

A Dash of History.   During the decade from 1665 to 1675, about a century before the American War of Independence, both Isaac Newton in England and Gottfried Leibniz in Continental Europe developed the ideas of calculus.

Newton began his work during years of turmoil. The years 1665–1666 were the years of the Great Plague, which wiped out about one quarter of the population of London. The Great Fire of London erupted in 1666, destroying almost half of London.[15] Cloistered in his small hometown, Newton developed calculus in order to understand physical phenomena. The language of fluxions he used to explain his ideas reflected his scientific perspective; his terminology is no longer in common usage. Newton used the notation $\dot{y}$ to denote a derivative. Although we will not adopt this notation, it is still used by some physicists and engineers.

Around the same time that Newton did his work Leibniz was also developing calculus. The language and notation he used in his work differed from Newton's; part of his genius was his introduction of very useful notation, notation still very much in use.

Leibniz's notation for the derivative, $\frac{dy}{dx}$, gives the derivative the appearance of a fraction. It is not a fraction; rather, it is the limit of the ratio $\frac{\Delta y}{\Delta x}$ as $\Delta x \to 0$. As a mental model, however, it can be useful to think of dx as an infinitesimally small change in x and dy as the corresponding change in y, where the d reminds us of delta. This notion is not, as stated, entirely accurate (what can be meant by an infinitesimally small change in x?) but it was essentially the model used by Leibniz; part of the genius of his notation is that the mental model generally does not lead us astray.

Well over a century after Newton and Leibniz did their work the word "derivative" was introduced by Joseph Lagrange (1736–1813). It was Lagrange who introduced the notation f' emphasizing that the derivative is a function.

---

[13] For our model we hold all other factors (such as the bicyclist, weather conditions, terrain, and the bicycle itself) constant.

[14] We say approximately because we are interpreting the rate of change over an *interval*, albeit a small one, and the function is probably not perfectly linear.

[15] These facts are from David Burton's *The History of Mathematics—An Introduction*, The McGraw-Hill Companies, Inc., 1997; p.349.

Answers to Selected Exercises

ANSWERS TO EXERCISE 5.5 a, b, c, and d are equal to $f'(1)$.

PROBLEMS FOR SECTION 5.4

1. Which of the following are equal to $g'(3)$? A sketch with labeled points will be useful.

(a) $\lim\limits_{h \to 0} \dfrac{g(3+h) - g(3)}{h}$

(b) $\lim\limits_{x \to 0} \dfrac{g(x) - g(3)}{x - 3}$

(c) $\lim\limits_{x \to 3} \dfrac{g(x) - g(3)}{x - 3}$

(d) $\lim\limits_{s \to 3} \dfrac{g(3) - g(s)}{3 - s}$

(e) $\lim\limits_{\Delta x \to 3} \dfrac{g(3 + \Delta x) - g(3)}{\Delta x}$

(f) $\lim\limits_{\Delta x \to 0} \dfrac{g(3 + \Delta x) - g(3)}{\Delta x}$

(g) $\lim\limits_{\Delta x \to 0} \dfrac{g(3) - g(3 + \Delta x)}{-\Delta x}$

2. An orange is growing on a tree. Assume that the orange is always spherical, and that it has not yet reached its mature size. Its current radius is r cm.

   (a) If the radius increases by 0.5 cm, what is the corresponding increase in volume? What is $\frac{\Delta V}{\Delta r}$?

   (b) If the radius of the orange increases by Δr, what is the corresponding increase in volume? What is $\frac{\Delta V}{\Delta r}$? (Please simplify your answer.)

   (c) Show that $\lim_{\Delta r \to 0} \frac{\Delta V}{\Delta r} = 4\pi r^2$. Conclude that for Δr very small $\Delta V \approx (4\pi r^2)\Delta r$.

   (d) The surface area of a sphere is $4\pi r^2$. Explain, in terms of an orange, why the approximation $\Delta V \approx (4\pi r^2)\Delta r$ make sense.

3. Let $F(t)$ be the number of fish in a pond at time t, where t is given in years. We'll denote by C the carrying capacity of the pond; C is a constant that tells us how many fish the pond can support. Suppose at time $t = 0$ the fish population is small. At first the fish population will grow at an increasing rate, but eventually the fish compete for limited resources and the number of fish levels out at the carrying capacity of the pond. A graph of $F(t)$ is drawn below.

   (a) Sketch the graph of $F'(t)$, the slope function, versus time.

   (b) Interpret the slope of F as a rate of change.

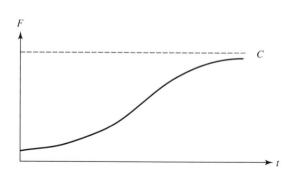

4. The curve below is an indifference curve, which shows combinations of food and clothing giving equal satisfaction, among which the household is indifferent. The slope of the tangent line T is called the marginal rate of substitution at the point b.

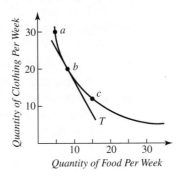

(a) Explain in terms of food and clothing what it would mean to know that the slope of the line T was -2.

(b) The slope of the tangent line at a is a negative number of larger magnitude than the slope of the tangent line at b. What does this mean in terms of food and clothing?

5. This problem deals with the effect of altitude on how far a batted ball will travel. The drag resistance on the ball is proportional to the density of the air, i.e., the barometric pressure if the temperature is held constant. Let us take as an example a 400-foot home run in Yankee Stadium, which is approximately at sea level. On average, an increase in altitude of 275 feet would increase the length of this drive by 2 feet. (Adair, Robert K. *The Physics of Baseball*. New York: Harper & Row, 1990.)

   Let $B(a)$ be the distance this ball would travel as a function of the altitude of the ballpark in which it is hit. Assume the relationship between altitude and distance is linear.

(a) What is the meaning of $\frac{dB}{da}$? What are its units?

(b) What is the numerical value of $\frac{dB}{da}$?

(c) Write an equation for $B(a)$.

(d) Prior to major league baseball's 1993 expansion into Denver, Atlanta, which has an altitude of 1050 feet, was the highest city in the majors. How far would this 400-foot Yankee Stadium drive travel in Atlanta?

(e) How far would it travel in Denver (altitude 5280 feet)?

6. Suppose that $A(p)$ gives the number of pounds of apples sold as a function of the price (in dollars) per pound.

(a) What are the units of $\frac{dA}{dp}$?

(b) Do you expect $\frac{dA}{dp}$ to be positive? Why or why not?

(c) Interpret the statement $A'(0.88) = -5$.

7. Between 1940 and 1995 the size of the average farm in America increased from 174 acres to 469 acres. (Facts from the *World Almanac and Book of Facts 1997*.) Suppose

that $A(t)$ gives the average number of acres of an American farm t years after 1940. $A(t)$ is an increasing function.

(a) What are the units of $\frac{dA}{dt}$?

(b) Average farm size increased much more dramatically in the 50s than in the 80s. Which is larger, $A(12)$ or $A(43)$?
Which do you think is larger, $A'(12)$ or $A'(43)$?

8. A baked apple is taken out of the oven and put into the refrigerator. The refrigerator is kept at a constant temperature. Newton's Law of Cooling says that the difference between the temperature of the apple and the temperature of the refrigerator decreases at a rate proportional to itself. That is, the apple cools down most rapidly at the outset of its stay in the refrigerator, and cools increasingly slowly as time goes by. You have the following pieces of information: At the moment the apple is put in the refrigerator its temperature is 110 degrees and is dropping at a rate of 4 degrees per minute. Twenty minutes later the temperature of the apple is 70 degrees.

(a) Let T be the temperature of the apple at time t, where t is measured in minutes and $t = 0$ is when the apple is put in the refrigerator. Express the three bits of information provided above in functional notation. Sketch a graph of T versus t.

(b) Using the same set of axes as you did in part (a), draw the cooling curve the baked apple would have if it were cooling linearly, with the initial temperature of 110 degrees and initial rate of cooling of 4 degrees per minute. What would the temperature of the apple be after 15 minutes using this linear model? In reality, is the temperature more or less?

(c) Since the apple's temperature dropped from 110 degrees to 70 degrees in twenty minutes, the average rate of change of temperature over the first twenty minutes is $\frac{-40 \text{ degrees}}{20 \text{ minutes}}$ or $-2 \frac{\text{degrees}}{\text{minute}}$. Using the same set of axes as you did in parts (a) and (b), draw the cooling curve the baked apple would have if it were cooling linearly, with the initial temperature of 110 degrees and rate of cooling of 2 degrees per minute. What would the temperature of the apple be after 15 minutes using this linear model? In reality, is the temperature more or less?

9. A hot-air balloonist is taking a balloon trip up a river valley. The trip begins at the mouth of the river. The balloon's altitude varies throughout the trip.
   Suppose that $A(t)$, is the function that gives the balloon's height (in feet) above the ground at time t, where t is the time from the start of the trip measured in hours.

(a) Suppose that at time $t = 4$ hours $A'(4)$ is 70. Interpret what $A'(4) = 70$ tells us in words.

(b) Let f be the function that takes as input x, where x is the balloon's horizontal distance from the mouth of the river (x measured in feet) and gives as output the time it has taken the balloon to make it from the mouth of the river to this point. In other words, if $f(1000) = 4$ then the balloon has taken 4 hours to travel 1000 feet up the river bank.

    i. Let $h(x) = A(f(x))$, where f and A are the functions given above. Describe in words the input and output of the function h.

    ii. Interpret the statement $h(700) = 100$ in words.

    iii. Interpret the statement $h'(700) = 60$ in words.

10. Suppose that $C(s)$ gives the number of calories that an average adult burns by walking at a steady speed of s miles per hour for one hour.

    (a) What are the units of $\frac{dC}{ds}$?

    (b) Do you expect $\frac{dC}{ds}$ to be positive? Why or why not?

    (c) Interpret the statement $C'(3) = 25$.

    Hint: If you are having difficulties with this problem, consider sketching a graph. What are the labels on the axes? (That is, what are the independent and dependent variables?) Thinking about these variables, what should the graph look like? How do your assumptions about the graph relate to the questions posed above?

C H A P T E R

6

The Quadratics: A Profile of a Prominent Family of Functions

6.1 A PROFILE OF QUADRATICS FROM A CALCULUS PERSPECTIVE

Face to Face with Quadratics

Quadratic functions arise naturally in a large variety of situations. For instance, if you are tossing a dart, firing a cannon, throwing a football, or contemplating bungee jumping you have to concern yourself with the physics of falling bodies, and the laws of physics are such that the height of a falling object can be modeled with a quadratic function of time. An economist looking at the relationship between revenue and price from the perspective of a monopolist can end up with a quadratic model. In our introductory example our goal is to construct a garden with the largest possible area, subject to certain constraints. This area can be expressed as the product of two linear functions, resulting in a quadratic. In many situations, whether planning a garden or running a business, we make decisions based on optimizing (maximizing or minimizing) some quantity and these optimization problems sometimes bring us face to face with quadratic functions. The quadratics form a family whose members have predictable behavior and characteristics. In Example 6.1 we investigate the behavior of one particular quadratic, but quadratics are such an elemental part of the mathematical landscape that we will find it useful to become very well acquainted with the whole family of quadratic functions so we know what kind of behavior to expect from its members.

◆ **EXAMPLE 6.1**   A gardener has 30 feet of fencing that she will use to fence off a rectangular tomato plot along the south wall of her house. What should the dimensions of the garden be in order for her to have the most area to grow her tomatoes?

SOLUTION   Let's begin by drawing a picture of the tomato plot. Because the plot is alongside the house, she needs fencing for three sides of the plot only. She wants to maximize the area of the plot. Our strategy is to express the area, A, as a function of one variable.

We know that area = (width)(length). Suppose we call the width of the garden w feet and the length ℓ. Can we express the area just in terms of w? If the width is w, why can't the length be anything the gardener likes? Because she has only 30 feet of fencing. The amount of fencing constrains her options. Once she chooses w, ℓ is determined and therefore the area is determined.

$$(width + width + length) = 30$$
$$w + w + \ell = 30$$
$$2w + \ell = 30$$
$$\ell = 30 - 2w$$

So $A(w) = (width)(length) = w(30 - 2w)$.

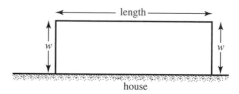

Figure 6.1

What value of w will make A as big as possible? We can investigate this by looking at the graph of $A(w)$. If we were to do this mindlessly using a graphing calculator with the default range and domain of $[-10, 10]$, we would get a picture that looks like Figure 6.2(a) and provides little information. To get useful information from the graph we need to pick a reasonable domain and range.

What values of w are reasonable? Certainly w, the width, can't be negative, and $2w$, twice the width, can't be greater than the 30 feet of fencing. Therefore, $0 < w < 15$. Simply adjusting the domain to $[0, 15]$ doesn't help much (see Figure 6.2b), but by also adjusting the range we can make some headway. Figure 6.2(c) shows the graph with the range of $[0, 150]$. Use of the trace key (or some more sophisticated calculator feature) indicates that the high point has coordinates of approximately $(7.5, 112.5)$; when $w \approx 7.5$ feet, then $A \approx 112.5$ feet. The gardener should use about 7.5 feet of fencing for the width of the plot and the remainder, $(30 - 2(7.5))$ feet = 15 feet, for the length. This will give a plot of 112.5 square feet for her tomatoes.

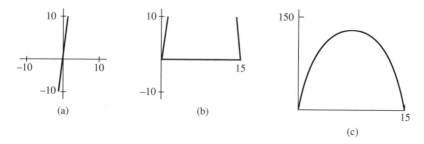

Figure 6.2   ◆

Notice that in the example just completed the area of the garden could be expressed as the product of two linear functions, as $w(30 - 2w)$ or $30w - 2w^2$. This expression is a quadratic, due to the w^2 term, and the graph that we looked at is a parabola. Similarly, there are other quantities that can be expressed as the product of two expressions; for example, revenue $=$ (price per item) $\cdot$ (number of items sold). If the two factors are linear functions of the same variable (or are modeled by linear functions), then the product will be a quadratic.

Definitions

A function $f(x)$ is **quadratic** if it can be expressed in the form $f(x) = ax^2 + bx + c$, where a, b, and c are constants and $a \neq 0$. The graph of a quadratic function is called a **parabola**.

A Calculus Perspective

There are several approaches we can take to studying quadratic functions. Let's begin by examining their rates of change. We compute the derivative of a quadratic using the limit definition of derivative.

$$
\begin{aligned}
f'(x) &= \lim_{h \to 0} \frac{f(x + h) - f(x)}{h} \\
&= \lim_{h \to 0} \frac{a(x + h)^2 + b(x + h) + c - (ax^2 + bx + c)}{h} \quad &&\text{Using the function} \\
& &&f(x) = ax^2 + bx + c \\
&= \lim_{h \to 0} \frac{a(x^2 + 2hx + h^2) + bx + bh + c - ax^2 - bx - c}{h} \quad &&\text{Expanding in order to simplify} \\
&= \lim_{h \to 0} \frac{ax^2 + 2ahx + ah^2 + bx + bh + c - ax^2 - bx - c}{h} \\
&= \lim_{h \to 0} \frac{2ahx + ah^2 + bh}{h} \\
&= \lim_{h \to 0} \frac{h(2ax + ah + b)}{h} \quad &&\text{Factoring out an } h \\
&= \lim_{h \to 0} 2ax + ah + b \quad &&\text{Because } h \neq 0 \text{ we can say } h/h = 1. \\
&= 2ax + b
\end{aligned}
$$

If $f(x) = ax^2 + bx + c$, then $f'(x) = 2ax + b$.

Observations

i. The derivative of a quadratic function is a linear function.

ii. This formula for f' is valid regardless of the values of a, b, and c. We can, for example, set $b = 0$ and $c = 0$ and conclude that if $f(x) = ax^2$ then $f'(x) = 2ax$. Similarly, we know that the derivative of bx is b, and the derivative of the constant c is 0.

iii. The value of c does not affect the value of the derivative. This makes geometric sense; c only shifts the graph vertically and therefore has no effect on the slope at any given x-value.

◆ **EXAMPLE 6.2**   Find the derivatives of the following functions.

$$\text{i. } f(x) = -3x^2 - 6x \qquad \text{ii. } g(x) = \frac{-x^2}{\sqrt{7}} + \frac{\pi^2}{3} \qquad \text{iii. } h(x) = \frac{(2x-1)(x-4)}{2}$$

ANSWER

i. $f'(x) = -6x - 6$

ii. $g'(x) = \frac{-2}{\sqrt{7}}x$

iii. $h'(x) = 2x - 4.5$ (Multiply out, then differentiate.) ◆

We have shown that if f is a quadratic function, then f' is linear. Let's use this information about f' to see what we can say about the graph of f itself. We'll refer back to the derivatives of the functions in Example 6.2.

◆ **EXAMPLE 6.3**   Use the derivatives f', g', and h' found in Example 6.2 to get information about the graphs of f, g, and h. In particular, determine the x-coordinate of any turning point.
The sign of the derivative function tells us where the original functions are increasing and where they are decreasing.

i. $f'(x) = -6x - 6$ ii. $g'(x) = \frac{-2}{\sqrt{7}}x$ iii. $h'(x) = 2x - 4.5$

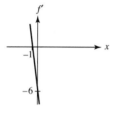

f is increasing for $x < -1$
f is decreasing for $x > -1$

graph of f
graph of f' + -1 $-$

parabolas with derivative f'

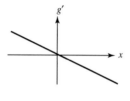

g is decreasing for $x > 0$
g is increasing for $x < 0$

graph of g
graph of g' + 0 $-$

parabolas with derivative g'

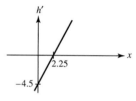

h is increasing for $x > 9/4$
h is decreasing for $x < 9/4$

graph of h
graph of h' $-$ 2.25 $+$

parabolas with derivative h'

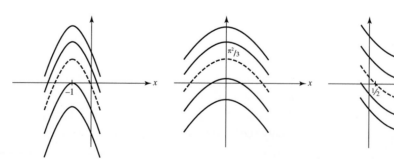

Figure 6.3

Notice that we have drawn *families* of parabolas. This is because shifting a graph vertically does not change its slope at a given point. The derivative determines a function only up to a constant. We have indicated by a dotted line the parabolas corresponding to those given in Example 6.2. Although the derivative gives us information about the shape of the parabola, it doesn't give us any information about the vertical positioning of the parabola. The derivative will not help us pick out any one member of the family of parabolas drawn.   ◆

The Graph of a Quadratic Function

The derivative of the quadratic function $f(x) = ax^2 + bx + c$ is the linear function $f'(x) = 2ax + b$. The linear function is not horizontal because $a \neq 0$.[1] Any nonhorizontal linear function has exactly one x-intercept; at this intercept the f' line cuts the x-axis and f' changes sign. This corresponds to a **turning point** of f; the quadratic function either changes from increasing to decreasing or *vice versa*.

The turning point of a parabola is called the **vertex** of the parabola. The derivative gives us easy access to the shape of the parabola and the x-coordinate of its vertex; we could have used derivatives to solve the problem in Example 6.1 efficiently, and exactly.

Question: How can we find the x-coordinate of the vertex of the parabola?

Answer:

The x-coordinate of the vertex is the zero of the derivative function, so $x = \frac{-b}{2a}$.
There is *no* need to memorize this. In practice, simply differentiate f and find the zero of f'.

Question: How can we determine whether the parabola opens upward or downward?

Answer:

If the graph of f' has a positive slope, ╱ , the parabola looks like ⌣
If the graph of f' has a negative slope, ╲ , the parabola looks like ⌢

If $f(x) = ax^2 + bx + c$, then the slope of the graph of f' is given by "a." Therefore we know that

if $a > 0$ then the parabola opens upward;

if $a < 0$ then the parabola opens downward.

Definition

The slope of f' is written f'' (read "f double prime") and is called the **second derivative of** f. In Leibniz' notation f'' is $\frac{d}{dx}\left(\frac{df}{dx}\right)$ or $\frac{d^2 f}{dx^2}$.

Note that

if $f'' > 0$, then the derivative is increasing; the graph of f is concave up;

if $f'' < 0$, then the derivative is decreasing; the graph of f is concave down.

This observation holds in general; there is nothing particular to quadratics here.

---

[1] If $a = 0$ then the function is no longer quadratic; it is linear. In this case its derivative *is* a horizontal line, because the rate of change of a linear function is constant.

PROBLEMS FOR SECTION 6.1

1. On the left below are graphs of the quadratic functions f, g, h, and j. Match f', g', h', and j' with the lines drawn on the right.

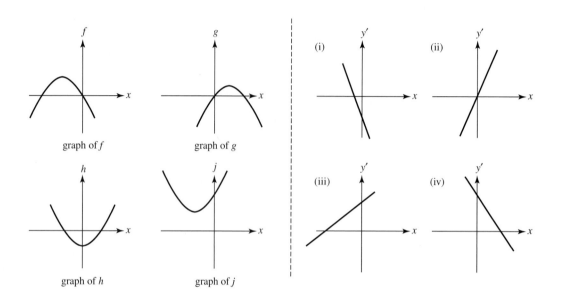

graph of f graph of g

graph of h graph of j

For each of the quadratics in Problems 2 through 4, identify the x- and y-coordinates of the vertex and determine whether the vertex is the highest point on the curve or the lowest point on the curve.

2. $y = 2x^2 - 3x + \pi$

3. $y = -\sqrt{3}x^2 + \sqrt{27}x + 15$

4. $y = -\frac{2x^2+7}{4} + \frac{3x-1}{3}$

For Problems 5 through 8, find a quadratic or linear function $f(x)$ whose derivative is the line specified and whose graph passes through:

(a) the origin,

(b) the point $(0, 2)$.

5. (a) $f'(x) = 3$ (b) $f'(x) = \pi$

6. (a) $f'(x) = 2x$ (b) $f'(x) = -2x + 8$

7. (a) $f'(x) = 6x - 2$ (b) $f'(x) = mx + b, m \neq 0$

8. (a) $f'(x) = \pi x$ (b) $f'(x) = -\frac{x}{3}$

6.2   QUADRATICS FROM A NONCALCULUS PERSPECTIVE

In this section we'll look at quadratics without using calculus. This is a different perspective on the same material to help you get a better intuitive feel for quadratics. In addition, we will take the opportunity to discuss the zeros of a quadratic, because the derivative gives us no information about the roots.

■   The graph of a quadratic function is a parabola opening up or down; conversely, a parabola opening up or down is the graph of a quadratic function. $y = x^2$ is an example of a parabola opening upward; $y = -x^2$ is an example of a parabola opening downward.

$x = y^2$ is an example of a parabola opening to the right, and $x = -y^2$ is an example of a parabola opening to the left.

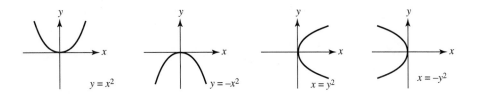

For the most part we are interested in the graphs of *functions of x*; therefore, when studying the latter two graphs, we will usually be restricting our interest to $y = \sqrt{x}$, $y = -\sqrt{x}$, $y = \sqrt{-x}$, and $y = -\sqrt{-x}$.

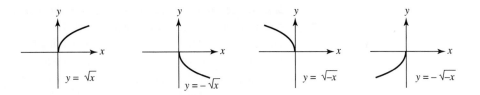

■   A parabola has one turning point, called the vertex of the parabola.

Figure 6.4

The graph of a quadratic function is symmetric about its longitudinal axis, the vertical line through its vertex. Verification is left as an exercise (see the Problems for Section 6.2). Look at the simplest quadratic function, $f(x) = x^2$, as an illustration of this symmetry.

■   We find the x-intercepts of a parabola by solving $ax^2 + bx + c = 0$. The solutions are given by the quadratic formula:[2]

---

[2] The quadratic formula is discussed in Reference Section A: Algebra, Part IIIB.

$$x = \frac{-b}{2a} \pm \frac{\sqrt{b^2 - 4ac}}{2a}.$$

The number of roots depends on the sign of $b^2 - 4ac$; there are two distinct roots if this expression is positive, one root if it is zero, and no real roots if it is negative.

If $ax^2 + bx + c = 0$ has no solutions, then the parabola $y = ax^2 + bx + c$ has no x-intercepts.

If $ax^2 + bx + c = 0$ has one solution, then the parabola $y = ax^2 + bx + c$ has one x-intercept.[3]

If $ax^2 + bx + c = 0$ has two solutions, then the parabola $y = ax^2 + bx + c$ has two x-intercepts.

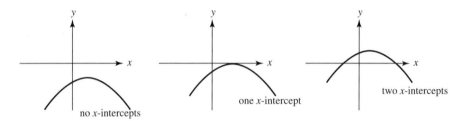

Figure 6.5

By thinking about the symmetry of the parabola we can see the following.

If there are two distinct x-intercepts, then the x-coordinate of the vertex lies midway between them. Compute the average of the two distinct roots of the corresponding quadratic equation by adding them and dividing by 2; you'll get $x = -b/2a$. Therefore the vertex of the parabola must have an x-coordinate of $-b/2a$.

Note that the number midway between $Q + R$ and $Q - R$ is Q;

$$\frac{(Q + R) + (Q - R)}{2} = \frac{2Q}{2} = Q.$$

If there is only one x-intercept, then that intercept, $x = -b/2a$, is the x-coordinate of the vertex. (Why? Think graphically.)

If there are no x-intercepts, then we use the fact that the x-coordinate of the vertex is midway between the points of intersection of the parabola and any horizontal line $y = d$ that cuts it in two points. This means the x-coordinate of the vertex is midway between the solutions to $ax^2 + bx + c = d$ for some d, and the assertion about the vertex being at $x = -b/2a$ follows. (Basically, the "c" in the quadratic formula is adjusted so that there are two roots.)

■   The parabola given by $f(x) = ax^2 + bx + c$ opens upward if $a > 0$ and downward if $a < 0$.

Reasoning: For x large enough in magnitude (think about $x \to \infty$ and $x \to -\infty$), the ax^2 term dominates the other two terms; by this we mean that it overpowers the

---

[3] This is a real root with multiplicity two. If r is a root of multiplicity two, then the polynomial has the factor $(x - r)^2$ as opposed to simply $(x - r)$.

other terms and determines the sign and behavior of f. x^2 is always positive, allowing us to reason as follows.

If $a > 0$, then ax^2 is positive. Therefore $f(x)$ is positive for $|x|$ large enough, and the parabola opens upward.

If $a < 0$, then ax^2 is negative. Therefore $f(x)$ is negative for $|x|$ large enough, and the parabola opens downward.

■ Three points determine a quadratic. If we know three points satisfying the equation $y = ax^2 + bx + c$, we can solve a system of three simultaneous linear equations[4] for the three unknown constants, a, b, and c.[5] If one of the points we know happens to be the vertex of the parabola (and we are aware of that), then the parabola is determined by just the vertex and one other point.[6]

---

[4] See Reference Section B on Simultaneous Equations.

[5] We know that two points determine a line. Essentially, the number of points needed to determine a particular type of equation boils down to the number of constants one can "play around with" in the equation's general form. For example, for a linear equation $y = mx + b$, we can adjust the m and the b; two points or independent pieces of information are needed. A quadratic equation $y = ax^2 + bx + c$ has three constants that we can modify, so we need three points to determine its equation.

[6] Why? Using the symmetry of the parabola over the vertical line through its vertex, from the second point we can find a third point.

Exploratory Problems for Chapter 6

Tossing Around Quadratics

1. Each graph below is the graph of a function of the form $f(x) = ax^2 + bx + c$. Determine the sign of $a, b,$ and c: positive, negative, or zero. Explain your criteria.

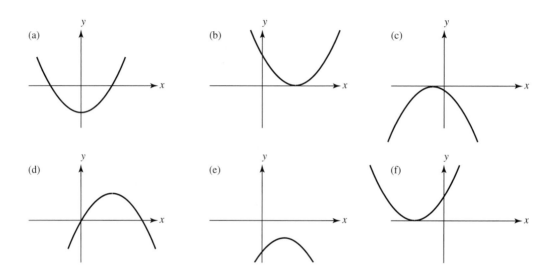

After having completed this problem, compare your strategy with those of your classmates. There are quite a few different ways to approach this problem.

2. Quadratics arise naturally in analyzing projectile motion due to the physics of falling objects. In this problem you will work on understanding why. Suppose an object is thrown straight upward from a height of 2 feet with an initial velocity of 25 feet per second.

   (a) When an object is in free fall, the only force acting upon it is the force of gravity. Gravity causes a downward acceleration of 32 feet per second squared.[7] You know that acceleration is the rate of change of velocity. Let $v(t)$ be the velocity of the falling object at time t, where t is measured in seconds and $t = 0$ corresponds to the time at which the object is thrown. The object is accelerating downward due to the force of gravity, so

$$\frac{dv}{dt} = -32.$$

---

[7] Acceleration is $\dfrac{\text{change in velocity}}{\text{change in time}}$ so we can measure it in $\dfrac{\text{feet/second}}{\text{second}}$ or ft/sec^2.

Find all possible functions $v(t)$ such that $\frac{dv}{dt} = -32$. You will get a family of functions. In other words, you are asked to find the family of functions with a constant slope of -32.

(b) Pick out the one function in the family of functions found in part (a) that is relevant to our situation by using the fact that at time $t = 0$ the velocity of the object was 25 feet per second.

(c) You should now have a velocity function for the object. Let $s(t)$ be the height of the object at time t. You know that velocity is the rate of change of position, and, because the object was thrown straight up, you can equate position with height. Therefore you can equate $v(t)$ with $\frac{ds}{dt}$. At this point, you should have $\frac{dv}{dt}$ set equal to a linear function determined in part (b). The work you've done in this chapter indicates that $s(t)$ could be a quadratic function. In fact, you should be able to find a family of quadratic functions whose derivative is this specified linear function. Use the fact that $s(0) = 2$ to pick out the quadratic that gives the object's height as a function of time.

PROBLEMS FOR SECTION 6.2

1. Let $f(x) = x + 3$, $g(x) = x - 5$, $h(x) = f(x)g(x)$, and $j(x) = \frac{f(x)}{g(x)}$.
   Solve the following equations. Find *all* x that satisfy the equation.

   (a) $h(x) = 0$

   (b) $h(x) = -7$

   (c) $h(x) = -15$

   (d) $h(x) = c$ c is a constant that will appear in your answer.

   (e) $j(x^2) - 2 = 0$

   (f) $[j(x)]^2 - 1 = 0$

   (g) $h(x) = j(x)$

2. Match each equation below with a possible graph. a, b, and c are all positive constants.

   a) $y = a(x - b)^2$ b) $y = -a(x + b)^2$ c) $y = (x + b)(x + c)$

   d) $y = (x + b)(x - c)(x + a)$ e) $y = ax - b^2$

(i) 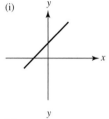   (ii) 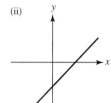   (iii) 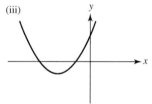   (iv)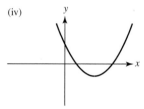

(v) 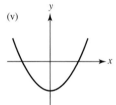   (vi) 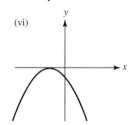   (vii)    (viii)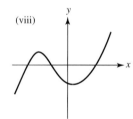

3. Solve for x. Work as efficiently as possible.

   (a) $x^2 - 7 = 0$ (b) $5x^2 = 125$ (c) $(x + 1)^2 = 25$

   (d) $x^2 + 2x + 1 = 25$ (e) $(2x + 3)^2 = 9$ (f) $(3x + 1)^2 = 7$

   (g) $(x + 3)(x - 1) = 0$ (h) $(x + 3)(x - 1) = 7$

4. Sketch the graphs of the following functions. Use what you know about the basic shapes plus shifting, flipping, and stretching to draw the graph without plotting lots of points. In each case tell us what "basic" function you are transforming.
   Label the x- and y-intercepts of each graph. Work as efficiently as possible.

   (a) $f(x) = 2(x - 3)^2 - 5$ (b) $g(x) = -4(x + 1)^2 + 3$

   (c) $h(x) = |x + 1| - 3$ (d) $j(x) = |x - 3|$

   (e) $k(x) = x^2 - 3$ (f) $l(x) = |x^2 - 3|$ (*Hint:* This has two sharp corners.)

5. Refer to Problem 4 for your answers to this question.

(a) How many solutions are there to $2(x-3)^2 - 5 = -6$?

(b) How many solutions are there to $-4(x+1)^2 + 3 = -6$?

(c) How many solutions are there to $|x+1| - 3 = -2$?

(d) Solve $x^2 - 3 \geq 1$.

(e) How many solutions are there to $|x^2 - 3| = 1$?

6. This question refers to Problem 4. True or false. (If the statement is false, give all counterexamples.)

(a) All the functions in Problem 4 are continuous everywhere.

(b) All the functions in Problem 4 have derivatives that are continuous everywhere.

7. Consider the first three functions from Problem 4. At what x-values does each function take on its maximum and minimum values on the interval $[0, 4]$ for the parabola in part (a)? In part (b)? In part (c)?

(This problem is equivalent to saying "Look at your graph only on the domain $[0, 4]$. Where does the function take on its biggest value? Its smallest value?")

8. Using functional notation, state the criterion for the function g to be symmetric about the vertical line $x = k$. The figure below should help you arrive at this criterion.

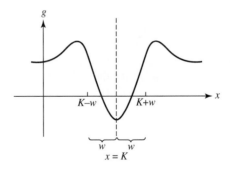

9. We have stated that the graph of a parabola is symmetric about the vertical line through its vertex. The goal of this problem is to prove this assertion. Since the vertex of the parabola $f(x) = ax^2 + bx + c$ is at $x = \frac{-b}{2a}$, we must show that the graph of $f(x)$ is symmetric about the vertical line $x = -\frac{b}{2a}$. This is equivalent to showing that

$$f\left(\frac{-b}{2a} + x\right) = f\left(\frac{-b}{2a} - x\right)$$

for all x. (To arrive at this criterion on your own, do Problem 8.)

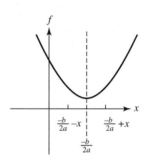

Show that if $f(x) = ax^2 + bx + c$, then $f\left(\frac{-b}{2a} + x\right) = f\left(\frac{-b}{2a} - x\right)$.

10. Find possible equations to fit the following graphs.

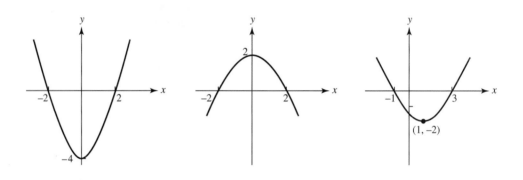

11. Solve for the indicated variable. (Your answers will be messy and will involve lots of letters. Don't let that faze you. In each problem, begin by determining whether the equation is linear or quadratic in the indicated variable. Then solve using the appropriate technique.)

a) $Q = b(j + 3) - jQ$ (j)   b) $\lambda(1 + 5\lambda) = 3\pi$ (λ)

c) $\frac{\Omega}{6} = \frac{1}{3}(\lambda + \Omega + 1) - \frac{1}{3}\lambda^2$ (Ω)   d) $\Omega = 3(\lambda + \Omega + 1) - 2\lambda^2$ (λ)

e) $R + \frac{R}{\Omega} = R + V$ (Ω)   f) $x(x + y) - yz - 1 = -ky$ (x)

g) $x = \frac{-y \pm \sqrt{y^2 - 4(ky - yz + 1)}}{2}$ (x)

12. Solve:

(a) $x^4 + x^2 = 6$.   (b) $x^4 - 5x^2 = -6$.

13. Solve: $2x^6 + 5x^3 - 3 = 0$.

14. Solve: $(x - 2)^4 - 2(x - 2)^2 = -1$.

6.3   QUADRATICS AND THEIR GRAPHS

Graphing Parabolas

Beginning with the basic parabola $y = x^2$ and using the results of Section 3.4 on shifting, stretching, shrinking, and flipping, we can obtain any other parabola in which we are interested.

EXAMPLE 6.4   Graph $f(x) = \frac{1}{2}(x - 4)^2 - 2$.

SOLUTION   This is the graph of $y = x^2$ shifted to the right 4 units, shrunk vertically to half the height at each x-value, and then shifted downward by 2 units. (Use the same order of operations graphically as you would algebraically).[8] In particular, the vertex of $y = x^2$ is shifted right 4 units, to (4, 0), shrunk vertically (no effect, because zero can't be "shrunk"), and then shifted down 2 units, to (4, −2). A graph is shown below.

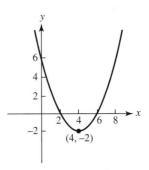

Figure 6.6   ◆

Using analogous lines of reasoning, we can graph any function of the form $f(x) = a(x - h)^2 + k$. In fact, by using the algebraic technique of completing the square, any quadratic can be expressed in this form. [9]

Suppose we are given a quadratic in the form $f(x) = a(x - r)(x - s)$, where a, r, and s are constants. This is also convenient to graph. We can immediately read off the zeros of the function; $f(x) = 0$ when $x = r$ or $x = s$. If $a > 0$, then the parabola opens upward, and if $a < 0$, then the parabola opens downward. Only parabolas that intersect the x-axis can be expressed in this form.

EXERCISE 6.1   Find the equation of the family of parabolas with vertex $(5, 0)$. Your answer should involve one parameter, that is, one unspecified constant.

Suppose the quadratic in Example 6.4 was not given in the form $f(x) = a(x - h)^2 + k$; suppose we were given this same function in the form

$$f(x) = \frac{1}{2}x^2 - 4x + 6.$$

To sketch this by hand there are a couple of options. One (which we won't use here) is to put this equation in the form $f(x) = a(x - h)^2 + k$, a process that involves what is called

---

[8] For additional work on order of operations, refer to Appendix A: Algebra.
[9] See the derivation of the quadratic formula in the Algebra Appendix to see how this is done.

"completing the square";[10] the other (given below) is to identify the vertex and one or two other points and draw. We can do this without calculus as follows.

Identify the x- and y-intercepts:

The y-intercept, $f(0)$, is straightforward to find: $f(0) = 6$.

Finding the x-intercept(s) is equivalent to the problem of solving[11] the quadratic equation $\frac{1}{2}x^2 - 4x + 6 = 0$.

$$\frac{1}{2}x^2 - 4x + 6 = 0$$

$$x^2 - 8x + 12 = 0 \qquad \text{Multiply both sides by 2.}$$

$$(x - 6)(x - 2) = 0 \qquad \text{Now it's easier to factor.}$$

$$x - 6 = 0 \quad \text{or} \quad x - 2 = 0$$

$$x = 6 \quad \text{or} \quad x = 2$$

The parabola has two x-intercepts, one at $x = 2$ and one at $x = 6$. Because a parabola is symmetric about its vertical axis, the x-coordinate of the vertex of the parabola lies midway between the two intercepts, at $x = 4$. The x-coordinate of the vertex is 4, so the y-coordinate must be

$$f(4) = \frac{1}{2}(4)^2 - 4(4) + 6 = 8 - 16 + 6 = -2.$$

Knowing that the vertex is $(4, -2)$ and that the x-intercepts are 2 and 6, we can sketch the graph. Knowing the y-intercept is simply a bonus.

Alternatively, we could use our knowledge of calculus to find the vertex. The x-coordinate of the vertex is the zero of $f'(x)$. We find the y-coordinate as above; identify one additional point on the graph by using symmetry, and sketch.

---

[10] See Appendix A: Algebra for details. Briefly:

$$ax^2 + bx + c = a\left(x^2 + \frac{b}{a}x + \frac{c}{a}\right)$$

$$= a\left[\left(x + \frac{b}{2a}\right)^2 - \frac{b^2}{4a^2} + \frac{c}{a}\right]$$

$$= a\left(x + \frac{b}{2a}\right)^2 - \frac{b^2 - 4ac}{4a}.$$

[11] Read Appendix A: Algebra, Part III, Solving Equations, and Section B: Quadratics, for a discussion of solving quadratic equations.

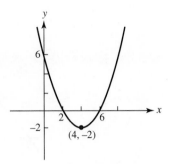

Figure 6.7

Notice that if we were handed the function in the form $f(x) = \frac{1}{2}(x-4)^2 - 2$ and wanted to find the x-intercepts, we would have to solve a quadratic equation. Rather than multiplying out, it is more efficient to take advantage of the perfect square as follows.

$$\frac{1}{2}(x-4)^2 = 2$$
$$(x-4)^2 = 4$$
$$x - 4 = \pm 2$$
$$x = 4 \pm 2$$
$$x = 4 + 2 \quad \text{or} \quad x = 4 - 2$$
$$x = 6 \quad \text{or} \quad x = 2$$

Finding a Function to Fit a Parabolic Graph

Three noncolinear points determine a parabola. We might know

  i. three arbitrary points,
 ii. the x-intercepts, r_1 and r_2, and a point, or
iii. the vertex (p, q) and a point.

Depending upon the information available, we can look for an equation of the form

  (i) $y = ax^2 + bx + c$ (ii) $y = k(x - r_1)(x - r_2)$ (iii) $y - q = k(x - p)^2$

respectively. From equation (i) and three arbitrary points we can obtain a system of three linear equations that we can solve for the constants a, b, and c. We can use the x-intercepts r_1 and r_2 in equation (ii) and use the other known point to solve for k. Similarly, if we know the vertex and use it in equation (iii) we can solve for k using the other known point.

◆ **EXAMPLE 6.5** Find a function to fit the parabola drawn below.

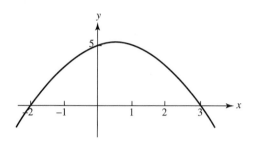

Figure 6.8

SOLUTION   Our strategy is to find an equation of the form

$$y = k(x - r)(x - s),$$

where r and s are the x-intercepts of the graph. First, let's nail down the x-intercepts; then we can adjust k as needed. (Recall that k can stretch or shrink the graph vertically, flipping vertically if k is negative. None of this action will affect the zeros, so we can nail them down with peace of mind.) The x-intercepts are -2 and 3, so -2 and 3 must be zeros of the function. The quadratic must be of the form

$$y = k(x + 2)(x - 3),$$

where k is a constant. We know that when $x = 0$, $y = 5$, so

$$5 = k(0 + 2)(0 - 3)$$
$$5 = k(-6)$$
$$k = \frac{-5}{6}.$$

Therefore, $y = \frac{-5}{6}(x + 2)(x - 3)$.   ◆

◆ **EXAMPLE 6.6** Find a function to fit the parabola drawn below.

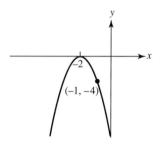

Figure 6.9

Approach I: Again, we nail down the x-intercept. It is $x = -2$, so the quadratic must be of the form

$$y = k(x + 2)^2.$$

The $(x + 2)$ term must be squared in order to obtain a quadratic.[12] Alternatively, $(-2, 0)$ is the vertex of the parabola, so $y - 0 = k(x - (-2))^2$.
When $x = -1$, $y = -4$, so

$$-4 = k(-1 + 2)^2$$
$$-4 = k$$
$$y = -4(x + 2)^2.$$

Approach II: This is our standard parabola $y = x^2$ shifted left 2 units and stretched and flipped. Therefore, it is of the form $y = k(x + 2)^2$. Find k as in Approach I. We expect k to be negative to flip the graph, and indeed it is.   ◆

PROBLEMS FOR SECTION 6.3

For Problems 1 through 8, graph the function. Label the x- and y-intercepts and the coordinates of the vertex.

1. $f(x) = -x^2 + 1$

2. $f(x) = (x + 3)(2x - 6)$

3. $f(x) = x^2 + 5x - 6$

4. $f(x) = 2(x - 1)^2 - 4$

5. $f(x) = |-x^2 + 1|$

6. $f(x) = -3(x + 2)^2 + 9$

7. $f(x) = (3 - x)(x + 1)$

8. $f(x) = x^2 + \pi x + 1$

9. Find possible equations for the following graphs assuming that the first three are parabolas and the last is the graph of some cubic.

---

[12] Without squaring we would be left with the equation of a line. Also, on either side of $x = -2$ the sign of y doesn't change, so $(x + 2)$ must be raised to an even power. This type of analysis will be examined in more detail later.

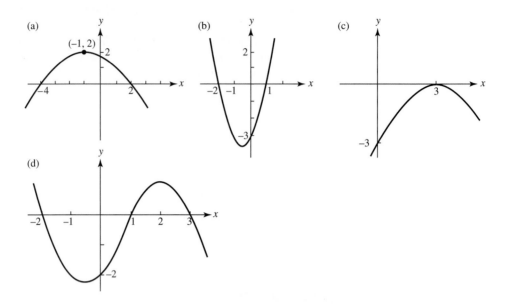

For Problems 10 through 16, give the set of functions all having the specified characteristics. Example: the set of nonvertical lines passing through (3, 2). Solution: $y - 2 = m(x - 3)$, so $f(x) = m(x - 3) + 2$, where m is any constant.

10. The set of nonvertical lines passing through $(0, \pi)$

11. The set of lines with slope 2

12. The set of parabolas with x-intercepts at $x = 0$ and $x = 3$

13. The set of parabolas with x-intercepts at $x = -5$ and $x = 1$

14. The set of parabolas with vertex $(2, 3)$

15. The set of parabolas with vertex $(-1, \pi)$

16. The set of parabolas passing through $(0, 3)$ with a slope of 2 at $(0, 3)$

Strategize. In the example above we used $y - y_1 = m(x - x_1)$ as the equation of our line, not $y = mx + b$. Similarly, when working with parabolas, different forms of the equation for a parabola are best suited to different situations.

In Problems 17 through 21, find the equation of the parabola with the specifications given.

17. x-intercepts of 3 and -2; maximum value of 1

18. x-intercepts of π and 3π; y-intercept of 6

19. x-intercepts of π and 3π; y-intercept of -2

20. Vertex at $(1, 5)$; y-intercept of 1

21. Vertex at $(-2, 3)$; passing through $(1, -1)$

22. Sketch the graphs of the following functions. Beneath the sketch of the function, sketch the graph of the derivative. If the graph is the graph of a quadratic, label the coordinates of the vertex of the corresponding parabola; if the graph has corners, label the coordinates of the corners.
    (a) $f(x) = -(x + 2)^2$
    (b) $f(x) = (x - 2)^2 + 3$
    (c) $f(x) = (x - 2)(x + 4)$
    (d) $f(x) = -2(x - 2)(x + 4)$
    (e) $f(x) = |x + 4| + 2$
    (f) $f(x) = |2x| - 3$

23. Find the equation of the parabola through the points $(0, 3)$, $(1, 0)$, and $(2, -1)$.

24. Find the coordinates of the vertex of the parabola passing through the points $(2, 0)$, $(-1, 9)$, and $(1, -5)$.
    Decide upon a strategy for doing this problem.

6.4   THE FREE FALL OF AN APPLE: A QUADRATIC MODEL

As mentioned in Section 6.1, quadratic functions arise naturally when modeling the motion of a falling body. We could be interested in the height of a tennis ball thrown in the air, a cannon ball shot from a cannon, a baseball hit by a bat, a football thrown by a quarterback; the list is endless, but the underlying laws of physics are the same. Below we'll look at the free fall of an urban apple.

◆ **EXAMPLE 6.7**   A fellow in a high-rise apartment building decides to get rid of a bad apple by depositing it in the open trash bin standing on the street directly below his window. He opens the window, leans out, and tosses the apple which lands at street level in the bin.

Let us assume the following. The apple travels vertically only. The only force acting on the apple after it leaves the fellow's hand is the force of gravity which gives the apple a downward acceleration of 32 ft/sec^2. We will ignore the effect of air resistance. The laws of physics dictate that the initial height and initial velocity of the apple completely determine its height as a function of time. In this case it turns out that

$$h(t) = -16t^2 + 20t + 130,$$

where $h(t)$ is the height of the apple (in feet) t seconds after it is tossed.[13]

(a) From what height is the apple tossed?

(b) i. What is the apple's initial velocity?
    ii. Is the apple dropped, tossed down, or tossed up?

---

[13] In the Exploratory Exercise you constructed such a function on your own when given this information plus some initial conditions.

(c) i. What is the maximum height of the apple?

ii. When does it reach this height?

(d) When does the apple land in the trash can?

(e) What is the velocity of the apple when it lands?

(f) Sketch $h(t)$.

(g) Sketch the graph of velocity versus time.

(h) How is the apple's acceleration reflected in your answer to part (g)?

SOLUTION   First go through the questions and determine what is being asked from a mathematical viewpoint. Rewrite each question in terms of the function $h(t)$. Try this on your own and compare your answers with those below.

The basic ideas upon which the problem rests are the following:

■ $h(t)$ gives the height of the apple at time t. The graph of $h(t)$ is a parabola opening downward.

■ $h'(t)$ gives the velocity (instantaneous rate of change of height with respect to time) of the apple at time t. The graph of $h'(t)$ is a line.

$|h'(t)|$ gives the speed of the apple at time t.

■   The derivative of $h'(t)$, written $h''(t)$, gives the **acceleration** (instantaneous rate of change of velocity with respect to time) at time t.

Translating the questions into mathematical language gives us the following.

(a) What is $h(0)$?

(b) i. What is $h'(0)$?

ii. What is the sign of $h'(0)$?

If $h'(0)$ is negative, the apple is thrown down.

If $h'(0)$ is zero, the apple is dropped.

If $h'(0)$ is positive, the apple is tossed up.

(c) If the apple is tossed up, then these questions translate to questions about the coordinates of the vertex.

i. What is the h-coordinate of the vertex of the parabola?

ii. What is the t-coordinate of the vertex?

If the apple is thrown down or dropped, then part (i) is simply the initial height, $h(0)$, and part (ii) is $t = 0$.

(d) Find t such that $h(t) = 0$. (We want t to be positive.)

(e) Let's denote by t_L the answer to part (d). Evaluate $h'(t_L)$.

(f) Graph the parabola $h(t) = -16t^2 + 20t + 130$. (Be careful to use the appropriate domain.)

(g) Graph $h'(t)$ and interpret it. (The domain should be as in part f.)

(h) Find $h''(t)$. This is the slope of the graph of $h'(t)$.

ANSWER

(a) $h(0) = 130$. The apple is tossed from a height of 130 feet.

(b) $h'(t) = -32t + 20$

i. $h'(0) = 20$. The apple's initial velocity is 20 ft/sec.

ii. $h'(0)$ is positive. The apple is tossed up.

(c) We can find the vertex with or without using calculus. Let's do the former. At the vertex $h'(t) = 0$, the velocity of the apple is zero and the tangent line to the parabola is horizontal. Solving the equation $h'(t) = -32t + 20 = 0$ gives $t = \frac{-20}{-32} = 5/8$.

i. The h-coordinate of the vertex of the parabola is

$$h(5/8) = -16\left(\frac{5}{8}\right)^2 + 20\left(\frac{5}{8}\right) + 130$$

$$= \frac{-16 \cdot 25}{8 \cdot 8} + \frac{100}{8} + 130$$

$$= \frac{-25}{4} + \frac{50}{4} + 130$$

$$= \frac{25}{4} + 130$$

$$= 136.25.$$

Thus, the apple reaches a maximum height of 136.25 feet.

ii. The t-coordinate of the vertex is 5/8. The apple reaches its maximum height 5/8 seconds after being thrown.

(d)
$$-16t^2 + 20t + 130 = 0$$

$$-8t^2 + 10t + 65 = 0$$

$$t = \frac{-10 \pm \sqrt{100 - 4(-8)(65)}}{2(-8)}$$

$$= \frac{-10 \pm \sqrt{2180}}{-16}$$

$$= \frac{-10 \pm 2\sqrt{545}}{-16}$$

$$= \frac{2[-5 \pm \sqrt{545}]}{-16}$$

$$= \frac{-5 \pm \sqrt{545}}{-8} \quad \text{(exact answers)}$$

$$\approx 3.54 \quad \text{or} \quad -2.29 \quad \text{(numerical approximations)}$$

The answer we are looking for is $t = \frac{5+\sqrt{545}}{8}$, or $t \approx 3.54$ seconds, because the apple hits the trash bin *after* it is thrown, not before.

(e) $h'(t) = -32t + 20$

$$h'\left(\frac{5+\sqrt{545}}{8}\right) = -32\left(\frac{5+\sqrt{545}}{8}\right) + 20$$

$$= -4(5 + \sqrt{545}) + 20$$

$$= -20 - 4\sqrt{545} + 20$$

$$= -4\sqrt{545}$$

$$\approx -93.38$$

The apple is falling so the velocity is negative ($-4\sqrt{545}$ ft/sec or ≈ -93.38 ft/sec), and the speed is positive. [14]

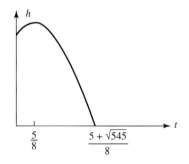

Figure 6.10

(f)

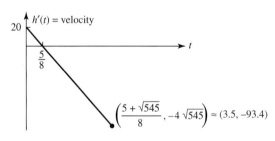

Figure 6.11

(g) Notice that the velocity is positive when the apple is going up, negative when the apple is traveling down, and zero at the instant it changes direction (at its maximum height).

(h) The acceleration, $h''(t)$, is -32 ft/sec^2, as expected. The acceleration is the slope of the velocity graph. ◆

---

[14] $h'(3.54) = -93.28$. The approximation -93.38 is more accurate than this; rounding off *after* doing all the calculations produces less error. When computing $h'(3.54) = -93.28$, the error is multiplied by 32.

PROBLEMS FOR SECTION 6.4

1. A seat on a round-trip charter flight to Cairo costs $720 plus a surcharge of $10 for every unsold seat on the airplane. (If there are 10 seats left unsold, the airline will charge each passenger $720 + $100 = $820 for the flight.) The plane seats 220 travelers and only round-trip tickets are sold on the charter flights.

   (a) Let x = the number of unsold seats on the flight. Express the revenue received for this charter flight as a function of the number of unsold seats. (*Hint:* Revenue = (price + surcharge)(number of people flying).)

   (b) Graph the revenue function. What, practically speaking, is the domain of the function?

   (c) Determine the number of unsold seats that will result in the maximum revenue for the flight. What is the maximum revenue for the flight?

2. Troy is interested in skunks and has purchased a bevy of them to study. He plans to keep the skunks in a rectangular skunk corral, as shown below. He will have a divider across the width in order to separate the males and females. He has 510 meters of fencing to make his corral.

   (a) Let x = the width of the corral and y = the length of the corral. Use the fact that Troy has only 510 meters of fencing to express y in terms of x.

   (b) Express A, the total area enclosed for the skunks, as a function of x.

   (c) Find the dimensions of the corral that maximize the total area inside the corral.

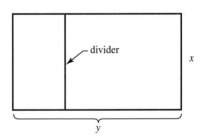

3. The height of a ball (in feet) t seconds after it is thrown is given by

$$h(t) = -16t^2 + 32t + 48 = -16(t + 1)(t - 3).$$

   (a) Graph $h(t)$ for the values of t for which it makes sense. Below it graph $v(t)$. Be sure that $v(t)$ looks like the derivative of $h(t)$.

   (b) From what height was the ball thrown?

   (c) What was the ball's initial velocity? Was it thrown up or down? How can you tell?

   (d) Was the ball's height increasing or decreasing at time $t = 2$?

   (e) At what time did the ball reach its maximum height? How high was it then? What was its velocity at that time?

   (f) How long was the ball in the air?

   (g) What is the ball's acceleration? Does this make physical sense?

4. We know that Revenue = (price) · (quantity). Suppose a certain company has a monopoly on a good. If the company wants to increase its revenue it can do so by raising its prices up to a certain point. However, at some point the price becomes so high that there are not enough buyers and the revenue actually goes down. Therefore, if a monopolist is attempting to maximize revenue, the monopolist must look at the demand curve. Suppose the demand curve for widgets is given by

$$p = 1000 - 4q,$$

where p is measured in dollars and q in hundreds of items.

(a) Express revenue as a function of price and determine the price that maximizes the monopolist's revenue.

(b) What price(s) gives half of the maximum revenue?

5. Suppose that q, the quantity of gas (in gallons) demanded for heating purposes, is given by $q = mp + b$, where m and b are constants (m negative and b positive) and p is the price of gas per gallon. The gas company is interested in its revenue function.

(a) Explain why m is negative.

(b) Express revenue as a function of price. (Note that m and b are constants, so if you express revenue in terms of m, b, and p, you have expressed revenue as a function of price.)

(c) Graph your revenue function, labeling all intercepts.

(d) From your graph, determine the price that maximizes revenue. (Your answer will be in terms of m and b.)

(e) Find R' and graph it.

6. A catering company is making elegant fruit tarts for a huge college graduation celebration. The caterer insists on high quality and will not accept shoddy-looking tarts. When there are 7 pastry chefs in the kitchen they can each turn out an average of 44 tarts per hour. The pastry kitchen is not very large; let us suppose that for each additional pastry chef put to the fruit-tart task the average number of tarts per chef decreases by 4 tarts per hour. (Assume that reducing the number of chefs will increase the average production by 4 tarts per hour, until the number of chefs has decreased to 3. At that point reducing the number of chefs no longer increases the productivity of each chef.)

(a) How many chefs will yield the optimum hourly fruit-tart production?

(b) What is the maximal hourly fruit-tart production?

(c) How many chefs are in the kitchen if the fruit-tart production is 320 tarts per hour?

7. The function $R(p) = 35p(75 - p)$ gives revenue as a function of price, p, where the price is given in dollars.

(a) Find the price at which the revenue is maximum.

(b) What is the maximum price?

8. David and Ben are sitting in a tree house sorting through a basket of apples. They find two that are rotten and decide to toss them into a garbage bin on the ground below. The height of Ben's apple t seconds after he throws it is given by

$$B(t) = -16t^2 + 4t + 10,$$

and the height of David's apple t seconds after he throws it is given by

$$D(t) = -16t^2 - 2t + 10.$$

(a) From what height are the apples thrown?

(b) What is the maximum height of Ben's apple? Explain your answer.

(c) What is the maximum height of David's apple?

(d) Does Ben toss the apple up or throw it down? Does David toss the apple up or throw it down? Explain your reasoning.

(e) How many seconds after he throws it does Ben's apple hit the ground?

9. Amelia is a production potter. If she prices her bowls at x dollars per bowl, then she can sell $120 - 5x$ bowls every week.

(a) For each dollar she increases her price how many fewer bowls does she sell?

(b) Express her weekly revenue as a function of the price she charges per bowl.

(c) Assuming that she can produce bowls more rapidly than people buy them, how much should she charge per bowl in order to maximize her weekly revenue?

(d) What is her maximum weekly revenue from bowls?

CHAPTER 7

The Theoretical Backbone: Limits and Continuity

7.1 INVESTIGATING LIMITS—METHODS OF INQUIRY AND A DEFINITION

The idea of taking a limit is at the heart of calculus. The limiting process allows us to move from calculating average rates of change to determining an instantaneous rate of change; in other words, it allows us to determine the slope of a tangent line. We will need it to tackle another fundamental problem in calculus, calculating the area under a curve. We've also seen that the language of limits is useful as a descriptive tool.

Roots of the limiting processes can be found in the work of the ancient Greek mathematicians. Archimedes essentially used a limiting process in studying the area of a circle, calling his technique involving successive approximations "the method of exhaustion." The limiting process was instrumental in the work of Isaac Newton and Gottfried Leibniz as they developed calculus and can also be found in work of their predecessors. It is interesting that these early practitioners of calculus in the late 1600s achieved many useful and valid results without carefully defining the notion of a limit. Not until substantially later were mathematicians able to put the work of their predecessors on solid ground with a rigorous definition of a limit. While we will not work much with the rigorous definition, we will use this chapter to put the notion of limit on more solid ground.

The challenge of computing instantaneous velocity from a displacement function, $s(t)$, led us to limits. To find the instantaneous velocity at, say, $t = 3$, we took successive approximations using average velocity over the interval $[3, 3 + h]$.

$$s'(3) \approx \frac{\Delta \text{ displacement}}{\Delta \text{ time}}$$

$$s'(3) \approx \frac{s(3+h) - s(3)}{h}$$

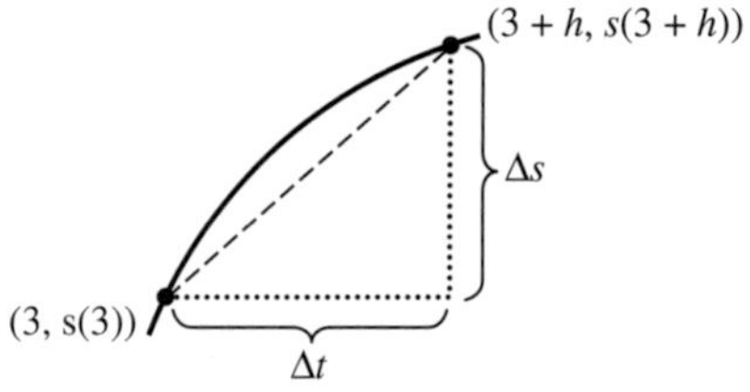

Figure 7.1

If $h = 0$ this expression is undefined; we have $\frac{0}{0}$. But the closer h is to zero the better the approximation becomes. We are interested in what happens to $\frac{s(3+h)-s(3)}{h}$ as h approaches zero but is not equal to zero. This problem motivates our definition of limit.

What Do We Mean by the Word "Limit"?

Below we give a short answer, which we will expand upon throughout the section.

- $\lim_{x\to 2} f(x) = 5$ means $f(x)$ stays arbitrarily close to 5 provided that x is sufficiently close to 2, but not equal to 2.

  It does *not* tell us that $f(2) = 5$. It gives us *no* information about $f(2)$; $f(2)$ could be 5, or $\sqrt{3}$, or undefined.

  However, we can guarantee that the difference between $f(x)$ and 5 is smaller than *any* positive number, no matter how miniscule, if x is close enough to 2 (but not equal to 2).

- $\lim_{x\to\infty} f(x) = 6$ means that the values of $f(x)$ stay arbitrarily close to 6 provided x is large enough.

- $\lim_{x\to 2} f(x) = \infty$ means that $f(x)$ increases without bound as x approaches 2.

We'll clarify the meaning of "arbitrarily close to" through the next two examples.

◆ **EXAMPLE 7.1** Argue convincingly that if $g(x) = \frac{1}{x}$, then $\lim_{x\to\infty} g(x) = 0$.

SOLUTION It "appears," from the graph of $g(x) = \frac{1}{x}$ (see Figure 7.2), that $\lim_{x\to\infty} \frac{1}{x} = 0$; numerical evidence suggests this hypothesis as well.

x	$g(x) = \frac{1}{x}$
10	.1
100	.01
1000	.001
10000	.0001
10^5	10^{-5}
10^6	10^{-6}

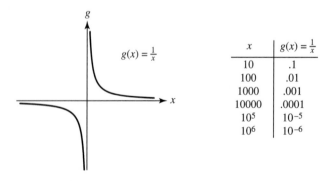

Figure 7.2

But appearances alone *can* be deceiving. For instance, suppose we graph $h(x) = \frac{1}{x} + 10^{-15}$. Adding 10^{-15} to $g(x)$ shifts the graph of $g(x)$ vertically up by 10^{-15}. But 10^{-15} is such a miniscule number that it is difficult to distinguish between h and g graphically (particularly using a graphing calculator). Yet if $\lim_{x \to \infty} \frac{1}{x} = 0$ then $\lim_{x \to \infty}(\frac{1}{x} + 10^{-15})$ ought to be 10^{-15}. So although the graph in Figure 7.2 is very useful, it isn't convincing evidence that $\lim_{x \to \infty} \frac{1}{x} = 0$.

The table of values suggests a way of nailing things down. Since $\frac{1}{x}$ is positive and decreasing, we see that

$g(x)$ is within 10^{-5} of zero provided $x > 10^5$,

$g(x)$ is within 10^{-6} of zero provided $x > 10^6$,

$g(x)$ is within 10^{-100} of zero provided $x > 10^{100}$.

Even this last statement is not enough to show that $\lim_{x \to \infty} g(x) = 0$. We must show that $g(x)$ will be *arbitrarily* close to zero for all x large enough. This means someone can issue a challenge with any miniscule little number. Let ϵ (read: epsilon) be any small positive number; ϵ can be excruciatingly small. We must show that for x large enough, $g(x)$ is within ϵ of 0.

$$g(x) = \frac{1}{x} \text{ is within } \epsilon \text{ of zero provided } x > \frac{1}{\epsilon}.$$

We're done.   ◆

In our next example we'll look at $\lim_{x \to \infty} \left(\frac{1}{2}\right)^x$.

Before launching in, first we note that we define $\left(\frac{1}{2}\right)^n$ for n a positive integer as $\frac{1}{2}$ multiplied by itself n times:

$$\underbrace{\frac{1}{2} \cdot \frac{1}{2} \cdot \frac{1}{2} \cdots \frac{1}{2}}_{n \text{ times}}.$$

Using the rules of exponent algebra, which can be reviewed in either the Algebra Appendix or Chapter 9 for any rational exponent x, we can define $\left(\frac{1}{2}\right)^x$ for x any rational number. For now we'll deal with irrational exponents simply by approximating the irrational number by a sequence of rational ones. Suppose x is irrational, r and s are rational, and $r < x < s$. Then $\left(\frac{1}{2}\right)^r < \left(\frac{1}{2}\right)^x < \left(\frac{1}{2}\right)^s$. We define b^x so as to make the graph of b^x continuous. A more satisfactory definition can be given after taking up logarithmic functions.

◆ **EXAMPLE 7.2**   Argue convincingly that $\lim_{x \to \infty} \left(\frac{1}{2}\right)^x = 0$.

SOLUTION   Let's begin by trying to get a feel for what happens to $\frac{1}{2^x}$ as x increases without bound. Consider the graph of $f(x) = \left(\frac{1}{2}\right)^x = \frac{1}{2^x}$ on the following page.

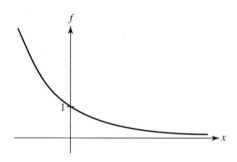

Figure 7.3

A second way of getting a feel for the limit is to look numerically at the outputs of $f(x) = \frac{1}{2^x}$ as x gets increasingly large.

x	$f(x) = \frac{1}{2^x} = (0.5)^x$ (approximate values)
20	$9.5 \times 10^{-7} = 0.00000095$
50	$8.8 \times 10^{-16} = 0.0000000000000088$
100	7.8×10^{-31}
200	6.2×10^{-61}
300	4.9×10^{-91}

It "appears" that $\lim_{x \to \infty} \frac{1}{2^x} = 0$. This is looking quite convincing; 4.9×10^{-91} is quite close to zero. But consider the following. When asked to compute 0.5^{328}, a TI-81 calculator gives approximately 1.8×10^{-99}. But when asked for 0.5^{329}, it gives the answer as 0. In fact, according to this calculator, for $x > 329$, $\frac{1}{2^x} = 0$. We know this is false; a fraction can be equal to zero only if its numerator is zero, and here the numerator is 1. It appears that the calculator rounds 10^{-100} off to zero. How can we be sure that $\lim_{x \to \infty} \frac{1}{2^x} = 0$, and not 10^{-120} for instance? It is not *quite* good enough to simply say $0 < f(x) < 10^{-99}$ provided $x > 330$.[1]

As in Example 7.1, we must show that $f(x)$ will be *arbitrarily* close to zero provided that x is large enough. Again, a challenge is issued with any excruciatingly small number ϵ; we must show that $f(x)$ is within ϵ of 0 for x big enough. To figure out how big is "big enough," let's compare $f(x) = \frac{1}{2^x}$ with $g(x) = \frac{1}{x}$.

---

[1] This statement is true because as x increases $f(x)$ decreases.

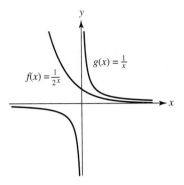

Figure 7.4

Notice that $\frac{1}{2^x} < \frac{1}{x}$ for all $x > 0$. (This is equivalent to the statement that if $x > 0$, then $2^x > x$.)

So for $x > 0$, if $\frac{1}{x}$ is within ϵ of zero, then $\frac{1}{2^x}$ is certainly within ϵ of zero.

And $\frac{1}{2^x}$ is within ϵ of zero provided $x > \frac{1}{\epsilon}$. (Notice by looking at the table of values that for large x, $\frac{1}{2^x}$ is *much, much* smaller than $\frac{1}{x}$, so requiring that $x > \frac{1}{\epsilon}$ is overkill—but that's all right.) ◆

The last two examples highlight a couple of important ideas about limits in general and $\lim_{x \to \infty} f(x) = L$ in particular:

1. Graphical and numerical investigations are both useful methods of inquiry that can provide compelling data from which to arrive at a conjecture about a limit. They cannot, however, be conclusive on their own.

2. When we say "$f(x)$ is arbitrarily close to L" we mean that the distance between $f(x)$ and L can be made arbitrarily small. To be arbitrarily small means that we can answer a challenge set out by *any* miniscule positive number ϵ, no matter how excruciatingly small ϵ may be, that the distance between $f(x)$ and L can be made less than ϵ given certain conditions on x.

◆ **EXAMPLE 7.3**   Find $\lim_{x \to 3} \frac{x}{2}$.

SOLUTION   Very loosely speaking, this problem asks "what does $f(x) = \frac{x}{2}$ approach as x gets closer and closer to 3 but is not equal to 3?" Intuitively, it should make sense that $\lim_{x \to 3} \frac{x}{2} = 1.5$. There is nothing particularly special happening to $\frac{x}{2}$ around $x = 3$.

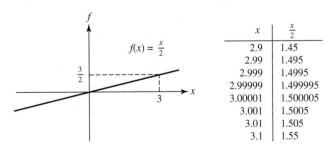

Figure 7.5

More rigorously, we can show that if x is close enough to 3 (but not equal to 3), then $\frac{x}{2}$ can be made to stay arbitrarily close to 1.5 as follows.

Look at Figure 7.6(a). We can see that if x is within 1 unit of 3, then $f(x)$ is within 0.5 units of 1.5. Similarly, we see in Figure 7.6(b) that if x is within 0.1 of 3, then $f(x)$ is within $0.05 = \frac{0.1}{2}$ of 1.5.

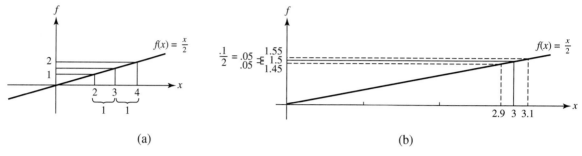

(a)   (b)

Figure 7.6

Because $f(x)$ is a straight line with slope 1/2, we can see that the ratio $\frac{\Delta f}{\Delta x}$ is always 0.5. We use this to ensure that the distance between $f(x)$ and 1.5 can be made arbitrarily small for x close enough to 3 but $x \neq 3$.

For ϵ any positive number, no matter how excruciatingly small, if x is within 2ϵ of 3 ($x \neq 3$), then $f(x)$ will be within ϵ of 1.5. Therefore $\lim_{x \to 3} \frac{x}{2} = 1.5$.   ◆

This last example illustrates what we mean by $\lim_{x \to 3} f(x) = L$. Showing that $f(x)$ stays arbitrarily close to L for x close enough to 3 (but not equal to 3) is equivalent to the following: Given the challenge of any excruciatingly small positive ϵ, we can guarantee that $f(x)$ will be within ϵ of L provided x is close enough to 3. We can write this more compactly: If $f(x)$ is within ϵ of L, then the distance between $f(x)$ and L is less than ϵ. Let's use absolute values to express this distance.

The distance between $f(x)$ and L is $|f(x) - L|$.

Similarly, the distance between x and 3 is $|x - 3|$.

To show that x is within some distance δ (delta for distance) of 3 but not equal to 3, we write $0 < |x - 3| < \delta$.

We arrive at the following definition, which we use in the examples which follow.

Definition

$\lim_{x \to 3} f(x) = L$ if, for every excruciatingly small positive number ϵ, we can come up with a distance δ so that

$$|f(x) - L| < \epsilon \text{ provided } 0 < |x - 3| < \delta.$$

This is the formal definition of a limit[2] (where we can substitute "a" for 3 to get $\lim_{x \to a} f(x) = L$). We applied this definition in the last example. As promised at the

---

[2] f must be defined on an open interval around 3, although not necessarily at $x = 3$.

beginning of this section, we will not be applying this definition much in practice; it is the *meaning* behind the definition that is of primary importance to us.

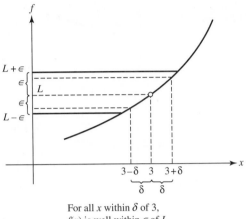 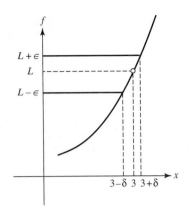

For all x within δ of 3,
$f(x)$ is well within ϵ of L.

(i)

For all x within δ of 3,
$f(x)$ is within ϵ of L.

(ii)

Figure 7.7

◆ **EXAMPLE 7.4**   Consider the function $f(x) = \begin{cases} \frac{x}{2}, & \text{for } x \neq 3, \\ \text{undefined}, & \text{for } x = 3. \end{cases}$

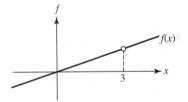

Figure 7.8

We can write this function compactly as $f(x) = \frac{x(x-3)}{2(x-3)}$, because $\frac{x(x-3)}{2(x-3)} = \frac{x}{2}$ for $x \neq 3$ and is undefined at $x = 3$. Find $\lim_{x \to 3} \frac{x(x-3)}{2(x-3)}$.

SOLUTION

$$\lim_{x \to 3} \frac{x(x-3)}{2(x-3)} = \lim_{x \to 3} \frac{x}{2} \quad (\text{provided } x \neq 3)$$

$$= \frac{3}{2}$$

The argument given in Example 7.3 holds without alteration since we always worked with the condition $x \neq 3$. A single hole in the graph makes no difference in the limit. In fact, inserting any finite number of holes in the graph of a function has no effect on the computation of limits.   ◆

◆ **EXAMPLE 7.5**   Let $f(x) = \begin{cases} \frac{x}{2}, & \text{for } x \neq 3, \\ 2, & \text{for } x = 3. \end{cases}$
Find $\lim_{x \to 3} f(x)$.

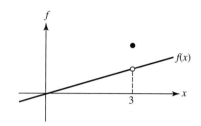

Figure 7.9

SOLUTION   Again $\lim_{x \to 3} f(x) = \frac{3}{2}$ since $f(x) = \frac{x}{2}$ for $x \neq 3$ and the arguments given above, both formal and informal, hold here as well.   ◆

Notice that we've established that $\lim_{x \to 3} f(x) = \frac{3}{2}$ for each of the functions $f(x)$ drawn below. The limit tells us about the behavior of f *near* $x = 3$ but not *at* $x = 3$.

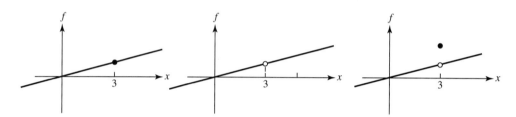

Figure 7.10

◆ **EXAMPLE 7.6**   Find $\lim_{x \to 0} \frac{4x + x^2}{x}$.

We're interested in the behavior of $f(x) = \frac{4x + x^2}{x} = \frac{(4+x)x}{x}$ for x near zero but not *at* $x = 0$. For $x \neq 0$, $\frac{(4+x)x}{x} = 4 + x$, so $\lim_{x \to 0} \frac{(4+x)x}{x} = \lim_{x \to 0} 4 + x = 4$.

More formally, we must show that for any positive ϵ, $|f(x) - 4| < \epsilon$ provided $|x|$ is small enough (and nonzero). For $x \neq 0$, we have:

$$|f(x) - 4| = \left| \frac{(4+x)x}{x} - 4 \right| = |4 + x - 4| = |x|,$$

so if $0 < |x| < \epsilon$, then $|f(x) - 4| < \epsilon$ as well.

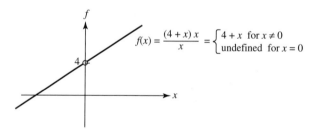

Figure 7.11   ◆

The limit in Example 7.6 is the limit that we would compute if we were computing the derivative of $g(x) = x^2$ at $x = 2$.

$$g'(2) = \lim_{h \to 0} \frac{g(2+h) - g(2)}{h}$$

$$= \lim_{h \to 0} \frac{(2+h)^2 - 4}{h}$$

$$= \lim_{h \to 0} \frac{4 + 4h + h^2 - 4}{h}$$

$$= \lim_{h \to 0} \frac{(4+h)h}{h} \qquad \text{This is the limit given in Example 7.6.}$$

$$= \lim_{h \to 0} 4 + h$$

$$= 4$$

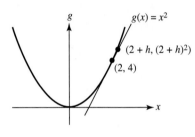

Figure 7.12

EXERCISE 7.1   Show that

$$\lim_{x \to 3} \frac{2x^2 - 6x}{x - 3} = 6$$

using the formal definition of a limit.

Try this first on your own. The answer is provided below.

Answer

Let $f(x) = \frac{2x^2 - 6x}{x - 3}$. We must show that for any positive ϵ, $|f(x) - 6| < \epsilon$ provided $|x - 3|$ is small enough but nonzero. If $|x - 3| \neq 0$ then

$$|f(x) - 6| = \left| \frac{2x(x-3)}{x-3} - 6 \right| = |2x - 6| = 2|x - 3|.$$

Therefore, $|f(x) - 6| < \epsilon$ provided $0 < |x - 3| < \frac{\epsilon}{2}$.

◆ **EXAMPLE 7.7**   Interpret $\lim_{x \to 3} \frac{x^2 - 9}{x - 3}$ as the derivative of some function $K(x)$ evaluated at $x = a$.

(a) Determine $K(x)$ and the value of a.

(b) Evaluate the limit.

SOLUTION   (a) A derivative is the limit of the slope of the secant lines, so we want to interpret $\frac{x^2-9}{x-3}$ as $\frac{\Delta K}{\Delta x}$, the slope of the line through the points (x, x^2) and $(3, 9)$.

We see that the limit is being taken as x approaches 3, so the point (x, x^2) is approaching $(3, 9)$ along the curve $K(x) = x^2$.

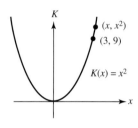

Figure 7.13

$K(x) = x^2$ and $a = 3$. That is, the derivative of x^2 at $x = 3$ is $\lim_{x \to 3} \frac{x^2-9}{x-3}$.

(b) There are several ways of evaluating this limit.

Method (i): In part (a) we showed that $\lim_{x \to 3} \frac{x^2-9}{x-3}$ is the derivative of x^2 at $x = 3$. In Chapter 6 we showed that the derivative of x^2 is $2x$. Evaluating this at $x = 3$ gives $\lim_{x \to 3} \frac{x^2-9}{x-3} = 6$.

Method (ii): Following the reasoning of the previous few examples, we can observe that

$$\frac{x^2 - 9}{x - 3} = \frac{(x - 3)(x + 3)}{x - 3} = \begin{cases} x + 3, & \text{for } x \neq 3, \\ \text{undefined}, & \text{for } x = 3. \end{cases}$$

So

$$\lim_{x \to 3} \frac{x^2 - 9}{x - 3} = \lim_{x \to 3} \frac{(x - 3)(x + 3)}{x - 3}$$

$$= \lim_{x \to 3} x + 3 \quad \text{since} \quad \frac{x - 3}{x - 3} = 1 \text{ for } x \neq 3$$

$$= 6.$$

Alternative ways of approaching the limit: Suppose you're staring at this limit and feel at a loss, or for some reason are feeling queasy about your answer. Or maybe you just want to double-check your work. You can investigate the limit graphically or investigate it numerically.

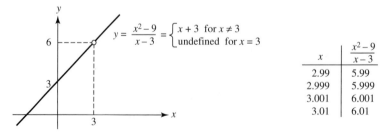

Figure 7.14

These methods lead you to conjecture that the limit is 6.   ◆

◆ **EXAMPLE 7.8**   Find $\lim_{x\to\infty} x^2$.

SOLUTION   We are being asked about the behavior of x^2 as x grows without bound. As x grows without bound, x^2 grows without bound; about these facts there should be no controversy. The problem is how we should answer the question. Reasonable people may differ. Some mathematicians don't like to write

$$\lim_{x\to\infty} x^2 = \infty.$$

After all, infinity is not something that $f(x) = x^2$ can snuggle up to; $f(x)$ can't get arbitrarily close to infinity. On the other hand, "as $x \to \infty$, $x^2 \to \infty$" is a reasonable shorthand for "as x grows without bound, x^2 grows without bound." Therefore in this text we use the convention that

$$\lim_{x\to\infty} x^2 = \infty$$

is a shorthand for "as x grows without bound, x^2 grows without bound." Similarly, $\lim_{x\to a} f(x) = \infty$ is shorthand for "as x approaches a, $f(x)$ grows without bound".   ◆

Before moving on from the definition of a limit, let's take a moment to debunk a common misconception with the next example.

◆ **EXAMPLE 7.9**   Find $\lim_{x\to 0} \frac{x}{x}$.

The graph of $\frac{x}{x}$ is given below. $\frac{x}{x} = 1$ for $x \neq 0$ and is undefined for $x = 0$.

$$\lim_{x\to 0} \frac{x}{x} = \lim_{x\to 0} 1$$
$$= 1$$

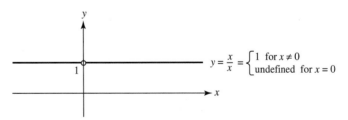

$$y = \frac{x}{x} = \begin{cases} 1 & \text{for } x \neq 0 \\ \text{undefined} & \text{for } x = 0 \end{cases}$$

Figure 7.15   ◆

From this example we see that $\lim_{x\to a} f(x) = L$ does *not* mean "as x gets closer and closer to a, $f(x)$ gets closer and closer to L but *never reaches* L"; in Example 7.9 $f(x)$ *does* reach $L = 1$ as x approaches 0. It is possible (but not required) that f will hit the value L any number of times.

EXERCISE 7.2   Show, using the methods of Example 7.3, that $\lim_{x\to 3} k = k$ for any constant k. Conclude that $\lim_{x\to c} k = k$ for c and k any constants.

EXERCISE 7.3   Show, using the methods of Example 7.3, that $\lim_{x\to c} x = c$, where c is any constant.

PROBLEMS FOR SECTION 7.1

1. Evaluate the following limits; then discuss $\lim_{x \to \infty} b^x$ for $b > 0$.

   (a) $\lim_{x \to \infty} (1.1)^x$
   (b) $\lim_{x \to \infty} (0.9)^x$
   (c) $\lim_{x \to 0} (1.1)^x$
   (d) $\lim_{x \to -\infty} (1.1)^x$
   (e) $\lim_{x \to -\infty} (0.9)^x$

2. We can define $\left(-\frac{1}{2}\right)^n$ for any positive integer n, but not for every real number. For instance, $\left(-\frac{1}{2}\right)^{1/2} = \sqrt{-\frac{1}{2}}$, which is not defined in real numbers. We'll write $\lim_{n \to \infty} \left(-\frac{1}{2}\right)^n = L$ if $\left(-\frac{1}{2}\right)^n$ can be made arbitrarily close to L for all positive integers n sufficiently large.

   (a) Find $\lim_{n \to \infty} \left(-\frac{1}{2}\right)^n$, where n takes on only positive integer values.

   (b) Which two of the following statements are true? Explain.
   i. $\lim_{n \to \infty} (-2)^n = \infty$
   ii. $\lim_{n \to \infty} (-2)^n = -\infty$
   iii. $\lim_{n \to \infty} (-2)^n$ does not exist
   iv. $\lim_{n \to \infty} -2^n = \infty$
   v. $\lim_{n \to \infty} -2^n = -\infty$

3. Each of the following limits are of the form $\lim_{x \to a} f(x)$. Evaluate the limit *and* sketch a graph of f on some interval including a. Make it clear from your sketch whether or not f is defined at a.

   (a) $\lim_{h \to -2} \dfrac{(h-3)(h+2)}{h+2}$
   (b) $\lim_{x \to 5} \dfrac{x^2 - 25}{x + 5}$
   (c) $\lim_{x \to -5} \dfrac{x^2 - 25}{x + 5}$
   (d) $\lim_{t \to 0} \dfrac{t^2 + \pi t}{t}$
   (e) $\lim_{h \to 0} \dfrac{hk + h^2}{h}$, where k is a constant
   (f) $\lim_{w \to 2} \dfrac{(w-3)(w+1)(w-2)}{3w - 6}$

4. *Discussion Question.* Consider $\lim_{x \to \infty} (1 + \frac{1}{x})^x$. This is a very important limit.

   (a) What is your guess for this limit? *You are not expected to guess the right answer. Once you complete part (b) you will see that the problem is subtle.*

   (b) Use a calculator or computer to investigate the limit, graphically and numerically. Give a revised estimate of this limit. *We will return to this limit in Chapter 15.*

   In Problems 5 through 13, graph f and evaluate the limit(s).

5. $f(x) = \dfrac{x}{2} + 3$; $\lim_{x \to 2} f(x)$

6. $f(x) = \pi x - 4$;
   (a) $\lim_{x \to 0} f(x)$
   (b) $\lim_{x \to 1} f(x)$
   (c) $\lim_{x \to \infty} f(x)$

7. $f(x) = |x - 2|$;
   (a) $\lim\limits_{x \to 0} f(x)$ (b) $\lim\limits_{x \to 2} f(x)$

8. $f(x) = \dfrac{x^2 - 2x}{x - 2}$;
   (a) $\lim\limits_{x \to 0} f(x)$ (b) $\lim\limits_{x \to 2} f(x)$

9. $f(x) = \begin{cases} 5x + 1, & x \neq 2 \\ 7, & x = 2 \end{cases}$;
   (a) $\lim\limits_{x \to 1} f(x)$ (b) $\lim\limits_{x \to 2} f(x)$

10. $f(x) = \dfrac{(x + 2)(x^2 - x)}{x(x - 1)}$;
    (a) $\lim\limits_{x \to 0} f(x)$ (b) $\lim\limits_{x \to 1} f(x)$

11. $f(x) = \dfrac{x^2 - 3x - 4}{x + 1}$;
    (a) $\lim\limits_{x \to 1} f(x)$ (b) $\lim\limits_{x \to -1} f(x)$

12. $f(x) = \begin{cases} |x|, & x \neq 3 \\ 0, & x = 3 \end{cases}$;
    (a) $\lim\limits_{x \to 0} f(x)$ (b) $\lim\limits_{x \to 3} f(x)$ (c) $\lim\limits_{x \to -\infty} f(x)$

13. $f(x) = \dfrac{x^2 - 9}{x + 3}$;
    (a) $\lim\limits_{x \to 3} f(x)$ (b) $\lim\limits_{x \to -3} f(x)$

In Problems 14 through 18, evaluate the limit by interpreting it as the derivative of some function f evaluated at x = a. Specify f and a, then calculate the limit.

14. $\lim\limits_{x \to 2} \dfrac{e^x - e^2}{x - 2}$

15. $\lim\limits_{h \to 0} \dfrac{\sqrt{9 + h} - 3}{h}$

16. $\lim\limits_{h \to 0} \dfrac{e^{1+h} - e}{h}$

17. $\lim\limits_{h \to 0} \dfrac{\sqrt{7 + h} - \sqrt{7}}{h}$

18. $\lim\limits_{h \to 0} \dfrac{e^h - 1}{h}$

In Problems 19 and 20, give an example of a function having the set of characteristics specified.

19. (a) $\lim_{x \to 5} f(x) = 7; f(5) = 7$

    (b) $\lim_{x \to 5} g(x) = 7; g(5) = 8$

20. (a) $\lim_{x \to \infty} f(x) = \infty; \lim_{x \to -\infty} f(x) = -\infty; \lim_{x \to 0} f(x) = 1$

    (b) $\lim_{x \to \infty} g(x) = \infty; \lim_{x \to -\infty} g(x) = \infty; \lim_{x \to 0} g(x) = -1$

21. Look back at Example 7.6. When approximating the slope of x^2 at $x = 2$, we end up with the expression $(4h + h^2)/h$. If we assume $h \neq 0$, then we can cancel the h's, arriving at $4 + h$. In this problem, we will investigate a function like $(4h + h^2)/h$.

    Consider the following pay scale for employees at the nepotistic Nelson Nattle Company. Let D be the date of hire (we use the start-up date of the Nattle Company as our benchmark time, $D = 0$, D measured in years) and let S be the starting annual salary.

$$S(D) = \begin{cases} \dfrac{(15000 + 200D)D(D-1)}{D(D-1)} \\ 50,0000 \text{ for } D = 0 \\ 35,000 \text{ for } D = 1 \end{cases}$$

    (a) Sketch the graph of $S(D)$. The graph looks a little weird until you find out that Nelson Nattle is the only person hired at $D = 0$ and that his son, Nelson Nattle, Jr., is due back from college exactly one year after the Nattle Company's start-up date. The expression $\frac{(15000+200D)D(D-1)}{D(D-1)}$ is equal to $15000 + 200D$ as long as $D \neq 0$ and $D \neq 1$. For $D = 0$ and $D = 1$, $\frac{(15000+200D)D(D-1)}{D(D-1)}$ is undefined.

    (b) Nelson has just received a letter from his brother Nathaniel asking for a position in the company. Nathaniel's projected date of hire is $D = 1.5$. Nelson is thinking of offering his brother a starting salary of \$40,000. Adjust $S(D)$ to define $S(1.5)$ appropriately.

22. Let $f(t) = \dfrac{t(3+t)}{t}$.

    (a) Sketch the graph of $f(t)$.

    (b) What are the domain and range of $f(t)$?

■ 7.2   LEFT- AND RIGHT-HANDED LIMITS; SOMETIMES THE APPROACH IS CRITICAL

◆ EXAMPLE 7.10   How does $f(x) = \frac{1}{x}$ behave as x approaches zero?

SOLUTION   Let's begin by investigating this graphically and numerically.

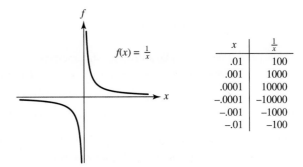

x	$\frac{1}{x}$
.01	100
.001	1000
.0001	10000
−.0001	−10000
−.001	−1000
−.01	−100

$f(x) = \frac{1}{x}$

Figure 7.16

As x approaches zero from the right (i.e., through numbers greater than zero), $\frac{1}{x}$ grows without bound. We can express this by writing

$$\lim_{x \to 0^+} \frac{1}{x} = \infty,$$

where "$x \to 0^+$" indicates that only numbers greater than zero are being considered. We read "$\lim_{x \to 0^+}$" as "the limit as x approaches 0 from the right."

As x approaches zero from the left (i.e., through numbers less than zero), $\frac{1}{x}$ decreases without bound. We can express this by writing

$$\lim_{x \to 0^-} \frac{1}{x} = -\infty,$$

where "$x \to 0^-$" indicates that only numbers less than zero are being considered.   ◆

We can define *finite* one-sided limits in general as follows.

Definitions

If the values of $f(x)$ stay arbitrarily close to L provided x is sufficiently close to a but greater than a, we write

$$\lim_{x \to a^+} f(x) = L$$

and read this as "the limit of $f(x)$ as x approaches a from the right is L."

Analogously, if the values of $f(x)$ can be made to stay arbitrarily close to L provided x is sufficiently close to a but less than a, then we write

$$\lim_{x \to a^-} f(x) = L.$$

These definitions can be made more precise in exactly the same manner the definition of $\lim_{x \to a} f(x) = L$ was made more precise. From that latter definition we see that in order to have $\lim_{x \to a} f(x) = L$ it is necessary that both the left- and right-hand limits be equal to L as well.

$$\boxed{\text{If } \lim_{x \to a^-} f(x) \neq \lim_{x \to a^+} f(x), \text{ then } \lim_{x \to a} f(x) \text{ does not exist.}}$$

(We will make a stronger statement about the relationship between one-sided and two-sided limits later in this section.) It follows that $\lim_{x \to 0} \frac{1}{x}$ does not exist.

EXAMPLE 7.11   Evaluate $\lim_{x \to 0^+} \frac{|x|}{x}$, $\lim_{x \to 0^-} \frac{|x|}{x}$, and $\lim_{x \to 0} \frac{|x|}{x}$.

SOLUTION   The basic strategy for dealing with absolute values is to split the problem into two cases, one where the quantity inside the absolute value is negative and the other where this quantity is nonnegative.

$$\begin{cases} \text{If } x > 0, & \dfrac{|x|}{x} = \dfrac{x}{x} = 1; \\[2mm] \text{if } x < 0, & \dfrac{|x|}{x} = \dfrac{-x}{x} = -1; \\[2mm] \text{if } x = 0, & \dfrac{|x|}{x} \text{ is undefined.} \end{cases}$$

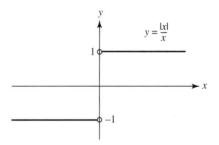

Figure 7.17

$\lim_{x \to 0^+} \frac{|x|}{x} = \lim_{x \to 0^+} 1 = 1,$

$\lim_{x \to 0^-} \frac{|x|}{x} = \lim_{x \to 0^-} -1 = -1,$

$\lim_{x \to 0} \frac{|x|}{x}$ does not exist because the left- and right-hand limits are unequal.   ◆

EXAMPLE 7.12   Evaluate $\lim_{x \to 0^+} |x|$, $\lim_{x \to 0^-} |x|$, and $\lim_{x \to 0} |x|$.

SOLUTION

$$\begin{cases} \text{If } x \geq 0, & |x| = x; \\ \text{if } x < 0, & |x| = -x. \end{cases}$$

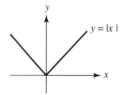

Figure 7.18

$$\lim_{x \to 0^+} |x| = \lim_{x \to 0^+} x = 0,$$

$$\lim_{x \to 0^-} |x| = \lim_{x \to 0^-} (-x) = 0,$$

$$\lim_{x \to 0} |x| = 0. \quad \blacklozenge$$

◆ **EXAMPLE 7.13**   Let $f(x) = |x|$.

(a) Sketch the graph of f and sketch the graph of f'.

(b) Evaluate $f'(1)$ and $f'(-2)$.

(c) Verify that $f'(0)$ is undefined.

SOLUTION

(a)   i. graph of $f(x) = |x|$

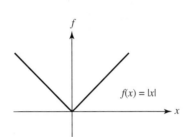

ii. graph of $f'(x) = \frac{d}{dx} |x|$

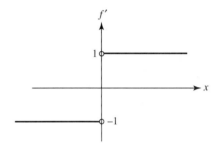

Figure 7.19

(b) For $x > 0$ $f(x) = x$, so $f'(1) = 1$.
    For $x < 0$ $f(x) = -x$, so $f'(-2) = -1$.

(c) Approaching the problem from a graphical viewpoint, we notice that the graph of f does not appear locally linear at $x = 0$. No matter how much the region around $x = 0$ is magnified the graph has a sharp corner there, so it will never look like a straight line. The slope of f is -1 if x approaches zero from the left and $+1$ if x approaches zero from the right.

Analytically we can argue similarly.

$$f'(0) = \lim_{h \to 0} \frac{f(0 + h) - f(0)}{h}$$

$$= \lim_{h \to 0} \frac{|h| - |0|}{h}$$

$$= \lim_{h \to 0} \frac{|h|}{h}$$

This is the limit we looked at in Example 7.11. We concluded that $\lim_{h \to 0} \frac{|h|}{h}$ does not exist. Therefore $\frac{d}{dx} |x| \Big|_{x=0}$, the derivative of $|x|$ evaluated at $x = 0$, does not exist.   ◆

◆ **EXAMPLE 7.14**   Find $\lim_{t \to 0} \frac{1}{t^2}$.

SOLUTION $\lim_{t \to 0^+} \frac{1}{t^2} = +\infty$; likewise, $\lim_{t \to 0^-} \frac{1}{t^2} = +\infty$.

Therefore we say $\lim_{t \to 0} \frac{1}{t^2} = +\infty$.

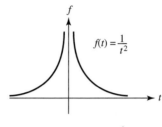

$f(t) = \frac{1}{t^2}$

Figure 7.20   ◆

The relationship between left- and right-hand limits and two-sided limits can be stated most succinctly as follows.

$$\lim_{x \to a} f(x) = L \text{ if and only if } \lim_{x \to a^+} f(x) = \lim_{x \to a^-} f(x) = L.$$

The limit as x approaches a of $f(x)$ is L if and only if the left- and right-hand limits are both equal to L.

In this section we've seen that if $\lim_{x \to a^-} f(x) = L_1$ and $\lim_{x \to a^+} f(x) = L_2$ but $L_1 \neq L_2$, then $\lim_{x \to a} f(x)$ does not exist. This is not the only situation in which a limit does not exist. Consider, for example, the function $g(x)$ that is 1 when x is rational and -1 when x is irrational. $\lim_{x \to 0} g(x)$ does not exist since in *any* open interval around zero, no matter how small, there will be both rational and irrational numbers. In this case neither the left- nor right-hand limit exists. Similarly, $\lim_{x \to \infty} g(x)$ does not exist, because for any N, no matter how big N is, there will always be both rational and irrational numbers larger than it.

PROBLEMS FOR SECTION 7.2

1. Let f be the first-class postage function for 1997 given in Example 2.3. The input of the function is W, where W is the weight, in ounces, of the item to be mailed and the output is the price of the postage. Return to Chapter 2 for details. Find the following.

(a) $\lim_{W \to 1^-} f(W)$ (b) $\lim_{W \to 1^+} f(W)$ (c) $\lim_{W \to 1} f(W)$

(d) $\lim_{W \to 1.5} f(W)$ (e) $\lim_{W \to 2} f(W)$ (f) $\lim_{W \to 12^+} f(W)$

(g) $\lim_{W \to 12^-} f(W)$ (h) $\lim_{W \to 12} f(W)$

In Problems 2 through 7, evaluate the limits. A graph may be useful.

2. (a) $\lim_{x \to 2^-} \frac{1}{x-2}$ (b) $\lim_{x \to 2^+} \frac{1}{x-2}$ (c) $\lim_{x \to 2} \frac{1}{x-2}$

3. (a) $\lim_{x \to 1^-} \frac{1}{(x-1)^2}$ (b) $\lim_{x \to 1^+} \frac{1}{(x-1)^2}$ (c) $\lim_{x \to 1} \frac{1}{(x-1)^2}$ (d) $\lim_{x \to -1} \frac{1}{(x-1)^2}$

4. $f(x) = \frac{|x-3|}{x-3}$

(a) $\lim\limits_{x \to 0} f(x)$ (b) $\lim\limits_{x \to 4} f(x)$ (c) $\lim\limits_{x \to 3^+} f(x)$

(d) $\lim\limits_{x \to 3^-} f(x)$ (e) $\lim\limits_{x \to 3} f(x)$

5. $f(x) = \begin{cases} 3x + 4, & x < 0 \\ 2x + 4, & x \geq 0 \end{cases}$

(a) $\lim\limits_{x \to 1} f(x)$ (b) $\lim\limits_{x \to -2} f(x)$ (c) $\lim\limits_{x \to 0^+} f(x)$

(d) $\lim\limits_{x \to 0^-} f(x)$ (e) $\lim\limits_{x \to 0} f(x)$

6. $f(x) = \begin{cases} \pi x + 1, & x > 0 \\ \pi x - 1, & x \leq 0 \end{cases}$

(a) $\lim\limits_{x \to \frac{1}{\pi}} f(x)$ (b) $\lim\limits_{x \to -\frac{1}{\pi}} f(x)$ (c) $\lim\limits_{x \to 0} f(x)$

7. $f(x) = \begin{cases} x^2 + 3, & x \geq 1 \\ 2x + 2, & x < 1 \end{cases}$

(a) $\lim\limits_{x \to 0} f(x)$ (b) $\lim\limits_{x \to 1} f(x)$ (c) $\lim\limits_{x \to 2} f(x)$

8. Let f be the function defined by $f(x) = \begin{cases} x + 1, & \text{for } x \text{ not an integer,} \\ 0, & \text{for } x \text{ an integer.} \end{cases}$

(a) Sketch f.

(b) Find the following limits.

   i. $\lim\limits_{x \to 1.5} f(x)$ ii. $\lim\limits_{x \to 2} f(x)$ iii. $\lim\limits_{x \to 0} f(x)$

(c) For what values of c is $\lim_{x \to c} f(x) = c + 1$? Have you excluded any values of c? If so, which ones and why? Explain.

9. Let $g(x)$ be the function defined by $g(x) = \begin{cases} x^2, & \text{for } x > 2, \\ x, & \text{for } x \leq 2. \end{cases}$

(a) Find the following.

   i. $\lim\limits_{x \to 2^+} g(x)$ ii. $\lim\limits_{x \to 2^-} g(x)$ iii. $\lim\limits_{x \to 2} g(x)$

(b) $h(x)$ is defined by $h(x) = \begin{cases} x^2, & \text{for } x > 2, \\ mx, & \text{for } x \leq 2. \end{cases}$

   Find the value of m that will make $f(x)$ a continuous function. (A graph is highly recommended.) Using the value of m you've found, determine $\lim_{x \to 2} h(x)$.

10. Let f be the function given by

$$f(x) = \begin{cases} \frac{1}{x+7}, & \text{for } x \leq -5, \\ x + 4, & \text{for } -5 < x < -2, \\ |x|, & \text{for } -2 \leq x \leq 2, \\ 2^{-x} + 1, & \text{for } x > 2. \end{cases}$$

A sketch of f is provided below.

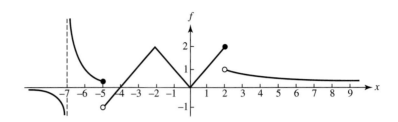

Evaluate the limits below.

(a) $\lim_{x \to -\infty} f(x)$ (b) $\lim_{x \to -7} f(x)$ (c) $\lim_{x \to -5^+} f(x)$

(d) $\lim_{x \to -5} f(x)$ (e) $\lim_{x \to -2} f(x)$ (f) $\lim_{x \to 2^+} f(x)$

(g) $\lim_{x \to \infty} f(x)$ (h) $\lim_{x \to 0} f(x)$

11. Let $f(x)$ be the function defined and graphed in Problem 10.
    (a) Sketch a graph of $f'(x)$.
    (b) Where is f' undefined?
    (c) Evaluate the following.
        i. $\lim_{x \to \infty} f'(x)$ ii. $\lim_{x \to -2^-} f'(x)$ iii. $\lim_{x \to -2^+} f'(x)$
        iv. $\lim_{x \to -2} f'(x)$ v. $\lim_{x \to -7^-} f'(x)$ vi. $\lim_{x \to -7^+} f'(x)$

12. (a) Sketch the graph of $f(x) = 2|x - 2|$.
    (b) What is $f'(0)$? What is $f'(4)$? What is $f'(2)$?
    (c) Find $\lim_{h \to 0^+} \frac{f(2+h)-f(2)}{h}$ and $\lim_{h \to 0^-} \frac{f(2+h)-f(2)}{h}$.
        Do your answers make sense to you?
    (d) On a separate set of axes graph $f'(x)$.

Each of the statements in Problems 13 through 16 is false. Produce a counterexample to show that the statement is false.

13. If $f(3) = 7$, then $\lim_{x \to 3} f(x) = 7$.

14. If $f(2)$ is undefined, then $\lim_{x \to 2} f(x)$ is undefined.

15. If $\lim_{x \to 0^+} f(x)$ and $\lim_{x \to 0^-} f(x)$ both exist, then $\lim_{x \to 0} f(x)$ exists.

16. If $\lim_{x \to 5} f(x) = 2$ then $f(5) = 2$.

For Problems 17 through 21, use the graph given to determine each of the following.

(a) $\lim_{x \to 1^+} f(x)$
(b) $\lim_{x \to 1^-} f(x)$
(c) $\lim_{x \to 1} f(x)$

17.

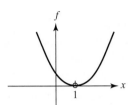

18.

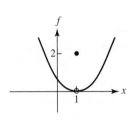

19.

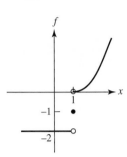

20.

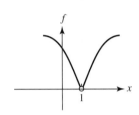

21.

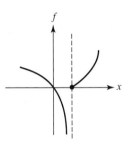

7.3   A STREETWISE APPROACH TO LIMITS

Suppose you and your trusty calculator are strolling down the street one night. As you pass a dark alley the Math Mugger jumps you, throwing you a nasty limit. A quick answer is demanded. Although normally cool and level-headed, you start to panic. You barely heard the question, but the words "zero," "one," "infinity," and "does not exist" are at the tip of your tongue. You try to calm yourself, asking boldly, "Could you repeat the question, please?"

Let's suppose the question thrown at you is "What is $\lim_{x \to 1} \frac{x^3 - 1}{x - 1}$?" First let's look at potholes to avoid. Then we'll supply some calculus street-survival tactics. Finally, we'll determine whether using these survival tactics would, in this particular instance, get you out of the bind you're in.

Potholes and Mudholes with Crocodiles

In your panic you might look at $\lim_{x \to 1} \frac{x^3 - 1}{x - 1}$ and think, "when $x \to 1$ the numerator tends toward 0, and if the numerator is 0 then the fraction is 0, so the answer is 0." Alternatively, you might think, "when $x \to 1$ the denominator tends toward 0, and when the denominator goes to 0 the limit does not exist or is infinite." If you've had these two thoughts in rapid succession, you might put them together and think, "so we have $\frac{0}{0}$, which is undefined whenever it's not 1."[3]

There are serious problems with these lines of thought. While it is true that both the numerator and the denominator of this fraction are approaching zero, we're interested in

---
[3] Notice that all this panicked thinking has put you in the same place you were before you heard the question.

what this *ratio* looks like as $x \to 1$. Think about this carefully; *every* derivative we calculate looks like $\lim_{\Delta x \to 0} \frac{\Delta f}{\Delta x}$ and therefore looks like $\frac{0}{0}$, yet this limit could work out to be *any* number, depending upon the function. To assert that every limit of the form $\frac{0}{0}$ is either 1 or does not exist is equivalent to asserting that *all* derivatives are 1 or undefined, an assertion you know is patently false!

Another approach you might take to $\lim_{x \to 1} \frac{x^3-1}{x-1}$ is to simplify the fraction $\frac{x^3-1}{x-1}$. That's a solid idea. If you can factor $x^3 - 1$, you're in great shape and don't need help. It's entirely possible, however, that you know how to factor the difference of two perfect squares but not of two perfect cubes. Do not resort to wishful-thinking algebra. If you're not sure of your algebra, you've got means at your disposal to check it.

- ■ If you're not sure of a factorization, multiply out to determine whether it works.

- ■ If you think you can simplify without factoring (you can't), do a spot-check on yourself. For example, whatever you simplify this to, when you evaluate at $x = 2$ you should get $\frac{2^3-1}{2-1} = 7$.

Suppose you're stuck; you can't figure out how to simplify the fraction $\frac{x^3-1}{x-1}$ and your panic level is mounting.[4] What you need are street-survival skills.

Street-Survival Tactics

The survival tactics are simple and can be applied in a large variety of situations, not just limit calculations in back alleys.

- ■ Investigate the situation graphically.

- ■ Investigate the situation numerically.

The potholes identified above are analytic and algebraic missteps. Don't put on analytic blinders; view the problem from different angles.

Graphical and Numerical Investigations

We're interested in $\lim_{x \to 1} \frac{x^3-1}{x-1}$, so we want the graph of $f(x) = \frac{x^3-1}{x-1}$ around $x = 1$. For starters we might graph on the interval $[0, 2]$ and then zoom in around $x = 1$. To make sure we have the graph in the viewing window, we can begin by letting the range of y-values be from $f(0) = \frac{0^3-1}{0-1} = 1$ to $f(2) = \frac{2^3-1}{2-1} = 7$. We can zoom in around $x = 1$ as many times as we like. (As pointed out in Chapter 2, although $f(x)$ is undefined at $x = 1$, the graph on the calculator may not reflect this.) Use the "trace" key to trace along the curve near $x = 1$.

---

[4] For help simplifying, see Section 11.2

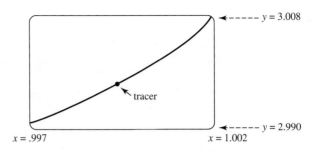

Figure 7.21

It is reasonable to conjecture that as $x \to 1$, $\frac{x^3-1}{x-1} \to 3$.

Another approach is to use your calculator to calculate values of $\frac{x^3-1}{x-1}$ for values of x approaching 1. (For instance, you could try 0.99, 0.9997, 1.0001, 1.00003, etc., or you could get your calculator to produce a table of values.) Again, your investigation would lead you to conjecture that as $x \to 1$, $\frac{x^3-1}{x-1} \to 3$.

So if the fellow in the back alley demands an answer immediately, tell him your best guess is 3.

Is this answer correct? Actually it is. $x^3 - 1$ can be factored into $(x-1)(x^2+x+1)$, allowing us to do the following.[5]

$$\lim_{x\to1} \frac{x^3-1}{x-1} = \lim_{x\to1} \frac{(x-1)(x^2+x+1)}{x-1}$$
$$= \lim_{x\to1} (x^2+x+1)$$
$$= 3$$

(Alternatively, this limit can be interpreted as the derivative of x^3 at $x = 1$ and calculated as $\lim_{h\to0} \frac{(1+h)^3-1^3}{h}$. Do this calculation as an exercise.)

Cautionary Notes

If you're going to use street-survival tactics, you're going to have to learn how to watch your back. The tactics recommended will often get you *near* the actual answer (*not* always, but frequently, if used wisely), but there is no reason to believe that they will lead you to the exact answer. For instance, in the last example at some point we simply guessed the answer was 3 (as opposed to some number very, very close to 3). If the actual answer was irrational, the calculator would probably not tell us this.

There are other limitations inherent in your trusty machine; sometimes the calculator can lead you astray. Suppose, for instance, we are interested in the derivative of 2^x at $x = 0$. It is given by

$$\lim_{h\to0} \frac{2^h-1}{h}.$$

[5] Chapter 11 will give you more insight into polynomials and their factorization.

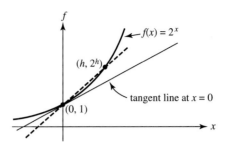

Figure 7.22

We can approximate this derivative by $\frac{2^h-1}{h}$ for h small. The approximation should be better and better the smaller h is. If h is positive, the value obtained should be greater than the actual limit. (See Figure 7.20.) The following are data collected from a calculator.

h	$\dfrac{2^h - 1}{h}$
$10^{-2} = 0.01$	0.6955550057
$10^{-3} = 0.001$	0.693387463
$10^{-4} = 0.0001$	0.6931712
$10^{-5} = 0.00001$	0.6931496
$10^{-6} = 0.000001$	0.693147
$10^{-7} = 0.0000001$	0.69315
$10^{-8} = 0.00000001$	0.6931
$10^{-9} = 0.000000001$	0.693
10^{-10}	0.69
10^{-11}	0.7
10^{-12}	1
10^{-13}	0
10^{-14}	0

Is the limit actually 0, not ≈ 0.693? The actual value of $\lim_{h\to 0}\frac{2^h-1}{h}$ is an irrational number. This irrational number is approximately 0.6931471806. What is going on here?

The information supplied by the calculator looks reasonable up until $h = 10^{-6}$. For the values of h smaller than this, the difference between 2^h and 1 is so tiny that the calculator is essentially losing one digit of information for each subsequent entry up through 10^{-12}. By $h = 10^{-13}$ and $h = 10^{-14}$, the difference $2^h - 1$ is so miniscule that the calculator just rounds the difference off to 0.

This phenomenon is not a quirk particular to this calculator, nor is it particular to the function 2^x. Similar issues arise whenever calculating a limit that can be interpreted as a derivative. So what should you do? A streetwise person is not an extremist. When investigating $\lim_{x\to a} f(x)$, try values of x close to a but not *too* close. And then make your guess and take your chances. You'll win some and you'll lose some. But frequently, thoughtful graphical and numerical investigation will put you in the vicinity of the actual answer.

Question: When is $\lim_{x \to a} f(x)$ simply equal to $f(a)$?

In the language of the streets, when can you just plug in $x = a$ and get the right answer even though taking the limit means x can't really be equal to a?[6]

This is a great question. Think about it for a minute. We'll recap Examples 7.3, 7.4, and 7.5 to give you food for thought.

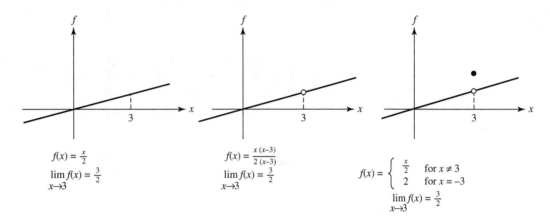

$$f(x) = \frac{x}{2}$$
$$\lim_{x \to 3} f(x) = \frac{3}{2}$$

$$f(x) = \frac{x\,(x-3)}{2\,(x-3)}$$
$$\lim_{x \to 3} f(x) = \frac{3}{2}$$

$$f(x) = \begin{cases} \frac{x}{2} & \text{for } x \neq 3 \\ 2 & \text{for } x = -3 \end{cases}$$
$$\lim_{x \to 3} f(x) = \frac{3}{2}$$

Figure 7.23

If f is continuous at $x = a$ then this 'method' will work. This brings us back from the streets to define continuity using limits.

PROBLEMS FOR SECTION 7.3

1. Is $\lim_{x \to \infty}(1 + \frac{1}{x})^x$ finite? If so, find two consecutive integers, one smaller than this limit and the other larger. We will return to this limit later in the course.

In Problems 2 and 3, find the limit.

2. $\lim_{x \to 3} \dfrac{2x^3 - 8x^2 + 5x + 3}{x - 3}$

3. $\lim_{x \to 0} \dfrac{\sqrt{4 + x} - 2}{x}$

4. (a) Suppose you are interested in finding $\lim_{x \to 2} f(x)$, where the function $f(x)$ is explicitly given by a formula. What approaches might you take to investigate this limit?

   (b) Suppose you're now interested in finding $\lim_{x \to \infty} f(x)$, where $f(x)$ again is explicitly given by a formula. What approaches might you take to investigate this limit?

---

[6] The more math you know the less you'll have to rely on street tactics.

7.4   CONTINUITY AND THE INTERMEDIATE AND EXTREME VALUE THEOREMS

Our intuitive notion is that a function is continuous if its graph can be drawn without lifting pencil from paper. For f to be continuous at $x = a$ the values of $f(x)$ must be close to $f(a)$ for x near a. To define continuity we'll begin by looking at the type of behavior we are trying to rule out. The function sketched below is discontinuous at $x = -3$, $x = -1$, $x = 0$, $x = 1$, and $x = 5.2$.

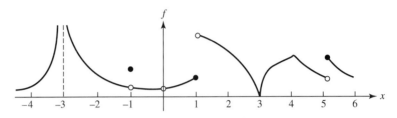

Figure 7.24

To begin with, we might insist that if f is to be continuous at $x = a$, then

$$\lim_{x \to a^-} f(x) = \lim_{x \to a^+} f(x),$$

where this limit is finite. For the function shown in Figure 2.24, this rules out the *jump* discontinuities at $x = -3$, $x = 1$, and $x = 5$ but doesn't take care of the *removable* discontinuities at $x = -1$ and $x = 0$. The latter reminds us that $f(a)$ must be defined and equal to the left- and right-hand limits. In other words, the following conditions must be satisfied:

i. f must be defined at $x = a$,

ii. left- and right-hand limits at $x = a$ must be equal,

iii. $f(a)$ must be equal to $\lim_{x \to a} f(x)$.

We can put this more succinctly.

Definition

The function f is **continuous at** $x = a$ if

$$\lim_{x \to a^-} f(x) = \lim_{x \to a^+} f(x) = f(a).$$

Equivalently, f is continuous at $x = a$ if

$$\lim_{x \to a} f(x) = f(a).$$

f is **continuous on an open interval** if f is continuous at every point in the interval.[7]

There are two theorems concerning continuous functions that we will use repeatedly.

---

[7] An open interval can be of the form (a, b), (a, ∞), $(-\infty, b)$ or $(-\infty, \infty)$.

Fact 1 (The Intermediate Value Theorem)

If f is continuous on the closed and bounded interval $[a, b]$ and $f(a) = A$, $f(b) = B$, then somewhere in the interval f attains every value between A and B. In particular, if a continuous function changes sign on an interval, it must be zero somewhere on that interval.

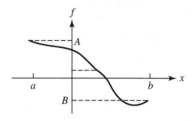

Figure 7.25

Fact 2 (The Extreme Value Theorem)

If a function f is continuous on a closed interval $[a, b]$, then f takes on both a maximum (high) and a minimum (low) value on $[a, b]$.[8]

For f to attain the maximum value of M on $[a, b]$ means that there is a number c in $[a, b]$ such that $f(c) = M$ and $f(x) \leq M$ for all x in $[a, b]$. Analogously, for f to attain the minimum value of m on $[a, b]$ means that there is a number c in $[a, b]$ such that $f(c) = m$ and $f(x) \geq m$ for all x in $[a, b]$.

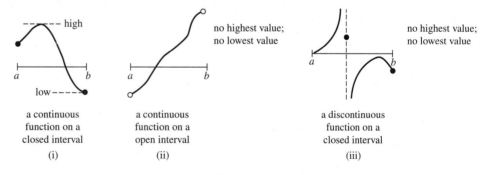

Figure 7.26

Studyinging parts (ii) and (iii) in Figure 7.26 should convince you that both conditions, the continuity of f and the interval being closed, are necessary in order for the statement to hold.

---

[8] For a a proof of either of these theorems, look in a more theoretical calculus book.

Note that even if $f(x) = k$, where k is a constant, the statement holds. Given the definitions of maximum and minimum presented above, if $f(x) = k$ on $[a, b]$, then for every $x \in [a, b]$ $f(x) = k$ is *both* a maximum value and a minimum value.

Principles for Working with Limits and Their Implications for Derivatives

The following general principles can be deduced from the definition of limits.

Suppose $\lim_{x \to a} f(x) = L_1$ and $\lim_{x \to a} g(x) = L_2$, where L_1 and L_2 are finite. Then:

1. $\lim_{x \to a}[f(x) \pm g(x)] = L_1 \pm L_2$ The limit of a sum (difference) is the sum (difference) of the limits.

2. $\lim_{x \to a}[f(x)g(x)] = L_1 \cdot L_2$ The limit of a product is the product of the limits (in particular, $g(x)$ may be constant: $\lim_{x \to a} kf(x) = kL_1$).

3. $\lim_{x \to a} \frac{f(x)}{g(x)} = \frac{L_1}{L_2}$, provided $L_2 \neq 0$.   The limit of a quotient is the quotient of the limits (provided the denominator has a nonzero limit).

4. If h is continuous at L_2, then $\lim_{x \to a} h(g(x)) = h(L_2) = h(\lim_{x \to a} g(x))$.

5. If $f(x) < g(x)$ for all x in the vicinity of a (although not necessarily at $x = a$), then $\lim_{x \to a} f(x) \leq \lim_{x \to a} g(x)$.

We'll also sometimes use what is known as the Sandwich Theorem, or the Squeeze Theorem.

Sandwich Theorem

If $f(x) \leq j(x) \leq g(x)$ for all x in the vicinity of a (although not necessarily at $x = a$) and

$$\lim_{x \to a} f(x) = \lim_{x \to a} g(x) = L,$$

then

$$\lim_{x \to a} j(x) = L.$$

The idea behind this theorem is that the functions f and g act as a vise, as lower and upper bounds for j in the vicinity of a. As x approaches a the lower and upper jaws of the vise get arbitrarily close together, trapping $\lim_{x \to a} j(x)$ between them.

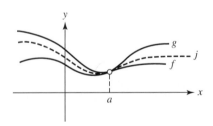

Figure 7.27

EXERCISE 7.4   Use the principles for working with limits, along with the conclusions of Examples 7.1, 7.2, and 7.3 and Exercises 7.1 and 7.2 to calculate the following.

$$\text{(a)} \ \lim_{x\to\infty} 3\cdot 2^{-x} + 4 \qquad \text{(b)} \ \lim_{x\to\sqrt{2}} (2x^2\sqrt{2}x + 3) \qquad \text{(c)} \ \lim_{x\to\infty} \frac{3 + 6x}{x}$$

We can use principles (1) and (2) given above to prove two very useful properties of derivatives.

Properties of Derivatives[9]

i. $\frac{d}{dx}kf(x) = k\frac{d}{dx}f(x)$, where k is any constant.
Multiplying f by a constant k multiplies its derivative by k.

ii. $\frac{d}{dx}[f(x) + g(x)] = \frac{d}{dx}f(x) + \frac{d}{dx}g(x)$
The derivative of a sum is the sum of the derivatives.

These are very important results. In the Exploratory Problems for this chapter you will be asked to prove these properties and to make sense out of them.

As an application of some of the principles for working with limits given in this section, we will verify the following fact.

Theorem: Differentiability Implies Continuity

If a function is differentiable at $x = a$ (that is, $f'(a)$ exists), then f is continuous at $x = a$.

Although the line of reasoning in the proof is easier to follow than to come up with, the conclusion should make sense intuitively. If f is differentiable at $x = a$, then f is locally linear at $x = a$; f looks like a line near $x = a$. It makes sense that f must be continuous at $x = a$.

PROOF OF THEOREM Suppose that a is a point in the domain of f, where f is defined on some open interval containing a. We will assume that $f'(a)$ exists and show that f is continuous at $x = a$.

According to our definition of continuity, f is continuous at $x = a$ if

$$\lim_{x\to a} f(x) = f(a).$$

We will show that this is true provided $f'(a)$ exists. Showing that $\lim_{x\to a} f(x) = f(a)$ is equivalent to showing that $\lim_{x\to a}[f(x) - f(a)] = 0$.

---

[9] Recall that $\frac{d}{dx}\ldots$ means "the derivative of $\ldots$"

$$\lim_{x \to a}[f(x) - f(a)] = \lim_{x \to a}\left[(f(x) - f(a))\frac{x - a}{x - a}\right] \qquad \text{We are multiplying by 1, since } x \neq a.$$

$$= \left[\lim_{x \to a}\frac{f(x) - f(a)}{x - a}\right] \cdot \left[\lim_{x \to a}(x - a)\right] \qquad \text{The limit of a product is the product of the limits if both exist and are finite.}$$

$$= f'(a) \cdot 0 \qquad \text{The first limit is } f'(a), \text{ which exists by assumption, and the second equals zero.}$$

$$= 0$$

We have shown that if $f'(a)$ exists, then $f(x)$ must be continuous at $x = a$.

Question: If f is continuous at $x = a$, is f necessarily differentiable (locally linear) at $x = a$?

Answer: No. Informally speaking, if f has a sharp corner at $x = a$, then it is not differentiable at $x = a$ because it is not locally linear there. A classic example of the latter situation is the function $f(x) = |x|$ at $x = 0$. $f(x) = |x|$ is continuous at $x = 0$ because $\lim_{x \to 0} f(x) = f(0) = 0$. However, as we saw in Example 7.13, $f'(0)$ does not exist; f is not differentiable at $x = 0$.

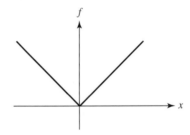

Figure 7.28

You are now prepared to read Appendix C. There we offer proofs of facts about derivatives stated without proof in Chapter 5.

Exploratory Problems for Chapter 7

Pushing the Limit

1. Let $h(t) = kf(t)$, where k is a constant and f is a differentiable function.

   (a) Use the principles of working with limits to show that $h'(t) = kf'(t)$. Begin with the limit definition of $h'(t)$ and then express h in terms of f.

   (b) Explain this result in graphical terms. Why is the slope of the tangent to h at t equal to k times the slope of the tangent to f at t?

   (c) Interpret this result in the case that $f(t)$ is a position function and $k = 2$. More specifically, consider the following example. Two women leave a Midwestern farmhouse and travel north on a straight road. One woman walks and the other woman runs. Suppose $f(t)$ gives the distance between the walker and the farmhouse at time t and $h(t) = 2f(t)$ gives the distance between the runner and the farmhouse. Interpret the result $h'(t) = 2f'(t)$ in this context. Generalize to the case $h'(t) = kf'(t)$.

2. Let $j(t) = f(t) + g(t)$, where f and g are differentiable functions.

   (a) Use the principles of working with limits to show that $j'(t) = f'(t) + g'(t)$. Begin with the limit definition of $j'(t)$ and then express j in terms of f and g.

   (b) Explain this result in graphical terms.

   (c) Interpret this result in the following scenario. A teacher has put retirement money in two accounts, TIAA and CREF. Let $f(t)$ be the value of his TIAA account at time t and let $g(t)$ be the value of his CREF account at time t. Interpret the result $j'(t) = f'(t) + g'(t)$ in this context.

3. (a) Use the properties of limits and the results of the previous problems to differentiate $f(x) = ax^3 + bx^2 + cx + d$ as efficiently as possible.

   (b) Along the way, you'll need to use the limit definition of derivative to differentiate x^3. Explain why your answer makes sense graphically by looking at the graphs of x^3 and $3x^2$.

4. *Looking for Patterns:* Use the results of Problem 3 above and formulas for derivatives of $\sqrt{x}$ and $\frac{1}{x}$ found in Chapter 5 to arrive at a formula for the derivative of x^n for $n = 0, 1, 2, 3, -1,$ and $\frac{1}{2}$. Try your formula on another value of n and see if it works.

PROBLEMS FOR SECTION 7.4

1. (a) Find the following limits. Illustrate your answers with graphs.

   i. $\displaystyle\lim_{x\to-\infty} -\frac{3}{x}$ ii. $\displaystyle\lim_{x\to\infty} -\frac{3}{x}$ iii. $\displaystyle\lim_{x\to\infty} -\frac{3}{x} - 3$

   iv. $\displaystyle\lim_{x\to\infty} \frac{x+1}{x}$ v. $\displaystyle\lim_{x\to\infty} \frac{2x+3}{x}$

   (b) In Section 7.1, Example 7.1, we showed $\lim_{x\to\infty} \frac{1}{x} = 0$. In Section 7.4 we listed some principles for working with limits. Show how your answers to all of the problems in part (a) can be deduced using

   $$\lim_{x\to\infty} \frac{1}{x} = 0 \quad \text{and} \quad \lim_{x\to\infty} k = k, \text{ for any constant } k$$

   and applying the principles listed in Section 7.4.

2. Find the following.

   (a) $\displaystyle\lim_{x\to4} \frac{1}{(x-4)^2}$ (b) $\displaystyle\lim_{x\to4} \frac{x+3}{(x-4)^2}$ (c) $\displaystyle\lim_{x\to4} \frac{x^2+16}{(x-4)^2}$

   (d) $\displaystyle\lim_{x\to4^+} \frac{1}{(x-4)}$ (e) $\displaystyle\lim_{x\to4^-} \frac{1}{(x-4)}$ (f) $\displaystyle\lim_{x\to4} \frac{x^2-16}{(x-4)}$

3. Find $\displaystyle\lim_{x\to4} \frac{\sqrt{x}-2}{x-4}$.

4. Suppose $|h(x)| \le 3$ for all x. Evaluate $\displaystyle\lim_{x\to\infty} \frac{h(x)}{x}$.

Each of the functions in Problems 5 through 10 is either continuous on $(-\infty, \infty)$ or has a point of discontinuity at some point(s) $x = a$. Determine any point(s) of discontinuity. Is the point of discontinuity removable? In other words, can the function be made continuous by defining or redefining the function at the point of discontinuity?

5. $f(x) = \frac{x^2-4}{x+2}$

6. $f(x) = \begin{cases} x^3, & x \ne 0, \\ 3, & x = 0. \end{cases}$

7. $f(x) = \frac{1}{x+2}$

8. $f(x) = \frac{1}{x^2+2}$

9. $f(x) = \begin{cases} -x^2 - 1, & x > 0, \\ 5x - 1, & x < 0. \end{cases}$

10. $f(x) = \begin{cases} -x^2 - x, & x > 0, \\ 5x - 1, & x < 0. \end{cases}$

11. Let $f(x) = \begin{cases} -x^2 + 1, & x > 0, \\ ax + b, & x \le 0. \end{cases}$

    What are the constraints on a and b in order for f to be continuous at $x = 0$?

12. Let $f(x) = \begin{cases} g(x), & x \geq a, \\ h(x), & x < a, \end{cases}$ where g is continuous on $[a, \infty)$ and h is continuous on $(-\infty, a)$.

What must be true about g and h in order for f to be continuous at $x = a$?

13. Let $f(x) = \begin{cases} -1, & \text{if } x \text{ is a rational number,} \\ 1, & \text{if } x \text{ is an irrational number.} \end{cases}$

   (a) Is $f(x)$ continuous at $x = 1$?

   (b) Is $f(x)$ continuous at $x = \pi$?

14. In this problem we'll look at a range of possible scenarios accompanying $\lim_{x \to a} f(x)$, where a is a finite number. Your job is to draw illustrations showing f in the vicinity of a with the specifications given.

   (a) Suppose $\lim_{x \to 2} f(x) = L$, where L is a finite number.

   Draw three qualitatively different pictures of what f could look like in the vicinity of $x = 2$. The first picture should show a continuous function f. The second should show a function discontinuous at $x = 2$ but defined at $x = 2$. The third should show f undefined at $x = 2$.

   (b) Illustrate the following two scenarios.

   i. $\lim_{x \to 2} f(x) = \infty$ ii. $\lim_{x \to 2} f(x) = -\infty$

   (c) Suppose $\lim_{x \to 2} f(x)$ is undefined.

   Draw three qualitatively different pictures of what f could look like in the vicinity of $x = 2$. The first picture should have f defined at 2. The second picture should have f undefined at 2 but the one-sided limits at 2 both finite. The third picture should have $\lim_{x \to 2+} f(x) = \infty$; $\lim_{x \to 2-} f(x)$ is left up to you.

   (Another possibility is that $f(x)$ does not approach any single finite value but also does not increase or decrease without bound.)

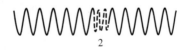

2

15. Let $f(x) = \begin{cases} x^2, & \text{for } x \geq 0, \\ -x^2, & \text{for } x < 0. \end{cases}$

   (a) Is f continuous at $x = 0$?

   (b) Is f differentiable at $x = 0$? If so, what is $f'(0)$?

16. Find the derivative of $f(x) = kx^4$, where k is a constant.

17. Let $g(x) = \begin{cases} x^2, & \text{for } x \geq 0, \\ x, & \text{for } x < 0. \end{cases}$

   (a) Is g continuous at $x = 0$?

   (b) Is g differentiable at $x = 0$? If so, what is $g'(0)$?

18. Let $f(x) = \frac{1}{x}$.

    (a) Draw the graph of $f(x)$ and $f'(x)$.

    (b) Use the graphs you've drawn in part (a) to do the following.

        i. Find $\displaystyle\lim_{x \to \infty} f'(x)$.

        ii. Find $\displaystyle\lim_{x \to 0^+} f'(x)$.

        iii. Find $\displaystyle\lim_{x \to 0^-} f'(x)$.

        iv. Find $\displaystyle\lim_{x \to -\infty} f'(x)$.

    (c) Use the limit definition of derivative to find $f'(x)$. Use your work to check your answers to parts (a) and (b).

19. The domain of a function f is all real numbers. The zeros of $f(x)$ are $x = -1$, $x = 2$, and $x = 6$. There are no other x-values such that $f(x) = 0$. Is it possible that $f(3) > 0$ and $f(4) < 0$? Explain.

20. The domain of a **continuous** function f is all real numbers. The zeros of f are $x = -1$, $x = 2$, and $x = 6$. There are no other x-values such that $f(x) = 0$. Is it possible that $f(3) > 0$ and $f(4) < 0$? Explain.

21. Sketch the graph of one function having *all* seven of the following characteristics.

    i. $f(x) > 0$ for all x,   ii. $\displaystyle\lim_{x \to 4} f(x) = 1$,   iii. $f(4) = 3$,   iv. $\displaystyle\lim_{x \to \infty} f(x) = 1$,

    v. $\displaystyle\lim_{x \to -\infty} f(x) = 1$,   vi. $\displaystyle\lim_{x \to 0^+} f(x) = 5$,   vii. $\displaystyle\lim_{x \to 0^-} f(x) = 2$.

22. Use the limit definition to differentiate $f(x) = \frac{1}{x^2}$.

CHAPTER 8

Fruits of Our Labor: Derivatives and Local Linearity Revisited

8.1 LOCAL LINEARITY AND THE DERIVATIVE

In Section 4.1 we discussed local linearity, but at that point we had not yet developed the concept of derivative. Therefore, in this section we revisit the idea. In Chapter 5 we pointed out that there are numerous forms in which the definition of derivative can be expressed, yet for the most part we've used $f'(x) = \lim_{h\to 0} \frac{f(x+h)-f(x)}{h}$. In this chapter, because we'll look at approximations that are good for a small range of the independent variable, we'll use notation that emphasizes the relative rates of change of the dependent and independent variables.

With that in mind, we re-establish the following conventions.

If $y = f(x)$, then $\frac{dy}{dx} = \lim_{\Delta x \to 0} \frac{\Delta y}{\Delta x}$, where $\Delta y = f(x + \Delta x) - f(x)$.

Equivalently, we can replace Δy by Δf,

$$\frac{df}{dx} = \lim_{\Delta x \to 0} \frac{\Delta f}{\Delta x} = \lim_{\Delta x \to 0} \frac{f(x + \Delta x) - f(x)}{\Delta x}.$$

(Notice that Δx and h play the same role.)

When we approximate a function near a point using local linearity, we might wonder whether the approximation is larger or smaller than the actual value. The answer depends on the second derivative, which was introduced in Chapter 6. There we saw that if $f(t)$ gives position as a function of time, then $f'(t)$ gives velocity, the rate of change of position with respect to time, and $f''(t)$ gives acceleration, the rate of change of velocity with respect to time. Let's now adopt a graphical perspective as well.

If $f'' > 0$, then f' is increasing and f is concave up.

If $f'' < 0$, then f' is decreasing and f is concave down.

279

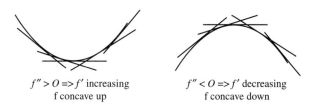

$$f'' > O \Rightarrow f' \text{ increasing}$$
f concave up

$$f'' < O \Rightarrow f' \text{ decreasing}$$
f concave down

Figure 8.1

◆ **EXAMPLE 8.1**   The weather pattern known as El Niño brought extreme weather conditions throughout the globe in 1997 and 1998. Some areas experienced severe drought, while others were beset by flooding. In a certain town, the amount of water in the reservoir has been decreasing for the past few weeks and there is no indication that rain is to be expected any time soon. An awareness of the crisis is spreading and the rate at which water is being consumed is dropping slightly. At present there are G_0 gallons of water in the reservoir and the level is dropping at a rate of 115 gallons per day.

(a) Let $W(t)$ be the amount of water in the reservoir, where t is measured in days and we choose $t = 0$ to be today. Translate the information given above into statements about W, W', and W''.

(b) Approximate the amount of water in the reservoir two days from now.

SOLUTION   (a) Today there are G_0 gallons of water in the reservoir, so $W(0) = G_0$.

$W(t)$ is decreasing and W' gives the rate of change of W, the amount of water in the reservoir, so $W'(t) = \frac{dW}{dt} < 0$.

$$\frac{dW}{dt}\bigg|_{t=0} = W'(0) = -115 \text{ gallons/day}$$

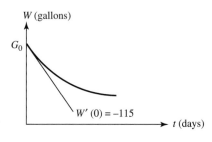

Figure 8.2

The rate of water consumption is decreasing, so W' is becoming less negative as t increases. The graph of W is concave up. $W''(t) \geq 0$.

(b)

$$\left(\begin{array}{c}\text{Amount of water in the}\\\text{reservoir 2 days from now}\end{array}\right) = \left(\begin{array}{c}\text{amount in the}\\\text{reservoir now}\end{array}\right) + \left(\begin{array}{c}\text{change in water}\\\text{in the next 2 days}\end{array}\right).$$

$$W(2) = G_0 + \left(\begin{array}{c}\text{change in water}\\\text{in the next two days}\end{array}\right).$$

We'll use the rate at which water is decreasing today to approximate the change in water level over the next two days.

$$\frac{dW}{dt} = \lim_{\Delta t \to 0} \frac{\Delta W}{\Delta t}, \quad \text{so}$$

$$\frac{dW}{dt} \approx \frac{\Delta W}{\Delta t} \quad \text{for } \Delta t \text{ small.}$$

$$\Delta W \approx \frac{dW}{dt} \Delta t$$

If we let $\Delta t = 2$ and use $\left.\frac{dW}{dt}\right|_{t=0} = -115$, we have

$$\Delta W \approx -115 \cdot 2 = -230.$$

This makes sense; the water is decreasing at a rate of 115 gallons/day so after two days we'd expect it to decrease by about 230 gallons.

$$W(2) \approx G_0 - 230$$

Because $W(t)$ is concave up, we expect the actual water level to be a bit higher than this.

◆

◆ EXAMPLE 8.2   Approximate $\sqrt{16.8}$. Use a first derivative to get a good approximation.

SOLUTION   Let's begin by sketching $\sqrt{x}$ and getting an off-the-cuff approximation of $\sqrt{16.8}$. This will help us see how a tangent line can be of use. We see that $\sqrt{16.8}$ is a bit larger than 4.

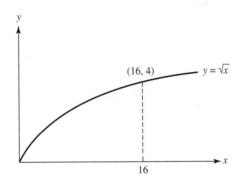

Figure 8.3

Question: How do we know this?
Answer: 16.8 is close to 16 and $\sqrt{16} = 4$.

Question: How do we know $\sqrt{16.8}$ is a tad *more* than 4?
Answer: We know that $\sqrt{x}$ is increasing between 16 and 16.8.

Question: How can we estimate *how* much to add to 4 to get a good approximation of $\sqrt{16.8}$?

Answer: This depends on the *rate* at which $\sqrt{x}$ is increasing near $x = 16$. This is where the derivative comes into the picture. The derivative of $\sqrt{x}$ at $x = 16$ gives the rate of increase.

$$\frac{d}{dx}\sqrt{x}\bigg|_{x=16} = \frac{1}{2\sqrt{x}}\bigg|_{x=16} = \frac{1}{2\sqrt{16}} = \frac{1}{8}.$$

Here we can adopt one of two equivalent viewpoints.

Viewpoint (i): The Derivative as a Tool for Adjustment. We begin with an off-the-cuff approximation of $\sqrt{16.8}$ grounded in a nearby value of $\sqrt{x}$ that we know with certainty. Knowing the rate of increase of $\sqrt{x}$ at $x = 16$ enables us to judge how to adjust our approximation (in this case, 4) to fit a nearby value of x. $\frac{dy}{dx} \approx \frac{\Delta y}{\Delta x}$ for Δx small (because $\frac{dy}{dx} = \lim_{\Delta x \to 0} \frac{\Delta y}{\Delta x}$). In this example $\Delta x = 16.8 - 16 = 0.8$.

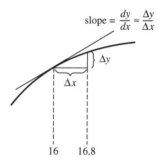

Figure 8.4

$$\frac{dy}{dx} = \frac{1}{8} \approx \frac{\Delta y}{0.8} \Rightarrow \Delta y \approx \frac{0.8}{8} = \frac{1}{10}$$

So $\Delta y \approx 0.1$ and we have the approximation $\sqrt{16.8} \approx 4 + 0.1 = 4.1$.

In other words, the "new" y-value is approximately the "old" y-value, 4, plus Δy, where we approximate Δy by $\frac{dy}{dx} \cdot \Delta x$.

Viewpoint (ii): Tangent Line Approximation. Again we begin with an off-the-cuff approximation grounded in a value of $\sqrt{x}$ we know with certainty. We choose $x = 16$. The tangent line to $\sqrt{x}$ at $x = 16$ is the best linear approximation of $\sqrt{x}$ around $x = 16$, so we use that tangent line to approximate $\sqrt{16.8}$.

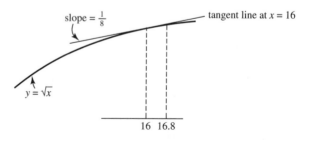

Figure 8.5

The y-value corresponding to $x = 16.8$ on the tangent line is very close to the corresponding y-value on the curve.

The tangent line has slope $\frac{1}{8}$ and passes through the point $(16, 4)$. Its equation is of the form $y - y_1 = m(x - x_1)$, from $\frac{y - y_1}{x - x_1} = m$ where $m = \frac{1}{8}$ and $(x_1, y_1) = (16, 4)$.

$$y - 4 = \frac{1}{8} \cdot (x - 16)$$

$$y = 4 + \frac{1}{8} \cdot (x - 16)$$

Notice that this just says that the "new" y-value is approximately the "old" y-value, 4, plus the adjustment term—as in the first approach.

The y-value on the tangent line when $x = 16.8$ is

$$y = 4 + \frac{1}{8} \cdot (16.8 - 16) = 4.1.$$

Is the approximation $\sqrt{16.8} \approx 4.1$ an overestimate or an underestimate? We know that $\frac{d}{dx} \sqrt{x} = \frac{1}{2\sqrt{x}}$; thus, as x increases the derivative decreases. Therefore $\sqrt{x}$ is concave down, the tangent line lies above the graph of the function, and this approximation is a bit too big.

Comparing our approximation, $\sqrt{16.8} \approx 4.1$, with the calculator approximation, $\sqrt{16.8} \approx 4.09878$, we see that the approximation is good. The error is about 0.00122; we are off by about 0.03%.   ◆

REMARK If we wanted to further refine our approximation, we could use information supplied by the second derivative to make the required adjustment.[1] Further successive refinements involve higher-order derivatives and give us what is called a *Taylor polynomial approximation* to the function at $x = 16$.

EXERCISE 8.1   Approximate $\sqrt{101}$ using a tangent line approximation. Compare your answer to that obtained using a calculator.

EXERCISE 8.2   Suppose you use a tangent line approximation to estimate $\sqrt{4.8}$. Do you expect the difference between your approximation and the actual answer to be greater or less than the analogous difference we found in Example 8.2? Explain, and then check your reasoning by computation and using a calculator.

EXERCISE 8.3   We argued that $f(x) = \sqrt{x}$ is concave down by reasoning that f' is decreasing. Show that f is concave down by showing that $f''(x) < 0$. Use the limit definition of derivative to compute $f''(x)$. If you have trouble, look back at Section 5.1 to see how the derivative of f was computed.

Answers to Exercises 8.1, 8.2, and 8.3 are supplied at the end of this section.

Generalizing Our Method

To approximate $f(\star)$ using a tangent line approximation, choose an x near $\star$ for which the value of f is known. Let's call this x-value a. Find the tangent line to f at $x = a$. The slope

---

[1] The larger the value of the second derivative at $x = 16$ the larger the rate of change of the slope near $x = 16$.

is $f'(a)$ and a point on the line is $(a, f(a))$. Therefore, the equation of the tangent line is

$$y - f(a) = f'(a) \cdot (x - a) \quad \text{or} \quad y = f(a) + f'(a) \cdot (x - a).$$

Evaluating at $x = \star$ gives the corresponding y-value on the tangent line.

Answers to Selected Exercises

Answer to Exercise 8.1

$$\sqrt{101} \approx 10 + \frac{1}{2\sqrt{100}} \cdot (101 - 100)$$

$$y \approx 10 + \frac{1}{20} \cdot (1) = 10.5$$

Answer to Exercise 8.2

The slope of $\sqrt{x}$ is changing more rapidly at $x = 4$ than at $x = 16$. In both cases we are using the tangent line to approximate the value of the function at a distance 0.8 from the known value. Therefore we expect the error to be greater for $\sqrt{4.8}$ than for $\sqrt{16.8}$.

$$\sqrt{4.8} \approx 2 + \frac{1}{2\sqrt{4}} \cdot (4.8 - 4)$$

$$\approx 2 + \frac{1}{4} \cdot (0.8) = 2.2$$

The difference between the approximation, 2.2, and the actual value, $2.19089\ldots$, is approximately 0.009.

Answer to Exercise 8.3

$$f''(x) = \frac{-1}{4(\sqrt{x})^3}$$

PROBLEMS FOR SECTION 8.1

1. Use the tangent line approximation (linear approximation) of $f(x) = \sqrt{x}$ at $x = 25$ to approximate the following. Use the graph of $\sqrt{x}$ and its tangent line at $x = 25$ to predict the relative accuracy of your approximations. Check the accuracy using a computer or calculator.

   (a) $\sqrt{23}$ (b) $\sqrt{24}$ (c) $\sqrt{24.9}$
   (d) $\sqrt{25.1}$ (e) $\sqrt{26}$ (f) $\sqrt{27}$

2. Suppose we want to use a tangent line approximation of $f(x) = \sqrt{x}$ at $x = a$ to approximate a particular square root numerically. Which values of a should we choose to approximate each of the following?

   (a) $\sqrt{102}$ (b) $\sqrt{8}$ (c) $\sqrt{18}$ (d) $\sqrt{115.5}$

3. Use a tangent line approximation to $f(x) = \frac{1}{x}$ at $x = 2$ to approximate $\frac{1}{1.9}$.

4. Approximate $\sqrt{98}$ using the appropriate first derivative to help you. Explain your reasoning.

5. Use the fact that $\frac{d}{dx}x^{1/3} = \left(\frac{1}{3}\right)x^{-2/3}$ to approximate $\sqrt[3]{30}$. Do you expect your answer to be an over-approximation or an under-approximation? Explain. Compare your answer to the approximation supplied by your calculator.

6. A steamboat is traveling down the Mississippi River. It is traveling south, making its way from point St. Paul, Minnesota, to Dubuque, Iowa. At noon it departs St. Paul traveling at 10 mph and is accelerating. It continues to accelerate over the next 10 minutes. Between noon and 12:10 P.M. it has covered 5 miles. Let $s(t)$ be the distance the steamboat has traveled from point St. Paul, where t is measured in minutes. We'll use noon as benchmark time of $t = 0$.

   (a) Determine the sign of $s(0)$, $s'(0)$, and $s''(0)$. Which of these expressions are you given enough information to specify numerically?

   (b) Sketch $s(t)$ over the time interval $[0, 10]$.

   (c) Find good upper and lower bounds for the distance the boat has traveled between noon and 12:05 P.M. Use a sketch to illustrate your reasoning graphically.

7. Consider the solid right cylinder with a fixed height of 10 inches and a variable radius. Let $V(r)$ be the volume of the cylinder as a function of r, the radius, given in inches. Interpret dV/dr geometrically. Explain why your answer makes sense by looking at ΔV geometrically.

Exploratory Problems for Chapter 8

Circles and Spheres

You know how to express the area and the circumference of a circle in terms of its radius; you also know how to express the volume and surface area of a sphere in terms of its radius. Having completed the Exploratory Problems for Chapter 7, you know how to take the derivative of both the area function for the circle and the volume function for the sphere. Have you observed the following?

$A(r) = \pi r^2$ is the area of a circle of radius r.
$\qquad \dfrac{dA}{dr} = 2\pi r$ is its circumference.

$V(r) = \frac{4}{3}\pi r^3$ is the volume of a sphere of radius r.
$\qquad \dfrac{dV}{dr} = 4\pi r^2$ is its surface area.

These relationships are not coincidental. The problems that follow ask you to make sense out of this observation. (Notice that the area of a square with sides of length s is given by $A(s) = s^2$ and its perimeter is $4s$, *not* $2s$.)

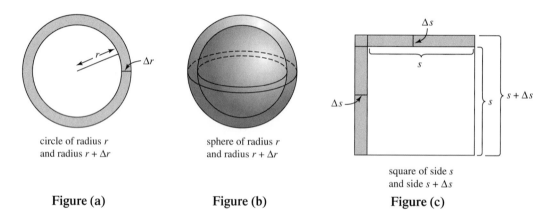

circle of radius r
and radius $r + \Delta r$

sphere of radius r
and radius $r + \Delta r$

square of side s
and side $s + \Delta s$

Figure (a)   **Figure (b)**   **Figure (c)**

1. Let $A(r)$ give the area of a circle as a function of its radius.

$$\frac{dA}{dr} = \lim_{\Delta r \to 0} \frac{\Delta A}{\Delta r} \quad \text{where} \quad \Delta A = A(r + \Delta r) - A(r)$$

Geometrically, ΔA corresponds to the shaded area in Figure (a) above.

(a) Imagine wrapping a piece of yarn around the outside of a circle of radius r, thereby increasing the radius of the circle by the thickness of the yarn. The thickness of the yarn corresponds to Δr. Let Δr be small in comparison to r. Approximate the additional area by clipping the yarn and laying it out straight.
   Explain why, for Δr very small, $\frac{\Delta A}{\Delta r} \approx 2\pi r = $ circumference of circle.

(b) Now approach the problem more algebraically. Instead of clipping a piece of yarn, look at the algebraic expression for

ΔA and show that it is $2\pi r \Delta r + \pi (\Delta r)^2$. Argue that for Δr very small in comparison to r, we can approximate ΔA by $2\pi r \Delta r$ and that as Δr tends toward zero this approximation gets better and better.

2. Investigate the relationship between the derivative of the volume of a sphere and the surface area of a sphere by approximating ΔV, when Δr is very small in comparison to r. As an aid in visualization imagine a grapefruit without the skin as the sphere of radius r and the same grapefruit with the skin as the sphere of radius $r + \Delta r$. ΔV corresponds to the volume of the skin of the grapefruit.

3. Let $A(s)$ be the area of a square with sides of length s. The derivative of the area of a square is *not* its perimeter; it is only half of the perimeter. Explain this by looking at ΔA geometrically and approximating it. (Refer to Figure (c).)

8.2   THE FIRST AND SECOND DERIVATIVES IN CONTEXT: MODELING USING DERIVATIVES

We've been interpreting the first derivative in context since its introduction but we have not yet used the second derivative extensively in context. Yet, according to a tongue-in-cheek article from *The Economist*, second-order derivatives were the rage in the 1990s. Below is an excerpt from the article "The Tyranny of Differential Calculus" published in *The Economist* on April 6, 1991.

> "The pace of change slows," said a headline on the *Financial Times*'s survey of world paints and coatings last week. Growth has been slowing in various countries—slowing quite quickly in some cases. Employers were invited recently to a conference on "techniques for improving performance enhancement." It's not enough to enhance your performance, Jones, you must improve your enhancement.
>
> Suddenly, everywhere, it is not the rate of change of things that matters, it is the rate of change of the rate of change. Nobody cares much about inflation; only whether it is going up or down. Or rather, whether it is going up fast or down fast. . . . No respectable budget director has talked about the national debt for decades; all talk sternly about the need to reduce the budget deficit, which is, after all, roughly the rate at which the national debt is increasing. Indeed, in recent years it is not the absolute size of the deficit that has mattered so much as the trend: Is the rate of change of the rate of change of the national debt positive or negative?

(© 1991 The Economist Newspaper Group, Inc. Reprinted with permission. Further reproduction prohibited. www.economist.com)

Consider the comment about the paint industry that reads "Growth has been slowing in various countries." If we let $P(t)$ be the size of the paint industry at time t, the graph of $P(t)$ is roughly given below. The graph is increasing because the industry is growing; it is concave down to reflect that the growth is slowing.

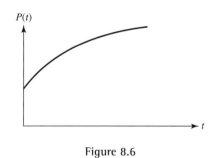

Figure 8.6

$\frac{dP}{dt} > 0$ because P is increasing with time; $\frac{d^2P}{dt^2} = \frac{d}{dt}\left(\frac{dP}{dt}\right) < 0$ because the growth is slowing.

EXERCISE 8.4   Suppose that in countries A and B the growth of the paint industry is slowing, and at time t_* growth is slowing more quickly in country A than in country B. Let $P_A(t)$ and $P_B(t)$ be the sizes of the paint industries in countries A and B, respectively, at time t.

(a) What are the signs of $P_A(t_*)$ and $P_B(t_*)$?
Can you determine which is larger?

(b) What are the signs of $P'_A(t_*)$ and $P'_B(t_*)$?
Can you determine which is larger?

(c) What are the signs of $P''_A(t_*)$ and $P''_B(t_*)$?
Can you determine which is larger?

Answers are provided at the end of this section.

◆ **EXAMPLE 8.3**   When politicians talk about the budget "deficit" they are referring to the difference between the government's revenues and expenditures for a given year. When they speak of "the national debt" they mean the *cumulative* deficit (or surplus) over all the years of the government's recorded transactions. Let $x(t)$ be the deficit for year t and let $y(t)$ be the national debt incurred up to and including year t. In our model $x(t)$ and $y(t)$ are both continuous and differentiable. $x(t)$ will be positive if there is a deficit, negative if there is a surplus.

(a) There has been considerable talk about a balanced budget. What would a balanced budget mean in terms of $x(t)$? In terms of $y(t)$?

(b) Politicians also often speak of deficit reduction plans. What would such a reduction mean in terms of $x(t)$ and $y(t)$ and their derivatives?

(c) The article from *The Economist* tells us that the budget deficit is roughly the rate at which the national debt is increasing. What, then, is the calculus-based relationship between $x(t)$ and $y(t)$?

SOLUTION   (a) A balanced budget would mean that $x(t) = 0$ and $y(t)$ is constant.

(b) Deficit reduction would mean $x(t)$ is positive and decreasing.

$$x(t) > 0 \text{ and } x'(t) < 0.$$

Deficit reduction would mean $y(t)$ is increasing at a decreasing rate,

$$y'(t) > 0 \text{ and } y''(t) < 0.$$

(c) $y'(t) = x(t)$.

This makes sense both practically speaking and in terms of the answers to parts (a) and (b).   ◆

Answers to Selected Exercises

Answers to Exercise 8.4

(a) $P_A(t_*)$ and $P_B(t_*)$ are both positive because the size of the paint industry won't be negative. We cannot determine which will be larger.

(b) $P'_A(t_*)$ and $P'_B(t_*)$ are both positive because both industries are growing. We cannot determine which rate of growth is larger.

(c) $P''_A(t_*)$ and $P''_B(t_*)$ are both negative, because growth is slowing.

Determining which is larger is semantically tricky. Since growth is slowing more rapidly in country A than in country B, $P''_A(t_*)$ is more negative than $P''_B(t_*)$. More negative means smaller, so $P''_A(t_*)$ is smaller than $P''_B(t_*)$.

■ PROBLEMS FOR SECTION 8.2

1. Water is being poured into a bucket at a steady rate. $h(t)$ gives the height of water at time t. Let t_* be the time when the bucket is half full. What can you say about the signs of $h'(t_*)$ and $h''(t_*)$. Explain your reasoning precisely in plain English.

2. Suppose that the revenue, R, brought in each month by the after-eight shows at a movie theater is a function of the price p of a ticket. Suppose that R is measured in thousands of dollars and that p is measured in dollars.

    Interpret the following statements in words.

    (a) $R'(3.5) = 50$

    (b) $R'(7.50) = -15$

3. A company is making industrial-size rolls of paper towels. A machine is wrapping paper around a roll at a steady rate. By this we mean that the same number of sheets of paper towels are added to the roll every minute. Let $D(t)$ be the diameter of the roll of paper towels at time t. Determine the sign of the following. Explain your answers using plain English.

    (a) $D(t)$ (b) $D'(t)$ (c) $D''(t)$

4. Let $Y(t)$ be the number of Japanese yen exchangeable for one U.S. dollar, where t is the number of days after January 1, 1996.

    (a) What is the practical significance of the values of t for which $Y'(t)$ is positive?

    (b) What is the practical significance of the values of t for which $Y'(t)$ is negative and $Y''(t)$ is negative?

    (c) What is the meaning of the statement $Y'(5) = 0.8$?

    (d) Interpret the quantity $\frac{Y(5) - Y(3)}{2}$.

■ 8.3 DERIVATIVES OF SUMS, PRODUCTS, QUOTIENTS, AND POWER FUNCTIONS

In this section we shift our focus from interpreting and using derivatives to actually computing them. Fruits of our labors with limits will be general procedures for differentiating sums, products, and quotients, and the ability to easily compute the derivative of any function of the form x^n, where n is an integer.

Sum Rule and Constant Multiple Rule

Section 7.4 dealt with principles for working with limits. We restate two of these principles here for reference.

Suppose $\lim_{x \to a} f(x) = L_1$ and $\lim_{x \to a} g(x) = L_2$, where L_1 and L_2 are finite. Then:

(1) $\displaystyle \lim_{x \to a} [f(x) \pm g(x)] = L_1 \pm L_2$
The limit of a sum/difference is the sum/difference of the limits.

(2) $\displaystyle \lim_{x \to a} [f(x) \cdot g(x)] = L_1 \cdot L_2$
The limit of a product is the product of the limits (in particular, $g(x)$ may be constant: $\lim_{x \to a} kf(x) = kL_1$).

We can use principles (1) and (2) given above to prove two properties of derivatives, the Constant Multiple Rule and the Sum Rule, stated in Section 7.4 and worked with in the Exploratory Problem for Chapter 7.

Properties of Derivatives

Constant Multiple Rule

$$\frac{d}{dx} kf(x) = k \frac{d}{dx} f(x) \quad \text{or} \quad (kf)'(x) = kf'(x), \text{ where } k \text{ is any constant.}$$

Multiplying f by a constant k multiplies its derivative by k.

Sum Rule

$$\frac{d}{dx} [f(x) + g(x)] = \frac{d}{dx} f(x) + \frac{d}{dx} g(x) \quad \text{or} \quad (f + g)'(x) = f'(x) + g'(x)$$

The derivative of a sum is the sum of the derivatives.

These properties are natural from a real-world point of view.

Sum Rule:.   Suppose a juice manufacturer is packaging cranapple juice by pouring cranberry and apple juice simultaneously into juice containers.

Let $c(t)$ be the number of liters of cranberry juice in a container at time t.

Let $a(t)$ be the number of liters of apple juice in a container at time t.

Then $c(t) + a(t)$ is the amount of juice in the container at time t.

It follows that $\frac{da}{dt}$ and $\frac{dc}{dt}$ are the rates at which apple and cranberry juice, respectively, are entering the container. $\frac{d}{dt}[a(t) + c(t)]$ is the rate at which juice is entering.

$$\frac{d}{dt}[a(t) + c(t)] = \frac{da}{dt} + \frac{dc}{dt}$$

Constant Multiple Rule:.   An athlete and a Sunday jogger start at the same place and run down a straight trail. Let $s(t)$ be the position of the Sunday jogger at time t. If the position of the athlete is always k times $s(t)$, then her velocity is k times that of the jogger.

$$\frac{d}{dt} ks(t) = k \frac{ds}{dt}$$

The Sum and Constant Multiple Rules are very important results and were the topic of the exploratory problems for the previous chapter. Proofs are provided at the end of the section. We will assume in the chapters that follow that you have completed the exercises below asking you to prove these rules and make sense out of them.

EXERCISE 8.5   Using the limit definition of derivative prove the two statements made above.

EXERCISE 8.6   Argue that the two properties above make sense from a graphical perspective.

EXERCISE 8.7   Differentiate the following.

$$\text{(a) } y = \frac{2 + 3x^3 + \sqrt{5}x}{x} \qquad \text{(b) } y = \frac{4\sqrt{x}}{\pi} + \frac{3\sqrt{\pi}x}{7}$$

Answers to these Exercises are provided at the end of the section.

The Product Rule: Differentiating $f(x) \cdot g(x)$

From the simplicity of the Sum Rule and the Constant Multiple Rule one might initially hope that the derivative of a product is the product of the derivatives and similarly for quotients. But this hope is quickly dashed by working through an example.[2]

Suppose, for instance, $f(x) = x$ and $g(x) = x^2$. Then $f'(x) = 1$ and $g'(x) = 2x$, so $f'(x) \cdot g'(x) = 1 \cdot 2x = 2x$. On the other hand, $f(x) \cdot g(x) = x \cdot x^2 = x^3$; we know that $(f(x) \cdot g(x))'$ should be $3x^2$, *not* $2x$.

CAVEAT   *The derivative of a product is* **not** *the product of the derivatives.*

However, if we know the derivatives of f and g, we *can* find the derivative of their product in terms of f, g, f', and g'. We'll figure out how to do this by working through the following example.

◆ **EXAMPLE 8.4**   A farmer growing a crop of tomatoes wants to know when he should harvest and sell them in order to collect the most revenue. Both the weight of the crop and the price of tomatoes are changing over time. Let $W(t)$ be the weight (in pounds) of his crop of tomatoes t days after the beginning of the harvest season, and let $P(t)$ be the going price (in dollars per pound) for tomatoes on day t. Then the revenue the farmer will receive for selling his crop on day t will be given by (weight in pounds) · ($/lb):

$$R(t) = W(t) \cdot P(t).$$

Suppose that $W(t)$ and $P(t)$ are both differentiable functions. As long as $R'(t) > 0$, waiting to harvest the crop will increase revenue. We want to compute $R'(t)$.

Let ΔW, ΔP, and ΔR be the change in weight, price, and revenue over the time interval Δt. There are two components contributing to the change in revenue:

$\Delta W \cdot P$ (the change in revenue due to increase in the weight of the tomatoes), and

$W \cdot \Delta P$ (the change in revenue due to increase in the price).

---

[2] Leibniz himself wondered about this in an unpublished manuscript dated November 1675 and concluded that the rules were not that simple. He arrived at the Product Rule ten days later. *(Burton, p.372)*

We are interested in $\frac{dR}{dt}$, the rate at which revenue is changing with respect to time.

$$\frac{dR}{dt} = \lim_{\Delta t \to 0} \frac{\Delta R}{\Delta t}.$$

The broad brushstroke of our argument looks like this.

$$\Delta R = \Delta W \cdot P + W \cdot \Delta P \qquad \text{Divide by } \Delta t.$$

$$\frac{\Delta R}{\Delta t} = \frac{\Delta W}{\Delta t} \cdot P + W \cdot \frac{\Delta P}{\Delta t} \qquad \text{Take the limit as } \Delta t \to 0.$$

$$\lim_{\Delta t \to 0} \frac{\Delta R}{\Delta t} = \lim_{\Delta t \to 0} \frac{\Delta W}{\Delta t} \cdot P + \lim_{\Delta t \to 0} W \cdot \frac{\Delta P}{\Delta t}$$

$$\frac{dR}{dt} = \frac{dW}{dt} \cdot P + W \cdot \frac{dP}{dt}$$

We will fill in more of the details while looking at the geometry of the problem. For simplicity's sake, suppose that W and P are both increasing with time.

$$\Delta W = W(t + \Delta t) - W(t) \text{ and } \Delta P = P(t + \Delta t) - P(t)$$

$$\Delta R = R(t + \Delta t) - R(t)$$

$$\Delta R = W(t + \Delta t)P(t + \Delta t) - W(t)P(t)$$

The drawings below will help us express ΔR in terms of ΔW and ΔP.

Revenue is represented by the area of a rectangle the length of whose sides represent the weight and price of tomatoes at a fixed time.

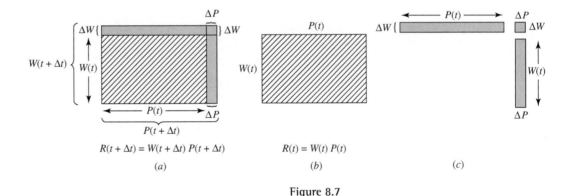

Figure 8.7

$$\Delta R = R(t + \Delta t) - R(t)$$

$$= \begin{pmatrix} \text{area of the entire} \\ \text{large rectangle} \end{pmatrix} - \begin{pmatrix} \text{area of the} \\ \text{hatched rectangle} \end{pmatrix}$$

ΔR can be represented by the three shaded rectangles shown in Figure 8.7(c).

$$\Delta R = (\Delta W)P(t) + (\Delta P)W(t) + (\Delta W)(\Delta P)$$

$$\frac{\Delta R}{\Delta t} = \frac{(\Delta W)P(t)}{\Delta t} + \frac{(\Delta P)W(t)}{\Delta t} + \frac{(\Delta W)(\Delta P)}{\Delta t}$$

$$\frac{dR}{dt} = \lim_{\Delta t \to 0} \frac{\Delta R}{\Delta t} = \lim_{\Delta t \to 0} \left[\frac{(\Delta W)P(t)}{\Delta t} + \frac{(\Delta P)W(t)}{\Delta t} + \frac{(\Delta W)(\Delta P)}{\Delta t} \right]$$

$$= \lim_{\Delta t \to 0} \frac{(\Delta W)P(t)}{\Delta t} + \lim_{\Delta t \to 0} \frac{(\Delta P)W(t)}{\Delta t} + \lim_{\Delta t \to 0} \frac{(\Delta W)(\Delta P)}{\Delta t}$$

Note that $P(t)$ and $W(t)$ are unaffected by letting Δt approach zero, so we obtain

$$\frac{dR}{dt} = P(t) \lim_{\Delta t \to 0} \frac{\Delta W}{\Delta t} + W(t) \lim_{\Delta t \to 0} \frac{\Delta P}{\Delta t} + \lim_{\Delta t \to 0} \frac{\Delta W}{\Delta t} \Delta P.$$

By definition

$$\lim_{\Delta t \to 0} \frac{\Delta W}{\Delta t} = \frac{dW}{dt} \quad \text{and} \quad \lim_{\Delta t \to 0} \frac{\Delta P}{\Delta t} = \frac{dP}{dt}.$$

Therefore,

$$\frac{dR}{dt} = P(t)\frac{dW}{dt} + W(t)\frac{dP}{dt} + \frac{dW}{dt} \lim_{\Delta t \to 0} \Delta P.$$

As $\Delta t \to 0$, we know that $\Delta P \to 0$,[3] so the last term is actually $\frac{dW}{dt} \cdot 0$. Thus,

$$\frac{dR}{dt} = P(t)\frac{dW}{dt} + W(t)\frac{dP}{dt}.$$

Another way of writing this is

$$R'(t) = P(t)W'(t) + W(t)P'(t). \quad \blacklozenge$$

We have constructed an argument that for increasing functions f and g,

$$\left[f(x)g(x) \right]' = f(x)g'(x) + g(x)f'(x).$$

This is known as the **Product Rule**. We will show below that this holds for the product of *any* two differentiable functions f and g. Before doing this, let's try an example.

◆ **EXAMPLE 8.5**   Find the derivative of each of the following functions using the Product Rule.

    i. $q(x) = x^3 = x \cdot x^2$
    ii. $h(x) = \sqrt{7x}(x^2 + 3x + 2)$

SOLUTION

    i. $q'(x) = x \cdot \frac{d}{dx}\left[x^2\right] + x^2 \cdot \frac{d}{dx}\left[x\right] = x \cdot 2x + x^2 \cdot 1 = 2x^2 + x^2 = 3x^2$
    ii. $h(x) = \sqrt{7} \cdot \sqrt{x}(x^2 + 3x + 2)$, so
        $h'(x) = \sqrt{7} \cdot \left(\frac{1}{2\sqrt{x}}(x^2 + 3x + 2) + \sqrt{x}(2x + 3) \right)$ ◆

---

[3] We know that the change in price approaches 0 as Δt approaches 0, because P is continuous. P must be continuous, because we have assumed it is differentiable.

Proof of the Product Rule

Let $j(x) = f(x)g(x)$, where f and g are differentiable functions.

$$\left[f(x)g(x)\right]' = j'(x) = \lim_{h \to 0} \frac{j(x+h) - j(x)}{h}$$

$$= \lim_{h \to 0} \frac{f(x+h)g(x+h) - f(x)g(x)}{h}$$

We'd love to see a $f(x+h)\left[g(x+h) - g(x)\right]$ in the numerator, but we're missing the $-f(x+h)g(x)$ term. However, $-f(x+h)g(x) + f(x+h)g(x) = 0$, so we can write

$$\left[f(x)g(x)\right]' = \lim_{h \to 0} \frac{f(x+h)g(x+h) - f(x+h)g(x) + f(x+h)g(x) - f(x)g(x)}{h}$$

$$= \lim_{h \to 0} \left[\frac{f(x+h)g(x+h) - f(x+h)g(x)}{h} + \frac{f(x+h)g(x) - f(x)g(x)}{h}\right]$$

$$= \lim_{h \to 0} \left[f(x+h)\frac{g(x+h) - g(x)}{h} + g(x)\frac{f(x+h) - f(x)}{h}\right]$$

$$= f(x)g'(x) + g(x)f'(x).$$

We've now proven the Product Rule.

The Product Rule: $\left[f(x)g(x)\right]' = f(x)g'(x) + g(x)f'(x)$

EXERCISE 8.8   Show that computing the derivative of $kf(x)$ is a special case of the Product Rule. In other words, apply the Product Rule to $kf(x)$ and show $\frac{d}{dx}kf(x) = k\frac{df}{dx}$.

The answer to this Exercise is provided at the end of the section.

The usefulness of the Product Rule will become increasingly apparent as the number of types of functions we can differentiate increases. Now we will see how the Product Rule enables us to differentiate a whole family of power functions in the blink of an eye, without having to return to the limit definition of the derivative.

The Derivative of x^n for n a Positive Integer

Mathematicians, like other scientists, look for patterns. Upon observing a pattern, a mathematician tries to generalize his observations and prove that the pattern holds true.

Let's gather our results from applying the limit definition of derivative to power functions, functions of the form x^n for n a constant. When looking for patterns, one must often play around with the form in which an expression is written. Results from Examples and Exercises are gathered in the table on the following page and written in a form so that the pattern is readily accessible. You can add results from other problems you have completed.

$f(x)$	$f'(x)$
x^3	$3x^2$
x^2	$2x$
x	1
$x^{\frac{1}{2}}$	$\frac{1}{2}x^{-\frac{1}{2}}$
$x^{\frac{-1}{2}}$	$\frac{-1}{2}x^{-\frac{3}{2}}$
x^{-1}	$-x^{-2}$

Looking at this table we might conjecture that the derivative of x^n is nx^{n-1}. Below we prove this conjecture for n any positive integer. The method we will use is called **mathematical induction**; it works on the domino principle. Suppose we arrange a collection of dominos in a line so that if any one domino falls, we are sure that the domino next to it will also fall. Then, if we knock over the first domino in the line, all the rest of the dominos will also be knocked down.

Proof by mathematical induction works similarly except that the dominos are the positive integers, so there are infinitely many of them.

i. First show that the statement $\frac{d}{dx}x^n = nx^{n-1}$ holds for $n = 1$.
   We show that the first domino will fall.

ii. Then we show that *if* $\frac{d}{dx}x^k = kx^{k-1}$ is true for some arbitrary k, where k is a positive integer, then it must also be true for $k + 1$. In other words, we must show that *if* $\frac{d}{dx}x^k = kx^{k-1}$, then $\frac{d}{dx}x^{k+1} = (k + 1)x^k$. This is called the *inductive step*.
   We show that if any one domino falls, it will knock down the domino after it.

Once this is shown our proof is complete. We will have shown that $\frac{d}{dx}x^n = nx^{n-1}$ holds true for $n = 1$, and that it therefore must hold true for $n = 2$; it holds true for $n = 2$, and therefore must hold true for $n = 3$, and so on, ad infinitum. It must hold true for all positive integers.

For more work with proof by induction, refer to Appendix D: Proof by Induction.

Proof Our hypothesis: $\frac{d}{dx}x^n = nx^{n-1}$.

i. Our hypothesis holds for $n = 1$.

$$\text{If } f(x) = x, \text{ then } f'(x) = 1 = 1 \cdot x^0.$$

ii. Let's suppose that our hypothesis holds for $n = k$.
   (This is known as the *induction hypothesis*.) We will show that it must then hold for $n = k + 1$. In other words, we show that $\frac{d}{dx}x^{k+1} = (k + 1)x^k$ assuming $\frac{d}{dx}x^k = kx^{k-1}$.
   Let $h(x) = x^{k+1} = x \cdot x^k$.
   By the Product Rule we know that

$$h'(x) = 1 \cdot x^k + x \cdot \frac{d}{dx}x^k.$$

But $\frac{d}{dx}x^k = kx^{k-1}$ by the induction hypothesis.

Therefore,

$$h'(x) = x^k + x \cdot kx^{k-1}$$
$$= x^k + kx^k$$
$$= (k+1)x^k.$$

Our proof is complete. The derivative of x^n is nx^{n-1} for $n = 1, 2, 3, \ldots$, that is, for n any positive integer.

EXERCISE 8.9   Differentiate $f(x) = \pi x^{19} - 5x^5 + \frac{3x^6}{2} + 7(\pi)^3$.

Answer

$f'(x) = 19\pi x^{18} - 25x^4 + 9x^5$. (Why is the derivative of $7(\pi)^3$ zero and not $21(\pi)^2$?)

The Quotient Rule

Although it is tempting to try to differentiate $\frac{f(x)}{g(x)}$ by dividing f' by g', this is incorrect. Below we use the Product Rule to arrive at a rule for differentiating a quotient.[4]

Let $h(x) = \frac{f(x)}{g(x)}$, where f and g are differentiable functions. We are looking for $h'(x)$.

$$f(x) = g(x) \cdot h(x) \qquad \text{Differentiate both sides of the equation.}$$

$$f'(x) = g'(x) \cdot h(x) + g(x) \cdot h'(x) \qquad \text{Solve for } h'(x).$$

$$g(x) \cdot h'(x) = f'(x) - g'(x) \cdot h(x)$$

$$h'(x) = \frac{f'(x) - g'(x) \cdot h(x)}{g(x)} \qquad \text{But } h(x) = \frac{f(x)}{g(x)}.$$

$$= \frac{f'(x) - g'(x) \cdot \frac{f(x)}{g(x)}}{g(x)}$$

$$= \frac{\frac{f'(x) \cdot g(x)}{g(x)} - \frac{g'(x) \cdot f(x)}{g(x)}}{g(x)}$$

$$= \left(\frac{f'(x) \cdot g(x) - g'(x) \cdot f(x)}{g(x)} \right) \cdot \frac{1}{g(x)}$$

$$= \frac{f'(x) \cdot g(x) - g'(x) \cdot f(x)}{[g(x)]^2}$$

$$\left(\frac{f}{g} \right)'(x) = \frac{f'(x) \cdot g(x) - g'(x) \cdot f(x)}{[g(x)]^2}$$

This is the Quotient Rule for taking the derivative of the quotient of two functions.

$$\frac{d}{dx}\left(\frac{f(x)}{g(x)} \right) = \frac{g(x) \cdot f'(x) - f(x) \cdot g'(x)}{[g(x)]^2} \qquad \text{or, without the } x\text{'s,} \qquad \left(\frac{f}{g} \right)' = \frac{g \cdot f' - f \cdot g'}{g^2}$$

---

[4] This is the argument that Leibniz used in the 1670s after realizing that the derivative of a quotient is not the quotient of the derivatives.

Notice that the order in which you do the subtraction matters. Here's a mnemonic to remember which term in the numerator comes first; take it or leave it. "Don't get fouled up" translates to gf' (g for "get," f for "fouled," and $'$ for "up"). Now that you have started, you're all set.

EXERCISE 8.10   Differentiate $f(x) = \frac{2x^7+3}{2x+1}$.

Answer

$$f'(x) = \frac{24x^7+14x^6-6}{(2x+1)^2}$$

Knowing the Quotient Rule allows us to differentiate x^n, where n is a negative integer. We'll differentiate $h(x) = \frac{1}{x^m}$, where m is positive. $\frac{1}{x^m} = x^{-m}$, so the exponent of x is negative.

If the exponent algebra in this proof is hard for you to follow, skip to Section 9.2 or the Algebra Appendix and, after studying that section, try this proof again.

$$h(x) = \frac{1}{x^m}$$

$$h'(x) = \frac{x^m \cdot 0 - mx^{m-1}}{(x^m)^2}$$

$$= \frac{-mx^{m-1}}{x^{2m}} \qquad \text{(because } (x^a)^b = x^{ab})$$

$$= -mx^{(m-1)-2m} \qquad \text{(because } \frac{x^a}{x^b} = x^{a-b})$$

$$= -mx^{-m-1}$$

This result fits our established pattern.

$$\frac{d}{dx}x^{-m} = -mx^{-m-1}$$

Therefore,

$$\boxed{\frac{d}{dx}x^n = nx^{n-1} \text{ for any integer } n.}$$

We've proven this result for n a positive integer using induction and for n a negative integer using the Quotient Rule. Verify that the result is true for $n = 0$ on your own. In fact, this result is true for *any* exponent, but we will not be able to prove this until we know more about exponential functions, their inverse functions, and taking the derivative of composite functions.

Answers to Selected Exercises

Answers to Exercise 8.5

i. If k is a constant, the derivative of $kf(x)$ is $kf'(x)$.

Let $g(x) = kf(x)$

$$g'(x) = \lim_{h \to 0} \frac{g(x+h) - g(x)}{h}$$

$$= \lim_{h \to 0} \frac{kf(x+h) - kf(x)}{h}$$

$$= \lim_{h \to 0} k \cdot \frac{f(x+h) - f(x)}{h}$$

$$= k \cdot \lim_{h \to 0} \frac{f(x+h) - f(x)}{h} \qquad \text{(limit property (2))}$$

$$= kf'(x)$$

ii. $\frac{d}{dx}[f(x) + g(x)] = \frac{d}{dx} f(x) + \frac{d}{dx} g(x)$

Let $j(x) = f(x) + g(x)$

$$j'(x) = \lim_{h \to 0} \frac{j(x+h) - j(x)}{h}$$

$$= \lim_{h \to 0} \frac{f(x+h) + g(x+h) - [f(x) + g(x)]}{h}$$

$$= \lim_{h \to 0} \frac{f(x+h) - f(x)}{h} + \frac{g(x+h) - g(x)}{h}$$

$$= \lim_{h \to 0} \frac{f(x+h) - f(x)}{h} + \lim_{h \to 0} \frac{g(x+h) - g(x)}{h} \qquad \text{(limit property (1))}$$

$$= f'(x) + g'(x)$$

Partial Answers to Exercise 8.6

i. This makes graphical sense because multiplying f by k rescales the height of the graph by a factor of k; therefore the ratio $\frac{\text{rise}}{\text{run}}$ is rescaled by a factor of k as well. (See Figure 8.8.)

ii. The height of the function $f + g$ is the height of f plus the height of g. Therefore,

$$\frac{\text{rise}}{\text{run}} = \frac{\text{rise due to } f + \text{ rise due to } g}{\text{run}}$$

$$= \frac{\Delta f}{\Delta x} + \frac{\Delta g}{\Delta x}.$$

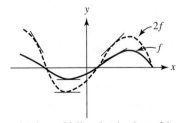

the slope of $2f$ is twice the slope of f.

Figure 8.8

Answers to Exercise 8.7

(a) $y = \frac{2}{x} + 3x^2 + \sqrt{5}$, so $y' = -\frac{2}{x^2} + 6x$

(b) $y = \frac{4}{\pi} \cdot \sqrt{x} + \frac{3\sqrt{\pi}}{7} \cdot x$, so $y' = \frac{4}{\pi} \cdot \frac{1}{2\sqrt{x}} + \frac{3\sqrt{\pi}}{7} \cdot 1 = \frac{2}{\pi\sqrt{x}} + \frac{3\sqrt{\pi}}{7}$

Answer to Exercise 8.8

Let $y = kf(x)$.

Then $y' = 0 \cdot f(x) + k \cdot f'(x) = kf'(x)$.

PROBLEMS FOR SECTION 8.3

For Problems 1 through 8, find $f'(x)$. Strategize to minimize your work. For example, $\frac{x^2+3}{3x}$ does not require the Quotient Rule. $\frac{x^2+3}{3x} = \frac{x}{3} + \frac{1}{x} = \frac{1}{3}x + x^{-1}$. This is simpler to differentiate.

1. $f(x) = 3x^2 + 3x + 3 + 3x^{-1} + 3x^{-2}$

2. $f(x) = \frac{x - 2x^2}{5}$

3. $f(x) = \pi(3x^2 + 7x + 1)(x - 2)$

4. $f(x) = \frac{1}{x^2+4}$

5. $f(x) = \frac{x}{x+2}$

6. $f(x) = \frac{x+2}{x}$

7. $f(x) = (\frac{5x^2}{2} + 7x^5 - 5x)x$

8. $f(x) = \frac{a}{x} + bx(c - dx)$, where a, b, c, and d are constants.

For Problems 9 through 11, rewrite each of these functions in a form that allows you to differentiate using the tools you have now, but with as little exertion as possible. Then differentiate. Work on strategy; none of these problems require the Quotient Rule.

9. (a) $f(x) = \frac{(x^2+2x)x}{3}$ (b) $f(x) = \frac{(x^2+1)^2}{x}$

10. (a) $f(x) = \frac{\frac{1}{x}+1}{\frac{x+1}{x^2+2x}}$ (b) $f(x) = \frac{(x^3+3x)}{2x^4}$

11. (a) $f(x) = (x + 1)(x - 1)x$ (b) $f(x) = \frac{x+\frac{1}{x}}{x}$

For each function in Problems 12 through 14:
(a) Sketch the graph of f.
(b) Find $f'(x)$. Are there any values of x for which f' is undefined?

12. $f(x) = x + |x|$

13. $f(x) = |x| - x$

14. $f(x) = \begin{cases} x^2, & x \geq 0 \\ x, & x < 0 \end{cases}$

15. The function $f(x) = x + |x|$ is continuous at $x = 0$ but not differentiable at $x = 0$. Explain, using the definitions of continuity at a point and differentiability at a point.

16. Let $f(x) = \frac{1}{x^2+1}$.

   (a) Sketch the graph of f. Do this by graphing $x^2 + 1$ and looking at the reciprocal. Check your answer with a computer or calculator.

   (b) Make a rough sketch of f' based on your graph of f.

   (c) Find $f'(x)$ analytically, using the Quotient Rule. Graph f'.

17. Let $f(x) = \begin{cases} x^3, & x \leq 1 \\ kx, & x > 1 \end{cases}$

   (a) What values of k makes $f(x)$ a continuous function?

   (b) If k is chosen so that f is continuous at $x = 1$, is f differentiable there?

18. Find the equation of the tangent line to $f(x) = x(x^2 + 2)$ at $x = 1$.

19. For what value(s) of x is the slope of the tangent line to $f(x) = \frac{1}{3}x^3$ equal to 1?

   For Problems 20 through 23, find the following:

20. $\dfrac{d}{dx}\left(\dfrac{x+1}{x^3 + 3x + 1}\right)$

21. $\dfrac{d}{dx}\left(\dfrac{\pi}{\pi x + \pi}\right)$

22. $\dfrac{d}{dx}\left(\dfrac{2x^2 + x + 1}{\sqrt{2x}}\right)$

23. $\dfrac{d}{dx}\left(\dfrac{x^2 + 5x}{2x^{10}}\right)$

C H A P T E R

Exponential Functions

■ 9.1 EXPONENTIAL GROWTH: GROWTH AT A RATE
PROPORTIONAL TO AMOUNT

Look around you at quantities in the world that change. While some functions are linear (the rate of change of output with respect to input is constant), you don't need to look far to see some very different patterns of growth and decay. Leave a piece of bread on a shelf for too long and mold will grow; not only will the amount of mold increase with time, but the rate at which the mold grows will also increase with time. Or observe the early stages of the spread of infectious disease in a large susceptible population; the rate at which the disease spreads increases as the number of infected people increases. Think about the rate of growth of money in an interest-earning bank account; the more money in the account, the faster additional money will accumulate. For instance, 3% of $1000 is greater than 3% of $100. The growth of bread mold, the spread of disease, and the accumulation of money in a bank account can all be modeled using exponential functions. Similarly, there are quantities that diminish at a rate proportional to the amount present. In particular, radioactive isotopes decay at a rate proportional to the amount present; modeling this decay by exponential functions enables scientists to date various events. We will see that if a quantity grows (or

303

decays) so that its rate of change at any moment is proportional to the amount of the quantity present at that moment, then the quantity grows (or decays) exponentially.

◆ **EXAMPLE 9.1**   Suppose you are a biologist growing a bacterial culture; you put some bacteria in a petri dish of agar where they will have everything they will need (food, space, etc.) to live and reproduce with abandon. At the start of the experiment, there are 5 bacteria in the petri dish. After an hour, there are 10 bacteria. One hour later, there are 20 bacteria, and the hour after that there are 40.

Think about the average rate of change of the bacteria population from one hour to the next. In the first hour, the average rate of change was 5 bacteria per hour, but in the second hour the average rate of change was 10 bacteria per hour, and in the third hour it was 20 bacteria per hour. The bacteria population at time t is clearly not increasing linearly, because the rate of change is not constant. Let's make a table of the data in order to come up with a model of the number of bacteria as a function of time. We'll measure time in hours, setting our benchmark time, $t = 0$, to be when we counted 5 bacteria in the dish. We'll denote the number of bacteria in the petri dish at time t by $B(t)$.

t (in hours)	$B(t)$	Change in B since previous hour
0	5	(not applicable)
1	10	5
2	20	10
3	40	20

Let's begin by looking at the hourly rate of change of the population. Notice that the average rate of growth each hour is equal to the population the previous hour. This is reasonable, as you would expect that the number of new bacteria produced would depend on the number of bacteria already living.[1] In this case it appears that on average each bacterium produces one new bacterium every hour. Looking at $B(t)$ we see that each hour the number of bacteria doubles. We'll use this observation to figure out an equation for $B(t)$. Let's rewrite the data in a way that makes the pattern more apparent.

t (in hours)	$B(t)$
0	5
1	$10 = 5 \cdot 2$
2	$20 = 5 \cdot 2 \cdot 2 = 5 \cdot 2^2$
3	$40 = 5 \cdot 2 \cdot 2 \cdot 2 = 5 \cdot 2^3$
$\vdots$	$\vdots$
t	$5 \cdot 2^t$

$B(t)$, the number of bacteria at time t, is given by $B(t) = 5 \cdot 2^t$. This equation makes sense;[2] we start with 5 bacteria and every hour the population is multiplied by 2. The function

---

[1] Most bacteria reproduce by binary fission; a new cell wall is formed and the one-cell organism splits in two.

[2] We should note here that when t is not an integer, this equation may give us a fractional part of a bacterium. For example, $B(0.5) = 7.07\ldots$. As we have seen before, scientists often model discrete phenomena with continuous models.

$B(t) = 5 \cdot 2^t$ is an example of an exponential function. In the expression b^a the "a" at the top right is called the **exponent** and the "b" is called the **base.**

Figure 9.1 is a graph of the number of bacteria versus t. Figure 9.2 illustrates the average rate of change of population over a one-hour time period by the slope of the secant line through $(t, B(t))$ and $(t + 1, B(t + 1))$ for two different values of t.

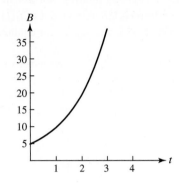

Figure 9.1

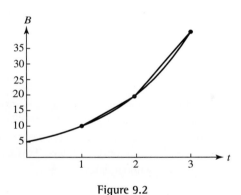

Figure 9.2

$$\begin{array}{l} \text{Average rate of change} \\ \text{of } B \text{ over } [t, t+1] \end{array} = \frac{B(t+1) - B(t)}{(t+1) - t} = \frac{5 \cdot 2^{t+1} - 5 \cdot 2^t}{1}$$

$$= 5 \cdot 2^t(2 - 1) = 5 \cdot 2^t = B(t). \quad \blacklozenge$$

Generally the rate of growth of a bacteria population under favorable conditions, while not equal to the population, is *proportional*[3] to the population, where the constant of proportionality depends on the organism. Below we give a real-life example.

◆ **EXAMPLE 9.2**   *Escherichia coli* is a bacterium that lives in the large intestine and helps the body with digestion. If *E. coli* enter the abdomen through a perforation in the intestine their presence can cause peritonitis, a potentially fatal disease if untreated. In order to understand the progress of the disease it is necessary to understand the growth of a population of *E. coli*.

---

[3] Recall that y is proportional to x means $y = kx$ for some constant k, called the constant of proportionality.

Microbiologists have found that under favorable conditions a population of *E. coli* doubles every 20 minutes. Such rapid growth rates are typical of bacteria. Suppose that at time $t = 0$, t measured in minutes, a person has 10 *E. coli* in his abdomen. How many *E. coli* are in this unfortunate individual's abdomen 12 hours later if the condition is untreated and there are no limiting factors?

SOLUTION   Let's begin by figuring out how many *E. coli* will be in his abdomen t minutes later if the condition is left untreated. Let $B(t)$ = the number of bacteria in his abdomen at time t. We'll make a table and search for a pattern.

t (in minutes)	$B(t)$
0	10
20	$10 \cdot 2$
40	$(10 \cdot 2) \cdot 2 = 10 \cdot 2 \cdot 2 = 10 \cdot 2^2$
60	$(10 \cdot 2^2) \cdot 2 = 10 \cdot 2^3$
80	$(10 \cdot 2^3) \cdot 2 = 10 \cdot 2^4$
$\vdots$	$\vdots$
t	$10 \cdot 2^{t/20}$

Notice that on the right-hand side, 2 is raised to a power that is always 1/20 of t. At $t = 20$, for example, 2 is raised to the $(1/20) \cdot 20 = 1$; at $t = 60$, 2 is raised to the $(1/20) \cdot 60 = 3$, and so on. So

$$B(t) = 10 \cdot 2^{t/20}.$$

Using this equation, we find that 12 hours later (12 hr $\cdot$ 60 min/hr = 720 min later) there are $10 \cdot 2^{720/20} = 10 \cdot 2^{36} \approx 10 \cdot (6.87 \cdot 10^{10}) = 6.87 \cdot 10^{11}$ bacteria. The number of *E. coli* in the poor fellow's abdomen has grown from a mere 10 to about 687,000,000,000, or 687 billion after only 12 hours. This example shows why it is important to treat bacterial infections without delay.[4] ◆

EXERCISE 9.1   Let $B(t) = 10 \cdot 2^{t/20}$. Show analytically that the average rate of change of $B(t)$ over the interval $[t, t + 20]$ is $\frac{B(t)}{20}$. That is, compute the average rate of change of B over the interval $[t, t + 20]$ as we did at the end of Example 9.1. If you have trouble with the algebra involved, read Section 9.2 and then try again.

Let's determine growth equations for other patterns of population growth; you'll see that we could have arrived at the last two growth equations bypassing the construction of a table of values. Let P_0 denote the size of a population at some benchmark time $t = 0$, t in years, and let $P(t)$ be the population at time t. For each of the examples below, try to arrive at the growth equation yourself; cover up the right-hand side of the page with a sheet of paper. (*Note:* The answers can be expressed in a myriad of ways. If your answer is not the one given, determine whether or not it is equivalent.)

---

[4] Another example of disease caused by bacterial infection is an equine disease manifested by the appearance of huge welts on a horse's skin. If treated immediately the disease is completely curable, but left untreated it can be fatal. When a groom at the Belmont Racetrack near New York informed his trainer that his horse had "gotten eaten real badly by mosquitos last night," the frenzy that ensued and the urgency with which the veterinarian worked made it clear that the groom had misdiagnosed the cause of the welts.

◆ **EXAMPLE 9.3**

(a) If the population doubles every year, then $P(t) = P_0 \cdot 2^t$.

(b) If the population triples every year, then $P(t) = P_0 \cdot 3^t$.

(c) If the population is cut in half every year, then $P(t) = P_0(\frac{1}{2})^t$.

(d) If the population increases by 30% every year, then $P(t) = P_0(1.30)^t$.

(e) If the population decreases by 10% every year, then $P(t) = P_0(1 - 0.10)^t = P_0(0.90)^t$.

(f) If the population doubles every 5 years, then $P(t) = P = P_0 \cdot 2^{t/5}$.

(g) If the population triples every 10 years, then $P(t) = P_0 \cdot 3^{t/10}$.

(h) If the population triples every half year, then $P(t) = P_0 \cdot 3^{2t}$.

(i) If the population increases by 10% every 3 years, then $P(t) = P_0(1.1)^{t/3}$.

(j) If the population is cut in half every 3 years, then $P(t) = P_0(\frac{1}{2})^{t/3}$.

SELECTED EXPLANATIONS

(d) To increase A by 30% gives $A + 30\%A = A + \frac{30}{100}A = A + 0.30A = A(1 + .030)$. In other words, to increase A by 30% is to get 130% of A, or $1.30A$. Therefore, each year the population is 130% of the previous year's population.

(e) To decrease A by 10% is to be left with 90% of A, or $0.90A$. Therefore each year the population is 90% of the previous year's population.

(f) The population must be multiplied by 2 not each year, but every 5 years. We want $P(5)$ to be $P_0 \cdot 2$ and $P(10)$ to be $P_0 \cdot 2^2$. In the expression $P(t) = P_0 \cdot 2^{t/5}$, each time t goes up by 5 the exponent increases by 1.

(h) Notice that when $t = \frac{1}{2}$ we have $P(\frac{1}{2}) = P_0 \cdot 3$ as specified. After one year the population triples twice: $P(1) = P_0 \cdot 3^2$.   ◆

COMMON ERROR Suppose a population doubles every three years. The temptation to write $P(t) = P_0 \cdot 2^{3t}$ can be thwarted by seeing what happens when $t = 3$. Using this incorrect equation we get $P(3) = P_0 \cdot 2^{3\cdot3} = P_0 \cdot 2^9$, and the error becomes glaringly clear. We can model the situation correctly using $P(t) = P_0 \cdot 2^{t/3}$.

PROBLEMS FOR SECTION 9.1

1. The number of bacteria in a certain culture is known to triple every day. Suppose that at noon today there are 200 bacteria.

(a) Construct a table of values to find a function that gives the number of bacteria after t days.

(b) Approximately what was the population count at noon yesterday? At noon 4 days ago?

(c) From now on, suppose the population at noon today is called B_0 rather than being specifically 200. Find a function that gives the number of bacteria after t days.

(d) Express the number of bacteria as a function of w, where w is time measured in weeks.

(e) How many bacteria will be present at noon one week from today?

2. As the story goes, a long time ago there was a Persian ruler who enjoyed the game of chess so much that in order to demonstrate his gratitude he offered its inventor any reward the man wanted. When the man was called in front of the ruler he requested 1 grain of rice for the first square of the chessboard, 2 for the second, 4 for the third, 8 for the fourth, and so on until rice was allocated for all 64 squares of the chessboard. At first the ruler thought that the request was quite modest—he would have been happy to give the man jewels instead of rice. Eventually he realized that the request was not modest at all and—according to some versions of this story—he ordered the man beheaded. What was all the fuss about? (This classic story is so popular that it has even been used on the TV show "I Love Lucy.")

(a) How many grains of rice were allocated to the 64th square?

(b) If we assume that a grain of rice is 0.02 g, approximately how many grams of rice are allocated to the 64th square? Compare this with the annual world production of rice ($\approx 4 \times 10^8$ tons in 1980). *Note:* 1 ton is 907.18 kilograms. For general edification, a table of commonly used prefixes is supplied below.

Symbol		Factor	
G	giga-	1 000 000 000	$= 10^9$
M	mega-	1 000 000	$= 10^6$
k	kilo-	1 000	$= 10^3$
m	milli-	0.001	$= 10^{-3}$
n	nano-	0.000 000 001	$= 10^{-9}$

(c) Challenge problem (optional): Find the total mass of the rice allocated to all 64 squares. *Note:* If you have to add 64 numbers without any shortcut, please skip this part of the problem!

(Facts about rice production are taken from J.C. Newby, *Mathematics for the Biological Sciences*, Oxford University Press, 1980. p. 53. The story itself is an old standard.)

3. Consider two strains of bacteria, one (*E. coli*) whose population doubles every 20 minutes and another, strain X, whose population doubles every 15 minutes. Suppose that at present the number of *E. coli* is 600 and the number of bacteria of strain X is 100.

(a) Express the number of E. coli as a function of h, the number of hours from now.

(b) Express the number of strain X bacteria as a function of h, the number of hours from now.

(c) After approximately how many hours will the populations be equal in number?

4. In the 1980s the small town of Old Bethpage, New York, made the front page of the *New York Times* magazine section as an illustration of what was termed a "dying suburb." In Old Bethpage schools are being converted to nursing homes as the population ages and the baby boomers move out. Suppose that the number of school-age children in 1980 was C_0, and was decreasing at a rate of 6% per year. Let's assume that the number of school-age children continues to drop at a rate of 6% each year. Let $C(t)$ be the number of school-age children in Old Bethpage t years after 1980.

(a) Find $C(t)$.

(b) Express the number of school-age children in Old Bethpage in 1994 as a percentage of the 1980 population.

(c) Use your calculator to estimate the year in which the population of school-age children in Old Bethpage will be half of its size in 1980.

5. In the middle of the 1994–95 academic year, in the middle of the week, in the middle of the day, there was a bank robbery and subsequent shootout in the middle of Harvard Square. Throughout the afternoon the news spread by word-of-mouth. Suppose that at the time of the occurrence 30 people know the story. Every 15 minutes each person who knows the news passes it along to one other person. Let $N(t)$ be the number of people who know at time t.

(a) Make a table with time in one column and $N(t)$ in the other. Identify the pattern and write N as a function of t.

(b) If you've written your equation for $N(t)$ with t in minutes, convert to hours. If you've done it in hours, convert to minutes. Make a table to check your answers. (It's easy to make a mistake the first time you do this.)

9.2   EXPONENTIAL: THE BARE BONES

You are going to be using exponential functions a great deal, so you need to be loose and limber with exponential algebra. You need to know exponential functions like the back of your hand. You will want to be able to picture an exponential function in your mind, with your eyes closed and your calculator behind your back. So without further ado, let's start from ground zero and rapidly work our way up.

The Algebra of Exponents

If n is a positive integer, then b^n is b multiplied by itself n times. In fact, b^a can be defined for any real exponent a and positive base b. How to do this will become clearer as we proceed.

The Basics of Exponential Algebra

Let x be a positive number. Then

$$\boxed{\begin{array}{c} \textbf{ExponentLaws} \\[4pt] \text{i) } x^a x^b = x^{a+b} \\[6pt] \text{ii) } \dfrac{x^a}{x^b} = x^{a-b} \\[6pt] \text{iii) } (x^a)^b = x^{ab} \end{array}}$$

These statements should make sense intuitively when you think of a and b as positive integers. (For (ii) think of $a > b$ for starters.) In actuality we will define x^a so that these statements are true for *any* exponents a and b. You have undoubtedly seen these laws before; you'll need to be able to recall and apply them without problem. Practice will aid application and concrete examples can supplement recall. Suppose, for instance, that you can't recall whether $x^a x^b$ is x^{a+b} or x^{ab}. Work through a simple concrete example to clarify your thoughts. Try an example where a and b are small positive integers, as done below.[5]

$$x^2 x^3 = (xx)(xxx) = x^5, \text{ so } x^a x^b \text{ must be } x^{a+b}, \text{ not } x^{ab}.$$

If the three exponent rules are to work for rational exponents, then the following must be true.

$$x^0 = 1$$
$$x^{-k} = \frac{1}{x^k}$$
$$x^{1/n} = \sqrt[n]{x}, \text{ i.e., the } n\text{th root of } x$$

These results can be deduced from the first three laws as follows.

$$1 = \frac{x^b}{x^b} = x^{b-b} = x^0, \quad \text{so} \quad x^0 = 1$$

$$\frac{1}{x^k} = \frac{x^0}{x^k} = x^{0-k} = x^{-k}, \quad \text{so} \quad x^{-k} = \frac{1}{x^k}$$

$(x^{1/n})^n = x^{n/n} = x$, so $x^{1/n}$ is a number that, when raised to the nth power, gives x. Therefore, $x^{1/n} = \sqrt[n]{x}$.

Notice the following:

$$\frac{1}{x^{-k}} = (x^{-k})^{-1} = x^k.$$

$$x^{k/n} = (x^k)^{1/n} = \sqrt[n]{x^k} \quad \text{and} \quad x^{k/n} = (x^{1/n})^k = (\sqrt[n]{x})^k;$$

consequently, $\sqrt[n]{x^k} = (\sqrt[n]{x})^k$.

At this point, we have defined b^a for any rational exponent a and positive number b. In fact, we can define b^t for any real number t. For now we'll deal with irrational exponents simply by approximating the irrational number by a sequence of rational ones. A more satisfactory definition can be given after taking up logarithmic functions. Suppose s is irrational, r and t are rational, and $r < s < t$. Then $b^r < b^s < b^t$. We define b^s so as to make the graph of b^x continuous.

Why do we restrict ourselves to b positive? If $b = 0$, then b^t is undefined for $t \leq 0$. If b is negative, then b^t is not a real number for many values of t. For example, $(-4)^{1/2} = \sqrt{-4}$ is not a real number; there is no real number whose square is -4. Try graphing $y = (-8)^t$ on your graphing calculator or computer and see what happens. We know that $(-8)^{1/3} = \sqrt[3]{-8} = -2$ and $(-8)^{-1} = \frac{1}{-8}$, but how does your calculator respond to the command to graph $(-8)^t$? How does it respond if you ask for $\sqrt{-8}$, that is, $(-8)^{1/2}$? Different calculators respond differently; can you make sense out of your calculator's response?

Keep the following points in mind:

■ An exponent refers only to the number immediately below it unless otherwise indicated by parentheses.

---

[5] Make sure you choose a and b so that $a + b \neq ab$; otherwise you have defeated the purpose of your experiment!

- Exponentiation does not distribute over addition or subtraction.

- Enough parentheses must be used when entering an exponential into your calculator or computer so that the typed expression is interpreted as you intended.

These points are illustrated below.

$$bc^2 = bcc \neq (bc)^2 \qquad \text{e.g., for } x \neq 0, \ -x^2 \text{ is negative while } (-x)^2 \text{ is positive.}$$

$$(ab)^n = a^n b^n \qquad \text{e.g., } (-2x)^2 = (-2)^2 \cdot x^2 = 4x^2 \neq -2x^2 = -(2x^2)$$

$$\left(\frac{a}{b}\right)^n = \frac{a^n}{b^n} \qquad \text{e.g., } \sqrt{\frac{4}{3}} = \left(\frac{4}{3}\right)^{1/2} = \frac{4^{1/2}}{3^{1/2}} = \frac{2}{3^{1/2}} = \frac{2}{\sqrt{3}}$$

But

$$(a+b)^n \neq a^n + b^n \quad \text{e.g., } (2+3)^2 = 5^2 = 25 \qquad\qquad \text{but } 2^2 + 3^2 = 4 + 9 = 13.$$

$$(a-b)^n \neq a^n - b^n \quad \text{e.g., } \sqrt{25-9} = (25-9)^{1/2} = 16^{1/2} = 4 \quad \text{but } 25^{1/2} - 9^{1/2} = 5 - 3 = 2.$$

Entering

$$\boxed{2}\ \boxed{*}\ \boxed{x}\ \boxed{\wedge}\ \boxed{5}\ \boxed{+}\ \boxed{\sqrt{\ }}\ \boxed{2}$$

in a calculator gives $2x^5 + \sqrt{2}$ not $2x^{5+\sqrt{2}}$ or $(2x)^{5+\sqrt{2}}$. To enter $(2x)^{5+\sqrt{2}}$ you can type

$$\boxed{(}\ \boxed{2}\ \boxed{x}\ \boxed{)}\ \boxed{\wedge}\ \boxed{(}\ \boxed{5}\ \boxed{+}\ \boxed{\sqrt{\ }}\ \boxed{2}\ \boxed{)}.$$

Different calculators may have slightly different conventions. Check yours out either by using the instruction manual or by playing around with numbers you know. Rounding off the bases of exponentials can cause substantial inaccuracy. Let your calculator work for you; do not round off early.

EXERCISE 9.2 Use a calculator or computer to evaluate the following to two decimal places.

(a) $\pi^3(1 + \frac{2}{3})^{50/3}$

(b) $100(1 + \frac{0.05}{12})^{12 \cdot 60}$

(c) $\frac{(3.001)^{3.001} - 3^3}{0.001}$

Answer

(a) 154500.10

(b) 1996.07

(c) 56.73

EXERCISE 9.3 Each statement that follows is *incorrect*. Find the incorrect step and correct it.

i) $\sqrt[3]{-8x^{-2}} = \sqrt[3]{-8}\sqrt[3]{x^{-2}} = \dfrac{1}{2}x^{-2/3}$

ii) $\left(\dfrac{B^3C^{-1}}{2B}\right)^{-3} = \left(\dfrac{B^2C^{-1}}{2}\right)^{-3} = \dfrac{B^{-6}C^3}{2^{-3}} = \dfrac{2^{-3}C^3}{B^6} = \dfrac{C^3}{8B^6}$

iii) $\sqrt{4x^2 + 16y^2} = \sqrt{4(x^2+4y^2)} = \sqrt{4}\sqrt{x^2+4y^2} = 2(x+2y) = 2x+4y$

iv) $\dfrac{A^n}{B^n} + \dfrac{C^n}{D^n} = \dfrac{A^nD^n + C^nB^n}{B^nD^n} = \dfrac{(AD)^n + (CB)^n}{(BD)^n} = \left(\dfrac{AD+CB}{BD}\right)^n$

v) $\left(\dfrac{1}{2}\right)^{-1}\left(\dfrac{1}{x}+\dfrac{1}{y}\right)^{-1} = 2\left(\dfrac{1}{x}+\dfrac{1}{y}\right)^{-1} = 2(x+y) = 2x+2y$

Answers are provided at the end of the chapter.

EXERCISE 9.4   Simplify the following.

i) $\dfrac{\sqrt{2}\sqrt{x^3}}{(2x)^{-1/2}}$ ii) $4(9y^{-x})^{1/2}$ iii) $\dfrac{b^{x+w} - b^x}{b^x}$ iv) $\dfrac{(\sqrt[3]{64x^3})^{1/2}}{\left(\frac{1}{2}\right)^{-2}}$ v) $\dfrac{Q^{3R} + Q^{R+1}}{Q^{2R}}$

Answers are provided at the end of the chapter.

The Exponential Function

The function f is called an **exponential function** if it can be written in the form $f(t) = Cb^t$, where C is a constant and b is a positive constant. This function is called exponential because the variable, t, is in the exponent. The domain of the exponential function is all real numbers, and the range is all positive real numbers. If a quantity Q **grows or decays exponentially** with time t, then $Q(t) = Cb^t$ for some constants C and b. There are two unknown constants, so two data points are necessary in order to find them. At $t = 0$, $Q(0) = Cb^0$ so $C = Q(0)$. Sometimes the quantity $Q(0)$ is denoted by Q_0, indicating the *initial quantity*; we write $Q(t) = Q_0b^t$.

In Figure 9.3(a) on the following page are the graphs of $f(t) = b^t$ for $b = 1, 2, 3,$ and 10. In Figure 9.3(b) are graphs of $f(t) = b^{-t}$ for $b = 2, 3,$ and 10 or, equivalently, of $f(t) = b^t$ for $b = \frac{1}{2}, \frac{1}{3},$ and $\frac{1}{10}$. The key notion is that numbers greater than 1 increase when squared, cubed, etc.

$$2^2 = 4; \quad 2^3 = 8; \quad 2^4 = 16; \quad 2^5 = 32; \ldots$$

while numbers between 0 and 1 decrease when squared, cubed, etc.

$$\left(\dfrac{1}{2}\right)^2 = \dfrac{1}{4}; \quad \left(\dfrac{1}{2}\right)^3 = \dfrac{1}{8}; \quad \left(\dfrac{1}{2}\right)^4 = \dfrac{1}{16}; \quad \left(\dfrac{1}{2}\right)^5 = \dfrac{1}{32}.$$

Recall that $x^{-k} = \frac{1}{x^k}$. Therefore, $\left(\frac{1}{2}\right)^t = \frac{1^t}{2^t} = \frac{1}{2^t} = 2^{-t}$; the graph of $\left(\frac{1}{2}\right)^t = 2^{-t}$ can be obtained by reflecting the graph of 2^t across the y-axis. Similarly, $\left(\frac{1}{3}\right)^t = 3^{-t}$ and $\left(\frac{1}{10}\right)^t = 10^{-t}$. Therefore, the functions graphed in Figure 9.3(b) can also be written as $f(t) = \left(\frac{1}{2}\right)^t$, $f(t) = \left(\frac{1}{3}\right)^t$, and $f(t) = \left(\frac{1}{10}\right)^t$, respectively.

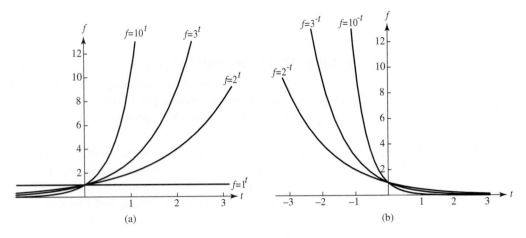

Figure 9.3

Observe that b^t is always positive.

■ Suppose $b > 1$.

  i. The graph of b^t is increasing and concave up. b^t increases at an ever-increasing rate.

  ii. The more negative t becomes the smaller b^t becomes. In fact, by taking t negative enough, b^t can be made arbitrarily close to zero. $\lim_{t \to -\infty} b^t = 0$.

■ Suppose $0 < b < 1$.

  i. The graph of b^t is decreasing and concave up.

  ii. As t increases without bound b^t gets arbitrarily close to 0. We can write this as $t \to \infty$, $b^t \to 0$; equivalently, $\lim_{t \to \infty} b^t = 0$.

The graph of b^t has a horizontal asymptote at $y = 0$.

EXERCISE 9.5

(a) Using your graphing calculator or computer, graph $f(t) = C \cdot 2^t$ for different values of C. (Be sure to include both positive and negative values of C.) How does the value of C affect the graph? Does it alter the y-intercept? The horizontal asymptote? If something changes, explain how it changes. Do your answers make sense to you?

(b) Using your graphing calculator or computer, graph $f(t) = 2^{-t} + D$. How does the value of D affect the graph? Does it alter the y-intercept? The horizontal asymptote? If something changes, explain how it changes. Do your answers make sense to you?

(c) Given your answers to parts (a) and (b), take a guess at what the graph of $f(x) = 3 - 2^x$ looks like. (*Note:* This function is the same as $g(t) = -2^t + 3$.)[6] Check your answer using your graphing calculator. Then try $h(t) = -3 - 2^{-t}$.

(d) Using your graphing calculator, graph $f(t) = 2^{kt}$ for different values of k. (Be sure to include both positive and negative values of k.) How does the value of k affect the

---

[6] Did changing the order in which the terms were written make the problem easier for you? If so, you need to practice fiddling around with the form in which expressions are written so you can help yourself by transforming an expression that looks unfamiliar into one that looks familiar. This is a standard problem-solving strategy.

graph? Does it alter the y-intercept? The horizontal asymptote? If something changes, explain how it changes. Rewriting $f(t)$ as $\left(2^k\right)^t$ helps in interpreting the results.

Manipulating Exponents: Taking Control

The three exponent laws given at the beginning of this section are identities; they can be read from left to right or from right to left. Flexibility and a sense of your own goals helps in applying them when working with exponential functions.

◆ **EXAMPLE 9.4**   In the beautiful desolate desert of the Sinai, garbage disposal is becoming a growing problem. An observer hypothesizes that in a certain region of the Sinai the number of plastic bags littering the desert is increasing exponentially. Two months after he arrived he estimates there are 100 plastic bags littering the region, and 7 months after he arrived he estimates there are 120 plastic bags. He sets the date of his arrival as a benchmark time of $t = 0$. Using the two data points and the observer's hypothesis of exponential increase, find an equation for $B(t)$, the number of bags in this region t months after the observer arrived.

Approach 1

The hypothesis of exponential growth implies

$$B(t) = Cb^t \quad \text{for constants } C \text{ and } b.$$

We know that

$$\text{when } t = 2, \ B = 100 \quad \text{and}$$
$$\text{when } t = 7, \ B = 120.$$

Therefore,

$$\begin{cases} 100 = Cb^2 \quad \text{and} \\ 120 = Cb^7. \end{cases}$$

We have a pair of simultaneous equations with two unknowns. Our goal is to eliminate a variable. One strategy is to divide one equation by the other, eliminating C. Equivalently, we could solve each equation for C and set them equal. We'll do this.

$$\begin{cases} C = \dfrac{100}{b^2} \quad \text{and} \\ C = \dfrac{120}{b^7} \end{cases}$$

So

$$\frac{100}{b^2} = \frac{120}{b^7} \qquad \text{Multiplying both sides by } b^7/100 \text{ leaves } b\text{'s on one side.}$$

$$\frac{b^7}{b^2} = \frac{120}{100}$$

$$b^5 = 1.2 \qquad \text{Raise both sides of the equation to 1/5 to solve for } b.$$

$$(b^5)^{1/5} = (1.2)^{1/5}$$

$$b = (1.2)^{1/5}.$$

We now have

$$B(t) = C[(1.2)^{1/5}]^t$$

$$B(t) = C(1.2)^{t/5}.$$

We can use either of the data points to solve for C.

$$100 = C(1.2)^{2/5}$$

$$\frac{100}{(1.2)^{2/5}} = C$$

$$C = \frac{100}{(6/5)^{0.4}} = 100(5/6)^{0.4}$$

Therefore,

$$B(t) = 100(5/6)^{0.4}(1.2)^{t/5}.$$

Approach 2

We have two data points: when $t = 2$, $B = 100$ and when $t = 7$, $B = 120$. In 5 months the number of bags has increased by 20. Because we're hypothesizing exponential growth, the *percent* increase is the important bit of information, not the actual increase itself. Twenty is 20% of 100, so every 5 months the number of bags increases by 20%.

$$B(t) = C(1.2)^{t/5}$$

Now find C as done above.

Note that if we were willing to adjust our benchmark time of $t = 0$ from the observer's day of arrival in the Sinai to the day he made the first bag count we would have

$$B(t) = 100(1.2)^{t/5}, \quad \text{where } t = 0 \text{ corresponds to 2 months after his arrival.} \quad \blacklozenge$$

◆ **EXAMPLE 9.5**   Suppose $f(x) = Ca^{2x}$, where C and a are constants, $a > 0$. The percent change in f over the interval $[x, x + 3]$ is given by

$$\frac{\text{final value} - \text{initial value}}{\text{initial value}} = \frac{f(x+3) - f(x)}{f(x)}, \quad \text{or} \quad \frac{Ca^{2(x+3)} - Ca^{2x}}{Ca^{2x}}.$$

Simplify this expression.

SOLUTION

$$\frac{C[a^{2(x+3)} - a^{2x}]}{Ca^{2x}} \qquad \text{Factor out the } C.$$

$$= \frac{a^{2x+6} - a^{2x}}{a^{2x}} \qquad \text{We want to factor out } a^{2x}, \text{ so we write } a^{2x+6} \text{ as } a^{2x} \cdot a^6.$$

$$= \frac{a^{2x} \cdot a^6 - a^{2x}}{a^{2x}} \qquad \text{Factor out } a^{2x}.$$

$$= \frac{a^{2x}[a^6 - 1]}{a^{2x}}$$

$$= a^6 - 1$$

Notice that the percent change in f over *any* interval 3 units in length is a constant. The percent change does not depend upon the particular endpoints of the interval. This is a defining characteristic of exponential functions.   ◆

▌ PROBLEMS FOR SECTION 9.2

For Problems 1 through 9, simplify the following expressions.

1. $\dfrac{x^{-1} + z^{-1}}{(z + x)x^{-2}}$

2. $\dfrac{(xy)^{-3}}{xy^{-3}}$

3. $\dfrac{\sqrt{2x^3} + \sqrt{12y^3x^4}}{\sqrt{6y^2x^5}}$

4. $\dfrac{x^{-1}}{zx^{-1} + z^{-1}}$

5. $\dfrac{z^0x^{-1}y^{-2}}{z^{-2}x^{-1}y^2}$

6. $\dfrac{x^{n+1}y^{2n}}{\left(\dfrac{x}{y}\right)^n}$

7. $\dfrac{(ab^x)^{-2}}{(ab)^{-x}}$

8. $\dfrac{(ab)^{-x}}{a^{-x} + b^{-y}}$

9. $\dfrac{(a^{-x+1}b)^3}{(a^2b^3)^x}$

For Problems 10 through 15, factor b^x out of the following expressions. Check your answer by multiplying out.

10. $b^{x+y} + b^x$

11. $b^{2x} + b^{x+1}$

12. $3b^{2x+1} - 4b^{2x-1}$

13. $(ab^2)^x + \left(\dfrac{a}{b}\right)^{-x}$

14. $b^{3x} - (2b)^{-1+2x}$

15. $\sqrt{ab^x} - a\sqrt{b^{3x}}$

For Problems 16 through 21, if the statement is always true, write "True;" if the statement is not always true, produce a counterexample. In these problems, a and b are positive constants.

16. $\sqrt{a^{2x}} = a^x$

17. $(a^x + b^x)^{\frac{1}{x}} = a + b$

18. $\sqrt{\dfrac{a^{2x}}{b^{-2x}}} = (ab)^x$

19. $a^{-x} + a^x = 1$

20. $a^x bc^x = (a^x b)(ac)^x$

21. $3^{2x} + 2^{3x} = 9^x + 8^x$

For Problems 22 through 24, simplify as much as possible.

22. $\dfrac{a^{x+y} - a^{2x}}{a^x}$

23. $\dfrac{a^{2x} - b^{4x}}{a^x + b^x}$

24. $\dfrac{A^{4+p} - A^{5p}}{A^{2+p} - A^{3p}}$

25. The graphs below are graphs of functions of the form $f(t) = Ca^t$. For each graph, determine the sign of C and whether $a \in (0, 1)$ or $a > 1$.

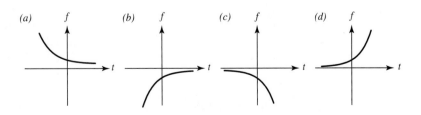

For Problems 26 through 29, find a function of the form $f(x) = Ca^x + D$ to fit the graph given.

26.

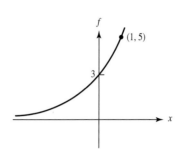

27.

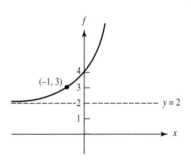

28.

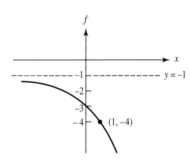

29.

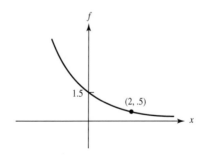

30. True or False: If the statement is always true, write "True;" if the statement is not always true, produce a counterexample.

(a) $(a^2 + b^2)^{1/2} = a + b$

(b) $(a + b)^{-1} = \frac{1}{a+b}$, $a, b \neq 0$

(c) $(a + b)^{-1} = \frac{1}{a} + \frac{1}{b}$, $a, b \neq 0$

(d) $R^{-1/2} = -\frac{1}{\sqrt{R}}$, $R > 0$

(e) $x^z + x^z = 2x^z$

(f) $x^z x^z = x^{(z^2)}$

(g) $x^z x^z = x^{2z}$

31. **Mix and Match:** Below are functions. To each function match the graph that best fits the function. Since there are no units on the graphs, you may match one graph to several different functions. *Note:* First do this problem without using a graphing calculator. You can then check your answers with the calculator. If you made a mismatch, figure out what your mistake was and learn from it.

(a) $f(x) = 3(\frac{3}{2})^x$

(b) $f(x) = -2(0.4)^x$

(c) $f(x) = 2 - 3^{-x}$

(d) $f(x) = 4(\frac{2}{3})^x$

(e) $f(x) = 4^{-x}$

(f) $f(x) = 1 + 2^x$

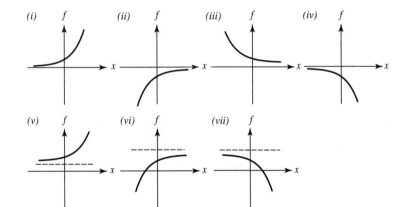

32. Simplify as much as possible:

(a) $\dfrac{x^{2y} + x^{y+2}}{x^y}$

(b) $\dfrac{\frac{\sqrt{x}}{x^{-1/2}y} - 1}{y - \frac{x^2}{y}}$

(c) $\dfrac{A^{B+4} - A^{3B}}{A^B(A^2 - A^B)}$

(d) $\dfrac{y^{3w} - y^{w+4}}{y^w(y^w + y^2)}$

33. Let $f(x) = a^x$. We know that replacing x by $x + k$ shifts the graph of f left k units. For exponential functions this is equivalent to vertical stretching or shrinking. Explain.

In Problems 34 through 37, evaluate the limits.

34. (a) $\lim\limits_{x \to \infty} -2(1.1)^{-x}$

(b) $\lim\limits_{x \to -\infty} -2(1.1)^x$

35. (a) $\lim\limits_{x \to \infty} (0.89^x - 1)$

(b) $\lim\limits_{x \to -\infty} (0.89^x - 1)$

36. (a) $\lim\limits_{x \to \infty} (\frac{3}{5})^{-x}$

(b) $\lim\limits_{x \to -\infty} (\frac{3}{5})^{-x}$

37. (a) $\lim\limits_{x\to\infty} -5(\frac{2}{3})^x + 7$ (b) $\lim\limits_{x\to-\infty} -5(\frac{2}{3})^x + 7$

38. Factor b^x out of each of the following expressions.

(a) $3b^x - b^{2x}$ (b) $(3b)^x - b^{x+2}$ (c) $b^{3x/2} - b^{2x-1}$

■ 9.3   APPLICATIONS OF THE EXPONENTIAL FUNCTION

In Section 9.2 we looked at exponential growth in the context of population growth under ideal conditions. In this section we'll look at the growth of money in an interest-earning account and the decay of radioactive materials.

The Growth of Money in a Bank Account

Exponential functions arise in questions dealing with money earning interest in a bank account and in calculations of loan paybacks. If money is deposited in a bank and left there (with no additional deposits), then the growth of the amount of money in the account is due to the interest earned. The amount of interest earned during each period is directly proportional to the amount of money in the account. Let's look at an example.

◆ **EXAMPLE 9.6**   Suppose we put $5000 in a bank account earning 4% interest per year compounded annually.

(a) Write a function that gives the amount of money in the account after t years.

(b) How long does it take for the money to double?

SOLUTION

(a) At the end of the first year, the interest added to the $5000 we already had is 4% of $5000, or $200 ($= 0.04 \cdot \$5000$). We start the next year with $5200, and at the end of that year get interest amounting to 4% of $5200, or $208, giving a total of $5408, and so on. Each year the amount in the account is 104% of the previous year's amount.

Recall that if a population grows at a rate of 4% per year, then $P(t) = P_0 \, (1.04)^t$, where P_0 is the population at $t = 0$. Similarly, if money in a bank account grows at a rate of 4% per year, then $M(t) = M_0 \, (1.04)^t$, where M_0 is the amount of money in the account at $t = 0$ and t is measured in years. Our function is

$$M(t) = 5000 \, (1.04)^t.$$

(b) To find out how long it takes for the money to double we want to solve $10,000 = 5000 \, (1.04)^t$, or, equivalently, $2 = (1.04)^t$. Observe that doubling time is independent of the amount of money we have in the account. Suppose the amount in the bank initially is denoted by B_0. The doubling time is the solution to the equation $2B_0 = B_0 \, (1.04)^t$. Notice that we are back to the equation $2 = (1.04)^t$.

This equation can be solved for t analytically using logarithms to obtain an exact answer.[7] The solution can be approximated using a graphing calculator. Solutions to an equation of the form $f(x) = k$, where k is a constant, can be approximated graphically in two ways.[8]

---

[7] We will take this up in Chapter 13.

[8] You might also have an "equation solver" on your calculator that allows you to approximate the solution.

i. Approximate the x-values where the graph of $y = f(x)$ intersects the horizontal line $y = k$. Equivalently, trace along the graph until the y-coordinate is k.

ii. Approximate the zeros of $y = f(x) - k$.

Let's use the second method to approximate the solution to $2 = (1.04)^t$. Estimate the zeros of the function $(1.04)^t - 2$. You should come up with approximately 17.67; the money doubles approximately every 17.67 years.[9]   ◆

Generalization

Suppose we put M_0 dollars in a bank at interest rate r per year compounded annually. (If the interest rate is 5%, then $r = 0.05$.) Assume the money is put in the bank at the beginning of the year and interest is compounded at the end of the year. Then each year the balance is multiplied by $1 + r$. The balance after t years is given by the function

$$M(t) = M_0(1 + r)^t. \tag{9.1}$$

In particular, if \$5000 is put in a bank paying 4% interest per year compounded annually, this formula says that $M(t) = 5000 \ (1.04)^t$, which agrees with our answer from Example 9.6. Notice that r is 0.04 and not 4; if we used $r = 4$ instead, then at the end of one year, the account would have grown from \$5000 to $\$5000 \cdot (1 + 4) = \$25{,}000$! (This is a nice deal if you can get it, but not a reasonable answer to this problem.)

COMMENT   When we write the equation $M(t) = 5000 \ (1.04)^t$ and let the domain be $t \geq 0$, we are modeling a discontinuous phenomenon with a continuous model. This equation will only mirror reality if t is an integer, i.e., if the bank has just paid the annual interest to the account. For instance, if $t \in (0, 1)$, then the interest has not yet been compounded so the balance should be exactly \$5000 over this interval. At $t = 1$ it should jump to \$5200. However, it is convenient to model this discrete process with a continuous function, as we have seen done in examples in previous chapters.

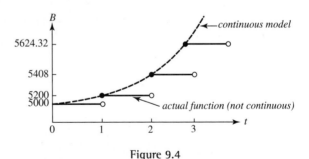

Figure 9.4

◆ **EXAMPLE 9.7**   Suppose we put the \$5000 in a bank paying interest at an interest rate of 4% per year but this time with interest compounded quarterly.

This does not mean that you get 4% interest each quarter.

---

[9] This illustrates a very useful rule of thumb for estimating doubling time: the Rule of 70. This rule of thumb says that to estimate the number of years it will take something growing at a rate of A% per year to double, we calculate $70/A$. In Example 9.6, we calculate $70/4 = 17.5$, and see that this is a very close estimate. Keep in mind that this is not a hard and fast rule but rather only an approximation. When we study logarithms we will see why this rule works, and for what values of A it works best.

Rather, your money earns 1%, $\left(\frac{4\%}{4}\right)$, interest each quarter of a year. The advantage is that interest is computed on the interest more frequently (four times per year, as opposed to just once per year in Example 9.6). The balance is growing by 1% every 1/4 of a year, so it grows according to

$$M(t) = 5000 \ (1.01)^{4t}.$$

(Refer to Example 9.3 if you need help with this.) Notice that the balance increases by 1% (is multiplied by 1.01) each quarter, so when $t = 1$, interest has been compounded four times, as expected. $M(1) = 5000(1.01)^4 \approx 5000(1.040604\ldots)$, so each year the interest is effectively $4.0604\ldots\%$. This is called the **effective** annual interest rate.

4% is called a **nominal** annual interest rate because the interest is 4% "in name only."   ◆

EXERCISE 9.6   Compare the amount of money you would have after 10 years in Example 9.6 with the amount in Example 9.7.

Answers are provided at the end of the chapter.

Generalization

Suppose you put $\$M_0$ in a bank at interest rate r per year compounded n times per year. Then the interest rate per compounding period is $\frac{r}{n}$. Each year there are n compounding periods, so in t years there are nt compounding periods. (The growth is $\frac{r}{n}$ every nth of a year.) Then

$$M(t) = M_0 \left(1 + \frac{r}{n} \right)^{nt}$$

where

$M(t) =$ the amount of money in the account at the end of t years,

$M_0 =$ the original amount of money,

$r =$ the annual interest rate (If the rate is 5%, then $r = 0.05$.),

$n =$ the number of compounding periods per year.

(If you have any trouble following this, make a table as was done in Example 9.2. Also, verify that Equation (9.1) is just a special case of the formula presented above.)

EXERCISE 9.7   Bank A gives a nominal interest rate of 6% per year compounded monthly. Bank B gives 6.1% interest per year compounded annually. Which is the preferable arrangement for an investor?

Radioactive Decay

Through experimentation scientists have found that unstable radioactive isotopes revert back to nonradioactive form at a rate proportional to the amount of the isotope present. The rate of decay of radioactive substances is typically indicated by identifying the *half-life* of the substance; this is the amount of time it takes for half of the original amount of radioactive substance to revert (decay) to its nonradioactive form.

◆ **EXAMPLE 9.8**   Tritium, the radioactive hydrogen isotope H_3, has a half-life of 12.3 years. (Tritium can be used for determining the age of wines in a process similar to carbon-dating, which we will look at in Exercise 9.8.) Let H_0 denote the amount of tritium at time $t = 0$. Find a formula for $H(t)$, the amount of tritium t years later.

Approach 1

Use the ideas of Example 9.3. This idea of radioactive half-life is very similar to the idea of doubling time we discussed when working with *E. coli*. When given information about doubling time we used 2 as the base; in the case of tritium it makes sense to use 1/2 as our base.

$$H(t) = H_0(1/2)^{t/12.3}$$

Approach 2

Make a table of values.

t	$H(t)$
0	H_0
12.3	$0.5 \; H_0$
24.6	$(0.5)(0.5) \; H_0$
36.9	$(0.5)^3 \; H_0$
⋮	⋮
t	$(0.5)^{t/12.3} \; H_0$

So $H(t) = H_0(0.5)^{t/12.3}$.

Approach 3

Radioactive material decays at a rate proportional to itself, so an exponential function models the process. Therefore, $H(t) = Cb^t$ for some constants C and b.

The "initial condition," the value of H when $t = 0$, helps us find C. At $t = 0$, $H = H_0$. Therefore, $H_0 = Cb^{(0)} = C$ and $H(t) = H_0 b^t$.

Now we must find b by using the fact that $H(12.3) = \frac{1}{2} H_0$.

$$0.5 H_0 = H_0 b^{12.3}, \text{ so } 0.5 = b^{12.3}.$$

How do we solve for b?

Recall: If $a^3 = 7$, we can solve for a by taking the cube root of both sides. That is,

$$(a^3)^{1/3} = 7^{1/3}, \text{ so } a = 7^{1/3}.$$

We can apply the same approach to solving for b in the equation $0.5 = b^{12.3}$.

$$0.5^{1/12.3} = (b^{12.3})^{1/12.3}$$

$$0.5^{1/12.3} = b$$

$$H(t) = H_0[0.5^{1/12.3}]^t \quad \text{or} \quad H_0(0.5)^{t/12.3}$$

A person using Approach 3 could decide that $0.5^{1/12.3}$ is a clumsy way to express the value of b and therefore replace the expression by a numerical approximation. $0.5^{1/12.3} \approx 0.945$. The answer can be expressed as: $H(t) \approx H_0(0.945)^t$. $H_0(0.945)^t$ is approximately the same as $(H_0)(0.5)^{t/12.3}$. Very large values of t will yield greater discrepancies than smaller ones. It is advisable to work with exact values and save any rounding off until the end of the problem. If you don't do this, then you'll need to think hard about error propagation.

We have two different forms in which we can express H as a function of t, t in years:

$$H(t) \approx H_0(0.945)^t \quad \text{and} \quad H = H_0 \left(\frac{1}{2}\right)^{t/12.3}.$$

From the former, we can quickly determine that every year there is 94.5% of the amount of tritium there was the previous year. (Another way of putting this is that each year the amount of tritium decreases by 5.5%, because $100\% - 94.5\% = 5.5\%$.) From the latter, we can read the information that every 12.3 years half the tritium remains.

We can take advantage of the fact that there are many (infinitely many, in fact) ways to write any given number. For example, $64 = 2^6 = 4096^{1/2} = 0.5^{-6} = 4^3$, and so on.   ◆

EXERCISE 9.8   *Carbon Dating*   Cosmic rays in the upper atmosphere convert some of the stable form of carbon, C_{12}, to the unstable radioactive isotope C_{14}, a form with two extra electrons. Through photosynthesis, plants take in carbon dioxide from the atmosphere and incorporate the carbon atoms; therefore plants contain the same ratio of C_{14} to C_{12} atoms as exists in the atmosphere. Animals eat plants and so have the same ratio of C_{14} to C_{12} as do the plants and the atmosphere. Thus, during the lifetime of a living organism the ratio of C_{14} to C_{12} in the organism is fixed; it is the same as the ratio in the earth's atmosphere.

When an organism dies it stops incorporating carbon atoms. The C_{14} that is in the organism decays exponentially (reverting to the stable C_{12} form), and the C_{14} supply is not replenished. Assuming that the ratio of C_{12} to C_{14} has remained constant in the atmosphere, scientists are able to use the unstable C_{14} isotope to date organisms.

The half-life of carbon-14 is approximately 5730 years.

(a) Let C_0 denote the amount of radioactive carbon in an organism right before it died. Find a formula for $C(t)$, the amount of radioactive carbon in this organism t years after it has died.

(b) A sample from a tree that perished in a volcanic eruption has been analyzed and found to have 70% of the C_{14} it would have had were it alive. Approximately how long ago did the volcanic eruption occur?

Answers are provided at the end of the chapter.

Writing and Rewriting Exponential Functions; Cheap Information, Yours for the Taking

Doubling Time in Minutes; Doubling Time in Hours.   We know that the number of organisms in a colony of *E. coli* under ideal reproductive conditions doubles in size every 20 minutes. The number of *E. coli* after t minutes is given by

$$B(t) = B_0 \cdot 2^{t/20}, \quad \text{where } B_0 = B(0) \text{ and } t \text{ is given in minutes.}$$

Doubling every 20 minutes is equivalent to doubling every 1/3 hour (or doubling three times each hour) so we can write

$$B(h) = B_0 \cdot 2^{3h}, \text{ where } h \text{ is given in hours.}$$

Percentage Change Per Minute, Percentage Change Per Hour. If we rewrite $B(t) = B_0 \cdot 2^{t/20}$ in the form $B_0 b^t$, we can readily read off the percent change in population each minute.

$$B(t) = B_0 \cdot 2^{t/20} = B_0(2^{1/20})^t \approx B_0(1.0353)^t = B_0(1 + 0.0353)^t,$$

so every minute the *E. coli* population increases by about 3.53%.[10]

For h measured in hours,

$$B(h) = B_0 2^{3h} = B_0(2^3)^h = B_0(8)^h = B_0(1 + 7)^h = B_0 \left(1 + \frac{700}{100}\right)^h,$$

so every hour the *E. coli* population increases by 700%.

Nominal Annual Interest and Effective Annual Interest. Suppose a bank has a nominal interest rate of 5.4% per year compounded monthly. If M_0 dollars are deposited in the account at time $t = 0$ and left for t years, the balance in the account will be

$$M(t) = M_0 \left(1 + \frac{0.054}{12}\right)^{12t}.$$

Effectively the annual interest will be slightly more than 5.4% due to the compounding. Rewriting the function above lets us read off the effective annual interest rate:

$$M(t) = M_0 \left(1 + \frac{0.054}{12}\right)^{12t} = M_0(1.0045)^{12t} = M_0[(1.0045)^{12}]^t \approx M_0(1.05536)^t.$$

We see that a nominal annual interest rate of 5.4% compounded monthly is equivalent to an effective annual growth rate of about 5.536%.

The idea behind the last few examples is that an exponential equation can be written in many different forms. Different forms make different bits of information (annual growth rate, or quarterly growth rate, or doubling time, or tripling time, etc.) available at our fingertips.

EXERCISE 9.9   A filtering system is used to purify water used in industrial processes. The longer the water remains in the filtering system the less contaminants remain in the water. The system removes contaminants at a rate proportional to the amount of contaminants in the water. Let $C(t)$ be the number of milligrams of contaminants in the water t hours after it enters the system. Suppose that when the water enters the system it contains C_0 milligrams of contaminant.

(a) Find $C(t)$ if

i. the system removes 20% of the contaminants every hour.

---

[10] Alternatively, we could have calculated $B(1)$ and computed the percent change from $B(0)$ to $B(1)$:

$$\frac{B(1) - B(0)}{B(0)} \approx 3.53\%.$$

ii. the system removes 10% of the contaminants every half hour.

iii. the system removes 4.9% of the contaminants every quarter of an hour.

iv. the system removes 60% of the contaminants every 4 hours.

(b) Write each of the functions from part (a) in the form $C(t) = C_0 b^t$. Use your answers to determine which of the systems is most efficient for water purification. Explain your reasoning.

Answers are provided at the end of the chapter.

Variations on a Theme—A Difference That Decays Exponentially

◆ **EXAMPLE 9.9** You've just taken a hot apple cobbler out of the oven and placed it on the kitchen counter to cool. The cobbler comes out of the oven at a piping hot 275°F. Assume the kitchen maintains a constant temperature of 65°F. How does the temperature of the cobbler change with time?

SOLUTION From experience we know that the cobbler will cool most rapidly when it is first taken from the oven, when the difference between its temperature and the temperature of the kitchen air is greatest. Newton's Law of Cooling tells us that the temperature difference between the cobbler and the air will decrease exponentially. In other words, if $A(t)$ is the temperature of the apple cobbler at time t, then the quantity $A(t) - 65$ decays exponentially with time.

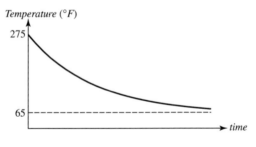

Figure 9.5

Recall that if y decays exponentially with time, then $y(t) = Cb^t$ for some $b \in (0, 1)$. Therefore our last statement translates to $A(t) - 65 = Cb^t$ for some $b \in (0, 1)$.

When $t = 0$, $A(t) = 275$ so at $t = 0$

$$275 - 65 = Cb^0$$
$$210 = C. \quad C \text{ is the initial temperature difference.}$$

We know $A(t) = 210b^t + 65$, where $b \in (0, 1)$. The graph of $y = b^t$ for $b \in (0, 1)$ looks like Figure 9.6(a). After doing Exercise 9.5, the graphs drawn in Figures 9.6(b) and (c) should make sense.

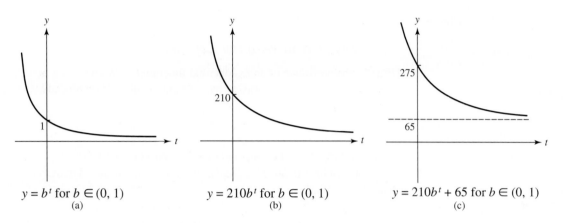

$y = b^t$ for $b \in (0, 1)$
(a)

$y = 210b^t$ for $b \in (0, 1)$
(b)

$y = 210b^t + 65$ for $b \in (0, 1)$
(c)

Figure 9.6

Notice that in this example $A(t)$ does not decay exponentially, $(A(t) \neq Cb^t)$, but the temperature difference, $(A(t) - 65)$, does decay exponentially with time.

In order to find b we would need to put a thermometer into the apple cobbler and find its temperature at some specified time t. This data point would allow us to solve for b. For instance, suppose that after 10 minutes the temperature of the cobbler is 175°F. As an exercise you can show that $A(t) = 210 \left(\frac{11}{21}\right)^{t/10} + 65$.   ◆

Exploratory Problems for Chapter 9

The Derivative of the Exponential Function

What is the derivative of an exponential function b^x? We will look at case studies to investigate this derivative numerically and graphically.

Case Studies.   Let $f(x) = 2^x$, $g(x) = 3^x$, and $h(x) = 10^x$.

You will investigate f', g', and h' from both graphical and numerical perspectives.

1. On the same set of axes, sketch the graphs of $f(x) = 2^x$, $g(x) = 3^x$, and $h(x) = 10^x$. Then, on another set of axes, draw a rough sketch of the graphs of $f'(x)$, $g'(x)$, and $h'(x)$. The graphs of f, g, and h all intersect the vertical axis at $(0, 1)$. On the other hand, the graphs of $f'(x)$, $g'(x)$, and $h'(x)$ do not intersect at the same point. Which has the largest vertical intercept? The smallest?

   This next problem is best done in groups. Each group chooses one of the functions f, g, or h. If there are many groups, other bases can be used in addition to 2, 3, and 10, but these constitute the minimum.

2. • Complete a table like the one below for your function. Approximate the slope of the tangent at a point by the slope of a secant line. For instance, to approximate the derivative of $f(x) = 2^x$ at $x = 1$, use the slope of the secant line between $(1, f(1))$ and $(1 + h, f(1 + h))$ for some very small h. Letting $f(x)$ be 2^x you get the two points $(1, 2^1)$ and $(1 + h, 2^{1+h})$. $f'(1) \approx m_{\text{sec}} = \frac{2^{1+h} - 2^1}{h}$ for h very small.

x	$f(x)$	Estimate of $f'(x)$
0		
1		
2		
3		
4		
5		

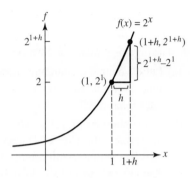

Post these completed tables so that everyone in the class can see them.

- In light of the data you have collected, make a conjecture about a formula for $f'(x)$ for your function. Post your conjecture under your completed table.

   If you and your group haven't been able to come up with a conjecture based on your data, look at other tables of data for inspiration.[11] Or see if you can predict $f'(6)$. Or collect more data until you see a pattern. Or look at the conjectures of other groups and see if you can adapt these conjectures to your own data.

3. After posting all the tables of data and all the conjectures from each of the groups, try to come up with a conjecture that might hold more generally, for any function of the form $f(x) = b^x$.

4. Do you think that there is a value of b for which the derivative of b^x is b^x? If so, about how big do you think this value is?

---

[11] Looking at the data for the function 10^x might help you out.

■ PROBLEMS FOR SECTION 9.3

1. *Carbon Dating:* Carbon-14, with a half-life of approximately 5730 years, can be used to date organic remains on earth. (See Exercise 9.8 for an explanation of carbon dating.)

   (a) Let C_0 denote the amount of radioactive carbon in an organism when it is alive. Find a formula for $C(t)$, the amount of radioactive carbon in this organism t years after it has died.

   (b) In the summer of 1993, while in Syria, I visited the ruins of Palmyra (Tadmor in Arabic—the City of Dates). Palmyra was mentioned in tablets dating as far back as the nineteenth century B.C., but it reached its heyday in the time of Queen Zenobia around 137 A.D. While exploring the site near the Funerary Towers, I came across something that looked disturbingly like that of a human leg bone lying in the sand. A Harvard medical student I met at the site confirmed that this was indeed a human bone—really. Assuming that the relevant person died around 137 A.D., what percentage of the original C_{14} remained in the bone?

   (c) Rewrite the equation you got in part (a) so it is in the form $f(t) = C_0 a^t$. From this, fill in the blank in the following statement: Each year the amount of carbon-14 in a deceased organism decreases by _____%.

2. On July 15, 1999, the Harvard University *Gazette* reported that workers gutting Holden Chapel in the course of renovations at the university unexpectedly came across some human bones. An archaeology concentrator, Rachel Sexton, who was observing the excavations, called in Professor Carole Mandryk of the Anthropology Department for her insight. Since some of the bones were sawed in half, Mandryk first hypothesized a horrible murder—but further investigation by Sexton revealed that before 1850 Holden Chapel had been used as a medical laboratory. The *Gazette* reports that

   > Whether the bones would be considered evidence or artifacts came down to a question of age. Under Massachusetts law, human remains less than 100 years old are the concern of the police and the forensics lab, while those more than 100 years old fall under the jurisdiction of the state archaeologist.[12]

   The bones were deemed to date back to around 1850. What percent of the original C_{14} remained in the bones in 1999?

3. A population of bacteria is growing exponentially. At 7:00 A.M. the mass of the population is 12 mg. Five hours later it is 14 mg.

   (a) What will be the mass of the bacteria after another 5 hours?

   (b) At 7:00 P.M. what do we expect the mass to be?

   (c) What was the mass of the population at 8:00 A.M.? Given your answer, by what percent is the mass of the population increasing each hour? By what percent is it increasing each day?

4. The population in a certain area of the country is increasing. In 1970 the population was 100,000, and by 1990 it was 200,000.

---

[12] Ken Gewertz, Harvard University *Gazette*, p. 4, July 15, 1999.

(a) If the population has been increasing linearly, was the population in 1980 equal to 150,000, greater than 150,000, or less than 150,000? Explain your reasoning.

(b) If the population has been increasing exponentially, was the population in 1980 equal to 150,000, greater than 150,000, or less than 150,000? Explain your reasoning. Note: Your answers to parts (a) and (b) should be different!

5. Let $D(t)$, $H(t)$, and $J(t)$ represent the annual salaries (in dollars) of David, Henry, and Jennifer, and suppose that these functions are given by the following formulas, where t is in years. $t = 0$ corresponds to this year's salary, $t = 1$ to the salary one year from now, and so on. The domain of each function is $t = 0, 1, 2, \ldots$ up to retirement.

$$D(t) = 40{,}000 + 2500t$$

$$H(t) = 50{,}000(0.97)^t$$

$$J(t) = 40{,}000(1.05)^t$$

(a) Describe in words how each employee's salary is changing.

(b) Suppose you are just four years away from retirement—you'll collect a salary for four years, including the present year. Which person's situation would you prefer to be your own?

(c) If you are in your early twenties and looking forward to a long future with the company, which would you prefer?

6. In Anton Chekov's play "Three Sisters," Lieutenant-Colonel Vershinin says the following in reply to Masha's complaint that much of her knowledge is unnecessary. "I don't think there can be a town so dull and dismal that intelligent and educated people are unnecessary in it. Let us suppose that of the hundred thousand people living in this town, which is, of course, uncultured and behind the times, there are only three of your sort. . . . Life will get the better of you, but you will not disappear without a trace. After you there may appear perhaps six like you, then twelve and so on until such as you form a majority. In two or three hundred years life on earth will be unimaginably beautiful, marvelous. Man needs such a life and, though he hasn't it yet, he must have a presentiment of it, expect it, dream it, prepare for it; for that he must know more than his father and grandfather. And you complain about knowing a great deal that is unnecessary."

Let us assume that Vershinin means that this doubling occurs every generation and take a generation to be 25 years. Suppose that the total population of the town remains unchanged.

(a) In approximately how many years will the people "such as [Masha] form a majority"?

(b) What percentage of the town will be "intelligent and educated" in the 200 years that Vershinin mentions?

(c) Now assume that the total population grows at a rate of 2% per year. Answer questions (a) and (b) with this new assumption.

7. Many trainers recommend that at the start of the season, a cyclist should increase his or her weekly mileage by not more than 15% each week.

(a) If a cyclist maintains a "base" of 50 miles per week during the winter, what is his or her maximum recommended weekly mileage for the fifth week of the season?

(b) Find a formula for $M(w)$, the maximum weekly recommended mileage w weeks into the season. Assume that initially the cyclist has a base of A miles per week.

8. Pasteurized milk is milk that has been heated enough to kill pathogenic bacteria. Pasteurization of milk is widespread because unpasteurized milk provides a good environment for bacterial growth. For example, tuberculosis can be transmitted from an infected cow to a human via unpasteurized milk. *Mycobacterium tuberculosis* has a doubling time of 12 to 16 hours. If a pail of milk contains 10 *M. tuberculosis* bacteria, after approximately how many hours should we expect there to be 1000 bacteria? Give a time interval.

(Facts from *The New Encylcopedia Britannica*, 1993, volume 14, p. 581.)

9. According to fire officials, a 1996 fire on the Warm Springs Reservation in central Oregon tripled in size to 65,000 acres in one day. A fire in Upper Lake, California, quintupled in size to 10,200 acres in one day.

(a) Assuming exponential growth, determine the doubling time for each fire.

(b) What was the hourly percentage growth of each fire?

10. During the decade from 1985 to 1995, Harvard's average return on financial investments in its endowment was 11.1% *per year*. Over the same period, Yale's *total* return on its investments was 287.3%. (*Boston Globe*, July 26, 1996.) Let's assume both Harvard and Yale's endowments are growing exponentially.

(a) What was Harvard's total return over this ten-year period?

(b) What was Yale's average annual rate of return?

(c) Which school got the higher return on its investments?

(d) What was the doubling time for each school's investments?

(e) In 1995, Harvard's endowment was approximately $8 billion. What was its instantaneous rate of growth (from investment only, ignoring new contributions)? Include units in your answer.

11. (a) If rabbits grow according to $R(t) = 1010(2)^{t/3}$, t in years, after how many years does the rabbit population double? What is the percent increase in growth each year?

(b) If the sheep population in Otrahonga, New Zealand, is growing according to $S(t) = 3162(1.065)^t$, t in years, after approximately how many years does the sheep population double? What is the percent increase in growth each year?

12. *Exploratory:* Which grows faster, 2^x or x^2?

(a) Using what you know about these two functions and experimenting numerically and graphically, guess the following limits:

i. $\lim\limits_{x \to \infty} x 2^{-x}$ ii. $\lim\limits_{x \to -\infty} \dfrac{x^2}{2^x}$ iii. $\lim\limits_{x \to \infty} \dfrac{x^2}{2^x}$ iv. $\lim\limits_{x \to \infty} \dfrac{2^x}{x^2}$

(b) For $|x|$ large, which function is dominant, 2^x or x^2? Would you have answered differently if we looked at 3^x and x^3 instead?

13. *Devaluation:* Due to inflation, a dollar loses its purchasing power with time. Suppose that the dollar loses its purchasing power at a rate of 2% per year.

   (a) Find a formula that gives us the purchasing power of $1 t years from now.

   (b) Use your calculator to approximate the number of years it will take for the purchasing power of the dollar to be cut in half.

14. A population of beavers is growing exponentially. In June 1993 (our benchmark year when $t = 0$) there were 100 beavers. In June 1994 ($t = 1$) there were 130 beavers.

   (a) Write a function $B(t)$ that gives the number of beavers at time t.

   (b) What is the percent increase in the beaver population from one year to the next?

15. We are given two data points for the cumulative number of people who have graduated from a newly established flying school, a school for training pilots. Our benchmark time, $t = 0$, is one year after the school opened.

   When $t = 0$, the number of people who have graduated $= 25$.

   When $t = 3$, the number of people who have graduated $= 127$.

   Find the cumulative number of people who have graduated at time $t = 5$ if

   (a) the cumulative number of people who have graduated is a linear function of time.

   (b) the cumulative number of people who have graduated is an exponential function of time.

16. According to a report from the General Accounting Office, during the 14-year period between the school year 1980–1981 and the school year 1994–1995, the average tuition at four-year public colleges increased by 234%. During the same period, average household income increased by 82%, and the Labor Department's Consumer Price Index (CPI) increased by 74%. (*Boston Globe*, August 16, 1996.)

   (a) Assuming exponential growth, determine the annual percentage increase for each of these three measures.

   (b) The average cost of tuition in 1994–1995 was $2865 for in-state students. What was it in 1980–1981?

   (c) Starting with an initial value of one unit for each of the three quantities, average tuition at four-year public colleges, average household income, and the Consumer Price Index, sketch on a single set of axes the graphs of the three functions over this 14-year period.

   (d) Suppose that a family has two children born 14 years apart. In 1980–1981, the tuition cost of sending the elder child to college represented 15% of the family's total income. Assuming that their income increased at the same pace as the average household, what percent of their income was needed to send the younger child to college in 1994–1995?

17. Suppose that in a certain scratch-ticket lottery game, the probability of winning with the purchase of one card is 1 in 500, or 0.2%; hence, the probability of losing is $100\% - 0.2\% = 99.8\%$. But what if you buy more than one ticket? One way to calculate the probability that you will win at least once if you buy n tickets is to subtract from 100% the probability that you will lose on all n cards. This is an easy calculation; the probability that you will lose two times in a row is $(99.8\%)(99.8\%) =$

99.6004%, so the probability that you will win at least once if you play two times is $1 - (99.8\%)(99.8\%) = 0.3996\%$.

(a) What is the probability that you will win at least once if you play three times?

(b) Find a formula for $P(n)$, the percentage chance of winning at least once if you play the game n times.

(c) How many tickets must you buy in order to have a 25% chance of winning? A 50% chance?

(d) Does doubling the number of tickets you buy also double your chances of winning?

(e) Sketch a graph of $P(n)$. Use $[0, 100]$ as the range of the graph. Explain the practical significance of any asymptotes.

9.4   THE DERIVATIVE OF AN EXPONENTIAL FUNCTION

As an exploratory problem you investigated the derivatives of 2^x, 3^x, and 10^x by numerically approximating the derivatives at various points. Below are tables of values of approximations to the derivatives of 2^x, 3^x, and 10^x.[13]

	$f(x) = 2^x$			$g(x) = 3^x$			$h(x) = 10^x$	
x	$f(x)$	$f'(x)$	x	$g(x)$	$g'(x)$	x	$h(x)$	$h'(x)$
0	1	0.693	0	1	1.099	0	1	2.305
1	2	1.386	1	3	3.298	1	10	23.052
2	4	2.774	2	9	9.893	2	100	230.524
3	8	5.547	3	27	29.679	3	1000	2305.238
4	16	11.094	4	81	89.036	4	10,000	23052.381

Based on the data gathered, we can conjecture that

$$f'(x) \approx (0.693)2^x, \qquad g'(x) \approx (1.099)3^x, \qquad h'(x) \approx (2.305)10^x.$$

Observation

In each case the derivative of the exponential function at any point appears to be proportional to the value of the function at that point. Furthermore, the proportionality constant appears to be the derivative of the function at $x = 0$.

$$f'(x) \approx f'(0) \cdot 2^x$$
$$g'(x) \approx g'(0) \cdot 3^x$$
$$h'(x) \approx h'(0) \cdot 10^x$$

---

[13] The approximations to $f'(c)$ were made with $\frac{f(c+h)-f(c)}{h}$ for $h = 0.001$. In each case the following was used:

$$\frac{b^{c+h} - b^c}{h} = b^c \left(\frac{b^h - 1}{h} \right).$$

The second term was calculated, multiplied by the first, and rounded off after three decimal places. You may have done this slightly differently, but our answers ought to be in the same ballpark.

This leads us to conjecture that, more generally, if $f(x) = b^x$, then $f'(x) = f'(0) \cdot b^x$.

Conjecture

If $f(x) = b^x$, then

$$f'(x) = (\text{the slope of the tangent line to } b^x \text{ at } x = 0) \cdot b^x.$$

All this is simply conjecture. The data we have gathered are based purely on numerical approximation; we have done only a few test cases with a few bases.

The conjecture agrees with what we know about exponential functions. For example, if a population is growing exponentially, then we expect its rate of growth (its derivative) to be proportional to the size of the population (the value of the function itself.)

Conjectures are wonderful; mathematicians are continually looking for patterns and making conjectures. After experimenting and conjecturing, a mathematician is interested in trying to prove his or her conjectures. Let's go back to the limit definition of derivative and see if we can prove our conjecture.

Proof of Conjecture

Let $f(x) = b^x$, where b is a positive constant. Consider $f'(c)$ for any real number c.

$$f'(c) = \lim_{h \to 0} \frac{f(c + h) - f(c)}{h}$$

$$f'(c) = \lim_{h \to 0} \frac{b^{c+h} - b^c}{h}$$

$$f'(c) = \lim_{h \to 0} \frac{b^c \cdot b^h - b^c}{h}$$

$$= \lim_{h \to 0} b^c \left[\frac{b^h - 1}{h} \right].$$

As h goes to zero b^c is unaffected, so

$$f'(c) = b^c \left[\lim_{h \to 0} \frac{(b^h - 1)}{h} \right].$$

Notice that

$$\lim_{h \to 0} \frac{(b^h - 1)}{h} = \lim_{h \to 0} \frac{(b^h - b^0)}{h} = \lim_{h \to 0} \frac{f(h) - f(0)}{h - 0}.$$

This is precisely the definition of $f'(0)$, the derivative of b^x at $x = 0$. How delightful! We have now proven our conjecture.

$$\boxed{\text{If } f(x) = b^x, \text{ then } f'(x) = f'(0) \cdot b^x = f'(0) \cdot f(x).}$$

We can approximate the derivatives at zero numerically, as done in the exploratory problem. Our results stand as

$$\frac{d}{dx} 2^x \approx (0.693) 2^x, \qquad \frac{d}{dx} 3^x \approx (1.098) 3^x, \qquad \frac{d}{dx} 10^x \approx (2.302) 10^x.$$

We have proven that $\frac{d}{dx}b^x = \alpha b^x$, where $\alpha =$ the slope of the tangent to the graph of b^x at $x = 0$. It follows that $\frac{d}{dx}Cb^x = \alpha Cb^x$. Regardless of whether we are looking at average or at instantaneous rates of change, we can state the following.

Exponential functions grow at a rate proportional to themselves.
Exponential functions have a constant percent change.

Question: Is there a base "b" such that the derivative of b^x is b^x?

This question asks us to look for a function whose derivative is, itself, a function f such that the slope of the graph of f at any point (x, y) is given simply by the y-coordinate. If we can find a base "b" such that the slope of the tangent line to b^x at $x = 0$ is 1, then we have found such a function. Since the slope of b^x at $x = 0$ increases as b increases, based on the data we've collected we posit the existence of such a number. Graphically and numerically it seems reasonable to believe such a number exists. We begin to look for such a base between 2 and 3 because we know that $\frac{d}{dx}2^x \approx (0.693)2^x$, and $\frac{d}{dx}3^x = (1.099) \cdot 3^x$. Some numerical experimentation allows us to narrow in on a value between 2.71 and 2.72.

Definition

We define the number e to be the base such that the slope of the tangent line to e^x at $x = 0$ is 1. In other words, we define e to be the number such that

$$\frac{d}{dx}e^x = e^x.$$

The number e is between 2.71 and 2.72. Later we will pin down the size of e more closely.

Given that $\frac{d}{dx}e^x = e^x$, we can find the derivative of e^{2x} and e^{3x} using the Product Rule. We know that $e^{2x} = e^x \cdot e^x$ and $e^{3x} = e^x \cdot e^{2x}$.

$$\left(e^{2x}\right)' = e^x \cdot e^x + e^x \cdot e^x = 2e^x \cdot e^x = 2e^{2x}$$

$$\left(e^{3x}\right)' = e^x \cdot 2e^{2x} + e^x \cdot e^{2x} = 3e^x \cdot e^{2x} = 3e^{3x}$$

Do you notice a pattern? The mathematician in you wants to generalize. In fact, using induction, we can show that this pattern holds for any integer k: $\frac{d}{dx}e^{kx} = ke^{kx}$. Partitioning the problem into cases facilitates this generalization.

Proof that $\frac{d}{dx}e^{kx} = ke^{kx}$ for any positive integer k

Our statement $(e^{kx})' = ke^{kx}$ holds for $k = 1$.

We need to show that *if* $\frac{d}{dx}e^{nx} = ne^{nx}$, then $\frac{d}{dx}e^{(n+1)x} = (n+1)e^{(n+1)x}$.

$$\frac{d}{dx}e^{(n+1)x} = \frac{d}{dx}\left(e^{nx} \cdot e^x\right)$$

$$= ne^{nx} \cdot e^x + e^{nx} \cdot e^x$$

$$= (n+1)e^{nx} \cdot e^x$$

$$= (n+1)e^{(n+1)x}$$

Therefore, our statement holds true for any positive integer and the proof is complete.

Check on your own that this statement holds true for the case $k = 0$. Using the Quotient Rule, we can show that this works for any negative integer as well. Consider the function $f(x) = e^{-kx}$, where k is positive. We can rewrite the function as $f(x) = \frac{1}{e^{kx}}$ and apply the Quotient Rule.

$$\frac{d}{dx}\left(\frac{1}{e^{kx}}\right) = \frac{e^{kx} \cdot 0 - 1 \cdot ke^{kx}}{\left(e^{kx}\right)^2}$$

$$= \frac{-ke^{kx}}{e^{2kx}}$$

$$= -ke^{kx-2kx}$$

$$= -ke^{-kx}$$

Therefore, $\frac{d}{dx}e^{kx} = ke^{kx}$ for any integer k.

Answers to Selected Exercises

Exercise 9.3 Answers

i. $\sqrt[3]{-8} \neq \frac{1}{2}$. Instead $\sqrt[3]{-8}\sqrt[3]{x^{-2}} = -2 \cdot x^{-2/3}$, not $\frac{1}{2}x^{-2/3}$.

ii. $\frac{1}{2^{-3}} \neq 2^{-3}$. We should have $\frac{B^{-6}C^3}{2^{-3}} = \frac{2^3 C^3}{B^6} = \frac{8C^3}{B^6}$.

iii. $\sqrt{4}\sqrt{x^2 + 4y^2} = 2\sqrt{x^2 + 4y^2}$. This cannot be simplified.
$\sqrt{x^2 + 4y^2} \neq x + 2y$. Try squaring the latter if you're unconvinced.

iv. $\frac{(AD)^n + (CB)^n}{BD^n}$ is as far as we can go.
$(AD)^n + (CB)^n \neq (AD + CB)^n$ (except for $n = 1$)

v. $2\left(\frac{1}{x} + \frac{1}{y}\right)^{-1} = 2\left(\frac{y}{xy} + \frac{x}{xy}\right)^{-1} = 2\left(\frac{y+x}{xy}\right)^{-1} = \frac{2xy}{y+x}$
The error is that $\left(\frac{1}{x} + \frac{1}{y}\right)^{-1} \neq x + y$. For instance, $\left(\frac{1}{2} + \frac{1}{2}\right)^{-1} = 1^{-1} = 1$, not $2 + 2 = 4$.

Exercise 9.4 Answers

i. $\frac{2^{.5}x^{1.5}}{2^{-.5}x^{-.5}} = 2^{.5-(-.5)} \cdot x^{1.5-(-.5)} = 2^{.5+.5} \cdot x^{1.5+.5} = 2x^2$

ii. $4(9y^{-x})^{1/2} = 4\sqrt{\frac{9}{y^x}} = \frac{4 \cdot 3}{\sqrt{y^x}} = \frac{12}{y^{x/2}}$ or $12y^{-x/2}$

iii. $\frac{b^{x+w} - b^x}{b^x} = \frac{b^x \cdot b^w - b^x}{b^x} = \frac{b^x(b^w - 1)}{b^x} = b^w - 1$.

iv. $\frac{(\sqrt[3]{64}x^3)^{1/2}}{\left(\frac{1}{2}\right)^{-1}} = \frac{(4x^3)^{1/2}}{(2^{-1})^{-2}} = \frac{2x^{3/2}}{2^2} = \frac{x^{3/2}}{2}$

v. $\frac{Q^{3R} + Q^{R+1}}{Q^{2R}} = \frac{Q^{R+2R} + Q^{R+1}}{Q^{R+R}} = \frac{Q^R(Q^{2R} + Q)}{Q^R Q^R} = \frac{Q^{2R} + Q}{Q^R}$ or $\frac{Q^{2R}}{Q^R} + \frac{Q}{Q^R} = Q^R + Q^{1-R}$

Exercise 9.6 Answers

About \$43.10 more if interest is compounded quarterly

Exercise 9.8 Answers

(a) $C(t) = C_0(0.5)^{t/5730}$

(b) ≈ 2948.5 years ago

Exercise 9.9 Answers

    i. $C(t) = C_0(0.8)^t = C_0 0.8^t$

    ii. $C(t) = C_0(0.9)^{2t} = C_0 0.81^t$

    iii. $C(t) = C_0(0.951)^{4t} \approx C_0 0.81794^t$

    iv. $C(t) = C_0(0.4)^{t/4} \approx C_0 0.795^t$

The most efficient system results in the smallest number of contaminants; 60% every 4 hours is therefore the most efficient.

PROBLEMS FOR SECTION 9.4

1. Let $g(t) = 3^{5t}$. Show that

$$\frac{g(t+h) - g(t)}{h} = g(t) \cdot \frac{g(h) - g(0)}{h}.$$

Some tips: (i) Write out the equation using the actual function. (ii) Now your job is to make the left and right sides look the same. Use the laws of exponents to do this.

2. Let $f(t) = 3^t$.

    (a) Sketch a graph of f.

    (b) Approximate $f'(1)$, the slope of the tangent line to the graph of $f(t) = 3^t$ at $t = 1$, by computing the slope of the secant line through $(1, f(1))$ and $(1.0001, f(1.0001))$.

    (c) Approximate $f'(0)$, the slope of the tangent line to the graph of $f(t) = 3^t$ at $t = 0$, by computing the slope of the secant line through $(0, f(0))$ and $(0.0001, f(0.0001))$.

    (d) Sketch a rough graph of the slope function f'.

3. The Exploratory Problems indicated that exponential functions grow at a rate proportional to themselves, i.e., if $f(x) = a^x$, then $f'(x) = ka^x$, for some constant k. Approximate the appropriate constant if $f(x) = 7^x$.

4. In the Exploratory Problems you approximated the derivatives of 2^x, 3^x, and 10^x for various values of x, and, after looking at your results, you conjectured about the patterns. Now, using the definition of the derivative of f at $x = a$, we return to this, focusing on the function $f(x) = 5^x$.

    (a) Using the definition of the derivative of f at $x = a$,

$$f'(a) = \lim_{h \to 0} \frac{f(a+h) - f(a)}{h},$$

    give an expression for $f'(0)$, the slope of the tangent line to the graph of at $x = 0$.

    (b) Show that for the function $f(x) = 5^x$, the difference quotient, $\frac{f(x+h)-f(x)}{h}$, is equal to $f(x) \cdot \frac{f(h)-f(0)}{h}$.

    (c) Using the definition of derivative,

$$f'(x) = \lim_{h \to 0} \frac{f(x+h) - f(x)}{h},$$

conclude that the derivative of $f(x) = 5^x$ is

$$f'(0) \cdot f(x).$$

Notice that you have now proven that the derivative of 5^x is proportional to 5^x, with the proportionality constant being the slope of the tangent line to 5^x at $x = 0$.

$$f'(x) = f'(0) \cdot f(x)$$

(d) Approximate the slope of the tangent line to 5^x at $x = 0$ numerically.

For Problems 5 through 9, differentiate the function given.

5. $f(x) = x^3 e^x$

6. $f(x) = \frac{e^{2x}}{x}$

7. $f(x) = 3e^{-x}$

8. $f(x) = \frac{x^2 + x}{e^x + 1}$

9. $f(x) = e^{2x}(x^2 + 2x + 2)$

10. Use the tangent line approximation of e^x at $x = 0$ to approximate e^{-1}. Is your answer larger than e^{-1} or smaller?

11. Find the equation of the line tangent to $f(x) = e^x$ at $x = 1$.

12. Differentiate the following.

  (a) $f(x) = \frac{x^2 e^x}{3}$

  (b) $f(x) = \frac{5x^2}{3e^x}$

  (c) $f(x) = \frac{1}{xe^{5x}}$

13. *Double, double, toil and trouble; Fire burn and caldron bubble.* Macbeth Act IV scene I.

    It is the eve of Halloween and the witches are emerging. As the evening progresses the number of witches grows exponentially with time. At the moment when the first star of the evening is sighted, there are 40 witches and the number of witches is growing at a rate of 10 witches per hour. Later, at the moment when there are 88 witches, at what rate is the number of witches increasing? Explain your reasoning.

14. Money in a bank account is growing exponentially. When there is $4000 in the account, the account is growing at a rate of $100 per year. How fast is the money growing when there is $5500 in the account? It is not necessary to find an equation for $M(t)$ in order to solve this problem. (In fact, you have not been given enough information to find $M(t)$.)

15. Consider the function

$$f(x) = \frac{x^2}{e^x}.$$

(a) Compute $f'(x)$. (The Quotient Rule is unnecessary.)

(b) For what values of x is $f'(x)$ positive? For what values of x is $f'(x)$ negative?

(c) For what values of x is $f(x)$ increasing? For which is it decreasing? Give exact answers.

(d) What is the smallest value ever taken on by $f(x)$? Explain your reasoning.

CHAPTER 10

Optimization

10.1 ANALYSIS OF EXTREMA

Introduction Via Examples

Many problems we deal with every day involve a search for some "optimal" arrangement. What route must be taken to travel the distance between two cities in the shortest amount of time? When should a farmer harvest his crop in order to maximize his profit? What dimensions will minimize the amount of material required to construct a can of a given volume? Example 6.1 was such an optimization problem; a gardener had a fixed amount of fencing and wanted to find out how to maximize the area of her garden.

These and many other questions require the optimization of some quantity. If the quantity we aim to optimize can be expressed as a function of one variable on a particular domain, then knowing about the graph of the function can help us locate the point at which the function achieves its maximum (or minimum, depending on the problem) value. Knowing the rate of change of the function can be very useful in finding this point, regardless of whether or not we actually produce a graph.

In this chapter, we will look at some basic types of optimization problems to get an idea of how to set them up and use both calculus-based and non-calculus-based methods to solve them. Throughout the course, as we study new kinds of functions and their derivatives, we will return to the topic of optimization.

◆ EXAMPLE 10.1   The Beta Shuttle flies passengers between Boston and New York City. Currently, it charges $150 for each one-way ticket. At this price, an average of 190 people buy tickets on the Shuttle, so the company receives ($150 per person)·(190 people) =$28,500 for each flight. The owners of the Beta Shuttle hire a consulting firm to help them figure out how the number of people buying tickets would change if the price were changed. The consultants conclude that for each one dollar increase in the price of the ticket, two fewer people would be willing to fly the Beta Shuttle; similarly, for each one dollar decrease in price, two more people would buy tickets. What should the price be to maximize the amount of revenue the Beta Shuttle owners receive? (Assume that the expenses of the flight are independent of the number of people on board because the large expenses like the cost of the plane, the salaries of the pilots, the rental space at the airports, and so forth, will be fixed, regardless of the number of passengers.) For now, let's assume that the plane's capacity is not an issue.

SOLUTION   Revenue, R, can be calculated by multiplying the price of the ticket by the number of people who buy the ticket,

$$R = (\text{price per passenger}) \cdot (\text{number of passengers}).$$

To express R as a function of one variable we first need formulas for the price and for the number of passengers. Since we know how many tickets will be sold in terms of the change in price from the current price, we let x be the change in the price of a ticket from its current level of $150.[1] If x is positive the price increases; if x is negative it decreases.

Let's express both price and number of passengers as functions of x. The price of a ticket now is $150, so changing the price by x dollars makes the new price $150 + x$.

$$\text{price} = 150 + x$$

At a price of $150, 190 people are willing to buy tickets; for each $1 increase in price, 2 fewer people will buy tickets.[2]

$$\text{number of passengers} = 190 - 2x$$

Taking the product of these two expressions, we obtain

$$R(x) = (150 + x)(190 - 2x).$$

We need to find the value of x that will make $R(x)$ as big as possible. There are several methods available for us to do this. Regardless of the method, we need to figure out the appropriate domain of the function. The relevant questions are: "What is the least possible value of x?" and "What is the greatest possible value of x?"

What is the least possible value for x? Because the price of a ticket must be a positive number, the price of the ticket is $(150 + x)$ dollars, and x must be greater than or equal to -150. What is the largest possible value for x? At first it may seem like the sky is the limit, but because the number of passengers is $(190 - 2x)$ and the number of passengers can't be negative, the largest possible value for x is the one that makes $190 - 2x = 0$; solving gives $x = 95$. Therefore, the domain is $[-150, 95]$.

One approach to the problem is to get a good graph of $R(x)$. $R(x)$ is a quadratic and the coefficient of the x^2 term is negative, so its graph is a parabola opening downward. (Multiplying out, we get $R(x) = -2x^2 - 110x + 28{,}500$.) To sketch the graph of the parabola, we'll first find its x-intercepts. These are the solutions to

---

[1] You can try letting x be the price itself; this makes things a bit more complicated.

[2] Our model is based on a completely linear demand curve. In reality this model will not be particularly good near the edges of the demand curve. For example, even if the price of a ticket gets so high that our model predicts that no one will buy a ticket, there is probably some rich fellow out there who will pick up one anyway. Our final answer is not going to be near this point, so this discrepancy with reality doesn't faze us.

$$R(x) = (150 + x)(190 - 2x) = 0$$

$$150 + x = 0, \quad \text{or} \quad 190 - 2x = 0$$

$$x = -150, \quad \text{or} \quad x = 95$$

The zeros of $R(x)$ are at $x = 95$ and $x = -150$. (Notice that we already found these zeros when we were finding the domain of R. They are the values of x for which the revenue will be zero. Either the price of a ticket is \$0 or there are no passengers.)

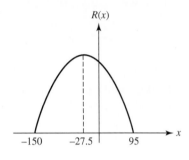

R(x)

−150   −27.5   95   x

Figure 10.1

Knowing that a parabola is symmetric about the vertical line through its vertex, we see that the maximum revenue occurs midway between the two zeros. Thus the value of x that makes $R(x)$ largest is the average of -150 and 95; $\frac{-150 + 95}{2} = -27.5$. If the Beta Shuttle reduces its fares by \$27.50 to a new fare of \$122.50, it will achieve its maximum amount of revenue of

$$(\$122.50 \text{ per person}) \cdot (190 - 2(-27.5) \text{ people}) = (\$122.50 \text{ per person}) \cdot (245 \text{ people})$$
$$= \$30{,}012.50.$$

This represents an increase of \$1,512.50 per flight over their current revenue of \$28,500 per flight.

A second approach to this optimization problem is to use R' to find the maximum. Either we can use R' to identify the x-coordinate of the vertex of the downward-facing parabola or, by looking at the derivative of $R(x)$, we can see where $R(x)$ is increasing and decreasing. (The latter will be useful if we fail to notice that we're dealing with a parabola.)

$$R(x) = -2x^2 - 110x + 28500, \text{ therefore } R'(x) = -4x - 110.$$

Let's draw a number line to indicate the sign of $R'(x)$ and what that tells us about $R(x)$. $R'(x)$ is continuous, so it can only change sign about a point at which it is zero. $R'(x) = 0$ when $x = -27.5$.

graph of R
sign of R' −150   +   −27.5   −   95

$R(x)$ is increasing for values of x less than -27.5 and decreasing for values of x greater than -27.5. Because $R(x)$ is continuous, we know the revenue will be maximized at $x = -27.5$.

The price of a ticket should be set at $150 − $27.50 = $122.50 in order to maximize the revenue taken in.   ◆

◆ EXAMPLE 10.2   Suppose that we are trying to solve the same problem as in Example 10.1, but now we are told that the Beta Shuttle flight has a maximum seating capacity of 210 people. We can't simply use our answer to Example 10.1, because we were figuring on 245 people purchasing tickets for the flight. Mathematically speaking, the difference between Example 10.1 and Example 10.2 is a change in the domain. Let's determine the new domain.

Recall that $R(x) = (150 + x)(190 − 2x) =$ (price)(number of passengers). Since the number of passengers must be between 0 and 210 inclusive,[3] we need to find x such that $0 \leq (190 − 2x) \leq 210$. We solve the following two equations.

$$190 − 2x = 0 \qquad \text{and} \qquad 190 − 2x = 210$$
$$-2x = -190 \qquad\qquad\qquad -2x = 20$$
$$x = 95 \qquad\qquad\qquad\quad x = -10$$

Looking at the number of passengers tells us that $-10 \leq x \leq 95$. None of these x-values pushes the price of the ticket under zero, so this is our domain.

Let's look at the graph of $R(x)$ restricted to the new domain. We're interested in the portion of the parabola between $x = -10$ and $x = 95$.

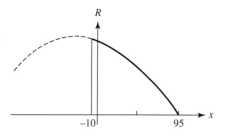

Figure 10.2

We can see that $R(-10)$ will be our maximum revenue given the plane size constraint in this example. The Shuttle should charge $140 per person to draw 210 passengers and obtain revenue of $29,400, an improvement of $900 per flight from their original revenue of $28,500.

Alternatively, looking at the sign of $R'(x)$ allows us to determine the maximum. $R'(x)$ is negative on the interval from -10 to 95, so $R(x)$ is decreasing on this interval. $R(-10)$ must be the highest value attained on this domain.

graph of R
sign of R' −10   −   95

◆

---

[3] The word "inclusive" tells us that the number of passengers can be *equal* to 0 and can be *equal* to 210.

◆ EXAMPLE 10.3   A machinist is making an open-topped box from a flat sheet of metal, 12 inches by 12 inches
in size, by cutting out equal-sized squares from each corner of the sheet and folding up the
remaining sides to form a box, as shown.

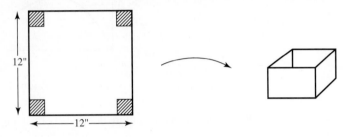

Figure 10.3

To maximize the volume of the box, what size squares should she cut from the corners?

SOLUTION   The goal is to maximize the volume of the box, so we'd like to get volume as a function
of one variable. Let's begin by using the variable x to label the lengths of the sides of the
square pieces she will cut out. Then the box formed will be $12 - 2x$ inches long, $12 - 2x$
inches wide, and x inches tall.

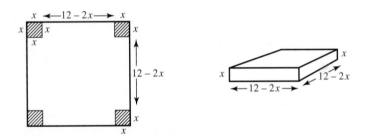

Figure 10.4

The volume, V, of the box is given by $V = $ (length)(width)(height), so we obtain the
equation

$$V(x) = (12 - 2x)(12 - 2x)(x).$$

Our goal is to maximize $V(x)$ for $0 \le x \le 6$.

   If we approach the problem without using derivatives, we find ourselves out of the
familiar realm of parabolas; the expression for V is a cubic. We will discuss the graphs
of cubics in Section 11.1, but because we have determined the domain it will be simple
to graph the function $V(x)$ using a graphing calculator. On the following page is a graph
on the appropriate domain. The maximum value of $V(x)$ on $[0, 6]$ appears to be around
$x = 2$. Notice that the beautiful symmetry of the parabola is, alas, lost on the cubic; the
maximum value of $V(x)$ does not lie midway between zeros of V. The machinist should
cut out squares with sides of length approximately 2 inches each.

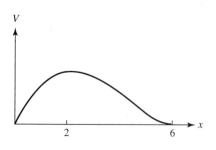

Figure 10.5

We can nail this down more definitively by using V' to find the maximum of V.

We are interested in the points where $V'(x)$ is either zero or undefined, because the sign of $V'(x)$, and hence the direction of V, can only change at these points.

$$V(x) = (12 - 2x)(12 - 2x)(x) = 4x^3 - 48x^2 + 144x$$

In the Exploratory Problems for Chapter 7 you found that if $f(x) = a_0 + a_1 x + a_2 x^2 + a_3 x^3$, then $f'(x) = a_1 + 2a_2 x + 3a_3 x^2$. Therefore,

$$V'(x) = 12x^2 - 96x + 144 = 12(x^2 - 8x + 12) = 12(x - 6)(x - 2).$$

$V'(x)$ is never undefined; it is zero at $x = 2$ and $x = 6$. On the interval [0, 6] the sign of V' can change only around $x = 2$. We need to check the sign of V' on the intervals [0, 2) and (2, 6). The simplest way to do this is to use a test point from each interval. We draw a number line to show important information about $V(x)$ and $V'(x)$.

graph of V
sign of V' 0   +   2   —   6

$V(x)$ is continuous and is increasing from $x = 0$ to $x = 2$ and decreasing from $x = 2$ to $x = 6$. So $x = 2$ gives the maximum volume. The dimensions of the box will be 8 inches × 8 inches × 2 inches with a total volume of 128 cubic inches.   ◆

Analysis of Extrema

In this section we use information about the derivative of f to identify where f takes on maximum and minimum values. We attempt to make arguments that are intuitively compelling. Our basic premise is, as previously stated,

$$f' > 0 \Rightarrow f \text{ is increasing,}$$

$$f' < 0 \Rightarrow f \text{ is decreasing.}$$

This makes sense; where the tangent line to f slopes upward, f is increasing. It actually takes a surprising amount of work to prove this. The proof rests on the Mean Value Theorem and is given in Appendix C.

We make a distinction between the absolute highest value of a function on its entire domain and a merely locally high value. This is analogous, for instance, to distinguishing between the point of highest altitude in the state of Massachusetts and the highest point of Winter Hill in Somerville, MA. We do similarly with low values.

Definitions

Let x_0 be a point in the domain of f.

x_0 is a **local maximum point** or **relative maximum point** of f if $f(x_0) \geq f(x)$ for all x in an open interval around x_0.[4] The number $f(x_0)$ is a **local maximum value**.

x_0 is a **local minimum point** or **relative minimum point** of f if $f(x_0) \leq f(x)$ for all x in an open interval around x_0. The number $f(x_0)$ is a **local minimum value**.

x_0 is a **global maximum point** or **absolute maximum point** of f if $f(x_0) \geq f(x)$ for all x in the domain of f. The number $f(x_0)$ is the **global** or **absolute maximum value**.

x_0 is a **global minimum point** or **absolute minimum point** of f if $f(x_0) \leq f(x)$ for all x in the domain of f. The number $f(x_0)$ is the **global** or **absolute minimum value**.

The term **extrema** is used to refer to local and to global maxima and minima.

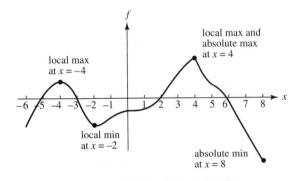

$x = 8$: absolute minimum point

$x = 4$: absolute maximum point

   local maximum point

$x = -2$: local minimum point

$x = -4$: local maximum point

$x = -6$: neither absolute nor local extremum

The graph of f with domain $[-6, 8]$.

Figure 10.6

Technical Conventions

■   In the language set out above, inputs to the function are called *points*; outputs are called *values*.

■   A maximum/minimum point may be just local or both local and global.

■   Because our definitions have "$\leq$" and "$\geq$" as opposed to $<$ and $>$, if f is a constant function, then every x-value is simultaneously a maximum point and a minimum point.

■   We will use the words "global" and "absolute" interchangeably when describing maxima and minima.

---

[4] Loosely speaking, "in an open interval around x_0" means "right around x_0, on both sides." More precisely, the statement above means there is an open interval (a, b), $a < x_0 < b$ such that $f(x_0) \geq f(x)$ for all $x \in (a, b)$.

■ We will use the words "local" and "relative" interchangeably when describing maxima and minima.

■ Endpoints of the domain can be **absolute** extrema but *not* local maxima and minima. This is because we have said that when at a local extrema you must be able to "look" on both sides of you and endpoints, by their nature, only allow you to "look" on one side.

With the exception of the last point, all the definitions given should fit in well with your intuitive notions of local and absolute maximum and minimum.

Where Should We Search for the Extrema of f?

■ Let's first look at the case in which f is continuous on an open connected domain. The domain could be of the form (a, b), (a, ∞), $(-\infty, b)$, or $(-\infty, \infty)$. We want to identify points at which the function changes from increasing to decreasing or vice versa; i.e., we are interested in the points where the derivative changes sign. The derivative can change sign only at a point where the derivative is either zero or undefined. Around these points the sign of f' *may* change or it *may not*.

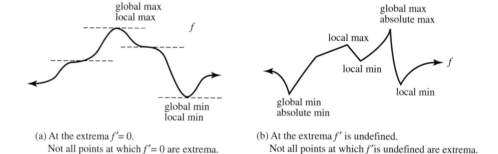

(a) At the extrema $f' = 0$.
    Not all points at which $f' = 0$ are extrema.

(b) At the extrema f' is undefined.
    Not all points at which f' is undefined are extrema.

Figure 10.7

If f is continuous on an open interval, it may be that f has no extrema. Consider, for example, $f(x) = x^3$ on the interval $(-1, 1)$. See Figure 10.8.

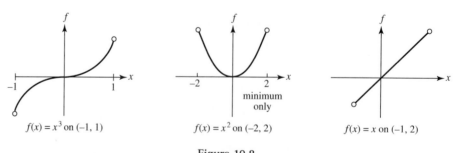

$f(x) = x^3$ on $(-1, 1)$ $f(x) = x^2$ on $(-2, 2)$ $f(x) = x$ on $(-1, 2)$

Figure 10.8

■ If f is continuous on a closed interval, then the Extreme Value Theorem guarantees that f attains both an absolute maximum and an absolute minimum value. The endpoints of the domain and points at which f' is zero or undefined are candidates for extrema. See Figure 10.9.

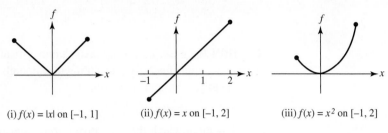

(i) $f(x) = |x|$ on $[-1, 1]$ (ii) $f(x) = x$ on $[-1, 2]$ (iii) $f(x) = x^2$ on $[-1, 2]$

Figure 10.9

■   If f is not continuous, we also need to look at points of discontinuity. (Keep in mind that f must be defined at x_0 in order for x_0 to be an extreme point.) At points of discontinuity f' is undefined. Notice that f' does not change sign at the extrema in the examples in Figure 10.10.

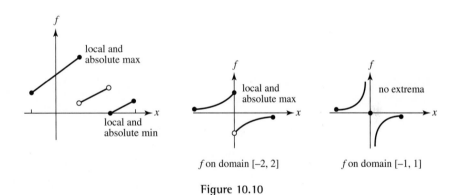

f on domain $[-2, 2]$ f on domain $[-1, 1]$

Figure 10.10

We'll refer to the set of points we need to sieve through in order to identify extrema as **candidates for extrema** or candidates for maxima and minima.

We can summarize our observations by identifying all candidates for extrema as follows.

Candidates for Maxima and Minima

Candidates for extrema are all x_0 in the domain of f for which

- $f'(x_0) = 0$, or
- $f'(x_0)$ is undefined, or   **Critical points of f**
- x_0 is an endpoint of the domain.

Definition

The **critical points of** f are the points x_0 in the domain of f such that either $f'(x_0) = 0$, $f'(x_0)$ is undefined, or x_0 is an endpoint of the domain.

CAUTION Be aware that many texts define critical points to be only points at which f' is zero or f' is undefined. We are adding in endpoints of the domain so that critical points and candidates for extrema are identical.

A point at which the derivative is zero is also sometimes referred to as a **stationary point**.[5]

Question: Suppose $f'(a) = 0$. Does that mean that $x = a$ is, necessarily, a maximum or a minimum?

Answer: No. $f'(a) = 0$ simply means that the slope of the graph at $x = a$ is zero. f' must change sign around $x = a$ if f has a local extremum there. It is possible that f has neither a maximum nor a minimum at $x = a$. See Figure 10.11.

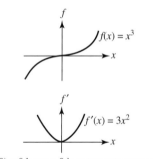

f

$f(x) = x^3$

x

f'

$f'(x) = 3x^2$

x

$f'(0) = 0$ but $x = 0$ is not an extremum of f.
The sign of f' does not change around $x = 0$.

Figure 10.11

Question: If f has a maximum or minimum at $x = b$, does it follow that $f'(b) = 0$?

Answer: No. It only follows that $x = b$ is a critical point. Think about the absolute value function, $f(x) = |x|$. f has a minimum at $x = 0$, and $f'(0)$ is undefined. Think also about the function $f(x) = x$ on the domain $[-1, 2]$. The absolute maximum occurs at $x = 2$ and the absolute minimum at $x = -1$. In neither case is f' zero. (See Figure 10.12.)

_____

[5] If f gives an object's displacement as a function of time, then this terminology makes a lot of sense. $f'(t_0) = 0$ means that at t_0 the object's velocity is zero. The object is stationary at this moment.

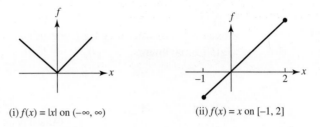

(i) $f(x) = |x|$ on $(-\infty, \infty)$ (ii) $f(x) = x$ on $[-1, 2]$

Figure 10.12

Local Extrema

Suppose $x = c$ is an interior critical point of a continuous function f. How can we tell if, at $x = c$, f has a local maximum, local minimum, or neither?

One approach is to look at the sign of f' to determine whether f changes from increasing to decreasing across $x = c$. This type of analysis is referred to as the **first derivative test.** If f is continuous, $x = c$ is an interior critical point of f, and f is differentiable on an open interval around c (even if f is not differentiable specifically at c), then:

■   if f' changes sign from negative to positive at $x = c$, then f has a local minimum at c;

■   if f' changes sign from positive to negative at $x = c$, then f has a local maximum at c;

■   if f' does not change sign across $x = c$, then f does not have a local extremum at c.

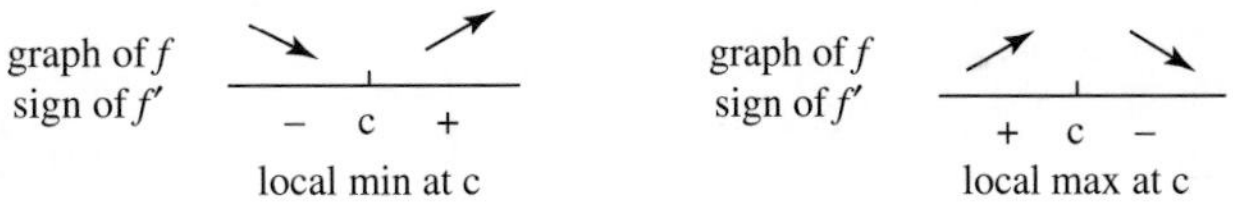

graph of f sign of f' $-$ c $+$ local min at c   graph of f sign of f' $+$ c $-$ local max at c

Figure 10.13

Suppose $x = c$ is an interior critical point of f but f is *not* continuous. You might wonder if there is a first derivative test we can apply. The answer is no. Look carefully at the graphs presented in Figure 10.14.

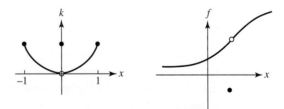

Figure 10.14

Global Extrema

Suppose we've rounded up the usual suspects for extrema; we've identified the critical points of f. Is there an easy way to identify the global maximum and minimum values? First we have to figure out whether or not the function *has* a global maximum. If we know it does, we can calculate the value of the function at each of the critical points. The largest value is the global maximum value. (The corresponding x gives you the absolute maximum point.) Sometimes you can exclude a few candidates. For instance, a local minimum will never be a global maximum.

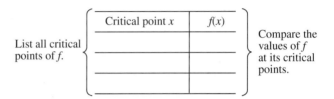

Some functions don't have absolute maximum and minimum values. Think about the functions $f(x) = 1/x$ and $h(x) = x^2$, each on its natural domain. The former has neither a maximum nor a minimum and the latter has only a minimum. Or consider $g(x) = x$ and $h(x) = x^2$ on the interval $(0, 1)$. On this open interval neither of these functions has a global maximum or a global minimum.

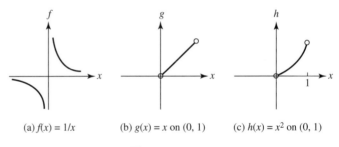

(a) $f(x) = 1/x$ (b) $g(x) = x$ on $(0, 1)$ (c) $h(x) = x^2$ on $(0, 1)$

Figure 10.15

Think about what went wrong in the cases above and see what criteria might guarantee a global extrema. To begin with, let's look at a function whose domain is a *closed* interval, $[a, b]$. This will fix the problems encountered in the case of functions f, g, and h above.

Now consider the functions $j(x) = \frac{1}{x}$ for $x \neq 0$ and 0 for $x = 0$ or $k(x) = (x^2)$ for $x \neq 0$ and 1 for $x = 0$, both restricted to the closed interval $[-1, 1]$. (See Figure 10.16.) To rule out these cases, we insist that the function be continuous. A continuous function on a closed interval is guaranteed to attain an absolute maximum value and an absolute minimum value. This is the Extreme Value Theorem discussed in Chapter 7. (It should seem reasonable, but its proof is difficult and beyond the scope of this text.)

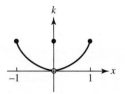

Figure 10.16

Summary

- If f has any local or global extrema, these will occur at critical points of f. x_0 is a critical point of f if x_0 is in the domain of f, and:

    $f'(x_0) = 0$, or

    $f'(x_0)$ is undefined, or

    x_0 is an endpoint of the domain of f.

- *Absolute extrema:* If f is continuous on a closed interval $[a, b]$, then f attains an absolute maximum and absolute minimum on $[a, b]$ (Extreme Value Theorem). These can be found by evaluating f at each of its critical points. Where the value of f is greatest, f has an absolute maximum; where the value of f is least, f has an absolute minimum. When looking for absolute extrema in general, you must treat the situations on a case-by-case basis. Stand back and take a bird's-eye view of the function to see if you expect global extrema. If f has any absolute maximum or minimum values, they will occur at the critical points.

- *Local extrema:* Suppose x_0 is a critical point of f. If f is continuous at x_0 and f' exists in an open interval around x_0, although not necessarily at x_0 itself, then we can apply the **first derivative test**. If the sign of f' changes at x_0, then f has a local extremum at x_0. Plot sign information on a number line to distinguish between maxima and minima. If the sign of f' does not change on either side of x_0, then f has neither a local maximum nor a local minimum at x_0.

◆ **EXAMPLE 10.4**   Let $f(x) = x^3 - 12x + 3$. Find all local extrema of f.

SOLUTION   Look for critical points.

$$f'(x) = 3x^2 - 12$$

f' is defined everywhere, and the function is defined for all real numbers, therefore the only critical points are the zeros of f'.

$$f'(x) = 3x^2 - 12 = 0$$
$$3(x^2 - 4) = 0$$
$$x = \pm 2.$$

To determine whether these points are extrema, we set up a number line and determine the sign of f' in each of the three intervals into which the critical points partition the line.

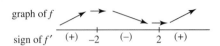

So f has a local minimum at $x = 2$ and a local maximum at $x = -2$. ◆

◆ **EXAMPLE 10.5**   Suppose f is continuous on $(-\infty, \infty)$. $f' = 0$ at $x = -1$ and at $x = 2$. f' does not exist at $x = -3$ and at $x = 0$. Classify the critical points of f as best you can given the information below regarding the sign of f'.

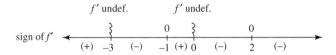

SOLUTION   Correlating the sign of the derivative with information about the slope of the function gives us the following number line.

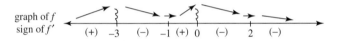

This gives us the following information about the extrema of f:

at $x = -3$, f has a local maximum;

at $x = -1$, f has a local minimum;

at $x = 0$, f has a local maximum;

at $x = 2$, f has neither a local maximum nor minimum.

There is not enough information to determine whether or not f has an absolute minimum value. It has an absolute maximum at either $x = -3$ or $x = 0$, or at both. There is not enough information to make a stronger statement. ◆

PROBLEMS FOR SECTION 10.1

In Problems 1 through 16, for each function:

(a) Find all critical points on the specified interval.

(b) Classify each critical point: Is it a local maximum, a local minimum, an absolute maximum, or an absolute minimum?

(c) If the function attains an absolute maximum and/or minimum on the specified interval, what is the maximum and/or minimum value?

1. $f(x) = x^3 - 3x + 2$ on $(-\infty, \infty)$

2. $f(x) = x^3 - 3x + 2$ on $[-5, 5]$

3. $f(x) = x^3 - 3x + 2$ on $[0, 3]$

4. $f(x) = x^3 - 3x + 2$ on $(0, 3)$

5. $f(x) = -2x^3 + 3x^2 + 12x + 5$ on $(-\infty, \infty)$

6. $f(x) = -2x^3 + 3x^2 + 12x + 5$ on $[-3, 4]$

7. $f(x) = x^5 - 20x + 5$ on $(-\infty, \infty)$

8. $f(x) = x^5 - 20x + 5$ on $[-2, 0]$

9. $f(x) = x^5 - 20x + 5$ on $[0, 2]$

10. $f(x) = 3x^4 - 8x^3 + 3$ on $(-\infty, \infty)$

11. $f(x) = 3x^4 - 8x^3 + 3$ on $[-1, 1]$

12. $f(x) = 3x^4 - 8x^3 + 3$ on $(0, 3)$

13. $f(x) = \frac{x^3}{3} + 2x + \frac{3}{x}$ on its natural domain. *Why is $x = 0$ not a critical point?*

14. $f(x) = \frac{x^3}{3} + 2x + \frac{3}{x}$ on $[-3, 0)$

15. $f(x) = \frac{x^3}{3} + 2x + \frac{3}{x}$ on $(0, 3]$

16. $f(x) = \frac{1}{x^2 + 4}$ on $(-\infty, \infty)$

17. Let $f(x) = \frac{e^x}{x}$.
    (a) Find all critical points of f.
    (b) Identify all local extrema.
    (c) Does f have an absolute maximum value? If so, where is it attained? What is its value?
    (d) Does f have an absolute minimum value? If so, where is it attained? What is its value?
    (e) Answer parts (c) and (d) if x is restricted to $(0, \infty)$.

18. Let $f(x) = x^2 e^{-x}$.
    (a) Find all critical points of f.
    (b) Classify the critical points.
    (c) Does f take on an absolute maximum value? If so, where? What is it?
    (d) Does f take on an absolute minimum value? If so, where? What is it?

In Problems 19 through 26, find and classify all critical points. Determine whether or not f attains an absolute maximum and absolute minimum value. If it does, determine the absolute maximum and/or minimum value.

19. $f(x) = (x^2 - 4)e^x$

20. $f(x) = \frac{10x}{x^2+1}$

21. $f(x) = \frac{10x^2}{x^2+1}$

22. $f(x) = \frac{x}{e^x}$

23. $f(x) = \frac{x-1}{x^2+3}$

24. $f(x) = \frac{4-x}{x^2+9}$

25. $f(x) = \frac{x^3}{x^2+1}$

26. $f(x) = \frac{e^x}{2x}$

10.2   CONCAVITY AND THE SECOND DERIVATIVE

In this section we will take another look at concavity to see what it can tell us when we are analyzing critical points. Recall that where f' is increasing the graph of f is concave up, and where f' is decreasing the graph of f is concave down. Let's see what bearing this has on optimization.

Suppose x_0 is a stationary point: $f'(x_0) = 0$. Then f has a horizontal tangent line at $x = x_0$.

If f' is increasing at x_0,   then locally the graph must look like Figure 10.17(a); f has a local minimum at x_0.

If f' is decreasing at x_0,   then locally the graph must look like Figure 10.17(b); f has a local maximum at x_0.

$(x_0, f(x_0))$

$(x_0, f(x_0))$

(a)   (b)

Figure 10.17

We know that

$$f'' > 0 \Rightarrow f' \text{ is increasing} \Rightarrow f \text{ is concave up;}$$
$$f'' < 0 \Rightarrow f' \text{ is decreasing} \Rightarrow f \text{ is concave down.}$$

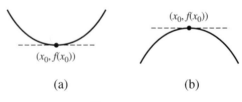

Therefore, if $f'(x_0) = 0$ and $f''(x_0)$ exists, we can look at the sign of $f''(x_0)$ and draw the following conclusions:

$$
\begin{cases}
\text{If } f''(x_0) > 0, & \text{then } f \text{ is concave up around } x_0 \text{ and therefore has a local minimum} \\
& \text{at } x_0. \\
\text{If } f''(x_0) < 0, & \text{then } f \text{ is concave down around } x_0 \text{ and therefore has a local maximum} \\
& \text{at } x_0. \\
\text{If } f''(x_0) = 0, & \text{we cannot draw any conclusions. The function could have a maximum} \\
& \text{at } x_0, \text{ a minimum at } x_0, \text{ or neither.}
\end{cases}
$$

This set of criteria is referred to as **the second derivative test**.

EXERCISE 10.1   Look at the case where $f'(x_0) = 0$ and $f''(x_0) = 0$. In the three examples below, determine whether the function has a maximum at $x_0 = 0$, a minimum at $x_0 = 0$, or neither a maximum nor a minimum at $x_0 = 0$. Sketch the graphs of these functions on the set of axes provided below as an illustration of why $f'(x_0) = 0$ and $f''(x_0) = 0$ does not allow us to classify x_0.

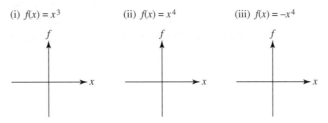

(i) $f(x) = x^3$ (ii) $f(x) = x^4$ (iii) $f(x) = -x^4$

Figure 10.18

Definition

A **point of inflection** is a point in the domain of f at which f changes concavity, i.e., a point at which f'' changes sign.

CAUTION The fact that $f''(x_0) = 0$ does *not necessarily* mean that $x = x_0$ is an inflection point. Look back at the graphs you've drawn in Exercise 10.1. $x = 0$ is a point of inflection for $f(x) = x^3$ but not for $f(x) = x^4$.

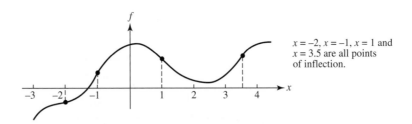

$x = -2, x = -1, x = 1$ and $x = 3.5$ are all points of inflection.

Figure 10.19

Putting It All Together

Suppose $x = 3$ is an interior critical point of a continuous function f and we are interested in classifying this critical point, that is, in determining whether $x = 3$ is a local maximum, a local minimum, or neither. What can we do?

Method (i) **First Derivative Test.**

We can look at the sign of f', the first derivative, on either side of 3. If f' changes sign around 3, then 3 is a local maximum or local minimum point.

Method (ii) **Second Derivative Test.**

If $f'(3) = 0$, then we can look at the sign of $f''(3)$.

If $f''(3) > 0$, then f has a local minimum at $x = 3$.

If $f''(3) < 0$, then f has a local maximum at $x = 3$.

If $f''(3) = 0$, we have insufficient information to draw a conclusion. In this case, turn to method (i).

If x_0 is an endpoint, then the only label it can get is absolute maximum or absolute minimum; otherwise it remains without a label. By looking at the sign of f' in the vicinity of x_0 we can determine whether it is a potential absolute maximum or potential absolute minimum. We need a bird's-eye view to determine whether there *is* an absolute extremum. If we expect one, we must compare the value of f at x_0 with its value at other candidates.

The first derivative test is widely applicable. It requires determining the sign of f' on intervals. Because the second derivative test only requires evaluating f'' at a point, it is simple to apply if f'' is easy to calculate. For instance, consider Example 10.3 where $V(x) = 4x^3 - 48x^2 + 144x$, $V'(x) = 12x^2 - 96x + 144$ and $V''(x) = 24x - 96$. The only interior critical point of V on $[0, 6]$ is $x = 2$, and $V''(2) = 24 \times 2 - 96 < 0$. We conclude that V has a local max at $x = 2$. The second derivative test, however, can only be applied at a point at which the first derivative is zero, i.e., at a stationary point. The graphs in Figure 10.20 illustrate why this is the case.

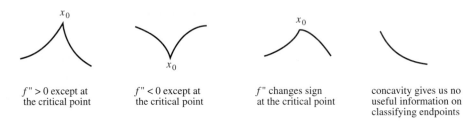

$f'' > 0$ except at the critical point　　$f'' < 0$ except at the critical point　　f'' changes sign at the critical point　　concavity gives us no useful information on classifying endpoints

Figure 10.20　The second derivative test *can't* be applied unless $f' = 0$ at the critical point.

f	positive / negative	increasing /decreasing	concave up / concave down
f'		positive / negative	increasing /decreasing
f''			positive / negative

PROBLEMS FOR SECTION 10.2

In Problems 1 through 12:

(a) Find all critical points.

(b) *Find f''. Use the second derivative, wherever possible, to determine which critical points are local maxima and which are local minima. If the second derivative test fails or is inapplicable, explain why and use an alternative method for classifying the critical point.*

1. $f(x) = x^3 - 6x + 1$

2. $f(x) = -x^3 + 3\pi^2 x$

3. $f(x) = x^3 + \frac{9}{2}x^2 - 12x + \frac{3}{2}$

4. $f(x) = x^5 - 5x$

5. $f(x) = 2x^4 + 64x$

6. $f(x) = x^6 + x^4$

7. $f(x) = x^4 + 4x^3 + 2$

8. $f(x) = 4x^{-1} + 2x^2$

9. $f(x) = e^x - x$

10. $f(x) = xe^x - e^x$

11. $f(x) = \frac{x^5}{5} - x^4 + \frac{4}{3}x^3 + 2$

12. $f(x) = 3x^4 - 8x^3 + 6x + 1$

13. Suppose that f is a continuous function and that $f(3) = 2$, $f'(3) = 0$, and $f''(3) = 3$. At $x = 3$, does f have a local maximum, a local minimum, neither a local maximum nor a local minimum, or is it impossible to determine? Explain your answer.

14. Suppose that f is a continuous function and that $f(4) = 1$, $f'(4) = 0$, and $f''(4) = 0$. At $x = 4$, f could have a local maximum, a local minimum, or neither. Sketch three graphs, all satisfying the conditions given, one in which f has a local minimum at $x = 4$, one in which f has a local maximum at $x = 4$, and one in which f has neither a local maximum nor a local minimum at $x = 4$.

15. Without using the graphing capabilities of your graphing calculator, sketch the following graphs. Label the x-coordinates of all peaks and valleys. Label exactly, not using a numerical approximation. (If the x-coordinate is $\sqrt{2}$, it should be labeled $\sqrt{2}$, not 1.41421.)

    Below the sketch of f, sketch $f'(x)$, labeling the x-intercepts of the graph of f'. (You can use your graphing calculator to check your answers.)

    (a) $f(x) = x(x - 9)(x - 3)$ (Start by looking at the x-intercepts. Then look at the sign of $f'(x)$ in order to determine where the graph of f is increasing and where it is decreasing.)

(b) $f(x) = -2x(x-9)(x-3)$ (Conserve your energy! Think!)

(c) $f(x) = -2x(x-9)(x-3) + 18$ (Conserve your energy! Think!)

16. (a) The function g with domain $(-\infty, \infty)$ is continuous everywhere. We are told that $g'(\sqrt{5}) = 0$. Some of the scenarios below would allow us to conclude that g has a local minimum at $x = \sqrt{5}$. Identify *all* such scenarios.

   i. $g(\sqrt{5}) = 0$, $g(2) = 1$, $g(3) = 1$
   ii. $g(\sqrt{5}) < 0$ and $g'(x) > 0$ for $x > \sqrt{5}$.
   iii. $g''(\sqrt{5}) > 0$
   iv. $g''(\sqrt{5}) < 0$
   v. $g'(x) > 0$ for $x < \sqrt{5}$ and $g'(x) < 0$ for $x > \sqrt{5}$
   vi. $g'(x) < 0$ for $x < \sqrt{5}$ and $g'(x) > 0$ for $x > \sqrt{5}$
   vii. $g'(\sqrt{5}) > 0$ and $g''(\sqrt{5}) = 0$

(b) The function h with domain $[-8, -3]$ has the following characteristics.

   h is continuous at every point in its domain.
   $h'(x) < 0$ for $-8 < x < -4$ and $h'(x) > 0$ for $(-4, -3)$.
   $h'(-4)$ is undefined.

   What can you conclude about the local and absolute extrema of h? Please say as much as you can given the information above.

17. The graph of f' (*not* f, but f') is a parabola with x-intercepts of $-\pi$ and 2π and a y-intercept of -2.

(a) Draw a graph of f'.

(b) Write an equation for f'. This equation should have no unknown constants.

(c) On the graph you drew in part (a), go back and label the x- and y-coordinates of the vertex.

(d) Find $f''(x)$.

(e) This part of the question asks about f, *not* f'.

   i. Where does f have a local maximum? Explain your reasoning clearly and briefly.
   ii. Where does f have a local minimum? Explain your reasoning clearly and briefly.
   iii. Does f have an absolute maximum or minimum value? Explain.
   iv. The function f has a single point of inflection. What is the x-coordinate of this point of inflection? Suppose you are told that the y-coordinate of the point of inflection is -1. Find the equation of the tangent line to the graph of f at its point of inflection.

18. Consider the function $f(x) = x^5 - 2x^4 - 7$ restricted to the domain $[-1, 1]$. Your reasoning for the questions below must be fully explained and be independent of a graphing calculator.

(a) Find the absolute maximum value of $f(x)$ on the interval $[-1, 1]$ or explain why this is not possible.

(b) Find the absolute minimum value of $f(x)$ on the interval $[-1, 1]$ or explain why this is not possible.

(c) Find the absolute minimum value of $f(x)$ on the open interval $(-1, 1)$ or explain why this is not possible.

10.3 PRINCIPLES IN ACTION

In this section we put the principles discussed in Sections 10.1 and 10.2 into action.

◆ **EXAMPLE 10.6** Below is a graph of f', the derivative of f. The questions that follow refer to f. The domain of f is all real numbers.

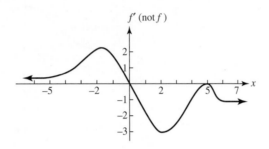

Figure 10.21

(a) Identify all critical points of f. Which of these critical points are also stationary points of f?

(b) On a number line plot all the critical points of f. Indicate the sign of f' and indicate where f is increasing and where f is decreasing. Is this enough information to classify all the extrema? Does f have a global maximum? If so, can we determine where it is attained? Does f have a global minimum? If so, can we determine where it is attained?

(c) Where is the graph of f increasing and concave down?
Where is the graph of f decreasing and concave up?

(d) Determine all points of inflection of f.

SOLUTION

(a) Critical points of f are points at which either $f' = 0$, f' is undefined, or endpoints of the domain. The critical points in this example are $x = 0$ and $x = 5$. Both are points at which $f' = 0$, so they are both stationary points.

(b)

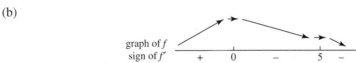

We can see that f has a local maximum at $x = 0$ because f changes from increasing to decreasing at $x = 0$. We can tell that $x = 0$ is also the global maximum point. (f is increasing for all negative x and decreasing for all positive x.)

$x = 5$ is not an extreme point, because f is decreasing both before and after $x = 5$.

$f' = -1$ for $x > 7$; consequently, f is decreasing with a slope of -1 on $(7, \infty)$. Therefore there is no global minimum.

(c) f is increasing and concave down on $(-2, 0)$, where f' is positive and decreasing. (Check the graph of f'.)

f is decreasing and concave up where f' is negative and increasing. That is, on $(2, 5)$.

(d) Points of inflection of f are points at which the concavity of f changes. That is, points of inflection of f are points at which f'' changes sign. f'' is the slope of f', so f'' changes sign at $x = -2$, $x = 2$, and $x = 5$.

Below is a sketch[6] of f. (It is drawn without the x-axis because we don't know where the x-axis is located.)

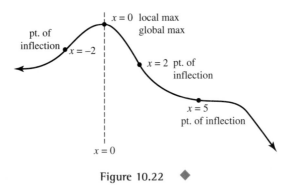

Figure 10.22   ◆

EXERCISE 10.2   Notice that in Example 10.6 one of the x-intercepts of f' corresponded to a local extrema of f and the other did not. Assuming f' is continuous, come up with a graphical criteria for determining which x-intercepts of f' correspond to local extrema of f.

EXERCISE 10.3   Explain why local extrema of f' correspond to points of inflection of f. (Assume f' is continuous.)

◆ EXAMPLE 10.7 f is a continuous function defined on $[-3, 3]$. Its derivative f' is given below. f' is an odd function, that is, $f'(-x) = -f'(x)$. Classify the critical points of f.

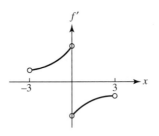

Figure 10.23

SOLUTION   The critical points of f are $x = 0$, $x = -3$, and $x = 3$. $x = 0$ is a critical point because f' is undefined there. $x = -3$ and $x = 3$ are endpoints of the domain.

graph of f
sign of f' -3 $+$ 0 $-$ 3

---

[6] Note that because f is differentiable we *know* f is continuous. We proved this in Chapter 7.

$x = 0$ is both a local maximum and the absolute maximum point.

$x = -3$ and $x = 3$ are both absolute minimum points, because f' is an odd function.

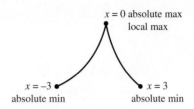

Figure 10.24   ◆

◆ **EXAMPLE 10.8**   Consider the function $f(x) = x^3 + 3x^2 - 9x + 3$. Identify the x-coordinates of all local maxima and minima. Is there a global maximum? A global minimum?

SOLUTION   Begin by identifying the critical points.

$$f'(x) = 3x^2 + 6x - 9$$
$$= 3(x^2 + 2x - 3)$$
$$= 3(x + 3)(x - 1)$$

f' is always defined, so the critical points are simply the stationary points.

$$f'(x) = 0 \text{ at } x = -3 \text{ and } x = 1$$

We test the sign of f' in each of the intervals $(-\infty, -3)$, $(-3, 1)$, and $(1, \infty)$. This is most efficiently done when f' is factored.

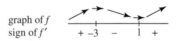

graph of f
sign of f' $+$ -3 $-$ 1 $+$

f has a relative maximum at $x = -3$ and a relative minimum at $x = 1$. Try showing this on your own using the second derivative test. Then see if you can convince yourself that there is neither a global maximum nor a global minimum. We will do this in the next chapter.
◆

◆ **EXAMPLE 10.9**   Find and classify the critical points of $f(x) = e^x(x^2 + x - 5)$. Find the x- and y- coordinates of all local extrema. Is there an absolute maximum value? An absolute minimum value?

SOLUTION   Begin by identifying critical points.

$$f'(x) = e^x(x^2 + x - 5) + e^x(2x + 1) \quad \text{By the Product Rule}$$
$$= e^x(x^2 + 3x - 4) \qquad\qquad\quad \text{Factor}$$
$$= e^x(x - 1)(x + 4)$$

f' is always defined, so the critical points are simply the stationary points.

$$f'(x) = 0 \quad \text{at } x = 1 \text{ and } x = -4$$

Both first and second derivative tests are simple to apply. We'll use the second derivative test.

$$f'(x) = e^x(x^2 + 3x - 4)$$
$$f''(x) = e^x(x^2 + 3x - 4) + e^x(2x + 3)$$
$$= e^x(x^2 + 5x - 1)$$

$f''(1) = e(1 + 5 - 1) = 5e > 0 \Rightarrow f$ has a local minimum at $x = 1$.
$f''(-4) = e^{-4}(16 - 20 - 1) = \frac{-5}{e^4} < 0 \Rightarrow f$ has a local maximum at $x = -4$.

When $x = 1$, $y = f(1) = e(1 + 1 - 5) = -3e$ local minimum at $(1, -3e)$
When $x = -4$, $y = f(-4) = e^{-4}(16 - 4 - 5) = \frac{7}{e^4}$ local maximum at $(-4, \frac{7}{e^4})$

Now we consider the global extrema.

From the local information we have gathered it is unclear what happens globally. A bird's-eye view will help. We need to look at

$$\lim_{x \to \infty} e^x(x^2 + x - 5) \quad \text{and} \quad \lim_{x \to -\infty} e^x(x^2 + x - 5).$$

As x grows without bound, both e^x and $(x^2 + x - 5)$ grow without bound. Therefore,

$$\lim_{x \to \infty} e^x(x^2 + x - 5) = \infty; \quad f(x) \text{ has no absolute maximum.}$$

As $x \to -\infty$, $e^x \to 0$ and $(x^2 + x - 5) \to \infty$. Therefore it is not immediately clear what happens to the product. But e^x is always positive and $(x^2 + x - 5)$ is positive for $|x|$ large. Therefore, the product is positive and $f(x)$ takes on an absolute minimum value of $-3e$. In fact, it can be shown that $\lim_{x \to -\infty} e^x(x^2 + x - 5) = 0$, the behavior of the exponential dominating that of the cubic.

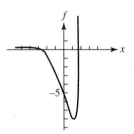

Figure 10.25

Look at the graph of $f(x)$ on your calculator or computer. Notice which features of the function are easy to discern from the graphs and which are more difficult. ◆

Exploratory Problems for Chapter 10

Optimization

1. Let $f(x) = x\, e^x$.

(a) Find and classify all critical points of f.

(b) Identify all points of inflection.

(c) Does f have an absolute maximum value? If so, what is it and where is it attained?

(d) Does f have an absolute minimum value? If so, what is it and where is it attained?

(e) Find $\lim_{x \to \infty} x\, e^x$.

(f) What is $\lim_{x \to -\infty} x\, e^x$?
 This one is harder. Why? Argue convincingly that you have found this limit with error less than 10^{-50}.

(g) Graph $f(x)$, labeling all critical points.

■ Optional variation on the theme: Answer the same questions for the function $f(x) = x^2\, e^x$ or $h(x) = x^3\, e^{-x}$.

2. Let $g(x) = x + \frac{4}{x}$.

(a) Find and classify all critical points of f. Why is $x = 0$ not a critical point?

(b) Identify all points of inflection.

(c) Does f have an absolute maximum value? If so, what is it and where is it attained?

(d) Does f have an absolute minimum value? If so, what is it and where is it attained?

(e) Find $\lim_{x \to \infty} g(x)$.

(f) What is $\lim_{x \to -\infty} g(x)$?

(g) Graph $g(x)$, labeling all critical points.

(h) Is g an even function? i.e., does $g(x) = g(-x)$?
 Is g an odd function? i.e., does $g(x) = -g(-x)$?

■ Optional variation on the theme: Answer the same questions for the function $g(x) = x + \frac{4}{x^2}$.

PROBLEMS FOR SECTION 10.3

1. Let $f(x) = \frac{e^x}{x^2+1}$. Find and classify the critical points.

2. A gardener has a fixed length of fence to fence off her rectangular chili pepper garden. Show that if she wants to maximize the area of her garden, then her garden should be square.

3. A gardener needs 90 square feet of land for her tomato plants. She will fence in a rectangular plot. The cost of the fencing increases with the length of the perimeter. Show that the cost of the fencing is minimum if she uses a square plot.

4. An open box with a rectangular base is to be constructed from a rectangular piece of cardboard 16 inches wide and 21 inches long by cutting a square from each corner and then bending up the resulting sides. Let x be the length of the sides of the corner squares. Our ultimate goal is to find the value of x that will maximize the volume of the box.

   (a) Express the volume V of the box as a function of x and determine the appropriate domain.

   (b) Use the sign V' to make a very rough sketch of the graph of V on $(-\infty, \infty)$. Identify the portion of the graph that is appropriate for the context of the problem.

   (c) Find the value of x that will maximize the volume of the box.

5. You want to cut a rectangular wooden beam from a cylindrical log 14 inches in diameter. The strength of the beam is proportional to the quantity h^2w, where h and w are the height and width of the cross section of the beam; the larger the quantity h^2w, the stronger the beam. Find the height and width of the strongest beam that can be cut from the log.

   (*Hint:* You will need to find a way of relating h and w. Sketch the circular cross section and sketch in a line denoting the diameter of the log. By placing the diameter line appropriately, you should be able to produce a right triangle made of w, h, and the diameter. This will enable you to relate the width and height by using the Pythagorean Theorem.)

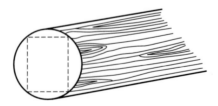

6. Let f be a function defined and continuous on the interval $[-5, 5]$. The graph of f' (NOT f) is given on the following page.

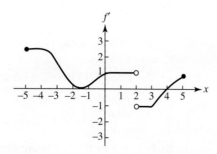

(a) Identify all critical points.

(b) For what value of x does f take on its maximum value?

(c) For what value of x does f take on its minimum value?

7. Graphs of $f(x)$ and its derivative $f'(x)$ are shown. Sketch graphs of the following and their derivatives. In each case

(a) Describe in words how the new function relates to $f(x)$.

(b) Describe in words how the new derivative relates to $f'(x)$.

(c) Identify the x-coordinates of all local minima.

    i. $g(x) = 2f(x)$

    ii. $j(x) = f(x) - 3$

    iii. $m(x) = |f(x)|$

    iv. $k(x) = f(x - 2)$

    v. $h(x) = f(2x)$

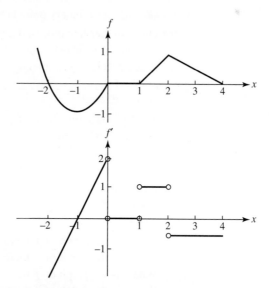

8. For each of the functions in parts (i)–(v) in the previous problem, consider the function restricted to the domain $[-2, 3]$ and answer the following.

(a) What are the critical points of the function on $[-2, 3]$?

(b) At what value of x does the function attain its maximum value on $[-2, 3]$?

(c) At what value of x does the function attain its minimum value on $[-2, 3]$?

9. (a) Use your knowledge of shifting, flipping, and stretching to graph the function
$f(x) = -2|x - 2| + 4$.

(b) At what value of x does $f(x)$ attain its maximum value? At this point, what is $f'(x)$?

(c) Does $f(x)$ have a minimum value?

(d) Where on the interval $3 \leq x \leq 8$ does f take on its maximum value? Its minimum value?

10. For each of the following functions, determine where the function is increasing and where it is decreasing. Find the x-coordinates of all local maxima and minima. (Give exact answers, not numerical approximations.)

(a) $f(x) = 2x^3 - 24x + 4$ (b) $f(x) = x^3 - 3x^2 - 9x + 2$

11. Below is a graph of $f'(x)$. The questions that follow are questions about f.

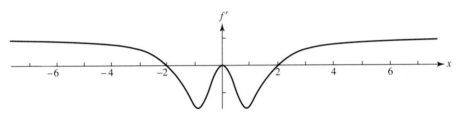

(a) Identify all critical points of f. Are these critical points also stationary points? (*Recall:* A stationary point is a point at which $f' = 0$.)

(b) On a number line plot all the critical points of f. Use the number line to indicate the sign of f'; above this indicate where the graph of f is increasing and where it is decreasing.

graph of f _____
sign of f'

(c) Some of the x-intercepts of f' correspond to the local extrema of f. How can you determine which ones do?

(d) On a number line plot all the critical points of f'. Use the number line to indicate the sign of f''; above this indicate where the graph of f is concave up and where it is concave down.

graph of f _____
sign of f''

(e) Where is the graph of f increasing and concave up?
Where is the graph of f increasing and concave down?
Where is the graph of f decreasing and concave up?
Where is the graph of f decreasing and concave down?

(f) Explain why the local extrema of f' correspond to the points of inflection of f.

(g) Suppose $f(0) = 0$. Draw a rough sketch of the graph of f.

12. For each of the two graphs given on the following page, answer questions (a)–(d) from Problem 11 above.

(a)

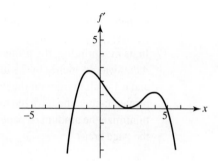

(b)

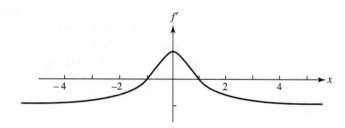

13. A tin can for garbanza beans is designed to be a cylinder with volume of 300 cubic centimeters. Denote the radius by r and the height by h. The top and bottom are thicker than the sides; for the purposes of our model, we'll assume that they are made with a double thickness of aluminum.

    (a) Give an expression for the volume of the can.
    (b) Give an expression for the amount of material used. (Remember that the top and bottom of the can are two layers thick.)
    (c) Make the expression from part (b) into a function of r alone.
    (d) What radius minimizes the material used?
    (e) What are the dimensions of the 300 cubic-centimeter can that require the least amount of material ?

14. A can for mandarin oranges is a cylinder with volume of 250 cubic centimeters. Denote the radius by r and the height by h. The material used for the top and bottom is stronger than that used for the sides. There is wasted material in constructing the top and bottom because they need to be cut from squares of metal and the scrap metal is not used. The manufacturers must pay for the material for the whole square from which the circle is cut. Suppose that the material for the top and bottom is three times as expensive as the material for the sides. What are the dimensions of the can that minimize the cost of the materials?

15. You are designing a wooden box for raspberries. It is to have the following specifications: a square base, no lid, and a volume of V cubic centimeters. What is the ratio of the height to the length of the base that minimizes the amount of wood required? Your answer should not involve V.

16. A box of Maine blueberries is packaged in a cardboard box with a square base and a see-through plastic top. The plastic costs three times as much per square inch as does

the cardboard. Assuming the volume of the box is fixed, what is the ratio of the height to the length of the base that minimizes the cost of the material for the box?

17. In its first printing, the printed material on a typical page of Frank McCourt's *Angela's Ashes* was $4\frac{1}{2}$ inches by $7\frac{1}{2}$ inches, with $\frac{1}{2}$-inch margins on the top and sides of the page and a 1-inch margin on the bottom. Assuming pages must hold 33.75 square inches of printed matter and have the margins specified, was this book laid out in such a way as to minimize the amount of paper per page? If not, what page dimensions would minimize the page area?

18. Q-Tips® are a brand of cotton swabs each 3 inches long. You can purchase a pack of 300 of them in a plastic rectangular container backed in cardboard. In other words, the plastic forms an open box and the "lid" is cardboard. The width of the box is 3 inches. What should the length and depth be if the goal is to minimize the amount of plastic used? In order to hold 300 Q-Tips the box must have a volume of 33.75 square inches.

   In reality, such a box is 7.5 inches long and 1.5 inches deep. Has the amount of plastic been minimized?

19. $f(x) = \frac{1}{3}x^3 - 2x - \frac{1}{x}$

   (a) Find f'.
   (b) Find f''.
   (c) Find all critical points. Which of these critical points are also stationary points?
   (d) Analyze the critical points. Are they local extrema? Global extrema? Points of inflection?
   (e) Is $x = 0$ the x-value of a critical point? Why or why not?
   (f) What is the absolute maximum value of the function? The absolute minimum value?

20. An artist wants to frame an 8-inch by 10-inch watercolor landscape with a mauve mat. She wants to use a total of 200 square inches of matting material, with x inches of matting above and below the painting and y inches of matting to the left and right of the painting. (*Note:* The matting is indicated by the shaded region in the accompanying figure; there is no matting underneath the painting.) The framing material is sold by the linear foot and is quite expensive. Therefore, she wants to minimize the perimeter of the frame.

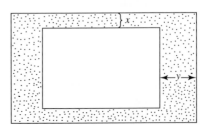

   (a) Express the perimeter of the frame as a function of x and y.
   (b) Express the perimeter of the frame as a function of x alone.
   (c) Find the dimensions of the frame that minimize its cost.

21. The graph below shows the total cost and total revenue curves for a certain firm. Both revenue and cost are functions of q, where q represents quantity.

$$\text{Profit} = \text{Total Revenue} - \text{Total Cost} = R(q) - C(q)$$

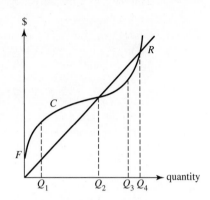

(a) Draw a graph of profit against quantity, labeling the points Q_1, Q_2, Q_3, and Q_4.

(b) What do the quantities Q_1, Q_2, Q_3, and Q_4 mean as far as the profits are concerned?

(c) Is the slope of the revenue curve constant or does it vary with q? Interpret the slope of the revenue curve in terms of the economic model. Economists call the slope of the revenue curve the *marginal revenue*.

(d) Is the slope of the cost curve constant or does it vary with q? Interpret the slope of the cost curve in words. Economists call the slope of the cost curve the *marginal cost*.

(e) Draw a set of axes with the vertical axis labeled $/item and the horizontal axis labeled q, for quantity (items produced and sold). On this set of axes, sketch both $R'(q)$ and $C'(q)$.

(f) Verify by looking at the accompanying graph that profit is a maximum (or a minimum) when the slope of the cost curve is equal to the slope of the revenue curve. How does this follow from the first derivative test?

(g) Bonus (especially for those interested in economics): Explain in words why it makes sense in economic terms that profit is a maximum (or a minimum) when the slope of the cost curve is equal to the slope of the revenue curve.

CHAPTER 11

A Portrait of Polynomials and Rational Functions

Linear functions, quadratics, and cubics are all members of the larger family of polynomial functions, a family whose members are functions that can be written in the form

$$f(x) = a_0 + a_1 x + a_2 x^2 + \cdots + a_n x^n,$$

where $a_0, a_1, a_2, \ldots, a_n$ are all constants and the exponents of the variable are nonnegative integers. We have already looked carefully at linear and quadratic functions; in this chapter we will look at characteristics of higher-degree polynomials so that we know what to expect of them. We'll begin with a case study of cubics both because we have run into cubics several times in the previous chapter and because familiarity with the behavior of cubics gives us insight into the behavior of the larger family of polynomials.

We will then turn briefly to look at rational functions, a larger class of functions that includes polynomials as well as some wilder relatives.

11.1 A PORTRAIT OF CUBICS FROM A CALCULUS PERSPECTIVE

A function $f(x)$ is **cubic** if it can be expressed in the form

$$f(x) = ax^3 + bx^2 + cx + d,$$

where a, b, c, and d are constants and $a \neq 0$. If $f(x) = ax^3 + bx^2 + cx + d$, then $f'(x) = 3ax^2 + 2bx + c$.

◆ **EXAMPLE 11.1**   Differentiate the functions below. Look at the relationship between the graphs of the derivatives and those of the corresponding function.

(a) $f(x) = \dfrac{x^3}{3} - 2x^2 + 3x + 2$ (b) $g(x) = \dfrac{-x^3 - 6x + \pi^2}{3}$

SOLUTION   (a) $f'(x) = x^2 - 4x + 3$

(b) Rewrite: $g(x) = -\frac{1}{3}x^3 - 2x + \frac{\pi^2}{3}$,   so $g'(x) = -x^2 - 2$.

The derivative of any cubic is a quadratic, so we can use our knowledge of quadratics to aid in sketching cubics.

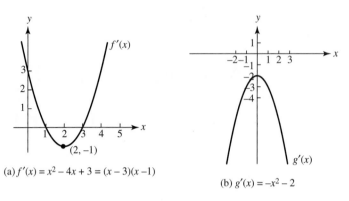

(a) $f'(x) = x^2 - 4x + 3 = (x-3)(x-1)$

(b) $g'(x) = -x^2 - 2$

Figure 11.1

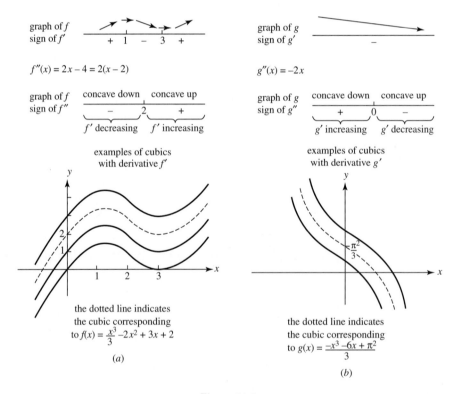

graph of f / sign of f'

$f''(x) = 2x - 4 = 2(x-2)$

graph of f / sign of f''

f' decreasing f' increasing

examples of cubics with derivative f'

the dotted line indicates the cubic corresponding to $f(x) = \frac{x^3}{3} - 2x^2 + 3x + 2$

(a)

graph of g / sign of g'

$g''(x) = -2x$

graph of g / sign of g''

g' increasing g' decreasing

examples of cubics with derivative g'

the dotted line indicates the cubic corresponding to $g(x) = \frac{-x^3 - 6x + \pi^2}{3}$

(b)

Figure 11.2

Notice that these are families of cubic graphs. To pick a particular cubic when we only know its derivative, we must also be given another identifying piece of information, such as a point on the graph of the cubic. ◆

Basic Characteristics of the Cubic Function f

Question: How many critical points can a cubic have?

Answer: The critical points of a cubic are the stationary points (i.e., wherever the derivative is zero). There are no other critical points, because f' is always defined and the natural domain of f is $(-\infty, \infty)$. The derivative of a cubic is a quadratic; a quadratic equation has either zero, one, or two real roots, so a cubic has either zero, one, or two critical points.

cubic with zero critical points   cubic with one critical point   cubic with two critical points

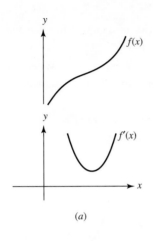

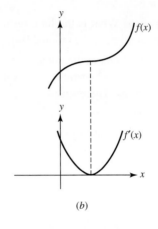

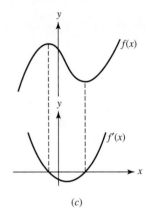

     (a)   (b)   (c)

Figure 11.3

Question: How many turning points (local extrema) can a cubic have?

Answer: A continuous function $f(x)$ has a turning point wherever $f'(x)$ changes sign. The derivative of a cubic is a quadratic; we need to ask how many times a quadratic can change sign. (Think about this. Draw some parabolas, or look at those in Figure 11.3.) The answer is zero times or two times. Therefore a cubic can have either no turning points (as in Figures 11.3a and 11.3b), or two turning points (as in Figure 11.3c), but *not* one turning point.

EXERCISE 11.1

(a) Find the point of inflection of $f(x) = \frac{x^3}{3} - 2x^2 + 3x + 2$. (This is the function from Example 11.1a.)

(b) Show that the point of inflection lies midway between the local maximum and local minimum points of f. (By "point" we are referring to the x-coordinate.)

Answers are provided at the end of the section.

EXERCISE 11.2

(a) How many inflection points (changes of concavity) does a cubic have?

(b) In Exercise 11.1 you found that the x-coordinate of the point of inflection lay midway between the x-coordinates of the turning points. If a cubic has two turning points, will this always be true? If yes, give a convincing argument. If no, give a counterexample.

Answers are provided at the end of the section.

Question: What is the long-term behavior of a cubic? In other words, if $f(x)$ is a cubic, what are $\lim_{x \to \infty} f(x)$ and $\lim_{x \to -\infty} f(x)$?

Answer: Look at the cubic $f(x) = ax^3 + bx^2 + cx + d$. When x is large enough in magnitude, the ax^3 term dominates the expression. By this we mean that it overpowers all the other terms. Therefore,

if $a > 0$, then $\lim_{x \to \infty} f(x) = \infty$ and $\lim_{x \to -\infty} f(x) = -\infty$, while

if $a < 0$, we have $\lim_{x \to \infty} f(x) = -\infty$ and $\lim_{x \to -\infty} f(x) = \infty$.

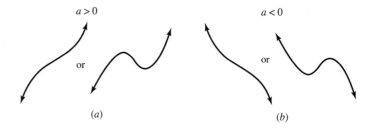

Figure 11.4

EXERCISE 11.3   Answer the question above basing your argument on $f'(x)$ rather than on the dominance of the x^3 term.

Question: How many real roots can a cubic equation have? Equivalently, how many x-intercepts can a cubic function have?

Answer: By looking at the long-term behavior of $f(x)$ in Figure 11.4, we see that a cubic must cross the x-axis somewhere because it is a continuous function. Therefore a cubic must always have at least one zero. Furthermore, a cubic can have at most two turning points, so it can have at most three x-intercepts, because it must turn at least once in between each two intercepts. A cubic can have either one, two, or three x-intercepts, as the graphs below illustrate.

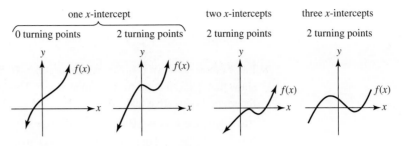

Figure 11.5

Question: What does the derivative tell us about *where* the zeros are?

Answer: Very little. The derivative tells us only about the *slope* of the graph of $f(x)$. For example, the functions in Figure 11.2(a) all have the same derivative, but the number and location of their zeros are completely different.

Answers to Selected Exercises

Answers to Exercise 1.1

(a) $x = 2$

(b) local maximum at $x = 1$; local minimum at $x = 3$

Answers to Exercise 1.2

(a) one

(b) The point of inflection corresponds to the vertex of the parabola given by f'. The local extrema correspond to the two x-intercepts of the parabola given by f'. The vertex of the parabola is midway between these two roots.

PROBLEMS FOR SECTION 11.1

For Problems 1 through 7, give an example of a cubic function $f(x)$ with the characteristic(s) specified. Your answer should be a formula, but a picture will be helpful. There may be many possible answers.

1. $f(x)$ has zeros at $x = -2$, $x = 3$, and $x = 0$.

2. $f(x)$ has zeros at $x = -1$ and $x = 2$ only. $f(0) = 1$.

3. $f(x)$ has only one zero. It is at $x = 1$. $\lim_{x \to \infty} f(x) = -\infty$.

4. f has a local maximum at $x = 0$ and a local minimum at $x = 2$.

5. f has a point of inflection at $x = 1$.

6. f is always increasing.

7. f is always decreasing. $f(3) = 0$ and $f(0) = 2$.

8. Let $f(x) = (x - a)(x - b)^2$, where $a > b > 0$. By looking at the sign of f you can show that f has a local maximum at $x = b$. This problem asks you to verify this using the second derivative test.

   (a) Using the Product Rule, show $f'(b) = 0$.

   (b) Use the second derivative test to show that f has a local maximum at $x = b$.

9. According to postal rules, the sum of the girth and the length of a parcel may not exceed 108 inches. What is the largest possible volume of a rectangular parcel with a square girth? ("Girth" means the distance around something. A person with a large girth needs a big belt.)

10. An open box with a rectangular base is to be constructed from a rectangular piece of cardboard 16 inches wide and 21 inches long by cutting a square from each corner and then bending up the resulting sides. Let x be the length of the sides of the corner squares. Find the value of x that will maximize the volume of the box.

11. Without using a graphing calculator, sketch the following graphs. Label all local maxima and minima. Beside the sketch of f, draw a rough sketch of $f'(x)$.

   (a) $f(x) = x(x - 3)(x + 5)$ (Start by looking at the x-intercepts. Then look at the critical points.)

   (b) $f(x) = -2x(x - 3)(x + 5)$ (Conserve your energy.)

   (c) $f(x) = x^3 + 3x^2 - 9x$ (Start by looking at the x-intercepts. Then look at the critical points.)

   (d) $f(x) = x^3 + 3x^2 - 9x + 1$ (This time the x-intercepts are difficult to find, so don't bother with them. Again, conserve your energy.)

In Problems 12 through 16, find and classify all critical points.

12. (a) $f(x) = x^3 - 3x + 1$
    (b) $f(x) = x^3 + 3x + 1$

13. (a) $f(x) = -x^3 - 3x^2 + 9x + 5$
    (b) $f(x) = x^3 + 3x^2 + 9x + 8$

14. $f(x) = x^3 + x^2 + x + 1$

15. $f(x) = -2x^3 + x^2 + 7$

16. $f(x) = x^3 + 2x^2 + 3x + 4$

In Problems 17 throough 20, graph the function given, labeling all x-intercepts, y-intercepts, and the x- and y-coordinates of any local maximum and minimum points.

17. $f(x) = x(x+2)(x-3)$

18. $f(x) = x(x-2)^2$

19. $f(x) = x^2(x-2)$

20. $f(x) = x^3 - x^2 - 6x$

21. Find the equation of the tangent line to $y = -2x^3 + 3x^2 + 6x - 2$ at its point of inflection.

11.2   POLYNOMIAL FUNCTIONS AND THEIR GRAPHS

Polynomials are quite well behaved, as functions go. They involve no operations on the variable other than addition, subtraction, and multiplication, so polynomials are defined everywhere; there is no need to worry about negatives under radicals or zeros in the denominators because polynomials, by definition, have no variables under square roots or in denominators.

Recall that a **polynomial function** is a sum of terms of the form $a_k x^k$, where k is a *nonnegative integer*. This means we can obtain a general polynomial by adding up functions of the form x^k for various integer values of k, giving them different weights by multiplying each x^k by some constant a_k. The result is an expression of the form

$$a_0 + a_1 x + a_2 x^2 + a_3 x^3 + \cdots + a_n x^n, \text{ where } a_0, a_1, a_2, \ldots, a_n$$

are constants, and $a_n \neq 0$.

We call the constant a_k the **coefficient** of the x^k term. (For instance, a_3 is the coefficient of the x^3 term.) The **degree** of a polynomial is the highest power to which x is raised. (The polynomial displayed above is of degree n, because it is specified that $a_n \neq 0$.) If a polynomial is of degree n, then a_n is called the **leading coefficient**.

◆ EXAMPLE 11.2   Decide whether or not the following are polynomials.

   i. $f(x) = 4x^3 - 3x + \frac{2}{x}$
   ii. $g(x) = 3x^2 - 7 + \sqrt{x}$
   iii. $h(x) = \frac{x^3}{\pi - 1} - \frac{2}{5}x^2 + x + \frac{1}{\sqrt{8}}$

SOLUTION

   i. $f(x)$ is not a polynomial because $\frac{2}{x} = 2x^{-1}$ has x raised to a negative power.

   ii. $g(x)$ is not a polynomial because $\sqrt{x} = x^{1/2}$ has x raised to a fractional power.

   iii. $h(x)$ is a polynomial because all the powers of x are nonnegative integers.   ◆

Let's start by looking at the graphs of some of the building block functions for polynomials, the power functions $y = x^k$, where $k = 0, 1, 2, 3, 4, 5, 6$.

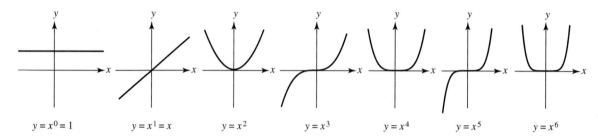

Figure 11.6

All of the building blocks are continuous functions, and since polynomials are constructed as weighted sums of these building blocks, they too must be continuous.

The Zeros of a Polynomial

> A polynomial equation of degree n can have at most n distinct real roots.

Suppose $P(x)$ is a polynomial and $P(x) = 0$ is the corresponding polynomial equation. If $x = c$ is a root of the equation, then $P(c) = 0$ and $(x - c)$ is a factor of $P(x)$. Consequently, if an nth degree polynomial equation has n distinct real roots $c_1, c_2, c_3, \ldots, c_n$, then $P(x)$ can be written in the form

$$P(x) = k(x - c_1)(x - c_2)(x - c_3) \cdots (x - c_n), \text{ where } k \text{ is a constant.}$$

The polynomial

$$P(x) = 6x^2(x - 2)(x - 5)^2(x + 1)^3$$

is an eighth degree polynomial. We can write

$$P(x) = 6(x - 0)(x - 0)(x - 2)(x - 5)(x - 5)(x + 1)(x + 1)(x + 1).$$

We say that $P(x) = 0$ has a **simple root** at $x = 2$, **double roots** or **roots of multiplicity 2** at $x = 0$ and $x = 5$, and a **root of multiplicity 3** at $x = -1$.

The polynomial

$$P(x) = -(x + 5)(x^2 + 1)(x - 2)^2$$

is a fifth degree polynomial with a zero of multiplicity 2 at $x = 2$ and a simple zero at $x = -5$. Notice that $x^2 + 1$ is positive for all real x. $x^2 + 1 = 0$ only if $x^2 = -1$, or $x = \pm\sqrt{-1} = \pm i$. If we include complex roots like these and count roots with their multiplicity, then we can say that

a polynomial equation of degree n has exactly n roots.

In this course we will mainly be working with the real number system as opposed to the complex number system, so unless otherwise stated, when we ask about the number of roots it is understood that we mean real roots.

EXERCISE 11.4   Construct a polynomial equation with the specification given. (The answers to this exercise are *not* unique!)

(a) a third degree polynomial equation with roots at $x = -2$, $x = 3$, and $x = 0$

(b) a second degree polynomial equation with a double root at $x = -1$

(c) a second degree polynomial equation with no roots

(d) a fifth degree polynomial equation with roots only at $x = -2$, $x = 3$, and $x = 0$

Answers are provided at the end of the section.

EXERCISE 11.5   Identify which of the following two polynomials is possible to construct, and construct it.

(a) a fifth degree polynomial with no real zeros

(b) a fourth degree polynomial with no real zeros

Answer is provided at the end of the section.

Finding the Zeros of a Polynomial

Finding the zero of a linear function is simple. Finding the zeros of a quadratic is no problem when you use the quadratic formula. It is considerably harder to find the roots of a general cubic equation. A systematic procedure for finding a cubic's roots does exist, but it is much more complicated than the quadratic formula and in practice is not often used. The story of the discovery of the cubic formula in Italy in the mid-1500s is one worth reading about in a math history book. It involves alleged lies about the discovery, a public competition to solve cubic equations being won by Nicolo Tartaglia and his "secret" method being passed on to Girolamo Cardano in confidence but then published anyway, and a final dispute from which one of the men is said to have been lucky to have escaped alive.[1] There is an algorithm for finding the zeros of fourth degree polynomials, but in the early 1800s the Norwegian mathematician Niels Abel (1802–1829) proved that a formula for finding the zeros of a fifth degree polynomial *does not exist*.

For a general nth degree polynomial it can be difficult to find the zeros, and as is clear from the last paragraph, we cannot give a recipe for pinning them down exactly. We can, however, take advantage of the continuity of polynomials; if $f(a) > 0$ and $f(b) < 0$, then there is a c somewhere between a and b such that $f(c) = 0$.

◆ **EXAMPLE 11.3**   Find the zeros of $f(x) = x^3 - 2x^2 - 8x$.

SOLUTION   We can factor out an x from each term.

$$x^3 - 2x^2 - 8x = x(x^2 - 2x - 8) \quad \text{Now we have a quadratic that we can factor.}$$
$$= x(x - 4)(x + 2)$$

So, the zeros are $x = 0$, $x = 4$, and $x = -2$.   ◆

---

[1] From Howard Eves, *An Introduction to the History of Mathematics*, and David Burton, *A History of Mathematics*.

◆ **EXAMPLE 11.4**   Find the zeros of $f(x) = x^3 + 3x^2 - 3x - 1$.

SOLUTION   In this case, it is not clear how to factor $f(x)$. Graph f on a graphing calculator.

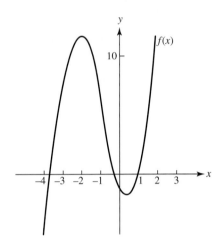

Figure 11.7

We see that there is a zero near $x = 1$. By zooming in, it looks like the zero might be exactly $x = 1$. To be sure, evaluate f at $x = 1$. $f(1) = (1)^3 + 3(1)^2 - 3(1) - 1 = 0$, so $x = 1$ is a zero. There are two other zeros. By zooming in, we estimate that they are at $x \approx -3.73$ and $x \approx -0.27$.

Let's try to find the exact values. $x = 1$ is a zero of $f(x)$; therefore, $(x - 1)$ is a factor of $f(x)$. In other words, we can write $x^3 + 3x^2 - 3x - 1 = (x - 1)p(x)$, where $p(x)$ is a new polynomial, a polynomial of degree 2.

To find $p(x)$, we do long division of polynomials. Long division of polynomials is analogous to long division for numbers. Let's review long division.

Find $\frac{732}{4}$.   Write denominator $\overline{)\text{numerator}}$.

$$
\begin{array}{r}
183 \\
4\overline{)732} \\
\underline{4} \\
33 \\
\underline{32} \\
12 \\
\underline{12} \\
0
\end{array}
$$

4 goes into 7 one time. Write the 1 on the top and 1 • 4 under the 7.

Subtract 4 from 7. Bring down the 3 from 732 to get 33.

4 goes into 33 eight times. Write the 8 on the top and 8 • 4 under the 33.

Subtract 32 from 33. Bring down the 2 from 732 to get 12.

4 goes into 12 three times. Write the 3 on the top and 3 • 4 under the 12.

Subtract 12 from 12. This gives zero. We've accounted for 732: 7 • 100 + 3 • 10 + 2, so we're done.

Find $\frac{x^3+3x^2-3x-1}{x-1}$.   Write denominator $\overline{)\text{numerator}}$.

$$
\begin{array}{r}
x^2 + 4x + 1 \\
x-1\overline{)x^3 + 3x^2 - 3x - 1} \\
\underline{x^3 - x^2} \\
4x^2 - 3x \\
\underline{4x^2 - 4x} \\
x - 1 \\
\underline{x - 1} \\
0
\end{array}
$$

x times x^2 gives x^3. Write the x^2 on the top and $x^2 \cdot (x-1)$ under the $x^3 + 3x^2$.

Subtract $x^3 - x^2$ from $x^3 + 3x^2$. Bring down the $-3x$ from above to get $4x^2 - 3x$.

x times $4x$ gives $4x^2$. Write the $4x$ on the top and $4x \cdot (x-1)$ under the $4x^2 - 3x$.

Subtract $4x^2 - 4x$ from $4x^2 - 3x$. Bring down the -1 from above to get $x - 1$.

x times 1 gives x. Write the 1 on the top and $1 \cdot (x-1)$ under the $x - 1$.

Subtract $x - 1$ from $x - 1$. This gives zero. We've accounted for $x^3 + 3x^2 - 3x - 1$, so we're done.

By doing long division as shown, we see that $p(x) = x^2 + 4x + 1$, so $f(x) = x^3 + 3x^2 - 3x - 1 = (x - 1)(x^2 + 4x + 1)$. We use the quadratic formula to find the roots of $x^2 + 4x + 1 = 0$ and find that

$$
x = \frac{-4 \pm \sqrt{16 - 4}}{2} = \frac{-4 \pm \sqrt{12}}{2} = \frac{-4 \pm 2\sqrt{3}}{2} = -2 \pm \sqrt{3}.
$$

Verify that the values of these two roots are approximately equal to our previous estimates from the graph.

The zeros of $f(x) = x^3 + 3x^2 - 3x - 1$ are $x = 1$, $x = -2 + \sqrt{3}$, and $x = -2 - \sqrt{3}$.

Notice that if we had not known an exact zero ($x = 1$ in this case), then we could not have used long division and would have needed to approximate all three zeros graphically without learning their exact values.[2]   ◆

EXERCISE 11.6   Find all the roots of $f(x) = x^3 - 1 = 0$. (This is the cubic discussed in Section 7.3.)

Characteristics of Polynomials and Differentiation of Polynomials

Let's return for a moment to the building blocks of polynomials, functions of the form $y = x^k$ for $k = 0, 1, 2, 3, \ldots$, and look at them more closely. They are graphed below on the same set of axes. Two scales are used, one allowing us to compare the graphs for x near zero, the other facilitating comparison for large values of x.

_____

[2] Another alternative is to use what is called "Newton's Method" for approximating roots. See the Appendix on Newton's Method for details.

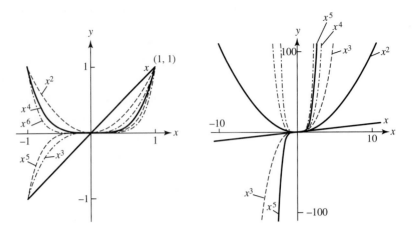

Figure 11.8

Behavior for small x (a bug's-eye view)

The larger k is, the flatter the graph of x^k is on the interval $-1 < x < 1$.

In other words, for $|x| < 1$, the higher the power to which x is raised, the smaller the magnitude of x^k.[3] For instance, $\left(\frac{1}{2}\right)^5$ is smaller than $\left(\frac{1}{2}\right)^2$.

Behavior for large x (a bird's-eye view)

For $|x| > 1$, the magnitude of x^k increases dramatically as k increases. For example, consider the values of y corresponding to $x = 10$ in each of the polynomials above. 10^5 is much greater than 10^2.

By summing up building block functions of the form x^k and weighting them with coefficients, we obtain a general nth degree polynomial $P(x) = a_0 + a_1x + a_2x^2 + \cdots + a_nx^n$. Characteristics of such a polynomial are discussed next.

A Bird's-Eye View of Polynomials

For x large enough in magnitude (zoom out a lot on your graphing calculator) the graph of an nth degree polynomial, $P(x) = a_0 + a_1x + a_2x^2 + a_3x^3 + \cdots + a_nx^n$, is dominated by a_nx^n and can be characterized as follows:

For n even, if a_n is positive, then the graph reaches upward;

if a_n is negative, then the graph reaches downward.

For n odd, if a_n is positive, then the graph reaches down on the left and up on the right;

if a_n is negative, the graph reaches up on the left and down on the right.

More precisely, let $P(x) = a_0 + a_1x + a_2x^2 + a_3x^3 + \cdots + a_nx^n$ be a polynomial of degree n.

---

[3] By "magnitude of x^k" we mean $|x^k|$. We are referring to the size, without regard to the sign.

Suppose n is even: If $a_n > 0$, then $\lim_{x \to \infty} P(x) = +\infty$ and $\lim_{x \to -\infty} P(x) = +\infty$.

If $a_n < 0$, then $\lim_{x \to \infty} P(x) = -\infty$ and $\lim_{x \to -\infty} P(x) = -\infty$.

Suppose n is odd: If $a_n > 0$, then $\lim_{x \to \infty} P(x) = +\infty$ and $\lim_{x \to -\infty} P(x) = -\infty$.

If $a_n < 0$, then $\lim_{x \to \infty} P(x) = -\infty$ and $\lim_{x \to -\infty} P(x) = +\infty$.

This bird's-eye view tells us that a polynomial of odd degree will have neither an absolute maximum value nor an absolute minimum value. Its range is $(-\infty, \infty)$. On the other hand, a polynomial of even degree will have a global maximum but no global minimum if the leading coefficient is negative and vice versa if the leading coefficient is positive.

A Bug's-Eye View of Polynomials

If we zoom in enough anywhere on the graph of any polynomial, the graph will eventually look like a straight line. Polynomial functions have the delightful property that they are locally linear everywhere. Another way to say this is that polynomials are differentiable everywhere. Their graphs are smooth; there are no sharp corners.

Polynomials are continuous, locally linear, and easy to evaluate. This makes polynomial functions very user-friendly. In fact, polynomials are so delightful to work with that people use them to approximate local behavior of functions that are more difficult to handle.[4] Another perk is that polynomial functions are a pleasure to differentiate. We can compute the derivative of x^n for n any positive integer, so we can differentiate any polynomial.

$$\frac{d}{dx}[x^n] = nx^{n-1}$$

We can conclude that the derivative of an nth degree polynomial is a polynomial of degree $n - 1$.

$$P(x) = a_0 + a_1x + a_2x^2 + a_3x^3 + \cdots + a_nx^n$$
$$P'(x) = a_1 + 2a_2x^2 + 3a_3x^3 + \cdots + na_nx^{n-1}$$

Critical Points of a Polynomial

The natural domain of polynomial functions is $(-\infty, \infty)$ and polynomials are differentiable everywhere; therefore, the critical points of a polynomial are simply the zeros of its derivative. The derivative of an nth degree polynomial is a polynomial equation of degree $(n - 1)$. A polynomial of degree $(n - 1)$ can have *at most* $(n - 1)$ real roots. We conclude that

a polynomial of degree n can have *at most* $(n - 1)$ turning points.

This makes sense graphically. If a polynomial of degree n has n zeros (the maximum possible), then between each of those zeros the polynomial must have a turning point so its graph can return to the x-axis. This means there would have to be $(n - 1)$ turning points. Notice also that a polynomial of degree n may have *fewer* than $n - 1$ turning points. Consider the graphs of $f(x) = x^k$ in Figure 11.6; each of them had either zero or one turning point.

---

[4] You might, for instance, try to approximate e^x near $x = 0$ by a third degree polynomial. One of the treats awaiting you is the amazing discovery that you can actually express functions such as e^x (and some of the trigonometric functions) as polynomials of infinite degree.

By looking at the behavior of polynomials for x large in magnitude (taking a bird's-eye view) we can argue that

a polynomial of odd degree must have at least one zero and

a polynomial of even degree must have at least one turning point.

Complete this argument as an exercise.

EXERCISE 11.7 Justify the assertion made above.

Symmetry. Recall that a function is said to be

even if $f(-x) = f(x)$ for all x in its domain (the graph is symmetric about the y-axis).

odd if $f(-x) = -f(x)$ for all x in its domain (the graph is symmetric about the origin)

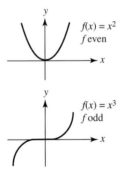

Figure 11.9

Looking back at the graphs of x^k in Figure 11.6, we can see why the names "even" and "odd" were chosen for these types of symmetry. Functions of the form x^k, where k is even, have even symmetry, and functions of the form x^k, where k is odd, have odd symmetry.

Any random polynomial may be even, odd, or neither. The following example is meant to illustrate criteria for a function to be even and to be odd.

◆ **EXAMPLE 11.5** For each of the following, determine whether the function is even, odd, or neither even nor odd.

(a) $g(x) = -5x^4 + 2x^2 + 1$

(b) $h(x) = 2x^3 - x$

(c) $f(x) = 2x^3 - x + 1$

SOLUTION Our strategy is to look at $g(-x)$ and see if it is equal to $g(x)$, $-g(x)$, or neither.

(a) $g(-x) = -5(-x)^4 + 2(-x)^2 + 8$

$\qquad = -5x^4 + 2x^2 + 8$

$\qquad = g(x)$

So $g(x)$ is even.

(b) $h(-x) = 2(-x)^3 - (-x)$

$$= -2x^3 + x$$
$$= -(2x^3 - x)$$
$$= -h(x)$$

So $h(x)$ is odd.

(c) $f(-x) = 2(-x)^3 - (-x) + 1$

$$= -2x^3 + x + 1$$

This is neither $f(x)$ nor $-f(x)$, so f is neither even nor odd.   ◆

We can deduce the following:

A polynomial with the variable raised to even powers only is even.

A polynomial with no constant term and the variable raised to odd powers only, with no constant terms, is odd.

Terminology: The polynomial $f(x) = x^2 + x$ is a polynomial of even degree, but it is *not* an even function. Notice that $f(x) = x(x + 1)$, so the zeros of the polynomial are at 0 and -1; the graph of f is not symmetric with respect to the y-axis. Similarly, the polynomial $g(x) = x^3 + 1$ is a polynomial of odd degree, but it is *not* an odd function. The graph of g has a y-intercept of 1, so it is not symmetric about the origin.

Answers to Selected Exercises

Answers to Exercise 11.4

(a) $y = x(x + 2)(x - 3)$ (b) $y = (x + 1)^2$

(c) $y = x^2 + 1$ (d) $y = (x + 2)x^2(x - 3)^2$

Answer to Exercise 11.5

(b) $y = x^4 + 3$

PROBLEMS FOR SECTION 11.2

1. Determine whether or not the expression given is a polynomial.

(a) $\frac{1}{\sqrt{2}}x + \sqrt{33}x^2 + \frac{19}{11}$

(b) $2x^2 + 3x^{-1} + 5x^3$

(c) $2x + x^{1/2} + 5x^5$

(d) $\frac{2}{x} + \frac{2x}{3} + 1$

(e) $5^{-1/2}x + 3^{-1}x^2 + \frac{1}{\pi^2 - 2} + 2$

(f) $(x^2 + 1)^{-1}$

In Problems 2 through 8, below, construct a polynomial $P(x)$ with the specified characteristics. Answers to these problems are not unique.

2. A second degree polynomial with zeros at $x = 1$ and $x = -3$.

3. A third degree polynomial with zeros at $x = -1, 0$, and 5.

4. A fourth degree polynomial with no zeros.

5. A fourth degree polynomial whose only zero is at $x = \sqrt{2}$.

6. A fifth degree polynomial with a zero of multiplicity two at $x = 9$ and zeros at $x = 0$, 3, and $-e$.

7. A third degree polynomial whose only zero is at $x = \pi + 1$, and whose y-intercept is 1.

8. A fourth degree polynomial whose only turning point is at $(-3, 2)$.

In Problems 9 through 14, construct a polynomial $P(x)$ with the specified characteristics. Determine whether or not the answer to the problem is unique. Explain and/or illustrate your answer.

9. A fourth degree polynomial with zeros of multiplicity two at $x = 2$ and $x = -3$, and a y-intercept of -2.

10. A fifth degree polynomial with zeros of multiplicity two at $x = 0$ and $x = \pi$, and a zero at $x = -2$; $\lim_{x \to \infty} P(x) = \infty$.

11. A third degree polynomial whose only zero is at $x = -1$ and such that $\lim_{x \to \infty} P(x) = \infty$.

12. The graph is a parabola with a vertex at $(\pi, 2)$ and a y-intercept of 0.

13. A fifth degree polynomial with a zero of multiplicity 3 at $x = 0$ and zeros at $x = 1$ and $x = -2$, and passing through the point $(-1, 2)$.

14. A third degree polynomial with zeros at $x = 1$ and $x = 2$, a turning point at $x = 1$, and a y-intercept of $\sqrt{e}$.

In Problems 15 through 21, find the (real) zeros of the polynomial given.

15. (a) $f(x) = 2x^3 + 2x^2 - 12x$
    (b) $g(x) = 2x^3 + 2x^2 + 12x$

16. (a) $f(x) = -x^3 - x^2 - 5x$
    (b) $g(x) = 0.5x^4 - 0.5$

17. (a) $P(x) = x^3 - x^2 - 4x + 4$

    (b) $Q(x) = x^3 - x^2 + 4x - 4$

Hint: Either guess a zero by observation or use a graphing calculator to guess a root; then use long division.

18. $f(x) = -x^5 + 16x^4$

19. $P(x) = x^4 - 2x^3 - 6x^2 + 12x$ (*Hint:* At a certain point, guess a root.)

20. $g(x) = 3x^3 + 3$

21. $h(x) = x^3 - 8$

In Problems 22 through 24, $P(x)$ is a polynomial with the characteristics specified. For each statement following the characteristic, determine whether the statement is definitely true; possibly true, but not necessarily true; or definitely false. Explain.

22. The graph of $P(x)$ is symmetric about the origin.
    (a) The degree of $P(x)$ is even.
    (b) The degree of $P(x)$ is odd.
    (c) $P(x)$ has no zeros.
    (d) $P(x)$ has at least one zero.
    (e) The number of turning points of $P(x)$ is of the form $2n$ where n is a nonnegative integer.

23. $P(x)$ is a polynomial whose degree is even and nonzero. $P(0) = 0$.
    (a) The graph of $P(x)$ is symmetric about the origin.
    (b) The graph of $P(x)$ is symmetric about the y-axis.
    (c) The graph of $P(x)$ has at least one turning point.
    (d) The graph of $P(x)$ has an even number of turning points.
    (e) The graph of $P(x)$ has an odd number of turning points.

24. $P(x)$ is a fifth degree polynomial. $\lim_{x \to \infty} P(x) = \infty$.
    (a) $P(x)$ has five distinct real roots.
    (b) $P(x)$ has no more than five roots.
    (c) $P(x)$ has five turning points.
    (d) $P(x)$ has four turning points.
    (e) $P(x)$ has no more than four turning points.
    (f) $P(x)$ has at least one real root.
    (g) $\lim_{x \to -\infty} P(x) = \infty$.

25. The graph of $P(x)$ is given below

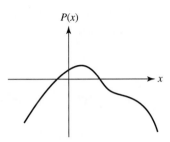

(a) The degree of $P(x)$ is at least _____.

(b) What is the sign of the leading coefficient of $P(x)$?

26. Below are graphs of polynomials, each with a zero at $x = a$.

(a) For each graph determine whether the zero is a simple zero, a zero of even multiplicity, or a zero of odd multiplicity.

(b) For each graph, determine whether the derivative at $x = a$ is positive, negative, or zero.

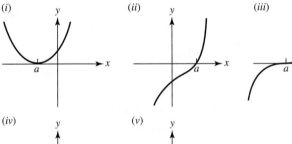

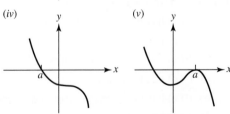

■ 11.3   POLYNOMIAL FUNCTIONS AND THEIR GRAPHS

Graphing Polynomials

We'll begin with a couple of examples and then arrive at a general strategy for understanding the graphs of polynomials.

◆ EXAMPLE 11.6   Graph $f(x) = x^3 - 2x^2 - 8x$, identifying the x-coordinates, all zeros, local extrema, and inflection points.

SOLUTION   *Zeros.*   This is the function from Example 11.2. In factored form $f(x) = x(x - 4)(x + 2)$; it has zeros at $x = -2$, $x = 0$, and $x = 4$. The graph of $f(x)$ does not cross the x-axis anywhere between these zeros. f is continuous, so its sign can't change between the zeros. Therefore, between $x = -2$ and $x = 0$, for example, $f(x)$ must be always positive or always negative. By determining the sign of $f(x)$ for just one test value of x on the interval $(-2, 0)$, we can determine the sign of $f(x)$ on the entire interval $(-2, 0)$. We draw a number line showing the zeros of $f(x)$ and the sign of $f(x)$ between those zeros.

Sign Computations—suggested method: Work with f in factored form: $f(x) = x(x - 4)(x + 2)$. We'll start with x very large and then look at the effect on the sign of the factors as x decreases and passes by each zero.

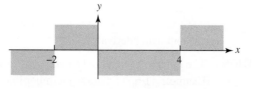

For x very large, all factors are positive: $(+)(+)(+) \Rightarrow (+)$

As x drops below 4, the factor $(x - 4)$ changes sign: $(+)(-)(+) \Rightarrow (-)$

As x drops below 0, the factor x changes sign: $(-)(-)(+) \Rightarrow (+)$

As x drops below -2, the factor $(x + 2)$ changes sign: $(-)(-)(-) \Rightarrow (-)$

A spot-check can be used to confirm this.

Figure 11.10

Local Extrema.   We need to look at the first derivative to determine critical points: $f'(x) = 3x^2 - 4x - 8$. We use the quadratic formula to determine where $f'(x) = 0$ and obtain $x = \frac{2 + 2\sqrt{7}}{3} \approx 2.43$, and $x = \frac{2 - 2\sqrt{7}}{3} \approx -1.10$. By drawing a number line showing these points where $f'(x) = 0$ and indicating the sign of $f'(x)$ between them, we can determine whether these stationary points are local minima, local maxima, or neither.

graph of $f(x)$
sign of $f'(x)$ $\quad$ (+) $\quad \dfrac{2-2\sqrt{7}}{3}$ $\quad$ (−) $\quad \dfrac{2+2\sqrt{7}}{3}$ $\quad$ (+)

Computations:

If x is large enough in magnitude, x^2 dominates and f' is positive.

For $x = 2/3$ (between the zeros) f' is negative.

So, $f(x)$ has a local maximum at $x = \dfrac{2-2\sqrt{7}}{3}$ and a local minimum at $x = \dfrac{2+2\sqrt{7}}{3}$.

We know that a cubic on a nonrestricted domain will have neither a global maximum value nor a global minimum value; when the domain of a cubic is $(-\infty, \infty)$, the range is also $(-\infty, \infty)$.

Inflection Points. Take the second derivative: $f''(x) = 6x - 4$. $f''(x) = 0$ when $x = 2/3$. We draw another number line to show the sign of $f''(x)$ and the corresponding concavity of $f(x)$.

concavity of $f(x)$ $\qquad$ concave down $\qquad$ concave up
sign of $f''(x)$ $\qquad\qquad$ (−) $\qquad \dfrac{2}{3} \qquad$ (+)

All this information taken together enables us to draw an accurate graph of $f(x)$.

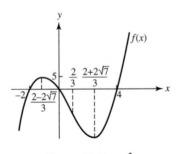

Figure 11.11 ◆

◆ **EXAMPLE 11.7**   Sketch a rough graph of $f(x) = -(x+3)(x-2)^2$, labeling all zeros and showing where $f(x)$ is positive and where $f(x)$ is negative. See how much information can be gotten without using derivatives.

SOLUTION   The cubic $f(x)$ has only two zeros: a zero at $x = -3$ and a double zero at $x = 2$. As in Example 11.6, let's draw a number line showing the zeros of $f(x)$ and the sign of $f(x)$ in between these zeros.

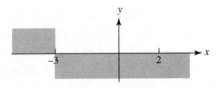

Figure 11.12

Notice that here $f(x)$ does *not* change sign when it passes through the zero at $x = 2$; it is negative on *both sides* of this zero. This is because the $(x-2)^2$ term in $f(x)$ doesn't change sign when x passes through $x = 2$.

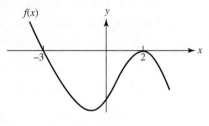

Figure 11.13

f has a local maximum at $x = 2$. To find the local minimum, we would need to calculate f'. Do this as an exercise.   ◆

Can you figure out in general what effect multiple (double, triple, or higher) roots have on the graph of $f(x)$? (*Hint:* Consider separately the cases where the factor $(x-a)$ is raised to even and odd powers and analyze the sign.)

Strategy for Graphing a Polynomial Function f

1. *Bird's-eye view*

   Look at the degree of the polynomial and the sign of the leading coefficient to anticipate what is ahead. From this information, determine whether the polynomial will have a global maximum or a global minimum or neither.

2. *Positive/negative*

   x-intercepts are simple to find *if* we can factor the polynomial into linear and quadratic factors. In theory this is always possible, but in practice it may be very difficult. Polynomials are continuous functions, so if $P(x)$ changes sign between $x = a$ and $x = b$ there must be at least one x-intercept between $x = a$ and $x = b$. If you find the x-intercepts, plot them. If you are certain you've identified all the zeros of $P(x)$, then check the sign of $P(x)$ in each interval.[5]

   Note: If the x-intercepts are symmetrically placed, check for even or odd symmetry. If x is only raised to even powers, the graph will be symmetric about the y-axis. If x is only raised to odd powers and there is no constant term, then the graph will be symmetric about the origin.

   y-intercept: $P(0)$

3. *Increasing/decreasing*

   Look at the sign of $P'(x)$. On a number line plot the points where $P'(x) = 0$ and then check the sign in every interval.

   $$P' > 0 \Rightarrow \text{the graph of } P \text{ is increasing}$$

   $$P' < 0 \Rightarrow \text{the graph of } P \text{ is decreasing}$$

---

[5] If the polynomial is in factored form, it's easy to check the sign of the polynomial at, say, $x = 3$, by checking the sign of each of the factors at $x = 3$ and then "multiplying the signs" together.

Since P' is continuous, $P' = 0$ at any turning point (local maximum or minimum) of the graph.

Note: If $P'(3) = 0$ we cannot conclude that the graph must have a local maximum or minimum at $x = 3$. We can only be sure that the graph has a horizontal tangent line at $x = 3$. P will have a local extrema at $x = 3$ only if P' *changes sign* at $x = 3$.

4. *Concavity—fine tuning* Look at the sign of $P''(x)$. On a number line plot the points where $P''(x) = 0$ and check the sign in every interval.

$$P'' > 0 \Rightarrow \quad \text{the graph of } P \text{ is concave up}$$

$$P'' < 0 \Rightarrow \quad \text{the graph of } P \text{ is concave down}$$

Note: P will have a point of inflection at $x = c$ only if P'' *changes sign* at $x = c$.

5. Look back at what you've done. Consider the degree of the polynomial and make sure that your graph makes sense.

A graphing calculator or computer can be useful for graphing polynomials. For instance, if zeros are hard to find a machine is useful in approximating them. It is up to you, however, to determine an appropriate viewing window.

◆ **EXAMPLE 11.8**   Graph[6] $P(x) = 100x^4 - 900x^2$.

SOLUTION

1. $P(x)$ is a fourth degree polynomial with a positive leading coefficient, so from a bird's-eye view the graph should look something like

At the extremities the graph reaches upward. The graph must have either one or three turning points.[7]

2. $P(x) = 100x^4 - 900x^2 = 100x^2(x^2 - 9) = 100x^2(x + 3)(x - 3)$

The x-intercepts are therefore at $x = 0$, $x = 3$, and $x = -3$. (Plot these zeros.)

The x-intercepts are symmetrically located, so we check for symmetry. All x's are raised to even powers, so $P(-x) = P(x)$. This means the graph has even symmetry; it is symmetric about the y-axis.

3. $P(x) = 100(x^4 - 9x^2)$, so $P'(x) = 100(4x^3 - 18x) = 200x(2x^2 - 9)$.

---

[6] Try to graph this on your graphing calculator. You'll probably find that it takes you some time to find a suitable viewing window.

[7] This follows from the previous statement. The derivative of a fourth degree polynomial is a cubic, which *can* have one, two, or three zeros, but the cubic will change sign either once, or three times.

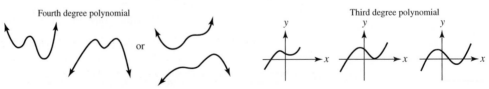

Fourth degree polynomial   or   Third degree polynomial

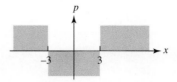

$P'(x) = 0$ at $x = 0$ and where $2x^2 - 9 = 0$, i.e., $x^2 = \frac{9}{2}$, or $x = \pm\frac{3}{\sqrt{2}}$.

graph of P						
sign of P'	$(-)$	$\frac{-3}{\sqrt{2}}$	$(+)$ 0 $(-)$	$\frac{3}{\sqrt{2}}$	$(+)$

There are local minimum points at $x = \frac{3}{\sqrt{2}}$ and at $x = -\frac{3}{\sqrt{2}}$ and a local maximum at $x = 0$.

$P(x) = 100x^2(x^2 - 9)$, so

$$P(0) = 0$$

$$P\left(\frac{3}{\sqrt{2}}\right) = 100\left(\frac{9}{2}\right)\left(\frac{9}{2} - 9\right)$$

$$= 450\left(-\frac{9}{2}\right)$$

$$= -2025.$$

Likewise,

$$P\left(-\frac{3}{\sqrt{2}}\right) = -2025.$$

4. $P''(x) = 100(12x^2 - 18) = 200(6x^2 - 9)$
   $P''(x) = 0$ provided that

$$6x^2 = 9$$

$$x^2 = \frac{9}{6}$$

$$x^2 = \frac{3}{2}$$

$$x = \pm\sqrt{\frac{3}{2}}.$$

graph of P	concave up		concave down		concave up
sign of P''	$(+)$	$-\sqrt{\frac{3}{2}}$	$(-)$	$\sqrt{\frac{3}{2}}$	$(+)$

First, using the information we have gathered, try to draw the graph yourself. Now use a graphing calculator or computer to check your answer. Notice that you have to be careful about your choice of range and domain, otherwise the machine will be of little help.

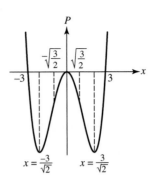

Figure 11.14   ◆

EXERCISE 11.8   Consider the polynomial $f(x) = 100x^4 - 900x^2 - 2$.

(a) How many zeros does $f(x)$ have? (Use the results of Example 11.8 to aid you.)

(b) Approximately where are the zeros? (Give your answers to within 0.01.)

(c) Show that 3 is not a zero of $f(x)$. Is the zero in the vicinity of 3 larger than 3 or smaller than 3? Explain your reasoning.

Answer

There are two zeros near $x = \pm 3$. In actuality, one is between 3.00031 and 3.00052, and the other is between -3.00031 and -3.00052.

EXERCISE 11.9   Graph $y = 100x^4 - 900x^2 + 800$. (*Hint:* Factor this first into two quadratics, then factor the quadratics themselves.)

Finding a Polynomial Function to Fit a Polynomial Graph

◆ EXAMPLE 11.9   Find a function to fit the cubic function $f(x)$ shown.

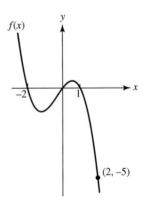

Figure 11.15

SOLUTION   We see that $f(x)$ has zeros at $x = -2$, $x = 0$, and $x = 1$, so $f(x)$ must have factors $(x + 2)$, x, and $(x - 1)$. Since it is a cubic, we can write

$$f(x) = k(x + 2)(x)(x - 1).$$

Multiplying by the constant k does not change the location of any of the x-intercepts (since it rescales vertically), but it does allow us to ensure that $f(x)$ passes through the point $(2, -5)$ as it does in the graph. Knowing that $x = 2$, $y = -5$ must satisfy the equation gives us

$$-5 = k(2 + 2)(2)(2 - 1)$$
$$-5 = 8k$$
$$k = -\frac{5}{8}.$$

The equation must be $f(x) = -\frac{5}{8}(x + 2)(x)(x - 1)$. You can verify this on your graphing calculator. ◆

◆ **EXAMPLE 11.10**   Find a possible equation for the polynomial function $g(x)$ shown.

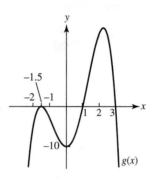

Figure 11.16

SOLUTION $g(x)$ has three zeros, at $x = -1.5$, $x = 1$, and $x = 3$. Notice that $g(x)$ does not change sign when it passes through $x = -1$; instead it "bounces off" the x-axis at this point. Therefore $x = -1$ must be a root of even multiplicity of $g(x) = 0$, as in Example 11.7 above. We write $g(x) = k(x + 1.5)^2(x - 1)(x - 3)$. A bird's-eye perspective confirms this is reasonable. $g(x)$ could be a fourth degree polynomial with a negative leading coefficient.

The graph passes through the point $(0, -10)$; we'll use that to find k.

$$g(x) = k\left(x + \frac{3}{2}\right)^2 (x - 1)(x - 3)$$

$$-10 = k\left(\frac{3}{2}\right)^2 (-1)(-3)$$

$$-10 = k \cdot \frac{27}{4}$$

$$-\frac{40}{27} = k$$

This gives us

$$g(x) = -\frac{40}{27}\left(x + \frac{3}{2}\right)^2 (x - 1)(x - 3).$$

You can verify that this is a reasonable fit for the graph by using a graphing calculator or computer. ◆

EXERCISE 11.10   In Example 11.10 we arrived at the function

$$g(x) = -\frac{40}{27}\left(x + \frac{3}{2}\right)^2 (x - 1)(x - 3).$$

Use your graphing calculator to observe the effect of the following variations.

$$h(x) = -\frac{40}{27}\left(x + \frac{3}{2}\right)^2 (x - 1)^3(x - 3)$$

$$j(x) = -\frac{160}{243}\left(x + \frac{3}{2}\right)^4 (x - 1)(x - 3).$$

Explain your observations by taking a numerical perspective.

EXERCISE 11.11   Find a possible equation for the cubic function $h(x)$ shown.

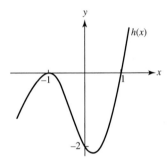

Figure 11.17

Basic Strategy for Fitting an Equation to a Polynomial Graph

■   Secure the x-intercepts: If they are at $x = a$, $x = b$, and $x = c$, start with $P(x) = k(x - a)(x - b)(x - c)$.

■   See whether or not the sign of y changes on either side of the x-intercept.

The factor $(x - a)$ changes sign as x increases from values less than a to those greater than a.

If the sign changes around $x = a$, then the factor $(x - a)$ must be raised to an odd power.

If the sign of y does not change around $x = a$, then the factor $(x - a)$ must be raised to some even power.[8]

■   If you know both the x- and y-coordinates of some point that is not a zero, use this information to find k. If this information is unavailable, check the sign somewhere to determine the sign of k.

---

[8] Generally it's wise to begin with the lowest power that has the appropriate effect. (Doing Exercise 11.10 will help you with this.) Below are figures illustrating the effect of $(x - a)$ versus $(x - a)^3$ in the neighborhood of $x = a$. Increasing the power of $(x - a)$ will flatten the graph in the immediate vicinity of $x = a$ and make it steeper far away from $x = a$.

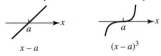

■   Take a bird's-eye view to make sure your answer is reasonable. Look at the degree and the leading coefficient.

REMARK This strategy is good for finding equations to fit graphs where nothing too exciting or extravagant is happening between the zeros. It won't help substantially in any of the cases sketched below.

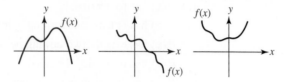

Figure 11.18

PROBLEMS FOR SECTION 11.3

1. Suppose you are given a polynomial expression in both factored and nonfactored form. When might you prefer one form over the other?

2. Suppose $P(x)$ is a polynomial whose derivative is $P'(x) = x^2(x + 2)^3$.
   (a) What degree is $P(x)$?
   (b) What are the critical points of $P(x)$?
   (c) Does $P(x)$ have an absolute minimum value? If so, where is it attained? Is it possible to find out what this minimum value is, if it exists? If yes, explain how; if no, explain why not.

3. Let $p(x)$ be a polynomial of degree n. What is the maximum number of points of inflection possible for the graph of $p(x)$?

4. A company is producing a single product. $P(x)$, the profit function, gives profit as a function of x, the number of hundreds of items produced. Suppose $P(0) = -200$ and $P'(x) = x^2(x - 1)^3$.
   Sketch the graph of P. Argue, using the sign of $P'(x)$, that the graph of P intersects the positive x-axis exactly once, i.e., for $x > 0$, that there is one and only one break-even point and that, if production levels are high enough, the profit will remain positive and increase with increasing x. The following questions will help guide you.
   (a) First, draw a rough sketch of the graph of $P'(x)$. (You need not determine precisely the position of the local minimum of P'; in other words, you need not take the second derivative—just use what you know about the intercepts and sign of $P'(x)$.)
   (b) Draw a number line and on it record the sign of P'. Above it indicate where P is increasing and decreasing.
   (c) Now, using the information that $P(0) = -200$ along with the information from part (b), make a rough sketch of P. You need not determine the positive x-intercept, just convincingly assert that it exists.

5. Suppose $P(x)$ with domain $(-\infty, \infty)$ is a polynomial of degree 4 whose leading coefficient is -3. For each statement given below, determine whether the statement is

■ necessarily true, or

■ possibly true, possibly false, or

■ definitely false.

Think carefully. This is a problem concerning both logic and polynomials.

(a) $\lim_{x \to \infty} P(x) = \infty$

(b) $\lim_{x \to -\infty} P(x) = -\infty$

(c) $P(x)$ has four zeros.

(d) $P(x)$ has at least one turning point.

(e) $P(x)$ has exactly two turning points.

(f) $P(x)$ has four critical points.

(g) $P(x)$ has an absolute maximum value.

(h) $P(x)$ has an absolute minimum value.

6. Suppose $P(x)$ is a polynomial of degree 7 whose leading coefficient is 2. For each statement given below, determine whether the statement is

■ necessarily true, or

■ possibly true, possibly false, or

■ definitely false.

If you decide the statement is *not* necessarily true, explain your reasoning!

(a) $P(x)$ has at least one zero and at most seven zeros.

(b) $P'(x)$ has no zeros.

(c) $P(x)$ has at least one point of inflection.

(d) $P(x)$ has five points of inflection.

7. Graph $f(x) = \frac{x^4}{4} + x^3 - 5x^2$. Indicate the x-coordinates of all local extrema and all points of inflection. What is the absolute minimum value of f? The absolute maximum value?

8. Find equations that could correspond to the graphs of the polynomials drawn below.

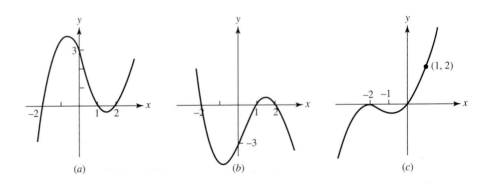

(a)   (b)   (c)

9. Find a possible equation to fit each polynomial graph below. Notice that a is a negative number.

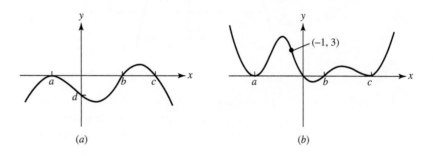

(a)   (b)

10. Find a polynomial to fit the graph below.

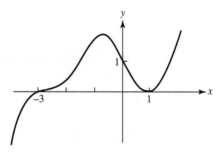

11. Each of the graphs on the following page is the graph of a polynomial $P(x)$. For each graph do the following.

(a) Determine whether the degree of $P(x)$ is even or odd.

(b) Despite the fact that you have just categorized each of the polynomials as being of either odd or even degree, *none* of the polynomials graphed are even functions and none are odd functions. Explain.

(c) Determine whether the leading coefficient is positive or negative.

(d) Determine a good lower bound for the degree of the polynomial. Explain your reasoning. (For example, the last graph on the right has one turning point, so it must be of degree 2 or more. It is not a parabola since it has a point of inflection; therefore we know the degree is higher than 2. It cannot be a polynomial of degree 3 because for $|x|$ large enough, $P(x)$ is positive. Therefore, it must be a polynomial of degree 4 or more.)

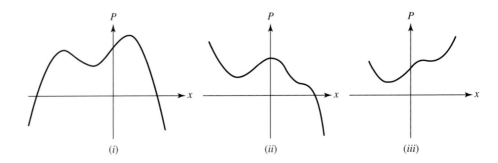

(i)   (ii)   (iii)

12. (a) Suppose $P(x)$ is a polynomial of degree 5. Which of the statements that follow must necessarily be true? If a statement is not necessarily true, provide a counterexample (an example for which the statement is false).

    i. $P(x)$ has at least one zero.

    ii. $P(x)$ has no more than four zeros.

    iii. The graph of $P(x)$ has at least one turning point.

    iv. The graph of $P(x)$ has at most four turning points.

(b) Suppose $P(x)$ is a polynomial of degree 5 with its natural domain $(-\infty, \infty)$. If $P'(\pi) = 0$ and $P''(\pi) = 5$, then which *one* of the following statements is true? Explain your answer.

    i. P has a local minimum at $x = \pi$ but this local minimum is not an absolute minimum.

    ii. P has a local minimum at $x = \pi$ and this local minimum *may* be an absolute minimum.

    iii. P has a local maximum at $x = \pi$ but this local maximum is not an absolute maximum.

    iv. P has a local maximum at $x = \pi$ and this local maximum *may* be an absolute maximum.

13. (a) Suppose $P(x)$ is a polynomial of degree 6. Which of the statements that follow must necessarily be true? If a statement is not necessarily true, provide a counterexample (an example for which the statement is false).

    i. $P(x)$ has at least one zero.

    ii. $P(x)$ has no more than five zeros.

    iii. The graph of $P(x)$ has at least one turning point.

    iv. The graph of $P(x)$ has at most five turning points.

(b) Suppose $P(x)$ is a polynomial of degree 6 with its natural domain $(-\infty, \infty)$. If $P'(2) = 0$ and $P''(2) = -1$, then which *one* of the following statements is true? Explain your answer.

    i. P has a local minimum at $x = 2$ but this local minimum is not an absolute minimum.

    ii. P has a local minimum at $x = 2$ and this local minimum *may* be an absolute minimum.

    iii. P has a local maximum at $x = 2$ but this local maximum is not an absolute maximum.

    iv. P has a local maximum at $x = 2$ and this local maximum *may* be an absolute maximum.

14. For each of the graphs below, all vertical and horizontal asymptotes are indicated with dotted lines. If there are no dotted lines there are no asymptotes.

   (a) Which of the following *could possibly* be the graph of a polynomial function? If the graph could be the graph of a polynomial, what can you say about the degree of the polynomial? Can you determine whether the degree is even or odd? Can you determine an n such that the degree of the polynomial is at least n?

   (b) Which *could possibly* be the graph of a function of the form $f(x) = Cb^x + D$, where C, b, and D are constants?

   (c) For each of the remaining graphs (graphs not listed as answers to the previous two questions), what characteristic of the graph made you rule it out?

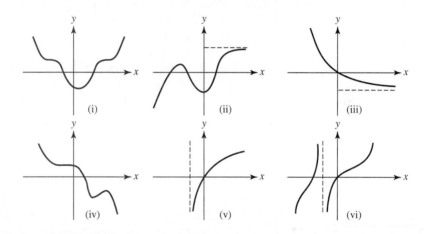

15. The functions that follow in this exercise are not polynomials. We ask you about their range, domain, and graphs with the goal of having you appreciate how nicely polynomial functions behave. For each of the following functions:

   (a) Determine the domain.

   (b) Determine the range.

   (c) Sketch a graph of the function. Do this using your knowledge of flipping, stretching, shrinking, shifting, and of graphing $\frac{1}{f(x)}$; check your graph with your graphing calculator.

   Your answers to parts (a) and (b) ought to agree with your answer to part (c). You can use your answers to parts (a) and (b) to select an appropriate viewing window in your calculator.

   i. $f(x) = \frac{5}{x+20}$ (The basic shape, before shifts and stretches, is $y = 1/x$.)

   ii. $g(x) = -2\sqrt{x} - 100$ (The basic shape, before shifts and stretches, is $y = \sqrt{x}$.

   iii. $h(x) = \frac{1}{\sqrt{x+40}}$ (Graph $y = \sqrt{x}$, shift, and then look at the reciprocal.)

   iv. $j(x) = \frac{2}{(x-20)(x+30)}$ (Graph $y = (x - 20)(x + 30)$, then look at the reciprocal.)

Exploratory Problems for Chapter 11

Functions and Their Graphs: Tinkering with Polynomials and Rational Functions

1. Find a polynomial function $P(x)$ that fits the graph drawn below. The x-intercepts should be at $x = -2$ and $x = 0$ and the function should have a global minimum of -6. It is not clear exactly where this minimum is attained.

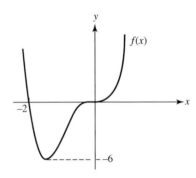

Figure 11.19

The goal of this problem is to encourage you to tinker with the equation using what you know about polynomials and their derivatives. The first few questions below are designed to steer you in the right direction.

(a) Is $P(x)$ a polynomial of even degree, or of odd degree?

(b) Give a lower bound for the degree of $P(x)$. Explain.

(c) What are the roots of $P(x)$? Take a first guess at the equation for $P(x)$ in factored form.

(d) What can you do (or what have you done) to make the graph of $P(x)$ flatten out at $x = 0$?

(e) Adjust your formula to assure that the minimum value of $P(x)$ will be -6.
    Do you want to stretch vertically or do you want to shift vertically? You don't want to uproot the x-intercepts you have so carefully nailed into place.

(f) Write a formula for $P(x)$.

(g) Given your answer to the last question, determine *where* $P(x)$ takes on its minimum value.
    Your answer should come from your function and be determined by analyzing its derivative; don't simply guess by reading off the graph.

The next set of problems asks you to think about rational functions, the topic of the next section of this chapter.

2. Graph each function f and under it, graph its reciprocal, $\frac{1}{f(x)}$. Then answer the following questions in as much generality as you can.

(a) How is the sign of $\frac{1}{f(x)}$ related to the sign of f?

(b) How is the magnitude (the absolute value) of $\frac{1}{f(x)}$ related to the magnitude of f?

(c) What characteristic(s) of f determines the location and type of vertical asymptote of $\frac{1}{f(x)}$?

i. $f(x) = x$ ii. $f(x) = x^2$

iii. $f(x) = x^2 - 1$ iv. $f(x) = x(x - 3)$

3. This problem should be done with the aid of a graphing calculator or computer. Give a very rough sketch of the graph of each of the following. If you like, a group can get together and split up the work, each person graphing a couple of these functions on his or her calculator. The important thing is for you to think about the relationships between the equations and their graphs.

(a) $y = \frac{1}{x-1}$ (b) $y = \frac{1}{(x-1)^2}$ (c) $y = \frac{1}{x(x-1)}$

(d) $y = \frac{1}{x^2(x-1)}$ (e) $y = \frac{1}{x(x-1)^2}$ (f) $y = \frac{1}{x^2(x-1)^2}$

(g) $y = \frac{x}{(x-1)}$ (h) $y = \frac{x^2}{(x-1)}$

Think about the relationships between these equations and their graphs, the effect of the factors in the denominators, and the effect of squaring certain factors. Present as many observations as you can come up with.

11.4   RATIONAL FUNCTIONS AND THEIR GRAPHS

An Introduction to Rational Functions

Rational functions are functions of the form $f(x) = \frac{\text{polynomial in } x}{\text{polynomial in } x}$. They are a class of functions that includes the polynomials[9] (a well-behaved family) as well as some more unruly relatives. Rational functions can exhibit much wilder and more varied behavior than polynomials. They may be undefined for certain values of x, and therefore may be discontinuous. Not only may they be discontinuous, but the magnitude of f may blow up around a point of discontinuity. In other words, it is possible that $\lim_{x \to c} |f(x)| = \infty$ for some finite number c. In this case, the rational function has a vertical asymptote. It is possible that $\lim_{x \to \infty} f(x) = k$ for some finite constant k, in which case $f(x)$ has a horizontal asymptote. Once you become accustomed to the behavior of rational functions and learn the relationship between the function and the behavior of its graph, you may very well find that rational functions are fun to work with. That alone could constitute a reason to get to know these functions. But there are more practical reasons as well.

An economist interested in the average cost per pound of producing q pounds of a good will divide the total cost function, $C(q)$, by the number of pounds produced. If $C(q)$ is modeled by a polynomial, then

$$\text{average cost per item} = \frac{C(q)}{q}$$

is a rational function.

Any time two variables are inversely proportional to one another, there is a functional relationship of the form $f(x) = k/x$, a form with which we have longstanding familiarity. Scientists observing naturally occurring phenomena have found such relationships ubiquitous. For instance, chemists use the combined gas laws relating the pressure, P, temperature, T, and volume, V, of a gas:

$$PV = kT \quad \text{or} \quad P = \frac{kT}{V}.$$

Physicists have found that the gravitational attraction between two objects is inversely proportional to the square of the distance between them. For example, a rocket on a journey in space will be subject to the gravitational force of the earth. The acceleration due to the gravitational attraction of the earth is given by

$$\frac{Gm_E}{r^2},$$

where G is the universal gravitational constant, m_E is the mass of the earth, and r is the distance from the rocket to the center of the earth. If the rocket is journeying from the earth to the moon, then the primary forces acting on it are the gravitational forces of the earth and the moon. The acceleration due to the gravitational attraction of the moon is given by

$$\frac{Gm_M}{R^2},$$

---

[9] If the polynomial in the denominator is a constant, then $f(x)$ is simply a polynomial.

where G is as above, m_M is the mass of the moon, and R is the distance from the rocket to the moon's center.

Let G be the acceleration of the rocket due to the combined gravitational forces of the earth and moon.

$$A = \left(\begin{array}{c} \text{acceleration due to} \\ \text{the earth's gravity} \end{array} \right) - \left(\begin{array}{c} \text{acceleration due to} \\ \text{the moon's gravity} \end{array} \right).$$

The two terms have opposite signs because, from the perspective of the rocket, the forces act in opposite directions. The distance between the center of the earth and the center of the moon is roughly 240,000 miles. We'll call this distance D. Then, when the rocket is a distance x from the center of the earth, its distance from the center of the moon is $D - x$.

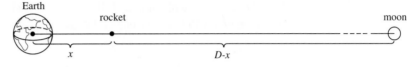

Figure 11.20

$$A(x) = \frac{Gm_E}{x^2} - \frac{Gm_M}{(D-x)^2}, \quad \text{or}$$

$$A(x) = G\left[\frac{m_E(D-x)^2 - m_M x^2}{x^2(D-x)^2} \right].$$

A is a rational function of the rocket's distance from the center of the earth.

Removable and Nonremovable Discontinuities and Asymptotes

The main ways in which rational functions can deviate from the behavior of polynomials are that they can have discontinuities (removable or nonremovable) and can have vertical and horizontal asymptotes.[10]

Points of Discontinuity

A rational function $f(x)$ will be undefined (and hence discontinuous) wherever the denominator is zero.

Removable Discontinuities: A bug's-eye view.   If the denominator and the numerator of a rational function are both zero at $x = b$, then the numerator and denominator have a common factor of $x - b$. If these factors occur with the same multiplicity,[11] then the graph has a pinhole at $x = b$. A pinhole, i.e., a situation in which $\lim_{x \to c^+} f(x) = \lim_{x \to c^-} f(x) = L$ where L is finite, but $f(c) \neq L$, is referred to as a removable discontinuity.

---

[10] The nonremovable discontinuities show up as vertical asymptotes.

[11] What is actually required is that the multiplicity in the numerator is greater than or equal to that in the denominator. For example, if $f(x) = \frac{(x^2+1)x^2}{x}$, then $f(x) = \begin{cases} (x^2+1)x & \text{for } x \neq 0 \\ \text{undefined} & \text{for } x = 0 \end{cases}$ and the graph of f has a pinhole at $x = 0$.

If $g(x) = \frac{(x^2+1)x}{x^2}$, then, $g(x) = \begin{cases} \frac{x^2+1}{x} & \text{for } x \neq 0 \\ \text{undefined} & \text{at } x = 0 \end{cases}$. The graph of $g(x)$ has a vertical asymptote at $x = 0$.

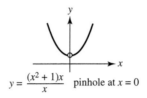

$$y = \frac{(x^2 + 1)x}{x} \quad \text{pinhole at } x = 0$$

Figure 11.21

Vertical Asymptotes: A bird's-eye view.   If the denominator of a rational function is zero at $x = b$ and the numerator is nonzero, then the graph has a vertical asymptote at $x = b$.[12] We need a bird's-eye view because as x approaches b, $f(x)$ will either increase without bound or decrease without bound. If f has a vertical asymptote at $x = b$, then near $x = b$ the graph of f will look like one of the graphs shown in Figure 11.22.

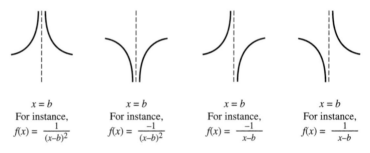

Figure 11.22

Simple sign information will distinguish between the four options. The base idea behind this is just what we discussed in Chapter 7 when looking at $\lim_{x \to 0^+} \frac{1}{x} = \infty$ and $\lim_{x \to 0^-} \frac{1}{x} = -\infty$. The graph will never cross its vertical asymptotes because the function is undefined there.

***Horizontal Asymptotes: *A bird's eye-view of rational functions.*[13]   A horizontal asymptote supplies information about f as the magnitude of x increases without bound; it indicates the behavior of the function toward the extremities of the graph in the case where these extremities look like horizontal lines. If $\lim_{x \to \infty} f(x) = K$ for some (finite) constant K, we say f has a horizontal asymptote at K, and similarly if $\lim_{x \to -\infty} f(x) = K$. Recall that polynomials never have horizontal asymptotes, and exponential functions have one-sided horizontal asymptotes. For any rational function f, *if* $\lim_{x \to \infty} f(x) = K$ where K is finite, then $\lim_{x \to -\infty} f(x) = K$ as well; the horizontal asymptotes are two-sided. In order to investigate whether or not the graph of a rational function $f(x)$ has a horizontal asymptote, we must look at the behavior of f as $x \to \infty$ and $x \to -\infty$. For x very large in magnitude, any polynomial is dominated by its term of highest degree; therefore, we will break down our investigations into cases in which we are concerned with the relative degrees of the

---

[12] What is actually required is that b is a zero of the denominator, and *if* it is also a zero of the numerator, the multiplicity of the root in the numerator is less than that in the denominator.

[13] A bird's-eye view would catch both horizontal and vertical asymptotes, but miss removable discontinuities.

numerator and the denominator of the rational function. We will use the fact that for any positive integer n and any constant k, $\lim_{x \to \infty} \frac{k}{x^n} = 0$.

Case I. Degree of Numerator Less Than Degree of Denominator

◆ **EXAMPLE 11.11** $f(x) = \frac{x-1}{2x^2-x-3}$. Calculate $\lim_{x \to \infty} \frac{x-1}{2x^2-x-3}$.

SOLUTION   Both the numerator and the denominator of this fraction are growing without bound, but the denominator grows much more rapidly than the numerator. (Try this out numerically on your calculator for very large x.) For x very large in magnitude the term of highest degree dominates any polynomial, so the numerator "looks like" x and the denominator like $2x^2$. Therefore, from a bird's-eye view, for x very large in magnitude $f(x)$ looks like $\frac{x}{2x^2} = \frac{1}{2x}$. $\lim_{x \to \infty} \frac{1}{2x} = 0$.

More formally, we can get a handle on this limit by dividing the numerator and denominator by the highest power of x occurring in the denominator:

$$\lim_{x \to \infty} \frac{x-1}{2x^2-x-3} = \lim_{x \to \infty} \frac{\frac{x}{x^2} - \frac{1}{x^2}}{\frac{2x^2}{x^2} - \frac{x}{x^2} - \frac{3}{x^2}} = \lim_{x \to \infty} \frac{\frac{1}{x} - \frac{1}{x^2}}{2 - \frac{1}{x} - \frac{3}{x^2}} = \frac{0}{2} = 0.$$

Similarly, we can show $\lim_{x \to -\infty} f(x) = 0$.   ◆

This argument can be generalized to show that if the degree of the numerator is less than the degree of the denominator, then the rational function has a horizontal asymptote at $y = 0$, the x-axis.

Case II. Degree of Numerator Equal to Degree of Denominator

◆ **EXAMPLE 11.12** $f(x) = \frac{x^2-1}{2x^2-x-3}$. Calculate $\lim_{x \to \infty} \frac{x^2-1}{2x^2-x-3}$.

SOLUTION   Both the numerator and the denominator of this fraction are growing without bound, but the denominator is growing about twice as rapidly as is the numerator. From a bird's-eye view, for x very large in magnitude the numerator "looks like" x^2 and the denominator like $2x^2$. Therefore, for x very large in magnitude $f(x)$ looks like $\frac{x^2}{2x^2} = \frac{1}{2}$. Again, you can test this out numerically on your calculator for very large x.

More formally, we can get a handle on this limit by dividing the numerator and denominator by the highest power of x occurring in the denominator:

$$\lim_{x \to \infty} \frac{x^2-1}{2x^2-x-3} = \lim_{x \to \infty} \frac{\frac{x^2}{x^2} - \frac{1}{x^2}}{\frac{2x^2}{x^2} - \frac{x}{x^2} - \frac{3}{x^2}} = \lim_{x \to \infty} \frac{1 - \frac{1}{x^2}}{2 - \frac{1}{x} - \frac{3}{x^2}} = \frac{1}{2}.$$

Similarly, we can show $\lim_{x \to -\infty} f(x) = \frac{1}{2}$.   ◆

This argument can be generalized to show that if the degree of the numerator is equal to the degree of the denominator, then the rational function has a horizontal asymptote at

$$y = \frac{\text{leading coefficient of numerator}}{\text{leading coefficient of denominator}}.$$

Case III. Degree of Numerator Greater Than Degree of Denominator

◆ **EXAMPLE 11.13** $f(x) = \frac{x^3-1}{2x^2-x-3}$. Calculate $\lim_{x \to \infty} \frac{x^3-1}{2x^2-x-3}$.

SOLUTION   Both the numerator and the denominator of this fraction are growing without bound, but the numerator is growing much more rapidly than the denominator. From a bird's-eye view, for x very large in magnitude the numerator looks like x^3 and the denominator like $2x^2$. Therefore, for x very large in magnitude $f(x)$ looks like $\frac{x^3}{2x^2} = \frac{x}{2}$. As $x \to \infty$, $\frac{x}{2} \to \infty$.

More formally, we can get a handle on this limit by dividing the numerator and denominator by the highest power of x occurring in the denominator:

$$\lim_{x \to \infty} \frac{x^3 - 1}{2x^2 - x - 3} = \lim_{x \to \infty} \frac{\frac{x^3}{x^2} - \frac{1}{x^2}}{\frac{2x^2}{x^2} - \frac{x}{x^2} - \frac{3}{x^2}} = \lim_{x \to \infty} \frac{x - \frac{1}{x^2}}{2 - \frac{1}{x} - \frac{3}{x^2}} = \lim_{x \to \infty} x = \infty.$$

We can show $\lim_{x \to -\infty} f(x) = \lim_{x \to -\infty} x = -\infty$.   ◆

This argument can be generalized to show that if the degree of the numerator is greater than the degree of the denominator, then the rational function has no horizontal asymptote.

Graphs from Equations/Equations from Graphs

A rational function may be discontinuous. At a point of discontinuity the function can change sign without passing through zero. If a function is discontinous its derivative will be discontinuous as well. (Recall that differentiability guarantees continuity, and therefore discontinuity guarantees a lack of differentiability.) Therefore the derivative can change sign without passing through zero and the function can change from increasing to decreasing or vice versa without having a horizontal tangent line.

Graphing a Rational Function $f(x) = \frac{\text{polynomial in } x}{\text{polynomial in } x}$

Simplify, if possible, and look for pinholes.   Factor the numerator and the denominator; some information is easiest to get when the expression is factored. If there is a common factor in the numerator and denominator, cancel with care. If the common factor is $(x - c)$, then the function is *undefined* at c, so its graph has a pinhole or a vertical asymptote at c.

$f(x) = \frac{(x+1)(x+2)}{(x+1)(x+3)} = \frac{(x+2)}{(x+3)}$ for $x \neq -1$. $f(x)$ is undefined when $x = -1$.

The graph has a pinhole at $x = -1$.

$f(x) = \frac{(x+1)}{(x+1)^2} = \frac{1}{x+1}$ for $x \neq -1$. $f(x)$ is undefined when $x = -1$.

The graph has a vertical asymptote at $x = -1$.

$f(x) = \frac{(x^2+1)(x+2)}{(x^2+1)(x+3)} = \frac{(x+2)}{(x+3)}$ The graph has no pinholes $(x^2 + 1) \neq 0$.

1. Identify all vertical asymptotes. Where is the denominator of the simplified expression zero?

$f(x)$ will be undefined at a vertical asymptote and nearby $|f(x)| \to \infty$.

The graph will look like one of the following.

Figure 11.23

2. **Find x-intercepts.** Where is the numerator of the simplified expression zero? (For a fraction to be zero, its numerator must be zero.)

3. **Positive/negative.** Figure out the sign of $f(x)$. The sign of *any* function can change only across a zero or a point of discontinuity. Draw a "sign of $f(x)$" number line. The sign can change only on either side of an x-intercept or vertical asymptote.

   Observe the following. If the factor $(x - a)$ occurs in either numerator of f with odd multiplicity, then f changes sign around $x = a$. If it occurs with even multiplicity, then f does not change sign around $x = a$. This is consistent with the behavior we observed in polynomials.

4. **Find the y-intercept.** Set $x = 0$. (This merely gives a point of reference.)

5. **Look for horizontal asymptotes.** A horizontal asymptote includes the behavior of $f(x)$ as $x \to \pm\infty$. A horizontal asymptote gives the behavior of $f(x)$ *only for x of very large magnitude*; therefore, it *can* be crossed. Find $\lim_{x\to\infty} f(x)$. (For rational functions if $\lim_{x\to\infty} f(x)$ is finite it will be equal to $\lim_{x\to-\infty} f(x)$.)

   degree top > degree bottom $\Rightarrow$ no horizontal asymptote $|f(x)| \to \infty$

   degree top < degree bottom $\Rightarrow$ horizontal asymptote at $y = 0$ (the x-axis)

   degree top = degree bottom $\Rightarrow$ horizontal asymptote at $y =$ fraction formed by leading coefficients

6. **Consider symmetry.** If the asymptotes and intercepts don't have the required symmetry, then the function is neither even nor odd. If they do, then consider whether the function has even symmetry, odd symmetry, or neither. You can look at the numerator and denominator separately first.

$$\text{even} = \frac{\text{even}}{\text{even}} \quad \text{or} \quad \frac{\text{odd}}{\text{odd}}; \quad \text{odd} = \frac{\text{even}}{\text{odd}} \quad \text{or} \quad \frac{\text{odd}}{\text{even}}$$

7. If you are interested in local extrema, then compute the derivative.

   The graphs of some rational functions can be built up from the sum or difference of other familiar rational functions. For example, $h(x) = \frac{x^2+1}{x} = x + \frac{1}{x}$. Refer to Figure 11.24 to see how the graph of $\frac{x^2+1}{x}$ can be built by summing the y-values of the graphs of x and $1/x$.

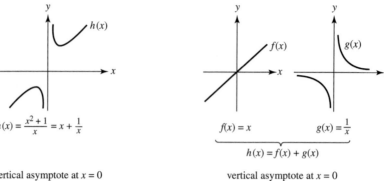

vertical asymptote at $x = 0$ vertical asymptote at $x = 0$

Figure 11.24

EXERCISE 11.12   Differentiate $h(x)$ in the example above to find the coordinates of the local extrema.

Answer

local minimum at $x = 1$, local maximum at $x = -1$.

Below are worked examples; rational functions are presented and graphed.

◆ **EXAMPLE 11.14**

(a) $f(x) = \frac{4x(x+2)}{x-3} = \frac{4x^2+8x}{x-3}$

x-intercepts: $x = 0$ and $x = -2$ **y-intercept:** $y = 0$

vertical asymptotes: $x = 3$ **symmetry?** None.

sign of y $\underset{-2}{-}$ $\underset{0}{+}$ $\underset{3}{-}$ $+$

horizontal asymptote?

$f(x) = \frac{4x^2+8x}{x-3}$, so as $x \to \pm\infty$, this looks like $\frac{4x^2}{x} = 4x$. No horizontal asymptote.

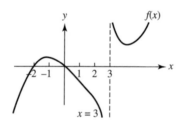

Figure 11.25

(b) $f(x) = \frac{x^2}{2x^2-8} = \frac{x^2}{2(x^2-4)} = \frac{x^2}{2(x+2)(x-2)}$

x-intercepts: $x = 0$ **y-intercept:** $y = 0$

vertical asymptotes: $x = 2, x = -2$ **symmetry?** Even, because $f(-x) = f(x)$.

Alternatively, $\frac{\text{even}}{\text{even}} = $ even.

sign of y $\begin{array}{ccccc} + & & - & & - & & + \\ \hline & -2 & & 0 & & 2 & \end{array}$

horizontal asymptote?

$y = \frac{x^2}{2x^2-8}$, so as $x \to \pm\infty$, this looks like $y = \frac{x^2}{2x^2} = \frac{1}{2}$. Horizontal asymptote at $y = \frac{1}{2}$.

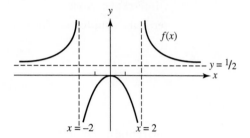

$x = -2$ $x = 2$

Figure 11.26

(c) $f(x) = \frac{(x+3)(x+1)^2}{x^3-3x^2+2x} = \frac{(x+3)(x+1)^2}{x(x-2)(x-1)}$

x-intercepts: $x = -3, x = -1$ **y-intercept:** None. (y-axis is asymptote)

vertical asymptotes: $x = 0, x = 2, x = 1$ **symmetry?** None.

sign of y $\begin{array}{ccccccc} (+) & & (-) & & (-) & (+) & (-) & & (+) \\ \hline & -3 & & -1 & & 0 & 1 & 2 & \end{array}$

horizontal asymptote?

$f(x) = \frac{(x-3)(x+1)^2}{x^3-3x^2+2x}$, so as $x \to \pm\infty$, this looks like $\frac{x^3}{x^3} = 1$. Horizontal asymptote at $y = 1$.

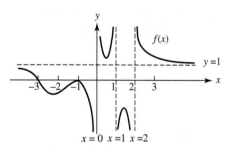

$x = 0$ $x = 1$ $x = 2$

Figure 11.27

Question. Does the graph cross the line $y = 1$?

$\frac{(x+3)(x+1)^2}{x^3-3x^2+2x} = 1$ or, multiplying out, $\frac{x^3+5x^2+7x+3}{x^3-3x^2+2x} = 1$.

$x^3 + 5x^2 + 7x + 3 = x^3 - 3x^2 + 2x$,   so $5x^2 + 7x + 3 = -3x^2 + 2x$,

or $8x^2 + 5x + 3 = 0$.

The discriminant is negative, $b^2 - 4ac = 25 - (4)(3)(8) < 0$, so there are no solutions.

The graph does not cross the line $y = 1$.

(d) $f(x) = \frac{x-2}{(x^2+1)(x-2)}$.

$$f(x) = \begin{cases} \frac{1}{x^2+1} & \text{for } x \neq 2, \\ \text{undefined} & \text{for } x = 2 \end{cases}$$

There is a pinhole at $x = 2$. There is just a pinhole because $\lim_{x \to 2^+} \frac{1}{x^2+1} = \lim_{x \to 2^-} \frac{1}{x^2+1} = \frac{1}{5}$. Both the left- and right-hand limits are equal and finite.

x-intercepts: None   **y-intercept:** 1

vertical asymptotes: None   **symmetry?** If the point of discontinuity at $x = 2$ were removed, then f would be even.

sign of y _____(+)_____

horizontal asymptote? As $x \to \pm\infty$, $f(x) = \frac{1}{x^2+1} \to 0$. Horizontal asymptote at $y = 0$.
$\frac{1}{x^2+1}$ is biggest when $x^2 + 1$ is smallest; the maximum value of $\frac{1}{x^2+1}$ is 1 and is attained at $x = 0$.

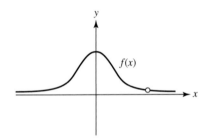

Figure 11.28

REMARK If you merely follow a recipe for doing mathematics problems you will not always do them in the most efficient way possible. Example (d) is most easily done as follows: $f(x) = \frac{x-2}{(x^2+1)(x-2)} = \frac{1}{x^2+1}$ for $x \neq 2$, and is undefined at $x = 2$. We can graph $\frac{1}{x^2+1}$ by beginning with the graph of $g(x) = x^2 + 1$ and then graphing its reciprocal, $\frac{1}{g(x)}$.

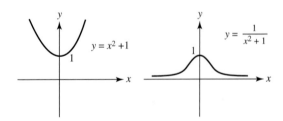

Figure 11.29

We insert a pinhole at $x = 2$ because $f(x)$ is undefined at $x = 2$.

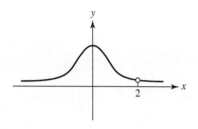

Figure 11.30   ◆

Fitting a Function to a Graph

We will use the graph below as reference.

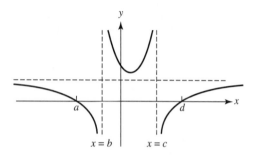

Figure 11.31

1. **Find all x-intercepts** $x = a$, $x = d$, ... and construct the factors $(x - a)$, $(x - d)$, ... for the numerator. Include a constant factor for stretching/flipping. Begin with $y = k\frac{(x-a)(x-d)}{?}$.

2. **Find all vertical asymptotes** $x = b$, $x = c$, ... and construct factors $(x - b)$, $(x - c)$, ... for the denominator.

$$y = k\frac{(x - a)(x - d)}{(x - b)(x - c)}$$

3. **Do preliminary sign analysis.** If the sign of y changes across an x-intercept or vertical asymptote at $x = q$, the factor $(x - q)$ should be raised to an *odd* power; if the sign does not change, the factor must be raised to an *even* power.

4. **Look at horizontal asymptotes.**

   If there is no horizontal asymptote, be sure that (degree of the numerator) > (degree of the denominator).

   If the x-axis is a horizontal asymptote, be sure that (degree of the numerator) < (degree of the denominator).

   For any other horizontal asymptote, the degree of the numerator and denominator must be equal; adjusting k adjusts the height of the horizontal asymptote.

Notice that introducing a factor like $(x^2 + 1)$ in the numerator or denominator changes its degree without introducing new zeros or vertical asymptotes because $(x^2 + 1)$ has no zeros.

EXERCISE 11.13   Find functions that fit the graphs below.

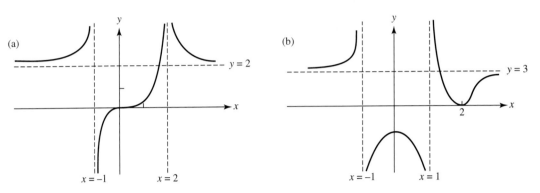

Figure 11.32

Possible answers

(a) $f(x) = \dfrac{2x^3}{(x+1)(x-2)^2}$ (b) $f(x) = \dfrac{3(x-2)^2}{(x-1)(x+1)}$

◆ EXAMPLE 11.15   What are the absolute maximum and minimum values taken on by

$$f(x) = \frac{x-2}{x^2+5}?$$

SOLUTION   First notice that $f(x)$ has no vertical asymptotes, as $x^2 + 5 > 0$. Observe that $\lim_{x\to\infty} f(x) = 0$, so we do expect to find absolute extrema for this function.

The extrema must occur at critical points of f. We find f' using the Quotient Rule.

$$f'(x) = \frac{(x^2+5)\cdot 1 - (x-2)\cdot 2x}{(x^2+5)^2}$$

$$= \frac{x^2+5-2x^2+4x}{(x^2+5)^2}$$

$$= \frac{-x^2+4x+5}{(x^2+5)^2}$$

$f'(x)$ is always defined; we look for x such that $f'(x) = 0$.

$$\frac{-x^2+4x+5}{(x^2+5)^2} = 0$$

$$x^2 - 4x - 5 = 0$$

$$(x-5)(x+1) = 0$$

$$x = 5 \quad \text{or} \quad x = -1$$

Critical points: $x = -1$ and $x = 5$.

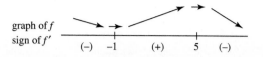

Figure 11.33

$$f'(x) - \frac{-(x-5)(x+1)}{(x^2+5)^2}$$

Minimum at $x = -1$. $f(-1) = \frac{-1-2}{(-1)^2+5} = \frac{-3}{6} = -\frac{1}{2}$. This is the absolute minimum of f.

Maximum at $x = 5$. $f(5) = \frac{5-2}{25+5} = \frac{3}{30} = \frac{1}{10}$. This is the absolute maximum value of f.   ◆

PROBLEMS FOR SECTION 11.4

1. Find an equation to fit the important features of the graph. For the purposes of this problem, the important features are the horizontal and vertical asymptotes, x-intercepts, and the sign of y.

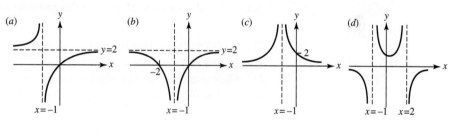

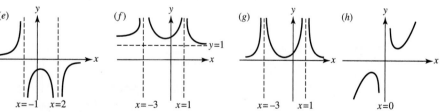

2. Suppose a distance function is given by $d(t) = 1/t$ for $0.5 \le t \le 20$.

   (a) What is the average velocity over the interval from $t = 1$ to $t = 5$?

   (b) Is there a time at which the instantaneous velocity is the same as the average velocity over the interval from $t = 1$ to $t = 5$? If so, find that time.

   (c) On the same set of axes, illustrate your answers to parts (a) and (b).

3. Find possible equations for the following graphs:

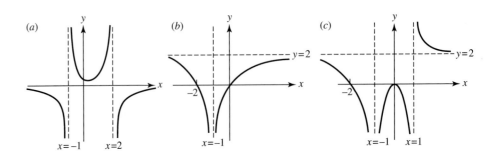

4. Let $f(x) = \frac{x^2+1}{x^2} = 1 + \frac{1}{x^2}$
   (a) Graph f.
   (b) Find the following.

   i. $\lim_{x \to \infty} f(x)$ ii. $\lim_{x \to 0} f(x)$ iii. $\lim_{x \to \infty} f'(x)$
   iv. $\lim_{x \to 0^+} f'(x)$ v. $\lim_{x \to 0^-} f'(x)$

   (c) Graph $f'(x)$.
   (d) Find $f''(x)$.
   (e) Are your answers to all parts of this problem consistent? (If not, find your errors.)

5. Graph $f(x) = \frac{x^3+x^2}{x^2-4}$. This function has four local extrema. One you can locate exactly. Where is it? Approximate the other three using your graphing calculator. (Notice that depending upon the viewing window you choose it may be very difficult to *realize* this function has four local extrema! When you use the calculator, use it carefully.)

6. Find a function whose graph matches the one drawn below.

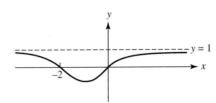

7. Graph the following, clearly labeling all x- and y-intercepts, vertical asymptotes, and horizontal asymptotes.
   (a) $y = \frac{x^2-4}{x^2-3x-4}$ (b) $y = \frac{3x^2}{(x-1)^2}$ (c) $y = \frac{(x-1)(x-2)}{x(x-1)(x-3)(x+1)}$

8. Find possible equations to match with the following graphs.

(a)

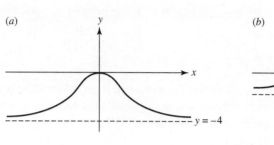

(b)

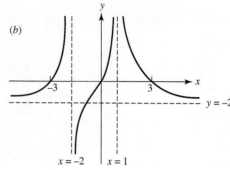

9. At one point in Leo Tolstoy's novella *The Death of Ivan Ilyich,* the title character states that the amount of blackness (the opposite of goodness in this context) in his life is in "inverse ratio to the square of the distance from death." Let $B(t)$ represent the amount of blackness in his life, where t measures the amount of time since his birth, and let $t = D$ represent the time of his death.

   (a) Write an equation for $B(t)$. (Your answer should include the constant D.)

   (b) Your equation for $B(t)$ should have an arbitrary constant in it. Can you determine the sign of this constant?

   (c) Sketch a graph of $B(t)$. Is it increasing or decreasing? Concave up or concave down? Label any t- or B-intercepts and any asymptotes.

10. What characteristics might the graph of a rational function (a polynomial divided by a polynomial) have that the graph of a polynomial will not have?

11. Graph $f(x) = \frac{4}{x} + x$.

   (a) Find $f'(x)$. Make a number line, marking all points at which f' is zero or undefined. Use the number line to indicate the sign of f'; above this indicate where the graph of f is increasing and where it is decreasing.

      Note: $x = 0$ is not a critical point, since f is undefined at $x = 0$. However, it is possible for the sign of f' to change on either side of a point at which f' is undefined, so $x = 0$ must be labeled on your number line.

   (b) Find $f''(x)$. Make a number line, marking all points at which f'' is zero and undefined. Use the number line to indicate the sign of f''; above this indicate where the graph of f is concave up and where it is concave down.

   (c) Graph $f(x)$. Label both the x- and y-coordinates of the local maxima and local minima.

   (d) Does $f(x)$ have an absolute maximum value? If so, what is it? Does $f(x)$ have an absolute minimum value? If so, what is it?

12. Graph each of the following equations without using calculus. Label the following.

   (a) The x-intercepts; the y-intercepts

   (b) The vertical asymptotes

   (c) The horizontal asymptotes

An analysis of where y is positive and where it is negative must be included. You need not find the coordinates of the local extrema. You need not look at y'.

i. $y = \frac{x}{(x-1)(x+1)}$ ii. $y = \frac{x^2(x-2)}{(x-1)(x+1)}$ iii. $y = \frac{x^2(x-2)}{(x-1)(x+1)}$

iv. $y = \frac{x^2(x-2)}{(x-1)^2(x+1)}$ v. $y = \frac{x(x-2)}{(x-1)^2(x+1)}$ vi. $y = \frac{(x-3)(x-2)}{(x-1)(x+1)}$

vii. $y = \frac{2}{x^2+1}$ viii. $y = \frac{-x^2}{x^2+1}$

13. Graph the following functions using the information provided by the derivatives for guidance. Indicate where the function is increasing, where it is decreasing, and the coordinates of all local extrema.

(a) $f(x) = x + \frac{1}{x}$

(b) $g(x) = x - \frac{1}{x}$

14. Suppose that $f(x)$ is a rational function with zeros at $x = 0$ and $x = 4$, vertical asymptotes at $x = -2$ and $x = 3$, and a horizontal asymptote at $y = 5$.
For each of the following functions, indicate the location of any

(a) zeros.   (b) vertical asymptotes.   (c) horizontal asymptotes.

If there is not enough information to answer part of any question, say so.

i. $g(x) = f(x - 3)$ ii. $h(x) = f(x) - 3$

iii. $j(x) = 2f(3x)$ iv. $k(x) = f(x^2)$

Inverse Functions: A Case Study of Exponential and Logarithmic Functions

C H A P T E R

12

Inverse Functions: Can What Is Done Be Undone?

What's done cannot be undone.
—*MacBeth*, Act V, scene 1
Now mark me, how I will undo myself.
—*Richard II*, Act IV, scene 1

12.1 WHAT DOES IT MEAN FOR F AND G TO BE INVERSE FUNCTIONS?

Some actions can be undone; other actions, once taken, can never be undone. If we think of a function as an action on an input variable to produce an output, we can make a similar observation. Let's begin by looking at functions whose actions can be undone. If a function acts on its input by adding 3, the action can be undone by subtracting 3. If a function acts by doubling its input, the action can be undone by halving. If f is a function whose action can be undone, we refer to the function that undoes the action of f as its *inverse function* and denote it by f^{-1}. We read f^{-1} as "f inverse."

CAUTION This notation, while quite standard, can be *very* misleading. "f inverse" is *not* $\frac{1}{f(x)}$. The *reciprocal* of $f(x)$ is written $\frac{1}{f(x)}$ or $[f(x)]^{-1}$.

◆ **EXAMPLE 12.1**   Below are some simple functions together with their inverse functions.

Function	Action done	Action undone	Inverse function
i. $f(x) = x + 3$	Add 3 to input	Subtract 3 from input	$f^{-1}(x) = x - 3$
ii. $g(x) = 2x$	Double input	Halve input	$g^{-1}(x) = x/2$
iii. $h(x) = x^3$	Cube the input	Take the cube root	$h^{-1}(x) = x^{1/3}$

Notice again that the inverse of f is *not* the reciprocal[1] of f.   ◆

What exactly do we mean when we say "the inverse function of f undoes the action of f"? If f assigns to the input "a" the output "b," then its inverse, f^{-1}, assigns to the input "b" the output "a." That is, if $f(a) = b$, then $f^{-1}(b) = a$. Equivalently, for every point (a, b) on the graph of f, the point (b, a) lies on the graph of f^{-1}.

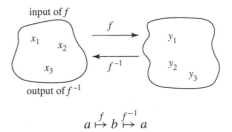

$$a \overset{f}{\mapsto} b \overset{f^{-1}}{\mapsto} a$$

If f and f^{-1} are inverse functions, f^{-1} undoes the action of f and *vice versa*, so

$$a \overset{f}{\mapsto} b \overset{f^{-1}}{\mapsto} a \quad \text{and} \quad b \overset{f^{-1}}{\mapsto} a \overset{f}{\mapsto} b$$

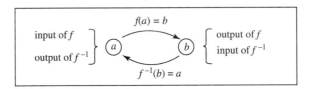

Figure 12.1

Look at Example 12.1(i), where $f(x) = x + 3$ and $f^{-1}(x) = x - 3$.

$$4 \overset{f}{\mapsto} 7 \overset{f^{-1}}{\mapsto} 4 \quad \text{and} \quad 7 \overset{f^{-1}}{\mapsto} 4 \overset{f}{\mapsto} 7$$

$$5 \overset{f^{-1}}{\mapsto} 2 \overset{f}{\mapsto} 5 \quad \text{and} \quad 2 \overset{f}{\mapsto} 5 \overset{f^{-1}}{\mapsto} 2$$

More generally, imagine sending x down an assembly line. The first machine on the line, let's call it f, acts on x. Its output, $f(x)$, is passed along to the second machine, f^{-1}, which undoes the work of the first. Therefore, the final output is x.

$$x \overset{f}{\mapsto} f(x) \overset{f^{-1}}{\mapsto} x \quad \text{and similarly} \quad x \overset{f^{-1}}{\mapsto} f^{-1}(x) \overset{f}{\mapsto} x.$$

---

[1] Do not confuse inverse *functions* with the statement "A is inversely proportional to B, $A = \frac{k}{B}$."

We can express this more succinctly using the composition of functions.[2]

$$f^{-1}(f(x)) = x \quad \text{and} \quad f(f^{-1}(x)) = x$$

Definition

The functions f and g are **inverse functions** if, for all x in the domain of f, $g(f(x)) = x$ and for all x in the domain of g, $f(g(x)) = x$.

EXERCISE 12.1   Verify that the pairs of functions in Example 12.1 actually *are* inverse functions. We'll do parts (i) and (ii) below. Part (iii) is left as an exercise.

i. Verify that $f(x) = x + 3$ and $f^{-1}(x) = x - 3$ are inverse functions.

$$x \overset{f}{\mapsto} (x+3) \overset{f^{-1}}{\mapsto} (x+3) - 3 = x; \quad f^{-1}(f(x)) = f^{-1}(x+3) = (x+3) - 3 = x.$$

$$x \overset{f^{-1}}{\mapsto} (x-3) \overset{f}{\mapsto} (x-3) + 3 = x; \quad f(f^{-1}(x)) = f(x-3) = (x-3) + 3 = x.$$

ii. Verify that $g(x) = 2x$ and $g^{-1}(x) = \frac{x}{2}$ are inverse functions.

$$x \overset{g}{\mapsto} 2x \overset{g^{-1}}{\mapsto} \frac{2x}{2} = x; \quad g^{-1}(g(x)) = g^{-1}(2x) = \frac{2x}{2} = x.$$

$$x \overset{g^{-1}}{\mapsto} \frac{x}{2} \overset{g}{\mapsto} 2\left(\frac{x}{2}\right) = x; \quad g(g^{-1}(x)) = g\left(\frac{x}{2}\right) = 2\left(\frac{x}{2}\right) = x.$$

Definition

A function is said to be **invertible** if it has an inverse function.

Let's review the characteristic of a function that makes it invertible, able to be undone. A function is an input/output relationship such that each input corresponds to a single output. To find the inverse of a function requires that we are able to begin with an output and trace it back to the corresponding input. Therefore no two inputs may share the same output. In other words, the function must be 1-to-1; there must be a 1-to-1 correspondence between inputs and outputs.

To illustrate this we'll reconstruct the soda machine example from Chapter 1. Both machine A and machine B can be modeled by functions, but only machine A is 1-to-1.

---

[2] Recall that $f(g(x))$ means find $g(x)$ and use $g(x)$ as the input of f. In other words, do f to $g(x)$. Therefore $f(f^{-1}(x))$ means do f^{-1} first and apply f to the output of f^{-1}. For a review of composition of functions, refer to Chapter 3.

Machine A		Machine B	
Button #	Output	Button #	Output
1	Coke	1	Coke
2	Diet Coke	2	Coke
3	Sprite	3	Diet Coke
4	Orange Crush	4	Diet Coke
5	Ginger Ale	5	Coke
		6	Coke

Soda machine B has six input buttons, but can give outputs of only Coke and Diet Coke; the function that models this machine is not 1-to-1. Knowing that the machine gave an output of Coke does not allow us to determine precisely which button was pressed. On the other hand, soda machine A, with five selection buttons, each one corresponding to a different type of soda, is modeled by a 1-to-1 function. To be invertible a function must be 1-to-1 because the inverse of f must map each output of f to the unique corresponding input; conversely, if f is 1-to-1, then f is invertible.

Let f be a function modeling machine A. f and its inverse function f^{-1} are shown below.

$$
\begin{array}{ccccc}
1 & \xrightarrow{f} & \text{Coke} & \xrightarrow{f^{-1}} & 1 \\
2 & \longmapsto & \text{Diet Coke} & \longmapsto & 2 \\
3 & \longmapsto & \text{Sprite} & \longmapsto & 3 \\
4 & \longmapsto & \text{Orange Crush} & \longmapsto & 4 \\
5 & \longmapsto & \text{Ginger Ale} & \longmapsto & 5
\end{array}
$$

Do we care whether a given function is invertible? It depends on the situation. For the soda machines above, it's probably not very important to be able to tell what button was pressed based on the type of soda that came out of the machine. It's merely important that the relationship between button and soda output be a function so that the input uniquely determines the output.

On the other hand, recall the bottle calibration problem from the beginning of the course. The entire exercise of calibrating a bottle is worthless unless this calibrated bottle can be used for measuring. We want to be able to pour water into the bottle and determine its volume based on the height of the water. It's important that the calibration function is invertible. The calibration function, C, takes volume as input and assigns height as output. For instance, one liter of water might fill the bottle to a height of 10 cm, in which case $C(1) = 10$. Once calibrated, the bottle can be used as a measuring device precisely because the calibration function is 1-to-1. C^{-1} turns the calibration procedure around; the height becomes the input and the volume the output. Because the function is 1-to-1 we know that a height of 10 cm corresponds to a volume of 1 liter, $C^{-1}(10) = 1$.

$$volume \xrightarrow{C} height \qquad height \xrightarrow{C^{-1}} volume$$

◆ **EXAMPLE 12.2**   Let's consider a particular bottle and its calibration function C. From the information about C given on the left in the following table we can construct a corresponding table for C^{-1}.

v (in liters)	$C(v) = h$ (in cm)		h (in cm)	$C^{-1}(h) = v$ (in liters)
0.25	4	$\Rightarrow$ corresponding table for C^{-1}	4	0.25
0.5	7		7	0.5
0.75	9		9	0.75
1	10		10	1

The graph of C is given in Figure 12.2(a) below. Since the function C^{-1} reverses the input and output of C, we can graph C^{-1} by reversing the coordinates of the points on the graph of C.

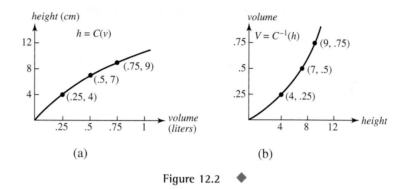

Figure 12.2 ◆

The Relationship Between the Graph of a Function and the Graph of Its Inverse

We have established that a function is invertible if and only if the function is 1-to-1.[3] How is this criterion reflected in the graph of $y = f(x)$? If f is invertible, then each y-value in the range must correspond to exactly one x-value in the domain.

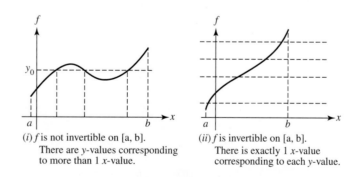

(i) f is not invertible on [a, b].
There are y-values corresponding
to more than 1 x-value.

(ii) f is invertible on [a, b].
There is exactly 1 x-value
corresponding to each y-value.

Figure 12.3

This gives us a graphical criterion for determining if a function is invertible.

---

[3] Recall that A if and only if B means that A and B are equivalent statements. If a function is invertible, then it is 1-to-1; if a function is 1-to-1, then it is invertible.

The Horizontal Line Test. A function f is invertible if and only if every horizontal line intersecting the graph of f intersects it in exactly one point. The horizontal line test is the reflection of the vertical line test about the line $y = x$. The vertical line test checks that there is at most one y for every x; that is, there is at most one output for each input. The horizontal line test checks that there is at most one x for every y; that is, there is at most one input for each output. Together, they check that the relationship is 1-to-1.

Consequence: If a *continuous* function f is invertible, then f must be either always increasing or always decreasing. A function with a turning point is not invertible; it cannot pass the horizontal line test. We can ascertain this information by calculating f'. If f is continuous on an open interval and f' is either always positive or always negative, then f has no turning points and therefore passes the horizontal line test and is invertible; if f is continuous and f' changes signs, then f has a turning point and therefore fails the horizontal line test and is therefore not invertible.[4]

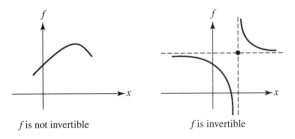

f is not invertible f is invertible

Figure 12.4

As illustrated in Example 12.2, the graph of f^{-1} is obtained by interchanging the coordinates of the points on the graph of f. If (a, b) lies on f, then (b, a) lies on the graph of f^{-1}. This is equivalent to reflecting the graph of $y = f(x)$ over the line $y = x$ as shown in the examples below. The graphs in Figure 12.5 correspond to the functions and inverse functions given in Example 12.1.

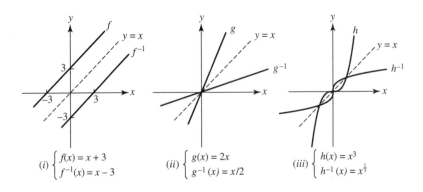

$(i) \begin{cases} f(x) = x + 3 \\ f^{-1}(x) = x - 3 \end{cases}$ $(ii) \begin{cases} g(x) = 2x \\ g^{-1}(x) = x/2 \end{cases}$ $(iii) \begin{cases} h(x) = x^3 \\ h^{-1}(x) = x^{\frac{1}{3}} \end{cases}$

Figure 12.5

---

[4] Note that it is not sufficient to check only if f' ever equals 0. For example, $f(x) = x^3$ is invertible since it has no turning points, although $f'(0) = 0$.

1. (a) Let S be the function that assigns to each living person a social security number. Is S 1-to-1? Is it invertible?

   (b) Let C be the counting function that allows a collection of 30 people to be put in six groups of five people each by "counting off" 1 to 6. Is C 1-to-1? Is it invertible?

   (c) Let A be the altitude function that assigns to each point in the White Mountains its altitude. Is A 1-to-1?

2. The identity function I is the function whose input equals its output: $I(x) = x$. If functions f and g have the property that $f(g(x)) = I(x)$ and $g(f(x)) = I(x)$, then f and g are inverse functions. For each function below, find the inverse function $g(x)$ and verify that $f(g(x)) = I(x)$ and $g(f(x)) = I(x)$.

   (a) $f(x) = 6x - 3$ (b) $f(x) = (x - 3)^3$

3. On the same set of axes, sketch the graphs of the following pairs of functions. In parts (a) and (b) find an expression for $f^{-1}(x)$. The graphs of f and $f^{-1}(x)$ are mirror images over the line $y = x$ since the roles of input and output are switched to obtain the inverse function. In other words, if $(1, 5)$ is a point on the graph of f, then $(5, 1)$ is a point on the graph of $f^{-1}(x)$.

   (a) $f(x) = 2x + 1$ and $f^{-1}(x)$ (b) $f(x) = x^2 - 2, x > 0$ and $f^{-1}(x)$

   (c) $f(x) = 10^x$ and $f^{-1}(x)$ (d) $f(x) = 2^{-x}$ and $f^{-1}(x)$

4. Suppose $f(v)$ is a calibration function for a bucket. f takes volumes (in liters) as inputs and gives heights (in inches) as outputs. Suppose $f(1) = 4$.

   (a) What is $f^{-1}(4)$?

   (b) What is the meaning of $f^{-1}(4)$ in physical terms?

   (c) Is $f^{-1}(4)$ greater than $f^{-1}(1)$? Explain in terms of the physical situation.

5. Which of the following functions are invertible on the domain given? Explain.

   (a) $P(w)$ is the price of mailing a package weighing w ounces; $w \in (0, 50]$.

   (b) $T(t)$ is the temperature at the top of the Prudential Center in Boston at time t, t measured in days, where $t = 0$ is February 1, 1998; $t \in [0, 365]$.

   (c) $C(w)$ is the cost of w pounds of ground coffee at a particular shop where coffee is sold by weight at a fixed price per pound; $w \in [0, 2]$.

   (d) $M(t)$ is the mileage on a car t days after it was purchased; $t \in [0, 365]$.

6. Let $f(x) = x^3 + 3x^2 + 6x + 12$.

   (a) Make a convincing argument that $f(x)$ is invertible. (It is not adequate to say it looks 1-to-1 on a calculator. How can you be absolutely sure it is 1-to-1 on $(-\infty, \infty)$?)

   (b) Find three points that lie on the graph of $f^{-1}(x)$. (Approximations are not adequate.) Explain your reasoning.

7. For each of the functions graphed below, determine whether or not the function is invertible. If it is not, restrict the domain to make it invertible. Then sketch f^{-1}, labeling any asymptotes and labeling two points on the graph of f^{-1}.

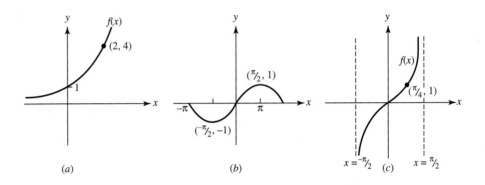

(a)   (b)   (c)

8. The graphs of $f(t)$, $g(t)$, and $h(t)$ are given below.

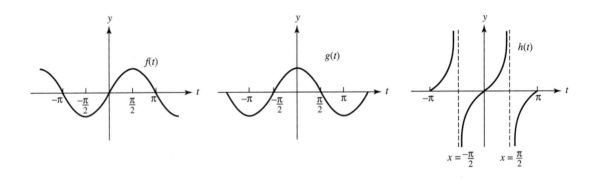

True or False?

(a) The function $f(t)$, restricted to the domain $[0, \pi]$, is invertible.

(b) The function $g(t)$, restricted to the domain $[0, \pi]$, is invertible.

(c) The function $h(t)$, restricted to the domain $(-\frac{\pi}{2}, \frac{\pi}{2})$, is invertible.

(d) The function $f(t)$, restricted to the domain $[-\frac{\pi}{2}, \frac{\pi}{2}]$, is invertible.

9. Which of the following functions are invertible?

(a) The function that assigns to each current senator the state he or she represents.

(b) The function $T(t)$ that gives the temperature in Moab at time t.

(c) The function $C(d)$, whose domain is the set of all performances of Broadway's *A Chorus Line*, and whose output is the cumulative number of people who have seen this show on Broadway.

(d) The function $L(d)$, whose domain is the set of all performances of Broadway's *The Lion King*, and whose output is the number of people seeing this Broadway show on the designated date.

■ 12.2 FINDING THE INVERSE OF A FUNCTION

If an invertible function is given by a table of values, the inverse function is constructed by interchanging the input and output columns. If an invertible function is presented graphically, the graph of f^{-1} is obtained by reflecting the graph of f over the line $y = x$. Suppose a function is given analytically. The subject of this section is how to arrive at an expression for its inverse function.

◆ **EXAMPLE 12.3** Suppose f is the function that doubles its input and then adds 3: $f(x) = 2x + 3$. What is its inverse function?

SOLUTION To undo the function f, do we subtract 3 and then divide by 2, or do we first divide by 2 and then subtract 3; is $f^{-1}(x) = \frac{x-3}{2}$ or is $f^{-1}(x) = \frac{x}{2} - 3$? We could try each out, since we know that $f^{-1}(f(x))$ should be x, or we can think about it in the following way.

Analogy. When getting dressed, you first put on your socks and then put on your shoes. To undo the process, you must first remove your shoes, then your socks. The last thing you did is the first thing you undo.

Accordingly, to undo $f(x)$, we first subtract 3 and then divide the result by 2: $f^{-1}(x) = \frac{x-3}{2}$.

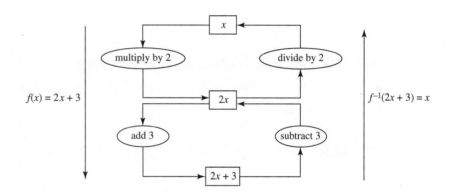

Check:

$$f^{-1}(f(x)) = f^{-1}(2x + 3) = \frac{(2x+3)-3}{2} = \frac{2x}{2} = x.$$

$$f(f^{-1}(x)) = f\left(\frac{x-3}{2}\right) = 2\left(\frac{x-3}{2}\right) + 3 = (x - 3) + 3 = x. \quad ◆$$

The three functions from Example 12.1 and the function from the example above had simple enough formulas that we could guess how to "undo" them in order to find formulas for their inverses. As this is not always the case, we need an analytic method for finding a formula for the inverse function.

◆ **EXAMPLE 12.4** Let $f(x) = \frac{x+1}{x-2}$. Find a formula for $f^{-1}(x)$.

SOLUTION In this example it is not very easy to figure out how to undo the action of f, so we'll use a different approach: f^{-1} reverses the output and input of f. Let's denote the output of f by y and the input by x. f is given by $y = \frac{x+1}{x-2}$. We interchange the input and output (x and y) to obtain f^{-1}.

$$x = \frac{y+1}{y-2},$$

where x is the input of f^{-1} and y is the output of f^{-1}. We solve for y to get the output of f^{-1} in terms of the input.

$$x = \frac{y+1}{y-2} \qquad \text{Get } y \text{ out of the denominator}$$
by multiplying both sides by $(y-2)$.

$$x(y-2) = y+1 \qquad \text{Multiply out.}$$

$$xy - 2x = y+1 \qquad \text{Gather the } y\text{'s together.}$$

$$xy - y = 2x+1 \qquad \text{Factor out } y.$$

$$y(x-1) = 2x+1 \qquad \text{Solve for } y.$$

$$y = \frac{2x+1}{x-1}$$

So $f^{-1}(x) = \frac{2x+1}{x-1}$. ◆

To summarize, we find a formula for f^{-1} by exploiting the fact that the inverse of f reverses the roles of input and output.

■   Write $y = f(x)$, and then interchange the variables x and y.

■   Solve for y in terms of x. We'll obtain $y = f^{-1}(x)$.

◆ **EXAMPLE 12.5**   Let $f(x) = 4x^3 + 2$. Find $f^{-1}(x)$ if f is invertible.

SOLUTION   We know that f is invertible because $f'(x) = 12x^2$ is positive for all $x \neq 0$. This indicates that f is a cubic polynomial with no turning points, so it's 1-to-1. Set $y = 4x^3 + 2$ and then interchange the roles of x and y to find the inverse relationship.

$$x = 4y^3 + 2$$

$$4y^3 = x - 2$$

$$y^3 = \frac{x-2}{4}$$

$$y = \left(\frac{x-2}{4}\right)^{1/3}$$

$$f^{-1}(x) = \left(\frac{x-2}{4}\right)^{1/3}$$

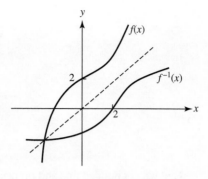

Figure 12.6

This makes sense: The function f cubes its input, multiplies the result by 4, and then adds 2. To undo this, we must subtract 2, divide the result by 4, and then take the cube root. See the diagram below for clarification.

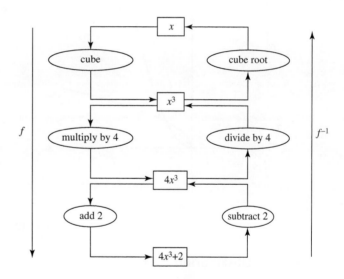

◆ **EXAMPLE 12.6**   Let $g(x) = x^2$. Find $g^{-1}(x)$ if g is invertible.

At first glance, it may seem that if g is the squaring function, its inverse must be the square root function. But we must be careful. g is not invertible because it is not 1-to-1. The problem is that given any positive output, say 4, it is impossible to determine uniquely the corresponding input. The input corresponding to 4 could be 2 or -2.

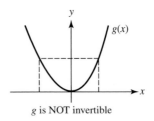

g is NOT invertible

Figure 12.7

Had we not noticed this and proceeded to look for the inverse analytically, we would soon realize that there was a problem.

Set $y = x^2$ and then interchange the roles of x and y to obtain an inverse relationship.

$$x = y^2 \qquad \text{Solve for } y.$$

$$y = \pm\sqrt{x} \qquad y \text{ is not a function of } x.$$

What we can do is restrict the domain of g to make g 1-to-1. If we restrict the domain to $[0, \infty)$, then $g^{-1}(x) = \sqrt{x}$. If we restrict the domain to $(-\infty, 0]$, then $g^{-1}(x) = -\sqrt{x}$.

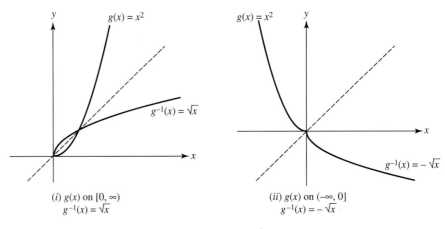

(i) $g(x)$ on $[0, \infty)$
$g^{-1}(x) = \sqrt{x}$

(ii) $g(x)$ on $(-\infty, 0]$
$g^{-1}(x) = -\sqrt{x}$

Figure 12.8 ◆

Observation

If a function is not 1-to-1 on its natural domain it is possible to restrict the domain in order to make the function invertible. Note that the domain of f is the range of f^{-1} and the range of f is the domain of f^{-1}.

PROBLEMS FOR SECTION 12.2

1. For each of the functions f, g, and h below, do the following.

   (a) Sketch the function and determine whether it is invertible.

   (b) If the function is invertible, sketch the inverse function on the same set of axes as the function and find a formula for the inverse function.

(c) Identify the domain and range of the function and its inverse.

    i. $f(x) = \sqrt{x-1}$ ii. $g(x) = \frac{1}{x}$ iii. $h(x) = \frac{x^3}{3} + 1$

2. For each of the functions below, find $f^{-1}(x)$.

    (a) $f(x) = 2 - \dfrac{x+1}{x}$ (b) $f(x) = \frac{x^5}{10} + 7$

3. Suppose f is an invertible function.

    (a) If f is increasing, is f^{-1} increasing, decreasing, or is there not enough information to determine?

    (b) If f is decreasing, is f^{-1} increasing, decreasing, or is there not enough information to determine?

    (c) Suppose f is increasing and concave up. Is f^{-1} concave up or concave down? (*Hint:* Let $y = f(x)$. What happens to the ratio $\frac{\Delta y}{\Delta x}$ as x increases? How does this translate into information about the inverse function? Check your conclusion with a concrete example.) We will be able to work this out analytically by Chapter 16.

4. Let

$$f(x) = \frac{2x-1}{3x+4}.$$

    Find $f^{-1}(x)$.

5. The function f is increasing and concave up on $(-\infty, \infty)$. $f'(x)$ is never zero. Denote by $g(x)$ the inverse of f.

    (a) What is the sign of g'?

    (b) What is the sign of g''?

    (c) If $f(3) = 5$ and $f'(3) = 10$, what is $g'(5)$?

The functions in Problems 6 through 10 are 1-to-1. Find $f^{-1}(x)$ and specify the domain of f^{-1}.

6. $f(x) = \frac{x}{x+3}$

7. $f(x) = \frac{2}{3-x}$

8. $f(x) = \sqrt{x+3}$

9. $f(x) = 2\sqrt{x-6}$

10. $f(x) = x^3 + 1$

For Problems 11 through 16, use the first derivative to determine whether the function given is 1-to-1. If it is, find its inverse function.

11. $f(x) = x^3 + 2x - 3$

12. $f(x) = x^3 - 2x + 3$

13. $f(x) = |x - 3|$

14. $f(x) = x^2 + 2x - 1$

15. $f(x) = 3 \cdot 2^x$

16. $f(x) = 5 \cdot 3^{-x}$

12.3   INTERPRETING THE MEANING OF INVERSE FUNCTIONS

In order to make sense of the information given by an inverse function in some real-world context it is helpful to clarify in words the input and output of both the function and its inverse. We illustrate this in the example below.

◆ **EXAMPLE 12.7**   Let $P(t)$ be the amount of money in a bank account at time t, where t is measured in years and $t = 0$ represents January 1, 1998. Suppose that P_0 dollars are originally deposited in the account.

Interpret the following expressions in words.

(a) $P(2)$ (b) $P^{-1}(5000)$ (c) $P^{-1}(2P_0)$

(d) $P^{-1}(2P_0 + 10)$ (e) $P^{-1}(2P_0) + 10$

SOLUTION   Begin by firmly establishing the input and output of P and P^{-1}.

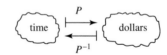

$P(\text{time}) = \text{dollars}$, so $P^{-1}(\text{dollars}) = \text{time}$

(a) $P(2)$ is the number of *dollars* in the account at $t = 2$ (January 1, 2000).

(b) $P^{-1}(5000)$ is the *time* when the balance will be $5000.

(c) $P^{-1}(2P_0)$ is the *time* when the balance will be twice P_0, i.e., when the balance will have doubled from its original amount.

(d) $P^{-1}(2P_0 + 10)$ is the *time* when the balance will be $10 more than twice the original balance.

(e) $P^{-1}(2P_0) + 10$ is the *time* ten years after the balance is twice P_0, i.e., ten years after the balance has doubled.   ◆

PROBLEMS FOR SECTION 12.3

1. Let $C(q)$ be the cost (in dollars) of producing q items. Translate the following equations into words.

    (a) $C(300) = 800$

    (b) $C^{-1}(1000) = 500$

    (c) $C'(200) = 1.5$

2. Apricots are sold by weight. In other words, the price is proportional to the weight. Let $C(w)$ be the cost of w pounds of apricots. Suppose that A pounds of apricots cost \$3.

    (a) Describe in words the practical meaning of each of the following and then evaluate the expression. (When evaluating, use the fact that price is proportional to weight. Your answers should be either a number or an expression in terms of A.)

        i. $C(3A)$

        ii. $C^{-1}(6)$

        iii. $C^{-1}(1)$

    (b) In this particular situation, which of the following statements are true?

        i. $C(3A) = 3C(A)$

        ii. $C^{-1}(2x) = 2C^{-1}(x)$

        iii. $C^{-1}(\frac{x}{2}) = \frac{C^{-1}(x)}{2}$

        iv. $C^{-1}(x + x) = C^{-1}(2x)$

    (c) Only *one* of the statements above is true for any invertible function C. Which statement is this?

3. Let $C(q)$ be the cost of producing q items. Suppose that right now A items have been produced at a cost of \$$B$. Interpret the following expressions in words. "A" and "B" should not appear in your answers; use words instead.

    (a) $C(400)$

    (b) $C^{-1}(3000)$

    (c) $C^{-1}(B + 100)$

    (d) $C(A + 10)$

    (e) $C^{-1}(2B)$

4. A typist's daily wages are determined by the number of words per minute he averages on his shift. Let $D(w)$ be his daily earnings (in dollars) as a function of w, the average number of words per minute he types. Suppose that yesterday he was paid \$$B$ for averaging C words per minute.

    Interpret each of the following equations or expressions in words. Your answer should be expressed in terms of pay and words per minute.

    (a) $D^{-1}(70) = 50$

    (b) $D(C + 5) = 1.1B$

    (c) $D^{-1}(B + 10)$

5. Let $R(d)$ be a function that models a company's annual revenue (the amount of money they receive from customers) in dollars as a function of the number of dollars they spend that year on advertising. Suppose that last year they spent \$$B$ on advertising and took in a total revenue of \$$C$.

    Interpret each of the following equations or expressions. Your answers should not contain \$$C$ or \$$B$, but words instead.

(a) $R(B/2) = C - 80{,}000$

(b) $R'(30{,}000) = 2.8$

(c) $R^{-1}(2C)$

6. Let $f(t) = 5(1.1)^{6t+2} + 1$.

   (a) The point $(\frac{1}{3}, 6)$ lies on the graph of f. What is $f^{-1}(6)$?

   (b) Find a formula for $f^{-1}(t)$.

   (c) Use your formula to find $f^{-1}(6)$. Does your answer agree with your answer to part (a)?

   (d) If $f(t)$ models the number of pounds of garbage in a garbage dump t days after the dump has officially opened, interpret $f^{-1}(30)$ in words.

7. A ball is thrown straight up into the air. t seconds after it is released, its height is given by $H(t) = -16t^2 + 96t$ feet.

   (a) Sketch a graph of $H(t)$.

   (b) What is the domain of $H(t)$? The range?

   (c) What is the ball's maximum height? When does it attain this height?

   (d) Sketch the inverse relation for $H(t)$. Is it a function? Explain.

   (e) How can you restrict the domain of $H(t)$ so that it will have an inverse?

   (f) Having restricted the domain so that $H(t)$ is invertible, evaluate $H^{-1}(80)$. What is its practical meaning?

Exploratory Problems for Chapter 12

Thinking About the Derivatives of Inverse Functions

1. Let $f(x) = x^2$ with the domain of f restricted to $x \geq 0$. Then $f^{-1}(x) = \sqrt{x}$.

   (a) Compare the derivative of x^2 at $(3, 9)$ with the derivative of $\sqrt{x}$ at $(9, 3)$. Accompany your answer by a sketch.

   (b) Compute the derivative of x^2 at (a, b).

   (c) Compute the derivative of $\sqrt{x}$ at (b, a).

   (d) Compare your answers to the previous two questions by either expressing both in terms of a or expressing both in terms of b. How are they related?

2. The function e^x is invertible. Denote its inverse function by $g(x)$.

   (a) On the same set of axes, graph e^x and its inverse function $g(x)$. What is the domain of g? The range of g?

   (b) The derivative of e^x at $(0, 1)$ is 1. What do you think the derivative of the inverse function is at $(1, 0)$?

   (c) What can you say about the sign of g' and of g''?

CHAPTER 13

Logarithmic Functions

Introductory Example

◆ **EXAMPLE 13.1**   A large lake has been serving as a reservoir for its nearby towns. Over the years, industries on the shore of the lake have contributed to the pollution of the lake. An awareness of the problem has caused community members to ban further pollution. Due to a combination of runoff and natural processes, the amount of pollutants in the lake is expected to decrease at a rate proportional to pollution levels. The number of grams of pollutant in the lake is now 1200; t years from now the number of grams is expected to be given by $1200(10)^{-t/8}$. If the water is deemed safe to drink only when the pollutant level has dropped to 400 grams, for how many years will the towns need to find an alternative source of drinking water?

SOLUTION   We must find t such that $400 = 1200(10)^{-t/8}$. This is equivalent to solving $\frac{400}{1200} = (10)^{-t/8}$, or $\frac{1}{3} = (10)^{-t/8}$. We can approximate the solution using a graphing calculator. One approach is to look for the root of $\frac{1}{3} - (10)^{-t/8}$. Another is to look for the point of intersection of $y = 1200(10)^{-t/8}$ and $y = 400$.

But suppose we would like an exact answer; we want to solve the equation analytically for t. We could simplify somewhat by converting $\frac{1}{3} = (10)^{-t/8}$ to $\frac{1}{3} = (10^t)^{-1/8}$ and raising both sides of this equation to the (-8) to get $\left(\frac{1}{3}\right)^{-8} = 10^t$. We know $\left(\frac{1}{3}\right)^{-8} = \left((3)^{-1}\right)^{-8} = (3)^8 = 6561$, so we must solve the equation

$$10^t = 6561.$$

t is the number we must raise 10 to in order to get 6561. Since $10^3 = 1000$ and $10^4 = 10,000$, we can be sure that t is a number between 3 and 4. Again we could revert to our calculator to get better and better estimates of the value of t. However, if we can find the inverse of the

439

function 10^t, then we can find the exact solution to the equation $10^t = 6561$. If $f(t) = 10^t$, then $t = f^{-1}(6561)$.

The Inverse of $f(x) = 10^x$

We know the function $f(x) = 10^x$ is invertible because it is 1-to-1. The inverse function, f^{-1}, is obtained by interchanging the input and output of f; the graph of f^{-1} can be drawn by reflecting the graph of 10^x over the line $y = x$.

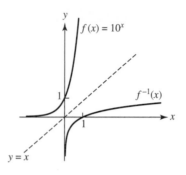

Figure 13.1

Suppose we go about looking for a formula for f^{-1} in the usual way, by interchanging the roles of x and y and solving for y. We write $x = 10^y$. What is y? y is the number we must raise 10 to in order to get x. We don't have an algebraic formula for this, but this function is quite useful, so we give it a name.

Definition

$\log_{10} x$ is the number we must raise 10 to in order to obtain x. $\log_{10} x$ is often written $\log x$. We read $\log_{10} x$ as "log base 10 of x."

$$y = \log_{10} x \text{ is equivalent to } 10^y = x.$$

The domain of f is $(-\infty, \infty)$ and the range is $(0, \infty)$. Therefore the domain of $\log x$ is $(0, \infty)$ and its range is $(-\infty, \infty)$. Note then that $\log x$ is defined only for $x > 0$. By examining the graph in Figure 13.1, we see that the graph of $\log x$ is increasing and concave down; it is increasing without bound, but it is increasing *very* slowly.

$$\lim_{x \to 0^+} \log_{10} x = -\infty \qquad \lim_{x \to \infty} \log_{10} x = +\infty$$

Note that although we now have a nice compact way of expressing the inverse function of 10^x, it might seem that all we have really accomplished so far is the introduction of a shorthand for writing "the number we must raise 10 to in order to obtain x." But there is a definite perk. A calculator will give a numerical estimate of the logarithm up to 10 digits with just a couple of key strokes.[1] In fact, many logarithms you'll work with are irrational, so a calculator will give an approximation of the value—but a nice, quick, accurate approximation it is!

Let's return to solving the equation $10^t = 6561$. Then $t = \log_{10} 6561$. This is the analytic solution of the equation. A calculator tells us that $\log_{10} 6561 \approx 3.817$, so people can start drinking the water from that polluted lake about 3.817 years from now.[2] ◆

Let's make sure the definition of $\log_{10} x$ is clear by looking at some examples.

◆ EXAMPLE 13.2

(a) $\log_{10} 100$ is the exponent to which we must raise 10 in order to get 100.
    $10^2 = 100$, so $\log_{10} 100 = 2$.

(b) $\log_{10} 0.1$ is the exponent to which we must raise 10 in order to get 0.1.
    $10^{-1} = 0.1$, so $\log_{10} = -1$.

(c) $\log_{10} 1$ is the exponent to which we must raise 10 in order to get 1.
    $10^0 = 1$, so $\log_{10} 1 = 0$.

(d) $\log_{10} 151$ is the exponent to which we must raise 10 in order to get 151.
    $10^2 = 100$ and $10^3 = 1000$, so we know that $\log_{10} 151$ is between 2 and 3. $\log_{10} 151$ is irrational; so we can't express it exactly using a decimal.

(e) $\log_{10} 0$ is the exponent to which we must raise 10 in order to get 0. There is no such number! $10^{\text{any number}}$ is always positive. Zero is not in the domain of the function $\log_{10} x$ because 0 is not in the range of 10^x, the inverse of $\log_{10} x$. See how your calculator responds when you enter log 0. (It should complain.) ◆

The Logarithm Defined

Let $f(x) = b^x$, where b is a positive number. Since f is a 1-to-1 function, it has an inverse function, f^{-1}. From the graph of $f(x) = b^x$ we can obtain the graph of its inverse function. The domain of f^{-1} is $(0, \infty)$ and the range is $(-\infty, \infty)$.

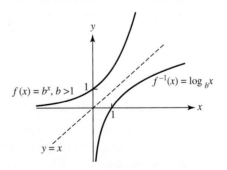

Figure 13.2

---

[1] Before there were calculators that could provide this information tables of logarithms were painstakingly constructed and used for reference.

[2] Keep in mind that 3.817 is an approximate solution to $10^t = 6561$. Try it out; $10^{3.817} = 6561.45 \ldots$. Even 3.816970038 is only an approximate solution. The exact solution is $\log_{10} 6561$.

We look for a formula for the inverse function by interchanging the roles of x and y and solving for y. We write $x = b^y$. What is y? y is the number we must raise b to in order to get x. As in the case of $f(x) = 10^x$ above, we haven't arrived at an algebraic formula for f^{-1}, but this function whose output is y is so useful that it is given its own name.

Definition

$\log_b x$ is the number we must raise b to in order to get x. We read $\log_b x$ as "log base b of x."

By definition, the following two statements are equivalent.[3]

$$\log_b x = y \qquad \Leftrightarrow \qquad b^y = x$$
logarithmic form   is equivalent to   exponential form

Test your understanding of this definition by working through the following exercises.

EXERCISE 13.1   If the statement is given in logarithmic form, write it in exponential form; if it is written in exponential form, write it in logarithmic form.

(a) $\log_3 81 = 4$ (b) $-0.5 = \log_2 \left(\dfrac{1}{\sqrt{2}} \right)$ (c) $A = W^b$

EXERCISE 13.2   Estimate the following logarithms by finding two consecutive integers, one smaller than the logarithm and the other larger.

(a) $\log_6 40$ (b) $\log_{10} 3789$ (c) $\log_2 40$

Answers to Exercises 1 and 2 are supplied at the end of the section.

EXERCISE 13.3   We know that $\log_{10} 10 = 1$ and $\log_{10} 100 = 2$. Therefore, $1 < \log_{10} 55 < 2$. However, although 55 is exactly halfway between 10 and 100, $\log_{10} 55$ is not midway between 1 and 2, since log is not a linear function. Is $\log_{10} 55$ greater than 1.5 or less than 1.5? Argue geometrically. Check your answer using your calculator.

Logarithms and Your Calculator.   A calculator will generally give numerical approximations of logs to two different bases, base 10 and base e.

$\log_{10} x$ is called the **common log of x** and is denoted by **log x**.

$\log_e x$ is called the **natural log of x** and is denoted by **ln x**.[4]

Recall that when we were studying the derivative of $f(x) = b^x$ we found that the derivative of b^x is kb^x, where k is the slope of the tangent to b^x at $x = a$. We searched for a base b such that the derivative of b^x is simply b^x. Such a base lies between the values of 2.71 and 2.72. We defined e to be this base. Just as the derivative of e^x is charming in its simplicity, so too will we find charm in the elegant simplicity of the derivative of ln x. It's something

---

[3] Recall that $\Leftrightarrow$ is a symbol that means "if and only if" or "is equivalent to." The symbol $A \Leftrightarrow B$ means that the statements A and B carry the same information (in different forms).

[4] Yes, the e is invisible in this notation. Soon you will get accustomed to this. In the meantime, if you get confused by the notation ln x you can write it as $\log_e x$ to clarify the meaning.

to look forward to in the next chapter! (Or try to discover it on your own!) Notice that on your calculator the pair of inverse functions 10^x and $\log x$ share the same key. Similarly, the pair of inverse functions e^x and $\ln x$ share the same key.[5]

$$y = \ln x \text{ is equivalent to } x = e^y.$$

Because $\ln x$ and e^x are inverse functions,

$$e^{\ln \star} = \star \text{ for all positive } \star, \text{ and } \ln e^\star = \star.$$

Answers to Selected Exercises

Answers to Exercise 13.1

1. (a) $3^4 = 81$ (b) $2^{-0.5} = \frac{1}{\sqrt{2}}$ (c) $\log_w A = b$

2. (a) $y = \log_6 40 \Leftrightarrow 6^y = 40$.
    Since $6^2 = 36$ and $6^3 = 216$, we know that $2 < \log_6 40 < 3$.

    (b) $y = \log_{10} 3789 \Leftrightarrow 10^y = 3789$.
    Since $10^3 = 1000$ and $10^4 = 10,000$, we know that $3 < \log_{10} 3789 < 4$.

    (c) $y = \log_2 40 \Leftrightarrow 2^y = 40$.
    Since $2^5 = 32$ and $2^6 = 64$, we know that $5 < \log_2 40 < 6$.

PROBLEMS FOR SECTION 13.1

1. Sketch the graphs of $f(x) = 2^x$ and the graph of $\log_2 x$ on the same set of axes. Label three points on each graph.

2. Fill in the blanks: When we write $\log_2 3$, we say "log base two of 3." We mean the power to which 2 must be raised in order to get 3.

    (a) When we write $\log_5 14$, we say "_____." We mean the power to which _____ must be raised in order to get _____.

    (b) When we say "log base 4 of 8," we write _____. We mean the power to which _____ must be raised in order to get _____.

    (c) We mean the power to which e must be raised in order to get 5, so we write _____ and we say _____.

    In Problems 3 through 5, approximate the values of the logarithms by giving two consecutive integers, one of which is a lower bound and the other an upper bound for the expressions given. Do this without a calculator. (You can use the calculator to check your answers, but the idea of the problem is to get you to think about what logarithms mean.) Explain your reasoning as in the example below.

    Example: $\log_{10} 113$ is between 2 and 3.

    Reasoning: $\log_{10} 113$ is the number we must raise 10 to in order to get 113.

---

[5] Notice too that the inverse functions x^2 and $\sqrt{x}$ share the same key. We will see that the trigonometric functions and their inverse functions also share a key. Your calculator's organization reflects the inverse function relation.

3. (a) $\log_7 50$

    (b) $\log_{10}(0.5)$

4. (a) $\log_{10}(0.05)$

    (b) $\log_3 29$

5. (a) $\log_2 \sqrt{30}$

    (b) $\log_5 \sqrt{30}$

    (c) $\log_{10} \sqrt{30}$

6. Simplify the following. (No calculators, except to check your answers if you like.)

   (a) $\log_2 \sqrt{8}$ (b) $\log_{10} 0.001$ (c) $\log_2 \left(\frac{4}{\sqrt{8}} \right)$ (d) $\log_3(1/9)$

   (e) $\log_k k^{3x}$ (f) $\log_k 1$ (g) $\log_k (k^x k^y)$

13.2   THE PROPERTIES OF LOGARITHMS

The properties of logarithms can be derived from the properties of exponentials and the inverse relation between logarithms and exponentials.

   We know that if f and g are inverse functions, then $f(g(x)) = x$ and $g(f(x)) = x$. By definition, the functions $\log_b x$ and b^x are inverse functions. Therefore we have the following inverse function identities:

$$\log_b b^\star = \star \qquad \text{and} \qquad b^{\log_b \star} = \star,$$

where $\star$ is any expression in the former identity, and any positive expression in the latter.

   We can also work our way through these by thinking about the meanings of the expressions.

$\log_b x$ is the number we must raise b to in order to get x.

Therefore,

$\log_b b^\star$ is the number we must raise b to in order to get $b^\star$. $\log_b b^\star = \star$.

On the other hand,

$b^{\log_b \star}$ is b raised to the number we must raise b to in order to get $\star$. $b^{\log_b \star} = \star$.

If we raise b to the power required to get $\star$ we ought to get $\star$!

Laws of Logarithmic and Exponential Functions

We have defined logarithms in terms of exponential functions, therefore we can deduce the laws of working with logarithms from those of exponents.

Exponent Laws	Logarithm Laws
(i) $b^x b^y = b^{x+y}$	(i) $\log_b QR = \log_b Q + \log_b R$
(ii) $\dfrac{b^x}{b^y} = b^{x-y}$	(ii) $\log_b \left(\dfrac{Q}{R}\right) = \log_b Q - \log_b R$
(iii) $(b^x)^y = b^{xy}$	(iii) $\log_b R^p = p \log_b R$

Derivation of Logarithmic Laws

Let $\log_b R = y$ and $\log_b Q = x$. Then $b^y = R$ and $b^x = Q$.

i. $\log_b QR = \log_b(b^x b^y) = \log_b(b^{x+y}) = x + y = \log_b Q + \log_b R$

ii. $\log_b \left(\dfrac{Q}{R}\right) = \log_b \left(\dfrac{b^x}{b^y}\right) = \log_b(b^{x-y}) = x - y = \log_b Q - \log_b R$

iii. $\log_b(R^p) = \log_b \left((b^y)^p\right) = \log_b(b^{py}) = py = p \log_b R$

Note:

$\log_b 1 = 0$ since $b^0 = 1$;

$\log_b \left(\dfrac{1}{R}\right) = \log_b 1 - \log_b R = 0 - \log_b R = -\log_b R$.

Alternatively, $\log_b(\dfrac{1}{R}) = \log_b R^{-1} = -\log_b R$.

CAUTION Resist the temptation to be sloppy with the log identities and laws we've just discussed. These laws mean precisely what they say, not more, not less. A novice might look at the law $\log AB = \log A + \log B$ and incorrectly think "multiplication and addition are the same for logs." But this is NOT what the law says.

$$(\log A)(\log B) \neq \log A + \log B.$$

Rather, the law says that the logarithm of a product is the sum of the logarithms. Similarly, many have succumbed to the temptation to look at

$$b^{\log_b \star} = \star$$

and draw incorrect conclusions. $b^{2 \log_b k}$ does not equal $2k$, despite the popular appeal. You can't just ignore that 2, use the inverse relation of logs and exponentials and then let the 2 rematerialize, lose altitude, and slip in right next to the k. Rather, you must rewrite $b^{2 \log_b k}$ so you can apply the identity above.

$$b^{2 \log_b k} = b^{\log_b k^2} \quad \text{by log law (iii)}$$

$$= k^2 \quad \text{by the inverse relation of logs and exponentials}$$

The examples and exercises that follow will provide practice in applying the log identities and laws. It might be useful to write the log laws out and be sure you can identify precisely which ones you are using as you work your way through the problems.

◆ **EXAMPLE 13.3**   Simplify the following.

  i. $7^{[\log_7(x^3) + \log_7 4]}$ ii. $7^{[2 \log_7 x + 3]}$

SOLUTION

    i. $7^{[\log_7(x^3)+\log_7 4]} = 7^{\log_7(x^3)} \cdot 7^{\log_7 4} = x^3 4 = 4x^3$

    ii. $7^{[2\log_7 x+3]} = 7^{2\log_7 x} \cdot 7^3 = 7^{\log_7 x^2} \cdot 7^3 = x^2 7^3 = 343x^2$ ◆

EXERCISE 13.4   Write the given expression in the form log( ) or ln( ).

    (a) $\log A - 3\log B + \frac{\log C}{2}$ (b) $\frac{-3}{5}\log 7 + \frac{1}{2}\log 49$

    (c) $\ln(x^4 - 4) - \ln(x^2 + 2)$ (d) $\ln 3x - 3\ln x$

EXERCISE 13.5   Simplify

    (a) $10^{3\log 2 - 2\log 3}$ (b) $5^{-0.5\log_5 3}$

    (c) $\frac{\log 4 - \log 1}{2}$ (d) $e^{-\ln \sqrt{x}}$

    (e) $\ln\left(\frac{1}{\sqrt{e}}\right)$ (f) $3\ln e^{\pi}$

    (g) $e^{2\ln x - \ln y}$

EXERCISE 13.6

    (a) What is the average rate of change of $\log x$ on the interval $[1, 10]$?

    (b) On the interval $[10, 100]$?

    (c) How many times bigger is the average rate of change of $\log x$ on $[1, 10]$ than on the interval $[10, 100]$?

Answers to Selected Exercises

Answers to Exercise 13.4

(a) $\log A - 3\log B + \frac{\log C}{2} = \log A - \log B^3 + \log C^{1/2}$
$= \log \frac{A}{B^3} + \log C^{1/2} = \log(\frac{AC^{1/2}}{B^3})$

(b) $-\frac{3}{5}\log 7 + \frac{1}{2}\log 49 = \log 7^{-3/5} + \log 49^{1/2}$
$= \log 7^{-3/5} + \log 7$
$= \log(7^{-3/5} \cdot 7)$
$= \log 7^{2/5}$

(c) $\ln(x^4 - 4) - \ln(x^2 + 2) = \ln \frac{x^4 - 4}{x^2 + 2}$
$= \ln \frac{(x^2+2)(x^2-2)}{x^2+2}$
$= \ln(x^2 - 2)$

(d) $\ln 3x - 3\ln x = \ln 3x - \ln x^3$
$= \ln \frac{3x}{x^3}$
$= \ln \frac{3}{x^2}$

Answers to Exercise 13.5

(a) $10^{3\log 2 - 2\log 3} = 10^{3\log 2} \cdot 10^{-2\log 3}$
$= (10^{\log 2})^3 \cdot (10^{\log 3})^{-2} = 2^3 \cdot 3^{-2} = 8 \cdot \frac{1}{9} = 8/9$

(b) $5^{-0.5\log_5 3} = (5^{\log_5 3})^{-0.5} = 3^{-0.5} = \frac{1}{\sqrt{3}}$

(c) $\frac{\log 4}{2} - \frac{\log 1}{2} = \frac{\log 4}{2}$

(d) $e^{-\ln \sqrt{x}} = \frac{1}{e^{\ln \sqrt{x}}} = \frac{1}{\sqrt{x}}$

(e) $\ln(\frac{1}{\sqrt{e}}) = \ln e^{-1/2} = -\frac{1}{2}\ln e = -\frac{1}{2}$

(f) $3\ln e^{\pi} = 3\pi \ln e = 3\pi$

(g) $e^{2\ln x - \ln y} = e^{2\ln x} \cdot e^{-\ln y} = e^{\ln x^2} \cdot e^{\ln \frac{1}{y}} = \frac{x^2}{y}$

Answers to Exercise 13.6

(a) $\frac{\log 10 - \log 1}{10 - 1} = \frac{1-0}{9} = \frac{1}{9}$

(b) $\frac{\log 100 - \log 10}{100 - 10} = \frac{2-1}{90} = \frac{1}{90}$

(c) The former is ten times larger.

An Historical Interlude on Logarithmic Functions

The Scotsman John Napier's invention of logarithms provided scientists and mathematicians of the seventeenth century with a revolutionary computational tool. (Today this is Napier's claim to fame, although in his own time he was also known as a widely published religious activist, a Protestant vehemently opposed to the Catholic Church and its Pope.) At the heart of Napier's logarithmic system was the ingenious idea of exploiting the laws of exponents in order to convert complicated multiplication, division, and exponentiation problems into substantially simpler addition, subtraction, and multiplication problems, respectively. Although his system did not exactly reflect the log system we use today, he associated numbers with exponents by creating extensive "log" tables. It is interesting to note that when Napier first began his work, fractional exponents were not in common use, so he had to choose a small enough base to make his system useful. The number that essentially played the "base role" was $1 - 10^{-7} = 0.9999999$.[6] The British mathematician Henry Briggs built upon Napier's work and by 1624 produced accurate log tables using a base of 10.

To add some historical perspective, all of this was going on at about the same time as Johannes Kepler (1571–1630) was painstakingly recording data and doing computations (all by hand, without a calculator) that led him to conclude that the earth's path around the sun is not a perfect circle, but rather an ellipse. Kepler began his computations without the help of logarithms. Historically logarithms arose as an invaluable computational tool, a role they have lost with the advent of calculators and computers.

How Do We Use Logarithms Today?   In modern-day mathematics the logarithm is important in modeling and in its role as the inverse of the exponential function.

■   **Logarithms help us solve for variables in exponents**
Look back at the introductory example. Logarithms helped us solve for the variable in the exponent. We'll return to this in the next section.

■   **Logarithms aid us in modeling and conveying information**
Many quantities grow exponentially. Try graphing $f(x) = 10^x$ for $0 \le x \le 6$. That's not a huge domain; it doesn't seem like an unreasonable request. And yet it's very impractical to convey the information graphically since the vertical scale must reach from 1 to 1,000,000! To graph on scales like this, sometimes a semilog plot is used. A semilog plot has one axis labeled in the usual linear way and the other axis, in order to accommodate the data, either labeled with the logarithm of the values or spaced

---

[6] From *e: The Story of a Number*, by Eli Maor, Princeton University Press, 1994, pp. 3–9.

according to the logarithms. For instance, 1, 10, 100, and 1000 would be spaced at equal intervals.

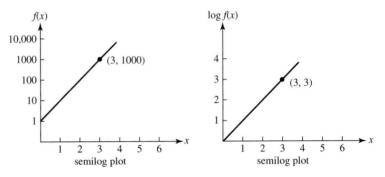

Figure 13.3

Many scales of measurements are logarithmic. For example, the Richter scale for earthquakes uses a logarithmic scale. An earthquake measuring 7 on the Richter scale is 10 times stronger[7] than one of magnitude 6, which in turn is 10 times stronger than one of magnitude 5.

PROBLEMS FOR SECTION 13.2

1. Simplify the following: $\log x$ is shorthand for $\log_{10} x$.

   (a) $3^{\log_3 2}$

   (b) $\log x + \log x^2 - 3 \log x$

   (c) $2 \log(x+3) - 3 \log(x+3) + \log(10^{\sqrt{7}})$ (d) $10^{\log x^2}$

   (e) $10^{3 \log x}$

   (f) $10^{-\log x}$

   (g) $10^{-0.5 \log x}$

   (h) $3^{-\log_3(x+y)}$

   (i) $2^{(\log_2 10 - \log_2 5)}$

   (j) $10^{\frac{\log x}{2}}$

2. If $\log_2 u = A$ and $\log_2 w = B$, express the following in terms of A and B (eliminating u and w).

   (a) $\log_2(u^2 w)$ (b) $\log_2(u^3/w^2)$

   (c) $\log_2(1/\sqrt{w})$ (d) $\log_2(\frac{2}{\sqrt{uw}})$

For Problems 3 through 9, simplify the expression given.

3. (a) $\sqrt{2} \cot 10^{\log 7}$ (b) $\pi e^{\ln 4}$

4. (a) $3^2 10^{2 \log 5}$ (b) $5 e^{-3 \ln 2}$

5. (a) $10^{\log 2 + 1}$ (b) $e^{3 - \ln 2}$

6. (a) $2^{\log_2 3 + 3}$ (b) $e^{2 \ln A + 1}$

7. (a) $10^{\log 2 - \log 3}$ (b) $e^{2 \ln 5 - \ln 2}$

8. (a) $10^{-\log \frac{1}{10}}$ (b) $e^{\frac{\ln 3}{2}}$

9. (a) $10^{\frac{\log 8 + 1}{2}}$ (b) $e^{-\frac{\ln 8}{3} + 2}$

In Problems 10 through 13, let $\log 2 = a$ *and* $\log 3 = b$. *Express each of the following in terms of a and b. There should be no logarithms explicitly in the expressions you give.*

10. $5 \log \frac{2}{3}$

11. $\frac{\log 12}{2}$

12. $5 \log \sqrt[3]{6}$

13. $\log(9\sqrt{2})$

14. (a) Evaluate the following limits. To do so rigorously, it is useful to apply L'Hôpital's rule (Appendix F). Otherwise, use a calculator to guess the answers.

   i. $\lim_{x \to \infty} \frac{\sqrt{x}}{\ln x}$ ii. $\lim_{x \to \infty} \frac{\ln x}{\sqrt{x}}$

   (b) Which grows faster as $x \to \infty$, $\ln x$ or $\sqrt{x}$?

In Problems 16 and 17, rewrite the expression given as a single logarithm.

15. $\ln \sqrt{x} - \frac{\ln x^3}{2} - 3 \ln x$

16. $a \ln(x + 3) - b \ln(\frac{1}{x}) - c \ln(x + 1)$

13.3   USING LOGARITHMS AND EXPONENTIATION TO SOLVE EQUATIONS

The fundamental ideas for solving equations using logarithms and exponentiation are these:

■   If $A = C$, then $b^A = b^C$.
   Exponentiating both sides of an equation preserves equality.

■   If $A = C$ where A and C are both positive, then $\log_b A = \log_b C$.
   Taking the logarithm of both sides of an equation preserves equality.

   Let's begin with two very simple examples. In both, our goal is to solve for x.

◆ **EXAMPLE 13.4**   Solve for x if $3^{5x} = 100$.

   SOLUTION   We need to "bring down the exponent" to solve for x, so we need to undo the exponentiation. Logarithms will help us to do this; we'll take the log of both sides of the equation. At first you might think that you *must* use $\log_3$. You *can* do this, but it actually carries a disadvantage. Let's try it and see.

$$3^{5x} = 100$$

$$\log_3(3^{5x}) = \log_3(100)$$

$$5x = \log_3(100)$$

$$x = \frac{\log_3 100}{5}$$

We've solved for x. But if we want a numerical approximation of the answer, it is not readily available because $\log_3$ is not one of the two log buttons on calculators. We'd have to convert the base 3 logarithm to something our calculators can handle. Let's do this problem again, using $\log x$ (i.e., $\log_{10} x$) and $\ln x$ (i.e., $\log_e x$). You'll see that the truly crucial property of logarithms that allows us to solve the equation is log property (iii), $\log_b(R^p) = p \log_b R$. Therefore we can use $\log_b$ with any base b.

$$3^{5x} = 100 \qquad\qquad\qquad 3^{5x} = 100$$

$$\log(3^{5x}) = \log(100) \qquad\qquad \ln(3^{5x}) = \ln(100)$$

$$5x \log 3 = \log(100) \qquad\qquad 5x \ln 3 = \ln(100)$$

$$x = \frac{\log 100}{5 \log 3} = \frac{2}{5 \log 3} \qquad\qquad x = \frac{\ln 100}{5 \ln 3}$$

Use a calculator to get a numerical approximation of the answer. Do this using both the natural logarithm and the common logarithm. In each case, you should come out with the answer of approximately 0.83836. Try it.

$$\boxed{\ln}\ \boxed{100}\ \boxed{\div}\ \boxed{(}\ \boxed{5}\ \boxed{\ln}\ \boxed{3}\ \boxed{)}\ \boxed{=}$$

Suppose we had the answer $x = \frac{\log_3 100}{5}$ and wanted to convert to base 10. Is it really necessary to begin the problem again? Of course not! Let's take a look at converting logarithms from one base to another. ◆

Converting Logarithms from One Base to Another

Goal: We want to convert $\log_b x$ to a logarithm with another base. Begin by writing $y = \log_b x$.

$$y = \log_b x \text{ is equivalent to } b^y = x.$$

We can convert $\log_b x$ to a log with *any* base by taking the appropriate logarithm of both sides of the exponential equation $b^y = x$ and solving for y.

Let's convert to the natural logarithm.

$$b^y = x$$

$$\ln(b^y) = \ln x$$

$$y \ln b = \ln x$$

$$y = \frac{\ln x}{\ln b}. \qquad \text{Since } b \text{ is a constant, } \ln b \text{ is simply a constant.}$$

Similarly, we could convert from log base b to log base k.

$$\boxed{\log_b x = \frac{\log_k x}{\log_k b}}$$

Returning to our original problem, we convert $\frac{\log_3 100}{5}$ to log base 10 as follows.

$$\frac{\log_3 100}{5} = \frac{1}{5}\frac{\log 100}{\log 3} = \frac{1}{5}\frac{2}{\log 3} = \frac{2}{5\log 3}$$

◆ EXAMPLE 13.5   Solve for x if $\log_7(4x) = 3$.

SOLUTION   In order to solve for x we need to free the x from the logarithm. There are two approaches we can take to this. The first is to undo $\log_7$ by exponentiating both sides of the equation with base 7.

$$\log_7(4x) = 3$$
$$7^{\log_7(4x)} = 7^3$$
$$4x = 7^3$$
$$x = \frac{7^3}{4} = \frac{343}{4}$$

Notice that if we use a base other than 7 to exponentiate, then we will get stuck. For instance, we could write $10^{\log_7(4x)} = 10^3$, but we cannot simplify the expression $10^{\log_7(4x)}$, so this course of action is unproductive.

An alternative mindset for solving the original problem is to think about what $\log_7(4x) = 3$ means and write the statement in exponential form. $\log_7(4x) = 3$ means "3 is the number we must raise 7 to in order to get $4x$."

So $7^3 = 4x$. Then $x = \frac{7^3}{4} = \frac{343}{4}$.   ◆

Theme and Variation.   Examples 13.4 and 13.5 are prototypical examples. Example 13.4 is of the form

$$B^{f(x)} = A,$$

and Example 13.5 is of the form

$$\log_b f(x) = A.$$

Knowing our way around equations of these forms will serve as a guideline for strategizing when we have more complicated examples. Below we show variations on these basic themes. Along the way, we'll try to point out some pitfalls so you can walk around them instead of falling into them. It's surprisingly easy either to get caught in a frenzy of unproductive manipulation or to plunge down a short, dead-end street when approaching exponential or logarithmic equations. Take time to strategize.

◆ EXAMPLE 13.6   Solve for x if $5^{2x+1} = 20$.

SOLUTION   We need to "bring down the exponent" to solve for x. Log property (iii) will help us do this. Again, although we can use $\log_b$ with any base b, if we want a numerical approximation then using common logarithms or natural logarithms simplifies the task.

$$5^{2x+1} = 20 \qquad \text{Take log of both sides.}[8]$$

$$\log 5^{2x+1} = \log 20 \qquad \text{Use log rule (iii); don't forget parentheses around } 2x + 1.$$

$$(2x + 1) \log 5 = \log 20$$

$$2x \log 5 + \log 5 = \log 20 \qquad \text{This is just a linear equation in } x. \text{ (This might not be immediately clear at first, but } \log 5 \text{ is just a constant, as is } \log 20.) \text{ Get all terms with } x \text{ on one side and all else on the other.}$$

$$2x \log 5 = \log 20 - \log 5 \qquad \text{Divide by the coefficient of } x.$$

$$x = \frac{\log 20 - \log 5}{2 \log 5}$$

At this point, you could leave the answer as is, simplify, or look for a numerical approximation. Even if you plan to do the latter, it still is helpful to simplify first.

$$\frac{\log 20 - \log 5}{2 \log 5} = \frac{\log(20/5)}{\log 25} = \frac{\log 4}{\log 25}.[9] \quad \blacklozenge$$

◆ **EXAMPLE 13.7**   Solve for x if $5^{2x+1} - 18 = 2$.

CAUTION We need a strategy here. Our goal is to "bring down the $2x + 1$," but taking logs right away doesn't help us. Although $\log(5^{2x+1} - 18) = \log 2$, we can't do much with the expression on the left because there is no log rule to simplify $\log(A + B)$. The x is trapped inside the log with no escape route. However, if we put the equation into the form $B^{f(x)} = A$ by adding 18 to both sides, then we're in business. In fact, we're back to $5^{2x+1} = 20$, as in Example 13.6.   ◆

REMARK

$$\text{If} \quad A = C + D, \quad \text{then}$$

$$\ln(A) = \ln(C + D), \quad \text{but}$$

$$\ln A \neq \ln C + \ln D.$$

You must take the log of each entire side of the equation.

As a concrete example,

$$2 = 1 + 1, \quad \text{but}$$

$$\ln 2 \neq \ln 1 + \ln 1,$$

$$\ln 2 \neq 0.$$

---

[8] An alternative strategy is to begin by dividing both sides of the equation by 5 to get $5^{2x} = 4$. Now take logs to get $\log(5^{2x}) = \log 4$, or $2x \log 5 = \log 4$. Consequently, $x = \frac{\log 4}{2 \log 5}$.

[9] Alternatively, $\frac{\log \frac{20}{5}}{2 \log 5} = \frac{\log 4}{2 \log 5} = \frac{\log 2^2}{2 \log 5} = \frac{2 \log 2}{2 \log 5} = \frac{\log 2}{\log 5}$.

Similarly,

$$\text{if}\quad A = C + D,\ \text{then}$$
$$e^A = e^{C+D} = e^C \cdot e^D,\ \text{but}$$
$$e^A \neq e^C + e^D.$$

As a concrete example,

$$2 = 1 + 1,$$
$$10^2 = 10^{(1+1)} = 100,\ \text{but}$$
$$10^2 \neq 10^1 + 10^1 = 20.$$

◆ **EXAMPLE 13.8**   Solve for x if $5^{x+1} - \frac{20}{5^x} = 0$.

SOLUTION   Let's strategize. We're trying to solve for a variable in the exponent, so perhaps we can get this into the form $B^{f(x)} = A$. First we'll clean it up a bit; we can clear the denominator by multiplying by 5^x.

$$5^x \left(5^{x+1} - \frac{20}{5^x} \right) = 5^x (0)$$
$$5^{2x+1} - 20 = 0$$
$$5^{2x+1} = 20,$$

and we're back to our familiar problem.   ◆

◆ **EXAMPLE 13.9**   Solve for x if $\log_7(x^3) + \log_7 4 = 2 \log_7(x) + 3$.

SOLUTION   We have plenty of options here. Below we'll work through a couple of approaches.

Approach 1.   Try to put this equation into the form $\log_b f(x) = A$.

$$\log_7(x^3) + \log_7 4 - 2 \log_7(x) = 3 \quad \text{Consolidate:}$$
$$\log_7(x^3) + \log_7 4 - 2 \log_7(x) =$$
$$\log_7(x^3) + \log_7 4 - \log_7(x^2) =$$
$$\log_7 \left(\frac{4x^3}{x^2} \right)$$

$\log_7 \left(\frac{4x^3}{x^2} \right) = 3$ But $\frac{4x^3}{x^2} = 4x$; $x = 0$ is not in the domain of $\log_7 x$.

$$\log_7(4x) = 3$$

Now we're back to Example 13.5.

Alternatively, we could have grouped terms as follows.

$$3 \log_7(x) - 2 \log_7(x) = -\log_7 4 + 3$$

$$\log_7(x) = -\log_7 4 + 3 \quad \text{Exponentiating now gives:}$$

$$7^{\log_7 x} = 7^{-\log_7 4 + 3} \qquad \text{Careful with the right-hand side!}$$

$$7^{A+B} = 7^A 7^B, \text{ not } 7^A + 7^B.$$

$$x = 7^{-\log_7 4} \cdot 7^3$$

$$x = 7^{\log_7(4^{-1})} \cdot 7^3$$

$$x = (4^{-1})7^3 = \tfrac{343}{4}$$

Approach 2.   Exponentiate immediately. This works, but not as neatly as Approach 1.

$$7^{[\log_7(x^3) + \log_7 4]} = 7^{[2\log_7 x + 3]}$$

Now use the laws of logs and exponents to simplify.

This is done in Example 13.3

$$4x^3 = x^2 7^3 \qquad \text{We can divide through by } x^2 \text{ provided } x \neq 0.$$

$$4x = 7^3 \text{ or } x = 0 \quad x \neq 0 \text{ because } 0 \text{ is not in the domain}$$
$$\qquad\qquad\qquad\quad \text{of the log function.}$$

$$x = \tfrac{7^3}{4} \qquad\qquad \text{Discard the extraneous root } x = 0. \quad \blacklozenge$$

Two Basic Principles

1. If you want to solve for a variable and it's caught up in an exponent, you need to bring down the variable. This requires taking the logarithm of both sides of the equation. If you can get the equation in the form

$$B^{f(x)} = A,$$

you are in great shape. Taking logs of both sides leaves you with

$$f(x) \log B = \log A$$

$$f(x) = \frac{\log A}{\log B}.$$

Since $\frac{\log A}{\log B}$ is independent of x, it is just a constant. Don't be fazed by its bulk; treat it as you would any other constant. You now have $f(x) = k$.

$CB^{f(x)} = A^{g(x)}$ is a nice form as well. Taking logs of both sides leaves you with

$$\log(C \cdot B^{f(x)}) = \log(A^{g(x)})$$

$$\log C + f(x) \log B = g(x) \log A.$$

Again, $\log A$, $\log B$, and $\log C$ are just constants, so the equation is analogous to $3 + 5f(x) = 7g(x)$.

2. If you want to solve for a variable and it is caught inside a logarithm, (i.e., in the argument, or input, of the log function), then you'll need to exponentiate to undo the log. Use the same base for exponentiation as is in the log. If you can get your equation in the form

$$\log_b f(x) = C,$$

then you can exponentiate both sides of the equation using b as a base to get

$$b^{\log_b f(x)} = b^C$$
$$f(x) = b^C.$$

Since b^C is just a constant, you're in good shape.

The Overarching Principle for Solving Equations of the Form $g(x) = $ Constant

Suppose you have an equation of the form $g(x) = k$, where k is an expression without x's. If g is invertible, you can solve for x by undoing g. $g^{-1}(g(x)) = g^{-1}$ (an expression without x's). Logarithms are undone through exponentiation, and exponentials are undone by taking logarithms, because logarithmic and exponential functions are inverses.[10]

◆ **EXAMPLE 13.10**   Consider the equation $\frac{1}{[\ln(x+2)]^3} + 7 = 34$. Solve for x.

SOLUTION   Think of what was done to x and undo it. Remember that because you put on your socks and then your shoes, your shoes must come off before your socks. To construct the left-hand side of the equation, we begin with x and do the following: Add 2, take the natural logarithm, cube the result, take the reciprocal, add 7. To undo this sequence, we can subtract 7, take the reciprocal, take the cube root, exponentiate, subtract 2.

$$\frac{1}{[\ln(x+2)]^3} + 7 = 34 \qquad \text{Subtract 7 from both sides.}$$

$$\frac{1}{[\ln(x+2)]^3} = 27 \qquad \text{Take the reciprocal of both sides.}$$
$$\qquad\qquad\qquad\qquad\text{(If } A = B, A \text{ and } B \neq 0, \text{ then } 1/A = 1/B.)$$

$$[\ln(x+2)]^3 = \tfrac{1}{27} \qquad \text{Take the cube root of both sides.}$$
$$\ln(x+2) = \tfrac{1}{3} \qquad \text{Exponentiate with base } e \text{ (since ln is } \log_e).$$
$$e^{\ln(x+2)} = e^{1/3}$$
$$x + 2 = e^{1/3} \qquad \text{Subtract 2.}$$
$$x = e^{1/3} - 2$$

◆

Worked Examples in Solving for x

(Try these on your own and then read the solutions. Notice that different strategies are adopted depending on the problem.)

---

[10] You may be wondering why when you solve the equation $B^{f(x)} = A$ you can use logs with any base to bring down $f(x)$ while to undo $\log_b$ you must exponentiate with base b. The reason is that $a^x = b^{kx}$ for the appropriate k, where b is any positive number, so a logarithm with any base will do. For example, $5^x = (10^{\log 5})^x = 10^{(\log 5)x} = 10^{kx}$, where k is the constant $\log 5$. Because $5^x = 10^{kx}$, log base 10 will help you out.

(a) $3^{2x+1} = 8^x$ Bring those exponents down; take ln of both sides.

     $\ln(3^{2x+1}) = \ln 8^x$

     $(2x + 1)\ln 3 = x \ln 8$ Multiply out to free the x.

     $2x \ln 3 + \ln 3 = x \ln 8$ This equation is linear in x.

     $2x \ln 3 - x \ln 8 = -\ln 3$

     $x(2\ln 3 - \ln 8) = -\ln 3$

     $x = \dfrac{-\ln 3}{2\ln 3 - \ln 8}$ or $x = \dfrac{-\ln 3}{\ln(9/8)}$

(b) $[\log(x^2 + 1)]^2 = 4$ Unpeel the problem as was done above in Example 13.10.

     $\log(x^2 + 1) = \pm\sqrt{4}$

     $\log(x^2 + 1) = 2$ or $\log(x^2 + 1) = -2$

     $10^{\log(x^2+1)} = 10^2$ $10^{\log(x^2+1)} = 10^{-2}$

     $(x^2 + 1) = 10^2$ $(x^2 + 1) = 10^{-2}$

     $x^2 = 100 - 1$ $x^2 = 0.01 - 1 = -0.99$

     $x = \pm\sqrt{99}$ x^2 can't be negative.

(c) $\ln \sqrt{x} = \ln x^5 - 7$ Rewrite this.

     $(1/2)\ln x = 5\ln x - 7$ This equation is linear in $\ln x$.

     $(1/2)\ln x - 5\ln x = -7$ Solve for $\ln x$ and then exponentiate.

     $-(9/2)\ln x = -7$

     $\ln x = 14/9$

     $x = e^{14/9}$

(d) $e^{2x} + 3e^x = 10$ This equation is quadratic in e^x.

     $(e^x)^2 + 3e^x - 10 = 0$ If you like, let $u = e^x$ and solve for u.

                                     Then return to e^x and solve for x.

     $u^2 + 3u - 10 = 0$

     $(u - 2)(u + 5) = 0$

     $u = 2$ or $u = -5$ But $u = e^x$.

     $e^x = 2$ or $e^x = -5$

     $x = \ln 2$ or $e^x = -5$ But e^x can never be negative, so $e^x \neq -5$.

     $x = \ln 2$

(e) $5^{\ln x} = 7x$ Bring down the exponent; take ln of both sides.

$\ln(5^{\ln x}) = \ln(7x)$ Separate out the ln x's.

$\ln x (\ln 5) = \ln 7 + \ln x$ This equation is linear in ln x.

Solve for ln x and then exponentiate.

$(\ln 5) \ln x - \ln x = \ln 7$

$\ln x (\ln 5 - 1) = \ln 7$

$\ln x = \frac{\ln 7}{\ln 5 - 1}$

$x = e^{\left(\frac{\ln 7}{\ln 5 - 1}\right)}$

or $x = \left(e^{\ln 7}\right)^{1/(\ln 5 - 1)} = 7^{1/(\ln 5 - 1)}$

Exploratory Problem for Chapter 13

Pollution Study

1. An environmental policy advisory board is studying information supplied to them concerning pollution levels in one of India's rivers. The data have been presented as follows.

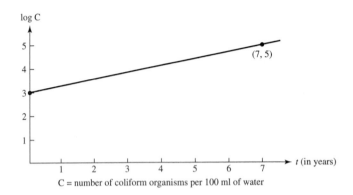

C = number of coliform organisms per 100 ml of water

   On the vertical axis is the log of the number of coliform organisms per 100 ml of water and on the horizontal axis is time, measured in years from the date the study of the river began. The data look more or less like a line passing through the points $(0, 3)$ and $(7, 5)$.

   This kind of presentation, plotting the logarithm of the pollution level against time, is a **semilog plot.** Find a formula for $C(t)$, the number of coliform organisms in 100 ml of water, as a function of time.

2. (a) Show that in general, if a quantity grows exponentially, then the semilog plot will be a straight line. In particular, show that if a quantity grows according to the equation $y = Ce^{kt}$, then $\ln y$ is a linear function of t. What is the significance of the slope of this line? The vertical intercept of the line?

   (b) Show that if $\ln y$ is a linear function of t, $\ln y = mt + b$, then y is an exponential function of t.

   Optional. Real-world data about pollution levels or population size won't follow a formula exactly. Nevertheless, we try to model their behavior by the most appropriate function. Go find some data about a quantity you think grows (or decays) exponentially, at least over a certain time period. Plot the data points on a graph using a semilog plot. Does it seem reasonable to fit a straight line to these points? If not, then the phenomenon must not actually be exponential. If so, find the line of best fit. If you know how to get this by doing linear regression with the help of a computer or calculator, do so. Otherwise, do it by eye. Then come up with an exponential function modeling the phenomenon you have chosen.

PROBLEMS FOR SECTION 13.3

1. Solve for t: $P = P_0 e^{kt}$.

2. Solve for x: (Don't expect "pretty" answers.)

   (a) $10^{2x} = 93$ (b) $10^{3x+2} = 1,000,000$

   (c) $2^{x+1} = 7$ (d) $3^x 3^{x^2} = 3$

   (e) $5B^x = (2C)^{x+1}$ (f) $\ln x + 2 = 5$

   (g) $\log_{10} x = 17$ (h) $\ln(5x - 40) = 3$

   (i) $\log_{10}(2x^2 + 4) = 2$ (j) $3 \cdot 2^{x/7} - 4 = 12$

3. (a) Approximate $\log_3 16$ (with error less than 0.005) using your calculator.
   (b) Rewrite $\log_3 16$ in terms of log base 10.
   (c) Rewrite $\log_3 16$ in terms of log base e.
   (d) Rewrite $\log_3 16$ in terms of log base 7.

4. Solve for x.
   (a) $3 \ln x + 5 = (\ln x) \ln 2$
   (b) $2 \left(7^{1+\log x}\right) = 8$
   (c) $Ke^x + K = Le^x - L$, where K and L are constants and $0 < K < L$
   (d) $R(1 + n)^{nx} = (Pn)^x$, where P, R, and n are constants
   (e) $3b^x = c^x 3^{2x}$, where b and c are constants

5. Solve for x.

   (a) $2^{x^2} 2^x = 3^x$ (b) $3^{x^2+2x} = 1$

   (c) $3 \ln(x^4) - 2 \ln 2x = 10$ (d) $e^{2x} + e^x - 6 = 0$

   (e) $e^x + 8e^{-x} = 6$ (f) $(\ln x)(\ln 5) = \ln 4x$

6. The Richter scale, introduced in the mid-1900s, measures the intensity of earthquakes. A measurement on the Richter scale is given by

$$M = \log \frac{I}{S},$$

where I is the intensity of the quake and S is some standard.

    Suppose we want to compare the intensity, I_1, of a particular earthquake with the intensity, I_2, of a less violent quake. The difference in their measurements on the Richter scale is

$$\log \frac{I_1}{S} - \log \frac{I_2}{S} = \log \left[\frac{\frac{I_1}{S}}{\frac{I_2}{S}} \right] = \log \frac{I_1}{I_2}.$$

In particular, suppose that one earthquake measures 7 on the Richter scale and another measures 4. Then

$$\log \frac{I_1}{I_2} = 7 - 4 = 3.$$

Therefore, $\frac{I_1}{I_2} = 10^3 = 1000$. The former earthquake has 1000 times the intensity of the latter.

(a) On August 20, 1999, there was an earthquake in Costa Rica (50 miles south of San Jose) measuring 6.7 on the Richter scale and another in Montana (near the Idaho border) measuring 5 on the Richter scale. How many times more intense was the Costa Rican earthquake?

(b) The 1989 earthquake in San Francisco measured 7.1 on the Richter scale. How many times more intense was the earthquake in Turkey on August 17, 1999, measuring 7.4 on the Richter scale?

In Problems 7 through 32, solve for x.

7. $5^{3x+2} = 2^{5x}$

8. $\frac{3}{2^{x-3}} = 7^{2x+1}$

9. $\pi \cdot 3^{1+2x} = \sqrt{\pi}5^x$

10. $e^{3x} = (\frac{5}{e})^{x+1}$

11. $e^2 e^x = \pi^{3x+3}$

12. $3^x \cdot \frac{5}{3^{x+1}} = 0$

13. $\frac{7+\pi 3^{x+2}}{2} = 3\pi$

14. $\log x - \log(x+1) = 2$

15. $\ln x^2 = 3 + \ln x$

16. $\ln \sqrt{x} + \ln x^2 = 1 - 2\ln x$

17. $[\ln(2x+3)]^2 - 9 = 0$

18. $\log x[\log(x+3) - 2] = 0$

19. $e^x(e^x - 5) = 0$

20. $e^x(e^x - 5) = 6$

21. $e^{2x} - 4e^x + 3 = 0$

22. $2e^{2x} + 6 = 11e^x$

23. $e^{-2x} - e^{-x} = 6$

24. $e^{-2x} = 2$

25. $e^x - 1 = e^{-x}$

26. $e^x - 2 = \frac{3}{e^x}$

27. $3^{\ln x} = 5x$

28. $\frac{4}{\ln(x+1)} + 5 = 13$

29. $\left[\frac{3}{\ln(2x+1)}\right]^2 - 1 = 10$

30. $\frac{3}{(e^x+1)^2} = 27$

31. $\frac{(5\pi)^{x+2}}{\pi} + \pi = 3$

32. $\ln(x - 3) - \ln(2x + 1) = 1$

In Problems 33 throuogh 36, solve for x; Q, R, and S are positive constants.

33. (a) $3^{5x+2} = 100$
    (b) $Q^{2x+1} = R$

34. (a) $2Q^{x+5} = R$
    (b) $(2Q)^{x+5} = R$

35. (a) $(Q + R)^x = S$
    (b) $(QR)^x = S$

36. (a) $R^{-x+1} - Q = S$
    (b) $\frac{R}{R^x} - S = R$

37. Acidity is determined by the concentration of hydrogen ions in a solution. The pH scale, proposed by Sorensen in the early 1900s, defines pH to be $-\log[H^+]$, where $[H^+]$ is the concentration of hydrogen ions given in moles per liter. A pH of 7 is considered neutral; a pH greater than 7 means the solution is basic, while a pH of less than 7 indicates acidity.

    (a) If the concentration of hydrogen ions in a solution is increased tenfold, what happens to the pH?
    (b) If a blood sample has a hydrogen ion concentration of 3.15 x 10^{-8}, what is the pH?
    (c) You'll find that the blood sample described in part (b) is mildly basic. Which has a higher concentration of hydrogen ions: the blood sample or something neutral? How many times greater is it?

In Problems 38 through 44 find all x for which each equation is true.

38. $[\log x]^3 = \log(x^3)$

39. $e^{x^3} = (e^x)^3$

40. $\ln x^{-1} = \frac{1}{\ln x}$

41. $\frac{\ln x}{\ln 2} = \ln x - \ln 2$

42. $e^{x+1} = e^x + e^1$

43. $10^{2x} = 10^2 10^x$

44. $\sqrt{\ln x} = \frac{1}{2} \ln x$

45. Suppose $\$M_0$ is put in a bank account where it grows according to:

$$M(t) = M_0 \left(1 + \frac{r}{12}\right)^{12t}, \quad \text{where } t \text{ is in years.}$$

   (a) If $r = 0.05$, how long will it take for the amount of money in the account to increase by 50%?
   (b) If the money doubles in exactly 8 years, what is r?

46. Find the equation of the straight line through the points $(2, \ln 2)$ and $(3, \ln 3)$.

47. Find the equation of the line through the points $(2, \ln 2)$ and $(2 + \epsilon, \ln(2 + \epsilon))$.

48. The "Rule of 70" says that if a quantity grows exponentially at a rate of $r\%$ per unit of time, then its doubling time is usually about $70/r$. This is merely a rule of thumb. Now we will determine how accurate an estimate this is and for what values of r it should be applied.

   Suppose that a quantity Q grows exponentially at $r\%$ per unit of time t. Thus, $Q(t) = Q_0 \left(1 + \frac{r}{100}\right)^t$.
   (a) Let $D(r)$ be the doubling time of Q as a function of r. Find an equation for $D(r)$.
   (b) On your graphing calculator, graph $D(r)$ and $70/r$. Take note of the values of r for which the latter is a good approximation of the former.

13.4   GRAPHS OF LOGARITHMIC FUNCTIONS: THEME AND VARIATIONS

In your mind's eye you should carry a picture of a pair of simple exponential and logarithmic functions (like 10^x and $\log x$, for instance), because a picture tells you a lot about how the functions behave. "Why clutter my mind? I can always consult my graphing calculator," you might be thinking. That's a bit like saying, "Why remember my mother's and my father's names? I can always ask my sister; she knows." The exponential and logarithmic functions behave so very differently that you want to be able to have identifiers for them. It's not difficult to reconstruct an exponential graph for yourself if necessary, and from that you can construct a logarithmic graph. (If you want to get a feel for e^x, try 3^x.)

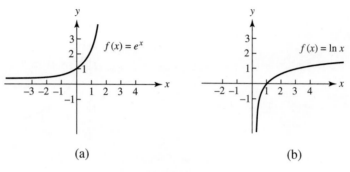

Figure 13.4

We briefly recap some important characteristics of the exponential and logarithmic functions.

Exponentials with a base greater than 1 grow increasingly rapidly. A log graph grows increasingly slowly, although the log function grows without bound: As $x \to \infty$, $\log x \to \infty$. Think of $y = \log_{10} x$ for a moment. If $x = 10$, then $y = 1$. To get to a height of 2, x must be 100. To reach a height of 3, x must increase to 1000; for a height of 6, x must be 1 million. This is sluggish growth.

Exponential functions are defined for all real numbers, but the range of b^x is only positive numbers. On the other hand, $\log_b x$ is defined only for positive numbers, while its range is all real numbers. These are not functions you want to confuse.

In this section we will play around a bit with some fancier variations on the basic logarithmic function and its graph.

◆ **EXAMPLE 13.11**   Sketch $f(x) = -\log_2 x$. Label at least three points on the graph.

SOLUTION $\log_2 x$ and 2^x are inverse functions. We are familiar with the graph of 2^x, so we'll begin with the graph of $y = 2^x$ and reflect it over the line $y = x$ (interchanging the x- and y-coordinates of the points) to obtain the graph of $y = \log_2 x$.

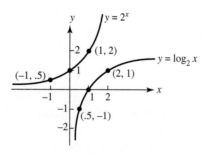

Figure 13.5

To obtain the graph of $f(x) = -\log_2 x$, we flip the graph of $\log_2 x$ over the x-axis.

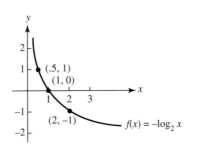

Figure 13.6   ◆

◆ EXAMPLE 13.12   Sketch the graph of $g(x) = -\log_2(-x)$. What is the domain of g?

SOLUTION   We can take the log of positive numbers only, so the domain of this function is $x < 0$. The graph of g can be obtained by reflecting the graph of f from the previous example over the y-axis.

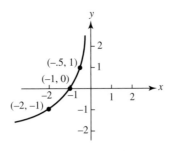

Figure 13.7   ◆

◆ EXAMPLE 13.13   Sketch the graph of $h(x) = \log_3 27x^2$.

SOLUTION   What is the domain of h? We can take the log of positive numbers only, so the domain of h is all nonzero x. Rewrite $h(x)$ so that it looks more familiar.

$$h(x) = \log_3 27x^2 = \log_3 27 + \log_3 x^2 = 3 + \log_3 x^2$$

$h(x)$ is an even function (because $h(-x) = h(x)$), so its graph is symmetric about the y-axis. Therefore, if we graph $y = 3 + 2\log_3 x$ and reflect this graph about the y-axis, we'll be all set.

How do we graph $y = 3 + 2\log_3 x$? We'll start with the graph of $\log_3 x$ and build it up from there. $\log_3 x$ and 3^x are inverse functions, so their graphs are reflections about the line $y = x$.

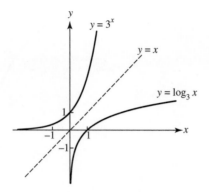

Figure 13.8

Multiplying $\log_3 x$ by 2 stretches the graph vertically, and adding 3 shifts the graph up 3 units. The graph of $y = 3 + 2\log_3 x$ is shown below.

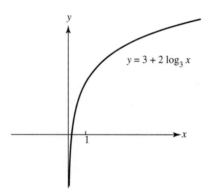

Figure 13.9

Should we want the exact value of the x-intercept, we can get it by solving the equation $0 = 3 + 2\log_3 x$. We obtain

$$\log_3 x = \frac{-3}{2} \qquad \text{Exponentiate.}$$

$$3^{\log_3 x} = 3^{-3/2}$$

$$x = \frac{1}{3^{3/2}} = \frac{1}{3\sqrt{3}}$$

The graph of $h(x)$ is given on the following page.

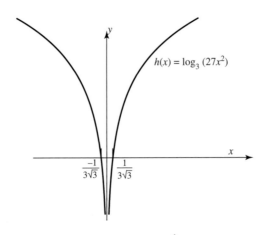

Figure 13.10   ◆

PROBLEMS FOR SECTION 13.4

In Problems 1 through 5, sketch the graph of the function without the use of a computer or graphing calculator.

1. $y = \ln(x + 1)$

2. $y = \ln(x^2)$

3. $y = |\ln x|$

4. $y = \ln |x|$

5. $y = \ln(\frac{1}{x})$

6. Sketch a rough graph of $y = \ln x - \ln(x^3) + 4\ln(x^2)$. (*Hint:* This will be straightforward after you have rewritten in the form $y = K \ln x$, where K is a constant.)

14

Differentiating Logarithmic and Exponential Functions

14.1 THE DERIVATIVE OF LOGARITHMIC FUNCTIONS

Why is ln x the most frequently used logarithm in calculus? Why is it known as the "natural" logarithm? What's so natural about it? You'll begin to understand as we investigate its derivative.

Exploratory Problem for Chapter 14

The Derivative of the Natural Logarithm

In this exploratory problem, you'll look at the derivative of $\ln x$ both qualitatively and numerically.

- ■   **Graphical Analysis:** Sketch the graph of $f(x) = \ln x$ and below that sketch its derivative.

- ■   **Numerical Analysis:** We know that $f'(b) = \lim_{h \to 0} \frac{f(b+h) - f(b)}{h}$, so $f'(b) \approx \frac{f(b+h) - f(b)}{h}$ for h very small. In a collaborative effort with a handful of your most trusted colleagues, calculate numerical approximations of $f'(b)$ for various b's by choosing a small h and calculating the slope of relevant secant lines. Complete the accompanying table using the results.

x	$f'(x)$ (approximated)
0.1	
0.5	
1	
2	
3	
4	
5	
6	

Conjecture: The derivative of $\ln x$ is _____.

The Derivative of $\ln x$

In the exploratory problem you probably conjectured that the derivative of $\ln x$ is $\frac{1}{x}$. This is only a conjecture; the numerical and graphical evidence is strong, and the result is beautiful enough to have a pull of its own, but we cannot *prove* the conjecture using numerical approximations and calculators. Mathematicians build on solid ground. We give a strong argument below.

Graphical Analysis

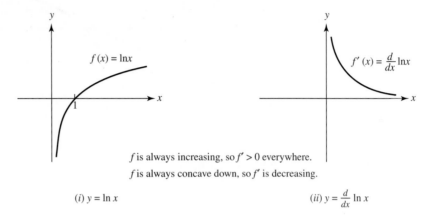

f is always increasing, so $f' > 0$ everywhere.
f is always concave down, so f' is decreasing.

(i) $y = \ln x$ (ii) $y = \frac{d}{dx} \ln x$

Figure 14.1

Solid Argument

We will show that $\frac{d}{dx} \ln x = \frac{1}{x}$ by using the inverse relation between $\ln x$ and e^x and the fact that $\frac{d}{dx} e^x = e^x$. (Because e was *defined* so that $\frac{d}{dx} e^x = e^x$ and $\ln x$ was *defined* to be the inverse of e^x, we are building our argument on a solid foundation.)

Below are graphs of e^x and $\ln x$ with an arbitrary point $(b, \ln b)$ labeled on the graph of $\ln x$. For any point (b, c) on the graph of $\ln x$ there is the point (c, b) on the graph of e^x because these functions are inverses.

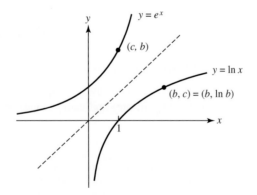

Figure 14.2

We are looking for the slope of the tangent line to $\ln x$ at $x = b$. Let L_1 be the tangent line to e^x at (c, b) and L_2 be the tangent line to $\ln x$ at (b, c).[1]

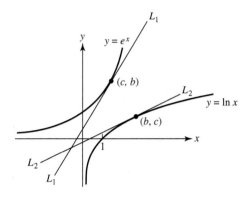

Figure 14.3

Because $\frac{d}{dx} e^x = e^x$ (i.e., the slope at any point on the graph of e^x is given by the y-coordinate of that point), the slope of the tangent line to e^x at (c, b) is b. Since the roles of x and y are interchanged on the graph of $\ln x$, the slope of the tangent line at (b, c) is $\frac{1}{b}$, as conjectured.

Let's make this argument more explicit. We assert that the slopes of L_1 and L_2 are reciprocals using the following reasoning. If (b, c) and (s, t) are two points on L_2, then (c, b) and (t, s) are two points on L_1. The slope of L_2 is $\frac{t-c}{s-b}$; the slope of L_1 is $\frac{s-b}{t-c}$.

$$\text{the slope of } L_2 = \frac{1}{\text{the slope of } L_1}$$

Conclusion:

$$\boxed{\frac{d}{dx} \ln x = \frac{1}{x}}$$

This result deserves some celebration! It's nice, neat, clean . . . . It's very beautiful.[2]

EXERCISE 14.1   Differentiate $f(x) = 3 \ln 2x$. (*Hint:* Pull apart the expression on the right so it is written as a sum. The two terms of the sum will be simple to differentiate.)

EXERCISE 14.2   Differentiate $f(x) = \log x$ by converting $\log x$ to log base e.

The Derivative of $\log_b x$

Let's find the derivative of $\log_b x$. We'll do this by converting $\log_b x$ to an expression using natural logarithms and then use what we've just learned about the derivative of $\ln x$.

---

[1] If an invertible function is differentiable at the point (c, b) with nonzero derivative, its inverse function will be differentiable at the point (b, c). This should make sense from a graphical point of view.

[2] There is even more to celebrate. Recall that we have shown that $\frac{d}{dx} x^n = nx^{n-1}$ for $n = -1, 0, 1, 2, \ldots$. It is also true that $\frac{d}{dx} x^n = nx^{n-1}$ for any constant n. We have not yet shown this, but we will. Using this fact, we can find a function whose derivative is x^k for any constant k, except $k = -1$. $\frac{d}{dx}\left(\frac{x^{k+1}}{k+1}\right) = \frac{k+1}{k+1} x^k = x^k$ for $k \neq -1$. Now we can also find a function whose derivative is x^{-1}.

$$y = \log_b x \quad \Longleftrightarrow \quad b^y = x$$

$$\ln b^y = \ln x$$

$$y \ln b = \ln x$$

$$y = \frac{\ln x}{\ln b}$$

$$\log_b x = \frac{\ln x}{\ln b} = \frac{1}{\ln b} \ln x$$

$\frac{1}{\ln b}$ is simply a constant. We know that the derivative of a constant times $f(x)$ is that constant times $f'(x)$, so

$$\frac{d}{dx} \left(\frac{1}{\ln b} \ln x \right) = \frac{1}{\ln b} \frac{d}{dx} (\ln x) = \frac{1}{\ln b} \frac{1}{x}.$$

Conclusion:

$$\boxed{\frac{d}{dx} \log_b x = \frac{1}{\ln b} \frac{1}{x}.}$$

The derivative of $\log_b x$ is just a constant times $\frac{1}{x}$. For example, $\frac{d}{dx} \log x = \frac{1}{\ln 10} \frac{1}{x}$, because $\log x = \frac{1}{\ln 10} \ln x$. Notice that there is no escaping the natural logarithm. It pops its head[3] right up into the derivative of the log base b of x no matter what the value of b!

◆ **EXAMPLE 14.1**   Find the equation of the line tangent to $f(x) = \log_3 27x^2$ at the point $(1, 3)$.

SOLUTION   The slope of the tangent line at $x = 1$ is given by $f'(1)$.

The equation of the line is $y - y_1 = m (x - x_1)$, or $y - 3 = f'(1) (x - 1)$.

$f(x) = \log_3 27 + \log_3 x^2 = 3 + \log_3 x^2$. For positive x, $f(x) = 3 + 2 \log_3 x$. We need to differentiate this and evaluate the derivative at $x = 1$.

$$Recall: \quad \frac{d}{dx} (3 + 2 \log_3 x) \Big|_{x=1} \quad \text{says "take the derivative of } 3 + 2 \log_3 x \text{ and evaluate it at } x = 1."$$

$$\frac{d}{dx} (3 + 2 \log_3 x) \Big|_{x=1} = \left(0 + 2 \left(\frac{1}{\ln 3} \frac{1}{x} \right) \right) \Big|_{x=1} = \frac{2}{\ln 3}.$$

Therefore, the equation of the tangent line is

$$y - 3 = \frac{2}{\ln 3} (x - 1) \quad \text{or} \quad y = \frac{2}{\ln 3} x - \frac{2}{\ln 3} + 3.$$

This is a linear equation; $\frac{2}{\ln 3}$ is a constant.   ◆

PROBLEMS FOR SECTION 14.1

In Problems 1 through 7, find y'.

---

[3] Of course we could write $\frac{\log b}{\log e}$ in place of $\ln b$. In particular, $\ln 10 = \frac{\log 10}{\log e} = \frac{1}{\log e}$. So $\frac{d}{dx} \log x = \frac{1}{\ln 10} \frac{1}{x} = (\log e) \frac{1}{x}$; but e is involved however we present the derivative.

1. $y = 2 \ln 5x$

2. $y = \pi \ln \sqrt{x}$

3. $y = \frac{\ln 3x}{5}$

4. $y = x \ln x$

5. $y = \frac{\ln \sqrt{2x}}{x}$

6. $y = 3 \log x$

7. $y = \frac{\log_2 x}{3}$

8. Show that $f(x) = \frac{\ln \sqrt{3x}}{2} + 3$ is invertible. Find $f^{-1}(x)$.

9. Show that $g(x) = \pi \log_2(\pi x) - \pi^2$ is invertible. Find $g^{-1}(x)$.

10. Find and classify the critical points of $f(x) = x \ln x$.

11. What is the lowest value taken on by the function $g(x) = x^2 \ln x$? Is there a highest value? Explain.

12. Use a tangent line approximation of $\ln x$ at $x = 1$ to approximate:
    (a) $\ln(0.9)$.
    (b) $\ln(1.1)$.

13. Graph $f(x) = \sqrt{x} - \ln x$, indicating all local maxima, minima, and points of inflection. Do this without your graphing calculator. (You can use your calculator to check your answer.) To aid in doing the graphing, do the following.
    (a) On a number line, indicate the sign of f'. Above this number line draw arrows indicating whether f is increasing or decreasing.
    (b) On a number line indicate the sign of f''. Above this number line indicate the concavity of f.
    (c) Find $\lim_{x \to 0^+} f(x)$ and $\lim_{x \to \infty} f(x)$ using all tools available to you. You should be able to give a strong argument supporting your answer to the former. The latter requires a bit more ingenuity, but you can do it.

14. Let $f(x) = \ln x - x$.
    (a) What is the domain of this function?
    (b) Find all the critical points of f. (The critical points must be in the domain of f.)
    (c) By looking at the sign of f', find all local maxima and minima. Give both the x- and y-coordinates of the extrema.
    (d) Find f''. Where is f concave up and where is f concave down?
    (e) Sketch the graph of $\ln x - x$ without using a calculator (except possibly to check your work).

14.2   THE DERIVATIVE OF b^x REVISITED

EXERCISE 14.3   Label the y-coordinates of the three points indicated in Figure 14.4 below. Use numerical approximations, rounding off your answers to three decimal places. Do these numbers look familiar?

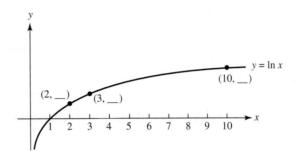

Figure 14.4

When we first investigated the derivative of b^x, we showed that $\frac{d}{dx}b^x = kb^x$, where k is a constant that depends on the base b. We know that k is the slope of b^x at $x = 0$, and we approximated k for various values of b. We found

$$\frac{d}{dx}2^x \approx (0.693)2^x \qquad \frac{d}{dx}3^x \approx (1.099)3^x \qquad \frac{d}{dx}10^x \approx (2.303)10^x,$$

and as a result of work on Exploratory Problems you may also know that

$$\frac{d}{dx}2.7^x \approx (0.993)2.7^x \qquad \text{and} \qquad \frac{d}{dx}2.8^x \approx (1.030)2.8^x.$$

None of this is completely satisfying. These results drive us to search for some structure to these constants. That search, or the exercise at the beginning of this section, might lead us to conjecture that $\frac{d}{dx}b^x = \ln b \cdot b^x$.

This conjecture we have is wonderful and delightful. If it is true, those constants that you've been memorizing or looking up in order to approximate the derivatives of 2^x, 3^x, and 10^x (the proportionality constants for the rates of growth of these functions), are just $\ln 2$, $\ln 3$, and $\ln 10$, respectively. The natural logarithm function just pops up—oh so naturally!

We will show that the conjecture $\frac{d}{dx}b^x = \ln b \cdot b^x$ holds. Our argument will rest on the fact that $\log_b x$ and b^x are inverse functions. For each point (c, d) on the graph of b^x, the point (d, c) lies on the graph of $\log_b x$. Notice that if (c, d) lies on the graph of b^x, then $d = b^c$.

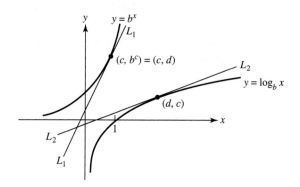

Figure 14.5

The slope of L_1, the line tangent to the graph of b^x at $x = c$, is $\dfrac{1}{\text{the slope of } L_2}$, where L_2 is the line tangent to the graph of $\log_b x$ at the point (d, c).

$$\text{The slope of } L_2 \text{ at the point } (d, c) = \frac{d}{dx} \log_b x \bigg|_{x=d} = \frac{1}{\ln b} \frac{1}{x} \bigg|_{x=d} = \frac{1}{\ln b} \frac{1}{d} = \frac{1}{d \ln b}.$$

Therefore,

$$\text{the slope of } L_1 = \frac{1}{\frac{1}{d \ln b}} = d \ln b = \ln b \cdot d.$$

But (c, d) lies on the graph of b^x so $d = b^c$. Consequently,

$$\text{the slope of } L_1 = \ln b \cdot b^c.$$

We have shown that the derivative of b^x at $x = c$ is $\ln b \cdot b^c$.
Conclusion:

$$\boxed{\frac{d}{dx} b^x = \ln b \cdot b^x}$$

An Alternative Approach to Finding the Derivative of b^x

We know that the derivative of b^x is

$$b^x \cdot \text{ (the slope of the tangent to } b^x \text{ at } x = 0).$$

We can concentrate our energy simply on finding the slope of the tangent line to b^x at $(0,1)$. It is the reciprocal of the slope of $\log_b x$ at the point $(1,0)$.

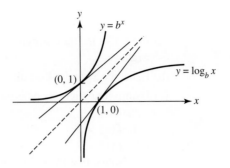

Figure 14.6

The slope of the line tangent to $\log_b x$ at $x = 1$ is $\frac{d}{dx} \log_b x \Big|_{x=1} = \frac{1}{\ln b} \frac{1}{x} \Big|_{x=1} = \frac{1}{\ln b}$.

Therefore, the slope of the tangent line to b^x at $x = 0$ is $\ln b$.

We conclude that $\frac{d}{dx} b^x = \ln b \cdot b^x$, as before.

EXERCISE 14.4   Differentiate $f(x) = 3 \cdot 2^x$.

Answer

$$f'(x) = 3 \ln 2 \cdot 2^x$$

PROBLEMS FOR SECTION 14.2

For Problems 1 through 3, find $\frac{dy}{dx}$.

1. $y = x^2 \cdot 2^x$

2. $y = \frac{5 \cdot 2^x}{3}$

3. $y = \frac{x^5 5^x}{5}$

4. Find the absolute minimum value of $f(x) = x \cdot 2^x$.

5. Find $f'(x)$ if
   (a) $f(x) = x^2 + e^x + x^e + e^2$.
   (b) $f(x) = \pi e^x - \frac{6e^x}{\sqrt{29}}$.
   (c) $f(x) = 3e^{x+3}$.   (*Hint:* Break this up into the product of e^x and a constant.)

6. Using the definition of derivative, show that the derivative of a^x is a^x times the slope of the tangent to a^x at $x = 0$. (We've done this, but refresh your memory.)

7. If your last name begins with A–J: Approximate the derivative of $(2.7)^x$ using the results of Problem 6 and $h = 0.000001$.

   If your last name begins with K–Z: Approximate the derivative of $(2.8)^x$ using the results of Problem 6 and $h = 0.000001$.

8. Graph $f(x) = e^x - x$.

    Only use a calculator to check your work after working on your own.

    (a) Find $f'(x)$. Draw a number line and indicate where f' is positive, zero, and negative.

    (b) Label the x- and y-coordinates of any local extrema (local maxima or minima).

    (c) Using your picture, determine how many solutions there are to the following equations.

       i. $f(x) = 5$
      ii. $f(x) = 0.5$

        Notice that these equations are "intractable"—try to solve $e^x - x = 5$ algebraically to see what this means. If we want to estimate the solutions, we can do so using a graphing calculator. At this point, we should know how many solutions to expect.

9. Find the quantity indicated.

    (a) $y = \ln 5x + \ln x^5 + \ln 5^x$
       i. Find y'.
      ii. Find the slope of the graph of the function at $x = 1$.

    (b) $f(x) = \log_{10} x$
       i. Find $f'(x)$.
      ii. Find $f'(100)$.

    (c) $P(x) = 7^x$
       i. Find $P'(x)$.
      ii. Find the instantaneous rate of change of P with respect to x when $x = 0$.

    (d) $y = e^{3x}$
       i. Find y'.
      ii. Find the slope of the graph of the function at $x = 0$.

    (e) $f(x) = 14e^{x/2}$
       i. Find f'.
      ii. Find $f'(\ln 9)$ and simplify your answer.

10. Differentiate $y = e^{3x} \ln(\frac{1}{\sqrt{5x}})$.

14.3   WORKED EXAMPLES INVOLVING DIFFERENTIATION

In Sections 14.1 and 14.2 we found some important derivative formulas. We've shown the following.

    i. $\frac{d}{dx} \ln x = \frac{1}{x}$

    ii. $\frac{d}{dx} \log_b x = \frac{d}{dx} \left(\frac{\ln x}{\ln b} \right) = \frac{d}{dx} \left[\frac{1}{\ln b} \ln x \right] = \frac{1}{\ln b} \cdot \frac{1}{x} = \frac{1}{x \ln b}$

    iii. $\frac{d}{dx} b^x = \ln b \cdot b^x$

    In this section we will work examples using a combination of these differentiation formulas and the laws of logs and exponents. The main piece of advice is to spend time putting the expression to be differentiated into a form that makes it simple to apply the differentiation formulas given above. Be sure to distinguish between constants and variables.

Worked Examples

In each of the following problems, find $\frac{dy}{dx}$.

◆ EXAMPLE 14.2 $y = \pi \ln \left(\frac{7x^6}{8} \right)$

SOLUTION   Rewrite:

$$y = \pi \left(\ln 7x^6 - \ln 8 \right)$$
$$= \pi \ln 7 + \pi \ln x^6 - \pi \ln 8$$
$$= \pi \ln 7 + 6\pi \ln x - \pi \ln 8$$
$$\text{(First and third terms are constant.)}$$

Differentiate:

$$\frac{dy}{dx} = \frac{6\pi}{x} \quad ◆$$

◆ EXAMPLE 14.3 $y = \frac{10 \log(bx^3)}{\sqrt{\pi}}$

SOLUTION   Rewrite:

$$y = \frac{10}{\sqrt{\pi}} \log \left(bx^3 \right)$$
$$= \frac{10}{\sqrt{\pi}} [\log b + 3 \log x]$$
$$= \frac{10}{\sqrt{\pi}} \log b + \frac{30}{\sqrt{\pi}} \frac{\ln x}{\ln 10}$$

Notice that $\frac{10}{\sqrt{\pi}} \log b$ is constant, as is $\frac{30}{\sqrt{\pi}} \frac{1}{\ln 10}$.

Differentiate:

$$\frac{dy}{dx} = \frac{30}{\sqrt{\pi}} \frac{1}{x \ln 10} \quad ◆$$

◆ EXAMPLE 14.4 $y = 17(3)^{t/5}$

SOLUTION   Rewrite:

$$y = 17 \left(3^{1/5} \right)^t$$

We'll treat $3^{1/5}$ as the base of $y = b^x$.

Differentiate:

$$\frac{dy}{dx} = 17 \left(\ln 3^{1/5} \right) \left(3^{1/5} \right)^t = \frac{17}{5} (\ln 3) \left(3^{t/5} \right) \quad ◆$$

◆ **EXAMPLE 14.5** $y = 5^{2x+3}$

SOLUTION   Rewrite:

$$y = 5^{2x} \cdot 5^3 = \left(5^3\right)\left(5^2\right)^x = 5^3\left(25^x\right)$$

Notice that 5^3 is a constant.

We'll treat 5^2, or 25, as the base of $y = b^x$.

Differentiate:

$$\frac{dy}{dx} = 5^3 \ln\left(5^2\right) \cdot 25^x$$

$$= 125 \ln 5^2 \cdot 5^{2x}$$

$$= 250 \ln 5 \cdot 5^{2x}   ◆$$

In the following examples, we generalize what we have done. Try them on your own before looking at the answers.

◆ **EXAMPLE 14.6** $y = A \ln\left(Bx^C\right)$, where A, B, and C are constants and $B > 0$.

SOLUTION   Rewrite:

$$y = A[\ln B + C \ln x]$$

$$= A \ln B + AC \ln x$$

Differentiate:

$$\frac{dy}{dx} = AC\frac{1}{x} = \frac{AC}{x}   ◆$$

◆ **EXAMPLE 14.7** $y = Ab^{kx}$, where A, b, and k are constants and $b > 0$.

SOLUTION   Rewrite:

$$y = A\left(b^k\right)^x$$

Differentiate:

$$\frac{dy}{dx} = A \ln\left(b^k\right)\left(b^k\right)^x$$

$$= Ak \ln b \cdot b^{kx}   ◆$$

In Chapter 16 we will acquire alternative means of finding these derivatives.

◆ **EXAMPLE 14.8**   Approximate ln 1.1 with the help of the first derivative.

SOLUTION   This problem is like the tangent-line approximations we discussed in Chapter 8. We begin by sketching ln x and getting an off-the-cuff approximation of ln 1.1.

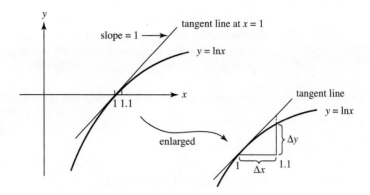

Figure 14.7

ln 1.1 is a bit larger than ln 1; how much larger can be approximated by looking at the *rate* at which ln x is increasing near $x = 1$. The derivative of ln x at $x = 1$ gives the rate of increase.

$$\frac{d}{dx} \ln x|_{x=1} = \frac{1}{x}\bigg|_{x=1} = 1.$$

Knowing the rate of increase of ln x at $x = 1$ enables us to judge how to adjust our approximation (in this case, 0) to fit a nearby value of x. $\frac{dy}{dx} \approx \frac{\Delta y}{\Delta x}$ for Δx small (because $\frac{dy}{dx} = \lim_{\Delta x \to 0} \frac{\Delta y}{\Delta x}$). In this problem, $\Delta x = 0.1$

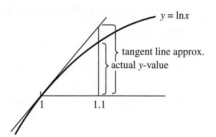

Figure 14.8

$$\frac{dy}{dx} = 1 \approx \frac{\Delta y}{.1} \Rightarrow \Delta y \approx 0.1$$

Therefore, $\Delta y \approx 0.1$. We obtain the approximation ln $1.1 \approx 0 + 0.1 = 0.1$.

Is the approximation ln $1.1 \approx 0.1$ an overestimate, or an underestimate? We know that ln x is concave down.

$$\frac{d^2}{dx^2} \ln x = \frac{d}{dx}\left(\frac{1}{x}\right) = -\frac{1}{x^2} < 0$$

Therefore the tangent line lies above the graph of the function, and this approximation is a bit too big.

Comparing our approximation, ln $1.1 \approx 0.1$, with the calculator approximation, ln $1.1 \approx 0.0953$, we see that the approximation is fairly good.

REMARK If we wanted to further refine our approximation, we could use information supplied by the second derivative to make the required adjustment.[4] This is equivalent to approximating the curve $y = \ln x$ at $x = 1$ by a parabola instead of a line. Further successive refinements involve higher order derivatives and give us what is called a *Taylor polynomial approximation* to the function at $x = 1$.   ◆

EXERCISE 14.5   Approximate ln 0.9 using a tangent-line approximation.

Answer

$$\ln 0.9 \approx -0.1$$

Logarithms: A Summary

$$\boxed{\log_b x = y \text{ is equivalent to } x = b^y.}$$

This equivalence follows from the definition of $\log_b x$. $\log_b x$ is the number we must raise b to in order to obtain x.

$\log_b x$ and b^x are inverse functions, so

$$\boxed{\log_b b^\star = \star} \quad \text{and} \quad \boxed{b^{\log_b \star} = \star.}$$

$\log x$ is the way we write $\log_{10} x$ (log base 10 of x, or the common log).

$\ln x$ is the way we write $\log_e x$ (log base e of x, or the natural log).

Thus,

$$\boxed{\ln x = y \text{ is equivalent to } x = e^y.}$$

It follows that

$$e^{\ln \star} = \star \quad \text{and} \quad \ln e^\star = \star.$$

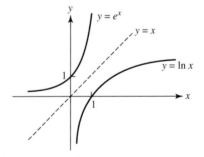

Figure 14.9

---

[4] The larger the value of the second derivative at $x = 1$ the larger the rate of change of the slope near $x = 1$.

Exponent Laws	*Logarithm Laws*	
		Logarithm laws
i. $b^x b^y = b^{x+y}$	i. $\log_b MN = \log_b M + \log_b N$	come from the exponent
ii. $\frac{b^x}{b^y} = b^{x-y}$	ii. $\log_b \left(\frac{M}{N} \right) = \log_b M - \log_b N$	laws because log and exponential
iii. $(b^x)^y = b^{xy}$	iii. $\log_b M^p = p \log_b M$	functions are inverse functions.

Logarithms are useful when we're trying to solve for a variable in the exponent.

If $A = B$ and A and B are positive, then $\ln A = \ln B$.

Similarly, if $A = B$, then $e^A = e^B$.

Caution: $1 + 1 = 2$ so $\ln(1 + 1) = \ln 2$,   *but* $\ln 1 + \ln 1 = 0 + 0 \neq \ln 2$.

$1 + 1 = 2$ so $10^{1+1} = 10^2$,   *but* $10^1 + 10^1 \neq 10^2$.

Changing the bases of logs: $y = \log_b M$

To change to natural logs, rewrite in exponential form and then take the natural logarithm of both sides.

$$y = \log_b M \text{ is equivalent to } b^y = M$$

$$y \ln b = \ln M$$

$$y = \frac{\ln M}{\ln b}$$

Derivatives: A Summary

1. $\frac{d}{dx}[f(x) + g(x)] = \frac{df}{dx} + \frac{dg}{dx}$

2. $\frac{d}{dx}[kf(x)] = k \frac{df}{dx}$, where k is any constant

3. $\frac{d}{dx} x^n = nx^{n-1}$, where n is a constant
(We've proven this for n any integer, but in fact it's true for any constant n.)

4. $\frac{d}{dx} \log_b x = \frac{d}{dx} \left(\frac{1}{\ln b} \ln x \right) = \frac{1}{x \ln b}$ (a) $\frac{d}{dx} \ln x = \frac{1}{x}$

5. $\frac{d}{dx} b^x = \ln b \cdot b^x$ (a) $\frac{d}{dx} e^x = e^x$ (b) $\frac{d}{dx} e^{kx} = k e^{kx}$

PROBLEMS FOR SECTION 14.3

1. If $f(x) = \frac{3 \ln \sqrt{x}}{x}$, what is $f'(e)$?

For Problems 2 through 5, compute y'.

2. $y = \ln(3x^2)$ (*Hint:* write $\ln(3x^2)$ as $\ln 3 + 2 \ln x$.)

3. $y = \frac{\ln x^2}{5e^{3x}}$

4. $y = \frac{\ln x}{x}$

5. $y = x \ln(\frac{1}{x})$

For Problems 6 through 13, differentiate the given function.

6. $f(x) = x \ln\left(\frac{1}{x}\right)$

7. $f(x) = \left(\frac{3 \ln(3^6 x^7)}{\pi}\right) + \frac{3 \ln(3^6)}{\pi}$

8. $f(x) = e^{5x} \ln\left(\frac{\pi}{\sqrt{x}}\right)$

9. $f(x) = 3^x(\log x)$

10. $f(x) = \frac{\ln(2x^3)}{3e^x}$

11. $f(x) = \frac{x + \ln(\frac{1}{x})}{x^2}$ (Conserve your energy.)

12. $f(x) = x^\pi + \pi^x + \ln\left(\frac{\pi}{x}\right)$

13. $f(x) = x^2 \ln\left(x\sqrt{\frac{61}{2x}}\right)$

14. Using what you know about the graph of $\ln x$, sketch the graphs of the following.

    (a) $y = |\ln x|$ (b) $y = \ln(|x|)$ (c) $y = |\ln(|x|)|$

    (d) For parts (a), (b), and (c), locate all critical points and identify all local maxima and minima.

15. Graph $f(x) = \ln x + 1/x$ for $x > 0$, indicating all local maxima and minima and any points of inflection. While answering this problem, do the following.

    (a) On a number line indicate the sign of f'. Above this number line draw arrows indicating whether f is increasing or f is decreasing.

    (b) On a number line indicate the sign of f''. Above this number line write "f is concave up" and "f is concave down" as appropriate.

    (c) Using all tools available, guess $\lim_{x \to 0^+}(\ln x + \frac{1}{x})$. Then make a convincing argument supporting your answer.

    (*Note:* Usually when graphing we would first look at f itself and determine where that is positive and negative if that information is easy to obtain. In this problem we didn't ask you to do that only because the information is *not* particularly easy to obtain. Now that we're done, we do in fact have that sign information.)

16. Differentiate. Spend time writing each of these in a form that makes the differentiation easy.

    (a) $y = 3 \ln 5x + 6 \ln\left(\frac{3}{x}\right)$ (b) $y = 20 \log\left(\frac{x}{100}\right)$

    (c) $y = \frac{ke^{2kx}}{\sqrt{k+1}}$ (d) $y = \left(\frac{(\ln 2)e^{5x}}{\ln 4}\right) + \frac{(\ln 2)e^{\ln 2}}{\ln 3}$

17. Suppose $f'(x) = x + 5(2^x)$. Find the equation of the tangent line to $f(x)$ at the point $(0, 3)$. Does the line lie above the graph of f or below the graph of f? How can you tell?

18. A report from the United Nations Food and Agriculture Organization (*Boston Globe*, July 22, 1996) made several projections about the growth of the world's population and need for food over the 55-year time period from 1995 to 2050. Among them were the following.

   I. The world's population is predicted to grow from 5.7 billion to 9.8 billion.

   II. North America will need to increase food production by a total of 30% to satisfy the demands of its population.

   III. Africa will need to increase food production by a total of 300% to satisfy the demands of its population.

   IV. The report also notes that sharing the world's food resources more fairly would "probably eliminate most cases of undernourishment."

   Given these projections, answer the following questions:

   (a) By what total percentage is the world's population predicted to increase over these 55 years?

   (b) Assuming that population growth is exponential, find an equation for the world's population as a function of time, with $t = 0$ in 1995. What is the annual percentage growth rate?

   (c) According to your equation, what is the instantaneous rate of growth at $t = 0$? At $t = 55$?

   (d) What is the average rate of growth between $t = 0$ and $t = 55$?

   (e) In what year is the instantaneous rate of growth equal to the average rate of growth? Use a sketch to illustrate whether this point should occur before or after $t = 27.5$ years.

   (f) Assuming the food production grows linearly, write an equation for North America's annual food production as a function of time, where N_0 is the amount produced in 1995 ($t = 0$). Do the same for Africa, where A_0 is the 1995 amount.

19. In a 1960s program to bring exotic animal species to the United States, 60 oryx (a 700-pound antelope with sharp 3-foot-long horns) were brought to the deserts of New Mexico. Thirty years later, the oryx population in New Mexico had grown to 2000 and was "destroying natural habitat." (*Boston Globe*, July 31, 1996.)

   (a) Assuming that the growth of the population was exponential, write an equation for $P(t)$, the number of oryx as a function of time, letting $t = 0$ when they were first imported.

   (b) What was their annual growth rate?

   (c) What was the doubling time for the population?

   (d) According to your equation, what was the instantaneous rate of growth at $t = 0$? At $t = 30$?

   (e) What was the average rate of change over the 30-year period?

   (f) Interpret each of the following in words.
       i. $P(10)$ ii. $P^{-1}(200)$

   (g) Estimate $P^{-1}(200)$.

20. A radioactive substance decays exponentially. Suppose its half-life is 5000 years and the initial amount of radioactive substance is denoted by R_0.

   (a) Write an equation of the form $R(t) = R_0 e^{kt}$ for $R(t)$, the amount of radioactive material left after t years.

   (b) If $R_0 = 3000$ mg, at what rate is the radioactive substance decaying at time $t = 0$?

21. Suppose we know a population grows exponentially; $P(2) = 1000$ and $P(4) = 1300$. Find the growth equation. (*Hint:* Write $P = P_0 e^{kt}$, or some other form of exponential growth. Put in the given information. Since you don't know P_0, divide one equation by the other so that the P_0's cancel.)

22. Suppose you put $500 in a bank account and your balance grows exponentially according to the equation

$$M = 500 e^{0.08t},$$

   where $M = M(t) = $ the amount of money in the account at time t.

   (a) Write the growth equation for the amount of money in the account in the form $M = 500 A^t$.

   (b) What is the annual growth rate of the money in the account? (Banks refer to this as the effective annual yield.) Please give your answer to the nearest tenth of a percent.

   (c) What is the instantaneous rate of change of money with respect to time?

   (d) When will you have enough money to buy a round-the-world plane ticket costing $1599?

23. Suppose you put M_0 dollars in a bank account with 6% interest compounded annually.

   (a) Write an equation for $M(t)$, the amount of money in the account at time t, t measured in years. Construct a continuous model.

   (b) Find $M'(t)$, the instantaneous rate of change of money with respect to time.

24. In early summer the fly population in Maine grows exponentially. The population at any time t can be given by $P(t) = P_0 e^{kt}$ for some constant k, where t is measured in days. Suppose that at some date, which we will designate as $t = 0$, there are 200 flies. Thirty days later there are 900 flies.

   (a) Find the constant k.

   (b) The mosquito population is also growing exponentially. At time $t = 0$ there are 100 mosquitoes, and the mosquito population doubles every 10 days. Write a function $M(t)$ that gives the number of mosquitoes at time t.

   (c) When will the number of flies and the number of mosquitoes be equal?

   (d) Find $P'(t)$.

   (e) Find $M'(t)$.

   (f) Find the rate at which each of the populations is growing when the populations are the same size. Which is growing more rapidly?

25. Suppose a population grows exponentially according to the equation

$$P = P_0 e^{0.4t}.$$

(a) Write the growth equation for the population in the form $P = P_0 A^t$.

(b) What is the annual growth rate of the population?

(c) What is the instantaneous rate of change of population with respect to time?

(d) How long does it take the population to double?

26. Let $f(x) = \frac{2 \ln x}{x}$. Does f have global extrema? Find the absolute maximum and minimum values taken on by f if these values exist. (In order to complete your argument, you will need to compute a limit. Make this argument explicit.)

27. Let $g(x) = x^2 \cdot 2^x$. Find all local extrema. Does $g(x)$ have a global maximum? A global minimum? If so, where? Explain your reasoning carefully.

28. Analyze the critical points of $h(x) = x^3 \cdot 3^x$. What is the absolute minimum value of h?

15

Take It to the Limit

15.1 AN INTERESTING LIMIT

In this chapter we begin with a concrete problem and after several iterations of the problem find ourselves hurtling toward a new and interesting perspective on something familiar. We'll begin with a brief look at inflation around the globe.

Many nations in the world suffer from runaway inflation. For instance, while the inflation rate in the United States in 1996 was about 3%, the inflation rate in Russia was about 22%. For Russia this was a considerable improvement over 1995 when its inflation rate was 131% and 1994 when its inflation rate was 2600%.[1] Brazil, Israel, and the former Yugoslavia have all had runaway inflation. For instance, Brazilians on a monthly salary habitually spend a large portion of their paychecks upon receipt because their buying power at the beginning of the month far exceeds that of the same sum at the end of the month. In the former Yugoslavia stores routinely used to shut down for a while during business hours to adjust prices in order to keep up with inflation! When inflation rates are high, interest rates on savings accounts become very high as well. In a country with runaway inflation it is not inconceivable to posit the existence of a bank offering 100% interest.

◆ **EXAMPLE 15.1** Let's suppose we put \$10,000 in a bank account with a nominal interest rate of 100%, and we leave the money in the bank untouched for exactly one year. We will compute the amount it will grow to by the end of the year under various compounding schemes. Compounding interest more than once a year will allow interest to generate more interest, yielding an *effective* annual interest rate that is greater than the *nominal* interest rate of 100%.

   (a) Suppose interest is compounded once a year. How much will be in the account by the end of the year?

---

[1] Data from the *Boston Globe*, November 2, 1996.

(b) BankBoston has savings accounts for which interest is compounded monthly. Suppose the bank in this problem compounded monthly as well. How much money will we have by the end of the year?

(c) The Cambridge Savings Bank has savings accounts for which interest is compounded daily. If the bank in this problem compounds interest daily, how much money will be in the account by the end of the year?

(d) Suppose interest is compounded hourly. How much money will be in the account by the end of the year?

(e) What if interest is compounded every minute?

SOLUTION   If $\$M_0$ is deposited in a bank account with a nominal interest rate of r compounded n times per year, then each compounding period the amount of money *increases* by $\frac{r}{n}$ times the amount of money present. In other words, each compounding period the money in the bank is multiplied by $(1 + \frac{r}{n})$.

For example, if the nominal interest is 8% per year ($r = 0.08$) and interest is compounded quarterly, then each quarter we will get $8\%/4 = 2\%$ interest. Each quarter the money in the account is multiplied by 1.02. In t years there are $n \cdot t$ compounding periods. Therefore, after t years the money grows to $M_0(1 + \frac{r}{n})^{nt}$. If $M_0 = 10,000$ and $r = 100\% = 1$, then after one year we have

$$\$10,000 \left(1 + \frac{1}{n}\right)^n.$$

(a) Interest compounded once a year: $\$10,000(1 + 1)^1 = \$20,000$

(b) Interest compounded monthly: $\$10,000(1 + \frac{1}{12})^{12} \approx \$26,130.35$

(c) Interest compounded daily: $\$10,000(1 + \frac{1}{365})^{365} \approx \$27,145.67$

(d) Interest compounded hourly: There are 24 hours in a day, so there are

$$\left(24 \, \frac{\text{hrs}}{\text{day}}\right) \left(365 \, \frac{\text{days}}{\text{yr}}\right) = 8760 \text{ hours in a year.}$$

$$\$10,000 \left(1 + \frac{1}{8760}\right)^{8760} \approx \$27,181.27$$

(e) Interest compounded every minute: There are 60 minutes in an hour, so there are

$$\left(60 \, \frac{\text{min}}{\text{hr}}\right) \left(8760 \, \frac{\text{hrs}}{\text{yr}}\right) = 525,600 \text{ minutes per year.}$$

$$\$10,000 \left(1 + \frac{1}{525,600}\right)^{525,600} \approx \$27,182.80$$

Notice that at an interest rate of 100% there is a substantial difference between compounding the interest once a year and compounding it 12 times per year. In our example it amounts to over \$6000 for the year. On the other hand, the difference between compounding daily and compounding hourly amounts to less than \$36, and the difference between compounding every minute instead of every hour only increases the yield by less than two dollars. Suppose we let n be the number of times we compound per year. If we let n grow without bound, will the amount of money in the account after one year also grow without bound? It appears not; there seems to be a limit.   ◆

Suppose M_0 is deposited in a bank account with a nominal interest rate of r compounded n times per year, and we take the limit as n grows without bound. Then $M(t) = M_0 \lim_{n\to\infty} \left(1 + \frac{r}{n}\right)^{nt}$, and we say that the account has an annual (nominal) interest rate of r **compounded continuously**.

We can pose the question of how much money we would have after one year if we put \$10,000 in a bank with an annual interest rate of 100% compounded continuously. In other words, what is $\lim_{n\to\infty} 10{,}000 \left(1 + \frac{1}{n}\right)^n$? Evidence suggests that this quantity *may* be heading toward some finite number, but this remains to be shown, and the exact number remains to be identified. Take a guess on your own.

Before tackling $10{,}000 \lim_{n\to\infty}(1 + \frac{1}{n})^n$, let's take a brief look at some simpler limits.

◆ **EXAMPLE 15.2**   Compute the following limits. In this problem, when we write $\lim_{n\to\infty}$ we mean n increases through *the integers* without bound. (Compare Figures 15.1 and 15.2; in the latter the functions are defined for integers only.)

(a) $\lim_{n\to\infty} 2^n$ (b) $\lim_{n\to\infty} \left(1 + \frac{1}{10}\right)^n$ (c) $\lim_{n\to\infty} \left(1 - \frac{1}{10}\right)^n$

(d) $\lim_{n\to\infty} \left(1 + \frac{1}{100{,}000}\right)^n$ (e) $\lim_{n\to\infty} a^n$ for some constant $a > 0$

SOLUTION   (a) $\lim_{n\to\infty} 2^n = \infty$ Each time n increases by 1 the expression doubles.

(b) $\lim_{n\to\infty}(1 + \frac{1}{10})^n = \infty$ Each time n increases by 1 the expression increases by 10%. (See Figure 15.2(i).)

(c) $\lim_{n\to\infty}(1 - \frac{1}{10})^n = 0$ Each time n increases by 1 the expression decreases by 10%.(See Figure 15.2(ii).)

(d) $\lim_{n\to\infty}(1 + \frac{1}{100{,}000})^n = \infty$ $1 + \frac{1}{100{,}000}$ is larger than 1; as n increases by 1 the expression increases by a positive percentage (0.001%) so the increase *increases* with n. $(1 + \frac{1}{100{,}000})^n$ is increasing at an increasing rate.

(e) $\lim_{n\to\infty} a^n = \begin{cases} \infty & \text{for } a > 1 \\ 1 & \text{for } a = 1 \\ 0 & \text{for } 0 < a < 1 \end{cases}$ If we know $\lim_{n\to\infty} b^n = \infty$ for $b > 1$ then it follows that $\lim_{n\to\infty} a^n = 0$ for $0 < a < 1$, since we can write a^n as $(\frac{1}{b})^n = \frac{1}{b^n}$ for some $b > 1$.

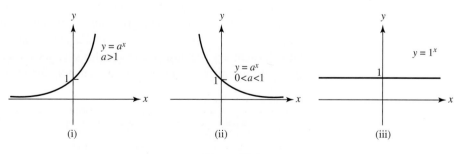

(i)   (ii)   (iii)

Figure 15.1

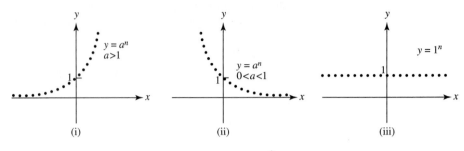

Figure 15.2 ◆

EXERCISE 15.1   Convince yourself that none of these answers would change if we took the limit as n increases without bound and n is *not* restricted to integer values.

Let's look at another set of examples.

◆ **EXAMPLE 15.3**   Compute the following limits.

(a) $\lim\limits_{n\to\infty} \left(1 + \dfrac{1}{n}\right)^2$ (b) $\lim\limits_{n\to\infty} \left(1 + \dfrac{1}{n}\right)^{10}$ (c) $\lim\limits_{n\to\infty} \left(1 + \dfrac{1}{n}\right)^{100,000}$

SOLUTION   For k any positive constant, x^k is a continuous function. Therefore we can apply limit principle (4) to get[2]

$$\lim_{n\to\infty} \left(1 + \frac{1}{n}\right)^k = \left[\lim_{n\to\infty} \left(1 + \frac{1}{n}\right)\right]^k = 1^k.$$

We apply this logic to each of the limits in this example.

(a) $\lim\limits_{n\to\infty} (1 + \frac{1}{n})^2 = 1^2 = 1$

(b) $\lim\limits_{n\to\infty} (1 + \frac{1}{n})^{10} = 1^{10} = 1$

(c) $\lim\limits_{n\to\infty} (1 + \frac{1}{n})^{100,000} = 1^{100,000} = 1$ ◆

Then what is $\lim_{n\to\infty}(1 + \frac{1}{n})^n$? Can our intuition get us anywhere? On one hand, the quantity $(1 + \frac{1}{n})$ gets arbitrarily close to 1 as $n \to \infty$, and 1 raised to any power is 1, so you might very well have the gut feeling that this limit will be 1. On the other hand, for any positive n the quantity $(1 + \frac{1}{n})$ is always bigger than 1; if we take any fixed number $a > 1$ and raise it to higher and higher powers, the resulting number grows without bound. Thinking along these lines might give someone the gut feeling that $(1 + \frac{1}{n})^n$ grows without bound as n grows without bound. If you think along both lines simultaneously the result is stomach churning. The upshot is that in this situation our intuition isn't going to get us too far; it pulls us rather strongly in two contradictory directions. The situation is too subtle.[3]

---

[2] Limit principle (4) tells us that if $\lim_{x\to a} g(x) = L$ and $h(x)$ is continuous at $x = L$, then $\lim_{x\to a} h(g(x)) = h(L) = h(\lim_{x\to a} g(x))$. It is not necessary for a to be finite for this principle to hold.

[3] Stomach-churning expressions such as this are called "indeterminate forms." In such cases our intuition pulls us in at least two very different directions simultaneously. A computational method that enables us to deal efficiently with indeterminate forms is introduced in Appendix F: l'Hôpital's Rule.

Let's return to Example 15.1. Recall that we have deposited \$10,000 into a bank with a nominal annual interest rate of 100% and left it for one year. If the interest were compounded n times a year we would have $\$10{,}000(1 + \frac{1}{n})^n$. The question is whether $(1 + \frac{1}{n})^n$ increases without bound or approaches some limiting value. (When looked at in this context, it seems unreasonable that the limit would be 1.) Let's experiment by returning to the numerical approach suggested by Example 15.1 and evaluating $(1 + \frac{1}{n})^n$ for large n. The largest value we looked at in Example 15.1 was $n = 525{,}600$. Below are the results of evaluating $(1 + \frac{1}{n})^n$ for various values of n using a TI-83. All the digits displayed by this calculator are recorded here.

For $n = 525{,}600$, the TI-83 gives 2.718279215.

For $n = 1{,}000{,}000$, the TI-83 gives 2.718280469.

For $n = 10^{10}$, the TI-83 gives 2.718281828.

For $n = 10^{15}$, the TI-83 gives 1.

What are we to make of this? For starters, look at the very last result. Do you honestly think that $(1 + \frac{1}{10^{15}})^{10^{15}} = 1$? No. The result *must* be larger than 1. As n increases, $(1 + \frac{1}{n})^n$ increases; we know this from the context of the problem. What in fact is happening is that the calculator has treated $(1 + \frac{1}{10^{15}})$ as 1 and then computed $1^{10^{15}}$ and arrived at 1. Therefore, when considering the numerical results from the calculator, we need to disregard this one. If we evaluate $(1 + \frac{1}{n})^n$ for n larger than 10^{10}, the TI-83 will keep giving us 2.718281828 until n gets so large that the TI-83 throws up its little calculator hands and gives us the number 1. The results of our numerical investigations might lead you to wonder whether the number 2.718281828 has some significance. Where have you seen this number before? If you make your calculator display (to the best of its ability) the number e, it will match up, decimal for decimal, with 2.718281828.[4] This might lead you to conjecture that $\lim_{n \to \infty}(1 + \frac{1}{n})^n = e$.

What Happens to the Limit as n Grows Without Bound? Does the Limit Equal e?

Looking at numerical data and conjecturing that $\lim_{n \to \infty}(1 + \frac{1}{n})^n = e$ is great, but if we want to verify this conjecture[5] we cannot simply try to match up decimal places. There are two logical difficulties. First of all, e is an irrational number; e cannot be written as a finite decimal or as an infinite repeating one, so we cannot ever hope to match all the decimal places for e. Second, and even more fundamental, is the question of where all of these decimal places for e are coming from to begin with. We have simply **defined** e to be the number "a" such that the derivative of the function a^x is a^x.[6] We have to rely exclusively on

---

[4] If you get hold of a few more digits for e, you'll get 2.718281828459 . . . . e is an irrational number; its decimal expansion is nonrepeating.

[5] We will show $\lim_{x \to \infty}(1 + \frac{1}{x})^x = e$ and then conclude that $\lim_{n \to \infty}(1 + \frac{1}{n})^n = e$. We must be careful. If $\lim_{x \to \infty} f(x) = L$ then $\lim_{n \to \infty} f(n) = L$, but the converse is not necessarily true. If $\lim_{n \to \infty} f(n) = L$ it is possible that $\lim_{x \to \infty} f(x) \neq L$. See the figure below.

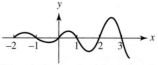

[6] The Swiss mathematician Leonhard Euler, who first introduced the number e in the mid-1700s, in fact *defined* e to be the limit in question (although a rigorous foundation for limits was not laid down until the 1800s). In this text we have not followed the actual historical development and must now show that e as we defined it is the same e that Euler defined.

this definition of e. While the calculator has been instrumental in suggesting the conjecture, it has no role in actually proving it.[7]

Knowing that the derivative of e^x is e^x, we concluded, after some work, that the derivative of its inverse function $\ln x$ is $1/x$; this information will turn out to be useful in verifying our conjecture. If we can show $\lim_{x \to \infty} \left(1 + \frac{1}{x}\right)^x = e$ we can conclude that $\lim_{n \to \infty} \left(1 + \frac{1}{n}\right)^n = e$.

Is $\lim_{x \to \infty} (1 + \frac{1}{x})^x$ Really e?

Let's use the longstanding tactic of naming what we are looking for. (What follows is valid if we assume this limit exists and is finite.)

$$\text{Let } B = \lim_{x \to \infty} \left(1 + \frac{1}{x}\right)^x.$$

We have a variable in the exponent; we'd like to "bring it down" in order to make the expression on the right more tractable, so we'll take the natural logarithm of both sides of the equation:

$$\ln B = \ln \left[\lim_{x \to \infty} \left(1 + \frac{1}{x}\right)^x \right].$$

Because the logarithm is a continuous function, this is equivalent to writing[8]

$$\ln B = \lim_{x \to \infty} \left[\ln \left(1 + \frac{1}{x}\right)^x \right] = \lim_{x \to \infty} x \, \ln \left(1 + \frac{1}{x}\right).$$

As $x \to \infty$ we have the product of two numbers, one of which is going toward zero (because $\ln 1 = 0$), and the other which is growing without bound.[9] $\ln(1 + \frac{1}{x})$ is approaching $\ln 1 = 0$, but it is being multiplied by x, which is approaching ∞. It is still hard to decipher what is going on. Our area of expertise is more along the lines of taking the limit as something tends toward zero, and this may tie in with our defining characteristic of e. Let's try a substitution in hopes of figuring out this limit. Substitution is a tool for transforming the unfamiliar into the familiar.

$$\text{Let } h = \frac{1}{x}. \quad \text{As } x \to \infty, h \to 0.$$

With this substitution, $\lim_{x \to \infty} x \ln(1 + \frac{1}{x})$ becomes $\lim_{h \to 0} \frac{1}{h} \ln(1 + h) = \lim_{h \to 0} \frac{\ln(1+h)}{h}$.

$$\ln B = \lim_{h \to 0} \frac{\ln(1 + h)}{h}$$

This latter limit is easier to evaluate. In fact, you may recognize that it is the definition of the derivative of $\ln x$ at $x = 1$. Verify, using Figure 15.3, that the slope of the secant line

---

[7] At this point, we don't know how the calculator has arrived at its decimal expansion of e. Numerically all *we* have shown is that e lies between 2.7 and 2.8. If we can prove our conjecture, then we have a way of numerically approximating e.

[8] If f is a continuous function, then $f(\lim_{x \to a} g(x)) = \lim_{x \to a} f(g(x))$. This is limit principle (4).

[9] Here's another "indeterminate form": $\infty \cdot 0$. On the one hand, any finite number multiplied by zero is zero. On the other hand, if a number is growing without bound and we multiply it by any positive number, the product ought to grow without bound. (Of course, the second factor is not identically zero; it is just tending toward zero. Neither is the first factor "∞"; it is just growing without bound.) Again, our gut reaction to the problem pulls us in two different directions and the result churns the stomach.

through $(1, 0)$ and $(1 + h, \ln(1 + h))$ is

$$\frac{\ln(1 + h) - \ln 1}{(1 + h) - 1} = \frac{\ln(1 + h)}{h}.$$

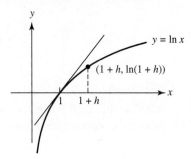

Figure 15.3

So $\displaystyle\lim_{h \to 0} \frac{\ln(1 + h)}{h} = \frac{d}{dx}(\ln x)\Big|_{x=1}.$

The derivative of $\ln x$ is $\frac{1}{x}$, and evaluating at $x = 1$ gives 1. Therefore, $\ln B = 1$. We were looking for B, not $\ln B$; exponentiating shows that $B = e^1 = e$. Eureka!

$$\boxed{\lim_{x \to \infty} \left(1 + \frac{1}{x}\right)^x = e}$$

Not only have we gained a whole new perspective on e, but we now have a method of approximating e numerically.

Having computed $\lim_{x \to \infty}(1 + \frac{1}{x})^x = e$, we can compute variations on this limit.

◆ **EXAMPLE 15.4**   Show that for any constant r, $\lim_{n \to \infty}(1 + \frac{r}{n})^n = e^r$.

SOLUTION   To evaluate this limit we will use the technique employed so successfully above; we will rename our variables. (Changing variables is a standard mathematical technique for converting something that looks a bit unfamiliar to something well known and having familiar structure.) We would like to replace $\frac{r}{n}$ by $\frac{1}{m}$ so we can utilize our previous result. If $\frac{1}{m} = \frac{r}{n}$, then $mr = n$, so $m = \frac{n}{r}$. Therefore, we use the substitution $m = \frac{n}{r}$. We can assume that $r \neq 0$, because the case $r = 0$ can be easily handled independently.

As $n \to \infty, \quad m = \frac{n}{r} \to \infty$ as well, because r is a constant.

Thus,

$$\lim_{n \to \infty} \left(1 + \frac{r}{n}\right)^n = \lim_{m \to \infty} \left(1 + \frac{1}{m}\right)^{mr} = \lim_{m \to \infty} \left[\left(1 + \frac{1}{m}\right)^m\right]^r.$$

But r is a constant, so this is equivalent to[10]

---

[10] We're using limit principle (4) again. If f is continuous, then $\lim_{x \to a} f(g(x)) = f(\lim_{x \to a} g(x))$.

$$\left[\lim_{m\to\infty}\left(1+\frac{1}{m}\right)^m\right]^r = e^r.$$

We conclude that

$$\boxed{\lim_{n\to\infty}\left(1+\frac{r}{n}\right)^n = e^r.}$$ ◆

◆ **EXAMPLE 15.5**   Find $\lim_{n\to\infty}(1+\frac{3}{n})^{4n}$ using the result of Example 15.4.

SOLUTION   CAUTION This limit business is subtle. We don't want to ad lib. Instead, we need to use exactly the results we have worked so hard to get.

$$\lim_{n\to\infty}\left(1+\frac{3}{n}\right)^{4n} = \lim_{n\to\infty}\left[\left(1+\frac{3}{n}\right)^n\right]^4 = [e^3]^4 = e^{12}$$ ◆

Implications of the Fact that $\lim_{n\to\infty}(1+\frac{r}{n})^n = e^r$

Recall: If you put $\$M_0$ in a bank account with nominal annual interest rate r compounded n times per year, then $M(t)$, the amount of money in the account after t years, is given by

$$\boxed{M(t) = M_0\left(1+\frac{r}{n}\right)^{nt}.}$$

If we let the number of compounding periods increase without bound, we obtain

$$M(t) = M_0\lim_{n\to\infty}\left(1+\frac{r}{n}\right)^{nt} = M_0\lim_{n\to\infty}\left[\left(1+\frac{r}{n}\right)^n\right]^t.$$

Having shown that $\lim_{n\to\infty}\left(1+\frac{r}{n}\right)^n = e^r$, we obtain $M(t) = M_0[e^r]^t = M_0e^{rt}$. Therefore, if a bank with nominal annual interest rate r compounds interest continuously,[11] then the money grows according to

$$\boxed{M(t) = M_0e^{rt}.}$$

EXERCISE 15.2   You plan to deposit a fixed sum of money into one of two bank accounts and leave it there for several years. Which is a better choice of accounts, an account with a nominal interest

---

[11] Do banks really do this? Certainly some banks compute interest on savings accounts every day. If the nominal annual interest rate is 5% and interest is compounded daily, then

$$M(t) = M_0\left(1+\frac{0.05}{365}\right)^{365t} \approx M_0(1.051267)^t.$$

If we modeled this situation using continuous compounding, we would have

$$M(t) = M_0e^{0.05t} \approx M_0(1.051271)^t.$$

Depending on the situation, the latter model may be quite reasonable and simpler to use. (Even using the equation $M(t) = M_0\left(1+\frac{0.05}{365}\right)^{365t}$ distorts reality a bit, because in reality if interest is compounded at the end of each day, $M(t)$ ought to be a step function with a point of discontinuity each time interest is computed, and yet this function is continuous.)

rate of 6% per year compounded monthly or an account with a nominal interest rate of $5\frac{1}{2}\%$ compounded continuously?

PROBLEMS FOR SECTION 15.1

1. Suppose you put $100,000 in a bank account with 6% interest and leave it for one year. How much money will there be in the account if the interest is compounded

   (a) annually?   (b) monthly?   (c) daily?   (d) hourly?

2. In 1996, inflation in Russia was 22%. This was a decline from the 131% inflation rate in 1995 and the 2600% inflation rate in 1994. By contrast, the inflation rate in the United States in 1996 was about 3%. (*Boston Globe*, November 2, 1996.)

   Compute the amount of time it would take for prices to double under each of the four inflation rates listed.

3. A *Boston Globe* article on January 1, 1997, said that the best stock of 1996 was Information Analysis, Incorporated, which closed the year at a price of $63 per share, an increase of 1525% during the year. The worst stock of 1996 was Mobilemedia Corporation, which closed the year at $7/16 per share, a decrease of 97.6%.

   What was the price of each of these stocks at the beginning of the year?

4. Compute the following limits. In each case stop to think of a strategy, and use whatever strategy seems simplest to you. For several of these limits there are different approaches.

   (a) $\lim_{x \to 3} \frac{3}{x-3}$

   (b) $\lim_{x \to 3} \frac{3}{(x-3)^2}$

   (c) $\lim_{x \to \infty} (1 - \frac{1}{3x})^{7x}$

   (d) $\lim_{t \to 0^+} (1 - 2t)^{1/t}$

5. Evaluate the following limits.

   (a) $\lim_{x \to \infty} \frac{e^{-x}}{x^2}$

   (b) $\lim_{x \to \infty} x^3 e^{-3x}$

6. Suppose you invest $10,000 in an account with a nominal annual interest rate of 5%. How much money will you have 10 years later if the interest is compounded

   (a) quarterly?   (b) daily?   (c) continuously?

7. (a) A certain amount of money is put in an account with a fixed nominal annual interest rate, and interest is compounded continuously. If 70 years later the money in the account has doubled, what is the nominal annual interest rate?

   (b) Answer the same question if the interest is compounded only once a year.

8. (a) Kevin has deposited money in a bank account that compounds interest quarterly. If the nominal interest rate is 5%, what is the effective interest rate?

   (b) Ama has deposited money in a bank account that compounds interest quarterly. If the effective interest rate is 5% per year, what is the nominal rate of interest?

9. Suppose that a person invests $10,000 in a venture that pays interest at a nominal rate of 8% per year compounded quarterly for the first 5 years and 3% per year compounded quarterly for the next 5 years.

(a) How much does the $10,000 grow to after 10 years?

(b) Suppose there were another investment option that paid interest quarterly at a constant interest rate r. What would r have to be for the two plans to be equivalent, ignoring taxes?

(c) If an investment scheme paid 3% interest compounded quarterly for the first 5 years and 8% interest compounded quarterly for the next 5 years, would it be better than, worse than, or equivalent to the first scheme?

10. Evaluate the following. Substitution may be helpful; these problems are variations on the theme $\lim_{n \to \infty} \left(1 + \frac{r}{n}\right)^n$.

(a) $\lim_{x \to 0^+} (1 + x)^{1/x}$

(b) $\lim_{w \to \infty} \left(\frac{w + 2}{w}\right)^w$

(c) $\lim_{x \to \infty} \left(\frac{x - 1}{x}\right)^{2n}$

(d) $\lim_{n \to \infty} \left(\frac{n}{n + 1}\right)^n$

(e) $\lim_{x \to 0^+} (1 + 2x)^{3/(2x)}$

11. Suppose you put $6000 in a bank account at 5% (nominal) annual interest compounded continuously.

(a) How much money do you have at the end of 7 years?

(b) How much money do you have at the end of t years?

(c) What is the instantaneous rate of change of money in the account with respect to time? (Find $\frac{dM}{dt}$.)

(d) True or False: $\frac{dM}{dt} = 0.05M$. Explain your reasoning!

(e) Write your answer to part (b) in the form $M = Ca^t$ and use your calculator to approximate the value of "a" numerically.

(f) Each year, by what percent does your money grow? (This is called the effective annual yield and, if interest is compounded more than once a year, it is always bigger than the nominal annual interest rate.)

12. Evaluate the following limits. Keep in mind that limit calculations can be subtle—don't ad lib, but instead keep the limits we looked at firmly in your mind and use substitution in order to make the transfer to the problems here. You can determine whether or not your answer is in the ballpark by using your calculator.

(a) $\lim_{x \to \infty} (1 + \frac{1}{x})^x$

(b) $\lim_{s \to \infty} (1 + \frac{1}{s})^{3s}$

(c) $\lim_{r \to \infty} (1 + \frac{0.3}{r})^r$

(d) $\lim_{w \to \infty} (1 + 2w^{-1})^w$

(e) $\lim_{w \to \infty} (1 + (2w)^{-1})^w$

13. Which is a better deal, an account offering 4% annual interest compounded continuously or an account offering 4.2% interest compounded annually? What is the effective annual yield of the former account?

14. If $M(t) = M_0 e^{rt}$, find $\frac{dM}{dt}$ and show that $\frac{dM}{dt} = rM$.

($\frac{dM}{dt} = rM$ is called a **differential equation** because it is an equation with a derivative in it. You have just shown that $M(t) = M_0 e^{rt}$ is a solution to this differential equation.)

15.2   INTRODUCING DIFFERENTIAL EQUATIONS

We have been characterizing exponential functions in two ways: If a quantity $Q = Q(t)$ grows (or decays) exponentially, then we have said either

    i. Q can be written in the form $Q(t) = Cb^t$, where C and b are constants; or

    ii. Q changes at a rate proportional to itself: $\frac{dQ}{dt} = kQ$ for some constant k.

The first statement gives an explicit formula for the amount, the second gives the rate of change. Let's express the statements in a form that makes their equivalence more transparent. In doing so we'll sidle up to the subject of differential equations, equations involving a rate (or rates) of change.

REMARK In the statements above we write Q and $Q(t)$ interchangeably, using the latter to emphasize that Q is a function of time.

We know that statement (i) implies statement (ii); if $Q(t) = Cb^t$, then

$$\frac{dQ}{dt} = C \cdot \ln b \cdot b^t$$
$$= \ln b \cdot Cb^t$$
$$= \ln b \cdot Q(t).$$

The constant "k" in $\frac{dQ}{dt} = kQ$ is equal to $\ln b$.

For a more aesthetically pleasing result, instead of writing $Q(t)$ as Cb^t, we can replace b by e^k and write $Q(t) = Ce^{kt}$. Now $\frac{dQ}{dt} = \ln e^k \cdot Q = kQ$.

We can rewrite statements (i) and (ii) as follows. If a quantity $Q = Q(t)$ grows (or decays) exponentially, then there is a constant k ($k > 0$ for growth, $k < 0$ for decay) such that

    i. $Q(t) = Ce^{kt}$ and

    ii. $\frac{dQ}{dt} = kQ$.

Let's put this in the framework of a bank account. If money in a bank account is growing at a nominal rate of 10% per year compounded continuously, then the instantaneous rate of change of money in the account is 10% of the amount in the account.

$$\left\{ \begin{array}{ll} \text{(i)} & M(t) = M_0 e^{0.1t} \\ \text{(ii)} & \frac{dM}{dt} = 0.1M \end{array} \right.$$

Equation (ii) really captures the situation in a simple way. The rate of change of the amount of money is 10% of the amount of money.

Frequently scientists have knowledge about the rate of change of a quantity and from this (and one piece of data about amount—like the initial amount) they try to find an amount equation. If you read scientific journals, you will find that the form $Q(t) = Ce^{kt}$ is often used to describe exponential growth or decay. If the scientist began with an equation of the form $\frac{dQ}{dt} = kQ$ describing the rate of change of the quantity in question, then that is the most natural form for the amount equation. The equation $\frac{dQ}{dt} = kQ$ is called a **differential**

equation. It is a differential equation that arises in many different disciplines, so we will take this opportunity to look at it more closely.

Differential Equations and Their Solutions

Definition

An equation that contains a derivative (or derivatives) is called a **differential equation**.

Some examples of differential equations are:

$$\frac{dy}{dx} = 3, \quad \frac{dy}{dt} = 2t, \quad \text{and} \quad \frac{dy}{dt} = y.$$

A function is a **solution** to a differential equation if it satisfies the differential equation; by this we mean that when the function and its derivative(s) are substituted in the appropriate places in the differential equation, the two sides of the equation are equal. A differential equation will actually have a whole family of solutions. A solution whose parameters have been determined is called a **particular solution**.

For example, the differential equation $\frac{dM}{dt} = 0.1$ has solutions $M(t) = Ce^{0.1t}$, where C can be any constant at all. $M(t) = Ce^{0.1t}$ gives a family of solutions. In fact, it can be proven that this is the **general solution** to the differential equation, meaning that *any* solution to the differential equation can be written in this form. $M(t) = 350e^{0.1t}$ is a particular solution to the differential equation.

Checking a solution to a differential equation is analogous to checking a solution to an algebraic equation in that if we are solving for y, then to check we must replace *all* occurrences of y (whether in the form y, y', or y'') by the proposed solution.

For example, to check whether $y = -1$ is a solution to the algebraic equation $y^2 - 3 = 2y$, we replace all occurrences of y by -1 (whether in the form y or y^2 or ...). We simply evaluate both sides of the equation at $y = -1$.

$$(-1)^2 - 3 \stackrel{?}{=} 2(-1)$$

$$1 - 3 \stackrel{?}{=} -2$$

$$-2 = -2 \quad \checkmark \quad y = -1 \text{ is a solution.}$$

Analogously, to check whether $y(t) = e^{-t}$ is a solution to the differential equation $\frac{d^2y}{dt^2} - 3y = 2\frac{dy}{dt}$, we evaluate both sides of the equation with $y = e^{-t}$.

$$\frac{d^2}{dt^2}[e^{-t}] - 3[e^{-t}] \stackrel{?}{=} 2\frac{d}{dt}[e^{-t}]$$

$$\frac{d}{dt}[-e^{-t}] - 3e^{-t} \stackrel{?}{=} -2e^{-t}$$

$$e^{-t} - 3e^{-t} \stackrel{?}{=} -2e^{-t}$$

$$-2e^{-t} = -2e^{-t} \quad \checkmark \quad y = e^{-t} \text{ is a solution.}$$

◆ **EXAMPLE 15.6**   Look for a solution to each of the differential equations given:
(a) $\frac{dy}{dx} = 3$ (b) $\frac{dy}{dt} = 2t$ (c) $\frac{dy}{dt} = y$.
Then look for a family of solutions. Verify that this family of functions actually solves the differential equation.

SOLUTION

(a) $\frac{dy}{dx} = 3$. We're looking for a function $y(x)$ whose derivative is 3.

$y = 3x$ will do. It's a line with slope 3.

$y = 3x + 2$ will work as well.

More generally, we have $y = 3x + C$ for any constant C.

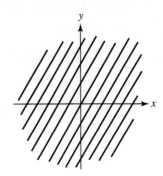

Figure 15.4   A family of lines with slope 3

Check: The differential equation is $\frac{dy}{dx} = 3$. Is $y = 3x + C$ a solution? Replace y by $3x + C$ and see if the two sides are equal.

$$\frac{d}{dx}[3x + C] \overset{?}{=} 3$$

$$3 = 3 \quad \checkmark$$

(b) $\frac{dy}{dt} = 2t$. We're looking for a function $y(t)$ whose derivative is $2t$.
$y = t^2$ will do. Its derivative is $2t$.

More generally, we can use $y = t^2 + C$ for any constant C.

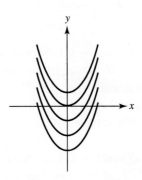

Figure 15.5   A family of parabolas with slope $2t$

Check: The differential equation is $\frac{dy}{dt} = 2t$. Is $y = t^2 + C$ a solution? Replace y by $t^2 + C$ and see if both sides are equal.

$$\frac{d}{dt}[t^2 + C] \stackrel{?}{=} 2t$$

$$2t = 2t \quad \checkmark$$

(c) $\frac{dy}{dt} = y$. We're looking for a function $y(t)$ whose derivative is itself. We know of one function whose derivative is itself: e^x. But we want a function of t, so $y(t) = e^t$.

Thinking back to what we were doing right before introducing the term "differential equation," we see that more generally $y = Ce^t$ for any constant C.

Check: The differential equation is $\frac{dy}{dx} = y$. Is $y = Ce^t$ a solution? Replace y by Ce^t and see if both sides are equal.

$$\frac{d}{dt}[Ce^t] \stackrel{?}{=} Ce^t$$

$$Ce^t = Ce^t \quad \checkmark$$

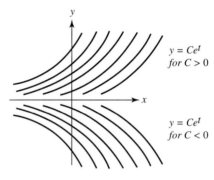

$y = Ce^t$
for $C > 0$

$y = Ce^t$
for $C < 0$

Figure 15.6   ◆

Suppose that, carried away by the success of adding C in parts (a) and (b), we thought that $y = e^t + C$ might be a solution. Let's check it.

Check: $\frac{dy}{dt} = y$. Is $y = e^t + C$ a solution? Replace y by $e^t + C$ and see if both sides are equal.

$$\frac{d}{dt}[e^t + C] \stackrel{?}{=} e^t + C$$

$$e^t \stackrel{?}{=} e^t + C \quad \text{No (unless } C = 0\text{)}.$$

So $y = e^t + C$ is not a solution to the differential equation.

We have verified that if $Q(t) = Ce^{kt}$, then $\frac{dQ}{dt} = kQ$; we know that the family of functions $Q(t) = Ce^{kt}$ is a solution to the differential equation $\frac{dQ}{dt} = kQ$. In fact, *any* solution *must* be of this form, so we refer to $Q(t) = Ce^{kt}$ as the **general solution** to the differential equation.

REMARK When we study differential equations in greater depth we will see that if $\frac{dy}{dt} = f(t)$, then, having found one solution, we can find the general solution by adding a constant C, whereas if $\frac{dy}{dt} = f(y)$, this is *not* the case. See if you can make sense of this graphically.

General and Particular Solutions to Differential Equations

The families of functions we found as solutions to the differential equations in Example 15.6 are not simply *more* general solutions but in fact are *the general solutions* to each of the differential equations. By this we mean that *any* particular solution to the differential equation can be expressed in this form.

Terminology: The graph of a solution is called a **solution curve**.

The table below gathers together examples of some differential equations and their solutions.

Differential Equation	Some Particular Solutions	The General Solution	Some Solution Curves
$\dfrac{dy}{dt} = 0$	$\begin{cases} y = 7 \\ y = -12 \end{cases}$	$y = C$	
$\dfrac{dy}{dx} = 3$	$\begin{cases} y = 3x \\ y = 3x + 2 \end{cases}$	$y = 3x + C$	
$\dfrac{dy}{dt} = 2t$	$\begin{cases} y = t^2 + 1 \\ y = t^2 - 4 \end{cases}$	$y = t^2 + C$	
$\dfrac{dy}{dt} = y$	$\begin{cases} y = e^t \\ y = -5e^t \\ y = 0 \end{cases}$	$y = Ce^t$	
$\dfrac{dy}{dt} = ky$	$\begin{cases} y = 5e^{kt} \\ y = -2e^{kt} \end{cases}$	$y = Ce^{kt}$	$k > 0$ $k < 0$

In this section we will focus primarily on the differential equation $\frac{dy}{dt} = ky$. It is ubiquitous, arising in fields as varied as economics, biology, and chemistry.

> The differential equation $\dfrac{dy}{dt} = ky$ has the general solution
>
> $$y(t) = Ce^{kt}, \text{ where } C \text{ is an arbitrary constant.}$$

In order to pick out a particular solution it is only necessary to know one point on the solution curve.[12] This is equivalent to being given exactly one data point, that is, one value of the independent variable along with the corresponding value of the dependent variable. Such a piece of information is called an **initial condition**.

EXERCISE 15.3   Let k be an arbitrary constant. Show that for any point P in the plane there is a value of C such that the curve $y = Ce^{kt}$ passes through P.

◆ EXAMPLE 15.7   For each of the differential equations below, find the particular solution corresponding to the initial condition given.

(a) $\frac{dQ}{dt} = -0.2Q$, where $Q(0) = 10$

(b) $\frac{dW}{dx} = 3W$, where $W(2) = -7$

SOLUTION

(a) The general solution is $Q(t) = Ce^{-0.2t}$. To find the particular solution, we use the information that $Q = 10$ when $t = 0$.

$$10 = Ce^{-0.2(0)}$$
$$10 = C$$

The particular solution is $Q(t) = 10e^{-0.2t}$.

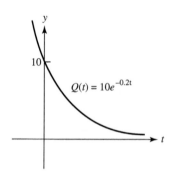

Figure 15.7

(b) This differential equation is basically of the same form as the one in part (a). The role of t is now played by x; the role of Q is now played by W. The general solution is

$$W = Ce^{3x}.$$

---

[12] This is true for any differential equation involving only first derivatives.

To find the particular solution, we use the information that $W = -7$ when $x = 2$.

$$-7 = Ce^{3(2)}$$

$$C = -7e^{-6} \approx -0.01735$$

So $W(x) = -7e^{-6}e^{3x}$.   ◆

Modeling a Situation with a Differential Equation

◆ **EXAMPLE 15.8**   For each of the scenarios given below, write a differential equation modeling the situation.

(a) **Economics.** Money is being kept in a savings account with a nominal interest rate of 5% per year compounded continuously. If no deposits and no withdrawals are made, then the *rate of change of the amount of money in the account* is proportional to the amount of money in the account at any given time.

(b) **Population Biology.** The *rate of growth of a population* under ideal circumstances (unlimited food, space, etc.) is proportional to the size of the population itself at any time.

(c) **Chemistry.** A radioactive substance *decays at a rate* proportional to the amount of the substance present at any given time.

SOLUTION

(a) Let $M = M(t)$ be the amount of money in the account as a function of time. Then $\frac{dM}{dt}$ is the rate of change of money in the account. *The rate of change of the amount of money is proportional to the amount of money* translates to

$$\frac{dM}{dt} = kM, \text{ where } k \text{ is a constant of proportionality.}$$

We know that $k = 0.05$ because that is the rate at which interest accumulates. Therefore

$$\frac{dM}{dt} = 0.05M.$$

Note that we've just shown this in Section 15.1; if money is compounded continuously at a rate of 5%, then the amount of money in the account after t years is given by $M(t) = \lim_{n \to \infty} M_0(1 + \frac{0.05}{n})^{nt} = M_0 e^{0.05t}$. If $M(t) = M_0 e^{0.05t}$, then $dM/dt = 0.05M(t)$. The proportionality constant is 0.05 *not* 1.05, because we are looking at the rate of *increase* of M; the *increase* is due to interest and the interest rate is 5% of $M(t)$, *not* 105% of $M(t)$.

(b) Let $P = P(t)$ represent the size of the population as a function of time. Then $\frac{dP}{dt}$ is the rate of growth of the population. Our task is really just a matter of translating prose into mathematical notation. *The population's rate of change is proportional to the population itself* translates to

$$\frac{dP}{dt} = kP, \text{ where } k \text{ is the constant of proportionality } (k > 0).$$

(c) Let $S(t)$ represent the number of milligrams of the substance present at time t. Then we can model the situation with the differential equation

$$\frac{dS}{dt} = kS.$$

Because S is decreasing over time, $\frac{dS}{dt}$ should be negative. That means that kS must also be negative. Because S (the amount of the substance) has to be positive, k must be a negative constant.[13]   ◆

EXERCISE 15.4   Suppose that the radioactive substance above decays at a continuous rate of 3%. Is the value of k 0.03, or -0.03, or 0.97, or -0.97?

The answer is supplied at the end of the section.

EXERCISE 15.5   Realistically speaking, most populations do not exhibit true exponential growth in the long run, because resources are not, in practice, unlimited. Consider, for example, the population of fish in a pond. In the short run, the number of fish may appear to grow exponentially, but in the long run, we don't expect this to continue. If it did, the pond would eventually be solidly packed with fish. Instead, in the long run, we expect the number of fish to level off at what ecologists refer to as the carrying capacity of the pond. Suppose the carrying capacity of this particular pond is 700 fish. In this problem we model the long-run behavior of the population. (When we return to differential equations, we will construct a model that reflects the situation *both* in the long run and at the beginning.)

I. We can model the situation by saying that due to limited resources, the difference between the number of fish and the carrying capacity of the pond decays exponentially.

(a) Use statement (i) in Section 15.2 characterizing exponential functions to write an equation for $D(t)$, where $D(t)$ is the difference between the number of fish at time t and the carrying capacity of the pond. Make all constants in your answer positive.

(b) Use the equation from part (a) to solve for $F(t)$, the number of fish at time t. Note that $D(t) = 700 - F(t)$, where $F(t)$ is the number of fish in the pond at time t.

(c) Graph $F(t)$, assuming that $F(0) < 700$.

II. Equivalently, we can model the situation by saying that due to limited resources, the difference $D(t)$ changes at a rate proportional to itself.

(a) Write a differential equation involving D and reflecting the situation. Again, use a positive proportionality constant k.

(b) Use the differential equation from part II(a) to write a differential equation involving $F(t)$. Write a differential equation reflecting the situation but this time put it in the form $dF/dt = \ldots$.

(c) Translate the differential equation you obtained in part II(b) to a sentence about the rate at which the fish population is growing.

III. Check that the function $F(t)$ obtained in part I(b) is a solution to the differential equation obtained in part II(b).   ◆

The answer is supplied at the end of the section.

◆ **EXAMPLE 15.9**   A population is growing at a rate proportional to itself. When the population is 11,500 it is growing at a rate of 225 per year. How fast is it growing when the population has reached 19,000?

---

[13] Alternatively, we could write $\frac{dS}{dt} = -kS$, where k is a positive constant. Often it is nice to keep all constants positive; $\frac{dS}{dt} = -kS, k > 0$ reminds us that the rate of change is negative.

SOLUTION   Let $P = P(t)$ represent the size of the population at time t. Notice that it is much simpler to write a differential equation modeling this scenario than it is to write an amount function.

$$\frac{dP}{dt} = kP$$

We are told that $\frac{dP}{dt} = 225$ when $P = 11{,}500$, so we can find k.

$$225 = k(11{,}500)$$

$$k = \frac{225}{11{,}500}$$

$$k = \frac{9}{460}$$

So $\frac{dP}{dt} = \frac{9}{460}P$. If $P = 19{,}000$ then $\frac{dP}{dt} = \frac{9}{460} \cdot 19{,}000 \approx 371.7$.

When the population is 19,000, it is growing at a rate of roughly 372 individuals per year.   ◆

Answer to Exercise 15.4

The value of k should be -0.03. k must be negative, because the quantity is decreasing. Why does $k = -0.03$? The rate of decay is 3% of itself, so the change in S is 3% of S, not 97% of S.

Answers to Exercise 15.5

I. (a) $D(t) = Ce^{-kt}$

(b) $700 - F(t) = Ce^{-kt}$

$$- F(t) = -700 + Ce^{-kt}$$
$$F(t) = 700 - Ce^{-kt}$$

(c)

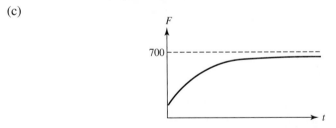

Figure 15.8

II. (a) $\frac{dD}{dt} = -kD$

(b) $\frac{d}{dt}(700 - F) = -k(700 - F)$

$$-\frac{dF}{dt} = -k(700 - F)$$
$$\frac{dF}{dt} = k(700 - F)$$

(c) The fish population is growing at a rate proportional to the difference between the carrying capacity of the pond and the number of fish in the pond.

III. We want to show that $F(t) = 700 - Ce^{-kt}$ is a solution to the differential equation $\frac{dF}{dt} = k(700 - F)$.

$$\frac{dF}{dt} = k(700 - F)$$

$$\frac{d}{dt}\left[700 - Ce^{-kt}\right] \overset{?}{=} k[700 - (700 - Ce^{-kt})]$$

$$Cke^{-kt} \overset{?}{=} kCe^{-kt}$$

$$kCe^{-kt} = kCe^{-kt} \qquad \checkmark$$

Exploratory Problems for Chapter 15

Population Studies

We've focused a lot of attention on the differential equation

$$\frac{dy}{dt} = ky.$$

In the problems that follow you will use substitution to extend your knowledge about this differential equation to a much larger class of differential equations.

Problem. In a laboratory, a colony of fruit flies are under scrutiny. Let $P = P(t)$ be the number of flies in the colony at time t.

Given ample food, the population would grow at a rate proportional to itself according to $\frac{dP}{dt} = kP$. But flies are continually being siphoned off to another lab at a constant rate of C flies per day.

The rate of change of the fly population is the rate at which it is increasing due to reproduction minus the rate at which flies are being siphoned off.

We model the situation as follows.

(Rate of change) = (rate of increase) − (rate of decrease)

$$\frac{dP}{dt} = kP - C$$

We want to solve this differential equation for $P(t)$.

1. Let's work on the equation

$$\frac{dP}{dt} = 0.02P - 2.$$

First we'll rewrite it as

$$\frac{dP}{dt} = 0.02(P - 100).$$

We'll convert this differential equation to the form that we are familiar with by making the substitution $y = P - 100$.

(a) Express the differential equation $\frac{dP}{dt} = 0.02(P - 100)$ in terms of y.

(b) Solve the differential equation you got in part (a).

(c) Knowing that $y = P - 100$, find $P(t)$. There will be an arbitrary constant in your answer because you will have found the general solution to the differential equation.

(d) Suppose that $P(0) = 3000$. Find the solution corresponding to this particular initial condition.

2. Use similar substitution techniques to solve the following differential equations.

(a) $\frac{dQ}{dt} = 2Q - 6$

(b) $\frac{dM}{dt} = 0.1M - 200$

PROBLEMS FOR SECTION 15.2

1. (a) Suppose a population grows at a rate of 5% per year: $P = P_0(1.05)^t$.
    i. Express this in the form $P = P_0 e^{rt}$.
    ii. Compute $\frac{dP}{dt}$.
    iii. Find the proportionality constant k so that $\frac{dP}{dt} = kP$.
   (b) Suppose a population grows according to $P = P_0 e^{0.05t}$
    i. Find the proportionality constant k so that $\frac{dP}{dt} = kP$.
    ii. By what percent does the population grow each year?
    Look back over this problem and *think* about it. Do your answers make sense to you?

2. Forestland in Borneo is being destroyed at a rapid rate. This is factual; the figures given in this problem are not. Assume that no new forestland is being created. Let $F(t)$ be the number of acres of forestland in Borneo.

   (a) Suppose that each year 10% of the existing forestland in Borneo is destroyed. What is $F(t)$? How many years will it take for half of the forestland in the country to be destroyed?

   (b) Suppose instead that forestland in Borneo is being destroyed at a continuous rate of 10% per year. What is $F(t)$? How many years will it take for half of the forestland in the country to be destroyed?

   (c) If forestland in Borneo is being destroyed at a continuous rate of 10% per year, what is $\frac{dF}{dt}$? Does the sign of your answer make sense? Write your answer in the form $\frac{dF}{dt} = kF$.

3. Solve the following differential equations. Give a general solution and then a particular solution corresponding to the initial condition given.
   (a) $\frac{dy}{dt} = 3y$ initial condition: $y(0) = 5$
   (b) $\frac{dy}{dx} = -0.01y$ initial condition: $y(2) = 1$
   (c) $\frac{dw}{ds} = w$ initial condition: $w(0) = \pi$

4. Solve the following differential equations. For each differential equation, find the general solution and then find a solution passing through the point $(0, \sqrt{2})$.
   (a) $\frac{dy}{dt} = -2y$ (b) $\frac{dy}{dt} = -2t$ (c) $\frac{dy}{dt} = -2$

5. (a) Show that $P = Ce^{2t}$ (where C is any constant) is a solution to the differential equation $\frac{dP}{dt} = 2P$. That is, show that if you compute $\frac{dP}{dt}$, you get $2P$.
   (b) Show that $P = e^{2t} + C$ is *not* a solution to the differential equation $\frac{dP}{dt} = 2P$.

6. Consider the differential equation $\frac{dy}{dt} = y - 2$. Which of the functions below are solutions? There could be more than one answer.
   (a) $y = e^t + 2$
   (b) $y = e^t + 3$
   (c) $y = Ce^t + 2$
   (d) $y = C(e^t + 2)$

7. (a) Is $y = e^t + \ln t$ a solution to the differential equation $\frac{dy}{dt} = y - \frac{y}{t}$?

   (b) Is $y = te^t$ a solution to the differential equation $\frac{dy}{dt} = y - \frac{y}{t}$?

8. A wet dish towel is put on the back of a kitchen chair to dry. It dries at a rate proportional to the difference in moisture content between the dishtowel and the kitchen air. Assume that the moisture content in the air is fixed and is given by M.

   (a) Set up the differential equation involving $W = W(t)$, the amount of water in the dish towel at time t.

   (b) Find and sketch the solution.

9. When a population has unlimited resources and is free from disease and strife, the rate at which the population grows is proportional to the population. Assume that both the bee and the mosquito populations described below behave according to this model.

   In both scenarios you are given enough information to find the proportionality constant k. In one case, the information allows you to find k solely using the differential equation, without requiring that you solve it. In the other scenario, you must actually solve the differential equation in order to find k.

   (a) Let $M = M(t)$ be the mosquito population at time t, t in weeks. At $t = 0$, there are 1000 mosquitoes. Suppose that when there are 5000 mosquitoes, the population is growing at a rate of 250 mosquitoes per week. Write a differential equation reflecting the situation. Include a value for k, the proportionality constant.

   (b) Let $B = B(t)$ be the bee population at time t, t in weeks. At $t = 0$, there are 600 bees. When $t = 10$, there are 800 bees. Write a differential equation reflecting the situation. Include a value for k, the proportionality constant.

10. Newton's law of cooling in its more general form tells us that the rate at which the temperature between an object and its environment changes is proportional to the difference in temperatures. In other words, if $D(t)$ is the temperature difference, then $\frac{dD}{dt} = kD$.

   (a) Solve the differential equation $\frac{dD}{dt} = kD$ for $D(t)$.

   (b) Suppose a hot object is placed in a room whose temperature is kept constant at R degrees. Let $T(t)$ be the temperature of the object. Newton's law says that the hot object will cool at a rate proportional to the difference in temperature between the object and its environment. Write a differential equation reflecting this statement and involving T. Explain why this differential equation is equivalent to the previous one.

   (c) What is the sign of the constant of proportionality in the equation you wrote in part (b)? Explain.

   (d) Suppose that instead of a hot object we now consider a cold object. Suppose that we are interested in the temperature of a cold cup of lemonade as it warms up to room temperature. Let $L(t)$ represent the temperature of the lemonade at time t and assume that it sits in a room that is kept at 65 degrees. At time $t = 0$, the lemonade is at 40 degrees. 15 minutes later it has warmed to 50 degrees.

      i. Sketch a graph of $L(t)$ using your intuition and the information given.

      ii. Is $L(t)$ increasing at an increasing rate, or a decreasing rate?

iii. Write a differential equation reflecting the situation. Indicate the sign of the proportionality constant.

iv. Find $L(t)$. Your final answer should have no undetermined constants. Does the graph of the function you got for $L(t)$ match the graph you drew?

v. How long will it take the lemonade to reach a temperature of 55 degrees?

11. Suppose that in a certain country the population grows at a rate proportional to itself with proportionality constant 0.02. Further suppose that due to a drought people are leaving the country at a constant rate of 1000 people per year. Let $P = P(t)$ be the population of the country at time t, where t is in years. Write a differential equation modeling the situation.

12. Money in a certain trust-fund account is earning 5% interest per year compounded continuously. Suppose money is being withdrawn from the account at a constant rate of $2000 per year. For the sake of our model, assume that money is being withdrawn continuously. The account begins with $30,000. Let $M = M(t)$ be the amount of money in the account at time t, where t is in years. Write a differential equation modeling the situation. What is the initial condition?

13. Solve the following differential equations. Use substitution to convert them to the form $\frac{dy}{dt} = ky$.

    (a) $\frac{dy}{dt} = 3y - 6$

    (b) $\frac{dy}{dt} = y + 1$

    (c) $\frac{dy}{dt} = 4 - 2y$

14. Solve $\frac{dy}{dt} = 2y - 6$ with the initial condition $y(0) = 2000$.

15. Return to Problem 12. Determine the amount of money in the account at time t by solving the differential equation and using the initial condition.

16. A boarding school with 800 students has been hit by a flu epidemic. If we assume that every student is either sick or healthy, that sick students will infect healthy ones, and that the disease is quite long in duration (it's a nasty flu) then we can model the epidemic using the following assumption.

    The rate at which students are getting sick is proportional to the product of the number of sick students and the number of healthy ones.

    (a) Let $S = S(t)$ be the number of sick students at time t. Translate the statement above into mathematical language.

    (b) $S(t)$ is an increasing function. The rate at which students are getting sick is a quadratic function of S. When the rate at which students are getting sick is highest, how many students are sick?

17. In the beginning of a chemical reaction there are 800 moles of substance A and none of substance B. Over the course of the reaction, the 800 moles of substance A are converted to 800 moles of substance B. (Each molecule of A is converted to a molecule of B via the reaction.) Suppose the rate at which A is turning into B is proportional to the product of the number of moles of A and the number of moles of B.

(a) Let $N = N(t)$ be the number of moles of substance A at time t. Translate the statement above into mathematical language. (*Note:* The number of moles of substance B should be expressed in terms of the number of moles of substance A.)

(b) $N(t)$ is a decreasing function. The rate at which N is changing is a function of N, the number of moles of substance A. When the rate at which A is being converted to B is highest, how many moles are there of substance A?

C H A P T E R

16

Taking the Derivative
of Composite Functions

16.1 THE CHAIN RULE

We can construct conglomerate functions in two different ways. One way is to combine the functions' outputs by taking, for example, their sum or product. We can differentiate a sum by summing the derivatives and differentiate a product by applying the Product Rule.

Another way to construct a conglomerate is to have functions operate in an assembly-line manner. In this configuration, the output of one function becomes the input of the next, creating a composite function. In this section we will look at the derivatives of composite functions. The work we do will have plentiful rewards, as the results have extensive application.

Our goal is to express the derivative of the composite function $f(g(x))$ in terms of f, g, and their derivatives. Our assumption throughout is that f and g are differentiable functions.

Many functions we can't yet differentiate can be decomposed and expressed as the composite of simpler functions whose derivatives we know. For example:

$h(x) = \sqrt{\ln x}$ can be decomposed into $f(g(x))$, where $f(x) = \sqrt{x}$ and $g(x) = \ln x$.

$h(x) = (x^2 + 1)^{30}$ can be decomposed into $f(g(x))$, where $f(x) = x^{30}$ and $g(x) = x^2 + 1$.

$h(x) = e^{x^2}$ can be decomposed into $f(g(x))$, where $f(x) = e^x$ and $g(x) = x^2$.

$h(x) = \ln(x + 8x^2)$ can be decomposed into $f(g(x))$, where $f(x) = \ln x$ and $g(x) = x + 8x^2$.

We'll look at the problem of differentiating a composite function in the context of the next example.

◆ **EXAMPLE 16.1**   The number of fish a lake can support varies with the water quality. The water quality is affected by industry around the lake; the level of grime in the lake varies with time. For the purposes of this example, we'll assume that the level of grime in the lake is always increasing. The number of fish decreases as the level of grime goes up.

Let $g(t)$ give the level of grime in the lake as a function of time t, measured in years. Let $f(g)$ be the fish population as a function of the amount of grime. Then the population of fish as a function of time is the composite function $f(g(t))$. Assume that f and g are differentiable functions. Find the rate at which the fish population is changing over time.

SOLUTION   We must compute $\frac{d}{dt} f(g(t))$, or $\frac{df}{dt}$. Essentially, we argue as follows.

$$\frac{\Delta f}{\Delta t} = \frac{\Delta f}{\Delta g} \cdot \frac{\Delta g}{\Delta t}.$$

From a purely algebraic standpoint, this must be true, provided Δg and Δt are not zero.

Let $\Delta g = g(t + \Delta t) - g(t)$ and $\Delta f = f(g + \Delta g) - f(g)$. As $\Delta t \to 0$, we know $\Delta g \to 0$ because g is a continuous function.

$$\lim_{\Delta t \to 0} \frac{\Delta f}{\Delta t} = \lim_{\Delta t \to 0} \left(\frac{\Delta f}{\Delta g} \cdot \frac{\Delta g}{\Delta t} \right)$$

$$\lim_{\Delta t \to 0} \frac{\Delta f}{\Delta t} = \left(\lim_{\Delta t \to 0} \frac{\Delta f}{\Delta g} \right) \cdot \left(\lim_{\Delta t \to 0} \frac{\Delta g}{\Delta t} \right)$$

$$\lim_{\Delta t \to 0} \frac{\Delta f}{\Delta t} = \left(\lim_{\Delta g \to 0} \frac{\Delta f}{\Delta g} \right) \cdot \left(\lim_{\Delta t \to 0} \frac{\Delta g}{\Delta t} \right)$$

$$\frac{df}{dt} = \frac{df}{dg} \cdot \frac{dg}{dt}$$

For the purposes of this problem, since we've asserted that g is increasing, we haven't run into trouble. In terms of generalizing, we have trouble only if $\Delta g = 0$ infinitely many times as $\Delta t \to 0$. But if this is the case, it can be shown that both $\frac{dg}{dt} = 0$ and $\frac{df}{dt} = 0$, so the equation $\frac{df}{dt} = \frac{df}{dg} \cdot \frac{dg}{dt}$ still holds.

Let's return to the fish and make things more concrete. Suppose we want to find the rate of change of the fish population with respect to time at $t = 3$, and suppose that at that time the grime level in the lake is 700 units. Let's look at $\frac{\Delta f}{\Delta t}$, the rate of change of the number of fish over the time interval $[3, 3 + \Delta t]$, where Δt is very small.

Over a given time interval, the level of grime varies, and this causes a fluctuation in the number of fish in the lake. We're interested in finding out how a change in time, Δt, affects the fish population. The change in time affects the fish population indirectly via the change in the grime level of the lake. There is a chain reaction; time affects grime, and grime affects fish. Therefore, we first look at the change in the level of grime produced by a Δt change in time.

How does $g(t)$ change during this small interval? We know that $\frac{dg}{dt} \approx \frac{\Delta g}{\Delta t}$ for Δt very small, so we can solve for Δg.

$$\Delta g \approx \frac{dg}{dt}\Delta t$$

Looking at the time interval $[3, 3 + \Delta t]$ gives us the following.

$$\Delta g = g(3 + \Delta t) - g(3)$$

$$\Delta g \approx \left[\frac{dg}{dt}\bigg|_{t=3} \right] \Delta t$$

$$\approx g'(3)\Delta t$$

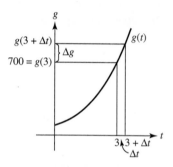

Figure 16.1

Now we need to determine how this change in the level of grime, Δg, will affect the fish population. (We've assumed in the problem that g is always increasing, so $\Delta g \neq 0$.) At $t = 3$, the grime level is 700. As the grime level changes from 700 to $700 + \Delta g$, how does the fish population change?

As above, we can write $\frac{df}{dg} \approx \frac{\Delta f}{\Delta g}$ for Δg very small. Because Δt is very small (and will approach zero), Δg is very small as well.[1] Solving for Δf gives

$$\Delta f \approx \frac{df}{dg}\Delta g.$$

Focusing on what is happening at $t = 3$ (when $g = 700$) leads to this equation.

$$\Delta f = f(g(3 + \Delta t)) - f(g(3))$$

$$= f(700 + \Delta g) - f(700)$$

$$\Delta f \approx \left[\frac{df}{dg}\bigg|_{g=700} \right] \cdot \Delta g$$

$$\approx f'(700)\Delta g$$

---

[1] This is because g is a continuous function. We know g is continuous because we are working under the assumption that g is differentiable.

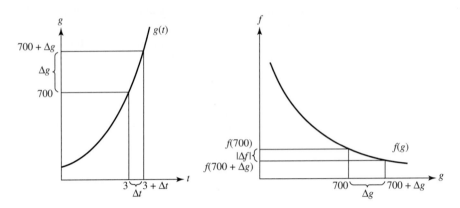

Figure 16.2

We can use the expression for Δg on the preceding page.

$$\Delta f \approx f'(700)\Delta g$$

$$\approx f'(700)g'(3)\Delta t$$

$$\approx f'(g(3)) \cdot g'(3)\Delta t$$

Now we can find an expression for $\frac{\Delta f}{\Delta t}$, the rate of change of the population with respect to time.

$$\frac{\Delta f}{\Delta t} \approx \frac{f'(g(3))g'(3)\Delta t}{\Delta t}$$

$$\approx f'(g(3))g'(3)$$

Let's look at the approximations we made in this discussion. As Δt gets closer and closer to zero, the approximation $\frac{dg}{dt} \approx \frac{\Delta g}{\Delta t}$ gets better and better, because

$$\lim_{\Delta t \to 0} \frac{\Delta g}{\Delta t} = \frac{dg}{dt}.$$

As Δt approaches zero, $\Delta g = g(t + \Delta t) - g(t)$ also approaches zero. As Δg gets closer and closer to zero, the approximation $\frac{df}{dg} \approx \frac{\Delta f}{\Delta g}$ gets better and better, because

$$\lim_{\Delta g \to 0} \frac{\Delta f}{\Delta g} = \frac{df}{dg}.$$

This leads us to conclude that

$$\frac{d}{dt}f(g(t))\bigg|_{t=3} = f'(g(3)) \cdot g'(3).$$

In fact, there was nothing special about the time $t = 3$, so this relation should hold for all values of t:

$$\left[f(g(t))\right]' = f'(g(t)) \cdot g'(t).$$

A unit analysis gives $\frac{\text{fish}}{\text{time}} = \frac{\text{fish}}{\text{grime}} \cdot \frac{\text{grime}}{\text{time}}$, which makes sense. The crucial characteristic of f and g is that they are differentiable. In our discussion, we need the assumption that $\Delta g \neq 0$, but, as previously mentioned, there are ways of getting around this restriction.

◆

The result can be stated for any two differentiable functions f and g; it goes by the name "the Chain Rule."

The Chain Rule:

$$\frac{d}{dt} f(g(t)) = f'(g(t)) \cdot g'(t) \qquad \text{or} \qquad \frac{df}{dt} = \frac{df}{dg} \cdot \frac{dg}{dt}$$

Interpreting the Chain Rule

■ Thoughts on the form $\frac{d}{dt} f(g(t)) = f'(g(t)) \cdot g'(t)$:
More informally, we can state the Chain Rule as

$$\frac{d}{dt} (f(mess)) = f'(mess) \cdot (mess)',$$

where *mess*, of course, is just a function of t; it's $g(t)$.

Notice that the derivative of f is evaluated at $g(t)$; $f'(g(t))$, **not** $f'(t)$. We obtain $f'(g(t))$ by calculating $f'(x)$ and replacing x by $g(t)$. Think about our example; f is a function of grime, so f' is also a function of grime, not of time. We evaluated f' at $g = 700$ (or $g(3)$), not at $g = 3$.

$$\left[f(g(t))\right]' = f'(g(t)) \cdot g'(t) = (\text{derivative of } f \text{ evaluated at } g(t)) \cdot (\text{derivative of } g)$$

■ Thoughts on the form $\frac{df}{dt} = \frac{df}{dg} \cdot \frac{dg}{dt}$: Leibniz's notation gives us a very nice way of expressing the Chain Rule. Although $\frac{df}{dg}$ and $\frac{dg}{dt}$ are *not* fractions, the notation works well; that is the genius of it.

Let's take a few more quick passes on interpreting the Chain Rule in this form.

If g changes three times as fast as t and f changes twice as fast as g, then f changes 6 times as fast as t. $\frac{df}{dt} = \frac{df}{dg} \cdot \frac{dg}{dt}$

Or, think of gears—either interlocking or connected by a chain as in bicycle gears. Suppose the little gear spins 12 times per minute and the big gear spins once for every three turns of the little gear. Then the big gear spins 4 times per minute.

$$\frac{\text{rotations of big gear}}{\text{rotations of little gear}} \cdot \frac{\text{rotations of little gear}}{\text{minute}} = \frac{\text{rotations of big gear}}{\text{minute}}.$$

Applying the Chain Rule

Let's apply the Chain Rule to the functions $h(x) = f(x + k)$ and $j(x) = f(kx)$, corresponding to a horizontal shift of f and a horizontal compression of f, respectively. If you haven't previously done so, first spend a minute trying to determine $h'(x)$ and $j'(x)$ by graphical means.

◆ **EXAMPLE 16.2** $f(x+k)$ can be thought of as the composite $f(g(x))$, where $g(x) = x + k$.

$$[f(x+k)]' = f'(g(x)) \cdot g'(x)$$
$$= f'(x+k) \cdot 1$$
$$= f'(x+k)$$

This is in agreement with our graphical intuition.   ◆

◆ **EXAMPLE 16.3** $f(kx)$ can be thought of as the composite $f(g(x))$, where $g(x) = kx$.

$$[f(kx)]' = f'(g(x)) \cdot g'(x)$$
$$= f'(kx) \cdot k$$
$$= kf'(kx)$$

This makes sense graphically. Suppose $k > 0$. Because the function's graph is compressed horizontally, its derivative will be as well. This horizontal compression results in steeper slopes; hence $f'(kx)$ is multiplied by k.   ◆

Some functions that we could differentiate by taking advantage of laws of logs and exponentials we can now differentiate using the Chain Rule. The next example illustrates these options.

◆ **EXAMPLE 16.4**   Find the derivatives of the following, where k is a constant.

(a) $q(x) = e^{kx+2}$ (b) $s(x) = \ln(kx)$

SOLUTION   (a) *Using Exponent Laws:* $q(x) = e^{kx}e^2$, so $q'(x) = e^2 k e^{kx} = k \cdot e^{kx+2}$.

Using the Chain Rule: Let the inside function, $g(x)$, be $kx + 2$ and the outside function, $f(u)$, be e^u. Then $g'(x) = k$, $f'(u) = e^u$, and $f'(g(x)) = e^{g(x)} = e^{kx+2}$. So

$$q'(x) = f'(g(x)) \cdot g'(x)$$
$$= e^{kx+2} \cdot k.$$

(b) *Using Log Laws:* $s(x) = \ln k + \ln x$. $\ln k$ is a constant, so $s'(x) = \frac{1}{x}$.

Using the Chain Rule: Let the inside function, $g(x)$, be kx and the outside function, $f(u)$, be $\ln u$. Then $g'(x) = k$, $f'(u) = \frac{1}{u}$, and $f'(g(x)) = \frac{1}{g(x)} = \frac{1}{kx}$. So

$$s'(x) = f'(g(x)) \cdot g'(x) = \frac{1}{kx} \cdot k = \frac{1}{x}.$$ ◆

Observe that, by using the Chain Rule, we can generalize the three basic derivative rules.

$$\begin{cases} \frac{d}{dx}[x^n] = nx^{n-1} & \text{can be generalized to} & \frac{d}{dx}\left[g(x)\right]^n = n\left[g(x)\right]^{n-1} \cdot \frac{dg}{dx} \\ \frac{d}{dx}[b^x] = \ln b \cdot b^x & \text{can be generalized to} & \frac{d}{dx}\left[b^{g(x)}\right] = \ln b \cdot b^{g(x)} \cdot \frac{dg}{dx} \\ \frac{d}{dx}\left[\log_b x\right] = \frac{1}{\ln b} \cdot \frac{1}{x} & \text{can be generalized to} & \frac{d}{dx}\left[\log_b\left[g(x)\right]\right] = \frac{1}{\ln b} \cdot \frac{1}{g(x)} \cdot \frac{dg}{dx} \end{cases}$$

◆ **EXAMPLE 16.5**   Decompose each of the following functions into $f(g(x))$ and then compute the derivative.

(a) $j(x) = (x^6 + 5x^3 + x^2)^8$ (b) $k(x) = \ln(x^3 + 3^x)$ (c) $h(x) = 3^{x^2}$

SOLUTION   (a) $j(x) = (x^6 + 5x^3 + x^2)^8$. Multiplying out would be terribly tiring and tedious. Instead, use the Chain Rule, where the inside function, $g(x)$, is $x^6 + 5x^3 + x^2$ and the outside function, $f(u)$, is u^8. (Check that $f(g(x)) = j(x)$.)

$$j'(x) = f'(g(x)) \cdot g'(x)$$

We know $f'(u) = 8u^7$, so

$$f'(g(x)) = 8\left[g(x)\right]^7 = 8(x^6 + 5x^3 + x^2)^7.$$

$$g'(x) = 6x^5 + 15x^2 + 2x.$$

So $j'(x) = 8(x^6 + 5x^3 + x^2)^7 \cdot (6x^5 + 15x^2 + 2x)$.

This is a lot easier than multiplying out in the beginning!

Basically, this function is of the form $(mess)^8$, so its derivative is $8(mess)^7 \cdot (mess)'$.

(b) $k(x) = \ln(x^3 + 3^x)$. Let the inside function, $g(x)$, be $x^3 + 3^x$ and the outside function be $f(u) = \ln u$. (Check that $f(g(x)) = k(x)$.)

$$k'(x) = f'(g(x)) \cdot g'(x)$$

We know $f'(u) = \frac{1}{u}$, so

$$f'(g(x)) = \frac{1}{g(x)} = \frac{1}{x^3 + 3^x}$$

$$g'(x) = 3x^2 + (\ln 3)3^x.$$

So $k'(x) = \frac{1}{x^3+3^x} \cdot (3x^2 + (\ln 3)3^x)$.

Basically, this function is of the form $\ln (mess)$, so its derivative is $\frac{1}{mess} \cdot (mess)'$.

(c) $h(x) = 3^{x^2}$. Let the inside function, $g(x)$, be x^2 and the outside function be $f(u) = 3^u$.

$$h'(x) = f'(g(x)) \cdot g'(x)$$

We know $f'(u) = (\ln 3)3^u$, so

$$f'(g(x)) = (\ln 3)3^{g(x)} = (\ln 3)3^{x^2}.$$

$$g'(x) = 2x$$

So $h'(x) = (\ln 3)3^{x^2}(2x) = (2 \ln 3)(x)(3^{x^2})$.

Basically, this function is of the form 3^{mess}, so its derivative is $(\ln 3)3^{mess} \cdot (mess)'$.   ◆

PROBLEMS FOR SECTION 16.1

1. (a) Which of the following are equal to $(\ln x)^2$?
      i. $(\ln x)(\ln x)$ ii. $\ln x^2$ iii. $\ln[(x)(x)]$ iv. $2 \ln x$

    (b) Which of the following are equal to $2 \ln x$?
      i. $(\ln x)(\ln x)$ ii. $\ln x^2$ iii. $\ln[(x)(x)]$

    (c) Differentiate $y = (\ln x)^2$. (Do this twice, first using the product rule and then using the Chain Rule.)

    (d) Differentiate $y = \ln x^2$. (Do this twice, first using the log rules and the derivative of $\ln x$ and then using the Chain Rule.)

In Problems 2 through 20, find $f'(x)$. Do these problems without using the Quotient Rule.

2. (a) $f(x) = 3(x + 2)^{-5}$ (b) $f(x) = 2(3x + 7)^{-8}$

3. $f(x) = \ln \sqrt{\pi x + 1} + \sqrt{\pi x} + (\pi x + \pi)^5 + \frac{1}{(\pi x^2 + 1)^3}$
   (*Hint:* Use log operations to simplify the first term.)

4. $f(x) = \frac{x}{(x^3 + 7x)^4}$

5. $f(x) = \frac{e^x}{3x^2 + 1}$

6. $f(x) = e^{5x}(1 + 2x)^6$

7. $f(x) = (1 - \frac{1}{x})e^{-x}$

8. $f(x) = \ln(\sqrt{x^3})e^{6x}$

9. $f(x) = 5 \ln(2x^2 + 3x)$

10. $f(x) = (3x^3 + 2x)^{13}$

11. $f(x) = \frac{e^{\pi x}}{(x + x^2)^3}$

12. $f(x) = \frac{\pi^2}{3(x^3 + 2)^6}$

13. $f(x) = \frac{1}{x^3 + 7x + 5}$

14. $f(x) = \frac{3^x}{2^x + 1}$

15. $f(x) = x 5^{\frac{x+1}{2}}$

16. $f(x) = \frac{4}{\sqrt{e^x + 1}}$

17. $f(x) = \sqrt{e^x + \ln(x + 1)^2}$

18. $f(x) = \frac{1}{\ln(x^2 + 2)}$

19. $f(x) = \ln(e^x + x^2)$

20. $f(x) = x \ln \left(\frac{x}{x^2 + 1} \right)$

21. Find a formula for $\frac{dy}{dx}$ if $y = f(g(h(x)))$, where f, g, and h are differentiable everywhere.

In Problems 22 through 25, graph $f(x)$, labeling the x-coordinates of all local extrema. Strategize. Is it more convenient to keep expressions factored?

22. $f(x) = x(x+3)^2$

23. $f(x) = (x-2)(x+1)^2$

24. $f(x) = (3-x)^2(x-1)$

25. $f(x) = e^x(x-3)^3$

26. Prove that if $f(x) = (x-a)^3(x-b)$ where $a, b > 0$ and $a \neq b$, then f has a point of inflection at $x = a$.

27. (a) Which of the following are equal to e^{-x^2}? Identify *all* correct answers.

   i. $e^{-(x)(x)}$ ii. $(e^{-x})^x$ iii. $\left(\frac{1}{e^x}\right)^x$ iv. $\left(\frac{1}{e^x}\right)^2$

   v. $(e^{-x})^2$ vi. $(e^{-2})^x$ vii. e^{-2x} viii. $e^{(-x)^2}$

   (b) Differentiate e^{-x^2}.

28. Graph $f(x) = e^{-x^2}$ and answer the following questions.

   (a) What is the domain of f? The range?
   (b) Is f an even function, an odd function, or neither?
   (c) For what values of x is f increasing? Decreasing?
   (d) Find all relative maximum and minimum points.
   (e) Does f have an absolute maximum value? An absolute minimum value? A greatest lower bound? If any of these exist, identify them.
   (f) Find the x-coordinates of the points of inflection.

   (*Note:* This function is a good one to keep in mind. It is extremely useful in both probability and statistics because, with some minor adjustments, it gives us a normal distribution curve.)

16.2   THE DERIVATIVE OF x^n WHERE n IS ANY REAL NUMBER

At this point we have proven that $\frac{d}{dx}x^n = nx^{n-1}$ for any integer n. In fact, this formula holds if n is *any* real number. We can use the Chain Rule to prove this fact.

Suppose that r is any real number and we want to find $\frac{d}{dx}x^r$. The key is to rewrite x^r so we can use derivative formulas that we already know.

$$x^r = e^{\ln x^r} \qquad e^x \text{ and } \ln x \text{ are inverses, so } e^{\ln z} = z.$$
$$= e^{r \ln x} \qquad \text{Use log rule (iii).}$$

Now we take the derivative.

$$\frac{d}{dx}x^r = \frac{d}{dx}e^{r \ln x}$$

$$= e^{r \ln x} \cdot \frac{d}{dx}(r \ln x) \qquad \text{Use the Chain Rule: } \frac{d}{dx}e^{mess} = e^{mess} \cdot (mess)'.$$

$$= e^{r \ln x} \cdot \frac{r}{x} \qquad \text{Keep in mind that } r \text{ is a constant.}$$

$$= x^r \cdot \frac{r}{x} \qquad \text{Rewrite } e^{r \ln x} \text{ as } x^r.$$

$$= rx^{r-1}$$

We now can take the derivative of x^n where n is any real number.

$$\boxed{\frac{d}{dx}x^n = nx^{n-1}}$$

EXERCISE 16.1   Find y' if $y = \frac{x^\pi \pi^x + (2x)^{2.1}}{\pi^2}$.

Answer

$$y' = \pi^{-2}[\pi x^{\pi-1}\pi^x + x^\pi (\ln \pi)\, \pi^x + (2.1)2^{2.1}x^{1.1}]$$

PROBLEMS FOR SECTION 16.2

For Problems 1 through 3, find y'.

1. $y = x^{2\pi} + 2\pi^x$

2. $y = (2x^2 + 1)^{\sqrt{3}}$

3. $y = (3x)^{\sqrt{2}+1} + \frac{1}{\sqrt{\pi x}}$

4. Differentiate the following, simplifying the expression first if useful.

   (a) $y = \pi e^{3t^2 + \pi}$ (b) $y = \ln(e^t + 1)$ (c) $y = \frac{\pi^2}{\sqrt{x^2+4}}$

   (d) $y = \frac{1}{(\ln x)^{2.6}}$ (e) $y = \frac{1}{(\ln x^2)^{1.5}}$ (f) $y = \sqrt[3]{\ln(e^t + 1)}$

5. Find y', simplifying the expression first where useful.

   (a) $y = e^x x^e$ (b) $y = e^{1/x}$ (c) $y = \sqrt{e^{-x}x}$

   (d) $y = \left[\ln \sqrt{1-x}\right]^{-3.5}$ (e) $y = \ln\left(\frac{x+1}{x-1}\right)$ (f) $y = (1 - \ln x)^{5/4}$

   (g) $y = \ln \sqrt{x(x+1)}$ (h) $y = \dfrac{5}{\sqrt{\frac{1}{e^{6x}}+x}}$

6. Identify and classify all critical points of the function $f(x) = (x^2 - 4)x^{\pi+1}$ for $x > 0$.

16.3   USING THE CHAIN RULE

In this section, we take a second look at applying the Chain Rule.

◆ **EXAMPLE 16.6**   Once we know the Chain Rule, there is no problem if we forget the Quotient Rule, because $\frac{f(x)}{g(x)}$ can be written as $f(x) \cdot [g(x)]^{-1}$. Derive the Quotient Rule from the Product Rule.

SOLUTION

$$\frac{d}{dx}\left[\frac{f(x)}{g(x)}\right] = \frac{d}{dx}\left[f(x) \cdot [g(x)]^{-1}\right]$$ Rewrite the quotient as a product.

$$= f(x) \cdot \frac{d}{dx}[g(x)^{-1}] + [g(x)^{-1}] \cdot \frac{d}{dx}f(x)$$ Use the Product Rule.

$$= f(x) \cdot (-1)[g(x)]^{-2}g'(x) + [g(x)]^{-1} \cdot f'(x)$$ Use the Chain Rule to find
$$\frac{d}{dx}[g(x)^{-1}] = -1[g(x)]^{-2} \cdot g'(x).$$

$$= \frac{-f(x) \cdot g'(x)}{[g(x)]^2} + \frac{f'(x)}{g(x)}$$ Rewrite negative exponents as fractions.

$$= \frac{-f(x) \cdot g'(x)}{[g(x)]^2} + \frac{g(x) \cdot f'(x)}{[g(x)]^2}$$ Get a common denominator of $[g(x)]^2$.

$$= \frac{g(x) \cdot f'(x) - f(x) \cdot g'(x)}{[g(x)]^2}$$ Combine into one fraction.   ◆

We can apply the Chain Rule to relate rates of change; as illustrated in the following example.

◆ **EXAMPLE 16.7**   In Nepal the daily kerosene consumption in restaurants and guesthouses throughout the country can be modeled by the function $K(x)$, where x is the number of tourists in the country. Tourism is seasonal; let $x(t)$ be the number of tourists at time t. In our model, $K(x)$ and $x(t)$ are continuous and differentiable. At a certain time, there are 5,000 tourists in Nepal and the number is decreasing at a rate of 40 per day. Write an expression for the rate at which kerosene consumption is changing with time at this moment.

SOLUTION   We're looking for $\frac{dK}{dt}$. We know that $\frac{dK}{dt} = \frac{dK}{dx} \cdot \frac{dx}{dt}$, and $\frac{dx}{dt} = -40$.
Therefore, $\frac{dK}{dt} = K'(5000) \cdot (-40) = -40K'(5000)$, where

$$K'(5000) = \left. \frac{dK}{dx} \right|_{x=5000}.$$ ◆

Before applying the Chain Rule to more complicated expressions and situations, we'll summarize our knowledge of differentiation, using the Chain Rule to express familiar differentiation formulas in a more general way.

Derivatives: A Summary

Graphical Interpretation: If $y = f(x)$, then the derivative of f is the slope function, giving the slope of the tangent line to f at the point $(x, f(x))$. For instance, $f'(2)$ gives the slope of f at $x = 2$.

Instantaneous Rate of Change: The derivative evaluated at $x = 2$ gives the instantaneous rate of change of y with respect to x at $x = 2$.

Notation: y', f', $f'(x)$, $y'(x)$, $\frac{dy}{dx}$, $\frac{df}{dx}$ all mean the same thing. $\frac{d}{dx}$ is an operator that means "take the derivative of what follows."

Limit Definition: The following expressions are equivalent and are equal to $f'(x)$.

$$f'(x) = \lim_{h \to 0} \frac{f(x+h) - f(x)}{h} = \lim_{\Delta x \to 0} \frac{\Delta y}{\Delta x} = \lim_{\Delta x \to 0} \frac{f(x + \Delta x) - f(x)}{\Delta x}$$

Differentiation Formulas

1. $y = f(x) + g(x)$		$y' = f'(x) + g'(x)$	*Sum Rule*
2. $y = kf(x)$	k constant	$y' = kf'(x)$	*Constant Multiple Rule*
3. $y = f(x)g(x)$		$y' = f'(x)g(x) + g'(x)f(x)$	*Product Rule*
			For the product of two functions
4. $y = \dfrac{f(x)}{g(x)}$		$y' = \dfrac{gf' - fg'}{g^2}$	*Quotient Rule*
5. $y = [g(x)]^n$	n constant	$y' = n[g(x)]^{n-1} \cdot g'(x)$	*Generalized Power Rule* (Variable in *base*)
6. $y = b^{g(x)}$	b constant	$y' = b^{g(x)}(\ln b)g'(x)$	*Exponential Function* (Variable in *exponent*)
$y = e^{g(x)}$		$y' = e^{g(x)}g'(x)$	
7. $y = \log_b\left[g(x)\right]$		$y' = \dfrac{1}{\ln b} \cdot \dfrac{1}{g(x)} g'(x)$	*Logarithmic Function*
$y = \ln[g(x)]$		$y' = \dfrac{1}{g(x)} g'(x)$	
8. $y = f(g(x))$		$y' = f'(g(x))g'(x)$	*Chain Rule*

The world of functions we can differentiate has broadened immensely! If faced with a function we cannot differentiate using these shortcuts,[2] we can return to the limit definition of derivative. If we cannot make headway obtaining an exact value for a derivative and the function we're working with is reasonably well behaved, we can still use numerical methods to approximate the derivative at a point, and we can obtain qualitative information using a graphical approach.

CAUTION Be sure you know the difference between an exponential function like $f(x) = (\pi + 1)^{2x}$ and a generalized power function like $f(x) = (x^2 + \pi)^{\pi^2}$. If the variable is in the base, then we differentiate using the generalized power rule; if the variable is in the exponent, we use the generalized rule for exponential functions.

Feel free to think of some of these differentiation formulas more informally. For instance,

the derivative of $(mess)^n$ is $n(mess)^{n-1} \cdot (mess)'$,

the derivative of $\ln(mess)$ is $\dfrac{1}{mess} \cdot (mess)'$, and

the derivative of e^{mess} is $e^{mess} \cdot (mess)'$.

---

[2] This will be the case when we first run into trigonometric functions.

These rules are all obtained by applying the Chain Rule to the differentiation rules we have been using. Then when you look at a complicated expression, you can categorize it. For instance, the expression

$$\frac{1}{\sqrt{\ln(x^2 + 1)}}$$

is basically $(mess)^{-1/2}$, meaning were we to construct the expression assembly-line style, the *last* worker on the line would take the mess coming along the line and raise it to the $-1/2$. The derivative of $(mess)^{-1/2}$ is $-\frac{1}{2}(mess)^{-3/2} \cdot (mess)'$. What's the mess? Basically, it is $\ln(stuff)$, so

$$(mess)' = \frac{1}{stuff} \cdot (stuff)'.$$

Putting this together,

$$\frac{d}{dx}\left[\frac{1}{\sqrt{\ln(x^2 + 1)}}\right] = \frac{d}{dx}\left[\ln(x^2 + 1)\right]^{-\frac{1}{2}}$$

$$= -\frac{1}{2}\left[\ln(x^2 + 1)\right]^{-\frac{3}{2}} \cdot \frac{1}{(x^2 + 1)} \cdot 2x.$$

What we've done—albeit informally—is to decompose $\frac{1}{\sqrt{\ln(x^2+1)}}$ into $f(g(h(x)))$, where

$$f(x) = x^{-1/2}, \quad g(x) = \ln x, \quad \text{and } h(x) = x^2 + 1.$$

$$x \xrightarrow{h} \left(x^2 + 1\right) \xrightarrow{g} \ln(x^2 + 1) \xrightarrow{f} \frac{1}{\sqrt{\ln(x^2 + 1)}}$$

So

$$\left[f(g(h(x)))\right]' = f'(g(h(x))) \cdot g'(h(x)) \cdot h'(x).$$

A WORD OF ADVICE Many students, having learned shortcuts to differentiation, approach problems by charging them, sleeves rolled up, brutally hacking away. There is no virtue in doing a problem in the most difficult way possible; it does *not* put you on higher moral ground. Rather, you increase the likelihood of making an error. Instead, when presented with a function to differentiate, take a moment to consider how to *prepare the function for differentiation*. Sometimes, a bit of thoughtful reformulation will save you time and energy. (Of course, there are times when it is necessary to dig in and get your hands all dirty. If it is necessary, do it with relish!) The meaning of this advice is clarified in the next example.

◆ **EXAMPLE 16.8**   Differentiate $y = \ln\sqrt{\frac{1+x}{(1-x)^3}}$.

SOLUTION   If you're enthusiastically charging the problem, you might informally say that basically this is $\ln(mess)$, where *mess* is $(stuff)^{1/2}$ and to differentiate *stuff*, the Quotient Rule and generalized power rule come into play. If you're formally charging headlong into this, you might valiantly decompose the function as follows:

$$h(x) = \frac{1+x}{(1-x)^3}$$

$$g(x) = \sqrt{x}$$

$$f(x) = \ln x,$$

where

$$\ln \sqrt{\frac{1+x}{(1-x)^3}} = f(g(h(x))).$$

Notice that h' alone is a bulky expression.

Alternatively, you can save yourself a lot of work by rewriting $\ln \sqrt{\frac{1+x}{(1-x)^3}}$:

$$\ln \sqrt{\frac{1+x}{(1-x)^3}} = \ln \left[\frac{1+x}{(1-x)^3} \right]^{\frac{1}{2}}$$

$$= \frac{1}{2} \ln \left[\frac{1+x}{(1-x)^3} \right]$$

$$= \frac{1}{2} \ln(1+x) - \frac{1}{2} \ln(1-x)^3$$

$$= \frac{1}{2} \ln(1+x) - \frac{3}{2} \ln(1-x).$$

Now the expression is ready for differentiation.

$$\frac{dy}{dx} = \left(\frac{1}{2} \right) \frac{1}{1+x} - \left(\frac{3}{2} \right) \frac{1}{1-x} \cdot -1$$

$$= \frac{1}{2(1+x)} + \frac{3}{2(1-x)}$$

$$= \frac{1}{2+2x} + \frac{3}{2-2x} \quad \blacklozenge$$

◆ **EXAMPLE 16.9**   Suppose f and g are differentiable functions. Find $h'(x)$ if

$$h(x) = \sqrt{[f(x)]^3 \cdot \ln(g(x))}.$$

SOLUTION   This is basically of the form $(mess)^{1/2}$ where $mess$ is the product $[f(x)]^3 \cdot \ln(g(x))$.

$$h'(x) = \frac{1}{2}(mess)^{-1/2}, (mess)'$$

$$= \frac{1}{2} \left[[f(x)]^3 \cdot \ln(g(x)) \right]^{-1/2} \cdot \left\{ \frac{d}{dx}[f(x)]^3 \cdot \ln(g(x)) + [f(x)]^3 \frac{d}{dx} \left[\ln(g(x)) \right] \right\}$$

$$= \frac{1}{2} \left[[f(x)]^3 \cdot \ln(g(x)) \right]^{-1/2} \cdot \left\{ 3[f(x)]^2 \cdot f'(x) \cdot \ln(g(x)) + [f(x)]^3 \cdot \frac{g'(x)}{g(x)} \right\}$$

Notice that the operator notation, $\frac{d}{dx}$, is useful for record-keeping. It can be read as "I plan to take the derivative of this, but I haven't done it yet." ◆

◆ **EXAMPLE 16.10**   Two ships leave from a port. The first one sails due east, while the second one sails due north. At 10:00 A.M. the first boat is 4 miles from the port and is traveling at 35 miles per hour. At this instant the second boat is 3 miles north of the port and is traveling at 15 miles per hour. At what rate is the distance between the ships increasing at 10:00 A.M.?

SOLUTION   Let $x(t)$ = the distance between the port and the first boat at time t. $y(t)$ = the distance between the port and the second boat at time t. Then the distance between the two boats at time t, t in hours, is given by

$$D(t) = \sqrt{[x(t)]^2 + [y(t)]^2}.$$

We want to find $\frac{dD}{dt}$ at 10:00 A.M. $D(t)$ is basically of the form $(mess)^{1/2}$.

$$D'(t) = \frac{1}{2}\left[[x(t)]^2 + [y(t)]^2\right]^{-1/2} \cdot \left[2x(t) \cdot x'(t) + 2y(t) \cdot y'(t)\right]$$

At 10:00A.M. $x(t) = 4$, $y(t) = 3$, $x'(t) = 35$, and $y'(t) = 15$, so

$$\frac{dD}{dt} = \frac{1}{2}(4^2 + 3^2)^{-1/2} \cdot [2 \cdot 4 \cdot 35 + 2 \cdot 3 \cdot 15]$$

$$= \frac{1}{2}\frac{1}{\sqrt{25}} \cdot \left[2(140 + 45)\right]$$

$$= \frac{1}{5} \cdot 185$$

$$= 37.$$

At 10:00 A.M. the distance between the ships is increasing at 37 miles per hour.   ◆

In Section 17.4 we will look at other ways of approaching this problem and problems similar. These other approaches depend, ultimately, on the Chain Rule.

Exploratory Problems for Chapter 16

Finding the Best Path

1. A pumping station that will service two towns is to be built on the shore of a straight river. The towns are 10 miles apart; one of them is 3 miles from the river and the other is 9 miles from the river. Pipes will run from the pumping station to the center of each town. Where should the pumping station be located in order to minimize the amount of piping used?

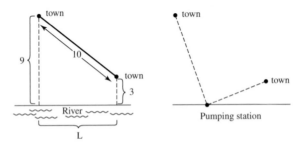

Hint: Begin by determining the distance L in the figure above. Then the amount of piping can be expressed in terms of x where x is the distance between the pumping station and the point on the river closest to one of the towns. What is the domain of the function you are minimizing?

2. A family lives on the western shore of a long, straight river in a rural area. The only store in the vicinity is located on the eastern shore of the river, directly across from a point 6 miles down the river. The river is 1 mile wide. They can get to the store by a combination of boat and foot. Where should they build their dock so the commute to the store takes the minimum amount of time?

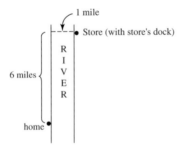

(a) Answer this question assuming that one can walk at a rate of 4 miles per hour and row at a rate of 3 miles per hour.

(b) Now assume that one can walk at a rate of 4 miles per hour, but with a new boat upgrade, the boating rate can be increased to 5 miles per hour. Where should the dock be built if they purchase a new boat?

PROBLEMS FOR SECTION 16.3

1. Just outside Newburgh, the New York State Thruway (I-87), running north-south, intersects Interstate 84, which runs east-west. At noon a car is at this intersection and traveling north at a constant speed of 55 miles per hour. At this moment a Greyhound bus is 150 miles west of the intersection and traveling east at a steady pace of 65 miles per hour.

   (a) When will the bus and the car be closest to one another?

   (b) What is the minimum distance between the two vehicles?

   (c) How far away from the intersection is the bus at this time?

   (*Hint:* Minimize the square of the distance rather than the distance itself. Explain why this strategy is valid.)

2. Draw a semicircle of radius 2. Inscribe a rectangle as shown. What are the dimensions of the rectangle of the largest area? What is the largest area?

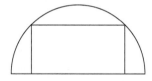

   (*Hint:* Strategize. What is your goal? What are you trying to maximize? Write an expression for the thing you're trying to maximize. Make sure you get it in one variable. Find and classify the critical points.)

3. What is the largest value of $\frac{\ln(x+1)}{x+1}$?

4. Differentiate.

   (a) $y = \frac{1}{x \ln 2 + 1}$

   (b) $y = \ln(5x^3 + 8x)$

   (c) $y = (2^x)(x^2 + x)^7$

   (d) $y = \sqrt{\ln(5x) + e^{6x}}$

   (e) $y = \frac{7}{\sqrt{\ln x}}$

   (f) $y = \left(4^{x/3}\right) \ln 3x$

5. Differentiate.

   (a) $y = \left(2^2\right)^x$

   (b) $y = 2^{2^x}$

   (c) $y = \frac{e^{\pi x}}{x}$

   (d) $y = \frac{x^3 + 1}{x^2 + 1}$

   (e) $y = 5 \ln\left(\frac{5x+3}{\sqrt{x}}\right)$

   (f) $y = \frac{3}{2 \ln(8x^2 + 1)}$

6. What is the global maximum value of the function $f(x) = \frac{3}{\sqrt{x^2 + 1}}$ and where is it attained?

   Instructions: First just look at this function. Without *any* calculus, try to figure out the answer. (It may be useful to check symmetry considerations.) Now use the first derivative to support your answer.

7. The graph of $h(x)$ is given below.

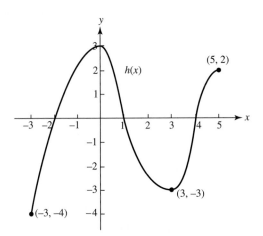

(a) For which values of x is $h'(x)$ negative?
(b) This part of the question concerns the function f given by

$$f(x) = [h(x)]^2,$$

where the graph of $h(x)$ is given above.
   i. Find all the points (x-values) at which the graph of $f(x)$ has a horizontal tangent line.
   ii. On a number line, indicate the sign of $f'(x)$.
   iii. Identify the x-coordinates of all the local maxima and minima of f.
   iv. On the interval $[-3, 5]$, what is the largest value taken on by f (i.e., what is the global or absolute maximum value of f for x in $[-3, 5]$)?
   v. On the interval $[-3, 5]$, what is the smallest value taken on by f?
(c) Let $g(x) = |h(x)|$.
   i. Sketch the graph of $g(x)$.
   ii. Is $g'(x)$ defined everywhere? If not, where is $g'(x)$ undefined?

8. Suppose that f and g are differentiable functions. We are given the following information:

x	$f(x)$	$f'(x)$	$g(x)$	$g'(x)$
0	3	5	2	−1
1	2	−1	0	2
2	0.3	2.5	4	−2
3	4	5	1	3
4	11	2	3	5

Evaluate the following. If there is not enough information to do so, indicate this.
(a) $y = f(g(x))$. What is $\frac{dy}{dx}\big|_{x=0}$?
(b) $y = g(f(x))$. What is $y(0)$?

(c) Find $\frac{d}{dx}\left(\frac{f}{g}\right)\Big|_{x=3}$.

(d) Let $y = \left[f(x)\right]^2 g(x^2)$. What is $\frac{dy}{dx}\Big|_{x=2}$?

(e) Let $y = \sqrt{f(x^2)}$. Find $y'(2)$.

9. Below is a graph of $f(x)$ on the interval $[-2, 3]$. $j(x)$ is given by $j(x) = \ln\left[f(x)\right]$ on the domain $[-2, 3]$.

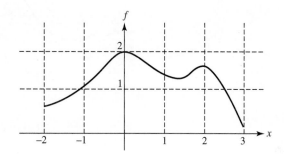

(a) How many zeros does $j(x)$ have?

(b) Find $j'(x)$.

(c) Approximate the critical points of j.

(d) Identify the local extrema of j. (Estimate the positions.)

(e) Where does j attain its absolute maximum value? Its absolute minimum value?

(f) Which function has a higher absolute maximum value, f or j? Which function has a lower absolute minimum value, f or j? Explain.

10. $g(x)$ is a continuous function with exactly two zeros, one at $x = 1$ and the other at $x = 4$. $g(x)$ has a local minimum at $x = 3$ and a local maximum at $x = 7$. These are the only local extrema of g. Let $f(x) = \left[g(x)\right]^4$.

(a) Find $f'(x)$ in terms of g and its derivatives.

(b) Can we determine (definitively) whether g has an absolute minimum value on $(-\infty, \infty)$? If we can, where is that absolute minimum value attained? Can we determine (definitively) whether g has an absolute maximum value? If we can, where is that absolute maximum value attained?

(c) What are the critical points of f?

(d) On what intervals is the graph of f increasing? On what intervals is it decreasing?

(e) Identify the local maximum and minimum points of f.

(f) Can we determine (definitively) whether f has an absolute minimum value? If so, can we determine what that value is?

If you haven't already done so, step back, take a good look at the problem (a bird's-eye view) and make sure your answers make sense.

11. Assume that f, g, and h are differentiable. Differentiate $p(x)$ where

(a) $p(x) = f(x)g(x)h(x)$. (*Hint:* Use the Product Rule twice.)

(b) $p(x) = \sqrt{g(x) + \ln f(x)}$.

12. Let $f(x)$ be the function whose graph is drawn on the axes below.

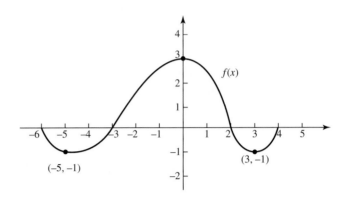

Let $a(x) = 2f(x),$
$$b(x) = f(x+2), \text{ and}$$
$$c(x) = f(2x).$$

(a) On three separate sets of axes, draw the graphs of the function $a(x) = 2f(x)$, $b(x) = f(x+2)$, and $c(x) = f(2x)$, labeling the x-intercepts, the y-intercept, and the x- and y-coordinates of the local extrema.

(b) Suppose we know that
$$f'(-4) = 1/2, \quad f'(-2) = 3/2, \quad f'(0) = 0.$$
Find the following, explaining your reasoning briefly.
   i. $a'(-2) = $ _____
   ii. $b'(-2) = $ _____
   iii. $c'(-2) = $ _____

In Problems 13 through 18, find $h'(x)$. Assume that f and g are differentiable on $(-\infty, \infty)$. Your answers may be in terms of f, g, f', and g'.

13. $h(x) = f(\ln x) - \ln(f(x))$

14. $h(x) = \sqrt{f(x)g(x)}$

15. $h(x) = \dfrac{1}{\sqrt{f(g(x))}}$

16. $h(x) = f(x^2)e^{g(x)}$

17. $h(x) = \dfrac{1}{[f(x)]^2} + f\left(\dfrac{1}{x^2}\right)$

18. $h(x) = [f(x)]^3 g(2x)$

In Problems 19 through 22, find $\frac{dy}{dx}$. Take the time to prepare the expression so that it is as simple as possible to differentiate.

19. $y = 3\ln\left(\dfrac{x^2-1}{x+2}\right)$

20. $y = \sqrt{(x^2 + 3)^5}$

21. $y = \frac{5x^\pi - x^3 - 1}{x^2}$

22. $y = 5 \ln \sqrt{\frac{3x}{x^2+1}}$

In Problems 23 through 29, differentiate. In Problems 23 through 25, assume f is differentiable. Your answers may be in terms of f and f'.

23. $y = \ln\left(\frac{x}{f(x^2)}\right)$

24. $y = [f(x)]^2 + 2^{f(x)}$

25. $y = \ln\left(\frac{x \cdot f(x)}{\sqrt{3x^3+2}}\right)$

26. $f(x) = \ln(e^{(x+5)^2})$

27. $f(x) = e^{(g(x))^2}$

28. $f(x) = (x^3 + e)^\pi$

29. Let $f(x) = x^x$.
    (a) Use numerical methods to approximate $f'(2)$.
    (b) Refer to your answer to part (a) to show that $f'(x) \neq x \cdot x^{x-1}$. What is it about f that makes it *not* a power function?
    (c) Refer to your answer to part (a) to show that $f'(x) \neq \ln x \cdot x^x$. What is it about f that makes it *not* an exponential function?
    (d) *Challenge:* Figure out how to rewrite x^x so you can use the Chain Rule to differentiate it.

30. Consider the function $f(x) = e^{-x^2}(-x^2 + 1)$. You must give exact answers for all of the following questions. Show your work. Your work must stand independent of your calculator.
    (a) Find all the x-intercepts.
    (b) Identify the local extrema of $f(x)$.
    (c) Sketch a graph of f, labeling the x-coordinates of all local and global extrema.
    (d) Now consider the function $g(x) = |f(x)|$.
        i. What are the critical points of g?
        ii. Classify the critical points of $g(x)$.

31. A craftsman is making a mobile consisting of hanging circles each with an inscribed triangle of stained glass. Each piece of stained glass will be an isosceles triangle. Show that if she wants to maximize the amount of stained glass used, the glass triangles should be equilateral. In other words, show that the isosceles triangle of maximum area that can be inscribed in a circle of radius R is an equilateral triangle.

32. A holiday ornament is being constructed by inscribing a right circular cone of brightly colored material in a transparent spherical ball of radius 2 inches. What is the maximum possible volume of such an inscribed cone?

33. A craftsman is making a ribbon ornament by inscribing an open hollow cylinder of colored ribbon in a transparent spherical ball of radius R. What is the maximum surface area of such a cylinder?

34. Phone cables are to be run from an island to a town on the shore. The island is 5 miles from the shore and 13 miles from the town. The cable will be run in a straight line from the island to the shore and then in another straight line along the shoreline. If it costs 60% more to run the cable under water than it does to run it under the ground, how far should the cable be run along the shore?

35. The volume of a cylindrical tree trunk varies with time. Let $r(t)$ give the radius of the trunk at time t and let $h(t)$ give the height of time t.

    (a) Express the rate of change of A, the cross-sectional area, with respect to time in terms of r and r'.

    (b) Express the rate of change of volume with respect to time in terms of r, r', h, and h'.

36. Suppose the amount of power generated by an energy generating system is a function of v, the volume of water flowing through the system. The function is given by $P = P(v)$. The volume of water in the sytem is determined by r, the radius of an adjustable valve; $v = v(r)$. The radius varies with time: $r = r(t)$.

    (a) Express $\frac{dP}{dr}$, the rate of change of the power with respect to a change in the valve's radius, in terms of the functions $P(v)$ and $v(r)$ and their derivatives.

    (b) Express $\frac{dP}{dt}$, the rate of change of the power with respect to time, in terms of the functions $P(v)$, $v(r)$, and $r(t)$ and their derivatives.

17

Implicit Differentiation and its Applications

17.1 INTRODUCTORY EXAMPLE

In this section you will add sophistication to your differentiation skills by applying two fundamental principles.

Principle (i): If two expressions are equal, their derivatives are equal.

Principle (ii): Whether y is given implicitly or explicitly in terms of x,[1] we can differentiate an expression in y with respect to x by treating y the same way we would treat $g(x)$ and using the Chain Rule.

Try Example 17.1 below. Do the problem on your own, writing down your solution so you can compare it with the discussion that follows.

◆ **EXAMPLE 17.1**   Let $f(x) = x^{x+1}$, where $x > 0$.

   (a) Approximate $f'(2)$ numerically.

   (b) Using appropriate rules of differentiation, find $f'(2)$ exactly.

   (c) Compare your answers to parts (a) and (b). If they are not very close to one another, identify your error.

SOLUTION   (a) To approximate the slope of the tangent line to $y = f(x)$ at $x = 2$, we find the slope of the secant line through $(2, f(2))$ and a nearby point on the graph of f, say $(2.0001, f(2.0001))$.

---

[1] For example, $xy^3 = 1$ is an equation that gives y implicitly in terms of x; $y = x^2 + 3x$ is an equation that gives y explicitly in terms of x.

$$f'(2) \approx \frac{f(2.0001) - f(2)}{2.0001 - 2}$$

(Equivalently, we could say $f'(2) = \lim_{h \to 0} \frac{f(2+h)-f(2)}{h}$, so $f'(2) \approx \frac{f(2+h)-f(2)}{h}$ for h very small. Then we choose a very small value for h, say $h = 0.0001$.)

$$f'(2) \approx \frac{(2.0001)^{2.0001+1} - 2^{2+1}}{0.0001}$$

$$f'(2) \approx \frac{(2.0001)^{3.0001} - 2^3}{0.0001}$$

$$\approx 17.547$$

REMARK When using a calculator to evaluate this, do not round off the numerator before dividing by 0.0001. To divide by 0.0001 is to multiply by 10,000, so any roundoff error in the numerator will be multiplied ten thousand-fold.

(b) There are two popular but fatally flawed approaches to differentiating x^{x+1}. We'll set them up and knock them down before proceeding.

Fatally flawed approach #1: We know that $\frac{d}{dx} x^n = n x^{n-1}$ for any constant n. It is easy to succumb to temptation and propose $(x+1)x^x$ as the derivative of x^{x+1}. But this differentiation rule *cannot* be applied because $(x+1)$ is **not** a constant; x^{x+1} is not a power function! (If you apply it anyway, comparing your answers to parts (a) and (b) (17.547 versus 24) ought to alert you to the error.)

Fatally flawed approach #2: We know that $\frac{d}{dx} b^{g(x)} = \ln b \cdot b^{g(x)} \cdot g'(x)$ for any positive constant b. Trying to apply this rule to x^{x+1} by letting $b = x$ leads to a derivative of $\ln x \cdot x^{x+1}$. The error is similar to that in the previous approach; b must be a constant, and x is **not** constant. Again, comparing your answers in parts (a) and (b) (17.547 versus 5.545) ought to alert you to an error if you use this approach.

A correct approach: In order to apply a differentiation rule we know, we must alter the form of $f(x)$. The problem is that x is in *both* the base and the exponent. A flash of inspiration can lead us to express $f(x)$ as follows.

$$f(x) = x^{x+1} = e^{\ln\left(x^{x+1}\right)} = e^{(x+1)\ln x}$$

Now we have an expression of the form $e^{g(x)}$ and can use the fact that $\frac{d}{dx} e^{g(x)} = e^{g(x)} \cdot g'(x)$ to obtain

$$f'(x) = e^{(x+1)\ln x} \frac{d}{dx}[(x+1)\ln x]$$

$$= e^{(x+1)\ln x}\left[\frac{x+1}{x} + 1 \cdot \ln x\right]$$

$$= x^{x+1}\left[\frac{x+1}{x} + \ln x\right].$$

Then $f'(2) = 2^3\left[\frac{3}{2} + \ln 2\right] = 12 + 8\ln 2$. This is $17.54517\ldots$, quite close to the approximation found in part (a).

From what source could this flash of inspiration, expressing x^{x+1} as $e^{(x+1)\ln x}$, have arisen? We know that $e^{\ln A} = A$ for any positive A, but what could make us think of applying that knowledge in this problem? Recall that in proving $\frac{d}{dx}x^n = nx^{n-1}$ for any real number n we used the same idea.[2] We rewrote x^n as $e^{\ln x^n}$ and then as $e^{n \ln x}$. (This is why we needed the Chain Rule before we could establish the proof.)

An alternative correct approach: In the approach taken above, to differentiate $f(x) = x^{x+1}$ we changed its form, converting the *expression* x^{x+1} to something we could more easily differentiate. Another approach is to deal instead with the entire *equation* $f(x) = x^{x+1}$. If we can bring the $(x + 1)$ down from the exponent before we differentiate, we'll be in good shape. Taking logs of *both sides* of the equation $f(x) = x^{x+1}$ accomplishes this while still preserving the balance of the equation. Keep in mind that taking the log of an expression changes it into a new and different expression; we are not proposing to take the logarithm of an *expression*, but rather of both sides of an *equation*.

$$f(x) = x^{x+1}$$

$\ln f(x) = \ln(x^{x+1})$ Take ln of each side. (If $A = B$ and A, $B > 0$, then $\ln A = \ln B$.)

$\ln f(x) = (x + 1)(\ln x)$ Use $\ln a^b = b \ln a$ to bring down the exponent.

Now we can differentiate both sides using principle (i), which says that if $g(x) = h(x)$, then $\frac{d}{dx}g(x) = \frac{d}{dx}h(x)$. Remember to use the Chain Rule on the left side (where we have $\ln(f(x))$) and the Product Rule on the right.

$\frac{d}{dx}[\ln f(x)] = \frac{d}{dx}[(x + 1)(\ln x)]$ Differentiate each side.

$\frac{1}{f(x)} f'(x) = (1)(\ln x) + (x + 1)\left(\frac{1}{x}\right)$

$f'(x) = \left[\ln x + \frac{x + 1}{x}\right] f(x)$ Solve for $f'(x)$.

$f'(x) = \left[\ln x + \frac{x + 1}{x}\right] x^{x+1}$ Replace $f(x)$ by x^{x+1}.

This gives us an expression for $f'(x)$, as desired.   ◆

Expressions Versus Equations

The distinction between an expression and an equation is critical to understanding the two approaches discussed above, so we'll reiterate the distinction.

■   The governing criterion for working with an expression is that its value remain unchanged; only its form may vary.

Zero can be added to an expression; an expression can be multiplied by 1; an expression can be factored. An expression A in the domain of f can be written as $f^{-1}(f(A))$. For example, $g(x) = \ln(e^{g(x)})$ and if $g(x) > 0$, then $g(x) = e^{\ln g(x)}$.

---

[2] Sometimes establishing proofs and doing practical problem solving seem to be disparate activities. Often, however, they are more intimately intertwined; a proof can provide problem-solving inspiration, and problem solving might suggest a more general theorem and sometimes even a proof.

- An equation, on the other hand, establishes an equality of two expressions; we can manipulate an equation in any way that does not alter the equality of the two sides. The governing criterion for working with equations is that balance be maintained.

Adding/subtracting the same thing to/from both sides of the equation maintains the balance. Multiplying/dividing both sides of the equation by the same nonzero quantity maintains the balance.

If $A = C$, where A and C are both positive, then $A^n = C^n$. For example, squaring both sides of an equation ($n = 2$) maintains balance (although it may introduce extraneous roots). Similarly, taking reciprocals of both sides of an equation ($n = -1$) maintains balance. If $J + K = C$, then $(J + K)^n = C^n$. In particular, if $J + K = C$, then $\frac{1}{J+K} = \frac{1}{C}$.

If $A = C$, where A and C are both positive, then $\ln A = \ln C$. For example, if $J + K = C$, then $\ln(J + K) = \ln C$.

If $A = C$, then $b^A = b^C$ for any positive constant b. For example, if $J + K = C$ then $e^{J+K} = e^C$.

We have presented two different successful approaches to tackling the problem presented in Example 17.1(b). The first dealt with the *expression* x^{x+1}; we used the fact that $e^{\ln A} = A$ for any positive A to convert the *expression* x^{x+1} to the *equivalent expression* $e^{(x+1)\ln x}$. The second approach dealt with the *equation* $f(x) = x^{x+1}$. We took the natural logarithm of both sides of the *equation* to obtain an *equivalent equation*, differentiated both sides of the *equation*, and solved for $f'(x)$. This latter technique of differentiation is called **logarithmic differentiation**.

PROBLEMS FOR SECTION 17.1

Differentiate the following.

1. (a) $y = 3^x$ (b) $y = x^3$ (c) $y = x^x$, where $x > 0$.

2. $y = (x + 1)^{(x+1)}$, where $x > -1$.

3. $y = (3x^2 + 2)^x$

4. $y = x^{x^2}$, where $x > 0$.

17.2   LOGARITHMIC DIFFERENTIATION

Logarithmic differentiation deals with the task of differentiating a positive function $f(x)$ by working with both sides of the equation $y = f(x)$ as follows.

Using Logarithmic Differentiation to Find y'

1. Begin with an equation $y = f(x)$, where $f(x) > 0$. Take the natural logarithm of *both sides* of the equation.

2. Use log rules to bring down exponents and/or simplify expressions.

3. Differentiate both sides of the equation. y is a function of x so the Chain Rule *must* be applied to differentiate $\ln y$ or $\ln f(x)$.

$$\frac{d}{dx}[\ln f(x)] = \frac{1}{f(x)} f'(x) \text{ or, equivalently, } \frac{d}{dx}[\ln y] = \frac{1}{y}\frac{dy}{dx}.$$

4. Solve for $\frac{dy}{dx}$ or $f'(x)$.

5. To express y' in terms of x, replace y or $f(x)$ with the equivalent expression in terms of x.

When to Use Logarithmic Differentiation

i. The technique is useful in differentiating a function that has the variable in both the base and the exponent.

ii. We may choose to use logarithmic differentiation to make the differentiation of quotients or products more palatable, provided that we only take the log of positive quantities.

◆ **EXAMPLE 17.2**   Let $y = 2x^{e^x}$, where $x > 0$. Find $\frac{dy}{dx}$.

SOLUTION   The variable is in the base and the exponent; logarithmic differentiation enables us to bring down the expression in the exponent.

$$y = 2x^{e^x}$$

$$\ln y = \ln\left(2x^{e^x}\right) \qquad \text{Take the natural logarithm of each side.}$$

$$\ln y = \ln 2 + e^x \ln x \qquad \text{Use log rules (i) and (iii) to bring down the exponent.}$$

$$\frac{d}{dx}[\ln y] = \frac{d}{dx}[\ln 2 + e^x \ln x] \qquad \text{Differentiate each side. } \ln 2 \text{ is a constant.}$$

$$\frac{1}{y}\frac{dy}{dx} = (e^x)(\ln x) + (e^x)\left(\frac{1}{x}\right) \qquad \text{Use the Chain Rule on left because } y \text{ is a function of } x.$$

$$\frac{dy}{dx} = e^x\left(\ln x + \frac{1}{x}\right) y \qquad \text{Solve for } \frac{dy}{dx}.$$

$$\frac{dy}{dx} = e^x\left(\ln x + \frac{1}{x}\right)\left(2x^{e^x}\right) \qquad \text{Replace } y \text{ by its expression in } x. \quad ◆$$

◆ **EXAMPLE 17.3**   Differentiate $y = \dfrac{(x+3)^5(x^2+7x)^8}{x(x^2+5)^3}$, where $x > 0$.[3]

---

[3] The condition $x > 0$ assures that $(x+3)$, (x^2+7x), and x are all positive.

$$y = \frac{(x+3)^5(x^2+7x)^8}{x(x^2+5)^3}$$

$$\ln y = \ln \frac{(x+3)^5(x^2+7x)^8}{x(x^2+5)^3} \qquad \text{Take the natural logarithm of each side.}$$

$$\ln y = \ln[(x+3)^5(x^2+7x)^8] - \ln[x(x^2+5)^3] \qquad \text{Use the fact that } \ln \frac{a}{b} = \ln a - \ln b.$$

$$\ln y = 5\ln(x+3) + 8\ln(x^2+7x) - \ln x - 3\ln(x^2+5) \qquad \text{Use } \ln(ab) = \ln a + \ln b \text{ and } \ln a^b = b\ln a.$$

$$\frac{d}{dx}[\ln y] = \frac{d}{dx}[5\ln(x+3) + 8\ln(x^2+7x) - \ln x - 3\ln(x^2+5)] \qquad \text{Differentiate each side.}$$

$$\frac{1}{y}\frac{dy}{dx} = \frac{5}{x+3} + \frac{8(2x+7)}{x^2+7x} - \frac{1}{x} - \frac{3(2x)}{x^2+5} \qquad \text{Apply the Chain Rule.}$$

$$\frac{dy}{dx} = \left[\frac{5}{x+3} + \frac{16x+56}{x^2+7x} - \frac{1}{x} - \frac{6x}{x^2+5}\right] y \qquad \text{Solve for } \frac{dy}{dx}.$$

$$\frac{dy}{dx} = \left[\frac{5}{x+3} + \frac{16x+56}{x^2+7x} - \frac{1}{x} - \frac{6x}{x^2+5}\right] \frac{(x+3)^5(x^2+7x)^8}{x(x^2+5)^3} \qquad \text{Replace } y \text{ by its expression in } x.$$

Ugly, but not as painful as differentiating this using the Quotient and Product Rules.[4]   ◆

PROBLEMS FOR SECTION 17.2

1. Find $f'(x)$.
   (a) $f(x) = 2x^x$, where $x > 0$
   (b) $f(x) = 5(x^2+1)^x$
   (c) $f(x) = (2x^4+5)^{3x+1}$

2. Find $f'(x)$.
   (a) $f(x) = 3 \cdot 2^x + 2 \cdot x^3 + 3 \cdot x^{2x+3}$, where $x > 0$
   (b) $f(x) = x(2x^3+1)^x + 5$, where $x > 0$

3. Find $g'(t)$.
   (a) $g(t) = \frac{2^t}{t^{2t}}$, where $t > 0$
   (b) $g(t) = \ln(t+1)^{t^2+1}$, where $t > -1$

4. Find $\frac{dy}{dx}$ using logarithmic differentiation. You need **not** simplify.
   (a) $y = x^{\ln \sqrt{x}}$, where $x > 0$
   (b) $y = \frac{xe^{5x}}{(x+1)^2\sqrt{x-2}}$, where $x > 0$
   (c) $y = (e^{2x})(x^2+3)^5(2x^2+1)^3$

5. Suppose $y = f(x)g(x)$, where $f(x)$ and $g(x)$ are positive for all x. Use logarithmic differentiation to find $\frac{dy}{dx}$. Verify that your result is simply the Product Rule.

---

[4] If you ever forget the Product or Quotient Rules but remember the derivative of $\ln x$, you can use logarithmic differentiation to reconstruct the other rules for yourself.

6. Suppose $y = \frac{f(x)}{g(x)}$, where $f(x)$ and $g(x)$ are positive for all x. Use logarithmic differentiation to find $\frac{dy}{dx}$. Verify that this is the same result you would get had you used the Quotient Rule.

7. If you felt so inclined, you could come up with a "rule" for taking the derivative of functions of the form $f(x)^{g(x)}$ where $f(x)$ is positive. You might call it "the Tower Rule" since you have a tower of functions, or you might think of a more descriptive name. In any case, what would this rule be?

17.3   IMPLICIT DIFFERENTIATION

When using the process of logarithmic differentiation, we differentiate an equation in which y is not explicitly expressed as a function of x. Logarithmic differentiation is a special case of the broader concept of **implicit differentiation**, a concept with far-reaching implications and applications. The basic idea is that we can find $\frac{dy}{dx}$ even when y is not given explicitly as a function of x. We differentiate both sides of the equation that relates x and y, applying the Chain Rule to differentiate terms involving y because y varies with x.

Implicit differentiation is an important concept; we'll begin with a very straightforward example to illustrate what is going on.

◆ **EXAMPLE 17.4**   Consider the circle of radius 2 centered at the origin.[5] It is given by $x^2 + y^2 = 4$. Find the slope of the line tangent to the circle at the following points.

(a) $(1, \sqrt{3})$

(b) $(1, -\sqrt{3})$

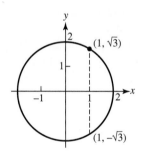

Figure 17.1

SOLUTION   Although y is not a function of x, it can be expressed as two different functions of x.

$$y = \sqrt{4 - x^2} \text{ (the top semicircle)} \quad \text{and} \quad y = -\sqrt{4 - x^2} \text{ (the bottom semicircle)}$$

Each of these functions gives y explicitly as a function of x. One approach is to differentiate the former expression to get information about the point $(1, \sqrt{3})$ and the latter to deal with

---

[5] This circle is the set of all points a distance 2 from the origin. If (x, y) is a point on this circle, then the distance formula tells us that $\sqrt{(x-0)^2 + (y-0)^2} = 2$. And conversely, if (x, y) satisfies the equation $\sqrt{x^2 + y^2} = 2$, then (x, y) is a point on the circle. Therefore, $x^2 + y^2 = 4$ is the equation of the circle.

$(1, -\sqrt{3})$. Do this on your own and compare your answers with those below. We'll take a more efficient approach.

We are looking for $\frac{dy}{dx}$, the rate of change of y with respect to x, at a point. The equation $x^2 + y^2 = 4$ implicitly gives a relationship between x and y. Because $\frac{dy}{dx}$ is a local measure of a rate of change, we are interested in a bug's-eye view rather than a bird's-eye view of the curve. Let's peer at the point $(1, \sqrt{3})$ through a magnifying glass. In the immediate vicinity of $(1, \sqrt{3})$ the curve looks like the graph of a differentiable function, despite the fact that a bird's-eye view tells us it is not.[6] Because a derivative is a local measure of a rate of change, this is all that is really important. We differentiate both sides of the equation treating y as if it were a function of x. We differentiate each side *with respect to x* because we are looking for $\frac{dy}{dx}$, the derivative of y with respect to x.

$$\frac{d}{dx}[x^2 + y^2] = \frac{d}{dx}[4]$$

Because we are treating y as if it were a function of x, when we differentiate y^2 we get $2y\frac{dy}{dx}$. Just as we needed to use the Chain Rule to find $\frac{d}{dx}[\ln y]$ when we did logarithmic differentiation in Example 17.2, we need to use the Chain Rule here to evaluate $\frac{d}{dx}y^2$. Because y depends on x, what we are really trying to find here is $\frac{d}{dx}(mess)^2$. The Chain Rule tells us that the derivative of $(mess)^2$ is $2(mess) \cdot (mess)'$. Therefore,

$$2x + 2y\frac{dy}{dx} = 0 \qquad \text{Now solve for } \tfrac{dy}{dx}.$$

$$2y\frac{dy}{dx} = -2x$$

$$\frac{dy}{dx} = \frac{-x}{y}.$$

You may initially be startled that the formula for $\frac{dy}{dx}$ involves not only x, but y also. On second thought, this should not be too surprising, because y is *not* a function of x. A given x-value may correspond to more than one y-value and therefore may have more than one slope associated with it. We need to know both coordinates of a point on the curve, not just the x-value, to determine the slope of the tangent line. For example, $x = 1$ at points $(1, \sqrt{3})$ and $(1, -\sqrt{3})$ on the circle; we need to specify a value for y to know which one is meant. It makes sense that our formula for $\frac{dy}{dx}$ should involve both x and y.

At $(1, \sqrt{3})$, the slope of the tangent line is $\frac{-1}{\sqrt{3}}$, while at $(1, -\sqrt{3})$ the slope of the tangent line is $\frac{1}{\sqrt{3}}$.

We've solved the original problem, but let's look back at it one more time. We found $\frac{dy}{dx} = \frac{-x}{y}$. If we are looking at a point on the top semicircle where $y = \sqrt{4 - x^2}$, we can write $\frac{dy}{dx} = \frac{-x}{\sqrt{4-x^2}}$; if we were looking at a point on the bottom semicircle where $y = -\sqrt{4 - x^2}$, we can write $\frac{dy}{dx} = \frac{-x}{-\sqrt{4-x^2}}$.

$\frac{dy}{dx}$ is undefined when $y = 0$. This corresponds to the points $(2, 0)$ and $(-2, 0)$ on the circle. Get out your magnifying glass again and take a good look at the curve in the immediate vicinity of each of these points. No matter how much the curve is magnified,

---

[6] The graph of the circle does not pass the vertical line test. Any bird with enough height can see that.

it does *not* look like the graph of a function. Intuitively speaking, $\frac{dy}{dx}$ is defined at a point P, and we can find its value using implicit differentiation only if, under magnification, the curve around P looks like the graph of a differentiable function. ◆

◆ **EXAMPLE 17.5**   The curve shown below is called the *folium of Descartes*. It is the set of all points satisfying the equation $x^3 + y^3 = 6xy$. The point $(3, 3)$ lies on this curve; when we substitute $x = 3$ and $y = 3$ into the equation, both sides are equal. Find the slope of a tangent line to $x^3 + y^3 = 6xy$, first in general, and then at the point $(3, 3)$.

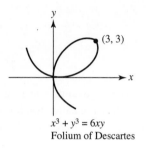

$$x^3 + y^3 = 6xy$$
Folium of Descartes

Figure 17.2

SOLUTION   We need to determine $\frac{dy}{dx}$. Notice that we do not have an explicit formula for y in terms of x. In fact, a glance at the graph shows that y is not a function of x. While it is possible to solve for y explicitly in terms of x using several functions, it is very difficult. However, if we magnify the curve right around the point $(3, 3)$, it does look like the graph of a function, so we can use implicit differentiation treating y as if it were a function of x and applying the Chain Rule where necessary.

$$\frac{d}{dx}(x^3 + y^3) = \frac{d}{dx}(6xy) \qquad \text{Differentiate each side with respect to } x, \text{ applying the Chain Rule.}$$

$$3x^2 + 3y^2\frac{dy}{dx} = 6y + 6x\frac{dy}{dx}$$

In differentiating the $6xy$ on the right side of the equation, we have used the Chain Rule in combination with the Product Rule. Now we need to solve for $\frac{dy}{dx}$. The equation we have is linear in $\frac{dy}{dx}$, so we use the standard strategy for solving linear equations.

$$3x^2 + 3y^2\frac{dy}{dx} = 6y + 6x\frac{dy}{dx} \qquad \text{Bring all terms involving } \tfrac{dy}{dx} \text{ to one side.}$$

$$3y^2\frac{dy}{dx} - 6x\frac{dy}{dx} = 6y - 3x^2 \qquad \text{Factor out a } \tfrac{dy}{dx}.$$

$$\frac{dy}{dx}(3y^2 - 6x) = 6y - 3x^2 \qquad \text{Divide through to isolate } \tfrac{dy}{dx}.$$

$$\frac{dy}{dx} = \frac{6y - 3x^2}{3y^2 - 6x}$$

$$\frac{dy}{dx} = \frac{2y - x^2}{y^2 - 2x}$$

We now have a formula for $\frac{dy}{dx}$; this expression makes sense, provided $y^2 \neq 2x$.[7]

To find the slope of the line tangent to the folium of Descartes at the point $(3, 3)$, substitute $x = 3$ and $y = 3$ into the expression for $\frac{dy}{dx}$, obtaining $\frac{dy}{dx} = -1$. This looks quite reasonable considering the symmetry of the graph around the line $y = x$. ◆

◆ **EXAMPLE 17.6**   Find the slope of the tangent line to the curve $3y^4 + 4xy^2 - 2x^2 = 9$ at the point $(2, 1)$.

SOLUTION   We take the derivative of both sides with respect to x, thinking of y as if it were a function of x.

$$\frac{d}{dx}(3y^4 + 4xy^2 - 2x^2) = \frac{d}{dx}(9)$$

$$12y^3\frac{dy}{dx} + 4y^2 + 8xy\frac{dy}{dx} - 4x = 0$$

Although we could solve for $\frac{dy}{dx}$ and then substitute in $x = 2$ and $y = 1$, because a general formula is not called for, it is algebraically much cleaner to substitute in the x- and y-values immediately after differentiation.[8]

$$12(1)^3\frac{dy}{dx} + 4(1)^2 + 8(2)(1)\frac{dy}{dx} - 4(2) = 0$$

$$28\frac{dy}{dx} - 4 = 0$$

$$\frac{dy}{dx} = \frac{1}{7}$$

Thus, the slope at the point $(2, 1)$ is $\frac{1}{7}$. ◆

◆ **EXAMPLE 17.7**   The equation $2x^2 + 4xy + 3y^2 = 6$ describes an ellipse[9] centered at the origin and rotated as shown below. Find the maximum and minimum values of y on this curve.

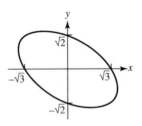

Figure 17.3

SOLUTION   We need to find the points at which the line tangent to the curve is horizontal, i.e., where $\frac{dy}{dx} = 0$. To find $\frac{dy}{dx}$ we use implicit differentiation.

---

[7] At the points on the folium of Descartes where $y^2 = 2x$ the derivative is undefined. y does not look like a differentiable function of x, no matter how much you magnify the area around the point in question. For example, look at the point $(0, 0)$.

[8] Don't substitute in values of x and y *before* differentiating; you would just be taking the derivative of constants.

[9] For information on conic sections, see Appendix E, Conic Sections.

$$\frac{d}{dx}(2x^2 + 4xy + 3y^2) = \frac{d}{dx}(6)$$

$$4x + 4y + 4x\frac{dy}{dx} + 6y\frac{dy}{dx} = 0$$

$$\frac{dy}{dx}(4x + 6y) = -4x - 4y$$

$$\frac{dy}{dx} = \frac{-2x - 2y}{2x + 3y}$$

$\frac{dy}{dx} = 0$ where the numerator of the fraction is zero and the denominator is not simultaneously zero. Simplifying $-2x - 2y = 0$ gives $y = -x$.

We must check that the denominator is *not* zero at the same time. Substituting $y = -x$ into the equation $2x + 3y = 0$ gives $x = 0$. Thus, when x and y are both zero, $\frac{dy}{dx}$ is not defined. But the point $(0, 0)$ is not on the ellipse, so we need not worry about it.

$y = -x$ is an entire line of points; we need to find out which of these points are also on the ellipse. We find the points of intersection of the ellipse and the line by solving the two equations simultaneously.

$$\begin{cases} 2x^2 + 4xy + 3y^2 = 6 \\ y = -x \end{cases}$$

The simplest way to do this is to substitute the second equation into the first to eliminate the variable y and then solve for x.

$$2x^2 + 4xy + 3y^2 = 6$$

$$2x^2 + 4x(-x) + 3(-x)^2 = 6$$

$$2x^2 - 4x^2 + 3x^2 = 6$$

$$x^2 = 6$$

$$x = \sqrt{6} \quad \text{or} \quad x = -\sqrt{6}$$

These are the two values of x at which the tangent line to the ellipse is horizontal. Our original goal is to find the maximum and minimum y-values, so we can substitute these x-values back into the original equation. Or, because we know that they will lie on the line $y = -x$, and this is a much simpler equation, we can say that at $x = \sqrt{6}$, the minimum y-value of $-\sqrt{6}$ is attained and that at $x = -\sqrt{6}$, the maximum y-value of $\sqrt{6}$ is attained.

◆

The Process of Implicit Differentiation

1. Decide which variable you want to differentiate with respect to (x if you want $\frac{dy}{dx}$, t if you want $\frac{dy}{dt}$, etc.).

2. Differentiate both sides of the equation with respect to that variable. Remember the Chain Rule!

   Suppose you are differentiating with respect to t. Distinguish between quantities that vary with t (treating them as functions of t) and those that are independent of t (treating them as constants). In particular, when looking for $\frac{dy}{dt}$, think of y as a function of t.

3. If necessary, solve to find a formula for the desired derivative. If you only want to know the derivative at a specific point, substitute in the coordinates of that point before solving for the derivative you're trying to find.

A Brief Recap of Differentiating "with respect to" a Particular Variable

In practice, because a functional relationship between variables is not explicitly spelled out when we use implicit differentiation, it is necessary to think about what quantities change with which other variables. Suppose, for instance, that we will differentiate both sides of an equation with respect to the variable time t; we must establish which quantities vary with respect to time and which do not. We'll clarify this by means of an example.

◆ **EXAMPLE 17.8**   Let P be the pressure under which a gas is kept, V be the volume of the gas, and T be temperature measured on the absolute (or Kelvin) scale. Then the combined gas law tells us that $\frac{PV}{T} = K$, where K is a constant.[10]

(a) Suppose temperature is kept constant. Express the rate of change of volume with respect to pressure.

(b) Suppose volume is kept constant. Express the rate of change of temperature with respect to pressure.[11]

(c) Suppose volume is kept constant but pressure and temperature change with time. What is the relationship between the change in pressure with respect to time and the change in temperature with respect to time?

(d) Suppose pressure, temperature, and volume all change with time. How are these rates of change related?

SOLUTION   (a) Here we are thinking of P as the independent variable and looking for $\frac{dV}{dP}$. V varies with P, so we treat it as a function of P. T is treated as a constant in this part of the problem. K is also a constant. We can emphasize this by writing

$$\frac{P}{T}V(P) = K.$$

We want to find $\frac{dV}{dP}$. One option is to solve for V in terms of P to get

$$V(P) = \frac{KT}{P} = KTP^{-1}.$$

To find out how V varies with P we differentiate with respect to P, obtaining

$$\frac{d}{dP}[V] = \frac{d}{dP}[KTP^{-1}]$$

---

[10] Here's a little history for the chemists among you. In 1660 Robert Boyle published his gas law stating that if temperature is kept constant, then the product of the pressure and the volume is a constant. Later Jacques Charles looked at the relationship between volume and temperature when pressure is kept constant. The absolute (Kelvin) temperature scale allows Charles's law to be written $V/T = C_1$, where T is the temperature on the absolute scale and the constant C_1 depends on the pressure and the mass of gas present. In 1802 Joseph Guy-Lussac's investigation of gases yielded the result that $P/T = C_2$, where C_2 is a constant. Combining Boyle's, Charles's, and Guy-Lussac's laws gives the more general gas law, $PV/T = C_3$, where the value of the constant C_3 depends on the amount of gas present and T is the absolute (or Kelvin) temperature. I found this law handy on Christmas day 1996 when baking a cake for a potluck dinner in the Ecuadoran Andes at an altitude of 2530 meters (about 1.5 miles above sea level).

[11] This was the goal in the cake baking mentioned in the previous footnote. The pressure had dropped due to the high altitude. I was trying to figure out how the baking temperature should change in order to counterbalance this.

$$\frac{dV}{dP} = KT(-1)P^{-2} = \frac{-KT}{P^2}.^{12}$$

Another option is to differentiate implicitly.

$$\frac{d}{dP}\left[\frac{P}{T}V(P)\right] = \frac{d}{dP}K, \text{ so}$$

$$\frac{1}{T}\left(P\frac{dV}{dP} + V(P)\right) = 0$$

$$\frac{P}{T}\frac{dV}{dP} = \frac{-V(P)}{T}.$$

Solving for $\frac{dV}{dP}$ gives $\frac{dV}{dP} = \frac{-V(P)}{P}$. This is equivalent to the previous answer, because $V(P) = \frac{KT}{P}$.

(b) Here we are thinking of P as the independent variable and looking for $\frac{dT}{dP}$. Now V is treated as a constant and T as a function of P. We can emphasize this by writing $\frac{PV}{T(P)} = K$. The simplest strategy here is to solve for T explicitly and then differentiate with respect to P.

$$T(P) = \frac{PV}{K} = \frac{V}{K}P, \text{ where } V \text{ and } K \text{ are both constant. Therefore, } \frac{dT}{dP} = \frac{V}{K}.$$

(c) Here we are thinking of time t as the independent variable; T and P vary with time but we consider both V and K as constants. We differentiate implicitly.

It may be helpful to write

$$\frac{P(t) \cdot V}{T(t)} = K$$

to emphasize that P and T vary with time.

We will differentiate the equation with respect to t. Our job will be easier if we rewrite the equation as

$$V \cdot P(t) = KT(t)$$

because we won't need the Product and the Quotient Rules.

Differentiating with respect to t gives

$$V\frac{dP}{dt} = K\frac{dT}{dt}$$

or

$$\frac{dP}{dt} = \frac{K}{V}\frac{dT}{dt}.$$

If we like, we can eliminate V entirely, because $V = K\frac{T}{P}$.

---

[12] *Some remarks about notation:*

$\frac{d}{dP}$ is an operator; it is a symbol that represents the operation of differentiating with respect to P. $\frac{dV}{dP}$ is an expression, in the same way that $P + V$ is an expression.

$\frac{dV}{dP} = \lim_{\Delta P \to 0} \frac{\Delta V}{\Delta P}$, where $\Delta V = V(P + \Delta P) - V(P)$. The notation $\frac{dV}{dP}$ reminds us that we are differentiating V with respect to P. Here it reminds us that we are looking at the change in volume induced by a change in pressure.

$$\frac{dP}{dt} = K \cdot \frac{P}{KT} \cdot \frac{dT}{dt}$$

or

$$\frac{dP}{dt} = \frac{P}{T} \frac{dT}{dt}.$$

(d) Again we think of time t as the independent variable. Here P, V, and T vary with time. It may be helpful to write

$$V(t) \cdot P(t) = K \cdot T(t).$$

We'll differentiate with respect to t, so we'll need to use the Product Rule on the left-hand side.

$$V(t) \cdot \frac{dP}{dt} + \frac{dV}{dt} \cdot P(t) = K \cdot \frac{dT}{dt}$$

We can eliminate K completely, replacing it by $\frac{PV}{T}$.

$$V(t) \cdot \frac{dP}{dt} + \frac{dV}{dt} \cdot P(t) = \frac{PV}{T} \cdot \frac{dT}{dt}.$$

Observe that if we know P,V, and T at a certain instant, then knowing two of the rates of change at that moment allows us to determine the third.

REMARK Notice that the notation V' can be ambiguous. Is it $\frac{dV}{dP}$ or $\frac{dV}{dt}$? We use it only when it is clear from context what is meant. The notation V'' can also be ambiguous. The second derivative of V with respect to P, $\frac{d}{dP}\left(\frac{dV}{dP}\right)$ can be written as $\frac{d^2V}{dP^2}$. The second derivative of V with respect to t, $\frac{d}{dt}\left(\frac{dV}{dt}\right)$ can be written as $\frac{d^2V}{dt^2}$. ◆

PROBLEMS FOR SECTION 17.3

1. The equation $x^2 + y^2 = 169$ describes a circle with radius 13 centered at the origin.
   (a) Solve explicitly for y in terms of x. Is y a function of x?
   (b) Differentiate your expression(s) from part (a) to find $\frac{dy}{dx}$.
   (c) Now use implicit differentiation on the original equation to find $\frac{dy}{dx}$.
   (d) Which method of differentiation (that used in part(a), or that used in part (b)) was easier? Why?
   (e) What is the slope of the tangent line to the circle at the point $(5, 12)$? At $(5, -12)$?

2. Which is larger, the slope of the tangent line to $x^2 + y^2 = 25$ at the point $(4, -3)$ or the slope of the tangent line to $x^2 + 4y^2 = 25$ at the point $(4, -3/2)$? You can answer this analytically (find the two slopes), or you can answer this by looking at the graphs of the ellipse and the circle and sketching the tangent to each at the designated points. (To sketch the ellipse, look at the x- and y-intercepts.)

3. Consider the equation $x^2y + xy^2 + x = 1$. Find $\frac{dy}{dx}$ at all points where $x = 1$.

4. Find the slopes of the tangent lines to $(x - 2)^2 + (y - 3)^2 = 25$ at the two points where $x = 6$.

5. Find the equation of the line tangent to the curve $x^3 + y^3 - 3x^2y^2 + 1 = 0$ at the point $(1, 1)$.

6. Find $\frac{dy}{dx}$ for the curve $x^3 + 3y + y^2 = 6$. What is the slope at $(2, -1)$? At what points is the slope zero?

7. At what points (if any) is the tangent line to the curve $3x^2 + 6xy + 8y^2 = 8$ vertical?

8. (a) Graph the ellipse $4(x - 1)^2 + 9(y - 3)^2 = 36$. (Sketch $4x^2 + 9y^2 = 36$ by looking for x- and y-intercepts and then shift the graph horizontally and vertically as appropriate.)

   (b) Find the slope of the line tangent to the ellipse at the point $(1, 5)$ in two ways. First solve for y explicitly (using the appropriate half of the ellipse) and find $\frac{dy}{dx}$. Then do the problem by differentiating implicitly.

9. Consider the circle given by $x^2 + y^2 = 4$.

   (a) Show that $\frac{dy}{dx} = -\frac{x}{y}$.

   (b) Show that $\frac{d^2y}{dx^2} = \frac{d}{dx}\left(-\frac{x}{y}\right) = -\frac{4}{y^3}$. Explain your reasoning completely.

10. The equation $2(x^2 + y^2)^2 = 25(x^2 - y^2)$ gives a curve that is known as a lemniscate. Find the slope of the tangent line to the lemniscate at the point $(-3, 1)$.

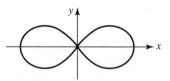

Lemniscate: $2(x^2 + y^2)^2 = 25(x^2 - y^2)$

11. Find an equation for the tangent line to $y^2 = x^3(3 - x)$ at the point $(1, 2)$. What can you say about the tangent line to this curve at the point $(3, 0)$?

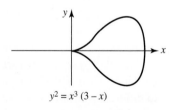

$y^2 = x^3 (3 - x)$

12. Find $\frac{dy}{dx}$.

   (a) $3x^2 + 6y^2 + 3xy = 10$ (b) $(x - 2)^3 \cdot (y - 2)^3 = 1$ (c) $xy^2 + 2y = x^2y + 1$

   (d) $(x^2y^3 + y)^2 = 3x$ (e) $e^{xy} = y^2$ (f) $x \ln(xy^3) = y^2$

   (g) $\ln(xy) = xy^2$

13. (a) Sketch a graph of $(x - 2)^2 - (y - 3)^2 = 25$. (Sketch $x^2 - y^2 = 25$ by looking for x- and y-intercepts and then shift your graph vertically and horizontally as appropriate.)

   (b) Find the equation of the tangent line to the graph at each of the following points.
       i. $(15, -9)$ ii. $(-3, 3)$

14. Find an equation for the tangent line to $x^{2/3} + y^{2/3} = 5$ at the point $(8, 1)$.

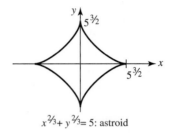

$x^{2/3} + y^{2/3} = 5$: astroid

15. Match the equations with the appropriate graphs. (These are graphs of conic sections. To learn more about conic sections, refer to Appendix E.)

(a) $\frac{x^2}{5} + \frac{y^2}{5} = 8$ (b) $3x^2 - 3y^2 = 27$ (c) $3x^2 + y^2 = 12$

(d) $x^2 + 3y^2 = 12$ (e) $-x^2 + 3y^2 = 12$

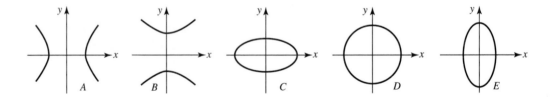

17.4 IMPLICIT DIFFERENTIATION IN CONTEXT: RELATED RATES OF CHANGE

In Sections 17.2 and 17.3 we introduced logarithmic and implicit differentiation. The former is particularly useful when we want to differentiate something of the form $[f(x)]^{g(x)}$ and the latter when it is either unappealing or impossible to solve for the variable we wish to differentiate. In fact, we can get a great deal more mileage from the two fundamental principles we used.

Principle (i): If two expressions are equal, their derivatives are equal.

Principle (ii): Whether y is given implicitly or explicitly in terms of x, we can differentiate an expression in y with respect to x by treating y the same way we would treat $g(x)$ and using the Chain Rule.

In this section we will demonstrate the power and versatility of implicit differentiation and the Chain Rule in context. The example that follows makes the transition from Section 17.3 to the applied problems of this section.

◆ EXAMPLE 17.9   Suppose that x and y are functions of t (they vary with time) and $x^2 + y^2 = 25$. In other words, the point (x, y) is moving on the circle of radius 5 centered at the origin. Think of a bug crawling around the circle; the coordinates of the bug at time t are given by (x, y) or $(x(t), y(t))$.

Suppose that $\frac{dy}{dt} = -6$ units/second, when $x = 3$ and $y > 0$. What is $\frac{dx}{dt}$ at this moment?

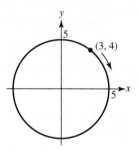

Figure 17.4

Let's think about the question in terms of the bug. We know that when $x = 3$, $y = 4$ (because $x^2 + y^2 = 25$ and $y > 0$). When the bug is at the point $(3, 4)$ its y-coordinate is decreasing at a rate of 6 units per second (the minus sign indicating a decrease). This tells us that the bug is traveling in a clockwise direction. Knowing that the bug stays on the circle, we want to find the rate of change of its x-coordinate. (By thinking about the bug's path, you can see that at $(3, 4)$, $\frac{dx}{dt}$ must be positive. Thinking harder can lead you to realize that $\frac{dx}{dt} > 6$.)

SOLUTION   Let's make the fact that x and y vary with t more explicit by writing

$$[x(t)]^2 + [y(t)]^2 = 25.$$

What We Know: $\frac{dy}{dt} = -6$.

What We Want: $\frac{dx}{dt}$, when $x = 3$ and $y > 0$.

Using principle (i) we differentiate both sides of this equation with respect to t.

$$\frac{d}{dt}\left\{[x(t)]^2 + [y(t)]^2\right\} = \frac{d}{dt}[25]$$

$$2x(t)\frac{dx}{dt} + 2y(t)\frac{dy}{dt} = 0$$

$$x\frac{dx}{dt} + y\frac{dy}{dt} = 0$$

We are interested in $\frac{dx}{dt}$ at the point where $x = 3$ and $y > 0$. When $x = 3$, $y = \sqrt{25 - 3^2} = \sqrt{25 - 9} = 4$.

$$3\frac{dx}{dt} + 4(-6) = 0$$

$$\frac{dx}{dt} = 24/3 = 8$$

REMARK At the moment we are interested in we know that $x = 3$ and $y = 4$, but because x isn't *always* 3 and y isn't *always* 4, we can't substitute $x = 3$ and $y = 4$ until *after* differentiating. (Similarly, if we are interested in $f'(2)$ we can't evaluate at 2 until *after* differentiating. If we were to evaluate f at 2 before differentiating, we would think that $f'(2)$ is always zero, regardless of f.) ◆

◆ EXAMPLE 17.10   A 5-foot ladder is leaning against a wall and begins to fall. Suppose the bottom of the ladder is being pulled away from the wall at the constant rate of 0.5 feet per second.

(a) How fast is the top of the ladder falling when the bottom of the ladder is 3 feet from the wall?

(b) Does the top of the ladder fall at a constant rate or does the rate depend on how far the bottom is from the wall? If the rate is not constant, how is it changing?

SOLUTION   We need to draw a picture of the situation and label variables and constants in order to express mathematically what we know and what we are trying to find. (It's crucial to determine what we know and what we're looking for; otherwise it is impossible to figure out a strategy for proceeding.)

Below are two pictures: Figure 17.5(a) represents the situation at time t and Figure 17.5(b) represents a snapshot at the moment in question.

Let $x =$ the distance from the foot of the ladder to the wall.

Let $y =$ the height of the top of the ladder.

The Pythagorean Theorem tells us that x and y are related by $x^2 + y^2 = 25$. Relating the variables x and y allows us to relate their rates of change. Because x and y vary with time, we can write $[x(t)]^2 + [y(t)]^2 = 25$.

(a) **What We Know:** $\frac{dx}{dt} = 0.5$ feet per second.
What We Want: $\frac{dy}{dt}$, when $x = 3$.

Differentiate $x^2 + y^2 = 25$ with respect to t to get

$$2x\frac{dx}{dt} + 2y\frac{dy}{dt} = 0.$$

Equivalently,

$$x\frac{dx}{dt} + y\frac{dy}{dt} = 0. \tag{17.1}$$

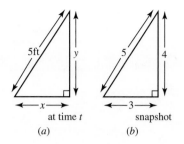

Figure 17.5

When $x = 3$ we know that $y = 4$, because x and y are related by $x^2 + y^2 = 25$. Evaluating (17.1) at $x = 3$, $y = 4$, and $\frac{dx}{dt} = 0.5$ gives

$$3(0.5) + 4\frac{dy}{dt} = 0$$

$$4\frac{dy}{dt} = -\frac{3}{2}$$

$$\frac{dy}{dt} = -\frac{3}{8}. \quad \frac{dy}{dt} \text{ is negative because } y \text{ is decreasing; the ladder is falling.}$$

So at this moment the top of the ladder is falling at a rate of 3/8 feet per second. Notice that the mathematics involved in Example 17.10(a) is just like that in Example 17.9 even though the situation is very different.

(b) We can start with the equation relating $\frac{dx}{dt}$ and $\frac{dy}{dt}$.

$$x\frac{dx}{dt} + y\frac{dy}{dt} = 0$$

The bottom of the ladder is being pulled away at a constant rate of 0.5 feet per second, so we know that $\frac{dx}{dt} = 0.5$. Therefore,

$$0.5x + y\frac{dy}{dt} = 0$$

$$\frac{dy}{dt} = \frac{-0.5x}{y} = -\frac{x}{2\sqrt{25 - x^2}}.$$

As x gets larger $\frac{x}{2\sqrt{25-x^2}}$ gets larger; the numerator increases and the denominator decreases, both contributing to the growth of the fraction. Therefore, the ladder starts falling slowly and speeds up as the bottom moves away from the wall. ◆

◆ EXAMPLE 17.11   At 10:00 A.M. a fishing boat leaves the dock in Vancouver and heads due south at 20 mph. At the same moment a ferry located 60 miles directly west of the dock is traveling toward the dock at 25 mph. Assume that the boats maintain their speeds for the next two hours. How fast is the distance between the boats changing at 11:00 A.M.? Are the boats getting farther apart or closer together?

SOLUTION   On the following page are two pictures of the situation. The picture on the left is a general picture; the one on the right is a snapshot taken at 11:00 A.M.

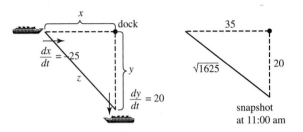

Figure 17.6

Let $x =$ the distance (in miles) between the ferry and the dock.

Let $y =$ the distance (in miles) between the fishing boat and the dock.

Let $z =$ the distance (in miles) between the two boats. x, y, and z are functions of time t, where t is measured in hours. $x = x(t)$, $y = y(t)$, and $z = z(t)$.

Approach 1

What We Know: $\frac{dx}{dt} = -25$ mph (negative because x is decreasing with time)

$\frac{dy}{dt} = 20$ mph (positive because y is increasing with time)

The Pythagorean Theorem tells us that x, y, and z are related by $x^2 + y^2 = z^2$.

What We Want: $\frac{dz}{dt}$, when $t = 1$.

$$x^2 + y^2 = z^2$$

Therefore,

$$2x\frac{dx}{dt} + 2y\frac{dy}{dt} = 2z\frac{dz}{dt}$$

$$x\frac{dx}{dt} + y\frac{dy}{dt} = z\frac{dz}{dt}.$$

When $t = 1$, $x = 60 - 25 = 35$, and $y = 20$. We use the relationship $x^2 + y^2 = z^2$ to find that $z = \sqrt{1625}$.

$$35(-25) + 20(20) = \sqrt{1625}\frac{dz}{dt}$$

$$\frac{dz}{dt} = \frac{-475}{\sqrt{1625}} \approx -11.78$$

The distance between the boats is *decreasing* at a rate of about 11.78 mph.

Approach 2

Express x, y, and z explicitly in terms of t and then find $\frac{dz}{dt}$ at $t = 1$. Let $t = 0$ correspond to 10:00 A.M.

$$y(t) = 20t \quad \text{(rate)} \cdot \text{(time)} = \text{distance}$$

$$x(t) = 60 - \text{(distance traveled)} = 60 - 25t$$

$$z(t) = \sqrt{(60 - 25t)^2 + (20t)^2}$$

We must find $\frac{dz}{dt}$ and then evaluate it at $t = 1$. This is left as an exercise for the reader.   ◆

PROBLEMS FOR SECTION 17.4

1. Suppose we toss a rock into a pond causing a circular ripple. If the radius is increasing at a rate of 3 feet per second when the diameter is 4, how fast is the area increasing?

2. The price and the demand for a certain item can be modeled by the equation $20p = -q + 200$.

   (a) Express the rate of change of quantity demanded with respect to price in terms of a derivative and evaluate it.

   (b) Suppose that price is determined by the world market, so that price and quantity can both be thought of as functions of time. At a certain instant the price is $20 and is increasing at a rate of $0.25 per week. At what rate is the quantity demanded changing at this instant? Why is your answer negative?

3. A bug is walking around the circle $x^2 + y^2 = 169$. At a certain instant the bug is at the point $(-5, 12)$ and its y-coordinate is decreasing at a rate of 3 units per second.

   (a) Is the bug traveling the circle in a clockwise direction or a counterclockwise direction?

   (b) How fast is its x-coordinate changing at this instant?

4. A spherical balloon is losing air at a steady rate of 0.5 cm^3/hour.

   (a) How fast is the radius decreasing when the diameter of the balloon is 30 cm?

   (b) How fast is the radius decreasing when the diameter of the balloon is 15 cm?

   (c) How fast is the diameter decreasing when the diameter of the balloon is 15 cm?

   (d) How fast is the radius decreasing when the circumference of the balloon is 12π centimeters?

5. A cylindrical barrel with radius 3 feet is filled with oil. The barrel stands upright and oil comes out a faucet at the base of the tank.

   (a) How are the volume of oil in the tank and the height of oil in the tank related?

   (b) What is the rate of change of volume with respect to height? Is it constant, or does it depend upon the height? Does your answer make sense to you?

   (c) As oil leaves the tank, both the volume and the height of oil in the tank change with respect to time.

      i. When the height is decreasing at a rate of 0.5 feet per hour, how fast is the volume of oil in the tank changing?

      ii. When oil is leaving the tank at a rate of 3 cubic feet per hour, how fast is the height of oil in the tank changing?

6. At 7:00 A.M. a truck is 100 miles due north of a car. The truck is traveling south at a constant speed of 40 mph, while the car is traveling east at 60 mph. How fast is the distance between the car and the truck changing at 7:30 A.M.?

7. As shown on the following page, a wheelbarrow is being wheeled down a perfectly straight 13-foot ramp. The top of the ramp is 5 feet high and the ramp covers a horizontal distance of 12 feet. When the wheelbarrow has moved a horizontal distance of 6 feet it is moving horizontally at a rate of 0.5 feet per second. At what rate is it moving vertically downward?

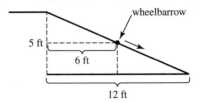

8. Valentina is sipping from a straw she has stuck in a conical cup of lemonade. The cup is an inverted circular cone of height 12 cm and radius 6 cm. When the height of the lemonade in the cup is 10 cm, Valentina is sipping lemonade at a rate of 2 cm^3/second. At this moment, how fast is the height of the lemonade in the cup decreasing?

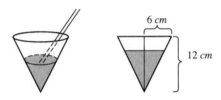

9. Two vehicles are on parallel roads that are 0.5 miles apart. At noon the vehicles are half a mile apart and traveling in the same direction. One is traveling at 30 mph and the other at 40 mph.

   (a) After 2 hours, how fast is the distance between the cars increasing?

   (b) When the cars are 40 miles apart, how fast is the distance between them increasing?

   (c) What would be the answer to part (a) if the cars were traveling in opposite directions?

10. In Barcelona there is a beautiful Spanish castle set 1/4 of a mile back from a straight road. A bicyclist rides by the castle at a velocity of 15 mph. Assuming that the biker maintains this speed, how fast is the distance between the biker and the castle increasing 20 minutes later?

11. A conical container is used to hold oil. It is positioned upright with the tip of the cone at the bottom. Oil comes out a faucet at the base of the container. We know that the volume of a cone is $V = \frac{1}{3}\pi r^2 h$.

    (a) As oil leaves the cone the height, radius, and volume of oil in the container change with time. Find $\frac{dV}{dt}$ in terms of r, h, $\frac{dr}{dt}$, and $\frac{dh}{dt}$.

    (b) Suppose the container has a height of 24 inches and a radius of 12 inches. Express the volume of oil in the container as a function of the height of oil in the container. (*Hint:* Use similar triangles to express the radius of the oil in terms of the height of the oil.)

    (c) Suppose oil is leaking out of the container at a rate of 5 cubic inches per hour. How fast is the height of the oil in the container decreasing when the height is 10 inches? When the height is 4 inches?
    Do the relative sizes of your answers make sense to you intuitively?

12. An airplane is flying at a speed of 600 mph at a constant altitude of 1 mile. It passes directly over an air traffic control tower. How fast is the distance between the control tower and the airplane increasing when the plane is 10 miles away from the tower?

13. Sand is being dumped into a conical pile whose height is 1/2 the radius of its base. Suppose sand is being dumped at a rate of 5 cubic meters per minute.
    (a) How fast is the height of the pile increasing when it is 9 meters high?
    (b) How fast is the area of the base increasing at this moment?
    (c) How fast is the circumference of the base increasing at this moment?
    (d) Will the height be increasing more slowly, more rapidly, or at a steady pace as time goes on?

14. It is evening and a woman 5 and a half feet tall is standing by a 14-foot-high street lamp on a cobbled road, waiting for her ride home from work. When she sees her husband pull his car up to the curb she begins to walk away from the light at a rate of 4 feet per second. How fast is the length of her shadow changing?

woman under streetlamp

C H A P T E R

18

Geometric Sums, Geometric Series

▉ 18.1 GEOMETRIC SUMS

An Introductory Example

Consider the modeling we've done with exponential functions: When modeling investments, we've restricted ourselves to the condition that no deposits and no withdrawals are made; when modeling the elimination of a drug from the body, we've restricted our model to an investigation of one dose of the drug, and one dose only. When you stop to think about this, we've been rather unrealistic about our restrictions. Why is this?

The reason for the restrictions we've put on the situations we have modeled is that we have been restricting ourselves to differentiable functions, functions whose graphs are continuous and smooth. If you withdraw $1000 from a bank account, the amount in the account plummets by $1000. The graph of $M(t)$, the amount of money in the account, will not be continuous and therefore will not be differentiable. Similarly, when a person on medication swallows a pill, the amount of medication in his body leaps up. In this chapter we will tackle these situations, situations in which a lump sum of some quantity is either introduced or removed at discrete intervals. This gives the chapter a different flavor from the work we have been doing so far.

◆ **EXAMPLE 18.1**   A doctor treating a patient who has had a heart attack may prescribe a medication, digitalis or digoxin, that helps the heart beat regularly. Digitalis is eliminated from the body at a rate proportional to the amount of digitalis in the body.[1] Suppose 30% of the drug present in the body at any moment is eliminated within 24 hours. If 0.04 mg of digitalis is taken at the same time every morning, how many milligrams are in the body one week later, immediately after the person has taken his eighth dose? How many milligrams will be in his body several years later, assuming he continues to take the drug regularly? (There are fluctuations, since there are 0.04 more mg of digitalis in the body immediately after taking a pill than there were right before taking the pill. Figure out the medication level immediately after the morning pill.)

SOLUTION   Immediately after the first pill of digitalis the person has 0.04 mg in his body. By the next morning 30% has been eliminated, so 70% of the 0.04 mg remains. Right after the second pill, the number of milligrams rises to

$$.04 + .7(.04).$$

The term 0.04 represents the medication from the second pill; the term 0.7(0.04) represents what remains of the first pill. By the following morning only 70% of what was previously in the body remains.

$$.7[.04 + .7(.04)], \text{ or equivalently, } (.7)(.04) + (.7)^2(.04) \text{ remains.}$$

Immediately after the third pill is taken the person has

$$.04 + (.7)(.04) + (.7)^2(.04) \text{ mg of digitalis,}$$

$$(\text{3rd pill}) + \binom{\text{what is left}}{\text{of 2nd pill}} + \binom{\text{what is left}}{\text{of 1st pill}}.$$

Day	Amount of digitalis in the person's system right after taking a pill
1	$.04$
2	$.04 + (.7)(.04)$
3	$.04 + (.7)(.04) + (.7)^2(.04)$
4	$.04 + (.7)(.04) + (.7)^2(.04) + (.7)^3(.04)$

Let's focus for a minute on an individual dosage. With each passing day the amount of digitalis remaining in the body from this individual pill decreases by 30%, or equivalently, is multiplied by 0.7.

After one week, medication from the first pill has been in the body for seven days, the second pill for six days, the third pill for five days, and so on. We are looking for the amount of medication in the body immediately after the eighth pill is taken, so we write a sum with eight terms, each term representing what is left of a pill that has been taken.

$$.04 \quad + (.7)(.04) \quad + (.7)^2(.04) \quad + (.7)^3(.04) \quad + (.7)^4(.04) \quad + (.7)^5(.04) \quad + (.7)^6(.04) \quad + (.7)^7(.04)$$

$$\binom{\text{8th}}{\text{pill}} + \binom{\text{what is left}}{\text{of 7th pill}} + \binom{\text{what is left}}{\text{of 6th pill}} + \binom{\text{what is left}}{\text{of 5th pill}} + \binom{\text{what is left}}{\text{of 4th pill}} + \binom{\text{what is left}}{\text{of 3rd pill}} + \binom{\text{what is left}}{\text{of 2nd pill}} + \binom{\text{what is left}}{\text{of 1st pill}}$$

---

[1] The half-life of digitalis in a patient with good renal function if 1.5–2 days. The figures given are consistent with that.

Calculating how many milligrams are in the body one week later requires simple addition. There are only eight terms, so entering them into a calculator is not too much work.

The second question, how many milligrams will be in the body after several years of taking a 0.04 mg pill every morning, is certainly an important one, but at first glance it looks like it will be a lot of work. Several years is not a well-defined period of time, but if we figured on about three years, with a pill taken every day of the year, we must sum $(365)(3) = 1095$ terms, 1096 terms if we look immediately after that 1096th pill is taken. The prospect of adding these terms up is not particularly appealing, nor is the prospect of writing them down.

But a mathematician would not become despondent at this prospect. Mathematicians look for patterns, and there is a pattern to this sum. The challenge is how to exploit this pattern to cut down on the workload. After three years, immediately after the 1096th pill is taken, the number of milligrams of digitalis in the body is

$$.04 + (.7)(.04) + (.7)^2(.04) + (.7)^3(.04) + \cdots + (.7)^{1095}(.04).$$

We can use a "$+ \cdots +$" precisely because there *is* a pattern. Each term (from the second one on) is (0.7) times the previous term. The ratio of one term to the previous term is constant. Sums with this characteristic are called **geometric sums**, and we will see that this characteristic makes it a real pleasure to add up a geometric sum.

A finite **geometric sum** is a sum of the form

$$a + ar + ar^2 + ar^3 + \cdots + ar^n,$$

where a denotes the first term and r is the constant ratio of any term to the previous term.

We'll begin by exploiting the pattern in order to find the amount of digitalis in the body immediately after the eighth pill. We use the strategy of giving a name to the thing we are trying to find.

Let $D = .04 + (.7)(.04) + (.7)^2(.04) + (.7)^3(.04) + (.7)^4(.04) + (.7)^5(.04)$
$\qquad + (.7)^6(.04) + (.7)^7(.04).$

Key Idea: If we multiply D by 0.7 we get something that looks very similar to D.

$.7D = (.7)(.04) + (.7)^2(.04) + (.7)^3(.04) + (.7)^4(.04) + (.7)^5(.04) + (.7)^6(.04)$
$\qquad + (.7)^7(.04) + (.7)^8(.04)$

Subtracting this equation from the previous one will make all but two terms disappear, and we can solve for D.

$$D = .04 + (.7)(.04) + (.7)^2(.04) + (.7)^3(.04) + (.7)^4(.04) + (.7)^5(.04) + (.7)^6(.04) + (.7)^7(.04)$$
$$.7D = \qquad (.7)(.04) + (.7)^2(.04) + (.7)^3(.04) + (.7)^4(.04) + (.7)^5(.04) + (.7)^6(.04) + (.7)^7(.04) + (.7)^8(.04)$$

$$D - .7D = .04 + 0 \qquad +0 \qquad +0 \qquad +0 \qquad +0 \qquad +0 \qquad +0 \qquad -(.7)^8(.04)$$
$$.3D = .04 - (.7)^8(.04)$$
$$D = \frac{.04 - (.7)^8(.04)}{.3} = \frac{.04[1 - (.7)^8]}{.3} \approx 0.1256 \text{ mg}$$

(This answer will agree with what you would obtain by adding up the eight terms individually.)

We can use the same technique to figure out the number of milligrams of digitalis in the body three years later, immediately after the 1096th pill. Let's call what we are trying to find S, multiply S by 0.7, subtract the latter equation from the former, and solve for S.

$$\text{Let } S = .04 + (.7)(.04) + (.7)^2(.04) + (.7)^3(.04) + \cdots + (.7)^{1095}(.04)$$

$$.7S = \qquad (.7)(.04) + (.7)^2(.04) + (.7)^3(.04) + \cdots + (.7)^{1095}(.04) + (.7)^{1096}(.04)$$

$$S - .7S = .04 \qquad\qquad\qquad\qquad\qquad\qquad\qquad\qquad - (.7)^{1096}(.04)$$

$$.3S = .04 - (.7)^{1096}(.04)$$

$$S = \frac{.04 - (.7)^{1096}(.04)}{.3} = \frac{.04[1-(.7)^{1096}]}{.3} \approx 0.1333$$

So three years later, immediately after taking his pill, the man has about 0.1333 mg of digitalis in his body. (Right before taking that 0.04-mg pill he must have had about 0.0933 mg of digitalis in his body.)

The question at the beginning of this section asked "How many milligrams will be in the body several years later, assuming that the drug continues to be taken regularly?" We figured out how many milligrams would be in the body three years later. Suppose we decided to look at the situation after five years of regular medication, immediately after the 1827th pill.[2] Let's name the sum F, then multiply F by 0.7, the ratio of the terms, and subtract the latter from the former.

$$F = .04 + (.7)(.04) + (.7)^2(.04) + (.7)^3(.04) + \cdots + (.7)^{1826}(.04)$$

$$.7F = \qquad (.7)(.04) + (.7)^2(.04) + (.7)^3(.04) + \cdots + (.7)^{1826}(.04) + (.7)^{1827}(.04)$$

$$.3F = .04 \qquad\qquad\qquad\qquad\qquad\qquad\qquad\qquad - (.7)^{1827}(.04)$$

$$F = \frac{.04 - (.7)^{1827}(.04)}{.3} = \frac{.04[1-(.7)^{1827}]}{.3} \approx 0.1333$$

As you can see, the issue of whether "several" years is closer to three years or five years is a moot one. In both cases a TI-83 calculator will give an answer of $0.1\overline{3}$. Why should this be? Let's look at the amount of digitalis left in the body from the very first pill three years after it is taken. The amount is $0.04(0.7)^{1095}$, a number so small that the TI-83 won't register it; it is negligible. After five years on digitalis the sum total of the medication left in the body from the first two years on the drug is less than $2 \cdot (365) \cdot (0.04)(0.7)^{1095}$ because each of the $2 \cdot (365)$ doses have been in the body for three or more years.[3] This too is negligible; it is less than 10^{-100}. ◆

EXERCISE 18.1   Sketch a rough graph of $D(t)$, the amount of digitalis in the man's bloodstream, as a function of time. You need not be overly concerned with the details but make sure that qualitatively the picture reflects the situation.

---

[2] (365 days/year)(5 yrs) = 1825, and there has to be at least one leap year in there, and then there is that last pill that he takes in the morning. In fact, we'll see that the difference between using 1825 and 1827 is negligible.

[3] An alternative approach is to compute the following sum: $.04(.7)^{1095} + .04(.7)^{1096} + \cdots + .04(.7)^{1095+1826}$ gives the amount of digitalis from the first three years that is still left at the end of five years.

Geometric sums are useful in many contexts. The mathematics is not limited to digitalis pills; it can be applied by physicians prescribing any medication that is administered at regular intervals and is eliminated by the body at a rate proportional to itself. Geometric sums can be used by environmental scientists studying pollutants whose potency diminishes exponentially over time and whose introduction into the environment occurs at regular intervals. Geometric sums are important to investors in analyzing and comparing investment schemes and to banks in calculating payments on loans and mortgages. And all this is in addition to the fact that they have a mathematical beauty in their own right.

Finite Geometric Sums

Definition

A finite geometric sum is a sum of the form

$$a + ar + ar^2 + ar^3 + \cdots + ar^n,$$

where r is the ratio of any term to the previous term and a is the first term. (Notice that there are $(n + 1)$ terms in the sum above; n is not the number of terms.)

The following are examples of geometric sums:

i) $\frac{1}{2} + \frac{1}{4} + \frac{1}{8} + \frac{1}{16} + \cdots + \left(\frac{1}{2}\right)^{50}$

$= \frac{1}{2} + \frac{1}{2}\left(\frac{1}{2}\right) + \frac{1}{2}\left(\frac{1}{2}\right)^2 + \frac{1}{2}\left(\frac{1}{2}\right)^3 + \cdots + \frac{1}{2}\left(\frac{1}{2}\right)^{49}$ $\qquad a = \frac{1}{2} \qquad r = \frac{1}{2} \qquad n = 49$

ii) $\frac{2}{3} - \frac{2}{9} + \frac{2}{27} - \cdots - 2\left(\frac{-1}{3}\right)^{12}$

$= \frac{2}{3} + \frac{2}{3}\left(-\frac{1}{3}\right) + \frac{2}{3}\left(-\frac{1}{3}\right)^2 + \frac{2}{3}\left(-\frac{1}{3}\right)^3 + \cdots + \frac{2}{3}\left(-\frac{1}{3}\right)^{11}$ $\qquad a = \frac{2}{3} \qquad r = -\frac{1}{3} \qquad n = 11$

iii) $3 + 6 + 12 + 24 + \cdots + 3072$

$= 3 + 3(2) + 3(2)^2 + 3(2)^3 + \cdots + 3(2)^{10}$ $\qquad a = 3 \qquad r = 2 \qquad n = 10$

(We find n by writing $3(2^n) = 3072$ so $2^n = 1024$.

Solve for n.)

iv) $.04 + (.9)(.04) + (.9)^2(.04) + \cdots + (.9)^7(.04)$ $\qquad a = .04 \qquad r = .9 \qquad n = 7$

These sums have 50, 12, 11, and 8 terms, respectively.

Consider a sum of the form $a + ar + ar^2 + \cdots + ar^{30}$. We can write this sum in a form that is computationally simpler to use by employing the strategy that worked so well in Example 18.1. Name the sum, multiply by r, subtract the latter equation from the former and solve for the original sum.

$$S = a + ar + ar^2 + \cdots + ar^{30}$$
$$rS = \quad\ ar + ar^2 + \cdots + ar^{30} + ar^{31} \qquad \text{subtracting gives}$$
$$\overline{}$$
$$S - rS = a \qquad\qquad\qquad\qquad - ar^{31}$$
$$(1 - r)S = a - ar^{31}$$
$$S = \frac{a - ar^{31}}{1 - r}.$$

The expression $\frac{a - ar^{31}}{1-r}$ is called the **closed form** of the sum $a + ar + ar^2 + \cdots + ar^{30}$. The closed form is computationally much easier to deal with than is adding 31 terms! One of the wonderful things about geometric sums is that they can be expressed in closed form.

To get any geometric sum into closed form, do the following.

1. Give the sum a name, like S, for instance.

2. Determine "r." (For a geometric sum this can be done simply by dividing any term by the previous one, the second by the first, for instance. Be sure to include signs.)

3. Subtract "r" times the sum from the original sum. All but two terms will drop out. That's the beauty of the thing.

4. Solve for S, and you're done.

The idea here is not to memorize a formula for closed form; rather it is to be able to put any geometric sum into closed form. Try this on the following example. Do the work on your own before looking at the solution provided.

◆ **EXAMPLE 18.2**   Write the following geometric sum in closed form: $-2 + \frac{8}{3} - \frac{32}{9} + \cdots - 2\left(\frac{4}{3}\right)^{100}$.

SOLUTION   Let's call the sum S. $r = \frac{8/3}{-2} = -\frac{4}{3}$ and $a = -2$.

$$S = -2 + \frac{8}{3} - \frac{32}{9} + \cdots - 2\left(\frac{4}{3}\right)^{100}$$

$$-\frac{4}{3}S = \qquad \frac{8}{3} - \frac{32}{9} + \cdots - 2\left(\frac{4}{3}\right)^{100} + 2\left(\frac{4}{3}\right)^{101} \qquad \text{(subtract)}$$

$$\left(1 + \frac{4}{3}\right)S = -2 \qquad\qquad\qquad\qquad\qquad - 2\left(\frac{4}{3}\right)^{101}$$

$$S = \frac{-2 - 2\left(\frac{4}{3}\right)^{101}}{1 + \left(\frac{4}{3}\right)}$$

$$S = \frac{-2 - 2\left(\frac{4}{3}\right)^{101}}{\frac{7}{3}} = \frac{-2\left(1 + \left(\frac{4}{3}\right)^{101}\right)}{\frac{7}{3}}$$

With the aid of a calculator, the right-hand side is easy to compute whereas the sum in its original form would be quite tedious.   ◆

PROBLEMS FOR SECTION 18.1

For Problems 1 through 17 determine whether or not the sum is geometric. Assume "$+\cdots$" indicates that the established pattern continues. If the sum is geometric, identify "a" and "r."

1. $1 - 10 + 100 - \cdots + 10^{10}$

2. $0.3 + 0.33 + 0.333 + 0.3333 + \cdots + 0.333333333$

3. $0.3 + 0.03 + 0.003 + 0.0003 + \cdots + 0.000000003$

4. $\frac{2}{3} + \frac{2}{9} + \frac{2}{27} + \cdots + \frac{2}{6561}$

5. $\frac{2}{3} + 2 + 6 + \cdots + 2(3)^{100}$

6. $1 + \frac{1}{2} + \frac{1}{3} + \cdots + \frac{1}{100}$

7. $1 - \frac{1}{2} + \frac{1}{4} - \frac{1}{8} + \cdots + \frac{1}{64}$

8. $\frac{2}{3} + 1 + \frac{3}{2} + \cdots + \frac{243}{32}$

9. $\frac{3}{2} - \frac{3}{4} + \frac{3}{8} - \cdots - \frac{3}{2^6}$

10. $\frac{3}{2} - \frac{4}{4} + \frac{5}{8} - \cdots - \frac{8}{2^6}$

11. $-a + a^2 - a^3 + \cdots + (-a)^{17}$

12. $\frac{1}{e} + \frac{2}{e^2} + \frac{4}{e^3} + \cdots + \frac{2^n}{e^{n+1}}$

13. $-2 + 4 - 6 + 8 - 10 + \cdots + 16$

14. $-2 + 4 - 8 + 16 - 32 + 64$

15. $0.2 + 0.06 + 0.018 + 0.0054 + 0.00162$

16. $10 + 9.5 + 9 + 8.5 + 8 + 7.5 + 7 + 6.5$

17. $p + p^3 + p^5 + p^7 + p^9$

Express each of the sums in Problems 18 through 32 in closed form. Wherever possible, give a numerical approximation of the sum, rounded off to 3 decimal places.

18. $1 - 10 + 100 - \cdots + 10^{10}$

19. $\frac{3}{2} - \frac{3}{4} + \frac{3}{8} - \cdots - \frac{3}{2^6}$

20. $\frac{2}{3} + 2 + 6 + \cdots + 2(3)^{100}$

21. $2e + 2e^2 + 2e^3 + \cdots + 2e^{12}$

22. $2e + (2e)^2 + (2e)^3 + \cdots + (2e)^n$

23. $-a + a^2 - a^3 + \cdots + (-a)^n$

24. $\frac{1}{e} + \frac{2}{e^2} + \frac{4}{e^3} + \cdots + \frac{2^n}{e^{n+1}}$

25. $5 + 15 + 45 + \cdots + 5 \cdot 3^{10}$

26. $\frac{2}{3} + \frac{1}{3} + \frac{1}{6} + \frac{1}{12} + \cdots + \frac{1}{3 \cdot 2^{100}}$

27. $1 + \frac{9}{10} + \left(\frac{9}{10}\right)^2 + \left(\frac{9}{10}\right)^3 + \cdots + \left(\frac{9}{10}\right)^n$

28. $1 + \frac{11}{10} + \left(\frac{11}{10}\right)^2 + \left(\frac{11}{10}\right)^3 + \cdots + \left(\frac{11}{10}\right)^{200}$

29. $-\frac{5}{2} + \frac{5}{v} - \frac{10}{v^2} + \frac{20}{v^3} - \frac{40}{v^4} + \cdots + (-1)^n \frac{5 \cdot 2^n}{v^{n+1}}$

30. $\frac{2}{p} + \frac{4}{p^3} + \frac{8}{p^5} + \frac{16}{p^7} + \cdots + \frac{2^{20}}{p^{39}}$

31. $mq + m^2 q^4 + m^3 q^7 + \cdots + m^{11} q^{31}$

32. $\frac{1}{x} - \frac{1}{x^2} + \frac{1}{x^3} - \frac{1}{x^4} + \cdots - \frac{1}{x^{10}}$ Simplify your answer.

33. A ball is dropped from a height of 10 feet. Each time the ball bounces it rises to 70% of its previous height. How far has the ball traveled when it hits the ground for the third time? For the 12th time?

18.2   INFINITE GEOMETRIC SERIES

We know that we can find $\frac{1}{2} + \frac{1}{4} + \frac{1}{8} + \cdots + \frac{1}{2^n}$ for any finite n. We can let n be a million; we can let n be a billion. What happens if we let n grow without bound? What sense can we make of $\frac{1}{2} + \frac{1}{4} + \frac{1}{8} + \cdots + \frac{1}{2^n} + \cdots$?

Definitions

The **terms** of a sum are the numbers being added. We denote by $a_1, a_2, a_3, \ldots a_n$, the first, second, third and nth term, respectively.

An **infinite series**, $a_1 + a_2 + a_3 + \cdots + a_n + \cdots$, is the sum of a sequentially ordered infinite set of terms;[4] it is defined to be $\lim_{n \to \infty}(a_1 + a_2 + a_3 + \cdots + a_n)$.

If $\lim_{n \to \infty}(a_1 + a_2 + a_3 + \cdots + a_n) = L$, where L is a finite number, then the series $a_1 + a_2 + \cdots + a_n + \cdots$ is said to **converge** to L.

If $\lim_{n \to \infty}(a_1 + a_2 + a_3 + \cdots + a_n)$ is unbounded or does not exist, then the series is said to **diverge.**

The **nth partial sum** of the series is the sum of the first n terms and is denoted by S_n. An infinite series converges to L if and only if $\lim_{n \to \infty} S_n = L$ for some finite number L.

A **geometric series** is an infinite series that can be written in the form $a + ar + ar^2 + ar^3 + \cdots + ar^n + \cdots$, where a and r are constants.

Notice that the "$+ \cdots +$" after a_n is shorthand for "and so on ad infinitum"; it is shorthand for an *infinite* number of terms. We often write some general term before the "$+ \cdots$"; this does not signify the end but simply a recipe for more terms.

---

[4] The adjective "infinite" in "infinite series" refers to the number of terms to be summed.

◆ **EXAMPLE 18.3**   What is $\frac{1}{2} + \frac{1}{4} + \frac{1}{8} + \cdots + \frac{1}{2^n} + \cdots$?

Your gut reaction to this may very well be "Hey, wait a minute! If we add an infinite number of positive terms the sum will be infinite." But think again, more carefully. A function can be always increasing and yet bounded. (Draw such a function for yourself.)

SOLUTION   *Take One:* Suppose I have a pie, a theoretical pie. I give you half of it, then half of what's left (I give you a quarter), then half of what's left (I give you an eighth), and so on ad infinitum. (We are assuming that a theoretical pie can be cut infinitely finely.) As I hand you pieces of pie you accumulate more and more of the pie. The more slices I hand you, the closer you will come to having the entire pie. You can come arbitrarily close to having the whole thing. However, you will never have more than one whole pie. We want to say that in some sense

$$\frac{1}{2} + \frac{1}{4} + \frac{1}{8} + \cdots + \frac{1}{2^n} + \cdots = 1.$$

Let's look at what is going on. After I hand you one piece, you have half the pie. After I give you two pieces, you have $\frac{1}{2} + \frac{1}{4} = \frac{3}{4}$ of the pie and $\frac{1}{4}$ remains; when you have three pieces, you have $\frac{1}{2} + \frac{1}{4} + \frac{1}{8} = \frac{7}{8}$ and $\frac{1}{8}$ remains. After I hand you the nth piece you have $\frac{1}{2} + \frac{1}{4} + \frac{1}{8} + \cdots + \frac{1}{2^n}$ of the pie and $\frac{1}{2^n}$ remains. Look at the pattern.

$$\frac{1}{2}$$

$$\frac{1}{2} + \frac{1}{4} = 1 - \frac{1}{4} = \frac{3}{4}$$

$$\frac{1}{2} + \frac{1}{4} + \frac{1}{8} = 1 - \frac{1}{8} = \frac{7}{8}$$

$$\frac{1}{2} + \frac{1}{4} + \frac{1}{8} + \frac{1}{1}6 = 1 - \frac{1}{16} = \frac{15}{16}$$

$$\frac{1}{2} + \frac{1}{4} + \frac{1}{8} + \frac{1}{16} + \frac{1}{32} = 1 - \frac{1}{32} = \frac{31}{32}$$

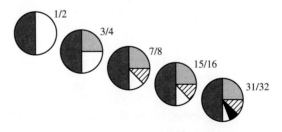

Notice that because I keep splitting what I have left evenly with you, then if you have $\frac{1}{2} + \frac{1}{4} + \frac{1}{8} + \frac{1}{16} + \frac{1}{32}$, I must have $\frac{1}{32}$. You have the rest; you have $1 - \frac{1}{32}$, the whole minus the slice I have.

More generally, we find

$$\frac{1}{2} + \frac{1}{4} + \frac{1}{8} + \cdots + \frac{1}{2^n} = 1 - \frac{1}{2^n}.$$

Then $\lim_{n\to\infty} \left(\frac{1}{2} + \frac{1}{4} + \frac{1}{8} + \cdots + \frac{1}{2^n} \right) = \lim_{n\to\infty} \left(1 - \frac{1}{2^n} \right)$.

$\lim_{n\to\infty} \left(1 - \frac{1}{2^n} \right) = 1$. So

$$\frac{1}{2} + \frac{1}{4} + \frac{1}{8} + \cdots + \frac{1}{2^n} + \cdots = \lim_{n\to\infty} \left(\frac{1}{2} + \frac{1}{4} + \frac{1}{8} + \cdots + \frac{1}{2^n} \right) = \lim_{n\to\infty} \left(1 - \frac{1}{2^n} \right) = 1.$$

We say that $\frac{1}{2} + \frac{1}{4} + \frac{1}{8} + \cdots + \frac{1}{2^n} + \cdots$ **converges** to 1.

Take Two: We define $\frac{1}{2} + \frac{1}{4} + \frac{1}{8} + \cdots + \frac{1}{2^n} + \cdots$ as $\lim_{n\to\infty} \left(\frac{1}{2} + \frac{1}{4} + \frac{1}{8} + \cdots + \frac{1}{2^n} \right)$.

This limit can be computed by putting $\frac{1}{2} + \frac{1}{4} + \frac{1}{8} + \cdots + \frac{1}{2^n}$ in closed form.

$$S_n = \frac{1}{2} + \frac{1}{4} + \frac{1}{8} + \cdots + \frac{1}{2^n}$$
$$\frac{1}{2} S_n = \qquad \frac{1}{4} + \frac{1}{8} + \cdots + \frac{1}{2^n} + \frac{1}{2^{n+1}}$$
$$\overline{\frac{1}{2} S_n = \frac{1}{2} \qquad\qquad\qquad\qquad - \frac{1}{2^{n+1}}}$$
$$S_n = 2 \left(\frac{1}{2} - \frac{1}{2^{n+1}} \right) = 1 - \frac{1}{2^n}$$

So $\lim_{n\to\infty} \left(\frac{1}{2} + \frac{1}{4} + \frac{1}{8} + \cdots + \frac{1}{2^n} \right) = \lim_{n\to\infty} \left(1 - \frac{1}{2^n} \right) = 1.$ ◆

◆ **EXAMPLE 18.4**   Does the series $2 + 4 + 8 + 16 + 32 + 64 + \cdots + 2^n + \cdots$ converge?

SOLUTION   Clearly the series diverges. The numbers that we are summing are themselves growing without bound, so the partial sums grow without bound.

Writing the nth partial sum in closed form we obtain

$$S_n = 2 + 4 + 8 + 16 + \cdots + 2^n$$
$$2 S_n = \qquad 4 + 8 + 16 + \cdots + 2^n + 2^{n+1}$$
$$\overline{-S_n = 2 \qquad\qquad\qquad\qquad - 2^{n+1}}$$
$$S_n = 2^{n+1} - 2.$$

$\lim_{n\to\infty}(2 + 4 + 8 + 16 + \cdots + 2^n) = \lim_{n\to\infty}(2^{n+1} - 2) = \infty$. This infinite series diverges. ◆

Question: How can we determine which infinite geometric series converge and which diverge?

Let's look at the problem in general. Recall our convention that the very first term of the series is "a" and the ratio of any one term to the previous term is "r".

$$a + ar + ar^2 + ar^3 + \cdots + ar^n + \cdots = \lim_{n\to\infty} (a + ar + ar^2 + ar^3 + \cdots + ar^n)$$

Our strategy for getting a handle on an infinite sum is to express the general partial sum in closed form and to compute the limit as n increases without bound. We'll give

$a + ar + ar^2 + ar^3 + \cdots + ar^n$ the name P (for partial sum) and express the partial sum P in closed form.

$$P = a + ar + ar^2 + \cdots + ar^n$$
$$rP = \quad\; ar + ar^2 + \cdots + ar^n + ar^{n+1} \quad \text{(subtract)}$$
$$\overline{\rule{0pt}{0pt}\hspace{11cm}}$$
$$(1-r)P = a \hspace{4.5cm} - ar^{n+1}$$

$P = \frac{a - ar^{n+1}}{1-r}$ for $r \neq 1$. We'll look at the case $r = 1$ later; for now assume $r \neq 1$.

$$\lim_{n \to \infty}(a + ar + ar^2 + ar^3 + \cdots + ar^n) = \lim_{n \to \infty}\frac{a - ar^{n+1}}{1-r}$$
$$= \lim_{n \to \infty}\frac{a}{1-r}(1 - r^{n+1})$$
$$= \frac{a}{1-r}\lim_{n \to \infty}(1 - r^{n+1})$$

What happens to r^{n+1} as $n \to \infty$? [5] The answer depends on r.

- If $|r| < 1$, then $\lim_{n \to \infty}r^{n+1} = 0$. (Think of $\lim_{n \to \infty}\left(\frac{1}{2}\right)^n$ or $\lim_{n \to \infty}\left(-\frac{1}{2}\right)^n$.) So $(1 - r^{n+1})$ is finite if $|r| < 1$, in which case the limit is 1.

- If $|r| > 1$, then $\lim_{n \to \infty}r^{n+1} = \infty$ for $r > 1$, and does not exist for $r < -1$. (Think of $\lim_{n \to \infty}2^n$ and of $\lim_{n \to \infty}(-2)^n$. The value of $(-2)^n$ oscillates wildly between large positive and large negative numbers as $n \to \infty$.)

- If $r = -1$, then $\lim_{n \to \infty}r^{n+1}$ does not exist since $(-1)^n$ oscillates between -1 and 1 as $n \to \infty$.[6]

- If $r = 1$, we cannot use the closed form for P given above. If $r = 1$, then we're looking at $a + a + a + a + \cdots$, so the closed form for the sum of the first $(n+1)$ terms is $a(n+1)$. $\lim_{n \to \infty}a(n+1)$ is finite only in the trivial case where $a = 0$. Therefore this series diverges.

We can summarize our results as follows.

$$a + ar + ar^2 + \cdots + ar^n + \cdots \begin{cases} \text{converges to } \frac{a}{1-r} & \text{for } |r| < 1 \text{ and} \\ \text{diverges} & \text{for } |r| \geq 1. \end{cases}$$

EXERCISE 18.2   Consider the series $1 + x + x^2 + x^3 + \cdots + x^n + \cdots$.

(a) For what values of x does this series converge? For these x-values, the series converges to $L(x)$. Find $L(x)$.

(b) Compare the values of $L(x)$, $1 + x$, $1 + x + x^2$, and $1 + x + x^2 + x^3$ for each of the following values of x.

      i) $x = 0.2$ ii) $x = 0.1$ iii) $x = 0$ iv) $x = -0.1$

---

[5] We are letting $n \to \infty$ through the integers.

[6] When $r = -1$, the geometric series looks like $a - a + a - a + \cdots$. If we tally just the first term, we get "a". If we tally the first two terms, we get zero; the first three terms sum to "a," etc. The partial sums oscillate between "a" and 0 indefinitely.

What do you observe about how well (or poorly) $1 + x + x^2$ and $1 + x + x^2 + x^3$ approximate $L(x)$ for each of these x-values?

◆ **EXAMPLE 18.5**   Let's return to the digitalis problem from Example 18.1. We are interested in how much digitalis is in the man's bloodstream in the long run if he takes 0.04 mg every morning for several years.

SOLUTION   Suppose we choose to model the situation with an infinite series. If we want to know how much digitalis is in the patient's body several years later, right after he takes his medicine we could calculate

$$.04 + (.7)(.04) + (.7)^2(.04) + \cdots + (.7)^n(.04) + \cdots .$$

This is a geometric series with $a = 0.04$ and $r = 0.7$. Since $|r| < 1$, the series converges to $\frac{a}{1-r} = \frac{.04}{1-.7} = \frac{.04}{.3} = 0.1\overline{3}$ mg. If we want to know how much digitalis is in the body several years later, right before taking a pill it would be about $0.1\overline{3}$mg $- 0.04$mg $= 0.09\overline{3}$ mg.

   Using an infinite series in this situation may seem suspect. First of all, no matter how effective digitalis is, no one will live forever, and secondly, "several" is a vague term. To put this objection to rest, let's look at the sum of all the "extra terms" we added. "Several" years implies more than one, so the difference between the actual amount, a finite geometric sum, and the infinite series we used is less than

$$(.7)^{365}(.04) + (.7)^{366}(.04) + \cdots .$$

The latter is a geometric series with $a = (.7)^{365}(.04)$ and $r = .7$. Its sum is $\frac{a}{1-r}$ or $\frac{(.7)^{365}(.04)}{1-.7} \approx$ 3.85×10^{-58}. This is extremely small, small enough in comparison to 0.4 to ignore.   ◆

Using Infinite Geometric Series to Compute Finite Geometric Sums

Notice that it is comparatively less work to compute the sum of a convergent infinite geometric series (identify a and r, verify that $|r| < 1$, and *voilà*, the sum is $\frac{a}{1-r}$), than it is to compute a finite geometric sum. If we have a finite geometric sum with $|r| < 1$, then we can capitalize on our ability to sum an infinite geometric series easily. We demonstrate this in the next example.

◆ **EXAMPLE 18.6**   Find the sum of $3 + 3(0.95)^2 + 3(0.95)^4 + 3(0.95)^6 + \cdots + 3(0.95)^{20}$.

SOLUTION   This sum is geometric, so we identify "a" and "r". $a = 3$ and $r = (0.95)^2$.

Method I.   Give the sum a name, like S for instance. Multiply S by r, subtract rS from S, and solve for S.

Method II. $|r| < 1$, so the infinite geometric series $3 + 3(.95)^2 + 3(.95)^4 + 3(.95)^6 + \cdots + 3(.95)^{20} + \cdots$ converges to $\frac{a}{1-r} = \frac{3}{1-(.95)^2} \approx 30.769$.

   We are interested in a finite sum, so we cut off the tail of the series and subtract it from 30.769. The infinite geometric series $3(0.95)^{22} + 3(0.95)^{24} + 3(0.95)^{26} + \cdots$ converges to $\frac{a}{1-r}$ where $a = 3(0.95)^{22}$ and $r = (0.95)^2$. It converges to $\frac{3(0.95)^{22}}{1-(0.95)^2} \approx 9.955$. Therefore, $3 + 3(0.95)^2 + 3(0.95)^4 + 3(0.95)^6 + \cdots + 3(0.95)^{20} \approx 30.769 - 9.955 \approx 20.814$.

   Note that this method of computing a finite geometric sum by looking at the difference of the sums of two infinite geometric series can only be employed in cases in which $|r| < 1$. If $|r| \geq 1$ we must use Method I because the related infinite series diverge.   ◆

PROBLEMS FOR SECTION 18.2

For Problems 1 through 11, determine whether the series converges or diverges. If it converges, find its sum.

1. $1 - 10 + 100 - \cdots + (-10)^n + \cdots$

2. $0.3 + 0.03 + 0.003 + 0.0003 + 0.00003 + \cdots$

3. $\frac{2}{3} + \frac{2}{9} + \frac{2}{27} + \cdots + \frac{2}{3^n} + \cdots$

4. $\frac{2}{3} + 2 + 6 + \cdots + 2(3)^n + \cdots$

5. $1 - \frac{1}{2} + \frac{1}{4} - \frac{1}{8} + \frac{1}{16} + \cdots$

6. $\frac{1}{4} - \frac{1}{8} + \frac{1}{16} - \frac{1}{32} + \cdots$

7. $\frac{2}{3} + 1 + \frac{3}{2} + \frac{9}{4} + \cdots$

8. $\frac{3}{2} - \frac{3}{4} + \frac{3}{8} - \cdots + \frac{(-1)^{n+1}3}{2^n} + \cdots$

9. $e + 1 + e^{-1} + e^{-2} + e^{-3} + \cdots$

10. $2e + 2e^2 + 2e^3 + \cdots + 2e^n + \cdots$

11. $(2e)^{-2} + (2e)^{-3} + (2e)^{-4} + \cdots + (2e)^{-n} + \cdots$

12. Find the sum of the following. (If there is no finite sum, say so.)
    - (a) $3 + 9 + 27 + \cdots + 3^{20}$
    - (b) $\frac{2}{3} + \left(\frac{2}{3}\right)^2 + \left(\frac{2}{3}\right)^3 + \cdots + \left(\frac{2}{3}\right)^n + \cdots$
    - (c) $(0.2)(10) + (0.2)(100) + (0.2)(1000) + \cdots$
    - (d) $3 + 3(0.8) + 3(0.8)^2 + 3(0.8)^3 + \cdots$
    - (e) $(0.2) + (0.2)(1.3) + (0.2)(1.3)^2 + (0.2)(1.3)^3 + \cdots$
    - (f) $1 + x^2 + x^4 + x^6 + \cdots$ for $-1 < x < 1$

13. Determine whether each of the following geometric series converges or diverges. If the series converges, determine to what it converges.
    - (a) $-\frac{4}{3} - \frac{1}{2} - \frac{3}{16} - \frac{9}{128} + \cdots$
    - (b) $-\frac{1}{100} + \frac{1.1}{(100)^2} - \frac{1.21}{(100)^3} + \frac{1.331}{(100)^4} - \cdots$
    - (c) $-\frac{7}{10000} + \frac{7}{11000} - \frac{7}{12100} + \frac{7}{13310} - \cdots$
    - (d) $1 - x + x^2 - x^3 + \cdots$ for $|x| < 1$

14. Write each of the following series first as a repeating decimal and then as a fraction.
    - (a) $2 + \frac{2}{10} + \frac{2}{100} + \frac{2}{1000} + \cdots$
    - (b) $3 + \frac{12}{10^2} + \frac{12}{10^4} + \frac{12}{10^6} + \cdots$

■ 18.3   A MORE GENERAL DISCUSSION OF INFINITE SERIES

In the previous four sections we focused on geometric sums and geometric series. In this section we broaden our discussion to investigate other infinite series. Our focus in this chapter is geometric series, but you will have a better appreciation of geometric series if you have some familiarity with series that are not geometric.

Given an infinite series

$$a_1 + a_2 + a_3 + \cdots + a_n + \cdots$$

the most basic question to consider is whether the series converges or diverges.

Suppose all the terms of the infinite series are positive. Then $S_n = S(n)$, the sum of the first n terms, is an increasing function. We know from our study of functions that an increasing function may increase without bound, or it may increase but be bounded, in which case it will be asymptotic to a horizontal line. In the latter case, the function must be increasing at a decreasing rate; in fact, if the function has a horizontal asymptote, its rate of increase must be approaching zero. Similarly, if $\lim_{n\to\infty} S_n = L$ for some finite constant L, then the rate at which $S(n)$ is increasing must be approaching zero. This translates to the observation that if an infinite series is to have any chance at converging, then its terms must be approaching zero, that is, $\lim_{n\to\infty} a_n$ must be 0.

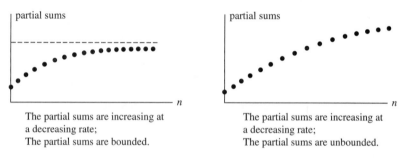

The partial sums are increasing at a decreasing rate;
The partial sums are bounded.

The partial sums are increasing at a decreasing rate;
The partial sums are unbounded.

Figure 18.1

If all of the terms of the series are negative, an analogous argument can be made. If some of the terms of an infinite series are positive and others are negative, it is still true that in order for the series to have any shot at converging the terms must be approaching zero. If $\lim_{n\to\infty} a_n = k$, where $k \neq 0$, then the partial sums will be eventually increasing without bound if $k > 0$ and eventually decreasing without bound if $k < 0$. [7]

Suppose the terms of a series are approaching zero; is this enough to guarantee convergence? In the case of geometric series the answer is "yes", but in the general case the answer is **NO!** The situation in general is much more subtle;[8] the next example will convince you of that fact.

---

[7] If $\lim_{n\to\infty} a_n$ does not exist, then the partial sums will be bouncing around and will not converge.

[8] We know that a function can be increasing at a decreasing rate and have a horizontal asymptote, but it can also be increasing at a rate tending toward zero and yet be unbounded. Consider, for example, $f(x) = \ln x$. Its rate of increase is given by $\frac{1}{x}$. $\lim_{n\to\infty} \frac{1}{x} = 0$, so the rate of increase of $\ln x$ tends toward zero as x increases without bound. Nevertheless, as $x \to \infty$ we know that $\ln x \to \infty$.

◆ **EXAMPLE 18.7**   Does the infinite series $\frac{1}{2} + \frac{1}{3} + \frac{1}{4} + \frac{1}{5} + \cdots + \frac{1}{n} + \cdots$ converge or diverge? This infinite series is called the **harmonic series**.

SOLUTION   The terms of this series are going toward zero: $\lim_{n\to\infty} \frac{1}{n} = 0$. The harmonic series is not a geometric series. Can we put S_n into closed form? Let's look at some partial sums.

$$\frac{1}{2}$$

$$\frac{1}{2} + \frac{1}{3} = \frac{5}{6}$$

$$\frac{1}{2} + \frac{1}{3} + \frac{1}{4} = \frac{13}{12}$$

$$\frac{1}{2} + \frac{1}{3} + \frac{1}{4} + \frac{1}{5} = \frac{77}{60}$$

$$\frac{1}{2} + \frac{1}{3} + \frac{1}{4} + \cdots + \frac{1}{n} = ??$$

Closed form allowed us to get a firm grip on something rather slippery. In this example our luck has run out. We cannot express the general partial sum in closed form. The expression $\lim_{n\to\infty} \left(\frac{1}{2} + \frac{1}{3} + \frac{1}{4} + \cdots + \frac{1}{n} \right)$ gives us no insight into the convergence or divergence of the series. We need to take a different perspective. We will compare this series to a familiar series.

$$\frac{1}{2} = \frac{1}{2}$$

$$\frac{1}{4} + \frac{1}{4} = \frac{1}{2} \qquad \text{so} \qquad \frac{1}{3} + \frac{1}{4} > \frac{1}{2}$$

$$\frac{1}{8} + \frac{1}{8} + \frac{1}{8} + \frac{1}{8} = \frac{1}{2} \qquad \text{so} \qquad \frac{1}{5} + \frac{1}{6} + \frac{1}{7} + \frac{1}{8} > \frac{1}{2}$$

$$8 \cdot \left(\frac{1}{16} \right) = \frac{1}{2} \qquad \text{so} \qquad \frac{1}{9} + \frac{1}{10} + \frac{1}{11} + \cdots + \frac{1}{16} > \frac{1}{2}$$

$$16 \cdot \left(\frac{1}{32} \right) = \frac{1}{2} \qquad \text{so} \qquad \frac{1}{17} + \frac{1}{18} + \frac{1}{19} + \cdots + \frac{1}{32} > \frac{1}{2}$$

and so on. . . . Think about this in terms of slices of pies. How many pies must we bake in order to give out slices as dictated by the harmonic series? If the series converges we need only bake some finite number of pies.

$$\frac{1}{2} + \frac{1}{3} + \frac{1}{4} + \frac{1}{5} + \frac{1}{6} + \frac{1}{7} + \frac{1}{8} + \frac{1}{9} + \cdots + \frac{1}{16} + \cdots \text{ equals}$$

$$\left(\frac{1}{2} \right) + \left(\frac{1}{3} + \frac{1}{4} \right) + \left(\frac{1}{5} + \frac{1}{6} + \frac{1}{7} + \frac{1}{8} \right) + \left(\frac{1}{9} + \cdots + \frac{1}{16} \right) + \cdots$$

$$\begin{pmatrix} \text{half a} \\ \text{pie} \end{pmatrix} + \begin{pmatrix} \text{more than} \\ \text{half a pie} \end{pmatrix} + \begin{pmatrix} \text{more than} \\ \text{half a pie} \end{pmatrix} + \begin{pmatrix} \text{more than} \\ \text{half a pie} \end{pmatrix} + \cdots,$$

which is greater than or equal to $\frac{1}{2} + \frac{1}{2} + \frac{1}{2} + \frac{1}{2} + \cdots$. But this latter series diverges; consequently, the same must be true of the harmonic series. By comparing the harmonic series with the divergent series $\frac{1}{2} + \frac{1}{2} + \frac{1}{2} + \frac{1}{2} + \cdots$, we see that the harmonic series must diverge. (No matter how many pies we bake, we will eventually run out and need more.)

$\blacklozenge$

We had previously observed that if an infinite series is to have any chance at converging, then the terms must be going toward zero. Although this is a necessary condition for convergence, a look at the harmonic series shows that it is not enough to guarantee convergence; the condition $\lim_{n \to \infty} a_n = 0$ is necessary for convergence but not sufficient.[9] It is important to realize that if the terms of an infinite series are going to zero, then the series *may* converge (as is true for all *geometric* series), yet on the other hand, the series may diverge (as in the example of the harmonic series).

A Summary of the Main Principles

■ *The Nth Term Test for Divergence:*

> If $\lim_{n \to \infty} a_n \neq 0$, then the series $a_1 + a_2 + a_3 + \cdots + a_n + \cdots$ diverges.[10]

Warning: This is a test for *divergence* only!

■ *Increasing and Bounded Partial Sums Test:* Suppose the terms of a series are all positive. Then S_n increases with n. If the partial sums are bounded, that is, there exists a constant M such that $S_n \leq M$ for all n, then it can be shown that $\lim_{n \to \infty} S_n$ exists and is finite. Therefore, the series converges.

In many cases the question of convergence or divergence of an infinite series is a very subtle one. Often, for instance, one can determine that a certain series converges without being able to say exactly what it converges to. We will return to infinite series in Chapter 30. Questions of convergence become much simpler if we focus on the special case of the geometric series, and this is our main focus in this chapter. In the case of geometric series, convergence and divergence are straightforward to establish.

$$a + ar + ar^2 + \cdots + ar^n + \cdots \begin{cases} \text{converges to } \frac{a}{1-r} & \text{for } |r| < 1 \\ \text{diverges} & \text{for } |r| \geq 1. \end{cases}$$

In Section 18.4 we introduce some convenient notation for working with series, and in Section 18.5 we apply geometric series to real-world situations.

PROBLEMS FOR SECTION 18.3

For Problems 1 through 9, determine whether the series converges or diverges. Explain your reasoning.

1. $1 - 2 + 3 - \cdots + (-1)^{(n+1)n} + \cdots$

2. $\frac{1}{1000} + \frac{2}{1000} + \frac{3}{1000} + \frac{4}{1000} + \cdots$

3. $\frac{2}{3} + \frac{2}{4} + \frac{2}{5} + \cdots + \frac{2}{n} + \cdots$

---

[9] "Necessary" versus "sufficient": In order for a polygon to be called a square it is necessary that it have four sides, but this alone is not sufficient to classify it as a square. In order to win a race it is necessary to finish it, but this alone is not sufficient.

[10] For an infinite series to diverge it is sufficient that $\lim_{n \to \infty} a_n \neq 0$. This, however, is not necessary. To lose a race it is sufficient not to finish it. That, however, is not the only way to lose a race!

4. $\frac{1}{2 \cdot 2^2} + \frac{1}{3 \cdot 2^3} + \frac{1}{4 \cdot 2^4} + \frac{1}{5 \cdot 2^5} + \cdots$

(*Hint:* Compare this term-by-term to a geometric series you know. Choose a convergent geometric series whose terms are larger than the terms of this series.)

5. $\frac{(\sin 1)^2}{3} + \frac{(\sin 2)^2}{3^2} + \frac{(\sin 3)^2}{3^3} + \cdots + \frac{(\sin n)^2}{3^n} + \cdots$

(*Hint:* Compare this term-by-term to a geometric series you know. Choose a convergent geometric series whose terms are larger than the terms of this series.)

6. $\frac{3^3}{3} + \frac{4^4}{4} + \frac{5^5}{5} + \cdots$

7. $\frac{-1}{2} - \frac{1}{3} - \frac{1}{4} - \frac{1}{5} - \cdots - \frac{1}{n} - \cdots$

8. $\frac{1}{2} - \frac{1}{2} + \frac{1}{2} - \frac{1}{2} + \cdots$

9. $(1 + 1)^1 + \left(1 + \frac{1}{2}\right)^2 + \left(1 + \frac{1}{3}\right)^3 + \cdots + \left(1 + \frac{1}{n}\right)^n + \cdots$

10. The sum $1 + 2 + 3 + 4 + 5 + \cdots + n = \sum_{k=1}^{n} k$ is not geometric, but we can express it in an easy-to-compute form.

    Let
    $$S = 1 + 2 + 3 + 4 + 5 + \cdots + n. \tag{18.1}$$

    Writing the terms from largest to smallest gives
    $$S = n + (n - 1) + (n - 2) + \cdots + 2 + 1. \tag{18.2}$$

    Add equations (18.1) and (18.2) and divide by two to show that
    $$1 + 2 + 3 + 4 + 5 + \cdots + n = \frac{n(n + 1)}{2}.$$

11. *Challenge:* Use the same line of reasoning as outlined in Problem 10 to show that
    $$1 + 3 + 5 + 7 + \cdots + (2n - 1) = n^2.$$

12. Give an example of each of the following.

    (a) An infinite series that converges and whose partial sums are always increasing
    (b) An infinite series that converges and whose partial sums oscillate around the sum of the series
    (c) An infinite series that diverges although its terms approach zero
    (d) An infinite series that diverges but whose partial sums do not grow without bound

18.4 SUMMATION NOTATION

Sums whose terms follow a consistent pattern can often be written in a more compact way by using summation notation. Summation notation is only notation; it is a compact shorthand for writing out a sum. We will introduce it via examples.

$$a_1 + a_2 + a_3 + \cdots + a_{28}$$

can be written in shorthand as

$$\sum_{i=1}^{28} a_i, \text{ where } \sum \text{ denotes the summation process.}[11]$$

The terms of the sum are all of the form a_i, where the "i"s form a sequence of consecutive integers starting with the integer indicated at the bottom of $\sum$ and ending with the integer at the top. In this example we successively substitute $i = 1, 2, 3, \ldots, 28$ in place of i in the expression a_i and add up these 28 terms. i is called the index. We can choose any letter we like for the index.

$$\sum_{i=1}^{28} a_i, \qquad \sum_{j=1}^{28} a_j, \qquad \text{and} \qquad \sum_{k=1}^{28} a_k$$

all are shorthand for $a_1 + a_2 + a_3 + \cdots + a_{28}$. In fact, $\sum_{q=0}^{27} a_{q+1}$ is also equivalent to $a_1 + a_2 + a_3 + \cdots + a_{28}$; there are infinitely many different ways of putting a sum into summation notation.

◆ **EXAMPLE 18.8**

i) $\sum_{k=0}^{14} 2^k$ is shorthand for $2^0 + 2^1 + 2^2 + \cdots + 2^{14}$.

ii) $\sum_{k=0}^{14} k^2$ is shorthand for $0^2 + 1^2 + 2^2 + \cdots + 14^2$.

iii) $\sum_{k=3}^{49} (-1)^k k$ is shorthand for $(-1)^3 3 + (-1)^4 4 + (-1)^5 5 + \cdots + (-1)^{49} 49$.

                or $-3 + 4 - 5 + 6 + \cdots - 49$.

iv) $a + ar + ar^2 + ar^3 + \cdots + ar^n + \cdots$ can be written as $\sum_{n=0}^{\infty} ar^n$.

v) $\frac{1}{2} + \frac{1}{2^2} + \frac{1}{2^3} + \cdots + \frac{1}{2^n} + \cdots$ can be written as $\sum_{k=1}^{\infty} \frac{1}{2^k}$, or, if we want this to look more

explicitly like the general geometric series in part (iv), it can be written $\sum_{k=0}^{\infty} \frac{1}{2} \left(\frac{1}{2}\right)^k$.

vi) $\frac{1}{2} - \frac{1}{2^2} + \frac{1}{2^3} + \cdots + \frac{1}{2^{33}}$ can be written $\sum_{k=0}^{32} \frac{1}{2} \left(\frac{-1}{2}\right)^k$.

vii) $f\left(\frac{1}{n}\right) \frac{1}{n} + f\left(\frac{2}{n}\right) \frac{1}{n} + f\left(\frac{3}{n}\right) \frac{1}{n} + \cdots + f\left(\frac{n}{n}\right) \frac{1}{n}$ can be written $\sum_{i=1}^{n} f\left(\frac{i}{n}\right) \frac{1}{n}$.

NOTE If a sum is geometric, we can begin by identifying "a" (the first term) and "r" (the ratio of any term to the previous one). Write $\sum_{k=0}^{w} ar^k$ and figure out what w ought to be. In part (vi) when we use this approach the upper index is 32 not 33. Writing $\sum_{k=1}^{33} -1 \left(-\frac{1}{2}\right)^k$ would also be correct.[12]   ◆

EXERCISE 18.3   Try these on your own and then check your answers with those given below.

(a) Express $\frac{2}{3} + \left(\frac{2}{3}\right)^2 + \left(\frac{2}{3}\right)^3 + \cdots + \left(\frac{2}{3}\right)^{40}$ in summation notation. Do this in two ways; with the index starting at 1 and with the index starting at 0.

(b) Express $\left(\frac{2}{3}\right)^2 - \left(\frac{2}{3}\right)^5 + \left(\frac{2}{3}\right)^8 - \left(\frac{2}{3}\right)^{11} + \cdots$ in summation notation.

Answers

(a) $\frac{2}{3} + \left(\frac{2}{3}\right)^2 + \left(\frac{2}{3}\right)^3 + \cdots + \left(\frac{2}{3}\right)^{40}$ can be written as $\sum_{n=1}^{40} \left(\frac{2}{3}\right)^n$ or as $\sum_{n=0}^{39} \frac{2}{3} \left(\frac{2}{3}\right)^n$.

This is a geometric sum with $a = \frac{2}{3}$ and $r = \frac{2}{3}$.

---

[11] $\sum$ is the uppercase Greek letter sigma; you can think of it as the letter S for sum. It denotes a process, much in the same way that $\frac{d}{dx}$ denotes a process or operation.

[12] To have the terms of this sum alternate sign, r must be negative.

(b) This is a geometric series. First identify r by finding the ratio of the second term to the first, $r = -\left(\frac{2}{3}\right)^3$. Once we know r and a, we're well on our way. The series can be written as $\sum_{k=0}^{\infty} \left(\frac{2}{3}\right)^2 \left(-\frac{2}{3}\right)^{3k}$, or, alternatively, $\sum_{k=2}^{\infty} \left(-\frac{2}{3}\right)^{3k+2}$.

EXERCISE 18.4   Determine whether each of the following statements is *always* true or not always true.

(a) $\sum_{i=1}^{n} c(1)^i = nc$ (b) $\sum_{i=1}^{n}(a_i + b_i) = \sum_{i=1}^{n} a_i + \sum_{i=1}^{n} b_i$

(c) $\sum_{i=1}^{n} a_i b_i = a_i \left(\sum_{i=1}^{n} b_i\right)$ (d) $\sum_{i=1}^{n} cb_i = c \sum_{i=1}^{n} b_i$

(e) $\sum_{i=1}^{n} a_i b_i = \left(\sum_{i=1}^{n} a_i\right)\left(\sum_{i=1}^{n} b_i\right)$ (f) If $a_i > b_i$ for $i = 1, 2, 3, \ldots, n$, then $\sum_{i=1}^{n} a_i > \sum_{i=1}^{n} b_i$.

(Four of the six statements are always true.)

◆ **EXAMPLE 18.9**   What rational number has the decimal expansion $5.123232323\ldots$?

SOLUTION

$$5.1232323\ldots = 5.1 + 0.0232323\ldots$$

$$= 5.1 + \frac{23}{10^3} + \frac{23}{10^5} + \frac{23}{10^7} + \cdots + \frac{23}{10^{2n+1}} + \cdots^{[13]}$$

Following 5.1 is a geometric series with $a = \frac{23}{10^3}$ and $r = \frac{1}{10^2}$. $|r| < 1$, so the series converges. We have

$$5.1 + \sum_{n=0}^{\infty} \frac{23}{10^3}\left(\frac{1}{100}\right)^n = \frac{51}{10} + \frac{\frac{23}{1000}}{1 - \frac{1}{100}} = \frac{51}{10} + \left(\frac{23}{1000} \cdot \frac{100}{99}\right)$$

$$= \frac{51}{10} + \frac{23}{990} = \frac{5072}{990} = \frac{2536}{495}. \quad ◆$$

PROBLEMS FOR SECTION 18.4

For Problems 1 through 10, write the sum using summation notation.

1. $2^3 + 3^3 + 4^3 + \cdots + 19^3$

2. $2 - 3 + 4 - 5 + 6 - \cdots + 100$

3. $2^3 + 3^4 + 4^5 + \cdots + 100^{101}$

4. (a) $4x^2 + 4x^3 + 4x^4 + 4x^5 + \cdots +$
   (b) $2x + 3x^2 + 4x^3 + 5x^4 + 6x^5 + \cdots +$

5. $1 - 10 + 100 - \cdots + (-10)^n + \cdots$

6. $0.3 + 0.03 + 0.003 + 0.0003 + 0.00003 + \cdots$

---

[13] In order to write a general term for this series we need to write $\frac{23}{10^{(\text{odd number})}}$. An even number can be written as $2n$ for n a positive integer. An odd number can be expressed as $2n + 1$ or $2n - 1$; hence the general term is given as $\frac{23}{10^{2n+1}}$.

7. (a) $\frac{2}{3} + \frac{2}{9} + \frac{2}{27} + \cdots + \frac{2}{3^n} + \cdots$

   (b) $\frac{2}{3} + 2 + 6 + \cdots + 2(3)^n + \cdots$

8. (a) $1 - \frac{1}{2} + \frac{1}{4} - \frac{1}{8} + \frac{1}{16} + \cdots$

   (b) $\frac{1}{4} - \frac{1}{8} + \frac{1}{16} - \frac{1}{32} + \cdots$

9. (a) $\frac{2}{3} + 1 + \frac{3}{2} + \frac{9}{4} + \cdots$

   (b) $\frac{3}{2} - \frac{3}{4} + \frac{3}{8} - \cdots + \frac{(-1)^n 3}{2^n}$

10. (a) $e + 1 + e^{-1} + e^{-2} + e^{-3} + \cdots$

   (b) $2e + 2e^2 + 2e^3 + \cdots + 2e^n + \cdots$

   (c) $(2e)^{-2} + (2e)^{-3} + (2e)^{-4} + \cdots + (2e)^{-n} + \cdots$

For Problems 11 through 16, do the following.

   i. Write out the first two terms of the series.

   ii. Determine whether or not the series converges.

   iii. If the series converges, determine its sum.

11. (a) $\sum_{k=1}^{\infty} \frac{(-1)^n}{3^n}$ (b) $\sum_{k=2}^{\infty} \frac{(-1)^n}{3^n}$

12. $\sum_{n=3}^{\infty} \frac{(-1)^n 3}{2^n}$

13. $\sum_{n=2}^{\infty} \frac{3^n}{4^{n-1}}$

14. $\sum_{n=1}^{\infty} \frac{(-1)^n}{3^n}$

15. $\sum_{n=100}^{\infty} \frac{10}{n}$

16. Does the series $\sum_{k=1}^{\infty} \frac{\ln(k+2)}{3k}$ converge or diverge? Explain.

17. Consider the sum

$$q^5 - q^7 + q^9 - q^{11} + \cdots + q^{41}.$$

   (a) Put the sum into closed form.

   (b) Put the sum into summation notation.

   (c) Now put $-q^5 + q^7 - q^9 + q^{11} - \cdots - q^{41}$ into summation notation.

18. For each of the following geometric sums, first write the sum using summation notation and then write the sum in closed form.

   (a) $\frac{2}{3^2} + \frac{2}{3^4} + \frac{2}{3^6} + \cdots \frac{2}{3^{18}}$

   (b) $1 - 2 + 2^2 - 2^3 + 2^4 - \cdots + 2^{46}$

   (c) $-\frac{1}{100} + \frac{1.1}{100} - \frac{1.21}{100} + \frac{1.331}{100} - \cdots - \frac{1.1^{100}}{100}$

   (d) $\frac{2}{3^2} + \frac{2^2}{3^3} + \frac{2^3}{3^4} + \cdots + \frac{2^{16}}{3^{17}}$

19. Write the following without using summation notation and answer the following questions.

(a) Is the series geometric?

(b) Does the series converge? If so, indicate to what it converges.

$$\text{i) } \sum_{n=0}^{\infty} \frac{3}{2}\left(\frac{5}{2}\right)^n \qquad \text{ii) } \sum_{n=0}^{\infty} \frac{15}{10^2}\left(\frac{1}{10}\right)^{3n} \qquad \text{iii) } \sum_{n=1}^{\infty} \frac{3}{2}\left(\frac{-2}{3}\right)^n \qquad \text{iv) } \sum_{n=1}^{\infty} \ln n$$

20. Write the following sums in summation notation. In each case, determine the sum; i.e., sum the first and determine what the second converges to.

   (a) $500 + 500e^{\cdot 1} + 500e^{\cdot 2} + 500e^{\cdot 3} + \cdots + 500e^2$

   (b) $\frac{5}{3} - \frac{5}{6} + \frac{5}{12} - \frac{5}{24} + \cdots$

18.5   APPLICATIONS OF GEOMETRIC SUMS AND SERIES

◆ **EXAMPLE 18.10**   Paulina is self-employed. On the first day of every month she puts $400 into an account she has set up for retirement. The account pays 0.5% per month, i.e., 6% per year compounded monthly. How much money will be in Paulina's retirement account five years after she sets up the account immediately, before her 61st payment? (Assume that interest is paid on the last day of every month.)

SOLUTION   *Making Estimates.*   If we ignored interest completely, there would be $400 \cdot 60 = \$24{,}000$ in the retirement account. Because there *is* interest, we expect the answer to be more than $24,000. Each payment is in the bank for a different amount of time. Money in the account grows according to

$$M(t) = M_0(1.005)^{12t} \qquad \text{or} \qquad M(t) = M_0(1.005)^m,$$

where t is time in years and m is the number of months the money has been in the bank. If *all* the money were in the bank for five years she would have $\$24{,}000(1.005)^{12 \cdot 5} = \$24{,}000(1.005)^{60} = \$32{,}372.40$. Thus, we expect the actual balance to be greater than $24,000 and less than $32,372.

Strategy: We will look at each of the 60 payments individually and determine how much each will grow to by the end of the five-year period. In other words, we will look at the future value of each of the 60 payments. Then we'll sum these future values.

   We can represent this diagramatically. We'll push each of the payments to the same point in the future.

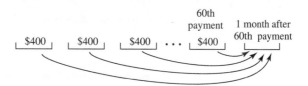

The first payment is in the bank for 60 months, the second for 59 months, the third for 58 months, and so on. The 60th payment is in the bank for 1 month.

future value of the 1st payment $= \$400(1.005)^{60}$

future value of the 2nd payment $= \$400(1.005)^{59}$

future value of the 3rd payment $= \$400(1.005)^{58}$

future value of the 4th payment $= \$400(1.005)^{57}$

$$\vdots$$

future value of the 59th payment $= \$400(1.005)^{2}$

future value of the 60th payment $= \$400(1.005)^{1}$

The total amount of money in the account after five years is given by[14]

$$\$400(1.005)^{1} + \$400(1.005)^{2} + \$400(1.005)^{3} + \cdots + \$400(1.005)^{59} + \$400(1.005)^{60}.$$

Denote this sum by S. "r" $= (1.005)$. Put this sum in closed form. You will get

$$\frac{400(1.005) - 400(1.005)^{61}}{1 - 1.005} \quad \text{or} \quad \$28,047.55.$$

Look back at our original expectations. This answer falls in the interval we expected.   ◆

◆ **EXAMPLE 18.11**   Marietta puts $200 into an account every year for four years in order to finance a long vacation in Greece. The bank pays 5% interest per year compounded annually. Her first payment is January 1, 2000. She estimates that her vacation will cost $900. Will she be able to go on the trip immediately after her fourth payment, on January 1, 2003? Will she be able to go on the trip one year after making her fourth payment?

SOLUTION   ***Making Estimates.***   Certainly if Marietta makes five payments she'll have over $1000 and the trip won't be a problem. After making only four payments she will have put a total of $800 into the vacation account. *If* that total were in the bank for 3 years it would grow to $800(1.05)^{3} = \$926.10$, but the total is not in the bank for three years.

Strategy: We'll find the amount of money in the account on January 1st, 2003. We can represent this diagramatically, pushing all the payments to this date.

January 1, 2003

On January 1, 2003

the first payment has grown to $\$200(1.05)^{3}$,

the second payment has grown to $\$200(1.05)^{2}$,

the third payment has grown to $\$200(1.05)^{1}$,

the fourth payment remains at $200.

---

[14] The order in which the addition is done does not matter. Had you chosen to write the sum as

$$\$400(1.005)^{60} + \$400(1.005)^{59} + \$400(1.005)^{58} + \cdots + \$400(1.005)^{2} + \$400(1.005)^{1}$$

you would have $r = (1.005)^{-1}$.

The sum of money in the account is

$$\$200(1.05)^3 + \$200(1.05)^2 + \$200(1.05) + \$200 = \$862.03.$$

Marietta cannot yet take her trip.
On January 1, 2004, before Marietta makes a fifth payment,

the first payment has grown to $\$200(1.05)^4$,

the second payment has grown to $\$200(1.05)^3$,

the third payment has grown to $\$200(1.05)^2$,

the fourth payment has grown to $\$200(1.05)$.

The sum of money in the account is

$$\$200(1.05)^4 + \$200(1.05)^3 + \$200(1.05)^2 + \$200(1.05).$$

This is just the previous sum multiplied by 1.05 (because the money has collected interest for another year). $\$862.03 \cdot (1.05) = \$905.12$. Marietta can be on her way, with $5.12 in her pocket as a slush fund.   ◆

◆ **EXAMPLE 18.12**   Mass Millions is a state-run lottery encouraging residents to support the state's public services by dangling the elusive prize of $1 million for the price of a winning lottery ticket. Instead of a million dollars, however, the winner actually receives 20 annual payments of $50,000. While this is a hefty sum, is this really a prize of $1 million? If you received $1 million today and put it into a bank account paying interest at a rate of 5% per year, just by taking the interest at the end of each year, you could pay yourself $50,000 per year starting one year hence and continuing on forever. The original million would stay in the bank generating interest. The 20 payments of $50,000 spread out over 20 years is not really the same as winning $1 million paid up front now.

Assume the first payment is made to you today and the 20th payment 19 years later. Let's compute the "up-front value" of the prize money of 20 annual payments of $50,000. The first payment of $50,000 received today is certainly worth $50,000 today. But how much is the second payment of $50,000 worth to you right now? How much is the 20th payment worth right now? Let's rephrase the question. Let's suppose that at the moment you win the prize, the state creates a bank account especially for you. The state puts a certain amount of money, P, into the account today and lets it earn interest. From this account the state doles out your 20 payments of $50,000; the final payment depletes your account. How much would the state have to deposit today to make all 20 payments, each at the allotted time? This sum depends on the interest rate in the bank account. Let's suppose that the account pays interest at 5% per year compounded annually.[15] Find P, the total amount of money in the account earmarked for you.

SOLUTION   Let's begin by breaking down the problem into manageable pieces.

The state must put away $50,000 for the very first payment. How much money must the state put away now in order to pay you $50,000 one year from now?

We know that money in this bank account grows according to $M(t) = M_0(1.05)^t$. In one year we want $50,000; we must solve for M_0.

---

[15] Assuming a fixed 5% interest rate over a 20-year period is a rather unrealistic assumption, but we make it to simplify our model. Assume that all payments after the original are made immediately after interest is credited.

$$50,000 = M_0(1.05)^1$$

$$M_0 = \frac{50,000}{1.05} \approx 47,619.05$$

The state must put \$47,619.05 in the account in order to pay you \$50,000 in one year.

The quantity \$47,619.05 is called the **present value** of \$50,000 in one year at an interest rate of 5% per year compounded annually.

Definition

> The present value of M dollars in t years at interest rate r per year compounded n times per year is the amount of money that must be put in an account (paying interest r per year compounded n times per year) *now* in order to have M dollars at the end of t years.

If an account has an interest rate of r per year compounded n times a year, then

$$M = M_0 \left(1 + \frac{r}{n} \right)^{nt}$$

$$\underset{\begin{pmatrix}\text{future}\\\text{value}\end{pmatrix}}{\nearrow} \quad \underset{\begin{pmatrix}\text{present}\\\text{value}\end{pmatrix}}{\nwarrow} \quad \underset{\begin{pmatrix}\frac{r}{n} = \text{interest per}\\\text{compounding period}\end{pmatrix}}{\nwarrow},$$

where M is the amount of money t years in the future and M_0 is the initial deposit, the present value. In this case the present value of M dollars in t years is $M_0 = \frac{M}{\left(1+\frac{r}{n}\right)^{nt}}$.

If an account has interest rate r and interest is compounded continuously, then $M = M_0 e^{rt}$ and the present value of M dollars in t years is $M_0 = \frac{M}{e^{rt}}$.

To continue with our lottery problem, we ask how much money the state must put away in order to pay you \$50,000 in two years. (With our new vocabulary, we can rephrase this: "What is the present value of \$50,000 in two years at an interest rate of 5% per year compounded annually?")

$$50,000 = M_0(1.05)^2 \text{ so } M_0 = \frac{50,000}{(1.05)^2} \approx 45,351.47$$

We continue in this vein.

The present value of the 1st payment is \$50,000.

The present value of the 2nd payment is $\frac{\$50,000}{1.05} \approx \$47,619.05$.

The present value of the 3rd payment is $\frac{\$50,000}{(1.05)^2} \approx \$45,351.47$.

$\vdots$

The present value of the 20th payment is $\frac{\$50,000}{(1.05)^{19}} \approx \$19,786.70$.

Notice that the amount the state must put away now in order to pay you \$50,000 in 19 years is only $\frac{\$50,000}{(1.05)^{19}} \approx \$19,786.70$.

The "up-front" value of the prize can be thought of as the sum of the present values of all of the payments. We can represent this diagramatically. Each of the 20 payments of \$50,000 is being pulled to the present.

$$\$50,000 + \frac{\$50,000}{1.05} + \frac{\$50,000}{(1.05)^2} + \cdots + \frac{\$50,000}{(1.05)^{19}}$$

This is a geometric sum with $r = \frac{1}{1.05}$. We can compute the sum by putting it in closed form.

$$P = \$50,000 + \frac{\$50,000}{1.05} + \frac{\$50,000}{(1.05)^2} + \cdots + \frac{\$50,000}{(1.05)^{19}}$$

$$\frac{1}{1.05} P = \qquad \frac{\$50,000}{1.05} + \frac{\$50,000}{(1.05)^2} + \cdots + \frac{\$50,000}{(1.05)^{19}} + \frac{\$50,000}{(1.05)^{20}}$$

$$\left(1 - \frac{1}{1.05}\right) P = \$50,000 \qquad\qquad\qquad\qquad\qquad\qquad - \frac{\$50,000}{(1.05)^{20}}$$

$$P = \frac{\$50,000 - \frac{\$50,000}{(1.05)^{20}}}{1 - \frac{1}{1.05}} \approx \$654,266.04. \qquad \blacklozenge$$

Work through the following two examples to make sure that the notion of present value makes sense. These examples are designed to emphasize the set-up of the problems, *not* the summation of a geometric series.

◆ **EXAMPLE 18.13**   Onnie has just won an award of $1000 per year for four years, with the first of the four payments being made to him today. Suppose that the money to finance this award is being kept in a bank account with 5% interest compounded annually. How much must be in the bank right now in order to pay for his award? In other words, what is the present value of his award?

SOLUTION   *Making Estimates.*   Although Onnie will receive a total of $4000 dollars, he is not getting it all right now. The bank account earmarked for this award needs *less* than $4000 in it because the money in the account will earn interest. Our goal is to figure out exactly what sum must be put in the account right now so that after Onnie's award has been paid the money in the account is depleted. We expect an answer slightly less than $4000.

Strategy: Treat each of the four payments separately. Calculate how much must be in the account to make the first payment, the second, the third, and the fourth. Then sum these four figures. In other words, sum the present values of the four payments. We can represent this diagramatically. We'll pull each of the payments back to the present.

We know that the present value of $1000 in t years at an interest rate of 5% per year is given by

$$\text{present value} = \frac{\$1000}{1.05^t}.$$

The present value of the first payment $=\$1000$.

The present value of the second payment $= \frac{\$1000}{1.05}$.

The present value of the third payment $= \frac{\$1000}{1.05^2}$.

The present value of the fourth payment $= \frac{\$1000}{1.05^3}$.

The present value of the award = the sum of the present values of the first, second, third and fourth payments.

$$\text{present value of award} = \$1000 + \frac{\$1000}{1.05} + \frac{\$1000}{1.05^2} + \frac{\$1000}{1.05^3}$$

There are only four terms here, so we'll just add them up. (This is a geometric sum with $a = 1000, r = \frac{1}{1.05}$.)

$$\text{present value of award} = \$1000 + \$952.38 + \$907.03 + \$863.84 = \$3723.25$$

The present value of the award is $3723.25. ◆

◆ **EXAMPLE 18.14** Julie has just won an award of $1000 per year for four years, with the first of the four payments being made to her three years from today. Suppose that the money to finance this award is being kept in a bank account with 5% interest compounded annually. How much must be in the bank right now in order to pay for her award? In other words, what is the present value of her award?

SOLUTION **Making Estimates.** Although Julie will receive a total of $4000, she is not getting any of it right now. The bank account earmarked for this award needs *less* than $4000 in it because the money in the account will earn interest. Our goal is to figure out exactly what sum must be put in the account right now so that after Julie's award has been paid the money in the account is depleted. We expect an answer substantially less than $3723.25.

Strategy: Treat each of the four payments separately. How much must be in the account to make the first payment? The second? The third? The fourth? Then sum these four figures.

We can represent this diagramatically. We'll pull each of the payments back to the present.

We know that the present value of $1000 in t years at an interest rate of 5% per year is given by

$$\text{present value} = \frac{\$1000}{1.05^t}.$$

The present value of the first payment $= \frac{\$1000}{1.05^3}$.

The present value of the second payment $= \frac{\$1000}{1.05^4}$.

The present value of the third payment $= \frac{\$1000}{1.05^5}$.

The present value of the fourth payment $= \frac{\$1000}{1.05^6}$.

The present value of the award = the present values of the first, second, third, and fourth payments.

$$\text{present value of award} = \frac{\$1000}{1.05^3} + \frac{\$1000}{1.05^4} + \frac{\$1000}{1.05^5} + \frac{\$1000}{1.05^6}$$

$$\text{present value of award} = \$863.84 + \$822.70 + \$783.53 + \$746.22 = \$3216.29$$

The present value of the award is \$3216.29. ◆

◆ **EXAMPLE 18.15**   Suppose a philanthropic organization wants to start a fund that will make payments of \$2000 each year to the American Cancer Society. The payments are to begin in five years and go on indefinitely. (They are setting up what is known as a *perpetual annuity*.) The fund will be kept in an account with a guaranteed 6% annual interest compounded continuously. How much money should be put in the fund today so the payments can begin five years from today?

SOLUTION   As is the case with many problems, there are several different constructive approaches to obtaining a solution. We'll look at two of these.

Approach 1.   Our strategy is to figure out how much the organization must put away now in order to make the first payment of \$2000, the second payment, and so on. We'll then add up these figures to find the total amount that must be invested today. In other words, we will look at the sum of the present values of the payments. We can represent this diagramatically.

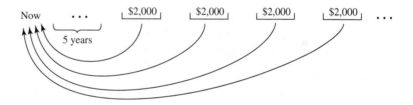

Interest is 6% compounded continuously, so $M(t) = M_0 e^{.06t}$. Solving for M_0 and denoting the future value by M gives $M_0 = \frac{M}{e^{.06t}}$.

The present value of the 1st payment $= \frac{\$2000}{e^{.06(5)}} \approx \$1481.64$.

The present value of the 2nd payment $= \frac{\$2000}{e^{.06(6)}} \approx \$1395.35$.

The present value of the 3rd payment $= \frac{\$2000}{e^{.06(7)}} \approx \$1314.09$.

The amount of money the foundation must put in the account is

$$\frac{\$2000}{e^{.06(5)}} + \frac{\$2000}{e^{.06(6)}} + \frac{\$2000}{e^{.06(7)}} + \cdots.$$

This is an infinite geometric series with $a = \frac{\$2000}{e^{.06(5)}}$ and $r = \frac{1}{e^{.06}}$. $|r| < 1$, so the series converges to $\frac{a}{1-r}$.

$$\frac{a}{1-r} = \frac{\frac{\$2000}{e^{.06(5)}}}{1 - \frac{1}{e^{.06}}} = \frac{\$2000 e^{-.06(5)}}{1 - e^{-.06}} \approx \$25{,}442.17$$

The organization must use \$25,442.17 to set up this perpetuity.

Approach 2. (This approach does not use geometric series.) Observe that in four years there should be enough money in the bank so that in each subsequent year the interest alone is $2000. Let F be the amount of money needed in four years. By the end of the next year this sum of F dollars ought to have earned $2000 in interest.

$$Fe^{.06} - F = 2000$$

$$F(e^{.06} - 1) = 2000$$

$$F = \frac{2000}{e^{.06} - 1}$$

After four years F dollars are needed. If we let M_0 be the amount of money that must be put in the account now, then

$$\frac{2000}{e^{.06} - 1} = M_0 e^{.06(4)}$$

$$M_0 = \frac{2000}{(e^{.06} - 1)e^{.06(4)}},$$

which is, to the nearest penny, $25,442.17. ◆

Time and Money From an economic standpoint, the essence of this whole discussion is that $1000 now, $1000 in 10 years, and $1000 in 20 years are not equivalent if the money could be earning interest. The $1000 now would grow to be much more than $1000 in 20 years if it were kept in an account paying interest. We can compare these three amounts by converting them to a common time frame. For instance, we can look at how much they are all worth in 20 years. If interest is 5% per year compounded monthly, then

$1000 now	is worth $2712.64	in 20 years.
$1000 in 10 years	is worth $1647.01	in 20 years.
$1000 in 20 years	is worth $1000.00	in 20 years.

Alternatively, we can compare them in the present. This is what calculating present value does.

$1000 now	is worth $1000.00	now.
$1000 in 10 years	is worth $607.16	now.
$1000 in 20 years	is worth $368.64	now.

REMARK If money were kept under a pillow instead of being kept in a bank paying interest, the present, future, and past values of $1000 would be the same in the context of our discussion. In this entire discussion we have concerned ourselves exclusively with interest, *not* inflation, the strength of the dollar in the world market, or other considerations. (Economists will often include other considerations.)

■ PROBLEMS FOR SECTION 18.5

1. You have the choice of two awards.

   Award 1: You will receive six yearly payments of $10,000, the first payment being made three years from today.

   Award 2: You will receive three payments of $20,000, the payments being made at two-year intervals, the first payment being made two years from today.

   Suppose that the interest rate at the bank is 4% per year compounded quarterly.

   (a) Find the present value of award 1 and the present value of award 2. Which present value is larger? Which award scheme would you choose?

   (b) Suppose you put each payment in the bank as soon as you receive it. How much money will be in the account eight years from today under the first award scheme? Under the second award scheme?

2. *Banking Basics:* We've been looking at banking because it ties in with the theoretical mathematics we are studying and because most of us have some sort of interface with a bank, be it only via an ATM card. The following questions are designed to direct your attention to the basics.

   (a) Suppose a bank compounds interest n times per year, $n > 1$. Which will be larger, the nominal interest rate, or the effective interest rate?

   (b) Suppose interest rates are fixed at $r\%$ per year compounded annually. Which is larger, the present value of $1000 in T years or $1000 in $(T + 1)$ years?

   (c) Which is larger, the present value of $1000 in T years at a rate of 4% compounded annually or the present value of $1000 in T years at a rate of 5% compounded annually?

3. (a) A friendly benefactor, impressed with Joselyn's enthusiasm for her mathematical studies, decides to award her a scholarship of $6000 to be paid to her five years from today. How much money must the benefactor put aside today in an account earning a nominal annual interest of 4% compounded continuously in order to cover Joselyn's award? (This question asks "what is the present value of $6000 in five years at an interest rate of 4% compounded continuously?")

   (b) Another benefactor, interested in Patrick's potential, promises Pat that if he continues his studies in mathematics he will be awarded a scholarship. The scholarship will given in three payments of $2000, the first payment being made in three years (when he graduates), the second in four years, and the last payment being made five years from today. The benefactor has put aside money for Pat's scholarship in an account earning 4% nominal annual interest compounded continuously in order to cover Pat's award. Pat says that he will promise to take mathematics, but he would like his scholarship money up front and immediately. Because the benefactor has not set aside the full $6000 now, he agrees to give Pat the present value of the award, i.e., the amount of money he has set aside in an account for Pat. How much money should Pat be expecting? (*Hint:* You need to do three separate calculations. Find the present value of the first payment, then of the second, and then of the third.)

(c) Which answer did you expect to be bigger, the answer to part (a) or the answer to part (b)? Why? Have your calculations matched your expectations?

4. A ball is thrown from the ground to a height of 16 feet. Each time the ball bounces it rises up to 60% of its previous height. What is the total distance traveled by the ball? (*Hint:* Keep in mind that between every bounce the ball is going up and then coming back down.)

In Problems 5 through 12, make initial estimates to be sure that the answers you get are in the right ballpark.

5. Suppose you borrow $10,000 at an interest rate of 7% compounded annually. You begin paying back money one year from today and make uniform payments of P annually. You pay back the entire debt after 10 payments. What are your annual payments? *Hint:* Pull each of the 10 payments of P back to the present. The sum should be $10,000.

   Ballpark figures: If no interest were charged, then you would pay $10,000/10 = $1000. If you had to pay interest on the entire $10,000 for 10 years, then you'd pay $10,000 \cdot 1.07^{10}/10$. The actual answer is somewhere between these two extremes.

6. Suppose you borrow $10,000 at an interest rate of 6% compounded annually. You begin paying back the loan one year from today and make uniform payments annually. You pay back the entire debt after 8 payments. What are your annual payments?

7. Suppose you borrow $10,000 at an interest rate of 7% compounded annually. You begin paying back money five years from today and make payments annually. You pay back the entire debt after 10 payments. What are your fixed annual payments?

8. Suppose you borrow some money at an interest rate of 6% compounded annually. You begin paying back one year from today and make payments annually. You pay back the entire debt after 10 payments of $1000 each. How much money did you borrow?

9. Suppose you borrow some money at an interest rate of 6% compounded monthly. You begin paying back one month from today and make payments monthly. You pay back the entire debt after 180 payments of $1000 each. (This is a 15-year mortgage.) How much money did you borrow?

10. Suppose you borrow some money at an interest rate of 6% compounded monthly. You begin paying back one year from today and make payments annually. You pay back the entire debt after 30 payments of $1000 each. How much money did you borrow?

11. Suppose you are saving for a big trip abroad. You estimate that you'll need $4000. You plan to put away a fixed amount of money every month for the next two years (24 deposits) so that immediately after the 24th deposit you have enough money for your trip. You put your money into an account paying interest of 4.5% per year compounded monthly. How much must you deposit every month?

12. Suppose you are saving to buy some cattle. You plan to put $200 into an account every month for the next three years (36 deposits) to pay for the cows. You put your money into an account paying interest of 4.5% per year compounded monthly. Immediately after the 36th deposit, how much money will you have in your cattle fund?

13. People who have slow metabolism due to a malfunctioning thyroid can take thyroid medication to alleviate their condition. For example, the boxer Muhammad Ali took Thyrolar 3, which is 3 grains of thyroid medication, every day. The amount of the drug in the bloodstream decays exponentially with time. The half-life of Thyrolar is 1 week.

   (a) Suppose one 3-grain pill of Thyrolar is taken. Write an equation for the amount of the drug in the bloodstream t days after it has been taken. (*Hint:* In part (a) you are dealing with one 3-grain pill of Thyrolar. Knowing the half-life of Thyrolar, you are asked to come up with a decay equation. This part of the problem has nothing to do with geometric sums.)

   (b) Suppose that Ali starts with none of the drug in his bloodstream. If he takes 3 grains of Thyrolar every day for five days, how much Thyrolar is in his bloodstream immediately after having taken the fifth pill?

   (c) Suppose Ali takes 3 grains of Thyrolar each day for one month. How much thyroid medication will be in his bloodstream right *before* he takes his 31st pill? Right after?

   (d) After taking this medicine for many years, what was the amount of the drug in his body immediately after taking a pill?

   Historical note: Before one of his last fights Muhammad Ali decided to up his dosage to 6 grains. In doing so he mimicked the symptoms of an overactive thyroid. The result in terms of the fight was dismal.

14. Amanda, at the young age of 9, has gotten it firmly into her mind that she wants to be a doctor when she grows up. Her father, panic-stricken, wonders how the family will finance her college and medical school education. To assuage his anxiety he decides to set aside enough money in a bank account right now so that they will be able to withdraw $10,000 every year for eight years beginning nine years from today. Amanda's mother computes how much money they will have to put into the account with an annual interest rate of 6% compounded quarterly. What figure should she arrive at?

15. You have found a calling! You have some burning questions about elephants and want desperately to go to Kenya for a year. In addition to the plane fare you'll need some equipment, a guide, a jeep ... You'll need some money. You figure that you'll need $7000. Each month beginning today you plan to put a fixed amount of money into an account paying 6% interest compounded monthly. How much must you deposit into the account each month if you plan to begin your field work in four years?

16. A woman takes out a loan of $100,000 in order to finance a home. The interest rate is 12% per year compounded monthly and she has a 30-year mortgage. She will pay back the loan by paying a fixed amount, M dollars, every month beginning one month from today and continuing for the next 30 years.

   (a) What is M? (*Hint:* The sum of the present values of her 360 payments, pulled back to the present using an interest rate of 12%, should equal her loan.)

   (b) How much could she save each month if she could borrow at an interest rate of 6.75% per year compounded monthly?

17. Mike L. and Mike C. have decided to establish the Mike and Mike Math Millenium Miracle Prize. The M&M M^3 prize is worth $2000 to the lucky winner. Due to limited

funds, Mike and Mike have decided to award the prize once every 4 years, starting 10 years from now and going on indefinitely. (It's like the Fields Medal in Math, only more accessible.) They have begun to go door-to-door to take collections in order to establish the fund. How much money should the M&M M^3 Prize Fund contain right now in order to start payments 10 years from today?

Assume a guaranteed interest rate of 5% per year compounded annually.

18. Barry is thinking about buying a vehicle. He hears on the radio that he can buy a truck with no money down for two years and then make monthly payments of $150. He thinks this sounds good. He asks Angie if he should buy it. Angie says that she thinks he needs to know about the interest rates and how many years he'll have to make the monthly payments. Barry listens to the radio again and discovers that the monthly payments must be made for 10 years. He decides to compute the present value of the truck payments using an interest rate of 6% per year compounded monthly—or 0.5% per month. What answer should he get?

19. Brent and Rob were working on their math homework when Rob got a headache. Because Rob was incapacitated, Brent went to take a nap. Due to the headache he is blaming on the homework, Rob takes two aspirin. In the body aspirin metabolizes into salicylic acid, which has a half-life of two to three hours. (*Source:* The pharmacist at a CVS Pharmacy.) Rob is a big fellow, so for the purposes of this problem we'll say three hours is the half-life of salicylic acid.

   (a) The math headache is haunting him, so three hours later Rob takes two more aspirin. In fact, the headache is so bad that every three hours he takes two more aspirin. If he keeps this up indefinitely, will the level of salicylic acid in his body ever reach the level equivalent to taking four aspirin all at once?

   (b) Brent wakes up from a deep sleep, looks at his math homework, gets a headache, looks at Rob, and decides that he's going to take two aspirin every two hours. If he keeps this up indefinitely, will the level of salicylic acid in his body every reach the level equivalent to taking three aspirin all at once? Four aspirin all at once? Five aspirin all at once? (Assume again that the half-life of salicylic acid is three hours.)

20. Matt is saving money for his wedding. Suppose that at the beginning of every month he puts $300 in his savings account. The savings account gives interest of 0.5% every month, for a nominal annual interest rate of 6% per year compounded monthly. Matt does this for three years. How much will be in his savings account right after he makes the 36th deposit?

21. Nadia is saving for a trip to Venezuela. She estimates that she'll need $3000. She plans to put away a fixed amount of money every month for the next 30 months so that immediately after the 30th deposit she will have enough money for her trip. She puts her money into an account paying interest of 4% per year compounded monthly. How much must she deposit every month?

   Before you begin calculations, do an estimate. Will she have to put aside more than $100 each month, or less?

22. Suppose you borrow $18,000 at an interest rate of 8% compounded annually. You begin paying back money four years from today and make fixed payments annually. You pay back the entire debt after six payments. What are your annual payments?

    Begin by figuring out the ballpark figures. Will you pay more than $3000 each year? What is an upper bound for the amount of money you will pay each year?

23. A prince takes out a loan of $200,000 in order to finance his castle. The interest rate is 12% per year compounded monthly and he has a 15-year mortgage. He will pay back the loan by paying a fixed amount, F dollars, every month beginning one month from today and continuing for the next 15 years. What is F?

    Note that the sum of the present values of his payments (pulled back to the present using an interest rate of 12%) should equal his loan.

24. Lithium, a drug that is used to treat manic depression, or bipolar disorder, has a half-life of 24 hours. Suppose a patient begins taking a pill of M mg every 12 hours.

    What is the level of the drug in the patient's body two weeks into treatment, immediately after taking the 28th pill?

25. A physician prescribes a pill to be taken daily. Suppose that the half-life of the medication in the patient's bloodstream is 10 days. How many milligrams of medicine should the doctor prescribe if she wants the maximum level of the drug in the bloodstream to reach L mg, but not to surpass it. Assume that the drug is to be taken indefinitely. Your answer will be in terms of L.

26. At the beginning of each month a medical research center buries its refuse in its refuse dump. The monthly refuse deposit contains 40 grams of radioactive material. The radioactive material decays at a rate proportional to itself, with proportionality constant -0.2.

    (a) How much of the radioactive material buried at the beginning of the month is radioactive t months later?

    (b) Immediately after the 60th monthly dump, how much radioactive material is in the refuse site?

    (c) If the situation goes on indefinitely, how much radioactive material will the site contain?

27. You take out a loan of $3000 at an interest rate of 6% compounded monthly. You start paying back the loan exactly one year later. How much should each payment be if the loan is paid off after 24 equal monthly payments? Give an exact answer and an approximation correct to the nearest penny.

C H A P T E R

19

Trigonometry: Introducing Periodic Functions

Transition to Trigonometry

Familiarity with a variety of families of functions provides us with tools necessary for modeling phenomena in the world around us. In the next few chapters we will work with functions that are particularly useful in modeling cyclic or repeating phenomena because the functions themselves are cyclic. Examples of cyclic behavior abound in nature; the rhythm of a heartbeat, the length of a day, the height of the sun in the sky, the path of a sound wave, and the motion of the planets all feature repeating patterns. Wheels are spinning all around us—wheels of bikes, trucks, cars—gears of vehicles, watches, and other machines. Think about the motion of a spot on a steadily rotating gear, or of a seat on a steadily spinning Ferris wheel. The height of the seat is a cyclic function of time; it rises and falls in a smooth, repeating manner. In this chapter we introduce trigonometric functions,[1] cyclic functions that exhibit and help us explore behaviors we observe around us.

---

[1] The word trigonometry refers to triangles, not circles. Trigonometry can be viewed in two distinct ways. Historically it developed in the context of triangles, and hence the name of this family of functions refers to triangles. We will take a triangle perspective in Chapter 20.

19.1   THE SINE AND COSINE FUNCTIONS: DEFINITIONS AND BASIC PROPERTIES

Definition

Below is a circle of radius 1 centered at the origin. This is referred to as **the unit circle.** We'll define trigonometric functions with reference to a point $P = (u, v)$ on the unit circle. We locate P using a real number x as follows.

Start at $(1, 0)$.

If $x \geq 0$, travel along the circle in a counterclockwise direction a distance x units to arrive at $P(x)$.

If $x < 0$, travel along the circle in a clockwise direction a distance $|x|$ units to arrive at $P(x)$.

In other words, x indicates a directed distance around the unit circle. Equivalently, x indicates a directed arc length from $(1, 0)$, where the term **arc length** means a distance along a circle. As x varies, the point P moves around the unit circle. As the position of P varies, so do its u- and v-coordinates.

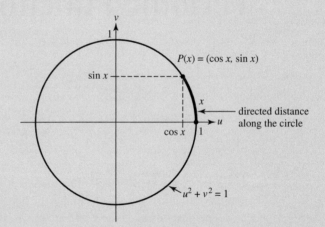

Figure 19.1

$\sin x$ = the v-coordinate of $P(x)$ (the vertical coordinate, the second coordinate, the (signed) height of P)

$\cos x$ = the u-coordinate of $P(x)$ (the horizontal coordinate, the first coordinate)[2]

These functions are called the **sine** and **cosine** functions, respectively.

From these definitions all the properties of the trigonometric functions follow.

_____

[2] Here's a way to remember which is which: put cosine and sine in alphabetical order. Cosine is first; it corresponds to the first coordinate of P. Sine is second, and corresponds to the second coordinate of P. We have labeled our coordinate axes u and v because we want x to determine the point P, making sine and cosine functions of x.

Our method for locating P essentially involves wrapping the real number line around the unit circle, with zero glued to the point $(1, 0)$ on the circle. Figure 19.2 shows the unit circle with the portion of the number line from 0 to 2π wrapped around it like measuring tape. We'll refer to this as the calibrated unit circle.

◆ **EXAMPLE 19.1** Use the calibrated unit circle to approximate $\sin(1.1)$ and $\cos(1.1)$.

SOLUTION Locate $P(1.1)$ by moving along the unit circle a distance 1.1 counterclockwise from the point (1.0). Approximate the coordinates of $P(1.1)$. $P(1.1) \approx (0.45, 0.9)$. Therefore $\sin(1.1) \approx 0.9$ and $\cos(1.1) \approx 0.45$.

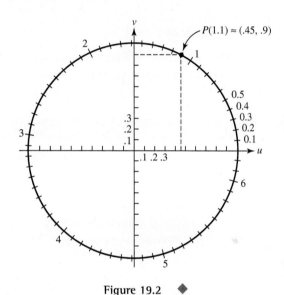

Figure 19.2 ◆

EXERCISE 19.1 (a) Find values of x between 0 and 2π such that the coordinates of $P(x)$ are

(i) $(1, 0)$. (ii) $(0, 1)$. (iii) $(-1, 0)$. (iv) $(0, -1)$.

(b) Use the definitions of sine and cosine to evaluate the following.

(i) $\sin \pi$ (ii) $\cos \pi$ (iii) $\sin \frac{\pi}{2}$ (iv) $\cos \frac{\pi}{2}$ (v) $\sin \frac{-3\pi}{2}$ (vi) $\cos \frac{3\pi}{2}$

EXERCISE 19.2 Use the calibrated unit circle shown to approximate the following. (You can check your answers using a calculator, but be sure that the calculator is in radian mode as opposed to degree mode.)[3]

(i) $\sin 0.3$ (ii) $\sin 2$ (iii) $\sin 3$ (iv) $\sin 4$ (v) $\sin 5$ (vi) $\cos 5$

EXERCISE 19.3 Use the calibrated unit circle shown to approximate all x-values between 0 and 2π such that

(i) $\sin x = 0.8$. (ii) $\sin x = -0.4$. (iii) $\cos x = 0.8$. (iv) $\cos x = -0.2$.

---

[3] Radians and degrees will be discussed in Section 19.4. To check whether your calculator is in radian mode, try to evaluate $\cos \pi$. You will get -1 only if your calculator is in radian mode.

(Again, use a calculator to check the accuracy of your answers. There are two answers to each question.)

The calibrated unit circle displayed in Figure 19.2 shows only the interval $[0, 2\pi]$ wrapped around the unit circle, but in fact we want to wrap the entire number line around the circle. The circumference of the unit circle is 2π, so every directed distance of 2π (positive *or* negative) brings us back to the same point P. Therefore there are infinitely many x-values corresponding to any point P. If x_0 corresponds to the point P, so do $x_0 + 2\pi$ and $x_0 - 2\pi$. In fact, $P(x_0) = P(x_0 + 2\pi n)$, where n is any integer, because circumnavigating the circle any integer number of times (in either direction) has no impact on the terminal point P.

Periodicity

One of the most striking characteristics of trigonometric functions is their cyclic nature. A periodic function is marked by repeated cycles.

Definition

A function f is **periodic** if there is a positive constant k such that for all x in the domain of f, $f(x + k) = f(x)$. The smallest of such constant k is called the **period** of f. If a function has period k, then we can select *any* single interval of length k in the domain, graph the function over this domain, and from this construct the entire function by horizontally shifting this fundamental block k units (left and right) repeatedly.

Below are graphs of some periodic functions.

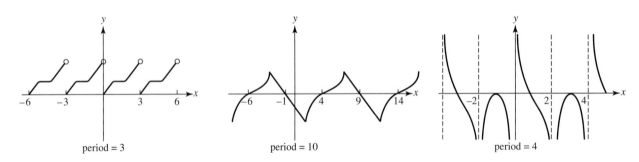

period = 3   period = 10   period = 4

Figure 19.3

From their definitions we observe that the values of the output of sine and cosine repeat every 2π units; $\sin x$ and $\cos x$ both are periodic with period 2π. Therefore, by graphing the functions on any interval of length 2π we know how they behave everywhere.

The Graphs of $\sin x$ and $\cos x$

Work through the following exercise.

EXERCISE 19.4   A steadily spinning Ferris wheel with a radius of 10 meters makes one counterclockwise revolution every 2 minutes. Placing the origin of a coordinate system at the center of the vertical wheel, consider the position of a seat that is at the point $(10, 0)$ at time $t = 0$.

(a) Plot the vertical position (height) of the seat as a function of time t, t in minutes.

(b) Plot the horizontal position of the seat as a function of time t, t in minutes.

The graphs of $\sin x$ and $\cos x$ are closely related to those of Exercise 19.4. To see the connection, think of a Ferris wheel with a radius of 1 unit and focus on the point on the rim that starts at position $(1, 0)$. Then, instead of plotting height versus time, plot height versus the distance the point travels. This will give the graph of $\sin x$. The graph of $\cos x$ is obtained by looking at the horizontal position of the same point. The graphs are sketched below.

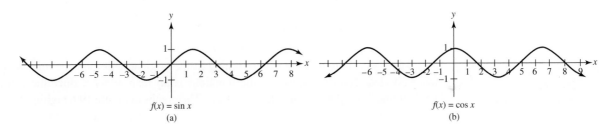

$f(x) = \sin x$ $f(x) = \cos x$

(a)   (b)

Figure 19.4

In Example 19.2 and Exercise 19.3 we *approximated* the values of $\sin x$ and $\cos x$ for various values of x. If x is any integer multiple of $\pi/2$, we can evaluate $\sin x$ and $\cos x$ *exactly* because we can find the coordinates of $P(x)$ exactly.

The circumference of the unit circle is 2π, therefore we know the following.

$P(x)$ is $(1, 0)$	for $x = 0, \pm 2\pi, \pm 4\pi, \ldots$, or $\ldots -4\pi, -2\pi, 0, 2\pi, 4\pi, \ldots$ i.e., for $x = 2\pi n$, where n is an integer.
$P(x)$ is $(-1, 0)$	for $x = \pi, \pi \pm 2\pi, \pi \pm 4\pi, \ldots$, or $-3\pi, -\pi, \pi, 3\pi, 5\pi, \ldots$ i.e., for $x = \pi + 2\pi n$, where n is an integer.
$P(x)$ is $(0, 1)$	for $x = \frac{\pi}{2}, \frac{\pi}{2} \pm 2\pi, \frac{\pi}{2} \pm 4\pi, \ldots$ i.e., for $x = \frac{\pi}{2} + 2\pi n$, where n is an integer.
$P(x)$ is $(0, -1)$	for $x = \frac{3\pi}{2}, \frac{3\pi}{2} \pm 2\pi, \frac{3\pi}{2} \pm 4\pi, \ldots$ i.e., for $x = \frac{3\pi}{2} + 2\pi n$, where n is an integer.

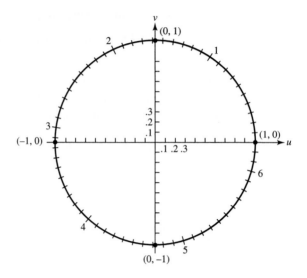

Figure 19.5

From the definitions of sine and cosine we can see that the zeros of these functions as well as all the local maxima and local minima occur at integer multiples of $\pi/2$. Therefore, when sketching $\sin x$ and $\cos x$, the x-axis is frequently labeled just in multiples of $\pi/2$. This labeling has been known to trap dozing students into assuming, *incorrectly,* that the trigonometric functions are defined only for x-values with the number π explicitly written as a factor. This notion is wrong! Returning to the original definitions of sine and cosine makes it clear that the domain of the functions is *all* real numbers since the entire real number line is wrapped around the unit circle. The graphs in Figure 19.6 highlight x-values for which the sine and cosine graphs have zeros, local maxima, and local minima.

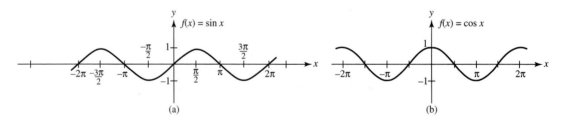

Figure 19.6

Domain and Range

The domain of $\sin x$ is $(-\infty, \infty)$. The range of $\sin x$ is $[-1, 1]$.

The domain of $\cos x$ is $(-\infty, \infty)$. The range of $\cos x$ is $[-1, 1]$.

Symmetry Properties of $\sin x$ and $\cos x$

EXERCISE 19.5   Using the unit circle, show that

$$\cos x = \cos(-x) \text{ and that } \sin x = -\sin(-x).$$

In Exercise 19.5 you have shown that

cos x is an even function; its graph is symmetric about the y-axis.

sin x is an odd function; its graph is symmetric about the origin.

Some Trigonometric Identities

(cos x, sin x) is a point on the unit circle $u^2 + v^2 = 1$; therefore,

$$(\sin x)^2 + (\cos x)^2 = 1$$

for all values of x. We can also see this by using the Pythagorean Theorem and the triangle drawn in Figure 19.7. For this reason it is sometimes called a Pythagorean identity. The equation $(\sin x)^2 + (\cos x)^2 = 1$ is called a trigonometric **identity** because the left- and right-hand sides of the equation are identically equal for *all* x.[4]

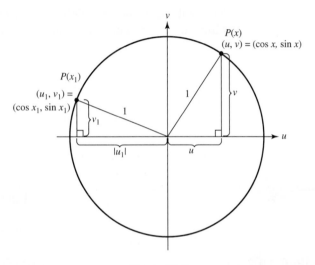

Figure 19.7

NOTATION   The conventional notation for $(\sin x)^2$ is $\sin^2 x$. In other words, $\sin^2 x$ *means* $(\sin x)^2$. By contrast, $\sin x^2$ *means* $\sin(x^2)$. The Pythagorean identity given above is usually written as

$$\sin^2 x + \cos^2 x = 1.$$

More generally,

$\cos^n x$ means $(\cos x)^n$ for all n except $n = -1$.

$\sin^n x$ means $(\sin x)^n$ for all n except $n = -1$.

The one important exception is $n = -1$. $\cos^{-1} x$ denotes the inverse of the cosine function, and *not* $(\cos x)^{-1}$. That's just the convention. To refer to the reciprocal of cos x, we write $(\cos x)^{-1} = \frac{1}{\cos x}$; this expression is generally referred to as sec x. In the next two chapters we will define trigonometric functions (like sec x) constructed from sine and cosine as

---

[4] An identity is an equation that holds for all possible values of the variable. Note the difference between an equation such as $x^2 = 4$, which holds only for $x = 2$ and $x = -2$, and the identity $(\sin x)^2 + (\cos x)^2 = 1$, which holds for all x, or the identity $(x + y)^2 = x^2 + 2xy + y^2$, which holds for all x and all y .

well as defining the inverse trigonometric functions. These notational conventions are very standard; in order to communicate in the language of mathematics you need to learn them.

A second set of trigonometric identities comes from the periodic nature of the functions.

$$\sin(x + 2\pi n) = \sin x \quad \text{for any integer } n$$

$$\cos(x + 2\pi n) = \cos x \quad \text{for any integer } n$$

A third set of trigonometric identities is suggested by looking at the relationship between the graphs of $\sin x$ and $\cos x$. These graphs are horizontal translates.[5] We can express this in innumerable ways. For instance, if the graph of $\cos x$ is shifted to the right $\pi/2$ units, we obtain the graph of $\sin x$. This observation is equivalent to

$$\sin x = \cos(x - \pi/2).$$

Shifting the graph of $\sin x$ horizontally π units produces the graph of $-\sin x$. Similarly, shifting the graph of $\cos x$ horizontally π units produces the graph of $-\cos x$.

$$\sin(x \pm \pi) = -\sin x$$

$$\cos(x \pm \pi) = -\cos x$$

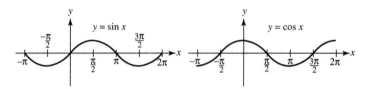

Figure 19.8

Trigonometric identities can be useful in simplifying expressions.

◆ **EXAMPLE 19.2**   Let $f(x) = \sqrt{4 - 4\cos^2(x - \pi)}$. For what values of x is f minimum? maximum?

SOLUTION   It is not necessary to differentiate f in order to answer this question. We'll begin by simplifying $f(x)$.

$$\text{We know } \cos(x - \pi) = -\cos x;$$

$$\text{therefore, } 4 - 4\cos^2(x - \pi) = 4 - 4[-\cos x]^2$$

$$= 4 - 4\cos^2 x$$

$$= 4(1 - \cos^2 x)$$

$$= 4\sin^2 x.$$

Thus,

$$f(x) = \sqrt{4\sin^2 x}.$$

You might, at first, think that $\sqrt{4\sin^2 x} = 2\sin x$, but that is not always so. $\sqrt{4\sin^2 x}$ must be nonnegative, and $\sin x$ can be negative; $\sqrt{4\sin^2 x} = 2|\sin x|$.

$$f(x) = 2|\sin x|$$

---

[5] This is true, but we have not proven it. The proof is left as an exercise in the next chapter, where it is more easily approached.

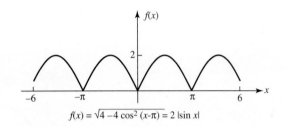

$$f(x) = \sqrt{4 - 4\cos^2(x-\pi)} = 2\,|\sin x|$$

Figure 19.9

$f(x)$ is minimum when $\sin x = 0$, that is, when $x = \pm \pi n$, where n is any integer. The minimum value of f is 0.

$f(x)$ is maximum where $\sin x$ takes on the values of ± 1. f is maximum at $x = \dots \frac{-\pi}{2}$, $\frac{\pi}{2}, \frac{3\pi}{2}, \frac{5\pi}{2}, \dots$. In other words, $f(x)$ is maximum at $x = \frac{\pi}{2} + \pi n$, where n is any integer. The maximum value of f is 2. ◆

Question: Let $f(x) = \sqrt{4 - 4\cos^2(x - \pi)}$, as in Example 19.2. Do you think that $f'(x) = 0$ at the local maxima of f? At the local minima of f?

PROBLEMS FOR SECTION 19.1

1. Use a straightedge and the calibrated unit circle drawn below to estimate each of the following values. (You can check your answers with a calculator set in radian mode.)

   (a) $\cos(1.1)$ (b) $\sin(1.1)$ (c) $\sin(3.5)$ (d) $\cos(4.2)$

   (e) $\cos(5.9)$ (f) $\sin(2.2)$ (g) $\sin(5.7)$

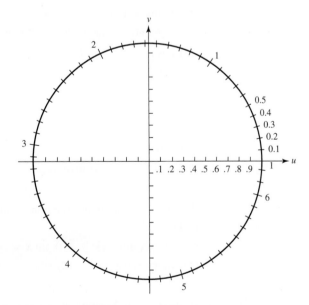

2. Use the calibrated unit circle to estimate all t-values between 0 and 6 such that

   (a) $\cos t = 0.3$.   (b) $\sin t = 0.7$.   (c) $\sin t = -0.7$.

3. $P(w)$ is indicated in the figure below.

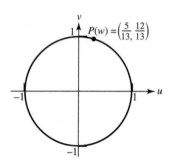

Find the following.

(a) $\sin w$

(b) $\cos w$

(c) $\sin(-w)$

(d) $\cos(-w)$

(e) $\sin(w + 6\pi)$

(f) $\cos(w - 2\pi)$

(g) Is $\cos(2w)$ positive, negative, or zero? Explain briefly.

4. Beginning at point $(1, 0)$ and traveling a distance t counterclockwise along the unit circle, we arrive at a point with coordinates $\left(\frac{-1}{3}, \frac{2\sqrt{2}}{3}\right)$. Find the following.

(a) $\cos t$

(b) $\sin t$

(c) $\sin(-t)$

(d) $\cos(-t)$

(e) $\sin(t - \pi)$

(f) $\sin(t - 10\pi)$

(g) Is $\sin(t + \frac{\pi}{2})$ positive, negative, or zero? Explain.

5. Which of the following equations hold for all x? Explain your answers in terms of the unit circle.

(a) $\sin x = \sin(-x)$ (b) $\sin x = -\sin(-x)$

(c) $\cos x = \cos(-x)$ (d) $\cos x = -\cos(-x)$

6. Evaluate the following limits. Explain your reasoning.

(a) $\lim_{x \to \infty} \sin x$ (b) $\lim_{x \to \infty} \frac{\sin x}{x}$

7. Evaluate the following limits.

(a) $\lim_{x \to \infty} \cos x$

(b) $\lim_{x \to 0} \sin \left(\frac{1}{x} \right)$

(c) $\lim_{x \to 0^+} \frac{1}{\sin x}$

(d) $\lim_{x \to \infty} \sin \left(\frac{x^2}{x+1} \right)$

(e) $\lim_{x \to \infty} \cos \left(\frac{\pi x^3 - 99}{x^3 - x^2 + 7} \right)$

19.2   MODIFYING THE GRAPHS OF SINE AND COSINE

We began our discussion of trigonometric functions by suggesting that they would be useful for modeling periodic phenomena. In order for the sine and cosine functions to be useful to us we must be able to alter their periods and adjust the levels of peaks and valleys. We introduce the following terminology.

Definitions

The **balance line** of a sine or cosine function (or modifications of them through shifting, flipping, stretching and shrinking) is the centrally located horizontal line about which the values of the function oscillate. The functions $f(x) = \sin x$ and $g(x) = \cos x$ have balance lines at $y = 0$. Shifting the graph vertically by k units shifts the balance line by k units. Stretching the $\sin x$ or $\cos x$ graphs, whether vertically or horizontally, does not alter the balance line. The **balance value** (also called the **average value** of the function over a complete cycle) is the y-value of the balance line.

The **amplitude** of a sine or cosine function (or modifications of them as specified above) is the positive number indicating the maximum vertical distance between the graph and its balance line.

We have already defined the **period** to be the smallest positive constant k such that $f(x + k) = f(x)$ for all x.

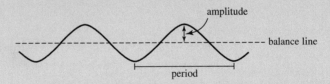

Figure 19.10

The term **sinusoidal function** is used to refer to sine functions, cosine functions, and modifications of them obtained by shifting, flipping, stretching, and shrinking.

If you are wondering why the word sinusoidal can refer to the cosine function, recall that $\cos x$ is simply a horizontal translate of $\sin x$.

Knowing the maximum and minimum values of a sinusoidal function, we can calculate its amplitude and balance value.

■ The amplitude is half of the vertical distance between the maximum and minimum values:

$$\text{amplitude} = \frac{\text{maximum value} - \text{minimum value}}{2}.$$

■   The balance value is the average of the maximum and minimum values; add them and divide by 2:

$$\text{balance value} = \frac{\text{maximum value} + \text{minimum value}}{2}.$$

Examples are given in Figure 19.11. From the pictures, determine the amplitude, period, and balance value. Check your answers with those given.

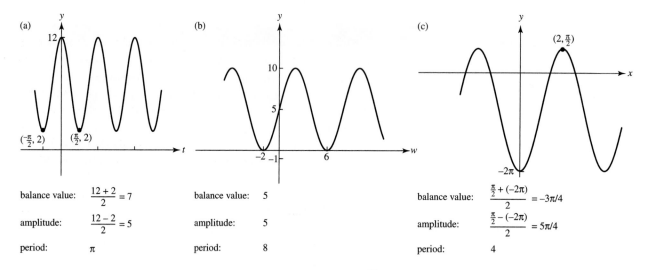

Figure 19.11

How Are Modifications of the Graphs Reflected by Modifications of the Functions, and Vice Versa?

■   To change the balance line of a sine or cosine function is to shift the function vertically (and vice versa).

■   To change the amplitude of a sine or cosine function is to stretch or shrink the function vertically (and vice versa).

■   To change the period of a sine or cosine function is to stretch or shrink the function horizontally (and vice versa).

We know how to shift a function vertically and how to stretch and shrink both horizontally and vertically, so we simply apply our knowledge to these new functions. We also know how to flip a graph about a horizontal line. For the time being we will try to write equations for graphs without involving horizontal shifts of the sine and cosine functions, if possible.

EXERCISE 19.6 Graph the following. Check your answers using a calculator (or computer), making sure the calculator is in radian mode. Do not rely completely on your calculator; blind faith in it can lead you *dreadfully* astray.[6]

(a) i. $y = 3 \sin x$ ii. $y = -2 \cos x + 2$ iii. $y = -(1/2) \sin x - 1$

(b) i. $y = \sin 2x$ ii. $y = \sin \frac{x}{2}$ iii. $y = 3 \cos(2\pi x)$

OBSERVATION After doing part (b) of Exercise 19.6, you'll see that

$\sin x$ has period 2π,

$\sin 2x$ has period π (one cycle is completed in π units) corresponding to horizontal compression by a factor of 2, and

$\sin(x/2)$ has period 4π (one cycle is completed in 4π units) corresponding to horizontal stretching by a factor of 2.

This is in keeping with what we know about horizontal compression and stretching; in the case of trigonometric functions it is simpler to understand why replacing x by $2x$ will compress the graph and replacing x by $\frac{x}{2}$ will stretch it. The point $P(x)$ makes a complete spin around the unit circle as x varies from 0 to 2π. If we replace x by $2x$, the point $P(2x)$ makes a complete spin around the unit circle as x varies from 0 to π, that is, in an interval of half the length. Analogously, the point $P(\frac{x}{2})$ makes one complete spin around the unit circle as x varies from 0 to 4π, that is, in an interval of double the original length. We can generalize to say that if we replace x by Bx, the point $P(Bx)$ makes a complete spin around the unit circle as x varies from 0 to $\frac{2\pi}{|B|}$, so the period of $\sin Bx$ is $\frac{2\pi}{|B|}$.

How must we modify the sine or cosine functions to reflect either the Ferris wheel problem from Exercise 19.4 or to model the height or horizontal position of a particular point on the rim of any spinning wheel as a function of time?

The balance value corresponds to the height of the center of the wheel,

the amplitude corresponds to the radius of the wheel, and

the period corresponds to the amount of time it takes to complete one full revolution.

Replacing t by $(2t)$ doubles the speed of the wheel, halving the time to complete one trip around. Similarly, replacing t by $\left(\frac{t}{2}\right)$ cuts the speed of the wheel in half, thereby doubling the time it takes for a point on the rim to return to its original position.

If time is measured in minutes, then the period gives a measure of minutes/revolution. The number of revolutions per unit time is called the **frequency**. The frequency is defined to be the reciprocal of the period.[7]

$$\text{frequency} = \frac{1}{\text{period}}$$

EXERCISE 19.7 (a) A Ferris wheel with diameter 30 meters makes one revolution every 2 minutes and rotates counterclockwise. Consider a seat that starts at ground level, a height of 0 meters at time $t = 0$. Express the vertical height of the seat as a function of time t.

---

[6] Your calculator can be *very* misleading when graphing trigonometric functions. For example, try graphing $y = \sin(50x)$ and $y = \sin(20x)$ with domain $[-10, 10]$ on your calculator. You will be appalled!

[7] When talking about radio waves, for instance, it is more common to talk of the frequency than it is to talk of the period.

(b) If the Ferris wheel slows down to one revolution every 3 minutes, what is the new equation?

(c) A Ferris wheel with diameter 25 meters makes one revolution every minute and rotates clockwise. Write an equation for the vertical height of a seat that starts at ground level at time $t = 0$.

Changing the Amplitude, Period, and Balance Value: A Summary

$$\text{If } y = A \sin Bx + K, \begin{cases} \text{the balance value is } K, \\ \text{the amplitude is } |A|, \\ \text{the period is } \frac{2\pi}{|B|}. \end{cases}$$

$$\text{If } y = A \cos Bx + K, \begin{cases} \text{the balance value is } K, \\ \text{the amplitude is } |A|, \\ \text{the period is } \frac{2\pi}{|B|}. \end{cases}$$

We can always write a trigonometric equation in such a way that B is positive since $\sin(-x) = -\sin x$ and $\cos(-x) = \cos x$. When constructing equations to match graphs in the examples below, we will choose B to be positive.

Finding an Equation to Fit a Sinusoidal Graph

Below we fit an equation to a sinusoidal graph that at $x = 0$ is either at an extreme value or at its balance value. If neither of these were the case, we could either shift the graph horizontally or establish a new benchmark for $x = 0$ to ensure one of these cases.

■ If the graph has a maximum or minimum at $x = 0$, it is simplest to use an equation of the form $y = A \cos Bx + K$, whereas if balance value is attained at $x = 0$, it is simplest to use an equation of the form $y = A \sin Bx + K$.

■ Determine the period. The period is $\frac{2\pi}{|B|}$ and we can choose B to be positive, so $B = \frac{2\pi}{\text{the period}}$.

■ Determine the amplitude. The amplitude is $|A|$. Choose A or $-A$ by comparing the graph in question to the standard sine and cosine graphs.

■ Determine the balance value. This is K.

◆ EXAMPLE 19.3   Find possible equations to match the graphs displayed earlier (in Figure 19.11) and reproduced on the next page.

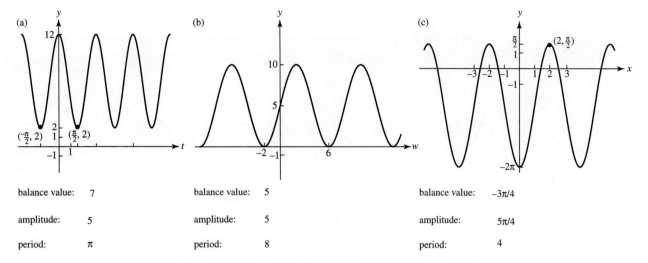

Figure 19.12

SOLUTION   (a) When $t = 0$ the graph has a maximum, so we model it with the cosine function.

$y = A \cos Bt + K$

The period is $\frac{2\pi}{|B|} = \pi$, so $\frac{2}{|B|} = 1$ and $|B| = 2$. We'll choose $B = 2$.

There is no vertical flipping, so A is positive. $A = 5$

Knowing the balance value is 7 gives $y = 5 \cos(2t) + 7$.

(b) When $w = 0$ this graph is at its balance value, so we model it with the sine function.

$y = A \sin Bw + K$

The period is $\frac{2\pi}{|B|} = 8$, so $8|B| = 2\pi$. Let $B = \pi/4$.

There is no vertical flipping, so A is positive. $A = 5$

Knowing the balance value is 5 gives us $y = 5 \sin\left(\frac{\pi}{4}w\right) + 5$.

(c) When $x = 0$ this graph is at a minimum, so we model it with the cosine function.

$y = A \cos Bx + K$

The period is $\frac{2\pi}{|B|} = 4$, so $4|B| = 2\pi$ and $|B| = \pi/2$. Choose $B = \pi/2$.

The cosine function must be flipped over a horizontal line, so A is negative. $A = \frac{-5\pi}{4}$

Knowing the balance value is $\frac{-3\pi}{4}$ gives us $y = -\frac{5\pi}{4} \cos\left(\frac{\pi}{2}x\right) - \frac{3\pi}{4}$.   ◆

Horizontal Shifting

We shift the sine and cosine graphs horizontally k units just as we would any other graph. For instance, replacing x by $x - 3$ shifts the graph to the right 3 units. We can express any cosine graph as a sine graph by making an appropriate horizontal shift. For instance, $\cos x = \sin(x + \pi/2)$. More generally,

$y = A \sin[B(x - C)]$ is the graph of $A \sin Bx$ shifted C units right.

$y = A \cos[B(x - C)]$ is the graph of $A \cos Bx$ shifted C units right.

A bit of care must be taken when simultaneously shifting horizontally *and* changing the period. Suppose we were to graph $y = \sin\left(\frac{t}{2} + \pi\right)$. Rewriting $\sin\left(\frac{t}{2} + \pi\right)$ as $\sin\left[\frac{1}{2}(t + 2\pi)\right]$ helps us see that this is the graph of $\sin\frac{t}{2}$ shifted 2π units to the left.

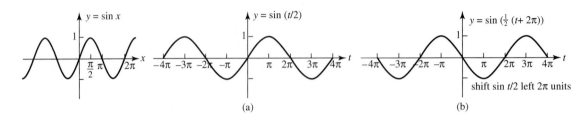

Figure 19.13

Notice that *first* changing the period of sine to 4π and *then* shifting 2π units to the left is different from shifting first (which in this case has no effect) and then changing the period. The order in which we execute the transformations matters: The transformation corresponding to multiplication/division takes precedent over that corresponding to addition/subtraction.

When faced with an expression like $\cos(2x + 3\pi)$, rewrite it as $\cos\left[2\left(x + \frac{3\pi}{2}\right)\right]$ to analyze the horizontal shift correctly. It is the graph of $\cos 2x$ shifted left $\frac{3\pi}{2}$ units.

EXERCISE 19.8   Graph the functions given below. Use a domain that allows you to display one full period. Check your answers with the help of a graphing calculator.

(a) $y = 5\cos 2\left(x + \frac{\pi}{2}\right)$

(b) $y = -3\cos\left(\frac{x}{2} + \pi\right)$

(c) $y = 5\sin(2x - \pi) - 1$

Testing the Limits of Your Technological Tools

When using a graphing calculator or computer to give the graph of a trigonometric function, a thoughtful choice of viewing window and a healthy dose of skepticism can be invaluable. Consider the function $f(x) = \sin 100x$. Use the technology available to you to graph this function on the intervals $[-5, 5]$, $[-4.5, 4.5]$, $[-3, 3]$, and $[-1, 1]$. We know that the graph is a sine function with a period of $\frac{2\pi}{100} = \frac{\pi}{50} \approx 0.063$. Therefore, on the interval $[-1, 1]$ there ought to be more than 30 cycles of the sine function. Depending upon the technology you have used to graph, the graphs will vary, but in all probability they will be highly misleading. A graphing calculator or computer generally plots points and connects the dots, but the oscillations of the function $f(x) = \sin 100x$ are so numerous these plotted points do not give a good reading of the behavior of the function overall.

Theme and Variation

We know that for constants A and B:

$$\text{If } y = A\cos Bx, \begin{cases} \text{the amplitude is } |A|, \\ \text{the period is } \frac{2\pi}{|B|}. \end{cases}$$

Suppose that the positions of A or of B were not filled by constants. The resulting functions would not be periodic, yet they would retain some of the oscillation of the cosine function.

EXERCISE 19.9   Match each of the five functions below with the appropriate graph from Figure 19.14. Use a graphing calculator or computer to check your work.

(a) $y = \cos(\pi x)$

(b) $y = e^{-.2x} \cos(\pi x)$

(c) $y = x \cos(\pi x)$

(d) $y = x^2 \cos(\pi x)$

(e) $y = \cos(x^2)$

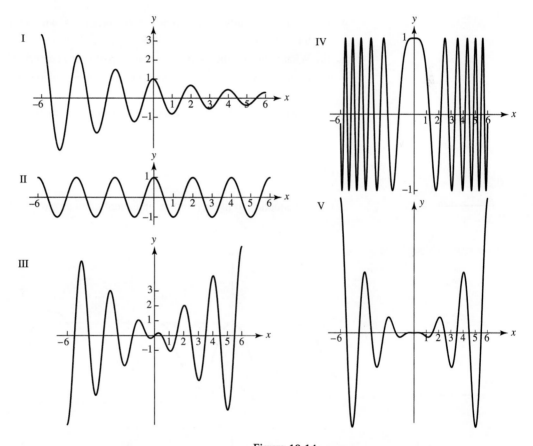

Figure 19.14

PROBLEMS FOR SECTION 19.2

1. (a) On the same set of axes graph the following. Set the domain to show at least one complete cycle of the function. (Colored pens/pencils can be helpful in identifying which graph is which.)

(i) $y = \sin x$ (ii) $y = 2 \sin x$ (iii) $y = -3 \sin x$

(b) Describe in words the effect of the parameter A in $y = A \sin(x)$.

2. (a) On the same set of axes graph the following.

   (i) $y = \sin x$ (ii) $y = \sin(2x)$ (iii) $y = \sin(x/2)$
   (iv) $y = \sin(4x)$ (v) $y = \sin(-2x)$

   (b) Describe in words the effect of the parameter B in $y = \sin(Bx)$.

3. (a) Describe in words the effect of the parameter C in $y = \sin(x) + C$.
   (b) Describe in words the effect of the parameter D in $y = \sin(x + D)$.

4. (a) Using the results of the previous three problems, without using a calculator graph the function $y = -3 \sin(2x) + 3$.
   (b) What is the amplitude of the curve in part (a)? The period?
   (c) Check your answer to part (a) using a graphing calculator or computer.

5. Find the domain and range of each of the following functions.

   (a) $f(x) = 3 \sin(2x + 1)$ (b) $g(x) = |2 \cos x|$ (c) $h(x) = \cos |x|$
   (d) $j(x) = 2 \cos x - 1$ (e) $k(x) = \sqrt{\sin x}$

6. Find an equation to fit each of the sinusoidal graphs below.

(a)

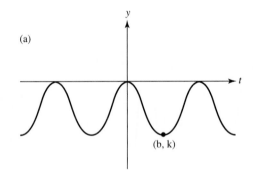

(b, k)

(b)

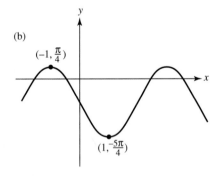

$(-1, \frac{\pi}{4})$

$(1, \frac{-5\pi}{4})$

(c)

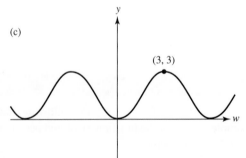

$(3, 3)$

(d)

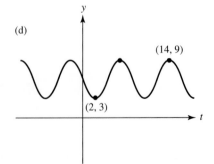

$(14, 9)$

$(2, 3)$

7. Find equations that fit the curves below:

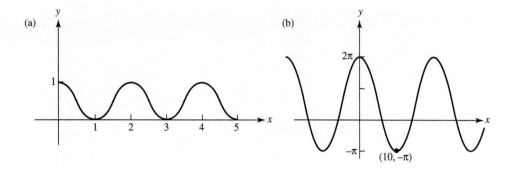

8. Graph the following.
   (a) $y = |\sin x|$ (b) $y = \sin |x|$

9. Find the period, amplitude, and balance value of each of the following functions.

   (a) $f(x) = 0.5 \sin(3x)$ (b) $g(x) = -4 \cos(x/3)$ (c) $h(x) = \frac{\sin(0.2x) + \pi}{\pi}$
   (d) $j(x) = 4[\sin(\pi x) - 1]$ (e) $k(x) = 4 \sin(\pi x - 1)$

10. (a) A Ferris wheel with diameter 20 feet makes one revolution every 8 minutes. Graph the height of a point on the Ferris wheel versus time, assuming that at $t = 0$ the point is at height 0. Give an equation whose graph is the picture you've drawn.
    (b) The Ferris wheel slows down so that it makes one revolution every 10 minutes. Adjust both your picture and your equation.

11. (a) What is the domain of $g(x) = \sqrt{\sin x}$?
    (b) On the same set of axes graph $f(x) = \sin x$ and $g(x) = \sqrt{\sin x}$. (Use different colors if you have them at your disposal.)

12. Suppose we have the equation of a sine curve, $y = \sin\left(\frac{\pi x}{2}\right)$, with period 4, amplitude 1.
    (a) We wish to shift the graph over 1 unit to the left. Which of the following will accomplish this?
        i. $y = \sin\left(\frac{\pi}{2}x - 1\right)$
        ii. $y = \sin\left(\frac{\pi}{2}(x - 1)\right) = \sin\left(\frac{\pi}{2}x - \frac{\pi}{2}\right)$
        iii. $y = \sin\left(\frac{\pi}{2}x + 1\right)$
        iv. $y = \sin\left(\frac{\pi}{2}(x + 1)\right) = \sin\left(\frac{\pi}{2}x + \frac{\pi}{2}\right)$
        Be sure that you get this right. Draw a picture of what you are aiming for, and then try a point or two to verify. For example, you might run a test on $x = 0$.
    (b) For each of the remaining three choices in part (a), describe in words what happens to the original graph.
    (c) In order to obtain the desired result in part (a), it is probably simplest to use a cosine function. What cosine function will give the desired result?
    (d) Suppose A, B, and C are positive constants and $y = A \sin(Bx + C)$. What are the period and amplitude of the sine graph? Describe the horizontal shift.

13. Match the equations with the graphs. Do this first without consulting a graphing calculator. (Think about spacing of intercepts, sign, relative heights of peaks, etc.)

(a) $y = x \sin x$ (b) $y = x^2 \sin x$ (c) $y = \sin(x^2)$

(d) $y = e^{-0.2x} \sin x$ (e) $y = e^{0.2x} \sin x$ (f) $y = 0.5x + \sin x$

(i)
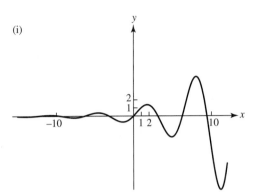

(ii)
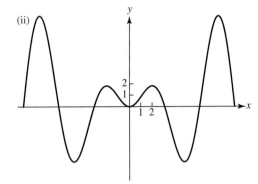

(iii)
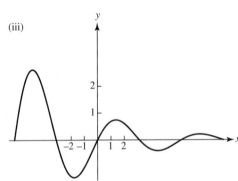

(iv)
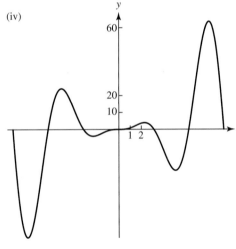

(v)
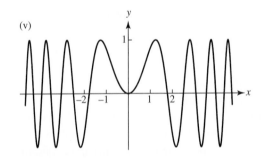

(vi)
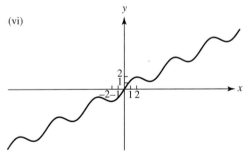

14. The region of Bogor in Java has a rainforest climate with 450 mm of rain falling in the rainiest month (February) and 230 mm of rain falling in the driest month. Model the amount of rain per month using a sinusoidal function. Let $R(t)$ be the number of millimeters of rain per month where $t = 0$ denotes the height of the rainy season and t is measured in months.

(a) Give an expression for $R(t)$.

(b) According to your model, on average how many millimeters of rain fall per month each year? Recall that the balance value of the function is the average value.

(c) How many millimeters of rain does your model predict each year? Compare this with the figure of 4370 mm recorded. What is the percent error involved in your model?

Facts from *Vegetation of the Earth in Relation to Climate and the Eco-physiological Conditions* by Heinrich Walter (translated from the second German edition by Joy Wieser), Springer-Verlag, 1973, p. 39.

15. Studies conducted over a nine-year period indicate that in the alpine belt of the tropics of Venezuela, in Páramo de Mucuchíes, the number of rainy days per month varies from an average low of 4 per month in the dry season to a high of 23 per month in the wet season, half a year later.

(a) Model the number of rainy days per month using a sinusoidal function. Let $t = 0$ correspond to the driest month.

(b) On average, how many rainy days does your model predict per year? Compare this with the recorded average number of rainy days per year: 181. (Your estimate will be a bit low, because in fact the rainy season is slightly longer than the dry season.)

Source: *Vegetation of the Earth in Relation to Climate and the Eco-physiological Conditions* by Heinrich Walter (translated from the second German edition by Joy Wieser), Springer-Verlag, 1973, p. 55.

16. A wave has amplitude 3 and frequency 10. Give a possible formula for the wave. (There are infinitely many correct answers.)

17. The gravitational pull of the sun and moon on large bodies of water produces tides. Tides generally rise and fall twice every 25 hours. (The length of a cycle is 12.5 hours as opposed to 12 hours due to the moon's revolution around the earth.) The range between high and low tide varies greatly with location. On the Pacific coast of America this range can be as much as 15 feet. The Bay of Fundy in New Brunswick has an extremely dramatic range of about 45 feet.

(a) Model the tidal fluctuations on the Pacific coast using a sinusoidal function. Let $H(t)$ give the height (in feet) above and below the average level of the ocean, where t is time in hours. Let $t = 0$ correspond to high tide.

(b) Model the tidal fluctuations in the Bay of Fundy using a sinusoidal function. Use the same conventions as in part (a).

Source: *Ecology* by Robert E. Ricklefs, Nelson, 1973, p. 124.

18. The average rental price for a two-bedroom apartment in Malden was $800 in 1990 and was $1000 in 2000. The price has been increasing over the past decade. We want to model the price of a two-bedroom apartment in Malden as a function of time and use our model to predict the price in the year 2020.

Alex thinks that prices are increasing at a constant rate, so he models the price with a linear function, $L(t)$.

Jamey thinks that the percent change in price is constant, so he models the price with an exponential function, $E(t)$.

Mike, an optimist who loves trigonometry, thinks that price is a sinusoidal function of time. He thinks that $800 is an all-time low and $1000 is an all-time high. He models the price with a sine or cosine function, $T(t)$.

(a) Suppose we let $t = 0$ correspond to the year 1990 and measure time in years. Find a formula for each of the following. Accompany your formula with a sketch.

   i. $L(t)$ ii. $E(t)$ iii. $T(t)$

(b) Which model predicts the highest price for the year 2003? Which model predicts the lowest price for the year 2003?

(c) What prices will Alex, Jamey, and Mike predict for the year 2020?

(d) Alex, Jamey, and Mike are combing the newspapers for information that might lend credence to one model over the other two. Which model, the linear, exponential, or trigonometric, is best supported by each of the following statements?

   i. "Prices in Malden have been growing at an increasing rate over the past decade."

   ii. "In the early 1990s prices in Malden were increasing at an increasing rate. After 1995 the rate of increase began to drop off."

   iii. "Prices of apartments in Malden have been increasing very steadily over the past decade."

19. Determine whether or not each function is periodic. If a function is periodic, determine its period.

   (a) $f(x) = \cos |x|$ (b) $g(x) = \sin |x|$ (c) $h(x) = |\cos x|$

   (d) $j(x) = |\sin x|$ (e) $k(x) = \sin(x^2)$ (f) $l(x) = \sin^2 x$

20. Graph $f(x) = \frac{1}{\cos x}$ on $[-\pi, 2\pi]$.

21. Let $f(x) = \frac{\sin x}{x}$. This function will be quite important when we are interested in the derivative of sine and cosine.

   (a) What is the domain of $f(x)$?

   (b) Use a graphing calculator or computer to help you sketch the graph of $f(x)$.

   (c) Although $f(x)$ is undefined at $x = 0$, $\lim_{x \to 0} f(x)$ exists. What do you think this limit might be? Check out your conjecture numerically. Observe that if $\lim_{x \to 0} \frac{\sin x}{x} = L$, then for x very close to zero, $\frac{\sin x}{x} \approx L$, or, equivalently,

$$\sin x \approx Lx \quad \text{for } x \text{ close to zero.}$$

22. A typical person might have a pulse of 70 heartbeats per minute and a blood pressure reading of 120 over 80, where 120 is the high pressure and 80 is the low. Model blood pressure as a function of time using a sinusoidal function $B(t)$, where t is time in minutes.

   (a) What is the amplitude of $B(t)$?

   (b) What is the period of $B(t)$? Notice that you have been given the frequency and from this must find the period.

   (c) Write a possible formula for $B(t)$.

19.3   THE FUNCTION $f(x) = \tan x$

Definition

The function **tan x**, called the tangent function, is defined as

$$\tan x = \frac{\sin x}{\cos x}.$$

Geometric Interpretation of $\tan x$

As before, locate the point $P(x) = (u, v) = (\cos x, \sin x)$ on the unit circle by traveling a directed distance x around the circle from the starting point $(1, 0)$. Draw a line from the origin to the point $P(x)$. The slope of this line is given by $\frac{\text{rise}}{\text{run}} = \frac{v}{u} = \frac{\sin x}{\cos x}$. Thus, $\tan x$ is the slope of the line from the origin to the point $P(x)$.

Let $\overline{OP}$ denote the line segment from the origin to $P(x)$. The slope of $\overline{OP}$ is 0 when $P = (1, 0)$ and $(-1, 0)$, that is, when $x = n\pi$, where n is any integer. The slope of $\overline{OP}$ is undefined when $P = (0, 1)$ and $(0, -1)$, that is, when $x = \pi/2 + n\pi$, where n is any integer.

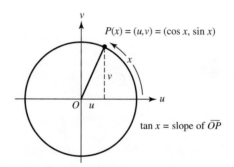

Figure 19.15

EXERCISE 19.10   Use your knowledge of the unit circle to show that the following are true.

(a) $\tan \frac{\pi}{4} = 1$ (b) $\tan \frac{3\pi}{4} = -1$ (c) $\tan 19\pi = 0$ (d) $\tan \frac{-19\pi}{2}$ is undefined

The Graph of $\tan x$

As x goes from 0 to $\pi/2$, the slope of $\overline{OP}$ increases without bound. As x increases from 0 to $\pi/4$, $\tan x$ increases from 0 to 1. As x increases from $\pi/4$ to $\pi/2$, $\tan x$ starts at 1 and increases without bound. For x in the interval $\left(\frac{\pi}{2}, \pi\right)$, the slope of $\overline{OP}$ is negative but gets less and less negative as x increases.

$$\lim_{x \to \pi/2^+} \tan x = -\infty \quad \text{and} \quad \lim_{x \to \pi/2^-} \tan x = \infty$$

We can see that $\tan x$ has no local extrema.

Figure 19.16, on the following pages, shows the graph of $\tan x$ on the interval $[0, \pi]$.

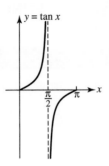

Figure 19.16

The Period of tan x

Notice that when x increases from π to $\frac{3\pi}{2}$, the line through the origin and $P(x)$ lies right on top of the lines we were looking at as x increased from 0 to $\frac{\pi}{2}$. We can conclude that the period of tan x is π. Do the exercise below to make a more rigorous argument.

EXERCISE 19.11   Show that tan x has period π by using the definition of tan x and your knowledge of sine and cosine. In other words, show that $\tan(x + \pi) = \tan x$ for all x by using the fact that $\tan(x + \pi) = \frac{\sin(x+\pi)}{\cos(x+\pi)}$.

Putting this information together gives a rough graph of tan x.

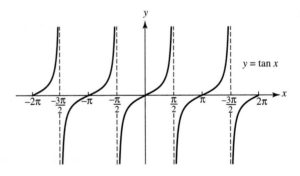

Figure 19.17

An Alternative Approach to Graphing tan x

To graph $\tan x = \frac{\sin x}{\cos x}$, we can use some of the same strategies used in graphing rational functions. If a function f is a fraction whose numerator and denominator are defined, continuous, and have no common zeros, then

■ f is zero wherever the numerator is zero;

■ f has a vertical asymptote wherever the denominator is zero;

■ f is positive wherever the numerator and denominator have the same signs, and negative wherever the signs are opposite.

The Zeros of tan x.   Since $\sin x$ and $\cos x$ have no common zeros, the zeros of $\tan x = \frac{\sin x}{\cos x}$ will occur at the zeros of $\sin x$. Thus, the zeros of $\tan x$ are at $x = n\pi$, where n is any integer. The zeros are $\ldots -3\pi, -2\pi, -\pi, 0, \pi, 2\pi, 3\pi \ldots$.

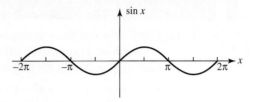

Figure 19.18

The Vertical Asymptotes of tan x. $\tan x$ is undefined where the denominator, $\cos x$, equals zero. $\tan x$ will have vertical asymptotes at $x = \frac{\pi}{2} + n\pi$, where n is any integer. In other words, the asymptotes are at $\ldots \frac{-5\pi}{2}, \frac{-3\pi}{2}, \frac{-\pi}{2}, \frac{\pi}{2}, \frac{3\pi}{2}, \frac{5\pi}{2} \ldots$.

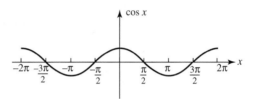

Figure 19.19

The Sign of tan x. $\tan x = \frac{\sin x}{\cos x}$, so $\tan x$ is positive where $\sin x$ and $\cos x$ have the same sign, and $\tan x$ is negative where they have opposite signs.

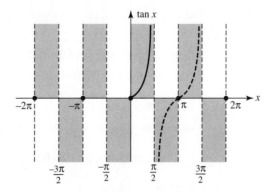

Figure 19.20

Between $x = 0$ and $x = \pi/2$, the numerator and denominator are positive. The numerator is increasing and the denominator is approaching zero, so $\tan x$ grows without bound.

$\tan(-x) = \frac{\sin(-x)}{\cos(-x)} = \frac{-\sin x}{\cos x} = -\tan x$; therefore $\tan x$ is an odd function. Thus, knowing the graph on $[0, \frac{\pi}{2})$, we can produce the graph on $(-\frac{\pi}{2}, 0]$. Now we've got an entire period graphed, so we are done.

Once we know the graph of $\tan x$, we can use our knowledge of shifts, stretches, shrinks, and flips to graph any function of the form $A \tan(Bx) + K$.

PROBLEMS FOR SECTION 19.3

1. (a) Using what you know about the properties of polynomial functions, explain how the graph of $f(x) = \sin x$ tells you that it is not a polynomial. (Think about the number of roots and the long-term behavior.)

    (b) Using what you know about the properties of rational functions, explain how the graph of $f(x) = \tan x$ tells you that it is not a rational function. (Think about the number of roots and vertical asymptotes.)

    (c) What are characteristics of trigonometric functions that distinguish them from other functions we've studied?

2. Evaluate the following limits.

   (a) $\lim_{x \to -\pi/2^+} \tan x$ (b) $\lim_{x \to -\pi/2^-} \tan x$ (c) $\lim_{x \to -\pi/2} \tan x$

3. Find equations to fit each of the periodic functions drawn below.

(a)

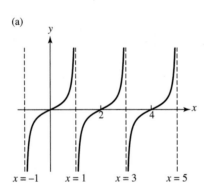

(b)

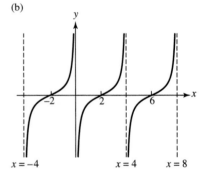

4. Suppose $\tan \alpha = b$. Find the following. Explain your reasoning.

   (a) $\tan(\alpha + \pi)$ (b) $\tan(-\alpha)$ (c) $\tan(\pi - \alpha)$

5. Graph the following.

   (a) $g(x) = -\tan x$ (b) $h(x) = |\tan x|$

6. Find all x such that $\tan x = 0$.

7. Find all x such that

   (a) $\tan x = 1$.   (b) $\tan x = -1$.

   Try to do this using the unit circle definitions.

8. Sketch the graph of $g(x) = \tan 2x$ on $[0, 2\pi]$.

9. Sketch the graph of $f(x) = 3 \tan(\frac{x}{2})$ on the interval $[-2\pi, 2\pi]$.

10. Consider the function $f(x) = \frac{\cos x}{\sin x}$.

(a) Where is f undefined?

(b) Where are the zeros of f?

(c) What is the period of f?

(d) Sketch the graph of f on the interval $[0, 2\pi]$.

11. Sketch the graph of $f(x) = \frac{1}{\sin x}$ on $[0, 2\pi]$. What is the period of f?

12. Suppose $\tan \beta = 7$.

(a) Find all x such that $\tan x = 7$.

(b) Find all x such that $\tan x = -7$.

13. Decide whether each of the following functions is even, odd, or neither.

(a) $f(x) = 1 + \cos x$ (b) $g(x) = 1 + \sin x$ (c) $h(x) = \sin 2x + \tan x$

(d) $j(x) = |\sin x|$ (e) $k(x) = \sin x + \cos x$

19.4   ANGLES AND ARC LENGTHS

The geometric interpretation of $\tan x$ as the slope of the line from the origin to $P(x)$ suggests an alternative way of thinking of the trigonometric functions. Rather than thinking of the point $P(x)$ as being determined by x, where x is a directed distance measured along the unit circle from the point $(1, 0)$, we can think of the point P as being determined by the intersection of the unit circle and a ray drawn from the origin and making a specific *angle* with the positive horizontal axis. Thus, we can think of the input of any trigonometric function as either a directed distance along the unit circle or as an angle.

How Should We Measure the Angle?

An angle can be thought of as being actively constructed by two rays (half-lines) sharing a common endpoint. Begin by having the rays coinciding. One of these rays is designated as the **initial side** of the angle. The other ray is rotated about its endpoint and in its terminal position is designated as the **terminal side** of the angle. The common endpoint of the rays is called the **vertex** of the angle. A positive angle measurement indicates that the angle was constructed by rotating the terminal side in a counterclockwise direction; a negative angle measurement indicates construction by rotating the terminal side in a clockwise direction. The Greek letter θ, pronounced "theta", is often used to refer to an angle.

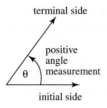

For the angle to determine a unique point P on the unit circle we place its initial side along the positive u-axis, the vertex at the origin, and let the terminal side (corresponding to $\overline{OP}$) cut the circle at point P. An angle positioned this way is said to be in **standard position**. Two angles in standard position are said to be **coterminal** if their sides coincide. The trigonometric functions of coterminal angles are identical because coterminal angles determine the same point P on the unit circle.

In ancient Babylonia, a convention was developed to measure angles in **degrees** (symbolized by °) with a full revolution corresponding to 360°, a quarter revolution (a **right angle**) corresponding to 90°, and so on. This way of measuring continues to be commonly used today.[8]

A very convenient alternative for measuring an angle is to put it in standard position and use the very same x that we've been using for the directed distance along the unit circle to measure the angle. The circumference of the unit circle is 2π units, so we use 2π to represent the angle of one complete revolution, $\pi/2$ for a quarter of a revolution, and so on. We give this angular measure the name *radians,* with a full revolution consisting of 2π radians. Many formulas are simplest when using this measure.[9] It is a good idea to put your calculator in radian mode and leave it there. Naturally, an angle need not be in standard position to be measured in radians. We give a general definition below.

Definition

An angle of x **radians** is the angle that subtends[10] an arc (a segment of a circle) of length x on the unit circle. Just as x indicates directed distance along the circle, so does it indicate direction for an angle, positive corresponding to counterclockwise revolution from initial to terminal side and negative corresponding to clockwise revolution.

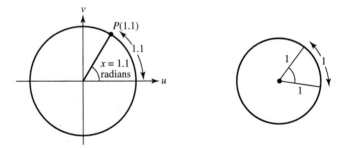

Figure 19.21

We can consider the point $P(x)$ on the unit circle as being determined by an angle of x radians measured from the positive horizontal axis or by a directed distance of x units along the unit circle from $(1, 0)$. For example, the point $(0, 1)$ is $\frac{\pi}{2}$ units along the unit circle from

---

[8] The choice of 360 may be due to there being close to 360 days in a year. There are other hypotheses as to the origin of the 360-degree set-up. For more information, see either Howard Eves's *An Introduction to the History of Mathematics, Sixth Edition,* Sauders College Publishing, 1990, p. 42, or Eli Maor's *Trigonometric Delights,* Princeton University Press, 1998, p. 15 and p. 18, footnote 2.

[9] It will make differentiation of trigonometric functions *much* cleaner later on.

[10] For an angle to subtend an arc of a circle, the vertex of the angle and the center of the circle should be made to coincide. The arc swept out from the initial ray to the terminal ray is the arc subtended by the angle.

(1, 0); the line drawn from the origin to (0, 1) makes an angle of $\frac{\pi}{2}$ radians with the positive horizontal axis.

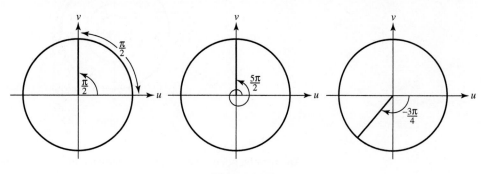

Figure 19.22

Radians are unitless. When we write sin(1.1) and want to think of 1.1 as an angle, then it is an angle in radians. In other words, sin(1.1) = sin(1.1 radians).

Converting Degrees to Radians and Radians to Degrees

You may be accustomed to measuring angles in degrees, but radians are much more convenient for calculus. We need a way to go back and forth between the two measures. The fact that one full revolution is 360°, or 2π radians, allows us to do this. Equivalently, half a revolution is 180°, or π radians.

$$180° = \pi \text{ radians} \qquad \pi \text{ radians} = 180 \text{ degrees}$$

$$1° = \frac{\pi}{180} \text{ radians} \qquad 1 \text{ radian} = \frac{180}{\pi} \text{ degrees } (\approx 57°)$$

EXERCISE 19.12   Convert from degrees to radians or radians to degrees.

(a) 45°   (b) −30°   (c) $\frac{3\pi}{2}$ radians   (d) −2 radians

Answers

(a) $\frac{\pi}{4}$ (b) $-\frac{\pi}{6}$ (c) 270°   (d) $-\frac{360}{\pi}° \approx -114.59°$

Arc Length

Arc lengths are usually described in terms of the angle that subtends the arc and the radius of the circle. We know that on a unit circle, an angle of x radians subtends an arc length of x units. What about on a circle of radius 3? The circumference of a circle of radius 3 is three times that of the unit circle, so the arc length subtended by an angle of x radians on the circle of radius 3 should be $3x$.

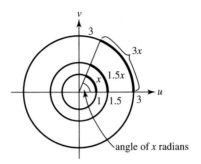

Figure 19.23

An arc subtended by an angle of θ radians on a circle of radius r will have

$$\textbf{arc length} = r\theta.$$

Notice that the formula for arc length is simple when the subtending angle is given in radian measure. The corresponding arc length formula for an angle of θ degrees is

$$r\theta° \cdot \frac{\pi}{180°} = \frac{r \cdot \theta\pi}{180}.$$

Trigonometric Functions of Angles

We can think of the input of any trigonometric function as either a directed distance along the unit circle or as an angle, because an angle in standard position will determine a point on the unit circle.

◆ **EXAMPLE 19.4**   Suppose angle θ is in standard position and the point (a, b) lies on the terminal side of θ but does not lie on the unit circle. Determine the sine and cosine of θ.

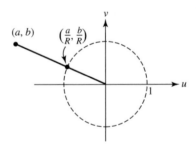

Figure 19.24

SOLUTION   Compute $\sqrt{a^2 + b^2}$ to determine the distance from point (a, b) to the origin. Let's denote this distance by R. The simplest approach is to just scale to a unit circle. If the point (a, b) lies on the terminal side of the angle, so does $(\frac{a}{R}, \frac{b}{R})$, and the latter point lies on the unit circle as well.

Therefore,

$$\cos \theta = \frac{a}{R} \quad \text{and} \quad \sin \theta = \frac{b}{R},$$

where $R = \sqrt{a^2 + b^2}$.   ◆

Tapping Circle Symmetry for Trigonometric Information

Suppose we know that $P = P(\theta) = (0.8, 0.6)$, where $\theta \in [0, 2\pi]$. Then not only do we know the sine and cosine of θ, but we know the trigonometric functions of any angle coterminal with θ, that is, $\theta \pm 2\pi n$ for n any integer. We actually know substantially more than this. The cosine and sine of θ are both positive, so θ must be an acute angle. Given the coordinates of point P, we also know the coordinates of points Q, R, and S (one in each of the remaining three quadrants), as shown in the Figure 19.25.

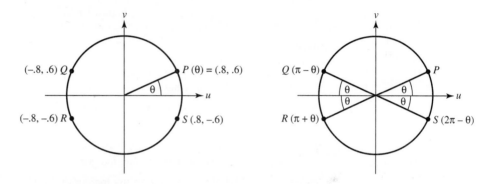

Figure 19.25

Not only do we know the trigonometric functions of θ and any angle coterminal to θ, but we also know the trigonometric functions of

$$\begin{cases} \pi - \theta \\ \pi + \theta, \quad \text{and any angle coterminal to any of these.}^{11} \\ 2\pi - \theta, \end{cases}$$

Use Figure 19.25 to verify the following.

$$\begin{array}{ll} \sin \theta = 0.6 & \cos \theta = 0.8 \\ \sin(\pi - \theta) = 0.6 & \cos(\pi - \theta) = -0.8 \\ \sin(\pi + \theta) = -0.6 & \cos(\pi + \theta) = -0.8 \\ \sin(2\pi - \theta) = -0.6 & \cos(2\pi - \theta) = 0.8 \end{array}$$

EXERCISE 19.13   Using only the symmetry of the unit circle and the information $\cos 1 \approx 0.54$ and $\sin 1 \approx 0.84$, approximate all solutions to the following equations.

(a) $\cos x = 0.54$

(b) $\cos x = -0.54$

---

[11] Actually we know more than this. By interchanging the u and v coordinates, we can obtain the trigonometric functions of $\frac{\pi}{2} - \theta$, $\frac{\pi}{2} + \theta$, $\frac{3\pi}{2} - \theta$, and $\frac{3\pi}{2} + \theta$.

(c) $\sin x = 0.84$

(d) $\sin x = -0.84$

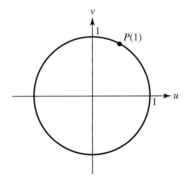

Figure 19.26

Answers

(a) $x \approx 1 + 2\pi n$, where n is any integer or
    $x \approx (2\pi - 1) + 2\pi n$, where n is any integer.

(b) $x \approx (\pi - 1) + 2\pi n$, where n is any integer or
    $x \approx (\pi + 1) + 2\pi n$, where n is any integer.

(c) $x \approx 1 + 2\pi n$, where n is any integer or
    $x \approx (\pi - 1) + 2\pi n$, where n is any integer.

(d) $x \approx (2\pi - 1) + 2\pi n$, where n is any integer or
    $x \approx (\pi + 1) + 2\pi n$, where n is any integer.

PROBLEMS FOR SECTION 19.4

1. (a) Convert the following to radians.
       (i) $60°$ (ii) $30°$ (iii) $45°$ (iv) $-120°$
   (b) Convert 2 radians to degrees.

2. Convert these angles to radian measure.

   (a) $-60°$ (b) $45°$ (c) $-270°$

   (d) $40°$ (e) $-120°$

3. Convert these angles given in radians to degrees.

   (a) $\frac{3\pi}{4}$ (b) $\frac{-3\pi}{4}$ (c) $\frac{5\pi}{6}$

   (d) $\frac{3\pi}{2}$ (e) $\frac{5\pi}{4}$ (f) -3.2 (g) 4

4. A second hand of a clock is 6 inches long.
   (a) How far does the pointer of the second hand travel in 20 seconds?
   (b) How far does the pointer of the second hand travel when the second hand travels through an angle of $70°$?

(c) In one hour the minute hand of the clock moves through an angle of 2π radians. In this amount of time, through what angle does the second hand travel? The hour hand? Give your answers in radians.

5. A bicycle gear with radius 4 inches is rotating with a frequency of 50 revolutions per minute. In 2 minutes what distance has been covered by a point on the corresponding chain?

6. A *nautical mile* is the distance along the surface of the earth subtended by an angle with vertex at the center of the earth and measuring $\frac{1}{60}^\circ$.

   (a) The radius of the earth is about 3960 miles. Use this to approximate a nautical mile. Give your answer in feet. (One mile is 5280 feet.)

   (b) The *Random House Dictionary* defines a nautical mile to be 6076 feet. Use this to get a more accurate estimate for the radius of the earth than that given in part (a).

7. The earth travels around the sun in an elliptical orbit, but the ellipse is very close to circular. The earth's distance from the sun varies between 147 million kilometers at *perihelion* (when the earth is closest to the sun) and 153 million kilometers at *aphelion* (when the earth is farthest from the sun).

   Use the following simplifying assumptions to give a rough estimate of how far the earth travels along its orbit each day.

   Simplifying assumptions:

   Model the earth's path around the sun as a circle with radius 150 million km.

   Assume that the earth completes a trip around the circle every 365 days.

8. A bicycle wheel is 26 inches in diameter. When the brakes are applied the bike wheel makes 2.2 revolutions before coming to a halt. How far has the bike traveled? (Assume the bike does not skid.)

9. A system of gears is set up as drawn.

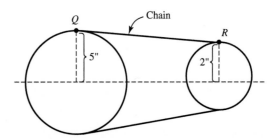

Consider the height of the point Q on the large gear. The height is measured as vertical position (in inches) with respect to the line through the center of the gears and is given as a function of time by

$$h(t) = 5\cos(\pi t),$$

where t is measured in seconds.

(a) How long does it take for Q to make a complete revolution?

(b) In 2 seconds how far does the chain travel?

(c) How many times does the large gear rotate in 2 seconds?

(d) How many times does the small gear rotate in 2 seconds? In 1 second?

(e) Write an equation for the height of point R on the small gear as a function of time (in seconds).

10. The point $(2, 3)$ lies on the terminal side of angle α when α is put in standard position. Label the coordinates of the points $P(\alpha)$; $P(-\alpha)$ and $P(\pi + \alpha)$ on the unit circle. Then find the following.

    (a) $\sin \alpha$ (b) $\cos(\alpha + 2\pi)$ (c) $\tan \alpha$

    (d) $\sin(-\alpha)$ (e) $\cos(\alpha + \pi)$ (f) $\sin(\pi - \alpha)$

    (g) $\tan(2\pi - \alpha)$

11. If $\sin \theta = \frac{5}{13}$ and $\cos \theta$ is negative, label the coordinates of the points $P(\theta)$, $P(-\theta)$ and $P(\pi - \theta)$ on the unit circle. Then find the following.

    (a) $\cos \theta$ (b) $\tan \theta$ (c) $\cos(-\theta)$

    (d) $\sin(\theta + \pi)$ (e) $\tan(\pi - \theta)$

CHAPTER 20

Trigonometry—Circles and Triangles

■ **20.1 RIGHT-TRIANGLE TRIGONOMETRY: THE DEFINITIONS**

An Historical Interlude

The word "trigonometry" has its origins in the words for "triangle" and "measurement," but as yet we have not given much reason for referring to $\sin x$, $\cos x$, and $\tan x$ as "trigonometric" functions. In ancient civilizations the development of trigonometry was spurred on by problems in astronomy, surveying, construction, and navigation, in which people were concerned with measuring sides of triangles. Astronomers from numerous civilizations spearheaded the field. On papyruses from ancient Egypt and cuneiforms from Babylonia one can find evidence of the early development of trigonometry. The Greek astronomer Hipparchus, born in Iznik, Turkey, around 190 B.C., made great strides in the development of trigonometry, calculating what were essentially trigonometric tables by inscribing a triangle in a circle.[1] These were passed down to us in the *Almagest*, a treatise written by the Greek scientist Ptolemy around 100 A.D. This work was later translated and studied by Arabic and Hindu astronomers who, in the Middle Ages, furthered the field. Astronomers in India made major contributions to trigonometry in their work; Jai Singh's observatories, such as Jantar Mantar in Jaipur, India, illustrate the remarkable degree of accuracy that had been obtained by the early 1700s.

While the approach taken in the previous chapter defines trigonometric functions in terms of the coordinates of a point on the unit circle and exploits the periodic nature of such functions, the classical development of trigonometry focused on right triangles and the ratios of the lengths of their sides. This triangle perspective is important in navigation, surveying,

---

[1] For more details, see *Trigonometric Delights* by Eli Maor, Princeton University Press, 1998, pp. 22–28.

optics, and astronomy. It is a valuable perspective whenever we analyze phenomena (such as force, velocity, or displacement) in terms of both magnitude and direction. The flexiblity of having two different viewpoints gives us added power in our use of trigonometric functions.

Definitions

Triangle trigonometry is based on the concept of similarity. **Similar triangles** have corresponding angles of equal measure and hence corresponding sides have proportional lengths; therefore the ratio of the lengths of pairs of corresponding sides of similar triangles is invariant. Consider a right triangle,[2] as drawn below. The side opposite the right angle is called the hypotenuse. The other two sides are referred to as the legs. The sum of the angles of any triangle is 180°, or π radians; therefore in a right triangle the measure of one acute angle determines the measure of the other.[3] It follows that all right triangles with a particular acute angle are similar.

Let θ be an acute angle in a right triangle.

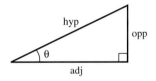

Figure 20.1

We'll denote

$$\begin{cases} \text{the length of the side of the triangle opposite } \theta \text{ by opp,} \\ \text{the length of the side of the triangle adjacent to } \theta \text{ by adj, and} \\ \text{the length of the hypotenuse by hyp.} \end{cases}$$

Trigonometry uses the fact that ratios of pairs of sides of right triangles are functions of θ. Definitions of $\sin \theta$, $\cos \theta$, and $\tan \theta$ are given below. They apply only to *right* triangles where θ is one of the acute angles.

$$\sin \theta = \frac{\text{opp}}{\text{hyp}} \qquad \cos \theta = \frac{\text{adj}}{\text{hyp}} \qquad \tan \theta = \frac{\text{opp}}{\text{adj}}$$

For any acute angle θ, these definitions are in complete agreement with the circle definitions given in the previous chapter, as illustrated in Figure 20.2.

---

[2] A right triangle is a triangle containing a right angle, an angle of $\pi/2$ radians or 90 degrees.
[3] If θ is acute, then $\theta \in (0, \frac{\pi}{2})$; if θ is obtuse, then $\theta \in (\frac{\pi}{2}, \pi)$.

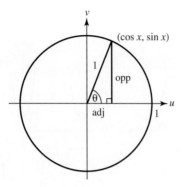

Figure 20.2

There are three more ratios that can be considered. Indeed, they have been not only considered but named. The functions cosecant, secant, and cotangent are the reciprocals of $\sin \theta$, $\cos \theta$, and $\tan \theta$, respectively.

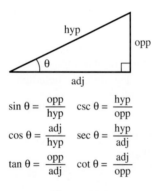

$$\sin \theta = \frac{\text{opp}}{\text{hyp}} \qquad \csc \theta = \frac{\text{hyp}}{\text{opp}}$$

$$\cos \theta = \frac{\text{adj}}{\text{hyp}} \qquad \sec \theta = \frac{\text{hyp}}{\text{adj}}$$

$$\tan \theta = \frac{\text{opp}}{\text{adj}} \qquad \cot \theta = \frac{\text{adj}}{\text{opp}}$$

Figure 20.3

We can sketch the graphs of $\csc x$, $\cos x$, and $\cot x$ by beginning with the graphs of $\sin x$, $\cos x$, and $\tan x$ and sketching the reciprocal of each function. We know about the sign of the reciprocal:

$$\frac{1}{n} \text{ is } \begin{cases} \text{positive} & \text{if } n \text{ is positive,} \\ \text{negative} & \text{if } n \text{ is negative,} \\ \text{undefined} & \text{if } n \text{ is zero.} \end{cases}$$

We know about the magnitude of the reciprocal:

$$\left| \frac{1}{n} \right| \text{ is } \begin{cases} > 1 \text{ if } |n| < 1 & \left(\text{the smaller } |n|, \text{ the larger } \left| \frac{1}{n} \right| \right), \\ < 1 \text{ if } |n| > 1 & \left(\text{the larger } |n|, \text{ the smaller } \left| \frac{1}{n} \right| \right), \\ = 1 \text{ if } |n| = 1 & \text{(that is, if } n = \pm 1). \end{cases}$$

We arrive at the graphs drawn below.

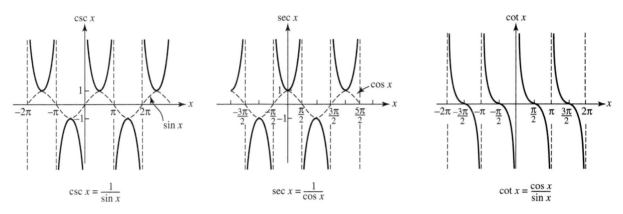

Figure 20.4

Notice that csc x and sec x have period 2π, as do their reciprocals. **The graphs of csc x and sec x hang off the graphs of sin x and cos x, attached at points where they take on a y-value of ± 1; cot x is periodic with period π, as is tan x.**

You may be wondering why trigonometric functions have names like sine and *co*sine, secant and *co*secant, tangent and *co*tangent. The "co" prefix refers to complementary angles. By definition, two angles are **complementary** if their sum is $\pi/2$. Therefore, the acute angles of any right triangle are complementary. Suppose we denote these acute angles by θ and β. The leg of the triangle that is opposite θ is adjacent to β. Therefore,

$$\sin \theta = \cos \beta, \qquad \tan \theta = \cot \beta, \qquad \sec \theta = \csc \beta.$$

In other words, the trigonometric function of θ is the "cofunction" of its complementary angle.

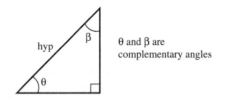

θ and β are complementary angles

Figure 20.5

EXERCISE 20.1   Show that $\sin x = \cos(\frac{\pi}{2} - x)$ for $x \in [0, \frac{\pi}{2})$. Conclude that $\sin x = \cos(x - \frac{\pi}{2})$; the graphs of sin x and cos x are horizontal translates.

The drawing that follows indicates how the secant and tangent functions got their names. The circle is a unit circle, so the tangent segment $\overline{AB}$ has length tan θ and the secant segment $\overline{OA}$ has length sec θ.[4]

---

[4] For an explanation of how the remaining trigonometric functions got their names, see Eli Maor's *Trigonometric Delights*, Princeton University Press, 1998, pp. 35–38.

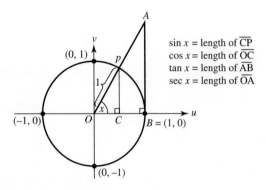

$\sin x = \text{length of } \overline{CP}$
$\cos x = \text{length of } \overline{OC}$
$\tan x = \text{length of } \overline{AB}$
$\sec x = \text{length of } \overline{OA}$

Figure 20.6

Trigonometry of Right Triangles

From times of antiquity, scientists and engineers have used triangle trigonometry to construct edifices and to estimate distances and heights—from estimating the height of the great Egyptian pyramids in Giza to estimating the sizes and distances of the sun and the moon. A triangle has three sides and three angles. To **solve a triangle** means to find measures for all three sides and all three angles from information given.

In the case of a right triangle, if we know the measure of one of the acute angles, then we know the measures of all the angles. This, together with the length of any one side, enables us to solve the triangle. If the lengths of any two particular sides of a right triangle are known, we can use the Pythagorean Theorem to determine the length of the third and trigonometry to determine the measure of the angles. Knowing the measures of all three angles does not determine the triangle, but knowing the length of all three sides does.[5] (Think about why this is so.)

Examples involving solving right triangles often include the terms "angle of elevation" and "angle of depression." **Angle of elevation** refers to the angle from the horizontal up to an object; **angle of depression** refers to the angle from the horizontal down to an object.

◆ EXAMPLE 20.1   A little girl flying a kite on a taut 350-foot string asks her father for the height of her kite. Her father estimates the angle of elevation of the kite to be 55°. Give an estimate for the height of the kite. Assume that the girl is holding the string 3 feet above the ground and her father is measuring the angle of elevation from this height.

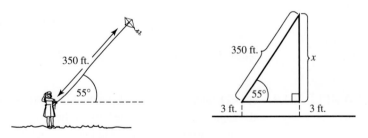

Figure 20.7

---
[5] To "determine" a triangle means that these specifications determine all three sides and all three angles.

SOLUTION   Let $x + 3 =$ the height of the kite.

We're looking for a trigonometric function that relates x and the known parts of the triangle. x is the side opposite the $55°$ angle and we know the length of the hypotenuse.

$$\sin 55° = \frac{x}{350}$$

$$\sin(55°) = \frac{x}{350}$$

$$x = 350 \cdot \sin(55°)$$

$$x \approx 286.7$$

The kite is flying at a height of approximately 290 feet.   ◆

◆ EXAMPLE 20.2   When the sun is $38°$ above the horizon, a totem pole casts a shadow of 25 feet. How tall is the totem pole?

SOLUTION   Let T be the height of the totem pole.

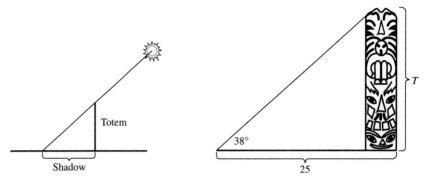

Figure 20.8

Using the diagram, we can write a trigonometric equation that relates the information we know with the information we want to find.

$$\tan 38° = \frac{T}{25}$$

Thus, $T = 25 \tan 38° \approx 19.5$ feet.   ◆

◆ EXAMPLE 20.3   A clock tower sits on top of a tall building. From a point 300 feet from the base of the building the angle of elevation to the base of the clock tower is $40°$ and the angle of elevation to the top of the tower is $49°$. How tall is the clock tower?

SOLUTION   Let $x =$ the height (in feet) of the clock tower.

Let $y =$ the height (in feet) of the building up to the base of the tower.

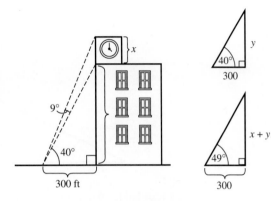

Figure 20.9

We'll use a trigonometric function that relates the lengths of the sides opposite and adjacent to the known angle.

Using the smaller triangle we can solve for y.

$$\tan 40° = \frac{y}{300}$$

$$y = 300 \tan 40°$$

Using the larger triangle we have

$$\tan 49° = \frac{x + y}{300}$$

$$x + y = 300 \tan 49°$$

$$x = 300 \tan 49° - y = 300 \tan 49° - 300 \tan 40°$$

$$x = 300(\tan 49° - \tan 40°)$$

$$x \approx 93.4.$$

The clock tower is approximately 93.4 feet tall.   ◆

PROBLEMS FOR SECTION 20.1

1. Given each triangle below, write expressions for the six trigonometric functions.

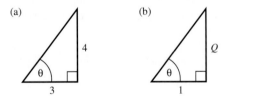
(a)   (b)

(c)

2. As illustrated on the following page, a tree casts a shadow of 20 feet when the sun's rays make an angle of 0.7 radians with the ground. How tall is the tree?

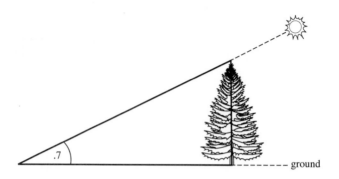

3. A 25-foot ladder is leaning against a straight wall. If the base of the ladder is 7 feet from the wall, what angle is the ladder making with the ground and how high up the wall does it go?

4. You're in an apartment looking across a 25-foot boulevard at a building across the way. The angle of depression to the foot of the building is 15° and the angle of elevation to the top of the building is 50°. How tall is the building?

5. You are standing on an overpass, 35 feet above street level, waving to a friend who is at the window of a high-rise dormitory. The angle of depression to the bottom of the dormitory (on street level) is 15°. The angle of elevation to your friend's window is 45°. What is your friend's elevation?

6. We are standing on flat ground in Monument Valley trying to estimate the height of the edifices. We have surveying equipment and take all of our measurements from a height of 5 feet. We find the angle of elevation to the top of one structure is 23°. We move 500 feet closer to the structure and find that the angle of elevation is now 29°. How tall is the structure?

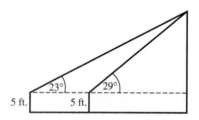

7. Graph each of the following. Be sure to display at least one full period of the function. For parts (a), (b), and (e), use what you know about graphing $\frac{1}{f(x)}$ when given the graph of $f(x)$.

(a) $y = 3 \sec x$ (b) $y = -\cot x$ (c) $y = \tan\left(\frac{x}{2}\right)$

(d) $y = -3 \tan x$ (e) $y = 2 \csc x$

8. In several ancient civilizations trigonometry was a highly developed field. This can in part be attributed to ancient astronomers. How could an ancient astronomer who could measure angles of elevation use trigonometry to estimate the height of the moon?

9. Peter is measuring the height of a church steeple. He stands on level ground 500 feet from the base of the church and determines that the angle of elevation from the ground to the base of the steeple is 23°. From the same spot he measures the angle of elevation to the highest point of the steeple and finds it is 29°.

   (a) How high is the church, from the base of the church at ground level to the tip of the steeple? Give an exact answer and then give a numerical approximation.

   (b) How high is the steeple? Give an exact answer and then give a numerical approximation.

20.2   TRIANGLES WE KNOW AND LOVE, AND THE INFORMATION THEY GIVE US

In Section 20.1 we used the ratios of the lengths of sides of a right triangle to define $\sin\theta$, $\cos\theta$, $\tan\theta$, and their reciprocals for any acute angle θ. Generally these definitions aren't computationally practical if our aim is very accurate computation of the values of trigonometric functions. We don't relish the thought of sitting down with a protractor and ruler, constructing some angle θ, drawing a right triangle, measuring the sides, and computing the ratio, do we? No! We'd much prefer to whip out a calculator and push some buttons.[6] In the absence of a calculator but in the presence of a calibrated unit circle we can use the circle to approximate the value of the trigonometric function, but the degree of accuracy we can expect is limited. There are, however, two special triangles that are very useful to us in obtaining the exact values of the trigonometric functions of the angles they contain, as well as the exact values of trigonometric functions of their close relatives. These two triangles are triangles all students of trigonometry come to know and love.

The 45°, 45° Right Triangle[7]

Draw an isosceles right triangle.[8] Its acute angles are equal; both are $\pi/4$ radians (or 45°). Let's label the legs. We can call them anything. Let's use 1 for simplicity. Using the Pythagorean Theorem we see that the hypotenuse has length $\sqrt{2}$. From the triangle we see that[9]

$$\sin(\pi/4) = \frac{1}{\sqrt{2}}, \qquad \cos(\pi/4) = \frac{1}{\sqrt{2}}, \qquad \tan(\pi/4) = 1.$$

---

[6] You may wonder "How does a calculator do it? How does it evaluate $\sin 0.1$, for instance?" This is a very good and interesting question. The short (and somewhat accurate) answer is that the calculator uses a polynomial to approximate $\sin x$ near $x = 0$ and evaluates that polynomial at $x = 0.1$. The concept is along the lines of what we do when we take tangent-line approximations, but it is done with much more sophistication. We will study this in Chapter 30.

[7] These triangles tend to be identified by their degree-angle measurements; they are easier to say aloud, and the study of these triangles is traditionally first undertaken in a trigonometry class as opposed to a calculus class. Radians show their advantage with calculus, as we will see in the next chapter.

[8] An isosceles triangle has two sides of equal length.

[9] In some places the dogma is that writing a fraction with a radical in the denominator, like $1/\sqrt{2}$, is a mortal sin. Relax. Writing $1/\sqrt{2}$ doesn't disturb us. Use $1/\sqrt{2}$ or $\sqrt{2}/2$, whichever makes you happiest.

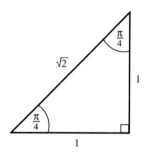

Figure 20.10

This information could also be obtained from the unit circle. $P(\pi/4)$ is the point of intersection of the line $u = v$ and the unit circle $u^2 + v^2 = 1$. Substituting $v = u$ into the equation of the circle we obtain

$$u^2 + u^2 = 1$$
$$2u^2 = 1$$
$$u^2 = \frac{1}{2}$$
$$u = \pm\frac{1}{\sqrt{2}}.$$

If $u = \frac{1}{\sqrt{2}}$ then $v = \frac{1}{\sqrt{2}}$; if $u = -\frac{1}{\sqrt{2}}$ then $v = -\frac{1}{\sqrt{2}}$.

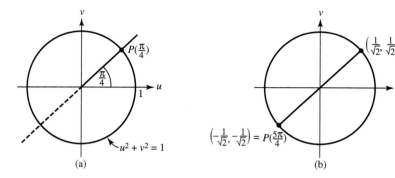

Figure 20.11

At no extra cost we obtain the coordinates of a second point on the unit circle. Using the symmetry properties of the circle we can identify the coordinates of two more. (See Figure 20.12.) This allows us to compute all the trigonometric functions for any x that takes us to one of these four points. Combined with the u- and v-intercepts, this gives us a total of eight points on the unit circle whose exact coordinates we can associate with $P(x)$. From Figure 20.12, we can read off the following information.

$$P(x) = \left(\frac{1}{\sqrt{2}}, \frac{1}{\sqrt{2}} \right) \text{ for } x = \pi/4 + 2\pi n, \qquad P(x) = \left(\frac{-1}{\sqrt{2}}, \frac{-1}{\sqrt{2}} \right) \text{ for } x = 5\pi/4 + 2\pi n,$$

$$P(x) = \left(\frac{-1}{\sqrt{2}}, \frac{1}{\sqrt{2}} \right) \text{ for } x = 3\pi/4 + 2\pi n, \qquad P(x) = \left(\frac{1}{\sqrt{2}}, \frac{-1}{\sqrt{2}} \right) \text{ for } x = 7\pi/4 + 2\pi n,$$

where n is any integer.

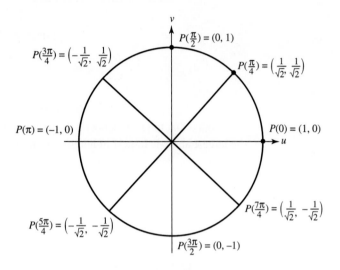

Figure 20.12

In general, if we know the coordinates of a point P on the unit circle, we can identify the coordinates of points Q, R, and S in the other three quadrants. Quadrants are labeled I, II, III, and IV, working counterclockwise from the first quadrant, where both coordinates are positive.

The 30°, 60° Right Triangle

Draw an equilateral triangle. All its sides are of equal length, so all of its angles are of equal measure; each angle is $\pi/3$ radians, or 60°. This time let's call the length of each side 2. (We'll chop one side in half right away.) Drop a perpendicular to create a right triangle with acute angles of $\pi/3$ and $\pi/6$. The perpendicular cuts one of the sides of the original equilateral triangle in two. We now have a right triangle in which the only missing piece of information is the length of the side opposite $\pi/3$. We find this using the Pythagorean Theorem.

$$1^2 + x^2 = 2^2$$
$$x^2 = 3$$
$$x = \pm\sqrt{3}$$
$$x = \sqrt{3} \qquad \text{because } x \geq 0.$$

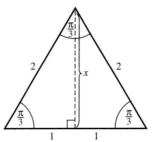
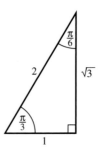

Figure 20.13

Notice that the side opposite the smallest angle is smallest in length; the side opposite the largest angle is largest in length. From the triangle we see that

$$\sin(\pi/3) = \frac{\sqrt{3}}{2} \qquad \cos(\pi/3) = \frac{1}{2} \qquad \tan(\pi/3) = \sqrt{3}$$

$$\sin(\pi/6) = \frac{1}{2} \qquad \cos(\pi/6) = \frac{\sqrt{3}}{2} \qquad \tan(\pi/6) = \frac{1}{\sqrt{3}}$$

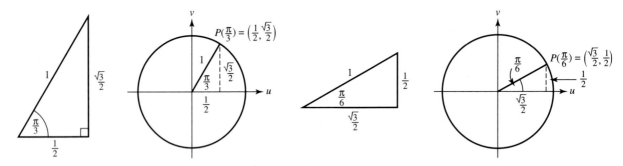

Figure 20.14

We can transport all this information to the unit circle. Scale down the triangle so its hypotenuse is 1 by cutting all sides in half. We use the symmetry of the circle to obtain the coordinates of eight points on the circle, as shown in Figure 20.15. On the diagram we have listed only x-values between 0 and 2π, but infinitely many others can be obtained by adding multiples of 2π.

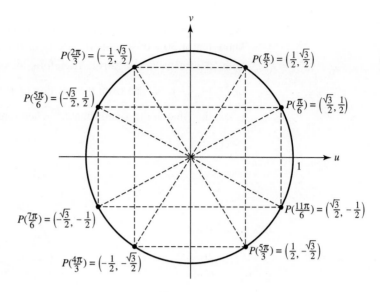

Figure 20.15

We can determine the coordinates of $P(x)$ for x any integer multiple of $\frac{\pi}{6}$, $x = \frac{n\pi}{6}$ where n is an integer. Now we have a host of angles for which we can evaluate trigonometric functions exactly.

◆ **EXAMPLE 20.4**   (a) Evaluate $\cos(-5\pi/6)$.   (b) Evaluate $\sin 3\pi/4$.

SOLUTIONS   (a) First locate $P(-5\pi/6)$ on the unit circle. We want the coordinates of $P = P(-5\pi/6)$.

Approach 1

Find the coordinates of P by using the symmetry of the circle to bounce into the first quadrant where we have an acute angle and can use the $\pi/6 - \pi/3$ right triangle.
    Then return to P.

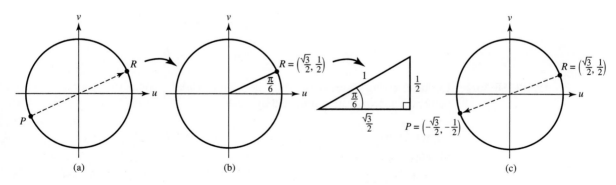

Figure 20.16

Shortcut

We're looking for the coordinates of $P(x)$ where x is an integer multiple of $\frac{\pi}{6}$. We know that a 30°, 60° right triangle is involved and, due to the position of P, that both coordinates are negative. Therefore, the coordinates of P are either $\left(\frac{-1}{2}, \frac{-\sqrt{3}}{2}\right)$ or $\left(\frac{-\sqrt{3}}{2}, \frac{-1}{2}\right)$. We see from the position of P that the magnitude of the v-coordinate is less than that of the u-coordinate.[10] Therefore, the coordinates of P are $\left(\frac{-\sqrt{3}}{2}, \frac{-1}{2}\right)$. $\cos\left(\frac{-5\pi}{6}\right) = \frac{-\sqrt{3}}{2}$.

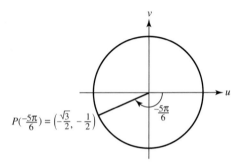

Figure 20.17

(b) To evaluate $\sin(3\pi/4)$, first locate $P(3\pi/4)$ on the unit circle, then find its coordinates. Due to its position, we know the coordinates of $P(3\pi/4)$ are (negative, positive). We know that a 45°, 45° right triangle is involved, so the coordinates are $\left(\frac{-1}{\sqrt{2}}, \frac{1}{\sqrt{2}}\right)$. Therefore, $\sin\left(\frac{3\pi}{4}\right) = \frac{1}{\sqrt{2}}$.

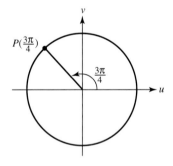

Figure 20.18   ◆

EXERCISE 20.2   Evaluate, giving an exact answer. Check your work by comparing your answer with the numerical approximation provided by a calculator.

(a) $\sin(3\pi/4)$ (b) $\sin(-3\pi/4)$ (c) $\tan(7\pi/6)$

(d) $\cos(-\pi/3)$ (e) $\tan(-\pi/4)$

---

[10] We know that $\sqrt{3} > 1$.

◆ **EXAMPLE 20.5**   You're interested in knowing the height of a very tall tree. You position yourself so that your line of sight to the top of the tree makes a 60° angle with the horizontal. You measure the distance from where you stand to the base of the tree to be 45 feet. How tall is the tree? (Assume that your eyes are five feet above the ground.)

SOLUTION   We'll begin with a sketch. It's simplest to call the height of the tree $x + 5$ feet and find x.

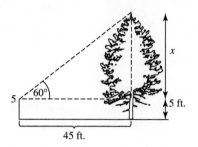

Figure 20.19

$$\tan(\pi/3) = \frac{x}{45}$$

$$\sqrt{3} = \frac{x}{45}$$

$$x = 45\sqrt{3}$$

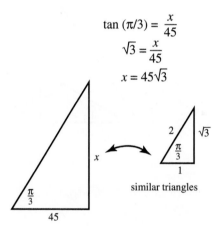

similar triangles

Figure 20.20

The tree is $(45\sqrt{3} + 5)$ feet tall, or approximately 82.9 feet tall.   ◆

◆ **EXAMPLE 20.6**

(a) Find all x such that $\tan x = -1$.

(b) Find all x such that $\sin x = 1/2$.

(c) Find all x on the interval $[0, 2\pi]$ such that $\cos x = -1/2$.

SOLUTIONS   (a) There are several approaches to solving $\tan x = -1$. One approach is to use the unit circle and interpret $\tan x$ as the slope of $\overline{OP}$. There are two points on the unit circle for which the slope of $\overline{OP}$ is -1. They are the points of intersection of the line $v = -u$ and the unit circle. The points correspond to $x = \frac{3\pi}{4}$ and $x = -\frac{\pi}{4}$. Therefore, $x = \frac{3\pi}{4} + 2\pi n$ or $x = -\frac{\pi}{4} + 2\pi n$, where n is any integer.

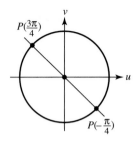

Figure 20.21

(b) We begin by looking for one x-value that will satisfy $\sin x = 1/2$. In our heads dance visions of triangles we know and love. We draw one of our all-time favorites and see that $x = \frac{\pi}{6}$ fits the bill.

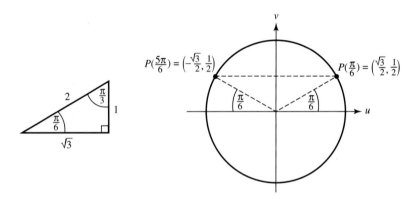

Figure 20.22

Draw $P\left(\frac{\pi}{6}\right)$ on the unit circle. Now use the unit circle to find all points P with v-coordinate $1/2$. Using symmetry we see that

$$P_1 = P(\pi/6) = \left(\frac{\sqrt{3}}{2}, \frac{1}{2}\right) \quad \text{and} \quad P_2 = P(5\pi/6) = \left(\frac{-\sqrt{3}}{2}, \frac{1}{2}\right).$$

Therefore, if $\sin x = 1/2$, then

$$x = \pi/6 + 2\pi n \quad \text{or} \quad x = 5\pi/6 + 2\pi n,$$

where n is an integer.

(c) We begin by looking for one x such that $\cos x = -1/2$. In fact, to begin with let's not even worry about the negative sign. For what x is $\cos x = 1/2$? Again, don't turn to strangers, turn to a triangle you know and love.

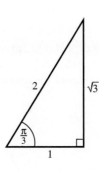 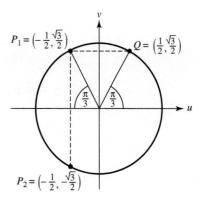

Figure 20.23

We see that $\cos(\pi/3) = 1/2$. Let's put that on the unit circle. We label it Q, because it is not what we're looking for. By symmetry considerations we can find two points, P_1 and P_2 , with u-coordinate of $-1/2$. What are the values of x between 0 and 2π that correspond to points P_1 and P_2? The angle between $\overline{OQ}$ and the positive u-axis is $\pi/3$; the angle between $\overline{OP}_1$ and the negative u-axis is $\pi/3$ as well. Therefore, $P_1 = P(\pi - \pi/3) = P(2\pi/3)$.

We have to go halfway around the unit circle (π radians) to get from Q to P_2. Therefore, $P_2 = P(\pi + \pi/3) = P(4\pi/3)$.

$$x = 2\pi/3 \text{ and } 4\pi/3 \quad \blacklozenge$$

Suppose we were interested in finding some x such that $\cos x = 0.3$. In this case our favorite triangles are useless. We do have a few tools at our disposal to approximate x; we could approximate x using a calibrated unit circle or we could graph $y = \cos x$ and $y = 0.3$ on a graphing calculator and approximate a point of intersection. It would be much more convenient, however, if we could just "undo" cosine; if we have the inverse function for $\cos x$ at our disposal, we can express a solution exactly. We turn to inverse trigonometric functions in Section 20.3.

PROBLEMS FOR SECTION 20.2

1. Find the following exactly.
   (a) $\sin(\pi/6)$ (b) $\sin(\pi/4)$ (c) $\sin(\pi/3)$ (d) $\sin(-\pi/3)$ (e) $\cos(\pi/3)$
   (f) $\cos(-\pi/3)$ (g) $\tan(\pi/4)$ (h) $\tan(3\pi/4)$ (i) $\tan(5\pi/3)$ (j) $\sin(-7\pi/6)$

2. (a) For what values of x is $\tan x = \sqrt{3}$?
   (b) For what values of x is $\tan(x) = -\sqrt{3}$?

3. Find the *exact* values of the following.
   (a) $\sin(2\pi/3)$ (b) $\cos(5\pi/4)$ (c) $\tan(7\pi/4)$ (d) $\sec(\pi/6)$
   (e) $\cot(-\pi/6)$ (f) $\csc(4\pi/3)$ (g) $\sin(801\pi)$ (h) $\cos(39\pi/4)$

4. (a) Find an acute angle x such that $\sin x = 0.5$.

   (b) Find all x between 0 and 2π such that $\sin x = 0.5$.

   (c) Find an acute angle x such that $\cos x = 0.5$.

   (d) Find all $x \in [0, 2\pi]$ such that $\cos(2x) = 0.5$.

5. Find the lengths of all the sides of the triangle drawn below.

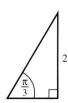

6. Fill in the following table.

θ in degrees	θ in radians	$\sin \theta$	$\cos \theta$	$\tan \theta$
30°				
45°				
60°				

7. Find exact values for each of the following. (No calculator—or use it only to check your answers.)

   (a) $\cos(\pi/4)$ (b) $\cos(5\pi/4)$ (c) $\cos(-3\pi/4)$

   (d) $\sin(5\pi/6)$ (e) $\sin(-13\pi/6)$ (f) $\cos(-2\pi/3)$

8. A cross-section of a feed trough is shown below. Find the area of the cross-section. Give an exact answer and then a numerical approximation.

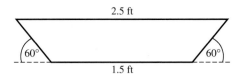

9. Some kids are sitting on their stoop wondering about the height of a tall street post. They estimate that the street post is casting a shadow 15 feet in length and that the angle of elevation of the sun (from the ground) is about 60°. Estimate the height of the street post.

For Problems 10 through 14, rewrite each of the following expressions in terms of a positive acute angle. This positive acute angle is sometimes referred to as a reference angle.

10. (a) $\sin(-48°)$ (b) $\cos(-48°)$

11. (a) $\sin 92°$ (b) $\sin(-92°)$

12. (a) cos 130°   (b) cos(−130°)

13. (a) cos 200°   (b) sin(200°)

14. (a) tan 200°   (b) tan(−200°)

20.3   INVERSE TRIGONOMETRIC FUNCTIONS

◆ **EXAMPLE 20.7**   A New York City lawyer is working on the case of an elderly woman who twisted her ankle while chasing a runaway shopping cart down a ramp outside a Manhattan grocery store. The curb is 6.5 inches high and the ground from the bottom of the curb to the end of the ramp measures 50 inches. New York City law specifies that ramps must have a slope of no more than 5 degrees. Was the construction of this ramp in accordance with the law?[11]

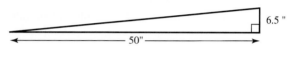

6.5 "

50"

Figure 20.24

We need to find the acute angle x such that $\tan x = \frac{6.5}{50}$, or $\tan x = 0.13$. To solve for x exactly we need to "undo" the tangent function. We can solve for x easily once we define inverse trigonometric functions. We'll do that and then return to the lawyer's case.   ◆

Inverse Functions Revisited

Given a function f, its inverse function, f^{-1}, undoes f. Recall the following.

■ f has an inverse function only if f is 1-to-1; that is, each output value of the function is taken on only once. If a function is not 1-to-1, we can restrict its domain so that it becomes 1-to-1.

■   If f is 1-to-1, then f^{-1} assigns to each output of f the unique associated input. We look for a formula for the inverse function by writing $y = f(x)$, interchanging x and y (input and output) and solving for y.

■   The graph of $f^{-1}(x)$ can be obtained by interchanging the x- and y-coordinates for every point on the graph of f. This is equivalent to reflecting the graph of $y = f(x)$ around the line $y = x$.

We want to define inverse functions for $\sin x$, $\cos x$, and $\tan x$. Our first obstacle is that none of these functions is 1-to-1. In fact, because they are all periodic, each output value in the range corresponds to an infinite number of input values. We must restrict the domain for each of them. We do so as follows.

---

[11] This is a true story, facts, measurements, and all. My brother-in-law, the lawyer, called with the question.

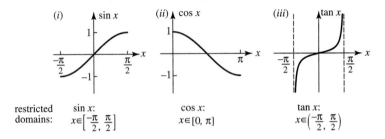

Figure 20.25

In every case we include the interval from 0 to $\pi/2$. For sine and cosine we use a continuous half-cycle; for $\tan x$ we use an entire continuous cycle. Each of these restricted domains includes all possible output values of the function. Think about the restricted domains given; they're fairly natural ones.

Once we have restricted the domains, we can obtain the inverse functions. Let's start with $y = \sin x$. Interchanging the roles of x and y gives $x = \sin y$. What is y? y is the angle between $-\pi/2$ and $\pi/2$ whose sine is x. We call the inverse function **inverse sine**, or **arcsine**, and denote it by $\sin^{-1} x$ or **arcsin** x. Notice that arcsine gives as its output an angle or a real number corresponding to a directed distance along the unit circle. In radians the measure of the angle and the corresponding arc on the unit circle are identical.

Recall the convention that $\sin^{-1} x$ refers to the inverse sine function, *not* to $(\sin x)^{-1}$. When discussing the reciprocal of $\sin x$ it is safest to write $\frac{1}{\sin x}$ or $\csc x$.

We construct the definitions of the other inverse trigonometric functions, arccosine and arctangent, similarly.

domain of sine $[-\pi/2, \pi/2]$ range of arcsine	$\sin x$ $\longrightarrow$ $\longleftarrow$ $\arcsin x$	range of sine $[-1, 1]$ domain of arcsine	domain of tangent $(-\pi/2, \pi/2)$ range of arctangent	$\tan x$ $\longrightarrow$ $\longleftarrow$ $\arctan x$	range of tangent $(-\infty, \infty)$ domain of arctangent

The Inverse Trigonometric Functions

Definition

$\sin^{-1} x$ is the angle between $-\pi/2$ and $\pi/2$ whose sine is x.   domain of $\sin^{-1} x$: $[-1, 1]$

$\tan^{-1} x$ is the angle between $-\pi/2$ and $\pi/2$ whose tangent is x.   domain of $\tan^{-1} x$: $(-\infty, \infty)$

$\cos^{-1} x$ is the angle between 0 and π whose cosine is x.   domain of $\cos^{-1} x$: $[-1, 1]$

Below are the graphs of the inverse trigonometric functions.

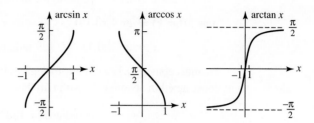

Figure 20.26

Observations

- $\sin^{-1}(\sin x) = x$ if $x \in [-\pi/2, \pi/2]$.

  If x is not in $[-\pi/2, \pi/2]$ this statement won't hold. For example, $\sin^{-1}(\sin(2\pi)) = \sin^{-1}(0) \neq 2\pi$. The *output* of arcsine is always an angle between $-\pi/2$ and $\pi/2$.
  $\sin(\sin^{-1} x) = x$ for all x in the domain of $\sin^{-1} x$.
  $\tan^{-1}(\tan x) = x$ if $x \in (-\pi/2, \pi/2)$.
  $\tan(\tan^{-1} x) = x$ for all x (the domain of $\tan^{-1} x$ is $(-\infty, \infty)$).
  $\cos^{-1}(\cos x) = x$ if $x \in [0, \pi]$.
  $\cos(\cos^{-1} x) = x$ for all x in the domain of $\cos^{-1} x$.

- On most calculators the arcsine and sine functions share a key, the arccosine and cosine functions share a key, and the arctangent and tangent functions share a key.[12]

- If A is positive, then

  $\arcsin(A)$, $\arccos(A)$, and $\arctan(A)$ are all angles between 0 and $\pi/2$ (provided A is in the function's domain), and

  $\arcsin(-A) = -\arcsin(A)$, $\arccos(-A) = \pi - \arccos(A)$, $\arctan(-A) = -\arctan(A)$.

  You can convince yourself of these last three statements by looking at a unit circle together with the appropriate domain restrictions.

- While it is possible to define inverses of $\sec x$, $\csc x$, and $\cot x$, we will not do so. After all, the equation $\sec x = A/B$ is equivalent to $\cos x = B/A$.

Example 20.7 (continued)

Let's return to Example 20.7, the problem of the grocery-store ramp and the runaway cart in New York City. We were looking for the acute angle whose tangent is $\frac{6.5}{50}$, or 0.13. We want to solve the equation $\tan x = 0.13$. We can think about this in either of the following ways:

x is the acute angle whose tangent is 0.13, so $x = \tan^{-1}(0.13) \approx 0.129$ radians,

---

[12] Just as you might wonder how a calculator computes sin 1, you may wonder how your calculator arrives at arcsin(0.3). Again, it is a good and interesting question. Again, the short and somewhat accurate answer is that it uses polynomial approximations of arcsin x.

or

$$\tan x = 0.13. \quad \text{To undo the tangent, take the arctangent of both sides.}^{13}$$

$$\tan^{-1}(\tan x) = \tan^{-1}(0.13)$$

$$x = \tan^{-1} 0.13 \approx 0.129 \text{ radians}$$

We must convert 0.129 radians to degrees in order to determine whether or not the ramp was constructed in accordance with City law.

$$\pi \text{ radians } = 180°$$

$$0.129 \text{ radians} = 0.129(180°/\pi) \approx 7.4°$$

The ramp was not constructed in accordance with the law since 7.4° is more than the allowed 5°.

◆ **EXAMPLE 20.8**   Simplify the following. Both a and b are positive constants.

(a) $\cos(\arcsin(0.2))$ (b) $\tan\left(\cos^{-1}\left(-\frac{a}{b}\right)\right)$ (c) $\sin(\cos^{-1}(-a))$

SOLUTION   (a) $\cos(\arcsin(0.2))$ is the cosine of the angle $\theta \in \left[\frac{-\pi}{2}, \frac{\pi}{2}\right]$ whose sine is 0.2.
Draw θ in a right triangle, as shown in Figure 20.27. We know $\theta \in \left(0, \frac{\pi}{2}\right)$ because $0.2 > 0$. $\cos \theta = \frac{x}{1}$; solve for x.

$$x^2 + (0.2)^2 = 1^2$$

$$x = \sqrt{1 - (0.2)^2} = \sqrt{0.96}$$

$$\cos \theta = \frac{\sqrt{0.96}}{1} = \sqrt{0.96}$$

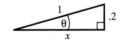

Figure 20.27

(b) $\tan\left(\cos^{-1}\left(\frac{-a}{b}\right)\right)$ is the tangent of the angle $\theta \in [0, \pi]$ whose cosine is $-\frac{a}{b}$.
We know $\theta \in \left[\frac{\pi}{2}, \pi\right]$ because $\cos \theta$ is negative. Therefore $\tan \theta$ is negative as well. See Figure 20.28.

$$\tan \theta = \frac{x}{-a} = \frac{\sqrt{b^2 - a^2}}{-a} = \frac{-\sqrt{b^2 - a^2}}{a}$$

---

[13] Don't even dream of getting rid of tan by dividing by tan. This would be mathematical blasphemy! tan x is NOT (tan) times x.

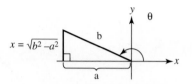

Figure 20.28

(c) $\sin(\cos^{-1}(-a))$ is the sine of the angle $\theta \in [0, \pi]$ whose cosine is $-a$.
We know $\theta \in \left[\frac{\pi}{2}, \pi\right]$ because $\cos\theta$ is negative . Therefore $\sin\theta$ is positive.

$$\sin^2\theta + \cos^2\theta = 1$$
$$\sin^2\theta + a^2 = 1$$
$$\sin\theta = \sqrt{1 - a^2}$$

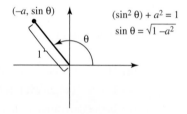

Figure 20.29

PROBLEMS FOR SECTION 20.3

1. Evaluate each of the following expressions exactly. Do not give numerical approximations.

   (a) $\sin^{-1}(1)$ (b) $\tan^{-1}(1)$ (c) $\sin^{-1}(-1)$

   (d) $\cos^{-1}(-1)$ (e) $\sin^{-1}(-0.5)$ (f) $\cos^{-1}(-0.5)$

   (g) $\tan^{-1}(\sqrt{3})$ (h) $\left[\cos^{-1}(-1)\right]^2 + \left[\tan^{-1}(-1)\right]^2$

2. Use the calibrated unit circle below to approximate the following. In each case, explain in words how you used the calibrated unit circle to make your estimate.

   (a) $\sin^{-1}(0.8)$ (b) $\cos^{-1}(0.6)$ (c) $\tan^{-1}(2)$

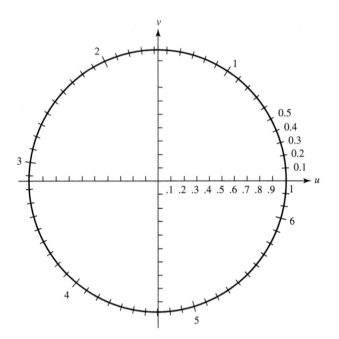

3. The function $f(x) = \tan x$ has an inverse function when its domain is restricted to $(-\pi/2, \pi/2)$.

   (a) Graph $y = \tan^{-1}(x)$. Where is the derivative positive? Negative?

   (b) Is $\tan^{-1}(x)$ an even function, an odd function, or neither?

   (c) Is the derivative of $\tan^{-1}(x)$ even, odd, or neither?

4. (a) Let $h(x) = \sin(\sin^{-1}(x))$. What is the domain of h? Is $h(x) = x$ for every x in its domain? If not, for what x is $\sin(\sin^{-1}(x)) \neq x$?

   (b) Let $j(x) = \sin^{-1}(\sin(x))$. What is the domain of j? For which values of x is $j(x) = x$? For which values of x is $j(x) \neq x$? (*Caution:* There are infinitely many values of x for which $j(x) \neq x$. Be sure to identify them all.)

5. On the same set of axes sketch the graphs of $f(x)$ and $f^{-1}(x)$.

   (a) $f(x) = \sin x$, where $x \in [-\pi/2, \pi/2]$.

   (b) $f(x) = \cos x$, where $x \in [0, \pi]$.

6. Find the angle between $\pi/2$ and π whose sine is

   (a) 0.5.

   (b) 0.2. (Give an exact answer and then a numerical approximation.)

7. Simplify the following.

   (a) $\sin\left(\arctan \frac{3}{4}\right)$

   (b) $\tan(\cos^{-1}(0.5))$

   (c) $\cos(\sin^{-1} x), \quad x < 0$

   (d) $\tan\left(\sin^{-1}\left(\frac{w}{r}\right)\right), \quad w, r > 0$

8. Let $f(x) = \cos x$ and $g(x) = \arctan x$. Find the following, where a and b are positive constants. Your answers should be exact and as simple as possible.

   (a) $g(f(\pi))$

   (b) $f\left(g\left(\frac{-a}{b}\right)\right)$

In Problems 9 through 11, simplify the expressions given that $x \in \left[0, \frac{\pi}{2}\right]$.

9. (a) $\sin^{-1}(\sin x)$ (b) $\cos^{-1}(\cos x)$

10. (a) $\sin^{-1}(\sin(-x))$ (b) $\cos^{-1}(\cos(-x))$

11. (a) $\tan^{-1}(\tan(x))$ (b) $\tan^{-1}(\tan(-x))$

In Problems 12 through 14, simplify the expressions given that $x \in \left(\frac{\pi}{2}, 2\pi\right)$.

12. (a) $\arcsin(\sin x)$ (b) $\arccos(\cos x)$

13. (a) $\arcsin(\sin(-x))$ (b) $\arccos(\cos(-x))$

14. (a) $\arctan(\tan(x))$ (b) $\arctan(\tan(-x))$

◼ 20.4   SOLVING TRIGONOMETRIC EQUATIONS

A trigonometric equation is an equation in which the variable to be solved for is the argument of a trigonometric function. If we can get the trigonometric equation into the form $\sin u = k$, $\cos u = k$, or $\tan u = k$, where k is a constant, then we can use the inverse trigonometric functions to help solve for u.

◆ **EXAMPLE 20.9**   Solve for x.

$$4\cos x + 1 = -1$$

SOLUTION

$$4\cos x + 1 = -1$$
$$4\cos x = -2$$
$$\cos x = -\frac{1}{2}$$

One solution to this equation is $\cos^{-1}\left(-\frac{1}{2}\right)$. But there are actually two solutions for $x \in [0, 2\pi]$ and infinitely many solutions due to the periodicity of the cosine function. For this problem we can turn to a triangle we know and love. We know $\cos(\pi/3) = \frac{1}{2}$. To have a negative cosine the angle must be in the second or third quadrant. So, the solutions are

$$x = \frac{2\pi}{3} + 2\pi n \quad \text{or} \quad x = \frac{4\pi}{3} + 2\pi n,$$

where n is any integer. See Figure 20.30 on the following page for illustration.

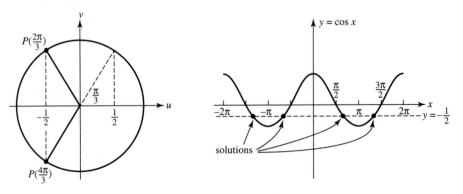

Figure 20.30   ◆

◆ **EXAMPLE 20.10**   Solve for θ.

$$2 \sin^2 \theta - \sin \theta = 1$$

SOLUTION   This is a quadratic equation in $\sin \theta$. We can either work with the $\sin \theta$ or we can begin the problem with the substitution $u = \sin \theta$, solve for u, and then return to $\sin \theta$. We'll take this latter approach.

$$2u^2 - u = 1$$
$$2u^2 - u - 1 = 0$$
$$(u - 1)(2u + 1) = 0$$
$$u - 1 = 0 \quad \text{or} \quad 2u + 1 = 0$$
$$u = 1 \quad \text{or} \quad u = -\frac{1}{2}$$
$$\sin \theta = 1 \quad \text{or} \quad \sin \theta = -\frac{1}{2}$$

So

$$\theta = \frac{\pi}{2} + 2\pi n \quad \text{or} \quad \theta = -\frac{\pi}{6} + 2\pi n \quad \text{or} \quad \theta = -\frac{5\pi}{6} + 2\pi n,$$

where n is an integer.

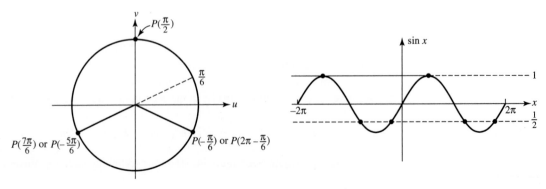

Figure 20.31   ◆

◆ EXAMPLE 20.11   Solve for x.

$$2 \cos^2 x + \sin x = 1$$

SOLUTION   We begin by converting the cosines into sines so we have a quadratic in $\sin x$.

$$2 \cos^2 x + \sin x = 1$$
$$2(1 - \sin^2 x) + \sin x = 1$$
$$2 - 2 \sin^2 x + \sin x - 1 = 0$$
$$-2 \sin^2 x + \sin x + 1 = 0$$
$$2 \sin^2 x - \sin x - 1 = 0$$

This is the equation we began with in the previous example, so the solutions are the same. ◆

◆ EXAMPLE 20.12   Find all x in the interval $[-2\pi, 2\pi]$ such that

$$\tan x \cdot \sec x = \tan x.$$

SOLUTION   CAUTION Don't divide both sides of the equation by $\tan x$ and leave yourself with only $\sec x = 1$. The equation is true if $\tan x = 0$. Lobbing off the $\tan x$ means you'll miss a couple of solutions.

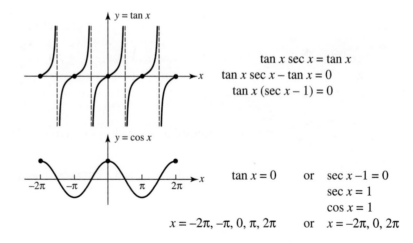

$$\tan x \sec x = \tan x$$
$$\tan x \sec x - \tan x = 0$$
$$\tan x (\sec x - 1) = 0$$

$\tan x = 0$ or $\sec x - 1 = 0$
                $\sec x = 1$
                $\cos x = 1$
$x = -2\pi, -\pi, 0, \pi, 2\pi$ or $x = -2\pi, 0, 2\pi$

Figure 20.32

Putting this all together, we get

$$x = -2\pi, \;\; -\pi, \;\; 0, \;\; \pi, \;\; 2\pi. \quad ◆$$

◆ EXAMPLE 20.13   Find all $x \in [0, 2\pi]$ such that $\cos(2x) = 0.3$. Give exact answers and then give numerical approximations.

SOLUTION   Let's begin by letting $w = 2x$. Then we can solve $\cos w = 0.3$.

We want $\;\; 0 \leq x \leq 2\pi$,

so $\;\;\;\; 2 \cdot 0 \leq 2x \leq 2 \cdot 2\pi$

and $\;\;\;\; 0 \leq w \leq 4\pi$.

We can get *one* solution to $\cos w = 0.3$ by using inverse cosine.

$$w = \arccos(0.3) \approx 1.266$$

From Figure 20.33 we see that there is a second point, Q, on the unit circle with a u-coordinate of 0.3. Let's locate Q with an angle between 0 and 2π.

$$Q = 2\pi - \arccos(0.3)$$

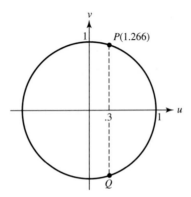

Figure 20.33

We want all $w \in [0, 4\pi]$ such that $\cos w = 0.3$.

$$w = \arccos(0.3), \quad \text{and} \quad w = \arccos(0.3) + 2\pi, \quad \begin{pmatrix} \text{corresponding} \\ \text{to point } P \end{pmatrix}$$

$$w = 2\pi - \arccos(0.3), \quad \text{and} \quad w = 2\pi - \arccos(0.3) + 2\pi. \quad \begin{pmatrix} \text{corresponding} \\ \text{to point } Q \end{pmatrix}$$

$x = \frac{w}{2}$, so

$$x = \frac{1}{2}\arccos(0.3), \qquad x = \frac{1}{2}\arccos(0.3) + \pi,$$

$$x = \pi - \frac{1}{2}\arccos(0.3), \qquad x = \pi - \frac{1}{2}\arccos(0.3) + \pi = 2\pi - \frac{1}{2}\arccos(0.3).$$

Numerical approximations give us

$$x \approx 0.63 \quad \text{or} \quad 3.77 \quad \text{or} \quad 2.51 \text{ to } 5.65.$$

A sketch of $y = \cos(2x)$ and $y = 0.3$ on $[0, 2\pi]$ supports our solutions (and assures us that we have found *all* the solutions requested).

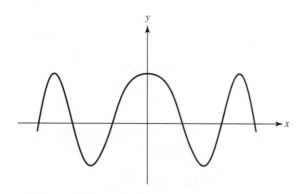

Figure 20.34   ◆

PROBLEMS FOR SECTION 20.4

1. Find all x between 0 and 2π such that
   (a) $4\cos^2 x = 3$.
   (b) $2\,\sin^2 x - \sin x - 1 = 0$.   (*Hint:* this is a quadratic in sin x.)
   (c) $\sin x = 2/3$. (Give exact answers, as well as numerical approximations.)

2. Find all solutions.
   (a) $\sin^2 x = 0.25$.
   (b) $\cos^2 x + 2\cos x + 1 = 0$. (*Hint:* Let $u = \cos x$ and first find u.)
   (c) $\cos^2 x + 4\cos x + 3 = 0$.

3. Find all solutions to the following equations:
   (a) $\sec^2 x = 2$.
   (b) $\cos^2 x = 0.2\cos x$. Why can't you cancel the cos x from both sides of the equation?
   (c) $\sin^2 x = 3\cos x + 1$.

4. Below is the graph of $f(x) = \cos(x^2)$ on $[-\pi, \pi]$. Find the exact coordinates of the x-intercepts.

For each of the equations in Problems 5 through 13, find all solutions in the interval
$[0, 2\pi]$. Give exact answers (as opposed to numerical approximations).

Use a graph to check that you have found all *solutions in this interval. (Check* $f(x) = 0.5$ *on* $[0, 2\pi]$ *by graphing* $y = f(x)$ *and* $y = 0.5$ *on* $[0, 2\pi]$ *and looking for points of intersection or by graphing* $y = f(x) - 0.5$ *on* $[0, 2\pi]$ *and looking for zeros.)*

5. $\sec x = -1$

6. $\cos(2x) = 1$

7. $2\sin^2(2x) = 1$

8. $3\tan^2 x - 1 = 0$

9. $\cos^2 x + 4\sin x = 4$

10. $\cos(3x) = 0.5$

11. $2\sin(3x) = -1$

12. $\tan^2 t - 4\tan t = 5$

13. $\cos x = 1 + \sin x$
    (*Hint:* Square both sides so you'll be able to convert cosines to sines. But be sure to check all your answers, because squaring both sides of an equation can introduce extraneous roots.)

14. Determine how many solutions each of the following equations has, and approximate the solutions. These equations involve both trigonometric and algebraic functions and therefore cannot be solved exactly using analytic methods. Instead, take a graphical approach.
    (a) $\sin x - x = 0$
    (b) $2\cos x = x$
    (c) $\sin x - \frac{x}{3} = 0$

15. Find all $t \in [0, \pi]$ such that $4\sec^2(2t) - 3 = 0$.

16. Find all $t \in [-\pi, \pi]$ such that $2\sin^2 t - 3\sin t + 1 = 0$.

17. Find all $x \in [0, 2\pi]$ such that

$$\cos 3x = -\frac{1}{\sqrt{2}}.$$

Give exact answers. Verify graphically that you have given the correct number of solutions.

18. Find all solutions to

$$5\cos(x) = 6\cos^3(x) - \sin(x)\cos(x)$$

in the interval $[0, 2\pi]$.

19. If x is a solution to $\cos x \sin x \cos(2x) \sin(2x) = 0$, find *all* possible values of $\cos(x)$.

20. Find all values of x in the interval $[0, 2\pi]$ such that

$$\sin(4x) = \frac{1}{\sqrt{2}}.$$

Explain how you know that you have all the values requested.

21. In the Alfred Hitchcock film "Rear Window," Jimmy Stewart plays a man who watches his neighbor's apartment through a camera with a telephoto lens because he believes that the man has committed a murder. Suppose that the neighbor's window is across a 50-foot courtyard and is at a height 10 feet below the location of Jimmy Stewart's camera as placed on the tripod.

At what angle should Jimmy Stewart aim his camera in order to see into his neighbor's window? Give an exact answer and a degree measurement correct to one decimal place.

22. A population of deer in a forest displays regular fluctuations in size. Scientists have chosen to model the population size with a sinusoidal function. At its height the deer population is 7000, while at its low it is 2000. The time between highs and lows is 6 months. The population at time $t = 0$ is 4500 and decreasing.

    (a) Model the deer population as a function of time t in months. A picture should accompany your answer.

    (b) If $t = 0$ is now, what is the deer population 3 months from now?

    (c) When is the first time in the future that the deer population will reach 3000? Give an exact answer and then a numerical approximation.

    (d) Call your answer to part (c) t_*. Give *any* one time *other* than your answer to part (c) at which the deer population is also 3000. (There are infinitely many correct answers.) Give an exact answer in terms of t_*.

20.5   APPLYING TRIGONOMETRY TO A GENERAL TRIANGLE: THE LAW OF COSINES AND THE LAW OF SINES

We began our study of triangle trigonometry by observing that if the measure of one acute angle of a right triangle is determined, then all the angles are determined and the ratio of the lengths of the sides are determined. This is the basis of trigonometry. We are able to apply trigonometry so that if we are given enough information to geometrically determine a triangle we can solve the triangle; that is, we can find the measure of all sides and all angles.

Suppose we are dealing with a triangle that is *not* a right triangle. Such a triangle is referred to as an **oblique triangle**. There are various configurations of information that will determine the triangle. The question we will look at in this section is how trigonometry can aid us in solving a generic triangle. Oblique triangles arise in many contexts, such as surveying, astronomy, and navigation.

In order to determine a generic triangle we need to know the length of at least one side of the triangle and we need at least two other measurements. This means we need one of the following four configurations of information.

1. Three sides (SSS)

2. Two sides and the included angle (SAS)

3. One side and any two angles (SAA)

4. Two sides and the angle opposite one of them (SSA)

In the first three cases, a unique triangle is determined by the information given. In the last case, the case of knowing the lengths of two sides and the angle opposite one of them, the triangle may not be uniquely determined. There may be a unique triangle, or two different triangles, or even no triangles corresponding to the specifications. Due to the ambiguities, we'll refer to this last case as the "ambiguous case."

There are two laws, the Law of Cosines and the Law of Sines, that allow us to export our knowledge from right triangles to oblique triangles. We will first state them, then give particular scenarios in which each can be applied, then prove them and discuss general applications of the laws.

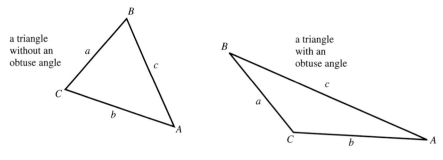

Figure 20.35

We'll denote the angles of the generic triangle by uppercase A, B, and C and denote by lowercase a, b, and c the lengths of the sides opposite A, B, and C, respectively.

The Law of Cosines

The Law of Cosines generalizes the Pythagorean Theorem; it says

$$c^2 = a^2 + b^2 - 2ab \cos C.$$

We'll do a reality check to make sure that this makes sense. We can think of the Law of Cosines as the Pythagorean Theorem with an adjustment term, $-2ab \cos C$, tagged on. First suppose that angle C is a right angle. Then the Law of Cosines just boils down to the Pythagorean Theorem, because $\cos C = 0$ if C is a right angle. Now, keeping the lengths of a and b fixed, let angle C change. As angle C decreases the length of c decreases. On the other hand, as angle C increases c increases as well.

If angle $C = \pi/2$,   then $\cos C = 0$ so the adjustment term, $-2ab \cos C$, is 0; we've got the Pythagorean Theorem.

If angle $C \in (0, \pi/2)$,   then $\cos C$ is positive, making the adjustment term, $-2ab \cos C$, negative. The smaller C, the more negative the adjustment term. This matches our intuition.

If angle $C \in (\pi/2, \pi)$,   then $\cos C$ is negative, making the adjustment term $-2ab \cos C$, positive. The larger C, the larger the adjustment term. Again, this matches our intuition.

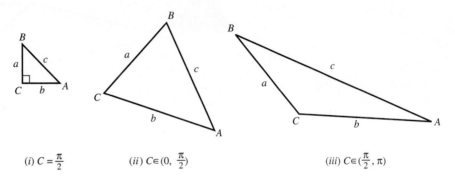

$(i)\ C = \dfrac{\pi}{2}$ $(ii)\ C \in (0, \dfrac{\pi}{2})$ $(iii)\ C \in (\dfrac{\pi}{2}, \pi)$

Figure 20.36

The Law of Sines

$$\frac{\sin A}{a} = \frac{\sin B}{b} = \frac{\sin C}{c}$$

We can rewrite $\frac{\sin A}{a} = \frac{\sin B}{b}$ as $\frac{\sin A}{\sin B} = \frac{a}{b}$. Certainly in the case of a right triangle this makes sense, because if C is a right angle then

$$\frac{\sin A}{\sin B} = \frac{\frac{a}{c}}{\frac{b}{c}} = \frac{a}{b}.$$

Why this is true in a generic triangle will be clarified in the proof.

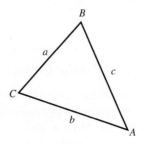

Figure 20.37

Scenario I

A ship is at sea and is heading toward port A. The captain knows the distance from his ship to port A, the distance from his ship to port B, and the distance between the two ports. Due to weather conditions he wants to change course and head to port B. Through what angle must the ship turn?

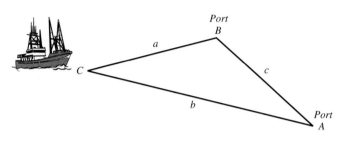

Figure 20.38

Geometrically, we are looking at a triangle, the two ports and the ship representing the vertices A, B, and C respectively. Let's call the three angles A, B, and C and denote by a, b, and c, respectively, the lengths of the sides opposite these angles. The captain knows a, b, and c. This is enough information to determine the triangle, that is, to determine all three of the angles. Since these angles are determined, we ought to be able to use our knowledge of trigonometry to find out what they are. The Law of Cosines allows us to do this. It tells us that $c^2 = a^2 + b^2 - 2ab \cos C$. We can solve for $\cos C$. Using arccosine we can find angle C. The range of arccosine is $[0, \pi]$ and angle C must be between 0 and π, so we will have solved the captain's problem.

EXERCISE 20.3   Suppose that in Scenario I given above the captain has the following data. His ship is 60 miles from port A and 40 miles from port B. Ports A and B are 30 miles apart. Through what angle must the ship turn?

ANSWER $\arccos\left(\frac{43}{48}\right) \approx 0.46$ radians or $\approx 26.38°$

Proof of the Law of Cosines

We know that the Law of Cosines, $c^2 = a^2 + b^2 - 2ab \cos C$, is true for $C = \pi/2$, so we need only look at the two cases that follow, the case where C is an acute angle and the case where C is an obtuse angle.

Place triangle ABC so that angle C is in standard position with its vertex at the origin and side b is lying along the positive x-axis as shown on the following page.

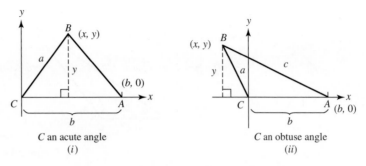

Figure 20.39

The coordinates of C are $(0, 0)$, the coordinates of A are $(b, 0)$, and we'll denote the coordinates of B by (x, y). Check that regardless of whether we use part (i) or part (ii) of Figure 20.39, we have

$$\cos C = \frac{x}{a}, \quad \text{or, equivalently,} \quad x = a \cos C$$

$$\sin C = \frac{y}{a}, \quad \text{or, equivalently,} \quad y = a \sin C.$$

The distance between points A and B is c.

$c =$ the distance between A and B

$$= \sqrt{(x - b)^2 + (y - 0)^2} \quad \text{(Using the formula for the distance between points)}$$

$$c = \sqrt{(a \cos C - b)^2 + (a \sin C - 0)^2}$$

Squaring both sides gives

$$c^2 = (a \cos C - b)^2 + (a \sin C)^2$$
$$c^2 = a^2 \cos^2 C - 2ab \cos C + b^2 + a^2 \sin^2 C$$
$$c^2 = a^2(\cos^2 C + \sin^2 C) - 2ab \cos C + b^2$$
$$c^2 = a^2 + b^2 - 2ab \cos C.$$

We have proven the Law of Cosines.

When Might We Use the Law of Cosines?

- When we know the lengths of all the sides of a triangle, we can use the Law of Cosines to find any angle (SSS).

- When we know the lengths of two sides and the angle between them, we can use the Law of Cosines to find the length of the third side of the triangle (SAS).

Scenario II

An astronomer sights a planet in her telescope. She measures the angle of elevation of her line of sight. She travels 5 miles and repeats the procedure. Can she determine how far away the planet is using the information she has gathered? Figure 20.40, on the following page, is not drawn to scale (and we are ignoring the curvature of the earth).

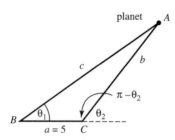

θ_1 and θ_2 are the angles of elevation of her line of sight from the two locations.

Figure 20.40

From her measurements of θ_1 and θ_2 (see Figure 20.40) she can determine two angles (and hence the third as well) of the triangle drawn. Knowing the length of the side between them determines the triangle. The Law of Cosines is not useful in this case, but the Law of Sines will be. The Law of Sines says $\frac{\sin A}{a} = \frac{\sin B}{b} = \frac{\sin C}{c}$. We can find A, B, and C and we know a, so we use $\frac{\sin A}{a} = \frac{\sin C}{c}$. This allows us to solve for c.

Before proving the Law of Sines we will look at an area formula for a triangle. It is interesting in its own right, and the Law of Sines follows from it with alacrity.

Obtuse Angles and the Area of an Oblique Triangle

Obtuse Angles

In Section 19.4 we looked at trigonometric functions of any angle by placing the angle in standard position. If the point (x, y) lies on the terminal side of the angle, we scaled to a unit circle, obtaining the point $\left(\frac{x}{\sqrt{x^2+y^2}}, \frac{y}{\sqrt{x^2+y^2}} \right)$, and from this we could read off the cosine and sine values. While this is completely adequate, it can be convenient, when we have an obtuse angle in the context of a triangle, not to have to refer back to a coordinate system. Suppose C is an obtuse angle in triangle ABC as shown in Figure 20.41.

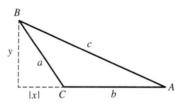
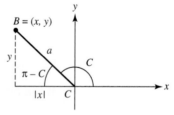

If C is obtuse, $(\pi - C)$ is acute

Figure 20.41

We see that $a = \sqrt{x^2 + y^2}$, so

$$\sin C = \frac{y}{a} = \sin(\pi - C) \quad \text{and} \quad \cos C = \frac{x}{a} = -\cos(\pi - C).$$

These are in fact identities, true for *any* value of C.

$$\sin x = \sin(\pi - x)$$

$$\cos x = -\cos(\pi - x)$$

EXERCISE 20.4   Verify the identities $\sin x = \sin(\pi - x)$ and $\cos x = -\cos(\pi - x)$ using either the unit circle or what you know about shifting and flipping the graphs of sine and cosine.

Area of a Triangle

The area of any triangle is given by

$$\text{area} = \frac{1}{2} \text{ base} \cdot \text{height}.$$

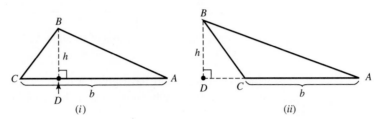

Figure 20.42

EXERCISE 20.5   Refer to Figure 20.42 above and verify that the formula area $= \frac{1}{2}b \cdot h$ for an oblique triangle can be derived using areas of right triangles. In Figure 20.42 part (i) verify that

$$\text{area of ABC} = \text{area of ABD} + \text{area of BCD} = \frac{1}{2}b \cdot h.$$

For Figure 20.42 part (ii) verify that

$$\text{area of ABC} = \text{area of BCD} - \text{area of ABD} = \frac{1}{2}b \cdot h.$$

Suppose that we know the measures of two sides of a triangle and the included angle. Let's say we know the measure of sides a and b and angle C. Place triangle ABC so that angle C is in standard position with its vertex at the origin and side b is lying along the positive x-axis as shown below.

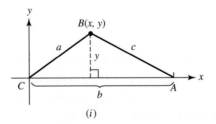

Figure 20.43

Denote the coordinates of B by (x, y).

$$\sin C = \frac{y}{a}, \quad \text{or, equivalently,} \quad y = a \sin C.$$

But $y =$ the height of the triangle. Therefore,

$$\text{area} = \frac{1}{2}b \cdot h$$

$$= \frac{1}{2}b \cdot a \sin C$$

$$\text{area} = \frac{1}{2}ab \sin C.$$

The area of a triangle with sides of length a and b and included angle C is given by

$$\boxed{\text{area of triangle ABC} = \frac{1}{2}ab \sin C.}$$

Notice that we could have given a coordinate-free proof using the result that $\sin x = \sin(\pi - x)$. We simply chose to keep the proof in line with that of the Law of Cosines.[14]

Proof of the Law of Sines

To prove the Law of Sines,

$$\frac{\sin A}{a} = \frac{\sin B}{b} = \frac{\sin C}{c},$$

we refer to Figure 20.43. Using the formula for the area of a triangle, we see that the area of triangle ABC is given by $\frac{1}{2}ab \sin C$. It is also given by $\frac{1}{2}ac \sin B$ and $\frac{1}{2}bc \sin A$. Therefore,

$$\frac{1}{2}ab \sin C = \frac{1}{2}ac \sin B = \frac{1}{2}bc \sin A.$$

Multiplying by 2 gives

$$ab \sin C = ac \sin B = bc \sin A,$$

and dividing by abc gives

$$\frac{\sin C}{c} = \frac{\sin B}{b} = \frac{\sin A}{a}.$$

When Might We Use the Law of Sines?

- When we know two angles of a triangle and the length of one side, we can use the Law of Sines to find the length of the other sides (AAS or ASA).

- When we know the lengths of two sides of the triangle and know the angle opposite one of these sides, we can try to find the length of the other side and other angles (SSA). Recall that this latter case is the ambiguous case.

Figure 20.44 illustrates why the ambiguous case has this name. Sides a and c are of given lengths and angle C is fixed. In cases (b) and (d) the triangle is completely determined. In case (c) there are two possible options for the triangle, and in case (a) the triangle is impossible to construct.

---

[14] The Law of Cosines can be proven in a coordinate-free manner.

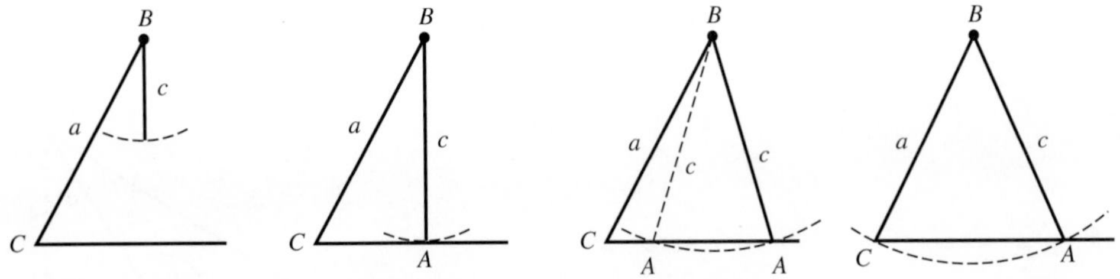

(a) c is too short
to make a triangle

(b) c is just long
enough for a right triangle

(c)   two possibilities

(d) $c \geq a$
only one possibility

The Ambiguous Case: two sides and the angle opposite one of them are known.
Measures of side a and c and angle C are known.

Figure 20.44

How is the ambiguity reflected in the mathematics when we apply the Law of Sines to the side-side-angle-opposite situation?

In the case in which the triangle is impossible to construct (like case a) we end up with an equation with no solution, such as $\sin A = 1.4$.

In the case in which the triangle is completely determined or there are two possible triangles, we will end up with an equation such as

$$\sin A = \frac{1}{\sqrt{3}}.$$

If we simply say $A = \arcsin(\frac{1}{\sqrt{3}})$, we end up with $A = \frac{\pi}{3} = 60°$. But we know that A could be $60°$ *or* $120°$. We must try both options. Either $A = 120°$ is a possibility or it is not.

In other words, we know that $\sin A = \frac{a}{c} \sin C$, but A is *not* necessarily $\arcsin\left(\frac{a}{c} \sin C\right)$. The problem is that $\left(\frac{a}{c} \sin C\right)$ is always positive, so $\arcsin\left(\frac{a}{c} \sin C\right)$ will give an angle between 0 and $\pi/2$, yet the angle in the triangle in question could be obtuse. We must consider this possibility independently. Try θ and $(\pi - \theta)$.

This complication did not arise when we used the Law of Cosines to solve for an angle because the cosine of an obtuse angle is negative.

PROBLEMS FOR SECTION 20.5

1. Find the area of the triangles below.

(a)
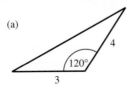

(b)
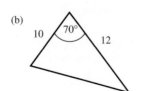

(c)
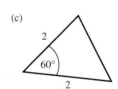

2. Find $\sin \theta$, $\cos \theta$, and $\tan \theta$.

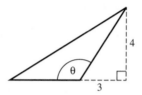

3. What methods would you employ to solve each of the triangles given.

(a)    (b) 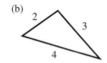   (c) 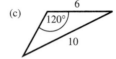   (d)

4. Solve the triangles in the previous problem. Label the lengths of the unidentified sides and the measures of the unidentified angles. Round off your answers to the nearest tenth.

5. After a hurricane a tree is left standing but makes an angle of 10° with its former upright position. Suppose the tree is tilting away from the sun and casting a shadow 25 feet long. If the angle of elevation of the sun is 30°, how long is the tree?

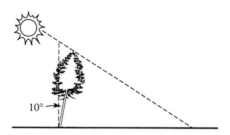

6. Three football players, Dante, Anthony, and Cliff, are on a playing field. There are 20 yards between Dante and Anthony, 30 yards between Dante and Cliff, and 40 yards between Anthony and Cliff. Dante fakes a pass to Anthony but throws the ball instead to Cliff. Through what angle must he have turned if he was first facing Anthony and is now facing Cliff?

7. A surveyor is interested in finding the distance across a lake from point P to point Q. He takes the measurements indicated in the figure on the following page. Find the distance from P to Q.

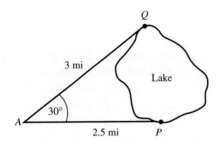

8. In a certain region land is valued at $25/square foot. What is the value of a triangular piece of property whose sides measure 90 feet, 140 feet, and 120 feet?

9. A boat is sailing up the Nile at a steady pace. The river is straight and the boat maintains a distance of 60 feet from the shoreline. At a given instant the boat's bearing to a temple on the shore is N 25° E (see picture). Fifteen minutes later the bearing is N 35° E. How fast is the boat going? Note: N 25° E means 25° East of due N.

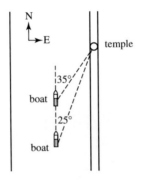
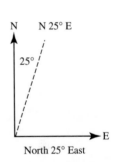

10. When designing the one-third-of-a-mile-long Georgia World Congress Center, the building that housed nearly one-fifth of the events of the 1996 Olympics, engineers had to take into account the curvature of the earth (*Sports Illustrated*, August 5, 1996). Assuming a constant curvature of the earth, how many feet would it curve in one-third of a mile? In other words, assume a cross-section of the earth is a perfect circle and draw a tangent line to the curve of this circle at one end of the building. How far away would the tangent line be from the circle itself at the other end of the building? (Use 3960 miles as the radius of the earth.)

20.6   TRIGONOMETRIC IDENTITIES

The six trigonometric functions are intimately related to one another. These relationships allow us to transform certain trigonometric expressions into equivalent expressions that may be easier for us to handle. Knowing some basic trigonometric identities can be useful when working with trigonometric expressions and solving trigonometric equations. Recall that an identity is an equation that is true for *all* values of x for which it makes sense. Some of these identities, like the Pythagorean identity discussed in Section 19.1, you should know off the top of your head. Others, which can be interpreted as horizontal shifts of trigonometric functions, can be quickly gleaned from the graphs of trigonometric functions. Other relationships are more complex. As a minimum you need to know of their existence and where you can look them up.

Among the more complex relationships are the addition formulas. It is crucial to realize that $\sin(A + B) \neq \sin A + \sin B$. The following example illustrates this.

◆ **EXAMPLE 20.14**   Show that $\sin(A + B) \neq \sin A + \sin B$.

SOLUTION   Any one counterexample will suffice to show that $\sin(A + B)$ is not equal to $\sin A + \sin B$. We could let A and B be almost anything. Let $A = \frac{\pi}{3}$ and $B = \frac{\pi}{3}$.

$$\sin\left(\frac{\pi}{3} + \frac{\pi}{3}\right) \stackrel{?}{=} \sin\left(\frac{\pi}{3}\right) + \sin\left(\frac{\pi}{3}\right)$$

$$\sin\left(\frac{2\pi}{3}\right) \stackrel{?}{=} \frac{\sqrt{3}}{2} + \frac{\sqrt{3}}{2}$$

$$\frac{\sqrt{3}}{2} \neq \sqrt{3}$$

so $$\sin(A + B) \neq \sin A + \sin B.$$

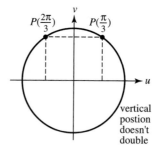

Figure 20.45   ◆

We can verify that $\sin 2A \neq 2 \sin A$ intuitively. Think about a Ferris wheel. If you ride twice the distance, does your vertical position double? No! Does your horizontal position double if you ride twice the distance? Again, the answer is no.

In fact, it can be shown that

$$\sin(A + B) = \sin A \cos B + \sin B \cos A,$$
$$\cos(A + B) = \cos A \cos B - \sin A \sin B.$$

These are called the **addition formulas** for sine and cosine. From the addition formulas we can derive subtraction formulas as well as important double-angle formulas and power-reducing formulas. Work carefully through Exercises 20.6–20.9 below.

EXERCISE 20.6   Use the addition formulas and the symmetry properties of sine and cosine to verify the following subtraction formulas.

$$\sin(A - B) = \sin A \cos B - \sin B \cos A$$
$$\cos(A - B) = \cos A \cos B + \sin A \sin B$$

EXERCISE 20.7   From the addition formulas, derive the following double-angle formulas.

$\qquad$ (a) $\sin 2x = 2 \sin x \cos x$

$\qquad$ (b) $\cos 2x = \cos^2 x - \sin^2 x = 2 \cos^2 x - 1 = 1 - 2 \sin^2 x$

EXERCISE 20.8   From the formula for $\cos 2x$, derive the following power-reducing formulas.

$\qquad$ (a) $\cos^2 x = \frac{1}{2}(1 + \cos 2x)$

$\qquad$ (b) $\sin^2 x = \frac{1}{2}(1 - \cos 2x)$

EXERCISE 20.9   From the Pythagorean identity $\sin^2 x + \cos^2 x = 1$ derive the following identities.

$\qquad$ (a) $\tan^2 x + 1 = \sec^2 x$

$\qquad$ (b) $1 + \cot^2 x = \csc^2 x$

Summary of Important Identities

Definitions	$\tan x = \frac{\sin x}{\cos x}$	$\cot x = \frac{\cos x}{\sin x}$
	$\sec x = \frac{1}{\cos x}$	$\csc x = \frac{1}{\sin x}$
Pythagorean	$\sin^2 x + \cos^2 x = 1 \qquad$ (20.1)	
identities	$\tan^2 x + 1 = \sec^2 x$	Divide both sides of (20.1) by $\cos^2 x$.
	$1 + \cot^2 x = \csc^2 x$	Divide both sides of (20.1) by $\sin^2 x$.
Symmetry-based	$\sin(-x) = -\sin x$	Sine is odd.
	$\cos(-x) = \cos x$	Cosine is even.
Addition/Subtraction	$\sin(A \pm B) = \sin A \cos B \pm \sin B \cos A$	
formulas	$\cos(A \pm B) = \cos A \cos B \mp \sin A \sin B$	
Double-angle	$\sin 2x = 2 \sin x \cos x$	Deduced from addition formulas above
formulas	$\cos 2x = \cos^2 x - \sin^2 x \qquad$ (20.2)	Now substitute $(1 - \cos^2 x)$ for $\sin^2 x$.
	$\cos 2x = 2 \cos^2 x - 1 \qquad$ (20.3)	
Power-reducing	$\cos^2 x = \frac{1}{2}(1 + \cos 2x)$	Solve (20.3) for $\cos^2 x$.
identities	$\sin^2 x = \frac{1}{2}(1 - \cos 2x)$	Substitute $(1 - \sin^2 x)$ for $\cos^2 x$ in (20.2) and solve for $\sin^2 x$.
Periodicity-based	$\sin x = \sin(x + k \cdot 2\pi)$, k any integer	Sine and cosine have period 2π.
	$\tan x = \tan(x + k\pi)$, k any integer	Tangent has period π.
Horizontal shift	Looking at the graphs of sine and cosine and knowing that they are horizontal translates can provide you with any of these.	
Cofunction identities	For complementary angles A and B	
	$\sin A = \cos B$	
	$\tan A = \cot B$	
	$\sec A = \csc B$	

There are also half-angle formulas for $\sin(x/2)$ and $\cos(x/2)$ and product identities for $\sin A \cos B$, $\sin A \sin B$, and the like. You can look these up in a reference book if you need them. Those listed above are enough for our purposes.

PROBLEMS FOR SECTION 20.6

1. True or false: If the equation is not always true, give a counterexample.

     (a) $\sin(A - B) = \sin(A) - \sin(B)$ (b) $\cos(A + B) = \cos A + \cos B$

2. Compute the following *exactly*. Do not use calculator approximations.

     (a) $\cos\left(\frac{\pi}{12}\right) = \cos\left(\frac{\pi}{4} - \frac{\pi}{6}\right)$ (b) $\sin\left(\frac{-\pi}{12}\right) = \sin\left(\frac{\pi}{6} - \frac{\pi}{4}\right)$ (c) $\tan\left(\frac{\pi}{12}\right)$

3. Using the addition formulas and what you know about even and odd trigonometric functions, find expressions for $\sin(x - 2y)$ and $\cos(x - 2y)$ in terms of $\cos x$, $\cos y$, $\sin x$ and $\sin y$.

4. Match with each of the expressions on the right with the equivalent expression on the left.

     (a) $\cos^2 x$ (i) $1 - \cos^2 x$

     (b) $\sin^2 x$ (ii) $1 - \sin^2 x$

     (c) $\tan^2 x$ (iii) $\frac{1}{2} - \frac{1}{2}\cos 2x$

     (d) $\sin x$ (iv) $-1 + \sec^2 x$

     (e) $\cos x$ (v) $1 - \sin^2(x + \pi/2)$

     (f) $-\cos x$ (vi) $\sin(x - \pi/2)$

                             (vii) $\frac{1}{2} + \frac{1}{2}\cos 2x$

                             (viii) $\cos(x - \pi/2)$

5. Use the addition formulas for $\sin x$ and $\cos x$ to derive the addition and subtraction formulas for $\tan x$.

$$\tan(A + B) = \frac{\tan A + \tan B}{1 - \tan A \tan B}$$

$$\tan(A - B) = \frac{\tan A - \tan B}{1 + \tan A \tan B}$$

6. Use the addition formula for $\tan(A + B)$ to show that

$$\tan 2x = \frac{2\tan x}{1 - \tan^2 x}.$$

7. Solve.

     (a) $\cos^2 x - \cos 2x = 0$ $x \in (-\infty, \infty)$

     (b) $\sin x \cos x = \sqrt{3}$ $x \in [0, 2\pi]$

     (c) $\sin^2 x - \cos 2x = 0$ $x \in [0, 2\pi]$

     (d) $-\cos^2 x + \frac{1}{2}\sin x + 1 = 0$ $x \in [0, 2\pi]$

8. Write a formula for $\cos 3x$ entirely in terms of sums and powers of $\cos x$.

9. Use the power-reducing formulas for $\sin^2 x$ and $\cos^2 x$ to show that

$$\tan^2 x = \frac{1 - \cos 2x}{1 + \cos 2x}.$$

20.7   A BRIEF INTRODUCTION TO VECTORS

Questions of displacement and velocity have figured prominently in our studies. When we talk about displacement, we ask

"how far?" and "in what direction?"

When we talk about velocity, we ask

"how fast?" and "in what direction?"

We've been answering the question "in what direction?" by using a plus or a minus sign. This has restricted the type of motion we describe. Our objects are going up or down; our boats are headed east or west; our cars travel on long straight highways; our trains are either inbound or outbound. If we wander along a meandering path, we can report our velocity only with reference to some benchmark point on the path; we're either heading toward it or away from it.

We can free ourselves from this world of dichotomy by using an arrow to indicate direction. In fact, we can have the arrow answer *both* the amount and direction questions by having the length of the arrow indicate "how far" or "how fast" and the direction the arrow is pointing indicate the direction of the displacement or motion. An arrow used in this way is called a vector.

Defining Vectors

Definition

A **vector** has two defining characteristics:

$$\begin{cases} \text{length (or magnitude) and} \\ \text{direction.} \end{cases}$$

A vector is often denoted by $\vec{v}$, a boldface letter with an arrow above it, and represented by an arrow. The length of $\vec{v}$ is denoted by $|\vec{v}|$.

Vectors can be used in a large variety of situations; for example, vectors can be used to model velocity, force, and displacement. A velocity vector is a vector whose length represents speed and whose direction indicates the direction of motion. A force vector is a vector whose magnitude corresponds to the magnitude of the force and whose direction indicates the direction of application of the force. We can give a similar interpretation for a displacement vector. We refer to the *head* of the vector (the tip of the arrow) and the *tail* of the vector (the other end). A vector represented by an arrow whose tail is at $A = (a_1, a_2)$ and whose head is at $B = (b_1, b_2)$ can be denoted by $\overrightarrow{AB}$.

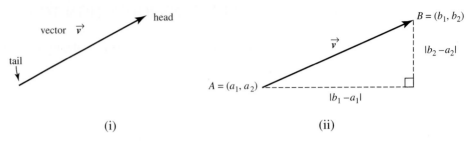

(i)   (ii)

Figure 20.46

Using the Pythagorean Theorem we calculate the length of $\overrightarrow{AB}$ as

$$\sqrt{(b_2 - a_2)^2 + (b_1 - a_1)^2}.$$

Two vectors, $\vec{u}$ and $\vec{v}$ are **equivalent** if they have the same magnitude and the same direction. For example, all the vectors drawn in Figure 20.47 are equivalent.

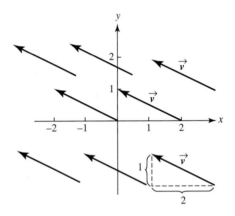

Figure 20.47

The simplest way of determining that these vectors are all equivalent is to notice that each is constructed (from tail to head) by a displacement of -2 units in the horizontal direction followed by a displacement of $+1$ unit in the vertical direction. We say that each of these vectors has a **horizontal component** of -2 and a **vertical component** of $+1$. The horizontal and vertical components of a vector together completely determine the vector. They determine its direction, and its magnitude is given by

$$\sqrt{(\text{horizontal component})^2 + (\text{vertical component})^2}.$$

Horizontal and Vertical Components of Vectors

Consider the vector $\overrightarrow{OP}$, where $O = (0, 0)$ and P is a point on the unit circle. The length of $\overrightarrow{OP}$ is 1. Let's suppose that $\overrightarrow{OP}$ makes an angle of θ with the positive x-axis, where $\theta \in [0, 2\pi]$ and θ is swept out counterclockwise from the positive x-axis. The coordinates of P are $(\cos \theta, \sin \theta)$, where $\cos \theta$ is the horizontal component of $\overrightarrow{OP}$ and $\sin \theta$ is the vertical component of $\overrightarrow{OP}$.

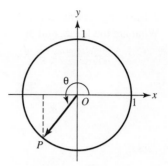

Figure 20.48

Given *any* vector $\vec{v}$ of length L we can position $\vec{v}$ so that its tail is at the origin. Its head lies somewhere on a circle of radius L centered at the origin, as shown in Figure 20.49. Let θ be the angle $\vec{v}$ makes with the positive x-axis, where $\theta \in [0, 2\pi]$.

$$\text{The horizontal component of } \vec{v} = L \cos \theta, \text{ and}$$

$$\text{the vertical component of } \vec{v} = L \sin \theta.$$

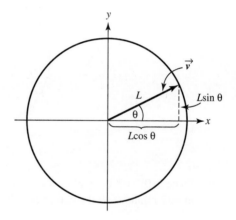

Figure 20.49

If the tail of a vector is at the origin, the horizontal component is the x-coordinate of the head and the vertical component is the y-coordinate of the head.

EXERCISE 20.10 D is the displacement vector for a turtle. Its horizontal component is $+3$ and its vertical component is -4.

(a) How far has the turtle moved?

(b) If the turtle started at $(1, 3)$, where did it end up?

ANSWERS

(a) 5 units   (b) $(4, -1)$

EXERCISE 20.11 $\vec{v}$ is the velocity vector at a certain instant for an object in motion. At this instant the object is traveling southwest at 50 miles per hour. Let due east correspond to the positive x-axis. What are the horizontal and vertical components of the velocity?

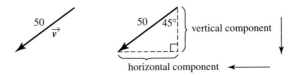

Figure 20.50

EXAMPLE 20.15   A wheel 60 inches in diameter is oriented vertically and spinning steadily at a rate of 1 revolution every $\pi/2$ minutes. There's a piece of gum stuck on the rim. Let $\overrightarrow{G(t)}$ be the position vector pointing from the center of the wheel to the wad of gum. At time $t = 0$, the vector $\overrightarrow{G(t)}$ is pointing in the same direction as the positive x-axis. Find a formula for $h(t)$, the vertical component of $\overrightarrow{G(t)}$.

SOLUTION   This is simply the Ferris wheel question formulated using different language.

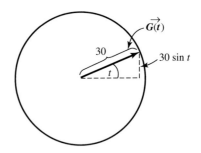

Figure 20.51

Since $h(0) = 0$, a formula of the form $30 \sin(Bt)$ can be used. One revolution is completed in $\pi/2$ minutes, so the period is $\pi/2$. $\frac{2\pi}{B} = \pi/2 \Rightarrow B = 4$.

$$h(t) = 30 \sin(4t) \quad \blacklozenge$$

The purpose of Example 20.15 is to show that we already know how to find the vertical and horizontal components of a vector. Given a vector (such as a velocity vector, force vector, or displacement vector), physicists, engineers, and architects often find it useful to resolve[15] the vector into its vertical and horizontal components. The next example asks for the resolution of a force vector into its horizontal and vertical components.

EXAMPLE 20.16   A suitcase is being pulled along a horizontal airport floor. A force of 50 pounds is being applied at an angle of 35° with the horizontal.

(a) What is the component of the force in the direction of motion?

(b) What is the component of the force perpendicular to the direction of motion?

---

[15] To "resolve" a vector is to separate it into its constituent parts.

SOLUTION   (a) Let $x =$ the component of the force in the direction of motion.

$$\cos 35° = \frac{x}{50}$$

$$x = 50 \cos 35°$$

$$x = 50 \cos \left(35° \frac{\pi}{180°}\right) \approx 40.96$$

Approximately 40.96 pounds of force are being applied in the direction of motion.

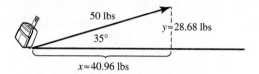

Figure 20.52

(b) Let $y =$ the component of the force perpendicular to the direction of motion.

$$\sin 35° = \frac{y}{50}$$

$$y = 50 \sin 35°$$

$$y \approx 28.68$$

Approximately 28.68 pounds of force are being applied perpendicular to the direction of motion.   ◆

What is the significance of what we have calculated? We've resolved the force into two perpendicular components. If a force of $(50 \sin 35°)$ pounds is applied upwards and simultaneously a force of $(50 \cos 35°)$ pounds is applied in the direction of motion, the sum of these two forces is equivalent to the original force.

◆ EXAMPLE 20.17   Consider the trajectory made by some thrown or launched object. The object could be a tennis ball, a javelin, a basketball, or a stone. The exact nature of the trajectory is of critical importance to athletes and marksmen alike. If we neglect air resistance and spin and consider only the force of gravity, then the trajectory of the object is determined by the initial velocity of the object, given by the velocity vector $\vec{v}_0$. Let's denote the length of $\vec{v}_0$ by v_0 (with no arrow above it). v_0 is the initial speed. We'll denote the angle of launch by θ, where θ is in the interval $(0, \pi/2)$. For simplicity's sake, we'll assume the object is launched from ground level. The only force acting on the object is the force of gravity, resulting in a downward acceleration of g meters per second squared.

(a) Find the vertical position of the object at time t.

(b) When will the object hit the ground?

(c) How far away will the object be from its launching spot when it hits the ground?

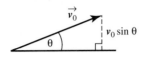

Figure 20.53

SOLUTION   (a) To analyze the vertical position of the object at time t, we can restrict our attention to the vertical component of velocity. We can express the vertical component of the initial velocity, $\vec{v_0}$, as $v_0 \sin \theta$.

Let $v(t)$ be the vertical velocity of the object (in meters per second) at time t and let $s(t)$ be the vertical position (in meters) of the object at time t.

Let $t = 0$ be the moment of launch.

The vertical acceleration of the object is $-g$, where g is the gravitational constant; therefore $v'(t) = -g$.

Our strategy is as follows:

Using information about acceleration, (the derivative of $v(t)$), find $v(t)$.

Using information about velocity, (the derivative of $s(t)$), find $s(t)$.

Let's guess a function whose derivative is $-g$; we know that any other such function differs from it by a constant.[16] The derivative of $-gt$ is $-g$ so

$v(t) = -gt + C$.   We know that $v(0) = v_0 \sin \theta$; this information is the initial condition that allows us to solve for the constant C.

$v_0 \sin \theta = -g \cdot 0 + C$

$v_0 \sin \theta = C$ (Keep in mind that $v_0 \sin \theta$ is a constant! Both v_0 and θ are fixed.)

$v(t) = -gt + v_0 \sin \theta$.

The velocity is the derivative of the displacement function; $v(t) = s'(t)$.

$$s'(t) = -gt + v_0 \sin \theta$$

Let's guess a function whose derivative is $-gt + v_0 \sin \theta$; we know that any other such function differs from it by a constant.

$s(t) = -g\frac{t^2}{2} + (v_0 \sin \theta)t + C_1$ We know that $s(0) = 0$; this information is the initial condition that allows us to solve for the constant C_1.

$0 = 0 + 0 + C_1$

$0 = C$

Therefore $s(t) = -g\frac{t^2}{2} + (v_0 \sin \theta)t$.

(b) The object hits the ground when $s(t) = 0$, so we must set $s(t)$ equal to zero.

$$-g\frac{t^2}{2} + (v_0 \sin \theta)t = 0 \quad \text{Solve for } t.$$

---

[16] See Appendix C for a proof of the Equal Derivatives Theorem.

$$t\left(-g\frac{t}{2} + v_0 \sin\theta\right) = 0$$

$$t = 0 \quad \text{or} \quad g\frac{t}{2} = v_0 \sin\theta$$

$$t = 0 \quad \text{or} \quad t = \frac{2v_0 \sin\theta}{g}$$

The object hits the ground $\frac{2v_0 \sin\theta}{g}$ seconds after it is launched.

(c) To find the horizontal distance the object will travel we direct our attention to the horizontal component of velocity. Unlike the vertical component of velocity (which is affected by the force of gravity), the horizontal component of velocity is constant. Therefore, the horizontal distance traveled is given by

(the horizontal component of velocity) · (time). The horizontal component of $\vec{v_0}$ is $v_0 \cos\theta$.

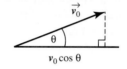

Figure 20.54

The time the object travels is the time between launch and when it hits the ground: $\frac{2v_0 \sin\theta}{g}$ seconds.

$$\text{horizontal distance traveled} = v_0 \cos\theta \cdot \frac{2v_0 \sin\theta}{g} = \frac{2v_0^2 \cos\theta \sin\theta}{g}$$

So, for instance, if an object is launched at ground level at an angle of 30° and with an initial speed of 28 meters per second it will hit the ground after $\frac{2(28 \text{ m/sec}) \sin(\pi/6)}{9.8\text{m/sec}^2} \approx 2.86$ sec. It travels a horizontal distance of $(28\text{m/sec}) \cos(\pi/6) \cdot \frac{2(28 \text{ m/sec}) \sin(\pi/6)}{9.8\text{m/sec}^2} \approx 69.28$ m. ◆

EXERCISE 20.12   Suppose an object is launched from the ground with an initial velocity of 96 feet per second at an angle of 40°. The gravitational constant is 32 ft/sec^2. Answer questions (a), (b), and (c) from the last example by working through all the steps using these concrete numbers.

ANSWERS   (a) $v(t) = -32t + 96 \sin 40°$, so $s(t) = -16t^2 + (96 \sin 40°)t + 5$.
$s(t) \approx -16t^2 + 61.7t + 5$ in ft/sec

(b) The object hits the ground after about 3.936 seconds.

(c) ≈ 289.463 feet

Finding the Component of a Vector in an Arbitrary Direction

In many situations it is useful to find the component of a vector not in the horizontal and vertical directions but in some other direction. For instance, a sailor might be interested in the component of the wind's velocity in the direction of his boat's motion. If an object is being pulled up an incline we are interested in the component of the force applied in the direction of the object's motion. A swimmer swimming in a current is interested in the component of the force of the water in her direction of motion.

The component of $\vec{v}$ in the direction of $\vec{u}$ gives a measure of how much of $\vec{v}$ is in the direction of $\vec{u}$. Below we give some examples. (Recall that $|\vec{v}|$ denotes the length or magnitude of $\vec{v}$.)

$$\begin{pmatrix} \text{The component of } \vec{v} \\ \text{in the direction of } \vec{u} \end{pmatrix} \text{ is } \begin{cases} |\vec{v}| \text{ if } \vec{u} \text{ and } \vec{v} \text{ point in the same direction} \\ -|\vec{v}| \text{ if } \vec{u} \text{ and } \vec{v} \text{ point in the opposite directions.} \\ 0 \text{ if } \vec{u} \text{ and } \vec{v} \text{ are perpendicular.} \end{cases}$$

We denote the component of $\vec{v}$ in the direction of $\vec{u}$ by $v_{\vec{u}}$. (It will be a number, not a vector, so we don't put an arrow over the v.) To visualize $v_{\vec{u}}$, begin by placing the tails of the vectors $\vec{v}$ and $\vec{u}$ together. Let θ be the angle between $\vec{u}$ and $\vec{v}$. We'll measure θ as an angle between 0 and π; there is no need to distinguish between its terminal and initial sides.

The *direction* of $\vec{u}$ matters but its length does not. If the angle between $\vec{u}$ and $\vec{v}$ is acute, then $v_{\vec{u}}$ is positive; if the angle is obtuse, then $\vec{v}$ is not in the direction of $\vec{u}$, and $v_{\vec{u}}$ is negative. You can visualize what is meant by the component of $\vec{v}$ in the direction of $\vec{u}$ in the examples shown in Figure 20.55 by tilting your head so that $\vec{u}$ looks horizontal and visualizing the horizontal component of $\vec{v}$.

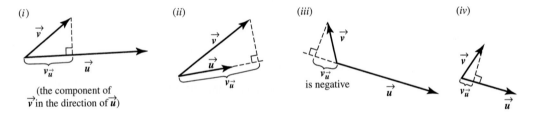

Figure 20.55

In the examples shown in Figure 20.56 it is easier to tilt your head so that you're letting $\vec{u}$ be vertical and visualize the vertical component of $\vec{v}$.

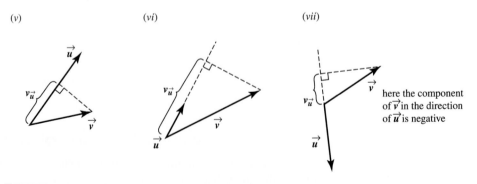

Figure 20.56

Defining and computing the component of $\vec{v}$ in the direction of $\vec{u}$

The component of $\vec{v}$ in the direction of $\vec{u}$, $v_{\vec{u}}$ is defined as $v_{\vec{u}} = |\vec{v}| \cos \theta$, where θ is the angle between $\vec{u}$ and $\vec{v}$.

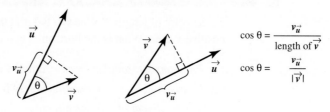

$$\cos \theta = \frac{v_{\vec{u}}}{\text{length of } \vec{v}}$$

$$\cos \theta = \frac{v_{\vec{u}}}{|\vec{v}|}$$

Figure 20.57

Observations

▪ The definition of $v_{\vec{u}}$ involves the length of $\vec{v}$ but *not* the length of $\vec{u}$. This is in line with our earlier remark that the length of $\vec{u}$ is irrelevant; only the direction of $\vec{u}$ matters.

▪ If $\theta \in [0, \pi/2)$, then $\cos \theta$ is positive; if $\theta \in (\pi/2, \pi]$, then $\cos \theta$ is negative. This is exactly what we want.

▪ $|\vec{v}| \cos \theta$
$$\begin{cases} \text{has a maximum value of } |\vec{v}| \text{ when } \theta = 0. & \left(\begin{smallmatrix}\vec{u} \text{ and } \vec{v} \text{ have the} \\ \text{same direction.}\end{smallmatrix}\right) \\ \text{has a minimum value of } -|\vec{v}| \text{ when } \theta = \pi. & \left(\begin{smallmatrix}\vec{u} \text{ and } \vec{v} \text{ have} \\ \text{opposite directions.}\end{smallmatrix}\right) \\ \text{is zero when } \theta = \pi/2. & (\vec{u} \text{ and } \vec{v} \text{ are perpendicular.}) \end{cases}$$

This is in agreement with our intuition.

◆ **EXAMPLE 20.18** A suitcase is being pulled up a ramp that makes a 10° angle with the horizontal. A force of 50 pounds is applied at an angle of 35° with the horizontal. What is the component of the force in the direction of motion?

SOLUTION Let $\vec{u}$ be a vector in the direction of motion and $\vec{F}$ be the force vector. $|\vec{F}| = 50$. The angle between $\vec{u}$ and $\vec{F}$ is 25°.

Let $x =$ the component of the force in the direction of motion.

$$x = 50 \cos 25°$$
$$\approx 45.3$$

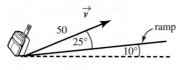

Figure 20.58

Approximately 45.3 pounds of force are being applied in the direction of motion. ◆

EXERCISE 20.13   The angle between vectors $\vec{u}$ and $\vec{v}$ is $2\pi/3$. $|\vec{u}| = 3$ and $|\vec{v}| = 8$.

(a) Find the component of $\vec{v}$ in the direction of $\vec{u}$. Give an exact answer. Accompany your answer by a sketch.

(b) Find the component of $\vec{u}$ in the direction of $\vec{v}$. Accompany your answer by a sketch.

PROBLEMS FOR SECTION 20.7

1. $\vec{v}$ is a vector of length 3. When its tail is at the origin, it makes an angle of $60°$ with the positive x-axis. What is the horizontal component of $\vec{v}$? The vertical component of $\vec{v}$?

For Problems 2 through 5, vectors $\vec{u}$ and $\vec{v}$ are described. Find
(a) the component of $\vec{v}$ in the direction of $\vec{u}$.
(b) the component of $\vec{u}$ in the direction of $\vec{v}$.

2. $|\vec{u}| = 5$, $|\vec{v}| = 7$, and the angle between $\vec{u}$ and $\vec{v}$ is $\frac{\pi}{6}$.

3. $|\vec{u}| = 5$, $|\vec{v}| = 7$, and the angle between $\vec{u}$ and $\vec{v}$ is $\frac{\pi}{3}$.

4. $|\vec{u}| = 5$, $|\vec{v}| = 7$, and the angle between $\vec{u}$ and $\vec{v}$ is $\frac{2\pi}{3}$.

5. $\vec{u}$ is a vector of length 2 directed due south. $\vec{v}$ is a vector of length 3 directed northeast.

6. Suppose $\vec{v}$ is a vector of length 8 and the component of $\vec{v}$ in the direction of $\vec{u}$ is 4.
   (a) Can the angle between $\vec{u}$ and $\vec{v}$ be determined? If so, what is it?
   (b) Can the direction of $\vec{u}$ be determined? If so, what is it?
   (c) Can the length of $\vec{u}$ be determined? If so, what is it?

7. You're pushing an object along a table, exerting a force of 10 pounds in the direction indicated below. What is the component of force in the direction of motion?

8. A plane is traveling 300 miles per hour. There is a 50-mph wind. The angle between the velocity vector of the wind and the velocity vector of the plane is $110°$.
   (a) What is the component of the direction of the plane's motion?
   (b) In the absence of the wind but all else remaining the same, how fast would the plane be traveling?

9. Suppose an object is launched from a height of 64 feet with an initial velocity of 96 feet per second at an angle of $\pi/6$ radians. Assume that the only force acting on the object is the force of gravity, which results in a downward acceleration of 32 ft/sec.
   (a) Find the vertical position of the object at time t.
   (b) When will the object hit the ground?

    (c) How far has the object traveled horizontally when it hits the ground? (In other words, what is the horizontal component of its displacement vector?)

    (d) When the object hits the ground, how far is it from where it was launched? (In other words, what is the length of its displacement vector?)

10. Force A has a horizontal component of 3 pounds and a vertical component of 4 pounds. Force B has a horizontal component of 5 pounds and a vertical component of 12 pounds.

    (a) What is the strength of force A? What angle does this force vector make with the horizontal? (Give a numerical approximation in degrees.)

    (b) What is the strength of force B? What angle does this force vector make with the horizontal? (Give a numerical approximation in degrees.)

    (c) What is the component of force A in the direction of force B?

11. There's a 10-mile-per-hour wind, but a bicyclist calculates that its component in his direction of motion is only 9 miles per hour. What is the angle between the velocity vector of the wind and that of the bicyclist?

12. Due to the scampering of a goat, a rock has been dislodged from a mountain and is sliding down an incline making a 70° angle with level ground. The weight of the rock exerts a downward force of 3 pounds. What is the component of this force in the direction of motion of the rock?

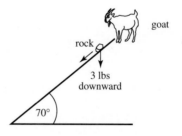

21

Differentiation of Trigonometric Functions

21.1 INVESTIGATING THE DERIVATIVE OF SIN X GRAPHICALLY, NUMERICALLY, AND USING PHYSICAL INTUITION

Graphical Analysis

Figure 21.1 presents the graphs of $f(x) = \sin x$, $g(x) = \cos x$, and sketches, qualitatively done, of their derivative functions.

Where the function is increasing, its derivative is positive or zero; where the function is decreasing, its derivative is negative or zero.

Where the function is concave up, its derivative is increasing; where the function is concave down, its derivative is decreasing.

Sine and cosine are periodic with period 2π. Similarly, their derivatives must be periodic with period 2π.[1] Notice that no vertical scale is given on the graphs of the derivative functions.

---

[1] A function with period k must have a derivative such that $f'(x + k) = f'(x)$. In general, it is possible for the period to be some number $s < k$ such that $ns = k$ for some positive integer n. For example, if the graph of f is as shown below, then the period of f' is half that of f.

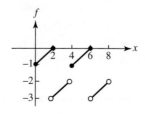

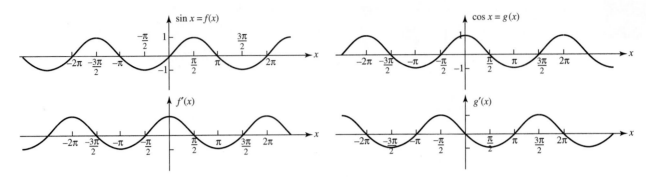

Figure 21.1

Let's focus on the derivative of $\sin x$. Its graph looks seductively like $\cos x$. However, simply from qualitative analysis we have no evidence leading us to believe, for instance, that the amplitude of its derivative is 1. In fact, the amplitude of the derivative will depend upon whether x is given in degrees or radians. Let's stick to radians and gather some numerical data by approximating the derivative of $\sin x$ at $x = 0$.

Numerical Investigation of the Slope of the Tangent Line to $\sin x$ at $x = 0$

The derivative of $\sin x$ at $x = 0$ is the slope of the tangent to $\sin x$ at $(0, 0)$. We can approximate the slope of the tangent line by finding the slope of the secant line between $(0, 0)$ and a nearby point on the sine curve. The slope of the secant line through $(0, \sin 0)$ and $(0 + h, \sin(0 + h))$ is

$$\frac{\sin h - \sin(0)}{h - 0} = \frac{\sin h}{h}.$$

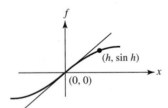

Figure 21.2

For h near 0 this is a good approximation to the slope of the tangent line. At $h = 0.0001$, the slope of this secant line is $\frac{\sin(0.0001)}{0.0001} \approx 0.9999999983$. If we use $h = 0.00001$ to approximate $\frac{d}{dx} \sin x|_{x=0}$, we compute $\frac{\sin(0.00001)}{0.00001}$. A TI-83 calculator gives this quotient as 1 (although in fact it is *not* exactly 1). The conjecture that the derivative of $\sin x$ at $x = 0$ is 1 looks quite reasonable,[2] and following closely upon its heels is the delightfully appealing conjecture that $\frac{d}{dx} \sin x = \cos x$. If we can prove this, it will make us happy.

---

[2] Notice that it appears (although we have not yet proven it) that $\lim_{x \to 0} \frac{\sin x}{x} = 1$.

EXERCISE 21.1   Graph $j(x) = \frac{\sin x}{x}$ on a graphing calculator or a computer. $j(x)$ is not defined at $x = 0$. What does the graph lead you to conjecture about $\lim_{x \to 0} \frac{\sin x}{x}$?

Radians versus Degrees.   The previous numerical analysis was done using radians. Let's see what happens if we measure the angle x in degrees and look at the slope of the line tangent to sin x at $x = 0$. Again look at

$$\frac{\sin h}{h} \quad \text{for } h \text{ very small}$$

but this time measure h in degrees. According to a TI-83 calculator, for $h = 0.001, 0.0001$, and 0.00001, $\frac{\sin h°}{h°} \approx 0.0174532925 \; \frac{1}{\text{degrees}}$.[3]

This is a far cry from the value of approximately 1 when using radians. Actually, the number makes sense *if* it is indeed true that $\frac{d}{dx} \sin x = \cos x$, when x is measured in radians. If w is an angle measured in degrees we can convert to radians and differentiate.

$$\frac{d}{dw} \left[\sin(w°) \right] = \frac{d}{dw} \left[\sin \left(w° \cdot \frac{\pi \text{ radians}}{180°} \right) \right]$$

$$= \frac{d}{dw} \left[\sin \left(\frac{w\pi}{180} \text{ radians} \right) \right]$$

$$= \left(\cos \frac{w\pi}{180} \right) \cdot \frac{\pi}{180} \qquad \text{Using the hypothesis } \frac{d}{dx} \sin x = \cos x$$
$$\text{and the Chain Rule}$$

$$= \cos w° \cdot \frac{\pi}{180}$$

$$\approx 0.0174532925 \cos w° \left(\text{ because } \frac{\pi}{180} \approx 0.0174532925 \right)$$

Spinning Wheels: Learning from Physical Intuition

We've been using the sine function to model the height of a marked point on a steadily spinning wheel, whether that point corresponds to position of a seat on a Ferris wheel, a mark on a gear, or gum on a suspended tire. Let's see what we can learn about the derivative of sin t by looking at the real-world situation we're modeling.

Suppose a mark is made on the rim of a vertically oriented wheel of radius 3 meters. For simplicity's sake, we'll assume the wheel is spinning counterclockwise at a steady rate of 1 revolution every 2π minutes so that the height of the mark at time t is given by

$$h(t) = 3 \sin t + 3,$$

where h is given in meters and t in minutes. Since the wheel is rotating steadily, we know the mark is moving at a constant speed. Its speed is given by

$$\frac{\text{circumference of wheel}}{2\pi \text{ seconds}} = \frac{6\pi \text{ meters}}{2\pi \text{ seconds}} = 3 \text{ meters per second.}$$

---

[3] Radians are unitless. When we write sin x with no units, we mean radians. We have the awkwardness of 1/degree because degrees are not unitless.

Although the *speed* of the mark is constant, the direction it is moving is not. At each time *t* the velocity can be resolved into its vertical and horizontal components.[4] The vertical component is the rate at which the height is changing, and the horizontal component is the rate at which the horizontal coordinate of the point is changing.

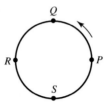

Figure 21.3

For now we'll focus on the vertical component of the velocity. $h(t) = 3 \sin t + 3$ gives the vertical position of the mark at time *t*, so $\frac{dh}{dt}$ must give the rate of change of height, or the vertical component of the velocity.

$$\frac{dh}{dt} = 3 \cdot \frac{d}{dt}[\sin t]$$

The point marked on the wheel has a vertical velocity of 3 m/sec when it is in position *P* (see Figure 21.4) because the entire velocity is directed upward. Its vertical velocity is decreasing but positive until it reaches position *Q*. At *Q* the velocity is directed horizontally, so the vertical component is zero.

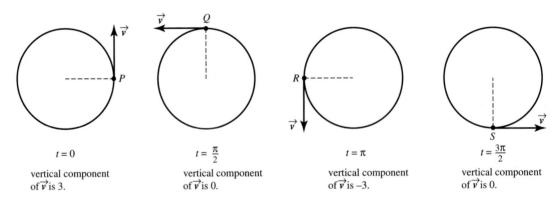

| $t = 0$ | $t = \frac{\pi}{2}$ | $t = \pi$ | $t = \frac{3\pi}{2}$ |
| vertical component of $\vec{v}$ is 3. | vertical component of $\vec{v}$ is 0. | vertical component of $\vec{v}$ is -3. | vertical component of $\vec{v}$ is 0. |

Figure 21.4

---

[4] The velocity can be modeled as a vector, $\vec{v}$ of length 3 tangent to the wheel. We are resolving this vector into its vertical and horizontal components.

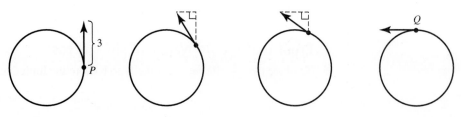

Figure 21.5

At position Q the point stops its ascent and is poised for its descent. Between Q and S (where the velocity again is entirely horizontal) the vertical component of velocity is negative. The most negative velocity is at position R where the entire velocity (3 m/sec) is directed downward.

Since $\frac{dh}{dt} = 3 \frac{d}{dt} \sin t$, this physical analysis indicates that the derivative of $\sin t$ has the following characteristics.

It has a maximum value of 1 at $t = 2\pi n$,

it has a minimum value of -1 at $t = \pi + 2\pi n$, and

it is zero at $\pi/2 + \pi n$.

The results of our physical analysis support our graphical and numerical work and the conjecture that the derivative of $\sin x$ is $\cos x$. We'll prove this conjecture in the next section.

PROBLEMS FOR SECTION 21.1

1. Estimate $\lim_{x \to \pi} \frac{\sin x}{x}$ both numerically and graphically.

2. Let $f(x) = \sin x$. Use the difference quotient with $h = 0.0001$ to estimate the value of $f'(\pi)$, the slope of the tangent line to $\sin x$ at $x = \pi$.

3. Let $f(x) = \sin(x)$.
   (a) Using a calculator, tabulate f at $x = 1.998, 1.999, 2.000, 2.001, 2.002$ (make a table with values of x and $f(x)$). Round values of f to six decimal places.
   (b) Estimate $f'(1.999)$, $f'(2)$, and $f'(2.001)$ using the tabulated values.
   (c) Estimate $f''(2)$ using the results from part (b).

4. (a) What is the limit definition of $\frac{d}{dx} \cos x \big|_{x=0}$?
   (b) Numerically approximate $\frac{d}{dx} \cos x \big|_{x=0}$.

5. Numerically approximate the derivative of $\cos x$ at $x = \pi$.

21.2   DIFFERENTIATING SIN X AND COS X

Proof that the Derivative of sin x is cos x

To prove that $\frac{d}{dx} \sin x = \cos x$ we'll go back to the limit definition of derivative.

$$\frac{d}{dx} \sin x = \lim_{h \to 0} \frac{\sin(x+h) - \sin(x)}{h}$$

We've already shown that $\sin(x + h) \neq \sin x + \sin h$ and asserted that $\sin(A + B) = \sin A \cos B + \sin B \cos A$. We'll use this fact below.

$$\frac{d}{dx} \sin x = \lim_{h \to 0} \frac{\sin(x+h) - \sin(x)}{h}$$

$$= \lim_{h \to 0} \frac{\sin(x) \cos(h) + \sin(h) \cos(x) - \sin(x)}{h}$$

$$= \lim_{h \to 0} \frac{\sin x[\cos(h) - 1] + \sin(h) \cos x}{h} \qquad \text{(gathering the } \sin x \text{ terms)}$$

$$= \lim_{h \to 0} \left(\frac{\sin x[\cos(h) - 1]}{h} + \frac{\sin(h) \cos x}{h} \right)$$

$$= \lim_{h \to 0} \left(\sin x \left[\frac{\cos(h) - 1}{h} \right] + \cos x \left[\frac{\sin(h)}{h} \right] \right)$$

$$= \lim_{h \to 0} \sin x \left[\frac{\cos(h) - 1}{h} \right] + \lim_{h \to 0} \cos x \left[\frac{\sin(h)}{h} \right] \qquad \begin{array}{l} (\sin x \text{ and } \cos x \\ \text{are independent of } h) \end{array}$$

$$= \sin x \left[\lim_{h \to 0} \frac{\cos(h) - 1}{h} \right] + \cos x \left[\lim_{h \to 0} \frac{\sin(h)}{h} \right]$$

There are two limits we need to evaluate: $\lim_{h \to 0} \frac{\cos(h) - 1}{h}$ and $\lim_{h \to 0} \frac{\sin(h)}{h}$. Let's take them on one by one. We already have evidence suggesting that $\lim_{h \to 0} \frac{\sin(h)}{h} = 1$; we'll prove this below.[5] Then we need only show the former limit is 0 to obtain the desired result.[6]

Proof that $\lim_{x \to 0} \frac{\sin x}{x} = 1$

First notice that $\frac{\sin x}{x}$ is an even function: $\frac{\sin(-x)}{-x} = \frac{-\sin x}{-x} = \frac{\sin x}{x}$. Therefore, it is enough to show that $\lim_{x \to 0^+} \frac{\sin x}{x} = 1$, because the limit as we approach zero from the left will be the same as the limit as we approach zero from the right.

---

[5] The proof given is the standard proof of this limit. When you first see this argument you may feel a bit bewildered. While each step follows logically from the previous one, chances are you will not see at the outset where you are being led. The proof is not laid out in the form in which someone thought of the argument. In contrast, the conclusion we want "drops out" at the end.

  Mathematicians spend a lot of time playing with ideas, poking, tugging, wrestling—and sometimes having a flash of inspiration. A mathematician may make the logical journey from point A to point B in a myriad of different ways. The proof he or she presents to the world may show a map of the journey or the proof may be polished and shined so that the marks of the hewing process are hidden from view. The aesthetics of a proof are very much part of the culture of mathematics. (A famous mathematician of our time, Paul Erdös (1913–1996), used to refer to what he called "The Book" of mathematical theorems in which each theorem exists along with its most beautiful and insightful proof.)

[6] The proof that we are in the midst of (that the derivative of sin x is cos x) began in a very straightforward manner. We have come up against two sticking points, and will resolve these two as separate problems, importing the result back to our original argument. These building blocks of the puzzle are called "lemmas."

When looking at the limit as x goes to 0 from the right, we can restrict our discussion to $0 < x < \pi/2$. That said, refer to Figure 21.6. B is the point $(1, 0)$ and $P = P(x)$ is a point on the unit circle in the first quadrant.

Look at the area of the pie-slice shaped sector of the circle delineated by OBP. For positive x, the area of this sector is greater than the area of the small shaded triangle OAP and less than the area of the large triangle OBQ.

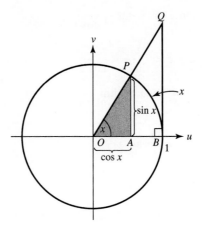

Figure 21.6

$$\begin{pmatrix} \text{area of} \\ \text{small triangle} \end{pmatrix} < \begin{pmatrix} \text{area of} \\ \text{sector} \end{pmatrix} < \begin{pmatrix} \text{area of} \\ \text{large triangle} \end{pmatrix}$$

The area of the small triangle $= \frac{1}{2}$ (length of $\overline{AP}$) $\cdot$ (length of $\overline{OA}$) $= \frac{1}{2}(\sin x)(\cos x)$

The area of the large triangle $= \frac{1}{2}$ (length of $\overline{BQ}$) $\cdot$ (length of $\overline{OB}$) $= \frac{1}{2}(\tan x)(1)$

$\qquad\qquad\qquad\qquad\qquad = \frac{1}{2} \tan x$

To find the area of the sector OBP look at the following ratio.

$$\frac{\text{area of the sector}}{\text{area of the circle}} = \frac{\text{arclength}}{\text{circumference}}$$

$$\frac{\text{area of the sector}}{\pi(1)^2} = \frac{x}{2\pi}$$

$$\text{area of the sector} = \frac{x}{2}$$

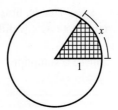

Figure 21.7

It follows that

$$\frac{1}{2}\sin x \cos x < \frac{x}{2} < \frac{1}{2}\tan x$$

$$\frac{\sin x \cos x}{2} < \frac{x}{2} < \frac{1}{2}\frac{\sin x}{\cos x}$$
If $A < B$ and k is any positive number, then $kA < kB.$[7] Multiply by 2.

$$\sin x \cos x < x < \frac{\sin x}{\cos x}$$
Because $\sin x > 0$ for $x \in (0, \pi/2)$, we can divide by $\sin x$.

$$\cos x < \frac{x}{\sin x} < \frac{1}{\cos x}$$
This inequality holds for $0 < x < \pi/2$.

$$\lim_{x\to 0^+}\cos x \le \lim_{x\to 0^+}\frac{x}{\sin x} \le \lim_{x\to 0^+}\frac{1}{\cos x}$$
In the limit the strict inequality "$<$" becomes "$\le$".[8]

$$1 \le \lim_{x\to 0^+}\frac{x}{\sin x} \le 1$$
The limit is squeezed in a vise until completely determined.

$$\text{so } \lim_{x\to 0^+}\frac{x}{\sin x} = 1$$

Therefore, $\lim_{x\to 0^+}\frac{\sin x}{x} = 1$. Because $\frac{\sin x}{x}$ is an even function, we have shown that $\lim_{x\to 0}\frac{\sin x}{x} = 1$.

We now have that

$$\frac{d}{dx}\sin x = \sin x \left[\lim_{h\to 0}\frac{\cos(h)-1}{h}\right] + \cos x \left[\lim_{h\to 0}\frac{\sin(h)}{h}\right]$$

$$= \sin x \left[\lim_{h\to 0}\frac{\cos(h)-1}{h}\right] + \cos x \cdot 1$$

$$= \sin x \left[\lim_{h\to 0}\frac{\cos(h)-1}{h}\right] + \cos x.$$

If we can show $\lim_{h\to 0}\frac{\cos(h)-1}{h} = 0$, then our proof that $\frac{d}{dx}\sin x = \cos x$ is complete.

WHAT IS $\lim_{h\to 0}\frac{\cos(h)-1}{h}$?

Looking at the graph of $\frac{\cos x - 1}{x}$ for x near zero and numerically investigating $\lim_{h\to 0}\frac{\cos(h)-1}{h}$ by evaluating $\frac{\cos(h)-1}{h}$ for h increasingly close to zero lead us to conjecture that $\lim_{h\to 0}\frac{\cos(h)-1}{h} = 0$. (Try this for yourself.)

Alternatively, we can reason that $\lim_{h\to 0}\frac{\cos(h)-1}{h} = \lim_{h\to 0}\frac{\cos(0+h)-\cos 0}{h} = \frac{d}{dx}\cos x \mid_{x=0}$. That is, $\lim_{h\to 0}\frac{\cos(h)-1}{h}$ is just the derivative of $\cos x$ evaluated at $x = 0$. If we assume that $\cos x$ is differentiable at $x = 0$ and has a local maximum at $x = 0$, then it follows that $\lim_{h\to 0}\frac{\cos(h)-1}{h} = \frac{d}{dx}\cos x \mid_{x=0} = 0$. We can prove that $\lim_{h\to 0}\frac{\cos(h)-1}{h} = 0$ without these assumptions as follows.

---

[7] If $A < B$ and k is negative, then $kA > kB$. For instance, $1 < 2$ but $(-5)(1) > (-5)(2)$.

[8] For example, $\frac{1}{x} < \frac{2}{x}$ for x positive but when we take the limit as $x \to \infty$ we must write $\lim_{x\to\infty}\frac{1}{x} \le \lim_{x\to\infty}\frac{2}{x}$; both these limits are zero.

Proof that $\lim_{h\to 0} \frac{\cos(h)-1}{h} = 0$

$$\lim_{h\to 0} \frac{\cos(h)-1}{h} = \lim_{h\to 0} \frac{\cos(h)-1}{h} \cdot \frac{(\cos(h)+1)}{(\cos(h)+1)}$$

$$= \lim_{h\to 0} \frac{\cos^2(h)-1}{h} \cdot \frac{1}{(\cos(h)+1)}$$

$$= \lim_{h\to 0} \frac{1-\cos^2(h)}{h} \cdot \frac{-1}{(\cos(h)+1)}$$

$$= \lim_{h\to 0} \frac{\sin^2(h)}{h} \cdot \frac{-1}{(\cos(h)+1)}$$

$$= \lim_{h\to 0} \frac{\sin(h)}{h} \cdot \frac{-\sin(h)}{(\cos(h)+1)}$$

But

$$\lim_{h\to 0} \frac{\sin(h)}{h} = 1 \quad \text{and} \quad \lim_{h\to 0} \frac{-\sin(h)}{(\cos(h)+1)} = \frac{0}{1+1} = 0,$$

so

$$\lim_{h\to 0} \frac{\cos(h)-1}{h} = (1)(0) = 0.$$

We have now shown that $\frac{d}{dx}\sin x = \cos x$, as conjectured!

The Chain Rule tells us that $\frac{d}{dx}\sin[g(x)] = \cos[g(x)]g'(x)$. Now that we have proven $\frac{d}{dx}\sin x = \cos x$, the derivative of $\cos x$ is easy to tackle by using the fact that $\cos x$ and $\sin x$ are related to one another by a horizontal shift. Looking back at the graphs of $\cos x$ and its derivative it is easy to speculate that the derivative of $\cos x$ is $-\sin x$.

Proof that the Derivative of cos x is $-\sin x$

Observe that $\sin\left(x + \frac{\pi}{2}\right) = \cos x$; replacing x by $\left(x + \frac{\pi}{2}\right)$ shifts the sine graph left $\frac{\pi}{2}$ units. Similarly, $\cos\left(x + \frac{\pi}{2}\right) = -\sin x$; replacing x by $\left(x + \frac{\pi}{2}\right)$ shifts the cosine graph left $\frac{\pi}{2}$ units.

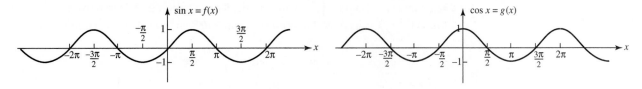

Figure 21.8

$$\frac{d}{dx}\cos x = \frac{d}{dx}\sin\left(x + \frac{\pi}{2}\right)$$

$$= \cos\left(x + \frac{\pi}{2}\right) \cdot 1 \qquad \text{(by the Chain Rule)}$$

$$= -\sin x$$

Combining this result with the Chain Rule gives us

$$\frac{d}{dx}\sin[g(x)] = \cos[g(x)] \cdot g'(x) \quad \text{or informally} \quad \frac{d}{dx}\sin[mess] = \cos[mess] \cdot [mess]'$$

$$\frac{d}{dx}\cos[g(x)] = -\sin[g(x)] \cdot g'(x) \qquad\qquad \frac{d}{dx}\cos[mess] = -\sin[mess] \cdot [mess]',$$

where *mess* is a function of x.

Using the Chain Rule and either the Product or Quotient Rule allows us to find the derivatives of $\tan x$, $\csc x$, $\sec x$, and $\cot x$.

EXERCISE 21.2   Show that $\boxed{\dfrac{d}{dx}\tan x = \sec^2 x.}$

EXERCISE 21.3   Show that $\boxed{\dfrac{d}{dx}\sec x = \sec x \tan x.}$

◆ **EXAMPLE 21.1**   Differentiate the following.

$$\text{(a) } y = 3x\sin(x^2) \qquad \text{(b) } y = 7\cos^2(3x + 5) \qquad \text{(c) } y = \sqrt{\tan(x^2)}$$

SOLUTIONS   (a) This is the product of $3x$ and $\sin[mess]$.

$$y' = 3\sin(x^2) + 3x\cos(x^2)(2x)$$

$$= 3\sin(x^2) + 6x^2\cos(x^2)$$

(b) $7\cos^2(3x + 5) = 7[\cos(3x + 5)]^2$, so basically this is $7[mess]^2$ and its derivative is $14[mess] \cdot [mess]'$, where the *mess* is $\cos(3x + 5)$. Then the Chain Rule must be applied to $\cos(3x + 5)$.

$$y' = 14\cos(3x + 5)(-\sin(3x + 5)) \cdot (3)$$

$$= -42\cos(3x + 5)\sin(3x + 5)$$

(c) $y = \sqrt{\tan(x^2)} = [\tan(x^2)]^{1/2}$. This is basically $[mess]^{1/2}$, so its derivative is $\frac{1}{2}[mess]^{-1/2} \cdot [mess]'$. We know $mess = \tan(stuff)$, so $[mess]' = \sec^2(stuff) \cdot (stuff)'$.

$$y' = (1/2)\left[\tan(x^2)\right]^{-1/2}\sec^2(x^2)2x$$

$$y' = \frac{x\sec^2(x^2)}{\sqrt{\tan(x^2)}} \qquad ◆$$

An Excursion: Polynomial Approximations of Trigonometric Functions

We spent a fair amount of energy proving that $\lim_{x\to 0}\frac{\sin x}{x}=1$. Now that this fact is ours, we can get some mileage out of it. From $\lim_{x\to 0}\frac{\sin x}{x}=1$ it follows that $\sin x \approx x$ for very small values of x.[9] This approximation is excellent for very small values of x, but as x gets increasingly far from zero the approximation becomes poor and eventually is useless. For example, $\sin 0.1 = 0.0998334\ldots \approx 0.1$ and $\sin 0.02 = 0.0199986\ldots \approx 0.02$, but $\sin 3 = 0.1411200\ldots$, which is not close to 3.

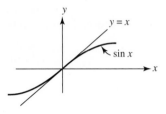

Figure 21.9

Historically the approximation $\sin x \approx x$ for very small values of x was used in ancient times, well before the development of calculus. Using it, $\sin 1°$ can be approximated with a great degree of accuracy. From there, using addition formulas and knowledge about specific triangles, trigonometric tables can be built up. The approximation $\sin x \approx x$, when used to estimate $\sin 1°$, gives

$$\sin 1° = \sin\left(\frac{\pi}{180}\right) \approx \frac{\pi}{180} = 0.01745329\ldots,$$

where the actual value of $\sin 1° \approx 0.0174524\ldots$.

The approximation $\sin x \approx x$, the tangent line approximation of $\sin x$ at $x = 0$, gives an estimate for $\sin x$ that is too large if x is positive and too small if x is negative. (See Figure 21.9). This approximation can be improved upon. The approximation $\sin x \approx x - x^3/6$, the best cubic approximation to $\sin x$ around $x = 0$, was used well before the seventeenth century. This gives a higher degree of accuracy for x near zero. Using it to approximate $\sin 1°$ gives $\sin 1° = \sin(\pi/180) \approx \pi/180 - \frac{1}{6}(\pi/180)^3$

$$\approx 0.0174524064.$$

This is identical to the numerical approximation supplied by a calculator.

The third degree polynomial approximation of $\sin x$ around $x = 0$ can be improved upon by using a fifth degree approximation, which can in turn be improved upon by using a seventh degree polynomial. (We use only polynomials of odd degree because sine is an odd function.) As the degree of the polynomial used to approximate sine increases, the accuracy of the approximation near $x = 0$ increases and the interval around $x = 0$ for which the approximation is reasonable enlarges as well. Amazingly enough, if we continue along in this way we can come up with an infinite "polynomial" that is exactly equal to $\sin x$ everywhere. These polynomial expansions are known as **Taylor series**, after the British mathematician Brook Taylor. (We take up Taylor series in Chapter 30).

---

[9] This statement is true only if x is given in radians.

By the seventeenth century mathematicians were using infinite polynomial expansions of functions, trigonometric and others. The polynomial expansion of $\sin x$ is given by

$$\sin x = x - \frac{x^3}{3 \cdot 2} + \frac{x^5}{5 \cdot 4 \cdot 3 \cdot 2} - \frac{x^7}{7 \cdot 6 \cdot 5 \cdot 4 \cdot 3 \cdot 2} + \cdots.$$

If we use *factorial* notation, letting $n! = n(n-1)(n-2) \ldots 3 \cdot 2 \cdot 1$, this can be written as

$$\sin x = x - \frac{x^3}{3!} + \frac{x^5}{5!} - \frac{x^7}{7!} + \cdots.$$

The first several terms can be used to approximate $\sin x$ for x small. For instance,

$$\sin(0.1) = 0.1 - \frac{(0.1)^3}{3!} + \frac{(0.1)^5}{5!} - \cdots,$$

so

$$\sin(0.1) \approx 0.1 - \frac{(0.1)^3}{3!} = 0.09983\overline{3} \ldots.$$

Compare this with $\sin(0.1) \approx 0.0998334166 \ldots.$ Using one more term of the polynomial gives

$$\sin(0.1) \approx 0.1 - \frac{(0.1)^3}{3!} + \frac{(0.1)^5}{5!} \approx 0.0998334167.$$

A good match!

PROBLEMS FOR SECTION 21.2

1. Using the derivatives of sine and cosine and either the Product Rule or the Quotient Rule, show that $\frac{d}{dx} \tan x = \sec^2 x$.

2. Show that $\frac{d}{dx} \sec x = \sec x \tan x$.

3. Find the first and second derivatives of the following.

   (a) $f(x) = 5 \cos x$ (b) $g(x) = -3 \sin(2x)$
   (c) $h(x) = 0.5 \tan x$ (d) $j(x) = 2 \sin x \cos x$

4. Differentiate the following.

   (a) $y = \cos^2 x$ (b) $y = \cos(x^2)$ (c) $y = x \tan^2 x$
   (d) $y = \sin^3(x^4)$ (e) $y = 7[\cos(5x) + 3]^x$

5. Consider the function $f(x) = e^{-0.3x} \sin x$.
   (a) For what values of x does $f(x)$ have its local maxima and local minima?
   (b) Is $f(x)$ a periodic function?
   (c) Sketch the graph of $f(x) = e^{-0.3x} \sin x$.

(d) What is the maximum value of for $e^{-0.3x} \sin x$ for $x \geq 0$? At what x-value is this maximum attained? Your answers must be exact, not numerical approximations from a calculator. Give justification that this value is indeed the maximum.

Evaluate the following derivatives. $u(x)$ is a differentiable function.

6. (a) $\frac{d}{dx} \sin (u(x))$ (b) $\frac{d}{dx} \cos(u(x))$ (c) $\frac{d}{dx} u(x)(\sin x)$

7. (a) $\frac{d}{dx} u(x)(\cos x)$ (b) $\frac{d}{dx} \tan(u(x))$ (c) $\frac{d}{dx} u(x)(\tan x)$

Evaluate.

8. $\frac{d}{dx} \sin(x^3 + \ln 3x)$

9. $\frac{d}{dx} \cos^2(\sin x)$

10. $\frac{d}{dx} \left[\frac{1}{\sin^3(\cos 2x)} \right]$

11. $\frac{d}{dx} \sqrt{\sin(2x^3)}$

12. $\frac{d}{dx} \frac{4}{\sqrt{2-\cos(x/7)}}$

13. $\frac{d}{dx} \left[e^{3x} \cos^2(7x) \right]$

14. Find dy/dx in terms of x and y.

$$\sin(xy) + y = y \cos x$$

15. Find y'.

     (a) $y = \frac{x}{\sin x}$ (b) $y = 3 \tan^3(x^2)$ (c) $y = \tan \left(\frac{x}{3}\right) \sec(3x)$

16. Why have we been telling you that radians are more appropriate than degrees when using calculus? Suppose x is measured in degrees. Then $\cos x° = \cos \left(\frac{x° \pi \, \text{radians}}{180°} \right) = \cos \left(\frac{\pi x}{180}\right)$ where the argument is now in radians. Find the derivative. Is the derivative $-\sin x°$?

21.3   APPLICATIONS

Optimization

◆ **EXAMPLE 21.2**   What angle of launch will propel an object (such as a cannonball or a baseball) farthest horizontally?

This question is vitally important to engineers and sportsmen alike. If we consider only the force of gravity (ignoring air resistance, the Coriolis effect, etc.), then it can be shown that the path the object will take is a parabola. In Section 20.7 we showed that if an object is launched at ground level at an angle θ and with an initial velocity of v_0, then the horizontal distance it will travel is given by

$$R(\theta) = \frac{2v_0^2 \cos\theta \sin\theta}{g},$$

where g is the acceleration due to gravity. We want to find θ such that $R(\theta)$ is maximum.

SOLUTION   Using the trigonometric identity $\sin 2\theta = 2\sin\theta\cos\theta$, $R(\theta)$ can be rewritten as

$$R(\theta) = \frac{v_0^2 \sin(2\theta)}{g} \quad \text{or} \quad \frac{v_0^2}{g}\sin(2\theta),$$

where g and v_0 are constants.

One approach to this problem is to find the critical points of $R(\theta)$. For our purposes θ must be between 0 and $\pi/2$. Therefore, the critical points of R are the endpoints $\theta = 0$ and $\theta = \pi/2$, both resulting in $R(\theta) = 0$, and the values of θ between 0 and $\pi/2$ such that $\frac{dR}{d\theta} = 0$.

$$\frac{dR}{d\theta} = \frac{v_0^2}{g}(2\cos(2\theta)) \quad \text{Setting } \frac{dR}{d\theta} \text{ equal to 0 gives}$$

$$0 = 2\frac{v_0^2}{g}\cos(2\theta) \quad 2\frac{v_0^2}{g} \text{ is a constant.}$$

$$0 = \cos(2\theta)$$

Let $u = 2\theta$. When $0 \leq \theta \leq \frac{\pi}{2}$, $2 \cdot 0 \leq 2\theta \leq 2 \cdot \frac{\pi}{2}$, so

$$0 \leq u \leq \pi.$$

$\cos u = 0$ when $u = \pi/2$. This is the only value of $u \in [0, \pi]$ that satisfies the equation.

$$2\theta = \pi/2 \quad (\text{Substitute } 2\theta = u.)$$

$$\theta = \pi/4$$

$\theta = \pi/4$ is a candidate for the maximum.

We can show it is actually the maximum by looking at $R''(\theta)$, or $\frac{d^2R}{d\theta^2}$.

$$R''(\theta) = \frac{2v_0^2}{g}\frac{d}{d\theta}(\cos 2\theta)$$

$$= \frac{2v_0^2}{g}(-2\sin 2\theta)$$

$$= \frac{-4v_0^2}{g}\sin 2\theta$$

$$R''(\pi/4) = \frac{-4v_0^2}{g}\sin(\pi/2)$$

$$= \frac{-4v_0^2}{g}$$

$R''(\pi/4) = \frac{-4v_0^2}{g} < 0$, so the graph of $R(\theta)$ is concave down at $\theta = \pi/4$ and $R(\theta)$ has a maximum at $\frac{\pi}{4}$. Therefore, an angle of $\frac{\pi}{4}$ radians, or $45°$, is the angle that will give the greatest horizontal distance.

NOTE An alternative approach would be to do this optimization entirely without calculus. $R(\theta) = \frac{v_0^2 \sin(2\theta)}{g}$; we need to find the angle $\theta \in \left[0, \frac{\pi}{2}\right]$ to make $\sin(2\theta)$ greatest. Either this can be calculated directly, or we can again let $u = 2\theta$.

If $\theta \in \left[0, \frac{\pi}{2}\right]$, then $u = 2\theta \in [0, \pi]$. On $[0, \pi]$ $\sin u$ is maximum at $u = \frac{\pi}{2}$, so $\theta = \frac{u}{2} = \frac{\pi}{4}$.

We have shown that for any fixed initial speed the projectile will travel the farthest horizontal distance if it is launched at a $45°$ angle.   ◆

EXAMPLE 21.3   A lighthouse is located 3 kilometers away from a long, straight beach wall. The beacon of light is rotating steadily at a rate of $1\frac{1}{2}$ revolutions per minute.

(a) A lone soul is sitting on the beach wall 5 kilometers from the lighthouse, staring into the sea and contemplating the universe. At what rate is the ray of light moving along the beach wall when it passes the thinker?

(b) At what point along the beach wall is the beam moving most slowly?

SOLUTION   (a) Begin with a picture. Do this on your own, thinking carefully about what is known and what you are trying to find.[10] Then compare your work with what is given below.

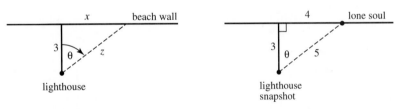

Figure 21.10

What We Want: $\frac{dx}{dt}$, the rate at which the ray of light is moving along the beach wall when $z = 5$.

What We Know: The light makes $1\frac{1}{2}$ revolutions per minute, so it goes through $\frac{3}{2}\frac{\text{revolutions}}{\text{minute}} \cdot \frac{2\pi \text{ radians}}{1 \text{ revolution}} = 3\pi \frac{\text{radians}}{\text{minute}}$.

$$\frac{d\theta}{dt} = 3\pi \text{ radians/minute}$$

Strategy   We know $\frac{d\theta}{dt}$ and we want to find $\frac{dx}{dt}$. Our strategy is to write an equation relating θ and x.[11] We then differentiate with respect to time to get a relationship between $\frac{d\theta}{dt}$ and

---

[10] A common error is to assume the beam of light has a fixed length.

[11] x varies. It is *not* always 4. It is only *after* we have differentiated that we can substitute in values for quantities that vary with time.

$\frac{dx}{dt}$. We try to find a trigonometric function involving only sides we know or are concerned about.

$$\frac{x}{3} = \tan \theta$$

$$\frac{d}{dt}\left[\frac{x}{3}\right] = \frac{d}{dt}(\tan \theta) \qquad \text{Differentiate each side with respect to time.}$$

$$\frac{1}{3}\frac{dx}{dt} = (\sec^2 \theta)\frac{d\theta}{dt} \qquad x \text{ and } \theta \text{ are functions of } t, \text{ so use the Chain Rule.}$$

$$\frac{dx}{dt} = 3(\sec^2 \theta)\frac{d\theta}{dt} \qquad \text{Solve for } \frac{dx}{dt}, \text{ the rate we want to find.}$$

Now we can substitute the values we know.

We know $\frac{d\theta}{dt} = 3\pi$. To find $\sec \theta$, we use the fact that $z = 5$ at the moment in question.

$$\sec \theta = \frac{\text{hyp}}{\text{adj}} = \frac{5}{3}$$

Therefore, at the moment when the beam passes the thinker, $z = 5$ and

$$\frac{dx}{dt} = 3(\sec^2 \theta)\frac{d\theta}{dt}$$

$$= 3\left[\frac{25}{9}\right] \cdot 3\pi$$

$$= 25\pi \text{ miles per minute.}$$

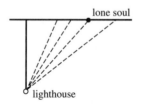

lone soul

lighthouse

Figure 21.11

(b) At what point along the cliffs is the spot of light moving most slowly?

For our purposes, the angle θ must be between $-\frac{\pi}{2}$ and $\frac{\pi}{2}$ or the light will not be shining on the cliffs. We know the rate at which the spot of light moves:

$$\frac{dx}{dt} = 3(\sec^2 \theta)\frac{d\theta}{dt} = 3 \cdot 3\pi (\sec^2 \theta) \text{ radians/minute.}$$

To find the place where the light moves most slowly, we want to find the angle θ that minimizes $\frac{dx}{dt}$. Since $\sec^2 \theta = \frac{1}{\cos^2 \theta}$, we want to make $\cos^2 \theta$ as large as possible. This will occur when $\cos \theta$ is at its maximum, which occurs at $\theta = 0$, or when the light is directly across from the lighthouse, when the beam is shortest.

Notice that the further away an object on the beach wall is from the lighthouse, the faster the beam will sweep past it. ◆

PROBLEMS FOR SECTION 21.3

1. Let $f(x) = x + 2 \sin x$.
   (a) Find all of the critical points.
   (b) Where is $f(x)$ increasing? Decreasing?
   (c) Where does $f(x)$ have local maxima? Local minima?
   (d) Does $f(x)$ have global maxima? Global minima? If so, what are the absolute maximum and minimum values?
   (e) Where is $f(x)$ concave up? Concave down?
   (f) Sketch a graph of $f(x)$.

2. Consider the function $f(x) = -\cos x + \frac{1}{2} \sin 2x$.
   (a) Explain how you can tell that f is periodic with period 2π.
   (b) Find and classify all the critical points of f on the interval $[0, 2\pi]$. Do the trigonometric "algebra" on your own, then check your answers using a graphing calculator. (*Hint:* You'll get a $\cos 2x$ that you'll need to rewrite.)

For Problems 3 through 6, graph f on the interval $[0, 2\pi]$ labeling the x-coordinates of all local extrema.

3. $f(x) = \cos x + \sqrt{3} \sin x$

4. $f(x) = \cos x - \sin x$

5. $f(x) = \cos 2x - 2 \cos x$

6. $f(x) = e^x \sin x$

7. Use a tangent line approximation to approximate the following. In each case, use concavity to determine whether the approximation is larger or smaller than the actual value. Then compare your results with the approximations given by a calculator or computer.
   (a) $\sin 0.2$
   (b) $\sin 0.1$
   (c) $\sin 0.01$
   (d) $\sin(-0.1)$

8. Creme Fraiche and Caveat are battling their way to the finish line in the last leg of horseracing's Triple Crown. At the finish line, 30 feet away from the track itself, is a camera that is focused on the leading horse who is moving down the stretch at a rate of 46 feet per second. At what rate is the camera rotating when the lead horse is 50 feet from the finish line?

9. Verify that $\sec x$ has local minima at $x = 2\pi k$ and local maxima at $x = \pi + 2\pi k$ (k an integer) by identifying its critical points and using the second derivative test for maxima and minima.

10. Verify that $\tan x$ has points of inflection at $x = \pi k$, k an integer, by showing that the sign of its second derivative changes at these points.

11. Let $f(x) = 3\cos x + 2\sin x$.

   (a) What is the period of f?

   (b) What are the maximum and minimum values of f?

12. If we ignore air resistance, a baseball thrown from shoulder level at an angle of θ radians with the ground and at an initial velocity of v_0 meters per second will be at shoulder level again when it is $\frac{v_0^2 \sin(2\theta)}{g}$ meters away. g is the acceleration due to gravity (9.8 m/sec^2).

   (a) Express the maximum distance the baseball can travel (from shoulder level to shoulder level) in terms of the initial velocity.

   (b) The fastest baseball pitchers can throw about 100 miles per hour. How far would such a ball travel if thrown at the optimal angle? (*Note:* 1 mile = 5280 feet and 1 meter $\approx$ 3.28 feet.)(*)

13. A policewoman is standing 80 feet away from a long, straight fence when she notices someone running along it. She points her flashlight at him and keeps it on him as he runs. When the distance between her and the runner is 100 feet he is running at 9 feet per second. At this moment, at what rate is she turning the flashlight to keep him illuminated? Include units in your answer.

14. A sewage gutter is to be constructed from a piece of sheet metal 8 feet long and 4 feet wide by folding up a 1-foot strip on each side. Denote by θ the angle between the sides and the vertical, as shown in the figure below. What angle θ will result in a sewage gutter of maximum volume?

15. A lookout tower is located 0.5 kilometers from a line of warehouses. A searchlight on the tower is rotating at a rate of 6 revolutions per minute. How fast is the beam of light moving along the wall of warehouses when it passes by a window located 1 kilometer from the tower?

16. Graph $f(x) = 2^{\cos x}$.

   (a) Is the function periodic? If so, what is its period?

   (b) What is its maximum value? Its minimum value? Give exact answers.

17. Let $f(x) = -\cos x$ and $g(x) = \sin x$.

   (a) What is the maximum distance between these two curves on the interval $[-\frac{\pi}{4}, \frac{3\pi}{4}]$?

   (b) What is the point of intersection of the tangent lines to these curves at the points from part (a) where the curves are farthest apart? Does this answer surprise you? Explain.

18. A wheel of radius 5 meters is oriented vertically and spinning counterclockwise at a rate of 7 revolutions per minute. If the origin is placed at the center of the wheel a point on the rim has a horizontal position of $(7, 0)$ at time $t = 0$. What is the horizontal component of the point's velocity at $t = 2$?

19. In this problem you will show that $y = C_1 \sin kx + C_2 \cos kx$ is a solution to the differential equation

$$y'' = -k^2 y.$$

Recall that a differential equation is an equation involving a derivative and a function is a solution to the differential equation if it satisfies the differential equation.

(a) Show that $y_1 = \sin kx$ is a solution to $y'' = -k^2 y$. To do this, first find y_1''. Then write

$$y_1'' \overset{?}{=} -k^2 y_1$$

and verify that the two sides are indeed equal.

(b) Show that $y_2 = C_1 \sin kx$ is a solution to $y'' = -k^2 y$.

(c) Show that if y_1 and y_2 are solutions to $y'' = -k^2 y$, then $y_3 = C_1 y_1 + C_2 y_2$ is a solution to $y'' = -k^2 y$ as well. Conclude that $y = C_1 \sin kx + C_2 \cos kx$ is a solution to the differential equation $y'' = -k^2 y$.

20. For each of the functions below, determine whether the function is a solution to differential equation (i), differential equation (ii), or neither. Differential equations (i) and (ii) are given below.

$$\text{i.} \, y'' = 16y \qquad \text{ii.} \, y'' = -16y$$

(a) $y_1(t) = \sin 16t$

(b) $y_2(t) = e^{4t}$

(c) $y_3(t) = 3 \cos 4t$

(d) $y_4(t) = \sin 4t + 1$

(e) $y_5(t) = e^{-16t}$

(f) $y_6(t) = -3e^{-4t}$

(g) $y_7(t) = e^{4t} + 3$

(h) $y_8(t) = -\sin 4t$

(*Hint:* Four of the eight functions given are solutions to *neither* differential equation.)

21. Each of the functions below is a solution to *one* of the differential equations below.

$$\text{i. } y'' = 9 \qquad \text{ii. } y'' = 9y \qquad \text{iii. } y'' = -9y$$

For each function, determine which of the three differential equations it satisfies.

(a) $y_1(t) = 5 \sin 3t$

(b) $y_2(t) = e^{3t}$

(c) $y_3(t) = 2 \cos 3t$

(d) $y_4(t) = 4.5t^2 + 3t + 8$

(e) $y_5(t) = 4e^{-3t}$

(f) $y_6(t) = 4.5t^2 - t + 2$

22. After having done the previous problem, make up a solution to each of the three differential equations below.

i. $y'' = 9$ ii. $y'' = 9y$ iii $y'' = -9y$

Your answers must be different from the solutions given in the preceding problem, but you can use those answers for inspiration.

Check that your answers are right by "plugging them back" into the differential equation. For instance, if you guess that $y = e^{3t} + 1$ is a solution to the differential equation $y'' = 9y$, test it out as follows. First calculate y''. Since $y' = e^{3t} \cdot 3$, $y'' = 9e^{3t}$. Now see if it satisfies the differential equation.

$$y'' \overset{?}{=} 9y$$

$$9e^{3t} \overset{?}{=} 9(e^{3t} + 1)$$

$$9e^{3t} \neq 9e^{3t} + 9$$

So $y = e^{3t} + 1$ is *not* a solution to $y'' = 9y$.

23. As you're riding up an elevator inside the Hyatt Hotel right next to the Charles River, you watch a duck swimming across the Charles, swimming straight toward the base of the elevator. The elevator is rising at a speed of 10 feet per second, and the duck is swimming at 5 feet per second toward the base of the elevator. As you pass the eighth floor, 100 feet up from the level of the river, the duck is 200 feet away from the base of the elevator.

(a) At this instant, is the distance between you and the duck increasing or decreasing? At what rate?

(b) As you're watching the duck, you have to look down at more and more of an angle to see it. At what rate is this angle of depression increasing at the instant when you are at a height of 100 feet? Include units in your answer.

(Problem written by Andrew Engelward)

24. *Approximating the function $f(x) = \sin x$ near $x = 0$ by using polynomials.*

The point of this problem is to show you how the values of $\sin x$ can be approximated numerically with a very high degree of accuracy. It is an introduction to Taylor polynomials.

(a) Find the equation of the line tangent to $f(x) = \sin x$ at $x = 0$.

(b) Find the equation of a quadratic $Q(x) = a + bx + cx^2$ such that the function $Q(x)$ and its nonzero derivatives match those of $\sin x$ at $x = 0$. In other words, $Q(0) = f(0)$, $Q'(0) = f'(0)$, and $Q''(0) = f''(0)$. The quadratic that you found is the quadratic that best "fits" the sine curve near $x = 0$. In fact, the "quadratic" turns out not to really be a quadratic at all. Sine is an odd function, so there is no parabola that "fits" the sine curve well at $x = 0$.

(c) Find the equation of a cubic $C(x) = a + bx + cx^2 + dx^3$ such that the function $C(x)$ and its nonzero derivatives match those of $\sin x$ at $x = 0$. In other words, $C(0) = f(0)$, $C'(0) = f'(0)$, $C''(0) = f''(0)$, and $C'''(0) = f'''(0)$. The cubic that you found is the cubic that best "fits" the sine curve near $x = 0$.

Using a calculator, on the same set of axes graph $\sin x$, the tangent line to $\sin x$ at $x = 0$, and $C(x)$, the cubic you found. Now "zoom in" around $x = 0$. Can you see that near $x = 0$ the line is a good fit to the sine curve but the cubic is an even better fit for small x and the cubic hugs the sine curve for longer? The next set of questions asks you to investigate how good the fit is.

(d) Use $C(x)$ from part (c) to estimate the following, and then compare with the actual value using a calculator.
$\sin(0.01)$ $\sin(0.1)$ $\sin(0.5)$ $\sin(1)$ $\sin(3)$

(e) (Challenge) Find the "best" fifth degree polynomial approximation of $\sin(x)$ for x near 0 by making sure that the first five derivatives of the polynomial match those of $\sin(x)$ when evaluated at $x = 0$. Graph $\sin x$ along with its first, third, and fifth degree polynomial approximations on your graphing calculator. The higher the degree of the polynomial, the better the fit to $\sin x$ near $x = 0$, right?

(f) Using a calculator, on the same set of axes graph $\sin x$ and the polynomial given below.

$$x - \frac{x^3}{6} + \frac{x^5}{120} - \frac{x^7}{5040} + \frac{x^9}{362880}$$

This polynomial is an even better fit than the last one, right?

Now graph the difference between $\sin x$ and this polynomial; in other words, graph

$$y = \sin x - \left[x - \frac{x^3}{6} + \frac{x^5}{120} - \frac{x^7}{5040} + \frac{x^9}{362880} \right].$$

On approximately what interval is the difference between $\sin x$ and this polynomial less than 0.005?

21.4 DERIVATIVES OF INVERSE TRIGONOMETRIC FUNCTIONS

We can find the derivatives of inverse trigonometric functions using implicit differentiation.

The Derivative of arcsin x

Let $y = \arcsin x$. Our goal is to find $\frac{dy}{dx}$.

$y = \arcsin x$	y is the angle between $-\dfrac{\pi}{2}$ and $\dfrac{\pi}{2}$ whose sine is x.
$\sin y = x$	Therefore, the sine of y is x. (sine and arcsine are inverse functions.)
$\dfrac{d}{dx} \sin y = \dfrac{d}{dx} x$	Differentiate each side with respect to x.
$(\cos y) \dfrac{dy}{dx} = 1$	Remember the Chain Rule on the left-hand side, since y is a function of x.
$\dfrac{dy}{dx} = \dfrac{1}{\cos y}$	Solve for $\dfrac{dy}{dx}$.

We'd like to express $\frac{dy}{dx}$ in terms of x, not y. We know that $x = \sin y$ and that $\sin y$ and $\cos y$ are related by $\sin^2 y + \cos^2 y = 1$. Solving for $\cos y$ gives us

$$\cos y = \pm\sqrt{1 - \sin^2 y}$$
$$= \pm\sqrt{1 - x^2}.$$

Do we want the positive square root, or the negative square root? y is an angle between $-\frac{\pi}{2}$ and $\frac{\pi}{2}$, so $\cos y$ is nonnegative; we choose $\sqrt{1 - x^2}$. Therefore, $\frac{dy}{dx} = \frac{1}{\sqrt{1-x^2}}$.

$$\boxed{\frac{d}{dx} \arcsin x = \frac{1}{\sqrt{1 - x^2}}}$$

Alternatively, once we know $\frac{dy}{dx} = \frac{1}{\cos y}$, we can replace y by $\arcsin x$ obtaining

$$\frac{dy}{dx} = \frac{1}{\cos(\arcsin x)}.$$

$\cos(\arcsin x)$ is the cosine of the angle between $-\pi/2$ and $\pi/2$ whose sine is x. We can draw a triangle with an acute angle θ whose sine is x, use the Pythagorean Theorem to solve for the length of the third side, and find $\cos(\arcsin x) = \sqrt{1 - x^2}$, as illustrated in Figure 21.12.[12]

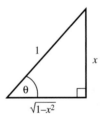

Figure 21.12

The Derivative of arctan x

We can use the same method to compute $\frac{d}{dx} \arctan x$.

$$y = \arctan x \qquad y \text{ is the angle between } -\frac{\pi}{2} \text{ and } \frac{\pi}{2} \text{ whose tangent is } x.$$

$$\tan y = x$$

$$\frac{d}{dx} \tan y = \frac{d}{dx} x \qquad \text{Use the Chain Rule on the left-hand side.}$$

$$(\sec^2 y)\frac{dy}{dx} = 1$$

$$\frac{dy}{dx} = \cos^2 y$$

---

[12] Knowing that $\theta \in [-\pi/2, \pi/2]$ assures us that $\cos \theta$ is nonnegative.

To express $\frac{dy}{dx}$ in terms of x we can replace y with arctan x. cos(arctan x) is the cosine of the angle between $-\pi/2$ and $\pi/2$ whose tangent is x. Notice that cosine is positive on this interval. We can draw a triangle with an acute angle θ whose tangent is x, use the Pythagorean Theorem to solve for the length of the third side, and find cos(arctan x) $=$ $\frac{1}{\sqrt{1+x^2}}$, as illustrated in Figure 21.13.

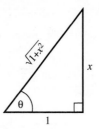

Figure 21.13

$\cos y = \frac{1}{\sqrt{1+x^2}}$, so $\frac{dy}{dx} = \cos^2 y = \frac{1}{1+x^2}$.

$$\boxed{\frac{d}{dx} \arctan x = \frac{1}{1+x^2}}$$

EXERCISE 21.4   Using the same method demonstrated in the previous two examples, show the following.

$$\boxed{\frac{d}{dx} \arccos x = \frac{-1}{\sqrt{1-x^2}}}$$

COMMENT Surprisingly, the derivatives of the inverse trigonometric functions involve neither trigonometric functions nor inverse trigonometric functions. We will find it handy to keep inverse trigonometric functions and their derivatives in mind later on, when searching for a function whose derivative is $\frac{1}{1+x^2}$ or $\frac{1}{\sqrt{1-x^2}}$. We mention this here because the inverse trigonometric functions are probably *not* the first things you might think of when looking for such functions!

◆ **EXAMPLE 21.4**   Differentiate.

   (a) $f(x) = x \arcsin(x^2)$ (b) $j(x) = 5 \tan^{-1} x^2$ (c) $g(x) = \tan \sqrt{x} \cdot \tan^{-1} \sqrt{x}$

SOLUTIONS   (a) Use the Product Rule. The Chain Rule tells us that
$\frac{d}{dx} \arcsin(mess) = \frac{1}{\sqrt{1-(mess)^2}} \cdot (mess)'$. Here $mess = x^2$.

$$f'(x) = x \cdot \frac{d}{dx}[\arcsin(x^2)] + \arcsin(x^2)$$

$$= x \cdot \frac{1}{\sqrt{1-(x^2)^2}} \cdot 2x + \arcsin(x^2)$$

$$= \frac{2x^2}{\sqrt{1-x^4}} + \arcsin(x^2)$$

(b) The Chain Rule tells us that $\frac{d}{dx} \tan^{-1}(mess) = \frac{1}{1+(mess)^2} \cdot (mess)'$, where $mess = x^2$.

$$j'(x) = \frac{5}{1 + (x^2)^2} \cdot 2x$$

$$= \frac{10x}{1 + x^4}$$

CAUTION $(mess)'$ is the derivative of x^2, not of x^4.

(c) $g(x) = \tan \sqrt{x} \cdot \tan^{-1} \sqrt{x}$. Again we need the Product Rule together with the Chain Rule.

$$g'(x) = \left[\sec^2 \sqrt{x} \cdot \frac{1}{2} \frac{1}{\sqrt{x}} \right] \cdot \tan^{-1} \sqrt{x} + \tan \sqrt{x} \cdot \frac{1}{1 + (\sqrt{x})^2} \cdot \frac{1}{2} \frac{1}{\sqrt{x}}$$

$$= \frac{1}{2\sqrt{x}} \left[\sec^2 \sqrt{x} \cdot \tan^{-1} \sqrt{x} + \frac{\tan \sqrt{x}}{1 + x} \right] \quad \blacklozenge$$

PROBLEMS FOR SECTION 21.4

For Problems 1 through 6, differentiate the function given.

1. $y = 3 \tan x - 4 \tan^{-1} x$

2. $f(x) = 3 \arctan \left(2\sqrt{x} \right)$

3. $y = \sin x \cdot \arcsin x$

4. $y = \sqrt{\tan^{-1} x}$

5. $y = x \tan^{-1} x$

6. $y = \frac{\arctan(e^x)}{e}$

7. (a) Show that the derivative of arccos x is $\frac{-1}{\sqrt{1-x^2}}$.
   (b) What is the domain of arccos x?
   (c) What is the range of arccos x?
   (d) Where is the graph of arccos x decreasing?
   (e) Where is the graph of arccos x concave up? Concave down? If there is a point of inflection, where is it?
   (f) Graph $f(x) = \arccos x$.

8. Differentiate $f(x) = 3 \cos \left(\frac{1}{x^2+1} \right) + x \arctan \left(\frac{1}{x} \right)$.

9. Compute $\frac{d}{dx} \frac{\sin^{-1} x}{\cos^{-1} x}$. Is it the same as $\frac{d}{dx} \tan^{-1} x$?

21.5   BRIEF TRIGONOMETRY SUMMARY

Unit Circle Definitions of Trigonometric Functions

$\sin x = v$

$\cos x = u$

$\tan x = \dfrac{\sin x}{\cos x} = \dfrac{v}{u}$

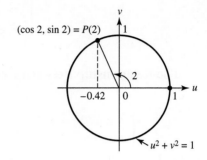

Right-Triangle Definitions of Trigonometric Functions

$\sin x = \dfrac{\text{opp}}{\text{hyp}}$ $\csc x = \dfrac{1}{\sin x}$

$\cos x = \dfrac{\text{adj}}{\text{hyp}}$ $\sec x = \dfrac{1}{\cos(x)}$

$\tan x = \dfrac{\text{opp}}{\text{adj}}$ $\cot x = \dfrac{1}{\tan(x)}$

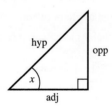

Radians and Degrees. π radians $= 180°$

Special triangles

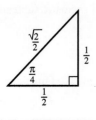

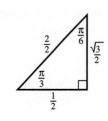

Note: The shortest side is opposite the smallest angle and the largest side is opposite the largest angle.

45°-45°-90°   30°-60°-90°

Graphs of the Trigonometric Functions

Graph of $f(x) = \sin(x)$

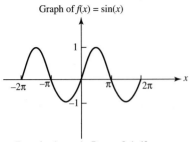

Domain: $(-\infty, \infty)$ Range: $[-1, 1]$

Graph of $f(x) = \cos(x)$

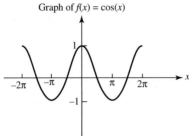

Domain: $(-\infty, \infty)$ Range: $[-1, 1]$

Graph of $f(x) = \tan(x)$

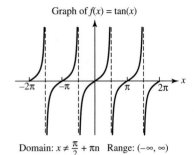

Domain: $x \neq \frac{\pi}{2} + \pi n$ Range: $(-\infty, \infty)$

The Relationship Between Graphs and Equations

$$y = \begin{cases} A\,\sin(Bx) + K \\ A\,\cos(Bx) + K \end{cases} \quad \text{amplitude} = |A|, \quad \text{period} = \frac{2\pi}{|B|}, \quad \text{balance value} = K$$

$y = A\,\sin(Bx + C)$ should be rewritten as $y = A\,\sin\left[B\left(x + \frac{C}{B}\right)\right]$; to read off

amplitude $= |A|$, period $= \frac{2\pi}{|B|}$, shift the graph of $y = A\,\sin(Bx)$ to the left $\frac{C}{B}$ units.

Inverse Trigonometric Functions

We must restrict the domain of sine, cosine, and tangent in order to construct inverse functions, because a function must be one-to-one to have an inverse function.

$\arcsin(x)$ is the angle between $-\pi/2$ and $\pi/2$ whose sine is x.   domain of arcsin x: $[-1, 1]$

$\arctan(x)$ is the angle between $-\pi/2$ and $\pi/2$ whose tangent is x.   domain of arctan x: $(-\infty, \infty)$

$\arccos(x)$ is the angle between 0 and π whose cosine is x.   domain of arccos x: $[-1, 1]$

NOTE If A is positive then $\arcsin(A)$, $\arccos(A)$, and $\arctan(A)$ are all angles between 0 and $\pi/2$ (provided A is in the function's domain).

If A is positive and in the function's domain, then

$\arcsin(-A) = -\arcsin(A)$,

$\arccos(-A) = \pi - \arccos(A)$, and

$\arctan(-A) = -\arctan(A)$.

Arc Length.   The arc length of a circle of radius R subtended by an angle of x radians is Rx.

Derivatives

Function	Derivative	Function	Derivative
$f(x) = \sin(x)$	$f'(x) = \cos(x)$	$f(x) = \arcsin(x)$	$f'(x) = \dfrac{1}{\sqrt{1-x^2}}$
$f(x) = \cos(x)$	$f'(x) = -\sin(x)$	$f(x) = \arccos(x)$	$f'(x) = -\dfrac{1}{\sqrt{1-x^2}}$
$f(x) = \tan(x)$	$f'(x) = \sec^2(x)$	$f(x) = \arctan(x)$	$f'(x) = \dfrac{1}{1+x^2}$

Remember the Chain Rule

$$\frac{d}{dx}\sin[g(x)] = \cos[g(x)] \cdot \frac{dg}{dx} \qquad \frac{d}{dx}\arcsin[g(x)] = \frac{1}{\sqrt{1-[g(x)]^2}} \cdot \frac{dg}{dx}$$

$$\frac{d}{dx}\cos[g(x)] = -\sin[g(x)] \cdot \frac{dg}{dx} \qquad \frac{d}{dx}\arccos[g(x)] = \frac{-1}{\sqrt{1-[g(x)]^2}} \cdot \frac{dg}{dx}$$

$$\frac{d}{dx}\tan[g(x)] = \sec^2[g(x)] \cdot \frac{dg}{dx} \qquad \frac{d}{dx}\arctan[g(x)] = \frac{1}{1+[g(x)]^2} \cdot \frac{dg}{dx}$$

Trigonometric Identities
Pythagorean Identity

$$\sin^2 A + \cos^2 A = 1$$

It follows that

$$\frac{\sin^2 A}{\cos^2 A} + \frac{\cos^2 A}{\cos^2 A} = \frac{1}{\cos^2 A} \quad \text{so } \tan^2 A + 1 = \sec^2 A$$

$$\frac{\sin^2 A}{\sin^2 A} + \frac{\cos^2 A}{\sin^2 A} = \frac{1}{\sin^2 A} \quad \text{so } 1 + \cot^2 A = \csc^2 A.$$

Addition Formulas and the Double-Angle Formulas

$$\sin(A + B) = \sin A \cos B + \sin B \cos A$$
$$\cos(A + B) = \cos A \cos B - \sin A \sin B$$

It follows that

$$\sin(2A) = 2 \sin A \cos A \text{ and } \cos(2A) = \cos^2 A - \sin^2 A.$$

General Triangles

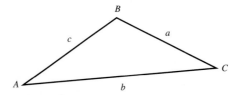

Law of Cosines

$$c^2 = a^2 + b^2 - 2ab \cos C$$

You can think of $-2ab \cos C$ as a "correction term" for the Pythagorean Theorem; it is positive when $C > \pi/2$ and negative when $C < \pi/2$.

Law of Sines

$$\frac{a}{\sin A} = \frac{b}{\sin B} = \frac{c}{\sin C}$$

C H A P T E R

22

Net Change in Amount and Area: Introducing the Definite Integral

■ **22.1 FINDING NET CHANGE IN AMOUNT: PHYSICAL AND GRAPHICAL INTERPLAY**

Introduction

The derivative allows us to answer two related problems.

■ How do we calculate the instantaneous rate of change of a quantity?

■ How do we calculate the slope of the line tangent to a curve at a point?

The physical and graphical questions are intertwined; the slope of the graph of an "amount" function can be interpreted as the instantaneous rate of change of the function.

Given an "amount" function, we can derive a "rate" function; we now shift our viewpoint and investigate the problem of how to recover an "amount" function when given a "rate" function.[1] If we know the rate of change of a quantity, how can we find the net change in the quantity over a certain time period? Suppose, for example, that we know an object's velocity over a specified time interval. Then we ought to be able to use that information to

---

[1] We did this when we looked at projectile motion in Section 20.7.

determine the object's net change in position during that time. To figure out how to do this we begin by looking at some simple examples where the rate of change is constant and then apply what we learn to cases where the rate of change is not constant.

Let's clarify what is meant by *net* change. If you take two steps forward and one step back, your net change in position is one step forward. If, over the course of a day, the stock market falls 100 points and then gains 40 points, the net change for the day is −60 points.

We now prepare to tackle the following two questions.

- ▪ Given a rate function, how do we calculate the net change in amount?

- ▪ How do we calculate the area under the graph of a function?

In this chapter we aim to convince you that the physical and graphical questions are closely related. We'll approach the questions using the strategy that served us so well in developing the derivative: the method of successive approximation followed by a limiting process.

Calculating Net Change When Rate of Change Is Constant

◆ EXAMPLE 22.1   Suppose that a moose is strolling along at a constant velocity. The distance she travels can be calculated by the familiar formula distance = (rate) · (time).

If the moose moves at 4 miles per hour between 1:15 P.M. and 1:30 P.M., the distance she has traveled (or her net change in position) is

$$4 \ \frac{\text{miles}}{\text{hour}} \cdot \frac{1}{4} \ \text{hours} = 1 \ \text{mile},$$

where the units cancel to give us an answer in the units we expect.

More generally, let $s(t)$ be the position of an object at time t and suppose that its velocity, $\frac{ds}{dt}$, is **constant** between times $t = a$ and $t = b$.

$$\text{net change in position} \ = \ (\text{rate of change of position}) \ \cdot \ (\text{time elapsed})$$
$$\Delta s \ = \ \frac{ds}{dt} \ \cdot \ \Delta t$$

Δt denotes the change in time, $\Delta t = (b - a)$, and $\Delta s = s(b) - s(a)$.

We can represent the net change in position graphically.

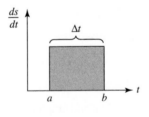

Figure 22.1

When velocity is constant, net change in position is calculated by multiplying rate of change, $\frac{ds}{dt}$, by time; geometrically, this corresponds to the area of the shaded rectangle in Figure 22.1.   ◆

◆ EXAMPLE 22.2   Let $W(t)$ be the amount of water in a swimming pool as a function of time, where time is measured in hours. Suppose the pool is being filled at a constant rate of $\frac{dW}{dt}$ gallons per hour between noon and 12:30 P.M. In that half-hour,

the net change in amount of water $\;=\;$ rate at which water is entering $\;\cdot\;$ (time elapsed)

$$\Delta W \qquad = \qquad \frac{dW}{dt} \qquad \cdot \qquad \Delta t,$$

where Δt is one-half of an hour.

Graphically, this net change can be represented as the area of the rectangle shaded in Figure 22.2.

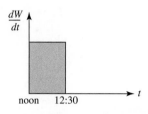

Figure 22.2   ◆

◆ **EXAMPLE 22.3**   Let $C(x)$ be the cost (in dollars) of producing x grams of feta cheese. Suppose that for each additional gram produced the cost increases by a constant amount; $\frac{dC}{dx}$ is constant.[2] Then,

the net change $\qquad=\qquad$ rate of change $\qquad\cdot\qquad$ change in number
in production cost $\qquad\qquad\qquad$ of cost per gram $\qquad\qquad$ of grams produced

$$\Delta C \qquad = \qquad \frac{dC}{dx} \qquad \cdot \qquad \Delta x$$

Notice that the independent variable (in this case x, the number of grams of cheese produced) does not have to represent time.

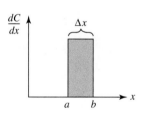

Figure 22.3   ◆

◆ **EXAMPLE 22.4**   On a certain day, the temperature in Kathmandu decreases at a constant rate of two degrees per hour from 6:00 P.M. to 10:00 P.M. If we let $T(t)$ be the temperature at time t, we can write the following.

net change in temperature $\;=\;$ rate of change of temperature $\;\cdot\;$ (time elapsed)

$$\Delta T \qquad = \qquad \frac{dT}{dt} \qquad \cdot \qquad \Delta t$$

$$\Delta T \qquad = \qquad (-2)\,\frac{degrees}{hour} \qquad \cdot \;(10-6)\;hours$$

$$= \qquad -8\;degrees$$

---

[2] Students of economics will recognize this as marginal cost.

When we represent this geometrically, we have a rectangle *under* the t-axis. We'd like to convey the information that ΔT is negative by attaching a sign to the area. The net change in temperature is negative because the temperature is decreasing; $\frac{dT}{dt}$ is negative. We can say that the "height" of the rectangle is -2 (since it lies below the t-axis) and assign the rectangle a signed area of -8.

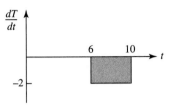

Figure 22.4

When using areas to represent net changes, we will use the idea of "signed area"—areas are positive or negative, depending on whether the region lies above or below the horizontal axis.

To illustrate further how signed area is used, suppose that we also know that the temperature in Kathmandu on this particular day increased at a rate of 1.5 degrees per hour between 3:00 P.M. and 6:00 P.M.

$$\Delta T \text{ from 3:00 P.M. to 6:00 P.M.} = 1.5 \frac{\text{degrees}}{\text{hour}} \cdot (6-3) \text{ hours}$$
$$= 4.5 \text{ degrees}$$

Putting the two time periods together gives a net change of $4.5° + (-8°) = -3.5°$ between 3:00 P.M. and 10:00 P.M. Graphically, this is represented by one rectangle above the t-axis (when $\frac{dT}{dt}$ is positive) and one rectangle below the t-axis (when $\frac{dT}{dt}$ is negative).

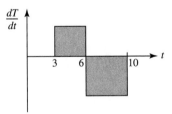

Figure 22.5

REMARK: Although we know that the temperature dropped 3.5 degrees between 3:00 P.M. and 10:00 P.M., we have no idea of what the temperature actually was at any time. We need to know one specific data point for the function $T(t)$ before we can determine any actual values of T. For example, if we knew that at 3:00 P.M. the temperature was 55 degrees, then we would know that $T(10) = 55 - 3.5 = 51.5$ degrees. In fact, we could calculate $T(t)$ for any time between 3:00 P.M. and 10:00 P.M. On the other hand, if the temperature at 3:00 P.M. was 40 degrees, then $T(10) = 40 - 3.5 = 36.5$. ◆

As illustrated by the preceding examples, finding a quantity's net change is straightforward provided that its rate of change is constant. However, the rate of change often is not constant. How can we approach these situations?

Approximating Net Change When Rate of Change is Not Constant

◆ **EXAMPLE 22.5a**   Suppose that a cheetah's velocity is given by $v(t) = 2t + 5$ meters per second on the interval $[1, 4]$. Approximate its net change in position over this interval by giving upper and lower bounds that differ by less than 0.1 meter.

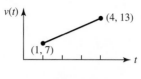

Figure 22.6

SOLUTION   The problem we face is analogous to the one confronting us when we first set out to find the slope of a curve. At that time, we knew how to find the slope of a straight line, but not the slope of a curve, so we approximated the slope of a curve at $x = a$ by the slope of a secant line through points $(a, f(a))$ and $(a + \Delta x, f(a + \Delta x))$. To get successively better approximations, we repeatedly shortened the interval Δx. Then, to obtain an exact value for the slope at $x = a$, we evaluated the limit as Δx approached zero.

To tackle the present problem, we will follow a similar procedure, that of successive approximations. If $r(t)$ is a nonconstant rate of change on an interval $[a, b]$, we divide the interval into smaller subintervals estimate the net change within each of those subintervals, and sum to find the accumulated change. On each of these subintervals we approximate $r(t)$ by a constant rate of change.

The Method of Successive Approximations

Take One.   The function $v(t)$ is increasing throughout the interval $[1, 4]$, so the velocity is at least $v(1) = 7$ m/sec and at most $v(4) = 13$ m/sec.

$$7 \leq v(t) \leq 13, \quad \text{for all } t \text{ in the interval } [1, 4].$$

If the cheetah were moving at a constant rate, then its net change in position would be given by (rate) · (time); therefore we know the following.

$$(7\text{m/sec}) \cdot (3 \text{ sec}) \leq \text{net change in position} \leq (13 \text{ m/sec})(3 \text{ sec})$$

$$21 \text{ meters} \leq \text{net change in position} \leq 39 \text{ meters}$$

Therefore, the cheetah moved between 21 and 39 meters.

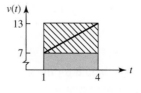

Figure 22.7

How can we improve on these bounds?

Take Two.   To get a better approximation, let's divide the interval $[1, 4]$ into smaller subintervals and approximate the rate of change by a constant function on each subinterval.

For the sake of convenience, let's divide the interval into three subintervals of equal width, [1, 2], [2, 3], and [3, 4], and call the width of each subinterval Δt.[3] The table below shows the velocity of the cheetah at the endpoints of each of these subintervals.

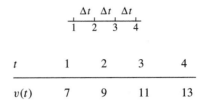

t	1	2	3	4
$v(t)$	7	9	11	13

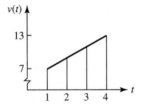

Figure 22.8

Because the velocity is increasing on the interval [1, 4], the velocity is least at the beginning of each subinterval (i.e., at the left endpoint) and greatest at the end of each subinterval (the right endpoint). During the time interval [1, 2] the velocity is greater than or equal to $v(1) = 7$ m/sec and less than or equal to $v(2) = 9$ m/sec. Again using the fact that if rate is constant, (rate) $\cdot$ (time) = net change, we deduce the following.

$v(1) \cdot (1 \text{ sec}) = (7 \text{ m/sec}) \cdot (1 \text{ sec}) = 7$ m is a **lower bound** for the distance traveled in the interval [1, 2]

$v(2) \cdot (1 \text{ sec}) = (9 \text{ m/sec}) \cdot (1 \text{ sec}) = 9$ m is an **upper bound** for the distance traveled in the interval [1, 2]

We can refer to the lower bounds and upper bounds as "underestimates," and "overestimates," respectively.

We use the same method to get lower and upper bounds for the distance the object travels in each of the other subintervals. Adding together the lower bounds for the three subintervals will give us a lower bound for the net change in the cheetah's position over the entire interval [1, 4]; similarly, summing the upper bounds for the subintervals will give an upper bound for the whole interval.

lower bound for the net change in position on [1, 4]	$v(1) \cdot 1 + v(2) \cdot 1 + v(3) \cdot 1 = 7 \cdot 1 + 9 \cdot 1 + 11 \cdot 1 =$ 27 meters
upper bound for the net change in position on [1, 4]	$v(2) \cdot 1 + v(3) \cdot 1 + v(4) \cdot 1 = 9 \cdot 1 + 11 \cdot 1 + 13 \cdot 1 =$ 33 meters

The net change in the cheetah's position on [1, 4] is more than 27 meters and less than 33 meters.

---

[3] Here $\Delta t = 1$, but we will soon subdivide further, and as we do so Δt will shrink.

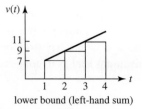
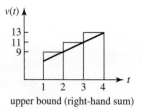

lower bound (left-hand sum)   upper bound (right-hand sum)

Figure 22.9

When we approximate the value of a function over each subinterval by using the value of the function at the left endpoint of each subinterval, then multiply by the length of the subinterval and sum the results, we call the sum a **left-hand sum**. It is denoted by L_n, where n denotes the number of subintervals. When we use the right endpoints of each subinterval, we call the sum a **right-hand sum** and denote it by R_n.

In this example, because the function $v(t)$ is increasing, the left-hand sum will be a lower bound and the right-hand sum will be an upper bound. In the graphical representation in Figure 22.9, the rectangles corresponding to the lower bound are inscribed rectangles and the rectangles corresponding to the upper bound are circumscribed rectangles.

Take Three.   We have now improved upon our first estimates and determined that the cheetah's net change in position is between 27 and 33 meters. To improve further on these estimates we can use even smaller subintervals. This time, let's divide the interval $[1, 4]$ into six subintervals, each of length $\Delta t = \frac{3}{6} = 0.5$ seconds. Then our estimates are the following.[4]

$$\text{lower bound} = L_6 = v(1.0)\Delta t + v(1.5)\Delta t + v(2.0)\Delta t + v(2.5)\Delta t + v(3.0)\Delta t + v(3.5)\Delta t$$
$$= 7(0.5) + 8(0.5) + 9(0.5) + 10(0.5) + 11(0.5) + 12(0.5)$$
$$= 28.5 \text{ meters}$$
$$\text{upper bound} = R_6 = v(1.5)\Delta t + v(2.0)\Delta t + v(2.5)\Delta t + v(3.0)\Delta t + v(3.5)\Delta t + v(4.0)\Delta t$$
$$= 8(0.5) + 9(0.5) + 10(0.5) + 11(0.5) + 12(0.5) + 13(0.5)$$
$$= 31.5 \text{ meters}$$

These bounds are closer together than in our previous estimates.

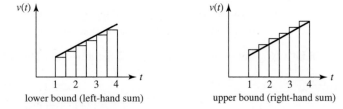

lower bound (left-hand sum)   upper bound (right-hand sum)

Figure 22.10

---

[4] These kinds of tedious sums are the sorts of tasks for which computers or programmable calculators are perfect. See if you can figure out how to construct such a program. Then learn how to use the technology available to you in order to compute left- and right-hand sums easily.

Notice that as we use more subdivisions, the graphical representation of the lower and upper bounds as the sum of areas of inscribed and circumscribed rectangles, respectively, is getting closer to the exact area under the rate of change function; as n increases the left-hand sum approaches the area under $v(t)$ from below and the right-hand sum approaches the area from above. This might lead us to conjecture that the exact value of the net change is the area under the curve.

Take Four.   Suppose we improve upon our estimates by partioning the interval $[1, 4]$ into 300 subintervals, each of length $\Delta t = \frac{3}{300} = \frac{1}{100}$ seconds. We need a convenient system for labeling the endpoints of these 300 subintervals. We label as indicated below.

The subintervals are $[t_0, t_1], [t_1, t_2], \ldots, [t_{299}, t_{300}]$, where $t_k = 1 + k(\Delta t) = 1 + \frac{k}{100}$ for $k = 0, 1, \ldots, 300$. Using this notation we have the following.

$$\text{lower bound} = L_{300} = v(t_0)\Delta t + v(t_1)\Delta t + \cdots + v(t_{299})\Delta t$$

$$= \sum_{i=0}^{299} v(t_i)\Delta t$$

$$= 29.97$$

$$\text{upper bound} = R_{300} = v(t_i), \Delta t + v(t_2)\Delta t + \cdots + v(t_{300})\Delta t$$

$$= \sum_{i=1}^{300} v(t_i)\Delta t$$

$$= 30.03$$

These bounds differ by less than 0.1 meter.   ◆

The Difference Between Left- and Right-Hand Sums

◆ EXAMPLE 22.5b   How many subdivisions were actually necessary in order to find the cheetah's net change in position with the specified degree of accuracy? Can we give an exact answer to the question of the cheetah's net change in position?

SOLUTION   Let's compare the terms of the left- and right-hand sums in Example 22.5a.

3 subdivisions:

$$L_3 = v(1) \cdot 1 + v(2) \cdot 1 + v(3) \cdot 1$$
$$R_3 = \qquad\quad v(2) \cdot 1 + v(3) \cdot 1 + v(4) \cdot 1$$

6 subdivisions:

$$L_6 = v(1)(.5) + v(1.5)(.5) + v(2)(.5) + v(2.5)(.5) + v(3)(.5) + v(3.5)(.5)$$
$$R_6 = \qquad\qquad v(1.5)(.5) + v(2)(.5) + v(2.5)(.5) + v(3)(.5) + v(3.5)(.5) + v(4)(.5)$$

300 subdivisions:

$$L_{300} = v(t_0)\Delta t + \sum_{i=1}^{299} v(t_i)\Delta t$$

$$= v(1)\Delta t + \sum_{i=1}^{299} v(t_i)\Delta t$$

$$R_{300} = \sum_{i=1}^{299} v(t_i)\Delta t + v(t_{300})\Delta t$$

$$= \sum_{i=1}^{299} v(t_i)\Delta t + v(4)\Delta t$$

In each case, except for the first term in the left-hand sum and the last term in the right-hand sum, all the terms are shared by both. Thus, the difference between the two estimates is merely the difference between these two terms.

3 subdivisions: $R_3 - L_3 = v(4) \cdot 1 - v(1) \cdot 1 = 13 - 7 = 6$

6 subdivisions: $R_6 - L_6 = v(4) \cdot (.5) - v(1) \cdot (.5) = (.5)[v(4) - v(1)] = .5(13 - 7) = 3$

300 subdivisions: $R_{300} - L_{300} = v(4)\Delta t - v(1)\Delta t$

$$= \Delta t \left[v(4) - v(1) \right] = \frac{1}{100}[13 - 7] = \frac{6}{100} = 0.06$$

No matter how many subintervals we use, the right endpoint of the first subinterval will be the left endpoint of the second, the right endpoint of the second interval will be the left endpoint of the third, and so on. In Example 22.5, if we use n subintervals, then the width of each subinterval is $\Delta t = \frac{3}{n}$, but the values of the velocity function at the endpoints of the interval remain the same. Because $v(t)$ is increasing on $[1, 4]$, R_n always gives an upper bound while L_n always gives a lower bound.

The difference between right- and left-hand sums for n subdivisions is given by

$$R_n - L_n = v(4) \cdot \Delta t - v(1) \cdot \Delta t$$

$$= \Delta t[v(4) - v(1)]$$

$$= \frac{3}{n}(13 - 7)$$

$$= \frac{18}{n}$$

For $Rn - Ln$ to be less than 0.1 we must have $\frac{18}{n} < \frac{1}{10}$, or $n > 180$.

If we let the number of subdivisions grow without bound, then the difference between the upper and lower bounds, $\frac{18}{n}$, will approach zero.

This has a nice graphical representation. (See Figure 22.11.) On each subinterval, the difference between the rectangles from R_n and L_n is a small "difference" rectangle. Sliding all these difference rectangles over to one side forms a total difference rectangle with height $v(4) - v(1) = 13 - 7 = 6$ and width $\Delta t = \frac{3}{n}$. As Δt goes to zero, the area of this rectangle representing the difference between the R_n and L_n must also go to zero.

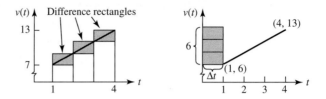

Figure 22.11

The rectangles for the upper bound lie above the velocity curve and the rectangles for the lower bound lie below the velocity curve. If the difference between the two bounds approaches zero, the area of each set of rectangles must be approaching the area under the velocity curve.

To summarize, because $v(t)$ is increasing, R_n (the right-hand sum) will always give an upper bound for the cheetah's net change in position and L_n (the left-hand sum) will always give a lower bound.

$$L_n < \text{net change in position} < R_n$$

Similarly, because $v(t)$ is increasing

$$L_n < \text{the area under } v(t) \text{ on } [1, 4] < R_n.$$

But $\lim_{n \to \infty} (R_n - L_n) = 0$, so $\lim_{n \to \infty} R_n = \lim_{n \to \infty} L_n = $ the area under $v(t)$ on $[1, 4]$

by the Squeeze Theorem.

This means that our conjecture is true. To compute the net change exactly, we need to find the area under the curve $v(t)$ between $t = 1$ and $t = 4$. In this particular example this region is a trapezoid so we can compute its area easily.

$$\text{area} = \frac{1}{2}(\text{height}_1 + \text{height}_2) \cdot (\text{base})$$

$$= \frac{1}{2}(7 + 13) \cdot (3)$$

$$= 30 \text{ meters}$$

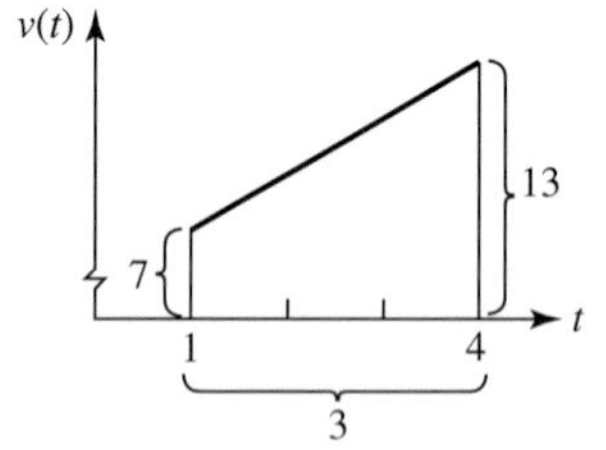

Figure 22.12

The cheetah's net change in position over the time interval $[1, 4]$ is 30 meters.

REMARK: In this example the function $v(t)$ is linear and therefore the exact area under $v(t)$ was midway between Ln and Rn. Draw a picture to convince yourself that for a nonlinear function this is generally not true.

Let's return for a moment to the problem of calculating the cheetah's net change in position and approach it with a different mindset. Let $s(t) =$ the cheetah's position at time t. Then $v(t) = s'(t)$. In other words, $\frac{ds}{dt} = 2t + 5$; $s(t)$ is a function whose derivative is $2t + 5$. Because $s'(t)$ is linear, we might suspect that $s(t)$ is quadratic.

If $s(t) = t^2 + 5t$, then $s'(t) = 2t + 5$. In fact, if two functions have the same derivative, then the functions must differ by an additive constant; if $f'(x) = g'(x)$, then $f(x) = g(x) + C$ for some constant C. Therefore,

$$s(t) = t^2 + 5t + C$$

for some constant C. Then the change in position from $t = 1$ to $t = 4$ is given by

$$s(4) - s(1) = 4^2 + 5(4) + C - [1^2 + 5(1) + C]$$
$$= 16 + 20 + C - 6 - C$$
$$= 30. \quad \blacklozenge$$

In Chapter 24 we will explore the relationship between the two mindsets presented in Example 22.5b and arrive at a wonderful theorem that unifies them. For the time being, there is a lot to be learned from the first mindset; we will stick with it for a while.

◆ **EXAMPLE 22.6**   A gazelle's velocity is given by the graph below. $v(t)$ is increasing on $[0, 3]$ and decreasing on $[3, 6]$. How can we find the net change in the gazelle's position over the interval $[0, 5]$?

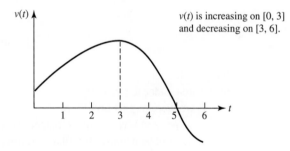

Figure 22.13

SOLUTION   Because v is increasing on the interval $[0, 3]$, we can find lower bounds for the gazelle's net change in position by using left-hand sums (inscribed rectangles) and upper bounds by using right-hand sums (circumscribed rectangles).

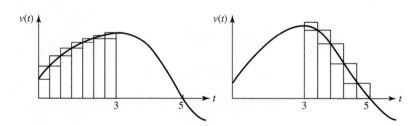

Figure 22.14

Because v is decreasing on the interval [3, 5], we can find lower bounds for the net change in position by using right-hand sums (inscribed rectangles) and upper bounds by using left-hand sums (circumscribed rectangles). See Figure 22.14.

We can obtain lower and upper bounds for the gazelle's net change in position on [0, 5] by treating the intervals [0, 3] and [3, 5] independently. For instance,

$$\mathcal{L} = (L_n \text{ on } [0, 3]) + (R_m \text{ on } [3, 5])$$

gives a lower bound, while

$$\mathcal{U} = (R_n \text{ on } [0, 3]) + (L_m \text{ on } [3, 5])$$

gives an upper bound.

If we compute the limit as n and m increase without bound, $\mathcal{L}$ and $\mathcal{U}$ will both approach the area under the velocity curve from $t = 0$ to $t = 5$. This area corresponds to the gazelle's net change in position on [0, 5].

If we did not treat the intervals [0, 3] and [3, 5] independently, we could still look at left- and right-hand sums to approximate the gazelle's net change in position. We could not, however, label them as under- or overestimates.

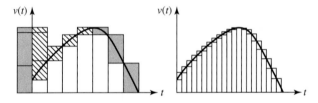

Figure 22.15

Nevertheless, $\lim_{n\to\infty}(R_n - L_n) = 0$; in fact, $\lim_{n\to\infty} R_n$ and $\lim_{n\to\infty} L_n$ both correspond to the area under the velocity curve.

If we were to partition [5, 6] into n equal pieces and compute L_n and R_n, they would give us upper and lower bounds, respectively, for the displacement. See Figure 22.16. Both L_n and R_n will be negative for all $n > 1$. $\lim_{n\to\infty} L_n = \lim_{n\to\infty} R_n = $ the signed area under the curve.

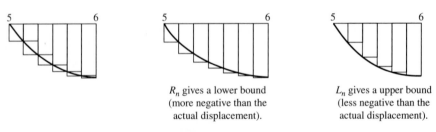

R_n gives a lower bound
(more negative than the
actual displacement).

L_n gives a upper bound
(less negative than the
actual displacement).

Figure 22.16   ◆

Examples 22.5 and 22.6 illustrate the interplay between the graphical and physical problems posed at the beginning of the section. The net change of a quantity can be represented as the signed area under the graph of the rate of change function. The question of how to find the area under the graph of a function is an important one, and is of interest on its own merits. It will be the focus of the next section.

Do Problem 1; it's a key problem and worthy of discussion.

1. It's flu season and a health clinic has set up a flu shot program for its patients. The clinic is open on Saturday from 8:00 A.M. to 4:00 P.M. (16:00) giving flu shots on a first-come, first-serve basis. The clinic has the capacity to serve 15 patients per hour. The function $r(t)$, whose graph is given below, gives the rate at which people are arriving at the clinic for shots.

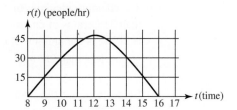

Explain your answers to the questions below by relating them to points, lengths, or areas on the graph.[5]

(a) At what time does a line start forming?

(b) At approximately what time is the length of the line increasing most rapidly?

(c) At approximately what time is the line the longest? (The answers to (b) and (c) are different. Explain why.)

(d) When the line is longest, approximately how many people are in line?

(e) Approximate the longest amount of time a person could wait for a shot. (Assume that doors close to new arrivals at 4:00 P.M. but everyone who has arrived by 4:00 P.M. is served.)

(f) Approximately how long is the line at 4:00 P.M.?

(g) Approximate the number of people who came to the clinic for a flu shot this day.

2. Maple syrup is being poured at a decreasing rate out of a tank. By taking readings from the valve on the tank, we have the following information on the rate at which the syrup is leaving the tank.

t (seconds)	0	2	4	6	8
rate (in cm^3/sec)	10	9	7	4	2

(a) Find a good upper bound for the amount of maple syrup that has been poured out between time $t = 0$ and $t = 8$.

(b) Find a good lower bound for this same amount.

3. An industrial chemist is making a mixture in a large container. A certain chemical, B, is being introduced into the mixture at an ever-increasing rate. Some of the rates have

---

[5] This problem, like Problem 8 in Section 22.2, was inspired by Peter Taylor's wicket problem, from *Calculus: The Analysis of Functions* by Peter Taylor, Wall and Emerson, Inc., 1992, p. 393.

been registered below. Time $t = 0$ marks the first introduction of this chemical into the mixture.

time (in minutes)	0	3	7	10	13	15
rate (in grams/min.)	10	12	20	23	25	29

(a) Determine reasonable upper and lower bounds for the number of grams of chemical B in the mixture at time $t = 13$.

(b) The chemist calculates that after a quarter of an hour more than 257 grams and less than 318 grams of chemical B are in the mixture. Midway between these bounds is $287\frac{1}{2}$. Is the amount of chemical B in the mixture at this time more than $287\frac{1}{2}$ grams, less than $287\frac{1}{2}$ grams, or exactly $287\frac{1}{2}$ grams?

4. (a) By partitioning the interval $[0, \pi/2]$ into four equal pieces and using the areas of inscribed and circumscribed rectangles as appropriate, find upper and lower bounds for the area between the graph of $\sin x$ and the x-axis for x in the interval $[0, \pi/2]$. Draw a picture illustrating what you have done. (*Note:* You will have to use your calculator to get some of the values of $\sin x$ and to get a numerical answer.)

(b) Using the work you did in part (a), find upper and lower bounds for the area under the graph of $\sin x$ between $x = 0$ and $x = \pi$. Explain what you have done using a picture.

(c) Using the work you did in part (b), give upper and lower bounds for the area under the graph of $\cos x$ between $x = -\pi/2$ and $\pi/2$.

5. Suppose velocity (in miles per hour) is given by $v(t) = 3t$, where t is measured in hours. We are interested in the distance traveled from $t = 0$ to $t = k$, where k is a constant.

(a) By solving the differential equation $ds/dt = 3t$ and using the initial condition $s(0) = s_0$, find the distance function $s(t)$. Using $s(t)$, find
   i. $s(0)$.
   ii. $s(k)$.
   iii. the distance traveled between $t = 0$ and $t = k$.

(b) Find the area under the graph of $v(t)$ from $t = 0$ to $t = k$. Verify that your answers to part (a) iii and (b) are the same.

6. Suppose velocity (in miles per hour) is given by $v(t) = mt + c$, where m and c are positive constants.

(a) Using your knowledge of the area of a trapezoid, find the area under the graph of $v(t)$ on the interval $[a, b]$, where a and b are positive constants.

(b) By solving the differential equation $ds/dt = mt + c$ and using the initial condition $s(0) = s_0$, find the distance function $s(t)$. Using $s(t)$ find
   i. $s(a)$.
   ii. $s(b)$.
   iii. the distance traveled between $t = a$ and $t = b$.
   Verify that your answers to parts (a) and (b) iii are the same.

7. Below is the graph of the velocity of a bee traveling in a straight line from a clover to a hive. Find the following.

   (a) the distance traveled between $t = 1$ and $t = 3$.

   (b) the distance traveled between $t = 0$ and $t = 8$.

   (c) the distance traveled between $t = 3$ and $t = 5$.

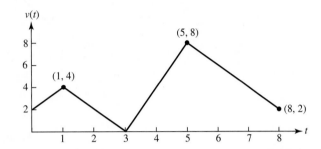

8. (a) The velocity of an object at time t is given by $v(t) = t^2$ ft/sec. Partition the time interval $[0, 3]$ into 3 equal pieces each of length 1 second. Find upper and lower bounds for the distance the object traveled between time $t = 0$ and $t = 3$.

   (b) Illustrate your work in part (a) by graphing $v(t)$ and using areas of inscribed and circumscribed rectangles. Draw two pictures, one illustrating the upper bound and the other the lower bound.

   (c) Repeat part (a), but this time partition the interval into 6 equal pieces, each of length 1/2. Make a sketch indicating the areas you have found.

   (d) What is the difference between R_n and L_n if the interval is partitioned into 50 equal pieces? 100 equal pieces?

   (e) Into how many equal pieces must we partition $[0, 3]$ to be sure that the difference between the right- and left-hand sums is less than or equal to 0.01?

9. Suppose $v(t)$ gives the velocity of a trekker on the time interval $[0, 3]$ and suppose that $v(t)$ is positive and decreasing over this interval. If we use a left-hand sum to approximate the distance she has covered over this time interval, will the approximation give a lower bound or an upper bound?

22.2   THE DEFINITE INTEGRAL

Suppose we want to find the signed area under the graph of a continuous function f on the interval $[a, b]$. We'll use the method of successive approximations and then apply a limit process. We divide the interval $[a, b]$ into n subintervals of equal width, each subinterval being of width $\Delta x = \frac{b-a}{n}$. We label as shown where $x_k = a + k\Delta x$ for $k = 0, 1, 2, \ldots, n$. The subintervals are $[x_0, x_1], [x_1, x_2], \ldots, [x_{n-1}, x_n]$[6].

---

[6] Notice that $x_n = a + n\Delta x = a + n\left(\frac{b-a}{n}\right) = a + b - a = b$.

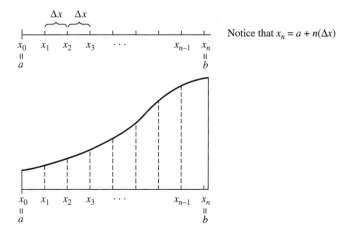

Notice that $x_n = a + n(\Delta x)$

Figure 22.17

On each subinterval, we approximate the function $f(x)$ by a constant function and approximate the (signed) area under f by the (signed) area of a rectangle.

For a left-hand sum, we approximate $f(x)$ on each subinterval by a constant function whose height is the value of f at the left endpoint of that subinterval. For example, on the second subinterval, $[x_1, x_2]$, the height of the rectangle is $f(x_1)$ because x_1 is the x-coordinate of the left endpoint of that subinterval. The width of every rectangle is Δx. Accordingly, we write the left-hand sum using n subintervals as follows.

$$L_n = f(x_0)\Delta x + f(x_1)\Delta x + \cdots + f(x_{n-1})\Delta x$$
$$= \sum_{i=0}^{n-1} f(x_i)\Delta x$$

We write the right-hand sum using n subintervals in a similar way, but this time beginning with x_1 and ending with x_n.

$$R_n = f(x_1)\Delta x + f(x_2)\Delta x + \cdots + f(x_n)\Delta x$$
$$= \sum_{i=1}^{n} f(x_i)\Delta x$$

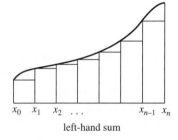

left-hand sum

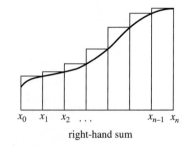

right-hand sum

Figure 22.18

Notice that we are not making any statement about these approximating sums being upper or lower bounds for the exact area. We are taking f to be a generic continuous bounded function, so we don't know where f is increasing and where f is decreasing.

The left- and right-hand sums are special cases of what is called a **Riemann sum**, after the mathematician Bernhard Riemann. We will focus our attention on left- and right-hand sums, partitioning $[a, b]$ into equal pieces, but Riemann sums are approximating sums that allow for more freedom of construction than R_n and L_n. To construct a Riemann sum for a bounded function f on $[a, b]$, chop the interval $[a, b]$, into n subintervals (not necessarily equal in width). Label the chops consecutively $a = x_0 < x_1 < x_2 < \cdots x_n = b$. From each of the subintervals choose an x-value in that interval. Call these x-values $x_1^*, x_2^*, \ldots, x_n^*$ where the subscript k indicates an x-value from the kth interval. Let Δx_k be the width of the kth subinterval. $\Delta x_k = x_k - x_{k-1}$ for $k = 1, 2, \ldots, n$. Any sum of the form $\sum_{k=1}^{n} f(x_k^*)\Delta x_k$ is referred to as a Riemann sum. It is more compact to use summation notation than to write out the individual terms, and this way of expressing the sums is suggestive of the notation we will soon introduce.

Finding the Exact Area

To get better approximations to the area, we increase n, the number of subdivisions. To get an exact value for the area, we look at the limit as n increases without bound. When we did so in Example 22.5b, L_n and R_n both approached the same value, this being the exact area under the curve. In the general case the left- and right-hand sums will approach the same limiting value provided that f is continuous.[7] The size of their difference is given by

$$|R_n - L_n| = |(f(x_1)\Delta x + f(x_2)\Delta x + \cdots + f(x_n)\Delta x) - (f(x_0)\Delta x + f(x_1)\Delta x$$
$$+ \cdots + f(x_{n-1})\Delta x)|$$
$$= |f(x_n)\Delta x - f(x_0)\Delta x|$$
$$= |f(b) - f(a)| \cdot \frac{b - a}{n}.$$

$f(b)$, $f(a)$, b, and a are all fixed quantities; they do not change as n grows without bound. Thus, the difference between the left- and right-hand sums approaches zero as n grows without bound.

$$\lim_{n \to \infty} R_n = \lim_{n \to \infty} L_n.$$

We define the **signed area** under f from a to b to be $\lim_{n \to \infty} R_n$ (or $\lim_{n \to \infty} L_n$) provided the limit exists. The signed area under f on the interval $[a, b]$ is called the definite integral of f from a to b. Signed area means that we consider the area between f and the horizontal axis to be positive where f is positive (where f lies above the horizontal axis) and negative where f is negative (where f lies below the horizontal axis).

---

[7] Being continuous is sufficient to guarantee that f is integrable, i.e., that the limit of the Riemann sums exists. This is proven in more advanced texts.

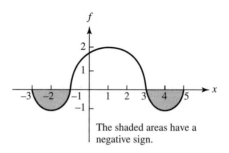

The shaded areas have a negative sign.

Figure 22.19

The definition given above agrees with our intuitive notion of signed areas in those instances in which we have such a notion. For f a reasonably well-behaved function, the interval $[a, b]$ can be partitioned into regions on which either f is increasing or f is decreasing.[8] By doing so, we can guarantee that within each region left- and right-hand sums will provide upper and lower bounds for the signed area within the region. $\lim_{n\to\infty} R_n$ equals $\lim_{n\to\infty} L_n$, so we can conclude that these quantities must also equal the exact value of the signed area. If f is not reasonably well-behaved, then the signed area under f on $[a, b]$ is *defined* to be $\lim_{n\to\infty} R_n$.

Definition

The **definite integral** of $f(x)$ from $x = a$ to $x = b$ is written $\int_a^b f(x)\, dx$ and read as "the integral from a to b of $f(x)$". We define it as follows. Subdivide the interval $[a, b]$ into n equal subintervals of width $\Delta x = \frac{b-a}{n}$ and label these subintervals $[x_0, x_1], [x_1, x_2], \ldots, [x_{n-1}, x_n]$, where $x_i = a + i\,\Delta x$, for $i = 0, 1, 2, \ldots n$.

$$\int_a^b f(x)\, dx = \lim_{n\to\infty} \sum_{i=1}^{n} f(x_i)\Delta x = \lim_{n\to\infty} \sum_{i=0}^{n-1} f(x_i)\Delta x, \text{ provided the limits exist.}$$

If f is continuous on $[a, b]$, then this limit is guaranteed to exist.[9]

The symbol $\int_a^b f(x)\, dx$ is laden with meaning. $\int$ is a script S, reminding us that the definite integral is the limit of a sum. It recalls the Greek letter Σ for summation in the Riemann sums approximating the area. We approximate the area with areas of rectangles; $f(x)$ represents *all* the heights $f(x_i)$, and dx represents the widths, the Δx, in the Riemann sums. You can think of the limiting process as an agent of metamorphoses from $\sum_{i=1}^{n} f(x_i)\Delta x$ to $\int_a^b f(x)dx$.

---

[8] More precisely, by "increasing" we really mean nondecreasing, and by "decreasing" we mean nonincreasing. Intervals on which f is constant can be included in one or the other.

[9] The definite integral $\int_a^b f(x)dx$ can be defined more generally as the limit of a general Riemann sum. When the limit is computed $|\Delta x_k|$ must approach zero for all Δx_k. If the limit of the Riemann sum as the width of the largest Δx_k goes to zero exists, it is equal to $\int_a^b f(x)\, dx$. Again, if f is continuous on $[a, b]$, then the limit is guaranteed to exist.

Notice that the numbers a and b, called the **endpoints of integration**, do not appear explicitly in the Riemann sums $\sum_{i=1}^{n} f(x_i)\Delta x$ and $\sum_{i=0}^{n-1} f(x_i)\Delta x$. At first this may surprise you, but on more careful examination you can see that $x_0 = a$ and $x_n = b$. As n increases without bound, x_1 gets arbitrarily close to a and x_{n-1} gets arbitrarily close to b.

Terminology

The function $f(x)$ being integrated in $\int_a^b f(x)dx$ is called the **integrand**.

The values a and b are called the **endpoints of integration** or the **lower and upper limits of integration**.

Interpretations of the Definite Integral

As we saw in the example of computing distance traveled, the definite integral, $\int_a^b f(x)\,dx$, can be interpreted as

- the signed area under the graph of $f(x)$ between $x = a$ and $x = b$, or

- the net change in the amount $A(x)$ between $x = a$ and $x = b$ if $f(x)$ is the rate of change of $A(x)$.

EXERCISE 22.1   Let f be the function graphed in Figure 22.19. Its graph is composed of three semicircles. Evaluate the following definite integrals by interpreting the definite integral as signed area.

(a) $\int_{-1}^{3} f(x)\,dx$ (b) $\int_{3}^{5} f(x)\,dx$ (c) $\int_{-1}^{5} f(x)\,dx$

Answers

(a) 2π (b) $\frac{-\pi}{2}$ (c) $2\pi - \frac{\pi}{2} = \frac{3\pi}{2}$

EXERCISE 22.2   Suppose that f is continuous on the interval $[-1, 3]$. We partition $[-1, 3]$ into n equal subintervals, each of length Δx, and form left- and right-hand Riemann sums, L_n and R_n, respectively. Explain and illustrate the following.

If $f' > 0$, then $L_n < \int_{-1}^{3} f(x)\,dx < R_n$; and

if $f' < 0$, then $R_n < \int_{-1}^{3} f(x)\,dx < L_n$.

These inequalities hold regardless of the sign of f.

PROBLEMS FOR SECTION 22.2

1. On the following page are the graphs of the velocities of three runners on the interval $[0, 5]$. Express the distance each runner has traveled in this interval in terms of a definite integral. Who has traveled the greatest distance in this time interval? Who has traveled the smallest distance in the time interval?

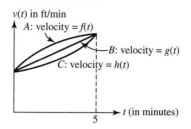

2. *Summation notation review*

(a) Write the following in summation notation.

   i. $3 - 4 + 5 - 6 + 7 - \cdots - 300$

   ii. $2 + 4 + 6 + \cdots + 1000$

   iii. $1 + 3 + 5 + \cdots + 999$

   iv. $\frac{2}{3} - \frac{2}{9} + \frac{2}{27} - \frac{2}{81} + \cdots + \frac{2}{3^{15}}$

   v. $x + x^2 + x^3 + x^4 + \cdots + x^{40}$

   vi. $1^2 + 2^2 + 3^2 + \cdots + 100^2$

   vii. $a_0 + a_1 x + a_2 x^2 + a_3 x^3 + \cdots + a_n x^n$

(b) Write out the following sums.

   i. $\sum_{i=2}^{5} i^2$ ii. $\sum_{k=0}^{4} 2^k$ iii. $\sum_{j=0}^{3} a_j x^j$

3. Find the following and express your answer as simply as possible.

(a) $\sum_{k=1}^{10} \left(\frac{k}{5}\right)^2 - \sum_{k=0}^{9} \left(\frac{k}{5}\right)^2$

(b) $\sum_{k=1}^{n} \left(\frac{k}{n}\right)^2 - \sum_{k=0}^{n-1} \left(\frac{k}{n}\right)^2$

4. Find upper and lower bounds for each of the following definite integrals by calculating left- and right-hand Riemann sums with the number of subdivisions indicated.

(a) $\int_0^2 x^3 dx$ $(n = 4)$ (b) $\int_1^3 \frac{1}{t} dt$ $(n = 6)$

5. Below are the velocity graphs for a chicken and a goat. Assume that at time $t = 0$ the goat and the chicken start out side by side and they both travel along the same straight dirt path. Answer the questions below. When the quantity corresponds to an area, describe this area on the graph and then give it using the appropriate definite integral or sums and differences of definite integrals. The graphs intersect at $t = 1.5$ and $t = 6.5$. The graph of $v_g(t)$ is maximum at $t = 4.5$.

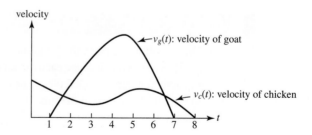

(a) The lead the chicken has when the goat begins to move

(b) The distance between the goat and the chicken at time $t = 1.5$

(c) The distance between the goat and the chicken at time $t = 3$

(d) The time at which the goat is farthest ahead of the chicken

(e) Approximate the time(s) at which the chicken and the goat pass one another on the path.

(f) Approximate the time at which the distance between the goat and the chicken is increasing most rapidly. At this time, what is the relationship between v'_g and v'_c?

6. Suppose we want to approximate $\int_0^2 f(x)\, dx$ by partitioning the interval $[0, 2]$ into n equal pieces and constructing left- and right-hand sums. Let $f(x) = x^3$.

(a) Put the following expressions in ascending order.

$$\int_0^2 f(x)\, dx, \quad L_4, \quad R_4, \quad L_{20}, \quad R_{20}, \quad L_{100}, \quad R_{100}$$

(b) Find $|R_4 - L_4|$.

(c) Find $|R_{100} - L_{100}|$.

(d) How large must n be to assure that $|R_n - L_n| < 0.05$?

(e) Write out R_4, once using summation notation, once without. Evaluate R_4.

7. Turn back to Problem 1 of Problems for Section 22.1 on page 723. Express the following quantities using a definite integral or the sum or difference of definite integrals.

(a) The length of the line at 10:00 A.M.

(b) The length of the line at its longest

(c) The number of people who came to the clinic for flu shots

(d) The number of people actually served by the clinic

(e) The length of the line at 4:00 P.M.

(f) The amount of time a person arriving at noon has to wait in line

(g) The amount of time by which the clinic must extend its hours in order to serve everyone who is in line before 4:00 P.M.

8. Repeat parts (a) through (e) of Problem 6, letting $f(x) = \frac{1}{x+1}$.

22.3   THE DEFINITE INTEGRAL: QUALITATIVE ANALYSIS AND SIGNED AREA

In Section 22.1 we established that the signed area under a rate graph is of particular interest to us; this area corresponds to the net change in amount. In Section 22.2 we established that if f is a reasonably well-behaved function,[10] we can, in theory, find the signed area under the graph of $f(t)$ on the interval $[a, b]$ as follows. Partition the interval $[a, b]$ into n equal pieces, each of length $\Delta t = \frac{b-a}{n}$, and label as shown on the following page.

---

[10] A continuous function on a closed interval is reasonably well-behaved.

The signed area is defined to be $\lim_{n\to\infty} \sum_{i=1}^{n} f(t_i)\Delta t$, provided this limit exists,[11] and is denoted by $\int_a^b f(t)\, dt$. In this section we interpret $\int_a^b f(t)\, dt$ as signed area (positive for $f > 0$ and negative for $f < 0$) and use our knowledge of functions and their graphs, along with a bit of basic geometry, to approximate some definite integrals and evaluate others.[12]

REMARK: The definite integrals $\int_a^b f(t)dt$, $\int_a^b f(x)\, dx$, $\int_a^b f(w)dw$, and $\int_a^b f(s)ds$ are all equivalent. Each gives the signed area between the graph of f and the horizontal axis from a to b. The variables t, x, w, and s are all dummy variables.

◆ **EXAMPLE 22.7**   The function $r(t)$ gives the rate at which water is flowing into (or out of) a backyard swimming pool. The graph of $r(t)$ is given below. At time $t = 0$ there are 200 gallons of water in the pool.

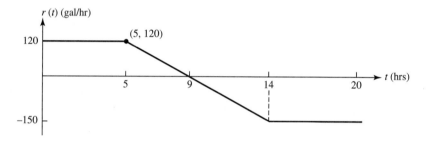

Figure 22.20

(If $w(t)$ is the number of gallons of water in the pool at time t, where t is given in hours, then $r(t) = \frac{dw}{dt}$.)

Express your answers to the questions below using definite integrals wherever appropriate and evaluate these integrals.

(a) When is the amount of water in the pool greatest? At that time, how many gallons of water are in the pool?

(b) What is the net flow in or out of the pool between $t = 4$ and $t = 12$?

(c) At what time is the pool's water level back at 200 gallons?

(d) Sketch a graph of the amount of water in the pool at time t.

SOLUTIONS   (a) The amount of water is maximum at $t = 9$ because water is entering the pool from $t = 0$ to $t = 9$ and leaving for t greater than 9.

---

[11] If f is continuous we can guarantee that this limit exists.

[12] We will rely on the following bits of geometry:

The area of a disk is πr^2.

The area of a trapezoid is $\frac{1}{2}(h_1 + h_2) \cdot b$, where $\frac{1}{2}(h_1 + h_2)$ is the average of the heights.

We can think of a triangle as a trapezoid with one "height" of length zero and a rectangle as a trapezoid with both heights of equal length.

The amount of water entering the pool on the interval $[0, 9]$ is $\int_0^9 r(t)dt$, the area under the graph from 0 to 9.

$$\int_0^9 r(t)dt = \text{the area of the rectangle} + \text{the area of the triangle}$$
$$= (5)(120) + (1/2)(4)(120)$$
$$= 600 + 240$$
$$= 840$$

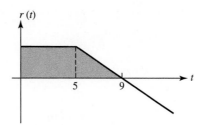

Figure 22.21

840 gallons of water have entered the pool in those 9 hours. Adding this to the original 200 gallons gives a total of 1040 gallons of water in the pool.

(b) The net flow in or out of the pool between $t = 4$ and $t = 12$ is given by $\int_4^{12} r(t)dt$. Compute this by summing the signed area under $r(t)$ from $t = 4$ to $t = 5$, from 5 to 9, and from 9 to 12. In other words, compute $\int_4^{12} r(t)dt$ by expressing it as follows.

$$\int_4^{12} r(t)dt = \int_4^5 r(t)dt + \int_5^9 r(t)dt + \int_9^{12} r(t)dt$$
$$= \text{area of rectangle} + \text{area of triangle} + (-\text{ area of triangle})$$
$$= 120(1) + (1/2)(120)(4) - (1/2)(3)(\text{height of triangle})^{13}$$
$$= 120 + 240 - (1/2)(3)(90)$$
$$= 360 - 135$$
$$= 225$$

The pool has a net gain of 225 gallons of water.

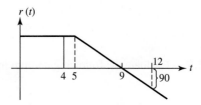

Figure 22.22

---

[13] The slope of the line forming the triangles is $(-120/4) = -30$, so between $t = 9$ and $t = 12$ the value of $r(t)$ drops by 90, making the height of this triangle 90.

(c) There are many ways to approach this question. From part (a), we know that by time $t = 9$ the pool has gained 840 gallons of water. After $t = 9$ the pool loses water. Between $t = 9$ and $t = 14$ the pool has lost $(1/2)(5)(150) = 375$ gallons of water, so it now has a net gain of 840 gallons $-$ 375 gallons $=$ 465 gallons. Subsequently water is flowing out at a rate of 150 gallons per hour. In t more hours the pool will have lost an additional $150t$ gallons of water. For what t will $150t = 465$? When $t = 465/150 = 3.1$. Therefore, at time $t = 14 + 3.1 = 17.1$, (17 hours and 6 minutes) the pool will be back at its original level of 200 gallons. Using integral notation, we have solved the following equation for T.

$$\int_0^T r(t)\,dt = 0$$

Or, equivalently,

$$\int_0^9 r(t)\,dt = -\int_9^T r(t)\,dt.$$

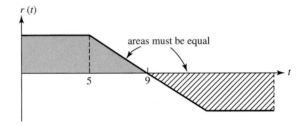

Figure 22.23

(d) Here is a graph of $w(t)$, where $w(t)$ is the amount of water in the pool at time t.

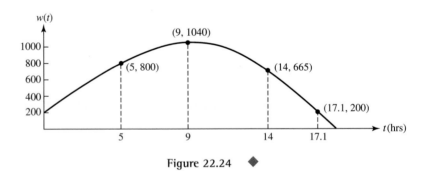

Figure 22.24   ◆

◆ EXAMPLE 22.8   Evaluate the following definite integrals by interpreting each as a signed area.

(a) $\int_0^{2\pi} \cos x\,dx$ (b) $\int_{-a}^a \sin x\,dx$

(c) $\int_{-7}^7 \frac{x}{x^4 + x^2 + 1}\,dx$ (d) $\int_{-2}^3 |x| - 2\,dx$

SOLUTIONS   (a) $\int_0^{2\pi} \cos x\,dx = 0$. The "positive" area and the "negative" area cancel.

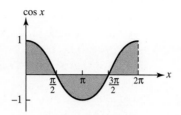

Figure 22.25

(b) $\int_{-a}^{a} \sin x \, dx = 0$. Sine is an odd function ($\sin(-x) = -\sin x$), so again the negative and positive areas cancel regardless of the value of a.

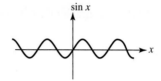

Figure 22.26

(c) The integrand is an odd function: $\frac{(-x)}{(-x)^4 + (-x)^2 + 1} = -\frac{x}{x^4 + x^2 + 1}$. Thus, the negative and positive areas cancel and we conclude that $\int_{-7}^{7} \frac{x}{x^4 + x^2 + 1} \, dx = 0$.

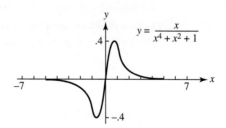

Figure 22.27

(d) Sketch the graph of $|x| - 2$. The area of the triangle lying below the x-axis is $(1/2)(4)(2) = 4$. Therefore, $\int_{-2}^{2}(|x| - 2) \, dx = -4$. The area of the triangle lying above the x-axis between $x = 2$ and $x = 3$ is $(1/2)(1)(1) = 1/2$. We conclude that

$$\int_{-2}^{3}(|x| - 2) \, dx = \int_{-2}^{2}(|x| - 2) \, dx + \int_{2}^{3}(|x| - 2) \, dx$$
$$= -4 + 0.5 = -3.5.$$

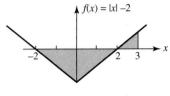

Figure 22.28   ◆

◆ **EXAMPLE 22.9**   Put the following expressions in descending order.

$$A = \int_{-2\pi}^{-\pi} e^{-0.1x} \sin x \, dx, \qquad B = \int_{-\pi}^{0} e^{-0.1x} \sin x \, dx, \qquad C = \int_{0}^{2\pi} e^{-0.1x} \sin x \, dx,$$

$$D = \int_{0}^{3\pi} e^{-0.1x} \sin x \, dx, \qquad E = \int_{\pi}^{2\pi} e^{-0.1x} \sin x \, dx, \qquad F = \int_{\pi}^{\pi} e^{-0.1x} \sin x \, dx$$

SOLUTION   Sketch the graph of $f(x) = e^{-0.1x} \sin x$.

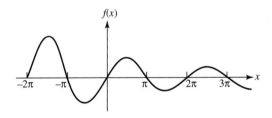

Figure 22.29

We begin by grouping the expressions above by sign: $F = \int_{\pi}^{\pi} e^{-0.1x} \sin x \, dx = 0$: $\int_{a}^{a} f(x) \, dx$ is always zero; there is no area contained between $x = a$ and $x = a$.
The following expressions are positive:

$$A = \int_{-2\pi}^{-\pi} e^{-0.1x} \sin x \, dx, \; C = \int_{0}^{2\pi} e^{-0.1x} \sin x \, dx, \; D = \int_{0}^{3\pi} e^{-0.1x} \sin x \, dx.$$

The following expressions are negative:

$$B = \int_{-\pi}^{0} e^{-0.1x} \sin x \, dx, \qquad E = \int_{\pi}^{2\pi} e^{-0.1x} \sin x \, dx.$$

Convince yourself, by considering the graphical interpretation of each of these definite integrals, that the integrals should be ordered as follows.

$$\int_{-2\pi}^{-\pi} e^{-0.1x} \sin x \, dx > \int_{0}^{3\pi} e^{-0.1x} \sin x \, dx > \int_{0}^{2\pi} e^{-0.1x} \sin x \, dx >$$

$$\int_{\pi}^{\pi} e^{-0.1x} \sin x \, dx > \int_{\pi}^{2\pi} e^{-0.1x} \sin x \, dx > \int_{-\pi}^{0} e^{-0.1x} \sin x \, dx$$

$$A > D > C > F > E > B \qquad ◆$$

PROBLEMS FOR SECTION 22.3

1. By interpreting the definite integral as signed area, calculate the following.

   (a) $\int_0^5 x\,dx$ (b) $\int_{-2}^2 x\,dx$ (c) $\int_{-2}^2 |x|\,dx$ (d) $\int_{-1}^3 3\,dx$

   (e) $\int_0^{2\pi} \sin t\,dt$ (f) $\int_{-\pi}^\pi \cos z\,dz$ (g) $\int_{-2}^2 (x+1)\,dx$ (h) $\int_{-2}^2 |x+1|\,dx$

2. Let f be the function whose graph is given below.

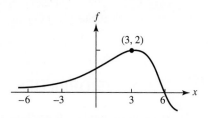

   $\int_{-6}^3 f(x)\,dx$ is closest to which of the following? Explain your reasoning.
   (a) -18 (b) -9 (c) -3 (d) 1.5 (e) 3 (f) 9 (g) 18

3. Put the following integrals in ascending order, placing "<" or "=" signs between them as appropriate. (*Strategy:* First determine which integrals are positive, which are negative, and which are zero.)

   (a) $\int_0^\pi \sin(t)\,dt$ (b) $\int_{-\pi}^{2\pi} \sin(t)\,dt$ (c) $\int_{-\pi}^{2\pi} \cos(t)\,dt$

   (d) $\int_0^{\pi/2} \cos(t)\,dt$ (e) $\int_0^\pi \cos(2t)\,dt$ (f) $\int_0^{3\pi/2} |\sin(t)|\,dt$

4. Put the following in ascending order, placing "<" or "=" signs between them as appropriate.

   (a) $\int_0^{\pi/2} \sin(t)\,dt$ (b) $\int_0^{\pi/2} t\,dt$ (c) $\int_{-\pi/2}^0 t\,dt$

   (d) $\int_{-\pi/2}^0 \sin(t)\,dt$ (e) $\int_0^{\pi/2} 1\,dt$ (f) $\pi/2$

Below is the graph of a function $f(x)$. Problems 5 and 6 refer to this function. The curve from $(0,0)$ to $(6,0)$ is a semicircle.

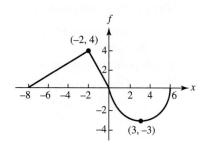

Use this graph to evaluate the following:

5. (a) $\int_{-8}^{-2} f(x)\, dx$ (b) $\int_{-8}^{0} f(x)\, dx$ (c) $\int_{0}^{6} f(x)\, dx$ (d) $\int_{3}^{6} f(x)\, dx$

6. (a) $\int_{-2}^{3} f(x)\, dx$ (b) $\int_{-8}^{6} f(x)\, dx$ (c) $\int_{0}^{6} |f(x)|\, dx$ (d) $\int_{-8}^{6} |f(x)|\, dx$

7. (a) What is the equation of a circle of radius 2 centered at the origin?

   (b) Write a function, complete with domain, that gives the equation of the top half of a circle of radius 2 centered at the origin.

   (c) Let $f(x) = \begin{cases} \sqrt{4 - x^2} & \text{for } -2 \leq x \leq 0, \\ 2x & \text{for } x > 0. \end{cases}$

   Evaluate $\int_{-2}^{2} f(x)\, dx$.

22.4   PROPERTIES OF THE DEFINITE INTEGRAL

From the definition of the definite integral and its interpretation as signed area, we obtain the following properties.[14]

1. **Constant Factor Property**
   For any constant k, $\int_{a}^{b} kf(t)\, dt = k \int_{a}^{b} f(t)\, dt$.
   A constant factor can be "pulled out" of a definite integral.

2. **Dominance Property**
   If $f \leq g$ on the interval $[a, b]$, then $\int_{a}^{b} f(t)\, dt \leq \int_{a}^{b} g(t)\, dt$.

3. **Endpoint Reversal Property**
   $\int_{b}^{a} f(t) = -\int_{a}^{b} f(t)\, dt$

4. **Additive Integrand Property**
   $\int_{a}^{b} [f(t) + g(t)]\, dt = \int_{a}^{b} f(t)\, dt + \int_{a}^{b} g(t)\, dt$

5. **Splitting Interval Property**
   $\int_{a}^{c} f(t)\, dt = \int_{a}^{b} f(t)\, dt + \int_{b}^{c} f(t)\, dt$
   (This is true regardless of the relative positions of a, b, and c.)

6. **Symmetry Property**
   $\int_{-a}^{a} f(t)\, dt = 0$ if f is odd; $\int_{-a}^{a} f(t)\, dt = 2 \int_{0}^{a} f(t)\, dt$ if f is even.

These properties are illustrated below.

Properties Illustrated

1. For any constant k, $\int_{a}^{b} kf(t)\, dt = k \int_{a}^{b} f(t)\, dt$.
   We can pull a constant multiple out of a definite integral. This is a property we use repeatedly in our work.

   (a) *Interpretation as the integral of a rate function*
       Suppose $f(t)$ is the rate that water is entering a pool. Doubling the rate at which water enters the pool over a certain time interval will double the amount of water

---

[14] Assume f is bounded on the intervals of integration and all integrals exist.

added to the pool in that time interval. Similarly, halving the rate will halve the net change in water.

(b) *Interpretation as area under f from t = a to t = b*

If the height of f is doubled, the area it bounds doubles. (If this doesn't make sense intuitively, consider partitioning the region and approximating its area with rectangles; the height of each rectangle will double, so the area of each rectangle will double.)

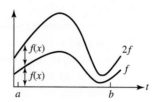

Figure 22.30

2. If $f \leq g$ on the interval $[a, b]$, then $\int_a^b f(t)\, dt \leq \int_a^b g(t)\, dt$.

(a) *Interpretation as the integral of a rate function*

Suppose $f(t)$ and $g(t)$ give the velocities of two cars. If the velocity of the second is greater than or equal to that of the first, then the second car will have the same or greater net change in position. In other words, if the rate of change of position of the second car is greater, the net change in position will also be greater.

(b) *Interpretation as area under f from t = a to t = b*

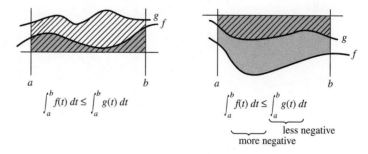

Figure 22.31

3. $\int_b^a f(t)\, dt = - \int_a^b f(t)\, dt$.

(a) *Interpretation as the integral of a rate function*

Suppose $f(t)$ is the rate at which water is entering a pool and $t = a$ corresponds to 10:00 A.M., while $t = b$ corresponds to noon. $\int_a^b f(t)dt$ is the net change in water in the pool between 10:00 A.M. and noon. Let's say this is 1000 gallons. Then the net change in water between noon and 10:00 A.M. (2 hours earlier) is -1000 because there were 1000 gallons less water at 10:00 A.M. than at noon. $\int_{12}^{10} f(t)\, dt = - \int_{10}^{12} f(t)\, dt$. When you see $\int_b^a f(t)\, dt$, where $a < b$, you can think of running a movie backward.

(b) *Interpretation as area under f from $t = a$ to $t = b$*

Suppose $a < b$. Think back to the approximating rectangles; each rectangle has base Δt, where $\Delta t = \frac{b-a}{n}$ is positive. If we switch the roles of a and b, then our measure of Δt becomes negative, so the value of the integral reverses sign.

Advice: Whenever you come across $\int_b^c f(t)\,dt$, where $b > c$, switch the endpoints and negate immediately. For instance, when dealing with $\int_0^{-2} f(t)\,dt$, write it as $-\int_{-2}^0 f(t)\,dt$.

4. $\int_a^b [f(t) + g(t)]\,dt = \int_a^b f(t)\,dt + \int_a^b g(t)\,dt$

(a) *Interpretation as the integral of a rate function*

A pool is being filled with two hoses. Suppose $f(t)$ is the rate at which water is entering through one hose and $g(t)$ is the rate that water is entering via another hose. Then the total amount of water that enters the pool is the water that enters through one hose plus the water that enters through the other.

(b) *Interpretation as area under f from $t = a$ to $t = b$*

Think about the approximating rectangles. In the bottom graph of Figure 22.32 the rectangles are positioned on top of one another; their positioning relative to one another does not affect their sum.

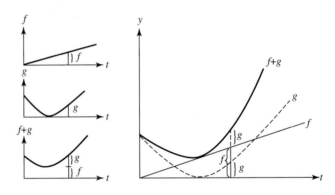

Figure 22.32

5. $\int_a^c f(t)\,dt = \int_a^b f(t)\,dt + \int_b^c f(t)\,dt$

(a) *Interpretation as the integral of a rate function*

Suppose $f(t)$ is the rate at which water is entering a pool. The total amount of water that enters the pool from 9:00 A.M. to 5:00 P.M. is the amount that enters from 9:00 A.M. to noon plus the amount that enters from noon to 5:00 P.M.

(b) *Interpretation as area under f from $t = a$ to $t = c$*

Using the endpoint reversal property, (3), you can convince yourself that the splitting interval property holds regardless of whether or not b lies between a and c.

The figures on the following page illustrate the statement for different relative positions of a, b, and c. Areas shaded with gray indicate a positive weighing in; areas shaded with hash marks indicate a negative weighting due to endpoint reversal. Work through each scenario slowly to assure yourself that in each case

$$\int_a^c f(t)\,dt = \int_a^b f(t)\,dt + \int_b^c f(t)\,dt.$$

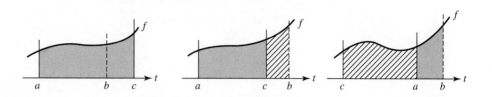

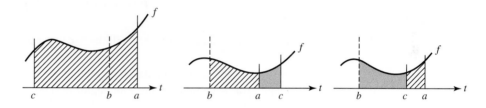

Figure 22.33

6. $\int_{-a}^{a} f(t)\,dt = 0$ if f is odd; $\int_{-a}^{a} f(t)\,dt = 2 \int_0^a f(t)\,dt$ if f is even.

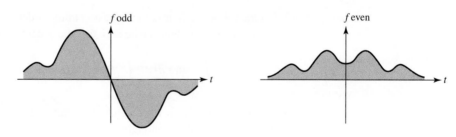

Figure 22.34

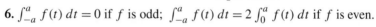

PROBLEMS FOR SECTION 22.4

For Problems 1 and 2, evaluate the following.

1. (a) $\int_2^0 x\,dx$ (b) $\int_4^{-1} (x+1)\,dx$

2. (a) $\int_0^1 \sqrt{1-x^2}\,dx$ (b) $\int_1^{-1} \sqrt{1-x^2}\,dx$
   (Hint: Think about circles.)

3. Choose the correct answer and explain your reasoning.
   (a) $\int_{-a}^{a} \frac{1}{1+x^2}\,dx =$

   (i) 0   (ii) $\int_0^a \frac{1}{1+x^2}\,dx$ (iii) $2\int_0^a \frac{1}{1+x^2}\,dx$ (iv) $-\int_0^a \frac{1}{1+x^2}\,dx$

(b) $\int_{-a}^{a} \frac{x}{1+x^2} \, dx =$

(i) 0   (ii) $\int_{0}^{a} \frac{x}{1+x^2} \, dx$ (iii) $2 \int_{0}^{a} \frac{x}{1+x^2} \, dx$ (iv) $-\int_{0}^{a} \frac{x}{1+x^2} \, dx$

4. (a) Explain why $\int_{0.5}^{3} \frac{1-\ln x}{x^2+1} \, dx > \int_{0.5}^{4} \frac{1-\ln x}{x^2+1} \, dx$.
   (*Hint:* Look at the sign of the integrand.)

   (b) Put in ascending order:

   $$0, \quad \int_{1/e}^{1} \frac{1-\ln x}{x^2+1} \, dx, \quad \int_{1/e}^{2} \frac{1-\ln x}{x^2+1} \, dx, \quad \int_{1/e}^{e} \frac{1-\ln x}{x^2+1} \, dx, \quad \int_{1/e}^{4} \frac{1-\ln x}{x^2+1} \, dx.$$

5. (a) Assume that $a < b$. Insert a ">" or "<" sign between the expressions below as appropriate. Explain your reasoning clearly.

   $$\int_{a}^{b} |f(t)| \, dt \quad \text{and} \quad \left| \int_{a}^{b} f(t) \, dt \right|$$

   (b) Under what circumstances will the two expressions be equal?

6. For each of the following, sketch a graph of the indicated region and write a definite integral (or, if you prefer, the sum and/or differences of definite integrals) that gives the area of the region.

   (a) The area between the horizontal line $y = 4$ and the parabola $y = x^2$

   (b) The area between the line $y = x + 1$ and the parabola $y = x^2 - 1$

7. Put the following four integrals in ascending order (from smallest to largest). Explain, using graphs, how you can be sure that the order you gave is correct.

   $$\int_{0}^{\pi} \sin t \, dt, \quad \int_{0}^{\pi} 2 \sin t \, dt, \quad \int_{0}^{\pi} \sin(2t) \, dt, \quad \int_{\pi}^{0} \sin t \, dt$$

8. Suppose $\int_{0}^{5} f(t)dt = 10$. Evaluate four out of the five expressions that follow. One of them you do not have enough information to evaluate.

   (a) $\int_{0}^{5} 7f(t) \, dt$ (b) $\int_{0}^{5} (f(t) + 7) \, dt$ (c) $\int_{0}^{5} f(t) \, dt + 7$

   (d) $\int_{0}^{5} 7f(t + 7) \, dt$ (e) $\int_{-7}^{-2} 7f(t + 7) \, dt$

9. If $\int_{0}^{5} f(t) \, dt = 10$ and $\int_{0}^{5} g(t) \, dt = 3$ does it follow that

   (a) $f(t) > g(t)$ for all t between 0 and 5? Explain.
   Mathematicians might write this statement in mathematical symbols as follows: $f(t) > g(t) \; \forall \, t$ in $[0, 5]$. The symbol "$\forall$" is read "for all."

   (b) $f(t) > g(t)$ for some t between 0 and 5? Explain.
   Mathematicians might write this statement in mathematical symbols as follows: $\exists \, t$ in $[0, 5]$ such that $f(t) > g(t)$. The symbol "$\exists$" is read "there exists."

C H A P T E R

23

The Area Function and Its Characteristics

The Big Picture

Our two fundamental interpretations of the definite integral $\int_a^b f(x)\,dx$ are:

■   the signed area between the graph of f and the horizontal axis on the interval $[a, b]$;

■   the net change in amount $A(x)$ between $x = a$ and $x = b$ if $f(x)$ is the rate of change of $A(x)$.

Note the parallel in construct between our interpretations of the definite integral and our interpretations of the derivative evaluated at a point. We have a geometric interpretation and a physical interpretation. (For the derivative at a point these were (i) the slope of a curve at a point and (ii) instantaneous rate of change of a quantity.) Not only did we study the derivative evaluated at a point, and the significance of this *number*, but we studied the derivative *function,* and interpreted it as the slope function or instantaneous rate of change function. The definite integral $\int_a^b f(x)\,dx$, where a and b are constants, is a *number*— whether interpreted as signed area or net gain (or loss) in amount. In this chapter we look at the area (or amount) *function.* We'll introduce this function via some examples.

23.1   AN INTRODUCTION TO THE AREA FUNCTION $\int_a^x f(t)\,dt$

◆ **EXAMPLE 23.1**   Consider $\int_{-1}^x f(t)\,dt$, where f is the constant function $f(t) = 4$. Is $\int_{-1}^x f(t)\,dt$ itself a function? If so, can it be expressed by an algebraic formula?

SOLUTION

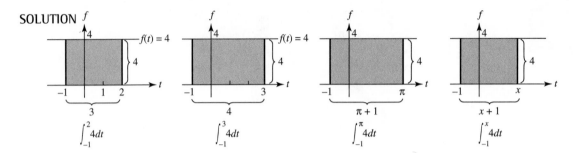

Figure 23.1

Indeed, $\int_{-1}^{x} 4\,dt$ *is* a function. It is the function that gives the signed area under $f(t) = 4$ between $t = -1$ and $t = x$. This area depends upon x and is uniquely determined by x. In fact, because the area is a rectangle, we can compute it by multiplying the height, 4, by the base, $x + 1$, to obtain the function $4 \cdot (x + 1) = 4x + 4$.

We could name this area function $A(x)$, but that wouldn't tell us that it's the area function for f with anchor point -1, so we dress it up a little more and write $_{-1}A_f(x)$.[1] The formula $_{-1}A_f(x) = 4x + 4$ holds for $x \le -1$ as well as $x > 1$.

For $x = -1$ we have $_{-1}A_f(-1) = \int_{-1}^{-1} 4\,dt = 0$. Any area function will be zero at its anchor point.

For $x < -1$ we use the endpoint reversal property of definite integrals to write

$$\int_{-1}^{x} 4\,dt = -\int_{x}^{-1} 4\,dt = -\big[4(-1 - x)\big]$$

$$= 4 + 4x.$$

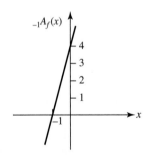

Figure 23.2

For example, we evaluate $_{-1}A_f(-2)$ as follows:

$$_{-1}A_f(-2) = \int_{-1}^{-2} 4\,dt = -\int_{-2}^{-1} 4\,dt = -(4)(1) = -4. \quad \blacklozenge$$

Computing Distances: General Principles

In the previous example, and in many applications of integration to come, you will be computing both vertical distances and horizontal distances.

---

[1] This chapter was inspired by Arnold Ostebee and Paul Zorn's treatment of the area function in their calculus text. *Calculus From Graphical, Numerical, and Symbolic Points of View*, Saunders College Publishing, 1997. The notation A_f is theirs.

vertical distance = (high y-value) − (low y-value)

horizontal distance = (right x-value) − (left x-value)

It is as simple as that. The location of the x- and y-axes is completely irrelevant. Be sure this makes sense to you as you look at Figure 23.3.

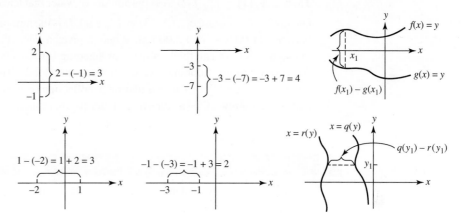

Figure 23.3

Let's return to the area function from Example 23.1 but change the anchor point from −1. Below we look at the area functions $_{-2}A_f(x)$, $_0A_f(x)$, and $_1A_f(x)$, where $f(t) = 4$ and the anchor points are −2, 0, and 1, respectively.

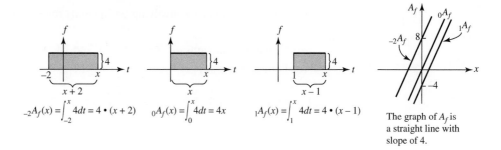

$$_{-2}A_f(x) = \int_{-2}^x 4\,dt = 4 \cdot (x+2) \qquad _0A_f(x) = \int_0^x 4\,dt = 4x \qquad _1A_f(x) = \int_1^x 4\,dt = 4 \cdot (x-1)$$

The graph of A_f is a straight line with slope of 4.

Figure 23.4

Notice that changing the anchor point only changes the area function by an additive constant.

EXERCISE 23.1   Argue that the formulas given for the functions $_{-2}A_f(x)$, $_0A_f(x)$, and $_1A_f(x)$ above are valid for x to the left of the anchor point.

Definition

For a continuous function f and a constant "a" in the domain of f we define $_aA_f(x)$ to be $\int_a^x f(t)\, dt$.

$_aA_f(x)$ gives the signed area between the graph of f and the t-axis from $t = a$ to $t = x$. Therefore, we will refer to it as the **area function**.

Amount Added, Accumulation, Accruement, . . .

In this chapter we will look at the characteristics of the function $_aA_f(x) = \int_a^x f(t)\,dt$. While our predominant interpretation of $_aA_f(x)$ will be as the area function, we could give it other interpretations as well. For example, suppose $t = 0$ corresponds to noon, t is measured in hours, and $f(t)$ gives the rate, in gallons per hour, at which water is leaving a storage tank. Then $_{-1}A_f(x) = \int_{-1}^x f(t)\,dt$ is the amount of water that leaves the tank between 11:00 A.M. and time x. For instance, $_{-1}A_f(3) = \int_{-1}^3 f(t)\,dt$ is the amount of water that has left the tank between 11:00 A.M. and 3:00 P.M. When f gives a rate, $A_f(x)$ represents the net change in amount. So, we can think of the A as standing for **A**rea (signed area), or as standing for **A**mount where by "amount" we mean "amount added." The "A" can stand for **a**ccretion (an increase by addition), **a**ccumulation (a collection over time), or **a**ccruement (an amount added). Certainly "A" is a convenient letter for this function.

PROBLEMS FOR SECTION 23.1

1. Let $f(t) = 7$. We'll define three area functions. The difference between their definitions is the anchor point. We'll denote them as follows.

$$_0A_f(x) = \int_0^x f(t)\,dt \qquad _2A_f(x) = \int_2^x f(t)\,dt \qquad _3A_f = \int_3^x f(t)\,dt$$

(a) Find a formula (not involving an integral) for $_0A_f(x)$, where $x \geq 0$. What is $_0A_f(-3)$? Does your formula work in general for negative x?

(b) Find a formula (not involving an integral) for $_2A_f(x)$, where $x \geq 2$. Does your formula work for $x < 2$ as well?

(c) Find a formula (not involving an integral) for $_3A_f(x)$, where $x \geq 3$. Does your formula work for $x < 3$ as well?

(d) Graph the functions $_0A_f(x)$, $_2A_f(x)$, and $_3A_f(x)$.

(e) Differentiate $_0A_f(x)$, $_2A_f(x)$, and $_3A_f(x)$.

2. Let $f(t) = 2t$. We'll define three area functions. The difference between their definitions is the anchor point. We'll denote them as follows:

$$_0A_f(x) = \int_0^x f(t)\,dt \qquad _2A_f(x) = \int_2^x f(t)\,dt \qquad _3A_f = \int_3^x f(t)\,dt$$

(a) Find a formula (not involving an integral) for $_0A_f(x)$, where $x \geq 0$. What is $_0A_f(-3)$? Does your formula work in general for negative x?

(b) Find a formula (not involving an integral) for $_2A_f(x)$, where $x \geq 2$. Does your formula work for $x < 2$ as well?

(c) Find a formula (not involving an integral) for $_3A_f(x)$, where $x \geq 3$. Does your formula work for $x < 3$ as well?

(d) Graph the functions $_0A_f(x)$, $_2A_f(x)$, and $_3A_f(x)$.

(e) Differentiate $_0A_f(x)$, $_2A_f(x)$, and $_3A_f(x)$.

3. Let $f(t) = \sin t$. Let $A_f(x) = \int_0^x \sin t\,dt$, where $x \geq 0$.

(a) Put the following in ascending order, with "<" or "=" signs between them.

$$A_f(0), \quad A_f\left(\frac{\pi}{2}\right), \quad A_f(\pi), \quad A_f\left(\frac{3\pi}{2}\right), \quad A_f(2\pi)$$

(b) For what values of x is $A_f(x) = 0$?

(c) For what values of x is $A_f(x)$ negative?

(d) Fow what values of x is $A_f(x)$ maximum?

4. The rate that water is entering a tank is given by $f(t) = 40 - 2t$ gallons/minute where t is measured in minutes past noon.

(a) Interpret $_0A_f(x)$ in words, for $x \geq 0$.

(b) For what values of x is $_0A_{f(x)}$ increasing?

(c) At what time is the water level in the tank the same as it was at noon? At this time, what is the value of $_0A_f(x)$?

23.2 CHARACTERISTICS OF THE AREA FUNCTION

◆ **EXAMPLE 23.2**   Let $_1A_f(x) = \int_1^x f(t)\,dt$, where the graph of f is drawn below and the domain of A_f is restricted to $1 \leq x \leq 12$.

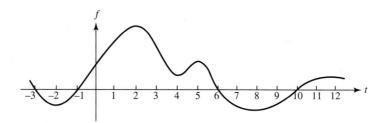

Figure 23.5

(a) On what interval(s) is the function $_1A_f(x)$ increasing? What characteristic of f ensures that $_1A_f(x)$ is increasing?

(b) On what interval(s) is the function $_1A_f(x)$ decreasing? What characteristic of f ensures that $_1A_f(x)$ is decreasing?

(c) Where on [1, 12] does $_1A_f(x)$ have its local maxima and minima?
Where on [1, 12] does $_1A_f(x)$ attain its absolute maximum and minimum values?

(d) On what interval(s) is the function $_1A_f(x)$ concave up? Concave down?
What characteristic of f ensures that $_1A_f(x)$ is concave up and what characteristic ensures that $_1A_f(x)$ is concave down?

SOLUTIONS   (a) $_1A_f(x)$ is increasing on [1, 6] and [10, 12]. Wherever f is positive, the area function increases with x, because a positive contribution to area is being made.

(b) $_1A_f(x)$ is decreasing on [6, 10]. Wherever f is negative, the accumulated area decreases with x, because a negative contribution is being made.

(c) On the number line below we indicate where $_1A_f(x)$ is increasing and where it is decreasing.

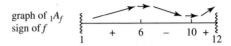

graph of $_1A_f$
sign of f

The local maximum occurs at $x = 6$; the local minimum at $x = 10$.

To identify where $_1A_f(x)$ takes on its absolute maximum and minimum values, we must also consider the endpoints of the domain, $x = 1$ and $x = 12$.

The absolute minimum occurs either at $x = 1$ or $x = 10$. We must compare the value of the function at these two points.

$_1A_f(1) = \int_1^1 f(t)\, dt = 0$. For $1 < x \leq 12$, $_1A_f(x)$ is always positive, since between $t = 1$ and any value greater than 1 there is always more positive accumulated area than there is negative accumulated area. Therefore the absolute minimum value of $_1A_f$ on $[1, 12]$ is 0 and it is attained at $x = 1$.

To find out where $_1A_f$ attains its absolute maximum value, we must compare the values at $x = 6$ and $x = 12$. The area between 6 and 10 gives a negative contribution and is larger in magnitude than the positive contribution between 10 and 12, so $_1A_f(12)$ is smaller than $_1A_f(6)$. The absolute maximum is attained at $x = 6$.

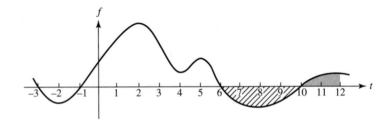

Figure 23.6

(d) We'll break this discussion into cases.

We begin by looking at $_1A_f$ on regions where f is positive. Where f is constant, area is accumulating at a constant rate; thus, $_1A_f$ is increasing at a constant rate and is neither concave up nor concave down.

Where f is positive and increasing, the area function is increasing at an increasing rate; $_1A_f$ is concave up.

Where f is positive and decreasing, the area function is increasing at a decreasing rate; $_1A_f$ is concave down.

Now let's look at $_1A_f$ on regions where f is negative. Where f is constant, the area function is decreasing at a constant rate and is neither concave up nor concave down.

Where f is negative and increasing (i.e., f is getting closer to zero), $_1A_f$ is decreasing by smaller and smaller amounts, so $_1A_f$ is concave up.

Where f is negative and decreasing, $_1A_f$ is decreasing more and more steeply, so $_1A_f$ is concave down.

We conclude that

■   where f is increasing, $_1A_f$ is concave up;

■   where f is decreasing, $_1A_f$ is concave down.

$_1A_f$ is concave up on $[1, 2), (4, 5), (8, 12]$ and concave down on $(2, 4), (5, 8)$.   ◆

◆ **EXAMPLE 23.3**   Let $_1A_f$ be the area function given by $\int_1^x f(t)$ as in Example 23.2, but this time let's enlarge the domain to $[-3, 12]$. We already know the behavior of $_1A_f$ on $[1, 12]$, so we'll concentrate on the interval $[-3, 1]$. We're interested in determining whether the conclusions we drew above when considering $_1A_f(x)$ for x to the right of the anchor point $x = 1$ also hold to the left of this anchor point. In particular, we'd like to answer the following questions.

(a) Where f is positive, is $_1A_f$ increasing? Where f is negative, is $_1A_f$ decreasing?

(b) Where f is increasing, is $_1A_f$ concave up? Where f is decreasing, is $_1A_f$ concave down?

The graph of f is reproduced below.

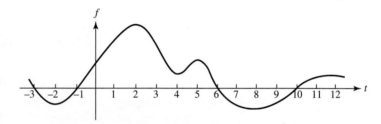

Figure 23.7

SOLUTION   Direct analysis of the area function to the left of the anchor point can sometimes make your head spin.[2] Instead, it will be simpler just to move the anchor point so that there are no values of the domain to the left of it. We'll look at the function $_{-3}A_f$ and then show that $_1A_f$ and $_{-3}A_f$ differ only by an additive constant; $_{-3}A_f$ and $_1A_f$ are vertical translates.

$_{-3}A_f$ is simple to analyze using the same logic as in Example 23.2.

Where f is positive, $_{-3}A_f$ is increasing; where f is negative, $_{-3}A_f$ is decreasing.

Where f is increasing, $_{-3}A_f$ is concave up; where f is decreasing, $_{-3}A_f$ is concave down.

How are the functions $_{-3}A_f$ and $_1A_f$ related? How can we apply our knowledge of the former to the latter?

$$_{-3}A_f(x) = \int_{-3}^x f(t) \text{ while } _1A_f(x) = \int_1^x f(t)\, dt.$$

From the splitting interval property we know that $\int_a^b f(t)\, dt = \int_a^c f(t)\, dt + \int_c^b f(t)\, dt$. Therefore,

---

[2] You need to use the fact that $_1A_f(x) = \int_1^x f(t)\, dt = -\int_x^1 f(t)\, dt$ in combination with the idea that if a sequence of numbers is increasing (e.g., 1, 2, 3), then when negated the sequence is decreasing $(-1, -2, -3)$.

$$\int_{-3}^{x} f(t)\, dt = \int_{-3}^{1} f(t)\, dt + \int_{1}^{x} f(t)\, dt$$

$$_{-3}A_f(x) = \int_{-3}^{1} f(t)\, dt + {_1}A_f(x)$$

$${_{-3}}A_f(x) - \int_{-3}^{1} f(t)\, dt = {_1}A_f(x).$$

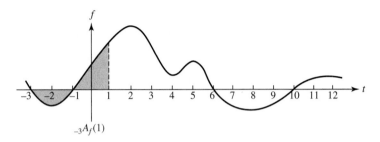

Figure 23.8

$\int_{-3}^{1} f(t)\, dt$ is a *constant*, so $_1A_f(x) = {_{-3}}A_f(x) + K$ for some constant K. $_1A_f(x)$ and $_{-3}A_f(x)$ differ by a constant, so their graphs are vertical translates of one another. Therefore, where $_{-3}A_f(x)$ is increasing, $_1A_f(x)$ is also increasing. Where $_{-3}A_f(x)$ is decreasing, $_1A_f(x)$ is also decreasing. Similarly, where $_{-3}A_f(x)$ is concave up, $_1A_f(x)$ is also concave up, and likewise for concave down. We summarize below.

> Where f is positive, $_1A_f(x)$ is increasing; where f is negative, $_1A_f(x)$ is decreasing.
> Where f is increasing, $_1A_f(x)$ is concave up; where f is decreasing, $_1A_f(x)$ is concave down.

In fact, any anchor point "a" in the interval $[-3, 12]$ could be chosen and $_aA_f(x)$ would simply be a vertical translate of $_{-3}A_f(x)$. For any constant "a" we know that

$$\int_{-3}^{x} f(t)\, dt = \int_{-3}^{a} f(t)\, dt + \int_{a}^{x} f(t)\, dt$$

$$\int_{-3}^{x} f(t)\, dt = K + \int_{a}^{x} f(t)\, dt. \quad \blacklozenge$$

Let's look at a case that's more complicated than the constant function of Example 23.1 but more concrete and computational than Example 23.2.

◆ **EXAMPLE 23.4**   Let $f(t) = 2 - t$. Find formulas for $_{-2}A_f(x)$, $_0A_f(x)$, and $_1A_f(x)$, the area functions anchored at -2, 0, and 1, respectively. How are these area functions related? Restrict the domains of these functions to the interval $[-2, 6]$.

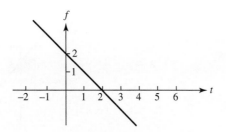

Figure 23.9

SOLUTION   Let's begin with $_{-2}A_f(x)$ because none of the domain is to the left of the anchor point.

$_{-2}A_f(x)$ is increasing on $[-2, 2]$ and decreasing on $[2, 6]$.

$_{-2}A_f(x)$ is concave down on $[-2, 6]$ because f is decreasing.

Finding a Formula for $_{-2}A_f(x)$**.**   First consider $_{-2}A_f(x)$ for x on the interval $[-2, 2]$. $_{-2}A_f(x)$ is the area of the trapezoid bounded by $f(t) = 2 - t$, the t-axis, $t = -2$, and $t = x$. (See Figure 23.10.) Its area is

$$\frac{1}{2}(h_1 + h_2) \cdot b = \frac{1}{2}[4 + (2 - x)] \cdot (x - (-2))$$

$$= \frac{1}{2}(6 - x) \cdot (x + 2)$$

$$= \frac{1}{2}(12 - x^2 + 4x)$$

$$= -\frac{x^2}{2} + 2x + 6.$$

$_{-2}A_f(x) = -\frac{x^2}{2} + 2x + 6$ on $[-2, 2]$.

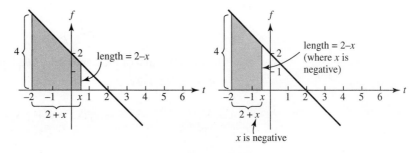

Figure 23.10

For x on $(2, 6]$ to find $_{-2}A_f(x)$ we must subtract from the area of the triangle above the t-axis the area of the appropriate triangle below. (See Figure 23.11.)

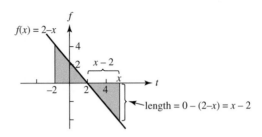

Figure 23.11

(area of triangle above the t-axis) $-$ (area of triangle below the t-axis)

$$= \left[\frac{1}{2} \cdot 4 \cdot 4 \right] - \left[\frac{1}{2} \cdot (x-2) \cdot (x-2) \right]$$

$$= 8 - \frac{1}{2}(x^2 - 4x + 4)$$

$$= 8 - \frac{x^2}{2} + 2x - 2$$

$$= -\frac{x^2}{2} + 2x + 6$$

So $_{-2}A_f(x) = -\frac{x^2}{2} + 2x + 6$ on the entire domain given.

Once we have found $_{-2}A_f(x)$, finding $_0A_f(x)$ and $_1A_f(x)$ is not difficult because the functions differ only by additive constants.

Finding a Formula for $_0A_f(x)$. $_0A_f(x) = \int_0^x f(t)\, dt$. To relate $_0A_f(x)$ to $_{-2}A_f(x)$ we use the splitting interval property, $\int_a^b f(t)\, dt = \int_a^c f(t)\, dt + \int_c^b f(t)\, dt$.

$$\int_{-2}^x f(t)\, dt = \int_{-2}^0 f(t)\, dt + \int_0^x f(t)\, dt$$

$$_{-2}A_f(x) = \int_{-2}^0 f(t)\, dt + {_0A_f(x)},$$

where $\int_{-2}^0 f(t)\, dt$ is the area of the trapezoid shaded in Figure 23.12.

$\int_{-2}^0 f(t)\, dt = \frac{1}{2} \cdot (4+2) \cdot 2 = 6$

$$_{-2}A_f(x) = 6 + {_0A_f(x)}$$

So

$$_0A_f(x) = {_{-2}A_f(x)} - 6.$$

$_{-2}A_f(x) = -\frac{x^2}{2} + 2x + 6$; therefore, $_0A_f(x) = -\frac{x^2}{2} + 2x$.

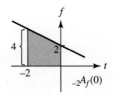

Figure 23.12

Finding a Formula for $_1A_f(x)$.   Similarly, $\int_{-2}^{x} f(t)\, dt = \int_{-2}^{1} f(t)\, dt + \int_{1}^{x} f(t)\, dt$.

$$_{-2}A_f(x) = \int_{-2}^{1} f(t)\, dt + {}_1A_f(x),$$

where $\int_{-2}^{1} f(t)\, dt$ is the area shaded in Figure 23.13.

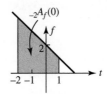

Figure 23.13

$$_{-2}A_f(1) = \tfrac{1}{2}(4+1)\cdot 3 = \tfrac{15}{2}$$

$$_{-2}A_f(x) = \frac{15}{2} + {}_1A_f(x)$$

So

$$_1A_f(x) = {}_{-2}A_f(x) - \frac{15}{2}.$$

$_{-2}A_f(x) = -\frac{x^2}{2} + 2x + 6$; therefore, $_1A_f(x) = -\frac{x^2}{2} + 2x - 1.5$.

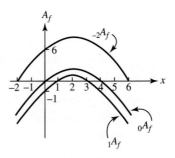

Figure 23.14

Alternatively, we could have found $_1A_f(x)$ by starting with $_1A_f(x) = -\frac{x^2}{2} + 2x + K$. Knowing that $_1A_f(1) = 0$ we can solve for K.

$$0 = -\frac{1}{2} + 2 + K$$

$$K = -\frac{3}{2}$$

Therefore,

$$_1A_f(x) = -\frac{x^2}{2} + 2x - 1.5.$$

Notice that

$$A_f(x) = -\frac{x^2}{2} + 2x + K,$$

where the value of K depends upon the anchor point. The area function graphs are vertical translates of one another. In fact,

$$\frac{d}{dx}A_f(x) = \frac{d}{dx}\left[-\frac{x^2}{2} + 2x + K\right] = -x + 2 = f(x). \quad \blacklozenge$$

Interesting . . . How much of this can we generalize?

We begin by generalizing the observation that area functions of f with different anchor points differ only by an additive constant. This can be shown using the splitting interval property of integrals. Let a and c be constants in the domain of f. Then

$$\int_a^x f(t)\,dt = \int_a^c f(t)\,dt + \int_c^x f(t)\,dt, \text{ so}$$

$$_aA_f(x) = \int_a^c f(t)\,dt + {}_cA_f(x),$$

where $\int_a^c f(t)\,dt$, the area under f from a to c, is a constant.[3]

How are A_f and f related?

Let's gather our observations from the previous examples and see what we can make of them.

Observations

■   Area functions for f are vertical translates of one another.

■   Where f is positive, A_f is increasing;
where f is negative, A_f is decreasing.

■   Where f is increasing, A_f is concave up;
where f is decreasing, A_f is concave down.

These statements are true regardless of the anchor point (so we've omitted it from the notation).

In addition, from the examples in which we were able to arrive at algebraic expressions for the area functions (Examples 23.1 and 23.4), we made the following intriguing observation.

---

[3] $\int_a^c f(t)\,dt$ can also be written as $_aA_f(c)$.

■ In two examples we find the derivative of the area function A_f is f itself.
For the time being this is only an observation *of a phenomenon that presented itself in two examples. It does not have the status of a fact.*

This last observation ties together the first three.

■ Where the derivative of a function is positive, the function is increasing; where the derivative is negative, the function is decreasing.

■ Where the derivative of a function is increasing, the function is concave up; where the derivative is decreasing, the function is concave down.

■ If two functions have the same derivative, then they differ by a constant.[4]

We have a conjecture in the making.

Conjecture

The derivative of the area function A_f is f itself.

We will prove this conjecture in the next section and find that it has beautiful and far-reaching consequences.

PROBLEMS FOR SECTION 23.2

1. Let $A_f(x)$ be the area function given by $A_f(x) = \int_0^x f(t)\, dt$, where $0 \le x \le 11$. The graph of f is given below.

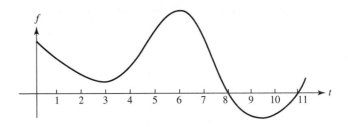

(a) On what interval(s) is the function $A_f(x)$ increasing?

(b) On what interval(s) is the function $A_f(x)$ decreasing?

(c) What is $A_f(0)$?

(d) Is $A_f(x)$ ever negative on the interval $[0, 11]$?

(e) On $[0, 11]$, where is $A_f(x)$ maximum? Minimum?

---

[4] This is the Equal Derivatives Theorem. For the proof, refer to Appendix C.

2. Below is the graph of $f(t)$. f is an odd function.

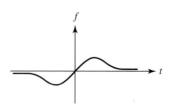

(a) Which of the graphs below could be the graph of $_0A_f(x) = \int_0^x f(t)\,dt$? Explain your reasoning.

(b) Which of the graphs below could be the graph of $_2A_f(x) = \int_2^x f(t)\,dt$? Explain.

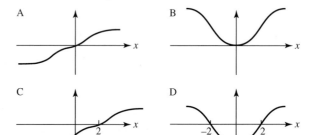

3. Below is the graph of $f(t)$.

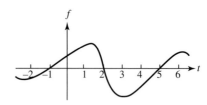

(a) If $F(x) = \int_0^x f(t)\,dt$, which of the graphs given below could be a graph of $F(x)$? Explain your criterion.

(b) If $G(x) = \int_{-2}^x f(t)\,dt$, which of the graphs given below could be a graph of $G(x)$?

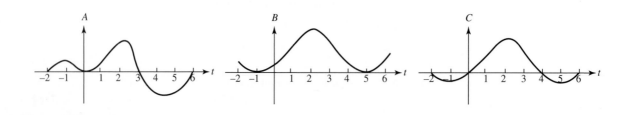

23.3   THE FUNDAMENTAL THEOREM OF CALCULUS

We are now on the threshold of obtaining a truly wonderful and important result. At the end of the previous section, we conjectured that the derivative of the area function A_f is f. In this section, we give a geometric justification of this conjecture. Then we'll see what an amazing and extremely useful result it is.

Proof that if f is a Continuous Function, then $\frac{d}{dx}A_f(x) = f(x)$

Let f be a continuous bounded function and c a constant in the domain of f. Consider the function $A_f(x) = \int_c^x f(t)\, dt$, the signed area between the graph of f and the t-axis between $t = c$ and $t = x$.

To calculate $\frac{d}{dx}A_f(x)$ let's begin by looking at

$$A_f(x + \Delta x) - A_f(x),$$

the numerator of the difference quotient. We will show that

$$\lim_{\Delta x \to 0} \frac{A_f(x + \Delta x) - A_f(x)}{\Delta x} = f(x)$$

by using the "Squeeze Theorem" or "Sandwich Theorem."

For simplicity's sake, let's begin by assuming that f is positive and increasing on the interval $[x, x + \Delta x]$, where Δx is positive. Then we can represent $A_f(x + \Delta x) - A_f(x)$ by the region shaded in Figure 23.15(b). (Later we will drop the requirements that f is positive and increasing; our argument can be adapted easily to the general case.)

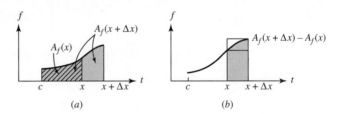

Figure 23.15

We can write the following inequalities. (Refer to Figure 23.15b.)

area of smaller rectangle	$\leq$	area under $f(t)$ on $[x, x + \Delta x]$	$\leq$	area of larger rectangle
$f(x)\Delta x$	$\leq$	$A_f(x + \Delta x) - A_f(x)$	$\leq$	$f(x + \Delta x)\Delta x$

It's legal to divide by Δx because Δx is positive.

$f(x)$	$\leq$	$\frac{A_f(x+\Delta x)-A_f(x)}{\Delta x}$	$\leq$	$f(x + \Delta x)$

Take the limit as Δx tends toward zero.

$\lim_{\Delta x \to 0} f(x)$	$\leq$	$\lim_{\Delta x \to 0} \frac{A_f(x+\Delta x)-A_f(x)}{\Delta x}$	$\leq$	$\lim_{\Delta x \to 0} f(x + \Delta x)$

The middle expression is the definition of derivative.

$\lim_{\Delta x \to 0} f(x)$	$\leq$	$\frac{d}{dx}A_f(x)$	$\leq$	$\lim_{\Delta x \to 0} f(x + \Delta x)$

Evaluate the limits.

$f(x)$	$\leq$	$\frac{d}{dx}A_f(x)$	$\leq$	$f(x)$

$\frac{d}{dx} A_f(x)$ is being "squeezed" on both sides by $f(x)$; it must be equal to $f(x)$.

This argument can be adapted to deal with functions that are not necessarily positive and increasing on $[x, x + \Delta x]$. The fact that $[x, x + \Delta x]$ is a closed, bounded interval and f is continuous means we can always find M and m, the absolute maximum and minimum values of f on $[x, x + \Delta x]$, respectively. We can set up the inequality

$$\left(\begin{array}{c} \text{minimum value} \\ \text{of } f \text{ on } [x, x + \Delta x] \end{array} \right) \Delta x \leq A_f(x + \Delta x) - A_f(x) \leq \left(\begin{array}{c} \text{maximum value} \\ \text{of } f \text{ on } [x, x + \Delta x] \end{array} \right) \Delta x$$

and proceed as above.[5] We obtain

$$\lim_{\Delta x \to 0} \left(\begin{array}{c} \text{minimum value} \\ \text{of } f \text{ on } [x, x + \Delta x] \end{array} \right) \leq \frac{d}{dx} A_f(x) \leq \lim_{\Delta x \to 0} \left(\begin{array}{c} \text{maximum value} \\ \text{of } f \text{ on } [x, x + \Delta x] \end{array} \right).$$

f is continuous, so as $\Delta x \to 0$ the minimum and maximum values of f on $[x + \Delta x]$ both approach $f(x)$.

The conclusion is that for any continuous function f, $\frac{d}{dx} A_f(x) = f(x)$. In words, this means that the rate of change of the area function at any point is the height of the function f at that point. Think about this geometrically for a minute; it makes sense.

The Fundamental Theorem of Calculus, version 1

If f is continuous on $[a, b]$ and $c \in [a, b]$, then the function

$$A_f(x) = \int_c^x f(t) \, dt \qquad x \in [a, b]$$

is differentiable on (a, b) and $\dfrac{d}{dx} A_f(x) = f(x)$.

We can rewrite the Fundamental Theorem of Calculus without explicit reference to the area function:

If f is continuous on $[a, b]$, then $\int_a^x f(t) \, dt$ is differentiable on (a, b) and

$$\frac{d}{dx} \int_a^x f(t) \, dt = d(x).$$

In the next chapter we will see that this result is *amazingly* useful.

---

[5] This inequality assumes $\Delta x > 0$. If $\Delta x < 0$, simply switch the inequalities to $\geq$ and switch again after dividing by Δx.

PROBLEMS FOR SECTION 23.3

1. The graph of $f(t)$ is given below. Let $F(x) = \int_{-6}^{x} f(t)\, dt$ and $G(x) = \int_{1}^{x} f(t)\, dt$.

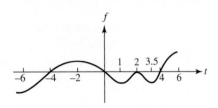

   (a) Where on $[-6, 6]$ is $F(x)$ increasing? Decreasing? Concave up? Concave down?
   (b) Where on $[-6, 6]$ is $G(x)$ increasing? Decreasing?
   (c) Where on $[-6, 6]$ does F have a local minimum?
   (d) Explain why the local extrema of F and G occur at the same values of x.
   (e) How are the graphs of F and G related?

2. Let $C(q)$ be the cost of producing q widgets. The graph below gives $\frac{dC}{dq}$, the marginal cost of producing widgets. Economists interpret the term "marginal cost" as the additional cost of producing an additional item.

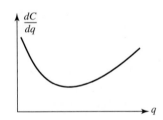

   Assume that fixed costs, the cost when no widgets are being produced, is F. Graph $C(x) = F + \int_{0}^{x} \frac{dC}{dq}\, dq$.

3. Below is a graph of $r(t)$, the rate of flow of liquid in and out of a tank. $t = 0$ corresponds to noon.

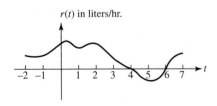

Suppose you know the following.

$$\int_{-2}^{7} r(t)\,dt = 4 \qquad \int_{2}^{6} r(t)\,dt = 0 \qquad \int_{2}^{7} r(t)\,dt = 1$$

(a) What is the net change in the amount of liquid in the tank between 10:00 A.M. and 6:00 P.M.?

(b) If there are 50 liters in the tank at 2:00 P.M., how many liters are in the tank at 7:00 P.M.?

4. A jug of cold lemonade is loaded into a cooler to be brought on a summer picnic. The lemonade is 40 degrees at the start of the trip. During the hour and a half drive to the park the lemonade gets steadily warmer at a rate of 1 degree every 15 minutes. Upon arriving at the park it is carried (in the cooler) to the river via a 40-minute hike. During the hike the lemonade gains 1 degree every 10 minutes. Once it reaches the river it is poured into cups. Now that it is out of the cooler the lemonade warms at a rate proportional to the difference between the temperature of the air and that of the liquid. Half an hour later it has warmed 12 degrees.

Let $L(t)$ be the temperature of the lemonade at time t, where t is the number of minutes into the trip.

(a) Sketch $\frac{dL}{dt}$ versus time. Label important points.

(b) Sketch $L(t)$ versus time. Label important points. Be sure your pictures of L and dL/dt are consistent.

24

The Fundamental Theorem of Calculus

24.1 DEFINITE INTEGRALS AND THE FUNDAMENTAL THEOREM

We concluded the previous chapter with the **Fundamental Theorem of Calculus**, version 1.

If f is continuous on $[a, b]$, then $\int_a^x f(t)\, dt$ is differentiable on (a, b) and

$$\frac{d}{dx} \int_a^x f(t)\, dt = f(x).$$

One reason this result is so exciting is that we can use it to obtain a simple and beautiful method for computing definite integrals. Let's look at how this result helps us compute $\int_a^b f(t)\, dt$, where a and b are constants.

Definition

A function F is an **antiderivative** of f if its derivative is f; that is, F is an antiderivative of f if $F' = f$.

Recall that if two functions have the same derivative, then they differ only by an additive constant. In other words, if F and G are both antiderivatives of f (i.e., if $F' = G' = f$), then $F(x) = G(x) + C$ for some constant C. Using this terminology, we can rephrase our last result as follows. Suppose c is between a and b.

761

$\int_c^x f(t)\,dt$ is an antiderivative of $f(x)$ because the derivative of $\int_c^x f(t)\,dt$ is $f(x)$.

Let $F(x)$ be any antiderivative of $f(x)$. Then $F(x) = \int_c^x f(t)\,dt + C$ for some constant C. (Any two antiderivatives of f differ only by an additive constant.) It follows that

$$F(b) = \int_c^b f(t)\,dt + C \quad \text{and} \quad F(a) = \int_c^a f(t)\,dt + C.$$

Suppose that we want to compute $\int_a^b f(t)\,dt$. We know that

$$\int_a^b f(t)\,dt = \int_a^c f(t)\,dt + \int_c^b f(t)\,dt \quad \text{(by the splitting interval property of definite integrals) Consequently,}$$

$$\int_a^b f(t)\,dt = \int_c^b f(t)\,dt - \int_c^a f(t)\,dt \quad \text{(using the endpoint reversal property of definite integrals)}$$

$$= F(b) + C - [F(a) + C]$$

$$= F(b) - F(a)$$

We've shown that $\int_a^b f(t)\,dt = F(b) - F(a)$.

This is the **Fundamental Theorem of Calculus**, version 2.

The Fundamental Theorem of Calculus, version 2

Let f be continuous on $[a, b]$. If F is an antiderivative of f, that is, $F' = f$, then

$$\int_a^b f(t)\,dt = F(b) - F(a).$$

The Fundamental Theorem tells us that to compute the signed area between the graph of f and the horizontal axis over the interval $[a, b]$ we need only find an antiderivative F of f and compute the difference $F(b) - F(a)$.

Recall that our working definition of $\int_a^b f(t)\,dt$ is $\lim_{n\to\infty}\sum_{i=1}^n f(x_i)\Delta x$, where we partition $[a, b]$ into n equal pieces, each of length Δx, and label $x_i = a + i\Delta x$, for $i = 1, \ldots, n$. Calculating $F(b) - F(a)$ is a wonderful alternative to computing $\lim_{n\to\infty}\sum_{i=1}^n f(x_i)\Delta x$!

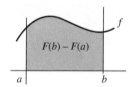

$$F(b) - F(a)$$

Figure 24.1

The Fundamental Theorem of Calculus gives us a fantastic amount of power when computing definite integrals. For example, to compute $\int_1^3 3t^2\,dt$, we need to find a function

$F(t)$ whose derivative is $f(t) = 3t^2$. The function $F(t) = t^3$ comes to mind.

$$\int_1^3 3t^2 \, dt = F(3) - F(1)$$
$$= 3^3 - 1^3$$
$$= 26$$

It's as easy as that! The task of evaluating definite integrals essentially amounts to reversing the process of taking derivatives. It's like the game show "Jeopardy"; our task is to find a function whose derivative is what we have been given.[1]

The Fundamental Theorem of Calculus is a truly amazing result. Stop and think about it for a minute. Back in Chapter 5, we set to work on the problem of finding the slope of a curve. This is an interesting problem all by itself and certainly worthy of a whole math course. Recently, we've been looking at the equally interesting problem of finding the area under a curve. This seems like a question worthy of another whole *separate* math course. But we have just found that these two questions are intimately related. The process of evaluating definite integrals involves the process of antidifferentiating—the process of finding derivatives in reverse! When we first started looking at areas, would you have guessed that such a marvelous relationship would exist? Probably not. But it does. Astounding!

Take a deep breath and think about this for a minute. Then go and explain this amazing result to someone—your roommate, your best friend, your grandmother, your goldfish, or all of them!

Using the Fundamental Theorem of Calculus

We'll begin by applying the Fundamental Theorem to a familiar example, an example we could do without the Theorem.

◆ **EXAMPLE 24.1**   Find the area under $v(t) = 2t + 5$ on the interval $t = 1$ to $t = 4$.

This is the example we looked at in Section 22.1, Example 22.5 in the following guise: $v(t) = 2t + 5$ is the velocity of a cheetah on the interval $[1, 4]$. How far has the cheetah traveled from $t = 1$ to $t = 4$?

SOLUTION   The area is $\int_1^4 (2t + 5) \, dt$. To evaluate this using the Fundamental Theorem of Calculus we need an antiderivative of $2t + 5$. The derivative of t^2 is $2t$ and the derivative of $5t$ is 5, so $F(t) = t^2 + 5t$ is an antiderivative of $2t + 5$.

$$\int_1^4 (2t + 5) \, dt = F(4) - F(1)$$
$$= (4^2 + 5 \cdot 4) - (1^2 + 5 \cdot 1)$$
$$= 30$$

This is the same answer we obtained by calculating the area of the trapezoid directly.

---

[1] It turns out that this game of "Jeopardy" is in fact harder to play than is the derivative game. In general it's simpler to differentiate than it is to find an antiderivative. Try finding a function whose derivative is e^{x^2} if you doubt this.

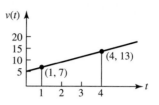

Figure 24.2

In fact, using the Fundamental Theorem of Calculus to solve this problem is the approach that was referred to as the "second mindset" in Chapter 22.1, Example 22.5b. The velocity of the cheetah is given by $v(t)$. Looking for a position function $s(t)$ is equivalent to looking for a function F whose derivative is $v(t)$. Finding the displacement by computing change in position, $s(4) - s(1)$, is essentially what we find when computing $F(4) - F(1)$. $F(t) = s(t) + C$, and the constants cancel when we compute $F(4) - F(1)$.   ◆

EXERCISE 24.1   Any antiderivative of $2t + 5$ can be written in the form $F(t) = t^2 + 5t + C$. Show that when using the Fundamental Theorem to evaluate the integral it doesn't matter *which* antiderivative is used.

NOTATION: The notation $F(x)\Big|_a^b$ is used as a shorthand for $F(b) - F(a)$. It is convenient because it allows us to write $F(x)$ explicitly before evaluating.

◆ EXAMPLE 24.2   Compute $\int_0^1 x^2 dx$.

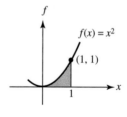

Figure 24.3

SOLUTION   We want an antiderivative of x^2, i.e., we're looking for a function whose derivative is x^2. $F(x) = \frac{x^3}{3}$ is such a function.

$$\int_0^1 x^2 dx = \frac{x^3}{3}\Big|_0^1$$

$$= \frac{1}{3} - 0$$

$$= \frac{1}{3}   ◆$$

◆ **EXAMPLE 24.3**   Compute $\int_e^5 \frac{7}{w}dw$.

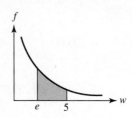

Figure 24.4

SOLUTION $\int_a^b kf(x)dx = k\int_a^b f(x)dx$ (the constant factor property), so we can rewrite this.

$$\int_e^5 \frac{7}{w}dw = 7\int_e^5 \frac{1}{w}dw \quad \text{We want a function whose derivative is } \frac{1}{w}. \; F(w) = \ln w \text{ works.}$$

$$= 7\left[\ln w \Big|_e^5\right]$$

$$= 7[\ln 5 - \ln e]$$

$$= 7[\ln 5 - 1] \quad ◆$$

Notice that calling the variable in the integrand w instead of t or x makes no difference; the name of the variable has no impact. Also note that the properties of definite integrals we found in Section 22.4 are coming in very handy here and that these properties agree with the properties of derivatives. The following are of particular computational importance.[2]

$$\int_a^b kf(x)dx = k\int_a^b f(x)\,dx \quad \text{and} \quad \frac{d}{dx}kf(x) = k\frac{d}{dx}f(x)$$

$$\int_a^b f(x) + g(x)\,dx = \int_a^b f(x)dx + \int_a^b g(x)\,dx \quad \text{and} \quad \frac{d}{dx}(f+g) = \frac{df}{dx} + \frac{dg}{dx}$$

◆ **EXAMPLE 24.4**   Suppose that water enters a reservoir at a rate of $r(t) = 40{,}000 + 60{,}000\cos t$ gallons per month, where t is measured in months. What is the net change in water level between $t = 0$ and $t = 2$?

SOLUTION   To find the net change we calculate the signed area under the rate of change function.

---

[2] When applicable, symmetry considerations can save us a lot of work.

$$\int_0^2 (40{,}000 + 60{,}000 \cos t)\, dt$$

$$= \int_0^2 10{,}000(4 + 6 \cos t)\, dt$$

$$= 10{,}000 \int_0^2 (4 + 6 \cos t)\, dt \qquad \text{constant factor property}$$

$$= 10{,}000 \left[\int_0^2 4\, dt + 6 \int_0^2 \cos t\, dt \right] \qquad \text{additive integrand property}$$

$$= 10{,}000 \left[4t \Big|_0^2 + 6(\sin t) \Big|_0^2 \right] \qquad \sin t \text{ is an antiderivative of } \cos t$$

$$= 10{,}000 \left[8 - 0 + 6 \sin 2 - 6 \sin 0 \right]$$

$$= 10{,}000 \left[8 + 6 \sin 2 \right]$$

Observe that $r(t)$ is the rate function for water entering the reservoir. The function giving the amount of water in the reservoir at time t is an antiderivative of $r(t)$. The net change is the difference between the final and initial amount. While the antiderivative that we use is not necessarily the amount function, it differs from the amount function by a constant, so the net change in amount is preserved.   ◆

◆ **EXAMPLE 24.5**   Compute $\int_0^1 \left(\frac{1}{1+x^2} - 3x^9 \right) dx$.

SOLUTION

$$\int_0^1 \left(\frac{1}{1+x^2} - 3x^9 \right) dx = \int_0^1 \frac{1}{1+x^2} dx - 3 \int_0^1 x^9 dx$$

For the first definite integral we are looking for a function whose derivative is $\frac{1}{1+x^2}$. arctan x is such a function. For the second we are looking for a function whose derivative is x^9. $\frac{x^{10}}{10}$ is such a function.

$$\int_0^1 \left(\frac{1}{1+x^2} - 3x^9 \right) dx = \arctan x \Big|_0^1 - 3 \cdot \frac{x^{10}}{10} \Big|_0^1$$

$$= [\arctan 1 - \arctan 0] - 3 \left[\frac{1}{10} - \frac{0}{10} \right]$$

$$= \frac{\pi}{4} - 0 - 3 \left(\frac{1}{10} \right)$$

$$= \frac{\pi}{4} - \frac{3}{10}$$

How did we know that an antiderivative of $\frac{1}{1+x^2}$ is arctan x? We had to remember the derivative of arctan x. (It's actually a bit surprising that *every* antiderivative of $\frac{1}{1+x^2}$ is of the form arctan $x + C$.)

How did we know that an antiderivative of x^9 was $\frac{x^{10}}{10}$? We might make a first guess of x^{10} but the derivative of x^{10} is $10x^9$. This is almost what we want, but we don't want the constant of 10 out front. To get rid of it, we divide by 10: $\frac{d}{dx}\left(\frac{1}{10}x^{10}\right) = \frac{1}{10} \cdot 10x^{10} = x^{10}$, as desired.   ◆

Appreciating the Fundamental Theorem

Before arriving at the Fundamental Theorem of Calculus we did not have a convenient, widely applicable method of evaluating definite integrals. In this section we tackle two problems, $\int_0^1 x^2 dx$ (Example 24.2), and $\int_0^\pi \sin x\, dx$, *without* using the Fundamental Theorem of Calculus. The purpose of this exercise is both to encourage you to recall the limit definition of the definite integral (essential in applications and in instances where we *can't find* an antiderivative) and to help you develop an appreciation for the power of the Fundamental Theorem.

◆ EXAMPLE 24.6   Evaluate $\int_0^1 x^2 dx$ using the definition of the definite integral but *not* using the Fundamental Theorem of Calculus.

SOLUTION   We begin by partitioning $[0, 1]$ into n equal subintervals labeled as shown.

Figure 24.5

Each subinterval has width $\Delta x = \frac{1}{n}$. $x_k = k\Delta x = \frac{k}{n}$, for $k = 0, 1, \ldots, n$, so

$$x_0 = 0, \quad x_1 = 1/n, \quad x_2 = 2/n, \ldots, \quad x_n = n/n = 1.$$

$\int_0^1 x^2 dx$ is defined to be $\lim_{n\to\infty} R_n$. We begin by constructing R_n, the general right-hand sum. Let $f(x) = x^2$ and refer to Figure 24.6.

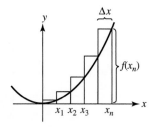

Figure 24.6

The height of the first rectangle is $f(x_1)$, the height of the second is $f(x_2)$, and so on. $\int_0^1 x^2 dx = \lim_{n\to\infty} R_n = \lim_{n\to\infty}\left(\sum_{i=1}^n f(x_i)\Delta x\right)$.

$$\int_0^1 x^2 dx = \lim_{n \to \infty} \left[f(x_1)\Delta x + f(x_2)\Delta x + f(x_3)\Delta x + \cdots + f(x_n)\Delta x \right]$$

$$= \lim_{n \to \infty} \left[f(x_1) + f(x_2) + f(x_3) + \cdots + f(x_n) \right] \Delta x$$

$$= \lim_{n \to \infty} \left[(x_1)^2 + (x_2)^2 + (x_3)^2 + \cdots + (x_n)^2 \right] \Delta x \quad \text{(because } f(x) = x^2\text{)}$$

$$= \lim_{n \to \infty} \left[\left(\frac{1}{n}\right)^2 + \left(\frac{2}{n}\right)^2 + \left(\frac{3}{n}\right)^2 + \cdots + \left(\frac{n}{n}\right)^2 \right] \frac{1}{n} \quad \text{(because } x_k = \frac{k}{n}, \ \Delta x = \frac{1}{n}\text{)}$$

$$= \lim_{n \to \infty} \left[\frac{1^2 + 2^2 + 3^2 + \cdots + n^2}{n^3} \right]$$

In order to evaluate this limit we must express $1^2 + 2^2 + 3^2 + \cdots + n^2$ in closed form. Although it is not a geometric sum, it *can* be expressed in closed form as follows.

$$1^2 + 2^2 + 3^2 + \cdots + n^2 = \frac{n(n+1)(2n+1)}{6}$$

This identity has been delivered to you like manna dropped from heaven. It can be proven by mathematical induction.[3] We use it to obtain $\int_0^1 x^2 dx = \lim_{n \to \infty} \frac{n(n+1)(2n+1)}{6n^3}$.

To evaluate this limit think back to the work we've done with rational functions and consider $\lim_{x \to \infty} \frac{x(x+1)(2x+1)}{6x^3}$. The degree of the numerator and denominator are equal (both are of degree 3), so the limit is given by the fraction formed by the leading coefficients of the numerator and denominator.[4]

$$\lim_{x \to \infty} \frac{x(x+1)(2x+1)}{6x^3} = \frac{2}{6} = \frac{1}{3}.$$

Therefore,

$$\lim_{n \to \infty} \frac{n(n+1)(2n+1)}{6n^3} = \frac{1}{3}.$$

We conclude that

$$\int_0^1 x^2 dx = \frac{1}{3}. \quad \blacklozenge$$

◆ **EXAMPLE 24.7**   Find the area under one arc of $\sin x$ without using the Fundamental Theorem of Calculus.

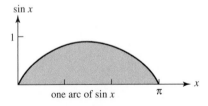

one arc of $\sin x$

Figure 24.7

---

[3] Refer to Appendix D: Proof by Induction.

[4] For large n this fraction "looks like" $\frac{2x^3}{6x^3}$, the dominant term of the numerator over the dominant term of the denominator.

SOLUTION   Consider $\int_0^\pi \sin x \, dx$. To use the limit definition of the definite integral we begin by chopping $[0, \pi]$ into n equal subintervals labeled as shown.

$$0 = x_0 \quad x_1 \quad x_2 \quad x_3 \cdots \qquad\qquad x_n = \pi$$

Each subinterval has width $\Delta x = \frac{\pi}{n}$. $x_k = k \Delta x = \frac{k\pi}{n}$, for $k = 0, \ldots, n$, so

$$x_0 = 0, \quad x_1 = \pi/n, \quad x_2 = 2\pi/n, \quad x_3 = 3\pi/n, \ldots, \quad x_n = n\pi/n = \pi.$$

Let $f(x) = \sin x$. We know

$$\int_0^\pi \sin x \, dx = \lim_{n \to \infty} R_n = \lim_{n \to \infty} \left(\sum_{i=1}^n f(x_i) \Delta x \right).$$

$$\int_0^\pi \sin x \, dx = \lim_{n \to \infty} \left[f(x_1) \Delta x + f(x_2) \Delta x + f(x_3) \Delta x + \cdots + f(x_n) \Delta x \right]$$

$$= \lim_{n \to \infty} \left[f(x_1) + f(x_2) + f(x_3) + \cdots + f(x_n) \right] \Delta x$$

$$= \lim_{n \to \infty} \left[\sin x_1 + \sin x_2 + \sin x_3 + \cdots + \sin x_n \right] \Delta x$$

$$\int_0^\pi \sin x \, dx = \lim_{n \to \infty} \left[\sin(\pi/n) + \sin(2\pi/n) + \sin(3\pi/n) + \cdots + \sin(n\pi/n) \right] (\pi/n)$$

We look up, but no manna drops down from heaven. Alas, we have no way of expressing $[\sin(\pi/n) + \sin(2\pi/n) + \sin(3\pi/n) + \cdots + \sin(n\pi/n)]$ in closed form; we have a problem evaluating this limit. Without using the Fundamental Theorem, we can only approximate $\int_0^\pi \sin x \, dx$. There are problems at the end of this section asking you to do this.   ◆

EXERCISE 24.2   Although we can write what we've done in Example 24.7 using summation notation, that is just shorthand, *not* a closed form; it is nice and compact, but it does not help us evaluate the limit. As an exercise, express what we've done using summation notation and compare your answer with what is given below.

$$\int_0^\pi \sin x \, dx = \lim_{n \to \infty} \sum_{k=1}^n \sin \left(\frac{k\pi}{n} \right) \frac{\pi}{n}$$

If we *do* use the Fundamental Theorem of Calculus, the problem of computing $\int_0^\pi \sin x \, dx$ becomes straightforward. We look for a function whose derivative is $\sin x$. We might first guess $\cos x$. $\frac{d}{dx} \cos x = -\sin x$, which is almost what we want. $\frac{d}{dx}(-\cos x) = \sin x$. Therefore

$$\int_0^\pi \sin x \, dx = -\cos x \Big|_0^\pi$$

$$= -\cos \pi - (-\cos 0)$$

$$= -(-1) + 1$$

$$= 2.$$

Pause for a moment to admire what we've discovered. The area under one arc of sin x is 2. Exactly. Remarkable!

Is the moral of Example 24.7 that the limit definition of the definite integral is merely academic? *No!* In practice, we use the Fundamental Theorem of Calculus to compute definite integrals exactly whenever we can, but sometimes it is difficult (or impossible) to find an antiderivative of the integrand. (For instance, try to find a function whose derivative is e^{-x^2}. But don't try for too long; it's impossible.[5]) In this case, the best we can do is to *approximate* the definite integral. The limit definition of derivative gives us a means of approximation. For example, since

$$\int_0^\pi \sin x \, dx = \lim_{n \to \infty} \sum_{k=1}^n \sin\left(\frac{k\pi}{n}\right) \frac{\pi}{n},$$

we know that

$$\int_0^\pi \sin x \, dx \approx \sum_{k=1}^n \sin\left(\frac{k\pi}{n}\right) \frac{\pi}{n}, \quad \text{for } n \text{ large.}$$

While this latter sum is tedious to compute by hand for n large, a computer or programmable calculator can deal with it easily. We'll return to numerical approximations in Chapter 26. In addition to providing a computational tool for approximating definite integrals, the limit definition allows us to apply the work we've done with definite integrals to new situations. This will be our focus in Chapter 27 and again in Chapter 29.

The Fundamental Theorem of Calculus, Take Two

The Fundamental Theorem of Calculus tells us that we can find the area under the graph of a continuous function f on the interval $[a, b]$ by finding any antiderivative F of f and evaluating the difference $F(b) - F(a)$. We've seen that this is an amazing and powerful result. We now show you (through examples) that *if* the integrand is thought of as a rate function, then the Fundamental Theorem of Calculus tells us something we've known for some time.

◆ EXAMPLE 24.8   A camel is walking along a straight desert road with velocity $f(t)$. How can we express the camel's displacement from time $t = a$ to time $t = b$?

SOLUTION   On the one hand we can partition the time interval $[a, b]$ into n equal pieces, approximate the camel's displacement on each subinterval by assuming constant velocity on the subinterval, sum the displacements, and take the limit as n grows without bound. The camel's net change in position is given by

$$\int_a^b f(t) \, dt.$$

On the other hand, $f(t)$ is the velocity function, so $f(t) = \frac{ds}{dt}$, where $s(t)$ gives the camel's position at time t. $s'(t) = f(t)$, so $s(t)$ is an antiderivative of $f(t)$. We know that if we can find $s(t)$, then the camel's net change in position must be given by

$$s(b) - s(a).$$

---

[5] It is impossible to find such a function unless you're willing to use an infinite polynomial.

So $\int_a^b f(t)\, dt = s(b) - s(a)$, where s is an antiderivative of f.

$$\begin{pmatrix} \text{net change in} \\ \text{position on } [a, b] \end{pmatrix} = \begin{pmatrix} \text{position} \\ \text{at } t = b \end{pmatrix} - \begin{pmatrix} \text{position} \\ \text{at } t = a \end{pmatrix}$$

When we interpret the integrand as a rate function the Fundamental Theorem of Calculus becomes transparent.

Because any two antiderivatives of f differ only by an additive constant and the constants will cancel in the course of subtraction, it follows that $s(t)$ could be any antiderivative of f.

$$\int_a^b \frac{ds}{dt}\, dt = s(t)\Big|_a^b = s(b) - s(a) \quad \blacklozenge$$

◆ **EXAMPLE 24.9** If $f(t)$ is the rate at which water is flowing into or out of a reservoir and we let $W(t)$ be the amount of water in the reservoir at time t, then $\frac{dW}{dt} = f(t)$. $W(t)$ is an antiderivative of $f(t)$. The net change in the amount of water in the reservoir over the time interval $[a, b]$ is given by

$$\int_a^b f(t)\, dt = \int_a^b \frac{dW}{dt}\, dt = W(b) - W(a).$$

Again, when we interpret the integrand as a rate function, the Fundamental Theorem of Calculus is transparent.

It is when the integrand, $f(t)$, is not thought of as a rate function that the Fundamental Theorem is most surprising. ◆

▌ PROBLEMS FOR SECTION 24.1

1. (a) Using a computer or programmable calculator, find upper and lower bounds for the area under one arc of $\cos x$ using Riemann sums. Explain how you can be sure your lower bound is indeed a lower bound and your upper bound is an upper bound. (Do not use the Fundamental Theorem of Calculus to do so.) Your upper and lower bounds should differ by no more than 0.01.

   (b) Use the Fundamental Theorem of Calculus to show that the area under one arc of the cosine curve is exactly 2.

2. An object's velocity at time t, t in seconds, is given by $v(t) = 10t + 3$ meters per second. Find the net distance traveled from time $t = 1$ to $t = 9$. Do this in two ways. First, look at the appropriate signed area and solve geometrically, without the Fundamental Theorem. Then calculate the definite integral

$$\int_1^9 (10t + 3)\, dt$$

   using the Fundamental Theorem of Calculus.

3. Use the Fundamental Theorem of Calculus to calculate $\int_1^2 t^3\, dt$.

4. Evaluate $\int_1^3 \frac{1}{t}\, dt$.

5. Find the area under the graph of $y = e^x$ between $x = 0$ and $x = 1$.

6. Find the area under the graph of $y = e^{-x}$ between

   (a) $x = -1$ and $x = 0$. (Why should the answer be the same as the answer to the previous problem?)

   (b) $x = 0$ and $x = 1$.

7. Aimee and Alexandra spent Friday afternoon eating hot cinnamon hearts. If they gobbled hearts at a rate of $1.5t + \sqrt{t}$ hearts per minute, then how many hearts did they consume between time $t = 0$ and $t = 9$, t given in minutes?

8. Suppose the temperature of an object is changing at a rate of $r(t) = -2e^{-t}$ degrees Celsius per hour, where t is given in hours.

   (a) Is the object heating, or cooling?

   (b) Between time $t = 0$ and $t = 1$, how much has the temperature changed?

   (c) Between $t = 1$ and $t = 2$, how much has the temperature changed?

   (d) If the object was 100 degrees Celsius at time $t = 0$, how hot is it at time $t = 1$?

9. Let $f(x) = \int_1^x \frac{1}{t}\, dt$, $x > 0$.

   (a) Find $f(1)$, $f(5)$, $f(10)$, and $f(1/2)$.

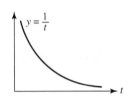

   (b) What is an alternative formula for $f(x)$?

   (c) Often mathematicians *define* the natural logarithm by $\ln x = \int_1^x \frac{1}{t}\, dt$ for $x > 0$. Suppose this was the definition you had been given. Use the Fundamental Theorem of Calculus to show that $\ln x$ is increasing and concave down for $x > 0$.

10. Find the value of $x > 1$ such that the area under the graph of $1/t$ from 1 to x is 1.

11. The rate of change of water level in a tank is given by $r(t) = 2 \sin\left(\frac{\pi}{4}t\right)$ gallons per hour, where t is measured in hours. At time $t = 0$ there are 30 gallons of water in the tank.

   (a) Between time $t = 0$ and $t = 8$, when will the water level in the tank be the highest?

   (b) What is the maximum amount of water that will ever be in the tank?

   (c) What is the minimum amount of water that will ever be in the tank?

   (d) Is the amount of water added to the tank between $t = 0$ and $t = 1$ less than, greater than, or equal to the amount lost between $t = 4$ and $t = 5$? (Try to answer without doing any computations.)

12. Let $g(x) = \int_0^x e^{-t^2}\,dt$.

   (a) Where is $g(x)$ zero? Positive? Negative?

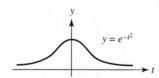

   (b) Where is $g(x)$ increasing? Decreasing?

   (c) Where is $g(x)$ concave up? Concave down?

   (d) Is $g(x)$ even, odd, or neither?

   (e) Although we *cannot* find an antiderivative for e^{-t^2}, we are able to get a lot of information about $g(x)$. Sketch $g(x)$.

   (f) Using a computer or programmable calculator, approximate $g(1)$, $g(-1)$, and $g(2)$.

   Go back and look at all your answers to this question; make sure that they are consistent with one another.

13. Let $h(x) = \int_0^x \sin(t^2)\,dt$.

   (a) Graph $\sin(t^2)$ on the domain $[-3, 3]$.

   (b) On $(0, \infty)$, where is $h(x)$ positive? Negative?

   (c) Is $h(x)$ even, odd, or neither? Explain.

   (d) On $[0, 3]$ where is $h(x)$ increasing? Decreasing? Give exact answers.

   (e) What is the absolute maximum value of $\sin(t^2)$? What is the absolute minimum value of $\sin(t^2)$?

   (f) Where on $[0, 3]$ does h attain its maximum and minimum values? Will your answers change if the domain is $(0, \infty)$? If the domain is $(-\infty, \infty)$?

   (g) Numerically approximate the maximum value of h on $[0, 3]$.

14. (a) Find $\int_{-1}^{1} |x|\,dx$ using the area interpretation of the definite integral.

   (b) Show that $\frac{|x^2|}{2}$ is *not* an antiderivative of $|x|$ on $[-1, 1]$ by showing that applying the Fundamental Theorem as if it were, gives the wrong answer.

   (c) Find an antiderivative of $|x|$ on $[-1, 0)$. Find an antiderivative of $|x|$ on $(0, 1]$.

15. Find the following two definite integrals *without* using the Fundamental Theorem of Calculus. Instead, use the area interpretation of the definite integral.

   (a) $\int_{-3}^{7} (\pi + 1)dx$ (b) $\int_{-3}^{7} |-2x - 4|dx$

16. Evaluate the following. (If you haven't done Problem 14, do that first.)

   (a) $\int_{-3}^{3} |x^2 - 4|dx$ (b) $\int_0^5 |(x + 3)(x - 1)|dx$

17. Calculate the following definite integrals by calculating the limit of Riemann sums. You'll need to use the formulas provided. Check your answers using the Fundamental Theorem of Calculus.

(a) $\displaystyle\int_0^5 x\,dx$ $1 + 2 + 3 + \cdots + n = \dfrac{n(n+1)}{2}$

(b) $\displaystyle\int_0^5 x^2\,dx$ $1^2 + 2^2 + 3^2 + \cdots n^2 = \dfrac{n(n+1)(2n+1)}{6}$

(c) $\displaystyle\int_0^2 x^3\,dx$ $1^3 + 2^3 + 3^3 + \cdots n^3 = \left[\dfrac{n(n+1)}{2}\right]^2$

18. Read Appendix D: Proof by Induction. Then prove that the formulas provided in the previous problem are valid.

19. Given the graph below, compute $\int_a^b [f'(x)g(x) + g'(x)f(x)]\,dx$. Assume that f and g are differentiable functions.

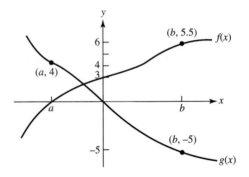

20. (a) Write a Riemann sum with 10 equal subdivisions that gives an overestimate for the area under $\ln x$ on $[1, 6]$. Write your answer in two ways, once with summation notation and one without summation notation (using $+ \cdots +$). State clearly and precisely the meaning of any notation used in your sum.

    (b) Consider the Riemann sum

    $$\sum_{k=0}^{49} \ln(3 + k \cdot (5/50)) \cdot \frac{5}{50}.$$

    This is an underestimate for what integral? (The answer to this question is not unique.)

21. Put the following in ascending order, using $<$ or $=$ as appropriate.

    $$\int_1^2 \ln x\,dx \qquad \int_{0.5}^2 \ln x\,dx \qquad \int_1^{2.5} \ln x\,dx,$$

22. Put the following in ascending order, using $<$ or $=$ as appropriate.

    $$0 \qquad \int_1^6 \ln x\,dx \qquad L_{50} \qquad R_{50} \qquad L_{10} \qquad R_{10}$$

    Explain your reasoning briefly.

23. (a) Which of the following is an antiderivative of $\ln x$?

i. $\frac{1}{x}$ ii. $x \ln x$ iii. $x \ln x - x$ iv. $\arctan(\ln x)$

(b) Evaluate $\int_1^6 \ln x \, dx$.

24.2 THE AVERAGE VALUE OF A FUNCTION: AN APPLICATION OF THE DEFINITE INTEGRAL

In the modern world, a data set is often presented in a way to succinctly convey basic information about the central tendency of the data. The average is one measure of central tendency of a data set. We're familiar with computing averages when given a discrete set of data points. For example, to determine the average score on a quiz, sum all quiz scores and divide by the number of quizzes. In this section we will look at the average value of a continuous function and find that the definite integral enables us to compute this. For example, we will look at how to determine the average velocity of an object given a continuous velocity function and the average temperature of an object given a continuous temperature function.

◆ **EXAMPLE 24.10** A dove flies at a velocity of $w(t) = 3t^2 + 4$ meters per second on the interval $[1, 3]$. A hawk flies at a velocity of $v(t) = \frac{51}{2}(-t^2 + 4t - 3)$ meters per second over the same interval.

(a) How far does each travel during this time?

(b) What is the average velocity of each bird on the interval $[1, 3]$?

SOLUTION (a) The net change in position (or distance traveled, as velocity is nonnegative here) is given by a definite integral.

$$\text{Dove:} \quad \int_1^3 (3t^2 + 4) \, dt \qquad = (t^3 + 4t)\Big|_1^3$$

$$= (27 + 12) - (1 + 4)$$

$$= 34 \text{ meters}$$

$$\text{Hawk:} \quad \int_1^3 \frac{51}{2}(-t^2 + 4t - 3) \, dt = \frac{51}{2}\left(\left(\frac{-t^3}{3}\right) + 2t^2 - 3t\right)\Big|_1^3$$

$$= \frac{51}{2}\left[(-9 + 18 - 9) - \left(\frac{-1}{3} + 2 - 3\right)\right]$$

$$= \frac{51}{2}\left(0 - \frac{-4}{3}\right)$$

$$= 34 \text{ meters}$$

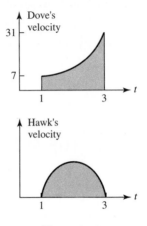

Figure 24.8

Both birds travel the same distance; even though their velocity functions look quite different, the areas under each on the interval $[1, 3]$ are the same.

(b) To compute average velocity, we divide the net change in position by the change in time. This will be the same for each bird.

$$\text{average velocity} = \frac{\text{net change in position}}{\text{time}}$$

$$= \frac{34 \text{ meters}}{2 \text{ seconds}}$$

$$= 17 \text{ meters per second}$$

Notice that the units (meters per second) are the units we expect for velocity. ◆

The key notion to grasp here is that in order to find the average value of the *velocity* function, we divide the net change in *position* by the change in time. We integrate the velocity function rather than use any specific values of it.

$$\text{average velocity on } [a, b] = \frac{\text{net change in position}}{\text{time}}$$

$$= \frac{\int_a^b v(t)\, dt}{b - a} = \frac{1}{b - a} \int_a^b v(t)\, dt$$

Notice that if we let $s(t)$ denote position, where $s'(t) = v(t)$, then

$$\frac{\int_a^b v(t)\, dt}{b - a} = \frac{s(b) - s(a)}{b - a} = \frac{\Delta s}{\Delta t}.$$

We can generalize the discussion. For instance, to find the average rate at which water enters a tank, we integrate the rate function over the time interval (finding the net change in the amount of water) and divide by the change in time. But not only can we generalize to other rate functions, we can generalize to find the average value of *any* integrable function. We define the average value of a function on a closed interval $[a, b]$ as follows.

Definition

$$\text{The average value of } f(x) \text{ on } [a, b] \text{ is } \frac{1}{b-a} \int_a^b f(x)\, dx.$$

We can interpret the average value of f graphically. Denote by f_{ave} the average value of $f(x)$ on $[a, b]$. Then $\int_a^b f(x)\, dx = f_{\text{ave}}(b - a)$. The latter expression can be interpreted as the area of a rectangle with base of length $b - a$ and height of f_{ave}. Thus the average value is the constant function with the same signed area under it on $[a, b]$ as f. For instance, an eagle traveling at a constant velocity of 17 meters per second would travel the same distance as the dove and the hawk over the interval $[1, 3]$.

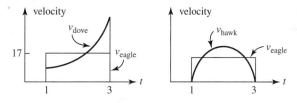

Figure 24.9

SOME COMMON MISCONCEPTIONS: There are several conceptual potholes to be avoided when thinking about average value. We point them out in the example below so you can bypass the potholes instead of falling into them.

The situation: A dove travels at $3t^2 + 4$ meters per second, a hawk at $\frac{51}{2}(-t^2 + 4t - 3)$ meters per second, and an eagle at 17 meters per second. We've established that on the interval $[1, 3]$ all three birds travel the same total distance and hence have the same average velocity.

Mistake I. The average velocity on $[1, 3]$ is *not* the average of the velocities at $t = 1$ and $t = 3$. We are interested in the average *over an interval*; the values of $v(t)$ *between* $t = 1$ and $t = 3$ are *vitally* important. Notice that the average of the final and initial velocities gives different results for each of the three birds. Especially alarming is the case of the hawk; because $v(3)$ and $v(1)$ are both zero, the average of the two is $\frac{0+0}{2} = 0$. But the hawk's velocity is positive for all $t \in (1, 3)$, so it is nonsensical to conclude that its average velocity on $[1, 3]$ is zero.

Mistake II. The average velocity on $[1, 3]$ is *not* half the difference between the velocities at $t = 1$ and $t = 3$. This computation also gives different results for the dove and the hawk and assigns both the hawk and the eagle the nonsensical value of zero as an average velocity. The eagle is flying at 17 meters per second on $[1, 3]$; certainly its average velocity is *not* zero.

◆ **EXAMPLE 24.11**   A hawk's velocity is given by $v(t) = \frac{51}{2}(-t^2 + 4t - 3)$ meters per second over the interval $[0, 3]$. $v(t)$ positive indicates the hawk is flying east. $v(t)$ negative indicates the hawk is flying west.

(a) What is the hawk's average velocity on $[0, 3]$?

(b) What is the hawk's average speed on $[0, 3]$?

SOLUTIONS   (a) The hawk's velocity on $[0, 3]$ is graphed in Figure 24.10. The average velocity on $[0, 3]$ is the average value of the velocity function $v(t)$:

$$v_{ave} = \frac{1}{3} \int_0^3 v(t) \, dt.$$

This is the *net* displacement divided by the time elapsed.

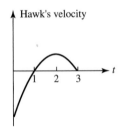

Hawk's velocity

Figure 24.10

$$v_{ave} = \frac{1}{3} \int_0^3 v(t) \, dt$$

$$= \frac{1}{3} \int_0^3 \frac{51}{2}(-t^2 + 4t - 3) \, dt$$

$$= \frac{1}{3} \cdot \frac{51}{2} \left[\frac{-t^3}{3} + 2t^2 - 3t \right] \Big|_0^3$$

$$= \frac{51}{6} [-9 + 18 - 9]$$

$$= \frac{17}{2} [0]$$

$$= 0$$

The average velocity of the hawk on $[0, 3]$ is zero because its net displacement on the interval $[0, 3]$ is zero. It flys the same distance west on the interval $[0, 1]$ as it does east on the interval $[1, 3]$.

(b) The average speed on $[0, 3]$ is the average value of the speed function $|v(t)|$:

$$|v|_{ave} = \frac{1}{3} \int_0^3 |v(t)| \, dt.$$

This is the *total* distance traveled divided by the time elapsed.

$$|v|_{\text{ave}} = \frac{1}{3} \int_0^3 |v(t)|\, dt$$

To integrate $|v(t)|$ we must take off the absolute value signs. This means we must split the interval of integration into pieces such that on each subinterval the sign of $v(t)$ does not change.

$$|v|_{\text{ave}} = \frac{1}{3} \left[\int_0^1 |v(t)|\, dt + \int_1^3 |v(t)|\, dt \right]$$

$$= \frac{1}{3} \int_0^1 - [v(t)]\, dt + \frac{1}{3} \int_1^3 v(t)\, dt$$

$$= \frac{1}{3} \int_0^1 - \left[\frac{51}{2}(-t^2 + 4t - 3) \right] dt + \frac{1}{3}[34]$$

$$= \frac{1}{3}\left(-\frac{51}{2}\right)\left(-\frac{t^3}{3} + 2t^2 - 3t\right) \Big|_0^1 + \frac{34}{3}$$

$$= \frac{1}{3}(34) + \frac{34}{3}$$

$$= \frac{1}{3}(34 + 34)$$

$$= \frac{68}{3}$$

$$= 22\frac{2}{3}$$

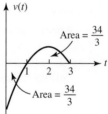

Figure 24.11

The hawk's average speed is $22\frac{2}{3}$ meters per second. ◆

EXERCISE 24.3   On the interval $[1, 3]$ the hawk's average speed and average velocity are equal, because $v(t) \geq 0$ on $[1, 3]$. On $[1, 3]$ the hawk's average speed is 17 meters per second. On $[0, 1]$ the hawk's average speed is $\frac{1}{1} \int_0^1 |(v(t)|\, dt = \ldots = 34$ meters per second. If we average 17 meters per second and 34 meters per second we get $25\frac{1}{2}$ meters per second, which is *not* the hawk's average speed. If, instead, we take a weighted average, giving the average speed on $[1, 3]$ twice the weight of the average speed on $[0, 1]$ we get

$$\frac{34 + 2 \cdot 17}{3} = 22\frac{2}{3}.$$

Explain.

◆ **EXAMPLE 24.12**   Suppose that the temperature in Toronto is given by $T(t) = 56 + \sqrt{t}$ degrees on the interval $[4, 9]$, where t is measured in hours. What is the average temperature over this time period?

SOLUTION

$$\text{average temperature} = \frac{1}{9 - 4} \int_4^9 \left(56 + \sqrt{t}\right) \, dt$$

$$= \frac{1}{5} \left(56t + \frac{2}{3}t^{3/2}\right) \Big|_4^9$$

$$= \frac{1}{5} \left[(504 + 18) - \left(224 + \frac{16}{3}\right)\right]$$

$$= \frac{1}{5} \frac{878}{3} \approx 58.53 \text{ degrees}$$

Notice that in this case the integral alone has no readily apparent meaning because the integrand is not a rate function. It is only after dividing by the time interval that we obtain a meaningful result.   ◆

PROBLEMS FOR SECTION 24.2

1. Find the average value of $3 \sin x + 5$ on the interval $[0, 2\pi]$. Do this in two ways, first geometrically and then using the Fundamental Theorem of Calculus.

2. (a) Suppose f is an odd function. Can you determine the average value of f on $[-a, a]$? If so, what is the average value?

   (b) Suppose f is an even function. Are the following equal? If not, can you determine which is largest? Explain your answer.
      i. the average value of f on $[-a, a]$
      ii. the average value of f on $[0, a]$
      iii. the average value of f on $[-a, 0]$

3. Find the average value of $\sin x$ on $[0, \pi]$.

4. The velocity of an object on $[0, 6]$ is given by $v(t) = -t^2 + 4t$.
   (a) Find the average velocity on $[0, 6]$.
   (b) Find the average speed on $[0, 6]$.

5. The function $v(t)$ gives the velocity (in miles per hour) of a bicyclist on a cross-country trip. We let $t = 0$ denote noon (t measured in hours) and designate $v(t)$ to be positive when the bicyclist is going east. Interpret the following in terms of distance, position, velocity, etc. Your interpretation should be intelligible to a twelve year old with no calculus background.

   (a) $v(5)$ (b) $\int_0^5 v(t) \, dt$ (c) $\int_0^5 |v(t)| \, dt$

   (d) $|v(5)|$ (e) $v'(5)$ (f) $\dfrac{\int_0^5 v(t) \, dt}{5}$

6. The amount of a certain chemical in a mixture varies with time. If $g(t) = 5e^{-t}$ is the number of grams of the chemical at time t, what is the average number of grams of the chemical in the mixture on the time interval $[0, 1]$?

7. The velocity of an object is given by $3\sin(\pi t)$.
   (a) What is the object's speed as a function of time?
   (b) What is the object's net displacement from $t = 0$ to $t = 2$?
   (c) How far has the object traveled from $t = 0$ to $t = 2$?
   (d) What is the object's average velocity on $[0, 2]$?
   (e) What is the object's average speed on $[0, 2]$?

8. The graphs of functions f, g, h, and k are given below.
   Let I denote the average value of f on $[0, 4]$.
   Let II denote the average value of g on $[0, 4]$.
   Let III denote the average value of h on $[0, 4]$.
   Let IV denote the average value of k on $[0, 4]$.
   Put I, II, III, and IV, in ascending order, with "<" or "≤" signs between them as appropriate.

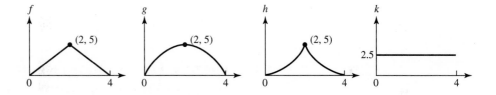

9. A bicycle speedometer will give the average velocity of a bicyclist over the time period the bicycle is moving. By pressing a button the bicylist can reset the average velocity counter. Suppose a long-distance cyclist has averaged 14 miles per hour for the first two hours of her trip. She resets the average velocity counter. For the next four hours her average velocity is 18 miles per hour.
   (a) What is the cyclist's average velocity for the six-hour trip?
   (b) How far has she traveled?

10. It takes a bicyclist 8 minutes to ride 1 mile uphill and then 2 minutes to ride 1 mile downhill. Explain how we know that the cyclist's average velocity for the hill is 12 miles per hour.

11. The temperature of a hotplate of radius 5 inches varies with the distance from the center of the plate. For the area within 2 inches of the center the average temperature is 100 degrees. For the area between 2 and 5 inches from the center the average temperature is 80 degrees. What is the average temperature of the plate?

12. Find the average value of $|\sin(3t)|$ on $[0, 2\pi]$. Explain your reasoning.

13. The graph below shows the birthrate, $B(t)$, and the death rate, $D(t)$, for a population of fish in a lake. t is measured in years, and $t = 0$ represents 1960. The population of fish in 1960 was 4500. (Assume that births and deaths are the only factors that affect the population—no fishing, no immigration, etc.)

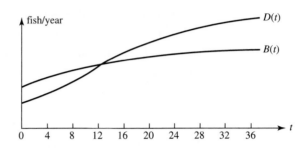

(a) Write an expression for the total number of births between 1960 and 1996.

(b) Write an expression for the average death rate between 1980 and 1990.

(c) Write an expression for the fish population in 1996.

(d) Approximate the year the population was greatest.

(e) Was the 1996 population greater or less than 4500? Explain.

Applications and Computation of the Integral

C H A P T E R

25

Finding Antiderivatives— An Introduction to Indefinite Integration

25.1   A LIST OF BASIC ANTIDERIVATIVES

In order to apply the Fundamental Theorem of Calculus we need to be able to find antiderivatives. This process can be challenging. Therefore, it is useful to have at our disposal a list of functions we can readily antidifferentiate. We obtain this list by thinking about functions we can readily differentiate and working backward.

Definition

The symbol $\int f(x)\, dx$ stands for the entire family of functions that are antiderivatives of $f(x)$. $\int f(x)\, dx$ is called the **indefinite integral of** $f(x)$. $f(x)$ is the **integrand**.

Recall that if $F(x)$ is an antiderivative of $f(x)$, then $F(x) + C$ is also an antiderivative of $f(x)$ for any constant C.

$$\frac{d}{dx}[F(x) + C] = \frac{d}{dx}F(x) + 0 = f(x)$$

Any two antiderivatives of f differ by an additive constant, so every antiderivative of f can be written in the form $F(x) + C$.

We know that

$$\frac{d}{dx} \sin x = \cos x; \quad \text{therefore} \quad \int \cos x \, dx = \sin x + C.$$

Equivalently, we can write

$$\frac{d}{dw} \sin w = \cos w; \quad \text{therefore} \quad \int \cos w \, dw = \sin w + C.$$

Or

$$\frac{d}{du} \sin u = \cos u; \quad \text{therefore} \quad \int \cos u \, du = \sin u + C.$$

The variables x, w, and u are sometimes referred to as "dummy variables," meaning that the statements we've made are equivalent, regardless of the variable we use. $\int \cos x \, dx = \sin x + C$, $\int \cos w \, dw = \sin w + C$, and $\int \cos u \, du = \sin u + C$ all say the same thing.

Derivatives	Corresponding Antiderivatives
$\frac{d}{dx} x^n = n x^{n-1}$	$\int x^n \, dx = \frac{x^{n+1}}{n+1} + C$ for $n \neq -1$
$\frac{d}{dx} \ln x = \frac{1}{x}$	$\int \frac{1}{x} \, dx = \ln|x| + C$ (An explanation of the need for $|x|$ will follow.)
$\frac{d}{dx} e^x = e^x$	$\int e^x \, dx = e^x + C$
$\frac{d}{dx} b^x = (\ln b) \cdot b^x$	$\int b^x \, dx = \frac{1}{\ln b} b^x + C$
$\frac{d}{dx} \sin x = \cos x$	$\int \cos x \, dx = \sin x + C$
$\frac{d}{dx} \cos x = -\sin x$	$\int \sin x \, dx = -\cos x + C$ $\left(\frac{d}{dx} \cos x = -\sin x, \text{ so } \frac{d}{dx}[-\cos x] = \sin x \right)$
$\frac{d}{dx} \tan x = \sec^2 x$	$\int \sec^2 x \, dx = \tan x + C$
$\frac{d}{dx} \arcsin x = \frac{1}{\sqrt{1-x^2}}$	$\int \frac{1}{\sqrt{1-x^2}} \, dx = \arcsin x + C$
$\frac{d}{dx} \arccos x = \frac{-1}{\sqrt{1-x^2}}$	$\int \frac{-1}{\sqrt{1-x^2}} \, dx = \arccos x + C$
$\frac{d}{dx} \arctan x = \frac{1}{1+x^2} \, dx$	$\int \frac{1}{1+x^2} \, dx = \arctan x + C$

Comments:

■ The "$+ C$," in each of these indefinite integrals is necessary; without the "$+ C$" we have given only one antiderivative as opposed to the entire family of antiderivatives. From a graphical perspective, the "$+ C$" indicates that all antiderivatives of the function are vertical translates of one another.

■ Why do we need the absolute value bars in $\int \frac{1}{x} \, dx = \ln|x| + C$?
We are looking for an antiderivative of $\frac{1}{x}$.

If $x > 0$ and $F(x) = \ln x$, then $F'(x) = \frac{1}{x}$. Therefore, $\ln x$ is an antiderivative of $\frac{1}{x}$ for $x > 0$.

If $x < 0$, then $\ln x$ is undefined. However, if we let $F(x) = \ln(-x)$, then $F'(x) = \frac{1}{-x} \cdot (-1) = \frac{1}{x}$. Therefore, $\ln(-x)$ is an antiderivative of $\frac{1}{x}$ for $x < 0$.

$$\ln |x| = \begin{cases} \ln x & \text{for } x > 0 \\ \ln(-x) & \text{for } x < 0 \end{cases}.$$

Therefore, we can use $F(x) = \ln |x|$ as an antiderivative of $\frac{1}{x}$ for all $x \neq 0$.

EXERCISE 25.1   On the one hand, $\int -\frac{1}{\sqrt{1-x^2}} \, dx = \arccos x + C$; on the other hand, $\int \frac{-1}{\sqrt{1-x^2}} \, dx = -\int \frac{1}{\sqrt{1-x^2}} \, dx = -\arcsin x + C$. Therefore, $\arccos x$ and $-\arcsin x$ must differ by a constant.

$$-\arcsin x = \arccos x + C$$

Find C.

In addition to the list of specific antiderivatives given in the table on page 784, we can deduce principles for integration of the sum of functions and integration of a constant multiple of a function from the corresponding principles for differentiation.

General Differentiation Rules

$\frac{d}{dx}[f(x) + g(x)] = \frac{d}{dx} f(x) + \frac{d}{dx} g(x)$

$\frac{d}{dx} \left[kf(x) \right] = k \frac{d}{dx} f(x)$

Corresponding General Integration Rules

$\int \left[f(x) + g(x) \right] \, dx = \int f(x) \, dx + \int g(x) \, dx$

$\int kf(x) \, dx = k \int f(x) \, dx$

The antiderivatives listed in the table on page 784 together with the two general integration rules above, enable us to evaluate a wide variety of definite integrals. Learning how to undo the Chain Rule will allow us to evaluate an even broader range of definite and indefinite integrals. We will do this, using a technique called substitution, in this chapter. We can further expand the type of functions we can integrate by using the Product Rule in reverse and arriving at a technique called integration by parts. This will be taken up in Section 29.1.

In the problems below we apply general integration rules in combination with the basic antiderivatives displayed in the table on page 784. Our strategy is to manipulate the integrand algebraically so that it can be expressed as a sum. Then the integral can be pulled apart into the sum of simpler integrals.

◆ **EXAMPLE 25.1**   Integrate the following. (Try these on your own first; then read the solutions.)

$$\text{(a) } \int \left(1 + x^2\right)^2 \, dx \quad \text{(b) } \int (x+7)\sqrt{x} \, dx \quad \text{(c) } \int \frac{2x+3}{x^2} \, dx$$

SOLUTIONS

$$\text{(a) } \int \left(1 + x^2\right)^2 \, dx = \int \left(x^4 + 2x^2 + 1\right) dx$$

$$= \frac{x^5}{5} + \frac{2x^3}{3} + x + C \quad \begin{pmatrix} \text{Note that there's no need to} \\ \text{use three separate constants.} \end{pmatrix}$$

(b) $\displaystyle\int (x+7)\sqrt{x}\,dx = \int x\sqrt{x}\,dx + \int 7\sqrt{x}\,dx$

$\displaystyle\qquad\qquad = \int x^{3/2}\,dx + 7\int x^{1/2} + C$

$\displaystyle\qquad\qquad = \frac{2}{5}x^{5/2} + 7\cdot\frac{2}{3}x^{3/2} + C$

$\displaystyle\qquad\qquad = \frac{2}{5}x^{5/2} + \frac{14}{3}x^{3/2} + C$

(c) $\displaystyle\int \frac{2x+3}{x^2}\,dx = \int \frac{2x}{x^2} + \frac{3}{x^2}\,dx$

$\displaystyle\qquad\qquad = \int \frac{2}{x}\,dx + \int \frac{3}{x^2}\,dx$

$\displaystyle\qquad\qquad = 2\int \frac{1}{x}\,dx + 3\int x^{-2}\,dx$

$\displaystyle\qquad\qquad = 2\ln|x| + 3\frac{x^{-2+1}}{-2+1} + C$

$\displaystyle\qquad\qquad = 2\ln|x| - \frac{3}{x} + C \quad\blacklozenge$

PROBLEMS FOR SECTION 25.1

In Problems 1 through 11, compute the integral.

1. $\int 3x^3 + 2x + \pi\,dx$

2. $\int At^n\,dt,$ where A and n are constants and $n \neq -1$.

3. $\int 3x^{-1}\,dx$

4. $\int \frac{dx}{2x}$

5. $\int 3\sin t - \frac{3}{1+t^2}\,dt$

6. $\int \frac{5\,dx}{7x}$

7. $\int \frac{2\cos w}{3}\,dw$

8. $\int \frac{1}{x^2+1}\,dx$

9. $\int \frac{e^p}{2}\,dp$

10. $\int \frac{\sec^2 t}{5}\,dt$

11. $\int \cos t + \sec t\tan t\,dt$

In Problems 12 through 14, find antiderivatives for the given functions. In other words, for each function f, find a function F such that $F' = f$. Check your answers.

12. (a) $f(x) = e^{3x}$ 　　　　　　 (b) $f(x) = \frac{3}{e^x}$

13. (a) $f(x) = \frac{-1}{2x}$ 　　　　　　 (b) $f(x) = \frac{4}{1+x^2}$

14. (a) $f(x) = \sin 2x$ 　　　　　　 (b) $f(x) = \cos(x/3)$

15. (a) Differentiate $f(x) = 5\tan(x^2) + \arctan 3x$.
    (b) Find $\int 10x \sec^2(x^2) + \frac{3}{1+9x^2}\,dx$

16. (a) Differentiate $y = \frac{-\pi \cos 3x}{3}$.
    (b) Find $\int A\sin Bx\,dx$, where A and B are constants.

17. (a) Suppose the velocity of an object is given by $v(t) = \frac{1}{1+t^2}$ miles per hour. Find its net change in position from $t = 0$ to $t = 1$, t measured in hours. What is the total distance traveled on $[0, 1]$?
    (b) Suppose that the velocity of an object is given by $v(t) = 5t(t-1)$ meters per second. What is the net change in position between $t = 0$ and $t = 2$, t measured in seconds? What is the total distance traveled on $[0, 2]$?

18. Find the following indefinite integrals.

    (a) $\int \frac{2+x}{x}\,dx$ 　　 (b) $\int \frac{3}{x^2}\,dx$ 　　 (c) $\int \frac{3}{1+x^2}\,dx$ 　　 (d) $\int \left(\frac{t^3}{4} + \frac{4}{\sqrt{t}}\right)dt$

19. Evaluate the following integrals.

    (a) $\int (x+\pi)x^2\,dx$ 　　　　 (b) $\int \frac{kx}{\sqrt{x}}\,dx$ 　　　　 (c) $\int \frac{3t^2+t}{6t^3}\,dt$

    (d) $\int \left(2 - \frac{1}{x}\right)\sqrt{x}\,dx$ 　　 (e) $\int (x+1)\sqrt{5x}\,dx$

25.2   SUBSTITUTION: THE CHAIN RULE IN REVERSE

In Section 25.1, we used our knowledge of the derivatives of basic building block functions (like $\sin x$, x^n, and $\ln x$) to create a list of "basic" or "familiar" integrals to keep at our fingertips. We used two general rules of differentiation, the rules for differentiating the sum of functions and the constant multiple of a function, to obtain two general rules of integration. In this section we will use the Chain Rule in reverse to arrive at an integration technique referred to as substitution. This will enable us to greatly expand the type of functions we can antidifferentiate.

From functions $h(x)$ and $u(x)$ we can build the composite function $h(u(x))$ in which the output of u is the input of h. The Chain Rule tells us how to differentiate the composite of differentiable functions:

$$\frac{d}{dx}h(u(x)) = h'(u(x)) \cdot u'(x).$$

The derivative of $h(u(x))$ is the product of the derivative of h evaluated at $u(x)$ and the derivative of u.

In the examples below we use the Chain Rule to differentiate composite functions and then give the corresponding antiderivative problem. After presenting these problems (all variations on a theme) we will generalize.

◆ **EXAMPLE 25.2**   ***Theme and Variations.***   In parts (a)–(c) we differentiate and then give the corresponding antiderivative problem. We apply the results of part (c) to parts (d)–(f).

(a) $\frac{d}{dx}\sin(5x) = \cos(5x) \cdot 5$ $\qquad$ $\int \cos(5x) \cdot 5\, dx = \sin(5x) + C$

(b) $\frac{d}{dx}\sin(x^2) = \cos(x^2) \cdot 2x$ $\qquad$ $\int \cos(x^2) \cdot 2x\, dx = \sin(x^2) + C$

(c) $\frac{d}{dx}\sin(u(x)) = \cos(u(x)) \cdot u'(x)$ $\qquad$ $\int \cos(u(x)) \cdot u'(x)\, dx = \sin(u(x)) + C$

or, equivalently,

$\frac{d}{dx}\sin(u(x)) = \cos(u(x)) \cdot \frac{du}{dx}$ $\qquad$ $\int \cos(u(x)) \cdot \frac{du}{dx}\, dx = \sin(u(x)) + C$

(d) $\int \cos(e^x) \cdot e^x\, dx$ has the underlying structure $\int \cos(u(x)) \cdot u'(x)\, dx$.

$$\int \underbrace{\cos(e^x)}\ \cdot\ \underbrace{e^x}\ dx = \underbrace{\sin(e^x)} + C$$

$$\int \cos(u(x)) \cdot u'(x)\, dx = \sin(u(x)) + C, \text{ where } u(x) = e^x \text{ and } u'(x) = e^x$$

(e) $\int \cos(x + 5)\, dx$ has the underlying structure $\int \cos(u(x)) \cdot u'(x)\, dx$.

$$\int \underbrace{\cos(x + 5)}\ \cdot\ \underbrace{1}\ dx = \sin(x + 5) + C$$

$$\int \cos(u(x)) \cdot u'(x)\, dx = \sin(u(x)) + C, \text{ where } u(x) = x + 5 \text{ and } u'(x) = 1$$

(f) $\int \frac{\cos(\sqrt{x})}{2\sqrt{x}}\, dx$ has the underlying structure $\int \cos(u(x)) \cdot u'(x)\, dx$.

$$\int \underbrace{\cos\sqrt{x}}\ \cdot\ \underbrace{\frac{1}{2\sqrt{x}}}\ dx = \underbrace{\sin(\sqrt{x})} + C$$

$$\int \cos(u(x)) \cdot u'(x)\, dx = \sin(u(x)) + C, \text{ where } u(x) = \sqrt{x} \text{ and } u'(x) = \frac{1}{2\sqrt{x}}\ ◆$$

There is a whole world of integrals accessible to us using the ideas of Example 25.2. Before we generalize, we distinguish between what we are able to do and what we are unable to do in the next example.

◆ **EXAMPLE 25.3**   Which of the following antiderivatives can be computed by using the Chain Rule in reverse? Compute those we can do in this way.

(a) $\int \cos 2x\, dx$ $\quad$ (b) $\int x \cos(x^2)\, dx$ $\quad$ (c) $\int 2 \cos(x^2)\, dx$

SOLUTIONS   (a) $\int \cos(2x)\, dx$

Can we put this integral into the form $\int \cos(u(x)) \cdot u'(x)\, dx$?

If so, $u(x) = 2x$. Therefore, $u'(x) = 2$. We'd like to see $u'(x)$ in the integrand; we're missing a 2.

$$\int \cos(2x)\,dx = \int \cos(2x) \cdot 2 \cdot \frac{1}{2}\,dx \qquad \text{Multiply the integrand by } \frac{2}{2} = 1.$$

$$= \frac{1}{2}\int \cos(2x) \cdot 2\,dx \qquad \int kf(x)\cdot dx = k\int f(x)\,dx$$

$$= \frac{1}{2}\int \cos(u(x)) \cdot u'(x)\,dx, \ \text{where } u(x) = 2x$$

$$= \frac{1}{2}\sin(u(x)) + C$$

$$= \frac{1}{2}\sin(2x) + C$$

(b) $\int x\cos(x^2)\,dx$

Can we put this integral into the form $\int \cos(u(x)) \cdot u'(x)\,dx$?
If so, $u(x) = x^2$. Therefore, $u'(x) = 2x$. We'd like to see $u'(x)$ in the integrand; we've got the x but are missing a 2.

$$\int x\cos(x^2)\,dx = \int \cos(x^2) \cdot x\,dx$$

$$= \frac{1}{2}\int \cos(x^2) \cdot 2x\,dx$$

$$= \frac{1}{2}\int \cos(u(x)) \cdot u'(x)\,dx, \ \text{where } u(x) = x^2$$

$$= \frac{1}{2}\sin(u(x)) + C$$

$$= \frac{1}{2}\sin(x^2) + C$$

(c) $\int 2\cos(x^2)\,dx$

Can we put this integral into the form $\int \cos(u(x)) \cdot u'(x)\,dx$?
If so, $u(x) = x^2$. Therefore, $u'(x) = 2x$. We'd like to see $u'(x)$ in the integrand; we're missing an x.

$$\int 2\cos(x^2)\,dx = \int \cos(x^2) \cdot 2\,dx$$

Now we're stuck. We can multiply by $\frac{x}{x}$ (for $x \neq 0$), but we can't pull the $\frac{1}{x}$ out of the integral.

$$\int kf(x)\,dx = k\int f(x)\,dx, \ \textit{only if } k \text{ is constant.}$$

A missing constant in $u'(x)$ need not worry us, but a missing variable makes all the difference in the world; we can't find an antiderivative using the Chain Rule in reverse.[1]
◆

Just as the Chain Rule allows us to generalize each shortcut to differentiation we know, e.g., $\frac{d}{dx}\ln x = \frac{1}{x}$ generalizing to $\frac{d}{dx}\ln(u(x)) = \frac{1}{u(x)} \cdot \frac{du}{dx}$, so too does it allow us to generalize each antiderivative in our table of antiderivatives.

---

[1] In fact, we can't find an antiderivative at all unless we're willing to express it as an infinite polynomial.

$\int x^n \, dx = \frac{x^{n+1}}{n+1} + C$ generalizes to $\int \left[u(x)\right]^n u'(x) \, dx = \frac{[u(x)]^{n+1}}{n+1} + C, n \neq -1$

$\int \frac{1}{x} \, dx = \ln |x| + C$ generalizes to $\int \frac{1}{u(x)} \cdot u'(x) \, dx = \ln |u(x)| + C$

$\int e^x \, dx = e^x + C$ generalizes to $\int e^{u(x)} \cdot u'(x) \, dx = e^{u(x)} + C$

$\int \sin x \, dx = -\cos x + C$ generalizes to $\int \sin(u(x)) \cdot u'(x) \, dx = -\cos(u(x)) + C$

$\int \cos x \, dx = \sin x + C$ generalizes to $\int \cos(u(x)) \cdot u'(x) \, dx = \sin(u(x)) + C$

$\int \frac{1}{1+x^2} \, dx = \arctan x + C$ generalizes to $\int \frac{1}{1+[u(x)]^2} \cdot u'(x) \, dx = \arctan(u(x)) + C$

and so on.

This generalized table of integrals incorporates the antidifferentiation analogue of the Chain Rule. This table is unnecessary, however, if we streamline our notation using the method of substitution. Essentially we will replace $u(x)$ by u and $u'(x) \, dx$ by du.

The Mechanics of Substitution

When faced with an unfamiliar integral, we can use the technique of substitution to attempt to transform it (altering its form but not its substance) into a familiar integral. In other words, some integrals that look intimidating are really sheep in wolves' clothing; structurally they are familiar integrals, but they are in disguise. Substitution is a long-standing method we use to uncover the underlying structure of a problem.[2]

For instance, in Example 25.3(b) we looked at $\int x \cos(x^2) \, dx$. In this form it is certainly not one of the integrals with which we are familiar. Substitution allows us to see that essentially this integral is $\int \cos u \, du$ in disguise. We make the substitution $u = x^2$ and transform the integral $\int x \cos(x^2) \, dx$ to an integral in u. $\cos(x^2)$ becomes $\cos u$. We must write $x \, dx$ in terms of u.

Since

$$u = x^2$$
$$\frac{du}{dx} = 2x.$$

We've differentiated both sides with respect to x. We can write

$$du = 2x \, dx.$$

In other words, $\frac{du}{dx} dx = 2x \, dx$ and we equate $\frac{du}{dx} dx$ with du.

Going from $\frac{du}{dx} = 2x$ to $du = 2x \, dx$ should strike you as a little underhanded (and conceivably illegal). We're treating $\frac{du}{dx}$ as if it were a fraction, but it is not. We have actually not defined du and dx independently. However, this abuse of notation turns out to be all right; we will justify it momentarily. First, let's finish the example so you appreciate the handiness of it.

Writing $du = 2x \, dx$ allows us to express our original integral entirely in terms of u.

---

[2] Similarly, when working with equations like $x^4 - 5x^2 + 4 = 0$ the substitution $u = x^2$ helped expose the underlying quadratic structure of the equation: $u^2 - 5u + 4 = 0$, so $(u - 4)(u - 1) = 0$. We solve for u and then return to the original variable.

$$u = x^2$$

$$du = 2x \, dx \quad \text{so} \quad x \, dx = \frac{1}{2} du$$

$$
\begin{aligned}
\int x \cos(x^2) \, dx &= \int \cos(x^2) \, x \, dx \\
&= \int \cos u \cdot \frac{1}{2} du && \text{Transform the integral using substitution.} \\
&= \frac{1}{2} \int \cos u \, du && \text{This is one of our basic integrals.} \\
&= \frac{1}{2} \sin u + C && \text{Now return to the original variable.} \\
&= \frac{1}{2} \sin(x^2) + C
\end{aligned}
$$

We can verify that this answer is correct by differentiating it.[3]

$$\frac{d}{dx}\left[\frac{1}{2} \sin(x^2) + C \right] = \frac{1}{2} \cos(x^2) \cdot 2x = x \cos x^2.$$

Why Is the Substitution $du = u'(x) \, dx$ Legitimate?

In the example above we went from $\frac{du}{dx} = u'(x)$ to $du = u'(x) \, dx$. Why is this valid?
   Suppose that

$$\int f(x) \, dx = F(x) + C.$$

That is,

$$F'(x) = f(x).$$

Then

$$\int f(u) \, du = F(u) + C.$$

Is it true that

$$\int f(u(x)) \cdot u'(x) \, dx = F(u(x)) + C?$$

Sure!

$$\frac{d}{dx} F(u(x)) = F'(u(x)) \cdot u'(x) = f(u(x)) \cdot u'(x)$$

Therefore, it "works" to replace $u(x)$ by u and $u'(x) \, dx$ by du. Replacing $\frac{du}{dx} dx$ by du is legitimate. This is part of the genius of Leibniz' notation $\frac{du}{dx}$ for the derivative. One generally doesn't run into trouble treating $\frac{du}{dx}$ as it appears, despite the fact that $\frac{du}{dx}$ is really not a fraction.
   Now we return to more examples.

---

[3] We can *always* check an answer to an integration problem by differentiation.

Using Substitution to Reverse the Chain Rule

The key idea when making a substitution in order to reverse the Chain Rule is to choose u so that the integral is of the form $\int f(u(x)) \cdot u'(x)\, dx$. We must choose u so that u' is already sitting by in the integrand, up to a constant multiple.

◆ **EXAMPLE 25.4**   Evaluate the following integrals.

(a) $\int \frac{5}{3x+2}\, dx$ (b) $\int \frac{1}{\sqrt{x}e^{\sqrt{x}}}\, dx$

(c) $\int \frac{3x}{1+x^2}\, dx$ (d) $\int \cos^2 x \sin x\, dx$

Try these problems on your own and then compare your answers with those below. (You can *always* check your answers by differentiating.)

SOLUTIONS   (a) $\int \frac{5}{3x+2}\, dx = 5\int \frac{1}{3x+2}\, dx$. We hope we essentially have $\int \frac{1}{u} du$.

$$\text{Let } u = 3x + 2. \quad \text{Then}$$

$$\frac{du}{dx} = 3 \quad \text{so}$$

$$\frac{1}{3} du = dx.$$

Rewrite the original integral in terms of u.

$$5\int \frac{1}{3x+2}\, dx = 5\int \frac{1}{u} \cdot \frac{1}{3}\, du$$

$$= \frac{5}{3}\int \frac{1}{u}\, du \qquad \text{This is a familiar integral.}$$

$$= \frac{5}{3}\ln|u| + C \qquad \text{Return to the original variable.}$$

$$= \frac{5}{3}\ln|3x + 2| + C$$

(b) $\int \frac{1}{\sqrt{x}e^{\sqrt{x}}}\, dx = \int \frac{e^{-\sqrt{x}}}{\sqrt{x}}\, dx$. We hope we essentially have $\int e^u du$.

$$\text{Let } u = -\sqrt{x}.$$

$$\frac{du}{dx} = -\frac{1}{2}\frac{1}{\sqrt{x}}$$

$$-2du = \frac{1}{\sqrt{x}}\, dx$$

Rewrite the original integral in terms of u.

$$\int e^{-\sqrt{x}}\frac{1}{\sqrt{x}}\, dx = \int e^u \cdot (-2)\, du$$

$$= -2\int e^u du \qquad \text{This is a familiar integral.}$$

$$= -2e^u + C \qquad \text{Return to the original variable.}$$

$$= -2e^{-\sqrt{x}} + C$$

(c) $\int \frac{3x}{1+x^2} \, dx = 3 \int \frac{x}{1+x^2} \, dx$

Depending upon how we look at this, we might make the substitution $u = 1 + x^2$, or we might let $u = x^2$. We can do the problem either way, although the latter choice is less efficient.

Option 1: Let $u = 1 + x^2$.   (We're hoping we essentially have $\int \frac{1}{u} \, du$.)

$$\frac{du}{dx} = 2x$$

$$\frac{1}{2} du = x \, dx$$

Rewrite the original integral in terms of u.

$$3 \int \frac{x}{1+x^2} \, dx = 3 \int \frac{1}{1+x^2} x \, dx$$

$$= 3 \int \frac{1}{u} \cdot \frac{1}{2} \, du$$

$$= \frac{3}{2} \int \frac{1}{u} \, du$$

$$= \frac{3}{2} \ln |u| + C$$

$$= \frac{3}{2} \ln(1 + x^2) + C$$

Option 2: Let $u = x^2$.   (We're trying to simplify.)

$$\frac{du}{dx} = 2x$$

$$\frac{1}{2} du = x \, dx$$

Rewrite the original integral in terms of u.

$$3 \int \frac{x}{1+x^2} \, dx = 3 \int \frac{1}{1+x^2} \cdot x \, dx$$

$$= 3 \int \frac{1}{1+u} \cdot \frac{1}{2} du$$

$$= \frac{3}{2} \int \frac{1}{1+u} \, du$$

You may be able to guess an antiderivative at this point. Alternatively, you can use substitution again.

$$\text{Let } v = 1 + u.$$

$$\frac{dv}{du} = 1$$

$$dv = du$$

Rewrite $\frac{3}{2} \int \frac{1}{1+u} \, du$ in terms of v.

$$\frac{3}{2} \int \frac{1}{1+u} du = \frac{3}{2} \int \frac{1}{v} dv$$

$$= \frac{3}{2} \ln |v| + C \qquad \text{But } v = 1 + u.$$

$$= \frac{3}{2} \ln |1 + u| + C \qquad u = x^2$$

$$= \frac{3}{2} \ln(1 + x^2) + C$$

(d) $\int \cos^2 x \sin x \, dx$. We're hoping we essentially have $\int u^2 \, du$.

$$\text{Let } u = \cos x.$$

$$\frac{du}{dx} = -\sin x$$

$$-du = \sin x \, dx$$

$$\int \cos^2 x \sin x \, dx = \int u^2 (-du)$$

$$= -\int u^2 \, du$$

$$= \frac{-u^3}{3} + C$$

$$= -\frac{1}{3} \cos^3 x + C \quad \blacklozenge$$

Substitution in Definite Integrals

When using substitution to evaluate a definite integral $\int_a^b f(x) \, dx$, we have two options.

Option 1: Change the limits of integration to the new variable and never look back.

Option 2: Leave the limits in terms of the original variable throughout, but rewrite the integral as $\int_{x=a}^{x=b} g(u) \, du$ to make it clear that we do **not** mean $u = a$ and $u = b$.[4] After antidifferentiating with respect to u, replace u by its original expression in terms of x. Then evaluate at $x = b$ and $x = a$, subtracting the latter from the former as usual.

◆ **EXAMPLE 25.5**   Evaluate $\int_0^{\pi/3} \cos^2 x \sin x \, dx$.

SOLUTION   This is the same integrand as in Example 25.3(d). We use the substitution $u = \cos x$ as before.

$$u = \cos x$$

$$\frac{du}{dx} = -\sin x$$

$$-du = \sin x \, dx$$

Option 1:

$$\begin{cases} \text{When } x = 0, & u = \cos 0 = 1; \\ \text{when } x = \frac{\pi}{3}, & u = \cos \frac{\pi}{3} = \frac{1}{2}. \end{cases}$$

---

[4] For instance, $\int_0^\pi \sin 2x \, dx = 0$. If we let $u = 2x$, then $du = 2dx$ and $dx = \frac{1}{2} du$. $\int_0^\pi \frac{1}{2} \sin u \, du = 1 \neq 0$, whereas $\int_0^{2\pi} \frac{1}{2} \sin u \, du$ does equal zero.

$$\int_0^{\pi/3} \cos^2 x \sin x \, dx = -\int_1^{1/2} u^2 \, du$$

$$= -\frac{u^3}{3}\Big|_1^{1/2}$$

$$= -\frac{1}{3}\left[\frac{1}{8} - 1\right]$$

$$= -\frac{1}{3}\left(-\frac{7}{8}\right)$$

$$= \frac{7}{24}$$

Option 2:

$$\int_0^{\pi/3} \cos^2 x \sin x \, dx = -\int_{x=0}^{x=\pi/3} u^2 \, du$$

$$= -\frac{1}{3}u^3\Big|_{x=0}^{x=\pi/3}$$

$$= -\frac{1}{3}\cos^3 x\Big|_0^{\pi/3}$$

$$= -\frac{1}{3}\left[\left(\cos\frac{\pi}{3}\right)^3 - (\cos 0)^3\right]$$

$$= -\frac{1}{3}\left[\frac{1}{8} - 1\right]$$

$$= \frac{7}{24}$$

CAUTION Writing $\int_0^{\pi/3} u^2 \, du$ is *not* an option! Unless you explicitly specify that the endpoints of integration are in terms of a different variable, it is understood that they are in terms of u if all else is in terms of u.   ◆

Working many problems will help you become proficient at using substitution. You will see that integrands that at first glance appear similar may in fact yield vastly different antiderivatives. Try the exercises below. Answers are provided, although you can always check your answers on your own by differentiating.

Fundamental goals of substitution:

i. Simplify the integrand.

ii. Obtain $\int f(u(x)) \cdot u'(x) \, dx$, which we write as $\int f(u) \, du$.

Therefore, when choosing a "u," be sure that you already have du up to a constant multiple. If you don't, chances are you've bitten off too much to call "u," so try a smaller bite. If you're stuck on any of the problems that follow, look at the suggested substitutions given below the answers.

EXERCISE 25.2   Evaluate the following integrals.[5]

(a) $\int \frac{dx}{1+2x}$

(b) $\int \frac{dx}{(1+2x)^2}$

(c) $\int_0^{\ln 2} \frac{e^x}{1+e^x} dx$

(d) $\int \frac{e^x}{1+e^{2x}} dx$

(e) $\int \frac{x \, dx}{\sqrt{2+3x^2}}$

(f) $\int \tan x \, dx$

Answers

(a) $\frac{1}{2} \ln |1+2x| + C$

(b) $\frac{-1}{2(1+2x)} + C$

(c) $\ln \left(\frac{3}{2} \right)$

(d) $\arctan(e^x) + C$

(e) $\frac{1}{3} \sqrt{2 + 3x^2} + C$

(f) $- \ln | \cos x| + C$

Suggested substitutions

(a) $u = 1 + 2x$

(b) $u = 1 + 2x$

(c) $u = 1 + e^x$ or $u = e^x$

(d) $u = e^x$

(e) $u = 2 + 3x^2$

(f) Express $\tan x$ as $\frac{\sin x}{\cos x}$. Then let $u = \cos x$. (Letting $u = \sin x$ won't work because du will be $\cos x \, dx$, not $\frac{1}{\cos x} \, dx$.)

PROBLEMS FOR SECTION 25.2

1. Evaluate the following indefinite integrals. (Let your mind be limber. Integrals that look very similar may require very different mindsets!)

(a) $\int \frac{1}{x} dx$

(b) $\int \frac{1}{x+1} dx$

(c) $\int \frac{1}{(x+1)^2} dx$

(d) $\int \frac{1}{x^2+1} dx$

(e) $\int \frac{x}{x^2+1} dx$

(f) $\int \frac{x^2+1}{x} dx$

(g) $\int (1+x)^5 dx$

(h) $\int \frac{1}{(1+x)^5} dx$

(i) $\int (1+x^2)^2 dx$

2. Find the following indefinite integrals. Check your answers.

(a) $\int 3 \sin(5t) \, dt$

(b) $\int \pi \cos(\pi t) \, dt$

(c) $\int \sqrt{3x+5} \, dx$

(d) $\int \frac{\pi}{e^x} dx$

(e) $\int e^{-3t} \, dt$

(f) $\int \sqrt{e^t} \, dt$

(g) $\int \frac{6}{\sqrt{t^3}} \, dt$

(h) $\int \frac{1}{3t+8} \, dt$

In Problems 3 through 8, find the indefinite integrals.

3. (a) $\int (2x+1)^3 dx$

(b) $\int \frac{1}{(2x+1)^2} dx$

(c) $\int \frac{1}{(2x+1)} dx$

(d) $\int \frac{1}{\sqrt{2x+1}} dx$

4. (a) $\int x \sqrt{2x^2+1} \, dx$

(b) $\int \frac{x}{\sqrt{2x^2+1}} dx$

(c) $\int \frac{\cos \sqrt{x}}{\sqrt{x}} dx$

(d) $\int \sqrt{\cos x} \sin x \, dx$

5. (a) $\int \frac{5}{1+9x^2} dx$

(b) $\int \frac{\ln x}{x} dx$

(c) $\int \frac{e^x}{e^{-x}} dx$

(d) $\int \frac{(\ln w)^2}{w} dw$

6. (a) $\int t^2 \sin(t^3) \, dt$

(b) $\int xe^{-x^2} \, dt$

(c) $\int \frac{1}{x+5} dx$

(d) $\int \frac{t}{2t^2+7} dt$

7. (a) $\int \frac{t}{(t^2+1)^2} dt$

(b) $\int \tan 2t \, dt$

(c) $\int \frac{e^w}{e^w+1} dw$

(d) $\int \frac{e^w}{e^{2w}+1} dw$

---

[5] $\int \frac{dx}{1+2x}$ and $\int \frac{1}{1+2x} dx$ are two ways of writing the same thing.

8. (a) $\int (x^2+3)^3 x\,dx$ (b) $\int x\sqrt{x^2+4}\,dx$ (c) $\int (2x+1)\sqrt{x^2+x}\,dx$

9. Evaluate.

(a) $\int_0^1 \frac{3}{1+4w^2}\,dw$ (b) $\int_0^1 \frac{1+4w^2}{3}\,dw$ (c) $\int_{\pi/2}^{3\pi} \cos\left(\frac{t}{2}\right)\,dt$

(d) $\int_1^3 \frac{4}{3x+2}\,dx$ (e) $\int_1^4 \frac{1}{(2x+1)^2}\,dx$

10. Find the following definite integrals.

(a) $\int_1^3 \frac{1}{(2x+4)^2} + \frac{1}{x+4}\,dx$ (b) $\int_1^e \frac{(\ln x)^2}{x}\,dx$

In Problems 11 through 23, compute the following integrals.

11. $\int \frac{\cos(e^{-x})}{e^x}\,dx$

12. $\int (e^x+x)\sqrt{2e^x+x^2}\,dx$

13. $\int \frac{\sec^2(\ln x)}{x}\,dx$

14. $\int \frac{\sin(x)}{\sqrt{\cos(x)}}\,dx$

15. $\int_0^{\pi/4} \tan x\,dx$

16. $\int e^{2x}\sqrt{e^x}\,dx$

17. $\int_0^{\pi/2} \frac{\cos(x)}{1+\sin^2 x}\,dx$

18. $\int_0^{\ln 5} \frac{3e^x}{\sqrt{e^x+4}}\,dx$

19. $\int_0^{\ln 3} \frac{e^{2x}}{e^{2x}+1}\,dx$

20. $\int_0^{\ln 2} \frac{e^x}{e^{2x}+1}\,dx$

21. $\int \frac{3^x}{3^x+1}\,dx$

22. $\int \sec^2 x \tan^2 x\,dx$

23. $\int \frac{\sec^2 \sqrt{x}\,\tan^2 \sqrt{x}}{\sqrt{x}}\,dx$

24. It is 10:00 A.M. and five ants have found their way into a picnic basket. Ants are notorious followers, so ants from all over the vicinity follow their five brethren into the basket. The culinary treat awaiting them is unsurpassed elsewhere, so once the ants find their way into the basket they choose not to leave. If the rate at which ants are climbing into the basket is well modeled by $100e^{-0.2t}$ ants per hour, where $t=0$ is the benchmark hour of 10:00 A.M., how many ants will be in the basket x hours after 10:00 A.M.?

   How many ants are in the basket at 1:00 P.M., when the picnic is supposed to begin? Give your answer to the nearest ant.

25. Consider the integral $\int \cos x \sin x \, dx$.

   (a) Using the substitution $u = \sin x$, show that $\int \cos x \sin x \, dx = \frac{1}{2} \sin^2 x + C$.

   (b) Using the substitution $u = \cos x$, show that $\int \cos x \sin x \, dx = -\frac{1}{2} \cos^2 x + C$.

   (c) Explain why although these answers look different they are both correct.

26. It is April and, with the arrival of warm weather a pile of snow has turned into a puddle and the puddle is drying up. At time $t = 0$ there are 4 gallons of water in the puddle. The rate of change of water in the puddle is given by $f(t) = -e^{0.2t}$ gallons per hour. How much water is in the puddle t hours from now? When will the puddle dry up?

27. A little rock rolls off a little cliff. It experiences an acceleration of -32 ft/sec^2.

   (a) The derivative of the velocity function $v(t)$ is acceleration. Therefore, the antiderivative (indefinite integral) of acceleration is velocity. $v(t) = \int a(t) \, dt$. Find $v(t)$ assuming that the rock's initial vertical velocity is zero. (You'll use this initial condition to find the constant of integration.)

   (b) Let $s(t)$ give the rock's height at any time t. We'll call s the position function. The derivative of the position function $s(t)$ is velocity. Therefore, the antiderivative (the indefinite integral) of velocity is the distance function. $s(t) = \int v(t) \, dt$. Find $s(t)$ assuming that the cliff has a height of 25 feet. (You'll use this initial condition to find the constant of integration.)

25.3   SUBSTITUTION TO ALTER THE FORM OF AN INTEGRAL

There are many ways of altering the form of an integral to make it more tractable. In this section we will look at some of them.

   We saw that one strategy for approaching an integral is to manipulate the integrand algebraically so it can be expressed as a sum and the integral can be pulled apart into the sum of simpler integrals. Sometimes substitution is useful in evaluating the resultant integrals and other times in actually altering the form of the integrand so that it can be expressed as a sum. Both uses of substitution are illustrated in the examples below. Your goal in each case is to write the integral as a sum of simpler integrals. Don't allow your eagerness to express an integrand as a sum interfere with your algebraic judgement; naturally, the standard rules of algebra apply!

◆ **EXAMPLE 25.6**   Evaluate the following.

   (a) $\int_0^1 \frac{3x+7}{1+x^2} \, dx$ (b) $\int \frac{x}{x+3} \, dx$ (c) $\int 3x\sqrt{x+5} \, dx$

Try these on your own before reading the detailed solutions. If you need help, first look at the hints supplied.

Hints

   (a) *Split this into the sum of two integrals. Then use substitution on one of them.*

   (b) *Let $u = x + 3$ so that you can split this integral.*

   (c) *Let $u = x + 5$ so that you can split this integral.*

SOLUTIONS   (a) $\int_0^1 \frac{3x+7}{1+x^2} \, dx = 3 \int_0^1 \frac{x}{1+x^2} dx + 7 \int_0^1 \frac{1}{1+x^2} \, dx$

We'll evaluate each integral independently and then sum the results. $3 \int_0^1 \frac{x}{1+x^2} \, dx$ looks essentially like $\int \frac{du}{u}$.

$$\text{Let } u = 1 + x^2.$$

$$du = 2x \, dx \qquad \text{so } x \, dx = \frac{1}{2} \, du$$

$$\text{when } x = 0, \quad u = 1 + 0^2 = 1$$

$$\text{when } x = 1, \quad u = 1 + 1^2 = 2$$

$$3 \int_0^1 \frac{x}{1+x^2} \, dx = 3 \int_1^2 \frac{1}{u} \cdot \frac{1}{2} \, du$$

$$= \frac{3}{2} \int_1^2 \frac{1}{u} \, du$$

$$= \frac{3}{2} \ln |u| \Big|_1^2$$

$$= \frac{3}{2} [\ln 2 - \ln 1]$$

$$= \frac{3}{2} \ln 2$$

$$7 \int_0^1 \frac{1}{1+x^2} \, dx = 7 \tan^{-1}(x) \Big|_0^1$$

$$= 7 \left[\tan^{-1}(1) - \tan^{-1}(0) \right]$$

$$= 7 \left(\frac{\pi}{4} - 0 \right)$$

$$= \frac{7\pi}{4}$$

$$\text{so } \int_0^1 \frac{3x+7}{1+x^2} \, dx = \frac{3}{2} \ln 2 + \frac{7\pi}{4}$$

(b) $\int \frac{x}{x+3} \, dx$

CAUTION $\frac{x}{x+3} \neq \frac{x}{x} + \frac{x}{3}.$

In order to express this integrand as a sum we need the numerator expressed as a sum, not the denominator. The substitution $u = x + 3$ transforms the integral so that we can break it up.

$$\text{Let } u = x + 3.$$
$$du = dx$$
$$\text{and } \quad x = u - 3$$

$$\int \frac{x}{x+3}\,dx = \int \frac{u-3}{u}\,du$$

$$= \int 1 - \frac{3}{u}\,du$$

$$= \int 1\,du - 3 \int \frac{1}{u}\,du$$

$$= u - 3\ln|u| + C \qquad \text{Now return to the original variable.}$$

$$= x + 3 - 3\ln|x+3| + C \quad \text{or} \quad x - 3\ln|x+3| + C_1 \text{ (where } C_1 = 3 + C)$$

(c) $\int 3x\sqrt{x+5}\,dx$

CAUTION $\sqrt{x+5} \neq \sqrt{x} + \sqrt{5}.$[6]

In order to express this integrand as a sum we need the quantity outside the square root to be a sum, not the quantity *under* the square root. The substitution $u = x + 5$ takes care of this.

$$\text{Let } u = x + 5.$$
$$du = dx$$
$$\text{and} \quad x = u - 5$$

$$\int 3x\sqrt{x+5}\,dx = \int 3(u-5)\sqrt{u}\,du$$

$$= 3\int u\sqrt{u}\,du - 15\int \sqrt{u}\,du$$

$$= 3\int u^{3/2}\,du - 15\int u^{1/2}\,du$$

$$= 3 \cdot \frac{2}{5}u^{5/2} - 15 \cdot \frac{2}{3} \cdot u^{3/2} + C \qquad \text{Now return to the original variable.}$$

$$= \frac{6}{5}(x+5)^{5/2} - 10(x+5)^{3/2} + C \quad \blacklozenge$$

Sometimes a substitution is used as a first step in simplifying a problem. It might be followed by a second substitution.

◆ **EXAMPLE 25.7**   Evaluate the following.

(a) $\int \frac{3\,dx}{4+9x^2}$ 　　　　(b) $\int \frac{\tan\sqrt{x}}{\sqrt{x}}\,dx$ 　　　　(c) $\int \frac{e^{2x}}{e^{2x}+2e^x+1}\,dx$

SOLUTIONS   (a) $\int \frac{3\,dx}{4+9x^2} = 3\int \frac{1}{4+9x^2}\,dx$

This looks vaguely like a variation on $\int \frac{1}{1+u^2}\,du$. To try to get $1 + u^2$ in the denominator we'll divide the numerator and denominator of the integrand by 4.

$$3\int \frac{\frac{1}{4}}{\frac{4}{4}+\frac{9x^2}{4}}\,dx = \frac{3}{4}\int \frac{1}{1+\left(\frac{3x}{2}\right)^2}\,dx$$

---

[6] We know $5 = \sqrt{25} = \sqrt{16+9} \neq \sqrt{16} + \sqrt{9} = 4 + 3 = 7.$

$$\text{Let } u = \frac{3x}{2}.$$

$$du = \frac{3}{2} dx \quad \text{so } dx = \frac{2}{3} du$$

$$\frac{3}{4} \int \frac{1}{1 + \left(\frac{3x}{2}\right)^2} dx = \frac{3}{4} \int \frac{1}{1 + u^2} \cdot \frac{2}{3} du$$

$$= \frac{3}{4} \cdot \frac{2}{3} \int \frac{1}{1 + u^2} du$$

$$= \frac{1}{2} \arctan u + C$$

$$= \frac{1}{2} \arctan \left(\frac{3}{2}x\right) + C$$

(b) $\int \frac{\tan \sqrt{x}}{\sqrt{x}} dx$

Use substitution to get rid of the $\sqrt{x}$.

$$\text{Let } u = \sqrt{x}.$$

$$du = \frac{1}{2}\frac{1}{\sqrt{x}} dx \quad \text{so } 2du = \frac{1}{\sqrt{x}} dx$$

$$\int \tan \sqrt{x} \cdot \frac{1}{\sqrt{x}} dx = \int \tan u \cdot 2du$$

$$= 2 \int \tan u \, du$$

$$= 2 \int \frac{\sin u}{\cos u} du \quad \text{(This looks essentially like } \int \frac{dW}{W}.)$$

Let $W = \cos u$.

$$dW = -\sin u \, du \quad \text{so } -dW = \sin u \, du$$

$$2 \int \frac{\sin u}{\cos u} du = 2 \int \frac{-dW}{W}$$

$$= -2 \ln |W| + C \qquad \text{Replace } W \text{ by } \cos u.$$

$$= -2 \ln |\cos u| + C \qquad \text{Replace } u \text{ by } \sqrt{x}.$$

$$= -2 \ln |\cos \sqrt{x}| + C$$

(c) $\int \frac{e^{2x}}{e^{2x}+2e^x+1} dx$

We want to use substitution to get rid of e^x and e^{2x}. If we let $u = e^x$, then $e^{2x} = (e^x)^2 = u^2$. If, instead, we were to let $u = e^{2x}$, then $e^x = (e^{2x})^{1/2} = \sqrt{u}$. The first option looks much more appealing. Let's try it.

$$\text{Let } u = e^x.$$

$$du = e^x \, dx$$

$$\int \frac{e^{2x}}{e^{2x} + 2e^x + 1} dx$$

Either we can express this as

$$\int \frac{e^x}{e^{2x} + 2e^x + 1} e^x \, dx = \int \frac{u}{u^2 + 2u + 1} du$$

or

$$\int \frac{(e^x)^2}{(e^x)^2 + 2e^x + 1} \, dx = \int \frac{u^2}{u^2 + 2u + 1} \cdot \frac{1}{u} du \quad \left(\text{because } dx = \frac{1}{e^x} du = \frac{1}{u} du \right)$$

The two are equivalent.

$$\int \frac{u}{u^2 + 2u + 1} \, du = \int \frac{u}{(u + 1)^2} du$$

This is a simpler integral than the one we began with, but it is awkward. If we could get the sum in the numerator as opposed to the denominator, we'd be happier. We'll accomplish this with another substitution.

$$\text{Let } W = u + 1 \quad \text{so } u = W - 1.$$
$$dW = du$$

$$\int \frac{u}{(u + 1)^2} \, du = \int \frac{W - 1}{W^2} \, dW$$

$$= \int \frac{W}{W^2} - \frac{1}{W^2} \, dW$$

$$= \int \frac{1}{W} \, dW - \int W^{-2} \, dW$$

$$= \ln |W| - (-W^{-1}) + C \qquad \text{Replace } W \text{ by } u + 1.$$

$$= \ln |u + 1| + \frac{1}{u + 1} + C \qquad \text{Replace } u \text{ by } e^x.$$

$$= \ln(e^x + 1) + \frac{1}{e^x + 1} + C \quad \blacklozenge$$

EXERCISE 25.3   Check the answers to Example 25.7 by differentiating.

PROBLEMS FOR SECTION 25.3

For Problems 1 through 11, find the given indefinite integral.

1. $\int \frac{x}{x+1} \, dx$

2. $\int \frac{x+3}{x-7} \, dx$

3. $\int x\sqrt{3x + 5} \, dx$

4. $\int \frac{2x}{3+x} \, dx$

5. $\int 3t\sqrt{t^2 + 5} \, dt$

6. $\int t\sqrt{2t + 5} \, dt$

7. $\int \frac{x^\pi}{5}\, dx$

8. $\int \cot(3x)\, dx$

9. $\int (1.5)^{(1-t)}\, dt$

10. $\int \sin^4 t \cos t\, dt$

11. $\int \frac{2t}{\sqrt{2t+6}}\, dt$

12. Suppose you've forgotten the antiderivative of $\frac{1}{\sqrt{1-x^2}}$. In this problem you will use a sophisticated substitution that will help you proceed. The goal is to find

$$\int_0^{1/2} \frac{1}{\sqrt{1-x^2}}\, dx.$$

We can't let $u = x^2$ or $1 - x^2$ because $u'(x)$ is nowhere in sight. Instead we take a different approach. We'll replace x by something that makes $1 - x^2$ a perfect square. We'll exploit the trigonometric identity $\sin^2 \theta + \cos^2 \theta = 1$. We know $1 - \sin^2 \theta = \cos^2 \theta$, which is a perfect square. Let $x = \sin \theta$. Rewrite the integral (including the endpoints) in terms of θ and evaluate.

Check your answer by using the antiderivative of $\frac{1}{\sqrt{1-x^2}}$. (Look it up in the table on page 784 if you have forgotten it.)

(This is an example of an integration technique known as **trigonometric substitution**. It involves exploiting trigonometric identities to simplify expressions such as $\sqrt{c^2 - x^2}$, $\sqrt{x^2 - c^2}$, and $x^2 + c^2$, where c is a constant.)

13. Evaluate $\int \frac{2}{x(x+2)}\, dx$ by rewriting the integrand in the form

$$\frac{A}{x} + \frac{B}{x+2},$$

where A and B are constants. In other words, find A and B such that

$$\frac{2}{x(x+2)} = \frac{A}{x} + \frac{B}{x+2}.$$

(This is an example of an integration technique known as **partial fractions**.)

14. Evaluate $\int \frac{1}{x^2-1}\, dx$ by factoring the denominator of the integrand and rewriting the integrand in the form

$$\frac{A}{x-1} + \frac{B}{x+1},$$

where A and B are constants.

26

Numerical Methods of Approximating Definite Integrals

Introduction

Not only can we differentiate all the basic functions we've encountered, polynomials, exponential and logarithmic functions, trigonometric and inverse trigonometric functions, and rational functions, but armed with the Product and Chain Rules, we can happily differentiate any new function constructed by multiplication, division, addition, subtraction, or composition of these functions. This gives us a sense of competence and satisfaction.

Although at this point we can integrate many functions, there are basic functions (such as $\ln x$ and $\sec x$) that we have not yet tackled.[1] From the Chain Rule for differentiation we get the technique of substitution for antidifferentiation; from the Product Rule for differentiation we get a technique of antidifferentiation known as Integration by Parts. (The latter is something to look forward to learning in Chapter 29.) Learning more sophisticated methods of substitution and algebraic manipulation will enlarge the collection of functions we can antidifferentiate. Tables of integrals and high-powered computer packages can provide assistance if we're in dire straits. Nevertheless, there are some very innocent-looking functions that cannot be dealt with easily. For instance, all the technical skill in the world won't help us find an antiderivative for e^{-x^2}, or $\sin(x^2)$, or $\sin(\frac{1}{x})$.[2] Knowing that there is no guarantee that we *can* antidifferentiate can be unnerving. This chapter will restore our sense of having things under control when we are faced with a definite integral.

Suppose we are interested in evaluating a definite integral and we have not found an antiderivative for the integrand. In any practical situation we'll need to evaluate the definite

---

[1] $\int \ln x \, dx = x \ln x - x + C$ (verify this for yourself). This can be found either by some serious guess-work, methods given in Section 27.3, or using the technique called Integration by Parts.

With some work we can integrate $\sec x$. $\int \sec x \, dx = \int \sec x \cdot \frac{\sec x + \tan x}{\sec x + \tan x} \, dx = \int \frac{\sec^2 x + \sec x \tan x}{\sec x + \tan x} \, dx = \ln | \sec x + \tan x | + C$.

[2] That is, if we want to obtain an antiderivative that is a *finite* sum, product, or composition of the elementary functions.

integral with a certain degree of accuracy. Provided that an approximation is satisfactory, it is not necessary to be able to find an antiderivative of the integrand. Instead, we return to the basic ideas that led us to the limit definition of the definite integral.

The theoretical underpinnings of calculus involve the method of successive approximations followed by a limiting process.

■ **Differential Calculus**

Let f be a differentiable function. We obtain numerical *approximations* of the slope of the tangent to the graph of f at point P by looking at the slope of secant lines through P and Q, where Q is a point on the graph of f very close to P. By taking the *limit* as Q approaches P we determine the exact slope of the tangent line at P.

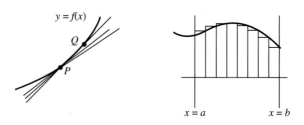

Figure 26.1

■ **Integral Calculus**

Let f be an integrable function. We obtain numerical *approximations* of the signed area between the graph of f and the x-axis on $[a, b]$ by partitioning the interval $[a, b]$ into many equal subintervals, treating f as if it is constant on each tiny subinterval, and approximating the signed area with a Riemann sum. By taking the *limit* as the number of subintervals increases without bound we determine the definite integral.

In this chapter we return to Riemann sums to obtain approximations of definite integrals. The numerical methods discussed here are often used in practice. Computers or programmable calculators are ideal for performing the otherwise tedious calculations.

Approximating Sums: L_n, R_n, T_n, and M_n

In the context of the following example we'll discuss left-hand, right-hand, midpoint, and trapezoidal sums, sums that can be used to approximate a definite integral. In order to be able to compare our approximations with the actual value, we'll look at an integral we can evaluate exactly .

◆ **EXAMPLE 26.1**   Approximate $\int_1^5 \frac{1}{x}\, dx$. Keep improving upon the approximation until you know the value to four decimal places.

SOLUTION   (i) To approximate the integral we chop up the interval of integration into n equal subintervals, (ii) approximate the area under the curve on each subinterval by the area of a rectangle, and (iii) sum the areas of these rectangles.

We'll construct three Riemann sums: the left-hand sum, denoted by L_n; the right-hand sum, denoted by R_n; and the midpoint sum, denoted by M_n. (We are already familiar with the first two.) We describe these sums as follows.

L_n: The height of the rectangle on each subinterval is given by the value of f at the left-hand endpoint of the subinterval.

R_n: The height of the rectangle on each subinterval is given by the value of f at the right-hand endpoint of the subinterval.

M_n: The height of the rectangle on each subinterval is given by the value of f at the midpoint of the subinterval. On the ith interval, $[x_{i-1}, x_i]$, the height of the rectangle is $f(c)$, where c is midway between x_{i-1} and x_i; $c = \frac{x_{i-1}+x_i}{2}$.

4 Subdivisions

Let $n = 4$; we chop $[1, 5]$ into 4 equal subintervals each of length $\Delta x = \frac{5-1}{4} = 1$.

Below are the left-hand sum, the right-hand sum, and the midpoint sum.

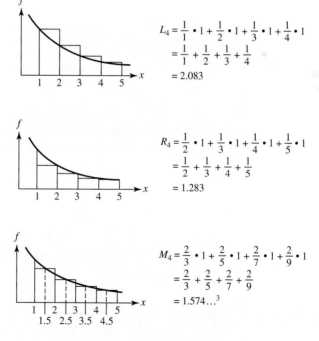

$$L_4 = \frac{1}{1} \cdot 1 + \frac{1}{2} \cdot 1 + \frac{1}{3} \cdot 1 + \frac{1}{4} \cdot 1$$
$$= \frac{1}{1} + \frac{1}{2} + \frac{1}{3} + \frac{1}{4}$$
$$= 2.083$$

$$R_4 = \frac{1}{2} \cdot 1 + \frac{1}{3} \cdot 1 + \frac{1}{4} \cdot 1 + \frac{1}{5} \cdot 1$$
$$= \frac{1}{2} + \frac{1}{3} + \frac{1}{4} + \frac{1}{5}$$
$$= 1.283$$

$$M_4 = \frac{2}{3} \cdot 1 + \frac{2}{5} \cdot 1 + \frac{2}{7} \cdot 1 + \frac{2}{9} \cdot 1$$
$$= \frac{2}{3} + \frac{2}{5} + \frac{2}{7} + \frac{2}{9}$$
$$= 1.574...[3]$$

Figure 26.2

---

[3] By "..." we mean there are more decimal places but we have stopped recording them.

The function $f(x) = \frac{1}{x}$ is decreasing; therefore the left-hand sums provide an upper bound and the right-hand sums a lower bound for $\int_1^5 \frac{1}{x}\, dx$.

$$1.283\ldots = R_4 < \int_1^5 \frac{1}{x}\, dx < L_4 = 2.083\ldots$$

Suppose we take the average of R_4 and L_4, $\frac{L_4 + R_4}{2}$. This is closer to the value of the integral than either the right- or left-hand sums. Geometrically this average is equivalent to approximating the area on each interval by a trapezoid instead of a rectangle, as shown below in Figure 26.3.

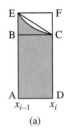
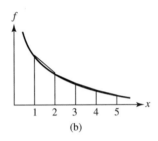

(a)   (b)

Figure 26.3

Averaging the area of rectangles of ABCD and AEFD in Figure 26.3(a) gives the area of the trapezoid AECD.

We refer to the average of the left- and right- hand sums, $\frac{L_n + R_n}{2}$, as the **trapezoidal sum** (or the Trapezoidal rule) and denote it by T_n.

$$T_4 = \frac{1.283\ldots + 2.083\ldots}{2} = 1.683\ldots$$

Because f is concave up[4] on $[1, 5]$ we know that on each subinterval the area under the trapezoid is larger than that under the curve. Thus T_4 gives an upper bound for the integral, a better upper bound than that provided by L_4.

$$R_4 < \int_1^5 \frac{1}{x}\, dx < T_4 < L_4$$

$$1.283\ldots < \frac{1}{x}\, dx < 1.683\ldots < 2.083\ldots$$

At this point the mind of the critical reader should be buzzing with questions. Perhaps they include the following.

What are the conditions under which T_n will be larger than the value of the integral? Smaller?

Where does the midpoint sum fit into the picture?

Let's investigate the first question by looking at some graphs.

---

[4] $f(t) = \frac{1}{t}$, so $f'(t) = -\frac{1}{t^2}$ and $f''(t) = \frac{2}{t^3}$. The latter is positive on $[1, 5]$, so f is concave up on the interval under discussion.

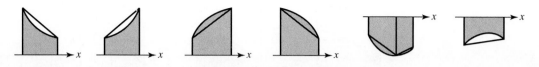

Figure 26.4

If f is concave up on $[a, b]$, then every secant line joining two points on the graph of f on $[a, b]$ lies *above* the graph.

If f is concave down on $[a, b]$, then every secant line joining two points on the graph of f on $[a, b]$ lies *below* the graph.

$$\text{We conclude that } \int_a^b f(t)\, dt < T_n \quad \text{if } f \text{ is concave up on } [a, b],$$

and

$$T_n < \int_a^b f(t)\, dt \quad \text{if } f \text{ is concave down on } [a, b].$$

Where does the midpoint sum fit into the picture? It turns out that T_n and M_n are a complementary pair. We will show that

$$\text{if } f \text{ is concave up on } [a, b], \text{ then } M_n < \int_a^b f(t)\, dt < T_n \quad \text{while}$$

$$\text{if } f \text{ is concave down on } [a, b], \text{ then } T_n < \int_a^b f(t)\, dt < M_n.$$

To do this we'll give an alternative graphical interpretation of the midpoint sum. Figure 26.5(i) illustrates the midpoint approximation, approximating the area under the graph of f on $[x_i, x_{i+1}]$ by the area of a rectangle whose height is the value of f at the midpoint of the interval. In Figure 26.5(ii) we approximate the area under f by the area of the trapezoid formed by the tangent line to the graph of f at the midpoint of the interval. We claim that these areas are identical; pivoting the line through A, B, and C about the midpoint C does not change the area bounded below.

Look at Figure 26.5(iii). The triangles ACD and BCE are congruent. We argue this as follows. Angles CAD and CBE are right angles. Angles ACD and BCE are equal. Therefore the two triangles in question are similar. But AC = CB because C is the midpoint of the interval $[x_i, x_{i+1}]$. We conclude that triangles ACD and BCE are congruent and hence have the same area. Therefore, rectangle $x_i AB x_{i+1}$ and trapezoid $x_i DE x_{i+1}$ (Figures 26.5(i) and (ii), respectively) have the same area; we can interpret the midpoint sum as the midpoint tangent sum.

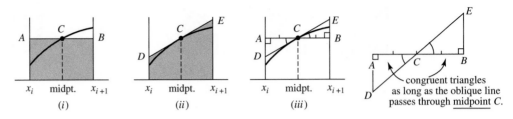

Figure 26.5

Below is a picture of M_4 using the midpoint tangent line interpretation. f is concave up, so the tangent lines lie below the curve. We know that on each subinterval the area under the midpoint tangent line is less than the area under the curve; therefore, M_4 gives a lower bound for the integral.

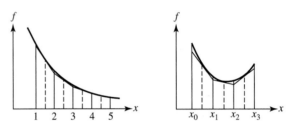

Figure 26.6

Where a function is concave up its tangent line lies below the curve; where a function is concave down its tangent line lies above the curve. It follows that if f is concave up on $[a, b]$, then $M_n < \int_a^b f(t)\, dt$; if f is concave down on $[a, b]$, then $\int_a^b f(t)\, dt < M_n$.

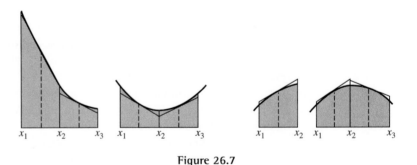

Figure 26.7

How do L_n, R_n, T_n, and M_n improve as we increase n?

8 Subdivisions

Let $n = 8$; we chop $[1, 5]$ into 8 equal pieces each of length $\Delta x = \frac{5-1}{8} = \frac{1}{2}$.

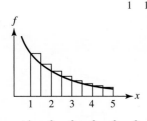

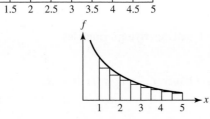

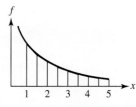

$$L_8 = \left(\frac{1}{1} + \frac{2}{3} + \frac{2}{4} + \frac{2}{5} + \frac{2}{6} + \frac{2}{7} + \frac{2}{8} + \frac{2}{9}\right) \cdot \frac{1}{2}$$
$$= 1.828\ldots$$

$$R_8 = \left(\frac{2}{3} + \frac{2}{4} + \frac{2}{5} + \frac{2}{6} + \frac{2}{7} + \frac{2}{8} + \frac{2}{9} + \frac{2}{10}\right) \cdot \frac{1}{2}$$
$$= 1.428\ldots$$

$$T_8 = \frac{L_8 + R_8}{2}$$
$$= 1.628\ldots$$

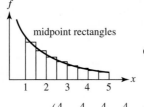

OR

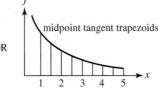

$$M_8 = \left(\frac{4}{5} + \frac{4}{7} + \frac{4}{9} + \frac{4}{11} + \frac{4}{13} + \frac{4}{15} + \frac{4}{17} + \frac{4}{19}\right) \cdot \frac{1}{2} = 1.599\ldots$$

We know that

$$1.428\ldots = R_8 < M_8 = 1.599\ldots < \int_1^5 \frac{1}{x}\,dx < T_8 = 1.628\ldots < L_8 = 1.828\ldots.$$

$f(x) = \frac{1}{x}$ is decreasing and concave up on $[1, 5]$. Therefore, we know that for any n

$$R_n < \int_1^5 \frac{1}{x}\,dx < L_n \quad \text{and} \quad M_n < \int_1^5 \frac{1}{x}\,dx < T_n.$$

If we are interested in more decimal places, we can simply choose larger values of n.

50 Subdivisions

Suppose $n = 50$; we chop $[1, 5]$ into 50 equal pieces each of length $\Delta x = \frac{5-1}{50} = 0.08$. We don't actually want to sum up 50 terms by hand. Work like this is painful to do by hand but it's child's play for a programmable calculator or computer. Get out your programmed calculator or computer and check the figures given below.[5]

$$R_{50} = 1.577\ldots \quad M_{50} = 1.60918\ldots \quad T_{50} = 1.60994\ldots \quad L_{50} = 1.641\ldots$$

400 Subdivisions

Suppose $n = 400$; we chop $[1, 5]$ into 400 equal pieces each of length $\Delta x = \frac{5-1}{400} = 0.01$. We obtain

$$R_{400} = 1.60544\ldots \quad M_{400} = 1.60943\ldots \quad T_{400} = 1.60944\ldots \quad L_{400} = 1.61344\ldots$$

Using T_{400} as an upper bound and M_{400} as a lower bound, we've nailed down the value of this integral to 4 decimal places.

---

[5] You'll have to enter the following information: the function (often as Y_1), the endpoints of integration, and the number of pieces into which you'd like to partition the interval. (And, if you're using a calculator without L_n, R_n, T_n, and M_n programmed, you'll have to enter the program.)

REMARK M_{50} and T_{50} give better approximations of $\int_1^5 \frac{1}{x}\,dx$ than do R_{400} and L_{400}. ◆

Summary of the Underlying Principles

◼ If f is increasing on $[a, b]$, then $L_n < \int_a^b f(t)\,dt < R_n$.

◼ If f is decreasing on $[a, b]$, then $R_n < \int_a^b f(t)\,dt < L_n$.

◼ If f is concave up on $[a, b]$, then $M_n < \int_a^b f(t)\,dt < T_n$.

◼ If f is concave down on $[a, b]$, then $T_n < \int_a^b f(t)\,dt < M_n$.

◼ For any given n, generally the trapezoidal and midpoint sums are much closer to the actual value of the definite integral than are the left- and right-hand sums.

◆ **EXAMPLE 26.2**   Approximate $\int_0^2 e^{-2x^2}\,dx$ by finding upper and lower bounds differing by no more than 0.001.[6]

SOLUTION   This is a great problem on which to practice, because it is impossible to find an antiderivative for $f(x) = e^{-2x^2}$ unless we resort to an infinite sum of terms. Look at the graph of $f(x)$. It is decreasing on the interval $[0, 2]$ and appears to have a point of inflection somewhere on this interval.[7]

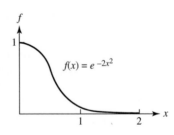

Figure 26.8

Suppose we were planning to use left- and right-hand sums only. As shown in Section 22.2, the difference between the left- and right-hand sums is given by

---

[6] The function e^{-kx^2} is of practical importance because, for the appropriate k, its graph gives the bell-shaped normal distribution curve. The area under the normal distribution curve over some given interval is of vital importance to probabilists and statisticians.

[7] How can we be *sure* f is decreasing? $f(x) = \frac{1}{e^{2x^2}}$. As x increases from 0 to 2, $2x^2$ increases, so e^{2x^2} is positive and increasing. Therefore, its reciprocal is decreasing. Alternatively, $\frac{d}{dx}e^{-2x^2} = e^{-2x^2}(-4x) = \frac{-4x}{e^{2x^2}} \le 0$ for $0 \le x \le 2$.

$$|R_n - L_n| = |f(b) - f(a)| \cdot \Delta x$$
$$= |f(b) - f(a)| \cdot \frac{b-a}{n}.$$

Therefore, in this example

$$|R_n - L_n| = |e^{-8} - e^0| \cdot \frac{2-0}{n} = |e^{-8} - 1| \cdot \frac{2}{n} < \frac{1.9994}{n}.$$

If we want $|R_n - L_n| < 0.001$, then we can solve $\frac{1.9994}{n} = 0.001$ for n and choose any integer larger than this.

$$n = \frac{1.9994}{0.001} = 1999.4,$$

so we can choose $n = 2000$.

If your calculator will accept a number this large, you're in good shape; the left-hand sum will be an upper bound and the right-hand sum will be a lower bound, because f is decreasing on $[0, 2]$. The computation may take some time for the machine to perform, depending upon its power. You should get $L_{2000} = 0.62711719\ldots$ and $R_{2000} = 0.62611753\ldots$.

REMARK Suppose f is monotonic[8] over the interval $[a, b]$, so the actual value of $\int_a^b f(x)\,dx$ is between L_n and R_n. Although it is possible simply to try larger and larger values of n until L_n and R_n are within the desired distance from one another, it is more efficient, if a high degree of accuracy is demanded, to use

$$|R_n - L_n| = |f(b) - f(a)|\Delta x$$

to find an appropriate value of n.

Generally, the midpoint and trapezoidal sums give us much better bounds for a particular n. If we find the point of inflection, we can use these sums to form a sandwich around the value of the integral.

$$f(x) = e^{-2x^2}$$
$$f'(x) = -4xe^{-2x^2}$$
$$f''(x) = -4[e^{-2x^2} + x(-4xe^{-2x^2})]$$
$$= 4e^{-2x^2}(-1 + 4x^2)$$

$f''(x) = 0$, where $-1 + 4x^2 = 0$, that is, where $x^2 = \frac{1}{4}$. On $[0, 2]$ $f''(x) = 0$ at $x = \frac{1}{2}$.

Looking back at $f''(x) = 4e^{-2x^2}(-1 + 4x^2)$, we can see that f'' changes sign (from negative to positive) at $x = \frac{1}{2}$, so $x = 0.5$ is a point of inflection. f is concave down on $[0, \frac{1}{2}]$ and concave up on $[\frac{1}{2}, 2]$. Therefore,

---

[8] f is monotonic on $[a, b]$ if it is either always increasing on $[a, b]$ or is always decreasing on $[a, b]$.

on $[0, \frac{1}{2}]$, T_n gives a lower bound for the integral and M_n gives an upper bound;

on $[\frac{1}{2}, 2]$, T_n gives an upper bound for the integral and M_n gives a lower bound.

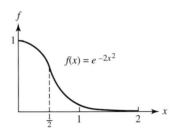

Figure 26.9

The combined error on $[0, \frac{1}{2}]$ and $[\frac{1}{2}, 2]$ must be no more than 0.001. We can split up the allowable error however we want. We can play around with the calculator or computer until the sum of $|T_n - M_n|$ on $[0, \frac{1}{2}]$ and $|T_m - M_m|$ on $[\frac{1}{2}, 2]$ is less than 0.001.

Here is the result of some playing around. If we try $n = 50$ on both intervals, we obtain the following.

On $[0, \frac{1}{2}]$ $T_{50} = 0.427802\ldots$ $M_{50} = 0.4278170\ldots$

On $[\frac{1}{2}, 2]$ $T_{50} = 0.198895\ldots$ $M_{50} = 0.198759\ldots$

On each interval the difference between T_{50} and M_{50} is substantially less than 0.0005. To get the final answer we need to add the lower bounds (T_{50} on $[0, \frac{1}{2}]$ and M_{50} on $[\frac{1}{2}, 2]$) to obtain a lower bound for the integral, and add the upper bounds to obtain an upper bound.

$$\int_0^2 e^{-2x^2}\, dx = \int_0^{.5} e^{-2x^2}\, dx + \int_{.5}^2 e^{-2x^2}\, dx$$

lower bound: $0.427802 + 0.198759 = 0.627065$

upper bound: $0.428170 + 0.198895 = 0.627065$

In the next section we will show an alternative and more efficient method of using the trapezoidal and midpoint sums to approximate this integral.   ◆

Numerical methods of integration, such as left- and right-hand sums and trapezoidal and midpoint sums, are very useful not only when we are trying to approximate $\int_a^b f(x)\, dx$ and can't find an antiderivative for f, but also in situations in which we don't even have a formula for f. A scientist may be taking periodic data readings, or a surveyor may be taking measurements at preset intervals, and these data sets may constitute the only information we have about f. Numerical methods of integration can be applied directly to the data sets. Exercise 26.1 below deals with information from a data set, and the results of Exercise 26.2 and Exercise 26.3 can be easily applied to data sets in which data have been collected at equally spaced intervals.

EXERCISE 26.1   Between noon and 5:00 P.M. water has been leaving a reservoir at an increasing rate. We do not have a rate function at our disposal, but we do have some measurements indicating the rate that water has been leaving at various times. The information is given below. These measurements were not taken at equally spaced time intervals.

Time	noon	1:00	1:30	2:00	3:00	3:45	4:15	5:00
Rate out (in gal/hr)	140	160	170	200	250	270	280	300

(a) Find good upper and lower bounds on the amount of water that has left the reservoir between noon and 5:00 P.M.

(b) Use a trapezoidal sum to approximate the amount of water that has left the reservoir between noon and 5:00 P.M.

Answers are provided at the end of the section.

EXERCISE 26.2   Our aim is to approximate $\int_a^b f(x)\,dx$ when we do not have a formula for $f(x)$. Our data consist of a collection of measurements of $f(x)$ taken at equally spaced intervals. Use a trapezoidal sum to approximate the definite integral.

Partition the interval $[a, b]$ into n equal subintervals, each of length $\Delta x = \frac{b-a}{n}$. $x_0, x_1, x_2, \ldots$, are as indicated. $x_0 = a$ and $x_k = a + k\Delta x$ for $k = 1, \ldots, n$. $y_k = f(x_k)$ for $k = 0, \ldots, n$.
Show that

> the trapezoidal sum can be computed as follows:
> $$T_n = \left(\tfrac{1}{2}\right)[y_0 + 2y_1 + 2y_3 + \cdots + 2y_{n-1} + y_n]\Delta x.$$

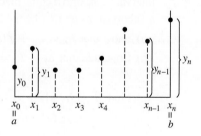

Figure 26.10

The answer is provided at the end of the section.

Answers to Selected Exercises

Exercise 1

(a) The rate is increasing, so the lower bound is given by the left-hand sum and the upper bound by the right-hand sum.

Lower bound: $140(1) + 160(0.5) + 170(0.5) + 200(1) + 250(0.75) + 270(0.5) + 280(0.25) = 897.5$

Upper bound: $160(1) + 170(0.5) + 200(0.5) + 250(1) + 270(0.75) + 280(0.5) + 300(0.25) = 1012.5$

Between 897.5 and 1012.5 gallons of water have left the reservoir.

(b) The trapezoidal approximation is the average of the left- and right-hand sums given above. The different lengths of the intervals do not change this basic principle. (Convince yourself of this.)

Trapezoidal sum $= \frac{1}{2}(897.5 + 1012.5) = 955$

Exercise 2.

$$T_n = \frac{1}{2}[(y_0 + y_1 + y_2 + \ldots + y_{n-1})\Delta x + (y_1 + y_2 + y_3 + \ldots + y_n)\Delta x]$$

$$= \frac{1}{2}(y_0 + 2y_1 + 2y_2 + \ldots + 2y_{n-1} + y_n)\Delta x$$

PROBLEMS FOR SECTION 26.1

Some of the problems in this problem set require the use of a programmable calculator or a computer. These are not highlighted in any way, but common sense should tell you that summing a couple of hundred terms is not a good use of your time. On the other hand, in order to make sure that you understand the numerical methods, some problems explicitly ask you to refrain from using a program to execute the computation.

1. Suppose we use right- and left-hand sums to approximate $\int_a^b f(t)\,dt$. We partition the interval $[a, b]$ into n equal pieces each of length Δt. Let R_n be the right-hand sum using n subdivisions and L_n be the left-hand sum using n subdivisions.

    (a) Show that $R_n = L_n + f(b)\Delta t - f(a)\Delta t$.

    (b) Conclude that $R_n - L_n = [f(b) - f(a)]\Delta t = [f(b) - f(a)]\frac{|b-a|}{n}$.

2. (a) Find $\int_1^x \frac{1}{t}\,dt$, where $x > 0$.

    (b) $\ln x$ can be defined to be $\int_1^x \frac{1}{t}\,dt$. Sketch the graph of $1/t$ and approximate $\ln 2$ numerically by partitioning the interval $[1, 2]$ into 4 equal pieces and computing L_4 and R_4. Write out the sum long-hand, without using a calculator program.

    (c) Into how many equal pieces would you have to subdivide the interval $[1, 2]$ in order to approximate $\ln 2$ so that R_n and L_n give upper and lower bounds that differ by no more than 0.01? Call this number p.

    (d) Compute M_p and T_p.

    (e) Into how many equal pieces would you have to subdivide the interval $[1, 3]$ in order to approximate $\ln 3$ so that R_n and L_n give upper and lower bounds that differ by no more than 0.001? That differ by no more than D?

3. Compute $\int_1^e \ln x\,dx$ with error less than 0.002. Try, by experimentation, to see how many subdivisions are required for M_n and T_n to differ by less than 0.002. How many are required for L_n and R_n to differ by less than 0.002?

4. (a) Let f be the function graphed below. f is increasing and concave down.

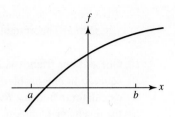

Let $A = \int_a^b f(x)\,dx$. Suppose that estimates of $f(x)\,dx$ are computed using the left, right, and trapezoid rules, each with the same number of subintervals. We'll denote these estimates by L_n, R_n, and T_n, respectively. Put the numbers A, L_n, R_n, and T_n in ascending order. Justify your answer and explain why your answer is independent of the value of n used.

   (b) Answer the same question as above if the graph of f is the one given below. f is decreasing and concave up on $[a, b]$.

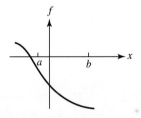

5. Consider $\int_1^4 \sqrt{x}\,dx$.

   (a) Find a value of n for which $\left| L_n - \int_a^b f(x)\,dx \right| \le 0.01$.

   (b) Use the value of n from part (a) to find L_n and R_n.

   (c) Is the average of the left- and right-hand sums larger than the integral, or smaller?

   (d) Compare your numerical approximations to the answer you get using the Fundamental Theorem of Calculus.

6. Approximate $\int_0^1 \sqrt{1 + x^4}\,dx$ with error less than 0.01.

7. Give upper and lower bounds for $\int_0^2 \frac{10}{2+x^5}\,dx$ such that the upper and lower bounds differ by less than 0.01.

8. Give upper and lower bounds for $\int_2^3 \frac{1}{\ln x}\,dx$ such that the two bounds differ from one another by less than 0.05. Explain how you know that the upper bound is indeed an upper bound and the lower bound is indeed a lower bound.

9. Consider $\int_4^9 \frac{1}{\sqrt{x}}\,dx$.

   (a) Find a value of n for which $|L_n - R_n| \le 0.01$.

   (b) Use that value of n to find L_n and R_n.

(c) If you take the average of the left- and right-hand sums, will your approximation be larger than the integral, or smaller?

(d) Compare your numerical approximations to the answers you get using the Fundamental Theorem of Calculus.

10. One of your friends is doing his mathematics homework on the run. He managed to squeeze in one problem between lunch and his expository writing class. He used his calculator to find L_n, R_n, T_n, and M_n to approximate a definite integral and jotted down the results on a napkin. They were

$$0.367617, \quad 0.3211885, \quad 0.341189, \quad \text{and} \quad 0.274760.$$

He didn't write down the problem, nor did he record the value of n he used. But he did sketch the function (over the relevant interval) on the napkin; it looked like this.

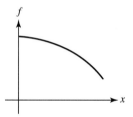

He wants your help in labeling the data he recorded as a left-hand sum, a right-hand sum, a midpoint sum, and a trapezoidal sum. Which is which? Explain your answer.

11. Measurements of the width of a pond are taken every 20 yards along its length. The measurements are: 0 yards, 60 yards, 50 yards, 70 yards, 50 yards, and 30 yards. Approximate the surface area of the pond using the Trapezoidal rule.

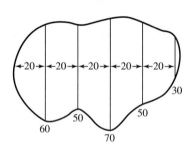

12. Approximate each of the following integrals with error less than 1/100.
    (*Note:* If you look at all the questions and think about a strategy in advance, you will only have to compute two integrals in order to answer all four questions with the desired degree of accuracy. There is no problem with being more accurate than is requested.) Briefly explain what you have done and how many subdivisions you used.
    (i) $\int_0^1 e^{-(\frac{1}{2})x^2}\, dx$ (ii) $\int_1^2 e^{-(\frac{1}{2})x^2}\, dx$ (iii) $\int_0^2 e^{-(\frac{1}{2})x^2}\, dx$ (iv) $\int_{-2}^2 e^{-(\frac{1}{2})x^2}\, dx$

13. Let $f(x) = \frac{3}{x^3+x}$. Let a and b be positive constants, $0 < a < b$. Below are two partitions of the interval $[a, b]$. One (given with w's) partitions $[a, b]$ into 8 equal subintervals, each of length Δw; the other (given with t's) partitions the interval into 12 equal pieces each of length Δt.

$$w_i = a + i\Delta w, \quad i = 0, 1, \ldots, 8$$
$$t_i = a + i\Delta t, \quad i = 0, 1, \ldots, 12$$

Put the following expressions in ascending order, with "<" or "=" signs between them. (Identify the sum with the letter preceding it.)

$$A = \sum_{i=0}^{7} f(w_i)\Delta w \qquad B = \sum_{i=1}^{8} f(w_i)\Delta w$$

$$C = \sum_{i=0}^{11} f(t_i)\Delta t \qquad D = \sum_{i=1}^{12} f(t_i)\Delta t$$

$$E = \int_{a}^{b} f(w)\, dw \qquad F = f(a) \cdot (b-a)$$

14. Approximate $\int_{-1}^{2} \arctan x \, dx$ using left- and right-hand sums to obtain an upper and lower bound for the integral with difference less than 0.05. Save time by graphing $y = \arctan x$ and using symmetry to simplify the problem.

15. The function $f(x)$ is decreasing and concave down on the interval $[3, 5]$. Suppose that you use a right-hand sum, R_{100}, a left-hand sum, L_{100}, a trapezoidal sum, T_{100}, and a midpoint sum, M_{100}, all with 100 subdivisions, to estimate $\int_{3}^{5} f(x)\, dx$. Select all of the following that must be true.

    (a) $L_{100} \geq R_{100}$
    (b) $\int_{3}^{5} f(x)\, dx \geq T_{100}$
    (c) $T_{100} \geq R_{100}$
    (d) $T_{100} \geq M_{100}$
    (e) $M_{100} \geq L_{100}$
    (f) $T_{100} = (L_{100} + R_{100})/2$

16. Suppose that g is a differentiable function whose derivative is $g'(x) = \frac{2}{x^2+3}$. Partition $[0, 2]$ into n equal pieces each of length Δx and let $x_k = k\Delta x$, where $k = 0, 1, \ldots, n$. Put the following expressions in ascending order (with "<" or "=" signs between them).

$$A = \sum_{i=1}^{n} g(x_i)\Delta x \qquad\qquad B = \sum_{i=0}^{n-1} g(x_i)\Delta x$$

$$C = \lim_{n\to\infty} \sum_{i=0}^{n-1} g(x_i)\Delta x \qquad D = \lim_{n\to\infty} \sum_{i=1}^{n} g(x_i)\Delta x$$

17. Suppose $L_{10} < \int_a^b f(x)\,dx < R_{10}$ and $|R_{10} - L_{10}| < 0.01$. Which of the following statements *must* be true? Circle *all* such statements.

(a) $|R_{10} - \int_a^b f(x)\,dx| < 0.01$ (b) $|T_{10} - \int_a^b f(x)\,dx| < 0.01$

(c) $|T_{10} - \int_a^b f(x)\,dx| < 0.005$ (d) $|L_{10} - \int_a^b f(x)\,dx| < 0.005$

18. Two of the following three integrals you can evaluate exactly. One you cannot, until learning integration by parts. Identify the one you cannot evaluate exactly and approximate it with an error under 0.05. Find exact answers for the other two integrals.

(a) $\int_1^e \frac{\ln x}{x}\,dx$ (b) $\int_0^1 xe^{x^2}\,dx$ (c) $\int_0^1 xe^x\,dx$

26.2   SIMPSON'S RULE AND ERROR ESTIMATES

Simpson's Rule

We began our discussion with left- and right-hand sums, sums that provide bounds for $\int_a^b f(x)\,dx$ if f is monotonic on $[a, b]$. Then we came up with the clever idea of using the average of L_n and R_n to approximate a definite integral. This led us to the trapezoidal sum, which generally gives a much better approximation to a definite integral than either the left or right sums for the same number of subdivisions. Encouraged by our success, we might start thinking about averaging the midpoint and trapezoidal sums. An even better idea would be to take a weighted average of these two sums. This is because the midpoint rule generally gives us a better fit than does the trapezoidal rule. In fact, analysis of numerical data indicates that the error in using the trapezoidal rule is about two times the size as that from using the midpoint rule. From the picture below this seems plausible. The weighting $\frac{M_n + M_n + T_n}{3} = \frac{2M_n + T_n}{3}$ gives us a method of approximating definite integrals known as **Simpson's rule.**

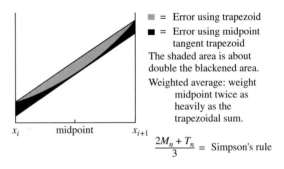

Figure 26.11

$$\text{Simpson's rule: } S_{2n} = \frac{2M_n + T_n}{3}$$

We call it S_{2n} because we need n data points for M_n and $(n+1)$ different data points for T_n. Sometimes it will be convenient to write $S_p = \frac{2M_n + T_n}{3}$ where $p = 2n$.

Simpson's rule generally gives a *substantially* better numerical estimate than either the midpoint or the trapezoidal sums. To get some appreciation for this, try it out on some integrals that you can compute exactly. Below we do this for the integral $\int_1^5 1/x \, dx$.

◆ **EXAMPLE 26.3** Compare M_n, T_n, and Simpson's rule (the weighted average) for $\int_1^5 \frac{1}{x} \, dx$, where $n = 4, 8,$ 50, and 100. Note that $\int_1^5 \frac{1}{x} \, dx = \ln |x| \Big|_1^5 = \ln 5 - \ln 1 = \ln 5 = 1.60943791 \ldots$.

SOLUTION

(a) $n = 4$: $M_4 = 1.57460\ldots$

$T_4 = 1.68333\ldots$

Simpson's rule $= \frac{2M_4 + T_4}{3} = 1.61084\ldots$

(b) $n = 8$: $M_8 = 1.59984\ldots$

$T_8 = 1.62896\ldots$

Simpson's rule $= \frac{2M_8 + T_8}{3} = 1.60955\ldots$

(c) $n = 50$: $M_{50} = 1.60918221\ldots$

$T_{50} = 1.60994957\ldots$

Simpson's rule $= \frac{2M_{50} + T_{50}}{3} = 1.60943799\ldots$

(d) $n = 100$: $M_{100} = 1.60937393\ldots$

$T_{100} = 1.60956589\ldots$

Simpson's rule $= \frac{2M_{100} + T_{100}}{3} = 1.60943791\ldots$ ◆

Like the trapezoidal sum, Simpson's rule is a convenient tool for approximating $\int_a^b f(x) \, dx$ even when a formula for f is not available. For instance, the values of $f(x)$ referred to in the discussion below could be measurements taken by a surveyor estimating the area of a plot of land or body of water.

Suppose our aim is to approximate $\int_a^b f(x) \, dx$. We can collect values of $f(x)$ at equally spaced intervals and use Simpson's rule to approximate the definite integral. Partition the interval $[a, b]$ into n equal subintervals, each of length $\Delta x = \frac{b-a}{n}$. We need to use a weighted average of the midpoint and trapezoidal sums; therefore, on each subinterval we need not only the value of f at each endpoint (for T_n) but also the value of f in the middle (for M_n). Although we are chopping $[a, b]$ into n equal subintervals, we will use $2n + 1$ values of f ($n + 1$ values for T_n and n values for M_n).

Let $y_k = f(x_k)$ for $k = 0, \ldots, 2n$, where $x_0, x_1, x_2, \ldots, x_{2n}$ are as indicated below.

$$x_k = a + k \left(\frac{\Delta x}{2} \right) \quad \text{for } k = 0.1, \ldots, 2n$$

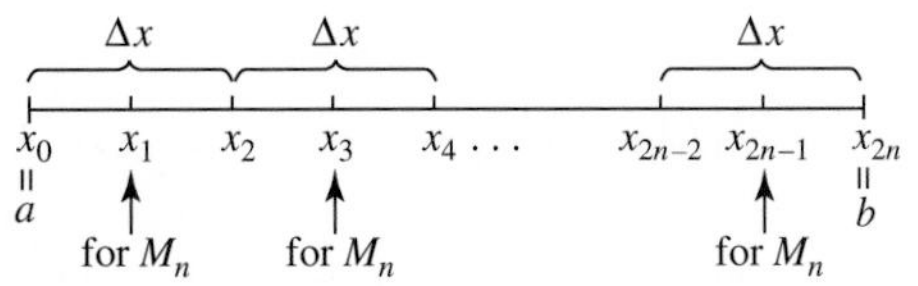

Figure 26.12

In the following exercise you will show that using this labeling convention, Simpson's rule can be used to approximate the definite integral using the following formula.

Simpson's rule:

$$S_{2n} = \left(\tfrac{1}{6}\right)[y_0 + 4y_1 + 2y_2 + 4y_3 + 2y_4 + \cdots + 2y_{2n-2} + 4y_{2n-1} + y_{2n}]\Delta x,$$

where $\Delta x = \dfrac{b - a}{n}$.

EXERCISE 26.3   In this exercise use the labeling system described above: Partition $[a, b]$ into n equal pieces of length $\Delta x = \frac{b-a}{n}$.

$$x_k = a + k\left(\tfrac{\Delta x}{2}\right) \text{ for } k = 0, 1, \ldots, 2n, \text{ where } \Delta x = \frac{b - a}{n}$$

Let $y_k = f(x_k)$ for $k = 0, \ldots, 2n$, where $x_0, x_1, x_2, \ldots, x_{2n}$.

(a) Show that using this labeling convention

$$T_n = \left(\tfrac{1}{2}\right)(y_0 + 2y_2 + 2y_4 + 2y_6 + \cdots + 2y_{2n-2} + y_{2n})\Delta x.$$

(b) Show that using this labeling convention

$$M_n = (y_1 + y_3 + y_5 + y_7 + \cdots + y_{2n-1})\Delta x.$$

(c) Show that using this labeling convention Simpson's rule is given by

$$\left(\tfrac{1}{6}\right)(y_0 + 4y_1 + 2y_2 + 4y_3 + 2y_4 + \cdots + 2y_{2n-2} + 4y_{2n-1} + y_{2n})\Delta x.$$

The answer to part (c) is given at the end of the section.

Notice that because Simpson's rule requires both the midpoint and trapezoidal rules and the relevant data points for each of these two rules are different, Simpson's rule requires approximately double the number of data points necessary for computing either M_n or T_n individually.

REMARKS

■   Simpson's rule can be presented without reference to the trapezoidal and the midpoint sums. Instead, it can be viewed as follows. On each subinterval $[x_{i-1}, x_i]$ approximate the area under f by the area under the parabola passing through the three points on the graph of f corresponding to the endpoints of the interval and the midpoint, or $(x_{i-1}, f(x_{i-1}))$, $(x_i, f(x_i))$, and $\left(\frac{x_{i-1}+x_i}{2}, f\left(\frac{x_{i-1}+x_i}{2}\right)\right)$.

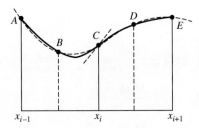

Find a parabola through A, B, and C.
Find another parabola through
C, D, and E.
Sum the areas under the parabolas to
approximate the area under the curve.

Figure 26.13

■   When Simpson's rule is applied to cubics, it will always give an exact answer, even
when using $\frac{2M_1+T_1}{3}$.

Error Bounds

In Section 26.1 we used approximating sums to provide upper and lower bounds for the
value of an integral. When we use Simpson's rule we have a "good" estimate, but without
getting a bound on the error involved we cannot be sure how "good" the estimate is. While
we cannot expect to know the exact error in using approximating sums, it would be quite
useful to get a bound on the error. Consider, for instance, Example 26.2. It would be useful
to be able to approximate $\int_0^2 e^{-2x^2}\, dx$ using the trapezoidal rule or Simpson's rule, having
a bound for the error but not worrying about the point of inflection.

Let's begin by thinking about the error involved in using left- and right-hand sums.

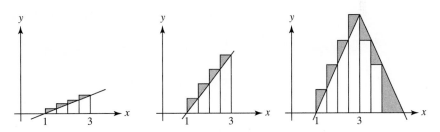

The steeper the slope of f the larger the error in approximating by rectangles.

Figure 26.14

As illustrated in Figure 26.14, the larger $|f'|$ is, the larger the magnitude of the error involved
in approximating the area under the curve by the area of a rectangle.

When approximating the area using the trapezoidal rule or the midpoint tangent trapezoid, the magnitude of f' is not an issue; the rate that f' is changing, (measured by f''), is
a factor controlling the error.

For L_n, R_n, T_n, and M_n, the larger n is, the smaller the expected error. Put another
way, the smaller Δx is, the smaller the expected error. Using the Mean Value Theorem, the
following error bounds can be proven.

Let $I = \int_a^b f(x)\,dx$. Let L_n, R_n, T_n, and M_n be left, right, trapezoidal, and midpoint approximations of I, respectively. Then

■ $\displaystyle |L_n - I| \le \frac{M_1(b-a)^2}{2n}$ where M_1 is the maximum value of $|f'|$ on $[a, b]$.

$\displaystyle |R_n - I| \le \frac{M_1(b-a)^2}{2n}$

■ $\displaystyle |T_n - I| \le \frac{M_2(b-a)^3}{12n^2}$ where M_2 is the maximum value of $|f''|$ on $[a, b]$.

$\displaystyle |M_n - I| \le \frac{M_2(b-a)^3}{24n^2}$

You might guess that the error bound for Simpson's rule involves an n^3 and the third derivative of f. In fact, Simpson's rule is much more accurate than you might expect. It can be shown that

■ $\displaystyle |S_{2n} - I| \le \frac{M_4(b-a)^5}{180(2n)^4}$ where M_4 is the maximum value of $|f^{(4)}|$ on $[a, b]$.

If p is an even number, then

$$|S_p - I| \le \frac{M_4(b-a)^5}{180p^4}.$$

◆ **EXAMPLE 26.4**   Find upper bounds for the error involved in using T_{10} and M_{10} to approximate $\int_0^2 e^{-2x^2}\,dx$.

SOLUTION   In order to use the error estimates for T_{10} and M_{10} we need to find an upper bound for f'' on $[0, 2]$. In Example 26.2 we computed $f''(x)$. If $f(x) = e^{-2x^2}$, then

$$f''(x) = 4e^{-2x^2}(-1 + 4x^2)$$
$$= \frac{4(-1 + 4x^2)}{e^{2x^2}}.$$

We can get a crude upper bound for f'' on $[0, 2]$ by noting that the numerator is no more than $4(-1 + 4 \cdot 2^2) = 4(-1 + 16) = 15 \cdot 4 = 60$.

$$|f''(x)| \le \frac{60}{e^{2x^2}} = 60e^{-2x^2}$$

$$e^{-2x^2} \le 1 \text{ on } [-0, 2] \text{ so}$$

$$|f''(x)| < 60.$$

NOTE We haven't found M_2, but we found a bound for M_2.

$\displaystyle |T_{10} - I| \le \frac{M_2(b-a)^3}{12n^2}$ where $a = 0, b = 2, n = 10$, and $M_2 < 60$

$\displaystyle |T_{10} - I| < \frac{60 \cdot 2^3}{12 \cdot 100} = \frac{480}{1200} = 0.4$ The error using T_{10} is guaranteed to be less than 0.4.

$\displaystyle |M_{10} - I| < \frac{60 \cdot 2^3}{24 \cdot 100} = \frac{0.4}{2} = 0.2$ The error using M_{10} is guaranteed to be less than 0.2.   ◆

◆ **EXAMPLE 26.5**   How big should n be to easily guarantee that M_n approximates $\int_0^2 e^{-2x^2} \, dx$ to within 0.001?

SOLUTION   From Example 26.4 we know that $|M_n - I| < \frac{480}{24n^2} = \frac{20}{n^2}$. We must find n such that $\frac{20}{n^2} < 0.001$. We'll solve $\frac{20}{n^2} = 0.001$ and pick the next higher integer.

$$n^2(0.001) = 20 \implies n^2 = 20{,}000 \quad \text{Then } n \approx 141.4.$$

So we choose $n = 142$.   ◆

◆ **EXAMPLE 26.6**   Get an upper bound for the error involved in using S_{10} to approximate $\ln 3 = \int_1^3 \frac{1}{x} \, dx$.

SOLUTION $f(x) = \frac{1}{x}$. Compute f', f'', f''', and $f^{(4)}$. Confirm that $f^{(4)}(x) = \frac{24}{x^5}$.

$$\text{On } [1, 3], \ \left| \frac{24}{x^5} \right| \leq 24.$$

$$|S_p - I| \leq \frac{M_4(b-a)^5}{180p^4} \quad \text{where } a = 1, b = 3, p = 10, \text{ and } M_4 = 24.$$

$$\leq \frac{24(3-1)^5}{180 \cdot 10^4}$$

$$\leq \frac{24 \cdot 2^5}{18 \cdot 10^5} < 4.27 \times 10^{-4} \approx 0.00043 \quad ◆$$

Answers to Selected Exercises

Exercise 26.3(c)

$$\text{Simpson's rule} = \frac{2M_n + T_n}{3}$$

$$= \left(\frac{1}{3}\right)[2(y_1 + y_3 + y_5 + \cdots + y_{2n-1})\Delta x + \left(\frac{1}{2}\right)(y_0 + 2y_2 + 2y_4 + 2y_6 + \cdots + 2y_{2n-2} + y_{2n})\Delta x]$$

$$= [2y_1 + 2y_3 + 2y_5 + + \cdots + 2y_{2n-1})] \cdot \Delta x \cdot \left(\frac{1}{3}\right) +$$

$$\qquad\qquad [y_0 + 2y_2 + 2y_4 + 2y_6 + \cdots + 2y_{2n-2} + y_{2n})]\left(\frac{1}{2}\right)\Delta x \cdot \left(\frac{1}{3}\right)$$

$$= [4y_1 + 4y_3 + 4y_5 + \cdots + 4y_{2n-1})] \cdot \left(\frac{1}{6}\right)\Delta x + [y_0 + 2y_2 + 2y_4 + 2y_6 + \cdots + 2y_{2n-2} + y_{2n})] \cdot \left(\frac{1}{6}\right) \cdot \Delta x$$

$$= [y_0 + 4y_1 + 2y_2 + 4y_3 + 2y_4 + 4y_5 + 2y_6 + 4y_7 + \cdots + 2y_{2n-2} + 4y_{2n-1} + y_{2n})] \cdot \left(\frac{1}{6}\right)\Delta x$$

PROBLEMS FOR SECTION 26.2

1. Approximate $\int_1^e \ln x \, dx$ using Simpson's rule with $p = 10$. $S_{10} = \frac{2M_5 + T_5}{3}$. Find an upper bound for the error. *At a certain point you'll use the fact that $e < 3$.*

2. Approximate the following integrals using the trapezoidal rule, the midpoint rule, and Simpson's rule for the specified number of subdivisions and compute error bounds using the formulas given in this section. Compare your answers to the exact answer.

   (a) $\int_0^2 x^2 \, dx$ $n = 4$ (b) $\int_0^1 \frac{1}{x+1} \, dx$ $n = 4$

3. Approximate the following integral using the trapezoidal rule, the midpoint rule, and Simpson's rule for the specified number of subdivisions. Look at T_n, M_n, and S_{2n}. Compute error bounds.

   $$\int_1^2 \ln x \, dx \quad \text{(a) } n = 4 \quad \text{(b) } n = 8 \quad \text{(c) } n = 10$$

4. (a) Approximate $\int_0^1 \cos(x^2) \, dx$ using M_{10}. Find an upper bound for the error.

   (b) Approximate $\int_0^1 \cos(x^2) \, dx$ using S_{20}. $S_{20} = \frac{2M_{10} + T_{10}}{3}$. Find an upper bound for the error.

5. A surveyor is measuring the cross-sectional area of a 30-foot-wide river beneath a bridge. He measures the river's depth every 5 feet. The data are given below.

   $$\overset{\text{5 ft}}{\frown} \quad \overset{\text{5 ft}}{\frown} \quad \overset{\text{5 ft}}{\frown} \quad \overset{\text{5 ft}}{\frown} \quad \overset{\text{5 ft}}{\frown} \quad \overset{\text{5 ft}}{\frown}$$
   Depth (in feet) 0.5   1.5   3   5   4   3   1

   (a) Use the trapezoidal sum to approximate the cross-sectional area of the river.

   (b) Use Simpson's rule to approximate the cross-sectional area of the river.

CHAPTER 27

Applying the Definite Integral: Slice and Conquer

■ **27.1 FINDING "MASS" WHEN DENSITY VARIES**

Up to this point we've used the definite integral to

■ find the signed area under the graph of f and

■ find the net change in amount when given a rate function.

In this chapter we will see how the idea of slicing that led us to the limit definition of the definite integral can help us apply our knowledge about definite integrals to other situations.

◆ **EXAMPLE 27.1** A geologist is working with a rectangular block-shaped chunk of sedimentary rock whose height is 3 meters, width is 4 meters, and length is 7 meters. A certain mineral is uniformly distributed throughout the rock sample with a density of 5 milligrams per cubic meter. How many milligrams of the mineral are in the sample?

SOLUTION To calculate the number of milligrams given the density we simply multiply volume by density. The volume of the rock is length · width · height $= (7 \text{ m}) \cdot (4 \text{ m}) \cdot (3 \text{ m}) = 84 \text{ m}^3$. Therefore,

$$\text{number of grams of the mineral} = 84 \text{ m}^3 \cdot 5 \frac{\text{mg}}{\text{m}^3} = 420 \text{ mg}.$$

Notice that the units of volume cancel to leave an answer in milligrams. ◆

◆ **EXAMPLE 27.2** The geologist is now looking at another mineral in the same sedimentary rock sample. This mineral had begun to settle as the rock was being formed, so its density varies with depth. The density is given by $\rho(h) = 1 + 0.1h^2$ milligrams per cubic meter, where h is the depth

827

below the top surface of the sample.[1] Notice that as the depth h increases, so does the density of the mineral in the rock sample. We want to find the number of milligrams of this mineral in the sample.

SOLUTION   Compare this problem to the first one. The difficulty is that the density is *not* constant; it varies from $\rho(0) = 1$ mg/m^3 at the top of the sample to $\rho(3) = 1.9$ mg/m^3 at the bottom. If we were to multiply the total volume of the rock sample by $\rho(0)$, we would get 84 mg; if we were to multiply the total volume of the rock sample by $\rho(3)$, we would get 159.6 mg. The former number is much too small and the latter is much too big. The actual number of grams of the mineral must be somewhere between these two extremes. How can we improve upon our under- and overestimates?

At any fixed depth the density *is* constant. Suppose we divide the sample into horizontal slabs, slices in which the density doesn't vary much. In each slice we can approximate the number of milligrams of the mineral; we'll add up the estimates in each slice to approximate the total number of milligrams. The finer the slices, the less the density will vary within a slice and the better our approximation will be.

Notice that this strategy is the same one we successfully applied to the problem of finding the area under a curve. In the area problem our dilemma was that the function f varied with x over the x-interval $[a, b]$. To deal with this we chopped the interval $[a, b]$ into n equal pieces, on each piece approximated the function by a constant, used that to approximate the area on the subinterval, and summed these areas. We found the area under the curve by letting the number of subintervals grow without bound. Here our dilemma is that the density varies with h, so we'll partition the h-interval $[0, 3]$ into pieces, thereby partitioning the rock sample into thin horizontal slabs in which density doesn't vary greatly.

Approximating the Number of Milligrams of Mineral in the Sample

Suppose we divide the h-interval $[0, 3]$ into six equal subintervals. Using the notation we developed for Riemann sums, these are $[h_0, h_1]$ through $[h_5, h_6]$, where the height of each subinterval is $\Delta h = 0.5$ m and $h_i = i\Delta h$ for $i = 0, 1, \ldots, 6$. This chops the entire rock sample into six slabs, each of which has height 1/2 meter, width 4 meters, and length 7 meters, as shown in Figure 27.1.

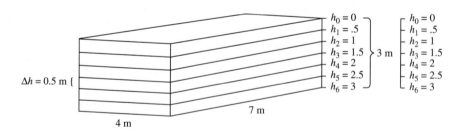

Figure 27.1

We can estimate the number of milligrams of mineral in each slice. For example, in the top slice, the density varies from $\rho(0) = 1 + 0.1(0)^2 = 1$ mg/m^3 at the top surface to $\rho(0.5) = 1 + 0.1(0.5)^2 = 1.025$ mg/m^3 at the bottom surface. The volume of the slice is

---

[1] The Greek letter ρ, pronounced "rho," is often used in the sciences to represent density.

$(4 \text{ m})(7 \text{ m})(\Delta h \text{ m}) = (4)(7)(0.5) \text{ m}^3 = 14 \text{ m}^3$. The density increases with h, so we have the following estimates for the number of milligrams of mineral in this top slice.

$$\text{underestimate: } \rho(0) \cdot \text{ (volume of slice)} = 1\frac{\text{mg}}{\text{m}^3} \cdot 14 \text{ m}^3 = 14 \text{ mg}$$

$$\text{overestimate: } \rho(0.5) \cdot \text{ (volume of slice)} = 1.025\frac{\text{mg}}{\text{m}^3} \cdot 14 \text{ m}^3 = 14.35 \text{ mg}$$

We follow the same procedure for each of the five other slices. Then, by adding all the underestimates for the separate slices, we obtain an underestimate for the total number of milligrams in the sample. Similarly, adding the overestimates on all six slices gives an overestimate for the total.

$$\text{underestimate} = \sum_{i=0}^{5} \rho(h_i) \cdot [28 \cdot \Delta h]$$

$$= \sum_{i=0}^{5} (1 + 0.1h_i^2) \cdot [28 \cdot \Delta h]$$

$$= 103.25 \text{ mg}$$

$$\text{overestimate} = \sum_{i=1}^{6} \rho(h_i) \cdot [28 \cdot \Delta h]$$

$$= \sum_{i=1}^{6} (1 + 0.1h_1^2) \cdot [28 \cdot \Delta h]$$

$$= 115.85 \text{ mg}$$

Notice that our method of slicing horizontally along the axis of the independent variable h ensured that the density would not vary as much within each slice as it had in the rock sample as a whole; this is what brought our estimates closer together. To slice vertically would have defeated the purpose of slicing, because the density in each slice would still have varied between $\rho(0)$ and $\rho(3)$, the same variance as in the entire rock sample.

KEY NOTION The choice of how to slice is at the heart of this kind of problem. This choice is determined by the density function. In this case, the density varies with height, so we need to slice in a way that keeps the height (and hence the density) approximately constant within each slice.

Computing the Exact Number of Milligrams in the Sample

We can improve upon our estimates by partitioning the h-interval $[0, 3]$ into finer and finer subintervals; this corresponds to cutting thinner and thinner horizontal slabs of the rock sample. As we let the number of subintervals grow without bound, we expect the difference between the overestimates and the underestimates to tend toward zero, just as it did in the area problem.

Chop the h-interval $[0, 3]$ into n equal subintervals each of height $\Delta h = \frac{3}{n}$ m and label $h_0, h_1, h_2, \ldots, h_n$ as shown. $h_i = (i)\frac{3}{n}$ for $i = 0, 1, \ldots, n$. This corresponds to slicing the rock sample horizontally into n slabs, each 7 meters long by 4 meters wide by Δh meters tall; the volume of each slice is (area) $\cdot$ (thickness) $= 28\Delta h \text{ m}^3$.

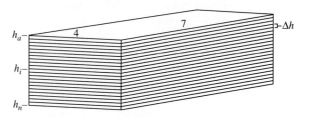

Figure 27.2

Let's look at a generic slice, say the ith slice.

$$\left(\begin{array}{c} \text{the \# of mg} \\ \text{in the } i\text{th slice} \end{array} \right) \approx \left(\begin{array}{c} \text{estimate of density of} \\ \text{the mineral in the slice} \end{array} \right) \cdot (\text{volume of the slice})$$

$$\approx \rho(h_i) \frac{\text{mg}}{\text{m}^3} \cdot (28 \Delta h \text{ m}^3) \qquad (\text{using } \rho(h_i) \text{ gives an overestimate})$$

$$\approx \rho(h_i) \cdot 28 \Delta h \text{ mg}$$

Thus, we obtain the following sums to estimate the total number of milligrams of the mineral in the sample.

$$\text{underestimate} = \sum_{i=0}^{n-1} \rho(h_i) \cdot [28 \cdot \Delta h] = \sum_{i=0}^{n-1} 28\rho(h_i) \cdot \Delta h$$

$$\text{overestimate} = \sum_{i=1}^{n} \rho(h_i) \cdot [28 \cdot \Delta h] = \sum_{i=1}^{n} 28\rho(h_i) \cdot \Delta h$$

These sums are Riemann sums. The difference between the overestimate and the underestimate is $28[\rho(h_n) - \rho(h_0)] \cdot \Delta h$. This can be written as $28[\rho(3) - \rho(0)] \cdot \Delta h$ or $28(0.9)\left(\frac{3}{n}\right) = \frac{75.6}{n}$. As n grows without bound this difference tends toward zero; the limit of the Riemann sums is the definite integral $\int_0^3 \rho(h) \cdot 28 \, dh$.

We now calculate.

$$\text{number of milligrams in sample} = \lim_{n \to \infty} \sum_{i=1}^{n} \rho(h_i) \cdot [28 \cdot \Delta h] \quad \left(= \lim_{n \to \infty} \sum_{i=0}^{n-1} \rho(h_i) \cdot [28 \cdot \Delta h]\right)$$

$$= \int_0^3 \rho(h) \cdot 28 \, dh \qquad \begin{array}{l} \text{(Both sums approach the} \\ \text{value of the integral.)} \end{array}$$

$$= \int_0^3 (1 + 0.1h^2) \cdot 28 \, dh$$

$$= 28 \left[h + 0.1\frac{h^3}{3} \right] \Big|_0^3$$

$$= 28 \left[(3 + 0.9) - (0) \right]$$

$$= 109.2 \text{ mg}$$

The strategy of slicing, approximating, summing, and then taking a limit of the sum allowed us to arrive at an exact answer. ◆

Observe the following.

■ We did *not* simply integrate $\rho(h)$. There was a 28 in the integrand as well. Let's parse the integral.

We are summing $\rho(h)$ ·   28   · dh from $h = 0$ to $h = 3$, corresponding to

$$\left(\begin{array}{c} \text{density} \\ \text{in mg/vol.} \end{array} \right) \cdot (\text{area of slice}) \cdot (\text{thickness of slice})$$

or

$$(\text{density in mg/vol.}) \cdot \text{volume of a slice.}$$

■ The limits of integration, while not explicit in the summation notation of the Riemann sum, are implicitly there. As $n \to \infty$, $h_1 \to 0$ and $h_{n-1} \to 3$. The limits of integration are always determined by the endpoints of the interval being chopped up.

◆ **EXAMPLE 27.3**   When a meteorite crashes into the earth, debris is scattered nearby. Suppose that the density of debris is modeled by $\rho(r) = \frac{1}{1+r^2}$ kilograms per square meter, where r is the number of meters from the center of the meteorite's impact. What is the mass of the debris that lies within 10 meters of the center of impact?

SOLUTION   In order to find mass, we need to multiply density by area. Density is not constant, so we will need to use a slicing approach to break the problem down into regions where the density is approximately constant.

KEY NOTION Just as in Example 27.2, we must slice in a way that keeps the density approximately constant within each slice. In this case density depends on r, the distance from the center of impact. Accordingly, we subdivide the r-interval [0, 10] into n equal subintervals each of length $\frac{10}{n} = \Delta r$ m; we label $r_0, r_1, r_2, \ldots, r_n$ as shown. The slices that result from this are concentric circular rings. Unlike in the rock sample example, in this situation the area of the slices varies.

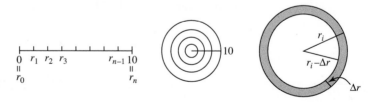

Figure 27.3

The mass in the ith ring $\approx$ (the approximate density in the ring) · (the area of the ith ring).

What is the area of the ith ring?

Intuitive Approach   Think of the ith ring as made of yarn. If the ring is thin enough, its inner and outer circumferences are approximately equal. To approximate the area, clip the yarn and unbend it, laying it out as a long narrow rectangle. The length of the rectangle is approximately the circumference of the ring and the width is Δr. The area of the ring is approximately the area of the rectangle, which is approximately $2\pi r_i \, \Delta r$.

Figure 27.4

Alternative Approach

The ith ring has an outer radius of r_i and an inner radius of $r_i - \Delta r$. We can compute the area of this ring by subtracting the area of a disk of radius $(r_i - \Delta r)$ from the area of a disk of radius r_i.

$$\text{area of } i\text{th ring} = \text{area of larger disk} - \text{area of smaller disk}$$

$$= \pi r_i^2 - \pi (r_i - \Delta r)^2$$

$$= \pi \left[r_i^2 - (r_i^2 - 2r_i \Delta r + (\Delta r)^2) \right]$$

$$= \pi \left[2r_i \Delta r - (\Delta r)^2 \right]$$

We will be constructing a Riemann sum and then taking the limit as the thickness of each slice (Δr) approaches zero. When Δr is very small, $(\Delta r)^2$ is much smaller still. (Consider for instance that if $\Delta r = 0.001$, then $(\Delta r)^2 = 0.000001$.) In the limit the $(\Delta r)^2$ term is insignificant when compared[2] with $r \Delta r$. We use the approximation: area of slice $\approx 2\pi r_i \Delta r$.

In Exercise 27.1 you will show that approximating the area of the ith ring in this way is valid by arriving at the same integral (using a different labeling system) without making any approximation in the area of the ith slice.

Now we estimate the mass of the meteorite debris that lies in each ring-shaped slice.

$$\text{the mass contained in the } i\text{th slice} \approx (\text{area of slice})(\text{approximate density in slice})$$

$$\approx (2\pi r_i \Delta r) \text{ m}^2 \cdot \left(\frac{1}{1 + r_i^2} \right) \frac{\text{kg}}{\text{m}^2}$$

$$\approx \frac{2\pi r_i \Delta r}{1 + r_i^2} \text{ kg}$$

The total mass contained within 10 meters of the center can be approximated by summing the mass lying in the individual slices; we'll use a right-hand sum.

$$\text{total mass} \approx \lim_{n \to \infty} \sum_{i=1}^{n} \frac{2\pi r_i}{1 + r_i^2} \Delta r$$

To get the exact mass, let n grow without bound. In the limit the Riemann sum becomes a definite integral in which the limits of integration are the endpoints of the r-interval being chopped up.

---

[2] This is the same line of reasoning we used in our discussion of the Product Rule.

$$\text{total mass} = \lim_{n \to \infty} \sum_{i=1}^{n} \frac{2\pi r_i}{1 + r_i^2} \Delta r$$

$$= \int_0^{10} \frac{2\pi r}{1 + r^2} \, dr$$

$$= \pi \int_0^{10} \frac{2r \, dr}{1 + r^2} \qquad \text{Substitute } u = 1 + r^2, \text{ so } du = 2r \, dr.$$

$$= \pi \int_{r=0}^{r=10} \frac{du}{u}$$

$$= \pi \left[\ln u \right] \Big|_{r=0}^{r=10}$$

$$= \pi \ln(1 + r^2) \Big|_{r=0}^{r=10}$$

$$= \pi \left[\ln 101 - \ln 1 \right]$$

$$= \pi \ln 101$$

$$\approx 14.50 \text{ kg} \quad \blacklozenge$$

EXERCISE 27.1   In this exercise you will verify that the approximation we made in Example 27.3 when looking at the area of the ith ring is valid. Partition the r-interval $[0, R]$ into n equal subintervals each of length $\Delta r = \frac{R}{n}$ centimeters. In this exercise, let r_i be the *midpoint* of the ith interval, $i = 1, \ldots, n$. That is, $r_1 = \frac{1}{2}\Delta r, r_2 = r_1 + \Delta r, r_3 = r_1 + 2\Delta r$, an so on, as shown in Figure 27.5.

$$r_i = \frac{1}{2}\Delta r + (i - 1)\Delta r \text{ for } i = 1, 2, \ldots, n$$

The ith ring has an outer radius of $r_i + \frac{1}{2}\Delta r$ and an inner radius of $r_i - \frac{1}{2}\Delta r$.

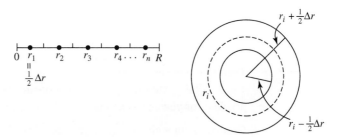

Figure 27.5

(a) Show that the area of the ith ring is *exactly* $2\pi r_i \Delta r$.

(b) Conclude that the mass of the debris described in Example 27.3 is given by $\int_0^{10} \rho(r)2\pi r \, dr$.

The solution is given at the end of the section.

EXERCISE 27.2   A potter is using a spray gun to apply a cobalt glaze to a plate of radius R centimeters.[3] The density of glaze varies with r, the distance from the center of the plate, and is given by $\rho(r)$ mg/cm^2. Write an expression that gives the amount of glaze on the plate.

The answer is given at the end of the section.

◆ EXAMPLE 27.4   A farmer plants a crop of wheat on a circular plot with radius 80 meters. A straight irrigation pipe 160 meters long runs down the center of the plot. Due to a drought, his yield varies with x, the distance from the irrigation pipe. Suppose his yield at a distance x from the pipe is given by $\rho(x)$ kg/m^2. What is the farmer's yield from the whole plot?

SOLUTION

KEY NOTION   We need to slice this disk so that $\rho(x)$ is approximately constant within each slice. Since the density of yield varies with x, the distance from the irrigation pipe, we need to keep x approximately constant within each slice. This means that we need to slice into long, thin strips, as shown in Figure 27.6.

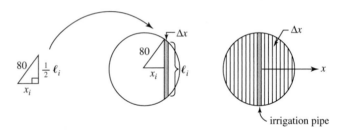

Figure 27.6

We can find the number of kilograms yield on one side of the pipe and double it to get the total yield. We'll assume for the sake of our model that the irrigation pipe itself has no thickness.

$$\text{yield in the } i\text{th strip} \approx \left(\begin{array}{c}\text{the approximate density of}\\ \text{yield in the strip}\end{array}\right) \cdot \left(\begin{array}{c}\text{the approximate area}\\ \text{of the strip}\end{array}\right)$$

$$\approx \rho(x_i) \cdot \ell_i \Delta x \quad \text{where } \ell_i \text{ is the length of the } i\text{th strip.}$$

This expression looks a bit different from the ones we have seen above because it has two variables in it. The Δx indicates that we will eventually integrate with respect to x. Therefore, we need to express ℓ_i in terms of x_i. From Figure 27.6 we see that 80, x_i, and $\frac{1}{2}\ell_i$, can be related using the Pythagorean Theorem. We'll leave the completion of this problem for you as an exercise.   ◆

Observe: Although the regions in Examples 27.3 and 27.4 are both circles, the shapes of the slices are different. The choice of how to slice is determined by the variable of the density function, *not* by the shape of the overall region.

EXERCISE 27.3   Show that the crop yield described in Example 27.4 is given by $4 \int_0^{80} \rho(x)\sqrt{6400 - x^2}\, dx$. If you are off by a factor of 2, be sure that you've looked at the whole field, not just half or a quarter of the field.

---

[3] Cobalt turns blue when fired. The shade of blue varies with the density of the application of cobalt.

The method of slicing allows us to calculate total "mass" in many different situations where the density is not constant. It is an example of the approach we have used repeatedly. To tackle a problem we start with a simpler situation and apply the strategy used there to approximate the quantity we are looking for. We construct a "generic" approximation and use a limiting process to arrive at an exact answer.

Divide and Conquer—Slicing Strategy

1. Determine the independent variable—the variable upon which the density depends.

2. Chop *along* the axis of this independent variable in order to keep density approximately constant within each slice. The endpoints of the interval you chop will determine the limits of integration.

3. Determine the shape and then the volume/area/length of a generic ith slice; this *may* vary with i.

4. Approximate the "mass" of each slice by multiplying volume/area/length by density.[4]

5. Write a Riemann sum by adding up the "masses" of the individual slices. There should be only one variable in this sum; the variable should be the same one determined in step 1.

6. Take the limit as the number of slices increases without bound in order to obtain a definite integral.

7. Evaluate the integral.

Common Errors of the Novice—Try to Avoid Them

■ The biggest calamity is to have sliced incorrectly. After establishing a way of slicing, check carefully to make sure that the density does not change much within a slice.

Alternatively, to get the slices to begin with, fix the independent variable in the density formula. This gives you a cut. Change the independent variable an iota (a very small amount); this gives you another cut. You've just carved out a slice for yourself.

■ Suppose, as in Example 27.2, that the relevant interval of the independent variable is $[0, 3]$. Beginners often have an intense desire to see the limits of integration appear explicitly in their Riemann sum. Thus, a typical error in writing the Riemann sum is to write $\sum_{i=0}^{3}$. When you write this you are summing up exactly four terms; to be useful to you in arriving at an exact answer, your Riemann sum should have n terms.

But the biggest logical problem is what comes next. Writing $\lim_{n \to \infty} \sum_{i=0}^{3}$ doesn't make sense since there is no 'n' in sight. The alternative, $\lim_{3 \to \infty} \sum_{i=0}^{3}$, is nonsense, because 3 is a constant and therefore glued in place on the number line; the number 3 can't pack up and march off "to infinity."

To arrive at an exact answer you'll want to write a **generic** Riemann sum with n terms:

$$\sum_{i=1}^{n}$$

(or, alternatively, $\sum_{i=0}^{n-1}$, a generic left-hand Riemann sum with n terms), so you can let n increase without bound.

---

[4] The units of the density will indicate whether to multiply by volume or area or length.

Try your hand at slicing and approximating the mass of the ith slice by working through the exercise below.

EXERCISE 27.4   Suppose the density of an object varies with height. Density $= \rho(h)$ grams per cubic centimeter. Your task is to write an integral giving the total mass of the object. For each of the objects listed, do the following.

(a) Describe a typical slice.

(b) Approximate the mass of the ith slice. Express the mass in terms of h_i.

(c) Write a generic Riemann sum approximating the total mass of the object.

(d) Write an integral giving the total mass.

Object 1: A right circular cylinder of height 20 centimeters and radius 5 centimeters

Object 2: A right circular cone of height 20 centimeters and base of radius 5 centimeters

Object 3: A sphere of radius 5 centimeters

Answers

Object 1

(a) Partition [0, 20] into n equal pieces. This slices the cylinder into coin-shaped disks of thickness Δh. Each disk has radius 5 and volume $25\pi \, \Delta h$.

(b) mass of ith disk $\approx$ density $\cdot$ volume $\approx \rho(h_i) \cdot 25\pi \, \Delta h$

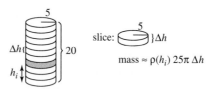

Figure 27.7

(c) total mass $\approx \sum_0^n \rho(h_i) \cdot 25\pi \, \Delta h$

(d) total mass $= \lim_{n \to \infty} \sum_0^n \rho(h_i) \cdot 25\pi \, \Delta h = \int_0^{20} \rho(h)25\pi \, dh$

Object 2

(a) Partition [0, 20] into n equal pieces. This slices the cone into approximately coin-shaped disks of thickness Δh. The radius of each disk varies with h. (The slices are not perfect disks, but they become increasingly disk-like as m increases.)

(b) mass of ith disk $\approx$ density $\cdot$ volume $\approx \rho(h_i) \cdot \pi(r_i)^2\Delta h$. We must get r_i in terms of h_i.

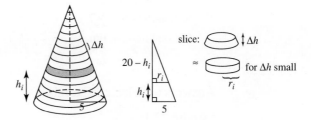

Figure 27.8

Using similar triangles we know that

$$\frac{20 - h_i}{20} = \frac{r_i}{5}$$

so

$$r_i = \frac{20 - h_i}{20} \cdot 5 = \frac{20 - h_i}{4} = 5 - \frac{1}{4}h_i.$$

mass of ith slice $\approx \rho(h_i)\pi(5 - \frac{1}{4}h_i)^2 \Delta h$

(c) total mass $\approx \sum_0^n \rho(h_i)\pi(5 - \frac{1}{4}h_i)^2 \Delta h$

(d) total mass $= \lim_{n\to\infty} \sum_0^n \rho(h_i)\pi(5 - \frac{1}{4}h_i)^2 \Delta h = \int_0^{20} \rho(h)\pi(5 - \frac{1}{4}h)^2 \, dh$

Object 3

(a) The height of the sphere is 10 centimeters. Partition $[0, 10]$ into n equal pieces. This slices the sphere into essentially coin-shaped disks of thickness Δh. The radius of each disk varies with h. The slices are not perfect disks, but they become increasingly disk-like as n increases.

(b) mass of ith disk $\approx$ density $\cdot$ volume $\approx \rho(h_i) \cdot \pi(r_i)^2 \Delta h$. We must get r_i in terms of h_i.

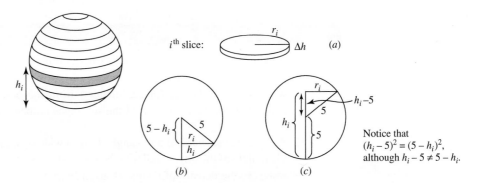

Figure 27.9

Using the Pythagorean Theorem (see Figure 27.9, parts b and c), we know

$$(5 - h_i)^2 + r_i^2 = 25$$

$$r_i^2 = 25 - (5 - h_i)^2.$$

the mass of the ith disk $\approx \rho(h_i) \cdot \pi [25 - (5 - h_i)^2] \Delta h$

(c) total mass $\approx \sum_0^n \rho(h_i) \cdot \pi [25 - (5 - h_i)^2] \Delta h$

(d) total mass $= \lim_{n \to \infty} \sum_0^n \rho(h_i) \cdot \pi [25 - (5 - h_i)^2] \Delta h$

$$= \int_0^{10} \rho(h) \cdot \pi [25 - (5 - h)^2] \, dh = \int_0^{10} \rho(h) \cdot \pi (10h - h^2) \, dh$$

Answers to Selected Exercises

Exercise 27.1

(a)

$$\text{area of } i\text{th ring} = \pi \left(r_i + \frac{\Delta r}{2} \right)^2 - \pi \left(r_i - \frac{\Delta r}{2} \right)^2$$

$$= \pi \left[r_i^2 + r_i \Delta r + \left(\frac{\Delta r}{2} \right)^2 \right] - \pi \left[r_i^2 - r_i \Delta r + \left(\frac{\Delta r}{2} \right)^2 \right]$$

$$= \pi \left[r_i^2 + r_i \Delta r + \left(\frac{\Delta r}{2} \right)^2 - r_i^2 + r_i \Delta r - \left(\frac{\Delta r}{2} \right)^2 \right]$$

$$= \pi (2 r_i \Delta r)$$

(b) In Example 27.3, $R = 10$. The amount of debris in the ith interval $\approx \rho(r_i) 2\pi r_i \Delta r$.

Therefore the total mass $\approx \sum_{i=1}^n \rho(r_i) 2\pi r_i \Delta r$.

Total mass $= \lim_{n \to \infty} \sum_{i=1}^n \rho(r_i) 2\pi r_i \Delta r = \int_0^{10} \rho(r) 2\pi r \, dr$.

Exercise 27.2

Partition the r-interval $[0, R]$ into n equal subintervals each of length $\Delta r = \frac{R}{n}$. Let $r_i = i \Delta r, i = 1, \ldots, n$.

The amount of glaze in the ith ring $\approx \rho(r_i) \cdot 2\pi r_i \Delta r$.

Therefore, the total amount of glaze $\approx \sum_{i=1}^n \rho(r_i) 2\pi r_i \Delta r$.

Total amount of glaze $= \lim_{n \to \infty} \sum_{i=1}^n \rho(r_i) 2\pi r_i \Delta r = \int_0^R \rho(r) 2\pi r \, dr$.

PROBLEMS FOR SECTION 27.1

1. A cylinder 80 centimeters tall with a 10-centimeter radius is filled with a compressible substance. The density of this substance is given by $\rho(h)$ grams per cubic centimeter, where h is the height (in centimeters) from the bottom of the cylinder. Write an expression for the total mass of the substance in the cylinder.

2. A city is in the shape of a rectangle 4 miles wide by 6 miles long. A river runs through the middle of the city, parallel to the 6 mile-long sides. People prefer to live nearer the water, so the density of people is given by $\rho(x) = 10,000 - 800x$ people per square mile, where x is the distance from the river. (You may ignore the width of the river in this problem.)

   (a) Show in a sketch how you will need to slice up the region.
   (b) What is the area of the ith slice?
   (c) What is the approximate population in the ith slice?
   (d) Write a Riemann sum to estimate the total population of the city.

(e) Calculate the exact population by taking the limit of the Riemann sum and evaluating the resulting definite integral.

3. Traditionally, when a college football team seems certain to receive a bid to play in the post-season Orange Bowl, fans begin to throw oranges onto the field. Suppose that at one point during a game, the number of oranges per square yard between the goal line and the 30-yard line is given by $\rho(x) = \frac{30-x}{3}$ oranges per square yard, where x is the number of yards from the goal line. If the field is 160 feet (160/3 yards) wide, how many oranges lie between the goal line and the 30-yard line?

4. (a) A farmer has planted corn on a rectangular plot of land 800 meters by 1000 meters. A straight stream runs alongside one of the long borders of the plot, and the farmer's irrigation system is such that his yield decreases with the distance from the stream. Suppose his yield is given by $f(x) = 50 - 0.3\sqrt{x}$ ears of corn per square meter, where x is the distance from the stream in meters. What is the farmer's yield from the plot?

(b) A second farmer plants his corn in a circular plot with radius 80 meters and he has a centralized irrigation system located in the middle of his field. His yield drops with the distance from the center of the field. Suppose his yield is also given by $f(x) = 50 - 0.3\sqrt{x}$ ears of corn per square meter, this time x being the distance from the center of the field. What is the farmer's yield from this plot?

5. Consider a box of cereal with raisins. The box is 5 centimeters deep, 25 centimeters tall, and 16 centimeters wide. The raisins tend to fall toward the bottom; assume their density is given by $\rho(h) = \frac{4}{h+10}$ raisins per cubic centimeter, where h is the height above the bottom of the box. How many raisins are in the box?

6. The density of dart holes on an old dartboard is given by $\rho(r) = \frac{1010}{\pi(r^2+1)^2}$ holes per square inch, where r is the distance, in inches, from the center of the board. If the board is a circle with diameter 20 inches, find the total number of holes in the board.

7. A beam of light is shining onto a screen creating a disk of radius 50 centimeters. The intensity of light is brightest at the center and diminishes away from the center. If the intensity of light at a distance r from the center of the beam is given by $f(r) = \frac{150}{20+r^2}$ watt/square cm, find the total wattage of the beam's image on the screen.

8. A coastal town is in the shape of a 7-mile by 2-mile rectangle, with one of the 7-mile sides along the coast. In this town people want to live near the beach and the population density at a distance x from the coast is given by $\delta(x) = 4000 - 2000x$ people per square mile.

(a) Write a general Riemann sum that approximates the total population of the town.

(b) Use your answer to part (a) to write a definite integral that represents the total population of the town and evaluate the integral.

9. In Example 27.3, the density within a circle depended on the distance from the center, so we sliced the circle into concentric circular rings and found the area of a slice by two different methods.

Now suppose that the density within a sphere is given by $\rho(r)$, where r is the distance from its center.

(a) Describe the slices that would be used when approximating the total mass.

(b) Find the volume of each slice by a method analogous to the first method used in Example 27.3. (You will need to know that the volume of a sphere of radius r is $\frac{4\pi r^3}{3}$.)

(c) Explain both geometrically and numerically why it is reasonable to approximate the volume of the ith slice by $4\pi r_i^2 \Delta r$.

10. A rectangular meadow is 100 meters. A straight irrigation pipe 400 meters long runs down the center of the meadow dividing it lengthwise in half. The density of wildflowers in the meadow varies with x, the distance from the irrigation pipe.

(a) If the density is given by $g(x)$ flowers per square meter, write an integral giving the number of wildflowers in the meadow. (*Hint:* Take advantage of symmetry.)

(b) If the density is given by $g(x) = \frac{50}{1+x^2}$ flowers per square meter, how many wildflowers are in the meadow?

11. (a) Suppose that the density of organisms in a certain petri dish varies with the distance from the center of the dish. The density at a distance x centimeters from the center is given by $f(x)$ organisms per square centimeter. The petri dish is 18 centimeters in diameter.

   i. Write an integral that gives the number of organisms in the dish.

   ii. Find the number of organisms in the dish if $f(x) = 100e^{-x^2}$ organisms per square centimeter.

(b) Suppose that the density of organisms in a certain petri dish varies with the distance from a strip of nutrients running along the diameter of the dish. The density at a distance x centimeters from the line of nutrients is given by $f(x)$ organisms per square centimeter. The petri dish is 18 centimeters in diameter.

   i. How will you slice up the petri dish?

   ii. Approximate the number of organisms in the ith slice.

   iii. Write a Riemann sum approximating the total number of organisms in the petri dish.

   iv. Write an integral that gives the number of organisms in the dish.

12. Suppose that the density of a planet of mass in a gaseous planet is given by the function $\rho(r) = \frac{40000}{1+.0001r^3}$ kilograms per cubic kilometer, where r is the number of kilometers from the center of the planet. Find the total mass of the planet if it has a radius of 8000 kilometers.

13. A chocolate truffle is a wonderfully decadent chocolate concoction. Truffles tend to be spherical or hemispherical.

(a) Consider a truffle made by dipping a round hazelnut into various chocolates, building up a delicious spherical delicacy. The number of calories per cubic millimeter varies with x, where x is the distance from the center of the hazelnut. If $\rho(x)$ gives the calories/mm^3 at a distance x millimeters from the center, write an integral that gives the number of calories in a truffle of radius R.

(b) Another truffle is made in a hemispherical mold with radius R. Layers of different types of chocolate are poured into the mold, one at a time, and allowed to set. The number of calories per cubic millimeter varies with x, where x is the depth from the top of the mold. The calorie density is given by $\delta(x)$ calories/mm^3. Write an integral that gives the number of calories in this hemispherical truffle.

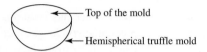

14. Liquid is being stored in a large spherical tank of radius 2 meters. The tank is completely full and has been left standing for a long time. A mineral suspended in the liquid is setting. Its density at a depth of h meters from the top is given by $5h$ milligrams per cubic meter. Determine the number of milligrams of the mineral contained in the tank.

15. A circular pond is 60 meters in diameter and has a bridge running along a diameter. At lunchtime people stand on the bridge and throw bread onto the water to feed the ducks. As a result, the density of ducks on the pond is given by a function $\rho(x)$ ducks per square meter, where x is the distance from the bridge. How many ducks are on the pond? (We will assume that the bridge itself is very thin so we can ignore its width.)

Notice that we cannot really say that the ducks are continuously distributed on the pond. Ducks, after all, are discrete. We are making a continuous model of a discrete phenomenon.

16. Let $W(t)$ be the amount of water in a pool at time t, t measured in hours and W measured in gallons. $t = 0$ corresponds to noon. Water is flowing in and out of the pool at a rate given by $\frac{dW}{dt} = 30 \cos\left(\frac{\pi}{2}t\right)$. During what time interval between noon and 5:00 P.M. ($0 \le t \le 5$) is water flowing *out* of the pool at a rate of 15 gallons an hour or more? How much water actually has left the pool in this time interval?

17. In the town of Lybonrehc there has been a nuclear reactor meltdown, which released radioactive iodine 131. Fortunately, the reactor has a containment building, which kept the iodine from being released into the air. The containment building is hemispherical with a radius of 100 feet. The density of iodine in the building was $6 \times 10^{-5}(200 - h)$ g/cubic feet, where h is the height from the floor (in feet). (It ranges from 12×10^{-3} g/cubic feet at the floor to 6×10^{-3} g/cubic feet near the top.)

    (a) Derive an integral that gives the amount of iodine in the building. You must explain your reasoning fully and clearly.

    (b) Calculate the amount of iodine in the building.

18. A spherical star has a radius of 90,000 kilometers. The density of matter in the star is given by $\rho(r) = \frac{K}{(r+1)^{3/2}}$ kilograms per cubic kilometer, where r is the distance (in kilometers) from the star's center and K is a positive constant.

    Write out (but do not evaluate) an expression for the total mass of the star. Your answer should contain the constant K.

19. A substance has been put in a centrifuge. We now have a cylindrical sample (radius 3 centimeters, height 4 centimeters) in which density varies with x, the distance (in centimeters) from the central axis. If the density is given by $\rho(x)$ mg/cm^3, write an integral that gives the total mass of the substance.

20. A very thin, lighted pole 10 feet tall is placed upright in a family's backyard to attract insects to it (where they are electrocuted). At one moment, the density of these insects is given by $\rho(r) = \frac{1.3}{\pi(r+1)}$ insects per cubic foot, where r measures the number of feet from the pole.

    (a) How many insects are within 5 feet of the pole at a height of 10 feet or less?

    (b) How many insects are within 5 feet of the pole at a height of 10 feet or more?

21. A circus tent has cylindrical symmetry about its center pole. The height a distance of x feet from the center pole is given by $h(x) = \frac{8}{1+\frac{x^2}{16}}$ feet. What is the volume enclosed by the tent of radius 4?

22. At the Three Aces pizzeria, the chef tosses lots of garlic on the pizza. The density of garlic varies with x, the distance from the center of the pizza, and is given by

$$g(x) = \frac{x}{(x^3 + 2)^2} \text{ ounces per square inch of pizza.}$$

If the pizza is 14 inches in diameter, and Three Aces cuts six slices from each pizza, how much garlic is on one slice of pizza? (Problem by Andrew Engelward)

23. (a) What is the present value of a single payment of $2000 three years in the future? Assume 5% interest compounded continuously.

    (b) What is the present value of a continuous stream of income at the rate of $100,000 per year over the next 20 years? Assume 5% interest compounded continuously. By "a continuous stream of income" we mean that we are modeling the situation by assuming that money is being generated continuously at a rate of $100,000 per year.

       Begin by partitioning the time interval [0, 20] into n equal pieces. Figure out the amount of money generated in the ith interval and pull it back to the present. Summing these pull-backs should approximate the present value of the entire income stream.

27.2   SLICING TO FIND THE AREA BETWEEN TWO CURVES

Section 27.1 is the heart of the "slicing" discussion. It allows us to apply what we know about integration to a very broad array of situations. In this section we'll use the same approach to enlarge the type of area problems we can deal with.

◆ **EXAMPLE 27.5**   Find an integral that gives the shaded area in Figure 27.10 the area bounded by $y = f(x)$, $y = g(x)$, and the vertical lines $x = a$ and $x = b$.

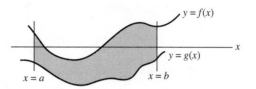

Figure 27.10

SOLUTION   Chop the area into tall, thin "rectangles" by partitioning the interval $[a, b]$ along the x-axis into n equal pieces as shown; each piece is of length $\Delta x = \frac{b-a}{n}$ and $x_i = a + i\Delta x$.[5]

Notice that the height of each rectangle is of the form $f(x_i) - g(x_i)$, regardless of where the x-axis is in relation to the graphs of f and g. To find any vertical distance, just subtract the bottom (lower) y-value from the top (higher) y-value. (See Figure 27.11.)

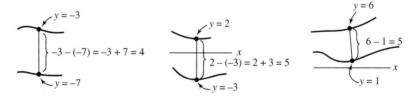

Figure 27.11

the area of the ith rectangle $\approx$ (height) $\cdot$ (length)
$$\approx [f(x_i) - g(x_i)] \cdot \Delta x$$

---

[5] When we chop up the area into many pieces we get a slew of pseudo-rectangles. They don't have flat tops and flat bottoms, so they are not really rectangles. But each one can be closely approximated by a genuine rectangle, one end of which lies on the curve $y = f(x)$ and the other end on the curve $y = g(x)$. For the remainder of this section, for ease of discussion we will say that we chop regions up into rectangles when more precisely what we mean is that we chop the region into pseudo-rectangles.

So,

$$\text{the area of the region} \approx \sum_{i=1}^{n} [f(x_i) - g(x_i)] \cdot \Delta x$$

$$\text{the area of the region} = \lim_{n \to \infty} \sum_{i=1}^{n} [f(x_i) - g(x_i)] \Delta x$$

$$= \int_{a}^{b} [f(x) - g(x)] \, dx.$$

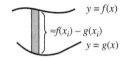

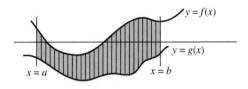

Figure 27.12   ◆

◆ **EXAMPLE 27.6**   Find an integral that gives the shaded area in Figure 27.13, the area bounded by $x = q(y)$, $x = r(y)$, and the horizontal lines $y = c$ and $y = d$.

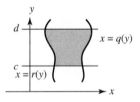

Figure 27.13

SOLUTION   Chop the area horizontally into long, narrow rectangles by partitioning the interval $[c, d]$ along the y-axis into n equal pieces as shown; each piece is of height $\Delta y = \frac{d-c}{n}$. Notice that the length of each rectangle is of the form $q(y_i) - r(y_i)$, regardless of where the y-axis is in relation to the graphs of q and r. To find any horizontal distance, we subtract the x-value on the left (the smaller one) from the x-value on the right (the larger one). (See Figure 27.14.)

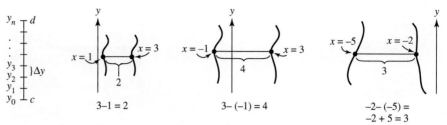

Figure 27.14

the area of the ith rectangle $\approx$ (length)· (height)

$$\approx [q(y_i) - r(y_i)] \cdot \Delta y$$

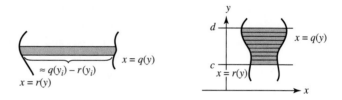

Figure 27.15

So,

the area of the region $\approx \displaystyle\sum_{i=1}^{n} [q(y_i) - r(y_i)] \cdot \Delta y$

the area of the region $= \displaystyle\lim_{n \to \infty} \sum_{i=1}^{n} [q(y_i) - r(y_i)] \cdot \Delta y$

$$= \int_{c}^{d} [q(y) - r(y)]\, dy. \quad \blacklozenge$$

EXERCISE 27.5 Find the area bounded above by $y = x(4 - x)$ and below by

(a) $y = 3$, (b) $y = -5$, (c) $y = -x - 6$.

Answers are given at the end of the section.

EXERCISE 27.6 Which of the following expressions gives the area bounded on the left by the y-axis, on the right by $y = \ln x$, below by the x-axis, and above by the line $y = 1$? (There is more than one correct answer.)

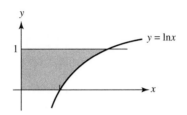

Figure 27.16

(a) $\int_0^1 (1 - \ln x)\, dx$ (b) $\int_0^1 1\, dx + \int_1^e (1 - \ln x)\, dx$

(c) $\int_0^1 e^y\, dy$ (d) $\int_0^e 1\, dx - \int_1^e \ln x\, dx$

(e) $\int_0^1 \ln y\, dy$ (f) $\int_0^e (1 - \ln x)\, dx$

Answers are given at the end of the section.

Using Slicing to Help with Some Definite Integrals—A Slick Trick

In this subsection we show you a nice byproduct of combining the idea of slicing with that of inverse functions.[6] By the end of this section you will be able to evaluate definite integrals such as $\int_1^e \ln x\, dx$, $\int_0^1 \arctan x\, dx$, and $\int_0^{0.5} \arcsin x\, dx$. These integrals look intractable because we don't know antiderivatives of $\ln x$, $\arctan x$, and $\arcsin x$. As we'll soon see, a change of perspective will work wonders for us.[7]

◆ **EXAMPLE 27.7**   Evaluate $\int_1^e \ln x\, dx$.

SOLUTION   Let's begin by drawing a picture of the region whose area is given by the integral above. Evaluating $\int_1^e \ln x\, dx$ is equivalent to finding the area of the shaded region. We're accustomed to slicing the area into vertical rectangles. However, we don't know an antiderivative of $\ln x$, so let's try a change of perspective. We'll look at the very same region but this time slice the region horizontally into "rectangles" as shown in Figure 27.18(b), chopping the interval $[0, 1]$ along the y-axis.

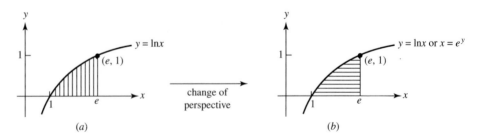

Figure 27.17

We partition the interval $[0, 1]$ into n equal pieces and label as shown. Each piece is of length Δy where $\Delta y = \frac{1-0}{n}$.

---

[6] It will be helpful to try Exercise 27.6 from the preceding page before proceeding.

[7] Generally these integrals are evaluated using a technique called integration by parts, a technique derived from the Product Rule and discussed in Section 29.1.

The area of the ith rectangle $\approx$ (length) $\cdot$ (height).

The height of each rectangle is Δy; the length is given by

$$e - \text{ (the } x\text{-value on the } y = \ln x \text{ curve).}$$

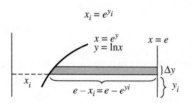

Figure 27.18

We want to write $y = \ln x$ in the form $x = r(y)$.

$$y_i = \ln x_i \quad \text{is equivalent to} \quad x_i = e^{y_i}.$$

Therefore,

$$\text{the area of the } i\text{th rectangle} \approx \text{(length)} \cdot \text{(height)}$$

$$\approx (e - e^{y_i}) \cdot \Delta y.$$

$$\text{the total area} \approx \sum_{i=1}^{n} (e - e^{y_i}) \cdot \Delta y$$

$$\text{the total area} = \lim_{n \to \infty} \sum_{i=1}^{n} (e - e^{y_i}) \cdot \Delta y$$

$$= \int_{0}^{1} (e - e^{y}) \, dy.$$

This integral is simple to evaluate.

$$\int_{0}^{1} (e - e^{y}) \, dy = (ey - e^{y}) \Big|_{0}^{1} = (e - e) - (0 - 1) = -(-1) = 1 \quad \blacklozenge$$

In Summary

The picture associated with the original integral in Example 27.7 enabled us to phrase the problem as "find the area under $y = \ln x$ from $x = 1$ to $x = e$." Slicing the area horizontally instead of vertically was the key to a new outlook. The problem then became "find the area between $x = e$ and $x = e^{y}$ from $y = 0$ to $y = 1$."

◆ **EXAMPLE 27.8**   Express the shaded area in Figure 27.19 on the following page as an integral (or sum or difference of integrals) and evaluate.

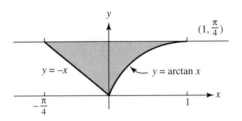

Figure 27.19

SOLUTION   If we chop vertically we must split up the area into two regions because the height of the rectangles requires two different descriptions. For x in the interval $[-\frac{\pi}{4}, 0]$ the rectangles have height $[1 - (-x)] = 1 + x$, while for x in the interval $[0, 1]$ they have height $(1 - \arctan x)$.

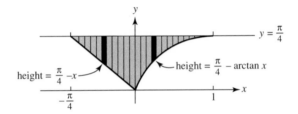

Figure 27.20

$$\text{the shaded area} = \int_{-\frac{\pi}{4}}^{0} (1 + x)\, dx + \int_{0}^{1} (1 - \arctan x)\, dx$$

The first of these integrals is not a problem for us to evaluate. (It's the area of a triangle.) Evaluating the second integral, however, requires finding an antiderivative of $\arctan x$; this information is not at our fingertips. Let's see if a change in perspective will help us out.

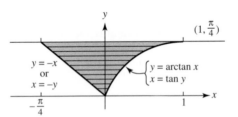

Figure 27.21

Suppose we chop up the region horizontally. We must express the equations $y = \arctan x$ and $y = -x$ in the form $x = q(y)$ and $x = r(y)$ respectively by solving each one for x.

$$y = \arctan x \quad \text{is equivalent to} \quad x = \tan y.$$

$$y = -x \quad \text{is equivalent to} \quad x = -y.$$

If we chop horizontally we can use one integral to describe the entire area, as the length of the rectangles is given by $(\tan y_i - (-y_i))$ throughout the region.

$$\text{the area} = \int_0^1 (\tan y - (-y))\ dy$$

This integral in y is easier to handle than the one above involving arctan x.

$$\int_0^1 (\tan y + y)\ dy = \int_0^1 \frac{\sin y}{\cos y}\ dy + \int_0^1 y\ dy$$

In the first integral, let $u = \cos y$, $du = -\sin y\ dy$. The first integral becomes

$$\int_{y=0}^{y=1} \frac{-du}{u} = -\ln|u|\Big|_{y=0}^{y=1} = -\ln|\cos y|\Big|_{y=0}^{y=1}.$$

Therefore,

$$\int_0^1 \frac{\sin y}{\cos y}\ dy + \int_0^1 y\ dy = -\ln|\cos y|\Big|_0^1 + \frac{y^2}{2}\Big|_0^1$$

$$= -\ln|\cos 1| + \ln|\cos 0| + \frac{1}{2} - 0$$

$$= -\ln|\cos 1| + \ln 1 + \frac{1}{2}$$

$$= -\ln|\cos 1| + \frac{1}{2}.$$

The area is $-\ln|\cos 1| + \frac{1}{2}$.   ◆

Answers to Selected Exercises

Exercise 27.5

Draw a picture for yourself.

(a) $\int_1^3 \left[x(4-x) - 3\right]\ dx = \cdots = \frac{4}{3}$

(b) $\int_{-1}^5 \left[x(4-x) - (-5)\right]\ dx = \cdots = 102\frac{2}{3}$

(c) Find the x-coordinates of the points of intersection. $x = -1$ and $x = 6$.
$\int_{-1}^6 x(4-x) - (-x - 6)\ dx = \int_{-1}^6 (-x^2 + 5x + 6)\ dx = \cdots = 57\frac{1}{6}$

Exercise 27.6

(b), (c), and (d) are correct. (a), (e), and (f) are incorrect.
(See Figure 27.22 on the following page.)

(b) corresponds to the sum of the two shaded areas

(c) uses the fact that $y = \ln x$ is equivalent to $x = e^y$

(d) corresponds to the area of the speckled rectangle minus that of the striped region

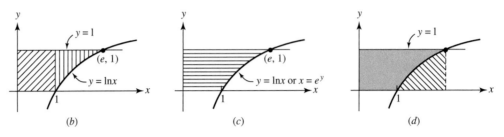

Figure 27.22

PROBLEMS FOR SECTION 27.2

1. Find the area bounded by the curves $y = e^x$, $y = 1 - x$, and $x = 1$.

2. Find the area bounded by $y = 2 - x^2$ and $y = x$.

3. Find the area bounded above by $y = -x + 6$, below by $y = x^2 + 1$, and on the left by the y-axis.

4. Find the area bounded below by the x-axis, on the left by the y-axis, and above by $y = x^2 + 1$ and $y = -x + 6$.

5. Find the area in the first quadrant bounded by $y = \arcsin x$, $y = \pi/2$, and $x = 0$. (*Hint:* To get an exact answer it will be simplest to integrate with respect to y.)

6. (a) Let A be the area between the cosine and the sine curves between $x = -\pi/4$ and $x = \pi/4$.
    i. Sketch the graphs of $\sin x$ and $\cos x$, shading the region A described above.
    ii. Write a definite integral giving the area A.
   (b) Let B be the area between the cosine and the sine curves between $x = 0$ and $x = 2\pi$.
    i. Sketch the graphs of $\sin x$ and $\cos x$, shading the region B described above.
    ii. Write a sum of the definite integrals giving the area B. Notice that over a certain interval the sine curve lies above the cosine curve, while on other intervals it lies below it. Therefore, it is necessary to split up the interval and write a sum of integrals.

7. Write an integral (or the sum or difference of integrals) that gives the area of the region shaded on the following page.

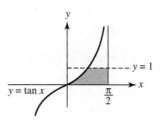

8. Write an integral (or the sum or difference of integrals) giving the area of the region shaded below. (You need not evaluate.)

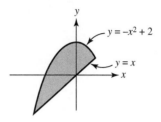

9. Which of the expressions below give the area of the shaded region? (Select *all* such expressions.)

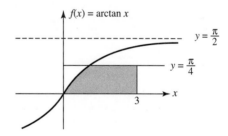

(a) $\int_0^{\pi/4} \arctan x \, dx + \int_{\pi/4}^3 \frac{\pi}{4} \, dx$

(b) $\int_0^1 \arctan x \, dx + \int_1^3 \frac{\pi}{4} \, dx$

(c) $\int_0^1 \tan y \, dy$

(d) $\int_0^{\pi/4} (3 - \tan y) \, dy$

(e) $\int_0^{\pi/4} 3 \, dy - \int_0^{\pi/4} \tan y \, dy$

(f) $\int_0^3 \arctan x \, dx - \int_1^3 (\arctan x - \frac{\pi}{4}) \, dx$

10. Find the area between the curve $y = \ln x$ and the x-axis for $1 \le x \le 10$. Get an exact answer. (*Hint:* Slice the area perpendicular to the y-axis so that the height of each slice is Δy. Use this to arrive at an integral that you can evaluate exactly.)

11. Find (exactly) the area bounded by $x = 1/e$, $y = \ln x$, and $y = 1$.

12. Find the area bounded below by the x-axis, and laterally by $y = \ln x$, and the line segment joining $(e, 1)$ to $(2e, 0)$.

13. Evaluate $\int_0^1 \arctan x \, dx$ by interpreting it as an area and slicing horizontally.

14. Evaluate $\int_0^{0.5} \arcsin x \, dx$.

15. The region A in the first quadrant is bounded by $y = 2x$, $y = -3x + 10$, and $y = -\frac{1}{9}(x^2 - 6x)$. It has corners at $(0, 0)$, $(2, 4)$, and $(3, 1)$. Express the area of A as the sum or difference of definite integrals. You need not evaluate.

CHAPTER 28

More Applications of Integration

28.1 COMPUTING VOLUMES

Volumes by Slicing

We compute the signed area of a region in the plane using a divide-and-conquer technique. To find the area under the graph of f we slice the region into n thin slices, each of width $\Delta x = \frac{b-a}{n}$ and approximate the area of each slice by the area of a rectangle. Let $x_i = a + i\Delta x$, for $i = 0, 1, \ldots, n$.

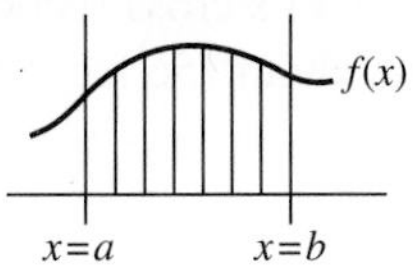

Area $\approx \sum_{i=1}^{n} f(x_i)\,\Delta x$ Where we approximate the height of the ith slice by $f(x_i)$.

Figure 28.1

Summing the areas of the slices and taking the limit as the number of slices increases without bound gives us the area in question.

$$\text{area} = \lim_{n \to \infty} \sum_{i=1}^{n} f(x_i)\Delta x = \int_a^b f(x)\,dx$$

We'll take a similar approach to calculating volume. Suppose we want to find the volume of a loaf of bread. It could be a plain shape, like a typical loaf of rye bread, or it could be a more complicated shape, like a braided loaf of challah. Whatever the loaf looks like, put the whole thing in a bread slicer and ask for thin slices. The volume of the ith slice, V_i, can be approximated by multiplying the area of one of the faces, $A(x_i)$, by the thickness of the slice, Δx. Suppose we have n slices each of thickness Δx.

853

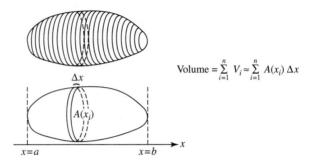

$$\text{Volume} = \sum_{i=1}^{n} V_i \approx \sum_{i=1}^{n} A(x_i)\,\Delta x$$

Figure 28.2

The thinner the slices, the less error is involved in approximating the volume of a slice by the area of a face times the thickness.

Let's attach coordinates to the problem. Set the loaf down along the x-axis and denote the positions of the ends of the loaf by $x = a$ and $x = b$ as shown in Figure 28.2. Slice perpendicular to the x-axis, partitioning $[a, b]$ into n equal pieces, each of length $\Delta x = \frac{b-a}{n}$. Let $x_i = a + i\,\Delta x$, so $a = x_0 < x_1 < x_2 < \cdots < x_n = b$. We'll refer to this as a standard partition of $[a, b]$. Let $A(x_i)$ denote the cross-sectional area cut by a plane perpendicular to the x-axis at x_i. Then

$$\text{volume} = \lim_{n \to \infty} \sum_{i=1}^{n} A(x_i)\,\Delta x = \int_a^b A(x)\,dx.$$

Naturally, this approach can be generalized from a loaf of bread to other solids.

EXAMPLE 28.1 The Egyptians built the Great Pyramid of Cheops in Giza around 2600 B.C. Its original height was about 481 feet and its cross sections are nearly perfect squares. The base is a square whose sides measure about 756 feet.[1] Find its volume.

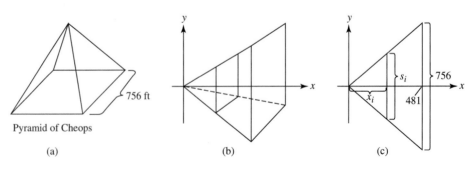

Figure 28.3

SOLUTION In your mind, pick up this massive pyramid and set it along the x-axis so the x-axis runs along the height of the pyramid and pierces the square base at its center. Start chopping at

---

[1] Facts from David Burton, *The History of Mathematics: An Introduction*, McGraw-Hill Companies, Inc., 1997, p 56.

$x = 0$ and stop at $x = 481$, slicing the interval $[0, 481]$ into n equal subintervals of length Δx. Let $x_i = i \Delta x$.

$$\text{volume} = \sum_{i=1}^{n} V_i \approx \sum_{i=1}^{n} A(x_i) \Delta x$$

Our job is to find $A(x_i)$. Let s_i be the side of the square cross section at x_i. Then $A(x_i) = (s_i)^2$. s_i is not constant; it varies with x. We use similar triangles to express s_i in terms of x_i.

$$\frac{s_i}{x_i} = \frac{756}{481}$$

$$s_i = \frac{756}{481} x_i \quad \text{so } A(x_i) \approx \left(\frac{756}{481} x_i \right)^2$$

$$\text{volume} = \lim_{n \to \infty} \sum_{i=1}^{n} A(x_i) \Delta x = \int_0^{481} A(x)\, dx$$

$$\text{volume} = \int_0^{481} \left(\frac{756}{481} x \right)^2 dx$$

$$= \left(\frac{756}{481} \right)^2 \int_0^{481} x^2 dx$$

$$= \left(\frac{756}{481} \right)^2 \frac{x^3}{3} \bigg|_0^{481}$$

$$= \left(\frac{756}{481} \right)^2 \frac{(481)^3}{3}$$

$$= (756)^2 \cdot \frac{481}{3}$$

$$\approx 91{,}636{,}272$$

The volume is approximately 91,636,272 cubic feet.   ◆

◆ **EXAMPLE 28.2**   A tent has a base that is an isosceles triangle. The mouth of the tent is 6 feet wide and 3 feet high; the length is 10 feet. The cross sections perpendicular to the base are all semicircles. What is the volume of the tent?

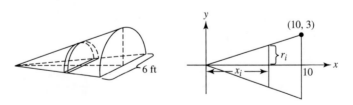

Figure 28.4

SOLUTION   Position the x-axis as shown in Figure 28.4. Volume $= \int_0^{10} A(x)\, dx$, where $A(x_i)$ is the cross-sectional area of a slice at x_i produced by a standard partition of $[0, 10]$ into n equal pieces . $A(x_i) = \frac{1}{2} \pi r_i^2$ because the cross sections are semicircles. We need to express r_i in terms of x_i, so we use similar triangles.

$$\frac{r_i}{x_i} = \frac{3}{10}$$

$$r_i = \frac{3}{10}x_i$$

$$A(x_i) = \frac{1}{2}\pi\left(\frac{3}{10}x_i\right)^2 = \frac{9\pi}{200}x_i^2$$

Therefore the volume is

$$\int_0^{10} \frac{9\pi}{200}x^2\,dx = \frac{9\pi}{200}\frac{x^3}{3}\Big|_0^{10} = \frac{3000\pi}{200} = 15\pi.$$

The volume is $15\pi\,\text{ft}^3$, or approximately 47.12 cubic feet.

REMARK Instead of taking slices we could have noticed that this tent is half of a right circular cone with height 10 and radius 3. Volume $= \frac{1}{2}\left(\frac{1}{3}\pi(3^2)\cdot 10\right) = \frac{1}{2}30\pi = 15\pi$. ◆

Knowing the area of a circle, we can derive the formulas for the volume of a right circular cone and the volume of a sphere using techniques similar to those employed in Examples 28.1 and 28.2. Alternatively, we can treat these objects as volumes of revolution, a viewpoint that sometimes amounts to cross-sectional slicing, as you will see.

Volumes of Revolution

Many familiar objects can be thought of as solids of revolution. A sphere, a cone, a bead, an ellipsoid, and a bagel all can be constructed geometrically by revolving a region in the plane about some axis of revolution. (See Figure 28.5.)

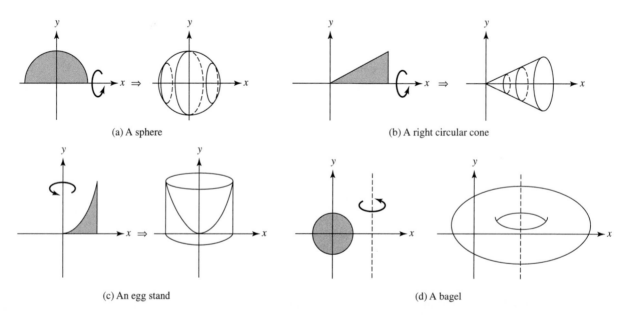

(a) A sphere

(b) A right circular cone

(c) An egg stand

(d) A bagel

Figure 28.5   Volumes of Revolution

In the case of the sphere or the right circular cone positioned as pictured above, we can think of slicing the object just like a loaf of bread, perpendicular to the x-axis . Each slice will have a circular cross-sectional area. The same is true in the case of the volume of the vase pictured in Figure 28.6.

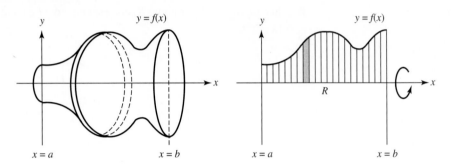

Figure 28.6

An alternative, but equivalent, point of view is to consider the volume as generated by revolving the region R bounded by $y = f(x)$ from $x = a$ to $x = b$ about the x-axis . We can chop up R into n rectangular-like strips as indicated in Figure 28.6, by rotating each strip about the x-axis, and then summing the resultant volumes. We'll see that this viewpoint, chopping the planar region, revolving each "rectangle" to get a volume, and summing the volumes, is quite versatile; it can be used in ways that are not equivalent to slicing bread.

Let's consider the different situations that can arise when rotating a rectangular strip around a vertical or horizontal axis. Three possibilities are given below. In Figure 28.7 the height of the shaded "rectangle" is approximated by $f(x_i)$ or $f(x_i) - g(x_i)$, as indicated, and the base is denoted by Δx.

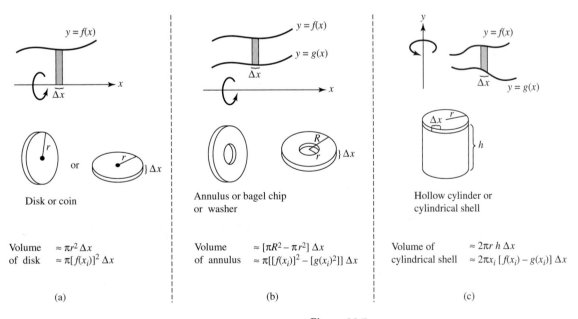

Disk or coin

Annulus or bagel chip
or washer

Hollow cylinder or
cylindrical shell

Volume $\approx \pi r^2 \, \Delta x$
of disk $\approx \pi [f(x_i)]^2 \, \Delta x$

Volume $\approx [\pi R^2 - \pi r^2] \, \Delta x$
of annulus $\approx \pi [[f(x_i)]^2 - [g(x_i)^2]] \, \Delta x$

Volume of $\approx 2\pi r \, h \, \Delta x$
cylindrical shell $\approx 2\pi x_i \, [f(x_i) - g(x_i)] \, \Delta x$

(a)

(b)

(c)

Figure 28.7

Observations

▪ Suppose a solid can be thought of as generated by revolving an area between the graph of $f(x)$ and the x-axis or between $f(x)$ and $g(x)$ around the x-axis. Cases (a) and (b) in Figure 28.7 are equivalent to slicing perpendicular to the x-axis.

▪ In case (b), the big radius, R, corresponds to $f(x_i)$ and the little radius, r, corresponds to $g(x_i)$.

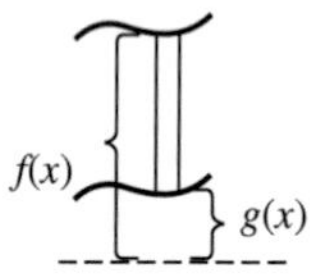

Figure 28.8

CAUTION $\pi(R^2 - r^2) \neq \pi(R - r)^2$

▪ In case (c), r corresponds to the distance from the y-axis and is therefore given by x_i; h corresponds to the height of the rectangle and is given by $f(x_i) - g(x_i)$. To see why the volume of the cylindrical shell is *approximately* $2\pi r h \Delta x$, picture constructing the shell from a sheet of paper. Then unroll the paper. We essentially want the volume of the sheet. Δx corresponds to the paper's thickness.

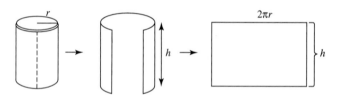

Figure 28.9

◆ **EXAMPLE 28.3**   Show that the volume of a sphere of radius R is $\frac{4}{3}\pi R^3$.

SOLUTION   Rotate the region in the first quadrant bounded by $y = \sqrt{R^2 - x^2}$ and the coordinate axes around the x-axis and double the result.

Chop along the x-axis. Partition $[0, R]$ into n equal pieces, creating a standard partition. Revolving a representative slice gives a "disk."

volume of ith "disk" $= V_i \approx \pi r_i^2 \Delta x$

$$\approx \pi \left[\sqrt{R^2 - x_i^2} \right]^2 \Delta x$$

$$\approx \pi \left(R^2 - x_i^2 \right) \Delta x$$

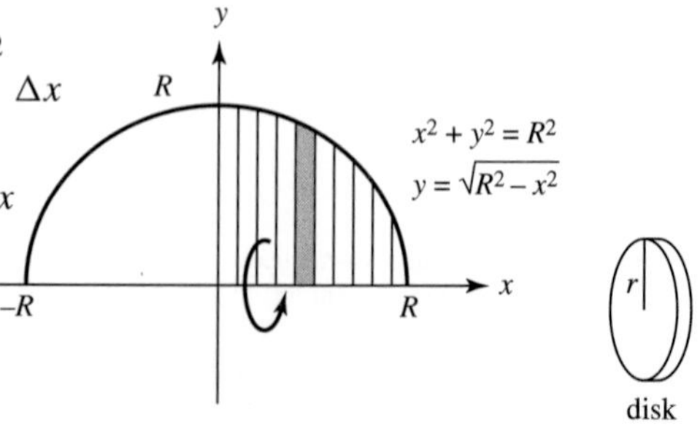

Figure 28.10

$$\text{volume} = 2 \lim_{n \to \infty} \sum_{i=1}^{n} \pi (R^2 - x_i^2) \Delta x$$

$$\text{volume} = 2 \int_0^R \pi (R^2 - x^2) \, dx$$

$$= 2\pi \left[R^2 x - \frac{x^3}{3} \right]_0^R$$

$$= 2\pi \left[R^3 - \frac{R^3}{3} \right]$$

$$= 2\pi \cdot \frac{2R^3}{3}$$

$$= \frac{4\pi R^3}{3}$$

This solution is like slicing a hemisphere as one would slice a bread loaf.   ◆

◆ **EXAMPLE 28.4**   Model a hard-boiled egg holder as follows. Let A be the region in the first quadrant bounded above by $y = x^2$, below by the x-axis, and laterally by $x = 2$. Revolve A about the y-axis to generate the egg holder. What is its volume?

SOLUTION   *Option 1:* Slice A along the x-axis as shown. Partition $[0, 2]$ into n equal pieces using a standard partition. Revolving a representative slice around the y-axis gives a cylindrical shell.[2]

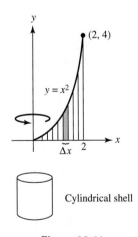

Figure 28.11

---

[2] From this point on we will refer to the geometric object obtained by revolving a "rectangle" about a vertical or horizontal line as a disk, cylindrical shell, or annulus as opposed to "disk," "cylindrical shell," and "annulus." In other words, in our language we will treat pseudorectangles as rectangles.

$$\text{volume of } i\text{th cylindrical shell } \approx 2\pi r_i h_i \Delta x$$

$$r_i = x_i = \text{distance from the } y\text{-axis}$$

$$h_i \approx y_i = x_i^2$$

$$\text{volume of } i\text{th cylindrical shell } \approx 2\pi x_i x_i^2 \Delta x = 2\pi x_i^3 \Delta x$$

So the volume $= \int_0^2 2\pi x^3 \, dx = 2\pi \frac{x^4}{4}\big|_0^2 = \frac{\pi}{2} \cdot 16 = 8\pi$.

Option 2: Slice A along the y-axis as shown. Partition $[0, 4]$ into n equal pieces using a standard partition. Revolving a representative slice gives an annulus.

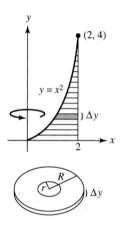

Figure 28.12

$$\text{volume of } i\text{th annulus } = \pi[R_i^2 - r_i^2]\Delta y$$

$$R_i = 2 \text{ (It is constant)}$$

r_i varies. It is given by the x-coordinate of the curve.

Because we have Δy, we want x in terms of y.

The curve is $x = \sqrt{y}$, so $x_i = \sqrt{y_i}$.

$$r_i = \sqrt{y_i}$$

$$\text{volume of } i\text{th annulus } \approx \pi \left[2^2 - (\sqrt{y_i})^2\right] \Delta y = \pi[4 - y_i]\Delta y$$

$$\text{volume} = \int_0^4 \pi[4 - y] \, dy = \pi \left[4y - \frac{y^2}{2}\right]\Big|_0^4 = \pi \left[16 - \frac{16}{2}\right] = 8\pi$$

Notice that we generally have two options on how to slice a region. How we slice the region is not predetermined by how we plan to rotate it. ◆

In Example 28.4, option (2) is equivalent to using a bread slicer along the y-axis. Option (1), however, does not correspond to taking cross sectional slices. It is more along the lines of coring an apple repeatedly, each time readjusting the size of the core.

◆ **EXAMPLE 28.5**   Let's model a bagel by revolving a disk of radius 1 centered at the origin about the vertical line $x = 2$. What is the volume of the bagel?

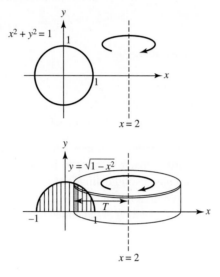

Figure 28.13

SOLUTION   We can slice the disk either along the x-axis or along the y-axis. The former will result in cylindrical shells and the latter in annuli or, more appropriately, bagel chips. We'll opt for slicing along the x-axis.

We can revolve the half-disk in the first two quadrants, obtaining the volume of the top half of the bagel, and doubling it to get the final answer. (Notice that if we revolved the half-disk in quadrants I and IV around the line $x = 2$ we would get less than half the total volume. Why?)

Partition $[-1, 1]$ into n equal pieces using a standard partition.

$$\text{volume of the } i\text{th cylindrical shell } \approx 2\pi r_i h_i \Delta x$$

$$h_i = \sqrt{1 - x_i^2}$$

$$r_i = (\text{the distance between } x = 2 \text{ and } x_i) = 2 - x_i$$

Note that this holds regardless of the sign of x_i.

$$\text{volume of } i\text{th cylindrical shell } \approx 2\pi (2 - x_i)\sqrt{1 - x_i^2}\Delta x$$

$$\text{volume of half of the bagel} = \int_{-1}^{1} 2\pi(2 - x)\sqrt{1 - x^2}\, dx$$

$$= 2\pi \int_{-1}^{1} 2\sqrt{1 - x^2} - x\sqrt{1 - x^2}\, dx$$

$$= 4\pi \left[\int_{-1}^{1} \sqrt{1 - x^2}\, dx \right] - 2\pi \int_{-1}^{1} x\sqrt{1 - x^2}\, dx$$

The first integral we can recognize as giving the area of a semicircle of radius 1.

$$4\pi \int_{-1}^{1} \sqrt{1 - x^2}\, dx = 4\pi \left[\frac{1}{2}\pi(1)^2\right] = 2\pi^2$$

The second integral has an integrand that is an odd function.

$$(-x)\sqrt{1(-x)^2} = -x\sqrt{1 - x^2}$$

Therefore,

$$\int_{-1}^{1} x\sqrt{1 - x^2}\, dx = 0$$

Volume of the whole bagel $= 2 \cdot 2\pi^2 = 4\pi^2$.

Observation.   Suppose this was not a real bagel but only a model made of clay. Suppose further that we chopped it open with a cleaver as shown in Figure 28.14 and opened it up into a solid cylinder of radius 1. The length of the cylinder should be neither the outer circumference of the bagel, nor the inner circumference of the bagel, but the circumference corresponding to the dotted line shown. The volume of this cylinder is the area of the circle, π, times the length, 4π, giving $4\pi^2$. This is precisely the volume of the bagel. In other words, as you unroll the clay there will be cracking on one side and buckling on the other, and they exactly cancel out.

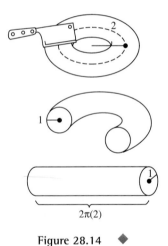

Figure 28.14   ◆

EXERCISE 28.1   Show that this result holds more generally. Rotating the disk $x^2 + y^2 \le a^2$ around the line $x = b$, where $b > a > 0$ generates a torus (the mathematical term for such a bagel-shape) whose volume is $\pi a^2 (2\pi b) = 2\pi^2 a^2 b$.

PROBLEMS FOR SECTION 28.1

1. A tent has a base that is an isosceles triangle. The mouth of the tent measures 8 feet and the length is 12 feet. The tent is constructed so that the cross sections perpendicular to the base are all equilateral triangles. Find the volume of the tent.

2. A Wisconsin cheese factory makes its cheese in solid cylinders of radius 2 inches. A wedge of cheese is cut from the cylinder by chopping through the diameter of the base at an angle of 45 degrees with the base. Find the volume of the wedge of cheese.

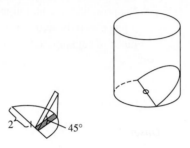

(*Hint:* The base of the wedge is a semicircle of radius 2. Attach a coordinate system to it. The cross-sections are right isosceles triangles.)

3. Find the volume generated when the region in the first quadrant bounded by $y = x^2$ and $y = 3x$ is rotated about

   (a) the x-axis,   (b) the y-axis,   (c) the line $x = -1$.

4. Find the volume generated when the region bounded by the y-axis, $y = x^2$, and $y = 4$ is rotated about the x-axis. Do this in three ways.

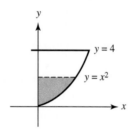

   (a) Chop the shaded region into vertical strips and rotate.
   (b) Chop the shaded region into horizontal strips and rotate.
   (c) Subtract volumes. Subtract the volume generated by rotating the region under $y = x^2$ from that generated by rotating the region under $y = 4$. (The latter is just a cylinder.)

5. Find the volume generated by rotating the region bounded by the x-axis and $y = \sqrt{\sin x}$ for $0 \leq x \leq \pi$ about the line $y = 0$.

6. The region bounded above by $y = \frac{1}{x}$, below by the x-axis, and laterally by $x = \frac{1}{2}$ and $x = 5$ is rotated about the x-axis. Find the volume of the funnel generated.

7. Let A be the region bounded by $y = x^2$ and $y = 4 - x^2$. Find the volume generated by rotating region A about

    (a) the y-axis,   (b) the x-axis.

8. Suppose that a cantaloupe is a perfect sphere with a radius of 5 inches. You slice off a piece as indicated. At its thickest your piece is 2 inches. What's the volume of your piece of cantaloupe?

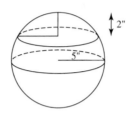

9. Find the volume generated by revolving the region bounded by $y = x^2$ and $y = 4$ about

    (a) the y-axis,   (b) the vertical line $x = 2$,

    (c) the horizontal line $y = 4$,   (d) the horizontal line $y = 5$.

10. A parfait cup is formed by revolving the curve $y = x^3, 0 \le x \le 2$, about the y-axis. The parfait cup is filled to the brim with hot chocolate. If you plan to drink exactly half the hot chocolate in the cup, at what height should the liquid be when you stop drinking?

11. The top 31 feet of the Great Pyramid of Cheops are now missing. What percentage of the structure remains? Refer to Example 28.1 for details.

12. We model a large, 9-inch-tall soup bowl by rotating a region A around the y-axis. A is the region bounded by the y-axis, $y = \frac{1}{4}x^2$, and $y = 9$.

    (a) What is the capacity of the bowl?

    (b) If the bowl was filled to the brim when set out on the table and an hour later was filled only to a height of 4 inches, how much soup was ladled out in the hour?

13. Find the volume of the ellipsoid generated by revolving the ellipse $\frac{x^2}{a^2} + \frac{y^2}{b^2} = 1$ about the x-axis.

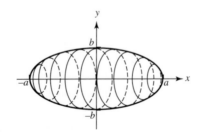

28.2   ARC LENGTH, WORK, AND FLUID PRESSURE: ADDITIONAL APPLICATIONS OF THE DEFINITE INTEGRAL

The strategy of getting a grip on something by partitioning, approximating, summing and computing a limit is applicable in a variety of situations. If we arrive at the limit of a Riemann sum, then we can solve the problem using a definite integral. In this section we apply this strategy to the problems of computing the length of a plane curve, computing the work done by a variable force acting on an object, and computing the pressure exerted by a fluid.

Arc Length

Not only is the definite integral useful in computing areas and volumes, but it is a tool for computing arc length as well.

Suppose f and f' are continuous on $[a, b]$. Our goal is to find the length of the curve $y = f(x)$ for $a \leq x \leq b$.

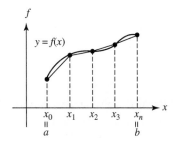

Figure 28.15

We partition $[a, b]$ into n equal pieces, each of length $\Delta x = \frac{b-a}{n}$. Let $x_i = a + i \Delta x$, for $i = 0, \ldots, n$ and $P_i = (x_i, y_i)$, where $y_i = f(x_i)$, $i = 0, \ldots, n$.

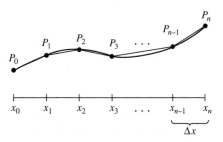

Figure 28.16

We approximate the length of the curve on the ith interval by s_i, the length of the line segment joining P_{i-1} and P_i.

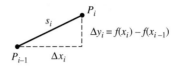

Figure 28.17

$$s_i = \sqrt{(\Delta x_i)^2 + (\Delta y_i)^2},$$

where $\Delta x_i = \Delta x$ and $\Delta y_i = y_i - y_{i-1}$. In order to arrive at a Riemann sum, we need an expression for s_i of the form $g(x_i)\Delta x$.

$$s_i = \sqrt{(\Delta x_i)^2 + (\Delta y_i)^2} \frac{\Delta x_i}{\Delta x_i}$$

$$= \sqrt{\frac{(\Delta x_i)^2 + (\Delta y_i)^2}{(\Delta x_i)^2}} \Delta x_i$$

$$= \sqrt{\left(\frac{\Delta x}{\Delta x}\right)^2 + \left(\frac{\Delta y_i}{\Delta x_i}\right)^2} \Delta x_i$$

$$= \sqrt{1 + \left(\frac{\Delta y_i}{\Delta x_i}\right)^2} \Delta x_i$$

Because we've made a uniform partition, all the Δx_i's are equal; the Δy_i's however are not. $\Delta y_i = f(x_i) - f(x_{i-1})$.

$$\text{As } n \to \infty, \Delta x \to 0 \quad \text{and} \quad \frac{\Delta y_i}{\Delta x_i} \to f'(x_i).$$

For Δx very small,[3] $s_i \approx \sqrt{1 + [f'(x_i)]^2}\Delta x$.

$$\text{arc length} \approx \sum_{i=1}^{n} s_i \approx \sum_{i=1}^{n} \sqrt{1 + [f'(x_i)]^2}\Delta x$$

Taking the limit as $n \to \infty$, we obtain

$$\text{arc length} = \int_a^b \sqrt{1 + [f'(x)]^2}\, dx.$$

Arc length:
We conclude that if f is differentiable on $[a, b]$ and f' is continuous on $[a, b]$, then the length of the graph of f from $x = a$ to $x = b$ is given by

$$\int_a^b \sqrt{1 + [f'(x)]^2}\, dx.$$

---

[3] Using the Mean Value Theorem it can be shown that $s_i = \sqrt{1 + [f'(x_i^*)]^2}\Delta x$ for some x_i^* in the ith interval even if Δx is not very tiny.

◆ **EXAMPLE 28.6**   Find the length of the curve given by $f(x) = \frac{1}{3}x^{3/2}$ on the interval [0, 2].

SOLUTION $f'(x) = \frac{1}{2}x^{1/2}$

$$\text{Arc length} = \int_0^2 \sqrt{1 + \left[\frac{x^{1/2}}{2}\right]^2}\, dx$$

$$= \int_0^2 \sqrt{1 + \frac{x}{4}}\, dx$$

$$= \frac{1}{2}\int_0^2 \sqrt{4 + x}\, dx$$

$$\text{Let } u = 4 + x \qquad \text{when } x = 0, u = 4$$
$$du = dx \qquad \text{when } x = 2, u = 6.$$

$$= \frac{1}{2}\int_4^6 u^{1/2}\, du$$

$$= \frac{1}{2}\frac{2u^{3/2}}{3}\Big|_4^6$$

$$= \frac{1}{3}\left(6^{3/2} - 4^{3/2}\right)$$

$$= \frac{1}{3}\left(6^{3/2} - 8\right) \qquad ◆$$

Example 28.6 was carefully chosen to give a straightforward integral. Usually the integrals obtained are not easy to evaluate exactly. Numerical methods together with a calculator or computer allow us to approximate them.

EXERCISE 28.2   Verify that the formula derived tells us that the length of $y = x$ from $x = 0$ to $x = 1$ is $\sqrt{2}$.

Work

How much work is needed to pump all the water out of a swimming pool? How much work is needed to compress a heavy spring? How much work is needed to launch a rocket? In order to make such calculations we need to clarify the meaning of work from the framework of a physicist.

If a constant force F acting in the direction of motion of an object causes a displacement d, then the work done by the force on the object is defined to be

$$\text{work} = \text{force} \cdot \text{displacement}$$
$$W = F \cdot d.$$

Notice that the physicists' definition of work differs from our colloquial use of the word. For example, if you wait at a bus stop for 15 minutes while holding your 25-pound nephew, you might feel tired but according to the physicist you've done no work. Your nephew hasn't moved. And if you give up on the bus and walk to the grocery store, carrying your nephew the whole way, your arms have still not done any work. Your arms have exerted upward force on your nephew; his motion, however, has been horizontal. Even worse, if you get tired and set him down after a few minutes, you've done negative work. If you lift him from his position around your waist to a perch on your shoulders, you and the physicist both agree that you have done some work. If your shoulder is 1.5 feet above your waist, you've done 1.5 feet × 25 pounds = 37.5 foot-pounds of work.

Units of Measure. Because work equals force times distance, in the English system it can be measured in foot-pounds, as above. Scientists generally measure force in newtons,[4] abbreviated N. A newton-meter is called a joule. To obtain newtons from kilograms (a unit of mass), apply Newton's second law:

$$\text{force} = \text{mass} \cdot \text{acceleration}$$
$$= m \cdot g,$$

where g is the acceleration due to gravity. Near the surface of the earth g is approximately 9.8 m/sec^2.

If a constant force is acting in the direction of motion, calculating work simply amounts to a multiplication problem. Consider a variable force acting in a straight line over a fixed distance. For example, consider compressing a spring. The applied force must increase as the spring becomes more compressed. Situate the x-axis so that the force is applied along it from $x = a$ to $x = b$. Let $F(x)$ be a continuous function giving the force applied at position x. To calculate the work done we partition $[a, b]$ into n equal pieces, each of length $\Delta x = \frac{b-a}{n}$ and let $x_i = a + i \Delta x$ for $i = 0, 1, \ldots, n$. On each subinterval we treat the force as approximately constant. This is reasonable because F is continuous.

$$\text{work done in } i\text{th subinterval } \approx F(x_i)\Delta x$$

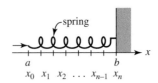

Figure 28.18

Total work done $\approx \sum_{i=1}^{n} F(x_i)\Delta x$. The approximation becomes better and better as n increases.

$$\lim_{n \to \infty} \sum_{i=1}^{n} F(x_i)\Delta x = \int_a^b F(x)\, dx$$

We conclude that if a continuous force F acts along the x-axis from $x = a$ to $x = b$ the work done by the force is given by

$$\int_a^b F(x)\, dx.$$

◆ **EXAMPLE 28.7**   A wheelbarrow is being pushed along a straight forest trail with a force of $F(x) = 10 + 9x - x^2$ pounds, where x is the distance from the trailhead. How much work is done in moving the wheelbarrow the first 9 feet?

---

[4]One newton corresponds to approximately 0.225 pound.

SOLUTION

$$\text{Work} = \int_0^9 (10 + 9x - x^2)\, dx$$

$$= 10x + \frac{9x^2}{2} - \frac{x^3}{3} \Big|_0^9$$

$$= 90 + 364.5 - 243 = 211.5 \text{ ft-lb.} \quad \blacklozenge$$

Let's return to the spring problem. Hooke's law, an experimentally derived law, says that the force necessary to compress or stretch a spring x units from its natural length is proportional to x.

$$F = kx,$$

where k is called the spring constant. Hooke's law is followed provided the elastic limit of the spring is not exceeded. Example 28.8 applies this law.

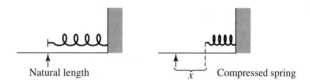

Natural length x Compressed spring

Figure 28.19

◆ **EXAMPLE 28.8**   An industrial-strength spring is being used to cushion the impact of packages as they come hurling off a conveyor belt. The spring has a spring constant of 1000 N/m. Find the work done by a package that compresses the spring 0.4 m.

SOLUTION

$$\text{Work} = \int_0^{0.4} 1000x\, dx$$

$$= \frac{1000x^2}{2} \Big|_0^{0.4} = 500 \cdot 0.16 = 80 \text{ N-m, or 80 joules.} \quad \blacklozenge$$

◆ **EXAMPLE 28.9**   A pulley system is being used to lift granite out of a quarry. If the cable used weighs 2 pounds per foot length, how much work is done lifting a 300-pound block of granite 150 feet to the level of the pulley? Assume friction is negligible.

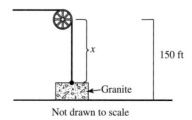

x 150 ft

Granite

Not drawn to scale

Figure 28.20

SOLUTION   Let $x =$ the length of cable between the granite and the pulley. x will vary between 150 feet and 0 feet. The force that must be applied to lift the granite varies.

$$\text{force} = (\text{weight of granite}) + (\text{weight of cable})$$
$$= 300 + 2x$$
$$\text{Work} = \int_0^{150} (300 + 2x)\, dx$$
$$= 300x + \frac{2x^2}{2}\Big|_0^{150}$$
$$= 300(150) + (150)^2 = 45{,}000 + 22{,}500 = 67{,}500$$

67,500 foot-pounds of work are required.   ◆

◆ **EXAMPLE 28.10**   A cylindrical above-ground swimming pool has side walls 1.5 m tall, a radius of 4 m, and is filled to a height of 1.4 m. How much work is required to pump water over the top and

(a) empty the pool?

(b) reduce the water level to 0.7 m?

SOLUTION   Our basic strategy is to slice along the vertical axis, chopping the water into thin solid disks. Think of pumping the water out one disk at a time. We'll calculate the work done to lift the ith disk to the top of the pool, sum over all the relevant slices, and, letting the number of slices increase without bound, get the limit of a Riemann sum.

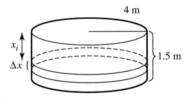

Figure 28.21

Let $x =$ the distance (in meters) between the top of the pool and the water to be lifted.

(a) To empty the pool, x must vary between 0.1 m and 1.5 m. Partition $[0.1, 1.5]$ into n equal pieces, where $\Delta x = \frac{1.4}{n}$ and $x_i = .1 + i\,\Delta x$ for $i = 0, \ldots, n$. The force required to lift a slice is the force acting against gravity; it is equal to the weight of the slice.

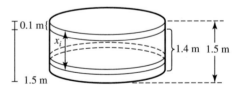

Figure 28.22

$$\text{amount of work to lift up the } i\text{th slice} \approx (\text{weight of } i\text{th slice}) \cdot (\text{distance to the top})$$

The volume of the ith slice $= \pi(4)^2 \Delta x = 16\pi \Delta x \text{ m}^3$.

The density of water is 1000 kg/m^3. Therefore

$$\text{the mass of the } i\text{th slice} = 1000 \cdot 16\pi \Delta x \text{ kg}.$$

We calculate the weight of the slice by multiplying by $g \approx 9.8 \text{ m/sec}^2$.

$$\text{The weight of the } i\text{th slice} \approx 1000 \cdot 16\pi \Delta x \cdot 9.8 \text{ newtons}$$
$$\approx 156{,}800\pi \Delta x \text{ newtons}$$

amt. of work to lift up the ith slice $\approx 156{,}800\pi \Delta x \cdot x_i$

$$\text{total work} = \lim_{n \to \infty} \sum_{i=1}^{n} 156{,}800\pi \cdot x_i \Delta x = \int_{0.1}^{1.5} 156{,}800\pi x \, dx$$

$$\text{total work} = 156{,}800\pi \left. \frac{x^2}{2} \right|_{0.1}^{1.5} = 156{,}800\pi \left[1.125 - 0.005 \right] = 175{,}616\pi$$

The work done is $175{,}616\pi$ joules, or $\approx 551{,}713.9$ joules.

(b) To reduce the water level to 0.7 m, to pump half the water out, requires partitioning the interval $[0.1, 0.8]$ into n equal pieces because the water slices must move between 0.1 and 0.8 m.

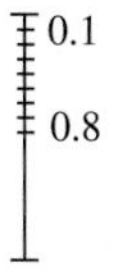

0.1

0.8

Figure 28.23

$$\int_{0.1}^{0.8} 156{,}800\pi x \, dx = 156{,}800\pi \left. \frac{x^2}{2} \right|_{0.1}^{0.8}$$
$$= 156{,}800\pi [0.32 - 0.005]$$
$$= 49{,}392\pi \text{ joules, or } \approx 155{,}169.5 \text{ joules}$$

Notice that it takes substantially less work to reduce the water depth from 1.4 m to 0.7 m than it does to reduce it from 0.7 m to 0 m. ◆

Fluid Pressure and Force

Water is an invaluable resource of the earth, and civilizations have devoted much scientific energy toward understanding it. We build bridges to travel over it, tunnels to travel under it, dams to contain it, plumbing, draining, and irrigation systems to direct it, and submarines to explore its depths. Understanding fluid pressure (force per unit area) is critical to the success of these projects.

> The fundamental principle is that fluid pressure is directly proportional to depth and, when a system is at equilibrium, it is the same in all directions. Pressure $= \left(\substack{\text{fluid} \\ \text{density}}\right) \cdot (\text{depth})$.

This means that given a submerged object in an equilbrium state, the fluid pressure down, the fluid pressure up, and the horizontal fluid pressures on the object are all the same.

> The hydrostatic force exerted by a fluid on a plane surface of area A situated horizontally at a depth d is given by
>
> $$\text{force} = (\text{pressure}) \cdot (\text{area}) = \rho d \cdot A,$$
>
> where ρ is the density of the liquid, its weight per unit volume.

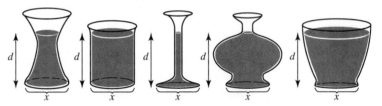

The hydrostatic force on the base of every bottle of water is the same

Figure 28.24

This means that if a bottle whose base has area A is filled with fluid to a height d, the fluid force on the base is unaffected by the shape of the bottle. (See Figure 28.24.)

NOTE In the metric system, water density is given as 1000 kg/m^3. To get weight per unit volume, multiply by g.

$$\text{water density} = (9.8)(1000) = 9800 \text{ newtons/m}^3.$$

In the English system,

$$\text{water density} = 62.4 \text{ lbs/ft}^3.$$

When building a dam or an underwater support for a bridge, the force of the water is more complicated to calculate because the structure is submerged vertically or at an angle as opposed to horizontally; the pressure increases with the depth.

◆ **EXAMPLE 28.11**   Suppose a particular dam has the shape of a parabola 40 feet wide at the top and 50 feet deep. Calculate the hydrostatic force on the dam when the water level is 5 feet from the top of the dam.

SOLUTION   We establish a coordinate system as shown in Figure 28.25 and take horizontal slices so that the depth does not vary much within a slice. Partition $[0, 45]$ into n equal pieces, each of height $\Delta y = \frac{45}{n}$. Let $y_i = i\,\Delta y$, for $i = 0, 1 \ldots, n$.

$$\begin{array}{c}\text{the force on the}\\ i\text{th strip}\end{array} \approx (\text{water density in lb/ft}^3) \cdot \left(\begin{array}{c}\text{depth of}\\ i\text{th strip}\end{array}\right) \cdot \left(\begin{array}{c}\text{area of}\\ i\text{th strip}\end{array}\right)$$

$$\approx \left(62.4 \text{ lbs/ft}^3\right) \cdot (45 - y_i) \text{ ft} \cdot \left(\begin{array}{c}\text{area of the}\\ i\text{th strip}\end{array}\right)$$

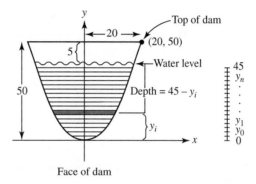

Figure 28.25

We can approximate the area of the horizontal slices by treating them as rectangles of height Δy and width $2x_i$. To express the width in terms of y we find the equation of the parabola. (This is where the dimensions of the dam come into play.)

$$y - kx^2 \quad \text{because we've put the vertex of the parabola at the origin.}$$

$$50 = k(20)^2 \Longrightarrow k = \frac{50}{400} = \frac{1}{8}$$

$$y = \frac{1}{8}x^2 \quad \text{and} \quad x = \pm\sqrt{8y}$$

Therefore, the area of the ith strip $\approx 2x_i \Delta y = 2\left(\sqrt{8y_i}\right)\Delta y = 4\sqrt{2}\sqrt{y_i}\,\Delta y$ ft^2

the force on the ith strip $\approx (62.4 \text{ lbs/ft}^3) \cdot (45 - y_i)\text{ft} \cdot 4\sqrt{2}\sqrt{y_i}\,\Delta y$ ft^2

$$\text{total force} = \lim_{n \to \infty} \sum_{i=1}^{n} 62.4(45 - y_i)4\sqrt{2}\sqrt{y_i}\,\Delta y$$

$$= \int_0^{45} \underbrace{62.4}_{} \underbrace{(45 - y)}_{} \underbrace{4\sqrt{2}\sqrt{y}\,dy}_{}$$

$$\text{fluid density} \cdot \text{depth} \cdot \text{area}$$

$$= 249.6\sqrt{2} \int_0^{45} (45y^{1/2} - y^{3/2})\,dy$$

$$= 249.6\sqrt{2} \left[45 \cdot \frac{2}{3}y^{3/2} - \frac{2}{5}y^{5/2} \right] \Big|_0^{45}$$

$$= 249.6\sqrt{2} \left[30 \cdot 45^{3/2} - \frac{2}{5} \cdot 45^{5/2} \right]$$

$$\approx 1{,}278{,}673 \text{ lb} \quad \blacklozenge$$

There are many other applications of integration. Integrals can arise in physics when computing moments, centroids, or centers of mass. They can arise in economics when computing consumer surplus and the present value of an income stream. They arise in biology when computing blood flow through a blood vessel. Integration is used to calculate the surface area of a surface of revolution. It is used extensively in statistics and probability. This is not an exhaustive list!

Pick an application of integration that interests you and learn about it. You have the basic tools you need in order to understand the exposition. You might try to read about it from a few sources—for instance, a calculus textbook and a book in the relevant discipline.

PROBLEMS FOR SECTION 28.2

1. Using the arc length formula, find the length of the line $f(x) = mx + b$ from $x = 0$ to $x = 3$. Confirm that this answer agrees with the distance formula.

   In Problems 2 through 4, approximate each length with error less than 0.05.

2. The length of one arc of the cosine curve, say from $x = \frac{-\pi}{2}$ to $x = \frac{\pi}{2}$

3. The length of $f(x) = x^2$ from $x = -1$ to $x = 1$

4. The length of $y = \ln x$ from $x = 1$ to $x = e$

5. Find the work done in pushing a stroller 200 feet along a path by applying a constant force of 12 pounds in the direction of motion.

6. A force of 5 pounds will stretch a certain spring 3 inches beyond its natural length.
   (a) What is the value of the spring constant?
   (b) How much work is done in stretching the spring from its natural length to 3 inches beyond its natural length?

(c) How much work is done in stretching the spring from 3 inches beyond its natural length to 6 inches beyond its natural length?

(d) Why is the answer to part (c) larger than the answer to part (b)?

7. Suppose 1.2 foot-pounds of work are required to compress a spring 2 inches.

(a) How much work is required to stretch this spring 2 inches from its equilibrium position?

(b) How much work is required to stretch this spring 4 inches from equilibrium?

(c) How much work is required to stretch the spring 5 inches from its equilibrium position?

8. A window washer weighing 160 pounds is attached to a rope hanging from the roof of the building whose windows he is washing. The rope weighs 0.6 lb/ft. Right now he is working 50 feet down from the rooftop.

(a) How much work is required to bring him to the windows that are 25 feet from the rooftop?

(b) How much work will it take to bring him from where he is to the roof?

9. Amelia and Beulah are city dwellers who have set up pulley systems to get their groceries delivered without walking the stairs. Amelia pulls her basket filled with 12 pounds of groceries up to her 40-foot-high balcony. Beulah pulls her basket filled with 16 pounds of cleaning supplies up to her 30-foot-high window. Assuming both women use ropes weighing 0.2 lb/ft, whose task requires more work? How much more work? (Assume friction is negligible.)

10. A rectangular swimming pool is 4 feet deep, 75 feet long, and 25 feet wide. It is completely filled with water.

(a) What is the hydrostatic force on the bottom of the pool?

(b) What is the hydrostatic force on the shorter wall of the pool?

(c) What is the hydrostatic force on the longer wall of the pool?

(d) How much work is required to pump all the water out of the pool (i.e., up to the top of the pool)?

11. A hemispherical tank with a radius of 7 feet is filled to a height of 6 feet with gasoline. How much work is required to pump all the gasoline over the top? The weight-density of gasoline is 42 lb/ft^3.

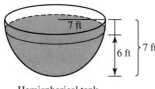

Hemispherical tank

12. As part of the pasteurization and homogenization process, milk is stored in a large tank as shown on the following page.

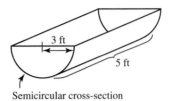

Semicircular cross-section

The next step in the processing requires the milk to be pumped out of this holding tank. If the tank is filled to the brim, how much work is required to empty it? The weight-density of milk is 64.5 lb/ft^3.

13. Refer to Problem 12. What is the force exerted by the milk on the semicircular side wall of the holding tank?

14. A dam in a canyon has the shape of an isosceles triangle. Its depth is 30 feet and its width is 15 feet. The water level is 2 feet below the top of the dam. What is the total hydrostatic force on the face of the dam?

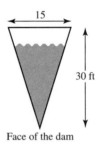

Face of the dam

15. A dam has the shape of an isosceles trapezoid. It is 40 feet deep. Its base is 30 feet long while on the surface it measures 60 feet across. (Its thickness is irrelevant to the problem.) If the water reaches the top of the dam, what is the total hydrostatic force on the face of the dam?

16. *Project* Pick an application of integration not explicitly discussed in this text and learn about it. Then either write an exposition of the application or prepare a short lesson on it. Include examples.

C H A P T E R

29

Computing Integrals

In this chapter we introduce techniques of integration that expand our computational ability. Section 29.1 presents integration by parts, the integration analogue of the Product Rule. Sections 29.2 and 29.3 introduce methods of transforming a difficult integrand into something more manageable. In the final section of this chapter we expand our notion of integration to encompass unbounded integrands and/or an unbounded interval of integration.

29.1 INTEGRATION BY PARTS—THE PRODUCT RULE IN REVERSE

Integration by parts is a method of integration derived from the Product Rule for differentiation. This technique will help us evaluate integrals such as $\int x \sin x \, dx$, $\int xe^x \, dx$, $\int x \ln x \, dx$, $\int \ln x \, dx$, and $\int \tan^{-1} x \, dx$. It is a useful technique for integrands involving the product of different types of functions as well as those involving a single intractable function, and it is a powerful tool for transforming an integral.

The Product Rule tells us that if f and g are differentiable functions, then

$$\frac{d}{dx}(f(x)g(x)) = f(x)g'(x) + g(x)f'(x).$$

Let's denote $f(x)$ by u and $g(x)$ by v. Then, using the differential notation introduced in the substitution section, we can write

$$du = f'(x)\,dx \quad \text{and} \quad dv = g'(x)\,dx.$$

The Product Rule becomes

877

$$d(uv) = u\,dv + v\,du.$$

$$\int d(uv) = \int u\,dv + \int v\,du$$

$$uv = \int u\,dv + \int v\,du$$

This can be rearranged to give what is referred to as the **formula for integration by parts.**

$$\boxed{\text{Integration by parts: } \int u\,dv = uv - \int v\,du}$$

This formula expresses the original integral in terms of an expression involving a new integral. Our aim is to arrive at a new integral that is easier to compute than the original.

Given an integral $\int h(x)\,dx$, the general idea is to choose u and dv such that

$$\int h(x)\,dv = \int u\,dv \qquad \text{and}$$

■   we can find $v = \int dv$,

■ du is simpler than u, and

■ $\int v\,du$ is simpler (or no more difficult) than $\int u\,dv$.

Indefinite Integrals Using Integration by Parts

◆ **EXAMPLE 29.1**   Find $\int xe^x\,dx$.

SOLUTION

$$\text{Let } u = x, \qquad dv = e^x\,dx$$

$$\text{Then } du = dx, \qquad v = \int e^x\,dx = e^x.$$

$$\int \underbrace{x}_{u}\,\underbrace{e^x\,dx}_{dv} = \underbrace{x}_{u}\,\underbrace{e^x}_{v} - \int \underbrace{e^x}_{v}\,\underbrace{dx}_{du} = xe^x - e^x + C$$

REMARKS

1. Notice that we wrote $v = \int e^x\,dx = e^x$, a particular antiderivative of e^x, as opposed to $e^x + C$, the family of antiderivatives. Try letting $v = e^x + C_1$ and verify that the answer remains the same. In Exercise 29.1 you'll verify that choosing any one antiderivative always suffices.

2. Suppose we try the following.

$$\text{Let } u = e^x, \qquad dv = x\,dx$$

$$du = e^x\,dx, \quad v = \frac{x^2}{2}.$$

We obtain

$$\int xe^x\,dx = e^x\left(\frac{x^2}{2}\right) - \frac{1}{2}\int x^2 e^x\,dx.$$

This new integral is more complex than the original integral, indicating a poor choice of u and dv. (Notice that these choices do not conform to the guidelines set out on the previous page.)

3. Given $\int xe^x \, dx$, we would like to get rid of the "extra x." Integration by parts is a mechanism that essentially allows us to do that. ◆

EXERCISE 29.1   Show that replacing v by $v + C$ in the formula for integration by parts does not alter the result. (You will end up getting $uC - uC$.)

◆ EXAMPLE 29.2   Find $\int x \ln x \, dx$.

SOLUTION

$$\text{Let } u = \ln x, \quad dv = x \, dx$$

$$du = \frac{1}{x} \, dx, \quad v = \frac{x^2}{2}.$$

$$\int x \ln x \, dx = \int \underbrace{\ln x}_{u} \cdot \underbrace{x \, dx}_{dv} = (\ln x) \cdot \frac{x^2}{2} - \frac{1}{2} \int x^2 \cdot \frac{1}{x} \, dx$$

$$= \frac{1}{2} x^2 \ln x - \frac{1}{2} \int x \, dx$$

$$= \frac{1}{2} x^2 \ln x - \frac{1}{4} x^2 + C$$

REMARK Both x and $\ln x$ become "simpler" when differentiated, but we choose $u = \ln x$ because x is much easier to integrate than $\ln x$. ◆

◆ EXAMPLE 29.3   Find $\int \ln x \, dx$.

SOLUTION

$$\text{Let } u = \ln x, \quad dv = dx$$

$$du = \frac{1}{x} \, dx, \quad v = x.$$

$$\int \ln x \, dx = x \ln x - \int x \cdot \frac{1}{x} \, dx$$

$$= x \ln x - \int 1 \, dx$$

$$= x \ln x - x + C$$

REMARK Notice that in this example the entire integrand is denoted by u. The approach when finding $\int \tan^{-1} \, dx$ or $\int \sin^{-1} \, dx$ or $\int (\ln x)^2 \, dx$ is similar. Integration by parts is amazingly effective in transforming these difficult integrals into simple ones. ◆

Repeated Use of Integration by Parts

Sometimes we have to use integration by parts multiple times. As long as the new integrals that arise are not more difficult than the original we may be on the right track.

◆ **EXAMPLE 29.4**   Find $\int x^2 e^x \, dx$.

SOLUTION

$$\text{Let } u = x^2, \qquad dv = e^x \, dx$$
$$du = 2x \, dx, \qquad v = e^x.$$

$$\int x^2 e^x \, dx = x^2 e^x - \int e^x \cdot 2x \, dx$$

$$= x^2 e^x - 2 \int x e^x \, dx$$

The integral $\int x e^x \, dx$ can be done by using integration by parts, as shown in Example 29.1. Therefore,

$$\int x^2 e^x \, dx = x^2 e^x - 2(x e^x - e^x) + C \quad \text{(by Example 29.1)}$$

$$= x^2 e^x - 2x e^x + 2e^x + C$$
$$= e^x (x^2 - 2x + 2) + C.$$

REMARKS

1. When doing integration by parts multiple times it is easy to make sign errors. Be cautious—use parentheses.

2. Sometimes you may end up with a constant of $-2C$, $\frac{1}{3}C$, etc. You may replace $-2C$ by C_1. Sometimes $-2C$ is simply replaced by C because C is an *arbitrary* constant.

3. To find $\int x^4 e^x \, dx$ we would do integration by parts four times.   ◆

EXERCISE 29.2   Use integration by parts to derive the reduction formula

$$\int x^n e^{ax} \, dx = \frac{1}{a} x^n e^{ax} - \frac{n}{a} \int x^{n-1} e^{ax} \, dx$$

where a and n are non-zero constants.

In the following example we use integration by parts twice, but the resulting integral is no simpler than the original. In fact, it is the same. You may at first think that we've gotten nowhere, but watch how we can solve algebraically for the integral in question.

◆ **EXAMPLE 29.5**   Find $\int e^{2x} \sin 3x \, dx$.

SOLUTION

$$\text{Let } u = e^{2x}, \qquad dv = \sin 3x \, dx$$
$$du = 2e^{2x} \, dx, \quad v = -\frac{1}{3} \cos 3x \, dx.$$

$$\int e^{2x} \sin 3x \, dx = -\frac{1}{3} e^{2x} \cos 3x - \int \left(-\frac{1}{3} \cos 3x \right) \cdot 2e^{2x} \, dx$$

$$= -\frac{1}{3} e^{2x} \cos 3x + \frac{2}{3} \int e^{2x} \cos 3x \, dx$$

The new integral is no simpler than the original, but it is no more difficult either. We do integration by parts again.

$$\text{Let } u = e^{2x}, \qquad dv = \cos 3x \, dx$$

$$du = 2e^{2x} \, dx, \quad v = \frac{1}{3} \sin 3x$$

$$\int e^{2x} \sin 3x \, dx = -\frac{1}{3} e^{2x} \cos 3x + \frac{2}{3} \left[\frac{1}{3} e^{2x} \sin 3x - \frac{1}{3} \int \sin 3x \cdot 2e^{2x} \, dx \right]$$

Then

$$\int \underbrace{e^{2x} \sin 3x \, dx}_{\text{I}} = -\frac{1}{3} e^{2x} \cos 3x + \frac{2}{9} e^{2x} \sin 3x - \frac{4}{9} \underbrace{\int e^{2x} \sin 3x \, dx}_{\text{I}} .$$

The original integral and the integral on the right are identical. We can solve algebraically for it. If you like you can denote the original integral by **I** and solve for **I**.

$$\frac{13}{9} \int e^{2x} \sin 3x \, dx = -\frac{1}{3} e^{2x} \cos 3x + \frac{2}{9} e^{2x} \sin 3x \qquad (29.1)$$

$$\int e^{2x} \sin 3x \, dx = -\frac{3}{13} e^{2x} \cos 3x + \frac{2}{13} e^{2x} \sin 3x + C$$

REMARKS

1. Notice that in Equation (29.1) we are missing a "+ C." This is because we never actually computed $\int v \, du$. We've got to insert a + C.

2. One of the key ingredients that makes this method work is that if $\sin x$ either is differentiated or integrated twice then we obtain $\sin x$ again, and similarly with e^x.

3. This integral could have been solved similarly by letting $u = \sin 3x$ and $du = e^{2x} \, dx$ and repeating this. It will not, however, be productive to let $u = \sin 3x$ the first time and then $u = e^{2x}$ the second time.   ◆

EXERCISE 29.3   Find $\int e^{2x} \sin 3x \, dx$ by letting u be the trigonometric function. Then try using integration by parts twice but letting $u = e^{2x}$ the first time and $u = \sin 3x$ the second time in order to verify that this is unproductive.

EXERCISE 29.4   The integral $\int \sin^2 x \, dx$ can be computed most easily using the trigonometric identity $\sin^2 x = \frac{1}{2}(1 - \cos 2x)$. However, it can also be solved using the techniques of Example 29.5. Do the latter. Check your answer by differentiating.

In Example 29.4 and Exercise 29.2 we saw that integration by parts can be used to reduce the complexity of an integral. There are numerous reduction formulas that can be derived by integration by parts. Exercise 29.5 below leads you through one such derivation.

EXERCISE 29.5   Derive the reduction formula

$$\int \sin^n x \, dx = -\frac{1}{n} \cos x \sin^{n-1} x + \frac{n-1}{n} \int \sin^{n-2} x \, dx,$$

where n is an integer, $n \geq 2$ as follows.

(a) Use integration by parts, letting $u = \sin^{n-1} x$ and $dv = \sin x \, dx$.

(b) In the resulting new integral, replace $\cos^2 x$ by $1 - \sin^2 x$ to obtain an equation with $\int \sin^n x \, dx$ on both sides.

(c) Solve algebraically for $\int \sin^n x \, dx$, as was done in Example 29.5.

Using Integration by Parts in Definite Integrals

To figure out how to adapt the integration by parts formula

$$\int u \, dv = uv - \int v \, du$$

to definite integrals, let's return to functions of x. Recall that $u = f(x)$ and $v = g(x)$. Then $du = f'(x) \, dx$ and $dv = g'(x) \, dx$ and the formula becomes

$$\int f(x)g'(x) \, dx = f(x)g(x) - \int g(x)f'(x) \, dx.$$

Suppose we want to compute the definite integral $\int_a^b f(x)g'(x) \, dx$. Assuming f' and g' are continuous, we simply evaluate between $x = a$ and $x = b$.

$$\int_a^b f(x)g'(x) \, dx = f(x)g(x) \Big]_a^b - \int_a^b g(x)f'(x) \, dx$$

In other words,

$$\int_{x=a}^{x=b} u \, dv = uv \Big]_{x=a}^{x=b} - \int_{x=a}^{x=b} v \, du.$$

This is equivalent to doing the indefinite integral and taking the difference between the antiderivative at $x = b$ and the antiderivative at $x = a$.

◆ **EXAMPLE 29.6**   Evaluate $\int_0^1 \tan^{-1} x \, dx$.

SOLUTION

$$\text{Let } u = \tan^{-1} x, \quad dv = dx$$

$$du = \frac{1}{1 + x^2} \, dx, \quad v = x.$$

$$\int_0^1 \tan^{-1} x \, dx = x \tan^{-1} x \Big]_0^1 - \int_0^1 \frac{x}{1 + x^2} \, dx$$

You may be able to do the latter integral in your head once you realize that the numerator is a constant multiple of the derivative of the denominator. Otherwise, use substitution.

$$\text{Let } w = 1 + x^2$$

$$dw = 2x \, dx$$

$$x \, dx = \frac{1}{2} \, dw.$$

$$\int \frac{x}{1+x^2}\,dx = \frac{1}{2}\int \frac{1}{w}\,dw = \frac{1}{2}\ln|w| + C$$

$$= \frac{1}{2}\ln(1+x^2) + C$$

$$\int_0^1 \tan^{-1}x = x\tan^{-1}x\Big]_0^1 - \frac{1}{2}\ln(1+x^2)\ \Big|_0^1$$

$$= \tan^{-1}1 - 0 - \left[\frac{1}{2}\ln(1+1) - \frac{1}{2}\ln(1)\right]$$

$$= \frac{\pi}{4} - \frac{1}{2}\ln 2 \quad \blacklozenge$$

The choice of u and dv sometimes requires a fair amount of thinking ahead. This is illustrated in the next example.

◆ **EXAMPLE 29.7**   Evaluate $\int_0^1 x^3 e^{x^2}\,dx$.

SOLUTION   Choosing $u = x^3$ won't work because we can't integrate e^{x^2}. Choosing $u = e^{x^2}$ is unproductive; the complexity of the integral will increase. Instead, we choose $u = x^2$; we can integrate xe^{x^2}.

$$\text{Let } u = x^2, \qquad dv = xe^{x^2}dx$$

$$du = 2x\,dx, \quad v = \int xe^{x^2}dx = \frac{1}{2}\int e^{x^2}\cdot 2x\,dx = \frac{1}{2}e^{x^2}.$$

$$\int_0^1 x^3 e^{x^2}dx = x^2\cdot\frac{1}{2}e^{x^2}\Big]_0^1 - \int_0^1 \frac{1}{2}e^{x^2}2x\,dx$$

$$= \frac{1}{2}x^2 e^{x^2}\Big]_0^1 - \frac{1}{2}\int_0^1 e^{x^2}2x\,dx$$

$$= \frac{1}{2}e - 0 - \frac{1}{2}\left[e^{x^2}\right]_0^1$$

$$= \frac{1}{2}e - \frac{1}{2}[e - 1]$$

$$= \frac{1}{2} \quad \blacklozenge$$

PROBLEMS FOR SECTION 29.1

In Problems 1 through 5, evaluate the integrals.

1. $\int x\sin x\,dx$

2. $\int x\cos x\,dx$

3. $\int 3xe^{-2x}dx$

4. $\int_1^e \ln x \, dx$

5. $\int_0^1 \cos^{-1} x \, dx$

6. $\int \sin^{-1} \left(\frac{x}{2} \right) dx$

7. $\int e^{-t} \sin(2t) \, dt$

8. $\int e^{-x} \cos x \, dx$

9. $\int x^2 \cos 3x \, dx$

10. $\int x \ln \frac{1}{x} \, dx$ (You may want to use substitution first.)

11. $\int x \sec^2 x \, dx$

12. $\int_0^1 t^3 e^{-t} dt$

13. $\int \sqrt{x} \ln x \, dx$

14. $\int \cos(\ln x) \, dx$

15. $\int (\ln x)^2 \, dx$

16. Find $\int \cos^2 x \, dx$ in two ways.

    (a) Use the trigonometric identity $\cos^2 x = \frac{1}{2}(1 + \cos 2x)$. *This is the most efficient way to do the problem.*

    (b) Use integration by parts. *You will solve algebraically for $\int \cos^2 x \, dx$.*

    (c) Check that your answers to parts (a) and (b) are correct by differentiating them.

    (d) Your answers to parts (a) and (b) are both antiderivatives of $\cos^2 x$; therefore they must differ by a constant (where the constant is possibly zero). What is the constant?

In Problems 17 through 29, evaluate the integral. In many cases it will be advantageous to begin by doing a substitution. For example, in Problem 19, let $w = \sqrt{x}$, $w^2 = x$; then $2w\,dw = dx$. This eliminates $\sqrt{x}$ by replacing x with a perfect square.
Not every integral in Problems 17 through 29 requires the technique of integration by parts.

17. $\int x^5 \cos x^3 dx$

18. $\int (\ln x)^3 dx$

19. $\int \sqrt{x} e^{\sqrt{x}} dx$

20. $\int_1^e \frac{\ln x}{x} \, dx$

21. $\int \frac{\ln x}{\sqrt{x}} \, dx$

22. $\int_0^{\frac{\pi}{4}} \sec^2 x \tan x \, dx$

23. $\int_1^e \ln \sqrt{w} \, dw$

24. $\int \sqrt{x} \ln x \, dx$

25. $\int \sin(\ln x) \, dx$

26. $\int_0^1 \cos \sqrt{x} \, dx$

27. $\int x^3 \ln x \, dx$

28. $\int_0^{\frac{\pi}{2}} x \sin x \cos x \, dx$ (*Hint:* Use a trigonometric identity.)

29. $\int_0^2 e^{\sqrt{x}} dx$

Problems 30 through 34 involve reduction formulas.

30. (a) Show that $\int \cos^n x \, dx = \frac{1}{n} \cos^{n-1} x \sin x + \frac{n-1}{n} \int \cos^{n-2} x \, dx$, for n an integer, $n \geq 2$. *If you need some hints, see Exercise 29.5.*

   (b) Use the results of part (a) to find $\int \cos^2 x \, dx$. Verify your answer using differentiation.

   (c) Use the results of parts (a) and (b) to find $\int \cos^6 x \, dx$.

31. (a) Use the reduction formula given in Exercise 29.5 to find $\int_0^{\frac{\pi}{2}} \sin^3 x \, dx$.

   (b) Find $\int_0^{\frac{\pi}{2}} \sin^5 x \, dx$.

   (c) Use the reduction formula given in Exercise 29.5 to show that for n a positive odd integer,

$$\int_0^{\frac{\pi}{2}} \sin^n x \, dx - \frac{2 \cdot 4 \cdot 6 \cdot \ldots \cdot (n-1)}{3 \cdot 5 \cdot 7 \cdot \ldots \cdot n}.$$

   This is called a *Wallis product*.

32. (a) Use the reduction formula given in Exercise 29.5 to find $\int_0^{\frac{\pi}{2}} \sin^4 x \, dx$.

   (b) Find $\int_0^{\frac{\pi}{2}} \sin^6 x \, dx$.

   (c) Show that for n a positive even integer,

$$\int_0^{\frac{\pi}{2}} \sin^n x \, dx - \frac{1 \cdot 3 \cdot 5 \cdot \ldots \cdot (n-1)}{2 \cdot 4 \cdot 6 \cdot \ldots \cdot n} \frac{\pi}{2}.$$

   Like the formula in part (c) of the previous problem, this is called a *Wallis product*.

33. Show that $\int (\ln x)^n \, dx = x(\ln x)^n - n \int (\ln x)^{n-1} dx$.

34. Show that $\int x^n e^x \, dx = x^n e^x - n \int x^{n-1} e^x \, dx$.

29.2   TRIGONOMETRIC INTEGRALS AND TRIGONOMETRIC SUBSTITUTION

The two fundamental techniques of integration involving calculus are

■   integration by substitution—coming from the Chain Rule—and

■   integration by parts—coming from the Product Rule.

The techniques discussed in this section are simply methods of transforming an integrand into something more manageable by exploiting trigonometric identities.

Trigonometric Integrals

Suppose we want to integrate the product of integer powers of trigonometric functions. We can convert the problem into one of the form $\int \sin^p x \cos^q x \, dx$ where p and q are integers. Integrals of this form can be evaluated. We'll demonstrate some methods below.

Integrals Involving $\sin x$ and $\cos x$

When dealing with trigonometric integrals, trigonometric identities can come in handy. Recall: $\cos(A + B) = \cos A \cos B - \sin A \sin B$

$$\text{so } \cos(2A) = \cos^2 A - \sin^2 A \nearrow \begin{array}{l} \cos 2A = 2\cos^2 A - 1 \\ \searrow \cos 2A = 1 - 2\sin^2 A \end{array} \Rightarrow \boxed{\begin{array}{l} \cos^2 A = \tfrac{1}{2}(1 + \cos 2A) \\ \sin^2 A = \tfrac{1}{2}(1 - \cos 2A) \end{array}}$$

These two identities, together with $\sin 2A = 2 \sin A \cos A$ and the Pythagorean identities, can be useful. We illustrate by example.

◆ **EXAMPLE 29.8**   Evaluate $\int_0^{2\pi} \sin^2 x \, dx$.

SOLUTION   As mentioned in Section 29.1, integration by parts can be applied to this integral. It's simpler, however, to use the trigonometric identity $\sin^2 x = \tfrac{1}{2}(1 - \cos 2x)$.

$$\int_0^{2\pi} \sin^2 x \, dx = \frac{1}{2} \int_0^{2\pi} (1 - \cos 2x) \, dx \quad \text{Integrate } \cos 2x \text{ by letting } u = 2x.$$

$$= \frac{1}{2} \left(x - \frac{1}{2} \sin 2x \right) \Big|_0^{2\pi}$$

$$= \frac{1}{2} \left(2\pi - \frac{1}{2} \sin(4\pi) \right) - \frac{1}{2} \left(0 - \frac{1}{2} \sin 0 \right)$$

$$= \pi \quad ◆$$

The trigonometric identities obtained from the formula for $\cos 2x$

$$\sin^2 x = \frac{1}{2}(1 - \cos 2x) \quad \text{and} \quad \cos^2 x = \frac{1}{2}(1 + \cos 2x)$$

are useful if the powers of sine and cosine are both positive and even. They are used to reduce powers. On the other hand, if the power of either sine or cosine is odd, the Pythagorean identities are useful.

EXERCISE 29.6   Evaluate $\int \sin^4 x \, dx$ by writing $\sin^4 x$ as $[\sin^2 x]^2$ and using the trigonometric identity $\sin^2 x = \frac{1}{2}(1 - \cos 2x)$. Along the way you'll have to use $\cos^2 u = \frac{1}{2}(1 + \cos 2u)$ as well.

Answer:

$$\tfrac{3}{8}x - \tfrac{1}{4} \sin 2x + \tfrac{1}{32} \sin 4x + C$$

Sometimes when integrating trigonometric functions we can use substitution to transform the integral into the form $\int u^n \, du$. This is the strategy we employ when facing $\int \sin^p x \cos^q x \, dx$ when at least one of p and q is odd. The Pythagorean identities can be useful in making this transformation.

◆ **EXAMPLE 29.9**   Evaluate $\int \cos^2 x \sin^3 x \, dx$.

SOLUTION   We'd like to convert $\sin^2 x$ to cosines using the Pythagorean identity and save the spare $\sin x$; it's just what we need when we make the substitution $u = \cos x$ and $du = -\sin x \, dx$.

$$\int \cos^2 x \sin^3 x \, dx = \int \cos^2 x (\sin^2 x) \sin x \, dx \qquad (\sin^2 x = 1 - \cos^2 x)$$

$$= \int \cos^2 x (1 - \cos^2 x) \sin x \, dx$$

$$= \int (\cos^2 x - \cos^4 x) \sin x \, dx$$

$$\text{Let } u = \cos x, \quad du = -\sin x \, dx$$
$$-du = \sin x \, dx.$$

$$= \int (u^2 - u^4)(-du)$$

$$= \int (-u^2 + u^4) \, du$$

$$= -\frac{u^3}{3} + \frac{u^5}{5} + C$$

$$= -\frac{\cos^3 x}{3} + \frac{\cos^5 x}{5} + C \quad ◆$$

The method used in Example 29.9 will work for integrals of the form

$$\int \sin^p x \cos^q x \, dx,$$

where p and q are positive and at least one of p and q is odd.

For instance, to evaluate $\int \cos^7 x \sin^2 x \, dx$, convert $\cos^6 x = [\cos^2 x]^3$ to sines using the Pythagorean identity and save the other $\cos x$ to use with dx after making the substitution $u = \sin x$. To evaluate $\int \cos^9 x \sin^3 x \, dx$, convert $\sin^2 x$ to cosines and save a $\sin x$ as part of du after making the substitution $u = \cos x$. This is much more efficient than converting $[\cos^2 x]^4$ to sines and letting $u = \sin x$.

EXERCISE 29.7   Find $\int \cos^3 x \, dx$. *Convert $\cos^2 x$ to sines and save a $\cos x$ for the subsequent substitution.*

Answer:

$$\sin x - \frac{\sin^3 x}{3} + C$$

NOTE Integrals of the form $\int \sin^n x \, dx$ or $\int \cos^n x \, dx$, where n is a positive integer, can be treated using a reduction formulas. (This is *not* always the most efficient strategy.)

$$\int \sin^n x \, dx = -\frac{\sin^{n-1} x \cos x}{n} + \frac{n-1}{n} \int \sin^{n-2} x \, dx, \quad n \neq 0$$

$$\int \cos^n x \, dx = \frac{\cos^{n-1} x \sin x}{n} + \frac{n-1}{n} \int \cos^{n-2} x \, dx, \quad n \neq 0.$$

See Exercise 29.5 for an outline of the derivation of these reduction formulas.

If the powers of $\sin x$ and $\cos x$ are not positive, it can be useful to express the integrand using $\tan x$ and $\sec x$.

Integrals Involving $\tan x$ and $\sec x$

Recall that

$\frac{d}{dx} \tan x = \sec^2 x,$

$\frac{d}{dx} \sec x = \sec x \tan x,$ and

$\tan^2 x + 1 = \sec^2 x.$ (The Pythagorean identity is obtained from dividing $\sin^2 x + \cos^2 x = 1$ by $\cos^2 x$.)

Again, we will proceed by example. Where possible, we attempt to transform the integral into the form $\int u^n \, du$.

◆ **EXAMPLE 29.10**   Evaluate $\int \tan^2 x \sec^4 x \, dx$.

SOLUTION   Convert $\sec^2 x$ to tangents, saving a $\sec^2 x \, dx$ for du when we let $u = \tan x$.

$$\int \tan^2 x \sec^4 x \, dx = \int \tan^2 x (\tan^2 x + 1) \sec^2 x \, dx$$

$$= \int (\tan^4 x + \tan^2 x) \sec^2 x \, dx$$

$$\text{Let } u = \tan x, \quad du = \sec^2 x \, dx.$$

$$= \int (u^4 + u^2) \, du$$

$$= \frac{u^5}{5} + \frac{u^3}{3} + C$$

$$= \frac{\tan^5 x}{5} + \frac{\tan^3 x}{3} + C \quad ◆$$

EXERCISE 29.8   Evaluate $\int \tan x \sec^3 x \, dx$. *Let* $u = \sec x$. *Use* $\tan x \sec x \, dx$ *as* du.

Answer

$\frac{\sec^3 x}{3} + C$

◆ **EXAMPLE 29.11**   Find $\int \tan^3 x \, dx$.

SOLUTION

$$\int \tan^3 x \, dx = \int \tan^2 x \tan x \, dx$$

$$= \int (\sec^2 x - 1) \tan x \, dx$$

$$= \int \tan x \sec^2 x \, dx - \int \tan x \, dx$$

$$= \int \tan x \, (\sec^2 x \, dx) + \int \frac{1}{\cos x} \sin x \, dx$$

In the first integral we let $u = \tan x$, $du = \sec^2 x \, dx$.
In the second, we let $w = \cos x$, $dw = -\sin x \, dx$.

$$\int \tan^3 x \, dx = \int u \, du - \int \frac{1}{w} \, dw$$

$$= \frac{u^2}{2} - \ln |w| + C$$

$$= \frac{1}{2} \tan^2 x - \ln |\cos x| + C \quad \blacklozenge$$

The next integral is hard to figure out unless you've seen the method. It might be worth remembering the answer or programming it into your calculator (if it is not already there).

◆ **EXAMPLE 29.12**   Evaluate $\int \sec x \, dx$.

SOLUTION

$$\int \sec x \, dx = \int \sec x \cdot \frac{\sec x + \tan x}{\sec x + \tan x} \, dx$$

$$= \int \frac{\sec^2 x + \sec x \tan x}{\sec x + \tan x} \, dx$$

Now the derivative of the denominator is sitting in the numerator.

Let $u = \sec x + \tan x$, $du = (\sec x \tan x + \sec^2 x) \, dx$.

$$\int \sec x \, dx = \int \frac{1}{u} \, du$$

$$= \ln |u| + C$$

$$= \ln |\sec x + \tan x| + C \quad \blacklozenge$$

Integrals of the form $\int \tan^n x \, dx$ or $\int \sec^n x \, dx$, where n is a positive integer, can be treated using reduction formulas.

$$\int \tan^n x \, dx = \frac{\tan^{n-1} x}{n-1} - \int \tan^{n-2} x \, dx, \quad n \geq 2$$

$$\int \sec^n x \, dx = \frac{\sec^{n-2} x \tan x}{n-1} + \frac{n-2}{n-1} \int \sec^{n-2} x \, dx, \quad n \geq 2.$$

The derivation of the first of these reduction formulas is given as Problem 46 at the end of this section.

EXERCISE 29.9   Use the reduction formula given above to find $\int \sec^3 x \, dx$. (Problem 45 at the end of this section leads the reader through this problem without reference to the reduction formula.)

Answer

$$\int \sec^3 x \, dx = \tfrac{1}{2}(\sec x \tan x + \ln | \sec x + \tan x|) + C$$

Other Trigonometric Integrals

If you are faced with integrals of the form $\int \sin Ax \cos Bx \, dx$, $\int \sin Ax \sin Bx \, dx$, or $\int \cos Ax \cos Bx \, dx$, you can use the appropriate product formula:

i. $\sin Ax \cos Bx = \tfrac{1}{2}[\sin(A - B)x + \sin(A + B)x]$

ii. $\sin Ax \sin Bx = \tfrac{1}{2}[\cos(A - B)x - \cos(A + B)x]$

iii. $\cos Ax \cos Bx = \tfrac{1}{2}[\cos(A - B)x + \cos(A + B)x]$

An alternative is to use integration by parts twice and solve algebraically for the original integral. These types of integrals will come up when you study Fourier series, a powerful tool in modern physics and engineering.

We derive formula (i) below. The other product formulas can be derived similarly.

$$\sin(A + B)x = \sin Ax \cos Bx + \sin Bx \cos Ax$$
$$+ \quad \sin(A - B)x = \sin Ax \cos Bx - \sin Bx \cos Ax$$

$$\sin(A + B)x + \sin(A - B)x = 2 \sin Ax \cos Bx$$

$$\frac{1}{2}[\sin(A + B)x + \sin(A - B)x] = \sin Ax \cos Bx$$

This is not meant to be an exhaustive discussion of trigonometric integrals. If you run into something you can't handle, see what a computer algebra system or a calculator can do with it.

Trigonometric Substitution

Trigonometric substitutions can be useful in transforming integrals whose integrands contain $\sqrt{c^2 - x^2}$, $\sqrt{x^2 - c^2}$, and $\sqrt{x^2 + c^2}$. We'll first illustrate why it can be useful to insert trigonometric functions into problems where they don't appear to belong, and then specify which substitutions to use.

◆ **EXAMPLE 29.13**   Find $\int \dfrac{1}{\sqrt{9-x^2}} \, dx$.

SOLUTION   Method 1: Recall that $\int \dfrac{1}{\sqrt{1-u^2}} \, du = \arcsin u + C$. We can transform the given integral into this form by dividing both numerator and denominator by 3, because $\tfrac{1}{3} = \tfrac{1}{\sqrt{9}}$.

$$\frac{1}{\sqrt{9-x^2}} = \frac{\frac{1}{3}}{\sqrt{\frac{9}{9} - \frac{x^2}{9}}} = \frac{1}{3}\frac{1}{\sqrt{1 - \left(\frac{x}{3}\right)^2}}$$

$$\int \frac{1}{\sqrt{9-x^2}}\,dx = \int \frac{1}{\sqrt{1 - \left(\frac{x}{3}\right)^2}}\frac{1}{3}\,dx \qquad \text{Let } u = \frac{x}{3}, \quad du = \frac{1}{3}\,dx,$$

$$= \arcsin\left(\frac{x}{3}\right) + C$$

Method 2: We'd love to get rid of that square root in the denominator. To do this, let's replace $9 - x^2$ by a perfect square. We can exploit the trigonometric identity

$$1 - \sin^2\theta = \cos^2\theta \quad \text{or} \quad 9 - 9\sin^2\theta = 9\cos^2\theta.$$

Let $x = 3\sin\theta$, where $\theta \in \left[-\frac{\pi}{2}, \frac{\pi}{2}\right]$, $\quad dx = 3\cos\theta\,d\theta.$

Then $\sqrt{9 - x^2} = \sqrt{9 - 9\sin^2\theta} = \sqrt{9\cos^2\theta} = 3\sqrt{\cos^2\theta}$

$$\int \frac{1}{\sqrt{9-x^2}}\,dx = \int \frac{1}{3\sqrt{\cos^2\theta}}\,3\cos\theta\,d\theta\,^1$$

$$= \int 1\,d\theta$$

$$= \theta + C$$

But $x = 3\sin\theta$, so $\frac{x}{3} = \sin\theta$ and $\theta = \arcsin\left(\frac{x}{3}\right)$

$$\int \frac{1}{\sqrt{9-x^2}}\,dx = \arcsin\left(\frac{x}{3}\right) + C$$

This second method will come in very handy, particularly when the first method cannot be applied. Notice that although the integrand involves no trigonometry, the antiderivative does. ◆

◆ **EXAMPLE 29.14**   Find $\int_0^1 \sqrt{4 - x^2}\,dx$.

SOLUTION   The strategy is to replace $4 - x^2$ by a perfect square in order to eliminate the square root.

Let $x = 2\sin\theta$, where $\theta \in \left[\frac{-\pi}{2}, \frac{\pi}{2}\right]$, $\quad dx = 2\cos\theta\,d\theta.$

$$\sqrt{4 - x^2} = \sqrt{4 - 4\sin^2\theta} = \sqrt{4\cos^2\theta} = 2\sqrt{\cos^2\theta}$$

What are the new endpoints of integration?

$$\begin{cases} \text{When } x = 0, & 0 = 2\sin\theta \Rightarrow \sin\theta = 0 \quad \Rightarrow \theta = \arcsin 0 = 0. \\ \text{When } x = 1, & 1 = 2\sin\theta \Rightarrow \sin\theta = \frac{1}{2} \quad \Rightarrow \theta = \arcsin\frac{1}{2} = \frac{\pi}{6}. \end{cases}$$

---

[1] $\sqrt{\cos^2\theta}$ is actually $|\cos\theta|$ as opposed to $\cos\theta$. In the context of this problem, however, $\sqrt{\cos^2\theta}$ can be replaced by $\cos\theta$ because θ has been restricted to $\left[-\frac{\pi}{2}, \frac{\pi}{2}\right]$ and $\cos\theta \geq 0$ on this interval.

$$\int_0^1 \sqrt{4-x^2}\, dx = \int_0^{\frac{\pi}{6}} 2\sqrt{\cos^2\theta}\, 2\cos\theta\, d\theta$$

$$= 4\int_0^{\frac{\pi}{6}} \cos^2\theta\, d\theta \qquad \text{Use } \cos^2\theta = \frac{1}{2}[1+\cos 2\theta].$$

$$= 2\int_0^{\frac{\pi}{6}} (1+\cos 2\theta)\, d\theta$$

$$= 2\left[\theta + \frac{1}{2}\sin 2\theta\right]\Big|_0^{\frac{\pi}{6}}$$

$$= 2\left[\frac{\pi}{6} + \frac{1}{2}\sin\frac{\pi}{3} - 0\right]$$

$$= \frac{\pi}{3} + \frac{1}{2}\frac{\sqrt{3}}{2}$$

Clarifications and Justifications

■ Notice that when we let $x = 2\sin\theta$, we are restricting x to lie between -2 and 2. That's fine, because $\sqrt{4-x^2}$ only makes sense for x in this interval.

■ In both of the previous examples we've said $\sqrt{\cos^2\theta} = \cos\theta$. Actually, $\sqrt{\cos^2\theta} = |\cos\theta|$, which is equal to $\cos\theta$ only if $\cos\theta$ is nonnegative. It is because we've restricted θ to $[-\frac{\pi}{2}\frac{\pi}{2}]$ that we can write $|\cos\theta| = \cos\theta$.

In fact, the substitution $x = g(\theta)$ is only valid if $g(\theta)$ is 1-to-1. In other words, when we say $x = 2\sin\theta$ we must restrict θ. In this case we choose $\theta \in \left[\frac{-\pi}{2}, \frac{\pi}{2}\right]$, the same restrictions used for $\sin^{-1} x$. This type of restriction will hold in all the examples that follow.

■ Notice that once we've established that $\theta \in \left[-\frac{\pi}{2}, \frac{\pi}{2}\right]$ we can say, as above, if $\sin\theta = 0$, then $\theta = 0$. ◆

Generalizing the Method of Trigonometric Substitution: k is a constant.

Original integrand contains	Substitution	The square root transformed	Restrictions on θ
$\sqrt{k^2 - u^2}$	$u = k\sin\theta$	$\sqrt{k^2 - u^2} = C\cos\theta$	$\theta \in \left[-\frac{\pi}{2}, \frac{\pi}{2}\right]$
$\sqrt{k^2 + u^2}$	$u = k\tan\theta$	$\sqrt{k^2 + u^2} = C\sec\theta$	$\theta \in \left(-\frac{\pi}{2}, \frac{\pi}{2}\right)$
$\sqrt{u^2 - k^2}$	$u = k\sec\theta$	$\sqrt{u^2 - k^2} = C\tan\theta$	$\theta \in \left[0, \frac{\pi}{2}\right), \left[\pi, \frac{3\pi}{2}\right)$

The restriction on θ for the substitution $u = k\sec\theta$ is designed to guarantee that $|\tan\theta| = \tan\theta$.[2] Restrictions on θ specified above will hold in the examples that follow.

CAUTION As always, try to make your work as simple as possible. The table supplied above does *not* mean you ought to use trigonometric substitution on $\int \frac{x}{\sqrt{x^2-4}}\, dx$, for instance. Use ordinary substitution, $u = x^2 - 4$, instead.

---

[2] In this particular case, care must be taken in converting an expression involving θ back to one involving u if $\sec\theta$ is negative.

◆ **EXAMPLE 29.15**   Find $\int \frac{1}{x^2\sqrt{9+x^2}}\, dx$.

SOLUTION   Let $x = 3\tan\theta$, $dx = 3\sec^2\theta\, d\theta$, $\theta \in \left(-\frac{\pi}{2}, \frac{\pi}{2}\right)$.

$$\sqrt{9+x^2} = \sqrt{9 + 9\tan^2\theta} = 3\sqrt{1 + \tan^2\theta} = 3\sqrt{\sec^2\theta} = 3\sec\theta$$

$$\int \frac{1}{x^2\sqrt{9+x^2}}\, dx = \int \frac{1}{9\tan^2\theta\, 3\sec\theta} \cdot 3\sec^2\theta\, d\theta$$

$$= \frac{1}{9}\int \frac{\sec\theta}{\tan^2\theta}\, d\theta$$

$$= \frac{1}{9}\int \frac{\cos^2\theta}{\cos\theta\,\sin^2\theta}\, d\theta$$

$$= \frac{1}{9}\int \frac{\cos\theta}{\sin^2\theta}\, d\theta$$

Let $u = \sin\theta$, $du = \cos\theta\, d\theta$.

$$= \frac{1}{9}\int \frac{1}{u^2}\, du$$

$$= \frac{1}{9}\int u^{-2}\, du$$

$$= -\frac{1}{9}u^{-1} + C$$

$$= -\frac{1}{9\sin\theta} + C$$

Now we must change from θ back to x. It is not appealing to write $\sin\theta = \sin\left(\tan^{-1}\frac{x}{3}\right)$ without simplifying. Instead, we can use a triangle to sort out the conversion to x as follows: $x = 3\tan\theta \Rightarrow \tan\theta = \frac{x}{3}$. Draw a right triangle, label θ, and label the sides so that $\tan\theta = \frac{x}{3}$. We determine the hypotenuse using the Pythagorean Theorem. (See Figure 29.1)

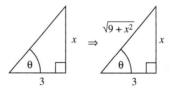

Figure 29.1

$$\sin\theta = \frac{x}{\sqrt{9+x^2}}$$

$$\int \frac{1}{x^2\sqrt{9+x^2}}\,dx = -\frac{1}{9}\frac{1}{\sin\theta} + C$$

$$= -\frac{1}{9}\frac{\sqrt{9+x^2}}{x} + C$$

$$= -\frac{\sqrt{9+x^2}}{9x} + C$$

Check this by differentiating.   ◆

◆ **EXAMPLE 29.16**   Find $\int \sqrt{4-x^2}\,dx$.

SOLUTION   This is the same integrand as in Example 29.14, so we use the same methods. We let $x = 2\sin\theta$ and find

$$\int \sqrt{4-x^2}\,dx = 2\left(\theta + \frac{1}{2}\sin 2\theta\right) + C$$

$$= 2\theta + \sin 2\theta + C.$$

The task now is to change from θ back to x. We'd like to use a triangle, as in the previous example, but we must first apply the trigonometric identity $\sin 2\theta = 2\sin\theta\cos\theta$.

$$\int \sqrt{4-x^2}\,dx = 2\theta + 2\sin\theta\cos\theta + C$$

$x = 2\sin\theta \Rightarrow \sin\theta = \frac{x}{2}$. Draw a right triangle, label θ, and label the sides so that $\sin\theta = \frac{x}{2}$. (See Figure 29.2.)

$$\int \sqrt{4-x^2}\,dx = 2\theta + 2\sin\theta\cos\theta + C$$

$$= 2\arcsin\left(\frac{x}{2}\right) + 2\left(\frac{x}{2}\right)\left(\frac{\sqrt{4-x^2}}{2}\right) + C$$

$$= 2\arcsin\left(\frac{x}{2}\right) + \frac{x\sqrt{4-x^2}}{2} + C$$

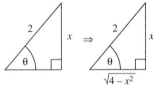

Figure 29.2   ◆

Can problems become more involved than this? Of course they can. Consider, for instance, $\int \sqrt{4x-x^2}\,dx$. $\sqrt{4x-x^2}$ can be expressed as $\sqrt{4-(x-2)^2}$ by completing the square. $\int \sqrt{4-(x-2)^2}\,dx$ is of the form $\int \sqrt{4-u^2}\,du$ and we continue as in the previous example.

At some point, when the level of complexity overpowers the level of intrigue, you'll turn to a computer or powerful calculator. Regardless, it is useful to have this method of trigonometric substitution in your array of tools.

PROBLEMS FOR SECTION 29.2

In Problems 1 through 18, evaluate the integral.

1. $\int \cos^2 x \, dx$

2. $\int_0^\pi \sin^3 x \, dx$

3. $\int \cos x \sin^2 x \, dx$

4. $\int \cos^3 x \sin^2 x \, dx$

5. $\int \cos^4 x \, dx$

6. $\int \cos^2 x \sin^2 x \, dx$

7. $\int \cos^4 x \sin^3 x \, dx$

8. $\int \cos^3 x \sin^{11} x \, dx$

9. $\int \cos^3(3x) \, dx$

10. $\int_0^{\frac{\pi}{2}} \cos^5 x \sqrt{\sin x} \, dx$

11. $\int \frac{\sin x}{\sqrt{\cos^3 x}} \, dx$

12. $\int \tan 3x \, dx$

13. $\int \tan 2x \sec 2x \, dx$

14. $\int_0^{\frac{\pi}{4}} \sqrt{\tan x} \sec^2 x \, dx$

15. $\int \tan x \sec^4 x \, dx$

16. $\int \tan^3 x \sec^4 x \, dx$

17. $\int \tan^3 x \sec^5 x \, dx$

18. $\int \tan^3 x \sec x \, dx$

19. $\int \frac{\tan^3 x}{\sec^4 x} \, dx$

20. $\int \tan^8 x \sec^4 x \, dx$

21. $\int \frac{\sin x}{\cos^2 x} \, dx$

Problems 22 through 24 refer to the information provided about Fourier series. A Fourier series expresses a function as a weighted infinite sum of terms of the form sin nx and cos nx, where n is a nonnegative integer. Fourier series are a very powerful tool. In order to construct such a series we need the following results. m and n are positive integers.

(a) $\int_{-\pi}^{\pi} \sin mx \cos nx \, dx = 0$

(b) $\int_{-\pi}^{\pi} \sin mx \sin nx \, dx = \begin{cases} \pi & \text{if } m = n \\ 0 & \text{if } m \neq n \end{cases}$

(c) $\int_{-\pi}^{\pi} \cos mx \cos nx \, dx = \begin{cases} \pi & \text{if } m = n \\ 0 & \text{if } m \neq n \end{cases}$

These results can be obtained using the product formulas given in this section.

22. Prove statement (a) above.

23. Prove statement (b) above.

24. Prove statement (c) above.

25. Find the volume generated by revolving the region under one arch of cos x about the x-axis.

26. Find the volume generated by revolving the region between the graphs of $y = \sin x$ and $y = \cos x$ from $x = \frac{\pi}{4}$ to $x = \frac{5\pi}{4}$ around the horizontal line $y = 2$.

27. Find $\int \frac{1}{\sin \theta} \, d\theta$.

28. Evaluate $\int \sec^3 x \, dx$ using a reduction formula.

In Problems 29 through 43, evaluate the integrals. Not all require a trigonometric substitution. Choose the simplest method of integration.

29. $\int x^3 \sqrt{4 - x^2} \, dx$

30. $\int \frac{x^3}{\sqrt{9-x^2}} \, dx$

31. $\int \frac{x}{\sqrt{4+x^2}} \, dx$

32. $\int \frac{x^3}{\sqrt{4+x^2}} \, dx$

33. $\int \frac{1}{\sqrt{9+x^2}} \, dx$

34. $\int x^2 \sqrt{x^2 - 9} \, dx$

35. $\int_0^{\frac{3}{4}} \sqrt{9 - 4x^2} \, dx$

36. $\int \sqrt{4 - 9x^2} \, dx$

37. $\int x\sqrt{4 - 9x^2} \, dx$

38. $\int_0^3 \frac{dx}{(4+x^2)^{\frac{3}{2}}}$

39. $\int \frac{\sqrt{x^2-4}}{x}\, dx$

40. $\int \frac{x}{\sqrt{x^2-1}}\, dx$

41. $\int_{\frac{2}{\sqrt{3}}}^2 \frac{\sqrt{x^2-1}}{x}\, dx$

42. $\int \frac{2x-3}{\sqrt{9-4x^2}}\, dx$

43. $\int \frac{x}{4-x^2}\, dx$

44. Find $\int \sqrt{k^2 - x^2}\, dx$ for any constant k.

45. Show that the area enclosed by the ellipse $\frac{x^2}{a^2} + \frac{y^2}{b^2} = 1$, where a and b are positive constants, is given by πab.

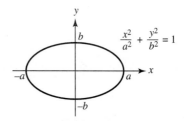

Hint: Find the area in the first quadrant and multiply by 4. Express y as a function of x. Notice the similarity between the formula for the area inside a circle and the formula for the area inside an ellipse.

46. Find $\int \sec^3 x\, dx$ as follows: Use integration by parts with $u = \sec x$. The resulting new integral will contain $\tan^2 x$. Replace $\tan^2 x$ by $\sec^2 x - 1$ and split the integral into the difference of two integrals, $\int \sec^3 x\, dx - \int \sec x\, dx$. Integrate the latter and solve algebraically for the former.

47. Derive the reduction formula

$$\int \tan^n x\, dx = \frac{\tan^{n-1} x}{n-1} - \tan^{n-2} x\, dx,$$

where n is an integer greater than or equal to 2. (*To do this, rewrite the integrand as $\tan^{n-2} x \cdot \tan^2 x$ and use a Pythagorean identity to convert $\tan^2 x$ into an expression involving $\sec x$.*)

48. Derive the formula $\sin Ax \sin Bx = \frac{1}{2}[\cos(A - B)x - \cos(A + B)x]$ given in this section. (*Begin with the addition formula for cosine.*)

29.3 INTEGRATION USING PARTIAL FRACTIONS

Practical motivation: Models commonly used in epidemiology and population biology are often based on what is known as a logistic growth model. For instance, while population growth may initially look exponential, due to limited resources growth generally levels off at what is known as the carrying capacity for the population. In order to derive an expression for $P(t)$, the population at time t, one computes $\int \frac{1}{kP(P-L)}\, dP$, where L is the carrying capacity and k is a constant determined by the population. The methods of this section enable us to calculate this integral.

Suppose an integral is of the form $\int \frac{P(x)}{Q(x)}\, dx$ where $P(x)$ and $Q(x)$ are polynomials and $Q(x)$ factors. The method of partial fractions is an *algebraic* technique enabling us to split up the integrand into the sum of simpler rational functions. For example, we know that

$$\frac{1}{x+2} - \frac{1}{x+3} = \frac{x+3-(x+2)}{(x+2)(x+3)} = \frac{1}{x^2+5x+6}.$$

Suppose we are struggling to find $\int \frac{1}{x^2+5x+6}\, dx$. If we replace the integrand by the equivalent expression $\frac{1}{x+2} - \frac{1}{x+3}$, our struggles are over.

$$\int \frac{1}{x^2+5x+6}\, dx = \int \left(\frac{1}{x+2} - \frac{1}{x+3} \right) dx$$

$$= \int \frac{1}{x+2}\, dx - \int \frac{1}{x+3}\, dx$$

$$= \ln|x+2| - \ln|x+3| + C$$

$$= \ln \left| \frac{x+2}{x+3} \right| + C$$

The method of **decomposition into partial fractions** is a method of integrating rational functions, functions of the form $\frac{P(x)}{Q(x)}$, where $P(x)$ and $Q(x)$ are polynomials by decomposing $\frac{P(x)}{Q(x)}$ into the sum of simpler rational functions.

The method of partial fraction decomposition works only if $\frac{P(x)}{Q(x)}$ is a proper fraction, that is, only if the degree of P is strictly less than the degree of $Q(x)$: $\deg(P) < \deg(Q)$. Suppose $\deg(P) \geq \deg(Q)$; the fraction is **improper**. Then we can do long division to split the fraction into a sum of a polynomial and a proper rational function. We can use the method of partial fractions on the proper part. We illustrate this below.

◆ **EXAMPLE 29.17**   Integrate $\int \frac{x^2}{x^2-1}\, dx$.

SOLUTION   Doing long division we get

$$\frac{x^2}{x^2-1} = 1 + \frac{1}{x^2-1} \text{ or } 1 + \frac{1}{(x-1)(x+1)}.$$

$\frac{1}{(x-1)(x+1)}$ can be decomposed into a sum of the form $\frac{A}{x-1} + \frac{B}{x+1}$ where A and B are constants.

$$\frac{1}{(x-1)(x+1)} = \frac{A}{x-1} + \frac{B}{x+1} \qquad \text{Clear denominators.}$$

$$1 = A(x+1) + B(x-1) \qquad \text{Gather } x \text{ terms; gather constants.}$$

$$0x + 1 = (A+B)x + (A-B)$$

This can be true only if $A + B = 0$ and $A - B = 1$. Solve these simultaneous equations to obtain $A = \frac{1}{2}$ and $B = -\frac{1}{2}$. Then

$$\frac{x^2}{x^2 - 1} = 1 + \frac{1}{2}\frac{1}{x-1} - \frac{1}{2}\frac{1}{x+1}.$$

$$\int \frac{x^2}{x^2 - 1}\,dx = \int \left[1 + \frac{1}{2}\frac{1}{x-1} - \frac{1}{2}\frac{1}{x+1} \right]\,dx$$

$$= x + \frac{1}{2}\ln|x-1| - \frac{1}{2}\ln|x+1| + C \quad \blacklozenge$$

The first step in partial fraction decomposition of a proper rational function is factoring the denominator. It can be proven that any polynomial can be factored into the product of linear and irreducible quadratic factors. A quadratic $ax^2 + bx + c$ is irreducible if it cannot be factored into the product of two linear factors, that is, if the discriminant, $b^2 - 4ac$, is negative. We'll look at polynomials that are not hard to factor. Computer algebra systems can take care of more difficult factoring jobs.

A proper rational function can be decomposed into the sum of proper rational functions. We'll look at some examples before giving the general procedure. Suppose the denominator factors into the product of distinct linear factors. For example, consider $\frac{-3}{(x+1)(x-2)}$. We can express $\frac{-3}{(x+1)(x-2)}$ as a sum of the form $\frac{A}{x+1} + \frac{B}{x-2}$. The task of determining A and B reduces to simultaneous equations.

$$\frac{-3}{(x+1)(x-2)} = \frac{A}{x+1} + \frac{B}{(x-2)} \qquad \text{Get a common denominator.}$$

$$\frac{-3}{(x+1)(x-2)} = \frac{Ax - 2A + Bx + B}{(x+1)(x-2)} \qquad \begin{array}{l}\text{Equate the numerators.} \\ \text{(Equivalently, clear denominators.)}\end{array}$$

$$-3 = Ax - 2A + Bx + B \qquad \text{Collect like powers of } x.$$

$$-3 = (A+B)x + (B - 2A)$$

$$0x - 3 = (A+B)x + (B - 2A) \qquad \begin{array}{l}\text{Equate the coefficients of } x; \\ \text{equate the constant terms.}\end{array}$$

$$\text{Then } \begin{cases} A + B = 0 \\ -3 = B - 2A. \end{cases} \qquad \text{simultaneous linear equations}$$

$$A = -B \text{ so } -3 = B - 2(-B)$$

$$-3 = 3B$$

$$-1 = B \text{ and } A = 1,$$

$$\text{Then } \frac{-3}{(x+1)(x-2)} = \frac{1}{x+1} - \frac{1}{x-2}. \qquad \text{Check by adding.}$$

Knowing the partial fraction decomposition of $\frac{-3}{x^2-x-2}$ makes finding its antiderivative a snap.

$$\int \frac{-3}{x^2 - x - 2}\, dx = \int \frac{1}{x + 1} - \frac{1}{x - 2}\, dx$$

$$= \ln |x + 1| - \ln |x - 2| + C$$

$$= \ln \left| \frac{x + 1}{x - 2} \right| + C$$

There's a shortcut to finding A and B. Given $\frac{-3}{(x+1)(x-2)} = \frac{A}{x+1} + \frac{B}{x-2}$, clear denominators and then evaluate both sides at values of x that *would* have made the denominator zero. It can be shown that this is legal; certainly it is quick.

$$-3 = A(x - 2) + B(x + 1)$$

Evaluate at $x = -1$: $-3 = A(-1 - 2)$ $\Rightarrow$ $A = 1$

Evaluate at $x = 2$: $-3 = B(2 + 1)$ $\Rightarrow$ $B = -1$

We'll apply this shortcut in the next example.

◆ **EXAMPLE 29.18**   Find $\int \frac{x+1}{x^2-3x^2+2x}\, dx$.

SOLUTION   The integrand is a proper fraction, so we factor the denominator.

$$\frac{x^2 + 1}{x^3 - 3x^2 + 2x} = \frac{x^2 + 1}{x(x^2 - 3x + 2)} = \frac{x^2 + 1}{x(x - 2)(x - 1)}$$

The denominator has distinct linear factors, so we set up partial fractions as follows.

$$\frac{x^2 + 1}{x(x - 2)(x - 1)} = \frac{A}{x} + \frac{B}{x - 2} + \frac{C}{x - 1}$$

$$x^2 + 1 = A(x - 2)(x - 1) + Bx(x - 1) + Cx(x - 2)$$

Evaluate at $x = 0$: $1 = A(-2)(-1) \Rightarrow A = \frac{1}{2}$

Evaluate at $x = 2$: $5 = B(2)(1) \Rightarrow B = \frac{5}{2}$

Evaluate at $x = 1$: $2 = C(1)(-1) \Rightarrow C = -2$

$$\frac{x^2 + 1}{x^3 - 3x^2 + 2x} = \frac{\frac{1}{2}}{x} + \frac{\frac{5}{2}}{x - 2} - \frac{2}{x - 1}$$

Therefore $\int \frac{x^2 + 1}{x^3 - 3x^2 + 2x}\, dx = \frac{1}{2} \int \frac{1}{x}\, dx + \frac{5}{2} \int \frac{1}{x - 2}\, dx - 2 \int \frac{1}{x - 1}\, dx$

$$= \frac{1}{2} \ln |x| + \frac{5}{2} \ln |x - 2| - 2 \ln |x - 1| + C \quad ◆$$

◆ **EXAMPLE 29.19**   Find $\int \frac{-2x-1}{x^2(x^2+1)}\, dx$.

SOLUTION   The integrand can be broken into a sum as follows,

$$\frac{-2x-1}{x^2(x^2+1)} = \frac{A}{x} + \frac{B}{x^2} + \frac{Cx+D}{x^2+1}$$

A repeated factor contributes multiple fractions, as shown.[3]

We can solve for A, B, C, and D. One method is to clear denominators and evaluate both sides at four different x-values. This leaves us with four equations and four unknowns.

We can show that

$$\frac{-2x-1}{x^2(x^2+1)} = \frac{-2}{x} + \frac{-1}{x^2} + \frac{2x+1}{x^2+1}.$$

$$\int \frac{-2x-1}{x^2(x^2+1)}\, dx = -2 \int \frac{1}{x}\, dx - \int \frac{1}{x^2}\, dx + \int \frac{2x+1}{x^2+1}\, dx$$

$$= -2\ln|x| - \frac{x^{-1}}{-1} + \int \frac{2x}{x^2+1}\, dx + \int \frac{1}{x^2+1}\, dx$$

$$= -2\ln|x| + \frac{1}{x} + \ln(x^2+1) + \arctan x + C \quad ◆$$

General Procedure for Partial Fraction Decomposition of $\frac{P(x)}{Q(x)}$

■   If $\frac{P(x)}{Q(x)}$ is improper, do long division to express it as the sum of a polynomial and a proper rational function.

■   Factor $Q(x)$ into the product of linear and irreducible quadratic factors.

$$(x - a) \text{ is a factor of } Q(x) \text{ if } Q(a) = 0$$

■   Rewrite the proper rational function as a sum of simpler rational functions as follows.
For every nonrepeated linear factor $(x - a)$ of Q there is a term $\frac{A}{x-a}$
For a repeated linear factor $(x - a)^n$ of Q there are n partial fractions. All numerators are constant; denominators consist of $(x - a)$ raised to successively higher powers. E.g., for $(x - a)^3$ a factor of Q there are fractions $\frac{A}{x-a} + \frac{B}{(x-a)^2} + \frac{C}{(x-a)^3}$.

For a nonrepeated *irreducible* quadratic factor of Q, $(x^2 + bx + c)$, there's a term $\frac{Ax+B}{x^2+bx+c}$.
If the quadratic factor of Q is repeated, there are multiple partial fractions, each with a linear expression in the numerator. E.g., $(x^2 + 4)^2$ as a factor of Q would result in the partial fractions $\frac{Ax+B}{x^2+4} + \frac{Cx+D}{(x^2+4)^2}$.

Finding numerical values for the constants is an algebra problem. Integrating the partial fractions coming from linear factors of Q is straightforward. Some quadratics and repeated quadratics are straightforward while others require both completing the square and trigonometric substitution.

---

[3] The denominators are x and x^2, not x and x.

PROBLEMS FOR SECTION 29.3

In Problems 1 through 6, write out the partial fraction decomposition of each rational function. You need not determine the coefficients; just set them up.

1. (a) $\frac{x^2+3}{x(x-1)(x+5)}$ (b) $\frac{x}{x^3+x}$

2. (a) $\frac{3}{x^3-4x}$ (b) $\frac{4}{x^3+2x}$

3. (a) $\frac{x^3}{x^3-4x}$ (b) $\frac{3x+1}{x^3+2x^2+1}$

4. (a) $\frac{x+5}{x^2+3x-4}$ (b) $\frac{x+5}{x^2-4}$

5. $\frac{3}{x^3+4x}$

6. $\frac{x^2+1}{(x^2+x+1)(x-1)}$

In Problems 7 through 17, evaluate the integrals.

7. $\int \frac{3x+9}{x^2-6x+5}\, dx$

8. $\int \frac{2}{x(x-1)^2}\, dx$

9. $\int_0^1 \frac{x^2}{2x+3}\, dx$

10. $\int \frac{2}{x^4-1}\, dx$

11. $\int \frac{3x^2+3}{(x^2-1)(x-2)}\, dx$

12. $\int \frac{1}{x(x-3)}\, dx$

13. $\int \frac{e^{2x}}{(e^x+2)(e^x-1)^2}\, dx$

    Hint: Use substitution. Do not attempt partial fractions until the integral is in the form $\frac{P(x)}{Q(x)}$ where $P(x)$ and $Q(x)$ are polynomials.

14. $\int x^{-2} \arctan x \, dx$

15. $\int x^2 \arctan x \, dx$

16. $\int \frac{x^3}{\sqrt{1-x^2}}\, dx$

17. $\int \ln(x^2 - 1)\, dx$

18. $\int \frac{2x^3-2x^2+4x+8}{(x-2)^2(x^2+3)}\, dx$

19. $\int \frac{x^3}{x^2+x-6}\, dx$

29.4   IMPROPER INTEGRALS

Mathematicians make a habit of testing and pushing the boundaries of their theorems and definitions. In this section we'll push the boundaries of the definition of the definite integral. The definition of the definite integral $\int_a^b f(x)\,dx$ requires that

■ the interval of integration, $[a, b]$, is finite and

■ the integrand, $f(x)$, is defined and bounded everywhere on $[a, b]$.

In this section we extend our notion of integral to include integrals for which one or both of these conditions break down. Such an integral is called an **improper integral**. The integral $\int_1^\infty \frac{1}{x^2}\,dx$ is improper because the interval of integration is unbounded; $\int_0^1 x^{-1/2}\,dx$ is improper because the integrand is unbounded at $x = 0$.

What motivates us to try to make sense of improper integrals? A perfectly good answer is "to see if we can." But in fact, physicists and statisticians work with such integrals on a regular basis. For example, probabilities are computed by calculating the area under a distribution $p(x)$, called a probability density function. The probability of a randomly chosen data point lying to the right of the marker "a" on the distribution is given by $\int_a^\infty p(x)\,dx$, where $p(x)$ is a nonnegative function.

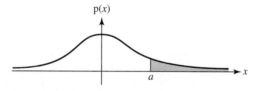

Figure 29.3

The probability of lying *somewhere* must be 1, (100 %), so if $p(x)$ is a probability density function, then $\int_{-\infty}^\infty p(x)\,dx = 1$.

Physicists use improper integrals in their models with great regularity, moving particles in "from infinity" in particle physics, and computing integrals such as $\int_0^1 \sqrt{\frac{1+x}{1-x}}\,dx$ when studying aerodynamics. The integral $\int_0^1 \sqrt{\frac{1+x}{1-x}}\,dx$ is improper because the integrand is unbounded, "blowing up" at $x = 1$. We'll make sense of improper integrals by taking up the types of improprieties, infinite interval and unbounded integrand, one at a time. Our method of dealing with an impropriety at an endpoint of integration is to excise it, constructing an ordinary integral with an endpoint that we allow to approach the trouble spot. If the appropriate limit exists and is finite, we define the improper integral to be that limit. If the impropriety does not occur at an endpoint, then we split up the integral so that it does. We'll begin with a few examples.

Infinite Interval of Integration

◆ **EXAMPLE 29.20**   Compute $\int_1^\infty \frac{1}{x^2}\,dx$, if possible.

SOLUTION $\int_1^\infty \frac{1}{x^2}\,dx$ corresponds graphically to the area under $\frac{1}{x^2}$ and is bounded on the left by $x = 1$. We approach this problem by looking at the proper integral $\int_1^b \frac{1}{x^2}\,dx$, and letting b increase without bound. As b increases the accumulated area increases; the question is whether or not the accumulated area increases without bound.

$$\int_1^b \frac{1}{x^2}\, dx = \frac{-1}{x}\, \bigg|_1^b = -\frac{1}{b} + 1$$

$$\lim_{b \to \infty} \int_1^b \frac{1}{x^2}\, dx = \lim_{b \to \infty} \left(-\frac{1}{b} + 1 \right) = 1$$

So the area of the shaded region approaches 1 as b increases without bound. We say $\int_1^\infty \frac{1}{x^2}\, dx$ is **convergent**, and converges to 1.

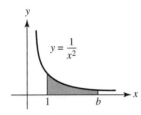

Figure 29.4   ◆

◆ **EXAMPLE 29.21**   Compute $\int_1^\infty \frac{1}{x}\, dx$, if possible.

SOLUTION   Again, we integrate from 1 to b (we compute a proper integral) and take the limit as $b \to \infty$.

$$\int_1^b \frac{1}{x}\, dx = \ln x\, \bigg|_1^b = \ln b - \ln 1 = \ln b$$

$$\lim_{b \to \infty} \int_1^b \frac{1}{x}\, dx = \lim_{b \to \infty} \ln b = \infty$$

We say $\int_1^\infty \frac{1}{x}\, dx$ **diverges**.

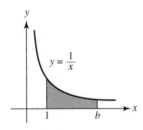

Figure 29.5   ◆

REMARK It is not at all clear from looking at the graphs of $\frac{1}{x^2}$ and $\frac{1}{x}$ that $\int_1^\infty \frac{1}{x^2}\, dx$ should converge to 1 and $\int_1^\infty \frac{1}{x}\, dx$ should be unbounded. In fact, it might surprise you. After all, as $x \to \infty$ both $\frac{1}{x}$ and $\frac{1}{x^2}$ approach 0. The crucial difference is that $\frac{1}{x^2}$ approaches 0 much faster. What should not surprise you is that $0 < \int_1^\infty \frac{1}{x^3}\, dx \le 1$, because $0 < \frac{1}{x^3} \le \frac{1}{x^2}$ for $x \ge 1$. More generally, if f is continuous and $f(x) \ge 0$ for all $x \ge a$, then for $b > a$ $\int_a^b f(x)\, dx$ increases with b. Therefore, $\lim_{b \to \infty} \int_a^b f(x)\, dx$ is either finite or increases without bound. In other words, if $\lim_{b \to \infty} \int_a^b f(x)\, dx$ is bounded, then $\int_a^\infty f(x)\, dx$ converges, provided f is nonnegative on $[a, \infty)$.

Similarly, we can argue that because

$$0 < \frac{1}{x} \leq \frac{1}{\sqrt{x}} \quad \text{for } x \geq 1,$$

$$0 < \int_1^b \frac{1}{x}\, dx \leq \int_1^b \frac{1}{\sqrt{x}}\, dx \text{ for } b \geq 1.$$

As b increases without bound $\int_1^b \frac{1}{x}\, dx$ increases without bound, so $\int_1^b \frac{1}{\sqrt{x}}\, dx$ increases without bound as well.

This type of comparison was not useful when dealing with $\frac{1}{x}$ and $\frac{1}{x^2}$. $0 < \frac{1}{x^2} \leq \frac{1}{x}$ for $x \geq 1$, but being larger than a convergent integral is an inconclusive characteristic, as is being smaller than a divergent integral.

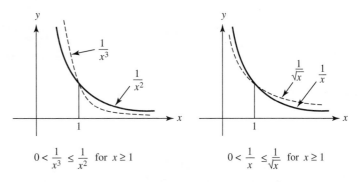

$$0 < \frac{1}{x^3} \leq \frac{1}{x^2} \text{ for } x \geq 1 \qquad\qquad 0 < \frac{1}{x} \leq \frac{1}{\sqrt{x}} \text{ for } x \geq 1$$

Figure 29.6

Definition

■ Consider $\int_a^\infty f(x)\, dx$, where $\int_a^b f(x)\, dx$ exists for all $b \geq a$. $\int_a^\infty f(x)\, dx = \lim_{b \to \infty} \int_a^b f(x)\, dx$ provided this limit exists and is finite. If the limit is finite, then $\int_a^\infty f(x)\, dx$ **converges**. Otherwise the integral **diverges**.

■ $\int_{-\infty}^a f(x)\, dx = \lim_{b \to -\infty} \int_b^a f(x)\, dx$ provided this limit exists and is finite, and $\int_b^a f(x)\, dx$ exists for all $b \leq a$.

■ If $\int_a^\infty f(x)\, dx$ and $\int_{-\infty}^a f(x)\, dx$ both converge, then

$$\int_{-\infty}^\infty f(x)\, dx = \int_{-\infty}^a f(x)\, dx + \int_a^\infty f(x)\, dx.$$

If one or both of $\int_a^\infty f(x)\, dx$ and $\int_{-\infty}^a f(x)\, dx$ diverges, then $\int_{-\infty}^\infty f(x)\, dx$ diverges.

◆ **EXAMPLE 29.22**   Evaluate $\int_{-\infty}^\infty \frac{1}{1+x^2}\, dx$ if this integral is convergent.

SOLUTION   Split this into two improper integrals. $x = 0$ makes a convenient breaking point.

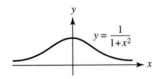

Figure 29.7

$\int_{-\infty}^{\infty} \frac{1}{1+x^2}\,dx = \int_{-\infty}^{0} \frac{1}{1+x^2}\,dx + \int_{0}^{\infty} \frac{1}{1+x^2}\,dx$, provided both these integrals converge.

$$\int_{0}^{\infty} \frac{1}{1+x^2}\,dx = \lim_{b \to \infty} \int_{0}^{b} \frac{1}{1+x^2}\,dx$$

$$= \lim_{b \to \infty} \arctan x \Big|_{0}^{b}$$

$$= \lim_{b \to \infty} \arctan b - \arctan 0$$

$$= \lim_{b \to \infty} \arctan b$$

$$= \frac{\pi}{2}$$

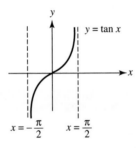

Figure 29.8

$\frac{1}{1+x^2}$ is an even function, so $\int_{-\infty}^{0} \frac{1}{1+x^2}\,dx = \frac{\pi}{2}$, and $\int_{-\infty}^{\infty} \frac{1}{1+x^2}\,dx = \frac{\pi}{2} + \frac{\pi}{2} = \pi$. ◆

So far the divergent improper integrals we've looked at diverge by growing without bound. The next example illustrates another way of diverging.

◆ **EXAMPLE 29.23**   Determine whether $\int_{0}^{\infty} \sin x\,dx$ converges or diverges.

SOLUTION

$$\int_{0}^{\infty} \sin x\,dx = \lim_{b \to \infty} \int_{0}^{b} \sin x\,dx$$

$$= \lim_{b \to \infty} -\cos x \Big|_{0}^{b}$$

$$= \lim_{b \to \infty} -\cos b + 1$$

This limit doesn't exist. The value of $-\cos b + 1$ oscillates between 0 and 1. This improper integral diverges by oscillation.   ◆

EXERCISE 29.10   Show that $\int_1^\infty e^{-x}\, dx = \frac{1}{e}$.
The solution is embedded in the solution to Example 29.28.

Observation

If $\lim_{x \to \infty} f(x) \neq 0$, then $\int_0^\infty f(x)\, dx$ diverges.

If $\lim_{x \to \infty} f(x) = 0$, then work must be done in order to determine whether or not $\int_a^\infty f(x)\, dx$ diverges or converges.

Unbounded and Discontinuous Integrands

We begin with an example.

◆ **EXAMPLE 29.24**   Find $\int_0^1 \frac{1}{\sqrt{x}}\, dx$.

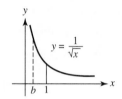

Figure 29.9

SOLUTION   This integral is improper because $\frac{1}{\sqrt{x}}$ blows up at $x = 0$. To get around this problem we compute

$$\int_b^1 \frac{1}{\sqrt{x}}\, dx \quad \text{for } b > 0 \text{ and take the limit as } b \to 0^+.$$

$$\lim_{b \to 0^+} \int_b^1 x^{-\frac{1}{2}}\, dx = \lim_{b \to 0^+} 2x^{\frac{1}{2}} \Big|_b^1$$

$$= \lim_{b \to 0^+} 2 \cdot 1 - 2\sqrt{b}$$

$$= 2 \quad ◆$$

◆ **EXAMPLE 29.25**   Show that $\int_0^1 \frac{1}{x}\, dx$ diverges.

SOLUTION

$$\int_0^1 \frac{1}{x}\, dx \Rightarrow \lim_{b \to 0^+} \int_b^1 \frac{1}{x}\, dx = \lim_{b \to 0^+} \ln |x| \Big|_b^1 = \lim_{b \to 0^+} 0 - \ln |b| = -(-\infty).$$

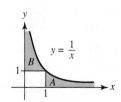

Figure 29.10

One could also argue that the integral diverges by the symmetry of the shaded regions in Figure 29.10. Area A is infinite, so area B is infinite. But $\int_0^1 \frac{1}{x}\, dx = 1 +$ (the area of B). Consequently $\int_0^1 \frac{1}{x}\, dx$ is divergent.   ◆

Definition

■ If f is continuous on $(c, d]$ and discontinuous at c, then

$$\int_c^d f(x)\, dx = \lim_{b \to c^+} \int_b^d f(x)\, dx$$

provided this limit exists and is finite.

■ If f is continuous on $[c, d)$ and discontinuous at d, then

$$\int_c^d f(x)\, dx = \lim_{b \to d^-} \int_c^b f(x)\, dx$$

provided this limit exists and is finite.

■ If f has a discontinuity at p, $c < p < d$, then

$$\int_c^d f(x)\, dx = \int_c^p f(x)\, dx + \int_p^d f(x)\, dx$$

provided both integrals are convergent.

It can be shown that

i. $\int_1^\infty \frac{1}{x^p}\, dx$ converges for $p > 1$ and diverges for $p \leq 1$,

ii. $\int_0^1 \frac{1}{x^p}\, dx$ converges for $p < 1$ and diverges for $p \geq 1$.

This is left as an exercise.

EXERCISE 29.11   Prove statement (i) above.

EXERCISE 29.12   Prove statement (ii) above.[4]

How to Approach an Improper Integral: An Informal Summary

1. Determine all improprieties: Infinite interval, discontinuous integrand, unbounded integrand.

2. If necessary, split up the integral into the sum of integrals so that
   (a) each integral has only one impropriety,
   (b) the impropriety is exposed; it occurs at the endpoint of an integral.

3. Compute each improper integral as the limit of a proper integral. Replace the endpoint at which the impropriety occurs and compute the appropriate one-sided limit.

4. The original integral converges only if each of the summands converges independently.

---

[4] Both exercises are also included as problems at the end of the chapter.

◆ **EXAMPLE 29.26**   Compute $\int_{-\infty}^{\infty} \frac{1}{x^3}\, dx$, if possible.

SOLUTION   This integral is improper at $x = 0$ and at the endpoints of integration. Split the interval up: $(-\infty, -1], [-1, 0], [0, 1], [1, \infty)$.

X ——— ⁻¹ ——— X ——— ⁰ ——— ¹ ——— X

$$\int_{-\infty}^{\infty} \frac{1}{x^3}\, dx = \int_{-\infty}^{-1} \frac{1}{x^3}\, dx + \int_{-1}^{0} \frac{1}{x^3}\, dx + \int_{0}^{1} \frac{1}{x^3}\, dx + \int_{1}^{\infty} \frac{1}{x^3}\, dx$$

We need to compute four different limits. We know $\int_{1}^{\infty} \frac{1}{x^3}\, dx$ will converge, either having completed Exercise 29.11 or because

$$0 < \frac{1}{x^3} \le \frac{1}{x^2} \quad \text{for } x \ge 1$$

and

$$\int_{1}^{\infty} \frac{1}{x^2}\, dx \text{ converges. So } \int_{1}^{\infty} \frac{1}{x^3} \text{ is increasing and bounded.}$$

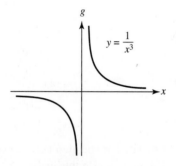

Figure 29.11

(By symmetry, we know $\int_{-\infty}^{-1} \frac{1}{x^3}\, dx$ is convergent, converging to $-\int_{1}^{\infty} \frac{1}{x^3}\, dx$.) If you've completed Exercise 29.12, you know $\int_{0}^{1} \frac{1}{x^3}\, dx$ diverges. We show this below.

$$\int_{0}^{1} \frac{1}{x^3}\, dx = \lim_{b \to 0^+} \int_{b}^{1} \frac{1}{x^3}\, dx$$

$$= \lim_{b \to 0^+} \left. \frac{x^{-2}}{-2} \right|_{b}^{1}$$

$$= \lim_{b \to 0^+} -\frac{1}{2} + \frac{2}{2b^2}$$

$$= \infty \quad \Rightarrow \text{ The integral } \int_{0}^{1} \frac{1}{x^3}\, dx \text{ diverges.}$$

If *any* one of the summands diverges the original integral is divergent, so we're done.

◆

◆ **EXAMPLE 29.27**   Compute $\int_0^3 \frac{1}{x-2}\,dx$, if possible.

SOLUTION   The integrand blows up at $x = 2$, and $2 \in [0, 3]$. Therefore, we must split the interval into $[0, 2]$ and $[2, 3]$.

$$\int_0^3 \frac{1}{x-2}\,dx = \int_0^2 \frac{1}{x-2}\,dx + \int_2^3 \frac{1}{x-2}\,dx$$

You can compute either of these summands independently. Both diverge, so the original integral diverges as well.

Alternatively, you can notice that

$$\int_2^3 \frac{1}{x-2}\,dx = \int_0^1 \frac{1}{u}\,du \quad \text{and}$$

this latter integral diverges. (This was shown in Example 29.25.)

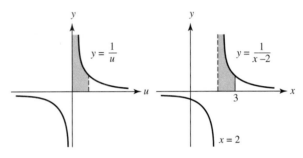

Figure 29.12

CAUTION   If you aren't careful you might not notice that $\int_0^3 \frac{1}{x-2}\,dx$ is an improper integral. Failure to notice this makes your work fatally flawed, as you would erroneously write

$$\int_0^3 \frac{1}{x-2}\,dx = \ln|x-2| \Big|_0^3 = \ln|1| - \ln|-2| = -\ln 2. \quad ◆$$

◆ **EXAMPLE 29.28**   Determine whether $\int_0^\infty e^{-x^2}\,dx$ converges or diverges.

SOLUTION   We have an interesting dilemma here because we can't find an antiderivative for e^{-x^2}. Therefore, we'll argue by comparison. e^{-x^2} approaches zero very quickly, so we expect the integral to converge. Therefore, we'll show that it is less than some convergent integral.

For $x \geq 1$,

$$0 < e^x \leq e^{x^2} \quad \text{so}$$

$$\frac{1}{e^x} \geq \frac{1}{e^{x^2}}$$

$$\int_0^\infty e^{-x^2}\,dx = \int_0^1 e^{-x^2}\,dx + \int_1^\infty e^{-x^2}\,dx.$$

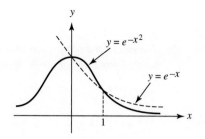

Figure 29.13

The first summand is proper. We'll concern ourselves with the second integral and compare with the convergent integral $\int_1^\infty e^{-x}\,dx$.

$$\int_1^\infty e^{-x^2}\,dx = \lim_{b\to\infty}\int_1^b e^{-x^2}\,dx$$

We will show that this limit exists and is finite.

We know $e^{-x^2} > 0$, so for $b > 1$, $\int_1^b e^{-x^2}\,dx$ increases with b. Therefore, as $b \to \infty$, $\int_1^b e^{-x}\,dx$ either grows without bound or is finite.

$$0 < \int_1^b e^{-x^2}\,dx < \int_1^b e^{-x}\,dx \quad \text{because } e^{-x^2} \le e^{-x} \text{ on } [1, b]$$

$$0 \le \lim_{b\to\infty}\int_1^b e^{-x^2}\,dx \le \lim_{b\to\infty}\int_1^b e^{-x}\,dx$$

$$\lim_{b\to\infty}\int_1^b e^{-x}\,dx = \lim_{b\to\infty} -e^{-x}\Big|_1^b = \lim_{b\to\infty} -\frac{1}{e^b} + \frac{1}{e} = \frac{1}{e}$$

so $0 \le \int_1^\infty e^{-x^2}\,dx \le \frac{1}{e}$. Therefore $\int_1^\infty e^{-x^2}\,dx$ is convergent. We conclude that $\int_0^\infty e^{-x^2}\,dx$ converges.   ◆

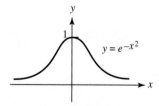

Figure 29.14

REMARK $\int_{-\infty}^\infty e^{-x^2}\,dx$ is an interesting integral. We've concluded that it is convergent; using more advanced methods it can be shown that its value is $\sqrt{\pi}$. If we wanted to approximate $\int_{-\infty}^\infty e^{-x^2}\,dx$, we could proceed as follows.

i. $\int_{-\infty}^\infty e^{-x^2}\,dx = 2\int_0^\infty e^{-x^2}\,dx$

ii. Cut the tail off of $\int_0^\infty e^{-x^2}\,dx$ and bound it. For instance, $\int_6^\infty e^{-x^2}\,dx < \int_6^\infty e^{-x}\,dx < 0.0025$. Even better, bound $\int_c^\infty e^{-x^2}\,dx$ by $\int_c^\infty xe^{-x^2}\,dx$. For instance,

$$\int_5^\infty e^{-x^2} \, dx < \int_5^\infty x e^{-x^2} \, dx = \frac{1}{2e^{25}} < 7 \times 10^{-12}$$

The interested reader has many details to fill in here. The claim is that xe^{-x^2} is a much better bound, a tighter fit, than is e^{-x}.

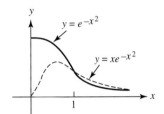

Figure 29.15

iii. Use numerical methods to approximate the proper integral $\int_0^k e^{-x^2} \, dx$ after the tail, $\int_k^\infty e^{-x^2} \, dx$, has been amputated.

Notice that the graph of e^{-x^2} is bell-shaped. As stands, e^{-x^2} cannot be a probability density function because the area under such a function must be exactly 1. A bit of tinkering takes care of this. A standard normal distribution in statistics is described mathematically by the formula

$$p(x) = \frac{1}{\sqrt{2\pi}} e^{-\frac{x^2}{2}}.$$

The Method of Comparison

We've used the method of comparison in several instances. We state the comparison theorem below. We omit the formal proof, but the statements should seem quite reasonable; an informal argument was provided earlier in this section.

Comparison Theorem

Let f and g be continuous functions with $0 \le g(x) \le f(x)$ for $x \ge a$.

- If $\int_a^\infty f(x) \, dx$ converges, then $\int_a^\infty g(x) \, dx$ converges.
- If $\int_a^\infty g(x) \, dx$ diverges, then $\int_a^\infty f(x) \, dx$ diverges.

Suppose $h(x)$ is positive and continuous.

To show $\int_a^\infty h(x) \, dx$ converges, we must produce a *larger* function whose improper integral converges.

To show $\int_a^\infty h(x) \, dx$ diverges, we must produce a *smaller* function whose improper integral diverges.

Naturally, we can't produce both, so we begin by taking a guess about whether or not $\int_a^\infty h(x)$ converges.

The integral $\int_1^\infty \frac{1}{x^p}\, dx$ (and constant multiples of this integral) can be useful for comparison. Recall that this integral converges for $p > 1$ and diverges for $p \le 1$.

EXAMPLE 29.29   Is $\int_1^\infty \frac{\sqrt{1+\sin^2 x}}{x}\, dx$ convergent?

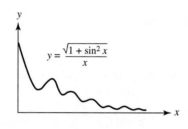

Figure 29.16

SOLUTION $0 \le \frac{1}{x} \le \frac{\sqrt{1+\sin^2 x}}{x}$

$\int_1^\infty \frac{1}{x}\, dx$ diverges, so $\int_1^\infty \frac{\sqrt{1+\sin^2 x}}{x}\, dx$ diverges by comparison.   ◆

REMARK Not only can we compare improper integrals with one another but we can compare improper integrals with infinite series. See Figure 29.17.

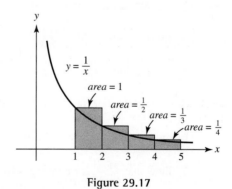

Figure 29.17

The shaded circumscribed rectangles have areas corresponding to the terms of the harmonic series. The rectangles lie above $\frac{1}{x}$. We can argue that the harmonic series $1 + \frac{1}{2} + \frac{1}{3} + \frac{1}{4} + \cdots$ diverges because $\sum_{k=1}^n \frac{1}{k}$ is larger than $\int_1^{n+1} \frac{1}{x}\, dx$ and $\int_1^\infty \frac{1}{x}\, dx$ is a divergent improper integral.

We can argue that the infinite series $\sum_{n=2}^\infty \frac{1}{n^2} = \frac{1}{2} + \frac{1}{3} + \frac{1}{4} + \cdots$ converges by comparing it to the convergent improper integral $\int_1^\infty \frac{1}{x^2}\, dx$.

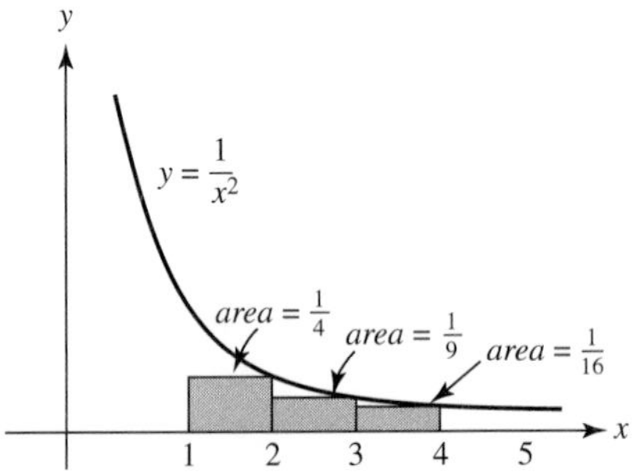

Figure 29.18

Similarly, after completing Exercise 29.11 we can argue that $\sum_{n=1}^{\infty} \frac{1}{n^p}$ converges for $p > 1$ and diverges for $p \leq 1$.

We will formalize the comparison between improper integrals and infinite series by the end of the next chapter.

PROBLEMS FOR SECTION 29.4

In Problems 1 through 5, pinpoint all the improprieties in the integral. If necessary rewrite the integral as a sum of integrals so that each impropriety occurs at an endpoint and there is only one impropriety per integral.

1. (a) $\int_0^\infty \frac{1}{x^2+4}\, dx$ (b) $\int_0^\infty \frac{1}{x^2-4}\, dx$

2. (a) $\int_0^\infty \frac{1}{x^2}\, dx$ (b) $\int_{-\infty}^\infty \frac{1}{x^2}\, dx$

3. (a) $\int_{-\infty}^\infty \frac{1}{x^2+4}\, dx$ (b) $\int_{-\infty}^\infty \frac{1}{x^2-4}\, dx$

4. (a) $\int_{-\frac{\pi}{2}}^{\frac{\pi}{2}} \tan x \, dx$ (b) $\int_0^\pi \tan x \, dx$

5. (a) $\int_0^\infty \tan^{-1} x \, dx$ (b) $\int_{-\infty}^\infty \tan^{-1} x \, dx$

6. Show that $\int_1^\infty \frac{1}{x^p}\, dx$ converges for $p > 1$ and diverges for $p \leq 1$.

7. Show that $\int_0^1 \frac{1}{x^p}\, dx$ converges for $p < 1$ and diverges for $p \geq 1$.

8. Show that $\int_{-1}^\infty \frac{1}{x^4}\, dx$ diverges.

9. (a) Evaluate $\int_0^\infty x e^{-x^2}\, dx$.   (b) Evaluate $\int_{-\infty}^\infty x e^{-x^2}\, dx$.

10. Show $\int_4^\infty e^{-x^2}\, dx < 0.0000001$. *Hint:* Compare it to $\int_4^\infty x e^{-x^2}\, dx$.

In Problems 11 through 36, determine whether the integral is convergent or divergent. Evaluate all convergent integrals. Be efficient. If $\lim_{x \to \infty} \neq 0$, then $\int_a^\infty f(x)\, dx$ is divergent.

11. $\int_0^\infty e^{-2x}\, dx$

12. $\int_0^\infty x\, dx$

13. $\int_0^\infty \cos x + 1\, dx$

14. $\int_0^\infty \cos x\, dx$

15. $\int_0^\infty xe^{-x}\, dx$

16. $\int_{-1}^1 \frac{1}{5x^2}\, dx$

17. $\int_{-\infty}^\infty \frac{1}{x^3}\, dx$

18. $\int_0^1 \ln x\, dx$

19. $\int_1^\infty \ln x\, dx$

20. $\int_1^\infty \frac{1}{x(x+1)}\, dx$

21. $\int_0^\infty \frac{1}{x(x+1)}\, dx$

22. $\int_e^\infty \frac{1}{x \ln x}\, dx$

23. $\int_{e^2}^\infty \frac{1}{x(\ln x)^2}\, dx$

24. $\int_0^\infty \frac{x}{2+x^2}\, dx$

25. $\int_0^\infty \frac{1}{\sqrt{x-1}}\, dx$

26. $\int_{-1}^1 \frac{1}{\sqrt{x+1}}\, dx$

27. $\int_1^5 \frac{1}{x-3}\, dx$

28. $\int_1^\infty \ln x\, dx$

29. $\int_1^\infty \frac{x}{\sqrt{3+x^2}}\, dx$

30. $\int_1^\infty \frac{\arctan x}{1+x^2}\, dx$

31. $\int_1^2 \frac{1}{x \ln x}\, dx$

32. $\int_1^\infty \arctan x\, dx$

33. $\int_0^\pi \tan x\, dx$

34. $\int_0^\infty \frac{x^2+3}{x+1}\,dx$

35. $\int_1^\infty \frac{1}{(x+1)^3}\,dx$

36. $\int_1^{e^2} \frac{dx}{x\sqrt{\ln x}}$

Use the comparison theorem to determine whether the integral is convergent or divergent.

37. $\int_1^\infty \frac{(\sin x)^2}{x^2}\,dx$

38. $\int_2^\infty \frac{1}{x(x+1)}\,dx$

39. $\int_2^\infty \frac{2}{x^2 \ln x}\,dx$

40. $\int_1^\infty \frac{1}{\sqrt{x^7+1}}\,dx$

41. $\int_0^\infty \sin x e^{-x}\,dx$

42. $\int_1^\infty \frac{\cos x}{x^2}\,dx$

43. (a) Show that $\int_1^\infty \frac{1}{1+x^4}\,dx$ converges.

   (b) Approximate $\int_1^\infty \frac{1}{1+x^4}\,dx$ with error < 0.01. This involves making some choices, but the gist should be as follows.

   i. Snip off the tail, $\int_c^\infty \frac{1}{1+x^4}\,dx$, for some constant c. Bound it using $\int_c^\infty \frac{1}{x^4}\,dx$.

   ii. Approximate $\int_1^c \frac{1}{1+x^4}\,dx$ using numerical methods.

   iii. Be sure the sum of the bound in part (i) and the error in part (ii) is less than 0.01.

44. The surface formed by revolving the graph of $y = \frac{1}{x}$ on $[1, \infty)$ about the x-axis is known as *Gabriel's horn*. Find the volume of the horn. *Curiously, you will find that the volume is finite even though the area under $y = \frac{1}{x}$ on $[1, \infty)$ is infinite.*

Probability Density Functions (Problems 45 through 48)

*As mentioned at the beginning of this section, statisticians use **probability density functions** to determine the probability of a random variable falling in a certain interval. If $p(x)$ is a probability density function, then $p(x) \geq 0$ for all x and $\int_{-\infty}^\infty p(x)\,dx = 1$.*

45. A probability density function of the form

$$p(x) = \begin{cases} \lambda e^{-\lambda x} & \text{for } x \geq 0, \\ 0 & \text{for } x < 0 \end{cases} \quad \text{where } \lambda \text{ is a positive constant}$$

describes what is known as an **exponential distribution**. Verify that

$$\int_{-\infty}^\infty p(x)\,dx = 1.$$

46. A **cumulative density function**, $C(x)$, gives the probability of a random variable taking on a value less than or equal to x. It is given by

$$C(x) = \int_{-\infty}^{x} p(y)\,dy.$$

Show that for an exponential distribution (refer to Problem 45), the cumulative density function is given by

$$C(x) = \begin{cases} 1 - e^{-\lambda x} & \text{for } x \geq 0, \\ 0 & \text{for } x < 0. \end{cases}$$

Find $\lim_{x \to \infty} C(x)$.

47. The mean of a probability distribution is given by

$$\mu = \int_{-\infty}^{\infty} xp(x)\,dx,$$

where $p(x)$ is the probability density function. Think of this as $p(x)$ giving a fractional weight to each value of x. Show that the mean of an exponential distribution (see Problem 45) is $\frac{1}{\lambda}$. (*Note:* You will need to use integration by parts to find μ.)

48. Suppose the number of minutes a caller spends on hold when calling a health clinic can be modeled using the probability density function.

$$p(x) = \begin{cases} 10e^{-10x} & \text{for } x \geq 0, \\ 0 & \text{for } x < 0. \end{cases}$$

The probability that a random caller will wait at least 5 minutes on hold is given by $\int_{5}^{\infty} p(x)\,dx$. Find this probability. *Note:* it is not necessary to compute an improper integral in order to answer this question.

49. Essay Question.

Two of your classmates are having some trouble with improper integrals.

Todd believes that improper integrals ought to diverge. He reasons that if f is positive, then the accumulated area keeps increasing, even if only by a little bit, so how can we get anything other than infinity?

Dylan, on the other hand, is convinced that if $\lim_{x \to \infty} f(x) = 0$, then $\int_{0}^{\infty} f(x)\,dx$ ought to converge. After all, he reasons, the rate at which area is accumulating is going to zero. Why isn't that enough to assure convergence?

Write an essay responding to Todd and Dylan's misconceptions. Your essay should be designed to help your classmates see the errors in their reasoning.

CHAPTER 30

Series

30.1 APPROXIMATING A FUNCTION BY A POLYNOMIAL

Preview

Addition and multiplication—these are our fundamental computational tools. A high-powered computer, for all its computational sophistication, ultimately relies on these basic operations. How then can a computer numerically approximate values of transcendental functions? How are values of exponential, logarithmic, and trigonometric functions computed?

Consider the sine function, for example. A calculator can approximate sin 0.1 with a high degree of accuracy, accuracy not readily accessible from unit circle or right triangle definitions of sin x. How can such a good approximation be obtained?

If we know the value of a differentiable function f at the point $x = b$, then we can use the tangent line to f at $x = b$ to approximate the function's values near $x = b$. The tangent line is the best linear approximation of f near $x = b$; higher degree polynomials offer the possibility of staying even closer to the values of f near $x = b$ and following the shape of f over a larger interval around b. In this section we will improve upon the tangent line approximation, obtaining quadratic, cubic, and higher degree polynomial approximations of f around $x = b$. We will generally find the "fit" improving with the degree of the polynomial.

Such polynomial approximations are convenient because they involve only the operations of addition and multiplication; they are easily evaluated, easily differentiated, and easily integrated.

The process of approximating an elusive quantity, successively refining the approximation, and using a limiting process to nail it down is at the heart of theoretical calculus. In this chapter we obtain successively better polynomial approximations of a function about a point by computing increasingly higher degree polynomial approximations. By computing the limit as the degree of the polynomial increases without bound, we will discover that, under certain conditions, we can represent a function as an infinite "polynomial" known as a power series. The fact that $\sin x$, $\cos x$, and e^x have representations as power series is remarkable in its own right. In addition, this alternative representation turns out to be computationally very useful. Power series representation of functions was known to Newton who used it as a computational aid, particularly for integrating functions lacking elementary antiderivatives. It was the subject of work published by the English mathematician Brook Taylor in 1712 and was popularized by the Scottish mathematician Colin Maclaurin in a textbook published in 1742. Although mathematicians had been using the ideas as early as the 1660s, the names of Taylor and Maclaurin have been associated with power series representations of functions.

Polynomial Approximations of $\sin x$ around $x = 0$

In this section we will use polynomials to numerically estimate values of some transcendental functions.

◆ **EXAMPLE 30.1**   A calculator or computer gives $\sin 0.1$ to ten decimal places, displaying 0.0998334166. Obtain this result by using a polynomial to approximate $\sin x$ near $x = 0$ and evaluating this polynomial at $x = 0.1$.

SOLUTION   We will approach this problem via a sequence of polynomial approximations to $f(x) = \sin x$ for x near zero until we arrive at the desired result. We denote by $P_k(x)$ the kth degree polynomial approximation. $P_k(x)$ is of the form $a_0 + a_1 x + a_2 x^2 + \cdots + a_k x^k$, where $a_0, a_1, \ldots, a_k$ are constants. We must determine the values of these constants so the $P_k(x)$ "fits" the graph of f well around $x = 0$.

Constant Approximation

Because $\sin x$ is continuous and 0.1 is near 0, we know $\sin 0.1 \approx \sin 0 = 0$.

$$P_0(x) = 0; \quad \sin 0.1 \approx P_0(0.1) = 0$$

Tangent Line Approximation

The tangent line passes through $(0, 0)$ and has a slope of $f'(0) = \cos 0 = 1$.

$$P_1(x) = x; \quad \sin 0.1 \approx P_1(0.1) = 0.1$$

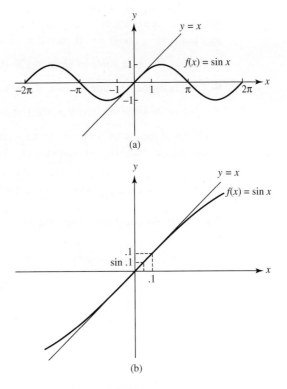

Figure 30.1

Refining the tangent line approximation: Any polynomial approximation, $P_k(x)$, of $\sin x$ about $x = 0$ certainly ought to be as good a local approximation as is the tangent line approximation, $P_1(x) = x$. Therefore, like $\sin x$ the graph of $P_k(x)$ must pass through $(0, 0)$ and must have a slope of 1 at $x = 0$. This means that

$$\begin{cases} P_k(0) = 0 & \text{and} \\ P_k'(0) = 1. \end{cases}$$

$$P_k(x) = a_0 + a_1 x + a_2 x^2 + \cdots + a_k x^k \qquad \text{so } P_k(0) = a_0 = 0.$$

$$P_k'(x) = a_1 + 2a_2 x + 3a_3 x^2 + \cdots + ka_k x^{k-1} \qquad \text{so } P_k'(0) = a_1 = 1.$$

Therefore, $P_k(x)$ is of the form $x + a_2 x^2 + a_3 x^3 + \cdots + a_k x^k$.

Note that $\sin x$ lies below the tangent line for $x > 0$ and above the tangent line for $x < 0$. Therefore the approximation $\sin x \approx x$ must be decreased for $x > 0$ and increased for $x < 0$ in order to improve upon it.

Second Degree Approximation

$P_2(x)$ must be of the form $x + a_2 x^2$, where a_2 is a constant. But $a_2 x^2$ cannot be negative for $x > 0$ and positive for $x < 0$, as required; we cannot improve upon the tangent line approximation by using a second degree polynomial. We need at least a third degree polynomial to improve upon the tangent line approximation.

Before moving on, let's look at the second degree polynomial approximation from a geometric viewpoint. If $P_2(x) = a_0 + a_1x + a_2x^2$ is to be the best parabolic approximation to $f(x) = \sin x$ about $x = 0$, then it must satisfy the following three conditions.

- It has the same value as $\sin x$ at $x = 0$. $\qquad\qquad P_2(0) = f(0)$

- It has the same slope as $\sin x$ at $x = 0$. $\qquad\qquad P_2'(0) = f'(0)$

- It has the same concavity as $\sin x$ at $x = 0$. $\qquad\quad P_2''(0) = f''(0)$

Each one of these conditions determines the value of one coefficient of $P_2(x)$. The first two result in $a_0 = 0$ and $a_1 = 1$, respectively. The second derivative of $\sin x$ at $x = 0$ is zero.

$$\frac{d^2}{dx^2} \sin x \Big|_{x=0} = \frac{d}{dx} \cos x \Big|_{x=0} = -\sin x \Big|_{x=0} = 0$$

The "parabola" must have a second derivative of zero; consequently, it is not a parabola at all.

Third Degree Approximation

To determine the coefficients of the third degree polynomial of best fit, we require that the polynomial, $P_3(x) = a_0 + a_1x + a_2x^2 + a_3x^3$, and $f(x) = \sin x$ agree at $x = 0$ and that each nonzero derivative of the polynomial is equal to the corresponding derivative of $\sin x$ at $x = 0$. These four conditions determine the four coefficients.

$$\begin{cases} P_3(0) = f(0) \\ P_3'(0) = f'(0) \\ P_3''(0) = f''(0) \\ P_3'''(0) = f'''(0) \end{cases}$$

We have already demonstrated that the first two conditions result in $a_0 = 0$, $a_1 = 1$. As an exercise, show that the third and fourth conditions require that $a_2 = 0$ and $a_3 = -\frac{1}{6}$, respectively.

$$P_3(x) = 0 + x + 0x^2 - \frac{1}{6}x^3 = x - \frac{1}{6}x^3.$$

Notice that the $-\frac{1}{6}x^3$ term is negative for $x > 0$ and positive for $x < 0$, providing an appropriate adjustment to the tangent line approximation. (See Figure 30.2.)

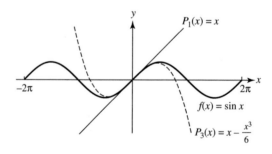

Figure 30.2

EXERCISE 30.1   Let $f(x) = \sin x$ and $P_3(x) = a_0 + a_1x + a_2x^2 + a_3x^3$. Calculate $P'(x)$, $P''(x)$, $P'''(x)$, $f'(x)$, $f''(x)$, $f'''(x)$, and evaluate each at $x = 0$. Show that the four conditions given above determine a_0, a_1, a_2, and a_3, respectively and that

$$a_0 = 0, \quad a_1 = 1, \quad a_2 = 0, \quad \text{and} \quad a_3 = -\frac{1}{6}.$$

Using the third degree polynomial to approximate $\sin 0.1$ gives

$$\sin 0.1 \approx P_3(0.1) = 0.1 - \frac{(0.1)^3}{6} = \frac{1}{10} - \frac{1}{6000} = 0.0998\overline{3}.$$

This matches the calculator estimate of $\sin 0.1$ to six decimal places. The actual value is a bit larger than $P_3(0.1)$.

Higher Degree Approximations

To find the kth degree polynomial approximation we require that $P_k(x)$ and $\sin x$ agree at $x = 0$ and each nonzero derivative of the polynomial matches that of $\sin x$.

The condition that the fourth derivatives agree ends up meaning that $a_4 = 0$, so we will proceed directly with the fifth degree polynomial.

Let $P_5(x) = a_0 + a_1 x + a_2 x^2 + a_3 x^3 + a_4 x^4 + a_5 x^5$. Requiring that all nonzero derivatives of $P_5(x)$ match the derivatives of $f(x) = \sin x$ at $x = 0$ means the following conditions must be satisfied.

$$\begin{cases} P_5(0) = f(0) \\ P_5'(0) = f'(0) \\ P_5''(0) = f''(0) \\ P_5'''(0) = f'''(0) \\ P_5^{(4)}(0) = f^{(4)}(0) \\ P_5^{(5)}(0) = f^{(5)}(0) \end{cases} \quad \text{where } f^{(p)} \text{ denotes the } p\text{th derivative of } f$$

As an exercise, show that these conditions determine a_0, a_1, a_2, a_3, a_4, and a_5, respectively, and that

$$P_5(x) = x - \frac{1}{6}x^3 + \frac{1}{120}x^5.$$

EXERCISE 30.2   Let $f(x) = \sin x$ and $P_5(x)$ be the fifth degree polynomial given above. Show that the six conditions stated mean that

$$a_0 = f(0), \quad a_1 = f'(0), \quad a_2 = \frac{f''(0)}{2!}, \quad a_3 = \frac{f'''(0)}{3!}, \quad a_4 = \frac{f^{(4)}(0)}{4!}, \quad a_5 = \frac{f^{(5)}(0)}{5!},$$

where $n! = n \cdot (n-1) \cdots 3 \cdot 2 \cdot 1$. Conclude that

$$a_0 = 0, \quad a_1 = 1, \quad a_2 = 0, \quad a_3 = \frac{-1}{3!} = -\frac{1}{6}, \quad a_4 = 0 \quad \text{and} \quad a_5 = \frac{1}{5!} = \frac{1}{120}.$$

Using the fifth degree polynomial approximation to $\sin x$ to approximate $\sin(0.1)$ gives

$$\sin 0.1 \approx P_5(0.1) = 0.1 - \frac{(0.1)^3}{6} + \frac{(0.1)^5}{120} = \frac{1}{10} - \frac{1}{6000} + \frac{1}{12 \cdot 10^6} \approx 0.0998334166.$$

This agrees with the 10 decimal places given for $\sin 0.1$.

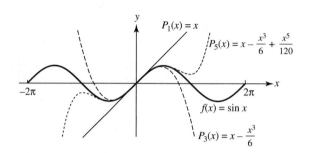

Figure 30.3

The graphs of $\sin x$, $P_1(x)$, $P_3(x)$, and $P_5(x)$ are given in Figure 30.3.   ◆

EXERCISE 30.3   Using a computer or graphing calculator, graph $f(x) = \sin x$, $P_1(x) = x$, $P_3(x) = x - \frac{x^3}{3!}$, and $P_5(x) = x - \frac{x^3}{3!} + \frac{x^5}{5!}$. Zoom in around $x = 0$ and observe how well each polynomial approximates the values of $\sin x$ near $x = 0$. Try to guess formulas for $P_7(x)$, $P_9(x)$, and $P_{11}(x)$. Graph these as well and decide how much confidence you have in your answers.

Below is a table of values given to 10 decimal places.

x	$\sin x$	$P_1(x)$	$P_3(x)$	$P_5(x)$
−0.2	−0.1986693308	−0.2	−0.1986666667	−0.1986693333
−0.1	−0.0998334166	−0.1	−0.0998333333	−0.0998334167
0	0	0	0	0
0.1	0.0998334166	0.1	0.0998333333	0.0998334167
0.5	0.4794255386	0.5	0.4791666667	0.4794270833
1	0.8414709848	1	0.8333333333	0.8416666667
2	0.9092974268	2	0.6666666667	0.9333333333
2.5	0.5984721441	2.5	−0.1041666667	0.7096354167

OBSERVATIONS From the graphical and numerical data gathered we observe that

   i. for a fixed x near zero, the higher the degree of the polynomial approximation the better its value approximates that of $\sin x$, and

   ii. the higher the degree of the polynomial, the further away from zero the approximation is reasonable.

NOTE In all the work we've done with $\sin x$, x *must* be in radians, *not* in degrees. $\frac{d}{dx} \sin x = \cos x$ only for x in radians.

Taylor Polynomial Approximations

In the previous example we constructed polynomial approximations to $f(x) = \sin x$ around $x = 0$ by choosing the coefficients of the polynomial such that the polynomial and all its nonzero derivatives matched f and its corresponding derivatives at $x = 0$. This method of constructing polynomial approximations to a function f about a number $x = b$ in its domain is remarkably useful.

Let f be a function whose first n derivatives exist at $x = b$. For the sake of simplicity, we begin with the case $b = 0$.

Definition

The nth degree polynomial, $P_n(x)$, that is equal to $f(0)$ when evaluated at $x = 0$ and whose first n derivatives are equal to those of $f(x)$ when evaluated at $x = 0$ is called **the nth degree Taylor polynomial generated by** f **at** $x = 0$. The polynomial is said to be **centered** at $x = 0$, or **expanded about** $x = 0$.

More generally, we can expand a function about $x = b$ using a polynomial in powers of $(x - b)$.

$$P_n(x) = a_0 + a_1(x - b) + a_2(x - b)^2 + a_3(x - b)^3 + \cdots + a_n(x - b)^n$$

Definition

The nth degree polynomial in powers of $(x - b)$ that is equal to $f(b)$ when evaluated at $x = b$ and whose first n derivatives match those of $f(x)$ at $x = b$ is called **the nth degree Taylor polynomial generated by** f **at** $x = b$. We refer to b as the **center** of the polynomial.

When evaluated at its center, a Taylor polynomial is equal to the value of its generating function. Our hope is that for x near the center the value of the polynomial is close to the value of the function.[1]

We now turn our attention to computing Taylor polynomials. In the next section we will look at the accuracy of Taylor polynomial approximations, and subsequently will see what we get by allowing the degree of the Taylor polynomial to increase without bound.

Computing a Taylor Polynomial Centered at $x = 0$

Suppose f and its first n derivatives exist at $x = 0$. We want to find constants $a_0, a_1, a_2, \ldots, a_n$ such that

$$P_n(x) = a_0 + a_1 x + a_2 x^2 + \cdots + a_n x^n$$

is the Taylor polynomial generated by f about $x = 0$.

We impose the following $(n + 1)$ conditions; each enables us to solve for one coefficient.

$$\begin{cases} P_n(0) = f(0) \\ P_n'(0) = f'(0) \\ P_n''(0) = f''(0) \\ P_n'''(0) = f'''(0) \\ \vdots \\ P_n^{(n)}(0) = f^{(n)}(0) \end{cases} \tag{30.1}$$

---

[1] While this is not *always* the case, often we will find it true.

In short, $P_n^{(k)}(0) = f^{(k)}(0)$ for $k = 0, 1, \ldots, n$.[2]

We begin by finding the first n derivatives of $P_n(x)$.

$$P_n(x) = a_0 + a_1 x + a_2 x^2 + a_3 x^3 \quad + a_4 x^4 \quad + \cdots + a_n x^n$$
$$P_n'(x) = \quad a_1 + 2a_2 x + 3a_3 x^2 \quad + 4a_4 x^3 \quad + \cdots + n a_n x^{n-1}$$
$$P_n''(x) = \quad 2 \cdot a_2 + 3 \cdot 2 \cdot a_3 x + 4 \cdot 3 \cdot a_4 x^2 \quad + \cdots + n(n-1) \cdot a_n x^{n-2}$$
$$P_n'''(x) = \quad 3 \cdot 2 \cdot a_3 \quad + 4 \cdot 3 \cdot 2 \cdot a_4 x + \cdots + n(n-1)(n-2) \cdot a_n x^{n-3}$$
$$P_n^{(4)}(x) = \quad 4 \cdot 3 \cdot 2 \cdot a_4 \quad + \cdots + n(n-1)(n-2)(n-3) \cdot a_n x^{n-4}$$
$$\vdots$$
$$P_n^{(n-1)}(x) = (n-1)(n-2)(n-3) \cdots 3 \cdot 2 \cdot a_{n-1} + n(n-1)(n-2) \cdots 3 \cdot 2 \cdot a_n x$$
$$P_n^{(n)}(x) = n(n-1)(n-2) \cdots 3 \cdot 2 \cdot a_n$$

Next we evaluate each expression at $x = 0$.

$$P_n(0) = a_0$$
$$P_n'(0) = a_1$$
$$P_n''(0) = 2 \cdot a_2$$
$$P_n'''(0) = 3 \cdot 2 \cdot a_3 = 3! a_3$$
$$P_n^{(4)}(0) = 4! a_4$$
$$\vdots$$
$$P_n^{(n-1)}(0) = (n-1)! a_{n-1}$$
$$P_n^{(n)}(0) = n! a_n$$

We summarize: $P_n^{(k)}(0) = k! a_k$ for $k = 0, 1, \ldots, n$. Returning to (30.1), the original $(n+1)$ conditions, we obtain[3]

$$k! a_k = f^{(k)}(0) \quad \text{for } k = 0, 1, \ldots, n.$$

Solving for a_k, the coefficient of x^k, we obtain $a_k = \frac{f^{(k)}(0)}{k!}$ for $k = 0, 1, \ldots, n$. We summarize our result.

The nth degree Taylor polynomial generated by $f(x)$ at $x = 0$ is given by

$$P_n(x) = f(0) + f'(0)x + \frac{f''(0)}{2!}x^2 + \frac{f'''(0)}{3!} + \cdots + \frac{f^{(n)}(0)}{n!}x^n.$$

That is, $P_n(x) = \displaystyle\sum_{k=0}^{n} \frac{f^{(k)}(0)}{k!} x^k.$

This work behind us, we compute the nth degree Taylor polynomial generated by f about $x = 0$ as follows.

---

[2] Here we use the convention that $P^0(x) = P(x)$.
[3] Recall: $0! = 1$ and $f^{(0)}(x) = f(x)$.

1. Compute the first n derivatives of f.

Be alert to the possibility of patterns emerging. You improve your chances of noticing patterns by not multiplying out. For instance, $5 \cdot 4 \cdot 3 \cdot 2$ is easier to recognize as $5!$ than is 120.

2. Evaluate f and each of its derivatives at $x = 0$.

3. The coefficient of x^k is the constant $\dfrac{f^{(k)}(0)}{k!}$.

EXAMPLE 30.2   Find the nth degree Taylor polynomial generated by e^x at $x = 0$.

SOLUTION

$$P_n(x) = f(0) + f'(0)x + \frac{f''(0)}{2!}x^2 + \cdots + \frac{f^{(n)}(0)}{n!}x^n$$

The derivative of e^x is e^x; therefore $f^{(k)}(0) = e^0 = 1$ for $k = 0, 1, 2, \ldots, n$. Thus

$$P_n(x) = 1 + x + \frac{x^2}{2!} + \frac{x^3}{3!} + \cdots + \frac{x^n}{n!}. \quad \blacklozenge$$

Graphs of e^x and several of its Taylor polynomials are shown in Figure 30.4. Note that $P_1(x) = 1 + x$ is simply the tangent line approximation to e^x at $x = 0$.

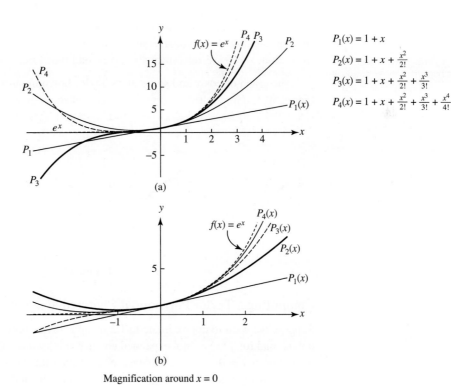

$$P_1(x) = 1 + x$$
$$P_2(x) = 1 + x + \frac{x^2}{2!}$$
$$P_3(x) = 1 + x + \frac{x^2}{2!} + \frac{x^3}{3!}$$
$$P_4(x) = 1 + x + \frac{x^2}{2!} + \frac{x^3}{3!} + \frac{x^4}{4!}$$

(a)

(b)

Magnification around $x = 0$

Figure 30.4

On the following page is a table of values produced using Taylor polynomials for e^x. Values are given to nine decimal places.

x	e^x	$P_1(x)$	$P_2(x)$	$P_3(x)$	$P_4(x)$	$P_5(x)$
0.1	1.105170918	1.1	1.105	1.105166666	1.105170833	1.105170917
0.2	1.221402758	1.2	1.22	1.221333333	1.221400000	1.221402667
0.5	1.648721271	1.5	1.625	1.645833333	1.6484375	1.648697917
1	2.718281828	2	2.5	2.66666666	2.708333333	2.716666666

◆ **EXAMPLE 30.3**   Find the 8th degree Taylor polynomial generated by $f(x) = \cos x$ about $x = 0$.

SOLUTION $P_8(x) = f(0) + f'(0)x + \frac{f''(0)}{2!}x^2 + \cdots + \frac{f^{(8)}(0)}{8!}x^8$

$f(x) = \cos x$	$f(0) = 1$	$f'(x) = -\sin x$	$f'(0) = 0$
$f''(x) = -\cos x$	$f''(0) = -1$	$f'''(x) = \sin x$	$f'''(0) = 0$
$f^{(4)}(x) = \cos x$	$f^{(4)}(0) = 1$	$f^{(5)}(x) = -\sin x$	$f^{(5)}(0) = 0$
$f^{(6)}(x) = -\cos x$	$f^{(6)}(0) = -1$	$f^{(7)}(x) = \sin x$	$f^{(7)}(0) = 0$
$f^{(8)}(x) = \cos x$	$f^{(8)}(0) = 1$		

$$P_8(x) = 1 - \frac{x^2}{2!} + \frac{x^4}{4!} - \frac{x^6}{6!} + \frac{x^8}{8!}$$

Notice that the coefficients of all the odd power terms are zero. This makes sense; cosine is an even function. Analogously, the coefficients of all even power terms in the expansion of $\sin x$ about $x = 0$ are zero since $\sin x$ is an odd function.   ◆

The graph of $\cos x$ and some of its Taylor polynomials centered at $x = 0$ are given in Figure 30.5. (Graph them yourself and you can zoom in around $x = 0$.)

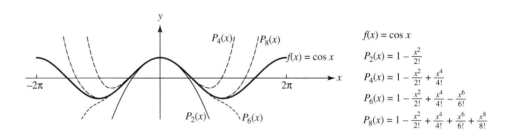

$f(x) = \cos x$

$P_2(x) = 1 - \frac{x^2}{2!}$

$P_4(x) = 1 - \frac{x^2}{2!} + \frac{x^4}{4!}$

$P_6(x) = 1 - \frac{x^2}{2!} + \frac{x^4}{4!} - \frac{x^6}{6!}$

$P_8(x) = 1 - \frac{x^2}{2!} + \frac{x^4}{4!} + \frac{x^6}{6!} + \frac{x^8}{8!}$

Figure 30.5

Computing a Taylor Polynomial Centered at $x = b$

Suppose we want to approximate $\ln 1.2$ using a Taylor polynomial. We can't use a Taylor polynomial for $f(x) = \ln x$ expanded about $x = 0$ because neither f nor any of its derivatives exist at $x = 0$. We can, however, either center the Taylor polynomial for $\ln(1 + x)$ at $x = 0$ or work with the Taylor polynomial for $\ln x$ expanded about $x = 1$. We will do the latter. First we will look at how to compute a Taylor polynomial centered at $x = b$.

Recall that the nth degree Taylor polynomial for $f(x)$ at $x = b$ is an nth degree polynomial in powers of $x - b$,

$$P_n(x) = a_0 + a_1(x - b) + a_2(x - b)^2 + \cdots + a_n(x - b)^n,$$

such that the values of $P_n(x)$ and its nonzero derivatives are equal to those of $f(x)$ when evaluated at $x = b$. That is, the coefficients $a_0, a_1, a_2, \ldots, a_n$ are determined by the conditions

$$P_n^{(k)}(b) = f^{(k)}(b) \quad \text{for } k = 0, 1, 2, \ldots, n. \tag{30.2}$$

As an exercise, calculate the first n derivatives of $P_n(x)$. (Don't multiply out $(x - b)^k$; use the Chain Rule.) Evaluating these derivatives at $a = b$, conclude that

$$P_n^{(k)}(b) = k!a_k.$$

This result, together with (30.2) enables us to solve for a_k, $k = 0, 1, 2, \ldots, n$.

$$a_k = \frac{f^{(k)}(b)}{k!}$$

The nth degree Taylor polynomial generated by $f(x)$ around $x = b$ is given by

$$P_n(x) = f(b) + f'(b)(x - b) + \frac{f''(b)}{2!}(x - b)^2 + \cdots + \frac{f^{(n)}(b)}{n!}(x - b)^n.$$

That is,[4] $P_n(x) = \displaystyle\sum_{k=0}^{n} \frac{f^{(k)}(b)}{k!}(x - b)^k.$

Given a particular function, f, and center, b, we compute the first n derivatives of f, evaluate each at $x = b$, and use $\frac{f^{(k)}(b)}{k!}$ as the coefficient of $(x - b)^k$.

Note:

- The equation of the line tangent to f at $x = b$ is of the form $y - y_1 = m(x - x_1)$, where $(x_1, y_1) = (b, f(b))$ and $m = f'(b)$. The equation is therefore $y = f(b) + f'(b)(x - b)$; this is $P_1(x)$, as expected.

- When $b = 0$ we're back to the Taylor polynomial centered at $x = 0$.

◆ **EXAMPLE 30.4**   (a) Find the nth degree Taylor polynomial for $\ln x$ centered at $x = 1$.

(b) Use P_5 to estimate $\ln(1.2)$.

SOLUTION   (a) $P_n(x) = f(1) + f'(1)(x - 1) + \frac{f''(1)}{2!}(x - 1)^2 + \frac{f'''(1)}{3!}(x - 1)^3 + \cdots + \frac{f^{(n)}(1)}{n!}(x - 1)^n$

---

[4] In order to use this summation notation we must adopt the convention that $(x - b)^0 = 1$ even if $x = b$.

Compute derivatives of $\ln x$, looking for a pattern.

$$
\begin{aligned}
f(x) &= \ln x & f(1) &= 0 \\
f'(x) &= \tfrac{1}{x} = x^{-1} & f'(1) &= 1 \\
f''(x) &= -x^{-2} & f''(1) &= -1 = -1! \\
f'''(x) &= 2 \cdot x^{-3} & f'''(1) &= 2 = 2! \\
f^{(4)}(x) &= -3 \cdot 2 \cdot x^{-4} & f^{(4)}(1) &= -3 \cdot 2 = -3! \\
f^{(5)}(x) &= 4 \cdot 3 \cdot 2 \cdot x^{-5} & f^{(5)}(1) &= 4 \cdot 3 \cdot 2 = 4! \\
&\vdots & &\vdots \\
f^{(n)}(x) &= (-1)^{n+1}(n-1)!x^{-n} & f^{(n)}(1) &= (-1)^{n+1}(n-1)!
\end{aligned}
$$

$$
P_n(x) = 0 + 1(x-1) + \frac{-1}{2!}(x-1)^2 + \frac{2!}{3!}(x-1)^3 + \frac{-3!}{4!}(x-1)^4 +
$$
$$
\cdots + \frac{(-1)^{n+1}(n-1)!}{n!}(x-1)^n
$$
$$
P_n(x) = (x-1) - \frac{(x-1)^2}{2} + \frac{(x-1)^3}{3} - \frac{(x-1)^4}{4} + \cdots + (-1)^{n+1}\frac{(x-1)^n}{n}.
$$

More compactly,

$$
P_n(x) = \sum_{k=0}^{n} \frac{f^{(k)}(1)}{k!}(x-1)^k = \sum_{k=0}^{n} \frac{(-1)^{k+1}(k-1)!}{k!}(x-1)^k
$$
$$
= \sum_{k=0}^{n}(-1)^{k+1}\frac{\cancel{(k-1)!}}{k \cdot \cancel{(k-1)!}}(x-1)^k
$$
$$
= \sum_{k=0}^{n}(-1)^{k+1}\frac{(x-1)^k}{k}.
$$

(b)
$$
\ln x \approx P_5(x) = (x-1) - \frac{(x-1)^2}{2} + \frac{(x-1)^3}{3} - \frac{(x-1)^4}{4} + \frac{(x-1)^5}{5}
$$
$$
\ln 1.2 \approx P_5(1.2) = 0.2 - \frac{(0.2)^2}{2} + \frac{(0.2)^3}{3} - \frac{(0.2)^4}{4} + \frac{(0.2)^5}{5} = 0.182330\overline{6}
$$

Compare this with the actual value of $\ln 1.2$; it matches for the first four decimal places.

◆

Aside: Dealing with Factorials and Alternating Signs

■ Factorials: Parentheses are important.
$$
(2n)! = (2n) \cdot (2n-1) \cdot (2n-2) \cdots 3 \cdot 2 \cdot 1
$$
$$
= 2n \cdot (2n-1)!
$$

On the other hand, $2n! = 2 \cdot n! = 2[n \cdot (n-1) \cdots 3 \cdot 2 \cdot 1]$. Similarly, $(2n+1)! \neq 2n+1! = 2n+1$.

■   Alternating signs:
$(-1)^k$ and $(-1)^{k+1}$ can be used to indicate alternating signs. Which is needed to do the job is determined by the notational system you happen to have chosen. The simplest way of determining which you need is by trial and error. Try $(-1)^k$ and check it with a particular k-value. If it doesn't work, switch to $(-1)^{k+1}$.

◆ **EXAMPLE 30.5**   Approximate $\sqrt[5]{34}$ using the appropriate second degree Taylor polynomial.

SOLUTION   Let $f(x) = x^{\frac{1}{5}}$. We must center the Taylor polynomial at a point near 34 at which the values of f, f', and f'' can be readily computed.

An off-the-cuff approximation of $\sqrt[5]{34}$ is $\sqrt[5]{34} \approx \sqrt[5]{32} = 2$; we know that $\sqrt[5]{34}$ is a bit more than 2. Center the Taylor polynomial at $x = 32$.

$$P_2(x) = f(32) + f'(32)(x-32) + \frac{f''(32)}{2!}(x-32)^2$$

$$f(x) = x^{\frac{1}{5}} \qquad\qquad f(32) = 2$$

$$f'(x) = \frac{1}{5}x^{-\frac{4}{5}} \qquad f'(32) = \frac{1}{5}\frac{1}{32^{\frac{4}{5}}} = \frac{1}{5}\frac{1}{2^4} = \frac{1}{80}$$

$$f''(x) = \frac{-4}{25}x^{-\frac{9}{5}} \qquad f''(32) = \frac{-4}{25}\frac{1}{2^9} = \frac{-1}{25\cdot 2^7} = \frac{-1}{3200}$$

Therefore,

$$P_2(x) = 2 + \frac{1}{80}(x-32) - \frac{1}{6400}(x-32)^2.$$

$$\sqrt[5]{34} \approx P_2(34) = 2 + \frac{1}{80}(2) - \frac{4}{6400} = 2 + \frac{1}{40} - \frac{1}{1600} = 2.024375$$

This agrees with the actual value of $\sqrt[5]{34}$ to four decimal places.   ◆

If you study closely the numerical data in this section you can start to get a sense of the magnitude of the error involved in a Taylor polynomial approximation. The size of the error can be estimated by graphing $f(x) - P_n(x)$ using a calculator or computer. In the next section we will state Taylor's Theorem, which will provide not only a method of estimating errors independent of a calculator, but also an invaluable theoretical tool.

PROBLEMS FOR SECTION 30.1

For Problems 1 through 7, do the following.

(a) Compute the fourth degree Taylor polynomial for $f(x)$ at $x = 0$.

(b) On the same set of axes, graph $f(x)$, $P_1(x)$, $P_2(x)$, $P_3(x)$, and $P_4(x)$.

(c) Use $P_1(x)$, $P_2(x)$, $P_3(x)$, and $P_4(x)$ to approximate $f(0.1)$ and $f(0.3)$. Compare these approximations to those given by a calculator.

1. $f(x) = e^{-x}$

2. $f(x) = \ln(1 + x)$

3. $f(x) = \tan^{-1} x$

4. $f(x) = (1 + x)^4$

5. $f(x) = \sqrt{1 + x}$

6. $f(x) = 2x^4 - 3x^2 + x - 1$

7. $f(x) = (1 + x)^{-2}$

8. Below is a graph of $f(x)$. For each quadratic given, explain why the quadratic could *not* be the second degree Taylor polynomial for $f(x)$ at $x = 0$.
   (a) $2 + 3x - \frac{1}{2}x^2$
   (b) $-1 - 5x + 2x^2$
   (c) $-2 + 2x - \frac{1}{3}x^2$

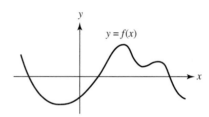

9. Let $P_2(x) = a_0 + a_1x + a_2x^2$ be the second degree Taylor polynomial generated by $f(x)$ at $x = 0$, where the graph of f is the one given in Problem 8 above. Use the graph to determine the signs of a_0, a_1, and a_2.

10. (a) Find the second degree Taylor polynomial generated by $\sec x$ at $x = 0$.
    (b) Graph $P_2(x)$ and $\sec x$ on the same set of axes.

11. (a) Compute the third degree Taylor polynomial for $\tan x$ about $x = 0$.
    (b) Why is it reasonable to expect the coefficient of the x^2 term to be zero?

12. The graph of a differentiable function $f(x)$ is given. Use the graph to determine the signs of the coefficients of the second degree Taylor polynomials indicated. f has a minimum at $x = 0$ and a point of inflection at $x = 2$.
    (a) $P_2(x) = a_0 + a_1x + a_2x^2$
    (b) $P_2(x) = a_0 + a_1(x - 1) + a_2(x - 1)^2$
    (c) $P_2(x) = a_0 + a_1(x - 2) + a_2(x - 2)^2$
    (d) $P_2(x) = a_0 + a_1(x - 3) + a_2(x - 3)^2$

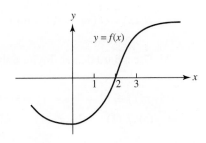

13. Let $f(x) = \ln(1 + x)$. Find the nth degree Taylor polynomial generated by f about $x = 0$.

14. Compute the nth degree Taylor polynomial expansion of $f(x) = \frac{1}{x}$ about $x = 1$. Graph f and P_1, P_2, P_3, and P_4 on a common set of axes.

In Problems 15 through 18, use a second degree Taylor polynomial centered appropriately to approximate the expression given.

15. $\sqrt[3]{8.3}$

16. $\sqrt{103}$

17. $\tan^{-1}(0.75)$

18. $\sqrt[3]{29}$

19. Compute the third degree Taylor polynomial generated by $\sin x$ at $x = \frac{\pi}{4}$.

20. Find the fifth degree Taylor polynomial for $\sqrt{x}$ centered at $x = 9$.

21. Write the third degree Taylor polynomial centered about $x = 0$ for $f(x) = \frac{1}{(1+x)^p}$, where p is constant.

Introduction to Error Analysis: Problems 22 and 23.

22. Let $f(x) = e^x$. Use the data given in the table on page 928 to compute the following.
   (a) $f(0.1) - P_k(0.1)$ for $k = 1, 2, \ldots, 5$
   (b) $f(0.2) - P_k(0.2)$ for $k = 1, 2, \ldots, 5$
   (c) $f(0.5) - P_k(0.5)$ for $k = 1, 2, \ldots, 5$
   (d) $f(1) - P_k(1)$ for $k = 1, 2, \ldots, 5$
   Compare the size of the difference between the actual value of the function and the polynomial approximation with that of the first "unused" term of the Taylor polynomial— that is, the last term of the next higher degree polynomial—and observe that they have the same order of magnitude.

23. Let $f(x) = e^x$ and let $P_k(x)$ be its kth degree Taylor polynomial about $x = 0$. Graph $R_k(x) = f(x) - P_k(x)$ for $k = 1, 2, \ldots, 5$.

24. Use a third degree Taylor polynomial to approximate $\ln 0.9$.

25. $f(x) = 12 + 3(x - 1) + 5(x - 1)^2 + 7(x - 1)^3$. Find the following.
    (a) $f'(1)$ (b) $f''(1)$ (c) $f'''(1)$ (d) $f(1)$

26. $f(x) = \sqrt{3} + 12(x - 5)^3 + 17(x - 5)^6$. Find the following.
    (a) $f(5)$ (b) $f''(5)$ (c) $f'''(5)$ (d) $f^{(6)}(5)$

27. Compute the sixth degree Taylor polynomial generated by $\sin x$ about $x = \pi$.

28. Compute the sixth degree Taylor polynomial generated by $\cos x$ about $x = -\frac{\pi}{2}$.

29. Let $f(x) = (1 + x)^p$, where p is a constant, $p \neq 0, 1, 2, 3, 4, 5$.
    (a) Compute the third degree Taylor polynomial for $f(x)$ around $x = 0$.
    (b) Compute the fifth degree Taylor polynomial for $f(x)$ around $x = 0$.

30. Using the results of Problem 29(a), approximate the following. Compare your results with the numerical approximations given by a calculator.
    (a) $\sqrt{1.002}$ (b) $\frac{1}{1.03}$ (c) $\sqrt[3]{1.001}$

30.2   ERROR ANALYSIS AND TAYLOR'S THEOREM

An approximation is of limited use unless we have a notion of the magnitude of the error involved. Every Taylor polynomial $P_n(x)$ has an associated error function, $R_n(x)$, defined by

$$f(x) = \qquad P_n(x) \qquad + \qquad R_n(x)$$

$$\text{function} = \begin{pmatrix} \text{polynomial} \\ \text{approximation} \end{pmatrix} + \begin{pmatrix} \text{associated} \\ \text{error} \end{pmatrix}$$

$R_n(x)$ is referred to as the **Taylor remainder**; $R_n(x) = f(x) - P_n(x)$.

For a Taylor polynomial centered at $x = b$ we expect the magnitude of the remainder to decrease as n increases and as x approaches b. Because each successive refinement of a Taylor polynomial involves a higher derivative, we might expect $R_n(x)$ to involve the $(n + 1)$st derivative of f. While Taylor's Theorem does not pin down the remainder precisely, it provides a means of putting an upper bound on the magnitude of the error.

Taylor's Theorem

Suppose f and all its derivatives exist in an open interval I centered at $x = b$. Then for each x in I

$$f(x) = f(b) + f'(b)(x - b) + \frac{f''(b)}{2!}(x - b)^2 + \frac{f'''(b)}{3!}(x - b)^3 + \cdots +$$

$$\frac{f^{(n)}(b)}{n!}(x - b)^n + R_n(x),$$

where

$$R_n(x) = \frac{f^{(n+1)}(c)}{(n + 1)!}(x - b)^{n+1} \quad \text{for some number } c \text{ in } I, c \text{ between } x \text{ and } b.$$

Note that $R_n(x)$ has the same form as the next term of a Taylor polynomial except that the $(n + 1)$st derivative is evaluated at some c between x and b instead of at b itself. Its form agrees with the expectations laid out before the statement of Taylor's Theorem. When applying the theorem we do not expect to be able to find c; if we could, an approximation wouldn't have been needed.

In practice, we look for a bound, M, such that $|f^{(n+1)}(c)| \leq M$ for all c between x and b and use the inequality

$$|R_n(x)| \leq \frac{M}{(n + 1)!}|x - b|^{n+1}.$$

This is referred to as Taylor's Inequality.

A sketch of the proof of Taylor's Theorem is given in Appendix H.

Let's revisit some of the problems from the previous section and see what information Taylor's remainder provides about the accuracy of approximations.

◆ **EXAMPLE 30.6** Give a good[5] upper bound for the error involved in estimating $\sin 0.1$ using the approximation $\sin x \approx x - \frac{x^3}{3!}$.

SOLUTION We can call $x - \frac{x^3}{3!}$ either $P_3(x)$ or $P_4(x)$, the two being equal. We'll call it $P_4(x)$ as this will give a better bound on the error.

$$f(0.1) = P_4(0.1) + R_4(0.1)$$

$$\sin(0.1) = \left[0.1 - \frac{(0.1)^3}{3!}\right] + R_4(0.1)$$

Taylor's Theorem says $R_n(x) = \frac{f^{(n+1)}(c)}{(n+1)!}(x - b)^{n+1}$ for some c between x and b. In this example $n = 4$, $f(x) = \sin x$, $b = 0$, and $x = 0.1$.

$$R_4(0.1) = \frac{f^{(5)}(c)}{5!}(0.1)^5$$

---

[5] We say "good" because 1 million, for instance, is an upper bound, but not what we are aiming for.

The derivatives of $\sin x$ are $\pm \sin x$ and $\pm \cos x$, so $|f^{(5)}(c)| \leq 1$.

$$0 \leq |R_4(0.1)| \leq \frac{1}{5!} \frac{1}{10^5} = \frac{1}{120 \cdot 10^5} = \frac{1}{1.2 \times 10^7} = 8.3 \times 10^{-8}. \quad \blacklozenge$$

◆ **EXAMPLE 30.7**   We want to use an nth degree Taylor polynomial for e^x centered at $x = 0$ to approximate e. How large must n be to assure that the answer differs from e by no more than 10^{-7}? *Assume we know $e < 3$.*

SOLUTION $|R_n(x)| = \frac{|f^{(n+1)}(c)|}{(n+1)!}|x - b|^{n+1}$ for some c between x and b. In this example $f(x) = e^x$, $b = 0$, and $x = 1$.

$$|f^{(n+1)}(c)| = |e^c| = e^c \text{ for } c \text{ between } 0 \text{ and } 1.$$

$$e^c \text{ increases with } c, \text{ so } e^c < e^1 < 3.$$

$$|R_n(1)| \leq \frac{3}{(n+1)!} \cdot 1 = \frac{3}{(n+1)!}$$

We must find an integer n such that $\frac{3}{(n+1)!} \leq \frac{1}{10^7}$, or, equivalently,

$$(n+1)! \geq 3 \cdot 10^7.$$

We find n by trial and error. $11! = 39,916,800 > 3 \cdot 10^7$, whereas $10!$ is not large enough.

$$n + 1 = 11, \text{ so } n = 10.$$

We must use the 10th degree Taylor polynomial: $P_{10}(x) = \sum_{k=0}^{10} \frac{x^n}{k!}$.

Checking, we see that $\sum_{k=0}^{10} \frac{1}{k!} = 1 + 1 + \frac{1}{2!} + \frac{1}{3!} + \cdots + \frac{1}{10!} \approx 2.718281801$, which differs from e by less than 10^{-7}. In fact, we solved the problem efficiently; had we used one less term of the expansion, the error would have been more than 10^{-7}. ◆

◆ **EXAMPLE 30.8**   In Example 30.5 we approximated $\sqrt[5]{34}$ using a second degree Taylor polynomial centered at $x = 32$. Find a reasonable upper bound for the magnitude of the error.

SOLUTION $|R_n(x)| = \frac{|f^{(n+1)}(c)|}{(n+1)!}|x - b|^{n+1}$ for some c between x and b. In this example $n = 2$, $f(x) = x^{\frac{1}{5}}$, $b = 32$, and $x = 34$.

$$|R_2(34)| = \frac{|f'''(c)|}{3!}|34 - 32|^3$$

$$|R_2(34)| = \frac{|f'''(c)|}{6} \cdot 8 \quad \text{for some } c \text{ between } 32 \text{ and } 34.$$

We must find M such that $|f'''(c)| \leq M$.

$$f'(x) = \frac{1}{5}x^{-\frac{4}{5}}; \quad f''(x) = \frac{-4}{25}x^{-\frac{9}{5}}; \quad f'''(x) = \frac{36}{125}x^{\frac{-14}{5}}$$

$$|f'''(c)| = \frac{36}{125}\frac{1}{(\sqrt[5]{c})^{14}} \quad \text{for some } c \text{ between } 32 \text{ and } 34.$$

The smaller c, the larger $|f'''(c)|$, so

$$0 \le |f'''(c)| \le \frac{36}{125}\frac{1}{(\sqrt[5]{32})^{14}} = \frac{36}{125}\frac{1}{2^{14}} = \frac{9}{125 \cdot 2^{12}}$$

$$0 \le |R_2(34)| \le \frac{9}{125 \cdot 2^{12}}\frac{1}{6}(8) = \frac{3}{125 \cdot 2^{10}} = \frac{3}{128000} \approx 2.344 \times 10^{-5}.$$

The error is less than 2.4×10^{-5}.

Taylor's Theorem gave a good estimate of the error; the actual error involved in Example 30.5 is approximately 2.24×10^{-5}.   ◆

If a computer or graphing calculator is at our disposal, error estimates can be readily available. Suppose, for example, that we plan to use the third degree Taylor polynomial for $\ln x$ centered at $x = 1$ in order to approximate $\ln x$ for $x \in [0.3, 1.7]$. We want an upper bound for the error involved in doing so. In other words, for $x \in [0.3, 1.7]$ we use the approximation

$$\ln x \approx (x - 1) - \frac{(x - 1)^2}{2} + \frac{(x - 1)^3}{3}$$

and want an estimate of $R_3(x) = \ln x - \left[(x - 1) - \frac{(x-1)^2}{2} + \frac{(x-1)^3}{3}\right]$. We can simply graph $|R_3(x)|$ on $[0.3, 1.7]$, obtaining the graph shown in Figure 30.6. Using the tracer we estimate that the magnitude of the error is less than 0.145.

As an exercise, use Taylor's Remainder to estimate the error.

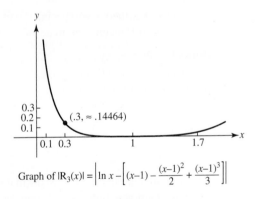

Graph of $|R_3(x)| = \left| \ln x - \left[(x-1) - \frac{(x-1)^2}{2} + \frac{(x-1)^3}{3}\right]\right|$

Figure 30.6

◆ **EXAMPLE 30.9**   Use graphical methods to find an upper bound for the error involved in using the tangent line approximation $1 - \frac{1}{2}x$ to approximate $\frac{1}{\sqrt{1+x}}$ for $|x| < 0.001$.

SOLUTION   Graph $R_2(x) = (1 + x)^{-\frac{1}{2}} - \left[1 - \frac{1}{2}x\right]$ on the domain $[-0.001, 001]$. (Play around with the range to obtain a useful graph.) The graph is given in Figure 30.7 on the following page.

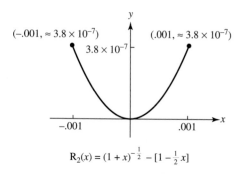

$$R_2(x) = (1 + x)^{-\frac{1}{2}} - [1 - \tfrac{1}{2}x]$$

Figure 30.7

For $|x| < 0.001$, the approximation $\frac{1}{\sqrt{1+x}} \approx 1 - \frac{1}{2}x$ produces an error of less than 4×10^{-7}. ◆

Any physicist will attest to the fact that physicists often use Taylor polynomials to simplify mathematical expressions. In fact, they often use only first or second degree polynomials. While this may at first strike you as a dubious strategy, the following example will demonstrate that in certain situations the error introduced is minimal.

◆ **EXAMPLE 30.10**   According to Newtonian physics, an object's kinetic energy, K, is given by

$$K = \frac{1}{2}m_0 v^2,$$

where m_0 is the mass of the object at rest and v is its velocity.

Einstein's theory of special relativity produces a more involved expression for K. According to Einstein, the mass of an object is a function of its velocity, $m = \frac{m_0}{\sqrt{1 - v^2/c^2}}$. Einstein's theory says energy, E, equals mc^2, where c is the speed of light. He concludes that an object's kinetic energy is given by the difference $mc^2 - m_0 c^2$. Using the expression for m, Einstein's theory says

$$K = m_0 c^2 \left[\frac{1}{\sqrt{1 - v^2/c^2}} - 1 \right] = m_0 c^2 \left[\left(1 - \frac{v^2}{c^2} \right)^{-\frac{1}{2}} - 1 \right]. \qquad (30.3)$$

Our goal in this example is to show that if an object is traveling much slower than the speed of light, then according to Einstein's theory, the error involved in using the Newtonian expression for K is small.

SOLUTION   We begin by noting that if v is substantially less than c, then $\frac{v}{c}$ is small, and $\left(\frac{v}{c}\right)^2$ is even smaller. From Example 30.9 we know that $(1 + x)^{-\frac{1}{2}}$ can be well approximated by its first degree Taylor polynomial, $1 - \frac{1}{2}x$, for $|x|$ small. Let $x = \frac{-v^2}{c^2}$. Using the approximation

$$\left(1 - \frac{v^2}{c^2} \right)^{-\frac{1}{2}} = \left(1 + \left(-\frac{v^2}{c^2} \right) \right)^{-\frac{1}{2}} \approx 1 - \frac{1}{2}\left(-\frac{v^2}{c^2} \right)$$

in Equation (30.3) we obtain

$$K = m_0 c^2 \left[\left(1 - \frac{v^2}{c^2} \right)^{-\frac{1}{2}} - 1 \right]$$

$$\approx m_0 c^2 \left[1 + \frac{1}{2}\frac{v^2}{c^2} - 1 \right] = m_0 c^2 \frac{1}{2}\frac{v^2}{c^2} = \frac{1}{2} m_0 v^2.$$

Let's estimate the size of the error introduced by using the Newtonian expression for K for an object traveling at speeds of 300 m/s or less. $c = 3 \cdot 10^8$ m/s.

We'll find an upper bound for the error in replacing $(1+x)^{-\frac{1}{2}}$ by $1 - \frac{1}{2}x$ for $|x| \leq \frac{300^2}{c^2}$ and multiply the answer by $m_0 c^2$.

$$|R_1(x)| = \frac{|f''(a)|}{2!}|x|^2 \qquad \text{for some } a \text{ between 0 and 300.}$$

$$f(x) = (1+x)^{-\frac{1}{2}}; \quad f'(x) = -\frac{1}{2}(1+x)^{-\frac{3}{2}}; \quad f''(x) = \frac{3}{4}(1+x)^{-\frac{5}{2}}$$

$$|R_1(x)| = \frac{3}{2 \cdot 4(\sqrt{1+a})^5}x^2 \leq \frac{3}{8\left(\sqrt{1 - \frac{300^2}{c^2}}\right)^5}\frac{300^4}{c^4} \approx 3.75 \times 10^{-25}$$

Multiplying by $m_0 c^2$ gives $m_0 3.375 \times 10^{-8}$.

Therefore, for speeds of up to 300 m/s, the error incurred in computing K using Newtonian physics is less than $3.4 \times 10^{-8} m_0$, where m_0 is the mass of the body at rest. ◆

PROBLEMS FOR SECTION 30.2

1. Find a good upper bound for the magnitude of the error involved in approximating $\cos x$ by $1 - \frac{x^2}{2!} + \frac{x^4}{4!}$ for $|x| \leq 0.2$. Do this using Taylor's Inequality; then check your answer by graphing the remainder function.

2. Use the third degree Taylor polynomial for e^x at $x = 0$ to estimate $\sqrt{e}$. Then use Taylor's Theorem to get a reasonable upper bound for the remainder.

3. We will use the nth degree Taylor polynomial for e^x, $\sum_{k=0}^{n} \frac{x^k}{k!}$, to approximate $\frac{1}{\sqrt{e}}$. What should n be in order to guarantee that the approximation is off by less than 10^{-5}?

4. Use the third degree Taylor polynomial for $\ln x$ centered at $x = 1$, $(x-1) - \frac{(x-1)^2}{2} + \frac{(x-1)^3}{3}$, to approximate $\ln(1.5)$. Then give an upper bound for the remainder using Taylor's Theorem.

5. The second degree Taylor polynomial for $f(x) = (1+x)^p$ is $1 + px + \frac{p(p-1)}{2!}x^2$. If the second degree Taylor polynomial is used to approximate $\sqrt{1+x}$ for $|x| \leq 0.2$, find an upper bound for the magnitude of the error. Use the Taylor Inequality; then check your answer by graphing $R_2(x)$.

6. For x near zero, $\cos x \approx 1 - \frac{x^2}{2!} + \frac{x^4}{4!} - \cdots + (-1)^n \frac{x^{2n}}{(2n)!}$. What degree Taylor polynomial must be used to approximate $\cos(0.2)$ with error less than $\frac{1}{10^8}$?

7. Approximate $\sqrt[3]{27.5}$ using an appropriate second degree Taylor polynomial. Find a good upper bound for the error by using Taylor's Inequality.

8. The second degree Taylor polynomial generated by $\ln(1 + x)$ about $x = 0$ is $x - \frac{x^2}{2}$. Use Taylor's Theorem to find a good upper bound on the error involved in using this polynomial to approximate the following.
   (a) $\ln(1.2)$ (b) $\ln(0.8)$

9. By graphing $R_2(x)$, estimate the values of x for which the approximation $\ln x \approx (x - 1) - \frac{(x-1)^2}{2!}$ can be used without producing an error of magnitude greater than 10^{-3}.

10. For x near zero, $e^x \approx 1 + x + \frac{x^2}{2!} + \frac{x^3}{3!}$. Find a reasonable upper bound for the magnitude of the error involved in using this approximation for $|x| < 0.5$. Use Taylor's Inequality and check your answer by graphing $R_3(x)$.

11. A hyena is loping down a straight path away from a stream. The hyena is 6 m from the stream, moving at a rate of 2 m/s and decelerating at a rate of 0.1 m/s². Use a second degree Taylor polynomial to estimate its distance from the stream 1 second later.

12. What degree Taylor polynomial for e^x about $x = 0$ must be used to approximate $e^{0.3}$ with error less than 10^{-5}?

13. (a) Find the nth degree Taylor polynomial for $f(x) = \frac{1}{1-x}$ centered at $x = 0$.
    (b) How many nonzero terms of the polynomial in part (a) must be used to approximate $f\left(\frac{1}{2}\right)$ with error less than 10^{-5}?

14. According to Einstein's theory of special relativity, the mass of an object moving with velocity v m/s is given by

$$m = \frac{m_0}{\sqrt{1 - \frac{v^2}{c^2}}},$$

where m_0 is the mass of the object at rest and c is the speed of light, $c = 3 \times 10^8$ m/s.
    (a) Use the first degree Taylor polynomial for $\frac{1}{\sqrt{1+x}}$ to arrive at the estimate

$$m \approx m_0 + \frac{m_0}{2} \frac{v^2}{c^2}.$$

    (b) If an object is moving at 100 m/s, find an upper bound for the error involved in using the approximation given in part (a).

30.3 TAYLOR SERIES

Defining Taylor Series

In many examples in this chapter we've observed that the higher the degree of the Taylor polynomial generated by f at $x = b$, the better it approximates $f(x)$ for x near b. For functions such as $\sin x$ and $\cos x$, the higher the degree of the Taylor polynomial the longer the interval over which the polynomial follows the undulations of the function's graph. Letting the degree of the polynomial increase without bound gives us the Taylor series for f.

Definition

If a function f has derivatives of all orders at $x = b$, then the **Taylor series of f at (or about)** $x = b$ is defined to be

$$f(b) + f'(b)(x - b) + \frac{f''(b)}{2!}(x - b)^2 + \cdots + \frac{f^{(n)}(b)}{n!}(x - b)^n + \cdots,$$

that is,

$$\sum_{k=0}^{\infty} \frac{f^{(k)}(b)}{k!}(x - b)^k.$$

We refer to this series as the Taylor *expansion* of f about $x = b$ or centered at $x = b$.

In the special case where $b = 0$, the series $\sum_{k=0}^{\infty} \frac{f^{(k)}(0)}{k!} x^k$ can be called the **Maclaurin series** for f.

From the work we've done with Taylor polynomials, we can easily find the Maclaurin series for e^x, $\sin x$, and $\cos x$.

◆ **EXAMPLE 30.11** Find the Maclaurin series for $f(x) = e^x$.

SOLUTION All derivatives of e^x are e^x. When evaluated at $x = 0$, e^x is 1. Maclaurin series for e^x:

$$1 + x + \frac{x^2}{2!} + \cdots + \frac{x^k}{k!} + \cdots = \sum_{k=0}^{\infty} \frac{x^k}{k!} \quad ◆$$

◆ **EXAMPLE 30.12** Find the Maclaurin series for $f(x) = \sin x$.

SOLUTION

Even order derivatives		**Odd order derivatives**	
$f(x) = \sin x$	$f(0) = 0$	$f'(x) = \cos x$	$f'(0) = 1$
$f''(x) = -\sin x$	$f''(0) = 0$	$f'''(x) = -\cos x$	$f'''(0) = -1$
$f^{(4)}(x) = \sin x$	$f^{(4)}(0) = 0$	$f^{(5)}(x) = \cos x$	$f^{(5)}(0) = 1$
$\vdots$	$\vdots$	$\vdots$	$\vdots$
$f^{(2k)}(x) = (-1)^k \sin x$	$f^{(2k)}(0) = 0$	$f^{(2k+1)}(x) = (-1)^k \cos x$	$f^{(2k+1)}(0) = (-1)^k$

Maclaurin series for $\sin x$:

$$x - \frac{x^3}{3!} + \frac{x^5}{5!} - \cdots + (-1)^k \frac{x^{2k+1}}{(2k + 1)!} + \cdots = \sum_{k=0}^{\infty} (-1)^k \frac{x^{2k+1}}{(2k + 1)!} \quad ◆$$

◆ **EXAMPLE 30.13**   Find the Maclaurin series for $f(x) = \frac{1}{1-x}$.

SOLUTION

$$f(x) = (1-x)^{-1} \qquad\qquad f(0) = 1$$
$$f'(x) = (1-x)^{-2} \qquad\qquad f'(0) = 1$$
$$f''(x) = 2(1-x)^{-3} \qquad\qquad f''(0) = 2$$
$$f'''(x) = 3\cdot 2(1-x)^{-4} \qquad\quad f'''(0) = 3!$$
$$\vdots \qquad\qquad\qquad\qquad \vdots$$
$$f^{(k)}(x) = k!(1-x)^{-(k+1)} \qquad f^{(k)}(0) = k!$$

Maclaurin series for $\frac{1}{1-x}$:

$$1 + x + \frac{2x^2}{2!} + \frac{3!x^3}{3!} + \frac{4!x^4}{4!} + \cdots + \frac{k!x^k}{k!} + \cdots$$

$$= 1 + x + x^2 + x^3 + \cdots + x^k + \cdots = \sum_{k=0}^{\infty} x^k$$

The Maclaurin series for $\frac{1}{1-x}$ should look familiar. $\sum_{k=0}^{\infty} x^k$ is a geometric series with $a = 1$ and $r = x$. Therefore we know that it converges to $\frac{1}{1-x}$ for $|x| < 1$ and diverges for $|x| \geq 1$. ◆

This observation at the end of Example 30.13 highlights the important question "What is the significance of the Taylor series for $f(x)$?" For instance, for what x does the Maclaurin series for $\sin x$ converge? When it converges, to *what* does it converge? In particular, does $\sin 0.1 = 0.1 - \frac{(0.1)^3}{3!} + \frac{(0.1)^5}{5!} - \cdots + (-1)^k \frac{(0.1)^{2k+1}}{(2k+1)!} + \cdots?$
Or, more generally, for *which* values of x is it true that

$$\sin x = x - \frac{x^3}{3!} + \frac{x^5}{5!} - \frac{x^7}{7!} + \cdots + (-1)^k \frac{x^{2k+1}}{(2k+1)!} + \cdots?$$

These latter questions can be answered using Taylor's Theorem.

$$f(x) = P_n(x) + R_n(x)$$

Taking the limit as n increases without bound gives

$$f(x) = \lim_{n\to\infty} P_n(x) + \lim_{n\to\infty} R_n(x).$$

Therefore, $f(x)$ is the sum of its Taylor series if and only if $\lim_{n\to\infty} R_n(x) = 0$. We state this more precisely below.

Theorem on Convergence of Taylor Series

If f is infinitely differentiable on an interval I centered around $x = b$, then the Taylor series for f at $x = b$ converges to $f(x)$ for all $x \in I$ if and only if $\lim_{n\to\infty} R_n(x) = 0$ for all $x \in I$, where $R_n(x)$ is the Taylor remainder.

In applying this theorem we frequently use the fact that $\lim_{n\to\infty} \frac{|x|^n}{n!} = 0$ for every x. Think about this; it should make sense that eventually $n!$ will be much larger than x^n for fixed x. We prove this below.

Fact: $\lim_{n\to\infty} \frac{|x|^n}{n!} = 0$ for every real number x.

Proof: $0 \leq \frac{|x|^n}{n!} = \frac{|x|}{1} \cdot \frac{|x|}{2} \cdot \frac{|x|}{3} \cdots \frac{|x|}{n}$

Let p be a positive constant integer such that $0 \leq \frac{|x|}{p} < 1$. Then

$$0 \leq \underbrace{\frac{|x|}{1} \cdot \frac{|x|}{2} \cdot \frac{|x|}{3} \cdots \frac{|x|}{p}}_{\substack{p \text{ positive terms, each} \\ \text{less than or equal to } |x|}} \cdot \underbrace{\frac{|x|}{p+1} \cdots \frac{|x|}{n}}_{\substack{n-p \text{ positive terms,} \\ \text{each less than } \frac{|x|}{p}}} \leq |x|^p \cdot \left(\frac{|x|}{p}\right)^{n-p}$$

$$\text{So } 0 \leq \frac{|x|^n}{n!} < |x|^p \cdot \left(\frac{|x|}{p}\right)^{n-p} \tag{30.4}$$

If $0 \leq r < 1$, then $\lim_{n\to\infty} r^n = 0$. Therefore $\lim_{n\to\infty} r^{n-p} = 0$ for $0 \leq r < 1$ and p constant.

$$\text{But } 0 \leq \frac{|x|}{p} < 1, \text{ so } \lim_{n\to\infty} \left(\frac{|x|}{p}\right)^{n-p} = 0.$$

Return to (30.4) and let n increase without bound.

$$\lim_{n\to\infty} 0 \leq \lim_{n\to\infty} \frac{|x|^n}{n!} \leq \lim_{n\to\infty} |x|^p \cdot \left(\frac{|x|}{p}\right)^{n-p}$$

$$0 \leq \lim_{n\to\infty} \frac{|x|^n}{n!} \leq |x|^p \cdot 0 = 0$$

$$\text{Therefore } \lim_{n\to\infty} \frac{|x|^n}{n!} = 0, \text{ by the Sandwich Theorem.}$$

We are now ready to show that $\sin x$ and e^x are equal to their respective Taylor series.

◆ **EXAMPLE 30.14**   Show that $\sin x = \sum_{k=0}^{\infty} (-1)^k \frac{x^{2k+1}}{(2k+1)!}$ for all x.

SOLUTION   For each x there exists a c between 0 and x such that

$$0 \leq |R_n(x)| = \frac{|f^{(n+1)}(c)|}{(n+1)!} |x|^{n+1}.$$

Therefore $|R_n(x)| \leq \frac{1 \cdot |x|^{n+1}}{(n+1)!}$.

The latter inequality holds because $|f^{(n+1)}(c)| = |\sin c|$ or $|\cos c|$ and both are bounded by 1.

$$\lim_{n\to\infty} 0 \leq \lim_{n\to\infty} |R_n(x)| \leq \lim_{n\to\infty} \frac{|x|^{n+1}}{(n+1)!}$$

$$0 \leq \lim_{n\to\infty} |R_n(x)| \leq 0$$

From the Sandwich Theorem we conclude that $\lim_{n\to\infty} |R_n(x)| = 0$ and therefore $\lim_{n\to\infty} R_n(x) = 0$ for all x. Thus, $\sin x$ is equal to its Taylor expansion about zero for all x. ◆

◆ **EXAMPLE 30.15**   Show that $e^x = \sum_{k=0}^{\infty} \frac{x^k}{k!}$ for all x.

SOLUTION   For each x there exists a c between 0 and x such that

$$0 \leq |R_n(x)| = \frac{|f^{(n+1)}(c)|}{(n+1)!} |x|^{n+1} = e^c \frac{|x|^{n+1}}{(n+1)!}$$

e^x is an increasing function, so $e^c \leq e^{|x|}$.

$$0 \leq |R_n(x)| \leq e^{|x|} \frac{|x|^{n+1}}{(n+1)!}$$

$$\lim_{n \to \infty} 0 \leq \lim_{n \to \infty} |R_n(x)| \leq \lim_{n \to \infty} e^{|x|} \frac{|x|^{n+1}}{(n+1)!}$$

But $\lim_{n \to \infty} e^{|x|} \frac{|x|^{n+1}}{(n+1)!} = e^{|x|} \lim_{n \to \infty} \frac{|x|^{n+1}}{(n+1)!} = e^{|x|} \cdot 0 = 0.$

$$0 \leq \lim_{n \to \infty} |R_n(x)| \leq 0$$

So $\lim_{n \to \infty} |R_n(x)| = 0$ by the Sandwich Theorem. Therefore, $\lim_{n \to \infty} R_n(x) = 0$.

We conclude that $e^x = 1 + x + \frac{x^2}{2!} + \frac{x^3}{3!} + \cdots$ for all x.   ◆

EXERCISE 30.4   Show that $\cos x$ is equal to its Maclaurin series for all x.

Take a moment to reflect upon the rather remarkable results we have accumulated. Not only *can* we express e^x, $\sin x$, and $\cos x$ as infinite "polynomials" (called power series), but we determined the coefficients using information about derivatives evaluated only at $x = 0$. We think of a derivative as giving local information, yet somehow information generating the *entire* function is encoded in the set of infinitely many derivatives. This is philosophically intriguing.

Let's take inventory on convergence issues.

■ A Taylor series might converge to its generating function for all x.
  For example, consider the Maclaurin series for e^x, $\sin x$, and $\cos x$.

■ A Taylor series might converge to its generating function only over a certain interval.
  For example, $\frac{1}{1-x} = \sum_{k=0}^{\infty} x^k$ only for $x \in (-1, 1)$.

■ At minimum a Taylor series will be equal to the value of its generating function at its center.[6]

Power Series

We'll put Taylor series in a broader context by discussing power series.

---

[6] It is possible for a Taylor series to converge, but not to its generating function, except at its center. This pathology is illustrated in Problem 35 at the end of this section.

Definition

A **power series in** x is an infinite series of the form $\sum_{k=0}^{\infty} a_k x^k$. A **power series in** $(x - b)$, or a **power series centered at** b, is a series of the form $\sum_{k=0}^{\infty} a_k (x - b)^k$.

Uniqueness Theorem for Power Series Expansions

If f has a power series expansion (or representation) at b, that is, if $f(x) = \sum_{k=0}^{\infty} a_k (x - v)^k$ for $|x - b| < R$, then that power series *is* the Taylor series for f at $x = b$.

The Uniqueness Theorem can be verified by repeatedly differentiating the power series expansion term by term and evaluating each successive derivative at $x = b$.

The Uniqueness Theorem carries with it computational power. For example, we could have avoided computing derivatives in Example 30.13 by using the fact that

$$\frac{1}{1 - x} = 1 + x + x^2 + \cdots \text{ for } x \in (-1, 1).$$

This is a power series expansion of $\frac{1}{1-x}$, and therefore it must be the Taylor series for $\frac{1}{1-x}$ at $x = 0$.

Convergence of a Power Series[7]

Theorem on the Convergence of a Power Series[8]

For a given power series $\sum_{k=0}^{\infty} a_k (x - b)^k$, one of the following is true:

i. The series converges for all x.

ii. The series converges only when $x = b$.

iii. There is a number R, $R > 0$ such that the series converges for all x such that $|x - b| < R$ (x is within R of the center) and diverges for all x such that $|x - b| > R$.

R is called the **radius of convergence**. If the series converges for all x, we say $R = \infty$; if the series converges only at its center, we say $R = 0$.

The set of all x for which a power series converges is called the **interval of converge** of the series. From the theorem stated above we see that a power series in $(x - b)$ will have an interval of convergence centered around $x = b$. At the endpoints of the interval the series could either converge or diverge; further investigation is necessary. In other words, if the radius of convergence is R, the interval of convergence will be one of the following:

---

[7] The student or instructor who prefers a thorough discussion of convergence before a discussion of the convergence of a power series can turn to page 964 (Section 30.5), and, after completing that section, return to this point.

[8] Justification is given in Appendix H.

$(b-R, b+R)$ $(b-R, b+R]$ $[b-R, b+R)$ $[b-R, b+R]$

The behavior of a power series at the points $b + R$ and $b - R$ can be tricky, but for $|x - b| < R$ we will find the behavior reassuringly like that of polynomials in many respects. We will use substitution, integration, and differentiation of power series on $|x - b| < R$ to obtain new Taylor series from familiar ones. Before moving in this direction we must add one more very important Taylor series to our list of "familiar" ones.

The Binomial Series

◆ **EXAMPLE 30.16**   THE BINOMIAL SERIES Find the Maclaurin series generated by $f(x) = (1 + x)^p$, where p is constant. This series is called the **binomial series.**

SOLUTION   The Maclaurin series is given by $\sum_{k=0}^{\infty} \frac{f^{(k)}(0)}{k!} x^k$.

$f(x) = (1 + x)^p$ $\hspace{3cm}$ $f(0) = 1$

$f'(x) = p(1 + x)^{p-1}$ $\hspace{3cm}$ $f'(0) = p$

$f''(x) = p(p - 1)(1 + x)^{p-2}$ $\hspace{2cm}$ $f''(0) = p(p - 1)$

$f'''(x) = p(p - 1)(p - 2)(1 + x)^{p-3}$ $\hspace{1.5cm}$ $f'''(0) = p(p - 1)(p - 2)$

$\vdots$ $\hspace{7cm}$ $\vdots$

$f^{(k)}(x) = p(p - 1)(p - 2) \cdots (p - k + 1)(1 + x)^{p-k}$ $\hspace{0.5cm}$ $f^{(k)}(0) = p(p - 1) \cdots (p - k + 1)$

$\vdots$ $\hspace{7cm}$ $\vdots$

Therefore the Maclaurin series is

$$1 + px + \frac{p(p - 1)}{2!}x^2 + \frac{p(p - 1)(p - 2)}{3!}x^3 + \cdots +$$
$$\frac{p(p - 1)(p - 2) \cdots (p - k + 1)}{k!}x^k + \cdots^9 \hspace{1cm} ◆$$

Fact: The Maclaurin series for $(1 + x)^p$ converges to $(1 + x)^p$ for $x \in (-1, 1)$ and diverges for $|x| > 1$.

$$(1 + x)^p = 1 + px + \frac{p(p - 1)}{2!}x^2 + \frac{p(p - 1)(p - 2)}{3!}x^3 + \cdots +$$
$$\frac{p(p - 1) \cdots (p - k + 1)}{k!}x^k + \cdots \text{ for } x \in (-1, 1)$$

Proving this fact by showing that $\lim_{n \to \infty} R_n(x) = 0$ is difficult, but possible. We omit the proof.[10]

REMARKS CONCERNING THE BINOMIAL SERIES

1. In the case that p is a positive integer the series terminates with the x^p term; subsequent coefficients all contain a factor $(p - p)$. We are left with an expansion of the polynomial

---

[9] The coefficients match those given by Pascal's Triangle.
[10] By the end of Section 30.5 you will be able to show that the radius of convergence of the binomial series is 1.

$(1 + x)^p$. As an exercise, show that if $p = 4$ the binomial series becomes $(1 + x)^4 = 1 + 4x + 6x^2 + 4x^3 + x^4$.

2. The notation $\binom{p}{k}$ is often used as an abbreviation for the binomial coefficients where $\binom{p}{k} = \frac{p(p-1)(p-2)\cdots(p-k+1)}{k!}$ for $k \geq 1$ and $\binom{p}{0} = 1$. Using this notation we can write

$$(1 + x)^p = \sum_{k=0}^{\infty} \binom{p}{k} x^k \quad \text{for } x \in (-1, 1).$$

3. The binomial expansion is valuable to know, as applications of it abound. Examples 30. 9, 30.10, and 30.13 all involve binomial expansions. Often one uses the first and second order approximations,

$$(1 + x)^p \approx 1 + px \quad \text{or} \quad (1 + x)^p \approx 1 + px + \frac{p(p-1)}{2!}x^2$$

for $|x|$ small, in computations in applied science.

EXERCISE 30.5 By letting $p = -1$, use the binomial series to find the Maclaurin series for $\frac{1}{1+u}$. Then let $u = -x$ to arrive at the Maclaurin series for $\frac{1}{1-x}$.

Below we list some commonly used Taylor expansions together with their intervals of convergence.

$$e^x = 1 + x + \frac{x^2}{2!} + \frac{x^3}{3!} + \cdots + \frac{x^k}{k!} + \cdots \qquad\qquad \text{for all } x$$

$$\sin x = x - \frac{x^3}{3!} + \frac{x^5}{5!} - \cdots + (-1)^k \frac{x^{2k+1}}{(2k+1)!} + \cdots \qquad\qquad \text{for all } x$$

$$\cos x = 1 - \frac{x^2}{2!} + \frac{x^4}{4!} - \cdots + (-1)^k \frac{x^{2k}}{(2k)!} + \cdots \qquad\qquad \text{for all } x$$

$$(1 + x)^p = 1 + px + \frac{p(p-1)}{2!}x^2 + \cdots + \frac{p(p-1)\cdots(p-k+1)}{k!}x^k + \cdots \quad \text{for } |x| < 1$$

$$\frac{1}{1-x} = 1 + x + x^2 + x^3 + \cdots + x^k + \cdots \qquad\qquad \text{for } |x| < 1$$

You will find it useful to know these series off the top of your head because other series can be derived directly from these.

Obtaining New Taylor Series From Familiar Ones: Substitution

EXAMPLE 30.17 Find the Taylor expansion for e^{-x^2} about $x = 0$.

SOLUTION Calculating this series by computing derivatives very quickly becomes unwieldy. Instead, we'll use substitution.

$$e^u = 1 + u + \frac{u^2}{2!} + \frac{u^3}{3!} + \cdots + \frac{u^k}{k!} + \cdots \qquad \text{for all } u. \quad \text{Let } u = -x^2.$$

$$e^{-x^2} = 1 + (-x^2) + \frac{(-x^2)^2}{2!} + \frac{(-x^2)^3}{3!} + \cdots + \frac{(-x^2)^k}{k!} + \cdots$$

$$e^{-x^2} = 1 = x^2 + \frac{x^4}{2!} - \frac{x^6}{3} + \cdots + (-1)^k \frac{x^{2k}}{k!} + \cdots \qquad \text{for all } x.$$

By the Uniqueness Theorem, this is the Taylor series for e^{-x^2} about $x = 0$. ◆

◆ **EXAMPLE 30.18**   Find the Maclaurin series for $f(x) = 2x \sin x \cos x$.

SOLUTION $f(x) = x \cdot 2 \sin x \cos x = x \cdot \sin(2x)$

$$\sin u = \sum_{k=0}^{\infty} (-1)^k \frac{u^{2k+1}}{(2k+1)!} \qquad \text{for all } u. \quad \text{Let } u = 2x.$$

$$\sin(2x) = \sum_{k=0}^{\infty} (-1)^k \frac{(2x)^{2k+1}}{(2k+1)!} = \sum_{k=0}^{\infty} (-1)^k \frac{2^{2k+1} x^{2k+1}}{(2k+1)!} \qquad \text{for all } x$$

$$x \cdot \sin(2x) = x \sum_{k=0}^{\infty} (-1)^k 2^{2k+1} \frac{x^{2k+1}}{(2k+1)!} = \sum_{k=0}^{\infty} (-1)^k \frac{2^{2k+1} x^{2k+2}}{(2k+1)!}$$

We can write this out as

$$2x^2 - \frac{2^3 x^4}{3!} + \frac{2^5 x^6}{5!} - \cdots + (-1)^k 2^{2k+1} \frac{x^{2k+2}}{(2k+1)!} + \cdots .$$

By the Uniqueness Theorem, this is the Maclaurin series for $2x \sin x \cos x$. Note the difference between substituting $2x$ for u in the first step and multiplying the whole series by x in the second step. ◆

◆ **EXAMPLE 30.19**   Find the fourth degree Taylor polynomial for $f(x) = \sqrt{9 - x^2}$ about $x = 0$. For what x-values does the Taylor series converge to f?

SOLUTION   Let's transform this function so that we can use the binomial series.

$$\sqrt{9 - x^2} = \sqrt{9 \left(1 - \frac{x^2}{9}\right)} = 3\sqrt{1 - \frac{x^2}{9}} = 3\left(1 - \frac{x^2}{9}\right)^{\frac{1}{2}}$$

From the binomial series we know

$$(1 + u)^p = 1 + pu + \frac{p(p-1)}{2!} u^2 + \cdots \qquad \text{for } |u| < 1.$$

so

$$(1 + u)^{\frac{1}{2}} = 1 + \frac{1}{2} u + \frac{\frac{1}{2}\left(-\frac{1}{2}\right)}{2!} u^2 + \cdots$$

$$(1 + u)^{\frac{1}{2}} = 1 + \frac{1}{2} u - \frac{1}{8} u^2 + \cdots \qquad \text{for } |u| < 1.$$

Let $u = -\frac{x^2}{9}$.

$$\left(1+\left(-\frac{x^2}{9}\right)\right)^{\frac{1}{2}} = 1 + \frac{1}{2}\left(-\frac{x^2}{9}\right) - \frac{1}{8}\left(-\frac{x^2}{9}\right)^2 + \cdots \quad \text{for} \ \left|-\frac{x^2}{9}\right| < 1$$

$$\sqrt{1-\frac{x^2}{9}} = 1 - \frac{1}{18}x^2 - \frac{1}{648}x^4 + \cdots \qquad\qquad \text{for} \ |x^2| < 9$$

$$3\sqrt{1-\frac{x^2}{9}} = 3 - \frac{1}{6}x^2 - \frac{1}{216}x^4 + \cdots \qquad\qquad \text{for} \ x \in (-3, 3)$$

Thus, the fourth degree Taylor polynomial is $3 - \frac{1}{6}x^2 - \frac{1}{216}x^4$. The Taylor series for $f(x)$ converges to $f(x)$ on $(-3, 3)$.   ◆

Note that in Examples 30.17 and 30.18 the "old" series being used converge for all real numbers. In Example 30.19 this was not the case; the new interval of convergence was obtained by substitution.

▮ PROBLEMS FOR SECTION 30.3

1. Find the Maclaurin series for $\cos x$ and show that it is equal to $\cos x$ for all x.

2. (a) Find the Maclaurin series for $\ln(1+x)$.
   (b) On the same set of axes, graph $\ln(1+x)$ and $P_6(x)$. Observe that the polynomial approximation to $\ln(1+x)$ is good for $|x| < 1$.
   (c) Graph $R_6(x) = \ln(1+x) - P_6(x)$. Observe that $R_6(x)$ is close to zero on $|x| < 1$.
   In the next section we will show that the radius of convergence of the Maclaurin series for $\ln(1+x)$ is 1.

3. The interval of convergence of the Maclaurin series for $\ln(1+u)$ is $u \in (-1, 1]$. On this interval the series converges to $\ln(1+u)$.
   (a) Find the Maclaurin series for $\ln(1+u)$.
   (b) By setting $u = x - 1$ in part (a), find the Taylor series for $\ln x$ centered at $x = 1$.
   (c) Find the Taylor series for $\ln x$ at $x = 1$ by taking derivatives. Make sure your answers to parts (b) and (c) agree.
   (d) What is the interval of convergence for the Taylor series for $\ln x$ centered at $x = 1$?
   (e) Graph $\ln x$ and several of its Taylor polynomials at $x = 1$ to be sure your answer to part (d) is reasonable.

In Problems 4 through 9, find the Taylor series for $f(x)$ centered at the indicated value of b.

4. $f(x) = \sin x, \quad b = \pi$

5. $f(x) = 2 \cos x, \quad b = \frac{\pi}{2}$

6. $f(x) = 10^x, \quad b = 0$

7. $f(x) = \frac{1}{\sqrt{x}}, \quad b = 1$

8. $f(x) = (3 + 2x)^3, \quad b = 0$

9. $f(x) = (1 + x)^5, \quad b = 0$

10. A power series centered at $b = 0$ has a radius of convergence of 5. For each value of x given below, determine whether the series converges, diverges, or there is not enough information available to determine.

(a) $x = 0$ (b) $x = 3$ (c) $x = 5$ (d) $x = 7$

(e) $x = -1.8$ (f) $x = -\sqrt{5}$ (g) $x = -5$ (h) $x = -6$

11. A power series of the form $\sum_{k=0}^{\infty} a_k (x - 2)^k$ has a radius of convergence of 3.
    (a) For what values of x can you say with confidence that the series converges?
    (b) For what values of x can you say with confidence that the series diverges?
    (c) For what values of x are you given inadequate information to determine convergence?

12. The interval of convergence of a power series is $(-2, 5]$.
    (a) What is the radius of convergence?
    (b) What is the center of the series?

13. A power series is of the form $\sum_{k=0}^{\infty} a_k (x + 3)^k$. Which of the intervals given below could conceivably be the interval of convergence of the series? For each option ruled out, explain the rationale.

(a) $(0, \infty)$ (b) $(2, 4)$ (c) $[-10, 4)$ (d) $[-3, 3]$

(e) $(-4, 2)$ (f) $(-5, -1]$ (g) $(-\infty, \infty)$

In Problems 14 through 21, use your knowledge of the binomial series to find the nth degree Taylor polynomial for $f(x)$ about $x = 0$. Give the radius of convergence of the corresponding Maclaurin series. One of these "series" converges for all x.

14. $f(x) = \sqrt{1 + 3x}, \qquad n = 3$

15. $f(x) = \frac{1}{\sqrt{1+x}}, \quad n = 2$

16. $f(x) = (1 - x)^{\frac{2}{3}}, \quad n = 3$

17. $f(x) = \sqrt[3]{1 + x^2}, \quad n = 5$

18. $f(x) = (1 + 3x)^5, \quad n = 6$

19. $f(x) = \frac{1}{(1+x)^2}, \quad n = 5$

20. $f(x) = 2(9 - x)^{\frac{1}{2}}, \quad n = 3$

21. $f(x) = \frac{x}{\sqrt{4+x}}, \quad n = 3$

22. (a) Expand $f(x) = (a + x)^4$ by multiplying out or by using Pascal's triangle.

   (b) Rewrite $f(x)$ as $[a(1 + \frac{x}{a})]^4 = a^4 \left(1 + \frac{x}{a}\right)^4$. Use the binomial series to expand $\left(1 + \frac{x}{a}\right)^4$, multiply by a^4, and demonstrate that the result is the same as in part (a).

23. Find the Maclaurin series for $\frac{1}{1+x^2}$. What is the radius of convergence?

24. Use the binomial series to find the Maclaurin series for $\frac{1}{\sqrt{1-x^2}}$. What is the radius of convergence?

In Problems 25 through 34, use any method to find the Maclaurin series for $f(x)$. (Strive for efficiency.) Determine the radius of convergence.

25. $f(x) = xe^{-x}$

26. $f(x) = \sin 3x$

27. $f(x) = \cos\left(\frac{x}{2}\right)$

28. $f(x) = 3e^{2x}$

29. $f(x) = \cos(x^2)$

30. $f(x) = 3^x$

31. $f(x) = x^2 \cos(-x)$

32. $f(x) = \cos^2 x$ (Hint: use a trigonometric identity)

33. $f(x) = (a + x)^p$, where "a" and "p" are constants and p is not a positive integer.

34. $f(x) = \frac{1}{2x+3}$

35. *Pathological Example:* Let $f(x) = \begin{cases} e^{-\frac{1}{x^2}} & \text{for } x \neq 0, \\ 0 & \text{for } x = 0. \end{cases}$

   (a) Graph $f(x)$ on the following domains: $[-20, 20]$, $[-2, 2]$, and $[-0.5, 0.5]$. (A graphing instrument can be used.)

   (b) It can be shown that f is infinitely differentiable at $x = 0$ and that $f^{(k)}(0) = 0$ for all k. Conclude that the Maclaurin series for $f(x)$ converges for all x but only converges *to* $f(x)$ at $x = 0$.

36. Find the Maclaurin series for $\frac{1}{\sqrt{e^x}}$. What is its radius of convergence?

37. For $x \in (-1, 1]$, $\ln(1 + x) = x - \frac{x^2}{2} + \frac{x^3}{3} - \frac{x^4}{4} + \cdots + (-1)^k \frac{x^{k+1}}{k+1} + \cdots$.

   (a) Find the Maclaurin series for $\ln(1 + 2x)$. What is its interval of convergence?

   (b) Find the Maclaurin series for $\ln(e + ex)$. What is its interval of convergence?

   (c) Find the Maclaurin series for $\log_{10}(1 + x)$.

38. *Discover something wonderful.* We know $e^u = 1 + u + \frac{u^2}{2!} + \frac{u^3}{3!} + \cdots + \frac{u^k}{k!} + \cdots$ for all real u. Now define e raised to a complex number, $a + bi$ where $i = \sqrt{-1}$, to be $e^a \cdot e^{bi}$ where $e^{bi} = 1 + (bi) + \frac{(bi)^2}{2!} + \frac{(bi)^3}{3!} + \cdots + \frac{(bi)^k}{k!} + \cdots$.

   (a) Use the fact that $i^2 = -1$, $i^3 = -i$, and $i^4 = 1$ to simplify the expression for e^{ib}. Gather together the real terms (the ones without i's) and the terms with a factor of i. Express e^{ib} as a sum of two familiar functions (one of them multiplied by i).

   (b) Use your answer to part (a) to evaluate $e^{\pi i}$.

39. The hyperbolic functions, hyperbolic cosine, abbreviated cosh, and hyperbolic sine, abbreviated sinh, are defined as follows.

$$\cosh x = \frac{e^x + e^{-x}}{2} \qquad \sinh x = \frac{e^x - e^{-x}}{2}$$

   (a) Graph $\cosh x$ and $\sinh x$, each on its own set of axes. Do this without using a computer or graphing calculator, except possibly to check your work.

   (b) Find the Maclaurin series for $\cosh x$.

   (c) Find the MacLaurin series for $\sinh x$.

   Remark: From the graphs of cosh x and sinh x one might be surprised by the choice of names for these functions. After finding their Maclaurin series the choice should seem more natural.

   (d) Do some research and find out how these functions, known as hyperbolic functions, are used. The arch in St. Louis, the shape of many pottery kilns, and the shape of a hanging cable are all connected to hyperbolic trigonometric functions.

30.4   WORKING WITH SERIES AND POWER SERIES

Absolute and Conditional Convergence

There are many ways in which power series can be treated very much as we treat polynomials, but there are ways in which they can behave differently and must be treated with caution. This makes sense; there are ways in which series and finite sums behave very differently. In order to sort this out a bit, not only do we need to steer clear of divergent series and power series outside their interval of convergence, but we need to refine our notion of convergence to distinguish between absolute and conditional convergence.

Definition

A series $\sum_{n=1}^{\infty} a_n$ is **absolutely convergent** if $\sum_{n=1}^{\infty} |a_n|$ converges.

Note that if the terms of a series are either all positive or all negative, then convergence implies absolute convergence. There is only an issue when some terms are positive and some terms are negative.[11]

---

[11] Actually, there is not an issue provided there exists a constant k such that a_n is either positive for all $n > k$ or negative for all $n > k$.

Fact: If a series converges absolutely, it converges. *This is proven in Appendix H.*

Definition

A series $\sum_{n=1}^{\infty} a_n$ is **conditionally convergent** if it is convergent but not absolutely convergent.

Why is this distinction handy? Well, one might hope that the order of the terms in a sum could be rearranged without altering the sum, yet for infinite series this is true *only* if the series converges absolutely. In fact, it can be proven that if $\sum_{k=0}^{\infty} a_n$ is conditionally convergent, then the order of the terms can be rearranged to produce *any* finite number. This unsettling fact is enough to make one wary of conditionally convergent series.

It's hard to be wary of something without a concrete example, so we will take this opportunity to look at alternating series. You will find that alternating series are fascinating in their own right, and that this excursion into the topic of alternating series will produce as a by-product an error estimate that will prove useful when dealing with many Taylor polynomials.

Alternating Series

Definition

A series whose successive terms alternate in sign is called an **alternating series.**

For any fixed x the Maclaurin series for $\sin x$ and $\cos x$ are alternating series. The Maclaurin series for e^x will alternate when x is negative and the one for $\ln(1 + x)$ will alternate when x is positive.

There is a simple convergence test, proved by Leibniz, that can be applied to alternating series. We know that for a general series, $\sum_{k=1}^{\infty} a_k$, the characteristic $\lim_{k \to \infty} a_k = 0$ is necessary but not sufficient for convergence. The divergence of the harmonic series $1 + \frac{1}{2} + \frac{1}{3} + \cdots + \frac{1}{n} + \cdots$ illustrates this fact. However, if a series is alternating, then if the magnitude of the terms decreases monotonically towards zero, this is enough to assure convergence.

Alternating Series Test

An alternating series, $\sum_{k=1}^{\infty} (-1)^k a_k$ or $\sum_{k=1}^{\infty} (-1)^{k+1} a_k$ for $a_k > 0$, converges if

   i. $a_{k+1} < a_k$,   *the terms are decreasing in magnitude* and

   ii. $\lim_{k \to \infty} a_k = 0$,   *the terms are approaching zero.*

The Basic Idea Behind the Alternating Series Test[12]

Consider the series $a_1 - a_2 + a_3 - a_4 + \cdots + (-1)^{k+1} a_k + \cdots$ for $a_k > 0$. Suppose that conditions (i) and (ii) are satisfied. In Figure 30.8 we plot partial sums.

---

[12] This is not a rigorous argument, but it can be made rigorous using the theorem that every bounded monotonic sequence is convergent.

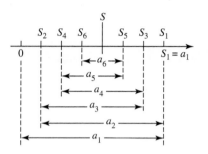

Figure 30.8

$S_1 = a_1$ is to the right of zero.

$S_2 = a_1 - a_2$ lies between 0 and S_1 because $a_2 < a_1$.

$S_3 = a_1 - a_2 + a_3$ lies between S_2 and S_1 because $a_3 < a_2$.

$\vdots$

Picture starting at the zero. Take a big step forward to S_1, then a smaller step backward to S_2, then an even smaller step forward to S_3, and so on. The partial sums oscillate; S_n is between S_{n-1} and S_{n-2} because $a_n < a_{n-1}$. The distance between S_{n-1} and S_n is a_n and $\lim_{n \to \infty} a_n = 0$. Therefore the sequence of partial sums is approaching a finite limit L, with successive partial sums alternately overshooting then undershooting L.

This argument can be made rigorous by considering the increasing but bounded sequence of partial sums, $S_2, S_4, S_6, \ldots$, and the decreasing but bounded sequence of partial sums $S_1, S_3, S_5, \ldots$, and showing that both sequences converge to the same limit.

Our analysis provides us with an easy-to-use error estimate. If an alternating series satisfies the two conditions of the Alternating Series Test and if we approximate the sum, L, using a partial sum S_n, then the magnitude of the error will be less than a_{n+1}, the magnitude of the first unused term of the series. Furthermore, if the last term of the partial sum is positive, then the partial sum is larger than L; if its last term is negative, then the partial sum is smaller than L. We refer to this as **the Alternating Series Error Estimate.**

◆ EXAMPLE 30.20   Consider the alternating harmonic series $1 - \frac{1}{2} + \frac{1}{3} - \frac{1}{4} + \frac{1}{5} - \cdots + (-1)^{k+1}\frac{1}{k} + \cdots$.

(a) Show that this series converges conditionally.

(b) It can be shown that $\sum_{k=1}^{\infty}(-1)^{k+1}\frac{1}{k}$ converges to $\ln 2$. How many terms of the series must be used in order to approximate $\ln 2$ with error less than 0.001?

SOLUTION   (a) The series $\sum_{k=1}^{\infty}(-1)^{k+1}\frac{1}{k}$ is alternating. It satisfies the conditions of the Alternating Series Test:

   i. The terms are decreasing in magnitude: $\frac{1}{k+1} < \frac{1}{k}$.

   ii. The terms approach zero: $\lim_{k \to \infty} a_k = \lim_{k \to \infty} \frac{1}{k} = 0$.

Therefore the series converges. But $\sum_{k=1}^{\infty}\left|(-1)^{k+1}\frac{1}{k}\right| = \sum_{k=1}^{\infty}\frac{1}{k}$ is the harmonic series, which diverges. Therefore the alternating harmonic series converges conditionally.

(b) By the Alternating Series Error Estimate we know that the magnitude of the error is less than the magnitude of the first omitted term. Therefore we use the estimate

$$\ln 2 \approx \sum_{k=1}^{999} (-1)^{k+1} \frac{1}{k};$$

we need 999 terms. This series for ln 2 converges *very* slowly! ◆

◆ **EXAMPLE 30.21**   Estimate $\frac{1}{\sqrt{e}}$ with error less than 10^{-3}.

SOLUTION $e^x = 1 + x + \frac{x^2}{2!} + \frac{x^3}{3!} + \cdots + \frac{x^k}{k!} + \cdots$ Thus

$$\frac{1}{\sqrt{e}} = e^{-\frac{1}{2}} = 1 - \frac{1}{2} + \frac{1}{2^2 \cdot 2!} - \frac{1}{2^3 \cdot 3!} + \cdots + (-1)^k \frac{1}{2^k \cdot k!} + \cdots.$$

This series is alternating, its terms are decreasing in magnitude, and its terms tend toward zero. Therefore, we can apply the Alternating Series Error Estimate. We must find k such that

$$\frac{1}{2^k \cdot k!} < \frac{1}{1000}, \qquad \text{or equivalently,} \quad 2^k \cdot k! > 1000.$$

We do this by trial and error. $2^4 \cdot 4! = 384$ but $2^5 \cdot 5! = 3840 > 1000$. $\frac{1}{2^5 \cdot 5!} < \frac{1}{1000}$, so we don't need to use this term.

$$e^{-\frac{1}{2}} \approx 1 - \frac{1}{2} + \frac{1}{2^2 \cdot 2!} - \frac{1}{2^3 \cdot 3!} + \frac{1}{2^4 \cdot 4!} = 1 - \frac{1}{2} + \frac{1}{8} - \frac{1}{48} + \frac{1}{384}$$

$$\frac{1}{\sqrt{e}} \approx .6068.$$

Notice that the Alternating Series Error Estimate is simpler to apply than Taylor's Remainder. ◆

Let's return to the disturbing remark made before introducing alternating series. The assertion was that if a series converges conditionally, then rearranging the *order* of the terms of the series can change the sum. We're now ready to demonstrate this.

$$1 - \frac{1}{2} + \frac{1}{3} - \frac{1}{4} + \frac{1}{5} - \frac{1}{6} + \frac{1}{7} - \frac{1}{8} + \frac{1}{9} - \frac{1}{10} + \frac{1}{11} - \cdots = \ln 2$$

Multiplying both sides by 2 gives

$$2 - \frac{2}{2} + \frac{2}{3} - \frac{2}{4} + \frac{2}{5} - \frac{2}{6} + \frac{2}{7} - \frac{2}{8} + \frac{2}{9} - \frac{2}{10} + \frac{2}{11} - \cdots = 2 \ln 2 = \ln 4 \quad (30.5)$$

Rearrange the order of the terms in Equation (30.5) so that after each positive term there are two negative terms as follows.

$$2 - 1 - \frac{2}{4} + \frac{2}{3} - \frac{2}{6} - \frac{2}{8} + \frac{2}{5} - \frac{2}{10} - \frac{2}{12} + \frac{2}{7} - \frac{2}{14} - \frac{2}{16} + \frac{2}{19} - \cdots$$

$$= (2 - 1) - \frac{2}{4} + \left(\frac{2}{3} - \frac{2}{6} \right) - \frac{2}{8} + \left(\frac{2}{5} - \frac{2}{10} \right) - \frac{2}{12} + \left(\frac{2}{7} - \frac{2}{14} \right) - \frac{2}{16} + \cdots$$

$$= 1 - \frac{1}{2} + \frac{1}{3} - \frac{1}{4} + \frac{1}{5} - \frac{1}{6} + \frac{1}{7} - \frac{1}{8} + \cdots$$

$$= \ln 2.$$

By rearranging the order of the terms we changed the sum from ln 4 to ln 2. Riemann proved that by rearranging the order of the terms we can actually get the sum to be any real number.

On the other hand, it can be proven that if a series converges absolutely to a sum of S, then any rearrangement of the terms has a sum of S as well. This is one of the reasons we prefer to work with absolutely convergent series whenever possible.

Manipulating Power Series

Having defined absolute convergence, we can return to the theorem on the convergence of a power series and state a stronger form. (See Appendix H for justification.)

Theorem on the Convergence of a Power Series

For a given power series $\sum_{k=0}^{\infty} a_n(x-b)^k$, one of the following is true:

i. The series converges absolutely for all x.

ii. The series converges only when $x = b$.

iii. There is a number R, $R > 0$, such that the series converges absolutely for all x such that $|x - b| < R$ and diverges for all x such that $|x - b| > R$.

The points $x = b + R$ and $x = b - R$ must be studied separately. At these endpoints the series could converge conditionally, converge absolutely, or diverge. For the sake of simplicity we will generally restrict our attention to the interval $(b - R, b + R)$ in which the power series converges absolutely.

Differentiation and Integration of Power Series

Differentiation and Integration of Power Series

Let $\sum_{k=0}^{\infty} a_k(x-b)^k$ be a power series with radius of convergence R, where $R > 0$, R possibly ∞. Then the function $f(x) = \sum_{k=0}^{\infty} a_k(x-b)^k$ can be differentiated term by term or integrated term by term on $(b - R, b + R)$. That is,

$$f'(x) = \sum_{k=0}^{\infty} \frac{d}{dx} a_k(x-b)^k = \sum_{k=1}^{\infty} k a_k(x-b)^{k-1} \qquad \text{with radius of convergence } R$$

and

$$\int f(x)\, dx = \sum_{k=0}^{\infty} \int a_k(x-b)^k\, dx = C + \sum_{n=0}^{\infty} a_k \frac{(x-b)^{k+1}}{k+1}$$

with radius of convergence R.

This result, whose proof is omitted, says that the radius of convergence remains the same after integration or differentiation; it gives no information about convergence or divergence at $x = b \pm R$.[13]

---

[13] The original series may diverge at an endpoint and yet converge once integrated, or vice versa.

This Theorem gives us convenient ways of generating new Taylor series from familiar ones and provides a tool for integrating functions that don't have elementary antiderivatives.

◆ **EXAMPLE 30.22**   Find the Maclaurin series for arctan x. What is the radius of convergence?

SOLUTION   This is unwieldy to compute by taking derivatives. Instead, we'll use the fact that

$$\int \frac{1}{1 + x^2} \, dx = \arctan x + C.$$

We know $\frac{1}{1-u} = 1 + u + u^2 + u^3 + \cdots + u^k + \cdots$ for $|u| < 1$.
Let $u = -x^2$.

$$\frac{1}{1 - (-x^2)} = 1 + (-x^2) + (-x^2)^2 + (-x^2)^3 + \cdots + (-x^2)^k + \cdots \qquad \text{for } |-x^2| < 1$$

$$\frac{1}{1 + x^2} = 1 - x^2 + x^4 - x^6 + \cdots + (-1)^k x^{2k} + \cdots \qquad \text{for } |x| < 1$$

$$\int \frac{1}{1 + x^2} \, dx = \int \left(1 - x^2 + x^4 - x^6 + \cdots + (-1)^k x^{2k} + \cdots \right) \, dx$$

$$\arctan x = C + x - \frac{x^3}{3} + \frac{x^5}{5} - \cdots + (-1)^k \frac{x^{2k+1}}{2k + 1} + \cdots$$

To determine C, evaluate both sides at $x = 0$. $\arctan 0 = C$, so $C = 0$.

$$\arctan x = x - \frac{x^3}{3} + \frac{x^5}{5} - \frac{x^7}{7} + \cdots + (-1)^k \frac{x^{2k+1}}{2k + 1} + \cdots$$

The radius of convergence is 1, so the series converges absolutely for $x \in (-1, 1)$ and diverges for $|x| > 1$.

In fact, although we have only shown convergence for $x \in (-1, 1)$, the series converges to arctan x for $x = \pm 1$ as well. When evaluated at $x = 1$, the series is

$$\frac{\pi}{4} = 1 - \frac{1}{3} + \frac{1}{5} - \frac{1}{7} + \cdots .^{14} \qquad ◆$$

◆ **EXAMPLE 30.23**   Find the Maclaurin series for $\ln(1 + x)$ by integrating the series for $\frac{1}{1+x}$. What advantage does this approach have over computing the series by taking derivatives?

SOLUTION   We know $\frac{1}{1-u} = 1 + u + u^2 + \cdots + u^k + \cdots$ for $|u| < 1$.
Let $u = -x$.

$$\frac{1}{1 + x} = \frac{1}{1 - (-x)} = 1 - x + x^2 - x^3 + \cdots + (-1)^k x^k + \cdots \qquad \text{for } |-x| < 1, \text{ i.e., } |x| < 1$$

$$\int \frac{1}{1 + x} \, dx = C + x - \frac{x^2}{2} + \frac{x^3}{3} - \frac{x^4}{4} + \cdots + (-1)^k \frac{x^{k+1}}{k + 1} + \cdots \text{ for } |x| < 1$$

$$\text{So } \ln(1 + x) = C + x - \frac{x^2}{x} + \frac{x^3}{3} - \frac{x^4}{4} + \cdots (-1)^k \frac{x^{k+1}}{k + 1} + \cdots .$$

---

[14] You will find this series is carved in stone at the entrance to Coimbra University's department of mathematics building in Coimbra, Portugal.

To determine C, evaluate at $x = 0$. $\ln(1 + 0) = C$, so $C = 0$.

$$\ln(1 + x) = x - \frac{x^2}{2} + \frac{x^3}{3} - \frac{x^4}{4} + \cdots + (-1)^k \frac{x^{k+1}}{k+1} + \cdots$$

An advantage of this method of arriving at the series is that we know the radius of convergence is 1, and that the series converges to $\ln(1 + x)$ for $|x| < 1$. ◆

Once we know $\ln(1 + x) = \sum_{k=0}^{\infty} (-1)^k \frac{x^{k+1}}{k+1}$ for $|x| < 1$ we can set $u = x + 1$, ($x = u - 1$) and find $\ln(u) = \sum_{k=0}^{\infty} (-1)^k \frac{(u-1)^{k+1}}{k+1} = (u - 1) - \frac{(u-1)^2}{2} + \frac{(u-1)^3}{3} - \cdots + (-1)^k \frac{(u-1)^{k+1}}{k+1} + \cdots$. When $-1 < x < 1$, we know that $0 < x + 1 < 2$, so the series for $\ln u$ about $u = 1$ must converge on $u \in (0, 2)$. In fact, it can be shown that both of these series converge at the right-hand endpoint of the respective interval of convergence.

$$\ln(1 + x) = x - \frac{x^2}{2} + \frac{x^3}{3} - \cdots + (-1)^k \frac{x^{k+1}}{k+1} + \cdots \qquad \text{for } x \in (-1, 1]$$

$$\ln u = (u - 1) - \frac{(u-1)^2}{2} + \frac{(u-1)^3}{3} - \cdots + (-1)^k \frac{(u-1)^{k+1}}{k+1} \qquad \text{for } u \in (0, 2]$$

REMARK We saw in Example 30.20 that the series $1 - \frac{1}{2} + \frac{1}{3} - \frac{1}{4} + \cdots$ converges *very* slowly. Similarly, observe that $1 - \frac{1}{3} + \frac{1}{5} - \frac{1}{7} + \cdots$ converges to $\frac{\pi}{4}$ *very* slowly. This series is aesthetically pleasing but computationally inefficient. For practical purposes the rate at which a series converges is important. For instance, it is more efficient to approximate $\ln 2$ by looking at the following:

$$-\ln 2 = \ln\left(\frac{1}{2}\right) = \ln\left(1 - \frac{1}{2}\right) = -\frac{1}{2} - \frac{1}{2^2 \cdot 2} - \frac{1}{2^3 \cdot 3} - \frac{1}{2^4 \cdot 4} + \cdots$$

$$= -\frac{1}{2} - \frac{1}{8} - \frac{1}{24} - \frac{1}{64} - \frac{1}{160} - \cdots.$$

$x = \frac{1}{2}$ is closer to the center of the series than is $x = 1$, so the series converges more rapidly at $\frac{1}{2}$ than at 1. For even more efficiency in approximating $\ln 2$ we can find the Maclaurin series for $\ln\left(\frac{1+x}{1-x}\right)$ and evaluate it at $x = \frac{1}{3}$. This is the topic of one of the problems at the end of this section.

One reason that it is so useful to be able to represent a function as a power series is that a power series is simple to integrate. The use of power series expansions as an integration tool figured prominently in Newton's work and continues to be important in the integration of otherwise intractable functions. Consider, for example, $f(x) = e^{-x^2}$, a function that has no elementary antiderivative. The graph of f is a bell-shaped curve which, with minor modifications, gives the standard normal distribution that plays such an important role in probability and statistics. It is crucial to know the area under the normal distribution, and for this we must compute a definite integral. The following example indicates how Taylor series can be used in such a computation.

◆ **EXAMPLE 30.24** Approximate $\int_0^{0.2} e^{-x^2}\, dx$ with error less than 10^{-8}.

SOLUTION   From Example 30.17 we know that

$$e^{-x^2} = 1 - x^2 + \frac{x^4}{2!} - \frac{x^6}{3!} + \cdots + (-1)^k \frac{x^{2k}}{k!} + \cdots \qquad \text{for all } x.$$

$$\int_0^{0.2} e^{-x^2}\, dx = \int_0^{0.2} \left(1 - x^2 + \frac{x^4}{2!} - \frac{x^6}{3!} + \cdots + (-1)^k \frac{x^{2k}}{k!} + \cdots \right) dx$$

$$= \left[x - \frac{x^3}{3} + \frac{x^5}{5 \cdot 2!} - \frac{x^7}{7 \cdot 3!} + \cdots + (-1)^k \frac{x^{2k+1}}{(2k+1)k!} + \cdots \right]_0^{0.2}$$

$$= 0.2 - \frac{(0.2)^3}{3} + \frac{(0.2)^5}{5 \cdot 2!} - \frac{(0.2)^7}{7 \cdot 3!} + \cdots + (-1)^k \frac{(0.2)^{2k+1}}{(2k+1)k!} + \cdots$$

We can apply the Alternating Series Error Estimate because the series above is alternating, its terms are decreasing in magnitude, and its terms tend toward zero. We look for a term whose magnitude is less than 10^{-8}

$$\frac{(0.2)^7}{7 \cdot 3!} = \frac{2^7}{7 \cdot 3! \cdot 10^7} \approx 3 \times 10^{-7} : \text{ not small enough}$$

$$\frac{(0.2)^9}{9 \cdot 4!} = \frac{2^9}{9 \cdot 4! \cdot 10^9} \approx 2.4 \times 10^{-9} < 10^{-8}$$

Therefore

$$\int_0^{0.2} e^{-x^2}\, dx \approx 0.2 - \frac{(0.2)^3}{3} + \frac{(0.2)^5}{5 \cdot 2} - \frac{(0.2)^7}{7 \cdot 6} \qquad \text{with error less than } 10^{-8}.$$

$$\int_0^{0.2} e^{-x^2}\, dx \approx 0.197365029 \qquad ◆$$

There are three main reasons for our interest in representing functions as power series. Such representations are useful in

■ approximating functions by polynomials and approximating function values numerically,

■ integrating functions that don't have elementary antiderivatives, and

■ solving differential equations.

Although we have illustrated the first two applications of power series, we have yet to give an example of the third. The Theorem on Differentiation of a Power Series plays the major role in this application.

Power Series and Differential Equations

The next example illustrates how power series can be used in solving differential equations.

◆ **EXAMPLE 30.25**   Use power series to solve the differential equation $y'' = -y$.

SOLUTION   Let $f(x)$ be a solution to the differential equation. Assume that $f(x)$ has a power series expansion.

$$f(x) = a_0 + a_1 x + a_2 x^2 + a_3 x^3 + \cdots + a_k x^k + \cdots$$

$$f'(x) = a_1 + 2a_2 x + 3a_3 x^2 + 4a_4 x^3 + \cdots + ka_k x^{k-1} + \cdots$$

$$f''(x) = 2a_2 + 3 \cdot 2 \cdot a_3 x + 4 \cdot 3 \cdot a_4 x^2 + \cdots + k(k-1)a_k x^{k-2} + \cdots$$

If $f(x)$ is a solution to $y'' = -y$, then $f''(x) = -f(x)$.

$$2a_2 + 3 \cdot 2 \cdot a_3 x + 4 \cdot 3 \cdot a_4 x^2 + \cdots + k(k-1)a_k x^{k-2} + \cdots = -a_0 - a_1 x - a_2 x^2 -$$
$$\cdots - a_k x^k \cdots$$

The key notion is that for these two polynomials to be equal the coefficients of corresponding powers of x must be equal. In other words, the constant terms must be equal, the coefficients of x must be equal, and so on.

$$2a_2 = -a_0$$
$$3 \cdot 2 \cdot a_3 = -a_1$$
$$4 \cdot 3 \cdot a_4 = -a_2$$
$$5 \cdot 4 \cdot a_5 = -a_3$$
$$\vdots$$
$$k(k-1) \cdot a_k = -a_{k-2}$$
$$\vdots$$

We can solve for all the coefficients in terms of a_0 and a_1.

Let $a_0 = C_0$, $a_1 = C_1$. We'll solve for a_k, $k = 2, 3, \cdots$ in terms of C_0 and C_1.

$$a_2 = -\frac{a_0}{2} = \frac{-C_0}{2!} \qquad\qquad a_3 = \frac{-a_1}{3 \cdot 2} = \frac{-C_1}{3!}$$

$$a_4 = \frac{-a_2}{4 \cdot 3} = \frac{C_0}{4 \cdot 3 \cdot 2!} = \frac{C_0}{4!} \qquad a_5 = \frac{-a_3}{5 \cdot 4} = \frac{C_1}{5 \cdot 4 \cdot 3!} = \frac{C_1}{5!}$$

$$a_6 = \frac{-a_4}{6 \cdot 5} = \frac{-C_0}{6 \cdot 5 \cdot 4!} = \frac{-C_0}{6!} \qquad a_7 = \frac{-a_5}{7 \cdot 6} = \frac{-C_1}{7 \cdot 6 \cdot 5!} = \frac{-C_1}{7!}$$

$$a_8 = \frac{-a_6}{8 \cdot 7} = \frac{C_0}{8 \cdot 7 \cdot 6!} = \frac{C_0}{8!} \qquad a_9 = \frac{-a_7}{9 \cdot 8} = \frac{C_1}{9!}$$

$$\vdots \qquad\qquad\qquad\qquad \vdots$$

$$a_{2n} = \frac{-a_{2n-2}}{(2n)(2n-1)} = (-1)^n \frac{C_0}{(2n)!} \qquad a_{2n+1} = \frac{-a_{2n-1}}{(2n+1)(2n)} = (-1)^n \frac{C_1}{(2n+1)!}$$

$$f(x) = C_0 + C_1 x - \frac{C_0}{2!} x^2 - \frac{C_1}{3!} x^3 + \frac{C_0}{4!} x^4 + \frac{C_1}{5!} x^5 - \frac{C_0}{6!} x^6 - \frac{C_1}{7!} x^7 + \cdots$$

$$f(x) = C_0 \underbrace{\left[1 - \frac{x^2}{2!} + \frac{x^4}{4!} - \cdots + (-1)^n \frac{x^{2n}}{(2n)!} + \cdots \right]}_{\cos x}$$

$$+ C_1 \underbrace{\left[x - \frac{x^3}{3!} + \frac{x^5}{5!} - \frac{x^7}{7!} + \cdots + (-1)^n \frac{x^{2n+1}}{(2n+1)!} + \cdots \right]}_{\sin x}$$

$$f(x) = C_0 \cos x + C_1 \sin x \quad \blacklozenge$$

EXERCISE 30.6   Verify that $f(x) = C_0 \cos x + C_1 \sin x$ is a solution to the differential equation $y'' = -y$. We have shown that if a solution to $y'' = -y$ has a power series representation, then that solution must be of the form $C_0 \cos x + C_1 \sin x$, where C_0 and C_1 are constants.

In the example just completed, we recognized the Maclaurin series for $\sin x$ and $\cos x$. It is entirely possible that we can solve for all the coefficients of a power series and simply have the solution expressed as and defined by the power series expansion. There are well-known functions defined by power series that arise in physics, astronomy, and other applied sciences. An example of such functions are the **Bessel functions**, named after the astronomer Bessel who came up with them in the early 1800s while working with Kepler's laws of planetary motion. The Bessel function $J_0(x)$ is defined by

$$J_o(x) = \sum_{k=0}^{\infty} (-1)^k \frac{x^{2k}}{(k!)^2 2^{2k}}.$$

As is often the case in mathematics, while Bessel functions arose in a particular astronomical problem they are now used in a wide array of situations. One such example is in studying the vibrations of a drumhead. A graph of the partial sum $J_0(x) \approx \sum_{k=0}^{13} (-1)^k \frac{x^{2k}}{(k!)^2 2^{2k}}$ is given in Figure 30.9.

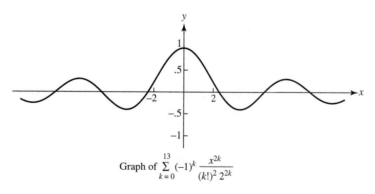

Graph of $\displaystyle\sum_{k=0}^{13} (-1)^k \frac{x^{2k}}{(k!)^2 \, 2^{2k}}$

Figure 30.9

Transition to Convergence Tests

Because this chapter began with Taylor polynomials, it was natural to move on to Taylor series directly, without the traditional lead-in of convergence tests for infinite series. Taylor's Theorem enables us to deal with some convergence issues quite efficiently. Not only are we able to show that the series for e^x, $\sin x$, and $\cos x$ converge, but we can determine that each converges to its generating function. Our previous work with geometric series allows us to conclude that the series for $\frac{1}{1-x}$ converges to its generating function on $(-1, 1)$. When we find a Taylor series by manipulating a known Taylor series, whether by substitution, differentiation, or integration, we can calculate the radius of convergence. But, faced with a generic power series, we have few tools at our disposal with which to determine convergence and divergence. More fundamentally, we have no systematic way of determining the convergence or divergence of an infinite series of the form Σa_k. The next section will remedy this situation.

PROBLEMS FOR SECTION 30.4

For each series in Problems 1 through 9, determine whether the series converges absolutely, converges conditionally, or diverges.

1. $\sum_{k=1}^{\infty} (-1)^k \frac{k!}{(k-1)!}$

2. $\sum_{k=1}^{\infty} (-1)^{k+1} \frac{k!}{(k+1)!}$

3. $\sum_{k=1}^{\infty} (-1)^k \frac{1}{3k}$

4. $\sum_{k=2}^{\infty} (-1)^n \frac{k}{\ln k}$

5. $\sum_{k=10}^{\infty} \frac{\cos(kn)}{10k}$

6. $\sum_{k=0}^{\infty} \left(-\frac{11}{12}\right)^k$

7. $\sum_{k=1}^{\infty} \frac{1}{100} \sin\left(\frac{k\pi}{2}\right)$

8. $\sum_{k=1}^{\infty} (-1)^k \frac{2^k}{k}$

9. $\sum_{k=0}^{\infty} (-1)^n \left(\frac{k^2-10}{2k^2+5k}\right)$

10. Is it possible for a geometric series to converge conditionally? If it is possible, produce an example.

11. How many nonzero terms of the Maclaurin series for $\ln(1 + x)$ are needed to approximate $\ln\left(\frac{3}{2}\right)$ with an error of less than 10^{-4}?

12. Approximate $\frac{1}{e}$ with error less than 10^{-5}.

13. Arrive at the series for $\cos x$ by differentiating the Maclaurin series for $\sin x$.

14. Find the Maclaurin series for $\arcsin x$ using the fact that $\int \frac{1}{\sqrt{1-x^2}} \, dx = \sin^{-1} x + C$. What is the radius of convergence of the series?

In Problems 15 through 17, write the given integral as a power series.

15. $\int \cos(x^2) \, dx$

16. $\int e^{x^3} \, dx$

17. $\int \frac{1}{1+x^5} \, dx$

18. Approximate $\int_0^{0.5} \sin(x^2) \, dx$ with error less than 10^{-8}. Is your approximation an overestimate, or an underestimate?

19. Approximate $\int_0^{0.1} \frac{x}{1+x^3} \, dx$ with error less than 10^{-10}.

20. Find the Maclaurin series for $\ln(2 + x)$ along with its radius of convergence.

21. (a) Find the Maclaurin series for $\ln \left(\frac{1+x}{1-x} \right)$ by subtracting the Maclaurin series for $\ln(1 - x)$ from that for $\ln(1 + x)$.
    (b) Show that when $x = \frac{1}{3}$, $\left(\frac{1+x}{1-x} \right) = 2$.
    (c) Use the first four nonzero terms of the series in part (a) to approximate $\ln 2$. Compare your answer with the approximation given by the first four terms of the series for $\ln(1 + x)$ evaluated at $x = 1$, and the value of $\ln 2$ given by a calculator or computer.

22. Show that $\sum_{k=0}^{\infty} \frac{(2x)^k}{k!}$ is a solution to the differential equation $f'(x) = 2f(x)$. What familiar function does this series represent?

23. Show that if $f(x) = \sum_{k=0}^{\infty} a_k x^k$ is a power series solution to $f'(x) = -f(x)$, then $f(x) = \sum_{k=0}^{\infty} (-1)^k \frac{x^k}{k!}$. What function does this series represent?

24. Use power series to solve the differential equation $f''(x) = 9f(x)$. What familiar function(s) does this series represent?

25. The Bessel function $J_0(x)$ is given by $J_0(x) = \sum_{k=0}^{\infty} (-1)^k \frac{x^{2k}}{(k!)^2 2^{2k}}$. It converges for all x.
    (a) If the first three nonzero terms of the series are used to approximate $J_0(0.1)$, will the approximation be too large, or too small? Give an upper bound for the magnitude of the error.
    (b) How many nonzero terms of the series for $J_0(1)$ must be used to approximate $J_0(1)$ with error less than 10^{-4}?

30.5   CONVERGENCE TESTS

In this section we focus on ways of determining whether or not a given series converges. We begin by looking at series of constants; in the last subsection we apply our results to the convergence of power series.

The Basic Principles

A series $\sum_{k=1}^{\infty} a_k$ converges to a sum S if the sequence of its partial sums converges to S, where S is a finite number. In other words, if $\lim_{n \to \infty} S_n = S$, where $S_n = \sum_{k=1}^{n} a_k$, then the infinite series converges to S. Otherwise, the series diverges. Note that if $\lim_{x \to \infty} f(x) = L$, then $\lim_{n \to \infty} f(n) = L$. The converse is not true.

Our first case study was geometric series. (Refer to Chapter 18.) For a geometric series we are able to express S_n in closed form and directly compute $\lim_{n \to \infty} S_n$. We find that

$$\sum_{k=0}^{\infty} ar^k \text{ converges to } \frac{a}{1-r} \text{ if } |r| < 1 \text{ and}$$

$$\text{diverges if } |r| \geq 1.$$

Once we leave the realm of geometric series it can be difficult or impossible to express S_n in closed form, so we generally can't compute $\lim_{n \to \infty} S_n$ directly. Instead, we might determine convergence or divergence by comparing the series in question to a geometric series or an improper integral.

We have already established one test for divergence; if the terms of the series don't tend toward zero then the series diverges.

nth Term Test for Divergence. If $\lim_{n \to \infty} a_n \neq 0$, then $\sum_{n=1}^{\infty} a_n$ diverges.

If $\lim_{n \to \infty} a_n = 0$, we have no information and must turn our attention back to the sequence of partial sums.

Definition

A sequence $\{s_n\}$ is **increasing** if $s_n \leq s_{n+1}$ for all $n \geq 1$. It is **decreasing** if $s_n \geq s_{n+1}$ for all $n \geq 1$. If a sequence is either increasing or it is decreasing it is said to be **monotonic.**

A sequence $\{s_n\}$ is **bounded above** if there is a constant M such that $s_n \leq M$ for all $n \geq 1$. It is **bounded below** if there is a constant m such that $m \leq s_n$ for all $n \geq 1$. A sequence is said to be **bounded** if it is bounded both above and below.

A bounded sequence may or may not converge. It could oscillate between the bounds like $\{(-1)^n\}$. However, if the sequence is bounded and increasing, then its terms must cluster about some number $L \leq M$. A similar statement can be made for a decreasing sequence. The following theorem will prove very useful.

Bounded Monotonic Convergence Theorem[15]

A monotonic sequence converges if it is bounded and diverges otherwise.

Suppose the terms of the series $\sum_{k=1}^{\infty} a_k$ are all positive. Then the sequence of partial sums is increasing: $s_n \leq s_{n+1} = s_n + a_{n+1}$. Because the terms are positive, the sequence of partial sums is bounded below by zero. Therefore, if $\{S_n\}$ is bounded above, then $\{S_n\}$ converges and consequently $\sum_{k=1}^{\infty} a_k$ converges; otherwise they diverge. We will use this line of reasoning repeatedly. We'll refer to it as the **Bounded Increasing Partial Sums Theorem**.

The Bounded Increasing Partial Sums Theorem

A series $\sum_{k=1}^{\infty} a_k$, where $a_k \geq 0$, converges if and only if its sequence of partial sums is bounded above.

Our focus in this section is on the question of convergence versus divergence and not on the sum of a convergent series. Therefore, the starting point of the series is not important; the first hundred or thousand terms of the infinite series can be chopped off without impacting convergence issues. Keep this in mind when applying the results of this section. For example, if the sequence of partial sums is *eventually* monotonic, then the Bounded Increasing Partial Sums Theorem can be applied.

In the next few subsections we will discuss convergence tests with the specification that the terms of the series are positive. From the observation made above, you can see that what is really required is that the terms a_k are positive for all k greater than some fixed number, or, more generally, have any of the required specifications in the long run.

The Integral Test

We revisit the idea of comparing an infinite series and an improper integral in the next example.[16]

◆ EXAMPLE 30.26   Determine whether the following series converge or diverge.

(a) $\sum_{k=1}^{\infty} \frac{1}{k^3} = \frac{1}{1^3} + \frac{1}{2^3} + \frac{1}{3^3} + \frac{1}{4^3} + \cdots$

(b) $\sum_{k=1}^{\infty} \frac{1}{\sqrt{k}} = \frac{1}{\sqrt{1}} + \frac{1}{\sqrt{2}} + \frac{1}{\sqrt{3}} + \cdots$

SOLUTION   In both of these series the terms are positive, decreasing, and going toward zero, but the terms of the series in part (b) are heading toward zero much more slowly than those in part (a). The values of some partial sums are given in the table on page 966. The information in the table is inconclusive, but it leads us to guess that $\sum_{k=1}^{\infty} \frac{1}{k^3}$ might converge and $\sum_{k=1}^{\infty} \frac{1}{\sqrt{k}}$ might diverge.

---

[15] A formal proof of this theorem rests on the Completeness Axiom for real numbers, which says that if a nonempty set of real numbers has an upper bound it must have a least upper bound.

[16] This was first introduced in Section 29.4.

n	$S_n = \sum_{k=1}^{n} \frac{1}{k^3}$	$S_n = \sum_{k=1}^{n} \frac{1}{\sqrt{k}}$
1	1.0	1
2	1.125	1.707106
3	1.162037	2.284457
4	1.177662	2.784457
5	1.185662	3.231670
6	1.190291	3.639918
7	1.193207	4.017883
8	1.195160	4.371436
9	1.196531	4.704770
10	1.197531	5.020997
11	1.198283	5.322509
12	1.198862	5.611184
13	1.199317	5.888534
14	1.199681	6.155795
15	1.199977	6.413994
16	1.204607	6.663994
17	1.204811	6.906530
18	1.204982	7.142232
19	1.205128	7.371648
20	1.205253	7.595255

(a)   (b)

Partial sums are recorded up to
six decimal planes.

(a) To prove that $\sum_{k=1}^{\infty} \frac{1}{k^3}$ converges it is enough to show that the increasing sequence of partial sums is bounded. We do this by comparing the partial sums to $\int_1^{\infty} \frac{1}{x^3}\, dx$, as shown in Figure 30.10.

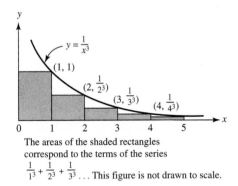

The areas of the shaded rectangles
correspond to the terms of the series
$\frac{1}{1^3} + \frac{1}{2^3} + \frac{1}{3^3} \dots$ This figure is not drawn to scale.

Figure 30.10

Each of the shaded rectangles has a base of length 1. The area = (base) · (height), so the areas of the rectangles, from left to right, are $\frac{1}{1^3}, \frac{1}{2^3}, \frac{1}{3^3}, \cdots$. The sum of the areas of the rectangles is $\sum_{k=1}^{\infty} \frac{1}{k^3}$. Chop off the first rectangle. $\sum_{k=2}^{n} \frac{1}{k^3} < \int_1^n \frac{1}{x^3}\, dx$; the rectangles lie under the graph of $\frac{1}{x^3}$. Consequently, $\lim_{n \to \infty} \sum_{k=2}^{n} \frac{1}{k^3} \leq \int_1^{\infty} \frac{1}{x^3}\, dx$. But $\int_1^{\infty} \frac{1}{x^3}\, dx$ converges.

$$\int_1^{\infty} \frac{1}{x^3}\, dx = \lim_{b \to \infty} \int_1^b x^{-3}\, dx = \lim_{b \to \infty} \frac{x^{-2}}{-2}\bigg|_1^b = \lim_{b \to \infty} \frac{-1}{2b^2} + \frac{1}{2} = \frac{1}{2}$$

Therefore, the partial sums of $\sum_{k=1}^{\infty} \frac{1}{k^3}$ are bounded by $1 + \frac{1}{2}$, and the series converges by the Bounded Increasing Partial Sums Theorem. The sum of the series is greater than 1 and less than 1.5.

(b) To prove that $\sum_{k=1}^{\infty} \frac{1}{\sqrt{k}}$ diverges it is enough to show that the sequence of partial sums is unbounded. We do this by comparing the partial sums to $\int_1^{\infty} \frac{1}{\sqrt{x}}\, dx$ as shown in Figure 30.11.

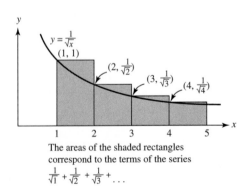

The areas of the shaded rectangles
correspond to the terms of the series
$\frac{1}{\sqrt{1}} + \frac{1}{\sqrt{2}} + \frac{1}{\sqrt{3}} + \dots$

Figure 30.11

$$\int_1^\infty \frac{1}{\sqrt{x}}\,dx = \lim_{b\to\infty} \int_1^b x^{-\frac{1}{2}}\,dx = \lim_{b\to\infty} 2x^{\frac{1}{2}}\,\Big|_1^b = \lim_{b\to\infty} 2\sqrt{b} - 2 = \infty$$

Because we want to show that the partial sums are unbounded, we draw rectangles that lie above the graph of $\frac{1}{\sqrt{x}}$. The areas of the shaded rectangles are, from left to right, $\frac{1}{\sqrt{1}}, \frac{1}{\sqrt{2}}, \frac{1}{\sqrt{3}}, \cdots$, so the sum of the areas of the shaded rectangles is $\sum_{k=1}^\infty \frac{1}{\sqrt{k}}$. We see that $\int_1^{n+1} \frac{1}{\sqrt{x}}\,dx < S_n$, so

$$\lim_{n\to\infty} \int_1^{n+1} \frac{1}{\sqrt{x}}\,dx \le \lim_{n\to\infty} S_n.$$

$$\infty \le \lim_{n\to\infty} S_n.$$

The series diverges. ◆

REMARKS

- To show that a series with positive terms converges we show that the increasing sequence of partial sums is bounded, that is, is less than some constant M. To show that it diverges, we show that $g(n) < S_n$ where $\lim_{n\to\infty} g(n) = \infty$.

- Suppose we compare the series $\sum_{k=1}^\infty f(k)$ with the improper integral $\int_1^\infty f(x)\,dx$. If $f(x)$ is positive, continuous, and decreasing on $[1, \infty)$, then by including or omitting the first term of the series, we can depict the area corresponding to the sum as lying above or below the area corresponding to the improper integral. (See Figures 30.12 and 30.13.)

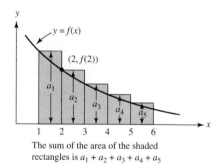

The sum of the area of the shaded
rectangles is $a_1 + a_2 + a_3 + a_4 + a_5$

Figure 30.12

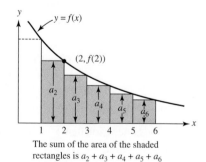

The sum of the area of the shaded
rectangles is $a_2 + a_3 + a_4 + a_5 + a_6$

Figure 30.13

Using the reasoning given we can obtain the Integral Test.

The Integral Test

Let $\sum_{k=1}^{\infty} a_k$ be a series such that $a_k = f(k)$ for $k = 1, 2, 3 \ldots$, where the function f is positive, continuous, and decreasing on $[1, \infty)$. Then

$$\sum_{k=1}^{\infty} a_k \quad \text{and} \quad \int_1^{\infty} f(x)\,dx$$

either both converge or both diverge.

The proof of the Integral Test is constructed along the lines of Example 30.26 and makes a nice exercise for the reader. Construct the proof for yourself. If you have difficulty, consult the proof in Appendix H.

EXERCISE 30.7   Use the Integral Test to show that the harmonic series diverges.

The harmonic series, $\sum_{k=1}^{\infty} \frac{1}{k}$, and the series from Example 30.26, $\sum_{k=1}^{\infty} \frac{1}{k^3}$ and $\sum_{k=1}^{\infty} \frac{1}{\sqrt{k}}$, are all examples of a class of series referred to as **p-series** because they can be written in the form $\sum_{k=1}^{\infty} \frac{1}{k^p}$, where p is a constant.

◆ **EXAMPLE 30.27**   Show that the p-series $\sum_{k=1}^{\infty} \frac{1}{k^p} = \frac{1}{1^p} + \frac{1}{2^p} + \frac{1}{3^p} + \frac{1}{4^p} + \cdots$ converges if $p > 1$ and diverges if $p \leq 1$.

SOLUTION   For $p \leq 0$ the nth term doesn't tend toward zero:

$$\text{For } p = 0, \ \lim_{n \to \infty} \frac{1}{n^p} = \lim_{n \to \infty} 1 = 1 \neq 0$$

$$\text{For } p < 0, \ \lim_{n \to \infty} \frac{1}{n^p} = \infty \neq 0$$

So for $p \leq 0$, $\sum_{k=1}^{\infty} \frac{1}{k^p}$ diverges by the nth Term Test for Divergence.

For $p > 0$, $f(x) = \frac{1}{x^p} = x^{-p}$ is positive, continuous, and decreasing on $[1, \infty)$.

For $p = 1$, we know $\sum_{k=1}^{\infty} \frac{1}{k}$ diverges. Let's consider $p \neq 1$ and apply the Integral Test. For $p > 0$, $p \neq 1$,

$$\int_1^{\infty} \frac{1}{x^p}\,dx = \lim_{b \to \infty} \int_1^{b} x^{-p}\,dx = \lim_{b \to \infty} \frac{x^{-p+1}}{-p+1} \Big|_1^{b} = \lim_{b \to \infty} \frac{b^{-p+1} - 1}{-p+1},$$

so

$$\int_1^{\infty} \frac{1}{x^p}\,dx \begin{cases} \text{converges to } \frac{1}{p-1} & \text{if } p > 1 \\ \text{diverges} & \text{if } p \in (0, 1). \end{cases}$$

Therefore $\sum_{k=1}^{\infty} \frac{1}{k^p}$ converges for $p > 1$ and diverges for $p \leq 1$.   ◆

The conclusion that $\sum_{k=1}^{\infty} \frac{1}{k^p}$ converges for $p > 1$ and diverges otherwise will be useful to keep in mind; we'll compare other series to p-series. Essentially, we've shown that for $p > 1$ the terms $\frac{1}{k^p}$ tend toward zero rapidly enough to make the series converge. If we run into an unfamiliar series whose terms go to zero more rapidly than those of a convergent p-series, we will find that this series converges as well.

EXERCISE 30.8   Which of the following series converge?

(a) $\sum_{n=1}^{\infty} n^{-1.1}$ (b) $\sum_{k=1}^{\infty} \frac{1}{\sqrt[5]{k^2}}$ (c) $\sum_{k=1}^{\infty} \frac{3}{k\sqrt{k}}$ (d) $\sum_{n=1}^{\infty} \frac{1}{e^n}$

Answers

(a), (c), and (d) converge. (a), (b), and (c) are p-series; (d) is not. (d) is a geometric series.

Comparison Tests

In Example 30.26(b) we showed $\sum_{k=1}^{\infty} \frac{1}{\sqrt{k}}$ diverges by comparing it to the divergent integral $\int_{1}^{\infty} \frac{1}{\sqrt{x}} \, dx$. This approach was taken to illustrate a tool that enabled us to treat all p-series easily. However, if we had been concerned only with $\sum_{k=1}^{\infty} \frac{1}{\sqrt{k}}$ it would have been simpler to compare this series with the harmonic series.

$$k \geq \sqrt{k} \text{ for } k \geq 1, \text{ so } \frac{1}{k} \leq \frac{1}{\sqrt{k}} \text{ for } k \geq 1.$$

Comparing $\sum_{k=1}^{\infty} \frac{1}{\sqrt{k}} = 1 + \frac{1}{\sqrt{2}} + \frac{1}{\sqrt{3}} + \frac{1}{\sqrt{4}} + \cdots$ with $\sum_{k=1}^{\infty} \frac{1}{k} = 1 + \frac{1}{2} + \frac{1}{3} + \frac{1}{4} + \cdots$, we see that each term of the former series is greater than or equal to the corresponding term of the harmonic series. The harmonic series diverges because the partial sums increase without bound. Therefore, $\sum_{k=1}^{\infty} \frac{1}{\sqrt{k}}$ must diverge as well.

Having shown that $\sum_{k=1}^{\infty} \frac{1}{\sqrt{k}}$ diverges because the partial sums increase without bound, we can reason that $\sum_{k=4}^{\infty} \frac{1}{\sqrt{k-1}}$ diverges as well. For $k \geq 4$ the latter is, term for term, larger.

$$0 < \frac{1}{\sqrt{k}} < \frac{1}{\sqrt{k} - 1} \quad \text{for } k \geq 4.$$

Similarly, knowing that the geometric series $\sum_{k=1}^{\infty} \frac{1}{2^k}$ converges, we can conclude that $\sum_{k=1}^{\infty} \frac{1}{2^k+1}$ converges as well. $0 < \frac{1}{2^k+1} < \frac{1}{2^k}$, so the partial sums of $\sum_{k=1}^{\infty} \frac{1}{2^k+1}$ are smaller than the corresponding partial sums of the geometric series. The partial sums of $\sum_{2^k=1}^{\infty} \frac{1}{2^k+1}$ are increasing and bounded. Therefore $\sum_{k=1}^{\infty} \frac{1}{2^k+1}$ converges. We can generalize this line of reasoning to compare pairs of series with positive terms.

The Comparison Test

Also known as the Direct Comparison Test

i. Let $0 \leq a_k \leq b_k$ for all k (or all $k \geq N$ for some constant N).

$$\text{If } \sum_{k=1}^{\infty} b_k \text{ converges, then } \sum_{k=1}^{\infty} a_k \text{ converges as well.}$$

ii. Let $0 \leq c_k \leq a_k$ for all k (or all $k \geq N$ for some constant N).

$$\text{If } \sum_{k=1}^{\infty} c_k \text{ diverges, then } \sum_{k=1}^{\infty} a_k \text{ diverges as well.}$$

Less formally, this says that we can compare series whose terms are positive as follows. We can show that a series is convergent by showing that its terms are (eventually) smaller than the terms of a series known to be convergent. We can show that a series is divergent by showing that its terms are (eventually) larger than the terms of a series known to be divergent.

Proof

    i. Let $S_n = \sum_{k=1}^{n} a_k$ and $\hat{S}_n = \sum_{k=1}^{n} b_k$ and suppose $0 \le a_k \le b_k$ for all k.
       Then $\{S_n\}$ and $\{\hat{S}_n\}$ and both increasing sequences, bounded below by 0. $S_n \le \hat{S}_n$ for all n. $\sum_{k=1}^{\infty} b_k$ converges, so $\{\hat{S}_n\}$ is bounded by some M. Then $\{S_n\}$ is likewise bounded by M, and hence $\sum_{k=1}^{\infty} a_k$ converges by the Bounded Increasing Partial Sums Theorem.
          Because the first N terms of a series don't affect whether or not it converges, we can draw the same conclusion if $0 \le a_k \le b_k$ for all $k \ge N$.

    ii. Let $\tilde{S}_n = \sum_{k=1}^{n} c_k$ and suppose $0 \le c_k \le a_k$ for all k.
       Then $\{S_n\}$ and $\{\tilde{S}_n\}$ are both increasing sequences and $\tilde{S}_n \le S_n$ for all n. But $\sum_{k=1}^{\infty} c_k$ diverges. By the Bounded Increasing Partial Sums Theorem $\{\tilde{S}_n\}$ is unbounded, and therefore $\{S_n\}$ is unbounded and $\sum_{k=1}^{\infty} a_k$ diverges.

◆ **EXAMPLE 30.28**   Does $\sum_{k=2}^{\infty} \frac{1}{\ln k}$ converge or diverge?

SOLUTION   Compare with the harmonic series.

$$k > \ln k \quad \text{for all } k \ge 2 \quad \text{(See Figure 30.14.)}$$

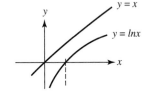

Therefore $0 \le \frac{1}{k} < \frac{1}{\ln k}$ for all $k \ge 2$. The series $\sum_{k=1}^{\infty} \frac{1}{k}$ diverges; therefore $\sum_{k=2}^{\infty} \frac{1}{\ln k}$ diverges by the Comparison Test.   ◆

Figure 30.14

To use the Comparison Test effectively we need to have some familiar series on call. The series most commonly used for comparison are the following:

■   The geometric series

$$\sum_{k=1}^{\infty} ar^k \begin{cases} \text{converges for } |r| < 1 \\ \text{diverges for } |r| \ge 1. \end{cases}$$

■   The *p*-series

$$\sum_{k=1}^{\infty} \frac{1}{k^p} \begin{cases} \text{converges for } p > 1 \\ \text{diverges for } p \le 1. \end{cases}$$

Suppose we're working with the series $\sum_{k=1}^{\infty} \frac{1}{10^k - 2}$. Basically this "looks like" $\sum_{k=1}^{\infty} \frac{1}{10^k} = \sum_{k=1}^{\infty} \left(\frac{1}{10}\right)^k$, which converges. Our series should behave similarly. But $10^k > 10^k - 2$, so $\frac{1}{10^k} < \frac{1}{10^k - 2}$. We can't apply the Direct Comparison Test. This is frustrating, because we know that the tail of the series is what really matters, and for large k, $10^k \approx 10^k - 2$. The Limit Comparison Test gets us out of this bind.

Limit Comparison Test

Suppose $\sum_{k=1}^{\infty} a_k$ and $\sum_{k=1}^{\infty} b_k$ are series whose terms are positive for all $k \geq N$ for some constant N.

i. If $\lim_{k\to\infty} \frac{a_k}{b_k} = L$ for $0 < L < \infty$, then both series converge or both series diverge.

ii. If $\lim_{k\to\infty} \frac{a_k}{b_k} = 0$ and $\sum_{k=1}^{\infty} b_k$ converges, then $\sum_{k=1}^{\infty} a_n$ converges as well.

iii. If $\lim_{k\to\infty} \frac{a_k}{b_k} = \infty$ and $\sum_{k=1}^{\infty} b_k$ diverges, then $\sum_{k=1}^{\infty} a_k$ diverges as well.

Proof

i. Choose positive constants α and β with $0 < \alpha < L < \beta$. Then for k sufficiently large

$$\alpha < \frac{a_k}{b_k} < \beta \quad \text{and} \quad \alpha b_k < a_k < \beta b_k.$$

Suppose $\sum_{k=1}^{\infty} b_k$ diverges. Then $\sum_{k=1}^{\infty} \alpha b_k$ diverges as well. $\sum_{k=1}^{\infty} a_k$ diverges by the Comparison Test.

Suppose $\sum_{k=1}^{\infty} b_k$ converges. Then $\sum_{k=1}^{\infty} \beta b_k$ converges as well. $\sum_{k=1}^{\infty} a_k$ converges by the Comparison Test.

The proofs of parts (ii) and (iii) are left as an exercise for the reader.

◆ **EXAMPLE 30.29**   Determine whether $\sum_{k=2}^{\infty} \frac{2k^2+3k}{k^4-k+1}$ converges or diverges.

SOLUTION   For k large the terms of the series "look like" $\frac{2k^2}{k^4} = \frac{2}{k^2}$. $\sum_{k=2}^{\infty} \frac{2}{k^2} = 2 \sum_{k=2}^{\infty} \frac{1}{k^2}$ converges. (It's a p-series with $p > 1$.) We use the Limit Comparison Test to show $\sum_{k=2}^{\infty} \frac{2k^2+3k}{k^4-k+1}$ converges as well.

$$\lim_{k\to\infty} \frac{a_k}{b_k} = \lim_{k\to\infty} \frac{\frac{2k^2+3k}{k^4-k+1}}{\frac{2}{k^2}} = \lim_{k\to\infty} \frac{2k^2+3k}{k^4-k+1} \cdot \frac{k^2}{2} = \lim_{k\to\infty} \frac{2k^4+3k^3}{2k^4-2k+2} = 1$$

Therefore $\sum_{k=2}^{\infty} \frac{2k^2+3k}{k^4-k+1}$ converges.   ◆

EXERCISE 30.9   Use the Limit Comparison Test to determine whether the following series are convergent or divergent.

(a) $\sum_{k=1}^{\infty} \frac{4}{10^k-2}$

(b) $\sum_{k=2}^{\infty} \frac{\sqrt{k^3+1}}{k(k^2-1)}$

Answers

(a) Convergent: Compare with $\sum_{k=1}^{\infty} \frac{4}{10^4}$ or $\sum_{k=1}^{\infty} \frac{1}{10^k}$, convergent geometric series.

(b) Convergent: "Looks like" $\sum_{k=2}^{\infty} \frac{k^{3/2}}{k^3} = \sum_{k=2}^{\infty} \frac{1}{k^{3/2}}$, a convergent p-series.

In the next example we tackle a more challenging problem.

◆ **EXAMPLE 30.30**   Determine whether $\sum_{k=2}^{\infty} \frac{\ln k}{k^2}$ converges or diverges.

SOLUTION   There are several options available for determining convergence.

- One option is to use the Integral Test. To do this, verify that $f(x) = \frac{\ln x}{x^2}$ is positive, decreasing, and continuous on $[2, \infty)$ and then compute $\int_2^{\infty} \frac{\ln x}{x^2} \, dx$. This improper integral requires integration by parts. As an exercise, verify that $\int_2^{\infty} \frac{\ln x}{x^2} \, dx = \frac{\ln 2 + 1}{2}$ and conclude that $\sum_{k=2}^{\infty} \frac{\ln k}{k^2}$ converges.

- An alternative is to use the Limit Comparison Test. The challenge is to make a good choice for comparison.

$$1 < \ln k < k \qquad \text{for } x \geq 3 \qquad \text{so}$$

$$\frac{1}{k^2} < \frac{\ln k}{k^2} < \frac{k}{k^2} = \frac{1}{k} \qquad \text{for } x \geq 3$$

The terms are greater than those of a known convergent series and less than those of a known divergent series! This doesn't clarify the convergence issues. If we try to use limit comparison by comparing with $\sum_{k=3}^{\infty} \frac{1}{k^2}$ we get

$$\lim_{k \to \infty} \frac{\frac{\ln k}{k^2}}{\frac{1}{k^2}} = \lim_{k \to \infty} \ln k = \infty, \text{ which is inconclusive.}$$

On the other hand, using limit comparison by comparing with $\sum_{k=3}^{\infty} \frac{1}{k}$ we get

$$\lim_{k \to \infty} \frac{\frac{\ln k}{k^2}}{\frac{1}{k}} = \lim_{k \to \infty} \frac{\ln k}{k^2} \cdot k = \lim_{k \to \infty} \frac{\ln k}{k} = 0, \text{ which is inconclusive.}$$

Let's compare with $\sum_{k=3}^{\infty} \frac{1}{k^{1.5}}$. This is a convergent p-series.

$$\lim_{k \to \infty} \frac{\frac{\ln k}{k^2}}{\frac{1}{k^{1.5}}} = \lim_{k \to \infty} \frac{\ln k}{k^2} \cdot k^{1.5} = \lim_{k \to \infty} \frac{\ln k}{k^{0.5}}$$

We can compute this limit by computing $\lim_{x \to \infty} \frac{\ln x}{x^{\frac{1}{2}}}$ using L'Hôpital's Rule.[17]

$$\lim_{x \to \infty} \frac{\ln x}{x^{\frac{1}{2}}} = \lim_{x \to \infty} \frac{\frac{1}{x}}{\frac{1}{2} \frac{1}{\sqrt{x}}} = \lim_{x \to \infty} \frac{1}{x} \cdot \frac{2\sqrt{x}}{1} = \lim_{x \to \infty} \frac{2}{\sqrt{x}} = 0$$

Therefore,

$$\lim_{k \to \infty} \frac{\ln k}{k^{0.5}} = 0 \text{ and } \sum_{k=2}^{\infty} \frac{\ln k}{k^2} \text{ converges.} \qquad ◆$$

The Ratio and the Root Test

In order for a series to converge its terms must be going toward zero sufficiently rapidly. Therefore, it is reasonable to think that valuable information can be obtained by looking at the ratio of successive terms, a_{k+1}/a_k, for k very large. The following convergence test, the

---

[17] See Appendix F for a discussion of L'Hôpital's Rule.

Ratio Test, does this. Its proof relies on our knowledge of geometric series, series for which the ratio of successive terms is fixed.

The Ratio Test

Let $\sum_{k=1}^{\infty} a_k$ be a series such that $a_k > 0$ for k sufficiently large. Suppose

$$\lim_{k \to \infty} \frac{a_{k+1}}{a_k} = L,$$

where L is a real number or ∞.

i. If $L < 1$, then $\sum_{k=1}^{\infty} a_k$ converges.

ii. If $L > 1$, then $\sum_{k=1}^{\infty} a_k$ diverges.

iii. If $L = 1$, the test is inconclusive.

Proof

i. Suppose $L < 1$. We will compare our series with a convergent geometric series. Since $0 \leq L < 1$, we can choose a constant r such that $0 \leq L < r < 1$. $\sum_{k=1}^{\infty} r^k$ converges.

$$\lim_{k \to \infty} \frac{a_{k+1}}{a_k} = L < r, \text{ so for } k \text{ sufficiently large, } \frac{a_{k+1}}{a_k} < r.$$

In other words, there exists an integer N such that

$$\frac{a_{k+1}}{a_k} < r \text{ for all } k \geq N.$$

Therefore, for $k \geq N$,

$$a_{k+1} < a_k r, \qquad \text{and}$$

$$a_{k+2} < a_{k+1} r < (a_k r) \cdot r = a_k r^2$$

$$a_{k+3} < a_{k+2} r < (a_k r^2) \cdot r = a_k r^3$$

$$\vdots$$

The series $\sum_{k=1}^{\infty} a_{N+k} = a_{N+1} + a_{N+2} + a_{N+3} + \cdots$ converges by comparison to

$$\sum_{k=1}^{\infty} a_N r^k = a_N r + a_N r^2 + a_N r^3 + \cdots.$$

This latter series is a geometric series with $|r| < 1$. Each of its terms is larger than the corresponding term in $\sum_{k=1}^{\infty} a_{N+k}$.

$$a_{N+i} < a_N r^i \quad \text{for } i = 1, 2, 3, \ldots.$$

Because $\sum_{k=N+1}^{\infty} a_k$ converges we conclude that $\sum_{k=1}^{\infty} a_k$ converges.

ii. Suppose $\lim_{k \to \infty} \frac{a_{k+1}}{a_k} = L$, $L > 1$, or $\lim_{k \to \infty} \frac{a_{k+1}}{a_k} = \infty$. Then there exists a number N such that $\frac{a_{k+1}}{a_k} > 1$ for all $k \geq N$; that is, $0 < a_k < a_{k+1}$ for all $k \geq N$. But then

$$\lim_{k \to \infty} a_k \neq 0;$$

the series $\sum_{k=1}^{\infty} a_k$ diverges by the nth Term Test for Divergence.

iii. To show that $\lim_{k \to \infty} \frac{a_{k+1}}{a_k} = 1$ is inconclusive, we need only produce both convergent and divergent series for which the limit of the ratio of successive terms is 1.

$$\sum_{k=1}^{\infty} \frac{1}{k} \text{ diverges. } \lim_{k \to \infty} \frac{a_{k+1}}{a_k} = \lim_{k \to \infty} \frac{\frac{1}{k+1}}{\frac{1}{k}} = \lim_{k \to \infty} \frac{k}{k+1} = 1.$$

$$\sum_{k=1}^{\infty} \frac{1}{k^2} \text{ converges. } \lim_{k \to \infty} \frac{a_{k+1}}{a_k} = \lim_{k \to \infty} \frac{\frac{1}{(k+1)^2}}{\frac{1}{k^2}} = \lim_{k \to \infty} \frac{k^2}{(k+1)^2} = 1.$$

As demonstrated, the Ratio Test is not useful for p-series or series whose terms look like polynomials. It is a great test to use for series whose terms involve factorials and/or exponentials.

◆ **EXAMPLE 30.31**   Test the following series for convergence.

(a) $\sum_{k=1}^{\infty} \frac{10^k}{k!}$ (b) $\sum_{k=1}^{\infty} \frac{k^k}{k!}$

SOLUTION   Both series have positive terms. We use the Ratio Test.

(a)

$$\frac{a_{k+1}}{a_k} = \frac{\frac{10^{k+1}}{(k+1)!}}{\frac{10^k}{k!}} = \frac{10^{k+1}}{(k+1)!} \cdot \frac{k!}{10^k} = \frac{10^k \cdot 10}{(k+1)k!} \cdot \frac{k!}{10^k} = \frac{10}{k+1}$$

Compute: $\lim_{k \to \infty} \frac{a_{k+1}}{a_k} = \lim_{k \to \infty} \frac{10}{k+1} = 0 < 1$. Therefore, $\sum_{k=1}^{\infty} \frac{10^k}{k!}$ converges.

(b)

$$\frac{a_{k+1}}{a_k} = \frac{\frac{(k+1)^{k+1}}{(k+1)!}}{\frac{k^k}{k!}} = \frac{(k+1)^{k+1}}{(k+1)!} \cdot \frac{k!}{k^k} = \frac{(k+1)^k (k+1) \cdot k!}{(k+1)k! k^k} = \left(\frac{k+1}{k}\right)^k = \left(1 + \frac{1}{k}\right)^k$$

Compute: $\lim_{k \to \infty} \left(1 + \frac{1}{k}\right)^k = e > 1$. Therefore, $\sum_{k=1}^{\infty} \frac{k^k}{k!}$ diverges.   ◆

REMARK In Example 30.31 part (b) we showed that for the series $\sum_{k=1}^{\infty} \frac{k^k}{k!}$, $\lim_{k \to \infty} \frac{a_{k+1}}{a_k} = e$. It follows that for the series $\sum_{k=1}^{\infty} \frac{k!}{k^k}$, $\lim_{k \to \infty} \frac{a_{k+1}}{a_k} = \frac{1}{e} < 1$. Thus $\sum_{k=1}^{\infty} \frac{k!}{k^k}$ converges and its terms must tend toward zero. We can conclude that k^k grows much more rapidly than does $k!$ From part (a) of Example 30.31 we can deduce that $k!$ in turn grows much more rapidly than does b^k for any constant b. The Ratio Test gives us an alternative way of proving the fact that $\lim_{k \to \infty} \frac{b^k}{k!} = 0$ for every real number b.

The following test gives us a convenient way of testing for convergence when the terms of a series are raised to the kth power.

The Root Test

Let $\sum_{k=1}^{\infty} a_k$ be a series such that $a_k > 0$ for k sufficiently large. Suppose that

$$\lim_{k \to \infty} \sqrt[k]{a_k} = L, \text{ where } L \text{ is a real number or } \infty.$$

i. If $L < 1$, then $\sum_{k=1}^{\infty} a_k$ converges.

ii. If $L > 1$, then $\sum_{k=1}^{\infty} a_k$ diverges.

iii. If $L = 1$, the test is inconclusive.

The proof of the Root Test runs along the same lines as that of the Ratio Test, so it is omitted.

EXERCISE 30.10 Use the Root Test to show that $\sum_{k=1}^{\infty} \left(\frac{2n^3+3n}{3n^3+1} \right)^n$ converges.

Extending Our Results

In this section we've presented several tests that can be used to determine the convergence or divergence of a series: The nth Term Test for Divergence, the Comparison and Limit Comparison Tests, the Integral Test, the Ratio Test, and the Root Test. Of these, all but the nth Term Test require that eventually all the terms of the series being tested must be positive. If, instead, all the terms of the series are negative, or eventually the terms are all negative, then we can factor out a (-1) and the tests can be applied. Otherwise, we can look at $\sum_{k=1}^{\infty} |a_k|$. If $\sum_{k=1}^{\infty} |a_k|$ converges, not only does $\sum_{k=1}^{\infty} a_k$ converge, but it **converges absolutely**. If $\sum_{k=1}^{\infty} |a_k|$ diverges, then $\sum_{k=1}^{\infty} a_k$ will either diverge or converge conditionally.[18] If the series is alternating we can try to apply the Alternating Series Test.

The Ratio Test and Root Test can be easily generalized to apply to arbitrary series.

Generalized Ratio Test

Let $\sum_{k=1}^{\infty} a_k$ be a series with $a_k \neq 0$ for k sufficiently large. Suppose

$$\lim_{k \to \infty} \frac{|a_{k+1}|}{|a_k|} = L, \text{ where } L \text{ is a real number or } \infty.$$

i. If $L < 1$, then $\sum_{k=1}^{\infty} a_n$ converges absolutely (and therefore converges).

ii. If $L > 1$, then $\sum_{k=1}^{\infty} a_n$ diverges.

iii. If $L = 1$, then the test is inconclusive.

The proof is left as a problem at the end of the section. The Root Test can be similarly generalized.

The generalized Ratio Test can be applied to a power series in order to find its radius of convergence and to show that in the interior of its interval of convergence (not including endpoints) the series converges absolutely.

Let $\sum_{k=0}^{\infty} a_k(x-b)^k$ be a power series centered at $x = b$. Let $w_k = a_k(x-b)^k$. Compute

---

[18] Conditional convergence, defined in Section 30.4, means $\sum a_k$ converges but $\sum |a_k|$ diverges.

$$\lim_{k\to\infty} \frac{|w_{k+1}|}{|w_k|} = \lim_{k\to\infty} \frac{|a_{k+1}(x-b)^{k+1}|}{|a_k(x-b)^k|} = \lim_{k\to\infty} \frac{|a_{k+1}|}{|a_k|}|x-b| = \left[\lim_{k\to\infty} \frac{|a_{k+1}|}{|a_k|}\right]|x-b|.$$

■ If $\lim_{k\to\infty} \frac{|a_{k+1}|}{|a_k|} = 0$ then by the generalized Ratio Test the series converges absolutely for all x.

■ If $\lim_{k\to\infty} \frac{|a_{k+1}|}{|a_k|} = Q$, where Q is finite and nonzero, then, the generalized Ratio Test says the series converges absolutely for all x such that $Q|x-b| < 1$, i.e., $|x-b| < \frac{1}{Q}$. The series diverges for $Q|x-b| > 1$, i.e., $|x-b| > \frac{1}{Q}$. Therefore, the radius of convergence is $\frac{1}{Q}$.

■ If $\lim_{k\to\infty} \frac{|a_{k+1}|}{|a_k|} = \infty$, then, by the generalized Ratio Test, the series converges only for $x = b$.

◆ **EXAMPLE 30.32**   The Bessel function $J_0(x) = \sum_{k=0}^{\infty} (-1)^k \frac{x^{2k}}{(k!)^2 2^{2k}}$ is a function *defined* as a power series. Find the set of all x for which the series converges absolutely.

SOLUTION   Apply the generalized Ratio Test.

$$\frac{|w_{k+1}|}{|w_k|} = \left|\frac{(-1)^{k+1}x^{2(k+1)}}{(k+1)!(k+1)!2^{2(k+1)}}\right| \cdot \left|\frac{(k!)(k!)2^{2k}}{(-1)^k x^{2k}}\right| = \frac{|x^{2k+2}|2^{2k}}{|x^{2k}|(k+1)(k+1)2^{2k+2}}$$

$$= \frac{|x^2|}{(k+1)^2 2^2} = \frac{x^2}{4}\frac{1}{(k+1)^2}$$

$$\lim_{k\to\infty} \frac{x^2}{4} \cdot \frac{1}{(k+1)^2} = 0 < 1.$$

The Bessel function $J_0(x)$ converges absolutely for all x.   ◆

◆ **EXAMPLE 30.33**   Find the interval of convergence of the power series $\sum_{k=2}^{\infty} \frac{(x+3)^k}{\ln k}$.

SOLUTION   Begin by applying the generalized Ratio Test. Compute $\frac{|w_{k+1}|}{|w_k|}$.

$$\frac{\left|\frac{(x+3)^{k+1}}{\ln(k+1)}\right|}{\left|\frac{(x+3)^k}{\ln k}\right|} = \frac{|x+3||\ln k|}{|\ln(k+1)|}$$

$$\lim_{k\to\infty} \frac{|w_{k+1}|}{|w_k|} = \lim_{k\to\infty} |x+3|\frac{\ln k}{\ln(k+1)} = |x+3|$$

The series converges for $|x+3| < 1$ and diverges for $|x+3| > 1$.

We must check the endpoints of the interval of convergence, $x = -4$ and $x = -2$, independently.

When $x = 4$ we have $\sum_{k=2}^{\infty} \frac{(-4+3)^k}{\ln k} = \sum_{k=2}^{\infty} \frac{(-1)^k}{\ln k}$. This converges by the Alternating Series Test, because the terms are decreasing in magnitude and tending toward zero.

When $x = -2$ we have $\sum_{k=2}^{\infty} \frac{(-2+3)^k}{\ln k} = \sum_{k=2}^{\infty} \frac{1}{\ln k}$. This series diverges by comparison with the harmonic series. (See Example 30.28 for details.)

The interval of convergence is centered at $x = -3$, the center of the series. The interval of convergence is $[-4, -2)$. The series converges absolutely on $(-4, -2)$ and converges conditionally at $x = -4$.   ◆

Summary of Convergence[19] Criteria for $\sum a_k$

Geometric Series: $\sum ar^k$ $\quad\begin{cases} \text{converges for } |r| < 1 \\ \text{diverges for } |r| \geq 1 \end{cases}$

p-series: $\sum \frac{1}{k^p}$ $\quad\begin{cases} \text{converges for } p > 1 \\ \text{diverges for } p \leq 1 \end{cases}$

nth Term Test for Divergence $\quad$ If $\lim_{k\to\infty} a_k \neq 0$, then the series diverges.

Comparison Test $\quad$ If $a_k \leq b_k$ for all k and $\sum b_k$ converges, so does $\sum a_k$.
terms all positive

$\quad$ If $b_k \leq a_k$ for all k and $\sum b_k$ diverges, so does $\sum a_k$.

Limit Comparison Test $\quad$ Suppose $\lim_{k\to\infty} \frac{a_k}{b_k} = L$, L possibly infinite.
terms all positive

$\quad\quad$ If $0 < L < \infty$, $\sum a_k$ and $\sum b_k$ either both converge or both diverge.

$\quad\quad$ If $L = 0$, and $\sum b_k$ converges, so does $\sum a_k$.

$\quad\quad$ If $L = \infty$, and $\sum b_k$ diverges, so does $\sum a_k$.

Integral Test $\quad$ If $f(k) = a_k$ and $f(x)$ is positive, continuous, and decreasing on $[1, \infty)$, then
terms positive $\quad \int_1^\infty f(x)\, dx$ and $\sum f(k)$ either both converge or both diverge.

Ratio Test $\quad$ Compute $\lim_{k\to\infty} \frac{|a_{k+1}|}{|a_k|} = L$, L possibly infinite.
particularly useful with factorials
and exponents $\quad\quad$ If $L < 1$, $\sum a_k$ converges absolutely.

$\quad\quad$ If $L > 1$, $\sum a_k$ diverges.

$\quad\quad$ If $L = 1$, the test is inconclusive.

Root Test $\quad$ Compute $\lim_{k\to\infty} \sqrt[k]{|a_k|} = L$, L possibly infinite.
particularly useful when a_k looks
like $(-)^k$ $\quad\quad$ If $L < 1$, $\sum a_k$ converges absolutely.

$\quad\quad$ If $L > 1$, $\sum a_k$ diverges.

$\quad\quad$ If $L = 1$, the test is inconclusive.

Alternating Series Test $\quad$ If the series is alternating, the terms are decreasing in magnitude, and the terms
terms alternate sign $\quad$ tend toward zero, then the series converges.

PROBLEMS FOR SECTION 30-5

For Problems 1 and 2, write out the first three terms of the series and then answer the question posed.

1. For what values of c will the series $\sum_{k=1}^{\infty} c^k$ converge? Explain.

2. For what values of w will the series $\sum_{k=1}^{\infty} k^w$ converge? Explain.

3. Suppose $0 \leq a_k \leq b_k \leq c_k$ for all k. Consider $\sum_{k=1}^{\infty} a_k$, $\sum_{k=1}^{\infty} b_k$ and $\sum_{k=1}^{\infty} c_k$. What conclusions can be drawn if you know that $\sum_{k=1}^{\infty} b_k$
   (a) converges. $\quad\quad$ (b) diverges.

---

[19] Here we use $\sum a_k$ to mean $\sum_{k=1}^{\infty} a_k$. The starting value of k is irrelevant since convergence is determined by the tail of the series.

In Problems 4 through 19, determine whether the series converges or diverges. It is possible to solve Problems 4 through 19 without the Limit Comparison, Ratio, and Root Tests.

4. $\sum_{k=2}^{\infty} \frac{3}{\sqrt{k}}$

5. $\sum_{k=10}^{\infty} \frac{10}{k\sqrt{k}}$

6. $\sum_{k=1}^{\infty} \frac{\ln k}{k}$

7. $\sum_{n=5}^{\infty} n^{-9/10}$

8. $\sum_{k=1}^{\infty} \frac{k}{e^k}$

9. $\sum_{k=2}^{\infty} 2k^{-10/9}$

10. $\sum_{k=1}^{\infty} e^{-2k}$

11. $\sum_{k=1}^{\infty} \frac{k+2}{3k^2}$

12. $\sum_{k=1}^{\infty} 2e^{-0.1k}$

13. $\sum_{k=2}^{\infty} \frac{1}{k \ln k}$

14. $\sum_{k=1}^{\infty} ke^{-k^2}$

15. $\sum_{k=1}^{\infty} \frac{k}{k^3+k+1}$

16. $\sum_{k=1}^{\infty} \frac{2}{3^k+1}$

17. $\sum_{k=2}^{\infty} \frac{5}{k-0.5}$

18. $\sum_{n=1}^{\infty} \frac{3^n}{2^n-1}$

19. $\sum_{n=1}^{\infty} \frac{1}{e^n+e}$

20. Suppose that $a_k = f(k)$ for $k = 1, 2, 3, \ldots,$ where $f(x)$ is positive, decreasing, and continuous on $[1, \infty)$. Put the following expressions in order, from smallest to largest. Explain your reasoning with a picture or two.

$$\sum_{k=2}^{n-1} a_k, \qquad \sum_{k=3}^{n} a_k, \qquad \int_2^n f(x)\, dx$$

21. Explain why the hypothesis that $f(x)$ is decreasing is important in the Integral Test.

22. Use your knowledge of improper integrals to give an upper and lower bound for $\sum_{k=1}^{\infty} \frac{1}{k^2}$.

23. Let $\sum_{k=1}^{\infty} a_k$ be a series and $S_k = a_1 + a_2 + \cdots + a_k$ its kth partial sum, where $k = 1, 2, 3, \ldots$. Let L be a constant, $0 < L < 1$.

(a) If $\lim_{k \to \infty} a_k = L$, what can you conclude about $\sum_{k=1}^{\infty} a_k$?

(b) If $\lim_{k \to \infty} S_k = L$, what can you conclude about $\sum_{k=1}^{\infty} a_k$?

24. Let $\sum_{k=1}^{\infty} a_k$ be a series and $S_n = \sum_{k=1}^{n} a_k$ its nth partial sum, where $n = 1, 2, 3, \ldots$. For each of the following, decide whether or not enough information is given to assure that $\sum_{k=1}^{\infty} a_k$ converges. M and m are constants. Explain your reasoning.

(a) $a_k > 0$ for all k and $S_n > m$ for all n.

(b) $a_k > 0$ for all k and $S_n < M$ for all n.

(c) $a_k < 0$ for all k and $S_n > m$ for all n.

(d) $m < S_n < M$ for all n.

In Problems 25 through 32, determine whether the series converges or diverges. In this set of problems knowledge of the Limit Comparison Test is assumed.

25. $\sum_{k=1}^{\infty} \frac{3}{2^k - 1}$

26. $\sum_{k=1}^{\infty} \frac{1}{e^k - 1}$

27. $\sum_{n=2}^{\infty} \frac{n-1}{2n^2 - n}$

28. $\sum_{k=1}^{\infty} \frac{2k^2 - k}{3k^4 + 1}$

29. $\sum_{k=3}^{\infty} \frac{k}{2k^3 - 2}$

30. $\sum_{n=2}^{\infty} \frac{1}{\sqrt{n^2 - n}}$

31. $\sum_{n=2}^{\infty} \frac{n+1}{\ln n}$

32. $\sum_{k=2}^{\infty} \frac{2^k}{5^k - 5}$

33. (a) Let $\sum_{k=1}^{\infty} a_k$ be a convergent series with $0 < a_k < 1$ for $k = 1, 2, 3, \ldots$.

   i. Show that $\sum_{k=1}^{\infty} a_k^2$ converges

   ii. Show that $\sum_{k=1}^{\infty} \frac{1}{a_k}$ diverges.

(b) Let $\sum_{k=1}^{\infty} b_k$ be a convergent series with $0 < b_k$ for $k = 1, 2, 3, \ldots$. Argue that $\sum_{k=1}^{\infty} b_k^2$ converges. (Use the results of part (a)).

In Problems 34 through 41, determine whether the series converges or diverges. In this set of problems knowledge of all the convergence tests from the chapter is assumed.

34. $\sum_{k=1}^{\infty} \frac{3k}{k!}$

35. $\sum_{n=1}^{\infty} \frac{n^3}{2^n}$

36. $\sum_{n=1}^{\infty} \frac{3^n}{2^n}$

37. $\sum_{k=1}^{\infty} \frac{k^3}{k!}$

38. $\sum_{k=1}^{\infty} \frac{k!}{k^3 3^k}$

39. $\sum_{k=2}^{\infty} \frac{2}{(\ln k)^k}$

40. $\sum_{k=1}^{\infty} \left(1 + \frac{1}{k}\right)^k$

41. $\sum_{k=1}^{\infty} \left(\frac{k^2 - 3k}{5k^2 + 1}\right)^k$

42. For what values of n, n a positive integer, does $\sum_{k=1}^{\infty} \frac{k^n}{k!}$ converge?

43. For what values of r does $\sum_{k=1}^{\infty} \frac{r^k}{k!}$ converge?

In Problems 44 through 51, determine whether the series converges absolutely, converges conditionally, or diverges. Explain your reasoning carefully.

44. $\sum_{k=1}^{\infty} \frac{(-1)^k k!}{(k+1)!}$

45. $\sum_{k=1}^{\infty} \frac{(-1)^{k+1}}{k\sqrt{2k}}$

46. $\sum_{k=1}^{\infty} \frac{\cos k}{k^3}$

47. $\sum_{k=1}^{\infty} \frac{\sin(2k)}{2^k}$

48. $\sum_{n=3}^{\infty} \frac{(-1)^n}{10\sqrt{n}}$

49. $\sum_{k=1}^{\infty} \frac{(-1)^k 5^k}{k!}$

50. $\sum_{k=1}^{\infty} \frac{(-k)^k}{k!}$

51. $\sum_{k=2}^{\infty} \frac{(-1)^k}{3k^3 + 3}$

52. Does the series $\sum_{k=2}^{\infty} \frac{(-1)^k \ln k}{k}$ converge absolutely, converge conditionally, or diverge? Explain your reasoning carefully and justify your assertions.

53. Prove the following version of the Integral Test. (It's a slightly weaker version than the one stated in this section.)

    Let $\sum_{k=1}^{\infty} a_k$ be a series such that $a_k = f(k)$ for $k = 1, 2, 3, \ldots$ where the function f is positive, continuous, and decreasing on $[1, \infty)$.

    (a) If $\int_1^{\infty} f(x)\,dx$ converges, then $\sum_{k=1}^{\infty} a_k$ converges.
    (b) If $\int_1^{\infty} f(x)\,dx$ diverges, then $\sum_{k=1}^{\infty} a_k$ diverges.

In Problems 54 through 59, use the Ratio Test or Root Test to find the radius of convergence of the power series given.

54. $\sum_{k=1}^{\infty} (-1)^k \frac{(2x)^k}{k!}$

55. $\sum_{k=1}^{\infty}(-1)^k \frac{(3x)^k}{k}$

56. $\sum_{k=1}^{\infty} k \left(\frac{x}{2}\right)^k$

57. $\sum_{n=1}^{\infty} \frac{n(x-5)^n}{(2n)!}$

58. $\sum_{k=3}^{\infty} \frac{(x-1)^{2k}}{(k-1)!}$

59. $\sum_{n=1}^{\infty} \frac{(x+2)^n}{n(2n+3)}$

In Problems 60 through 71, find the interval of convergence of the series. Explain your reasoning fully.

60. $\sum_{k=1}^{\infty} \frac{(3x)^k}{2^k}$

61. $\sum_{k=1}^{\infty} 2^k (x-3)^k$

62. $\sum_{k=1}^{\infty}(-1)^k \frac{(x+1)^k}{3k}$

63. $\sum_{k=1}^{\infty}(-1)^k \frac{(x+1)^k}{k}$

64. $\sum_{k=2}^{\infty} \frac{(x-3)^k}{2^k}$

65. $\sum_{k=1}^{\infty} \frac{(x-3)^k}{k!}$

66. $\sum_{k=1}^{\infty}(kx)^k$

67. $\sum_{k=1}^{\infty} \frac{(x-2)^k}{k5^k}$

68. $\sum_{k=1}^{\infty} \frac{(x-1)^k}{k^5}$

69. $\sum_{n=0}^{\infty} \frac{(-1)^n x^n}{(2n+1)}$

70. $\sum_{k=1}^{\infty} \frac{2^k x^k}{k!}$

71. $\sum_{k=2}^{\infty}(-1)^k \frac{x^k}{2 \ln k}$

72. Suppose $\lim_{k\to\infty} \sqrt[k]{|a_k|} = \frac{1}{3}$.
   (a) What is the radius of convergence of $\sum_{k=1}^{\infty} a_k(x-1)^k$?
   (b) For each value of x listed below, determine whether the series converges absolutely, converges conditionally, or diverges.
       i. $x = 0$ ii. $x = -3$ iii. $x = -1.5$ iv. $x = 5$

73. Give an example of each of the following.

    (a) a series that converges only at $x = 4$

    (b) a series that converges for $x \in (3, 5)$ and diverges otherwise

    (c) a series that converges for all x

    (d) a series that converges for $x \in (2, 6)$ and diverges otherwise

74. Show that if a power series $\sum_{k=0}^{\infty} a_k x^k$ has radius of convergence R, then $\sum_{k=0}^{\infty} a_k (x - b)^k$ also has a radius of convergence of R.

75. If R is the radius of convergence of $\sum_{k=0}^{\infty} a_k x^k$, determine the radius of convergence of the following.

    (a) $\sum_{k=100}^{\infty} a_k x^k$ (b) $\sum_{k=0}^{\infty} a_k (2x)^k$ (c) $\sum_{k=0}^{\infty} a_k \left(\frac{x}{2}\right)^k$

76. Show that the radius of convergence of the binomial series is 1.

77. Justify the generalized Ratio Test.

31

Differential Equations

Change is intrinsic in the universe and in the world around us; the world is in motion. Attempts to understand and predict change often involve creating models reflecting rates of change. Translating information about rates of change into the language of mathematics using a continuous model means setting up an equation containing a derivative. As discussed in Chapter 15, such an equation is called a differential equation. We set up this differential equation by interpreting the derivative as an instantaneous rate of change; subsequently, we can use the slope interpretation of the derivative to gain more information about the function itself.

In Section 15.2 we first introduced differential equations, highlighting the differential equation $\frac{dy}{dt} = ky$ reflecting exponential growth and decay. Here, in Section 31.1, we construct differential equations reflecting various situations. In Section 31.2 we discuss what it means to solve a differential equation. We turn our attention to qualitative analysis in Section 31.3 before returning to analytic solutions in Section 31.4. In Section 31.5 we look at systems of first order differential equations, using them to model epidemics and population interactions. In Section 31.6 we solve a special type of second order differential equation.

31.1 INTRODUCTION TO MODELING WITH DIFFERENTIAL EQUATIONS

Generally, real-life situations are complex and involve many interrelated variables. The way the situations change can often be modeled using differential equations, but generally only after some simplifying assumptions are made. A good model ought to identify the factors that are of primary importance to the situation and state clearly the simplifying assumptions. Problems must be formulated so that it can be translated into mathematics. The mathematical problem is analyzed using mathematical tools and the results of the analysis then considered in terms of their implications to the original problem.

Differential equations are used by scientists and social scientists in a wide variety of disciplines, such as physics, chemistry, medicine, agronomy, population biology, epidemiology, astronomy, and economics. Change is the unifying concept. Situations in disparate fields may be modeled by structurally similar differential equations. For instance, the rate of change of a quantity may be proportional to the quantity, or to the difference of some quantities, or to a product of quantities. In the examples below we use differential equations to model a variety of situations.

◆ **EXAMPLE 31.1**   *Population biology.* If a population is flourishing under ideal circumstances and with unlimited resources, its rate of growth is proportional to itself.

SOLUTION   Let $P = P(t)$ be the size of the population at time t. Then

$$\frac{dP}{dt} = kP, \quad \text{where } k \text{ is the constant of proportionality, } k > 0. \quad ◆$$

REMARK If we know $\frac{dP}{dt}$ for a certain value of P, then we can use the differential equation to determine k. On the other hand, if we know $\frac{dP}{dt}$ for a certain value of t, then we must solve the differential equation in order to determine k.

◆ **EXAMPLE 31.2**   *Cellular biology.* The concentration of a certain nutrient in a cell changes at a rate proportional to the difference between the concentration of the nutrient inside the cell and the concentration in the surrounding environment. Suppose that the concentration in the surrounding environment is kept constant and is given by N. If the concentration of the nutrient in the cell is greater than N, then the concentration in the cell decreases; if the concentration in the cell is less than N, then the concentration increases. Let $C = C(t)$ be the concentration of the nutrient within the cell. Write a differential equation involving the rate of change of C.

SOLUTION   We translate the following sentence into a mathematical equation.
The rate of change of C is proportional to the difference between C and N.

$$\frac{dC}{dt} = k(C - N),$$

where k is the constant of proportionality.

What is the sign of k? When $C(t) > N$, we know that $C(t)$ decreases, so $\frac{dC}{dt}$ is negative.

$$\frac{dC}{dt} = k(C - N) \Rightarrow (-) = k(+) \quad \text{so } k \text{ must be negative.}$$

Let's make sure this makes sense for $C(t) < N$.

When $C(t) < N$, we know that $C(t)$ increases, so $\frac{dC}{dt}$ is positive.

$$\frac{dC}{dt} = k(C - N) \Rightarrow (+) = (-)(-) \quad \text{This works as desired.}$$

So $\frac{dC}{dt} = k(C - N)$, where k is a negative constant. Alternatively, $\frac{dC}{dt} = -k(C - N)$, where $k > 0$. ◆

◆ **EXAMPLE 31.3**   *Newton's law of cooling.* We know from experience that hot objects cool down to the temperature of their surroundings and cold objects warm up to the temperature of their surroundings. Newton's law of cooling says more specifically that the temperature difference between and object and its surroundings changes at a rate proportional to the difference between the temperature of the object and that of its surroundings.

(a) Let $T(t)$ be the temperature of an object and $R(t)$ the temperature of the room in which it resides, where t is measured in minutes since the object was placed in the room. Write a differential equation that models the temperature dynamics.

The rate of change of the difference between T and R is proportional to the difference between T and R.

$$\frac{d(T-R)}{dt} = k(T-R)$$

(b) Suppose the temperature of the room is held constant at 65° and a cup of hot mulled apple cider with a temperature of 180° is placed in the room. Write a differential equation involving $T(t)$, the temperature of the cider as a function of time.

The room is kept at a constant 65°, so $R(t) = 65$, and $\frac{dR}{dt} = 0$.

$$\frac{d(T-65)}{dt} = k(T-65)$$

$$\frac{dT}{dt} = k(T-65)$$

Notice that the initial temperature of the cider does *not* appear in the differential equation. The temperature of the cider does not remain at 180°; $T(0) = 180$ is the initial condition. However, because the temperature of the room is constant (always 65°), we can replace R by the value 65.

What is the sign of the constant k? The cider is cooling down, so $\frac{dT}{dt}$ is negative. Because the cider's temperature is greater than 65°, the difference $T - 65$ is positive.

$$\frac{dT}{dt} = k(T-65)$$

$$(-) = k(+)$$

So k must be negative.

(c) Now suppose that the object in the 65° room is a glass of cold apple juice with an initial temperature of 40°. Does this change the differential equation we obtained in part (b)?

$$\frac{dT}{dt} = k(T-65)$$

This differential equation is exactly the same as the one for the hot mulled cider. The fact that the initial temperature of the object is different does not change the differential equation, only the initial condition.

Is the sign of k the same as in part (b)? The juice will warm up, so $\frac{dT}{dt}$ will now be positive, but $T - 65$ will now be negative.

$$\frac{dT}{dt} = k(T-65)$$

$$(+) = k(-) \quad \text{Yes, } k \text{ is negative.}$$

The differential equation holds regardless of the relative temperatures of the object and the room.[1] ◆

◆ **EXAMPLE 31.4**   *Epidemiology.*   The flu is spreading throughout a college dormitory of 300 students. It is highly contagious and long in duration. Assume that during the time period we are modeling no student has recovered and all sick students are still contagious. It is reasonable to assume that the rate at which students are getting ill is proportional to the product of the number of sick students and the number of healthy ones because there must be an interaction between a healthy and a sick student to pass along the disease. Let $S = S(t)$ be the number of sick students at time t. Write a differential equation reflecting the situation.

SOLUTION   We translate the following sentence into a differential equation.
The rate at which students are getting sick is proportional to the product of the number of sick students and the number of healthy ones.

$$\frac{dS}{dt} = kS(300 - S),$$

where k is the positive constant of proportionality.   ◆

◆ **EXAMPLE 31.5**   *Medicine.*   The rate at which a certain drug is eliminated from the bloodstream is proportional to the amount of the drug in the bloodstream. A patient now has 45 mg of the drug in his bloodstream. The drug is being administered to the patient intravenously at a constant rate of 5 milligrams per hour. Write a differential equation modeling the situation.

SOLUTION   Let $A = A(t)$ be the amount of the drug in the patient's bloodstream at time t.
The amount of drug is being affected in two ways; it is increasing due to intravenous injections and decreasing due to biological processes. We know the rate of increase and the rate of decrease.
Our basic strategy for dealing with multiple rates is to use the framework

total rate of change =	rate in	− rate out, or equivalently,
total rate of change = rate of increase −		rate of decrease.

$$\frac{dA}{dt} = 5 - kA, \qquad A(0) = 45$$

Notice that the initial amount of the drug in the bloodstream does *not* appear in the differential equation. 45 mg corresponds to $A(0)$, the initial condition that specifies a particular solution to the differential equation, not to a rate. k is the proportionality constant determined by the drug (and perhaps the weight or condition of the patient).   ◆

◆ **EXAMPLE 31.6**   *Economics.*   Ten thousand dollars is deposited in a bank account with a nominal annual interest rate of 5% compounded continuously. No further deposits are made. Write a differential equation reflecting the situation if money is withdrawn continuously at a rate of $4000 per year.[2]

---

[1] Instead of writing the temperature differences as $T - 65$, we could just as well have used $65 - T$ to get the differential equation $\frac{dT}{dt} = k(65 - T)$. Check that in this case the sign of k must be positive for both the hot cider and the cold juice.
[2] Of course, money is not actually being withdrawn continuously; we are using a continuous model for a discrete reality.

SOLUTION Let $M = M(t)$ be the amount of money in the account at time t. Consider the factors affecting the rate of change of the amount of money. Interest is responsible for the growth rate, and the withdrawals are responsible for the rate of decrease.

The rate at which interest is earned is proportional to the balance itself with constant of proportionality 0.05. The rate of withdrawals does *not* depend on either the balance or the time; the rate out is constant at $4000 per year.

$$\text{rate of change} = \quad \text{rate of growth} \quad - \quad \text{rate of decrease}$$

$$\text{rate of change} = \text{rate of earning interest} - \text{rate of withdrawals}$$

$$\frac{dM}{dt} \quad = \quad 0.05M \quad - \quad 4000$$

REMARK This differential equation often causes trouble, so heads up! The *rate of increase* is the rate of earning interest, not the accumulated money. Similarly, the *rate out* is the *rate at which withdrawals are made*.

■ One common error is to write the "rate out" as $4000t$ instead of 4000. The quantity $4000t$ is the total *amount* withdrawn during the first t years; it does not represent the *rate* of withdrawal.

■ A second common error is to make some attempt to write a formula for the amount of money in the account and to insert that into the differential equation. This attempt to compute $M(t)$ on the fly is invariably futile, as the situation is quite complex. Interest is not being earned on money that is withdrawn from the account; interest is only being earned on the money that is still there. But the beauty of modeling the situation with a differential equation is that we have given the amount of money in the account at any time the name $M(t)$ and this is all we need in order to set up the differential equation. It is then possible to solve this differential equation to obtain a formula for $M(t)$. ◆

◆ EXAMPLE 31.7 *Physics.* An object (football, tennis ball, soccer ball, rock . . .) falling through the air undergoes a constant downward acceleration of 32 feet per second per second due to the force of gravity.

(a) Write a differential equation involving $v(t)$, where $v(t)$ is the vertical velocity of the object at time t.

(b) Write a differential equation involving $s(t)$, where $s(t)$ is the object's height above the ground.

SOLUTION (a) Downward acceleration is the rate of change of vertical velocity. We translate the following sentence into a mathematical equation.

The rate of change of velocity is -32 *feet per second per second.*

$$\frac{dv}{dt} = -32.$$

(b) Downward acceleration is the second derivative of the height function.

$$v(t) = \frac{ds}{dt}, \quad \text{so} \quad \frac{dv}{dt} = \frac{d}{dt}\left(\frac{ds}{dt}\right) = \frac{d^2s}{dt^2}.$$

$$\frac{d^2s}{dt^2} = -32 \quad ◆$$

Let's establish some terminology.

The **order** of a differential equation is the order of the highest order derivative in the equation.

A differential equation involving only first derivatives is called a **first order differential equation.** A differential equation involving second derivatives but no higher order derivatives is called a **second order differential equation.**

In Example 31.7 the differential equation $\frac{dv}{dt} = -32$ is a first order differential equation; $\frac{d^2s}{dt^2} = -32$ is a second order differential equation conveying the same information.

A **solution** to a differential equation is a *function* that satisfies the equation.[3] The graph of a solution is called a **solution curve.** A differential equation has a family of solutions. An **initial condition** will specify a particular solution to a first order differential equation. The initial condition specifies a value of the dependent variable for a specific value of the independent variable. A second order differential equation will require two initial conditions. This should make sense in light of Example 31.7. If we start with the second order differential equation $\frac{d^2x}{dt^2} = -32$, we need to know two pieces of information, for instance, the initial position and the initial velocity, in order to determine the position at time t.

Let's look at another example, a variation on Example 31.6.

◆ **EXAMPLE 31.8**   Let $M = M(t)$ be the amount of money in a bank account at time t, given in years. Suppose

$$\frac{dM}{dt} = 3000 + 0.04M - 100t.$$

What scenario could be modeled by this differential equation?

SOLUTION   There are two terms contributing to the rate of increase of M. Because the rate of increase due to interest is proportional to M, the term $0.04M$ must reflect the rate of growth due to interest. It is possible that the interest is nominally 4% per year compounded continuously.

The term 3000 reflects deposits into the account at a constant rate of 3000 dollars per year.

The term $-100t$ accounts for the contribution to the decrease of money. It represents the rate of withdrawal. The rate at which money is being withdrawn is increasing with time. At $t = 1$ money is being withdrawn at a rate of $100 per year while at $t = 9$ it is being withdrawn at a rate of $900 per year. While deposits are being made at a constant rate, withdrawals are being made at an increasing rate.   ◆

Obtaining Information from a Differential Equation

In this section we have been writing differential equations without solving them or graphing their solutions. Yet in the differential equation itself there is a wealth of information. For example, consider the differential equation from Example 31.3: $\frac{dP}{dt} = kP$, where $k > 0$ and $P(0) > 0$. From the differential equation we see that $\frac{dP}{dt}$ is positive and increasing; consequently $P(t)$ is increasing and concave up; $P(t)$ is increasing at an increasing rate.

---

[3] Sometimes the solution to a differential equation can be a *relation* instead of a function. For instance, $x^2 + y^2 = C$ is a solution to the differential equation $\frac{dy}{dx} = \frac{-x}{y}$.

$\frac{d^2P}{dt^2}$ can be computed directly from the differential equation. $\frac{d^2P}{dt^2} = \frac{d}{dt}\left(\frac{dP}{dt}\right) = \frac{d}{dt}(kP) = k\frac{dP}{dt} = k(kP) = k^2P$. A more involved example is analyzed below.

◆ **EXAMPLE 31.9**   Consider the differential equation

$$\frac{dM}{dt} = 0.04M - 4000,$$

where $M(t)$ is the amount of money in a bank account at time t, where t is given in years. The differential equation reflects the situation in which interest is being paid at a rate of 4% per year compounded continuously and money is being withdrawn at a constant rate of $4000 per year.

(a) Suppose the initial deposit is $50000. Will the account be depleted?

(b) If money is to be withdrawn at a rate of $4000 per year, what is the minimum initial investment that assures the account is not depleted?

(c) If this is a trust fund that is being set up with $50000 and the idea is that the account should not be depleted, what should the restriction be on the rate of withdrawal? Assume money will be withdrawn at a constant rate.

SOLUTION   (a) If the initial deposit is $50,000, then at $t = 0$ we have $\frac{dM}{dt} = 0.04(50,000) - 4000 = -2000$, so the amount of money is decreasing. As M decreases, $\frac{dM}{dt}$ becomes increasingly negative. In other words, if the initial deposit is $50,000, then the account will lose money (more money each year) and be depleted.

(b) If $\frac{dM}{dt}$ is ever negative, then it will become increasingly negative. We need an initial deposit such that $\frac{dM}{dt} = 0.04M - 4000 = 0$. Solving for M we find that $100,000 is needed as an initial deposit in order to maintain the account indefinitely. If the initial deposit is $100,000, then money is being withdrawn at exactly the rate that it is growing; when $\frac{dM}{dt} = 0$ an equilibrium is maintained. $M(t) = 100,000$ is called the **equilibrium solution.**

(c) We need to find K such that $\frac{dM}{dt} = 0.04M - K$ is nonnegative if the initial deposit is $50,000. We'll solve

$$\frac{dM}{dt} = 0.04(50,000) - K = 0.$$

$K = 2000$. Money can be withdrawn at a rate of $2000 per year.   ◆

The analysis used in Example 31.9 can be applied to a myriad of other situations as well. For example, consider demographic changes that might require a demographer to take into account immigration and/or emigration. For instance, when the British relinquished their hold on India, and India and Pakistan became two independent nations, there was a massive movement of populations, Hindus from Pakistan immigrating to India and Moslems from India immigrating to Pakistan. There have been many other mass movements of populations: in the mid-1800s there was a mass emigration out of Ireland due to the potato famine, and in the early 1900s there was a population swap between Greece and Turkey. One could model migrations in Ethiopia during the recent civil war, or in the former Yugoslavia; the list is endless. If the net immigration/emigration rate is constant, then the basic equation that can be used to model many of these situations is of the form $\frac{dP}{dt} = kP + (I - E)$, where k is the proportionality constant due to population growth, E represents the rate of emigration, and I represents the rate of immigration. This is structurally identical to the equation used in the previous example.

PROBLEMS FOR SECTION 31.1

1. Money is deposited in a bank account with a nominal annual interest rate of 4% compounded continuously. Let $M = M(t)$ be the amount of money in the account at time t.

   (a) Write a differential equation whose solution is $M(t)$. Assume there are no additional deposits and no withdrawals.

   (b) Suppose money is being added to the account continuously at a rate of $1000 per year and no withdrawals are made. Write a differential equation whose solution is $M(t)$.

2. We can construct a model for the spread of a disease by assuming that people are being infected at a rate proportional to the product of the number of people who have already been infected and the number of those who have not. Let $P(t)$ denote the number of infected people at time t and N denote the total population affected by the epidemic. Assume N is fixed throughout the time period we are considering.

   We are assuming that every member of the population is susceptible to the disease and the disease is long in duration (there are no recoveries during the time period we are analyzing) but not fatal (no deaths during this period). The assumption that people are being infected at a rate proportional to the product of those who are infected and those who are not could reflect a contagious disease where the sick are not isolated.

   Write a differential equation whose solution is $P(t)$.

3. Solutes in the bloodstream enter cells through osmosis, the diffusion of fluid through a semipermeable membrane until the concentration of fluid on both sides of the membrane is equal. Suppose that the concentration of a certain solute in the bloodstream is maintained at a constant level of K mg/cubic cm. Let's consider $f(t)$, the concentration of the solute inside a certain cell at time t. The rate at which the concentration of the solute inside the cell is changing is proportional to the difference between the concentration of the solute in the bloodstream and its concentration inside the cell.

   (a) Set up the differential equation whose solution is $y = f(t)$.

   (b) Sketch a solution assuming that $f(0) < K$.

4. A large garbage dump sits on the outskirts of Cairo. Garbage is being deposited at the dump at a rate of T tons per month. Scavengers and salvagers frequent the dump and haul off refuse from the site. The rate at which garbage is being hauled off is proportional to the tonnage at the site. Let $G(t)$ be the number of tons of garbage in the dump.

   Write a differential equation whose solution is $G(t)$. The basic framework is

   rate of change of G = rate of increase of G − rate of decrease of G.

5. Elmer takes out a $100,000 loan for a house. He pays money back at a rate of $12,000 per year. The bank charges him interest at a rate of 8.5% per year compounded continuously. Make a continuous model of his economic situation. Write a differential equation whose solution is $B(t)$, the balance he owes the bank at time t.

6. Let's suppose that the population in a certain country has a growth rate of 2% and a population of 9 million at a time we'll designate as $t = 0$. Due to the political and economic situation, there is a massive rearrangement of populations in the region. The immigration and emigration rates are both constant, with people entering the country at a rate of 100,000 per year and leaving at a rate of 300,000 per year. Let $P = P(t)$ be the population in millions at time t.

   (a) Write a differential equation reflecting the situation. *Keep in mind that P is in millions.*

   (b) If this situation goes on indefinitely, what will happen to the country's population?

   (c) What initial population would support a net emigration of 200,000 per year?

7. In the beginning of a chemical reaction there are 600 moles of substance A and none of substance B. Over the course of the reaction, the 600 moles of substance A are converted to 600 moles of substance B. (Each molecule of A is converted to a molecule of B via the reaction.) Suppose the rate at which A is turning into B is proportional to the product of the number of moles of A and the number of moles of B.

   (a) Let $N = N(t)$ be the number of moles of substance A at time t. Translate the statement above into mathematical language. (*Note:* The number of moles of substance B should be expressed in terms of the number of moles of substance A.)

   (b) Using your answer to part (a), find $\frac{d^2N}{dt^2}$. Your answer will involve the proportionality constant used in part (a).

   (c) $N(t)$ is a decreasing function. The rate at which N is changing is a function of N, the number of moles of substance A. When the rate at which A is being converted to B is highest, how many moles are there of substance A?

8. There are many places in the world where populations are changing and immigration and/or emigration play a big role. People may move to find food, or to find jobs, or to flee political or religious persecution. Pick a situation that interests you. You could look at the number of Tibetans in Tibet, or the number of Tibetans in India, or the number of lions in the Serengeti, or the number of tourists in Nepal. Get some data and try to model the population dynamics using a differential equation. What simplifying assumptions have you made?

31.2   SOLUTIONS TO DIFFERENTIAL EQUATIONS: AN INTRODUCTION

Although knowing about the rate of change of a quantity is useful, often what we really want to know about is the actual *amount* of that quantity. After all, it's nice to know that the money in your bank account is growing at an instantaneous rate of 5%, but what you really want to know is when you'll finally have enough money to buy that new motorcycle or whatever it is that you're saving for. In this section we turn our attention to the solutions of differential equations.

What Does It Mean to Be a Solution to a Differential Equation?

We have said that a function f is a **solution to a differential equation** if it satisfies the differential equation. By this we mean that when the function and its derivative(s) are substituted in the appropriate places in the differential equation, the two sides of the equation are equal. We review work presented in Section 15.2.

◆ **EXAMPLE 31.10**   Is $y = x^3$ a solution to the differential equation $\frac{dy}{dx} = \frac{3y}{x}$?

SOLUTION   To determine whether $y = x^3$ is a solution, we need to substitute it into the differential equation.

$$\frac{dy}{dx} = \frac{3y}{x} \qquad \text{Replace } y \text{ by } x^3 \text{ wherever it appears.}$$

$$\frac{d}{dx}x^3 \overset{?}{=} \frac{3x^3}{x}$$

$$3x^2 \overset{?}{=} \frac{3x^3}{x}$$

$$3x^2 = 3x^2 \qquad \text{True.}$$

$y = x^3$ is a solution to $\frac{dy}{dx} = \frac{3y}{x}$ because it satisfies the equation.   ◆

◆ **EXAMPLE 31.11**   Is $y = xe^{3x}$ a solution to the differential equation $\frac{dy}{dx} = \frac{3y}{x}$?

SOLUTION

$$\frac{dy}{dx} = \frac{3y}{x} \qquad \text{Replace } y \text{ by } xe^{3x} \text{ wherever it appears.}$$

$$\frac{d}{dx}[xe^{3x}] \overset{?}{=} \frac{3xe^{3x}}{x}$$

$$3xe^{3x} + e^{3x} \overset{?}{=} 3e^{3x}$$

$$e^{3x}[3x + 1] \neq e^{3x}(3)$$

The equation is not satisfied, so $y = xe^{3x}$ is *not* a solution to the differential equation.   ◆

Differential equations have families of solutions. In Section 15.2 we looked at the family of solutions to each of the three differential equations given below. The discussion is summarized here.

i. $\frac{dy}{dt} = 2$ ii. $\frac{dy}{dt} = 2t$ iii. $\frac{dy}{dt} = 2y$

Graphical perspective: Solving a differential equation that involves $\frac{dy}{dt}$ means finding y as a function of t; therefore, on our graph we will label the vertical axis y and the horizontal axis t.

At any point P in the ty-plane we can use the differential equation to find the slope of the tangent line to the solution curve through P. We'll draw a short line segment through P indicating the slope of the solution curve there. The resulting diagram is called a **slope field**. (See Figure 31.1 on the following page.)

Observations

In part (i), $\frac{dy}{dt} = 2$, the slope is independent of the point P chosen.

In part (ii), $\frac{dy}{dt} = 2t$, the slope depends only on the t-coordinate of P.
This is true whenever we have a differential equation of the form $\frac{dy}{dt} = f(t)$.

In part (iii), $\frac{dy}{dt} = 2y$, the slope depends only on the y-coordinate of P.
This is true whenever we have a differential equation of the form $\frac{dy}{dt} = f(y)$.

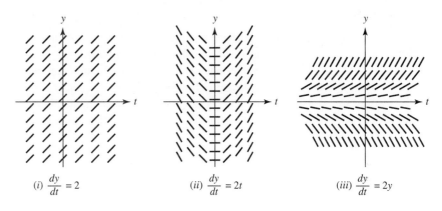

(i) $\dfrac{dy}{dt} = 2$ (ii) $\dfrac{dy}{dt} = 2t$ (iii) $\dfrac{dy}{dt} = 2y$

Figure 31.1

In Figure 31.1 (ii) the slope is positive whenever t is positive, negative whenever t is negative, and zero at $t = 0$. At $(2, 3)$ the slope is 4; at $(1, 5)$ the slope is 2.

In Figure 31.1 (iii) the slope is positive whenever y is positive, negative whenever y is negative, and zero at $y = 0$. At $(2, 3)$ the slope is 6; at $(1, 5)$ the slope is 10.

By following the slope fields, we can get a rough idea of the shapes of the solution curves. (See Figure 31.2.)

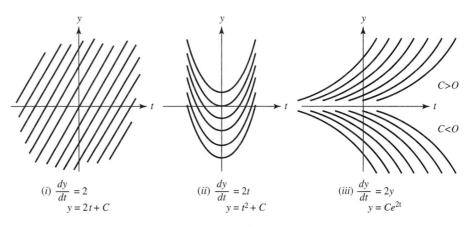

(i) $\dfrac{dy}{dt} = 2$ (ii) $\dfrac{dy}{dt} = 2t$ (iii) $\dfrac{dy}{dt} = 2y$
$\quad\;\; y = 2t + C$ $\quad\;\; y = t^2 + C$ $\quad\;\; y = Ce^{2t}$

Figure 31.2

We state, without proof, the following fact (known as the Existence and Uniqueness Theorem).[4]

---

[4] The statements you'll see in differential equations texts are actually much stronger than this.

Existence and Uniqueness Theorem (Weak form)

Let (a, b) be a point in the plane. Any differential equation of the form $\frac{dy}{dt} = g(t)$ where g is continuous, or of the form $\frac{dy}{dt} = f(y)$ where f and f' are continuous, has a solution passing through (a, b). The solution exists, and it is unique.

In particular the Existence and Uniqueness Theorem tells us given any point $P = (a, b)$ in the plane, a differential equation of the form $\frac{dy}{dt} = k$, $\frac{dy}{dt} = kt$, or $\frac{dy}{dt} = ky$ where k is constant has exactly one solution passing through the point P. This makes sense in our examples above; because the slope at P is completely determined by the coordinates of P, no two solution curves can cross. We can combine this graphical analysis with our knowledge of analytic solutions to these differential equations.[5]

$$y = 2t + C \quad \text{is a solution to part (i) for any constant } C.$$

$$y = t^2 + C \quad \text{is a solution to part (ii) for any constant } C.$$

$$y = Ce^{2t} \quad \text{is a solution to part (iii) for any constant } C.$$

The Existence and Uniqueness Theorem tells us that we have written general solutions to each of these differential equations. In other words, any solution to $\frac{dy}{dt} = 2$ can be expressed in the form $y = 2t + C$; any solution to $\frac{dy}{dt} = 2t$ can be expressed in the form $y = t^2 + C$; any solution to $\frac{dy}{dt} = 2y$ can be expressed in the form $y = Ce^{2t}$.

EXERCISE 31.1   Let k be an arbitrary constant. Show that for any point P in the plane there is a unique value of C such that the curve $y = Ce^{kt}$ passes through P.

EXERCISE 31.2   Let C be an arbitrary constant. Show that for any point P in the plane there is a unique value of C such that the curve $y = t^2 + C$ passes through P.

Suppose we know the general solution of a first order differential equation. We can determine a particular solution if an initial condition is specified. Geometrically this is equivalent to knowing one point through which the solution curve passes.

The conclusions drawn from these specific examples can be generalized.

1. **Suppose a differential equations is of the form $\frac{dy}{dt} = f(t)$.**
   - The slope of the solution curve at P is determined completely by the t-coordinate.
   - If F is an antiderivative of f, then $F(t) + C$ is the general solution to $\frac{dy}{dt} = f(t)$. Solving $\frac{dy}{dt} = f(t)$ is equivalent to finding $\int f(t)\, dt$.
   - The solutions to $\frac{dy}{dt} = f(t)$ are **vertical translates** of one another. (They differ from one another by an additive constant.)

2. **Suppose a differential equation is of the form $\frac{dy}{dt} = f(y)$.**
   - The slope of the solution curve at P is determined completely by the y-coordinate.
   - Solutions to $\frac{dy}{dt} = f(y)$ are **horizontal translates** of one another.

---

[5] See Section 15.2 for a review of solutions to differential equations of the form $\frac{dy}{dt} = ky$.

Solving Differential Equations: Analytic Solutions

Solving a differential equation can be quite difficult. However, there are several types of differential equations that we can already solve. In this section we will look at differential equations of the form $\frac{dy}{dt} = f(t)$ and $\frac{dy}{dt} = ky$. In Section 31.3 we'll look qualitatively at solutions to differential equations of the form $\frac{dy}{dt} = f(y)$. Then in Section 31.4 we will look at solutions to a larger class of differential equations, differential equations of the form $\frac{dy}{dt} = f(y)g(t)$. The type of differential equations we look at here are special cases of those we will look at in Section 31.4.

Differential Equations of the Form $\frac{dy}{dt} = f(t)$

Differential equations of the form $\frac{dy}{dt} = f(t)$ can be solved by integration. To solve such an equation is to find a function y whose derivative is $f(t)$, i.e., to find an antiderivative of $f(t)$. If $\frac{dy}{dt} = f(t)$, then $y = \int f(t)\, dt$.

◆ **EXAMPLE 31.12**   The differential equation governing the path of a projectile can be solved by finding antiderivatives. In Example 31.7 we considered an object falling through the air. Ignoring air resistance, we say that the object undergoes a constant downward acceleration of 32 feet per second due to the force of gravity. The object's initial vertical velocity, $v(0)$, is denoted by the constant v_0 and its initial height, $s(0)$, by s_0. Solve the differential equations modeling this situation.

SOLUTION   Let $v(t)$ be the vertical component of the velocity of the object at time t.

$$\frac{dv}{dt} = -32$$

$$v(t) = \int -32\, dt$$

$$v(t) = -32t + C_1$$

We solve for C_1 using $v(0) = v_0$ to get

$$v_0 = -32(0) + C_1, \quad \text{so } C_1 = v_0$$
$$v(t) = -32t + v_0.$$

We know that $v(t) = \frac{ds}{dt}$, where $s(t)$ is the height (vertical position) of the object at time t.

$$\frac{ds}{dt} = -32t + v_0$$

$$s(t) = \int (-32t + v_0)\, dt$$

$$\text{Then} \quad s(t) = -32\frac{t^2}{2} + v_0 t + C_2.$$

Using $s(0) = s_0$, we can solve for C_2 to get

$$s_0 = -32 \cdot \frac{0}{2} + v_0 \cdot 0 + C_2, \quad \text{so } C_2 = s_0.$$

$$s(t) = -16t^2 + v_0 t + s_0. \quad ◆$$

Review: Differential Equations of the Form $\frac{dy}{dt} = ky$

In Section 15.2 we looked at the differential equation $\frac{dy}{dt} = ky$, an equation that models any situation in which a quantity grows or decays at a rate proportional to the amount of the quantity itself and conclude the following.

> The general solution to $\frac{dy}{dt} = ky$ is $y(t) = Ce^{kt}$, where C is an arbitrary constant.

The general solution to $\frac{dy}{dt} = ky$, together with an initial condition, enables us to determine a particular solution. Using substitution, we are able to get quite a bit of mileage out of knowing how to find a solution to differential equations of this form.

◆ **EXAMPLE 31.13**   Solve the following differential equations.

(a) $\frac{dP}{dt} = -2P$

(b) $\frac{d}{dt}(700 - F) = -0.01(700 - F)$

(c) $\frac{dT}{dt} = -k(T - 65)$

SOLUTION   Our strategy is to put each of these equations in the form $\frac{dy}{dt} = ky$. Once in this form we know $y(t) = Ce^{kt}$.

(a) $\frac{dP}{dt} = -2P$. The general solution is $P(t) = Ce^{-2t}$.

(b) $\frac{d}{dt}(700 - F) = -0.01(700 - F)$

   $700 - F(t)$ plays the role of the dependent variable, y; the constant of proportionality is -0.01.

   The general solution is $700 - F(t) = Ce^{-0.01t}$.

   We can write $F(t) = 700 - Ce^{0.01t}$.

(c) $\frac{dT}{dt} = -k(T - 65)$. We need to rewrite this to get it into the form $\frac{dy}{dt} = ky$.

   Let $y = T - 65$. Then $\frac{dy}{dt} = \frac{dT}{dt}$ and the original equation becomes $\frac{dy}{dt} = -ky$.

$$y = Ce^{kt}$$

   The general solution is $T - 65 = Ce^{-kt}$, or $T(t) = 65 + Ce^{-kt}$.   ◆

We see that knowing how to solve any differential equation of the form

$$\frac{dy}{dt} = ky$$

allows us to solve $\frac{dT}{dt} = kT - b$ because we can use substitution to transform it into the form $\frac{dy}{dt} = ky$. Using substitution, we can solve several of the other differential equations that arose from examples in Section 31.1, namely,

(a) $\frac{dC}{dt} = k(C - N)$, where k and N are constants (Example 31.2);

(b) $\frac{dA}{dt} = 5 - kA$, where k is a constant (Example 31.5); and

(c) $\frac{dM}{dt} = 0.05M - 4000$ (Example 31.6).

We solve by using substitution to alter the form to be $\frac{dy}{dt} = ky$.

(a) $\frac{dC}{dt} = k(C - N)$ can be solved using the substitution $y = C - N$, where N is constant.

(b) $\frac{dA}{dt} = 5 - kA$ can be solved either by writing $\frac{dA}{dt} = -kA + 5 = -k\left(A - \frac{5}{k}\right)$ and using the substitution $y = A - \frac{5}{k}$ or simply by using the substitution $y = 5 - kA$.

(c) $\frac{dM}{dt} = 0.05M - 4000$ can be solved by writing $\frac{dM}{dt} = 0.05\left(M - \frac{4000}{0.05}\right)$ and using the substitution $y = M - \frac{4000}{0.05}$ or by letting $y = 0.05M - 4000$.

We can solve any differential equation of the form $\frac{dP}{dt} = aP + b$ either by rewriting it first as $\frac{dP}{dt} = a\left(P + \frac{b}{a}\right)$ and then using the substitution $y = P + \frac{b}{a}$ to get it into the form $\frac{dy}{dt} = ay$ or by using the substitution $u(t) = aP + b$, $P = \frac{1}{a}[u(t) - b]$.

EXERCISE 31.3   Solve the differential equation $\frac{dM}{dt} = 0.05M - 4000$ with the initial condition $M(0) = 10,000$. (This is the differential equation from Example 31.6.) Check your answer to make sure it works.

The differential equation $\frac{dP}{dt} = aP + b$ is a special case of differential equations of the form $\frac{dy}{dt} = f(y)$. Differential equations of the form $\frac{dy}{dt} = f(y)$ are called **autonomous differential equations.** The word autonomous means "not controlled by outside forces"; in an autonomous differential equation the rate of change of y is controlled (determined) only by y, not by other outside "forces." In other words, $\frac{dy}{dt}$ can be described in terms of y only, without reference to t or any other variables. In the next section we will study the behavior of solutions to autonomous differential equations (such as $\frac{dS}{dt} = kS(300 - S)$) from a qualitative perspective.

PROBLEMS FOR SECTION 31.2

1. Which one of the following is a solution to the differential equation $y'(t) = -5y$?
   (a) $y = e^{5t}$ (b) $y = t^2$ (c) $y = e^{-5t}$ (d) $y = \sin(5t)$

2. Which one of the following is a solution to the differential equation $y''(t) = -25y$?
   (a) $y = e^{5t}$ (b) $y = t^2$ (c) $y = e^{-5t}$ (d) $y = \sin(5t)$

3. (a) Verify that $y(t) = Ce^{kt}$ is a solution to the differential equation $\frac{dy}{dt} = ky$.
   (b) Verify that $y = ke^t$ is *not* a solution to $\frac{dy}{dt} = ky$.
   (c) Verify that $y = e^{kt} + C$ is *not* a solution to $\frac{dy}{dt} = ky$.

4. Which of the following is a solution to $\frac{dy}{dx} = \frac{-y}{x} + \frac{y^2}{(\ln x)^2}$?
   (a) $y = x + 1$ (b) $y = 1 + 1/x$ (c) $y = \frac{\ln x}{x}$ (d) $y = \ln x$

5. Determine which of the following functions are solutions to each of the differential equations below. (A given differential equation may have more than one solution.)
   Differential Equations:
   i. $\frac{dy}{dt} = t$ ii. $\frac{dy}{dt} = y$ iii. $\frac{dy}{dt} = e^t$ iv. $\frac{d^2y}{dt^2} = 4y$

Solution choices:

(a) $y = \frac{t^2}{2}$ (b) $y = \frac{t^2}{2} + 5$ (c) $y = e^{-2t}$ (d) $y = e^t + 5$

(e) $y = 2e^t$ (f) $y = e^{2t}$ (g) $y = 5e^{2t}$ (h) $y = e^{2t} + 5$

6. Which of the following is a solution to the differential equation

$$y'' - y' - 6y = 0?$$

(a) $y = Ce^t$ (b) $y = \sin 2t$ (c) $y = 5e^{3t} + e^{-2t}$ (d) $y = e^{3t} - 2$

7. Which of the following is a solution to the differential equation

$$y'' + 9y = 0?$$

(a) $y = e^{3t} + e^{-3t}$ (b) $y = Ce^t - t$ (c) $y = C(t^2 + t)$

(d) $y = \sin 3t + 6$ (e) $y = 5 \cos 3t$

8. Which of the following is a solution to the differential equation

$$\frac{dy}{dt} = y + 1?$$

(a) $y = Ce^t$ (b) $y = Ce^t - t$ (c) $y = C(t^2 + t)$

(d) $y = Ce^t - 1$ (e) $y = Ce^{-t} + 1$

9. Is $y = \frac{xe^x}{2} + \frac{e^x}{3x}$ a solution to the differential equation $x\frac{dy}{dx} + (1 - x)y = xe^x$? Justify your answer.

10. Solve the following differential equations by using the method of substitution to put them into the form $\frac{dy}{dt} = ky$.

(a) $\frac{dP}{dt} = 0.3(1000 - P)$ (b) $\frac{dM}{dt} = 0.4M - 2000$

11. Solve the following.

(a) $\frac{dx}{dt} = 6 - 2x$. Do this using substitution in two ways.

     i. Factor out a -2 from the right-hand side and let $u = x - 3$. Then solve.

     ii. Let $v = 6 - 2x$. Express $\frac{dx}{dt}$ in terms of v and then convert the equation $\frac{dx}{dt} = 6 - 2x$ to an equation in v and $\frac{dv}{dt}$ and solve.

(b) $\frac{dx}{dt} = 3x - 7$

(c) $\frac{dy}{dt} = ky + B$

12. When a population has unlimited resources and is free from disease and strife, the rate at which the population grows in often modeled as being proportional to the population. Assume that both the bee and the mosquito populations described below behave according to this model.

In both scenarios described below you are given enough information to find the proportionality constant k. In one case the information allows you to find k solely using the differential equation, without requiring that you solve it. In the other scenario you must actually solve the differential equation in order to find k.

(a) Let $M = M(t)$ be the mosquito population at time t, t in weeks. At $t = 0$ there are 1000 mosquitoes. Suppose that when there are 5000 mosquitoes the population is growing at a rate of 250 mosquitoes per week. Write a differential equation reflecting the situation. Include a value for k, the proportionality constant.

(b) Let $B = B(t)$ be the bee population at time t, t in weeks. At $t = 0$ there are 600 bees. When $t = 10$ there are 800 bees. Write a differential equation reflecting the situation. Include a value for k, the proportionality constant.

13. The population in a certain country grows at a rate proportional to the population at time t, with a proportionality constant of 0.03. Due to political turmoil, people are leaving the country at a constant rate of 6000 people per year. Assume that there is no immigration into the country. Let $P = P(t)$ denote the population at time t.

(a) Write a differential equation reflecting the situation.

(b) Solve the differential equation for $P(t)$ given the information that at time $t = 0$ there are 3 million people in the country. In other words, find $P(t)$, the number of people in the country at time t.

14. A population of otters is declining. New otters are born at a rate proportional to the population with constant of proportionality 0.04, but otters die at a rate proportional to the population with constant 0.09. Today, the population is 1000. A group of people wants to try to prevent the otter population from dying out, so they plan to bring in otters from elsewhere at a rate of 40 otters per year. We'll model the situation with continuous functions. Let $P(t)$ be the population of the otters t years after today.

(a) Write a differential equation whose solution is $P(t)$.

(b) Solve this differential equation. Your answer should include no unknown constants.

(c) According to this model, will the attempt to save the otter population work? Explain your answer. If it won't work, at what rate must otters be brought in to ensure the population's survival? If it will work, for how many years must the importation of otters continue?

15. (a) Suppose a hot object is placed in a room whose temperature is kept fixed at F degrees. Let $T(t)$ be the temperature of the object. Newton's law of cooling says that the hot object will cool at a rate proportional to the difference in temperature between the object and its environment. Write a differential equation reflecting this statement and involving T. Explain why this differential equation is a special case of the differential equation in Example 31.3 part (a).

(b) What is the sign of the constant of proportionality in the equation you wrote above? Explain.

(c) Suppose that we are interested in the temperature of a cold cup of lemonade as it warms up to room temperature. Let $L(t)$ represent the temperature of the lemonade at time t and assume that it sits in a room that is kept at 65 degrees. At time $t = 0$ the lemonade is at 40 degrees. Fifteen minutes later it has warmed to 50 degrees.

  i. Sketch a graph of $L(t)$ using your intuition and the information given.

  ii. Is $L(t)$ increasing at an increasing rate or a decreasing rate?

  iii. Write a differential equation reflecting the situation. Indicate the sign of the proportionality constant.

iv. Find $L(t)$. Your final answer should have no undetermined constants and should be consistent with the answer to part (i).

v. How long will it take the lemonade to reach a temperature of 55 degrees?

16. Money in a bank account is earning interest at a nominal rate of 4% per year compounded continuously. Withdrawals are made at a rate of $8000 per year. Assume that withdrawals are made continuously.

(a) Write a differential equation modeling the situation.

(b) Depending on the initial deposit, the amount of money in the account will either increase, decrease, or remain constant. Explain this in words; refer to the differential equation.

(c) Suppose the money in the account remains constant. What was the initial deposit? For what initial deposits will the amount of money in the account actually continue to grow?

(d) Show that $M(t) = M_0 e^{0.04t} - 8000t$ is *not* a solution to the differential equation you got in part (a).

17. Suppose a population is changing according to the equation $\frac{dP}{dt} = kP - E$, where E is the rate at which people are emigrating from the country. As established in part (d) of the previous problem, $P(t) = P_0 e^{kt} - Et$ is *not* a solution to this differential equation.

(a) Use substitution to solve $\frac{dP}{dt} = kP - E$. (Your answer ought to agree with that given in part (b).)

(b) Verify that $P(t) = Ce^{kt} + \frac{E}{k}$, where C is a constant, is a solution to the differential equation $\frac{dP}{dt} = kP - E$.

18. $P(t) = Ce^{kt} + \frac{E}{k}$, where C is a constant, is the general solution to the differential equation $\frac{dP}{dt} = kP - E$. Below is the slope field for $\frac{dP}{dt} = 2P - 6$.

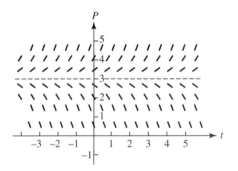

(a)   i. Find the particular solution that corresponds to the initial condition $P(0) = 2$.

ii. Sketch the solution curve through $(0, 2)$.

(b)   i. Find the particular solution that corresponds to the initial condition $P(0) = 3$.

ii. Sketch the solution curve through $(0, 3)$.

(c)   i. Find the particular solution that corresponds to the initial condition $P(0) = 4$.

ii. Sketch the solution curve through $(0, 4)$.

19. Let $M = M(t)$ be the amount of money in a trust fund earning interest at an annual interest rate of r compounded continuously. Suppose money is withdrawn at a rate of w dollars per year. Assuming the withdrawals are being made continuously, we can use a differential equation to model the situation. Analysis of this differential equation shows that the rate of change of money in the account at $t = 0$ could be either positive or negative, depending on the size of $M(0) = M_0$. Find the threshold value of M_0. Then, by analyzing the sign of the first and second derivatives, argue that if M_0 is less than this threshold value, then $M(t)$ is decreasing and concave down, and if M_0 is greater than this threshold value, then $M(t)$ is increasing and concave up.

20. A miser spends money at a rate proportional to the amount he has. Suppose that right now he has \$100,000 stowed under his mattress; he does not pay any taxes and does not earn any return on his money. Assume that this is all the money he has and that he has no other source of income. At the moment he is spending the money at a rate of \$10,000 per year.

    (a) At what rate will he be spending money when he has \$50,000?

    (b) At what time will the amount of money be down to \$10,000?

21. A drosophila colony (a colony of fruit flies) is being kept in a laboratory for study. It is being provided with essentially unlimited resources, so if left to grow, the colony will grow at a rate proportional to its size. If we let $N(t)$ be the number of drosophila in the colony at time t, t given in weeks, then the proportionality constant is k.

    (a) Write a differential equation reflecting the situation.

    (b) Solve the differential equation using N_0 to represent $N(0)$.

    (c) Suppose the drosophila are being cultivated to provide a source for genetic study, and therefore drosophila are being siphoned off at a rate of S drosophila per week. Modify the differential equation given in part (a) to reflect the siphoning off.

    (d) One of your classmates is convinced that the solution to the differential equation in part (c) is given by

    $$N(t) = N_0 e^{kt} - St.$$

    Show him that this is not a solution to the differential equation.

    (e) Your classmate is having a hard time giving up the solution he brought up in part (d). He sees that it does not satisfy the differential equation, but he still has a strong gut feeling that it ought to be right. Convince him that it is wrong by using a more intuitive argument. Use words and talk about fruit flies.

22. Solve the differential equations below. Find the general solution.

    (a) $\frac{dy}{dt} = \sin 3t$ (b) $\frac{dy}{dt} = 5 \cdot 2^t$ (c) $\frac{dx}{dt} = \frac{t+1}{t}$ (d) $\frac{dx}{dt} = \frac{t+1}{t^2}$

23. Solve the differential equations below. Find the general solution.

    (a) $\frac{dy}{dt} = 3t + 5$ (b) $\frac{dy}{dt} = 3y$ (c) $\frac{dy}{dt} = -y$

    (d) $\frac{dy}{dt} = 0$ (e) $\frac{dy}{dt} = 3y - 6$

24. For each differential equation below, sketch the slope field and find the general solution.

(a) $\frac{dy}{dt} = -y$ (b) $\frac{dy}{dt} = -t$ (c) $\frac{dy}{dt} = e^{-t}$

25. Each function below is a solution to one of the second order differential equations listed. To each function match the appropriate differential equation. C_1 and C_2 are constants.

Differential Equations

I. $\frac{d^2x}{dt^2} - 9x = 0$ II. $\frac{d^2x}{dt^2} + 9x = 0$ III. $\frac{d^2x}{dt^2} = 3x$

Solution Functions

(a) $x(t) = 5e^{3t}$ (b) $x(t) = -2e^{\sqrt{3}t}$ (c) $x(t) = 7\sin 3t$

(d) $x(t) = C_1 \sin 3t + C_2 \cos 3t$ (e) $x(t) = C_1 e^{\sqrt{3}t} + C_2 e^{-\sqrt{3}t}$

26. For what value(s) of β, if any, is
   (a) $y = C_1 \sin \beta t$ a solution to $y'' = 16y$?
   (b) $y = C_2 \cos \beta t$ a solution to $y'' = 16y$?
   (c) $y = C_3 e^{\beta t}$ a solution to $y'' = 16y$?

27. For what value(s) of β, if any, is
   (a) $y = C_1 \sin \beta t$ a solution to $y'' = -16y$?
   (b) $y = C_2 \cos \beta t$ a solution to $y'' = -16y$?
   (c) $y = C_3 e^{\beta t}$ a solution to $y'' = -16y$?

28. (a) There are two values of λ such that $y = e^{\lambda t}$ is a solution to $y'' + 7y' + 12y = 0$. Find them and label them λ_1 and λ_2.
   (b) Let $y = C_1 e^{\lambda_1 t} + C_2 e^{\lambda_2 t}$, where C_1 and C_2 are arbitrary constants. Verify that $y(t)$ is a solution to $y'' + 7y' + 12y = 0$.

29. (a) Find λ such that $y = e^{\lambda t}$ is a solution to $y'' + 4y' + 4y = 0$.
   (b) Verify that $y = te^{\lambda t}$ is also a solution to $y'' + 4y' + 4y + 0$.

31.3   QUALITATIVE ANALYSIS OF SOLUTIONS TO AUTONOMOUS DIFFERENTIAL EQUATIONS

One way of getting information about the behavior of solutions to a differential equation is to use the differential equation itself to sketch a picture of the solution curves. This will not give us a formula for the solutions; however, a graphical approach will often give us enough qualitative information about the solutions to answer some important questions.

In this section we will focus exclusively on autonomous differential equations, differential equations of the form $\frac{dy}{dt} = f(y)$.

◆ **EXAMPLE 31.14**   *Newton's law of cooling, revisited.*   Suppose a hot or cold beverage is put in a room that is kept at 65 degrees. Then the rate of change of the temperature of the beverage is proportional to the difference between the temperature of the room and the temperature of the drink.

$$\frac{dT}{dt} = k(65 - T),$$

where k is a positive constant and $T = T(t)$ is the temperature of the beverage at time t. Answer the following questions by taking a graphical perspective.

(a) What must the temperature of the beverage be in order for its temperature to remain constant?

(b) For what temperatures is the beverage cooling down? In other words, where is $\frac{dT}{dt}$ negative?

(c) Sketch representative solution curves corresponding to a variety of initial conditions.

SOLUTION   Solving a differential equation involving $\frac{dT}{dt}$ means finding temperature, T, as a function of time, t, so we label the vertical axis T and the horizontal axis t.

(a) If a quantity is not changing, then its derivative must be zero. $\frac{dT}{dt} = k(65 - T)$ is zero only if $T = 65$. The beverage's initial temperature must be $65°$ in order to remain constant. This agrees perfectly with our intuition.

(b) Because the equation $\frac{dT}{dt} = k(65 - T)$ is a continuous function of T, the sign of $\frac{dT}{dt}$ can only change around the zeros of $\frac{dT}{dt}$; $\frac{dT}{dt} = 0$ only at $T = 65$.

We draw the T number line vertically because temperature, T, is plotted on the vertical axis.

When $T > 65$, $\frac{dT}{dt}$ is negative; this makes sense because a hot beverage will cool off.

When $T < 65$, $\frac{dT}{dt}$ is positive; this makes sense because a cold beverage will warm up.

(c) We use the information contained in Figure 31.3. $\frac{dT}{dt} = k(65 - T)$, so the further T is from 65 the greater the magnitude of the slope and the closer T is to 65 the more gentle the slope.

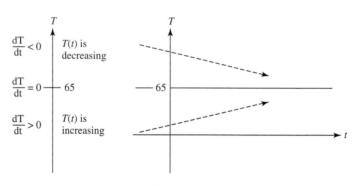

Figure 31.3

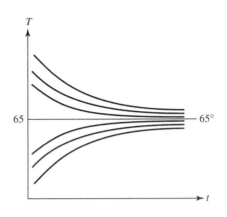

Figure 31.4

The nonconstant solutions are asymptotic to the constant solution, $T = 65$, meaning that the temperature of the drink will approach the temperature of the room. The constant solution $T(t) = 65$ is called the equilibrium solution.   ◆

Definition

An **equilibrium solution** to a differential equation is a solution that is constant for all values of the independent variable (often $t = $ time). If $\frac{dy}{dt} = f(y)$, then the equilibrium solutions can be found by setting $\frac{dy}{dt} = 0$ and solving for y. Equilibrium solutions are also referred to as **constant solutions**.

The next example will serve as a case study of differential equations of the form $\frac{dx}{dt} = f(y)$, where $f(y)$ is continuous.

◆ **EXAMPLE 31.15**   Do a qualitative analysis of the solutions to the differential equation

$$\frac{dy}{dt} = (y - 1)(y - 3).$$

Sketch representatives of the family of solutions.

SOLUTION   Solving the differential equation means finding y as a function of t, so we label the vertical axis y and the horizontal axis t.

First we identify the equilibrium, or constant solutions, solutions for which $\frac{dy}{dt} = 0$. These correspond to horizontal lines in the ty-plane. The constant solutions are $y = 1$ and $y = 3$.

Because $\frac{dy}{dt}$ is continuous, the sign of $\frac{dy}{dt}$ can only change around the zeros of $\frac{dy}{dt}$.

$$\text{For } y > 3, \quad \frac{dy}{dt} > 0 \Rightarrow \quad y(t) \text{ is increasing.}$$

$$\text{For } 1 < y < 3, \quad \frac{dy}{dt} < 0 \Rightarrow \quad y(t) \text{ is decreasing.}$$

$$\text{For } y < 1, \quad \frac{dy}{dt} > 0 \Rightarrow \quad y(t) \text{ is increasing.}$$

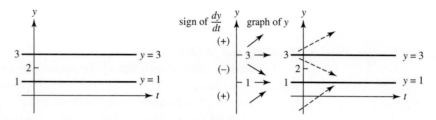

Figure 31.5

The equilibrium solutions of $\frac{dy}{dt} = f(y)$ divide the plane into horizontal strips. $f(y)$ is continuous so two distinct solutions to the differential equation cannot intersect. Therefore, each nonconstant solution lies completely within one strip. We check the sign of $\frac{dy}{dt}$ in each of these strips. Because $f(y)$ is a continuous function, the sign of $\frac{dy}{dt}$ within each strip does not change. Therefore each nonconstant solution is either strictly increasing or strictly decreasing.

Representative solution curves are given in Figure 31.6. Note that within each horizontal band the solutions are of the same form. (In each band one arbitrary solution has been highlighted; the other solutions within the band can be obtained by shifting the highlighted solution horizontally.)

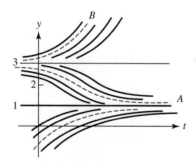

Figure 31.6

Each solution in a bounded strip is asymptotic to a constant solution. (Why? The solutions are horizontal translates and there will not be an "empty" horizontal band.) Each solution in an unbounded strip is either asymptotic to a constant solution or increases or decreases without bound. ◆

In summary, given a differential equation of the form $\frac{dy}{dt} = f(y)$, where $f(y)$ is continuous we know the following:

- The solution curves will not intersect.

- The constant solutions (equilibrium solutions) partition the ty-plane into horizontal strips in which each solution is of the same type (i.e., the solutions are horizontal translates of one another).

- Within each horizontal strip the solution curve is either strictly increasing or strictly decreasing because the sign of $\frac{dy}{dt}$ can only change around the zeros of $f(y)$.

■ Every solution curve is either asymptotic to a constant solution or increases or decreases without bound.

Equilibria and Stability

Equilibrium solutions to differential equations can be classified as stable, unstable, or semistable. In Example 31.5, $y = 1$ and $y = 3$ are the equilibrium solutions. The equilibrium at $y = 1$ is referred to as a **stable equilibrium**; under slight perturbation the system will tend back toward the equilibrium. For instance, if y is slightly less than 1, as $t \to \infty$, y increases toward 1. Similarly, if y is a bit more than 1 (i.e., anything under 3), then as $t \to \infty$, y decreases toward 1. The equilibrium at $y = 3$ is an **unstable equilibrium**; under slight perturbation the system does not return to this equilibrium. If y is slightly greater than 3, then as $t \to \infty$, y increases without bound. On the other hand, if y is slightly less than 3, then as $t \to \infty$, y tends away from 3 and toward the stable equilibrium of 1.

As illustration, consider two stationary coins, one lying flat and the other balancing on its edge. Since their positions are not changing with time, both coins are at equilibrium. The former configuration is stable under slight perturbations. The latter, however, is unstable because the coin balancing on its edge will topple under small perturbations.

In terms of modeling real-world situations, stable equilibrium solutions are states that systems would naturally gravitate toward, while unstable equilibrium solutions are thresholds between two qualitatively different types of outcomes. In any modeling situation knowing whether an equilibrium solution is stable or unstable is of tremendous importance, because real life is chock full of perturbations.

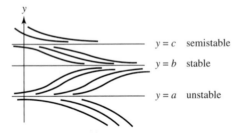
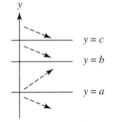

Figure 31.7

◆ **EXAMPLE 31.16**   Find and classify the equilibrium solutions of

$$\frac{dx}{dt} = x^2 - x.$$

SOLUTION   To find the equilibrium solutions, set $\frac{dx}{dt} = 0$.

$$\frac{dx}{dt} = x(x - 1) = 0$$

$$x = 0 \quad \text{or} \quad x = 1$$

Now look at the sign of $\frac{dx}{dt}$ on either side of the equilibrium values.

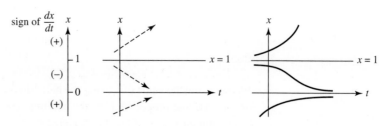

Figure 31.8

$x(t) = 0$ is a stable equilibrium solution, while $x(t) = 1$ is an unstable equilibrium solution.

◆

EXERCISE 31.4   Do a qualitative analysis of the solutions of the differential equation

$$\frac{dS}{dt} = 0.01S(300 - S).$$

(This is the differential equation that arose in modeling the spread of a flu in a college dormitory.) Sketch some representative solution curves and any constant solutions. Classify the equilibria. After analyzing the solutions in the abstract, determine what this says about the spread of the flu.

EXERCISE 31.5   Write a differential equation whose solution curves look like those sketched below. (Begin by constructing a differential equation with equilibrium solutions at $y = 3$ and $y = -2$.)

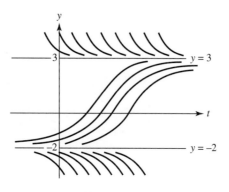

Figure 31.9

◆ **EXAMPLE 31.17**   An industrial plant produces radioactive material at a constant rate of 4 kilograms per year. The radioactive material decays at a rate proportional to the amount present and has a half-life of 20 years.

(a) Write a differential equation whose solution is $R(t)$, the amount of material present t years after this practice begins.

(b) Sketch some representative solutions corresponding to different initial values of R. Include the equilibrium solution. Can we predict the level of radioactive material in the long run?

SOLUTION   (a) The rate of production is constant at 4 kilograms per year regardless of the time and regardless of the amount present.

The rate out is proportional to the amount present, so it is kR, where k is a constant.

$$\text{rate of change} = \text{rate of production} - \text{rate of decay}$$
$$\frac{dR}{dt} \quad = \quad 4 \quad - \quad kR$$

We must find the value of k using the half-life information. We know that if a substance decays at a rate proportional to the amount present (and there are no further additions to the amount), then it can be modeled by $\frac{dS}{dt} = kS$. The solution to this differential equation is $S(t) = Ce^{kt}$. We know the half-life to be 20 years, so we can compute k.

$$S(20) = \frac{1}{2}Ce^{0t}$$
$$Ce^{20k} = 0.5C$$
$$e^{20k} = 0.5$$
$$20k = \ln 0.5$$
$$k = \frac{-\ln 2}{20} \approx -0.0347$$

In our case, since $\frac{dR}{dt} = 4 - kR$ already incorporates the negative sign indicating decay, $k = \frac{\ln 2}{20} \approx 0.0347$.
The differential equation is $\frac{dR}{dt} = 4 - \frac{\ln 2}{20}R$.

(b) The equilibrium solution occurs where $\frac{dR}{dt} = 0$. $4 = \frac{\ln 2}{20}R \Rightarrow R = \frac{80}{\ln 2} \approx 115.42$ kg. This is a stable equilibrium.

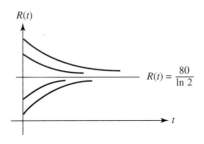

Figure 31.10

In the long run, the amount of radioactive material present approaches $\frac{80}{\ln 2}$ kg, or approximately 115.42 kg. ◆

Logistic Population Growth

Because unlimited resources are not observed in the real world, population growth is not exponential indefinitely. At a certain point members of the population begin to compete with one another for limited resources and the growth rate slows. The larger the population, the more prominent the role of competition. Ecologists and population biologists call the number of animals a particular environment can support the *carrying capacity* of that environment for that animal. Suppose a Tanzanian savannah has the carrying capacity L for lions, where L is a fixed constant. Let $P = P(t)$ be the population of lions at time t. We expect the graph of P versus t to look something like the graph drawn in Figure 31.11 below.

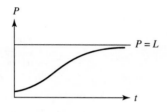

Figure 31.11

For P small the graph should be concave up . As P gets closer to the carrying capacity we expect the population growth to slow down and hence the graph of P to be concave down.

The most basic observation is that there are constant solutions at $P = 0$ and $P = L$; if there are no lions we don't expect lions to be spontaneously generated, and if the population is at the carrying capacity we expect it to stay there (barring natural or unnatural disasters). To formulate a mathematical model for the situation we'll write a differential equation that has equilibrium solutions at $P = 0$ and $P = L$ and is increasing for $P \in (0, L)$.

$$\frac{dP}{dt} = kP(L - P)$$

will work, provided k is a positive constant.

NOTE This equation, $\frac{dP}{dt} = kLP - kP^2$, can be thought of as

$$\frac{dP}{dt} = k_1 P - k_2 P^2, \text{ where } k_1 \text{ and } k_2 \text{ are constant.}$$

Without the $-k_2 P^2$ term this would just be the familiar exponential growth differential equation. But because lions are competing with one another for limited resources we need to incorporate a braking factor. This braking factor cannot be of the form kP, as this would still give exponential growth; only the proportionality constant would have changed. A braking factor of the form $k(P \cdot P)$ is reasonable because the competition is proportional to "interactions" between lions and lions.

A population growth model of the form

$$\frac{dP}{dt} = kP(L - P)$$

is known as a **logistic growth model**. The logistic model was first used in the early 1800s by the Belgian demographer Pierre-François Verthust to model human world population. The population figures he predicted for 100 years into the future were off by less than 1%.

◆ EXAMPLE 31.18   Suppose the number of fish in a lake grows according to the equation

$$\frac{dP}{dt} = 0.45P - 0.0005P^2.$$

(a) What is the lake's carrying capacity for fish? Is it a stable equilibrium?

(b) What size is the fish population when it is growing most rapidly?

SOLUTION   (a) Let's sketch representative solution curves. We begin by identifying the equilibrium solutions.

$$\frac{dP}{dt} = P(0.45 - 0.0005P) = 0$$

$$P = 0 \quad \text{or} \quad 0.0005P = 0.45$$

$$P = \frac{0.45}{0.0005} = 900$$

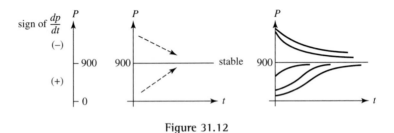

Figure 31.12

From this analysis we see that the carrying capacity is 900 fish and that this is a stable equilibrium.

(b) The fish population is growing most rapidly at the point of inflection of the curves lying in the strip between $P = 0$ and $P = 900$. Our analysis so far tells us that there ought to be at least one such point for each curve because the curve is asymptotic to both $P = 0$ and $P = 900$. We will show that for each solution curve in this interval there is only one such point of inflection and determine for what value of P this occurs.

There are two different approaches we can take to this problem.

Approach 1. We must determine the value of P in the interval $(0, 900)$ such that $\frac{dP}{dt}$ is maximum. We want to know where

$$0.45P - 0.0005P^2$$

is maximum. But the expression $0.45P - 0.0005P^2$ is quadratic, corresponding to a parabola with P-intercepts of 0 and 900, a parabola opening downward; hence its maximum value is at $P = \frac{-0.45}{2(-0.0005)} = \frac{0.45}{0.01} = 450$. The point of inflection is at $P = 450$.

Approach 2. $\frac{dP}{dt}$ will be maximum at the point at which the solution curve changes concavity. The point of inflection of $P(t)$ is the point at which the sign of $\frac{d^2P}{dt^2}$ changes.

$$\frac{dP}{dt} = 0.45P - 0.0005P^2$$

Differentiate both sides of the equation above *with respect to t* keeping in mind that P itself is a function of t.

$$\frac{dP}{dt} = 0.45 P(t) - 0.0005[P(t)]^2$$

$$\frac{d}{dt}\left[\frac{dP}{dt}\right] = 0.45\frac{dP}{dt} - 0.001 P \frac{dP}{dt}$$

$$\frac{d^2P}{dt^2} = (0.45 - 0.001 P)\frac{dP}{dt}$$

We can replace $\frac{dP}{dt}$ by $0.45P - 0.0005P^2$, or $P(0.45 - 0.0005P)$, obtaining

$$\frac{d^2P}{dt^2} = (0.45 - 0.001 P) P(0.45 - 0.0005 P) \cdot P$$

$\frac{d^2P}{dt^2}$ is continuous, so it can change sign only on either side of a zero. The zeros are where

$$0.45 - 0.001 P = 0, \qquad P = 0, \qquad \text{and} \qquad 0.45 - 0.0005 P = 0.$$

$$P = 450, \qquad P = 0, \qquad \text{and} \qquad P = 900.$$

$P = 450$ is what we're interested in.

$$\text{For } P \in (0, 450) \quad \frac{d^2P}{dt^2} > 0;$$

$$\text{for } P \in (450, 900) \quad \frac{d^2P}{dt^2} < 0.$$

Therefore the point of inflection is at $P = 450$, halfway between the two equilibrium solutions. The fish population is growing most rapidly when there are 450 fish.   ◆

Compartmental Analysis

◆ **EXAMPLE 31.19**   A 20-quart juice dispenser in a cafeteria is filled with a juice mixture that is 10% mango juice and 90% cranberry juice. An orange-mango blend is entering the dispenser at a rate of 4 quarts an hour and the well-stirred mixture leaves at the same rate. The orange-mango blend is 50% orange and 50% mango.

(a) Model the situation with a differential equation whose solution is $M = M(t)$, the number of quarts of mango juice in the container at time t. Include the initial condition. Sketch the solution curve, indicating the long-run behavior of M.

(b) Model the situation with a differential equation whose solution is $C(t)$, the number of quarts of cranberry juice in the container at time t. Include an initial condition.

(c) Suppose that once the percentage of mango juice in the mixture reaches 30%, high enough to be irresistible, the rate of juice consumption increases to 5 quarts per hour while the rate at which the container is being filled remains at 4 quarts an hour. Revise the differential equation in part (a) to reflect the situation. Reset the clock to denote by $t = 0$ the moment at which the consumption rate increases.

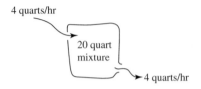

Figure 31.13

SOLUTION   (a) The basic structure for our model is *rate of change = rate in − rate out.* More specifically,

$$\left(\begin{array}{c}\text{rate of change}\\\text{of mango juice}\end{array}\right)=\left(\begin{array}{c}\text{rate at which}\\\text{mango juice enters}\end{array}\right)-\left(\begin{array}{c}\text{rate at which}\\\text{mango juice leaves}\end{array}\right).$$

The left-hand side of the equation is $\frac{dM}{dt}$ with units $\frac{\text{quarts of mango juice}}{\text{hour}}$. The rate in and rate out must also be measured in $\frac{\text{quarts of mango juice}}{\text{hour}}$.

Rate in: A mixture is entering at a rate of 4 $\frac{\text{quarts of mixture}}{\text{hour}}$. Half of this is mango juice, so mango juice is entering at a rate of 2 $\frac{\text{quarts of mango juice}}{\text{hour}}$.

$$\left(\frac{\text{quarts of mixture}}{\text{hour}}\right)\cdot\left(\frac{\text{quarts of mango juice}}{\text{quarts of mixture}}\right)=\left(\frac{\text{quarts of mango juice}}{\text{hour}}\right)$$

Rate out: The mixture is leaving at a rate of 4 $\frac{\text{quarts of mixture}}{\text{hour}}$, but what fraction of this exiting juice is mango juice? Because the juice in the container is well mixed, the fraction is the same as the ratio of mango juice to mixture in the container. The amount of mango juice in the container changes with time; it is given by $M(t)$. The amount of mixture in the dispenser remains constant at 20 quarts (because the rate at which juice is entering is equal to the rate that juice is leaving).

$$\text{rate out}=4\frac{\text{quarts of mixture}}{\text{hour}}\cdot\frac{M(t)}{20}\frac{\text{quarts of mango juice}}{\text{quarts of mixture}}$$

$$=\frac{4M(t)}{20}\frac{\text{quarts of mango juice}}{\text{hour}}$$

Now we can set up the differential equation.

$$\left(\begin{array}{c}\text{rate of change}\\\text{of mango juice}\end{array}\right)=\left(\begin{array}{c}\text{rate in}\\\text{of mango juice}\end{array}\right)-\left(\begin{array}{c}\text{rate out of}\\\text{mango juice}\end{array}\right)$$

$$\frac{dM}{dt}=2-4\cdot\frac{M}{20}$$

$$\frac{dM}{dt}=2-\frac{1}{5}M\quad\text{where }M(0)=2$$

The equilibrium solution is the value of M for which $\frac{dM}{dt}=0$.

$$2 - \frac{1}{5}M = 0$$

$$\frac{1}{5}M = 2$$

$$M = 10$$

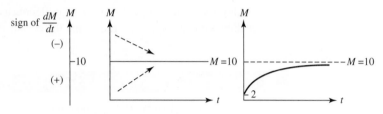

Figure 31.14

We know that the solution curve through $(0, 2)$ is concave down because its slope, given by $2 - \frac{1}{5}M$, decreases for $M \in (0, 10)$.

In the long run the amount of mango juice in the container will approach 10 quarts. This makes sense, because the juice in the container will eventually look like the incoming juice; it will eventually be half mango juice and half orange juice.

(b) Let $C = C(t)$ be the number of quarts of cranberry juice in the container at time t.

$$\text{rate of change} = \text{rate in} - \text{rate out}$$

$$\begin{pmatrix} \text{rate of change} \\ \text{of cranberry juice} \end{pmatrix} = \begin{pmatrix} \text{rate in of} \\ \text{cranberry juice} \end{pmatrix} - \begin{pmatrix} \text{rate out of} \\ \text{cranberry juice} \end{pmatrix}$$

$$\frac{dC}{dt} = 0 - 4 \left(\frac{\text{quarts of mixture}}{\text{hour}} \right) \cdot \left(\frac{C(t) \text{ quarts of cranberry juice}}{20 \text{ quarts of mixture}} \right)$$

$$\frac{dC}{dt} = -4 \frac{C(t)}{20}$$

$$\frac{dC}{dt} = -\frac{1}{5}C(t)$$

(c) Notice that from $t = 0$ on the number of quarts of juice in the dispenser goes down by 1 quart every hour.

$$\text{rate of change} = \text{rate in} - \text{rate out}$$

The rate in remains at $2 \frac{\text{quarts of mango juice}}{\text{hour}}$ but the rate out must be recalculated.

Rate out: $\left(5 \frac{\text{quarts of mixture}}{\text{hour}} \right) \cdot \left(\frac{M(t)}{20-t} \frac{\text{quarts of mango juice}}{\text{quarts of mixture}} \right) = 5 \frac{M(t)}{20-t}$

Therefore $\frac{dM}{dt} = 2 - \frac{5M}{20-t}$.[6] The initial condition is $M(0) = .3(20) = 6$.   ◆

While the amount of mango juice in a juice dispenser is not itself an issue of vital importance, the ideas involved in solving problems of this type are directly applicable to interesting problems in chemistry, ecology, and geology. For example, a lake may be fed by an underground

---

[6] This differential equation gives $\frac{dM}{dt}$ in terms of *both* M and t. We have not discussed how to solve it.

spring and other tributaries and may itself feed an outlet stream. The water quality may vary; assuming a well-mixed lake, an environmentalist can analyze the spring/lake/stream system and the level of pollutants over time. Similarly, an ecologist looking at Mono Lake, where the diversion of water sources has led to a change in salination levels affecting brine shrimp populations, can analyze the salt concentration in terms of

$$\text{rate of change} = \text{rate in} - \text{rate out.}$$

Geologists trying to date events across millennia use the same setup. The time spans they are dealing with can be so enormous that the oceans themselves are considered "well-mixed."

PROBLEMS FOR SECTION 31.3

1. Do a qualitative analysis of the family of solutions to each of the differential equations below. Then, in another color pen or pencil, highlight the graphs of the solutions corresponding to the given initial conditions. (If the solution is asymptotic to a horizontal line, draw and label that line.)

   (a) $\frac{dy}{dt} = 4y - 8$; $y(0) = 0,$ $y(0) = -1,$ $y(0) = 3$

   (b) $\frac{dy}{dt} = y^2 - 4$; $y(0) = -1,$ $y(0) = -3,$ $y(0) = 4$

   (c) $\frac{dy}{dt} = (y - 1)(y - 2)(y + 1)$; $y(0) = 0,$ $y(0) = 3$

   (d) $\frac{dy}{dt} = y^2 + 5y - 6$; $y(0) = -5,$ $y(0) = -7,$ $y(0) = 2$

2. For each of the following families of graphs, write a differential equation for which the graphs drawn could be solutions. (There are infinitely many correct answers; produce one.)

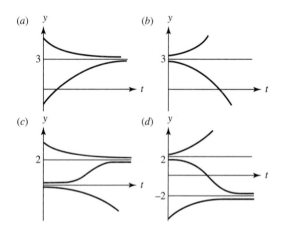

3. In each of Problems 2 (a)–(d), identify the equilibria and classify them as stable or unstable.

4. Consider the differential equation $\frac{dP}{dt} = kP(L - P)$, where k and L are positive constants.

(a) For what values of P is $\frac{dP}{dt}$ zero?

(b) Show that $P(t) = \frac{L}{1+Ce^{-kLt}}$, where C is a constant, is a solution of the logistic equation above.

5. The population of a town in the south of Bangladesh has been growing exponentially. However, recent flooding has alarmed residents and people are leaving the town at a rate of N thousand people per year, where N is a constant. The rate of change of the population of the town can be modeled by the differential equation

$$\frac{dP}{dt} = 0.02P - N,$$

where $P = P(t)$ is the number of people in the town in thousands.

(a) If $P(0) = 100$, what is the largest yearly exodus rate the town can support in the long run?

(b) How big must the population of the town be in order to support the loss of 1000 people per year?

6. Which of the graphs below could be the graph of a solution to $\frac{dy}{dt} = y^2 - 1$, where $y(0) = 0$? Explain why you eliminated each of the other options.

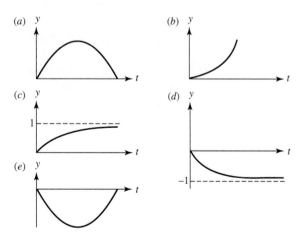

7. Which of the graphs on the following page could be a solution to the differential equation $\frac{dx}{dt} = x^2(2 - x)$ with initial condition $x(0) = 0.5$? Explain why you eliminated each of the other options.

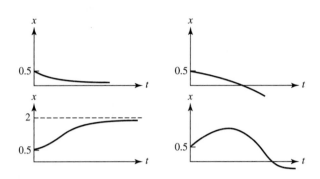

8. Essay question: One of your classmates is puzzled by what it means to solve a differential equation. He has two questions for you. Answer them in plain English.

   (a) How can you tell if something is a solution to a differential equation?

   (b) What is the difference between a general and a particular solution? Is the difference always that the general solution just has a "plus C" at the end? Are there some cases where that's the only difference?

9. Let $P(t)$ be the number of crocodiles in a mud hole at time t. Suppose $\frac{dP}{dt} = 0.01P - 0.0025P^2$.

   (a) What is the carrying capacity of the mud hole?

   (b) Find $\frac{d^2P}{dt^2}$. Remember: You are differentiating with respect to t, so the derivative of P is *not* 1.

   (c) Use your answer to part (b) to determine how many crocodiles are in the mud hole when the number of crocodiles is increasing most rapidly.

   (d) Sketch a solution curve if the number of crocodiles in the mud hole at time $t = 0$ is 3. (Label the vertical axis. You need not calibrate the t-axis.)

10. Sketch a representative family of solutions to the following differential equations. You need not take a second derivative.

   (a) $\frac{dy}{dt} = y(y^2 - 4)$ (b) $\frac{dy}{dt} = y^2(y - 2)$

In Problems 11 through 16 sketch a representative family of solutions for each of the following differential equations.

11. (a) $\frac{dy}{dt} = 2y - 6$ (b) $\frac{dy}{dt} = 6 - 2y$

12. (a) $\frac{dy}{dt} = \sin t$ (b) $\frac{dy}{dt} = \sin y$

13. $\frac{dy}{dt} = \tan y$

14. (a) $\frac{dy}{dt} = t^2$ (b) $\frac{dy}{dt} = y^2$

15. (a) $\frac{dy}{dt} = t^2 - 1$ (b) $\frac{dy}{dt} = y^2 - 1$

16. $\frac{dx}{dt} = (1 - x)(x + 2)(x - 3)$

17. Give an example of a differential equation with constant solutions at $y = -1$ and $y = 4$ with the characteristics specified.

    (a) The equilibrium at $y = -1$ is stable; the equilibrium at $y = 4$ is unstable.

    (b) The equilibrium at $y = -1$ is unstable; the equilibrium at $y = 4$ is stable.

    (c) Neither equilibrium solution is stable.

18. Give an example of a differential equation of the form $\frac{dv}{dt} = f(v)$ and whose solutions depend upon $v(0)$ as described below.

    If $v(0) > 5$, then $v(t)$ is increasing.

    If $v(0) = 5$, then $v(t) = 5$.

    If $2 < v(0) < 5$, then $v(t)$ is decreasing.

    If $v(0) = 2$, then $v(t) = 2$.

    If $v(0) < 2$, then $v(t)$ is increasing.

19. A canister contains 10 liters of blue paint. Paint is being used at a rate of 2 liters per hour and the canister is being replenished at a rate of 2 liters per hour by a pale blue paint that is 80% blue and 20% white. Assuming the canister is well-mixed, write a differential equation whose solution is $w(t)$, the amount of white paint in the canister at time t. Specify the initial condition.

20. A 5-gallon urn is filled with chai, a milky spicy tea. The chai in the urn is 90% tea and 10% milk. Chai is being consumed at a rate of 1/2 gallon per hour and the urn is kept full by adding a mixture that is 80% tea and 20% milk. Assume that the chai is well-mixed.

    (a) Write a differential equation whose solution is $M(t)$, the number of gallons of milk in the urn at time t. Specify the initial condition.

    (b) Use qualitative analysis to sketch the solution to the differential equation in part (a).

    (c) How much milk is in the urn after 2 hours?

21. The population of wildebeest in the Serengeti was decimated by a rinderpest plague in the 1950s. In 1961 the Serengeti supported a population of a quarter of a million wildebeest. By 1978 the wildebeest population was 1.5 million and by 1991 it had reached 2 million. (Craig Packer *Into Africa*, Chicago, The University of Chicago Press, 1996 p. 250.)

    Given this data, would you be more inclined to model the growth of the wildebeest population using an exponential growth model or using a logistic growth model? Explain your reasoning.

22. A lake contains 10^{10} liters of water. Acid rain containing 0.02 milligrams of pollutant per liter of rain falls into the lake at a rate of 10^3 liters per week. An outlet stream drains away 10^3 liters of water per week. Assume that the pollutant is always evenly distributed throughout the lake, so the runoff into the stream has the same concentration of pollutant as the lake as a whole. The volume of the lake stays constant at 10^{10} liters because the water lost from the runoff balances exactly the water gained from the rain.

(a) Write a differential equation whose solution is $P(t)$, the number of milligrams of pollutant in the lake as a function of t measured in weeks.

(b) Find any equilibrium solutions.

(c) Sketch some representative solution curves.

(d) How would you alter the differential equation if there was a dry spell and rain was falling into the lake at a rate of only 10^2 liters per week.

23. *Challenge Problem:*

(a) You plan to save money starting today at a rate of $4000 per year over the next 30 years. You will deposit this money at a nearly continuous rate (a constant amount each day) into a bank account that earns 5% interest compounded continuously.

Let $B(t)$ be the balance of money in the account t years from now, where $0 \leq t \leq 30$.

i. Write a differential equation whose solution is $B(t)$.

ii. Write an integral that is equal to $B(30)$, the amount in the account at the end of 30 years.

(b) Now assume that instead of making deposits continuously, you decide to make a deposit of $4000 once a year, starting today and continuing until you have made a total of 30 deposits. Suppose the bank account pays 5% interest compounded annually.

i. Write a geometric sum equal to the balance immediately after the final deposit.

ii. Find a closed form expression (no $+ \cdots +$, no summation notation) for this sum.

31.4   SOLVING SEPARABLE FIRST ORDER DIFFERENTIAL EQUATIONS

Differential equations of the forms $\frac{dy}{dt} = g(t)$ and $\frac{dy}{dt} = f(y)$ are special cases of a larger class of first order differential equations called **separable differential equations**. Separable differential equations can be written in the form

$$\frac{dy}{dt} = g(t)f(y).$$

Equivalently, we can write $\frac{dy}{dt} = \frac{g(t)}{h(y)}$. We can "separate" variables as follows:

$$h(y)\frac{dy}{dt} = g(t)$$

Integrating both sides with respect to t gives

$$\int h(y)\frac{dy}{dt}\, dt = \int g(t)\, dt$$

or

$$\int h(y)\, dy = \int g(t)\, dt.$$

Generally this is accomplished by separating the variables into the differential form

$$h(y)\, dy = g(t)\, dt$$

and integrating both sides.

Let's use this method on a familiar example.

◆ **EXAMPLE 31.20**   Solve $\frac{dy}{dt} = ky$ using separation of variables.

SOLUTION

$$\frac{dy}{y} = k\, dt \qquad \text{Assuming } y \neq 0 \text{, we can divide by } y.$$

$$\int \frac{dy}{y} = \int k\, dt$$

$$\ln |y| = kt + C$$

$$|y| = e^{kt+C} \qquad \text{Solve for } y \text{ explicitly in terms of } t.$$

$$|y| = e^C \cdot e^{kt} = Ae^{kt}$$

$$y = \pm Ae^{kt} \quad \text{where } A > 0.$$

In other words, $y = C_1 e^{kt}$, C_1 any constant.

There's one missing detail here. According to what we've done, C_1 shouldn't be zero because the separation of variables assumed $y \neq 0$. Treating the case $y = 0$ independently, we see $y = 0$ is a solution. Hence C_1 may be zero. ◆

◆ **EXAMPLE 31.21**   Solve $\frac{dy}{dx} = \frac{-x}{y}$.

SOLUTION

$$y\, dy = -x\, dx$$

$$\int y\, dy = \int -x\, dx$$

$$\frac{1}{2}y^2 = -\frac{1}{2}x^2 + C_1$$

$$x^2 + y^2 = C$$

Observations

■ y is not a function of x. The solution is a relationship between x and y.

■   You might remember this particular problem from Chapter 17. We began with the equation of the circle and, using implicit differentiation, arrived at $\frac{dy}{dx} = \frac{-x}{y}$. ◆

In the previous section we discussed the logistic growth model, both in the context of lions on a Tanzanian savannah and fish in a lake. Using separation of variables and partial fractions, we can solve the logistic differential equation.

◆ **EXAMPLE 31.22**   Solve $\frac{dP}{dt} = 0.5P - 0.001P^2$.

SOLUTION

$$\frac{dP}{dt} = P(0.5 - 0.001P) = 0.001P(500 - P)$$

Separating variables, we obtain

$$\frac{dP}{P(500-P)} = 0.001 \, dt \text{ for } P \neq 500, \, P \neq 0.$$

Integrating gives

$$\int \frac{dP}{P(500-P)} = 0.001t + C_1. \tag{31.1}$$

To integrate the left-hand side, we algebraically split the fraction into the sum of two fractions each of which is easier to integrate. We look for constants A and B such that

$$\frac{1}{P(500-P)} = \frac{A}{P} + \frac{B}{500-P}.$$

Clearing denominators gives

$$1 = A(500-P) + BP$$

and evaluating at $P = 0$ yields

$$A = 0.002.$$

If $A = 0.002$, then

$$1 = 1 - (0.002)P + BP, \quad \text{so } B = 0.002.$$

Therefore

$$\frac{1}{P(500-P)} = \frac{0.002}{P} + \frac{0.002}{500-P},$$

i.e.,

$$\frac{1}{P(500-P)} = \frac{1}{500}\left(\frac{1}{P} + \frac{1}{500-P}\right).$$

Hence

$$\int \frac{1}{P(500-P)} dP = \frac{1}{500} \int \frac{1}{P} + \frac{1}{500-P} \, dP.$$

Returning to Equation (31.1) we can now write

$$\frac{1}{500} \int \left(\frac{1}{P} + \frac{1}{500-P}\right) dP = 0.001t + C_1.$$

Then

$$\frac{1}{500}[\ln|P| - \ln|500-P|] = 0.001t + C_1$$

or

$$\ln\left|\frac{P}{500-P}\right| = 0.5t + C_2.$$

Exponentiating, we obtain

$$\left|\frac{P}{500-P}\right| = C_3 e^{0.5t}, \quad C_3 > 0.$$

Let us assume P is greater than zero and less than the carrying capacity 500. Then

$$\frac{P}{500 - P} = C_3 e^{0.5t}, \quad C_3 > 0.$$

Solve for P.

$$P = 500 C_3 e^{0.5t} - P C_3 e^{0.5t}$$

$$P + P C_3 e^{0.5t} = 500 C_3 e^{0.5t}$$

$$P(1 + C_3 e^{0.5t}) = 500 C_3 e^{0.5t}$$

$$P = \frac{500 C_3 e^{0.5t}}{1 + C_3 e^{0.5t}}$$

$$P(t) = \frac{500}{C e^{-0.5t} + 1}$$

This, together with $P = 0$ and $P = 500$, is the solution to the differential equation $P(0.5 - 0.001P)$ for $P(0)$ between 0 and 500.

Let's check that the answer is reasonable.

$$\lim_{t \to \infty} P(t) = \lim_{t \to \infty} \frac{500}{C e^{-0.5t} + 1} = 500$$

because

$$\lim_{t \to \infty} C e^{-0.5t} = 0.$$

This makes sense. For $0 < P(0) < 500$, the solution curves are asymptotic to the constant solutions. Verify that $\lim_{t \to -\infty} P(t) = 0$. ◆

◆ **EXAMPLE 31.23**   Is $2\frac{dy}{dt} + 3y = 4$ separable?

SOLUTION

$$2\frac{dy}{dt} + 3y = 4$$

$$2\frac{dy}{dt} = -3y + 4$$

$$2dy = (-3y + 4)\, dt$$

$$\frac{2dy}{-3y + 4} = dt$$

Yes, the differential equation is separable. It can be solved either by separation of variables or by using substitution to put it into the form $\frac{du}{dt} = ku$. ◆

EXERCISE 31.6   Solve $2\frac{dy}{dt} + 3y = 4$. Do it twice; use each of the methods suggested above.

Answer

$$y = \frac{4}{3} + C e^{-3/2t}$$

Taking Inventory

If a differential equation is separable and we can compute the relevant integrals, then we can find a relationship that is a solution to the differential equation. Differential equations of the forms $\frac{dy}{dt} = f(t)$ and $\frac{dy}{dt} = f(y)$ are special cases of separable differential equations.

- If $\frac{dy}{dt} = f(t)$ then $\int dy = \int f(t)\, dt$ so $y = \int f(t)\, dt$. Solutions are vertical translates of one another.

- If $\frac{dy}{dt} = f(y)$ then $\int \frac{dy}{f(y)} = \int dt$. Solutions are horizontal translates of one another.

What Can We Do With a Nonseparable First Order Differential Equation?

Suppose the differential equation $\frac{dy}{dt} = f(t, y)$ with initial condition $f(t_0) = y_0$ is nonseparable. One approach to approximating the solution is to start at the point (t_0, y_0) and to follow the flow line from this point by following the slope field for the differential equation. Imagine placing a bug in the ty-plane at the point (t_0, y_0). Evaluating the differential equation at (t_0, y_0) we find the slope there. Let the bug take one step in the direction indicated, arriving at a new point (t_1, y_1) where $t_1 = t_0 + \Delta t$ and $y_1 = y_0 + \left(\frac{dy}{dt} \big|_{(t_0, y_0)} \right) \cdot \Delta t$, and Δt is small. Evaluating the differential equation at (t_1, y_1) gives the bug its direction for its next step. $t_2 = t_1 + \Delta t$ and $y_2 = y_1 + \left(\frac{dy}{dt} \big|_{(t_1, y_1)} \right) \cdot \Delta t$. And so the bug proceeds, one step at a time, consulting the differential equation after each step in order to get directions for the next step.

$$t_{k+1} = t_k + \Delta t \quad \text{and} \quad y_{k+1} = y_k + \left[\frac{dy}{dt} \bigg|_{(t_k, y_k)} \right] \Delta t$$

This method of approximating a solution to a differential equation is known as **Euler's Method** (pronounced "oiler's" method).

In order to use Euler's Method, one must choose a "step size," that is, a value for Δt. We expect the approximation to improve as the size of Δt decreases. While Euler's Method is tedious to perform by hand, computers and programmable graphing calculators are well suited to carry out the task. You need to input the differential equation, the initial condition, and the step size.

Drawbacks to Euler's Method: Suppose we apply Euler's Method to Example 31.21, $\frac{dy}{dx} = -\frac{x}{y}$, with initial condition $t = 3$, $y = 4$. From the work we did using separation of variables, we know the solution curve is a circle of radius 5 centered at the origin. Using Euler's method, we step out along the tangent lines, so regardless of how small we program the step size, the path taken will spiral outward instead of giving the closed curve. To correct for such problems, more sophisticated variations of Euler's Method have been developed, but the underlying line of reasoning is similar. Improvements upon Euler's Method basically involve using more information about the slope in the vicinity of a particular point so that a more informed choice of direction can be made. One commonly used improvement upon Euler's Method is called the Runge-Kutta Method.

PROBLEMS FOR SECTION 31.4

In Problems 1 through 10, solve the given differential equation.

1. $\frac{dy}{dx} = \frac{x}{2y}$

2. $\frac{dy}{dx} = x^2 y$

3. $\frac{dy}{dx} = xy^2$

4. $\frac{dy}{dx} = \frac{y}{x}$

5. $\frac{dy}{dx} = \frac{x-1}{2y+1}$

6. $2y' - y = 1$

7. $\frac{dy}{dx} - y^2 = 1$

8. $\frac{dy}{dt} = t \cos^2 y$

9. $2\frac{dy}{dx} - 3xy = 0$

10. $\frac{dy}{dx} = \frac{\cos x}{-\sin y}$

11. Suppose that a freighter moving in a straight line with velocity v encounters water and air resistance proportional to its velocity. If the freighter is traveling with velocity v_0 when the motor is cut off, how far does it travel in time t? (*Hint:* Force = mass · acceleration. The only force acting on the freighter is the resistance from air and water.)

12. A very large container of juice contains four gallons of apple juice and one gallon of cranberry juice. Cranberry-apple juice (60% apple, 40% cranberry) is entering the container at a rate of three gallons per hour. The well-stirred mixture is leaving the container at three gallons per hour.

    (a) Write a differential equation whose solution is $C(t)$, the number of gallons of cranberry juice in the container at time t. Solve, using the initial condition.

    (b) Write a differential equation whose solution is $A(t)$, the number of gallons of apple juice in the container at time t. Solve the equation.

13. Suppose we change the previous problem so that the mixture is leaving the container at two gallons per hour.

    (a) Write a differential equation whose solution is $C(t)$, the number of gallons of cranberry juice in the container at time t. (This equation should be good as long as the container does not overflow.)

    (b) Write a differential equation whose solution is $A(t)$, the number of gallons of apple juice in the container at time t.

    (c) Are the differential equations you got in parts (a) and (b) separable? If so, solve them.

31.5   SYSTEMS OF DIFFERENTIAL EQUATIONS

Adjusting the simple exponential population growth model to reflect competition between the members of a population for limited resources led us to the logistic population growth model. In this section we will develop tools enabling us to deal with an added level of complexity, the relationship between two distinct populations. We can model competitive relationships, symbiotic relationships, and predator/prey relationships. These same tools enable us to develop alternative models for the spread of an epidemic, reflecting the nature of the disease and the way it is spread. Our mathematical focus in this section is on systems of differential equations.

Epidemic Models

Epidemic Model A: Contagious, Nonfatal Disease of Long Duration

Earlier in this chapter we considered a model for the spread of disease assuming that people are being infected by the disease at a rate proportional to the product of the number of people who have already been infected and the number of those who have not. This could reflect a contagious disease where the sick are not isolated. Letting $P(t)$ denote the number of infected people at time t and N denote the total population affected by the epidemic, we wrote $\frac{dP}{dt} = kP(N - P)$. In this model we assume that the population is fixed and that every member is susceptible to the disease. The disease is long in duration (there are no recoveries during the time period we are analyzing) but not fatal (no deaths during this period). A solution, with $0 < P_0 < N$, is sketched below. According to this model, eventually everyone will become infected.

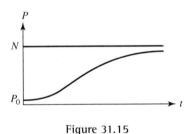

Figure 31.15

REMARK   This differential equation is logistic; it is of the same form as the fish-in-a-pond example. N is a limiting factor because $P(t) \leq N$. As P gets close to N, there are very few healthy people to infect, so $\frac{dP}{dt}$ will be small.

Epidemic Model B: A Contagious, Fatal Disease

Consider the situation in which a small group of people having a fatal infectious disease reside in a large population capable of catching the disease. Let us assume the disease has a negligibly short incubation time, so an infected individual can immediately infect others. We partition the population into two groups: the infected and the susceptible.[7] Assume that the population remains fixed during the time interval in question except for deaths due to

---

[7] In mathematics, to partition a population into classes is to assign every member of the population to one and only one class. In this example, every person is either susceptible or infected.

the disease; assume there are no births and no deaths apart from those deaths due to the disease.

Let $I(t) =$ the number of infected people at time t.

Let $S(t) =$ the number of susceptible people at time t.

Consider $\frac{dS}{dt}$. The susceptible group is not gaining new members. It is losing members to the infected group at a rate proportional to interactions between infected and susceptible individuals. We can write

$$\frac{dS}{dt} = -rSI, \quad \text{where } r \text{ is a positive constant.}$$

As the susceptible class loses members, the infected class gains members, the rate being rSI. The infected class loses people due to death from the disease. It is reasonable to suppose that the rate at which people die is proportional to the number of sick people. Then

$$\frac{dI}{dt} = rSI - kI, \quad \text{where } r \text{ and } k \text{ are positive constants.}$$

We must consider the *system* of equations

$$\begin{cases} \frac{dS}{dt} = -rSI \\ \frac{dI}{dt} = rSI - kI \end{cases}.$$

This is the first time we have run into a *system* of differential equations, so we will take a step back to analyze some simple systems of equations before returning to this particular example.

Analyzing Systems of Differential Equations: Some Examples[8]

◆ **EXAMPLE 31.24**   *Newton's Law of Cooling.*

Let $y = y(t) =$ the temperature of a large room at time t.

Let $x = x(t) =$ the difference in temperature between the room and a small object placed in it.

Let x be positive if the object is hotter than the room. The differential equations given below could reflect the room-object system.

$$\frac{dx}{dt} = -kx, \quad k > 0 \qquad \frac{dy}{dt} = 0$$

For the sake of simplicity, let $k = 1$.

(a)   i. Solve for y in terms of t and sketch the solution in the ty-plane.

   ii. Solve for x in terms of t and sketch the solution in the tx-plane.

(b) Graph the solutions in the xy-plane as follows: Pick any initial temperatures $y(0) = y_0$ and $x(0) = x_0$ for the room and the small object, respectively. Think about what will happen to x and y as t increases and sketch the appropriate path, drawing an arrow to indicate the direction the path is traveled. (t will not appear explicitly in the sketch.)

---

[8] The three examples that follow were originally suggested to me by Stephen DiPippo when we taught together.

Select various other starting points and sketch each path traced out by x and y as t increases. All paths should have arrows to indicate how x and y change with time.

(c) If the system is at equilibrium, then *both* x and y will remain constant as t varies, that is,

$$\frac{dy}{dt} = 0 \quad \text{and} \quad \frac{dx}{dt} = 0.$$

Under what conditions is the system at equilibrium? How do these equilibrium points relate to the picture drawn in part (a)?

SOLUTION   (a) Each differential equation can be solved independently.

i. $\frac{dy}{dt} = 0 \Rightarrow y = C$
If we let y_0 denote $y(0)$, then $y(t) = y_0$. In the ty-plane, solutions can be represented as shown in Figure 31.16.

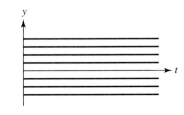

Figure 31.16

ii. $\frac{dx}{dt} = -x \Rightarrow x(t) = x_0 e^{-t}$
In the tx-plane, solutions can be represented as shown in Figure 31.17.

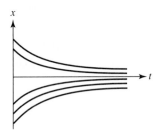

Figure 31.17

(b) Pick a point (x_0, y_0) as a starting point. The value of y will remain constant. If x_0 is positive, then as t increases x will remain positive but tend toward zero; if x is negative, then as t increases x will remain negative but tend toward zero. (This information can be read from Figure 31.17.) On the following page is a sketch in the xy-plane. Various starting points (corresponding to $t = 0$) are chosen, each designated by an "X." Arrows indicate the direction the path is traveled as t increases.

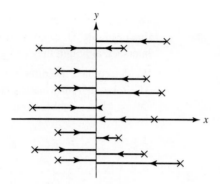

Figure 31.18

(c) If the system is at equilibrium, then

$$\frac{dy}{dt} = 0 \quad \text{and} \quad \frac{dx}{dt} = 0.$$

$\frac{dy}{dt}$ is identically 0. $\frac{dx}{dt} = -x$, therefore $\frac{dx}{dt} = 0$ if and only if $x = 0$. In other words, the system is at equilibrium as long as x, the temperature difference, is zero.

When we represent solutions to the system of differential equations in the xy-plane (with no t-axis), an equilibrium solution is simply a point. As t varies, x and y remain constant. (Contrast this with an equilibrium solution in the ty-plane or the tx-plane in which an equilibrium solution is a horizontal line.) In this example, every point along the y-axis is an equilibrium point.

Below representative solution curves are drawn. These curves, or **trajectories**, are traced by $(x(t), y(t))$ as t varies. Arrows indicate the direction the curve is traced as t increases. The xy-plane is often referred to as the **phase-plane** of the solutions. Initial points are usually not explicitly indicated.

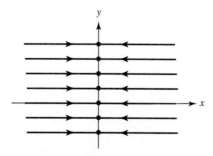

Figure 31.19

REMARKS

1. Trajectories on the right-hand side of the y-axis and the left-hand side of the y-axis are distinct. If x is initially positive, it will always be positive (see Figure 31.17). Likewise, if x is initially negative, it is always negative. In other words, if one starts at a point to the right of the y-axis, the trajectory through that point will never cross the y-axis. In fact, although the trajectory gets arbitrarily close to the y-axis, x never becomes zero

(again, refer to Figure 31.17). Each point on the y-axis is a distinct equilibrium point. (We can think of an equilibrium point as a degenerative trajectory.) Each horizontal line in Figure 31.19 consists of three distinct solution trajectories. Trajectories do not intersect, although they can get arbitrarily close to one another.

2. We determined the direction of the arrows on the trajectories by referring to Figure 31.17, but that information is directly accessible from the differential equations themselves. $\frac{dx}{dt} = -x$, therefore

$$\text{for } x > 0, \quad \frac{dx}{dt} < 0; \quad x \text{ decreases as } t \text{ increases.}$$

$$\text{For } x < 0, \quad \frac{dx}{dt} > 0; \quad x \text{ increases as } t \text{ increases.}$$

We can determine the "shape" of the trajectories in the xy-plane by using the original differential equations and looking at the relative rates of change of y and x as t varies. The slope of the trajectory is $\frac{dy}{dx} = \frac{dy/dt}{dx/dt}$, where equality follows from the Chain Rule, provided $\frac{dx}{dt} \neq 0$. Therefore

$$\frac{dy}{dx} = \frac{\frac{dy}{dt}}{\frac{dx}{dt}} = \frac{0}{-x} = 0 \Rightarrow \frac{dy}{dx} = 0 \Rightarrow y = C.$$

These are horizontal lines in the xy-plane. In Remark 1 we point out that each of these horizontal lines actually consists of three distinct trajectories, one to the right of the y-axis, one to the left of the y-axis, and an equilibrium point on the y-axis.   ◆

Note: $\frac{dy}{dx} = \frac{dy/dt}{dx/dt}$ provided that $\frac{dx}{dt}$ is nonzero. If, at a certain point P, $\frac{dx}{dt} = 0$ and $\frac{dy}{dt} = 0$, then neither x nor y is changing with t, so P is an equilibrium point. If, at a certain point Q, $\frac{dx}{dt} = 0$ and $\frac{dy}{dt} \neq 0$, then y is changing with t but x is not, so the trajectory is vertical at Q.

◆ **EXAMPLE 31.25**   Analyze the system of differential equations below, sketching solution trajectories in the phase-plane.

$$\begin{cases} \frac{dx}{dt} = -x \\ \frac{dy}{dt} = -y \end{cases}$$

SOLUTION   Each of these differential equations can be solved independently. They describe exponential decay.

$$\frac{dx}{dt} = -x \Rightarrow x(t) = x_0 e^{-t}$$

$$\frac{dy}{dt} = -y \Rightarrow y(t) = y_0 e^{-t}$$

In the tx-plane and the ty-plane, solutions can be represented as shown in Figure 31.20.

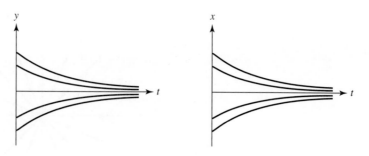

Figure 31.20

Now consider the trajectories in the xy-plane. As t changes, $(x(t), y(t))$ traces out a curve in the xy-plane. Wherever we start out in the plane, as t increases, both the x- and y-coordinates go toward zero. At the equilibrium

$$\frac{dx}{dt} = -x = 0 \quad \text{and} \quad \frac{dy}{dt} = -y = 0$$

simultaneously. So $x = 0$ and $y = 0$. The origin is the only equilibrium point.

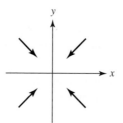

Figure 31.21

To determine the shape of the trajectories in the xy-plane, observe that for $x_0 \neq 0$

$$\frac{y(t)}{x(t)} = \frac{y_0 e^{-t}}{x_0 e^{-t}} = \frac{y_0}{x_0} = \text{constant.}$$

Thus, the trajectory starting at (x_0, y_0) when $t = 0$ satisfies $\frac{y}{x} = k$, where k is the constant $\frac{y_0}{x_0}$. We obtain $y = kx$, the equation of a straight line through the origin. The coordinates of the starting point determine the slope of the line. In Figure 31.22 arrows indicate the direction to travel along these lines as t increases.

We need to treat the case $x_0 = 0$ separately. If $x_0 = 0$ then $x(t) = 0$. For $y_0 > 0$, $y(t)$ decreases as t increases; for $y_0 < 0$, $y(t)$ increases as t increases. This gives two trajectories along the y-axis. If both x_0 and y_0 are zero, the system is at equilibrium.

Putting everything together, we obtain the picture sketched on the following page. Apart from the origin, each trajectory is a ray approaching the origin.

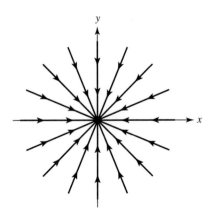

Figure 31.22   ◆

EXERCISE 31.7   In Example 31.25, we can find the shape of the trajectories by looking at $\frac{dy}{dx}$.

$$\frac{dy}{dx} = \frac{\frac{dy}{dt}}{\frac{dx}{dt}} = \frac{-y}{-x}, \text{ or}$$

$$\frac{dy}{dx} = \frac{y}{x}.$$

This is a separable differential equation. Solve for y, showing that $y = kx$, where k is a constant.

Examples 31.24 and 31.25 are special in that we can solve each of the original differential equations independently to get x and y explicitly in terms of t and we can see precisely what the trajectories look like in the xy- phase-plane. In general, the situation is more complex.

General Strategy

Consider a system of differential equations of the form

$$\begin{cases} \frac{dx}{dt} = f(x, y) \\ \frac{dy}{dt} = g(x, y) \end{cases}$$

where f and g are continuous functions.[9] Suppose we want to analyze the behavior of solution trajectories in the xy-phase-plane. The general direction of a trajectory at any point can be determined by knowing the signs of $\frac{dx}{dt}$ and $\frac{dy}{dt}$. Therefore, to get a handle on the phase-plane picture we begin by determining where $\frac{dx}{dt} = 0$ and where $\frac{dy}{dt} = 0$. The curves along which $\frac{dx}{dt} = 0$ or $\frac{dy}{dt} = 0$ are called **nullclines**. Nullclines are not, in general, trajectories. They are the set of points along which trajectories have either vertical or horizontal tangent lines.

---

[9] We haven't defined continuity for functions of two variables, but if $f(x, y)$ is continuous then for fixed y, f is a continuous function of x and for fixed x, f is a continuous function of y. The crucial consequence of continuity for our analysis is that if $\frac{dx}{dt} = f(x, y)$ where f is continuous, then $\frac{dx}{dt}$ can change sign only by passing through zero.

- Where $\frac{dx}{dt} = 0$ but $\frac{dy}{dt} \neq 0$, only y is changing with t so the trajectory's tangent line is vertical. The sign of $\frac{dy}{dt}$ indicates how y changes with t; we display this information by orienting the vertical tangents up or down.

- Where $\frac{dy}{dt} = 0$ but $\frac{dx}{dt} \neq 0$, only x is changing with t so the trajectory's tangent line is horizontal. The sign of $\frac{dx}{dt}$ indicates how x changes with t; we display this information by orienting the horizontal tangents right or left.

- Where $\frac{dx}{dt} = 0$ and $\frac{dy}{dt} = 0$ simultaneously, the system is at equilibrium. Equilibrium points are the points of intersection of nullclines for which $\frac{dx}{dt} = 0$ and those for which $\frac{dy}{dt} = 0$.

- The nullclines partition the phase plane into regions in which neither $\frac{dx}{dt}$ nor $\frac{dy}{dt}$ changes sign. In each region we'll have one of the following cases:

$$\underbrace{\frac{dx}{dt} > 0, \frac{dy}{dt} > 0}_{\substack{\text{as } t \text{ increases,} \\ \text{both } x \text{ and } y \text{ increase}}} \qquad \underbrace{\frac{dx}{dt} > 0, \frac{dy}{dt} < 0}_{\substack{\text{as } t \text{ increases,} \\ x \text{ increases and } y \text{ decreases}}}$$

$$\underbrace{\frac{dx}{dt} < 0, \frac{dy}{dt} > 0}_{\substack{\text{as } t \text{ increases,} \\ x \text{ decreases and } y \text{ increases}}} \qquad \underbrace{\frac{dx}{dt} < 0, \frac{dy}{dt} < 0}_{\substack{\text{as } t \text{ increases,} \\ \text{both } x \text{ and } y \text{ decrease}}}$$

- Sketch possible trajectories using the information gathered.
  Fact The limit point of a trajectory must be an equilibrium point. By this we mean the following. Suppose $\lim_{t \to \infty} x(t)$ and $\lim_{t \to \infty} y(t)$ both exist and are finite. Denote the limits by A and B, respectively. Then (A, B) must be an equilibrium point. (Verify that this is indeed the case in Examples 31.24 and 31.25.)

  Note the analogy to the case of solutions to autonomous differential equations. If $y_1(t)$ is a solution to an autonomous equation and $\lim_{t \to \infty} y_1(t) = L$, where L is finite, then $y(t) = L$ is an equilibrium solution.

- The slope of a trajectory at any point is $\frac{dy}{dx} = \frac{dy/dt}{dx/dt} = \frac{g(x,y)}{f(x,y)}$ evaluated at that point (provided $f(x, y) \neq 0$). Sometimes explicitly solving for a relationship between x and y is enlightening.

◆ **EXAMPLE 31.26**   Analyze the following system of differential equations, sketching solution trajectories in the xy-plane.

$$\begin{cases} \frac{dx}{dt} = y \\ \frac{dy}{dt} = -x \end{cases}$$

SOLUTION   We begin by finding the nullclines.

The trajectories are horizontal where $\frac{dy}{dt} = -x = 0$, i.e., along the y-axis.

The trajectories are vertical where $\frac{dx}{dt} = y = 0$, i.e., along the x-axis.

We can figure out the direction in which the trajectories are traveled along the nullclines by looking back at the original equations. For example, $\frac{dx}{dt} = y$, so on the part of the y-axis with $y > 0$ we know $\frac{dx}{dt} > 0$; the trajectories are traveled from left to right, x increasing with t. On the section of the y-axis with $y < 0$, $\frac{dx}{dt} < 0$ so the trajectories are traveled from right to left, x decreasing with t. Vertical and horizontal tangents are oriented as shown in Figure 31.23.

The only equilibrium point is the origin.

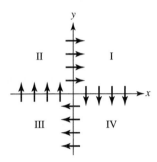

Figure 31.23

The nullclines partition the plane into the four regions labeled. In each region we look at the signs of $\frac{dy}{dt}$ and $\frac{dx}{dt}$ to determine the basic direction of the trajectory.

Region I: $\underbrace{\frac{dx}{dt} = y > 0}_{x \text{ increases}}, \underbrace{\frac{dy}{dt} = -x < 0}_{y \text{ decreases}}$ Region III: $\underbrace{\frac{dx}{dt} = y < 0}_{x \text{ decreases}}, \underbrace{\frac{dy}{dt} = -x > 0}_{y \text{ increases}}$

Region II: $\underbrace{\frac{dx}{dt} = y > 0}_{x \text{ increases}}, \underbrace{\frac{dy}{dt} = -x > 0}_{y \text{ increases}}$ Region IV: $\underbrace{\frac{dx}{dt} = y < 0}_{x \text{ decreases}}, \underbrace{\frac{dy}{dt} = -x < 0}_{y \text{ decreases}}$

This information is collected in Figure 31.24.

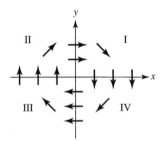

Figure 31.24

At this point we have some idea of what the trajectories look like. However, we are not sure, for instance, if they form closed circles, or ellipses, or if they spiral in or spiral out. We can look at $\frac{dy}{dx}$ and try to solve the resulting differential equation to uncover the shapes of the trajectories.

$$\frac{dy}{dx} = \frac{\frac{dy}{dt}}{\frac{dx}{dt}} = \frac{-x}{y}, \quad y \neq 0$$

so

$$\frac{dy}{dx} = \frac{-x}{y}.$$

Separate variables and solve.

$$y\, dy = -x\, dx$$

$$\int y\, dy = \int -x\, dx$$

$$\frac{y^2}{2} = -\frac{x^2}{2} + C_1 \Rightarrow x^2 + y^2 = C$$

Therefore, the trajectories in the xy-plane are circles centered at the origin; the picture in the phase plane is given in Figure 31.25.

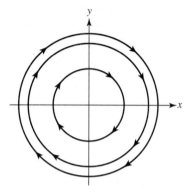

Figure 31.25

Through every point in the xy-plane there is one trajectory passing through that point. Each trajectory is a circle, except for the origin, which can be thought of as a degenerate circle with radius 0. As t increases, the point $(x(t), y(t))$ traces out a circle in the clockwise direction.   ◆

REMARK The system of equations in Example 31.26 can arise when modeling the vibrations of a frictionless spring. Consider a spring and attached block positioned as shown in Figure 31.26. Let x give the position of the block (of negligible weight) attached to the end of the spring. Suppose we stretch the spring by pulling the block from its original position ($x = 0$) out to x_0 and release the block.

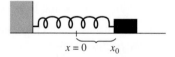

Figure 31.26

The block will oscillate back and forth about $x = 0$ and, in the absence of friction, will repeat its motion *ad infinitum*. We expect the graph of x versus t to look like Figure 31.27 on the following page.

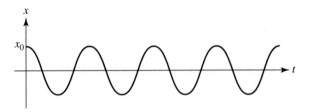

Figure 31.27

Let $y =$ the velocity of the block. Then $y = \frac{dx}{dt}$; velocity is the derivative of position with respect to time. By Newton's second law we know that force $=$ (mass) $\cdot$ (acceleration). We can denote the mass of the block by m and its acceleration by $\frac{d^2x}{dt^2}$ or $\frac{dy}{dt}$. Hooke's law, an experimentally derived law, tells us that the force exerted by the spring is proportional to the displacement from its equilibrium length. Putting the two laws together gives $-\propto x = m\frac{dy}{dt}$ or $\frac{dy}{dt} = -kx$ for some $k > 0$. Acceleration, $\frac{dy}{dt}$, is proportional to x but opposite in sign. The proportionality constant is determined by the mass of the block and the nature of the spring. (Think through some special cases to assure yourself that the equation $\frac{dy}{dt} = -kx$ makes sense in terms of the spring.) The frictionless block and spring system can therefore be modeled by the system of differential equations

$$\begin{cases} \frac{dx}{dt} = y \\ \frac{dy}{dt} = -kx. \end{cases}$$

For convenience let's assume that $k = 1$. Then the equations describing the motion become

$$\frac{dx}{dt} = y \quad \text{and} \quad \frac{dy}{dt} = -x.$$

Look back at the phase-plane diagram (Figure 31.25) to see how the motion of the spring is reflected in this picture. If we pull the block out a distance x_0 and simply let it go, giving it no initial velocity, we should look at the point $(x_0, 0)$ in the phase plane. Follow the trajectory around in the direction of the arrow and note how the position of the block (the x-coordinate) and the velocity of the block (the y-coordinate) change with time.

EXERCISE 31.8   It is unrealistic to ignore friction. Let $x(t)$ be the position of the block.

$$\begin{cases} x(0) = x_0 \\ x'(0) = 0 \end{cases}$$

(a) Sketch a possible graph of x versus t for the spring example, considering the force of friction and assuming the block vibrates back and forth several times.

(b) Now, letting $y = \frac{dx}{dt} =$ velocity, sketch the trajectory in the xy-plane. (*In the next section we will deal analytically with the friction issue.*)

The answer to Exercise 31.8 appears at the end of the section.

Returning to Epidemic Model B: A Contagious, Fatal Disease

◆ EXAMPLE 31.27   Consider the epidemic model that was used to motivate this discussion. Residing in a large susceptible population is a small group of people with a fatal infectious disease. An infected individual can immediately infect others, and any person not infected is susceptible. We make the assumption that the population remains fixed during the time interval in question except for deaths due to the disease. $I(t) =$ the number of infected people at time t, and $S(t) =$ the number of susceptible people at time t.

$$\frac{dS}{dt} = -rSI, \quad \text{where } r > 0$$

As the susceptible class loses members, the infected class gains members but it also loses people due to death from the disease.

$$\frac{dI}{dt} = rSI - kI, \quad \text{where } r, k > 0$$

Analyze the system of equations

$$\begin{cases} \frac{dS}{dt} = -rSI \\ \frac{dI}{dt} = I(rS - k) \quad r, k > 0, \quad \text{sketching solution trajectories in the } SI\text{-plane.} \end{cases}$$

SOLUTION   Look for the nullclines and equilibrium points. Let's plot S on the horizontal axis and I on the vertical axis. Then the slopes of the trajectories are given by $\frac{dI}{dS} = \frac{dI/dt}{dS/dt}$.

The trajectories are horizontal where $\frac{dI}{dt} = I(rS - k) = 0$, that is, for $I = 0$ or $S = \frac{k}{r}$.

The trajectories are vertical where $\frac{dS}{dt} = -rSI = 0$, that is, for $S = 0$ or $I = 0$.

The system is at equilibrium if and only if $\frac{dI}{dt} = 0$ and $\frac{dS}{dt} = 0$ simultaneously. S cannot simultaneously be 0 and $\frac{k}{r}$, so the equilibria for the system are at $I = 0$. Every point on the S-axis is an equilibrium point.

The nullclines are $S = 0$, $I = 0$, and $S = \frac{k}{r}$. These partition the first quadrant into two regions, I and II, as indicated in Figure 31.28.

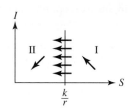

Figure 31.28

	Region I	Region II
sign of $\frac{dS}{dt}$:	−	−
sign of $\frac{dI}{dt}$:	+	−

We know that the solution curves go in the direction of the arrows drawn in Figure 31.28. Further information about the shape of the trajectories in the SI-plane can be obtained by looking at $\frac{dI}{dS}$.

$$\frac{dI}{dS} = \frac{dI/dt}{dS/dt} = \frac{I(rS - k)}{-rSI} = \frac{rS - k}{-rS} = -1 + \frac{k}{rS}.$$

The slope is a function of S; the trajectories are vertical translates. (As an exercise, solve for I in terms of S.) The second derivative of I with respect to S is given by $\frac{d^2I}{dS^2} = -\frac{k}{rS^2}$, where r and k are positive constants. $\frac{d^2I}{dS^2} < 0$, so the trajectories are concave down. See Figure 31.29.

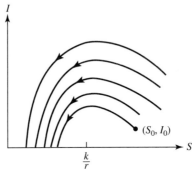

Figure 31.29

REMARKS Observe that if $S_0 < \frac{k}{r}$, then I immediately decreases toward zero; the disease leaves the population without becoming an epidemic. On the other hand, if $S_0 > \frac{k}{r}$, then the disease will spread and the number of infected people will increase until S drops to $\frac{k}{r}$. Only when S falls below the threshold value of $\frac{k}{r}$ does I begin to decrease. The disease does not die out completely for lack of a susceptible population but rather for lack of infected people. Generally, some individuals will remain who have not caught the disease. ◆

The same system of differential equations that were set up in the previous example can also be used to model an epidemic such as measles or chicken pox, where instead of the disease being fatal, sickness confers immunity upon those who recover. Once an individual is infected he cannot become reinfected, so after leaving the "infected" class he does not rejoin the "susceptible" class but belongs to a "recovered" class.

◆ EXAMPLE 31.28   In an isolated community of 800 susceptible children, one child is diagnosed with chicken pox. Suppose the spread of the disease can be modeled by the system of differential equations

$$\begin{cases} \frac{dS}{dt} = -0.001SI \\ \frac{dI}{dt} = 0.001SI - 0.3I. \end{cases}$$

(a) What is the maximum number of children sick at any one moment?

(b) According to our model, how many susceptible children will avoid getting chicken pox while the epidemic runs its course?

SOLUTION   (a) From the qualitative analysis done in Example 31.27 we know that I is maximum when $S = \frac{k}{r}$, that is, when $\frac{dI}{dS} = 0$. So I is maximum at $S = \frac{0.3}{.001} = 300$. To find I when $S = 300$ we need an explicit relationship between I and S. $I \neq 800 - S$ because some infected children have already recovered.

$$\frac{dI}{dS} = \frac{dI/dt}{dS/dt} = \frac{0.001SI - 0.3I}{-0.001SI} = -1 + \frac{0.3}{0.001S} = -1 + \frac{300}{S}$$

Separate variables and integrate to obtain I as a function of S.

$$dI = \left(-1 + \frac{300}{S}\right) dS$$

$$\int dI = \int (-1)dS + 300 \int \frac{dS}{S}$$

$$I = -S + 300 \ln S + C, \qquad S > 0, \text{ so } |S| = S.$$

Use the initial conditions, $I = 1$ and $S = 800$ when $t = 0$, to solve for C.

$$1 = -800 + 300 \ln(800) + C, \quad \text{so } C = 801 - 300 \ln(800) \approx -1204.38$$

Then

$$I(S) \approx -S + 300 \ln S - 1204, \text{ and}$$

$$I(300) \approx -300 + 300 \ln 300 - 1204 \approx 207.$$

There were at most about 207 children sick at one time. This is just over a quarter of the susceptible population.

(b) As pointed out in Example 31.27, the epidemic ends not for lack of susceptible children but for lack of infected children. Therefore, we set $I = 0$ and solve for S. Because the equation involves both S and $\ln S$ we can only approximate the solution. Using a calculator, computer, or numerical methods, we find that $S \approx 70$. At the end of the epidemic about 70 children will still be susceptible to chicken pox; 730 children will have caught the disease. ◆

In some sense, in the previous two examples it is the ratio $\frac{k}{r}$ that governs the course of the epidemic. The likelihood of the disease being passed from one individual to the next is reflected in r. For a disease such as chicken pox, communities sometimes attempt to raise the value of r (to confer immunity before adulthood) whereas for a fatal infectious disease, like AIDs, communities sometimes educate to lower the value of r. When dangerous epidemics break out in livestock populations, farmers and ranchers often remove sick animals, effectively raising the value of k, in order to end up with more uninfected animals.

Recurrent Epidemics

Many diseases recur in various populations with some regularity. For instance, in London in the early 1900s, measles epidemics recurred approximately every two years. In 1929, when the mathematical biologist H. E. Soper was attempting to model this cyclic measles epidemic, he dropped the assumption that the population remains fixed for the time interval being observed. Instead he assumed that the population of susceptibles grows at a constant rate μ and arrived at this system of differential equations:

$$\frac{dS}{dt} = -rSI + \mu$$

$$\frac{dI}{dt} = rSI - kI \quad \text{for } r, k, \text{ and } \mu \text{ positive constants}$$

In fact this system of equations does not predict recurrent outbreaks of the epidemic. Instead it predicts that the disease will reach a steady state level since the cycles are heavily damped. The problem with this model is the assumption that the susceptible population grows at a *constant* rate. Try to modify this system of equations. If you assume the population grows at a rate proportional to itself, then the solutions to the equations are cyclic and undamped, as desired. Think about the repeating cycle of measles epidemics again after working through Example 31.29.

Modeling Population Interactions

The more closely we look at the world the more clearly we see its interconnected nature. We can use the tools we've developed in this section to model interactions between distinct populations. We can model symbiotic interactions, such as the interaction between sea anemones and clown fish, or competitive interactions, such as the interaction between lion and hyena populations competing for small prey. We can model predator-prey interactions, such as that between hyenas and the Thomson's gazelle, lions and water buffalo, or cats and mice. In the examples in this section we'll simplify our models to focus on just two populations.

◆ **EXAMPLE 31.29**   Consider the predator-prey relationship between hyenas and the Thomson's gazelle, a small gazelle native to Africa. Let's make the following simplifying assumptions.

  i. Assume hyenas are the gazelles' major predator and that in the hyenas' absence the gazelle population would grow exponentially.

  ii. Assume gazelles are the major food source for hyenas; with no gazelles the hyenas would die off.

Model this interaction with a system of differential equations.

SOLUTION   Let $h(t) = $ the number of hundreds of hyenas at time t.

Let $g(t) = $ the number of hundreds of gazelles at time t.

We can model the interaction by a system of differential equations of the form

$$\begin{cases} \frac{dh}{dt} = -k_1 h + k_2 hg \\ \frac{dg}{dt} = k_3 g - k_4 hg \end{cases}$$

where k_1, k_2, k_3, and k_4 are positive constants. Just as the rate of transmission of disease is proportional to interactions between the infected and the susceptible, which is in turn proportional to the product of their numbers, so too is the rate of nourishing/fatal interaction between hyena and gazelle proportional to the product of their population sizes. Observe that if $h = 0$ then $\frac{dg}{dt} = k_3 g$ and if $g = 0$ then $\frac{dh}{dt} = -k_1 h$. For the sake of concreteness, we'll work with the values of k_1, k_2, k_3, and k_4 given below and analyze solutions in the gh-plane.

$$\frac{dh}{dt} = -0.3h + 0.1gh = 0.1h(-3 + g)$$

$$\frac{dg}{dt} = +0.4g - 0.4gh = 0.4g(1 - h)$$

Nullclines: $\frac{dh}{dt} = 0$ when $h = 0$ or $g = 3$. In the gh-plane this is where trajectories have horizontal tangent lines.

$\frac{dg}{dt} = 0$ when $g = 0$ or $h = 1$. In the gh-plane this is where trajectories have vertical tangent lines.

Equilibrium points: $\frac{dh}{dt} = 0$ at $h = 0$ or $g = 3$, so at an equilibrium point either $h = 0$ or $g = 3$.

Suppose $h = 0$. Then in order for $\frac{dg}{dt}$ to be zero we must have $g = 0$. $(0, 0)$ is an equilibrium point.

Suppose $g = 3$. Then in order for $\frac{dg}{dt}$ to be zero we must have $h = 1$. The point $(3, 1)$ is an equilibrium point.

Use the differential equations to orient the vertical and horizontal tangents. For instance, when $g = 3$ we know that $\frac{dg}{dt} > 0$ for $0 < h < 1$ and $\frac{dg}{dt} < 0$ for $h > 1$.

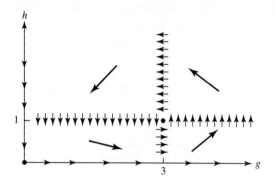

Figure 31.30

The h and g axes are oriented as shown in Figure 31.30. The nullclines divide the relevant first quadrant region into four subregions. The direction of trajectories in each region is indicated in Figure 31.30.

From Figure 31.30 we see that for $g(0) > 0$ and $h(0) > 0$ the trajectories either spiral in toward $(3, 1)$ or spiral outward, or are closed curves. From the slope field it appears that the curves are closed. Looking at $\frac{dh}{dg} = \frac{0.1h(-3+0.1g)}{0.4g(1-h)}$ enables us to distinguish between these options. Separating variables, we find that

$$\ln(h) - h = -3 \ln(g) + g + C.$$

Suppose we start at the point $(5, 1)$. We can solve for C and show that the trajectory through $(5, 1)$ intersects the line $g = 3$ exactly twice, once for $h > 1$ and once for $h < 1$. Similarly we can show that it intersects the line $h = 1$ exactly twice, once for $g < 3$ and once for $g > 3$. The trajectory through $(5, 1)$ is a closed curve. In fact, all the trajectories in the first quadrant are closed.

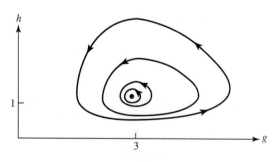

Figure 31.31

Our model predicts that the hyena and gazelle populations will oscillate cyclically. When there are few gazelle the hyena population decreases due to lack of food. The decrease in the number of hyenas allows the gazelle population to thrive, but as the gazelle population increases the hyenas' food source is replenished, allowing the hyena population to flourish. This flourishing takes its toll on the gazelles, and the cycle repeats.   ◆

To the extent that a model reflects observed population dynamics, it is a good model. Models that don't reflect observed behavior must be modified. There are various ways this predator-prey model can be modified. For instance, there may be competition among gazelle for limited grazing land so that in the absence of hyena the gazelle population exhibits logistic growth. This can be reflected in the system of differential equations by inserting a $-k_5 g^2$ term as shown.

$$\begin{cases} \frac{dh}{dt} = -k_1 h + k_2 hg \\ \frac{dg}{dt} = k_3 g - k_4 hg - k_5 g^2, \end{cases}$$

where the $-k_5 g^2$ term ($k_5 > 0$) reflects competition among gazelles.

The predator-prey system of differential equations given in Example 31.29 is sometimes referred to as Volterra's model after the Italian mathematician Vito Volterra (1860-1940) who was encouraged to analyze the predator-prey relationship between sharks and the fish they prey upon by his son-in-law, the biologist Humberto D'Ancona.[10] By inserting a term in each equation to account for fishing, Volterra was able to explain why fishing conducted in the Adriatic Sea was raising the average number of prey and lowering the average number of predators over any cycle. Similar analysis has been successfully used to analyze unexpected results of introducing DDT into a predator-prey system, leaving predator populations lowered and prey populations elevated.

Competition between species can be modeled by differential equations of the form

$$\begin{cases} \frac{dx}{dt} = k_1 x - k_2 xy \\ \frac{dy}{dt} = k_4 y - k_5 xy \end{cases} \quad \text{or} \quad \begin{cases} \frac{dx}{dt} = k_1 x - k_2 xy - k_3 x^2 \\ \frac{dy}{dt} = k_4 y - k_5 xy - k_6 y^2 \end{cases}$$

where $k_1, k_2, \ldots, k_6$ are positive constants. The latter set of differential equations takes into account competition between members of the same species in addition to competition between species.

Answers to Selected Exercises

Answers to Exercise 31.8

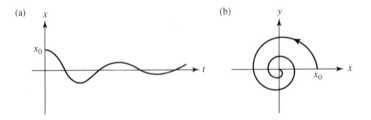

(a) x

(b) y

---

[10] For more information on a very interesting story see Martin Braun, *Differential Equations and their Applications*, 2nd ed. New York. Springer Verlag, 1978 or G. F. Gause, *The Struggle for Existence*, New York. Hafner, 1964, or Umberto D'Ancona, *The Struggle for Existence*, Leiden, Brill, 1954.

PROBLEMS FOR SECTION 31.5

1. Let $x = x(t)$ be the number of hundreds of animals of species A at time t.

   Let $y = y(t)$ be the number of hundreds of animals of species B at time t.

   For each system of differential equations, describe the nature of the interaction between the two species. What happens to each species in the absence of the other?

   (a) $\begin{cases} \frac{dx}{dt} = 0.02x - 0.001x^2 - 0.002xy \\ \frac{dy}{dt} = 0.008y - 0.004y^2 - 0.001xy \end{cases}$ (b) $\begin{cases} \frac{dx}{dt} = 0.02x - 0.01xy \\ \frac{dy}{dt} = -0.01y + 0.08xy \end{cases}$

   (c) $\begin{cases} \frac{dx}{dt} = 0.02x - 0.001x^2 + 0.002xy \\ \frac{dy}{dt} = 0.03y - 0.006y^2 + 0.001xy \end{cases}$

   For each system of differential equations in Problems 2 through 4, find the nullclines and identify the equilibrium solutions.

2. $\begin{cases} \frac{dx}{dt} = 0.02x - 0.001x^2 - 0.002xy \\ \frac{dy}{dx} = 0.008y - 0.004y^2 - 0.001xy \end{cases}$

3. $\begin{cases} \frac{dx}{dt} = 0.02x - 0.01xy \\ \frac{dy}{dt} = -0.01y + 0.08xy \end{cases}$

4. $\begin{cases} \frac{dx}{dt} = 0.02x - 0.001x^2 + 0.002xy \\ \frac{dy}{dt} = 0.03y - 0.006y^2 + 0.001xy \end{cases}$

5. Let $x = x(t)$ be the number of thousands of animals of species A at time t.

   Let $y = y(t)$ be the number of thousands of animals of species B at time t.

   Suppose $\begin{cases} \frac{dx}{dt} = x - 0.5xy \\ \frac{dy}{dt} = y - 0.5xy. \end{cases}$

   (a) Is the interaction between species A and B symbiotic, competitive, or a predator-prey relationship?

   (b) What are the equilibrium populations?

   (c) Find the nullclines and draw directed horizontal and vertical tangent lines in the phase-plane (as in Figures 31.28 and 31.30).

   (d) The nullclines divide the first quadrant of the phase-plane into four regions. In each region determine the general direction of the trajectories.

   (e) If $x = 0$, what happens to $y(t)$? How is this indicated in the phase-plane? If $y = 0$, what happens to $x(t)$? How is this indicated in the phase-plane?

   (f) Use the information gathered in parts (b) through (e) to sketch representative solution trajectories in the phase-plane. Include arrows indicating the direction the trajectories are traveled.

   (g) For each of the initial conditions given below, describe how the number of species of A and B change with time and what the situation will look like in the long run.

   i. $x(0) = 2 \quad y(0) = 1.8$

   ii. $x(0) = 2 \quad y(0) = 2.3$

   iii. $x(0) = 2.2 \quad y(0) = 2$

(h) Does this particular model support or challenge Charles Darwin's principle of competitive exclusion?

Problems 6 through 8 give systems of differential equations modeling competition between two species. In each problem find the nullclines. The nullclines will divide the phase-plane into regions; find the direction of the trajectories in each region. Use this information to sketch a phase-plane portrait. Then interpret the implications of your portrait for the long-term outcome of the competition. $x(t)$ and $y(t)$ give the number of thousands of animals of species A and B, respectively.

6. $\begin{cases} \frac{dx}{dt} = 0.03x - 0.01x^2 - 0.01xy \\ \frac{dy}{dt} = 0.05y - 0.01y^2 - 0.01xy \end{cases}$

7. $\begin{cases} \frac{dx}{dt} = 0.04x - 0.02x^2 - 0.01xy \\ \frac{dy}{dt} = 0.04y - 0.01y^2 - 0.01xy \end{cases}$

8. $\begin{cases} \frac{dx}{dt} = 0.01x(2 - 2x - y) \\ \frac{dy}{dt} = 0.01y(1 - y - 0.25x) \end{cases}$

9. Suppose we modify the Volterra predator-prey equations to reflect competition among prey for limited resources and competition among predators for limited resources. The equations would be of the form

$$\begin{cases} \frac{dx}{dt} = k_1x - k_2x^2 - k_3xy \\ \frac{dy}{dt} = -k_4y - k_5y^2 + k_6xy \end{cases}$$

where $k_1, k_2, \ldots, k_6$ are positive constants. Consider the system

$$\begin{cases} \frac{dx}{dt} = x(1 - 0.5x - y) \\ \frac{dy}{dt} = y(-1 - 0.5y + x). \end{cases}$$

(a) Find the equilibrium points.
(b) Do a qualitative phase-plane analysis. (In fact, solution trajectories will spiral in toward the non-trivial equilibrium point.)

For Problems 10 and 11, suppose that for a given system of differential equations

$$\frac{dy}{dt} = f(x, y) \qquad \frac{dx}{dt} = g(x, y)$$

a trajectory with initial conditions $y(0) = y_0$ and $x(0) = x_0$ is as drawn. On the same set of axes, sketch possible graphs of $x(t)$ and $y(t)$ corresponding to the trajectory shown. (There are infinitely many correct answers. What characteristics must they all share?)

10.

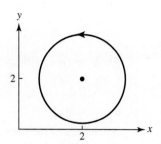

11.

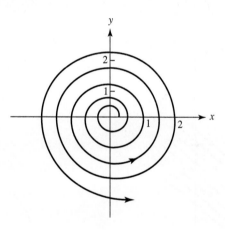

12.

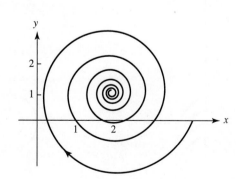

13. Consider the following systems of differential equations.

(a) $\frac{dx}{dt} = -2$ (b) $\frac{dx}{dt} = +3y$ (c) $\frac{dx}{dt} = +10x$

$\frac{dy}{dt} = -4x$ $\frac{dy}{dt} = -3x$ $\frac{dy}{dt} = +10y$

Match each of the pairs of differential equations with the correct figure below. Explain your rationale briefly.

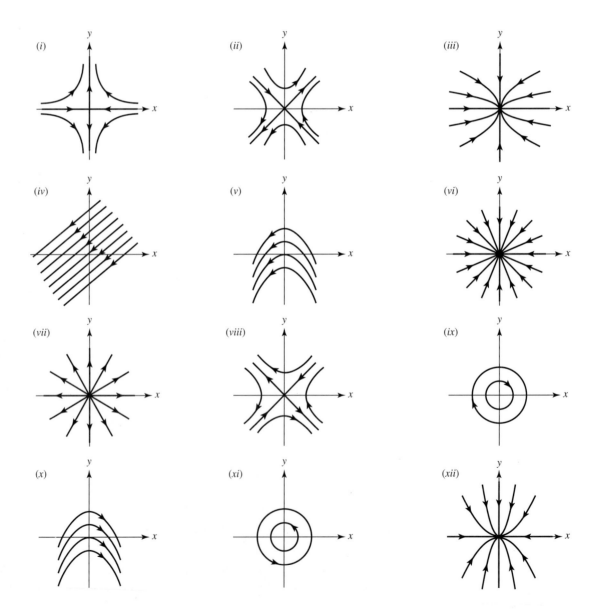

14. Match the phase-plane diagram with one of the systems of differential equations. Explain your reasoning.

(a) $\frac{dx}{dt} = -4x + y$ (b) $\frac{dx}{dt} = x - 4y$ (c) $\frac{dx}{dt} = -x + 4y$ (d) $\frac{dx}{dt} = x - 4y$

   $\frac{dy}{dt} = x - 4y$ $\frac{dy}{dt} = 4x - y$ $\frac{dy}{dt} = -4x - y$ $\frac{dy}{dt} = -4x - y$

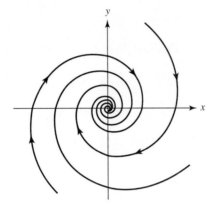

31.6   SECOND ORDER HOMOGENEOUS DIFFERENTIAL EQUATIONS WITH CONSTANT COEFFICIENTS

In the last section we looked at a frictionless model of a vibrating spring from the perspective of a system of differential equations. We sketched solution curves in the position-velocity phase-plane, but we did not find position as a function of time. In this section we will revisit this problem from the perspective of a second order differential equation, solve explicitly for position in terms of time, and tackle the issue of friction. By the end of the section we will be able to solve any second order homogeneous differential equation with constant coefficients and will have uncovered a surprising and beautiful mathematical result as a byproduct of our efforts. This result ties together differential equations, series, exponential functions and trigonometric functions.

Consider the block-spring system analyzed after Example 31.26 in Section 31.5. We let $x(t)$ be the position of the block at time t, where $x = 0$ corresponds to the equilibrium position of the system. Newton's second law says force = mass · acceleration.

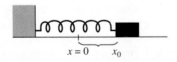

Figure 31.32

The forces on the system are the force of the spring, given by $-kx$ in accordance with Hooke's law, and the force of friction. The force of friction acts opposite the direction of motion; it is reasonable to model friction as proportional to velocity. Velocity is given by $\frac{dx}{dt}$ and acceleration by $\frac{d^2x}{dt^2}$. Therefore, we can deduce from Newton's law that

$$-kx - \lambda \frac{dx}{dt} = m \frac{d^2x}{dt^2}$$

where k and λ are positive constants and m is the mass of the block.

$$m \frac{d^2x}{dt^2} + \lambda \frac{dx}{dt} = 0$$

Equivalently,

$$\frac{d^2x}{dt^2} + \frac{\lambda}{m} \frac{dx}{dt} + \frac{k}{m} x = 0.$$

This is a second order differential equation of the form

$$\frac{d^2x}{dt^2} + b \frac{dx}{dt} + cx = 0.$$

If $x(0) = x_0 > 0$ and $x'(0) = 0$ then we might expect solutions as shown in Figure 31.33, with part (a) corresponding to the frictionless case and part (b) corresponding to some friction.

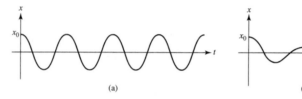

(a)   (b)

Figure 31.33

Definition

A differential equation of the form $y'' + by' + cy = 0$ is called a **second order homogeneous differential equation with constant coefficients**.

"Second order" refers to the second derivative, "constant coefficients" means the coefficients of y'', y', and y must be constants, and "homogeneous" refers to the zero on the right-hand side of the equation.

EXERCISE 31.9   Show that if y_1 and y_2 are solutions to the differential equation $y'' + by' + cy = 0$ then $y = C_1 y_1 + C_2 y_2$ is also a solution, for any constants C_1 and C_2.

FACT If y_1 and y_2 are solutions to $y'' + by' + cy = 0$ and y_1 and y_2 are not constant multiples of one another, then the general solution can be written in the form $y = C_1 y_1 + C_2 y_2$ where C_1 and C_2 are constants. (If $y_1 \neq C y_2$, C constant, we say y_1 and y_2 are *linearly independent*.)

In light of the fact presented, we would like to guess two solutions to $y'' + by' + cy = 0$. For inspiration, we'll look at the simpler case with $b = 0$. As a warm-up, do Exercise 31.10.

EXERCISE 31.10   For each differential equation, find two linearly independent solutions.

   i. $y'' = y$ ii. $y'' = 0$ iii. $y'' = -y$

◆ **EXAMPLE 31.30**   Find the general solution to each of the following differential equations.

   (a) $y'' = k^2 y$ (b) $y'' = 0$ (c) $y'' = -k^2 y$

SOLUTION   (a) $y_1 = e^{kt}$ and $y_2 = e^{-kt}$ are solutions, so the general solution is

$$y = C_1 e^{kt} + C_2 e^{-kt}.$$

(b) $y_1 = 1$ and $y_2 = t$ are solutions, so the general solution is

$$y = C_1 + C_2 t.$$

(c) $y_1 = \sin(kt)$ and $y_2 = \cos(kt)$ are solutions, so the general solution is

$$y = C_1 \sin(kt) + C_2 \cos(kt).$$

Recall that in Chapter 30 we looked at the differential equation $y'' = -y$ and showed that if the solution had a power series expansion it must be of the form $y = C_1 \sin t + C_2 \cos t$.

                                  ◆

Now let's return to the general case $y'' + by' + cy = 0$. Is it possible for a solution to be of the form e^{rt}? Suppose $y = e^{rt}$. Then $y' = re^{rt}$ and $y'' = r^2 e^{rt}$.

$$y'' + by + cy \overset{?}{=} 0$$
$$r^2 e^{rt} + br e^{rt} + c e^{rt} \overset{?}{=} 0$$
$$e^{rt}(r^2 + br + c) \overset{?}{=} 0$$

r is a solution to $r^2 + br + c = 0$. $y = e^{rt}$ is a solution.

Given the differential equation $y'' + by' + cy = 0$, we call the quadratic equation $r^2 + br + c = 0$ the *characteristic equation*. We'll break our analysis into three cases.

   Case 1. Suppose the characteristic equation has two distinct real roots, r_1 and r_2. Then $y_1 = e^{r_1 t}$ and $y_2 = e^{r_2 t}$ are solutions to $y'' + by' + cy = 0$. The general solution is $y = c_1 e^{r_1 t} + c_2 e^{r_2 t}$.

   Case 2. Suppose the characteristic equation has a double root: $r_1 = \frac{-b}{2}$. Then $y_1 = e^{r_1 t}$ is a solution. We need another solution.

   Claim: $y_2 = t e^{r_1 t}$ is a solution.

If $y_2 = t e^{r_1 t}$ then $y_2' = t r_1 e^{r_1 t} + e^{r_1 t}$ and $y_2'' = r_1 e^{r_1 t} + t r_1^2 e^{r_1 t} + r_1 e^{r_1 t}$.

$$y_2'' + by_2' + cy_2 \overset{?}{=} 0$$

$$e^{r_1 t}(r_1 + tr_1^2 + r_1) + b(e^{r_1 t})(tr_1 + 1) + cte^{r_1 t} \overset{?}{=} 0$$

$$e^{r_1 t}[2r_1 + tr_1^2 + btr_1 + b + tc] \overset{?}{=} 0$$

$$t\underbrace{(r_1^2 + br_1 + c)}_{0} + 2r_1 + b \overset{?}{=} 0$$

$$2r_1 + b \overset{?}{=} 0 \quad \text{We know } r_1 = \frac{-b}{2}.$$

$$2\left(\frac{-b}{2}\right) + b \overset{?}{=} 0$$

$$-b + b = 0 \quad \checkmark$$

$y_2 = te^{r_1 t}$ is a solution. Therefore, the general solution is of the form $y = C_1 e^{r_1 t} + C_2 te^{r_1 t}$.

Case 3. Suppose the characteristic equation has complex roots $\alpha \pm \beta i$ where $i = \sqrt{-1}$. We might first be tempted to turn away from this case, but a look back at Figure 31.33 should push us forward. Both curves in Figure 31.33 cross the t-axis multiple times whereas the functions $y = C_1 e^{r_1 t} + C_2 e^{r_2 t}$ and $y = (C_1 t + C_2)e^{r_1 t}$ have at most one zero.

If $\alpha \pm \beta i$ are roots of the characteristic equation then $y_1 = e^{(\alpha + \beta i)t}$ and $y_2 = e^{(\alpha - \beta i)t}$ are solutions to $y'' + by' + cy = 0$. $e^{(\alpha + \beta i))t} = e^{\alpha t} \cdot e^{\beta it}$. How can we make sense of $e^{\beta it}$? Let's turn to Taylor series. We know $e^u = 1 + u + \frac{u^2}{2!} + \frac{u^3}{3!} + \frac{u^4}{4!} + \cdots$.
Let $u = \beta it$. Then

$$e^{\beta it} = 1 + \beta it + \frac{(\beta it)^2}{2!} + \frac{(\beta it)^3}{3!} + \frac{(\beta it)^4}{4!} + \frac{(\beta it)^5}{5!} + \cdots$$

$$i = \sqrt{-1}, \text{ so } i^2 = -1, i^3 = -i, i^4 = 1, i^5 = i, \text{ etc.}$$

$$e^{\beta it} = 1 + i\beta t - \frac{(\beta t)^2}{2!} - i\frac{(\beta t)^3}{3!} + \frac{(\beta t)^4}{4!} + i\frac{(\beta t)^5}{5!} - \frac{(\beta t)^6}{6!} + \cdots$$

Gather together all terms without factors of i appearing.

$$e^{\beta it} = \underbrace{\left(1 - \frac{(\beta t)^2}{2!} + \frac{(\beta t)^4}{4!} - \frac{(\beta t)^6}{6!} + \cdots\right)}_{\cos(\beta t)} + i\underbrace{\left(\beta t - \frac{(\beta t)^3}{3!} + \frac{(\beta t)^5}{5!} - \cdots\right)}_{\sin(\beta t)}$$

We obtain what is known as *Euler's Formula*: $e^{i\beta t} = \cos(\beta t) + i\sin(\beta t)$. We define $e^{(\alpha + \beta i)t}$ to be

$$e^{(\alpha + \beta i)t} = e^{\alpha t}(\cos(\beta t) + i\sin(\beta t)).$$

The general solution to $y'' + by' + cy$ is therefore given by

$$y = C_1 e^{\alpha t} \cdot e^{\beta it} + C_2 e^{\alpha t} \cdot e^{-\beta it}$$
$$= e^{\alpha t}\left[C_1 \cos(\beta i) + C_1 i \sin(\beta t) + C_2 \cos(-\beta t) + C_2 i \sin(-\beta t)\right].$$

Cosine is an even function and sine is an odd function, so

$$y = e^{\alpha t}\big[(C_1 + C_2)\cos(\beta t) + (C_1 - C_2)i\,\sin(\beta t)\big] \text{ or}$$
$$y = e^{\alpha t}\big[C_3\cos(\beta t) + C_4\sin(\beta t)\big].$$

Let's return to the vibrating spring example. $x'' + \frac{\lambda}{m}x' + \frac{k}{m}x = 0$. The characteristic equation $r^2 + \frac{\lambda}{m}r + \frac{k}{m} = 0$, has roots given by

$$r = \frac{\frac{-\lambda}{m} \pm \sqrt{\frac{\lambda^2}{m^2} - 4\frac{k}{m}}}{2} = \frac{-\lambda}{2m} \pm \frac{1}{2m}\sqrt{\lambda^2 - 4km}.$$

EXERCISE 31.11   Argue that if the characteristic equation $r^2 + \frac{\lambda}{m}r + \frac{k}{m} = 0$ has two real roots then they are both negative. What physical situation does this case correspond to? It is referred to as "overdamped".

EXERCISE 31.12   The situation corresponding to one real root, $r = \frac{-\lambda}{2m}$, is called "critically damped." Find the solution given that the spring is compressed 0.5 inches and released and is critically damped. Your answer will be in terms of λ and m. Sketch this solution.

EXERCISE 31.13   The situation corresponding to complex roots is called "underdamped". Show that in all three situations, overdamped, critically damped, and underdamped, $\lim_{t\to\infty} x(t) = 0$ provided $\lambda > 0$. Explain why this makes sense.

EXERCISE 31.14   Use Euler's Formula to compute $e^{\pi i}$.

PROBLEMS FOR SECTION 31.6

Solve the differential equations in Problems 1 through 8.

1. $y'' - 4y' = -3$

2. $y'' - 2y' + y = 0$

3. $y'' + 25y = 0$

4. $y'' + 5y = 0$

5. $y'' + 5y' = 0$

6. $y'' = y' + 2y$

7. $3y'' + 3y' + 3y = 0$

8. $\frac{d^2x}{dt^2} + 25 = 3\frac{dx}{dt}$

In Problems 9 through 11, find the particular solution corresponding to the initial conditions given.

9. $\frac{d^2x}{dt^2} + \frac{dx}{dt} = 2x, \quad x(0) = -1, \quad x'(0) = 0$

10. $2x'' + 6x = 0,   x(0) = 0,   x'(0) = 4$

11. $4\frac{d^2x}{dt^2} - 4\frac{dx}{dt} = -x,   x(0) = 1,   x'(0) = 2$

Problems 12 through 15 refer to the equation

$$x'' + bx' + cx = 0.$$

12. What condition(s) must be satisfied to have a periodic solution? If the solution is periodic, what will its period be?

13. Find constants b and c such that if $x(0) = 5$ and $x'(0) = 0$ then $x(n) = 5$ for any integer n. Is this pair of constants unique?

14. Suppose $b > 0$ and $c > 0$. Is $\lim_{t \to \infty} x(t)$ necessarily zero? Explain. Interpret your response in terms of vibrating springs.

15. Suppose $b < 0$ and $c > 0$. For $x(0)$ and $x'(0)$ not both zero, is it possible that $\lim_{t \to \infty} |x(t)| = L$, where L is finite? Explain.

16. A spring with a 2-kg mass has a natural length of 0.6 m. A 10 N force is required to compress it to a length of 0.5 m. If the spring is compressed to 0.4 m and released, find the position of the mass at time t. Assume a frictionless system.

17. Write a second order homogeneous differential equation that is satisfied by $y(t) = e^t \sin t$. (The answer is not unique.)

18. Compute the following.
    (a) $e^{2\pi i}$ (b) $e^{-\pi i}$

Appendices

APPENDIX

Algebra

A.1 INTRODUCTION TO ALGEBRA: EXPRESSIONS AND EQUATIONS

Algebra is rooted in arithmetic. If you are confused about algebra your knowledge of arithmetic can provide guidance. Algebra is more abstract than arithmetic, but a constructive approach to understanding an abstract idea is to work through a concrete example. If you feel confused or amnesic try a very concrete example to get yourself grounded. This is a fundamental problem solving strategy; you will find ample opportunity to apply it—in mathematics as well as in other disciplines.

For example, suppose you're debating whether or not $\sqrt{x^2 + y^2}$ is equal to $x + y$. Instead of trying to "remember" or simply guessing, check it out with numbers. For example, if $x = 2$ and $y = 3$, $\sqrt{x^2 + y^2} = \sqrt{2^2 + 3^2} = \sqrt{4 + 9} = \sqrt{13}$, while $x + y = 2 + 3 = 5$. Therefore, $\sqrt{x^2 + y^2} \neq x + y$.

CAUTION From one example that does not "work" we can conclude that the statement does not hold in general. On the other hand, one example that does "work" is inconclusive. For instance, suppose the question of the hour is whether or not x^2 is equal to $2x$. You decide to check it out letting $x = 2$: $2^2 = 4 = 2 \cdot 2$. So far so good. You then double-check letting

1051

$x = 0$. Once again all goes smoothly: $0^2 = 0 = 2 \cdot 0$. You *cannot* conclude that squaring a number is the same as doubling it. *One* counterexample (try any number apart from 0 and 2) puts that idea to rest.

In order to keep your bearings algebraically, you must distinguish clearly between expressions and equations.

Expressions

$\frac{x^3-x}{2x^2+2x}$ is an expression, not an equation, just as "an automobile" is an expression, not a sentence. You *cannot* change the value of an expression without changing its meaning and obtaining a different (nonequivalent) expression. Replacing the expression "a car" by "an automobile" gives an equivalent expression, while replacing it by "a cat" does not. Mathematically, the only thing you can do with an expression is write the *same* thing in a different way. You can change its form (a stylistic change), but not its content. You *can* do anything that doesn't change its value. You can multiply by 1; you can add 0. Generally, we try to simplify expressions, making them look less complicated and easier to manipulate.

> The Fundamental Principle for working with expressions is preserve the value of the expression. An expression can only be transformed into an equivalent expression.

◆ **EXAMPLE A.1**   Simplify the following.

(a) $\frac{2x}{x}$

(b) $\frac{x^3-x}{2x^2+x}$

SOLUTION   (a) $\frac{2x}{x} = 2$ for $x \neq 0$ and is undefined for $x = 0$.

$$\frac{2x}{x} = \begin{cases} 2 & \text{if } x \neq 0; \\ \text{undefined} & \text{if } x = 0. \end{cases}$$

(b)

$$\frac{x^3 - x}{2x^2 + 2x} = \frac{x(x^2 - 1)}{2x(x + 1)} \qquad \text{(Factoring changes form, not value.)}$$

$$= \frac{x(x - 1)(x + 1)}{2x(x + 1)} \qquad \text{(Factor further.)}$$

$$= \frac{x(x + 1)}{x(x + 1)} \cdot \frac{x - 1}{2} \qquad \text{(We know } \frac{A}{A} = 1 \text{ for } A \neq 0.\text{)}$$

$$= \begin{cases} \frac{x-1}{2} & \text{for } x \neq 0,\, x \neq 1; \\ \text{undefined} & \text{for } x = 0 \text{ and } x = -1. \end{cases}$$

In other words, $\frac{x^3-x}{2x^2+2x}$ is equivalent to the expression $\frac{x-1}{2}$ except that the former is undefined at $x = 0$ and $x = -1$, a fraction being undefined when the denominator is zero.   ◆

> Division by zero is undefined.

Why is division by zero undefined? Let's begin by establishing that

$$\frac{a}{b} = c \text{ only if } a = bc.$$

Suppose we attempt to define $\frac{a}{0}$ to be some unique number c; then $a = 0 \cdot c = 0$. But then a must be zero; we cannot define $\frac{a}{0}$ for $a \neq 0$. If $a = 0$, then the equation $0 = 0 \cdot c$ holds for *any* c, not for some unique c, so $\frac{0}{0}$ is also undefinable.

Equations

An equation sets two expressions equal to one another.

$$\frac{x^3 - x}{2x^2 + 2x} = x$$

is an equation. We try to *solve* equations; in other words, we try to find out what values of x make the equation hold true.

> The Fundamental Principle for working with equations is to preserve the balance.

Therefore

- you can add or subtract anything from both sides of the equation,

- you can multiply or divide both sides of the equation by any nonzero quantity.

To solve for x means to get $x =$(an expression with no x's). Our goal, therefore, is to isolate x. How we proceed depends on what type of equation we have. If $x = a$ is a solution to an equation, then when both sides of the equation are evaluated at $x = a$, they are equal.

Meaning What You Say and Saying What You Mean

Math is a language, and a very precise one at that. When we write an expression it has a specified meaning. If you are opening up a new area of inquiry you can "invent" new words, but it is crucial to know the common conventions because that is what allows us to communicate. Sometimes a student will say "when I write '$3x^2$' what I mean is to square $3x$." This is as zany as writing an essay and saying "when I write 'yellow' what *I* mean is 'magenta.'" Below we review a few conventions worth knowing.

Exponents

$3x^2$ means $3 \cdot x^2 = 3 \cdot x \cdot x$. If you mean $(3x) \cdot (3x)$, then you must write $(3x)^2$, or $9x^2$. In the expression $3x^2$, 3 is called the **coefficient** of x^2. We say that 2 is an **exponent**; it is the power to which x is raised. The **base** is x.

> An exponent (power) refers to whatever is directly below it.

As you see, we can use parentheses to expand the sphere of influence of the exponent. For instance, $-5^2 = -5 \cdot 5 = -25$, whereas $(-5)^2 = (-5) \cdot (-5) = 25$. This means that if $f(x) = 3x^2$, then $f(-5) = +75 = f(5)$; if $g(x) = -3x^2$, then $g(5) = g(-5) = -75$.

Order of Operations

> In the absence of parentheses, the convention about order of operations is: First deal with exponents, then multiplication and division, and lastly addition and subtraction.

Parentheses can be used to override this order. In the case of nested parentheses, do what is in the innermost set first. Use parentheses whenever they make the meaning more obvious. *Note:* Multiplication and division are given equal status because they are really the same operation in two different guises; $A/B = A \cdot \frac{1}{B}$. Similarly, addition and subtraction are given equal status because $A - B = A + (-B)$. You need to understand how your calculator deals with orders of operation, or make copious use of parentheses to make your intended meaning perfectly clear. For example

$$\boxed{-}\boxed{3}\boxed{^\wedge}\boxed{2}\boxed{+}\boxed{2}\boxed{*}\boxed{9}\boxed{-}\boxed{6}\boxed{/}\boxed{3}\boxed{+}\boxed{1}$$

must, according to the conventions of order of operations, be interpreted as $-(3^2) + 2 \cdot 9 - \frac{6}{3} + 1 = -9 + 18 - 2 + 1 = 8$. Try this on your calculator. (*Note:* Many calculators have two separate keys, one for subtraction, as in $6 - 3$, and another for negative numbers, as in -3. The calculator will often give you an error if you mix up these two keys.)

EXERCISE A.1   Evaluate the following functions at the values indicated.

(a) $f(x) = x - 2x^2(-x + 4) - 6$; find $f(-5)$.

(b) $g(x) = x - 2x^2 - x + 4(-6)$; find $g(-2)$.

(c) $h(x, y) = xy[xy^3 - y^3x^2(x + y)]$; find $h(2, -1)$.

ANSWERS

(a) $f(-5) = (-5) - 2(25)(9) - 6 = -5 - 450 - 6 = -461$

(b) $g(-2) = -2 - (2)(4) + 2 - 24 = -2 - 8 + 2 - 24 = -10 - 22 = -32$

(c) $h(2, -1) = (2)(-1)[(2)(-1) - (-1)(4)(1)] = -2[-2 - (-4)] = -2[-2 + 4] = -2[2] = -4$

Square Roots

The square root of a nonnegative number A is defined to be the *nonnegative* number whose square is A. $\sqrt{9} = 3$ because $3^2 = 9$.

The solutions to the equation $x^2 = 9$ are $x = 3$ and $x = -3$. They can be expressed as $\sqrt{9}$ and $-\sqrt{9}$. When we write $\sqrt{9}$ we mean only positive 3. *Note:* $-\sqrt{9} \neq \sqrt{-9}$, the latter being undefined in the real number system because there is no real number whose square is -9. Try to take the square root of -9 on your calculator and see what you get.[1]

PROBLEMS FOR SECTION A.1

1. For what values of x is each expression undefined?

   (a) $\frac{3}{x^2}$

   (b) $\sqrt{x-1}$

   (c) $\frac{5}{x+3}$

   (d) $\frac{x-2}{2x-2}$

2. Are the expressions below equivalent? If the answer is no, show this with a concrete example or by simplifying the two expressions.

   Note: Some expressions are not equivalent, but are equal provided that a certain condition holds. For example, the expressions in part (a) are equal provided $x \neq 0$.

   (a) $\frac{x}{3x}$ and $\frac{1}{3}$

   (b) $\frac{x+1}{x(x+1)}$ and $\frac{1}{x}$

   (c) $-\frac{A}{B}$ and $\frac{-A}{B}$ and $\frac{A}{-B}$

   (d) -3^2 and $(-3)^2$

   (e) $\frac{A}{B}$ and $\frac{A^2}{B^2}$

   (f) $\sqrt{x^2}$ and x

   (g) xy^2 and y^2x

   (h) xy^2 and $(xy)^2$

3. Answer the questions below. If the answer is no, illustrate this with a concrete example.

   (a) If we square the numerator (top) and denominator (bottom) of an expression, will we obtain an equivalent expression? Is $\frac{A}{B}$ equivalent to $\frac{A^2}{B^2}$?

   (b) If we square both sides of an equation, will the balance be preserved? If $A = B$ will $A^2 = B^2$? If $A^2 = B^2$ does it follow that $A = B$?

   (c) If $A = B + C$, will $A^2 = B^2 + C^2$?

   (d) Suppose that $A = B$ and $A, B \neq 0$. Is $\frac{1}{A} = \frac{1}{B}$?

   (e) Suppose that $A = B + C$ and $A, B, C \neq 0$. Is $\frac{1}{A} = \frac{1}{B} + \frac{1}{C}$?

4. Write each of the following as an algebraic expression.

   (a) Square x and then multiply by 3.

   (b) Multiply x by 3 and square the result.

   (c) Subtract twice x from y and then take twice the square root of the result.

   (d) The cubed root of 3 times the reciprocal of x.

5. Simplify the following expressions.

---

[1] Some calculators will give an error message. Others give an ordered pair of numbers, indicating a complex number. $(0, 3)$ is shorthand for $0 + 3i$, where i stands for the complex number $\sqrt{-1}$.

(a) $-2^2 - 3\left[2 - 8 \cdot 4^{1/2} + (-3)^2\right]$

(b) $-x^2 - (-x^2) + (-x^2) + 3x^2$

(c) $\frac{-x}{y} - \frac{x}{y} + \frac{2x}{-y}$

(d) $3\sqrt{y} - \frac{2}{3}\sqrt{y} - \sqrt{y}$

6. (a) If we double an expression, will we have an equivalent expression?

(b) If we double the numerator and denominator of an expression, will we have an equivalent expression?

(c) If we double both sides of an equation, will the balance be preserved?

7. Evaluate the functions at the values indicated.

(a) $f(x) = -x^3 - 2x^2 + (-x)^2 + x$; find $f(1)$ and $f(-1)$.

(b) $f(x) = -\frac{1}{x} - \frac{2}{x^2} + \frac{-3}{-x^3}$; find $f(2)$ and $f(-2)$.

A.2   WORKING WITH EXPRESSIONS

In this section we will work with expressions. Keep in mind that the Fundamental Principle for working with expressions is to preserve the value of the expression. An expression can be transformed only into an entirely equivalent expression. We can change only the form of an expression.

Multiplying Out and Factoring: A Brief Introduction

One way of changing the form of an expression is to multiply out or factor. The expression $x(x + 1)$ is factored; $x^2 + x$ is the equivalent expression multiplied out.

The Distributive Law.   Multiplication distributes over addition:

$$(a + b) \cdot d = a \cdot d + b \cdot d.$$

Similarly,

$$(a + b + c) \cdot d = a \cdot d + b \cdot d + c \cdot d.$$

REMARK Multiplication does *not* distribute over itself. We know that

$$2 \cdot 3 \cdot 4 = (2 \cdot 3) \cdot 4 = 2 \cdot (3 \cdot 4) = 24$$
$$2 \cdot (3 \cdot 4) \neq 2 \cdot 3 \cdot 2 \cdot 4 = 48.$$

◆ EXAMPLE A.2   Multiply out the following expression, $(2x - 1)(x + 1)$

SOLUTION

$$(2x - 1) \cdot \underbrace{(x + 1)}_{\text{plays the role of } d} = 2x(x + 1) - 1(x + 1)$$

$$= 2x^2 + \underbrace{2x - x}_{\text{like terms}} - 1$$

$$= 2x^2 + x - 1.$$

Generally the first step is not written. We think of multiplying $2x$ by $(x + 1)$, and simply write out the result, and add to it the product of -1 and $(x + 1)$. We write $(2x - 1)(x + 1) = 2x^2 + 2x - x - 1$.   ◆

To factor an expression is to write it as a product. It is the reverse of multiplying out. (You can always check your factoring by multiplying out.) To factor out a common factor is to use the distributive law in reverse.

◆ **EXAMPLE A.3**   Factor $x^3 - x^2 + 4x^7$.

SOLUTION $x^3 - x^2 + 4x^7 = x^2(x - 1 + 4x^5)$ ◆

CAUTION Although $x^3 - x^2 + 4x^7 + 3 = x^2(x - 1 + 4x^5) - 3$, this expression is *not* factored. Similarly, $x^2 - x - 2 = x(x - 1) - 2$, but this expression is not factored. $x^2 - x - 2 = (x + 1)(x - 2)$ and this *is* factored. (See below.)

◆ **EXAMPLE A.4**   Factor $x^3 - x^2$.

SOLUTION $x^3 - x^2 = x^2(x - 1)$ or $x^2(-1)(-x + 1) = -x^2(1 - x)$, whichever suits us. All are factored forms.   ◆

We will take up the topic of factoring more systematically later in Section A.2.

Multiplication and Working with Exponents

> **ExponentLaws**
>
> (1) $c^m c^n = c^{m+n}$
>
> (2) $\dfrac{c^m}{c^n} = c^{m-n}$
>
> (3) $(c^m)^n = c^{mn}$

From these fundamental laws for working with exponents, it follows that

$$c^0 = 1,$$

because we know that $1 = \frac{c^n}{c^n} = c^{n-n} = c^0$;

$$c^{-n} = \frac{1}{c^n},$$

because $\frac{1}{c^n} = \frac{c^0}{c^n} = c^{0-n} = c^{-n}$;

$$c^{1/p} = \text{the } p\text{th root of } c,$$

because we know that $(c^{1/p})^p = c^1 = c$.

Make sure that you can use your calculator to compute quantities along the lines of, say, the fifth root of 32. On many calculators this would be along the lines of

$$\boxed{3}\,\boxed{2}\,\boxed{x^y}\,\boxed{(}\,\boxed{1}\,\boxed{/}\,\boxed{5}\,\boxed{)}\,\boxed{=}$$

or

$$\boxed{3}\,\boxed{2}\,\boxed{\sqrt[x]{x}}\,\boxed{5}\,\boxed{=}.$$

(You ought to get an answer of 2.)

Arithmetic Underpinnings[2]

◆ **EXAMPLE A.5** Simplify the expressions below.

(a) $3 \cdot \frac{1}{4} \cdot 1\frac{6}{12}$

$$3 \cdot \frac{1}{4} \cdot 1\frac{6}{12} = \frac{3}{1} \cdot \frac{1}{4} \cdot \frac{3}{2}$$
$$= \frac{3 \cdot 3}{4 \cdot 2} = \frac{9}{8}$$

(b) $\frac{2^3 \cdot 2^4}{2^6}$

$$\frac{2^3 \cdot 2^4}{2^6} = \frac{2^7}{2^6} = 2 \quad ◆$$

Extracting the Fundamentals

Multiplication of fractions:

$$\frac{a}{b} \cdot \frac{c}{d} = \frac{ac}{bd}.$$

Simplification of fractions:

$$\frac{ac}{cd} = \frac{a}{d}.$$

$(\frac{ac}{cd} = \frac{ca}{cd} = \frac{c}{c} \cdot \frac{a}{d} = 1 \cdot \frac{a}{d} = \frac{a}{d})$.

CAUTION $\frac{A+B}{C+B} \neq \frac{A}{C}$; e.g., $\frac{5}{4} = \frac{2+3}{1+3} \neq 2$. You can't "cancel over addition."

Canceling is based on the idea that multiplying by one doesn't change the value of the expression. You can cancel only when the numerator (top) and denominator (bottom) are each factored (expressed as a product) and have a common factor.

Using the fundamentals extracted from arithmetic, we are ready to do some algebra.

◆ **EXAMPLE A.6** Simplify $\frac{x-5}{x^2-1} \cdot \frac{x^2+1}{5-x}$.

SOLUTION

$$\frac{x-5}{x^2-1} \cdot \frac{x^2+1}{5-x} = \frac{(x-5)(x^2+1)}{(x^2-1)(5-x)}$$
$$= \frac{x^2+1}{(x^2-1)} \frac{(x-5)}{(5-x)}$$
$$= \frac{x^2+1}{(x^2-1)} \frac{(x-5)}{(-1)(x-5)} = \frac{x^2+1}{(x^2-1)(-1)}$$
$$= \frac{x^2+1}{-x^2+1} \quad ◆$$

---

[2] You might be thinking, "Why arithmetic? I know arithmetic." We start with arithmetic to emphasize that when you know arithmetic, algebra nicely follows without much ado.

Note: $\dfrac{-a}{b} = \dfrac{a}{-b} = -\dfrac{a}{b}$.

Similarly, $\dfrac{a-b}{b-a} = -1$, because $\dfrac{a-b}{b-a} = \dfrac{a-b}{(-1)(-b+a)} = \dfrac{a-b}{(-1)(a-b)} = \dfrac{1}{-1} = -1$.

◆ **EXAMPLE A.7**   Simplify $\dfrac{y^4\sqrt{y}}{(\sqrt{x}y^2)^3} \cdot \dfrac{x^3 y^{-1}}{6\sqrt{xy}}$.

SOLUTION

$$\frac{y^4\sqrt{y}}{(\sqrt{x}y^2)^3} \cdot \frac{x^3 y^{-1}}{6\sqrt{xy}} = \frac{y^{4+1/2-1}x^3}{x^{3/2}y^6 \cdot 6x^{1/2}y^{1/2}}$$

$$= \frac{y^{7/2}x^3}{x^2 y^{13/2}}$$

$$= y^{7/2-13/2}x^{3-2} = y^{-3}x, \text{ or, equivalently, } \frac{x}{y^3}. \quad ◆$$

Addition and Subtraction

Arithmetic Underpinnings

◆ **EXAMPLE A.8**   Add and simplify.

$$3 \cdot \frac{1}{4} \cdot \frac{18}{12} + 0.5 - \frac{5}{6} = \frac{9}{8} + \frac{1}{2} - \frac{5}{6}.$$

SOLUTION   We can only add "like" things, so to add these three terms we must write them with a common denominator. We could use $8 \cdot 2 \cdot 6$ as a common denominator but it is less cumbersome to factor the denominators completely and find the **least common denominator**.

$$\frac{9}{2 \cdot 2 \cdot 2} + \frac{1}{2} - \frac{5}{2 \cdot 3}.$$

The least common denominator is $2 \cdot 2 \cdot 2 \cdot 3 = 24$.

Finding the Least Common Denominator.   Every factor appearing in a denominator must be made a factor of the LCD (least common denominator) and must be raised to the highest power to which it appears in any one denominator.

Each term must now be written with the LCD without changing its value.

$$\frac{9}{8} \cdot \frac{3}{3} = \frac{27}{24}; \quad \frac{1}{2} \cdot \frac{12}{12} = \frac{12}{24}; \quad \frac{5}{6} \cdot \frac{4}{4} = \frac{20}{24}$$

$$\frac{9}{8} + \frac{1}{2} - \frac{5}{6} = \frac{27}{24} + \frac{12}{24} - \frac{20}{24} = \frac{27 + 12 - 20}{24} = \frac{19}{24} \quad ◆$$

Extracting the Fundamentals

Addition: We can only add like quantities. Therefore, we can't add fractions until we have written them with a common denominator. We can't change the *value* of a fraction in our eagerness to get a common denominator.

$$\frac{a}{b} + \frac{c}{d} = \frac{ad}{bd} + \frac{cb}{db} = \frac{ad + cb}{bd}$$

The product of all denominators is always a common denominator. It is generally most efficient to find the least common denominator. Rewrite each fraction with the LCD; then add or subtract.

Algebra

◆ **EXAMPLE A.9**   Simplify $\frac{x}{x+1} - \frac{2}{x+2}$.

SOLUTION   The terms are already simple. The LCD is $(x+1)(x+2)$. Write each fraction with the LCD and then add.

$$\frac{x}{x+1} = \frac{x}{x+1}\frac{x+2}{x+2} = \frac{(x)(x+2)}{(x+1)(x+2)};$$

$$\frac{2}{x+2} = \frac{2}{x+2} \cdot \frac{x+1}{x+1} = \frac{(2)(x+1)}{(x+2)(x+1)}$$

$$\frac{(x)(x+2)}{(x+1)(x+2)} - \frac{(2)(x+1)}{(x+2)(x+1)} = \frac{(x)(x+2) - (2)(x+1)}{(x+2)(x+1)}$$

$$= \frac{x^2 + 2x - 2x - 2}{(x+2)(x+1)}$$

$$= \frac{x^2 - 2}{(x+2)(x+1)}.$$

Here we multiplied out the terms of the numerator in hopes that the numerator would simplify (it did), and perhaps even share a common factor with the denominator (it didn't). We will not benefit by multiplying out the denominator.

This last remark deserves to be highlighted. There are *very* few situations in which it's helpful to have the denominator of such an expression expanded. Generally, it's handy to keep denominators in factored form. Then it's simpler to find an LCD if adding, or to cancel when multiplying. ◆

◆ **EXAMPLE A.10**   Simplify $\frac{x^2+4x-5}{2x^2-2x} - \frac{x}{4x+6}$.

SOLUTION   See if the terms themselves can be simplified; then, with denominators written in factored form, find the LCD. Write each fraction with the LCD and then add.

First term:

$$\frac{x^2 + 4x - 5}{2x^2 - 2x} = \frac{(x+5)(x-1)}{2x(x-1)} = \frac{x+5}{2x}.$$

Second term:

$$\frac{x}{4x+6} = \frac{x}{2(2x+3)}.$$

$$\frac{x^2 + 4x - 5}{2x^2 - 2x} - \frac{x}{4x+6} = \frac{x+5}{2x} - \frac{x}{2(2x+3)}$$

The LCD is $2x(2x+3)$. Express each term with the LCD:

$$\frac{x+5}{2x} = \frac{x+5}{2x} \cdot \frac{2x+3}{2x+3} = \frac{(x+5)(2x+3)}{(2x)(2x+3)}$$

$$\frac{x}{(2)(2x+3)} = \frac{x}{(2)(2x+3)} \cdot \frac{x}{x} = \frac{x^2}{(2x)(2x+3)}.$$

Now subtract:

$$\frac{(x+5)(2x+3)}{(2x)(2x+3)} - \frac{x^2}{(2x)(2x+3)} = \frac{(x+5)(2x+3) - x^2}{(2x)(2x+3)}.$$

We can multiply out the numerator (top) in hopes that we can add and then factor, but there is *no reason* to multiply out the denominator. (Note that after addition/subtraction the numerator is *not* in factored form.)

$$\frac{(x+5)(2x+3) - x^2}{(2x)(2x+3)} = \frac{2x^2 + 13x + 15 - x^2}{(2x)(2x+3)} = \frac{x^2 + 13x + 15}{(2x)(2x+3)}$$

The numerator does not factor, so this cannot be simplified.   ◆

Advice: One common pitfall for a novice is the loss of direction. Keep sight of your goal. Just because a certain operation is "legal" doesn't mean you benefit from doing it; make sure that what you are doing is leading you in the direction of your goal. For example, many times a student will correctly determine the LCD, express each term using the LCD, and then, after multiplying out the numerator of each term, proceed to simplify each term. If you do this accurately, then you are right back at the beginning of the problem, feeling a little dizzy from having completed a tight circle . . . .

> **It is important to have a strategy. Break the problem solving process into a sequence of manageable steps and keep track of where you are in your game plan.**

Another fairly common pitfall occurs when you're working with an expression once you've written something along the lines of $\frac{x}{x+1} = \frac{(x)(x+2)}{(x+1)(x+2)}$. You are still working with an expression, so the rules for expressions hold. You can't, for instance, simply decide to wipe out the denominator by "multiplying both sides by $(x+1)(x+2)$" because you really don't have two "sides." Clarifying your goal and game plan will help you avoid these mistakes.

Division and Complex Fractions

Arithmetic Underpinnings
To divide a quantity by 2 is to take half of it.

$$Q \div 2 = Q \cdot \frac{1}{2} = \frac{Q}{2}$$

To divide a quantity by 10 is to take a tenth of it.

$$Q \div 10 = Q \cdot \frac{1}{10} = \frac{Q}{10}$$

To divide a quantity by $\frac{2}{3}$ is $Q \div \frac{2}{3}$ or $\frac{Q}{2/3}$. We can multiply this expression by 1 in the form $\frac{\frac{3}{2}}{\frac{3}{2}}$ in order to eliminate the $\frac{2}{3}$ from the denominator:

$$\frac{Q}{\frac{2}{3}} = \frac{Q \cdot \frac{3}{2}}{\frac{2}{3} \cdot \frac{3}{2}} = \frac{Q \cdot \frac{3}{2}}{\frac{6}{6}} = Q \cdot \frac{3}{2}.$$

So, to divide by $\frac{2}{3}$ is to multiply by $\frac{3}{2}$. To divide Q by R is to multiply Q by $\frac{1}{R}$. To divide Q by $\frac{r}{s}$ is to multiply Q by $\frac{s}{r}$.

$$\frac{Q}{\frac{r}{s}} \cdot \frac{\frac{s}{r}}{\frac{s}{r}} = \frac{Q \cdot \frac{s}{r}}{1} = Q \cdot \frac{s}{r}$$

◆ **EXAMPLE A.11**   Simplify.   (a) $\frac{1/3}{2/9}$ (b) $\frac{\frac{1}{3}+1}{\frac{1}{3}+\frac{1}{2}}$

SOLUTION   (a)

$$\frac{\frac{1}{3}}{\frac{2}{9}} = \frac{1}{3} \cdot \frac{9}{2} = \frac{3}{2}$$

(b)

$$\frac{\frac{1}{3}+1}{\frac{1}{3}+\frac{1}{2}} = \frac{\frac{1}{3}+\frac{3}{3}}{\frac{2}{6}+\frac{3}{6}} = \frac{\frac{4}{3}}{\frac{5}{6}} = \frac{4}{3} \cdot \frac{6}{5} = \frac{8}{5}$$ ◆

Extracting the Fundamentals

$$\boxed{\frac{\frac{A}{B}}{\frac{C}{D}} = \frac{A}{B} \cdot \frac{D}{C}}$$

Why?

$$\frac{\frac{A}{B}}{\frac{C}{D}} = \frac{\frac{A}{B}}{\frac{C}{D}} \cdot \frac{\frac{D}{C}}{\frac{D}{C}} = \frac{\frac{A}{B} \cdot \frac{D}{C}}{1} = \frac{A}{B} \cdot \frac{D}{C}.$$

Algebra

◆ **EXAMPLE A.12**   Simplify $\frac{\frac{1}{x}-x}{1-\frac{1}{x^2}}$.

SOLUTION   Get the complex fraction into the form $\frac{\frac{A}{B}}{\frac{C}{D}}$. You can work on each part, $\frac{1}{x} - x$ and $1 - \frac{1}{x^2}$, individually or work both at once. We've done the latter.

$$\frac{\frac{1}{x}-x}{1-\frac{1}{x^2}} = \frac{\frac{1}{x}-\frac{x^2}{x}}{\frac{x^2}{x^2}-\frac{1}{x^2}}$$

$$= \frac{\frac{1-x^2}{x}}{\frac{x^2-1}{x^2}}$$

$$= \frac{1-x^2}{x} \cdot \frac{x^2}{x^2-1}$$

$$= \frac{(1-x)(1+x)}{x} \cdot \frac{x^2}{(x-1)(x+1)}$$

$$= \frac{(1-x)x}{x-1}$$

$$= \frac{(-1)(x-1)x}{(x-1)} = -x. \quad \blacklozenge$$

Talking the Talk—A Bit of Terminology

Most of our examples have dealt with polynomials or quotients of polynomials. Polynomials are sums of products of constants and nonnegative integer powers of variables. A polynomial can be written in the form

$$a_0 + a_1 x + a_2 x^2 + a_3 x^3 + \cdots + a_n x^n,$$

where $a_0, a_1, a_2, \ldots, a_n$ are constants. Since the powers to which the variable is raised are nonnegative integers $(0,1,2,\ldots)$, the variable is never in the denominator or under a radical. $4x^3 - 3x + 8$ is a polynomial, as are $\frac{4}{11}x^3 - \sqrt{3}x + \frac{8}{\sqrt{5}}$ and $\frac{abx}{\sqrt{c^2+1}}$ for a, b, and c constants. $-3\sqrt{x} + 5$ is not a polynomial.

The degree of a polynomial is the highest power to which x is raised. A polynomial of degree 2 is called a quadratic. We call a_k the coefficient of the term x^k. If a polynomial is of degree n, a_n is called the leading coefficient. In the term $\frac{4}{11}x^3$ we call $\frac{4}{11}$ the coefficient of the term. 3 is the exponent, or power, of x.

Integers, Rationals, Irrationals, and Real Numbers

Real Numbers:

$$\mathbb{R} = \left\{\text{all } x \text{ such that } x \text{ corresponds to a point on the number line}\right\}$$

Integers:

$$\mathbb{Z} = \{\ldots, -3, -2, -1, 0, 1, 2, 3, \ldots\}$$

Positive Integers:

$$\mathbb{Z}^+ = \{1, 2, 3, 4, \ldots\}$$

Rational Numbers: $\mathbb{Q}$, all numbers that can be written in the form $\frac{a}{b}$, where a and b are integers, $b \neq 0$. Equivalently, the rationals are all numbers that either repeat a pattern infinitely or terminate when written in decimal notation. For example:

$$\frac{13}{20} = 0.65$$

$$\frac{1}{3} = 0.333\ldots = 0.\bar{3}$$

$$\frac{2}{7} = 0.285714285714\ldots.$$

Irrational Numbers: Real numbers that cannot be written in the form $\frac{a}{b}$. Decimals that are nonterminating and nonrepeating are irrational numbers, and conversely. There are infinitely many irrational (and rational) numbers between any two rational numbers. Examples of irrational numbers are $\sqrt{2}$, $\sqrt{10}$, and π.

EXERCISE A.2   Let $f(x) = \frac{x}{x+1}$ and let $g(x) = \frac{1}{x^2-1}$.

Find the following. If your answer involves the sum or difference of fractions, add the fractions; do not leave complex fractions in your answers.

(a) $g(x) - f(x)$

(b) $\frac{f(x)}{g(x)}$

(c) $\frac{f(x+1)}{x+1}$

(d) $\frac{x}{f(x)} + \frac{1}{g(x)}$

(e) Find $f\big(g(x)\big)$.

(f) For what values of x is $f\big(g(x)\big) = 1$?

Answers:

(a) $\frac{-x^2+x+1}{(x+1)(x-1)}$ (b) $x(x-1)$ (c) $\frac{1}{x+2}$

(d) $x^2 + x$ (e) $\frac{1}{x^2}$ (f) $x = \pm 1$

Factoring

Why factor? From what we have already seen, factoring permits us to simplify expressions. We can find least common denominators only if we can factor. We will also see that factoring helps us solve equations.

What is factoring? To factor an expression is to write it as a product. $(x+3)(x-1)$, $x(-1)(+7)(x-5)$, and $\left(\frac{8}{x}\right)(x+4)(x)$ are all factored. $(x^2+1)(x^2-1)$ is factored, but it can be further factored into $(x^2+1)(x-1)(x+1)$. $(x+1)(x+2)+2$ is *not* factored, nor is $(x+1)(x+2)+3(x+4)$.

$$\underbrace{(x+3)(x-1)}_{\text{factored form}} = \underbrace{x^2+2x-3}_{\text{multiplied out}}$$

◆ **EXAMPLE A.13**   Multiply out. $(x+3)^2 = (x+3)(x+3) = x^2 + 6x + 9$

Note: A common error is to write $x^2 + 9$ as an answer and forget about the $6x$. Perhaps the picture in Figure A.1 can steer you away from this.

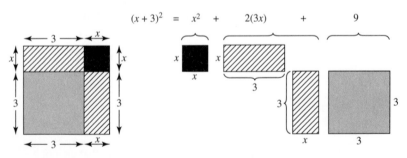

Figure A.1   ◆

◆ **EXAMPLE A.14**   Multiply out. $(x-2)(x+2) = x^2 + 2x - 2x - 4 = x^2 - 4$ ◆

Factoring and expanding, or multiplying out, are complementary procedures. We discussed multiplying out and the distributive law in Section A.2. Below we give a couple of more involved illustrations before we take up factoring.

◆ **EXAMPLE A.15**   Expand the following products.

(a) $(x^3 - 3x^2 + 1)(-x + 2x^3)$ (b) $(x+3)(x-1)(2x-1)$

SOLUTION   (a) $(x^3 - 3x^2 + 1)(-x + 2x^3) = x^3(-x + 2x^3) - 3x^2(-x + 2x^3) + 1(-x + 2x^3)$
$$= -x^4 + 2x^6 + 3x^3 - 6x^5 - x + 2x^3$$

We've multiplied; now we get organized.
$$= 2x^6 - 6x^5 - x^4 + 3x^3 + 2x^3 - x$$

We add like terms,
$$= 2x^6 - 6x^5 - x^4 + 5x^3 - x.$$

(b) $(x+3)(x-1)(2x-1)$. Multiply any two of these and then multiply the result by the third.

$$(x+3)(x-1)(2x-1) = [(x+3)(x-1)](2x+1)$$
$$= [x^2 + 2x - 3](2x+1)$$
$$= 2x^3 + 4x^2 - 6x + x^2 + 2x - 3$$
$$= 2x^3 + 5x^2 - 4x - 3.   ◆$$

Factoring an expression is to write it as a product. It is the reverse of multiplying out. You can always check your factoring by multiplying out.

Factoring Out a Common Factor: The Distributive Law in Reverse

◆ **EXAMPLE A.16**   Factor the following as much as possible.

(a) $(\lambda w)^2 + w\lambda^2 + \lambda w^2$

$$(\lambda w)^2 + w\lambda^2 + \lambda w^2 = \lambda^2 w^2 + w\lambda^2 + \lambda w^2 = \lambda w(\lambda w + \lambda + w)$$

(b) $(\lambda + C)^3 - (\lambda + C)^2 + 4(\lambda + C)^7$

$$(\lambda + C)^3 - (\lambda + C)^2 + 4(\lambda + C)^7 = (\lambda + C)^2 \left[(\lambda + C) - 1 + 4(\lambda + C)^5 \right]$$

(c) $(A+B)x^2 + 2(A+B)x = (A+B)(x)(x+2)$
   This final expression is factored completely.[3]

(d) $2x(x+1) + 3(x+1) = (2x+3)(x+1)$

(e) $2x(x-1) + 3(1-x) = 2x(x-1) - 3(x-1) = (2x-3)(x-1)$
   Notice that $1 - x$ could be rewritten as $-(x-1)$ so the common factor of $x - 1$ could be extracted.   ◆

---

[3] An expression is said to be *factored completely* if it is factored and each factor cannot be further factored.

Factoring Quadratics of the Form $x^2 + bx + c$

We want to express $x^2 + bx + c$ in the form $(x + \square)(x + \triangle)$, where $\square$ and $\triangle$ are constants.

$$(x + \square)(x + \triangle) = x^2 + \square x + \triangle x + \square\triangle = x^2 + (\square + \triangle)x + \square\triangle.$$

This has to be equal to $x^2 + bx + c$, so we must find two numbers whose product is c and sum is b. Replace $\square$ by one of them and $\triangle$ by the other.

◆ EXAMPLE A.17   Factor the expressions below.

(a) $x^2 - x - 2$

(b) $x^2 - 14x + 24$

(c) $x^2 - 10x + 24$

(d) $x^2 - 9x + 24$

SOLUTION

(a) Look for two numbers whose product is -2 and whose sum is -1. Answer: -2 and 1.

$$x^2 - x - 2 = (x + \underline{\quad})(x + \underline{\quad})$$
$$= (x - 2)(x + 1)$$

(b) Look for two numbers whose product is 24 and whose sum is -14. If they are going to add to -14, they will both be negative. Our options are $-1, -24$, or $-2, -12$, or $-3, -8$, or $-4, -6$. Answer: -12 and -2.

$$x^2 - 14x + 24 = (x + \underline{\quad})(x + \underline{\quad})$$
$$= (x - 12)(x - 2)$$

(c) Look for two numbers whose product is 24 and whose sum is -10. Answer: -4 and -6.

$$x^2 - 10x + 24 = (x + \underline{\quad})(x + \underline{\quad})$$
$$= (x - 4)(x - 6)$$

(d) Look for two numbers whose product is 24 and whose sum is -9. Our options are as in part , but we can't get a sum of 9. This expression *can't* be factored; it's irreducible.

◆

Factoring Quadratics of the Form $ax^2 + bx + c$

Option 1:   Guess and check:

$$ax^2 + bx + c = (ex + f)(gx + h) = egx + (fg + eh)x + fh.$$

To have equality we must find e, f, g, and h such that $a = ef$, $c = fh$, and $b = fg + eh$. This approach isn't as formidable as it looks provided that a and c have few factors.

◆ EXAMPLE A.18 $2x^2 + 5x - 3 = (\underline{\quad}x + \underline{\quad})(\underline{\quad}x + \underline{\quad})$

SOLUTION   We know that we must have $(2x + \underline{\quad})(x + \underline{\quad})$ and the only question is whether to use -1 and 3 or 1 and -3 and where to place these numbers.[4] If b is to be 5, we'll want to multiply the 2 by 3, so we have

$$2x^2 + 5x - 3 = (2x - 1)(x + 3).$$

Check this by multiplying out.   ◆

Try the next exercise on your own.

EXERCISE A.3   Factor.

(a) $3x^2 - 14x - 5$

(b) $3x^2 - 2x - 5$

ANSWER

(a) We know that we must have $(3x + \underline{\quad})(x + \underline{\quad})$.

$$3x^2 - 14x - 5 = (3x + 1)(x - 5)$$

(b)

$$3x^2 - 2x - 5 = (3x - 5)(x + 1)$$

Option 2: A systematic method for factoring $ax^2 + bx + c$: Look for two numbers whose product is ac and sum is b. If you can find such numbers, you can factor. Otherwise, you can't. We'll work an example to show you what to do with these two numbers and then describe the general procedure.

◆ **EXAMPLE A.19**   Factor $2x^2 + 5x - 3$.

SOLUTION   Look for two numbers whose product is -6 and sum is 5. Answer: 6 and -1. These numbers add to 5, so we can use them to split the $5x$ term.

$$2x^2 + 5x - 3 = 2x^2 + 6x - x - 3$$

Now we'll factor the first two terms and the last two terms independently. We will be able to arrange this so that the resulting two terms have a common factor.

$$2x^2 + 6x - x - 3 = 2x(x + 3) - (x + 3)$$

Lo and behold, we can factor out a common factor of $(x + 3)$.

$$2x(x + 3) - (x + 3)$$
$$= (x + 3)(2x - 1)$$

You might wonder what would happen if we split $5x$ into $-x + 6x$ instead of $6x - x$. No problem. $2x^2 + 5x - 3 = 2x^2 - x + 6x - 3 = x(2x - 1) + 3(2x - 1)$. We can factor out the common factor of $(2x - 1)$.

$$2x^2 + 5x - 3 = (2x - 1)(x + 3),$$

as before.   ◆

---

[4] Starting with $(-2x + \underline{\quad})(-x + \underline{\quad})$ is equivalent (factor -1 out of both factors) so we don't deal with it separately.

General Procedure

Given $ax^2 + bx + c$:

1. Look for two numbers whose product is ac and sum is b. Let's suppose q and r work.
2. Using these two numbers, split the bx term. Then group the first two terms together and the last two terms together. $(ax^2 + qx) + (rx + c)$
3. Factor each expression in parentheses so that a factor of the second group looks the same as a factor of the first group.
4. You will now have an expression out of which you can factor a common factor.

◆ **EXAMPLE A.20**   Factor.   (a) $-6x^2 + 19x - 15$ (b) $6x^2 + 5x - 6$

SOLUTION   (a) Look for two numbers that multiply to 90 and add to 19. Answers: 10 and 9.

$$-6x^2 + 19x - 15 = -6x^2 + 10x + 9x - 15$$
$$= (-6x^2 + 10x) + (9x - 15)$$
$$= 2x(-3x + 5) - 3(-3x + 5)$$

Note: We factored out -3 instead of 3 from the second group in order to make it match a factor from the first group.

$$= (-3x + 5)(2x - 3)$$

(b) Look for two numbers that multiply to -36 and add to 5. Answers: 9 and -4.

$$6x^2 + 5x - 6 = 6x^2 + 9x - 4x - 6$$
$$= 3x(2x + 3) - 2(2x + 3)$$
$$= (2x + 3)(3x - 2) \quad ◆$$

Factoring the Difference of Two Perfect Squares

$$(A^2 - B^2) = (A - B)(A + B)$$

Be on the lookout for the difference of perfect squares that might be "in disguise." An expression like $A^2 - B^2$ should jump out at you, screaming to be factored, but there can be less obvious examples, as illustrated below.

◆ **EXAMPLE A.21**   Factor.   (a) $9x^2 - 4z^6$ (b) $8x^8 - 18(x - y)^2$ (c) $x^4 - 1$

SOLUTION   (a)

$$9x^2 - 4z^6 = (3x)^2 - (2z^3)^2 = (3x - 2z^3)(3x + 2z^3)$$

(b)

$$8x^8 - 18(x - y)^2 = 2(4x^8 - 9(x - y)^2)$$
$$= 2\left[(2x^4)^2 - (3(x - y))^2\right]$$
$$= 2[2x^4 - 3(x - y)][2x^4 + 3(x - y)]$$
$$= 2(2x^4 - 3x + 3y)(2x^4 + 3x - 3y)$$

(c)
$$x^4 - 1 = (x^2 - 1)(x^2 + 1) = (x - 1)(x + 1)(x^2 + 1)$$

The sum of two perfect squares cannot be factored.   ◆

Variations on the Theme

Once you've learned about factoring out common factors and factoring quadratics (including the difference of two perfect squares), you can in practice do a bit more. Let's take the example $2x^2 - x - 1 = (2x + 1)(x - 1)$ and see how it can be dressed up a little.

◆ **EXAMPLE A.22**   Factor $2x^6 - x^4 - x^2$ completely.

SOLUTION   First factor out any factors common to all terms.

$$2x^6 - x^4 - x^2 = x^2(2x^4 - x^2 - 1)$$

Now look for a quadratic in masquerade.

$$x^2 \left[2(x^2)^2 - (x^2) - 1 \right]$$

The expression in brackets is of the form $2u^2 - u - 1$, where $u = x^2$.

$$2u^2 - u - 1 = (2u + 1)(u - 1).$$

Therefore we have $x^2(2x^2 + 1)(x^2 - 1)$, but this is still not factored completely.

$$2x^6 - x^4 - x^2 = x^2(2x^2 + 1)(x + 1)(x - 1)   ◆$$

◆ **EXAMPLE A.23**   Simplify $\dfrac{\frac{2x^6 - x^4 - x^2}{x - 2x^2}}{\frac{x - x^3}{-2x^2 - x + 1}}$.

SOLUTION   We'll use the result of Example A.22 to factor the numerator.

$$\frac{\frac{2x^6 - x^4 - x^2}{x - 2x^2}}{\frac{x - x^3}{-2x^2 - x + 1}} = \frac{x^2(2x^2 + 1)(x + 1)(x - 1)}{x(1 - 2x)} \cdot \frac{(x + 1)(-2x + 1)}{x(1 - x)(1 + x)}$$

$$= \frac{(2x^2 + 1)(x - 1)(x + 1)}{1 - 2x} \cdot \frac{\cancel{(-2x + 1)}}{1 - x}$$

$$= (2x^2 + 1)(x - 1)(x + 1) \cdot \frac{-1}{-1 + x}$$

$$= (2x^2 + 1)(-x - 1)   ◆$$

We have by no means exhausted the topic of factoring—one can learn to factor the difference of two perfect cubes (and also the sum) and more, but for now we're moderately well equipped to go on. As we discuss solving equations, we'll touch again on factoring.

PROBLEMS FOR SECTION A.2

1. Compute the following sums and simplify your answer.

   (a) $\frac{1}{x^2} + \frac{3-x}{x} + \frac{x}{3+x}$

   (b) $\frac{\frac{1}{x+w} - \frac{1}{w}}{w}$

   (c) $\left[\frac{2}{(y+z)^2} - \frac{2}{y^2} \right] \cdot \frac{1}{y}$

2. Simplify.

   (a) $\frac{\frac{2x^2-8}{x}}{-x-2}$

   (b) $\frac{y+\frac{1}{xy}}{\frac{x}{y}+x}$

   (c) $\frac{1}{\frac{1}{x}+\frac{1}{y}}$

   (d) $\frac{(\sqrt{x}+\sqrt{y})^2}{2xy} - \frac{\sqrt{xy}x^{-1}}{y}$

   (e) $\frac{x^2y^3-xy^4}{y^3x^2-x^3y^2}$

3. Factor as much as possible.

   (a) $x^2 - x - 6$

   (b) $2x^2 - x - 3$

   (c) $6x^3 + 6x^2 - x$

   (d) $16x^2y^4 - 1$

   (e) $(x-1)xy + 3(1-x)$

   (f) $x^4 - 3x^2 - 10$

   (g) $x^4 + 3x^2 - 4x^3$

4. Factor as much as possible.

   (a) $b^{w+2} - b^w$

   (b) $b^{2w} - b^w$

   (c) $x^3b^{x+2} - x^3b^x$

   (d) $b^{2x} - 4$

   (e) $b^{2w} - b^w - 6$

5. Simplify as much as possible $\dfrac{\frac{3x^2-x-2}{9x^2-4}}{\frac{x^4-9}{x^4+x^2}}$.

A.3   SOLVING EQUATIONS

The strategy one adopts for solving an equation depends upon what type of equation one has. In particular, to solve an equation for x one has to determine what type of equation it is *in x*.

Linear Equations

An equation is **linear in** x if it can be put into the form $ax + b = 0$, where a and b are constants or expressions *not* containing x, and where $a \neq 0$.

To determine whether an equation is linear in x, clear all x's from the denominators; if x is inside parentheses, remove the parentheses and multiply out.

◆ EXAMPLE A.24

(a) $3x - 2y = 7$ is linear in x and linear in y.

(b) $3y^2 + 7xy = 6xz$ is linear in x and in z, but *not* linear in y.

(c) $(x + 2)(x - 3) = x$ is not linear in x. It's quadratic, because x is raised to the second degree.   ◆

The graph of a linear function $f(x) = ax + b$ is a line. If $a \neq 0$, then $f(x) = K$ (where K is a constant) must have exactly one solution.

Solving.   Suppose an equation is linear in x or can be converted to such an equation. Solve for x as follows.

1. Clean up the equation if necessary; if x is inside parentheses, remove the parentheses. If x is in the denominator, clear the denominator.

2. Get all terms with x's on one side of the equation and all terms without x's on the other side: _____ $x +$ _____ $x +$ _____ $x =$ _____ .

3. Isolate x: Factor out the x and then divide both sides by the coefficient of x.

If x was in the denominator of the original equation, the equation is not really linear; however, it appears linear after clearing denominators. Be sure to check that your solution doesn't make the denominator zero.

◆ EXAMPLE A.25

Solve the following equations for x.

(a) $2\beta x - 5(x - \beta) = \gamma + 3x\,\alpha^2$

(b) $\sqrt{y}x + z^2(x - 1) = xy^3 + 2$

(c) $\frac{2}{x+1} = \frac{1}{x}$

(d) $\frac{2}{x} + \frac{m}{2} + \frac{m}{x}(2 - m) = 3m$

SOLUTION   (a) Solve for x in terms of α, β, and γ.

$$2\beta x - 5(x - \beta) = \gamma + 3x\alpha^2$$
$$2\beta x - 5x + 5\beta = \gamma + 3x\alpha^2$$
$$2\beta x - 5x - 3x\alpha^2 = -5\beta + \gamma$$
$$x(2\beta - 5 - 3\alpha^2) = -5\beta + \gamma$$
$$x = \frac{-5\beta + \gamma}{2\beta - 5 - 3\alpha^2}$$

(b) Solve for x in terms of y and z.

$$\sqrt{y}x + z^2(x-1) = xy^3 + 2$$
$$\sqrt{y}x + z^2x - z^2 = xy^3 + 2$$
$$\sqrt{y}x + z^2x - y^3x = 2 + z^2$$
$$x(\sqrt{y} + z^2 - y^3) = 2 + z^2$$
$$x = \frac{2 + z^2}{\sqrt{y} + z^2 - y^3}$$

(c) Multiply both sides of the equation by $x(x+1)$ to clear the denominators. This is fine as long as $x \neq 0$ and $x \neq -1$.

$$\frac{x(x+1)}{1} \cdot \frac{2}{x+1} = \frac{1}{x} \cdot \frac{x(x+1)}{1}$$
$$2x = x + 1$$
$$x = 1$$

(d) Solve for x in terms of m: $\frac{2}{x} + \frac{m}{2} + \frac{m}{x}(2-m) = 3m$.

First we'll multiply both sides by $2x$ to clear all denominators. This is fine as long as $x \neq 0$.

$$2x\left[\frac{2}{x} + \frac{m}{2} + \frac{m(2-m)}{x}\right] = 3m \cdot 2x$$
$$4 + mx + 2m(2-m) = 6mx$$
$$mx - 6mx = -4 - 2m(2-m)$$
$$x(m - 6m) = -4 - 2m(2-m)$$
$$x = \frac{-4 - 4m + 2m^2}{m - 6m}$$
$$= \frac{-4 - 4m + 2m^2}{-5m} \quad \blacklozenge$$

◆ **EXAMPLE A.26**   Solve for x: $\frac{4}{x-2} + 2 = \frac{5x-6}{x-2}$.

SOLUTION   Clear the denominator by multiplying both sides of the equation by $(x-2)$. This is fine provided $x \neq 2$.

$$(x-2)\left[\frac{4}{x-2} + 2\right] = \left[\frac{5x-6}{x-2}\right](x-2)$$
$$4 + 2(x-2) = 5x - 6$$
$$4 + 2x - 4 = 5x - 6$$
$$2x = 5x - 6$$
$$-3x = -6$$
$$x = 2$$

You might be lulled into thinking that the solution to this equation is $x = 2$, but if you check it by substituting $x = 2$ into the original equation you see that we have undefined terms, so $x = 2$ is *not* a solution.

What happened here? When we multiplied both sides of the equation by $x - 2$ we made the provision that $x \neq 2$ (otherwise we are multiplying both sides of an equation by zero). When we come up with $x = 2$ we are left to conclude that there are no solutions to the original equation.[5] ◆

Quadratic Equations

An equation is **quadratic in** x if it can be put into the form $ax^2 + bx + c = 0$, where a and b are constants or expressions not containing x, and $a \neq 0$.

$x^2 + bx + c = a$, $(x + 1)(x + 2) = 3x^2 + 5$, and $x^2 y^3 + 7xy = x$ are all quadratic in x.

The methods that worked for solving linear equations fall apart with quadratics. For example, suppose we want to solve $x^2 = -2x - 1$. We can get all x's on one side,

$$x^2 + 2x = -1,$$

but we don't have a constant times x on the left-hand side. If we factor out at x, then

$$x(x + 2) = -1$$

$$x = \frac{-1}{x + 2}.$$

This is *not* solved. We've got x on both sides.

How many solutions do we expect for the equation $ax^2 + bx + c = 0$? The graph of the quadratic function $f(x) = ax^2 + bx + c$ is a parabola, opening upward if $a > 0$, downward if $a < 0$; the location of the turning point [vertex] and the steepness are determined by a, b, and c.

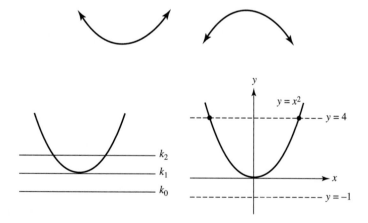

Figure A.2

---

[5] Here's another common way of understanding why there are no solutions to this equation. Getting a common denominator on the left-hand side, we have $\frac{4 + 2x - 4}{x + 2}$, or $\frac{2x}{x - 2}$. On the right-hand side, we have $\frac{5x - 6}{x - 2}$, or $\frac{2x}{x - 2} + \frac{3x - 6}{x - 2}$, or $\frac{2x}{x - 2} + 3 \frac{(x - 2)}{x - 2}$. The original equation can be written $\frac{2x}{x - 2} = \frac{2x}{x - 2} + 3 \frac{(x - 2)}{(x - 2)}$. This equation can only hold if $3 \frac{(x - 2)}{(x - 2)} = 0$. But this is impossible; $3 \frac{(x - 2)}{(x - 2)}$, if it is defined, must be equal to 3.

We see that unlike the case of $ax + b = k$, the equation $ax^2 + bx + c = k$ may have 0, 1, or 2 solutions.

As a concrete example look at $g(x) = x^2$. Solve $g(x) = -1$, $g(x) = 0$, and $g(x) = 9$.

(a) $x^2 = -1$. No solutions (x^2 is always nonnegative).

(b) $x^2 = 0 \Longrightarrow x = 0$. One solution.

(c) $x^2 = 9 \Longrightarrow x = \pm 3$. Two solutions.

Below is a sequence of examples leading to the quadratic formula.[6]

◆ **EXAMPLE A.27**   Solve.   (a) $x^2 = 5$ (b) $x^2 + 6x + 9 = 5$ (c) $x^2 + 6x + 4 = 0$

SOLUTION   (a)
$$x^2 = 5$$
$$x = \pm\sqrt{5}$$

(b) Express the left-hand side as a perfect square.

$$x^2 + 6x + 9 = (x + 3)^2$$
$$(x + 3)^2 = 5$$
$$x + 3 = \pm\sqrt{5}$$
$$x = -3 \pm \sqrt{5}$$

(c)
$$x^2 + 6x + 4 = 0$$
$$x^2 + 6x = -4$$

This is actually equivalent to part (b), presented in a different form. We want to write the left-hand side as a perfect square so that we can then solve for x by taking the square root of both sides as in Examples A.27(a) and (b). In order to achieve this, we will add an appropriate constant to both sides of the equation. This process is called *completing the square*.

How can we write $x^2 + 6x +$ _____ as a perfect square? k below is a constant. We'll solve for k; this will help us to figure out what to add to both sides of the equation.

$$x^2 + 6x + \underline{\quad} = (x + k)^2$$
$$x^2 + 6x + \underline{\quad} = x^2 + 2kx + k^2$$

So $2k$ must be equal to 6.

$$k = 3$$
$$(x + 3)^2 = x^2 + 6x + 9$$

So add 9 to both sides of the equation.

$$x^2 + 6x + 9 = -4 + 9$$
$$(x + 3)^2 = 5$$
$$x = -3 \pm \sqrt{5} \quad ◆$$

---

[6] Recall that $\sqrt{5}$ denotes, *by definition*, the positive square root of 5 ($\sqrt{5} = 2.236\ldots$). To denote the negative square root of 5, we must write $-\sqrt{5}$. A common error is to use $\sqrt{5}$ to denote both the positive and the negative square roots; this is *not* valid.

General Case: Deriving the Quadratic Formula   Solve for x: $ax^2 + bx + c = 0$

$$ax^2 + bx = -c$$

$$x^2 + \frac{b}{a}x = -\frac{c}{a}$$

We want to write the left-hand side as a perfect square and are perfectly willing to add the appropriate constant to both sides in order to accomplish this feat.

$$x^2 + \frac{b}{a}x + \underline{\quad} = (x + k)^2$$

$$= x^2 + 2kx + k^2$$

$$2k = \frac{b}{a}$$

$$k = \frac{b}{2a}$$

$$\left(x + \frac{b}{2a}\right)^2 = x^2 + \frac{b}{a}x + \frac{b^2}{4a^2}.$$

So we must add $\frac{b^2}{4a^2}$ to both sides of the equation.

$$x^2 + \frac{b}{a}x + \frac{b^2}{4a^2} = -\frac{c}{a} + \frac{b^2}{4a^2}$$

$$\left(x + \frac{b}{2a}\right)^2 = \frac{b^2 - 4ac}{4a^2}$$

$$x + \frac{b}{2a} = \pm\sqrt{\frac{b^2 - 4ac}{4a^2}}$$

$$x + \frac{b}{2a} = \pm\frac{\sqrt{b^2 - 4ac}}{2a}$$

$$x = \frac{-b}{2a} \pm \frac{\sqrt{b^2 - 4ac}}{2a}$$

$$x = \frac{-b \pm \sqrt{b^2 - 4ac}}{2a}$$

We've derived the quadratic formula by completing the square for a general quadratic equation.

We've shown that if $ax^2 + bx + c = 0$, then

$$\boxed{x = \frac{-b \pm \sqrt{b^2 - 4ac}}{2a}}$$

If $b^2 - 4ac > 0$, we've got a positive and a negative square root so the equation has two solutions.

If $b^2 - 4ac = 0$, there's one solution: $\left(x = -\frac{b}{2a}\right)$.

If $b^2 - 4ac < 0$, we have the square root of a negative number, so there are no real solutions.[7]

Using the Quadratic Formula to Solve Quadratics

Key step: Write the equation in the form $ax^2 + bx + c = 0$.

$$a = \text{coefficient of } x^2 \text{ term}$$
$$b = \text{coefficient of } x \text{ term}$$
$$c = \text{constant (i.e., any term or terms with no } x\text{'s)}$$

Then use the quadratic formula.

◆ **EXAMPLE A.28**   Solve for x.

(a) $x^2 + 1 = 2x^2 - 2x$ (b) $xy + x + 3yx^2 = 7y$ (c) $(x - 5)^2 - 1 = 0$

SOLUTION   (a) $x^2 + 1 = 2x^2 - 2x$.

$$x^2 + 1 = 2x^2 - 2x$$
$$0 = x^2 - 2x - 1$$

Then $a = 1$, $b = -2$, and $c = -1$, so

$$x = \frac{+2 \pm \sqrt{4 - 4(1)(-1)}}{2}$$
$$= \frac{2 \pm \sqrt{8}}{2} = \frac{2 \pm 2\sqrt{2}}{2} = \frac{2(1 \pm \sqrt{2})}{2}$$
$$= 1 \pm \sqrt{2}.$$

(b) Solve for x in terms of y. Treat y as a constant.

$$xy + x + 3yx^2 = 7y$$
$$3yx^2 + xy + x - 7y = 0$$
$$3yx^2 + (y + 1)x - 7y = 0$$

Then $a = 3y$, $b = y + 1$, and $c = -7y$, so

$$x = \frac{-(y + 1) \pm \sqrt{(y + 1)^2 - 4 \cdot 3y \cdot (-7y)}}{2 \cdot 3y}$$
$$= \frac{-y - 1 \pm \sqrt{y^2 + 2y + 1 + 84y^2}}{6y}$$
$$= \frac{-y - 1 \pm \sqrt{85y^2 + 2y + 1}}{6y}.$$

(c) Solve for x: $(x - 5)^2 - 1 = 0$.

---

[7] The square of any real number, x^2, is nonnegative.

Easier than using the quadratic formula is the following:

$$(x - 5)^2 = 1$$
$$x - 5 = \pm 1$$
$$x = 5 \pm 1 = 4 \text{ or } 6.$$

Alternatively,

$$x^2 - 10x + 25 - 1 = 0$$
$$x^2 - 10x + 24 = 0$$
$$(x - 6)(x - 4) = 0.$$

So $x = 6$ or $x = 4$.   ◆

Solving a Quadratic Equation by Factoring

Given $ax^2 + bx + c = 0$, if the left-hand side factors, we can use the fact that

> If $A \cdot B = 0$, then $A = 0$ and/or $B = 0$.[8]

◆ **EXAMPLE A.29**   Solve for x.   (a) $3x^2 = x$ (b) $x^2 - 5x = -4$

SOLUTION   (a) *Note:* We can't divide by x because x could be zero.

$$3x^2 = x$$
$$3x^2 - x = 0$$
$$x(3x - 1) = 0$$
$$x = 0 \quad \text{or} \quad 3x = 1$$
$$x = 0 \quad \text{or} \quad x = \frac{1}{3}$$

Notice that we would have lost the solution $x = 0$ had we made the mistake of dividing by x.

(b) Instead of using the quadratic formula we can factor.

$$x^2 - 5x + 4 = 0$$
$$(x - 4)(x - 1) = 0$$
$$x = 4 \quad \text{or} \quad x = 1.$$ ◆

How to Solve a Quadratic Equation.   If the equation is quadratic in x, we put the equation in the form $ax^2 + bx + c = 0$ and

---

[8] *Note:* $AB = 2$ does *not* imply $A = 2$ or $B = 2$; for example, we could have $\frac{1}{2} \cdot 4 = 2$ or $(-1)(-2) = 2$. Only if the product of numbers is zero can we draw a conclusion.

1. use the quadratic formula,

$$x = \frac{-b \pm \sqrt{b^2 - 4ac}}{2a}, \qquad \text{or}$$

2. factor into two linear factors (if possible). If $A \cdot B = 0$, then $A = 0$ and/or $B = 0$.[9]

Higher Degree Equations

Higher degree equations are usually much harder to solve algebraically. There exist formulas for solving cubic and fourth degree equations—but they are quite messy. Amazingly enough, it can be *proven* that such a formula for equations of degree five *does not exist*.[10]

If we are working with a higher order polynomial equation presented to us in the form of a product of linear or quadratic factors, then we can use the approach that if $A \cdot B \cdot C = 0$, then $A = 0$ or $B = 0$ or $C = 0$.

EXAMPLE A.30   Solve $(x^6 - x^4)(x^2 + 2x + 1)(2x^2 - x - 1) = 0$.

SOLUTION   First factor completely.

$$(x^6 - x^4)(x^2 + 2x + 1)(2x^2 - x - 1) = 0$$
$$x^4(x^2 - 1)(x + 1)^2(2x + 1)(x - 1) = 0$$
$$x^4(x - 1)(x + 1)(x + 1)^2(2x + 1)(x - 1) = 0$$
$$x^4(x - 1)^2(x + 1)^3(2x + 1) = 0$$

So

$x^4 = 0$		$(x - 1)^2 = 0$		$2x + 1 = 0$		$(x + 1)^3 = 0$
$x = 0$	or	$x - 1 = 0$	or	$2x = -1$	or	$x + 1 = 0$
$x = 0$		$x = 1$		$x = -\frac{1}{2}$		$x = -1$

Suppose we are not handed our equation in factored form. Then the situation is more difficult—but we can sometimes make progress.

EXAMPLE A.31   Solve $x^3 - 3x = -x + 1$.

SOLUTION $x^3 - 2x + 1 = 0$

There are at most three solutions to any cubic and at least one. Think about the graph of $y = x^3 - 2x + 1$; for x large enough y is positive, and for x negative enough y is negative. A cubic is continuous, therefore the graph must cut the x-axis at least once. On the other hand, a third degree equation can have at most three roots. We could try some numbers in an educated hunt-and-peck, hope-we-luck-out manner.

Let's first look at positive x. If x is large enough, x^3 outweighs $-2x$ and $x^3 - 2x + 1$ is positive. If $x = 0$, $x^3 - 2x + 1 = 1$, again positive. Perhaps if x is small enough, $-2x$

---

can outweigh $x^3 + 1$. Try $x = 1$:

$$1^3 - 2(1) + 1 = 1 - 2 + 1 = 0.$$

Eureka![11] We've been lucky. Now we're in business—either we can consider this graphically to see if there are two more roots or we can use the following fact:

> If $P(a) = 0$, then $(x - a)$ is a factor of $P(x)$.

So $x^3 - 2x + 1 = (x - 1) \cdot q(x)$, where $q(x)$ is some quadratic. We can find $q(x)$ by long division:

$$q(x) = \frac{x^3 - 2x + 1}{x - 1} = x^2 + x - 1.$$

(See the section on long division of polynomials for this.)

So $x^3 - 2x + 1 = (x - 1)(x^2 + x - 1)$. We must solve

$$x^3 - 2x + 1 = 0$$
$$(x - 1)(x^2 + x - 1) = 0$$
$$x = 1 \text{ or } x^2 + x - 1 = 0$$

Use the quadratic formula.

$$x = \frac{-1 \pm \sqrt{1 - 4(1)(-1)}}{2} = -\frac{1 \pm \sqrt{5}}{2}$$

$x = 1$, or $x = \frac{-1+\sqrt{5}}{2}$, or $x = \frac{-1-\sqrt{5}}{2}$.

An alternative initial attack: Given a graphing calculator, you can use that to help you. A cubic looks like one of the figures below with the turning points possibly taken out. Using a graphing calculator you can estimate the roots. You might even guess that one of the roots is exactly 1; you couldn't be sure until you tried it and saw that it actually worked exactly. As for the other two roots—if you simply want numerical approximations, a calculator is great. By zooming in you can get increasingly more accurate numerical estimates. Depending upon the sophistication of your calculator, you may get exact answers. Try using your equation solver to test this out.

---

[11] Legend has it that this phrase was coined by Archimedes of Syracuse, 287–212 B.C.E. While sitting in the bathtub, he realized that the amount of weight lost by a floating body was precisely equal to the weight of the amount of water displaced, solving a problem he had been given by the king. He was so excited he ran through the streets of Syracuse naked and dripping wet, shouting "*Εuρήκα!*" which means "I have found it!" We do not recommend such displays of exuberance.

Disguised Quadratics

◆ **EXAMPLE A.32** Solve for x in the following equations.

(a) $x^2(x^2 + 3) = 4$ (b) $x^2 + \frac{6}{x^2} = 5$

SOLUTION (a) Having the left-hand side factored when the right-hand side is not zero is not useful! Multiply it out:

$$x^4 + 3x^2 = 4.$$

Let $u = x^2$. Then

$$u^2 + 3u - 4 = 0$$

or equivalently, $x^4 + 3x^2 - 4 = 0$.

$$(u + 4)(u - 1) = 0$$

or $(x^2 + 4)(x^2 - 1) = 0$. Therefore,

$$
\begin{array}{ccc}
u = -4 & \text{or} & u = 1 \\
x^2 = -4 & \text{or} & x^2 = 1 \\
\text{impossible} & & x = \pm 1
\end{array}
$$

(b) Clear the denominator; multiply by x^2 on both sides. Make sure that $x = 0$ is not a solution.

$$x^2 + \frac{6}{x^2} = 5$$
$$x^4 + 6 = 5x^2$$
$$x^4 - 5x^2 + 6 = 0$$

Let $u = x^2$. Then

$$u^2 - 5u + 6 = 0$$
$$(u - 3)(u - 2) = 0$$

or equivalently, $(x^2 - 3)(x^2 - 2) = 0$.

$$
\begin{array}{ccc}
u = 3 & \text{or} & u = 2 \\
x^2 = 3 & \text{or} & x^2 = 2 \\
x = \pm\sqrt{3} & \text{or} & x = \pm\sqrt{2}
\end{array} \quad ◆
$$

Equations with Radicals

The procedure for solving an equation with radicals (e.g., square roots) is to isolate and then eliminate the radical. Consider the equation $\sqrt{x + 1} = x - 5$. To eliminate the square root, we square both sides of the equation. Then we'll have a quadratic and can solve as in the previous section.

◆ **EXAMPLE A.33** Solve for x: $x - \sqrt{x + 1} - 5 = 0$.

SOLUTION

$$x - \sqrt{x + 1} - 5 = 0$$
$$-\sqrt{x + 1} = -x + 5$$
$$\sqrt{x + 1} = x - 5$$

Notice that we need $x \geq 5$ for this equation to have a solution.

The radical is isolated. Square both sides. (This gets rid of the radical but may introduce extraneous roots. We'll have to check our answers.)

$$\left(\sqrt{x+1}\right)^2 = (x-5)^2$$

$$x + 1 = x^2 - 10x + 25$$

We have obtained a quadratic.

$$x^2 - 11x + 24 = 0$$

$$(x - 8)(x - 3) = 0$$

$$x = 8 \text{ or } x = 3$$

Check: $x = 8$:

$$\sqrt{8+1} \stackrel{?}{=} 8 - 5$$

$$\sqrt{9} \stackrel{?}{=} 3$$

$$3 = 3.$$

Yes, $x = 8$ is a solution. Check: $x = 3$:

$$\sqrt{3+1} \stackrel{?}{=} 3 - 5$$

$$\sqrt{4} \stackrel{?}{=} -2$$

$$2 \neq -2.$$

No, $x = 3$ is *not* a solution.

$x = 8$ is the only solution. ◆

As so often happens in science and mathematics, one question leads to another question. In this last example the question is, "Why did we get an answer that didn't work?" Notice that if we didn't check we wouldn't have realized that $x = 3$ did not work. To see what has gone on, we want to investigate squaring both sides with a simple example.

Demonstration:

$$x = -5$$

$$x^2 = (-5)^2$$

$$x^2 = 25$$

$$x = \pm\sqrt{25} = \pm 5.$$

Whoops. We've obtained an answer that doesn't work in the original equation.

The process of squaring both sides of an equation can give "answers" that don't work. These are called **extraneous roots**. The reason we obtain an extraneous root in the problem is that the squaring process destroys sign information. The antidote is to check your answers.

◆ **EXAMPLE A.34**   Solve $2\sqrt{x} - \sqrt{x+1} = 1$.

SOLUTION   We can't isolate both radicals simultaneously, so we isolate one of them.

$$2\sqrt{x} - 1 = \sqrt{x+1}$$

Square both sides to eliminate the radical on the right.

$$(2\sqrt{x} - 1)(2\sqrt{x} - 1) = \left(\sqrt{x+1}\right)^2$$

$$4x - 2\sqrt{x} - 2\sqrt{x} + 1 = x + 1$$

$$4x - 4\sqrt{x} = x$$

$$3x = 4\sqrt{x}$$

Square both sides again to eliminate the second radical.

$$9x^2 = 16x$$

Now we have a quadratic equation.

CAUTION We can't divide by x because x may be zero. Instead, solve by factoring.

$$9x^2 - 16x = 0$$

$$x(9x - 16) = 0$$

$$x = 0 \quad \text{or} \quad x = \frac{16}{9}$$

Check: $x = 0$:

$$2\sqrt{0} - 1 \stackrel{?}{=} \sqrt{0+1}$$

$$-1 \neq 1.$$

$x = 0$ is *not* a solution.

$x = \frac{16}{9}$:

$$2\sqrt{\frac{16}{9}} - 1 \stackrel{?}{=} \sqrt{\frac{16}{9} + 1}$$

$$2\left(\frac{4}{3}\right) - 1 \stackrel{?}{=} \sqrt{\frac{16}{9} + \frac{9}{9}}$$

$$\frac{8}{3} - 1 \stackrel{?}{=} \sqrt{\frac{25}{9}}$$

$$\frac{8}{3} - \frac{3}{3} \stackrel{?}{=} \frac{5}{3}$$

$$\frac{5}{3} = \frac{5}{3}.$$

$x = \frac{16}{9}$ is a solution.

$x = 16/9$ is the only solution.   ◆

MORAL When solving equations:

■   You can multiply/divide both sides of the equation by anything *except zero*.

If you multiply both sides of an equation by an expression that could be zero, you will introduce extraneous roots. This necessitates checking your answer(s). See Example A.26.

If you divide both sides of an equation by an expression that could be zero you will lose that as a root.

■   If you square both sides of an equation,[12] you may introduce extraneous roots. Therefore you must check your answers (either numerically or graphically). See Example A.33.

Solving an Equation in x:

Can you identify the equation as linear, quadratic, or radical in x?

If it is *linear* in x, get all terms with x's on one side; all else on the other. Factor out x and divide both sides by the coefficient of x.

If it is *quadratic* in x, get everything on one side and zero on the other. Either use the quadratic formula or factor into two linear factors and set each factor equal to zero.

If it is *radical* in x, isolate the radical and then eliminate it by squaring. Check for extraneous roots.

If it is a *higher degree polynomial*, you may be lucky or you may not. Can you factor? Guess a root? Is it a quadratic in disguise? The number of roots is at most the degree of the equation.

PROBLEMS FOR SECTION A.3

1. Find *all* y for which $y^3 = y^2$.

2. Solve the equation

$$\frac{\lambda}{1+x} + \frac{2}{\lambda} = \frac{1}{\lambda\beta}$$

for the variable indicated.

   (a) β

   (b) x

   (c) λ

3. Let $f(x) = x^2$ and $g(x) = 1 + x$. Find all w such that $f\big(g(w)\big) - 2g(w) = 0$. (You may find it simplest to solve for $u = 1 + w$ and then find w.)

4. Solve for y. Find *all* y satisfying the equation.

   (a) $y(2y - 3) = 5$

   (b) $y(2y - 3) = -5$

   (c) $y(y - 6) = 9$

   (d) $y(y - 6) = -9$

---

[12] Raise both sides of an equation to any even power can introduce extraneous roots.

(e) $\frac{y}{y-6} = -9$

(f) $\frac{1}{y-1} = \frac{3}{y}$

5. Let $f(x) = \sqrt{x}$ and $g(x) = x + 1$. Solve the following equations.

(a) $f(g(w)) = g(w)$

(b) $f(x^2 + 1) - g(x + 1) = 0$

(c) $\frac{1}{f(x)} + 3x = \frac{2}{f(x)}$

6. Solve for the indicated variables.

(a) $w(z + w) = z$ (z)

(b) $w(z + w) = z$ (w)

(c) $\frac{z+w}{w} = z$ (z)

(d) $\frac{z+w}{w} = z$ (w)

(e) $z^3 + 3z^2 + 2z = 0$ (z)

(f) $z^4 + 3z^2 + 2 = 0$ (z)

(g) $z^5 = 16z$ (z)

B

Geometric Formulas

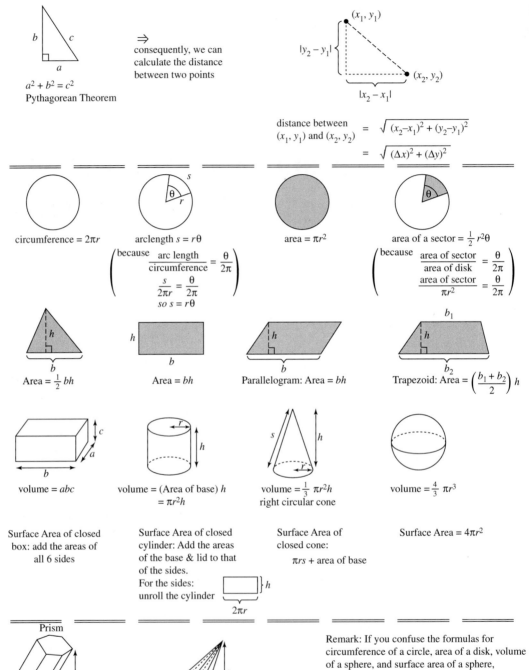

$a^2 + b^2 = c^2$
Pythagorean Theorem

$\Rightarrow$ consequently, we can calculate the distance between two points

$$\text{distance between} \atop (x_1, y_1) \text{ and } (x_2, y_2) = \sqrt{(x_2-x_1)^2 + (y_2-y_1)^2}$$

$$= \sqrt{(\Delta x)^2 + (\Delta y)^2}$$

circumference $= 2\pi r$

arclength $s = r\theta$

$$\left(\text{because} \quad \frac{\text{arc length}}{\text{circumference}} = \frac{\theta}{2\pi}\right.$$
$$\frac{s}{2\pi r} = \frac{\theta}{2\pi}$$
$$\left. so \; s = r\theta\right)$$

area $= \pi r^2$

area of a sector $= \frac{1}{2} r^2 \theta$

$$\left(\text{because} \quad \frac{\text{area of sector}}{\text{area of disk}} = \frac{\theta}{2\pi}\right.$$
$$\left. \frac{\text{area of sector}}{\pi r^2} = \frac{\theta}{2\pi}\right)$$

Area $= \frac{1}{2} bh$

Area $= bh$

Parallelogram: Area $= bh$

Trapezoid: Area $= \left(\frac{b_1 + b_2}{2}\right) h$

volume $= abc$

volume $=$ (Area of base) h
$\quad = \pi r^2 h$

volume $= \frac{1}{3} \pi r^2 h$
right circular cone

volume $= \frac{4}{3} \pi r^3$

Surface Area of closed box: add the areas of all 6 sides

Surface Area of closed cylinder: Add the areas of the base & lid to that of the sides.
For the sides: unroll the cylinder

$2\pi r$

Surface Area of closed cone:

$\pi r s +$ area of base

Surface Area $= 4\pi r^2$

Prism

volume $=$ (Area of base)h

volume $= \frac{1}{3}$ (Area of base)h

Remark: If you confuse the formulas for circumference of a circle, area of a disk, volume of a sphere, and surface area of a sphere, it may be useful to notice that the 1-dimensional measure has r^1, the two-dimensional measure has r^2, and the 3-dimensional measure has r^3.

Figure B.1

C

The Theoretical Basis of Applications of the Derivative

In this section we will give the theoretical underpinnings of many of the statements we have been taking as fact without proof. We give proofs of statements 1 and 3–7 that follow. They have been numbered, as the proof of each rests on the veracity of previous statements. (Statements 5, 6, and 7 follow from the Mean Value Theorem (statement 4), whose proof rests on statement 3, whose proof depends on statements 1 and 2.

Theorem 1 deals with an arbitrary function; Theorem 2 deals with continuous functions. Theorems 3 through 7 deal with functions that are

(a) continuous on $[a, b]$, and

(b) differentiable on (a, b).

These five theorems refer to functions whose graphs have no holes, no jumps, and no sharp corners.

1. **Local Extremum Theorem.** If a function f has a local maximum or local minimum at $x = c$, then either $f'(c) = 0$ or $f'(c)$ does not exist.

2. **Extreme Value Theorem.** If f is continuous on a closed interval $[a, b]$, then f attains an absolute maximum and absolute minimum value on $[a, b]$.

3. **Rolle's Theorem.** Suppose f is continuous on $[a, b]$ and differentiable on (a, b). Then if $f(a) = f(b)$, there exists some number $c \in (a, b)$ such that

$$f'(c) = 0.$$

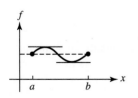

4. **Mean Value Theorem.** Suppose f is continuous on $[a, b]$ and differentiable on (a, b). Then there exists some number $c \in (a, b)$ such that

$$f'(c) = \frac{f(b) - f(a)}{b - a}.$$

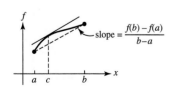

slope $= \frac{f(b) - f(a)}{b - a}$

1087

5. **Zero Derivative Theorem.** Suppose f is continuous on $[a, b]$ and differentiable on (a, b). If $f'(x) = 0$ for all $x \in (a, b)$, then f is constant on (a, b).

6. **Equal Derivatives Theorem.** Suppose f is continuous on $[a, b]$ and differentiable on (a, b). If $f'(x) = g'(x)$ for all $x \in (a, b)$, then $f(x)$ and $g(x)$ differ by a constant on (a, b). That is, there exists a constant C such that

$$f(x) = g(x) + C$$

for all $x \in (a, b)$.

7. **Increasing / Decreasing Function Theorem.** Suppose f is continuous on $[a, b]$ and differentiable on (a, b).

If $f'(x) > 0$ for all $x \in (a, b)$, then f is increasing on $[a, b]$;

if $f'(x) < 0$ for all $x \in (a, b)$, then f is decreasing on $[a, b]$.

Now let's take up the job of proving these theorems.

1. *Local Extremum Theorem:* If a function f has a local maximum or local minimum at $x = c$, then either $f'(c) = 0$ or $f'(c)$ does not exist.

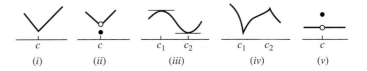

Figure C.1

Proof.   Suppose that f has a local maximum at $x = c$. Then for all x sufficiently close to c, $f(x) \le f(c)$.

Therefore, for h sufficiently small,

$$f(c + h) \le f(c)$$

and

$$f(c + h) - f(c) \le 0.$$

Now we want to divide both sides of the inequality by h for h sufficiently small but nonzero. We must treat the cases $h > 0$ and $h < 0$ separately; dividing by a positive number preserves the inequality while dividing by a negative number reverses it.[1]

Case 1: $h > 0$

$$\frac{f(c + h) - f(c)}{h} \le 0$$ On both sides of the inequality compute the limit as h approaches zero from the right.

$$\lim_{h \to 0^+} \frac{f(c + h) - f(c)}{h} \le \lim_{h \to 0^+} 0$$

$$\lim_{h \to 0^+} \frac{f(c + h) - f(c)}{h} \le 0$$

---

[1] $3 < 4$ but $-3 > -4$. Multiplying (or dividing) both sides of an inequality by a negative number reverses the inequality.

Case 2: $h < 0$

$$\frac{f(c+h) - f(c)}{h} \geq 0$$

On both sides of the inequality compute the limit as h approaches zero from the left.

$$\lim_{h \to 0^-} \frac{f(c+h) - f(c)}{h} \geq \lim_{h \to 0^-} 0$$

$$\lim_{h \to 0^-} \frac{f(c+h) - f(c)}{h} \geq 0$$

If $f'(c)$ exists, then

$$f'(c) = \lim_{h \to 0^+} \frac{f(c+h) - f(c)}{h} = \lim_{h \to 0^-} \frac{f(c+h) - f(c)}{h}.$$

$f'(c) \geq 0$ *and* $f'(c) \leq 0$, so we conclude that $f'(c) = 0$ if $f'(c)$ exists.

We have shown that the Local Extremum Theorem is true if f has a local maximum at $x = c$. The argument in the case of a local minimum at $x = c$ is analogous.

REMARK We have seen that neither the condition that $f'(c) = 0$ nor that $f'(c)$ is undefined is alone enough to guarantee a local extremum at $x = c$.

2. The *Extreme Value Theorem* is stated without proof. The proof is involved; please consult a more theoretical text for the proof.

3. *Rolle's Theorem*:[2] Suppose f is continuous on $[a, b]$ and differentiable on (a, b). Then if $f(a) = f(b)$, there exists some number $c \in (a, b)$ such that

$$f'(c) = 0.$$

(See Figure C.2 below for illustrations. Basically, either f is constant on $[a, b]$, in which case $f'(x) = 0$ on (a, b), or the graph of f must turn at least once in order to end up at the same height at which it began. At any turning point c, $f'(c)$ will be zero.

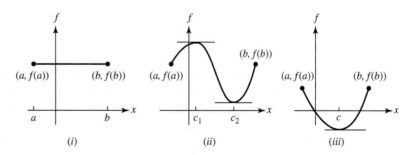

(i) $\qquad\qquad$ (ii) $\qquad\qquad$ (iii)

Figure C.2

Proof. Either f is constant on $[a, b]$ or it is not.

If f is constant on $[a, b]$, then $f'(x) = 0$ on (a, b) so c can be chosen to be anywhere on the interval (a, b).

---

[2] Rolle's Theorem is named after the French mathematician Michel Rolle, who published the result around 1697.

Suppose f is not constant on $[a, b]$. The Extreme Value Theorem says that if f is continuous on a closed interval $[a, b]$, then f attains an absolute maximum value and an absolute minimum value on $[a, b]$. Since $f(a) = f(b)$, either the absolute maximum or the absolute minimum or both are attained on the open interval (a, b). Therefore there is some $c \in (a, b)$ such that f has a local extremum at $x = c$. Since f is differentiable on (a, b), by the Local Extremum Theorem $f'(c) = 0$.

4. *Mean Value Theorem:* If the function f is continuous on $[a, b]$ and differentiable on (a, b), then there exists a number $c \in (a, b)$ such that

$$f'(c) = \frac{f(b) - f(a)}{b - a}.$$

The Mean Value Theorem was first stated in the late 1700s by the French/Italian mathematician Joseph Louis Lagrange more than a century after Newton and Leibniz did the bulk of their work.[3]

Geometrically, the Mean Value Theorem can be interpreted as telling us that between points $A = (a, f(a))$ and $B = (b, f(b))$ there is at least one point $C = (c, f(c))$ such that the slope of the tangent line through C is parallel to the secant line through A and B. See Figure C.3 below.

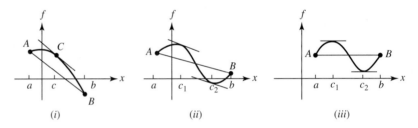

$$(i) \qquad\qquad (ii) \qquad\qquad (iii)$$

Figure C.3

Notice that Rolle's Theorem is a special case of the Mean Value Theorem.

For a practical interpretation, think of the word "mean" as the average. Then, if we let $f(x)$ be a displacement function, the Mean Value Theorem can be interpreted as telling us that there must be at least one point in the interval (a, b) such that the instantaneous velocity $f'(c)$ is the same as the average velocity on the interval $[a, b]$. This should strike you as quite reasonable; if a car averages 60 mph on the Massachusetts Turnpike (where its displacement function is both continuous and differentiable!), then at some instant the car must have been traveling at 60 mph.

Proof. We prove the Mean Value Theorem by concocting a new function, v, from f such that the values of v at a and b are equal and Rolle's Theorem can be applied to v. The function v is the vertical displacement function that gives the (signed) vertical displacement between a point $(x, f(x))$ on the graph of f and the secant line through $A = (a, f(a))$ and $B = (b, f(b))$.

---

[3] It was Lagrange who introduced the prime notation for derivatives.

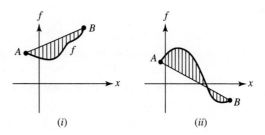

Figure C.4

The equation of the secant line through A and B is

$$y - f(a) = m(x - a), \quad \text{where} \quad m = \frac{f(b) - f(a)}{b - a}$$

or

$$y = \frac{f(b) - f(a)}{b - a}(x - a) + f(a).$$

Therefore the vertical displacement v is given by

$$v(x) = f(x) - \left[\frac{f(b) - f(a)}{b - a}(x - a) + f(a) \right] \quad \text{for} \quad x \in [a, b]$$

$$v(a) = f(a) - \left[\frac{f(b) - f(a)}{b - a}(a - a) + f(a) \right] = f(a) - f(a) = 0$$

$$v(b) = f(b) - \left[\frac{f(b) - f(a)}{b - a}(b - a) + f(a) \right] = f(b) - f(b) + f(a) - f(a) = 0.$$

Because f is continuous on $[a, b]$ and differentiable on (a, b), we know that v is continuous on $[a, b]$ and differentiable on (a, b). Therefore, we can apply Rolle's Theorem to v; there is a number $c \in (a, b)$ such that

$$v'(c) = 0.$$

We know that the derivative of a linear function is just the slope of the line, so

$$v'(x) = f'(x) - \frac{f(b) - f(a)}{b - a}.$$

Therefore,

$$v'(c) = f'(c) - \frac{f(b) - f(a)}{b - a}.$$

The statement $v'(c) = 0$ is equivalent to

$$f'(c) = \frac{f(b) - f(a)}{b - a}.$$

5. *Zero Derivative Theorem:* Suppose f is continuous on $[a, b]$ and differentiable on (a, b). If $f'(x) = 0$ for all $x \in (a, b)$, then f is constant on (a, b).

Proof. Label as x_1 and x_2 any two distinct numbers in the interval (a, b) where $x_1 < x_2$. f is differentiable on (a, b) so it must be differentiable on (x_1, x_2) and continuous on

$[x_1, x_2]$. Therefore, we can apply the Mean Value Theorem to f on $[x_1, x_2]$ to say that there exists $c \in (x_1, x_2)$ such that

$$f'(c) = \frac{f(x_2) - f(x_1)}{x_2 - x_1}.$$

But $f'(c) = 0$ by hypothesis ($f'(x) = 0$ for all $x \in (a, b)$), so

$$f(x_2) - f(x_1) = 0$$

$$f(x_1) = f(x_2).$$

Since x_1 and x_2 were chosen arbitrarily in (a, b), we conclude that $f(x)$ is constant throughout the interval (a, b).

6. *Equal Derivatives Theorem:* Suppose f is continuous on $[a, b]$ and differentiable on (a, b). If $f'(x) = g'(x)$ for all $x \in (a, b)$, then $f(x)$ and $g(x)$ differ by a constant on (a, b). That is, there exists a constant C such that

$$f(x) = g(x) + C$$

for all $x \in (a, b)$.

   Proof. Let $j(x) = f(x) - g(x)$.

$$j'(x) = f'(x) - g'(x) = 0 \quad \text{for all} \quad x \in (a, b).$$

Therefore, by the Zero Derivative Theorem, $j(x) = C$ for some constant C for all $x \in (a, b)$.

$$f(x) - g(x) = C$$

or

$$f(x) = g(x) + C \quad \text{for all} \quad x \in (a, b).$$

7. *Increasing/Decreasing Function Theorem:* Suppose f is continuous on $[a, b]$ and differentiable on (a, b).

   If $f'(x) > 0$ for all $x \in (a, b)$, then f is increasing on (a, b);
   if $f'(x) < 0$ for all $x \in (a, b)$, then f is decreasing on (a, b).

   Proof. We will show that if $f'(x) > 0$ for all $x \in (a, b)$, then f is increasing on (a, b).
   Label as x_1 and x_2 any two distinct numbers in the interval (a, b), where $x_1 < x_2$. Our goal is to show that

$$f(x_1) < f(x_2),$$

or equivalently,

$$f(x_2) - f(x_1) > 0.$$

To do this, we apply the Mean Value Theorem to f on the interval $[x_1, x_2]$. There exists a number $c \in (x_1, x_2)$ such that

$$f'(c) = \frac{f(x_2) - f(x_1)}{x_2 - x_1} \quad \text{or} \quad f'(c)(x_2 - x_1) = f(x_2) - f(x_1).$$

By hypothesis $f'(x) > 0$ on (a, b); therefore, $f'(c) > 0$. We know that $x_2 - x_1 > 0$ because $x_1 < x_2$. From sign analysis it follows that

$$f(x_2) - f(x_1) > 0.$$

f is increasing.

We leave it as an exercise to show that if $f'(x) < 0$ for all $x \in (a, b)$, then f is decreasing on (a, b).

D

Proof by Induction

Suppose we arrange a collection of dominos in a line so that if any one domino falls, we are sure that the domino next to it will also fall. Then, if we knock over the first domino in the line, all the rest of the dominos will also be knocked down. Proof by mathematical induction works under very much the same principle, except that it works with an infinite number of dominos.

Suppose we do some calculations and notice a pattern. We may wonder whether this pattern will hold indefinitely. For example, consider the sum of the first n odd numbers.

$$1 = 1$$
$$1 + 3 = 4$$
$$1 + 3 + 5 = 9$$
$$1 + 3 + 5 + 7 = 16$$
$$1 + 3 + 5 + 7 + 9 = 25$$
$$1 + 3 + 5 + 7 + 9 + 11 = 36$$

Do you notice a pattern? The numbers on the right are $1^2, 2^2, 3^2, 4^2, 5^2$, and 6^2. We might conjecture that the sum of the first n odd integers is n^2. But how can we prove this?

Let's call our conjecture C. This conjecture encompasses infinitely many assertions. We'll name them C_1, C_2, C_3, *et cetera*.

C_1: $1 = 1^2$

C_2: $1 + 3 = 2^2$

C_3: $1 + 3 + 5 = 3^2$

et cetera . . .

We will show that the conjecture is true using mathematical induction, working on the domino principle.

■ First we show that the statement holds for $n = 1$.

That is, we show C_1 is true.

We show that the first domino will fall.

■ Then we show that *if* C_k is true, where k is a positive integer, then C_{k+1} must be true as well.

We show that if any one domino falls, it will knock down the domino after it.

Then we have completed our proof. C_k holds true for all positive integers.

Let's apply it to our conjecture that the sum of the first n odd integers is n^2.

Proof. First, we show that C_1 holds. $1 = 1^2$. (Not hard to show!)

Now we show that *if* C_k is true, then C_{k+1} must be true.

Suppose that C_k is true. Then

$$1 + 3 + 5 + 7 + 9 + \cdots + (2k - 1) = k^2.$$

We must show that

$$1 + 3 + 5 + 7 + 9 + \cdots + (2k - 1) + (2(k + 1) - 1) = (k + 1)^2.$$

$$1 + 3 + 5 + 7 + 9 + \cdots + (2k - 1) + (2(k + 1) - 1)$$
$$= [1 + 3 + 5 + 7 + 9 + \cdots + (2k - 1)] + (2k + 2 - 1)$$
$$= [k^2] + (2k + 1) \quad \text{(by the induction hypothesis)}$$
$$= k^2 + 2k + 1$$
$$= (k + 1)^2$$

Therefore, if C_k is true, then C_{k+1} must be true. This completes the induction proof. The sum of the first n odd integers is n^2.

Below we give a second example.

◆ **EXAMPLE D.1**   Prove that

$$1^2 + 2^2 + 3^2 + 4^2 + \cdots + n^2 = \frac{n(n + 1)(2n + 1)}{6}.$$

Proof. We will show that this statement is true using mathematical induction.

First we'll show that the statement holds for $n = 1$.

$$1 \stackrel{?}{=} \frac{1(1 + 1)(2 \cdot 1 + 1)}{6}$$

$$1 \stackrel{?}{=} \frac{6}{6}$$

$$1 = 1 \quad \text{yes, indeed!}$$

Now we want to show that *if* the statement holds for $n = k$, then it must hold for $n = k + 1$. If we show this, then we know that because the statement holds for $n = 1$, it must hold for $n = 2$, and because it holds for $n = 2$, it must hold for $n = 3$, and so on, ad infinitum.

Therefore, we need to show that *if* $1^2 + 2^2 + 3^2 + 4^2 + \cdots + k^2 = \frac{k(k+1)(2k+1)}{6}$, then it follows that

$$1^2 + 2^2 + 3^2 + 4^2 + \cdots + k^2 + (k+1)^2 = \frac{(k+1)[(k+1)+1][2(k+1)+1]}{6}.$$

The right-hand side of this equation can be written as

$$\frac{(k+1)(k+2)(2k+3)}{6}.$$

By assumption,

$$1^2 + 2^2 + 3^2 + 4^2 + \cdots + k^2 = \frac{k(k+1)(2k+1)}{6}.$$

From this it follows that

$$
\begin{aligned}
1^2 + 2^2 + 3^2 + 4^2 + \cdots + k^2 + (k+1)^2 &= \frac{k(k+1)(2k+1)}{6} + (k+1)^2 \\
&= \frac{k(k+1)(2k+1)}{6} + \frac{6(k+1)^2}{6} \\
&= \frac{k(k+1)(2k+1) + 6(k+1)^2}{6} \\
&= \frac{(k+1)[k(2k+1) + 6(k+1)]}{6} \\
&= \frac{(k+1)[2k^2 + k + 6k + 6]}{6} \\
&= \frac{(k+1)[2k^2 + 7k + 6]}{6} \\
&= \frac{(k+1)(k+2)(2k+3)}{6}.
\end{aligned}
$$

We have proven what is known as the induction step. The statement holds for $n = 1$, so it holds for $n = 2$, and because it holds for $n = 2$, it holds for $n = 3$, and so on, $n = 1, 2, 3, 4 \ldots$. We have shown that

$$1^2 + 2^2 + 3^2 + 4^2 + \cdots + n^2 = \frac{n(n+1)(2n+1)}{6}$$

holds for n any positive integer. ◆

Later in our study of calculus we will find ourselves trying to calculate the area under the graph of $y = x^2$ on the interval $[0, 1]$ by evaluating the limit of a Riemann sum without using the Fundamental Theorem of Calculus. Just when we need it, the identity

$$1^2 + 2^2 + 3^2 + 4^2 + \cdots + n^2 = \frac{n(n+1)(2n+1)}{6}$$

will drop like manna from heaven. Below are some other bits of manna.

$$1 + 2 + 3 + 4 + \cdots + n = \frac{n(n+1)}{2}$$

$$1^3 + 2^3 + 3^3 + 4^3 + \cdots + n^3 = \frac{n^2(n+1)^2}{4}$$

They too can be proven by mathematical induction.

For the problems below, use mathematical induction to prove that each statement is true for all positive integers.

1. Prove

$$1 + 2 + 3 + 4 + \cdots + n = \frac{n(n+1)}{2}$$

for n any positive integer.

2. Prove

$$1^3 + 2^3 + 3^3 + 4^3 + \cdots + n^3 = \frac{n^2(n+1)^2}{4}$$

for n any positive integer.

3. Prove that the sum of the first n nonzero even integers is $n(n-1)$.

4. Prove

$$1 + 2 + 2^2 + 2^3 + \cdots + 2^n = 2^{n+1} - 1$$

for n any positive integer.

5. Prove

$$2(1 + 3 + 3^2 + 3^3 + \cdots + 3^n) = 3^{n+1} - 1$$

for n any positive integer.

6. Prove

$$\left(1 + \frac{1}{1}\right) \cdot \left(1 + \frac{1}{2}\right) \cdot \left(1 + \frac{1}{3}\right) \cdots \left(1 + \frac{1}{n}\right) = n + 1$$

for n any positive integer.

Conic Sections

Conic sections get their name from the fact that they can be thought of as the intersection of a plane with a pair of circular cones with vertices joined in an hourglass configuration. Depending upon the orientation of the plane relative to the cones we obtain a **circle**, an **ellipse**, a **parabola**, or a **hyperbola**.

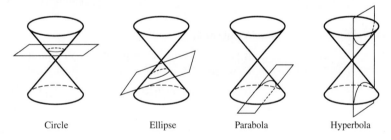

| Circle | Ellipse | Parabola | Hyperbola |

Figure E.1   Conic sections

If the plane passes through the vertex of the cones, then the resulting figure, a point, a line, or a set of two intersecting lines, is a "degenerate conic."

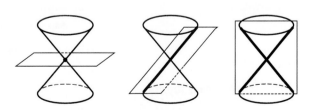

Figure E.2   Degenerate conics

We can describe conics in several ways: as "slices" of the cones, as a set of points with a given geometric property, or as the graph of the second degree equation in x and y of the form

$$Ax^2 + Bxy + Cy^2 + Dx + Ey + F = 0,$$

where A, B, C, D, E, and F are constants.

Below we present the characteristics of the conics from a geometric point of view.

E.1   CHARACTERIZING CONICS FROM A GEOMETRIC VIEWPOINT

■ A circle

A circle is the set of points (x, y) equidistant from a fixed point C (the center).

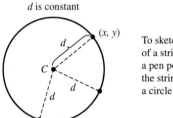

d is constant

To sketch a circle, tack one end of a string at C, and attach a pen point to the other. Holding the string taut, you can trace out a circle with the pen.

Figure E.3   Circle

■ A parabola

A parabola is the set of points (x, y) equidistant from a fixed line L and a fixed point F, where F does not lie on L.

L is called the **directrix**, and F is called the **focus.**

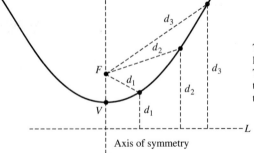

The vertex, V, of the parabola lies midway between F and L. The axis of symmetry runs through F and perpendicular to L.

Axis of symmetry

Figure E.4

■ An Ellipse

An ellipse is the set of points (x, y), the sum of whose distances from two fixed points F_1 and F_2 is constant.

F_1 and F_2 are called the **foci** of the ellipse.

$d_1 + d_2$ is constant

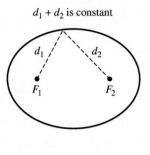

To sketch an ellipse, you can tack one end of a string to each focus and, with the point of a pen, hold the string taut. by moving around the foci you'll trace out an ellipse.

Pen

Figure E.5   Ellipse

The farther apart the two foci, the more elongated the ellipse. The closer they are to one another, the more circular the ellipse looks. If the two foci merge to one point, then the figure is a circle.

◼ **A hyperbola**

A hyperbola is the set of points (x, y) the difference of whose distances from two distinct fixed points F_1 and F_2 is constant.

F_1 and F_2 are called the **foci** of the hyperbola.

$|d_2 - d_1|$ is constant

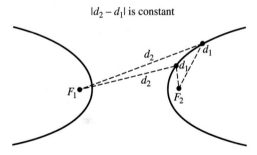

Figure E.6

◼ ## E.2   DEFINING CONICS ALGEBRAICALLY

Next we relate geometric and algebraic representations of conics. First the conclusions are presented. Then a road map is given. The actual trips from geometric to algebraic respresentations are left as problems at the end of the section. Signposts are provided. To find these problems, look under the appropriate conic section; they lead the section of problems for the particular conic.

Relating the Geometric and Algebraic Representations of Conics

■   If a circle has center $(0, 0)$ and the distance from the center is denoted by r, then the circle is given by the equation

$$x^2 + y^2 = r^2.$$

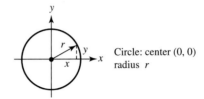

Circle: center $(0, 0)$
radius r

Figure E.7

■   If a parabola has focus $(0, c)$ and directrix $y = -c$, then it will have a vertex at $(0, 0)$ and be given by the equation

$$y = \frac{1}{4c}x^2.$$

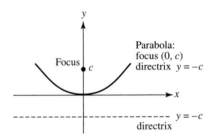

Parabola:
focus $(0, c)$
directrix $y = -c$

Figure E.8

■   If an ellipse has foci at $(c, 0)$ and $(-c, 0)$ and the sum of the distances from any point on the ellipse to the foci is denoted by $2a$, then the ellipse will have x-intercepts of $(a, 0)$ and $(-a, 0)$ and be given by the equation

$$\frac{x^2}{a^2} + \frac{y^2}{b^2} = 1,$$

where $b^2 = a^2 - c^2$. (See Figure E.9(a).)
If an ellipse has foci at $(0, c)$ and $(0, -c)$ and the sum of the distances from any point on the ellipse to the foci is denoted by $2a$, then the ellipse will have y-intercepts of $(0, a)$ and $(0, -a)$ and be given by the equation

$$\frac{y^2}{a^2} + \frac{x^2}{b^2} = 1,$$

where $b^2 = a^2 - c^2$. (See Figure E.9(b).)

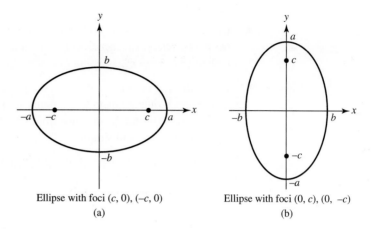

Ellipse with foci $(c, 0)$, $(-c, 0)$
(a)

Ellipse with foci $(0, c)$, $(0, -c)$
(b)

Figure E.9

■   If a hyperbola has foci at $(c, 0)$ and $(-c, 0)$ and the magnitude of the difference of the distances from any point on the hyperbola to the foci is denoted by $2a$, then the hyperbola will have x-intercepts of $(a, 0)$ and $(-a, 0)$ and be given by the equation

$$\frac{x^2}{a^2} - \frac{y^2}{b^2} = 1,$$

where $b^2 = c^2 - a^2$. (See Figure E.10(a).)

If a hyperbola has foci at $(0, c)$ and $(0, -c)$ and the magnitude of the difference of the distances from any point on the hyperbola to the foci is denoted by $2a$, then the hyperbola will have y-intercepts of $(0, a)$ and $(0, -a)$ and be given by the equation

$$\frac{y^2}{a^2} - \frac{x^2}{b^2} = 1,$$

where $b^2 = c^2 - a^2$. (See Figure E.10(b).)

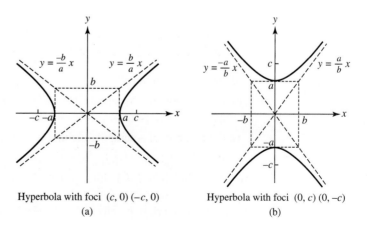

Hyperbola with foci $(c, 0)$ $(-c, 0)$
(a)

Hyperbola with foci $(0, c)$ $(0, -c)$
(b)

Figure E.10

Roadmaps

- *Circle:* Let (x, y) be a point on the circle with center $(0, 0)$. Use the distance formula (or the Pythagorean Theorem) to conclude that if the distance from (x, y) to $(0, 0)$ is r, then $x^2 + y^2 = r^2$.

- *Parabola:* Let (x, y) be a point on the parabola with focus F at $(0, c)$ and directrix L at $y = -c$. Using the distance formula, we know that the distance from (x, y) to F is $\sqrt{x^2 + (y - c)^2}$ and the distance from (x, y) to L is $\sqrt{(y + c)^2}$. Equate the distances, square both sides, and simplify.

- *Ellipse:* Let (x, y) be a point on the ellipse with foci at $(c, 0)$ and $(-c, 0)$. Denote the sum of the distances from any point on the ellipse to the foci by $2a$. (This distance is some positive number, and denoting it by $2a$ is convenient.) Using the distance formula we have that

$$\sqrt{(x + c)^2 + y^2} + \sqrt{(x - c)^2 + y^2} = 2a.$$

Isolate one radical and square both sides to eliminate it. Then isolate the other radical and square both sides to eliminate the second radical. Simplifying and denoting the quantity $a^2 - c^2$ by b^2 gives the desired result.

- *Hyperbola:* Let (x, y) be a point on the hyperbola with foci at $(0, c)$ and $(0, -c)$. Denote the magnitude (absolute value) of the differences of the distances from any point on the hyperbola to the foci by $2a$. (Denoting this difference by $2a$ is convenient.) Using the distance formula we obtain

$$\sqrt{(x + c)^2 + y^2} - \sqrt{(x - c)^2 + y^2} = \pm 2a.$$

Isolate one radical and square both sides to eliminate it. Then isolate the other radical and square both sides to eliminate the second radical. Simplifying and denoting the quantity $c^2 - a^2$ by b^2 gives the desired result.

Conics Given as $Ax^2 + Bxy + Cy^2 + Dx + Ey + F = 0$

Consider the equation

$$Ax^2 + Cy^2 = P,$$

where A, C, and P are constants. This is

i. the equation of a circle if $A = C$, and the signs of A, C, and P agree;

ii. the equation of an ellipse if A, C, and P have the same signs;

iii. the equation of a hyperbola if A and C have opposite signs (to find out which way the hyperbola opens, look for the x- and y-intercepts—you'll find only one pair);

iv. the equation of a parabola if either $A = 0$ or $C = 0$.

Consider the graph of $4(x - 2)^2 + 9(y - 3)^2 = 36$. This is the ellipse $4x^2 + 9y^2 = 36$ shifted 2 units right and 3 units up. This example can be generalized.

The graph of $A(x - h)^2 + C(y - k)^2 = P$ is the graph of the conic section $Ax^2 + Cy^2 = P$ shifted right h units and up k units. Multiplying out gives an equation of the form

$$Ax^2 + Cy^2 + Dx + Ey + F = 0,$$

where A, C, D, E, and F are constants.

Conversely, an equation of the form $Ax^2 + Cy^2 + Dx + Ey + F = 0$ can, by completing the square twice, be put in the form $A(x - h)^2 + C(y - k)^2 = P$. By looking at the signs of A and C as described above, one can determine the nature of the conic section.

◆ **EXAMPLE E.1**   Put the conic section

$$2x^2 - y^2 + 4x + 6y - 8 = 0$$

into the form $A(x - h)^2 + C(y - k)^2 = P$ and determine what it looks like.

SOLUTION

$$2x^2 - y^2 + 4x + 6y - 8 = 0$$
$$2x^2 + 4x - y^2 + 6y = 8$$
$$2(x^2 + 2x) - (y^2 - 6y) = 8$$
$$2\left[(x + 1)^2 - 1\right] - \left[(y - 3)^2 - 9\right] = 8$$
$$2(x + 1)^2 - 2 - (y - 3)^2 + 9 = 8$$
$$2(x + 1)^2 - (y - 3)^2 = 1$$

Therefore, this conic is the hyperbola $2x^2 - y^2 = 1$ (a hyperbola with x-intercepts at $\pm\frac{1}{\sqrt{2}}$) shifted left 1 unit and up 3 units.   ◆

In addition to shifting the conic $Ax^2 + Cy^2 = P$ horizontally and vertically, we can rotate the conic. The resulting algebraic equation is of the form

$$Ax^2 + Bxy + Cy^2 + Dx + Ey + F = 0,$$

where A, B, C, D, E, and F are constants.

It can be shown[1] that this equation gives

a parabola when $B^2 - 4AC = 0$,

an ellipse when $B^2 - 4AC < 0$ (a circle if, in addition, $A = C$),

a hyperbola when $B^2 - 4AC > 0$.

Notice that the sign of the expression $B^2 - 4AC$ discriminates between the three types of conics. This is not coincidental . . .

In addition to having fascinating geometric characteristics, conic sections have many practical applications.

---

[1] This can be shown most elegantly using linear algebra; therefore the more accessible but less appealing argument is not given here. Both arguments involve introducing a new, rotated coordinate system.

◼ E.3   THE PRACTICAL IMPORTANCE OF CONIC SECTIONS

Reflection Properties

The reflective properties of parabolas, hyperbolas, and ellipses are most useful when considering surfaces formed by rotating the two-dimensional figures around the line through the two foci in the latter two cases and through the focus and perpendicular to the directrix in the former.

◼ Reflecting properties of parabolas: A light source placed at the focus will be reflected parallel to the axis of symmetry of the parabola. Conversely, sound or light waves coming in parallel to the axis of symmetry of the parabola are reflected through the focus, thereby concentrating them there.

This property makes parabolas a useful shape in the design of headlights and search-lights, reflecting mirrors in telescopes, satellite dishes, radio antennas, and micro-phones designed to pick up a conversation far away.

Reflective property
of a parabola

Parabolic headlight with
the bulb at the focus

Figure E.11

◼ Reflecting property of ellipses: Sound or light emanating from one focus and reflecting off the ellipse will pass through the other focus.

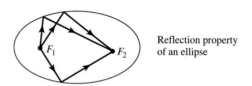

Reflection property
of an ellipse

Figure E.12

There are what are known as "whispering galleries" under the elliptic dome in the Capitol building of the United States as well as in the Mormon Tabernacle in Salt Lake City, Utah. A person standing at one focus and whispering to his neighbor can be heard quite clearly by an individual located at the other focus. As you can imagine, this can be a potential problem for those unaware of the reflection properties of the ellipse.

In 1980 an ingenious medical treatment for kidney stones, called lithotripsy (from the Greek word for "stone breaking"), was introduced. The treatment is based on the reflective properties of the ellipse. Intense sound waves generated outside the body are focused on the kidney stone, bombarding and thereby destroying it without invasive surgery. Over 80% of patients with kidney stones can be treated with lithotripsy; the recovery time is substantially less and the mortality rate much lower than with the traditional surgery .[2]

- ◼ Reflecting property of hyperbolas: Light traveling along a line through one focus of a hyperbola will be reflected off the surface of the hyperbola along the line through the point of reflection and the other focus.

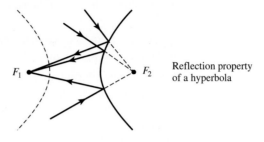

Reflection property of a hyperbola

Figure E.13

The reflective property of hyperbolas is used in the construction of camera and telescope lenses as well as radio navigation systems, such as the long-range navigation system (LORAN).

Conic Sections and Astronomy

The path of any projectile under the force of gravity is a parabolic arc. Around 1600 Kepler found that planets have elliptical orbits around the sun. This discovery involved painstaking calculations using measurements collected by Tycho Brahe in the late 1500s over a period of more than two decades. Kepler's conclusions (and the accuracy of Tycho Brahe's measurements) are rather astounding, given that the foci of most of the planets are very close together, making their paths very nearly circular.[3] Newton later put Kepler's observations on solid theoretical grounds.

Comets can travel in elliptical orbits (like Halley's comet) or in parabolic orbits. The moon has an elliptical orbit with the earth at one focus.

PROBLEMS FOR APPENDIX E

Parabolas

1. Begin with the geometric characterization of a parabola. Suppose that the focus F is at $(0, c)$ and the directrix L is at $y = -c$. You will show that the set of points equidistant from F and L is the parabola $y = \frac{1}{4c}x^2$.

   (a) Use the fact that the axis of symmetry is perpendicular to the directrix and passes through the focus to deduce that the vertex of the parabola is at $(0, 0)$.

---

[2] The story of the development of this treatment is fascinating. It came out of scientists looking at aircraft. You can get more information on the scientific aspects at http://www.hvlitho.com/lithotripsy and on the mathematics of it at http://www.math.iupui.edu:edu:80/m261vis/litho.

[3] For more information on Kepler, consult a history of mathematics book, such as *A History of Mathematics An Introduction* by Victor Katz, Addison Wesley 1998, pp. 409–416.

(b) Let (x, y) be a point on the parabola described above. Equating the distance between the point (x, y) and F with the distance between the point (x, y) and L gives the equation

$$\sqrt{x^2 + (y - c)^2} = \sqrt{(y + c)^2}.$$

Explain.

(c) Square both sides of the equation in part (b) to solve for y.

2. In Problem 1 you showed that a parabola with focus F at $(0, c)$ and the directrix L at $y = -c$ can be written in the form

$$y = \frac{1}{4c}x^2.$$

(a) Find the equation of the parabola with vertex $(0, 0)$ and focus $(0, 1)$. Sketch its graph. In the sketch show the location of the focus and directrix.

(b) Find the equation of the parabola with vertex $(0, 0)$ and directrix $y = 2$. Sketch its graph. In the sketch show the location of the focus and directrix.

3. In Problem 1 you showed that a parabola with focus F at $(0, c)$ and the directrix L at $y = -c$ can be written in the form

$$y = \frac{1}{4c}x^2.$$

Find the focus and directrix of the parabola $y = 2x^2$.

4. If the focus and directrix of a parabola are moved farther apart, does the parabola get narrower or does it get wider? Explain.

5. Find the equation of a parabola with its vertex at the origin and its focus at $(0, 6)$.

6. Find the equation of a parabola with vertex at $(0, 0)$ and directrix $y = 5$.

7. A headlight is made of reflective material in the shape of a parabolic dish. In order to take advantage of the fact that light emanating from the focus will be reflected by the parabolic bowl in the direction of the axis of symmetry, where should the light source be placed if the dish is 5 inches in diameter and 2 inches in depth?

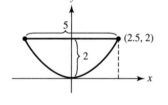

(*Hint: Look at a cross section of the reflector and introduce a coordinate system. Let the y-axis lie along the axis of symmetry of the reflector and let the origin lie at the vertex of the reflector.*)

8. The light filament in the bulb of a floodlight is 1.5 inches from the vertex of the paraboloid reflector. (The cross sections of the reflector taken through the vertex are all identical parabolas.)

(a) Find the equation of the parabolic cross section.

(b) If the reflector dish is 10 inches in diameter, how deep is it?

Ellipses

9. Let (x, y) be a point on the ellipse with foci at $(0, c)$ and $(0, -c)$. You will arrive at the formula

$$\frac{x^2}{a^2} + \frac{y^2}{b^2} = 1,$$

where $a^2 - C^2 = b^2$.

(a) Denote the sum of the distances from any point (x, y) on the ellipse to the foci by $2a$. (This distance is some positive number; denoting it by $2a$ is convenient.) Using the distance formula we know that

$$\sqrt{(x + c)^2 + y^2} + \sqrt{(x - c)^2 + y^2} = 2a.$$

Isolate $\sqrt{(x + c)^2 + y^2}$ and square both sides of the equation to eliminate it. After multiplying out you should be able to simplify the result to

$$a\sqrt{(x + c)^2 + y^2} = a^2 + cx.$$

(b) Square both sides of the equation $a\sqrt{(x + c)^2 + y^2} = a^2 + cx$ to eliminate the radical. Show that the result can be expressed in the form

$$(a^2 - c^2)x^2 + a^2 y^2 = a^2(a^2 - c^2). \qquad (\text{E.1})$$

(c) The distance between the two foci is $2c$. This must be less than the sum of the distance between (x, y) and the foci. Thus $2a > 2c$, or $a > c$. Consequently, $(a^2 - c^2) > 0$ and we can divide both sides of Equation E.1 by $a^2(a^2 - c^2)$ to get

$$\frac{x^2}{a^2} + \frac{y^2}{a^2 - c^2} = 1.$$

Denote $a^2 - c^2$ by b^2 to get the desired result.

10. (a) In Problem 9 you showed that the ellipse with foci at $(0, c)$ and $(0, -c)$ is given algebraically by

$$\frac{x^2}{a^2} + \frac{y^2}{b^2} = 1.$$

Explain from an algebraic point of view how you know that $b < a$.

(b) What is the geometric significance of the constants a and b?

Hyperbolas

11. Let (x, y) be a point on the hyperbola with foci at $(0, c)$ and $(0, -c)$. Denote the magnitude (absolute value) of the differences of the distances from any point on the hyperbola to the foci by $2a$.

(a) Using the distance formula show that

$$\sqrt{(x + c)^2 + y^2} - \sqrt{(x - c)^2 + y^2} = \pm 2a.$$

(b) Carry on as in Problem 9. Isolate one radical and square both sides to eliminate it. Then isolate the other radical and square both sides to eliminate the second radical. Show that you obtain

$$(c^2 - a^2)x^2 - a^2y^2 = a^2(c^2 - a^2).$$

(c) Argue that $c^2 - a^2 > 0$. Once this is done, we can set $b^2 = c^2 - a^2$. Simplify to obtain

$$\frac{x^2}{a^2} - \frac{y^2}{b^2} = 1.$$

L'Hôpital's Rule: Using Relative Rates of Change to Evaluate Limits

F.1 INDETERMINATE FORMS

In evaluating limits, we sometimes encounter situations in which our intuition might pull us in two different directions simultaneously. Many limits of this sort are called indeterminate forms; below are a few familiar examples.

i. $\lim_{x \to \infty} \frac{3x^2 + 1}{x^2}$ ii. $\lim_{h \to \infty} \frac{f(a+h) - f(a)}{h}$

iii. $\lim_{n \to \infty} \left(1 + \frac{1}{n}\right)^n$ iv. $\lim_{x \to 0^+} x \ln x$

We take these up one by one below.

i. $\lim_{x \to \infty} \frac{3x^2 + 1}{x^2}$

In the section on rational functions, we encountered limits like this. The numerator and the denominator both grow without bound. We will refer to this as a limit of the form $\frac{\infty}{\infty}$. Because the numerator is growing without bound we might expect the fraction to grow without bound; on the other hand, because the denominator is growing without bound we might expect the limit to be zero. We learned that in general the answer to $\lim_{x \to \infty} \frac{f(x)}{g(x)}$ depends on the degree of the polynomials involved; the limit could be zero (degree of numerator < degree of denominator), a nonzero finite number (degrees equal), or unbounded (degree of numerator > degree of denominator). In the case of $\lim_{x \to \infty} \frac{3x^2 + 1}{x^2}$, we know that in fact the limit is 3. The answer depends on the *relative rates* at which the numerator and the denominator are growing.

ii. $\lim_{h \to 0} \frac{f(a+h) - f(a)}{h}$

Every time that we use the limit definition of the derivative we also have intuition pulling us in two directions. In evaluating $f'(a) = \lim_{\Delta x \to 0} \frac{\Delta y}{\Delta x} = \lim_{h \to 0} \frac{f(a+h) - f(a)}{h}$, both the numerator and the denominator are approaching zero. The former fact might make us think that the limit should be zero, while the latter could seem to imply that the expression should be unbounded. We will refer to this as a limit of the form $\frac{0}{0}$. In fact, we know from our experience evaluating derivatives that $f'(a)$ might be any real number at all. Actually the value of $f'(a)$ depends on the *relative rates* at which Δy and Δx are approaching zero; the derivative is the ratio of these rates. We can evaluate the limit by comparing their rates of growth.

iii. $\lim_{n \to \infty} \left(1 + \frac{1}{n}\right)^n$

In Chapter 15 we worked on this limit. In trying to evaluate it, we noticed how our intuition pulled us in two opposite directions. On the one hand, the base is approaching 1, and we know that 1 raised to any power is still 1, making us think that the limit should be 1. But, at the same time, the exponent is growing without bound, leading us to think that $\left(1 + \frac{1}{n}\right)^n$ should grow without bounded as well, as $\lim_{n \to \infty} A^n$ is infinite if A is a constant greater than 1. As it turned out, the answer was somewhere in between;
$$\lim_{n \to \infty} \left(1 + \frac{1}{n}\right)^n = e = 2.71828\ldots.$$

iv. $\lim_{x \to 0^+} x \ln x$

This limit is of the form $0 \cdot \infty$. The two factors, $(-\infty)$ and 0 pull us in different directions.

Each of the examples above is an indeterminate form. As you probably remember, we put a fair amount of work into evaluating the limits listed above. Will we need to do so much work to evaluate all limits like this? Fortunately, the answer is no; there is a quick and simple method for evaluating many indeterminate forms. This method, called l'Hôpital's Rule, depends heavily on the work we have already done in learning how to compute the derivatives of many different types of functions; it would not be possible to use this method had we not done that previous work.

Indeterminate Forms of Type $\frac{0}{0}$ and $\frac{\infty}{\infty}$

The expression $\lim_{x \to a} \frac{f(x)}{g(x)}$ is an indeterminate form of type $\frac{0}{0}$ if $\lim_{x \to a} f(x) = 0$ and $\lim_{x \to a} g(x) = 0$. The expression $\lim_{x \to a} \frac{f(x)}{g(x)}$ is an indeterminate form of type $\frac{\infty}{\infty}$ if $\lim_{x \to a} f(x) = \pm\infty$ and $\lim_{x \to a} g(x) = \pm\infty$. These statements also hold if a is replaced by ∞, $-\infty$, or a one-sided limit.

L'Hôpital's Rule

If $\lim_{x \to a} \frac{f(x)}{g(x)}$ is of the form $\frac{0}{0}$ or $\frac{\infty}{\infty}$, and $\lim_{x \to a} \frac{f'(x)}{g'(x)}$ exists, then

$$\lim_{x \to a} \frac{f(x)}{g(x)} = \lim_{x \to a} \frac{f'(x)}{g'(x)},$$

if the latter limit exists. This result also holds if a is replaced by ∞, $-\infty$, or a one-sided limit.

The proof of a special case of this rule is given on page 1114, but it is not very enlightening in terms of showing why the rule should work; so, before we give this proof, we will look at a more illuminating (but not rigorous) "explanation" of the rule. First, though, we'll get our feet wet by showing how l'Hôpital's Rule is used in the following examples.

EXAMPLE F.1 Compute $\lim_{x \to \infty} \frac{e^x}{x}$.

SOLUTION $\lim_{x \to \infty} \frac{e^x}{x}$ is of the form $\frac{\infty}{\infty}$, so l'Hôpital's Rule tells us

$$\lim_{x \to \infty} \frac{e^x}{x} = \lim_{x \to \infty} \frac{e^x}{1} = \infty.$$

This reconfirms a fact we've asserted: e^x grows considerably faster than does x. ◆

EXAMPLE F.2 Compute $\lim_{x \to \infty} \frac{e^x}{x^3}$.

SOLUTION $\lim_{x \to \infty} \frac{e^x}{x}$ is of the form $\frac{\infty}{\infty}$, so l'Hôpital's Rule tells us

$$\lim_{x \to \infty} \frac{e^x}{x^3} = \lim_{x \to \infty} \frac{e^x}{3x^2}$$

if the latter exists. But $\lim_{x \to \infty} \frac{e^x}{3x^2}$ is of the form $\frac{\infty}{\infty}$, so l'Hôpital's Rule can be applied again.

$$\lim_{x \to \infty} \frac{e^x}{3x^2} = \lim_{x \to \infty} \frac{e^x}{6x}$$

Here we can either apply l'Hôpital's Rule again or use the result of Example F.1: $\lim_{x \to \infty} \frac{e^x}{6x} = \infty$. ◆

In the next example we compare the growth rates of exponentials as compared with polynomials.

EXAMPLE F.3 Evaluate $\lim_{x \to \infty} \frac{P(x)}{b^x}$, where $P(x)$ is any polynomial of degree n and b is a constant greater than 1.

SOLUTION This limit is of the form $\frac{\infty}{\infty}$. When we apply l'Hôpital's Rule, we differentiate the numerator, obtaining a new polynomial of degree $n - 1$, and we differentiate the denominator, obtaining $(\ln b)b^x$. If the degree of the new polynomial is 1 or greater, then as x tends toward infinity this polynomial will still tend to (positive or negative) infinity; the denominator will also still tend to infinity because it contains the b^x term. Thus, we will need to apply l'Hôpital's Rule again. But no matter how many times we differentiate the denominator, it will always be a constant times b^x and hence tend to infinity. On the other hand, once we have differentiated the numerator n times, it will be just a constant. So, we will be evaluating $\lim_{x \to \infty} \frac{K}{(\ln b)^n b^x} = 0$. ◆

The moral of Example F.3 is that, in the long run, exponentials dominate polynomials.

EXAMPLE F.4 Evaluate $\lim_{x \to 1} \frac{x-1}{\ln x}$.

SOLUTION This limit is of the form $\frac{0}{0}$. Before using l'Hôpital's Rule, we can look at this problem from a graphical perspective. The function $f(x) = x - 1$ and the function $g(x) = \ln x$ have equal slopes at $x = 1$. Because the rates at which $f(x)$ and $g(x)$ are changing are equal, we can conjecture that $\lim_{x \to 1} \frac{x-1}{\ln x}$ should be 1.

Numerical evidence also supports this, as choosing values of x near 1 gives values of $\frac{x-1}{\ln x}$ very nearly equal to 1. Now we use l'Hôpital's Rule to confirm our conjecture.

$$\lim_{x \to 1} \frac{x-1}{\ln x} = \lim_{x \to 1} \frac{\frac{d}{dx}(x-1)}{\frac{d}{dx}(\ln x)} \qquad \text{Apply l'Hôpital's Rule.}$$

$$= \frac{1|_{x=1}}{(1/x)|_{x=1}} \qquad \text{Compute the derivatives of the top and the bottom.}$$

$$= \frac{1}{1} \qquad \text{Evaluate them at } x = 1.$$

$$= 1$$

L'Hôpital's Rule provides us with a confirmation of our conjecture that this limit should be 1.   ◆

"Explanation" (nonrigorous) of why l'Hôpital's Rule works in the special case that $f(a) = 0$ **and** $g(a) = 0$ **and** $f'(a)$ **and** $g'(a)$ **exist; are finite, and** $g'(a) \neq 0$.

If f is differentiable at $x = a$, then it can be approximated at $x = a$ by the line through the point $(a, f(a))$ with slope $f'(a)$.

$$y - f(a) = f'(a)(x - a) \text{ or } y = f(a) + f'(a)(x - a)$$

Similarly, near $x = a$, $g(x)$ is approximated by the line $y = g(a) + g'(a)(x - a)$.

$$\lim_{x \to a} \frac{f(x)}{g(x)} \approx \lim_{x \to a} \frac{f(a) + f'(a)(x-a)}{g(a) + g'(a)(x-a)} \qquad \text{Use the linear approximations to } f(x) \text{ and } g(x).$$

$$\approx \lim_{x \to a} \frac{f'(a)(x-a)}{g'(a)(x-a)} \qquad f(a) \text{ and } g(a) \text{ are both zero by assumption.}$$

$$\approx \lim_{x \to a} \frac{f'(a)}{g'(a)} \qquad \text{Cancel the } (x - a) \text{ terms.}$$

$$\approx \frac{f'(a)}{g'(a)} \qquad \text{There are no } x\text{'s in the limit anymore.}$$

This result agrees with l'Hôpital's Rule as stated above. We see how the idea of local linearity means that it is reasonable that the rule should work; the value of the limit is the ratio of the rates of change of the numerator and denominator.

Proof of l'Hôpital's Rule for the Indeterminate Form $\frac{0}{0}$ **in the special case that** $f(a) = g(a) = 0$, f' **and** g' **are continuous,** $f'(a)$ **and** $g'(a)$ **exist and are finite, and** $g'(a)$ **is nonzero.**

$$\lim_{x \to a} \frac{f(x)}{g(x)} = \lim_{x \to a} \frac{f(x) - 0}{g(x) - 0}$$ Subtracting zero doesn't change anything.

$$= \lim_{x \to a} \frac{f(x) - f(a)}{g(x) - g(a)}$$ $f(a)$ and $g(a)$ are zero by assumption.

$$= \lim_{x \to a} \frac{\frac{f(x) - f(a)}{x - a}}{\frac{g(x) - g(a)}{x - a}}$$ Multiply top and bottom by $\frac{1}{x-a}$.

$$= \frac{\lim_{x \to a} \frac{f(x) - f(a)}{x - a}}{\lim_{x \to a} \frac{g(x) - g(a)}{x - a}}$$

$$= \frac{f'(a)}{g'(a)}$$ Recognize that $\lim_{x \to a} \dfrac{f(x) - f(a)}{x - a} = f'(a)$ and that $\lim_{x \to a} \dfrac{g(x) - g(a)}{x - a} = g'(a)$.

The proof without the added assumptions is more difficult, and the proof for the form $\frac{\infty}{\infty}$ more difficult yet. We omit these proofs.

CAUTION

- L'Hôpital's Rule does *not* say to take the derivative of $\frac{f(x)}{g(x)}$; for that you would need to use the Quotient Rule. Instead, it says to take the derivatives of the numerator and denominator *individually* and compute the limit as x approaches a.

- L'Hôpital's Rule must only be applied when its conditions are met. In particular, if the limit does not have the form $\frac{\infty}{\infty}$ or $\frac{0}{0}$, l'Hôpital's Rule will generally give a wrong answer.

◆ **EXAMPLE F.5** Evaluate $\lim_{x \to 0} \frac{e^{3x} - 1}{e^x - 1}$.

SOLUTION This is of the indeterminate form $\frac{0}{0}$. The slope of $e^{3x} - 1$ at $x = 0$ is three times the slope of $e^x - 1$ at $x = 0$, so the value of the limit should be 3.

$$\lim_{x \to 0} \frac{e^x - 1}{e^{3x} - 1} = \lim_{x \to 0} \frac{\frac{d}{dx}(e^{3x} - 1)}{\frac{d}{dx}(e^x - 1)}$$ Apply l'Hôpital's Rule.

$$= \frac{3e^{3x}|_{x=0}}{e^x|_{x=0}}$$ Compute the derivatives individually.

$$= \frac{3}{1}$$ Evaluate.

$$= 3 \quad ◆$$

◆ **EXAMPLE F.6** Evaluate $\lim_{x \to \infty} \frac{3x+1}{4x+17}$.

SOLUTION This is of the form $\frac{\infty}{\infty}$. Because this is a rational function where the degrees of the numerator and denominator are equal, it should have an asymptote at $y = 3/4$ (the ratio of the leading coefficients). We'll use l'Hôpital's Rule to confirm this.

$$\lim_{x \to \infty} \frac{3x + 1}{4x + 17} = \lim_{x \to \infty} \frac{3}{4}$$ Differentiate top and bottom and look at limit as x tends to infinity.

$$= \frac{3}{4}$$ Evaluate.   ◆

◆ EXAMPLE F.7   Evaluate $\lim_{x \to 0^+} \frac{\ln x}{1/x}$.

SOLUTION   This is of the form $\frac{\infty}{\infty}$.

$$\lim_{x \to 0^+} \frac{\ln x}{1/x} = \lim_{x \to 0^+} \frac{1/x}{-1/x^2}$$ Apply l'Hôpital's Rule.

$$= \lim_{x \to 0^+} -x$$ Simplify the fraction.

$$= 0$$ Evaluate.   ◆

The Indeterminate Form $0 \cdot \infty$.

If one quantity is tending to zero and another is tending to infinity, then their product might be zero or infinity or anywhere in between. Consider the following examples. (Note that we are not using l'Hôpital's Rule, but rather just algebra to evaluate them.)

$\lim_{x \to \infty} \frac{1}{x^2} \cdot x$ is of the form $0 \cdot \infty$, but $\lim_{x \to \infty} \frac{1}{x^2} \cdot x = \lim_{x \to \infty} \frac{1}{x} = 0$.

$\lim_{x \to \infty} \frac{1}{x} \cdot x^2$ is of the form $0 \cdot \infty$, but $\lim_{x \to \infty} \frac{1}{x} \cdot x^2 = \lim_{x \to \infty} x = \infty$.

$\lim_{x \to \infty} \frac{39}{x} \cdot x$ is of the form $0 \cdot \infty$, but $\lim_{x \to \infty} \frac{39}{x} \cdot x = \lim_{x \to \infty} 39 = 39$.

Most limits that take the indeterminate form $0 \cdot \infty$ are harder to evaluate than the three just given. Fortunately, we can often convert the indeterminate form $0 \cdot \infty$ into the indeterminate form $\frac{0}{0}$ or $\frac{\infty}{\infty}$ with a little algebraic manipulation.

◆ EXAMPLE F.8   Evaluate $\lim_{x \to 0^+} x \ln x$.

SOLUTION $\lim_{x \to 0^+} x \ln x = \lim_{x \to 0^+} \lim_{x \to 0^+} \frac{\ln x}{1/x}$. This latter limit is the one computed in Example F.7. The limit is 0.   ◆

◆ EXAMPLE F.9   Evaluate $\lim_{x \to \infty} xe^{-x}$. This is of the form $\infty \cdot 0$; we can convert it into the form $\frac{\infty}{\infty}$.

$$\lim_{x \to \infty} xe^{-x} = \lim_{x \to \infty} \frac{x}{e^x}$$ Convert $\infty \cdot 0$ into $\frac{\infty}{\infty}$.

$$= \lim_{x \to \infty} \frac{1}{e^x}$$ Apply l'Hôpital's Rule.

$$= 0$$ ◆

◆ EXAMPLE F.10   Evaluate $\lim_{x \to 2} (x - 2)^2 \ln(x - 2)$. This is of the form $0 \cdot (-\infty)$; algebra converts it to the form $\frac{-\infty}{\infty}$.

$$\lim_{x \to 2^+} (x - 2)^2 \ln(x - 2) = \lim_{x \to 2^+} \frac{\ln(x - 2)}{(x - 2)^{-2}} \qquad \text{Convert } 0 \cdot (-\infty) \text{ into } \tfrac{\infty}{\infty}.$$

$$= \lim_{x \to 2^+} \frac{\frac{1}{x-2}}{\frac{-2}{(x-2)^3}} \qquad \text{Apply l'Hôpital's Rule.}$$

$$= \lim_{x \to 2^+} \frac{-(x - 2)^2}{2} \qquad \text{Simplify.}$$

$$= 0 \qquad \text{Evaluate.} \quad \blacklozenge$$

The Indeterminate Forms 1^∞, ∞^0, and 0^0

In Chapter 15, when we were trying to evaluate the limit $L = \lim_{n \to \infty} \left(1 + \frac{1}{n}\right)^n$ in which the base was tending to 1 and the exponent to infinity, we took the natural logarithm of each side, evaluated $\ln L$, and then used that to determine L itself. We can use the same technique in conjunction with l'Hôpital's Rule to evaluate other indeterminate forms of the types 1^∞, 0^0, or ∞^0.

Notation:

1^∞: We use the notation 1^∞ to denote the situation in which the base is tending toward 1 (not when the base *is* 1) and the exponent is increasing without bound. If the base actually is exactly 1, then the limit must be 1 no matter what the exponent is.

0^0: We use the notation 0^0 to denote the situation in which both the base and the exponent are approaching zero. This form is indeterminate because, on the one hand, 0 raised to any power is zero, but on the other hand, any nonzero base raised to the power of zero is one; intuition pulls us in two different directions.

∞^0: We use the notation ∞^0 to denote the situation in which the base is growing without bound while the exponent is approaching zero. This is an indeterminate form because any nonzero number raised to the power of zero is one, but a base growing unboundedly raised to a positive power should also be unbounded.

◆ **EXAMPLE F.11**   Evaluate $\lim_{x \to 0^+} x^x$.

SOLUTION   This is the indeterminate form 0^0. Assuming the limit exists, we let $L = \lim_{x \to 0^+} x^x$ and take natural logs.

$$L = \lim_{x \to 0^+} x^x$$

$$\ln L = \lim_{x \to 0^+} x \ln x \qquad \text{This is a new indeterminate form, } 0 \cdot (-\infty).$$

$$= \lim_{x \to 0^+} \frac{\ln x}{1/x} \qquad \text{Convert to form } \tfrac{-\infty}{\infty} \text{ so we can apply l'Hôpital's Rule.}$$

$$= \lim_{x \to 0^+} \frac{1/x}{-1/x^2} \qquad \text{Take derivatives of top and bottom.}$$

$$= \lim_{x \to 0^+} -x \qquad \text{Simplify.}$$

$$= 0$$

We are not done yet. We have determined that $\ln L = 0$, so L must equal 1. Thus, $\lim_{x \to 0^+} x^x = 1$. ◆

◆ **EXAMPLE F.12**   Evaluate $\lim_{x \to 0^+} (1 + x)^{1/x}$ (assuming the limit exists.)

SOLUTION   This is of the form 1^∞, so we let $L = \lim_{x \to 0^+} (1 + x)^{1/x}$ and take the natural log of each side.

$$L = \lim_{x \to 0^+} (1 + x)^{1/x}$$

$$\ln L = \lim_{x \to 0^+} \frac{1}{x} \ln(1 + x) \qquad \text{This is a new indeterminate form } 0 \cdot (-\infty).$$

$$\ln L = \lim_{x \to 0^+} \frac{\ln(1 + x)}{x} \qquad \text{Convert to the form } \frac{0}{0}.$$

$$\ln L = \lim_{x \to 0^+} \frac{1/(1 + x)}{1} \qquad \text{Take derivatives according to l'Hôpital's Rule.}$$

$$\ln L = \lim_{x \to 0^+} \frac{1}{1 + x} \qquad \text{Simplify.}$$

$$\ln L = 1$$

Because $\ln L = 1$, L must equal e. Therefore, $\lim_{x \to 0^+} (1 + x)^{1/x} = e$. (Notice that this is the same as the $\lim_{n \to \infty} \left(1 + \frac{1}{n}\right)^n$ but with the change of variable $x = 1/n$.)   ◆

PROBLEMS FOR APPENDIX F

For Problems 1 through 15, evaluate the following limits. (Note: L'Hôpital's Rule is a fine tool, but it is not applicable to every problem.)

1. $\lim_{x \to \infty} \frac{\ln x}{x}$

2. $\lim_{x \to 0^+} \frac{\ln x}{x}$

3. $\lim_{x \to \infty} \frac{x}{e^x}$

4. $\lim_{x \to 0} \frac{x}{e^{-x}}$

5. $\lim_{x \to \infty} \frac{100x^3}{e^x}$

6. $\lim_{x \to \infty} \frac{\ln(5 + e^x)}{3x}$

7. (a) $\lim_{t \to 0} \frac{t^2 + 3}{2t^3 + 100t + 1}$ (b) $\lim_{t \to \infty} \frac{t^2 + 3}{2t^3 + 100t + 1}$

8. $\lim_{t \to 0} \frac{2t}{t^3}$

9. $\lim_{x \to \infty} 2x \cdot e^{-x}$

10. $\lim_{x \to -\infty} \frac{e^x + 2}{x}$

11. $\lim_{x \to \infty} \frac{r^x}{x}$, where $0 < r < 1$.

12. $\lim_{x \to 0.5} \frac{\ln(1-2x)}{2x-1}$

13. $\lim_{x \to \infty} \frac{e^{2x}+7}{5e^{3x}-10}$

14. $\lim_{t \to 3} \frac{\ln(t-3)}{t^2-t-6}$

15. Evaluate the following limits. n is a positive integer.

    (a) $\lim_{x \to \infty} \frac{\ln x}{x^n}$

    (b) $\lim_{x \to \infty} \frac{e^{nx}}{x^n}$

16. Does the series $\sum_{k=1}^{\infty} \left(1 + \frac{1}{k}\right)^k$ converge or diverge? Explain clearly and carefully.

17. (a) Evaluate $\lim_{x \to \infty} \frac{x}{\ln x}$.

    (b) Does the series $\sum_{k=1}^{\infty} \frac{k}{\ln k}$ converge or diverge? Explain.

In Problems 18 through 21, evaluate the limit.

18. $\lim_{x \to \infty} \frac{x^3+10000x}{3^x}$

19. $\lim_{x \to 0^+} e^x \ln x$

20. $\lim_{x \to \infty} \left(1 + \frac{5}{x}\right)^{3x}$

21. $\lim_{x \to \infty} e^{-x} \ln x$

22. Graph $f(x) = x^2 \ln x$, indicating the (exact) coordinates of all extrema. What is $\lim_{x \to 0^+} x^2 \ln x$? How is this indicated on the graph of f?

APPENDIX G

Newton's Method: Using Derivatives to Approximate Roots

Many problems involve finding roots of equations, and many equations we are unable to solve exactly. Consider for instance, finding the solutions to $x^4 = 2x + 2$. By graphing $y = x^4$ and $y = 2x + 2$ we find the graphs intersect at two points. The x-coordinates of the points correspond to the roots of $f(x) = x^4 - 2x - 2$.

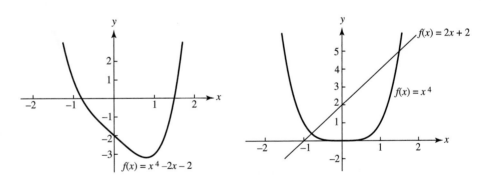

Figure G.1

The roots appear to be in the vicinity of $x = -0.8$ and $x = 1.5$. Zooming in on the graphs would help us get more accurate answers, but the method we are about to present, Newton's method, is more efficient.

Newton's method is a method of obtaining successive approximations to a root of a function. It uses the derivative to hone in on the root and generally produces a high degree of accuracy within a few iterations. The key notions are that

■ it is simple to find the root of a linear equation, and

■ the tangent line is a good linear approximation of a function about a point.

1121

Let's begin by finding the root of a linear equation. We will use the result of the next example repeatedly in applying Newton's method.

◆ **EXAMPLE G.1** Find the x-intercept of the line through (x_k, y_k) with slope m, where $m \neq 0$.

SOLUTION The equation of the line is $y - y_k = m(x - x_k)$. Set $y = 0$ and solve for x.

$$mx - mx_k = -y_k$$
$$mx = mx_k - y_k$$

$$\boxed{x = x_k - \frac{y_k}{m}}$$ ◆

We can use this result to refine an approximation of the root, r, of an arbitrary differentiable function, $f(x)$, as follows.

■ Approximate the root. Call this initial guess x_0. For instance, in our example $f(x) = x^4 - 2x - 2$, one of the roots appears to be around 1.5, so let $x_0 = 1.5$.

■ Find out where the tangent line to f at x_0 intersects the x-axis. Call this point x_1. Generally[1] the point x_1 will be closer to r than was x_0.

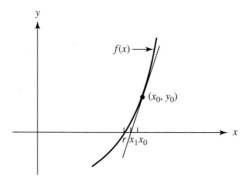

Figure G.2

From Example 2 we know $x = x_0 - \frac{y_0}{m}$, where $y_0 = f(x_0)$ and $m =$ the slope of the tangent line at $x_0 = f'(x_0)$. So

$$x_1 = x_0 - \frac{f(x_0)}{f'(x_0)}.$$

■ x_1 is our new guess. Repeat the process. x_2 is the x-intercept of the tangent line at x_1.

$$x_2 = x_1 - \frac{f(x_1)}{f'(x_1)}$$

---

[1] If x_1 is not closer to r than x_0, use another first guess.

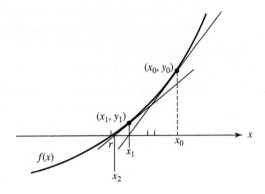

Figure G.3

By repeating the process we can produce a sequence $x_0, x_1, x_2, \ldots$ of successive approximations of r, where

$$x_{k+1} = x_k - \frac{f(x_k)}{f'(x_k)}.$$

REMARK

If, at any stage, $f(x_k)$ is zero, then we have found the root exactly.

If, at any stage, $f'(x_k)$ is zero, then we're stuck and must begin again.

Newton's Method for Approximating a Root of f

i. Take an initial guess and call it x_0.

ii. Successive approximations $x_1, x_2, x_3, \ldots$ are found using

$$x_{k+1} = x_k - \frac{f(x_k)}{f'(x_k)}, \quad f'(x_k) \neq 0.$$

Generally, the sequence of approximations will approach the root r as n increases, provided the initial guess was close enough to the root.

◆ **EXAMPLE G.2**   Use Newton's method to approximate the larger of the two roots of $f(x) = x^4 - 2x - 2$.

SOLUTION   Our first guess is $x_0 = 1.5$, $f(x_0) = 0.0625$.

$$f'(x) = 4x^3 - 2$$
$$f'(1.5) = 4(1.5)^3 - 2 = 11.5$$

$$x_1 = x_0 - \frac{f(x_0)}{f'(x_0)} = 1.5 - \frac{f(1.5)}{f'(1.5)} = 1.5 - \frac{0.0625}{11.5} \approx 1.494565$$

$$f(1.494565) \approx 3.9 \times 10^{-4}$$

We have a high degree of accuracy in just one step. To get more accuracy repeat the process.

◆

Notice that a computer or programmable calculator can be programmed to perform iterations and arrive at a sequence of approximations with ease.

Ways to be Led Astray

We've already remarked that Newton's method only makes sense when $f'(x_k) \neq 0$. If $f'(x_k) = 0$, then the tangent line is horizontal and does not intersect the x-axis. If this happens, start the process again. There are a few other problems one can run into, as illustrated in the figures below.

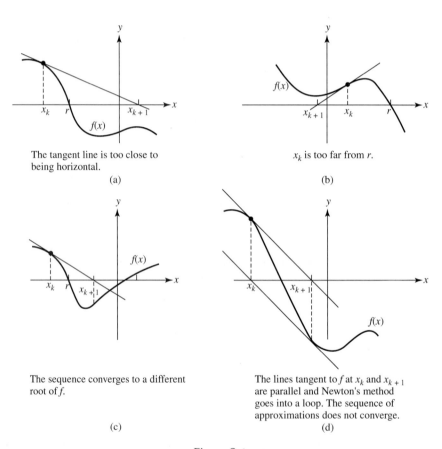

The tangent line is too close to being horizontal.

(a)

x_k is too far from r.

(b)

The sequence converges to a different root of f.

(c)

The lines tangent to f at x_k and x_{k+1} are parallel and Newton's method goes into a loop. The sequence of approximations does not converge.

(d)

Figure G.4

When can we be *guaranteed* that Newton's method won't lead us astray? It can be shown that if x_0 is chosen in an interval around r over which f is monotonic and doesn't change concavity, then the successive approximations will converge to r.

In fact, the criteria can be weakened. Newton's method will work if, in the interval between x_0 and r, the graph of f protrudes toward the x-axis, that is, f is concave up when $f(x_0) > 0$ or concave down when $f(x_0) < 0$. But even when the criteria do not hold, Newton's method may still converge to the root.

EXERCISE G.1   Verify for yourself that the graphs in Figure G.4 fail to meet these criteria.

◆ **EXAMPLE G.3**   Approximate $\sqrt[3]{17}$ using Newton's method, continuing until two successive approximations agree to five decimal places.

SOLUTION   This question is equivalent to finding the root $x^3 - 17 = 0$.

$$f(x) = x^3 - 17$$
$$f'(x) = 3x^2$$

We know $2 < \sqrt[3]{17} < 3$ so we'll let $x_0 = 2.5$.

$$x_1 = x_0 - \frac{f(x_0)}{f'(x_0)} = 2.5 - \frac{f(2.5)}{f'(2.5)} = 2.57\overline{3}$$

$$x_2 = x_1 - \frac{f(x_1)}{f'(x_1)} = 2.57\overline{3} - \frac{f(2.57\overline{3})}{f'(2.57\overline{3})} \approx 2.571283226$$

$$x_3 = x_2 - \frac{f(x_2)}{f'(x_2)} = 2.571283226 - \frac{f(2.571283226)}{f'(2.571283226)} \approx 2.57128591$$

x_3 and x_2 agree to five decimal places.
   Check: $2.57128^3 = 16.99996\ldots$ and $(2.571281591)^3 = 17.000085.\ldots$ ◆

PROBLEMS FOR APPENDIX G

In Problems 1 through 5, approximate the quantity indicated. Use Newton's method until two successive approximations agree to three decimal places. You are expected to use a calculator or computer.

1. The solution to $\ln x + x = 0$.

2. The positive root of $e^x - x = 3$.

3. The negative root of $e^x - x = 3$.

4. $\sqrt[3]{20}$

5. Write a computer program to approximate roots using Newton's method.

6. Approximate the negative root of $x^4 - 3x^2 + 2$. Use Newton's Method until two successive approximations agree to five decimal places.

7. Approximate the largest root of $x^3 - 3x^2 = -2$. Use Newton's method until two successive approximations agree to three decimal places.

8. Approximate the solution(s) to $\cos x = x$. Use Newton's method until two successive approximations agree to 5 decimal places.

9. Approximate the solution(s) to $\sin x = x^2$. Use Newton's method until two successive approximations agree to four decimal places.

H

Proofs to Accompany Chapter 30, Series

We begin this Appendix with a proof of Taylor's Theorem. The proof rests on Rolle's Theorem, which says that if f is continuous on $[a, b]$ and differentiable on (a, b) and $f(a) = f(b)$, then there exists a number $c \in (a, b)$ such that $f'(c) = 0$. Although the proof of Taylor's Theorem will not seem natural, it gets the job done.

Taylor's Theorem

Suppose f and all its derivatives exist in an open interval I centered at $x = b$. Then for each x in I,

$$f(x) = f(b) + f'(b)(x - b) + \frac{f''(b)}{2!}(x - b)^2 + \cdots + \frac{f^{(n)}(b)}{n!}(x - b)^n + R_n(x),$$

where $R_n(x) = \frac{f^{(n+1)}(c)}{(n+1)!}(x - b)^{n+1}$, for some number $c \in I$, c between a and b.

Proof

Let $a \in I$. The function F given below is constructed to satisfy the conditions of Rolle's Theorem.

$$F(x) = (a - b)^{n+1}\left[f(a) - f(x) - f'(x)(a - x) - \frac{f''(x)}{2!}(a - x)^2 - \cdots - \frac{f^{(n)}(x)}{n!}(a - x)^n \right] -$$

$$(a - x)^{n+1}\underbrace{\left[f(a) - f(b) - f'(b)(a - b) - \frac{f''(b)}{2!}(a - b)^2 - \cdots - \frac{f^{(n)}(b)}{n!}(a - b)^n \right.}_{\text{This is constant; denote it by } k.}$$

$F(a) = 0$ and $F(b) = 0$.

By Rolle's Theorem, there exists c between a and b such that $F'(c) = 0$. Compute $F'(x)$.

$$F'(x) = (a-b)^{n+1} \left[-f'(x) + f'(x) - f''(x)(a-x) + f''(x) - \frac{f'''(x)}{2!}(a-x)^2 \right.$$
$$\left. + \frac{f'''(x)}{2!}(a-x)^2 - \cdots - \frac{f^{(n+1)}(x)}{n!}(a-x)^n \right] + (n+1)k(a-x)^n$$

$$F'(x) = (a-b)^{n+1} \left[-\frac{f^{(n+1)}(x)}{n!}(a-x)^n \right] + (n+1)k(a-x)^n \qquad \text{Factor out } (a-x)^n.$$

$$F'(x) = (a-x)^n \left[-\frac{f^{(n+1)}(x)}{n!}(a-b)^{n+1} + (n+1)k \right]$$

There exists c between a and b such that $F'(c) = 0$. For such c we have

$$(a-c)^n \left[-\frac{f^{(n+1)}(c)}{n!}(a-b)^{n+1} + (n+1)k \right] = 0$$

$$k = \frac{f^{(n+1)}(c)}{(n+1)n!}(a-b)^{n+1}$$

$$f(a) - f(b) - f'(b)(a-b) - \frac{f''(b)}{2!}(a-b)^2 - \cdots - \frac{f^{(n)}(b)}{n!}(a-b)^n$$
$$= \frac{f^{(n+1)}(c)}{(n+1)!}(a-b)^{n+1}$$

Solving for $f(a)$ gives

$$f(a) = f(b) + f'(b)(a-b) + \frac{f''(b)}{2!}(a-b)^2 + \cdots + \frac{f^{(n)}(b)}{n!}(a-b)^n$$
$$+ \frac{f^{(n+1)}(c)}{(n+1)!}(a-b)^{n+1}.$$

Since a can be any number in I, we have completed the proof.

Absolute Convergence Implies Convergence

If $\sum_{k=1}^{\infty} |a_k|$ converges, then $\sum_{k=1}^{\infty} a_k$ converges.

Proof

Suppose $\sum_{k=1}^{\infty} |a_k|$ converges, and $\sum_{k=1}^{\infty} |a_k| = S$. Then

$$\sum_{k=1}^{\infty} 2|a_k| = 2S.$$

$|a_k|$ is either a_k or $-a_k$, so $0 \le a_k + |a_k| \le 2|a_k|$.

Therefore $\sum_{k=1}^{\infty}(a_k + |a_k|)$ is convergent; its sum is $\leq 2S$.

$$\sum_{k=1}^{\infty} a_k = \sum_{k=1}^{\infty}(a_k + |a_k|) - \sum_{k=1}^{\infty} |a_k|, \quad \text{the difference of two convergent series.}$$

We conclude that $\sum_{k=1}^{\infty} a_k$ converges.

A Convergence Theorem for Power Series

If a power series $\sum_{k=0}^{\infty} a_k x^k$ converges at $x = c, c \neq 0$ and $|\alpha| < |c|$, then the series converges absolutely for $x = \alpha$.

Proof

We assume $a_0 + a_1 c + a_2 c^2 + \cdots + a_k c^k + \cdots$ converges. Therefore

$$\lim_{k \to \infty} a_k c^k = 0.$$

Then there exists a positive constant M such that $|a_k c^k| < M$ for all k, $k = 0, 1, 2 \ldots$

$$0 \leq |a_0| + |a_1 \alpha| + |a_2 \alpha^2| + |a_3 \alpha^3| + \cdots + |a_n \alpha^n|$$

$$= |a_0| + |a_1 c| \left|\frac{\alpha}{c}\right| + |a_2 c^2| \left|\frac{\alpha}{c}\right|^2 + \cdots + |a_n c^n| \left|\frac{\alpha}{c}\right|^n$$

$$\leq M + M \left|\frac{\alpha}{c}\right| + M \left|\frac{\alpha}{c}\right|^2 + \cdots + M \left|\frac{\alpha}{c}\right|^n$$

But this is a geometric sum with $|r| < 1$, so

$$0 \leq \sum_{k=0}^{n} M \left|\frac{\alpha}{c}\right|^k = \frac{M}{1 - \left|\frac{\alpha}{c}\right|}.$$

Hence the sequence of partial sums of $\sum_{k=0}^{\infty} |a_k \alpha^k|$ is increasing (all terms are positive) and bounded $\left(\text{below by zero and above by } \frac{M}{1 - \left|\frac{\alpha}{c}\right|}\right)$. Therefore the sequence of partial sums converges and consequently $\sum_{k=0}^{\infty} |a_k \alpha^k|$ converges. $\sum_{k=0}^{\infty} a_k x^k$ converges absolutely at $x = \alpha$.

The Integral Test for the Convergence of a Series

Let $\sum_{k=1}^{\infty} a_k$ be a series such that $a_k = f(k)$ for $k = 1, 2, 3, \ldots$ where the function f is positive, continuous, and decreasing on $[1, \infty)$. Then

$$\sum_{k=1}^{\infty} a_k \text{ and } \int_1^{\infty} f(x)\, dx$$

either both converge or both diverge.

Proof

Given the hypotheses of the theorem, we can illustrate the relationship between the series and the integral in two ways, shown in Figure H.1 (with circumscribed rectangles) and Figure H.2 (with inscribed rectangles).

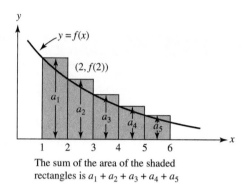

The sum of the area of the shaded rectangles is $a_1 + a_2 + a_3 + a_4 + a_5$

Figure H.1

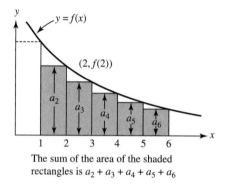

The sum of the area of the shaded rectangles is $a_2 + a_3 + a_4 + a_5 + a_6$

Figure H.2

Each rectangle has a base of 1, so its area corresponds to its height. In Figure H.1 the kth shaded rectangle has area $= a_k$; in Figure H.2 the kth shaded rectangle has area $= a_{k+1}$. Figure H.2 can be obtained from Figure H.1 by shifting the rectangles left one unit and dropping the rectangle corresponding to a_1, from consideration.

$$a_2 + a_3 + \cdots + a_n \leq \int_1^n f(x)\, dx \leq a_1 + a_2 + \cdots + a_{n-1} \qquad \text{(H.1)}$$

We begin by demonstrating that the behavior of the series can be determined from that of the integral. The terms of the series are positive; therefore the sequence of partial sums is increasing.

■   Suppose $\int_1^\infty f(x)\, dx$ converges. Denote $\int_1^\infty f(x)\, dx$ by A.

$$S_n = a_1 + a_2 + a_3 + \cdots + a_n \leq a_1 + \int_1^n f(x)\, dx$$

$$\lim_{n \to \infty} S_n \le \lim_{n \to \infty} a_1 + \int_1^n f(x)\, dx = a_1 + A$$

The sequence of partial sums is bounded and increasing. Therefore, by the Bounded Increasing Partial Sums Theorem, $\sum_{k=1}^{\infty} a_k$ converges.

■ Suppose $\int_1^{\infty} f(x)\, dx = \infty$. Refer to Equation (H.1) to obtain

$$\int_1^{n-1} f(x)\, dx \le a_1 + a_2 + \cdots + a_n = S_n$$

Taking the limit as $n \to \infty$ gives

$$\infty \le \lim_{n \to \infty} S_n.$$

So S_n grows without bound and $\sum_{k=1}^{\infty} a_k$ diverges.

Now we show that the behavior of the integral can be determined by that of the series.

Because $f(x) > 0$, decreasing, and continuous on $[1, \infty)$, $\lim_{b \to \infty} \int_1^b f(x)\, dx$ is either finite or grows without bound. Therefore, if we can find an upper bound, the integral converges. If it has no upper bound, it diverges.

■ Suppose $\sum_{k=1}^{\infty} a_k$ converges. Denote its sum by S. From Equation (H.1) we know

$$\int_1^n f(x)\, dx \le a_1 + a_2 + \cdots + a_{n-1}$$

$$\lim_{n \to \infty} \int_1^n f(x)\, dx \le \lim_{n \to \infty} \sum_{k=1}^{n-1} a_k = S.$$

If $\lim_{n \to \infty} \int_1^n f(x)\, dx$ is bounded, so too is $\lim_{b \to \infty} \int_1^b f(x)\, dx$ (given the hypotheses).

■ Suppose $\sum_{k=1}^{\infty} a_k$ diverges. Because the terms are all positive, we know $\lim_{n \to \infty} \sum_{k=1}^n a_k = \infty$. From Equation (H.1) we know

$$a_2 + \cdots + a_n \le \int_1^n f(x)\, dx$$

$$\lim_{n \to \infty} \sum_{k=2}^n a_k \le \lim_{n \to \infty} \int_1^n f(x)\, dx.$$

We conclude that $\lim_{b \to \infty} \int_1^b f(x)\, dx = \infty$; the improper integral diverges.

INDEX